Space Sciences Series of ISSI

Volume 47

For further volumes:
www.springer.com/series/6592

André Balogh • Andrei Bykov • Peter Cargill •
Richard Dendy • Thierry Dudok de Wit •
John Raymond
Editors

Microphysics of Cosmic Plasmas

Previously published in *Space Science Reviews* Volume 178,
Issues 2–4, 2013

Springer

Editors
André Balogh
Imperial College London
London, UK

Andrei Bykov
Russian Academy of Sciences
St. Petersburg, Russia

Peter Cargill
Imperial College London
London, UK

Richard Dendy
Euratom/UKAEA Fusion Association
Abingdon, UK

Thierry Dudok de Wit
LPC2E/CNRS-University of Orleans
Orleans Cedex 2, France

John Raymond
Harvard-Smithsonian Center for Astrophysics
Cambridge, MA, USA

ISSN 1385-7525 Space Sciences Series of ISSI
ISBN 978-1-4899-7412-9 ISBN 978-1-4899-7413-6 (eBook)
DOI 10.1007/978-1-4899-7413-6
Springer New York Heidelberg Dordrecht London

Library of Congress Control Number: 2013957365

Cover illustration: A combined optical (Hubble Space Telescope) and X-ray (Chandra Observatory) image of supernova remnant 0509-67.5 in the Large Magellanic Cloud galaxy. The soft blue and green colors indicate the X-ray emission detected by Chandra from the type Ia supernova shell, while the optical emission imaged with HST and shown in pink arises from a collisionless shock wave. This is an example of dynamical processes in collisionless plasmas produced by extreme events such as supernova explosions. Image credit: NASA, ESA, and B. Schaefer and A. Pagnotta (Louisiana State University, Baton Rouge); Image Credit: NASA, ESA, CXC, SAO, the Hubble Heritage Team (STScI/AURA), J. Hughes (Rutgers University).

Printed on acid-free paper

Springer is part of Springer Science+Business Media (www.springer.com)

Contents

Microphysics of Cosmic Plasmas: Background, Motivation and Objectives
A. Balogh · A. Bykov · P. Cargill · R. Dendy · T. Dudok de Wit · J. Raymond **1**

Microphysics in Astrophysical Plasmas
S.J. Schwartz · E.G. Zweibel · M. Goldman **5**

Solar Wind Turbulence and the Role of Ion Instabilities
O. Alexandrova · C.H.K. Chen · L. Sorriso-Valvo · T.S. Horbury · S.D. Bale **25**

Solar Surface and Atmospheric Dynamics · The Photosphere
V. Martínez Pillet **65**

Astrophysical Hydromagnetic Turbulence
A. Brandenburg · A. Lazarian **87**

Microphysics of Cosmic Ray Driven Plasma Instabilities
A.M. Bykov · A. Brandenburg · M.A. Malkov · S.M. Osipov **125**

Nonclassical Transport and Particle-Field Coupling: from Laboratory Plasmas to the Solar Wind
D. Perrone · R.O. Dendy · I. Furno · R. Sanchez · G. Zimbardo · A. Bovet · A. Fasoli · K. Gustafson · S. Perri · P. Ricci · F. Valentini **157**

Collisional and Radiative Processes in Optically Thin Plasmas
S.J. Bradshaw · J. Raymond **195**

Recent Evolution in the Theory of Magnetic Reconnection and Its Connection with Turbulence
H. Karimabadi · V. Roytershteyn · W. Daughton · Y.-H. Liu **231**

Notes on Magnetohydrodynamics of Magnetic Reconnection in Turbulent Media
P. Browning · A. Lazarian **249**

Microphysics of Cosmic Plasmas: Hierarchies of Plasma Instabilities from MHD to Kinetic
M.R. Brown · P.K. Browning · M.E. Dieckmann · I. Furno · T.P. Intrator **281**

In-Situ Observations of Reconnection in Space
G. Paschmann · M. Øieroset · T. Phan **309**

Kinetic Structure of Current Sheets in the Earth Magnetotail
A. Artemyev · L. Zelenyi **343**

Mechanisms of Spontaneous Reconnection: From Magnetospheric to Fusion Plasma
L. Zelenyi · A. Artemyev **365**

Topics in Microphysics of Relativistic Plasmas
M. Lyutikov · A. Lazarian **383**

Plasma Diagnostics of the Interstellar Medium with Radio Astronomy
M. Haverkorn · S.R. Spangler **407**

Microphysics of Quasi-parallel Shocks in Collisionless Plasmas
D. Burgess · M. Scholer **437**

The Dynamic Quasiperpendicular Shock: Cluster Discoveries
V. Krasnoselskikh · M. Balikhin · S.N. Walker · S. Schwartz · D. Sundkvist · V. Lobzin · M. Gedalin · S.D. Bale · F. Mozer · J. Soucek · Y. Hobara · H. Comisel **459**

Collisionless Shocks in Partly Ionized Plasma with Cosmic Rays: Microphysics of Non-thermal Components
A.M. Bykov · M.A. Malkov · J.C. Raymond · A.M. Krassilchtchikov · A.E. Vladimirov **523**

Electron-Ion Temperature Equilibration in Collisionless Shocks: The Supernova Remnant-Solar Wind Connection
P. Ghavamian · S.J. Schwartz · J. Mitchell · A. Masters · J.M. Laming **557**

Methods for Characterising Microphysical Processes in Plasmas
T. Dudok de Wit · O. Alexandrova · I. Furno · L. Sorriso-Valvo · G. Zimbardo **589**

Laboratory Astrophysics: Investigation of Planetary and Astrophysical Maser Emission
R. Bingham · D.C. Speirs · B.J. Kellett · I. Vorgul · S.L. McConville · R.A. Cairns · A.W. Cross · A.D.R. Phelps · K. Ronald **619**

DOI 10.1007/978-1-4899-7413-6_1
Reprinted from *Space Science Reviews* Journal, DOI 10.1007/s11214-013-0027-1

Microphysics of Cosmic Plasmas: Background, Motivation and Objectives

André Balogh · Andrei Bykov · Peter Cargill · Richard Dendy · Thierry Dudok de Wit · John Raymond

Received: 30 September 2013 / Accepted: 4 October 2013 / Published online: 17 October 2013

Abstract With the maturing of space plasma research in the solar system, a more general approach to plasma physics in general, applied to cosmic plasmas, has become appropriate. There are both similarities and important differences in describing the phenomenology of space plasmas on scales from the Earth's magnetosphere to galactic and inter-galactic scales. However, there are important aspects in common, related to the microphysics of plasma processes. This introduction to a coordinated collection of papers that address the several aspects of the microphysics of cosmic plasmas that have unifying themes sets out the scope and ambition of the broad sweep of topics covered in the volume, together with an enumeration of the detailed objectives of the coverage.

The space around and between astrophysical objects is occupied by plasma: that is, by matter which is fully ionised, or nearly so. These plasmas exist on all astrophysical scales, from the Earth's neighbourhood to clusters of galaxies. The density of plasma in these widely different environments is usually so low that binary collisions between plasma ions, or between

A. Balogh (✉) · P. Cargill
Imperial College London, Prince Consort Road, London SW7 2AZ, UK
e-mail: a.balogh@imperial.ac.uk

A. Bykov
Ioffe Institute of Physics and Technology, Russian Academy of Sciences, St. Petersburg, Russia

A. Bykov
Saint Petersburg State Polytechnical University, Politechnicheskaya, 26, St. Petersburg 194021, Russia

R. Dendy
Euratom/UKAEA Fusion Association, Culham Science Centre, Abingdon, Oxfordshire, OX14 3DB, UK

T. Dudok de Wit
LPC2E/CNRS 3A, Avenue de la Recherche Scientifique, 45071 Orléans cedex 2, France

J. Raymond
Harvard-Smithsonian Center for Astrophysics, 60 Garden Street, Cambridge, MA 02138, USA

ions and neutral atoms, are so rare that they can usually be neglected. While collisionless fully-ionised plasmas predominate in astrophysical environments, a complete picture also includes partially ionised plasmas and weakly collisional populations of particles. Plasmas are electrically quasi-neutral in the sense that, in most locations and on most timescales, the total charge density of electrons is equal and opposite to that of ions.

The Earth's magnetosphere has been a subject of intense research for the whole of the space age, and remains so today. It has proved to be a highly instructive laboratory for plasma physics under a wide range of conditions. Research into the Sun's atmosphere and the heliosphere has broadened the parameter space in which plasma phenomena are addressed. Plasmas in the solar system, from the Sun outwards through the heliosphere and on to the diverse planetary magnetospheres, are accessible to very detailed observations and measurements. In consequence, the physical processes that arise in space plasmas are generally well known and are certainly extensively documented. While our knowledge of the space plasma environment is extensive, large gaps remain in both conceptual and quantitative understanding of some basic phenomena: notably plasma turbulence, magnetic reconnection, and plasma kinetic processes.

Plasma research in the solar system, and specifically in the magnetosphere, has thus become a mature branch of space science. In March 2009 this prompted the International Space Science Institute (ISSI) in Bern, Switzerland, to organise a brainstorming Forum with the provocative title "*Is there a future for magnetospheric research*?" The answer was nuanced, as could be expected. It was agreed that many aspects of space plasma research have reached a level of maturity that qualifies them to be included in undergraduate textbooks, with the implication that both the phenomenology and the underlying theoretical models have been conclusively understood. However some very basic and essential aspects of space plasma phenomena are necessarily subject to continuing research. It was also recognised that there are many conceptual links between accessible space plasmas within the heliosphere, and the magnetospheric and astrophysical plasmas that exist throughout the universe. Such links need to be carefully analysed, interpreted and understood across many orders of magnitude in spatial and temporal scales.

Following the Forum, three broad topics were identified, in which links should be explored between plasmas on solar system and cosmic scales. ISSI has therefore organised three Workshops, on the topics of "*Particle Acceleration in Cosmic Plasmas*", "*The Microphysics of Cosmic Plasmas*" (the subject of the current volume) and "*Structure Formation and Dynamics in Cosmic Plasmas*". The three Workshops were held, respectively, in April each year from 2011 to 2013. As with all Workshops organised by ISSI, these three Workshops have resulted in coordinated collections of review papers. The first collection was published in the journal Space Science Reviews, Vol. 173, Nos. 1 to 4, 2012, followed by publication in the Space Science Series of ISSI, Vol. 45, edited by A. Balogh, A. Bykov, R.P. Lin, J. Raymond and M. Scholer. The present collection of papers in Space Science Reviews will also be published in the same series, as Volume 47. The third and final collection of papers, on structure formation and dynamics in cosmic plasmas, will be published in 2014.

The ISSI Workshop on "*The Microphysics of Cosmic Plasmas*" took place on from 16th to 20th April 2012. It was convened by André Balogh, Andrei Bykov, Peter Cargill, Richard Dendy, Thierry Dudok de Wit and John Raymond. The objective of the Workshop was to address the physical processes that underlie the observed large-scale properties, structure and dynamics of cosmic plasmas. As noted, these fill interplanetary, interstellar and intergalactic space, as well as the solar atmosphere and the Earth's magnetosphere. We aimed to review

the status of understanding of microscale processes in collisionless astrophysical plasmas, and also to consider the lessons that can be adapted from the extensive existing knowledge of laboratory plasmas. The topics covered in the Workshop were as follows:
Turbulence as a phenomenological description of the properties of plasmas on all scales

- General description of turbulence phenomena in space plasmas: the turbulent cascade, driving and dissipation processes
- Turbulence in the solar wind plasma
- Experimental and theoretical studies of dissipation in turbulent plasmas in the solar photosphere, chromosphere and corona
- Turbulence in astrophysical plasmas on all scales: supernovae remnants, interstellar medium, and intergalactic medium, together with accretion processes

A review and assessment of microprocesses in plasmas

- Hierarchies of plasma instabilities
- Non-local, non-diffusive transport processes on the scales of laboratory, space and astrophysical plasmas
- Ionisation and radiation processes

Magnetic reconnection

- Collisionless reconnection: conceptual problems and solutions
- Magnetohydrodynamic reconnection
- Experimental magnetic reconnection in laboratory plasmas
- Reconnection in solar system plasmas including magnetospheres
- The role of magnetic reconnection in astrophysical plasmas

Shock waves in cosmic plasmas

- Plasma kinetics of shocks
- 3D structures and shock reformation
- Interaction between turbulence and nonlinear structures and shocks
- Electron and ion heating at shocks
- Relativistic shocks

Techniques of plasma description

- Remote sensing of astrophysical plasmas
- Lessons from laser-plasma interactions and from magnetically confined laboratory plasmas

The 23 review papers in this volume cover these topics, providing a comprehensive and authoritative account of space plasma processes on all scales. The Editors are grateful for the extensive scientific interactions among the participants in this Workshop, which have led to the collaborations represented by the joint author teams for these reviews. Thanks are also due to the referees for their painstaking efforts that have contributed to the high quality of the papers. The Editors also thank the editorial and production staff of Space Science Reviews for their sensitive and patient work.

It is impossible to overemphasise the value and significance of ISSI's role in promoting and supporting the important task of reviewing and taking stock of key topics in space sciences. Acknowledgements are due for the Institute's role in the present example of ISSI's continuing task. In particular, we thank for their support: Roger-Maurice Bonnet, Executive

Director during the gestation of the workshops project on Cosmic Plasmas; the current Executive Director Rafael Rodrigo; and his fellow director Rudolf von Steiger. We are grateful also for the indispensible and always good-humoured support of the ISSI staff: in alphabetical order, Maurizio Falanga, Andrea Fischer, Saliba Saliba, Irmela Schweitzer, Silvia Wenger, Jennifer Zaugg and Danielle Zemp. Without their dedication and professionalism, space science would be definitely poorer.

DOI 10.1007/978-1-4899-7413-6_2
Reprinted from *Space Science Reviews* Journal, DOI 10.1007/s11214-013-9975-8

Microphysics in Astrophysical Plasmas

Steven J. Schwartz · Ellen G. Zweibel · Martin Goldman

Received: 25 January 2013 / Accepted: 8 March 2013 / Published online: 18 May 2013

Abstract Although macroscale features dominate astrophysical images and energetics, the physics is controlled through microscale transport processes (conduction, diffusion) that mediate the flow of mass, momentum, energy, and charge. These microphysical processes manifest themselves in key (all) boundary layers and also operate within the body of the plasma. Crucially, most plasmas of interest are rarefied to the extent that classical particle collision length- and time-scales are long. Collective plasma kinetic phenomena then serve to scatter or otherwise modify the particle distribution functions and in so-doing govern the transport at the microscale level. Thus collisionless plasmas are capable of supporting thin shocks, current sheets which may be prone to magnetic reconnection, and the dissipation of turbulence cascades at kinetic scales. This paper lays the foundation for the accompanying collection that explores the current state of knowledge in this subject. The richness of plasma kinetic phenomena brings with it a rich diversity of microphysics that does not always, if ever, simply mimic classical collision-dominated transport. This can couple the macro- and microscale physics in profound ways, and in ways which thus depend on the astrophysical context.

Keywords Microphysics · Plasmas · Astrophysics · Space plasmas

S.J. Schwartz (✉)
Blackett Laboratory, Imperial College London, South Kensington, London SW7 2AZ, UK
e-mail: s.schwartz@imperial.ac.uk

E.G. Zweibel
Astronomy Department, University of Wisconsin, Madison, WI 53706, USA
e-mail: zweibel@astro.wisc.edu

M. Goldman
Department of Physics, University of Colorado, Boulder, CO 80309-0390, USA
e-mail: Martin.Goldman@Colorado.EDU

1 Introduction

The astrophysical world is filled with plasmas, from the solar atmosphere through supernova remnants to distant galaxies. Despite these diverse environments, there are common underlying physical mechanisms at work. Shock waves form at flow interaction regions, current layers breakdown to release bottled-up magnetic energy, and turbulence tangles magnetic fields and cascades energy to small scales where it is dissipated.

In the classical view, these and many more phenomena are controlled by transport processes (diffusion, conduction, etc.) that are mediated by inter-particle collisions. The resulting collision frequencies and transport coefficients are then used in a fluid approach to the physics. In such an approach, these coefficients depend only on the local state parameters (e.g., density, temperature) independent of the large-scale region of interest. In this view, the media never stray far from Maxwellian thermodynamic equilibrium. If we lived in a fluid Universe, there would be no solar flares, no ultra-relativistic cosmic rays, and no Aurora Borealis.

However, most astrophysical plasmas are too rarefied for binary Coulomb collisions to be effective. That is, the characteristic length- and time-scales are too short for the infrequent collisions to maintain thermodynamic equilibrium and control the transport processes. Instead, the collective interaction of the plasma particles and fields results in a range of phenomena that dominate the microphysics. The absence of collisions opens up the possibility for some particles to be accelerated to become the highest energy cosmic rays while their neighbours participate in a less spectacular background plasma.

Thus the collective interactions operate selectively on the plasma particles, depending on their species, their location in velocity space (e.g., through resonant wave-particle interactions), or the pre-existence of fluctuations in electromagnetic fields. None of these are necessarily prescribed by the gross macroscopic conditions such as density or temperature or their gradients, and so the resulting description of the transport processes probably bears little resemblance to the classical collision-dominated one.

Early attempts to attribute the action of wave-particle interactions as some "anomalous resistivity η," for example Papadopoulos (1977), in which

$$\mathbf{j} = \mathbf{E}/\eta$$

while convenient are probably rarely accurate. In the first instance, even if this linear relationship holds, η could be a function of details in the particle distribution functions $f_i(\mathbf{v})$ and not simply density and temperature. And secondly, the nonlinear relationship between current and electric field could involve the global context including the total current, the contribution of particles traversing macroscopic scales to the local population, or DC electric or magnetic fields unrelated to the current which nonetheless influence the collective behaviour.

Because there is no single, general description, these non-classical forms are best illustrated through specific examples. In this paper, we will describe the typical applications to shock waves, magnetic reconnection, and turbulence. Those applications will be developed in the accompanying papers in this volume. We will see that some aspects of the microphysics are very specific to the parameter regimes involved while others are more universal, at least qualitatively. Accordingly, the juxtaposition of laboratory, solar system, and astrophysical applications should lead to important cross-fertilisation. We will also see that the microphysics couples into the macrophysics in profound ways that have no counterparts in classical transport theory.

2 Common Themes

Before discussing specific applications, we lay out a few common themes. These illustrate the origins of the departure from classical transport and some key features to consider in any context.

2.1 Plasma Hierarchy

It is instructive to re-examine the basis of transport and dissipation in a plasma. We begin with the Boltzmann equation for the one-particle distribution function $f_i(\mathbf{x}, \mathbf{v}, t)$ for a species i:

$$\frac{\partial f_i}{\partial t} + \mathbf{v} \cdot \frac{\partial f_i}{\partial \mathbf{x}} + \mathbf{a} \cdot \frac{\partial f_i}{\partial \mathbf{v}} = \left(\frac{\partial f_i}{\partial t} \right)_C \tag{1}$$

where the right hand side is a placeholder for collisions and other processes which are not represented within the particle acceleration **a** and which may alter the phase space density in an instantaneous or discontinuous manner not representable in this differential formulation. Such processes include interactions with other species, ionisation, etc.

Taking velocity-space moments of (1) results in an infinite hierarchy of equations for the density, velocity, pressure tensor, heat flux, ..., associated with the species. For example, multiplying by $m_i \mathbf{v}$ and integrating over velocities yields the momentum equation:

$$\frac{\partial (\rho_i \mathbf{V}_i)}{\partial t} + \nabla \cdot \underline{\underline{\mathsf{P}}}_i - \mathbf{F}_i = \left\langle m\mathbf{v} \left(\frac{\partial f_i}{\partial t} \right)_C \right\rangle \tag{2}$$

One obvious problem is that this requires knowledge of the next higher moment (the stress tensor $\underline{\underline{\mathsf{P}}}_i$) so that ultimately some closure assumption is required. The microphysical processes responsible for momentum transport manifest themselves in two terms of (2). Firstly, the right hand side contains all the interactions with other species, collisions, etc. This gives rise to the normal collisional coupling between species, and the viscosity, when particular forms of the collision operator are employed, as in the case of near-equilibrium collisional plasmas (Braginskii 1965).

Additionally, the force density $\mathbf{F}_i$ on the species as a whole can be thought of as arising from two contributions. One is derived from the acceleration **a** in (1) due to macroscopic, quasi-steady fields and particle distributions. The second is the correlation of any fluctuating acceleration $\delta\mathbf{a}$ with fluctuations δf_i. This nonlinear convolution has a non-zero average. In instability analyses this captures the nonlinear wave-particle interaction contributions to the momentum transport. Similar correlations appear when the stress-tensor $\underline{\underline{\mathsf{P}}}_i$ is recast in terms of the mean velocity (Che et al. 2011). There is no reason, of course, for these nonlinear contributions, even when spatially averaged or smoothed, to resemble the viscous or collisional species coupling terms. Wave-particle interactions act differently upon, e.g., resonant and non-resonant particles, shaping f_i in ways that cannot be determined by the macroscopic fluid parameters but which nonetheless contribute to the transport of, in this example, the momentum.

Another fundamental aspect of the microphysics is revealed in the $mv^2/2$ moment of (1) which yields an energy equation. In addition to dissipative processes which heat the species and clearly depend on the microphysics, the hierarchy of moment equations is often truncated at this level by some closure assumption, e.g., that the heat flux is zero, which turns this moment equation into an equation of state of the form

$$\left(\frac{\partial}{\partial t} + \mathbf{V} \cdot \frac{\partial}{\partial \mathbf{x}} \right) \left(P_i \rho_i^{-\gamma} \right) = \text{dissipative term} \tag{3}$$

where the term on the right hand side can be identified with the right hand side of (1) together with any nonlinear response to fluctuations in the plasma and fields. Now much of the unknown microphysics is collapsed into the unknown, and presumed constant, adiabatic index γ reflecting the equation of state for species i or, in the case of single fluid descriptions, for the plasma as a whole. This is a convenient but rarely appropriate prescription for the influence of the microphysics in transporting energy. Similar concerns arise in the derivation of the double-adiabatic equations for an anisotropic plasma (Chew et al. 1956), which rely on macroscopic conditions related to the first and second adiabatic invariants of particles in a magnetic field; conditions which are rarely realised in practise.

2.2 Coupling of Scales

All dynamical media involve disparate scales which couple to one another (Schwartz et al. 2009a, 2009b). In viscous fluids, the thin boundary layers are driven by, and strongly influence, the macroscopic flow field. Collisionless plasmas contain many more scales, ranging from the macroscopic context through energetic particle scales to thermal ion and electron scales. Additionally, there are magnetic scales linked to particle gyromotion and electric scales linked to charge separation, collisionless inertial lengths or skin depths (e.g., $c/\omega_{pi,e}$ where $\omega_{pi,e}$ is the plasma frequency), and Debye lengths. Fluctuations and turbulence can lead to broad probability distribution functions of pitch angle scattering times and lengths. These imply a nonlocal random walk, which can lead to nondiffusive transport.

Small-scale breakdown of thin current layers in the solar atmosphere, for example, can lead to the eruption of solar flares and the reconfiguration of the magnetic field topology over vast distances. Electromagnetic fields confined to thin layers at a collisionless shock can inject energetic particles into the unshocked medium where locally-driven turbulence can scatter and further accelerate those particles, leading to a feedback process by which large-scale shocks are mediated by self-generated cosmic rays that account for a significant fraction of the shock energetics.

2.3 Micro-instabilities

The non-equilibrium aspects of collisionless plasmas opens up the possibility for differential flows (currents or beams), temperature anisotropies in which the kinetic temperature along and perpendicular to the magnetic field, say, are different, and other kinetic features (e.g., “bumps on tail”, loss cones, ring distributions). If the removal of such features would lead to a lower energy state, they represent sources of “free energy” capable of driving plasma micro-instabilities (Gary 2005). This is another example of scale coupling, since the free energy is usually driven by macroscopic interactions or sources of some kind.

In a micro-unstable plasma, particles in resonant regions of velocity-space coherently interact with electrostatic or electromagnetic perturbations which grow at the expense of the resonant particles. In the nonlinear stages one expectation is that the unstable features (beam, anisotropy or whatever) will diminish toward marginal instability, with the free energy released residing in the fluctuating fields and non-resonant populations. Eventually that fluctuation energy would damp heating the plasma. Alternatively, nonlinear wave cascades can remove fluctuation energy before it can modify the free-energy regions of phase space; such a competition, as well as particle trapping, can yield a steady state that is not well-described as being marginally stable against the original free energy source.

Two heuristic scenarios are commonly invoked. In “anomalous transport” theory, the nonlinear interactions are cast into the form of a collision term with the collision frequency

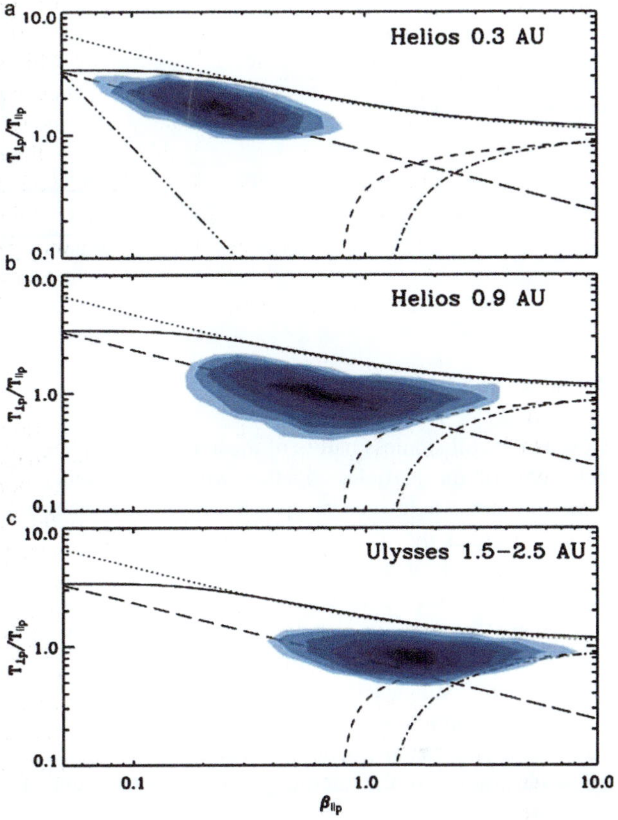

Fig. 1 Occurrence frequency (color scale) of solar wind plasma in proton temperature anisotropy ($T_\perp/T_\parallel$) vs. parallel plasma beta ($\beta_\parallel$) from Matteini et al. (2007). The *curves* show thresholds for small, positive growth rates near marginal stability for the ion-cyclotron/mirror mode (*upper curves*) and firehose instability (*lower curves*). The steeply falling *dash-dot-dot-dot* curve in (a) is the prediction based on conservation of particle magnetic moments. This figure suggests that these instabilities constrain the thermal properties of the expanding solar wind, which would otherwise be driven, e.g., to small anisotropy values through the action of adiabatic particle invariants

dependent on the intensity of the fluctuations. Thus, for example, the influence of an instability driven by an electric current is made to look in form like that due to classical resistivity, with the classical collision frequency replaced by the nonlinear wave-particle scattering rate. Similarly, the transport of streaming cosmic rays is influenced by effective wave-particle scattering arising from self-generated Alfvén waves due to the cosmic ray streaming instability (Skilling 1975), or by scattering due to pre-existing levels of turbulence.

Since kinetic instabilities grow on short, kinetic timescales, another approach when considering the larger scale consequences of micro-instabilities is to assume that the plasma can never be grossly unstable. In this case the plasma parameters (beam speed, anisotropy, or other free energy source) should be close to, or below, values corresponding to marginal stability. Figure 1 shows that the solar wind temperature anisotropy and plasma beta appear to be constrained within the near-marginally stable limits of the mirror and firehose instabilities (Matteini et al. 2007).

Such constraints can be employed directly in macroscopic models rather than attempts to predict the instability nonlinear wave intensities and corresponding effective collision frequencies. This prescription can only assume that when the macroscopic conditions bring the plasma into the stable regime the microphysical processes cease to operate. While both approaches to incorporate the action of micro-instabilities into a macroscopic description of the plasma are useful, both fall far short of a self-consistent approach to the micro-macroscale problem which remains largely unsolved.

Fig. 2 The formation of a DC electric field (and current) in the boundary between a flowing plasma and a vacuum magnetic field (after Cowley 1995 Fig. 1)

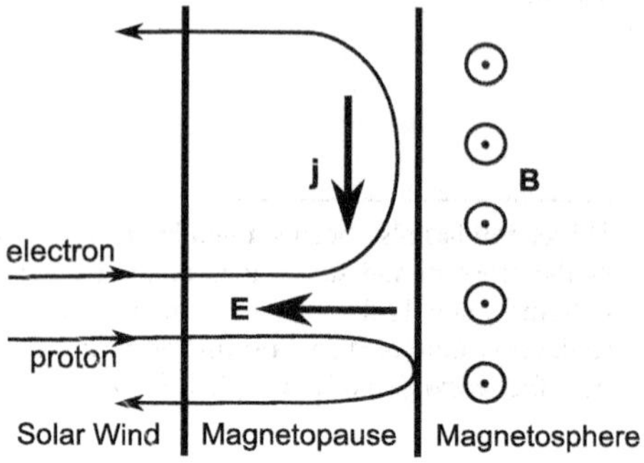

3 Shocks and Discontinuities

Despite the collisionless nature of most astrophysical plasmas, the collective self-consistent behaviour of the particles together with the influence of large-scale magnetic fields in inhibiting cross-field motion prevents wholesale intermingling of plasmas from different sources. Instead, the "Plasma Universe" (Alfvén 1986) is divided into cells of plasma separated by relatively sharp boundaries or discontinuities. Supersonic flows also drive nonlinear macroscopic shock waves, examples of which range from the relatively weak shocks driven by high speed streams or planetary obstacles in the solar wind flow (Schwartz 2006) through shocks driven by supernova explosions in the interstellar medium to shocks in galaxy clusters driven by extragalactic jets or large scale accretion flows.

These discontinuities are maintained through the role of microphysics in supporting the currents, DC electromagnetic fields, and transport processes within them. Here we provide a few illustrations of the microphysical processes found at such discontinuities.

3.1 DC Fields

It can be easy to overlook the DC electric fields at discontinuities and to underestimate the subtle microphysical processes and balances required to support them. Consider, for example, an unmagnetised plasma impinging transversely on a vacuum magnetic field as first studied by Chapman and Ferraro (1931) (see Cowley 1995 for a review) in the context of the Earth's magnetopause. While both ions and electrons will be turned around by the magnetic field, the ions will attempt to penetrate further owing to their larger momentum and hence gyroradius. That would lead to a large charge separation over the turnaround layer. Since the ion gyroradius is typically much larger than a Debye length, this charge separation must be inhibited through the establishment of an electric field which opposes the ion motion, enhances the electron penetration (and local gyroradii), and results in a layer of thickness c/ω_{pe} carrying an appreciable electric current as sketched in Fig. 2.

Similarly, collisionless shocks need to dissipate the incident kinetic energy through some microphysical process. Treumann (2009) provides a very comprehensive review of shock physics. Above a critical Mach number, which is only 2.8 or less depending on the shock parameters, the ions and electrons are heated by different mechanisms both of which rely on the DC fields within the shock layer. An electrostatic field is maintained by the electron temperature gradient. This, in concert with the magnetic field profile, reflects a fraction of the incident ions which gyrate back into the unshocked medium, but are convected back into and ultimately downstream. There they form a partial ring gyrating about the directly transmitted ions as sketched in Fig. 3 (see, e.g., Burgess 1995). This distribution of particles,

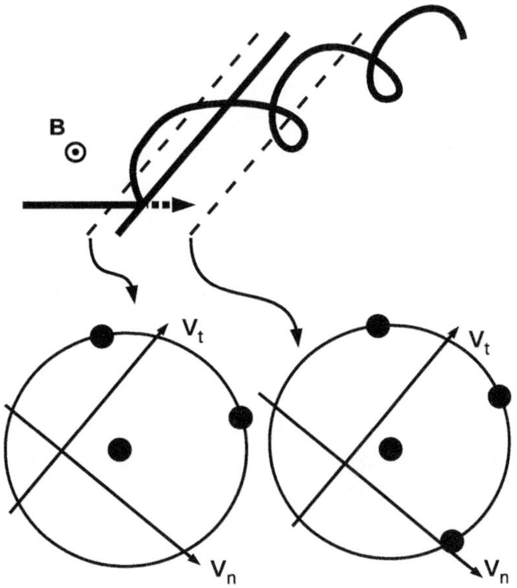

Fig. 3 Positional space (*top*) and velocity space (*bottom*) signatures of ions reflected at a shock surface (after Sckopke et al. 1983). The combination of directly transmitted and initially reflected particles downstream of the shock location results in a distribution that has slower net bulk flow speed and larger velocity spread (and hence kinetic temperature)

though far from equilibrium, has a lower bulk velocity and higher velocity spread than the incident plasma; that is, it has been slowed and "heated" entirely by the action of the DC fields without any scattering or collisions.

At the same time, this electric field causes the electron population to bifurcate into an incident population, which gains energy, and an escaping heated population, which loses energy during that escape and therefore originates downstream from a higher energy region in velocity space (Scudder et al. 1986). Thus the action of the DC field on the electrons results in a wide distribution downstream with a 'hole' at low energies. The consequence is that the DC fields are able to account for the inflation in velocity space expected through a heating mechanism.

Recently, Mitchell et al. (2012) investigated the added complication of electron heating at curved shocks, such as the bow shock formed by the impingement of the solar wind on the Earth's magnetosphere. Under collisionless conditions, electrons encountering the shock DC fields at one location traverse the region of shocked plasma to re-encounter the shock at a different location. This "cross-talk" connects portions of the curved shock at which the local conditions (e.g., Mach number, shock geometry) and hence total heating requirements are very different. Mitchell et al. found that the collisionless transport results in the entire shocked electron population being nearly isothermal, implying that the ion heating must vary greatly in order to balance the total energy budget around the shock surface. This work reveals a complex interplay amongst the DC fields (supported by the near-isothermal shocked electrons), local shock conditions, and global shock shape. In particular, for such shocks knowing the local shock parameters is insufficient to predict the energy partition amongst plasma species. Once again, there is strong coupling between different scales in the plasma.

3.2 Instabilities

In the shock example above, both the ion and electron distributions, as described, contain considerable free energy capable of driving micro-instabilities which would be expected to fill the electron 'hole' (together with electrons trapped downstream by the cross-shock electrostatic potential) and mix the transmitted/reflected ion components. Thus micro-instabilities finish the thermalisation process. See, e.g., Schwartz et al. (1996) and McKean et al. (1992) for discussion of ion waves and instabilities, of which the Alfvén ion cyclotron and mirror modes are the most commonly invoked. Electrons interact with pre-existing waves driven by the ion kinetics (e.g., the lower hybrid drift instability) or drive whistler or other electron-scale waves directly (e.g., Wu et al. 1984; Tsurutani et al. 1982; Masood et al. 2006).

There are many more examples of micro-instabilities associated with shocks and discontinuity layers. Some of these participate in acceleration processes.

3.3 Acceleration

While microinstabilities can thermalize the core of the particle distribution function, they can also extract a high energy tail. Galactic cosmic rays, which represent 10^{-9}–10^{-10} of interstellar particles but carry about as much energy as the thermal gas, dramatically exemplify this. Although the association between supernovae and cosmic rays had long been suspected, a viable mechanism was not suggested until the 1970's, when Bell (1978) and Blandford and Ostriker (1978) independently developed a theory for first order Fermi acceleration of cosmic rays by strong interstellar shocks driven by supernovae. In the original theory, the shock was assumed to be quasi-parallel, and was idealized as a discontinuity with the upstream and downstream flow properties connected by the Rankine-Hugoniot relations. In the shock frame, unshocked fluid streams into the shock at speed V_s and streams out at speed V_s/R, where R is the compression ratio of the shock. A particle with momentum p traveling upstream from the shock gains energy $2pV_s$ if scattered back downstream, and loses energy $2pV_s/R$ if scattered back upstream, resulting in a net energy gain of $2pV_s(1-1/R)$ per loop. The resulting spectrum in momentum space is a power law; $p^{-3R/(R-1)}$. For strong non-relativistic shocks, $R \sim 4$ leading to a power law $\propto p^{-4}$. Propagation through the interstellar medium steepens this spectrum, bringing it closer to the observed $p^{-4.7}$ spectrum.

Electromagnetic fluctuations generated by microinstabilities play two roles in this scenario: scattering particles back and forth across the shock, and keeping the distribution of particle pitch angles nearly isotropic. The fluctuations are thought to be Alfvén waves which interact with the cosmic rays through gyroresonance, and their primary source is thought to be an instability driven by super-Alfvénic streaming of the cosmic rays themselves (Wentzel 1968; Kulsrud and Pearce 1969). Later it was realized that including the momentum and energy of cosmic rays, the stresses on the fluid exerted by the waves (Dewar 1970), and the energy deposited in the fluid by wave dissipation modifies the Rankine-Hugoniot relations and broadens the shock by creating a cosmic ray precursor (Voelk et al. 1984). When these effects, and the increase in scattering mean free path with cosmic ray energy (which causes higher energy particles to "see" a larger velocity jump and gain more energy per loop), are accounted for, the power law expected from the simple theory is replaced by a more complicated distribution (Blasi 2012; Reville et al. 2009).

This Fermi acceleration process has been observed, though at non-relativistic energies, in situ e.g., at the Earth's bow shock. Figure 4 shows an experimentally determined upstream e-folding distance (related to the wave-scattering mean free path) as a function of particle energy as deduced by Kis et al. (2004). A long-standing feature of this theory is the requirement to "inject" suprathermal particles into this accelerator. It would appear, at least for low Mach number shocks, that processes within the shock layer are able to select, energise, and inject particles directly from the incident thermal population, as shown in Fig. 5. In astrophysical environs, any pre-existing suprathermal population would similarly inject particles into the Fermi process. Injected particles must be able to travel upstream against the incident flow to participate in the acceleration mechanism.

The gyroresonant streaming instability dominates when the cosmic ray energy density U_{cr}, background magnetic field energy density U_B, and cosmic ray drift speed v_D satisfy the inequality $U_{cr}/U_B < c/v_D$. When the inequality is reversed, a nonresonant instability driven by the thermal electron return current grows faster and drives electromagnetic fluctuations which differ significantly from Alfvén waves (Zweibel 2003; Bell 2004). PIC simulations of the instability suggest that it amplifies the magnetic field significantly, increasing the rate of

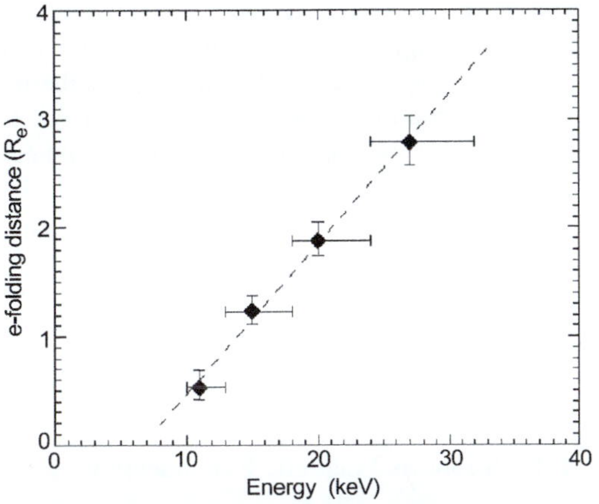

Fig. 4 Observationally determined exponential scale for Fermi-accelerated ions at the Earth's bow shock, from Kis et al. (2004)

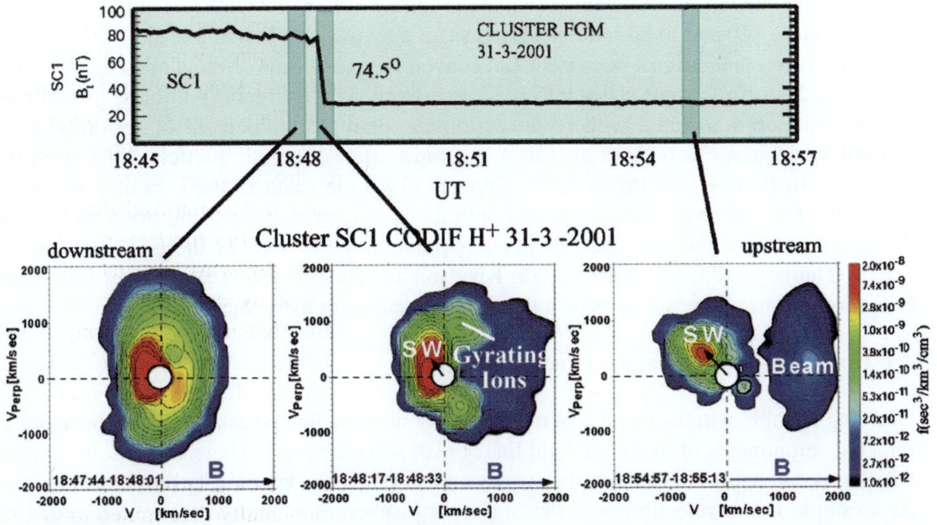

Fig. 5 Formation of ion beams via coherent process at the Earth's bow shock, from Kucharek et al. (2004). Note that the beam occupies regions of velocity space that are empty in the downstream region, thus establishing that the beam must originate at the shock itself. This is at least a partial answer to the "injection problem" for cosmic rays

shock acceleration and the maximum energy to which it can accelerate particles (Riquelme and Spitkovsky 2010; Ohira et al. 2009).

The cosmic ray acceleration picture cannot be tested with *in situ* probes the way solar system acceleration can, but there are predictions that can be checked remotely. Among them are the cosmic ray spectrum itself, the modification of the Rankine-Hugoniot relations (especially the postshock temperatures), the magnetic field strengths, and the cosmic ray energy densities and maximum energies in the shock itself.

Electron acceleration (as opposed to heating) is harder due to lack of efficient wave-particle scattering and more rapid energy losses. These effects must come into play in Galactic cosmic ray acceleration; electrons comprise only 1–2 % of primary cosmic rays. It is important to be able to estimate the relative efficiencies of cosmic ray electron and ion acceleration, as the electron component is far easier to detect remotely through its synchrotron and inverse Compton emission. There is evidence for coherent reflection at shocks which can lead to significant one-step energisation (Wu 1984; Leroy and Mangeney 1984). These processes exploit the fact that the electron thermal distribution is much wider in velocity due to the electron mass. As a result, simple magnetic mirroring in the frame in which the shock is at rest and the incident flow is field-aligned (removing the $-\mathbf{V} \times \mathbf{B}$ electric field), the deHoffmann-Teller frame, yields an energetic beam with appreciable density. There are also recent reports of Fermi-accelerated relativistic electrons at the Saturn's bow shock (Masters et al. 2013). Within the solar system, the Fermi process is limited by the finite size of planetary bow shocks which results in particle escape. Thus the relatively large size of the Saturnian bow shock is more favorable for acceleration to higher energies given appropriate solar wind conditions.

4 Collisionless Magnetic Reconnection

Magnetic reconnection is a ubiquitous energetic plasma process commonly found in planetary magnetospheres, in the solar corona, in the solar wind, in astrophysical objects (Zweibel and Yamada 2009) and in laboratory plasmas (for a review, see Yamada et al. 2010). Reconnection is of special interest because it can convert large amounts of magnetic energy stored on both sides of a current sheet into particle energy. The energetic particles, instabilities and/or radiation associated with reconnection can disrupt confinement of fusion plasmas in toroidal devices (Yamada et al. 1994) and can initiate energetic particle flows from the Sun to Earth (Forbes and Priest 1995). Dungey (1961) introduced the idea that reconnection at the Earth's magnetopause (and within the geomagnetic tail) could drive the aurora. We now know that the energised particles populate large regions of the Earth's magnetosphere (Dungey 1995; Vasyliunas 1975; Kivelson and Russell 1995) where they may affect communications and possibly pose a threat to spacecraft or astronauts.

4.1 How Does Topological Reconnection Begin?

One major problem in the physics of magnetic reconnection is to understand how oppositely-directed components of magnetic field lines break and reconnect. There are various ways in which this can happen. In dense plasmas reconnection can be enabled by particle collisions. An example is reconnection near the Sun's photosphere. Collisional reconnection can be modeled by *resistive* MHD (but not ideal MHD). However in Earth's magnetopause and magnetotail, and in many planetary and astrophysical environments, the plasma is collisionless, entropy is conserved, and modeling reconnection often requires *kinetic simulations*, such as *Particle In Cell (PIC)* simulations.

Collisionless reconnection in Earth's magnetopause is *asymmetric*—with different ambient conditions on the two sides of the current sheet. In Earth's magnetotail it is generally *symmetric* in the two lobes on either side of the plasmasheet. A simple initial condition often employed in kinetic simulations of symmetric reconnection consists of a uniform thin current sheet separating oppositely-directed straight magnetic field lines which asymptote to a constant magnetic field, $\mathbf{B}_0$, far from the current sheet. A common example is the kinetic Harris equilibrium (Harris 1962; Yamada et al. 2000), with 1-D spatial variation in the direction orthogonal to the current sheet plane (e.g., the plasmasheet in the magnetotail). A uniform out-of-plane guide field, $\mathbf{B}_g$, and uniform background (lobe) populations may be added without affecting the self-consistency of the equilibrium.

When the current sheet thickness is on the order of the ion skin depth (c/ω_i) or less, the Harris equilibrium is unstable to spontaneous reconnection through collisionless tearing instabilities (Drake and Lee 1977) that create x-points. The magnetic tension in newly reconnected flux tubes makes them snap away from the x-point towards the outflow. The out-of-plane induction (reconnection) electric field causes inflow of unreconnected field lines which are topologically converted into reconnected field lines at the x-point.

Tearing instabilities are usually very slow (growth rate on the order of 100's of ion inverse cyclotron frequencies, Ω_{ci}^{-1}). Kinetic simulations of spontaneous reconnection initiated solely by this instability have been performed (Pritchett 2005), but they can be time-consuming. To speed things up, simulations of kinetic reconnection are usually *driven*, either by an initial perturbation which effectively introduces a very small local magnetic field across the thin current sheet or by an $\mathbf{E} \times \mathbf{B}$ drift in the inflow, which thins the current sheet and speeds up the tearing. As an alternative to Harris sheet initialization, force free equilibrium current sheets are sometimes employed (Drake et al. 2003). The initial density can then be uniform because pressure is not necessary to balance magnetic forces. Another strategy for simulating reconnection is to begin with an already-reconnected field-line configuration (Pritchett 2007). Still another is to employ kinetic simulations with open boundary conditions allowing boundary inflow of magnetic flux and boundary outflow of reconnected magnetic flux; this can lead to steady state reconnection (Daughton et al. 2006). The rate of *steady state reconnection* as dictated by geometric properties of the current sheet were analysed early-on (Sweet 1958; Parker 1957; Petschek 1964).

4.2 Hall Reconnection

A major advance in modeling collisionless reconnection using MHD and fluid models occurred when it was shown that the Hall term in the generalized Ohm's law was essential to expedite the breaking of field lines (i.e., the breaking of the frozen-in condition in the absence of collisions). This term effectively separates electron physics near the x-point from ion physics further away, as sketched in Fig. 6. Work is performed on electrons in the small so-called electron diffusion region, where the electrons are not frozen-in. The electron diffusion region extends from the x-point to 10's of electron inertial lengths, c/ω_e, downstream on either side of the x-point. Beyond the electron diffusion region the electrons can be frozen-in but the ions are not, forming the so-called ion diffusion region, which is many ion inertial lengths long. Parallel electron currents that form around the separatrices due to flux-tube-widening (Uzdensky and Kulsrud 2006) act as Hall currents which produce a quadrupolar Hall magnetic field, $\mathbf{B}_H$, and a Hall electric field, $\mathbf{E}_H$, orthogonal to both the $\mathbf{B}_H$ and the Hall current. The necessity of including Hall physics has led to the development of Hall MHD (Shay et al. 1999;

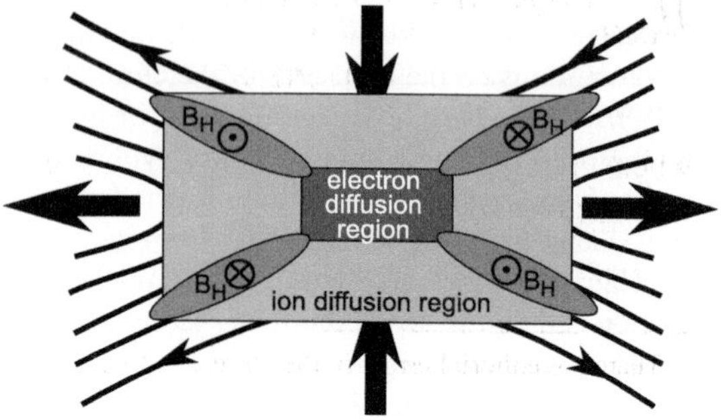

Fig. 6 Current systems and fields found in the Hall reconnection scenario

Huba 2005) and to the use of two-fluid and hybrid models (Karimabadi et al. 2004) in reconnection simulations. However it is important to realize that Hall processes are *always automatically included in kinetic simulations* of reconnection.

4.3 Kinetic Processes in Reconnection

PIC simulations and satellite measurements have also revealed a number of intrinsically kinetic processes that occur after reconnection has begun. Electron two-stream instabilities form near the separatrices that separate incoming flux tubes from outflowing reconnected flux tubes. These instabilities saturate by trapping the electron streams, thereby forming electron phase space holes propagating towards and away from the x-point. Electron holes have been seen in reconnection simulations (Lapenta et al. 2011) and observed in the tail through their associated bipolar electric fields (Cattell et al. 2002, 2005). Another (different) kind of trapping is associated with electron bounce motion due to magnetic mirror points and bipolar electric fields along flux tubes in the inflow and near-outflow. Analysis of these motions together with temperature anisotropy considerations have led to useful equations of state in the presence of a guide field (Le et al. 2009). There is still another electric field—the electrostatic Hall electric field, which is typically as large as 100 mV/m in the tail. This electric field has been shown to accelerate ions to multi-keV energies in the direction perpendicular to magnetic field lines (Wygant et al. 2005).

Kinetic Alfvén wave physics has been invoked in explaining the fast outflowing Poynting flux corresponding to the Hall electric and magnetic fields (Shay et al. 2011). Electron whistler waves have been identified radiating from the separatrices into the inflowing plasma (Goldman et al. 2012). Efficient electron acceleration at physical ion to electron mass ratios has been studied in kinetic simulations by Ricci et al. (2003), and has been associated with the magnetic stress in contracting curved flux tubes (magnetic islands) (Drake et al. 2006).

4.4 Guide Field Effects

Magnetic guide fields, B_g, can be as large as or larger than B_0 in Earth's magnetopause. Together with the density gradient across the magnetopause current sheet they can produce diamagnetic drifts that can move x-points in the outflow direction and even prevent reconnection (Swisdak et al. 2003). Che et al. (2011) have shown that the thin current sheet of background electrons which forms after the initial Harris current sheet is torn can be unstable to a shear instability in the presence of a moderate B_g. The instability creates a right-circularly polarized electromagnetic wave along $\mathbf{B}_g$ that saturates by thickening the current sheet, thereby removing its shear free-energy. The unstable waves lead to anomalous viscosity which can speed up reconnection.

Guide fields of order B_0 or less can change the flow of electrons along the separatrices, producing a higher density outflow along one axis of the separatrix and a lower density inflow on the other axis, where electron holes become more robust at early times. Even weak magnetic guide fields can have important effects in Earth's magnetotail. A guide field of $B_g = 0.1\, B_0$ or less can still create a strong asymmetry in the Hall quadrupole magnetic field (Eastwood et al. 2010). Such a weak guide field can also deflect towards the separatrix the outflowing collimated *electron jets* found in antiparallel reconnection simulations (i.e., with $B_g = 0$) (Goldman et al. 2011). Simulations of reconnection with small guide field are quite sensitive to the assumed ion to electron mass ratio. Guide field jet deflections and other effects are much more pronounced when the mass ratio is physical (e.g. 1836) (Goldman et al. 2011).

4.5 Flux Pile-up and Dipolarization Processes Associated with Collisionless Reconnection

PIC simulations of magnetotail reconnection initiated with a Harris sheet representing the plasmasheet and lower density background particles representing the lobes can display *pileup* of reconnected field lines as well as other features of measured dipolarization events (Runov et al. 2009). In the simplest case a single x-point arises from an initial perturbation in the simulation. As the tension in the reconnected field lines splits open the Harris sheet, leaving behind low density background plasma, a strong pressure force develops at the boundary of the opening higher density Harris sheet. This pressure force opposes the magnetic stress, thereby causing a pileup of reconnected field lines moving with the outflow as the Harris sheet continues to open. This is (roughly) a moving dipolarization front, although dipolarization physics effects are missing from this simple model (e.g., the shape, extent and motions of the plasmasheet, the connectivity of the magnetic field to Earth's magnetic field, etc.). Dipolarization fronts have been shown to be an important mechanism for energizing the plasmasheet (Hamrin et al. 2012). Simulations have shown that the moving pileup front is a critical boundary that separates the ion-diffusion region from the plasmasheet (Goldman et al. 2012).

4.6 Multiple Islands, x-Points and Flux Tubes in Collisionless Reconnection

Measurements in the solar wind have provided evidence for very long x-lines out of the reconnection plane (Phan et al. 2006). In driven or long-time PIC simulations multiple x-points can develop. In two-dimensions these are separated by multiple islands (centered on O-points). Island chains have been studied in simulations (Markidis et al. 2012) and have been found in the solar wind and elsewhere. Multiple islands can be found either on the thin background electron current sheets which form during reconnection or along the separatrix of a primary island. Multiple islands can be unstable to *merging instabilities* (Pritchett 2007). In pioneering recent massive 3D PIC simulations the islands are found to extend out of the reconnection plane. There they become wiggling flux tubes that can touch each other at different out-of-plane points and even produce secondary reconnections where they touch (Daughton et al. 2011).

5 Turbulence

Turbulence is the nonlinear transfer of energy in fluctuations from one scale to another. Typically that transfer proceeds from large scales, where macroscopic motion and interactions

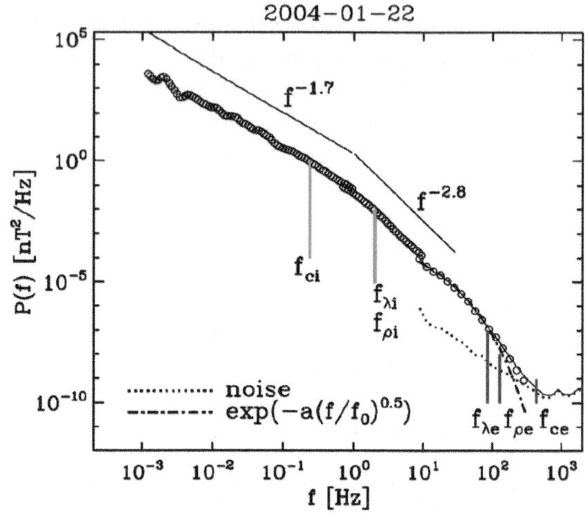

Fig. 7 Magnetic power spectrum of solar wind turbulence, illustrating the inertial regime in which the power falls off as $f^{-5/3}$ and successive steepenings at ion and electron scales (from Alexandrova et al. 2009)

"stir" the medium, toward small scales where eventually the energy is dissipated. Given our earlier discussions about dissipative processes in collisionless plasmas being mediated by microphysics, it should come as no surprise that plasma microphysics is expected to be involved in the turbulent cascade.

However, two aspects complicate plasma turbulence. The first is that there is more than one kinetic scale, and hence one would expect to see signatures of those scales and to examine the contribution each might make to the energy dissipation. Figure 7 shows a typical turbulence spectrum measured in the solar wind (Alexandrova et al. 2009; Sahraoui et al. 2009). Note the Kolmogorov-like $-5/3$ spectral slope at lower frequencies. This corresponds to the "inertial range" where energy is cascaded to shorter scales (higher frequencies) without loss in a self-similar fashion. The spectrum steepens beyond frequencies corresponding to typical ion scales and is believed to steepen further or roll-over at electron scales.

The second aspect of the dissipation range of collisionless plasmas is more subtle. Kinetic plasma waves are dispersive, unlike acoustic waves in a fluid. Thus the nonlinear interaction of two waves of similar frequencies will not, in general, lead to a mode at the sum of those frequencies that is a normal mode of the system. The orderly transfer of energy through interactions that are local in frequency (or wave-vector) space which features in the inertial range cannot operate. The nature of the turbulence at these kinetic scales, the local or non-local nature of cascaded energy transfer, and the actual dissipation mechanism(s) for that energy are all areas of ongoing research.

One possibility is that the turbulence generates coherent current structures at small scales. These structures, identified through the intermittency properties of the turbulence (Osman et al. 2011), then dissipate through local current-driven instabilities or reconnection. Observations in the solar wind (Phan et al. 2006) and in the turbulent magnetosheath behind the terrestrial bow shock (Retinò et al. 2007; Sundkvist et al. 2007) all provide direct evidence for reconnection events embedded in a turbulent plasma. These events are weak compared to the violent events responsible for solar flares and geomagnetic storms, but may play important roles in plasma heating.

On the larger scale, there must be a fast reconnection mechanism to remove small-scale tangles in the interstellar magnetic field so that the large scale component dominates, as observed (Kulsrud and Zweibel 2008). Possibilities include formation of tiny scale kinetic structures in an overall turbulent background (Servidio et al. 2011), enhancement of the reconnection rate by small scale turbulence in large scale current sheets (Lazarian and Vishniac 1999), breakup of current sheets through instabilities (Loureiro et al. 2007; Huang and Bhattacharjee 2013), and formation of thin current sheets by ion-neutral friction (Brandenburg and Zweibel 1994).

6 Concluding Remarks

In this paper we have given an overview of the role microphysical processes play in both solar system and astrophysical plasmas; although we have not discussed laboratory plasmas directly, under collisionless regimes the same remarks would and do apply there. Such processes arise due to the nonequilibrium nature of collisionless media. Although microphysics then controls the transport properties, energy partition (heating, acceleration, etc.), and hence effective "equation of state," the consequences of microphysical control cannot often be captured by retaining the mathematical form of collision-dominated conduction or diffusion. Instead, microphysics acts on discrete portions of velocity space and participates in physics that couples the largest scales to the smallest ones.

The papers that follow in this special edition explore specific applications of microphysics to collisionless plasmas. Many of them also attempt to extrapolate lessons learned in one regime or application to another. Reference to the underlying principles in the present paper will prove helpful in setting those papers in context.

Acknowledgements SJS and EGZ are grateful for the hospitality provided by ISSI during the workshop from which this special issue emerged. EGZ acknowledges support by NSF Grants NSF PHY-0821899 and NSF AST-0909900 to the University of Wisconsin, and the support of CMSO. SJS acknowledges the support of the UK STFC.

References

O. Alexandrova, J. Saur, C. Lacombe, A. Mangeney, J. Mitchell, S.J. Schwartz, P. Robert, Universality of solar-wind turbulent spectrum from MHD to electron scales. Phys. Rev. Lett. **103**(16), 165003 (2009). doi:10.1103/PhysRevLett.103.165003

H. Alfvén, Plasma Universe, in *Space Technology Plasma Issues in 2001*, ed. by H. Garrett, J. Feynman, S. Gabriel (1986), pp. 373–407

A.R. Bell, The acceleration of cosmic rays in shock fronts. II. Mon. Not. R. Astron. Soc. **182**, 443–455 (1978)

A.R. Bell, Turbulent amplification of magnetic field and diffusive shock acceleration of cosmic rays. Mon. Not. R. Astron. Soc. **353**, 550–558 (2004). doi:10.1111/j.1365-2966.2004.08097.x

R.D. Blandford, J.P. Ostriker, Particle acceleration by astrophysical shocks. Astrophys. J. Lett. **221**, 29–32 (1978). doi:10.1086/182658

P. Blasi, Origin of galactic cosmic rays, arXiv:1211.4799 (2012)

S. Braginskii, Transport processes in a plasma. Rev. Plasma Phys. **1**, 205 (1965)

A. Brandenburg, E.G. Zweibel, The formation of sharp structures by ambipolar diffusion. Astrophys. J. Lett. **427**, 91–94 (1994). doi:10.1086/187372

D. Burgess, Collisionless shocks, in *Introduction to Space Physics*, ed. by M.G. Kivelson and C.T. Russell (Cambridge University Press, Cambridge, 1995), pp. 129–163. Chap. 5

C. Cattell, J. Crumley, J. Dombeck, J.R. Wygant, F.S. Mozer, Polar observations of solitary waves at the Earth's magnetopause. Geophys. Res. Lett. **29**(5), 050000-1 (2002). doi:10.1029/2001GL014046

C. Cattell, J. Dombeck, J. Wygant, J.F. Drake, M. Swisdak, M.L. Goldstein, W. Keith, A. Fazakerley, M. André, E. Lucek, A. Balogh, Cluster observations of electron holes in association with magnetotail reconnection and comparison to simulations. J. Geophys. Res. **110**(A9), 1211 (2005). doi:10.1029/2004JA010519

S. Chapman, V.C.A. Ferraro, A new theory of magnetic storms. Terr. Magn. Atmos. Electr. **36**, 171 (1931). doi:10.1029/TE036i003p00171

H. Che, J.F. Drake, M. Swisdak, A current filamentation mechanism for breaking magnetic field lines during reconnection. Nature **474**, 184–187 (2011). doi:10.1038/nature10091

G.F. Chew, M.L. Goldberger, F.E. Low, The Boltzmann equation and the one-fluid hydromagnetic equations in the absence of particle collisions. Proc. R. Soc. Lond. Ser. A **236**, 112–118 (1956). doi:10.1098/rspa.1956.0116

S.W.H. Cowley, Theoretical perspectives of the magnetopause: a tutorial review, in *Physics of the Magnetopause*, ed. by P. Song, B.U.O. Sonnerup, M.F. Thomsen. Geophysical Monograph Series, vol. 90 (American Geophysical Union, Washington, 1995), p. 29

W. Daughton, J. Scudder, H. Karimabadi, Fully kinetic simulations of undriven magnetic reconnection with open boundary conditions. Phys. Plasmas **13**(7), 072101 (2006). doi:10.1063/1.2218817

W. Daughton, V. Roytershteyn, H. Karimabadi, L. Yin, B.J. Albright, B. Bergen, K.J. Bowers, Role of electron physics in the development of turbulent magnetic reconnection in collisionless plasmas. Nat. Phys. **7**, 539–542 (2011). doi:10.1038/nphys1965

R.L. Dewar, Interaction between hydromagnetic waves and a time-dependent, inhomogeneous medium. Phys. Fluids **13**, 2710–2720 (1970). doi:10.1063/1.1692854

J.F. Drake, Y.C. Lee, Kinetic theory of tearing instabilities. Phys. Fluids **20**, 1341–1353 (1977). doi:10.1063/1.862017

J.F. Drake, M. Swisdak, C. Cattell, M.A. Shay, B.N. Rogers, A. Zeiler, Formation of electron holes and particle energization during magnetic reconnection. Science **299**, 873–877 (2003). doi:10.1126/science.1080333

J.F. Drake, M. Swisdak, H. Che, M.A. Shay, Electron acceleration from contracting magnetic islands during reconnection. Nature **443**, 553–556 (2006). doi:10.1038/nature05116

J.W. Dungey, Interplanetary magnetic field and the auroral zones. Phys. Rev. Lett. **6**, 47–48 (1961). doi:10.1103/PhysRevLett.6.47

J.W. Dungey, Origins of the concept of reconnection and its application to the magnetopause: a historical view, in *Physics of the Magnetopause*, ed. by P. Song, B.U.O. Sonnerup, M.F. Thomsen. Geophysical Monograph Series, vol. 90 (American Geophysical Union, Washington, 1995), p. 17

J.P. Eastwood, M.A. Shay, T.D. Phan, M. Øieroset, Asymmetry of the ion diffusion region hall electric and magnetic fields during guide field reconnection: observations and comparison with simulations. Phys. Rev. Lett. **104**(20), 205001 (2010). doi:10.1103/PhysRevLett.104.205001

T.G. Forbes, E.R. Priest, Photospheric magnetic field evolution and eruptive flares. Astrophys. J. **446**, 377 (1995). doi:10.1086/175797

S.P. Gary, *Theory of Space Plasma Microinstabilities* (Cambridge University Press, Cambridge, 2005)

M.V. Goldman, G. Lapenta, D.L. Newman, S. Markidis, H. Che, Jet deflection by very weak guide fields during magnetic reconnection. Phys. Rev. Lett. **107**(13), 135001 (2011). doi:10.1103/PhysRevLett.107.135001

M.V. Goldman, D.L. Newman, G. Lapenta, S. Markidis, J.T. Gosling, L. Andersson, S. Eriksson, Particle and wave energy transport during magnetic reconnection, AGU Fall Meeting Abstracts, 24-05 (2012)

M. Hamrin, O. Marghitu, P. Norqvist, S. Buchert, M. André, B. Klecker, L.M. Kistler, I. Dandouras, The role of the inner tail to midtail plasma sheet in channeling solar wind power to the ionosphere. J. Geophys. Res. **117**(A16), 6310 (2012). doi:10.1029/2012JA017707

E. Harris, On a plasma sheath separating regions of oppositely directed magnetic field. Nuovo Cim. **23**(1), 115–121 (1962)

Y.-M. Huang, A. Bhattacharjee, Plasmoid instability in high-Lundquist-number magnetic reconnection, arXiv:1301.0331 (2013)

J.D. Huba, Hall magnetic reconnection: guide field dependence. Phys. Plasmas **12**(1), 012322 (2005). doi:10.1063/1.1834592

H. Karimabadi, D. Krauss-Varban, J.D. Huba, H.X. Vu, On magnetic reconnection regimes and associated three-dimensional asymmetries: hybrid, hall-less hybrid, and hall-MHD simulations. J. Geophys. Res. **109**(A18), 9205 (2004). doi:10.1029/2004JA010478

A. Kis, M. Scholer, B. Klecker, E. Möbius, E.A. Lucek, H. Rème, J.M. Bosqued, L.M. Kistler, H. Kucharek, Multi-spacecraft observations of diffuse ions upstream of Earth's bow shock. Geophys. Res. Lett. **31**, 20801 (2004). doi:10.1029/2004GL020759

M.G. Kivelson, C.T. Russell, *Introduction to Space Physics* (Cambridge University Press, Cambridge, 1995)

H. Kucharek, E. Möbius, M. Scholer, C. Mouikis, L. Kistler, T. Horbury, A. Balogh, H. Réme, J. Bosqued, On the origin of field-aligned beams at the quasi-perpendicular bow shock: multi-spacecraft observations by cluster. Ann. Geophys. **22**, 2301–2308 (2004). doi:10.5194/angeo-22-2301-2004

R. Kulsrud, W.P. Pearce, The effect of wave-particle interactions on the propagation of cosmic rays. Astrophys. J. **156**, 445 (1969). doi:10.1086/149981

R.M. Kulsrud, E.G. Zweibel, On the origin of cosmic magnetic fields. Rep. Prog. Phys. **71**(4), 046901 (2008). doi:10.1088/0034-4885/71/4/046901

G. Lapenta, S. Markidis, A. Divin, M.V. Goldman, D.L. Newman, Bipolar electric field signatures of reconnection separatrices for a hydrogen plasma at realistic guide fields. Geophys. Res. Lett. **38**, 17104 (2011). doi:10.1029/2011GL048572

A. Lazarian, E.T. Vishniac, Reconnection in a weakly stochastic field. Astrophys. J. **517**, 700–718 (1999). doi:10.1086/307233

A. Le, J. Egedal, W. Daughton, W. Fox, N. Katz, Equations of state for collisionless guide-field reconnection. Phys. Rev. Lett. **102**(8), 085001 (2009). doi:10.1103/PhysRevLett.102.085001

M.M. Leroy, A. Mangeney, A theory of energization of solar wind electrons by the Earth's bow shock. Ann. Geophys. **2**, 449–456 (1984)

N.F. Loureiro, A.A. Schekochihin, S.C. Cowley, Instability of current sheets and formation of plasmoid chains. Phys. Plasmas **14**(10), 100703 (2007). doi:10.1063/1.2783986

S. Markidis, P. Henri, G. Lapenta, A. Divin, M.V. Goldman, D. Newman, S. Eriksson, Collisionless magnetic reconnection in a plasmoid chain. Nonlinear Process. Geophys. **19**, 145–153 (2012). doi:10.5194/npg-19-145-2012

W. Masood, S.J. Schwartz, M. Maksimovic, A.N. Fazakerley, Electron velocity distribution and lion roars in the magnetosheath. Ann. Geophys. **24**, 1725–1735 (2006). doi:10.5194/angeo-24-1725-2006

A. Masters, L. Stawarz, M. Fujimoto, S.J. Schwartz, N. Sergis, M.F. Thomsen, A. Retinò, H. Hasegawa, B. Zieger, G.R. Lewis, A.J. Coates, P. Canu, M.K. Dougherty, Electron acceleration to relativistic energies at a strong quasi-parallel shock wave. Nat. Phys. **9**, 164–167 (2013). doi:10.1038/nphys2541

L. Matteini, S. Landi, P. Hellinger, F. Pantellini, M. Maksimovic, M. Velli, B.E. Goldstein, E. Marsch, Evolution of the solar wind proton temperature anisotropy from 0.3 to 2.5 AU. Geophys. Res. Lett. **34**, 20105 (2007). doi:10.1029/2007GL030920

M.E. McKean, D. Winske, S.P. Gary, Mirror and ion cyclotron anisotropy instabilities in the magnetosheath. J. Geophys. Res. **97**, 19421 (1992). doi:10.1029/92JA01842

J.J. Mitchell, S.J. Schwartz, U. Auster, Electron cross talk and asymmetric electron distributions near the earth's bowshock. Ann. Geophys. **30**, 503–513 (2012). doi:10.5194/angeo-30-503-2012

Y. Ohira, B. Reville, J.G. Kirk, F. Takahara, Two-dimensional particle-in-cell simulations of the nonresonant, cosmic-ray-driven instability in supernova remnant shocks. Astrophys. J. **698**, 445–450 (2009). doi:10.1088/0004-637X/698/1/445

K.T. Osman, W.H. Matthaeus, A. Greco, S. Servidio, Evidence for inhomogeneous heating in the solar wind. Astrophys. J. Lett. **727**, 11 (2011). doi:10.1088/2041-8205/727/1/L11

K. Papadopoulos, A review of anomalous resistivity for the ionosphere. Rev. Geophys. Space Phys. **15**, 113–127 (1977). doi:10.1029/RG015i001p00113

E.N. Parker, Acceleration of cosmic rays in solar flares. Phys. Rev. **107**, 830–836 (1957). doi:10.1103/PhysRev.107.830

H.E. Petschek, Magnetic field annihilation, in *Physics of Solar Flares*, ed. by W.N. Ness. NASA SP, vol. 50, 1964, p. 425

T.D. Phan, J.T. Gosling, M.S. Davis, R.M. Skoug, M. Øieroset, R.P. Lin, R.P. Lepping, D.J. McComas, C.W. Smith, H. Reme, A. Balogh, A magnetic reconnection X-line extending more than 390 Earth radii in the solar wind. Nature **439**, 175–178 (2006). doi:10.1038/nature04393

P.L. Pritchett, Onset and saturation of guide-field magnetic reconnection. Phys. Plasmas **12**(6), 062301 (2005). doi:10.1063/1.1914309

P.L. Pritchett, Kinetic properties of magnetic merging in the coalescence process. Phys. Plasmas **14**(5), 052102 (2007). doi:10.1063/1.2727458

A. Retinò, D. Sundkvist, A. Vaivads, F. Mozer, M. André, C.J. Owen, In situ evidence of magnetic reconnection in turbulent plasma. Nat. Phys. **3**, 236–238 (2007). doi:10.1038/nphys574

B. Reville, J.G. Kirk, P. Duffy, Steady-state solutions in nonlinear diffusive shock acceleration. Astrophys. J. **694**, 951–958 (2009). doi:10.1088/0004-637X/694/2/951

P. Ricci, G. Lapenta, J.U. Brackbill, Electron acceleration and heating in collisionless magnetic reconnection. Phys. Plasmas **10**, 3554–3560 (2003). doi:10.1063/1.1598207

M.A. Riquelme, A. Spitkovsky, Magnetic amplification by magnetized cosmic rays in supernova remnant shocks. Astrophys. J. **717**, 1054–1066 (2010). doi:10.1088/0004-637X/717/2/1054

A. Runov, V. Angelopoulos, M.I. Sitnov, V.A. Sergeev, J. Bonnell, J.P. McFadden, D. Larson, K.-H. Glassmeier, U. Auster, THEMIS observations of an earthward-propagating dipolarization front. Geophys. Res. Lett. **36**, 14106 (2009). doi:10.1029/2009GL038980

F. Sahraoui, M.L. Goldstein, P. Robert, Y.V. Khotyaintsev, Evidence of a cascade and dissipation of solar-wind turbulence at the electron gyroscale. Phys. Rev. Lett. **102**(23), 231102 (2009). doi:10.1103/PhysRevLett.102.231102

S.J. Schwartz, Shocks: commonalities in solar-terrestrial chains. Space Sci. Rev. **124**, 333–344 (2006). doi:10.1007/s11214-006-9093-y

S.J. Schwartz, D. Burgess, J.J. Moses, Low-frequency waves in the Earth's magnetosheath: present status. Ann. Geophys. **14**, 1134–1150 (1996). doi:10.1007/s00585-996-1134-z

S.J. Schwartz, T. Horbury, C. Owen, W. Baumjohann, R. Nakamura, P. Canu, A. Roux, F. Sahraoui, P. Louarn, J.-A. Sauvaud, J.-L. Pinçon, A. Vaivads, M.F. Marcucci, A. Anastasiadis, M. Fujimoto, P. Escoubet, M. Taylor, S. Eckersley, E. Allouis, M.-C. Perkinson, Cross-scale: multi-scale coupling in space plasmas. Exp. Astron. **23**, 1001–1015 (2009a). doi:10.1007/s10686-008-9085-x

S. Schwartz, S.D. Bale, M. Fujimoto, P. Hellinger, M. Kessel, G. Le, W. Liu, P. Louarn, I. Mann, R. Nakamura, C. Owen, J.-L. Pinçon, L. Sorriso-Valvo, A. Vaivads, R.F. Wimmer-Schweingruber, Cross-scale: multi-scale coupling in space plasma, Assessment Study Report, arXiv:0912.0856 (2009b)

N. Sckopke, G. Paschmann, S.J. Bame, J.T. Gosling, C.T. Russell, Evolution of ion distributions across the nearly perpendicular bow shock—specularly and non-specularly reflected-gyrating ions. J. Geophys. Res. **88**, 6121–6136 (1983). doi:10.1029/JA088iA08p06121

J.D. Scudder, A. Mangeney, C. Lacombe, C.C. Harvey, C.S. Wu, The resolved layer of a collisionless, high beta, supercritical, quasi-perpendicular shock wave. III—Vlasov electrodynamics. J. Geophys. Res. **91**, 11075–11097 (1986). doi:10.1029/JA091iA10p11075

S. Servidio, A. Greco, W.H. Matthaeus, K.T. Osman, P. Dmitruk, Statistical association of discontinuities and reconnection in magnetohydrodynamic turbulence. J. Geophys. Res. **116**(A15), 9102 (2011). doi:10.1029/2011JA016569

M.A. Shay, J.F. Drake, B.N. Rogers, R.E. Denton, The scaling of collisionless, magnetic reconnection for large systems. Geophys. Res. Lett. **26**, 2163–2166 (1999). doi:10.1029/1999GL900481

M.A. Shay, J.F. Drake, J.P. Eastwood, T.D. Phan, Super-Alfvénic propagation of substorm reconnection signatures and Poynting flux. Phys. Rev. Lett. **107**(6), 065001 (2011). doi:10.1103/PhysRevLett.107.065001

J. Skilling, Cosmic ray streaming. III—Self-consistent solutions. Mon. Not. R. Astron. Soc. **173**, 255–269 (1975)

D. Sundkvist, A. Retinò, A. Vaivads, S.D. Bale, Dissipation in turbulent plasma due to reconnection in thin current sheets. Phys. Rev. Lett. **99**(2), 025004 (2007). doi:10.1103/PhysRevLett.99.025004

P.A. Sweet, The neutral point theory of solar flares, in *Electromagnetic Phenomena in Cosmical Physics*, ed. by B. Lehnert. IAU Symposium, vol. 6, 1958, p. 123

M. Swisdak, B.N. Rogers, J.F. Drake, M.A. Shay, Diamagnetic suppression of component magnetic reconnection at the magnetopause. J. Geophys. Res. **108**, 1218 (2003). doi:10.1029/2002JA009726

R.A. Treumann, Fundamentals of collisionless shocks for astrophysical application, 1. Non-relativistic shocks. Astron. Astrophys. Rev. **17**, 409–535 (2009). doi:10.1007/s00159-009-0024-2

B.T. Tsurutani, E.J. Smith, R.R. Anderson, K.W. Ogilvie, J.D. Scudder, D.N. Baker, S.J. Bame, Lion roars and nonoscillatory drift mirror waves in the magnetosheath. J. Geophys. Res. **87**, 6060–6072 (1982). doi:10.1029/JA087iA08p06060

D.A. Uzdensky, R.M. Kulsrud, Physical origin of the quadrupole out-of-plane magnetic field in hall-magnetohydrodynamic reconnection. Phys. Plasmas **13**, 062305 (2006)

V.M. Vasyliunas, Theoretical models of magnetic field line merging. I. Rev. Geophys. Space Phys. **13**, 303–336 (1975). doi:10.1029/RG013i001p00303

H.J. Voelk, L.O. Drury, J.F. McKenzie, Hydrodynamic estimates of cosmic ray acceleration efficiencies in shock waves. Astron. Astrophys. **130**, 19–28 (1984)

D.G. Wentzel, Hydromagnetic waves excited by slowly streaming cosmic rays. Astrophys. J. **152**, 987 (1968). doi:10.1086/149611

C.S. Wu, A fast Fermi process—energetic electrons accelerated by a nearly perpendicular bow shock. J. Geophys. Res. **89**, 8857–8862 (1984). doi:10.1029/JA089iA10p08857

C.S. Wu, D. Winske, M. Tanaka, K. Papadopoulos, K. Akimoto, C.C. Goodrich, Y.M. Zhou, S.T. Tsai, P. Rodriguez, C.S. Lin, Microinstabilities associated with a high Mach number, perpendicular bow shock. Space Sci. Rev. **37**, 63–109 (1984). doi:10.1007/BF00213958

J.R. Wygant, C.A. Cattell, R. Lysak, Y. Song, J. Dombeck, J. McFadden, F.S. Mozer, C.W. Carlson, G. Parks, E.A. Lucek, A. Balogh, M. Andre, H. Reme, M. Hesse, C. Mouikis, Cluster observations of an intense normal component of the electric field at a thin reconnecting current sheet in the tail and its role in the shock-like acceleration of the ion fluid into the separatrix region. J. Geophys. Res. **110**(A9), 9206 (2005). doi:10.1029/2004JA010708

M. Yamada, R. Kulsrud, H. Ji, Magnetic reconnection. Rev. Mod. Phys. **82**, 603–664 (2010). doi:10.1103/RevModPhys.82.603

M. Yamada, F.M. Levinton, N. Pomphrey, R. Budny, J. Manickam, Y. Nagayama, Investigation of magnetic reconnection during a sawtooth crash in a high-temperature tokamak plasma. Phys. Plasmas **1**, 3269–3276 (1994). doi:10.1063/1.870479

M. Yamada, H. Ji, S. Hsu, T. Carter, R. Kulsrud, F. Trintchouk, Experimental investigation of the neutral sheet profile during magnetic reconnection. Phys. Plasmas **7**, 1781–1787 (2000). doi:10.1063/1.873999

E.G. Zweibel, Cosmic-ray history and its implications for galactic magnetic fields. Astrophys. J. **587**, 625–637 (2003). doi:10.1086/368256

E.G. Zweibel, M. Yamada, Magnetic reconnection in astrophysical and laboratory plasmas. Annu. Rev. Astron. Astrophys. **47**, 291–332 (2009). doi:10.1146/annurev-astro-082708-101726

DOI 10.1007/978-1-4899-7413-6_3
Reprinted from *Space Science Reviews* Journal, DOI 10.1007/s11214-013-0004-8

Solar Wind Turbulence and the Role of Ion Instabilities

**O. Alexandrova · C.H.K. Chen · L. Sorriso-Valvo ·
T.S. Horbury · S.D. Bale**

Received: 15 February 2013 / Accepted: 18 June 2013 / Published online: 31 August 2013

Abstract Solar wind is probably the best laboratory to study turbulence in astrophysical plasmas. In addition to the presence of magnetic field, the differences with neutral fluid isotropic turbulence are: (i) weakness of collisional dissipation and (ii) presence of several characteristic space and time scales. In this paper we discuss observational properties of solar wind turbulence in a large range from the MHD to the electron scales. At MHD scales, within the inertial range, turbulence cascade of magnetic fluctuations develops mostly in the plane perpendicular to the mean field, with the Kolmogorov scaling $k_\perp^{-5/3}$ for the perpendicular cascade and $k_\parallel^{-2}$ for the parallel one. Solar wind turbulence is compressible in nature: density fluctuations at MHD scales have the Kolmogorov spectrum. Velocity fluctuations do not follow magnetic field ones: their spectrum is a power-law with a $-3/2$ spectral index. Probability distribution functions of different plasma parameters are not Gaussian, indicating presence of intermittency. At the moment there is no global model taking into account all these observed properties of the inertial range. At ion scales, turbulent spectra have a break, compressibility increases and the density fluctuation spectrum has a local flattening. Around ion scales, magnetic spectra are variable and ion instabilities occur as a function of the local plasma parameters. Between ion and electron scales, a small scale turbulent cascade seems to be established. It is characterized by a well defined power-law spectrum in magnetic and density fluctuations with a spectral index close to -2.8. Approaching electron scales, the fluctuations are no more self-similar: an exponential cut-off is usually observed

O. Alexandrova (✉)
LESIA, Observatoire de Paris, 5, Place Jules Janssen, 92190 Meudon, France
e-mail: olga.alexandrova@obspm.fr

C.H.K. Chen · L. Sorriso-Valvo · S.D. Bale
Space Sciences Laboratory, University of California, Berkeley, CA 94720, USA

L. Sorriso-Valvo
IPCF/CNR, UOS di Cosenza, 87036 Rende, CS, Italy
e-mail: sorriso@fis.unical.it

T.S. Horbury
The Blackett Laboratory, Imperial College London, London SW7 2AZ, UK

(for time intervals without quasi-parallel whistlers) indicating an onset of dissipation. The small scale inertial range between ion and electron scales and the electron dissipation range can be together described by $\sim k_\perp^{-\alpha}\exp(-k_\perp\ell_d)$, with $\alpha \simeq 8/3$ and the dissipation scale ℓ_d close to the electron Larmor radius $\ell_d \simeq \rho_e$. The nature of this small scale cascade and a possible dissipation mechanism are still under debate.

Keywords Plasma turbulence · Solar wind · Kinetic scales · Ion instabilities

1 Introduction

Natural plasmas are frequently in a turbulent state characterized by large, irregular fluctuations of the physical parameters. The spatial and temporal scales of these fluctuations cover a large range, usually extending down to the smallest scales resolved by the observations. Well known examples are provided by the solar wind, the magnetosheath of planetary magnetospheres, the interstellar medium, etc.

Is there a certain degree of generality in the physics of the various astrophysical situations where turbulent states are observed? If this is the case, is it of the same nature as what happens in incompressible neutral (or magnetized) fluid turbulence, which is a non-linear process, non-reproducible locally but with some "universal" statistical properties? These "universal" statistical properties are thought to result from the combination of (1) an infinite number of degrees of freedom, each characterized by its spatial and temporal scale; (2) the absence of characteristic spatial and temporal scales, which implies some sort of equivalence between all of the degrees of freedom; (3) a nonlinear transfer of energy between these degrees of freedom, often called a cascade of energy.

To be more specific, the incompressible fluid turbulence occurs at large Reynolds numbers $Re = LV_L/\eta \gg 1$ (where L is the scale at which the energy is injected in the system, that is of the order of the correlation length of the largest turbulent eddy, V_L the typical value of velocity fluctuations at scale L and η the kinematic viscosity). This is verified when the energy injection scale is sufficiently far from the dissipation scale ℓ_d ($L \gg \ell_d$). Thanks to a number of observations, numerical simulations and theoretical works, the following universal properties of a turbulent system have been firmly established:

- In Fourier space, at intermediate scales $L^{-1} \ll k \ll \ell_d^{-1}$ (k being a wave-number), within the so called *inertial range*, the power spectrum of the velocity fluctuations is observed to follow a $k^{-5/3}$ law, independently of how the energy is injected in the system, and of how it is dissipated at small scales. A power-law spectrum suggests scale invariance, i.e., at each scale the same physical description is valid (the Navier-Stokes equation for fluids and the magnetohydrodymanic equations for magnetized plasmas are scale invariant and describe well self-similar turbulent fluctuations).
- Intermittency, due to spatial nonuniformity of the energy transfer across scales, manifests itself as a scale dependent departure from Gaussian distributions of the probability distribution functions of the turbulent fluctuations.

To date, 3D fluid turbulence is far from being understood, and there is no satisfactory theory, based on first principles, that fully describes it in a sufficiently general frame. Therefore one has to rely on "phenomenologies" which attempt to provide a framework for the interpretation of experimental results; for example the empirical $k^{-5/3}$ law is well described by the Kolmogorov's phenomenology (hereafter K41) (Kolmogorov 1941a; Frisch 1995). In this simple model of turbulence, kinetic energy E_c is supposed to cascade from large scales to small scales and the *cascade rate* (an energy per unit time) is constant over the inertial range $\varepsilon = \partial E_c/\partial t =$ const. Since the only timescale that appears in the system is the time of the energy exchange between the fluctuations (the *eddies*), also called the *non-linear* or *eddy turnover time* $\tau_{nl} = \ell/\delta v$, the cascade rate can be approximated by $\varepsilon \approx (\delta v)^2/\tau_{nl} =$ const. It follows that the velocity field fluctuations $\delta v \approx (\varepsilon\ell)^{1/3}$ so that the power spectrum $(\delta v)^2/k$ goes like $\ell^{5/3}$ or $k^{-5/3}$.

Intermittency is beyond the Kolmogorov phenomenology but it has been observed that in neutral fluids it appears in the form of coherent structures as filaments of vorticity. Their characteristic length can be of the order of the energy injection scale L but their cross-section is of the order of the dissipation scale ℓ_d (see the references of Sect. 8.9 in Frisch 1995). Thus, in Fourier space, these filaments occupy all scales including the edges of the inertial range.

As we have said, in the phenomenological framework of turbulence, the majority of the results are based on the interpretation of experimental results. However, one important theoretical result was obtained from the Navier-Stokes equation, independently of K41 phenomenology: it is known as Kolmogorov's 4/5 law (hereafter K4/5). The K4/5 law prescribes that, for fully developed incompressible turbulence in a stationary state[1], under conditions of isotropy, local homogeneity, and vanishing dissipation (i.e., in the inertial range), the third order moment of the longitudinal (i.e. along the bulk flow) velocity fluctuations δv scales linearly with the separation ℓ (or with the time scale $\tau = \ell/V$, with V being a bulk flow speed):

$$Y(\ell) = \langle \delta v^3 \rangle = -4/5\varepsilon\ell, \tag{1}$$

the proportionality factor ε being the mean energy transfer rate and dissipation rate of the turbulent cascade (see Frisch 1995, Sect. 6.2, and references therein). This law has been indeed observed in the neutral fluid turbulence, e.g. Danaila et al. (2001). Note that Kolmogorov 4/5 law can be obtained from the more general Yaglom (1949) law in case of Navier-Stokes isotropic turbulence.

When the energy cascade "arrives" to the spatial (or time) scale of the order of the dissipation scale ℓ_d, the spectrum becomes curved (Grant et al. 1962), indicating a lack of self-similarity. This spectrum is also universal (see, e.g., Fig. 8.14 in Frisch 1995) and can be described by $\sim k^3 \exp(-ck\ell_d)$ with $c \simeq 7$ (Chen et al. 1993). In neutral fluids the dissipation sets in usually at scales of the order of the collisional mean free path.

We shall restrict ourselves here to the solar wind turbulence, which is perhaps our best laboratory for studying astrophysical plasma turbulence (Tu and Marsch 1995; Bruno and Carbone 2005; Horbury et al. 2005; Matthaeus and Velli 2011). Does the solar wind turbulence share the above universal characteristics, such as power-law spectra, intermittency and linear dependence between the third order moment of the fluctuations and the energy transfer rate? How does the dissipation set in? and is its spectrum universal?

The solar wind expands radially but not with spherical symmetry. Fast, rather steady wind at around 700 km/s flows from coronal holes, generally at high solar latitudes. More variable slow wind (200–500 km/s) is thought to have its source around coronal hole boundaries or in transiently open regions. In general, the properties of fluctuations within fast and slow wind at 1 AU are rather different, with fast wind turbulence appearing less developed than that

[1] In a stationary state, the energy injection rate ε_{inj} at large scales is equal to the energy transfer rate within the inertial range $\varepsilon = (\delta v)^2/\tau_{nl}$ and to the energy dissipation rate within the dissipation range of scales $\varepsilon_{dis} = \eta\langle(\partial_x v(x))^2\rangle$, where η is the kinematic viscosity: $\varepsilon_{inj} = \varepsilon_{dis} = \varepsilon$.

in slow wind, indicating different "age of turbulence". Interactions between fast and slow wind, as well as transient events, produce compressions, rarefactions and shocks. When considering the innate properties of plasma turbulence, it is usually easier to treat steady, statistically homogeneous intervals of data from individual streams.

In situ spacecraft measurements in the solar wind provide time series of local plasma parameters. Therefore, in Fourier space, we have a direct access to frequency spectra. When the flow speed of the solar wind V_{sw} is much larger than the characteristic plasma speeds, one can invoke the Taylor's hypothesis (Taylor 1938; Perri and Balogh 2010) and convert a spacecraft-frame frequency f to a flow-parallel wavenumber k in the plasma frame $k = \frac{2\pi f}{V_{sw}}$. At scales larger than the proton characteristic scales, we can largely treat the solar wind fluctuations using magnetohydrodynamics (MHD) (Marsch and Mangeney 1987; Biskamp 1993; Schekochihin et al. 2009). The flow speed V_{sw} is typically much larger than the Alfvén speed $V_A = B/\sqrt{4\pi\rho} \simeq 50$ km/s (B being the magnetic field and ρ the mass density) and far faster than spacecraft motions, so that one can use Taylor's hypothesis. At plasma kinetic scales, the Taylor hypothesis can be used in the absence of quasi-parallel propagating whistler waves, which have a phase speed higher than V_{sw}.

The solar wind is pervaded with fluctuations on all measured scales. These fluctuations form energy spectra following power laws as expected for developed turbulence. For example, for magnetic fluctuations, at very large scales (for the spacecraft-frame frequencies $f < 10^{-4}$ Hz) the power spectrum goes as $\sim f^{-1}$. This spectrum can be interpreted in terms of uncorrelated large scale Alfvén waves (Matthaeus and Goldstein 1986; Horbury et al. 2005). A recent work proposes that it originates due to the nonlinear coupling in the corona between outgoing and ingoing Alfvén waves with the help of multiple reflections on the non-homogeneous transition region (Verdini et al. 2012). The corresponding frequency range is usually called the energy injection scales (Bruno and Carbone 2005). The maximal frequency f_0 of this range, or *outer scale* of the turbulent cascade, is close to 10^{-4} Hz at 1 AU. It was proposed by Mangeney et al. (1991), Salem (2000), Meyer-Vernet (2007), that at the outer scale there is a balance between the solar wind expansion time $\tau_{exp} = R/V_{sw}$ at a radial distance R and the eddy-turnover time τ_{nl}; and the turbulent cascade can develop at scales where $\tau_{nl} < \tau_{exp}$. Estimations at 1 AU for $V_{sw} = 600$ km/s give $\tau_{exp} \simeq 70$ h. The characteristic non-linear time at f_0 is of the order of $\tau_{nl} \simeq 70$ h as well[2]. At smaller scales, i.e. at higher frequencies $f > 10^{-4}$ Hz, the non-linear time becomes smaller than the expansion time and turbulent cascade develops. As τ_{exp} increases with R, the outer scale increases, i.e. f_0 shifts towards lower frequencies. This is indeed observed in the solar wind (Bruno and Carbone 2005). It will be interesting to verify the relationship between the outer-scale and τ_{exp} with solar wind observations for different turbulence levels and at different heliospheric distances.

Within the $\sim [10^{-4}, 10^{-1}]$ Hz range, magnetic spectrum is usually observed to follow the K41 scaling, interpreted as the inertial range (the details on the spectral slope of the inertial range will be discussed in Sect. 2). The spectrum undergoes new changes at the proton characteristic scales (appearing in the measured spectra at $\sim [0.1, 1]$ Hz) and at the electron scales $\sim [50, 100]$ Hz (see details in Sect. 3).

One of the important differences of the solar wind turbulence with the isotropic neutral fluid turbulence is the presence of the mean magnetic field $\mathbf{B}$, which introduces a privileged direction and so imposes an anisotropy of turbulent fluctuations. In the inertial range, the

[2]This is estimated using the Taylor hypothesis $\ell = V_{sw}/f_0 \simeq 6 \cdot 10^6$ km and a typical value of $\delta v \simeq 25$ km/s/$\sqrt{\text{Hz}}$ at $f_0 = 10^{-4}$ Hz.

observed magnetic fluctuations $\delta B_{\|}$ along the mean field are usually much smaller than the transverse Alfvénic fluctuations $\delta B_{\perp}$. The wave vector distributions are not isotropic either, $k_{\perp} > k_{\|}$. In Sect. 2 we will discuss in more details how this k-anisotropy has been detected within the inertial range of the solar wind turbulence and its possible interpretations. We will discuss as well intermittency in the solar wind and show recent verification of the K4/5 law.

Another important difference between neutral fluid turbulence and solar wind turbulence is the weakness of collisional dissipation in the solar wind, as for most of the space plasmas. The dissipation process at work and the dissipation length are not known precisely. There are observational indications and theoretical considerations that characteristic plasma scales may be good candidates to replace, in some sense, the dissipation scale of fluid turbulent cascade. The characteristic plasma scales are the ion Larmor radius $\rho_i = \sqrt{2k_B T_{i\perp}/m_i}/(2\pi f_{ci})$ (with k_B being the Boltzmann's constant, $T_{i\perp}$ being the ion temperature perpendicular to the magnetic field $\mathbf{B}$, m_i being the ion mass), the ion inertia length $\lambda_i = c/\omega_{pi}$ (with c the speed of light and ω_{pi} the ion plasma frequency), the corresponding electron scales ρ_e, λ_e, and the ion and electron cyclotron frequencies $f_{ci,e} = qB/(2\pi m_{i,e})$ (with q being the charge of the particle). At these scales different kinetic effects may take place. However, the precise mechanism (or mechanisms) which dissipates electromagnetic turbulent energy in the solar wind and the corresponding spatial and/or temporal scale(s) are still under debate. The details of the observations of solar wind turbulence around plasma kinetic scales will be discussed in Sect. 3. In particular, in Sect. 3.2 we discuss the ion temperature anisotropy instabilities which may control turbulent fluctuations around ion scales. Conclusions are found in Sect. 4.

2 The MHD Scale Cascade

An MHD theory of cascading turbulence similar to Kolmogorov, but carried by Alfvénic fluctuations propagating in the large-scale magnetic field $\mathbf{B}$ was proposed independently by Iroshnikov (1963) and Kraichnan (1965) (IK hereafter). In this model, the fluctuations are still assumed to be isotropic but most of the energy transfer is due to interactions between Alfvénic fluctuations moving in opposite direction along $\mathbf{B}$ with the Alfvén speed V_A. This limits the time during which two eddies interact, which is of the order of an Alfvén time $\tau_A \sim \ell/V_A$. It is also assumed that the interactions are weak such that $\tau_A \ll \tau_{nl}$, and thus a number of interactions proportional to τ_{nl}/τ_A is needed to transfer the energy (Dobrowolny et al. 1980). Following the argument of Kolmogorov and under the assumption of equipartition between magnetic and kinetic energies, for incompressible fluctuations and random interactions between the Alfvén wave packets, the velocity and magnetic turbulent spectra follow a $\sim k^{-3/2}$ scaling.[3]

However, the assumption of isotropy in IK model for the magnetized plasma is quite strong. Goldreich and Sridhar (1995) proposed an MHD model for anisotropic Alfvénic fluctuations. In that theory, the cascade energy is carried by perpendicular fluctuations $v_{\perp}$ with wavelength $\ell_{\perp} = 2\pi/k_{\perp}$. The Alfvén time is the time scale along $\mathbf{B}$, $\tau_A = \ell_{\|}/V_A$, and the eddie turnover time $\tau_{nl} \approx \ell_{\perp}/v_{\perp}$ governs the energy exchange in the plane perpendicular to $\mathbf{B}$.[4] Goldreich and Sridhar proposed that the turbulence is *strong*, so that these timescales are comparable, $\tau_{nl} \approx \tau_A$. This condition, called *critical balance*, implies that the nonlinear

[3]For the detailed demonstration we refer to the problem 6.6.4 in the book of Meyer-Vernet (2007).

[4]All over the paper, $\|$ ($\perp$) denotes direction parallel (perpendicular) to the mean magnetic field $\mathbf{B}$.

interaction occurs over a single Alfvén wave period. Using the argument of Kolmogorov, one can show that the perpendicular energy transfer rate is $\varepsilon(k_\perp) \sim v_\perp^3/\ell_\perp$. Under the assumption of $\varepsilon(k_\perp) = \text{const}$, the power spectral density of $k_\perp$-fluctuations goes therefore like $\sim k_\perp^{-5/3}$. For the parallel energy transfer rate $\varepsilon(k_\parallel)$ one gets $v_\perp^2 V_A/\ell_\parallel$ and a spectrum $v_\perp^2/k_\parallel \sim k_\parallel^{-2}$. An interesting consequence of the Goldreich-Sridhar model is the following: since the cascade is carried by the perpendicular fluctuation spectrum (and indeed this property is reinforced as the energy arrives at larger wavenumbers, where the k-anisotropy becomes important $k_\perp \gg k_\parallel$), the energy in the spectrum reaches dissipation scales (or characteristic plasma scales) in the perpendicular spectrum long before it does so in the parallel spectrum. This implies that relatively little energy of $k_\parallel$-fluctuations reaches the characteristic plasma scales due to the nonlinear cascade.

It should be pointed out that the model of Goldreich and Sridhar (1995) describes Alfvénic turbulence, i.e., the perpendicular magnetic $\delta B_\perp$ and velocity $\delta v_\perp$ fluctuations. This model has been extended to include the passive mixing of the compressive fluctuations by the Alfvénic turbulence (Goldreich and Sridhar 1995, 1997; Lithwick and Goldreich 2001; Schekochihin et al. 2009). However, the nature of compressible fluctuations observed in the solar wind, i.e. a passive scalar or an active turbulence ingredient, remains under debate.

Some theoretical results and solar wind observations suggest that ion cyclotron wave-particle interactions are an important source of heating for solar wind ions (Marsch and Tu 2001; Isenberg et al. 2001; Kasper et al. 2008, 2013; Bourouaine et al. 2010, 2011; Marsch and Bourouaine 2011; He et al. 2011b). However, this interpretation requires substantial turbulent energy at $k_\parallel \rho_i \approx 1$, that is in apparent contradiction to the Goldreich-Sridhar model and to the solar wind measurements described in the following section (Horbury et al. 2008; Podesta 2009; Luo and Wu 2010; Wicks et al. 2010; 2011; Chen et al. 2011a). This is another puzzle that has important ramifications for the coronal heating problem.

2.1 Scaling and Anisotropy as Observed in the Solar Wind

2.1.1 Magnetic Fluctuations

It has long been known that in the inertial range the power spectrum of magnetic field fluctuations in the solar wind is $P(f) \propto f^{-5/3}$, i.e. the same spectrum as for the velocity fluctuations in hydrodynamics turbulence (Kolmogorov 1941b; Frisch 1995). One might conclude that the turbulence in the solar wind is similar to that in a neutral fluid, like air. However, turbulence in a magnetofluid is radically different to that in a neutral fluid, due to the presence of a magnetic field which breaks the isotropy of the turbulence (Shebalin et al. 1983), leading to a correlation length parallel to the field longer than that across it, $\ell_\parallel > \ell_\perp$ (Matthaeus et al. 1990)—crudely, we can think of the turbulent eddies as being shorter perpendicular to the magnetic field than parallel to it, and more formally as having a dominance of turbulent power at wavevectors at large angles to the field, $k_\perp > k_\parallel$.

Measurements of the wave-vector anisotropy and of the corresponding spectra in the solar wind with one satellite are not trivial. A satellite provides time series measurements along its orbit; therefore, applying the Fourier (or wavelet) transform we obtain directly frequency spectra and not k-spectra. As we have discussed in the introduction, the Taylor hypothesis can be used, i.e. we can easily estimate k along the bulk flow through $k = \frac{2\pi f}{V_{sw}}$. Thus, if $\mathbf{V}_{\mathbf{sw}}$ is parallel to the mean field, the fluctuations with parallel wave vectors $k_\parallel$ will be measured, and if $\mathbf{V}_{\mathbf{sw}}$ is perpendicular to $\mathbf{B}$, the satellite resolves well fluctuations with $k_\perp$. We denote the local flow-to-field angle as θ_{BV}. Figure 1 shows magnetic spectra in the fast high latitude solar wind measured by the *Ulysses* spacecraft (at distance of 1.38 to 1.93

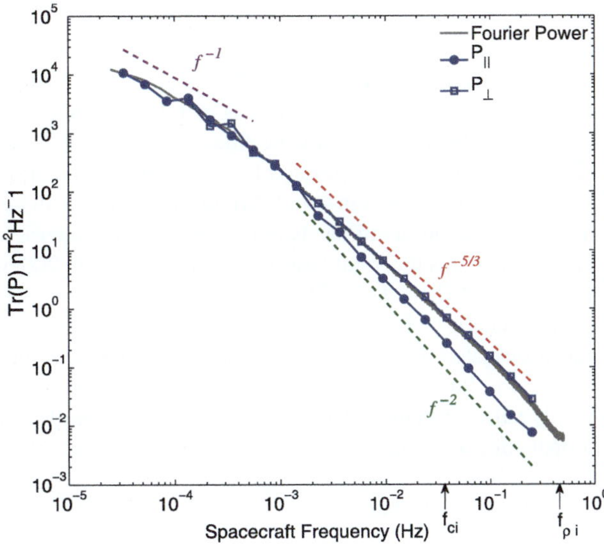

Fig. 1 Trace of the spectral matrix of magnetic field corresponding to the field being parallel ($\theta_{BV} \in [0, 10]°$) and perpendicular ($\theta_{BV} \in [80, 90]°$) to the plasma flow are shown by *blue lines*, the total Fourier spectrum is shown in *gray*. The field-perpendicular spectrum $P_\perp$ dominates turbulence within the inertial range, it follows a power-law with the spectral index $-5/3$. The field-parallel spectrum $P_\parallel$ has lower power, is steeper and has the spectral slope -2. At the energy injection scales $f < 5 \cdot 10^{-4}$ Hz ($k\rho_i < 2 \cdot 10^{-3}$) the fluctuations are isotropic and their spectrum follows $\sim f^{-1}$. Courtesy of R. Wicks. The same figure as a function of $k\rho_i$ can be found in Wicks et al. (2010)

AU from the Sun). As the spacecraft only measures wave vectors **k** parallel to $\mathbf{V}_{\mathbf{sw}}$, for small flow-to-field angles $\theta_{BV} \in [0, 10]°$, $P_\parallel$ (nT^2/Hz) represents an $E(k_\parallel)$ spectrum, and for quasi-perpendicular angles $\theta_{BV} \in [80, 90]°$, $P_\perp$ (nT^2/Hz), is the proxy of $E(k_\perp)$. The total Fourier power, without separation into different angles is also shown. Within the energy injection range, the fluctuations are found to be isotropic, $P_\parallel \simeq P_\perp$, and both spectra follow an $\sim f^{-1}$ power-law in agreement with previous observations (Bruno and Carbone 2005). In the inertial range one observes a bifurcation of the two spectra: the perpendicular spectrum follows the Kolmogorov's slope, $E(k_\perp) \sim k_\perp^{-5/3}$, while the parallel spectrum is steeper, $E(k_\parallel) \sim k_\parallel^{-2}$. This result, initially seen in fast wind measured by *Ulysses* (Horbury et al. 2008) has been confirmed by several other studies (Podesta 2009; Luo and Wu 2010; Wicks et al. 2010, 2011; Chen et al. 2011a). These magnetic field spectral scaling observations provide an intriguing, if not unequivocal, connection to the Goldreich-Sridhar theory (Higdon 1984; Goldreich and Sridhar 1995). It is important to notice that the spectral anisotropy, shown in Fig. 1, is only observed while the local anisotropy analyses is used (Horbury et al. 2008). Such analysis consists in following the magnetic field direction as it varies in space and scale, which may cause the measured spectra to contain higher order correlations (Matthaeus et al. 2012).

The importance of the local field for the turbulence anisotropy analysis has been pointed out already in Cho and Vishniac (2000), Maron and Goldreich (2001), Milano et al. (2001). The method proposed by Horbury et al. (2008), and used by Wicks et al. (2010) in Fig. 1, is equivalent in some sense to the one used in Milano et al. (2001) for numerical simulations, but can appear contradictory with the requirement of the ergodic theorem (equivalence between space and time averaging).[5] However, there are practical implications that have to be considered: an individual packet of plasma passes a spacecraft once and never returns, meaning that the average magnetic field direction over many correlation lengths measured from a time series is not necessarily representative of the actual magnetic field direction at any

[5] In order to insure the equivalence between space and time averaging, the average should be taken over several correlation lengths, i.e. several energy injection lengths.

point. Rather than taking simple time averages, here the local magnetic field direction (and local θ_{BV}) to each fluctuation is measured, and then fluctuations that have similar directions are averaged. Precisely, in Fig. 1, Wicks et al. (2010) used many hundreds of observations in each direction, so the ergodicity is met, but in a non-conventional way.

Beyond the anisotropy of the fluctuations with respect to the magnetic field direction, (Boldyrev 2006) also suggested that the turbulence can be anisotropic with respect to the local fluctuation direction – and that this anisotropy will be scale dependent. Remarkably, in the solar wind observations there is some recent evidence for the scale-dependent alignment predicted by this theory at large scales (Podesta et al. 2009b) and for the local 3D anisotropy small scales (Chen et al. 2012b).

The nature of imbalanced turbulence is also a topic of current interest. Alfvén waves can propagate parallel or anti-parallel to the magnetic field. Without the presence of both senses, the fluctuations are stable and will not decay. However, the level of imbalance is highly variable in the solar wind (fast wind is typically dominated by Alfvénic fluctuations propagating anti-sunward).

2.1.2 Velocity Fluctuations

Velocity fluctuations in the solar wind appear to have a spectrum significantly shallower than the magnetic field, with a spectral index near $-3/2$ (Grappin et al. 1991; Salem 2000; Mangeney et al. 2001; Podesta et al. 2007; Salem et al. 2009; Chen et al. 2011b, 2013b; Boldyrev et al. 2011; Borovsky 2012b). Figure 2 shows (a) a velocity spectrum and (b) a compensated spectrum with the $f^{3/2}$ function obtained from *Wind* measurements using the Haar wavelet technique (Salem et al. 2009). Such a spectrum was predicted by the IK phenomenology for Alfvénic fluctuations propagating in opposite directions along **B**. However, in this model, both magnetic field and velocity spectra are expected to follow the $\sim k^{-3/2}$ power-law. The difference of the solar wind inertial range with a pure Alfvénic turbulence described in the IK model (and with the anisotropic Goldreich-Sridhar model) is also an excess of magnetic energy with respect to the kinetic energy, see Fig. 8 in Salem et al. (2009). How can the difference between the velocity and the magnetic spectra, and the excess of magnetic energy in the solar wind, be explained? Direct simulations of incompressible MHD usually show an excess of magnetic energy as well. It has been attributed to a local dynamo effect which balances the linear Alfvén effect (Grappin et al. 1983). The difference between magnetic and kinetic energies is usually called in the literature "residual energy". The residual energy has been shown to follow a definite scaling which is related to the scaling of the total energy spectrum (Grappin et al. 1983; Müller and Grappin 2005), see also (Boldyrev and Perez 2009; Boldyrev et al. 2012; Chen et al. 2013b).

Another possible explanation of the difference between the observed magnetic and velocity spectra can be related to the presence of compressible fluctuations, not negligible for the energy exchange between scales.

2.1.3 Density fluctuations

Turbulent fluctuations within the inertial range are not only anisotropic in space (or in **k**), but as well in their amplitudes with respect to **B**. As we have discussed in the introduction, the non-compressive, Alfvénic turbulence dominates the solar wind at MHD scales, $\delta B_\perp \gg \delta B_\parallel$. Nevertheless, there is a sub-dominant population of $\delta B_\parallel$ and density $\delta\rho$ fluctuations always present, with scaling properties suggestive of a turbulent cascade (Celnikier et al.

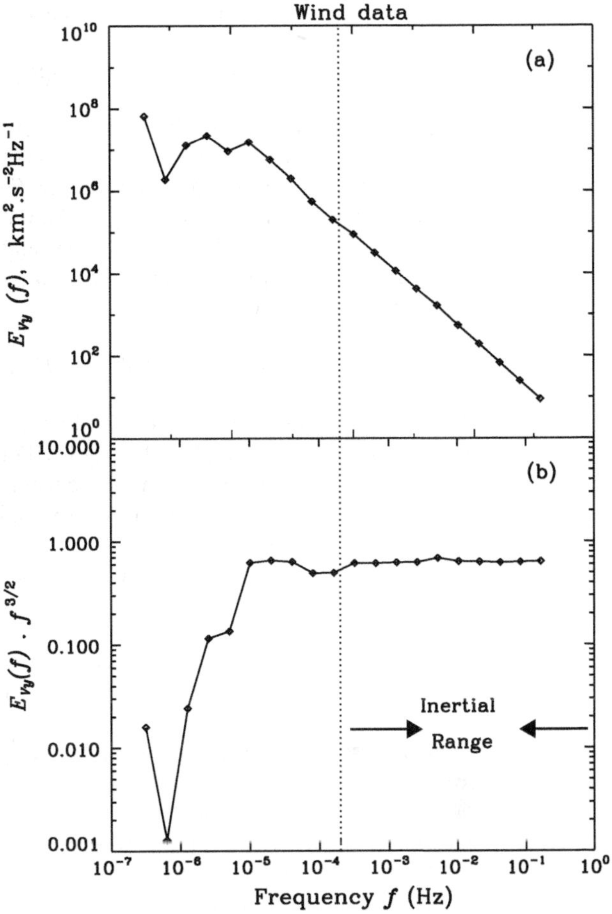

Fig. 2 (**a**) Spectrum of velocity fluctuations of V_y (GSE) component, measured by *Wind* as a function of the frequency in the spacecraft frame, the data have been published in Salem et al. (2009). (**b**) Compensated spectrum by an $f^{3/2}$ law: the resulting function is flat for $f > 10^{-4}$ Hz. Courtesy of C. Salem

1983; Marsch and Tu 1990; Manoharan et al. 1994; Kellogg and Horbury 2005; Hnat et al. 2005; Issautier et al. 2010; Chen et al. 2011b). Figure 3 shows an example of an electron density spectrum measured by the *ISEE 1-2* satellites in the $[6 \cdot 10^{-4}, 5]$ Hz frequency range (Celnikier et al. 1983). At MHD scales, $f < 10^{-1}$ Hz, the K41 scaling is observed. At higher frequencies, i.e. around ion scales, one observes a spectrum flattening and then another steep spectrum. These high-frequency features will be discussed in more details in Sect. 3.

The origin of the compressible fluctuations in the solar wind is not clear, as far as fast and slow mode waves are strongly damped at most propagation angles. Howes et al. (2012a) have recently argued, based on the dependence of the $\delta B_{\|}$-$\delta\rho$ correlation on the plasma beta β (ratio of thermal to magnetic pressure), that these fluctuations are slow mode and they appear to be anisotropic in wave-vectors (He et al. 2011a). Chen et al. (2012b) measured the $\delta B_{\|}$ fluctuations to be more anisotropic than the Alfvénic component in the fast solar wind, suggesting this as a possible reason why they are not heavily damped (Schekochihin et al. 2009). Yao et al. (2011) observe a clear anti-correlation between electron density and the magnetic field strength at different time scales (from 20 s to 1 h): the authors interpret their observations as multi-scale pressure-balanced structures which may be stable in the solar wind. This interpretation is consistent with the observation of intermittency in electron density fluctuations by the *Ulysses* spacecraft (Issautier et al. 2010).

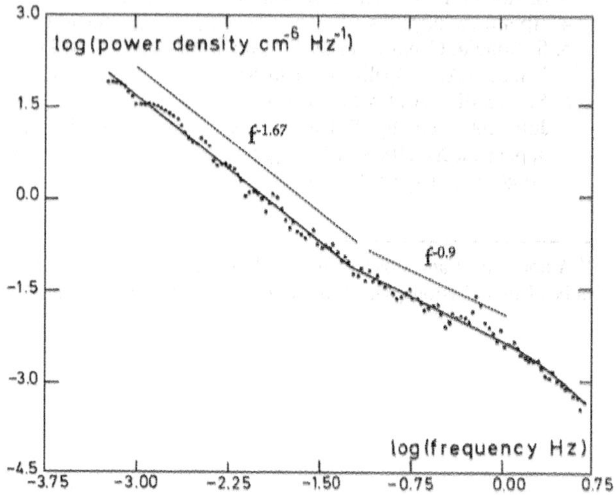

Fig. 3 Spectrum of electron density fluctuations n_e measured by the *ISEE 1–2* spacecraft: two distinct power-laws are observed, the spectrum follows $\sim f^{-1.67\pm0.05}$ within the frequency range $[10^{-3}, 6\cdot10^{-2}]$ Hz, the spectrum is about $f^{-0.9\pm0.2}$ at $f > 6\cdot10^{-2}$ Hz. Around 1–2 Hz the spectrum seems to change again, however, this high frequency range is too narrow to make any firm conclusion (the maximal measured frequency is 5 Hz). Figure from Celnikier et al. (1983)

2.2 Intermittency

In hydrodynamics, the amplitude of the fluctuations at a given scale—and hence the local energy transfer rate—is variable, a property known as intermittency, i.e. turbulence and its dissipation are non-uniform in space (Frisch 1995). This results in the turbulence being bursty, which can be easily seen from the test of regularity of turbulent fluctuations (Mangeney 2012). Usually, turbulent fluctuations at different time scales τ are approximated by increments calculated at these scales, $\delta y_\tau = y(t+\tau) - y(t)$. The time averages of these increments are called "structure functions" (for more details see the paper by Dudok de Wit et al. 2013 in this book). In the presence of intermittency, the scaling of higher order moments of the structure functions diverges from the simple linear behavior expected for non-intermittent, Gaussian fluctuations: in essence, at smaller scales, there are progressively more large jumps, as the turbulence generates small scale structures. This behavior is also observed in the solar wind on MHD scales (Burlaga 1991; Tu and Marsch 1995; Carbone et al. 1995; Sorriso-Valvo et al. 1999; Veltri and Mangeney 1999; Veltri 1999; Salem 2000; Mangeney et al. 2001; Bruno et al. 2001; Sorriso-Valvo et al. 2001; Hnat et al. 2003; Veltri et al. 2005; Bruno and Carbone 2005; Leubner and Voros 2005; Jankovicova et al. 2008; Greco et al. 2009, 2010; Sorriso-Valvo et al. 2010). Figure 4 shows probability distribution functions (PDF) of the tangential component of the standardized magnetic field fluctuations $\Delta B_y = \delta B_y/\sigma(\delta B_y)$, σ being the standard deviation of δB_y (in RTN coordinates[6]) computed for three different time scales τ. Intermittency results in the change of shape, from the large scale Gaussian to the small scale Kappa functions.

Intermittency is a crucial ingredient of turbulence. Being related to the full statistical properties of the fields, its characterization can give an important insight on the nature of turbulence and on possible dissipation mechanisms of turbulent energy.

Note, as well, that as far as the third-order moment of fluctuations is related to the energy dissipation rate and is different from zero (see the K4/5 law, Eq. (1)), turbulence must shows some non-Gaussian features.

Solar wind observations have shown that the intermittency of different fields can be remarkably different. In particular, it has been observed in several instances that the magnetic

[6] R is the radial direction, N is the normal to the ecliptic plane and T completes the direct frame.

Fig. 4 Probability distribution functions (PDFs) of the tangential component of the standardized magnetic field fluctuations ΔB_y (in RTN coordinates) computed for three different time lags, as indicated in the legend. PDFs were estimated using 6 second *Helios 2* data recorded in a stationary slow wind stream near 0.3 AU on days 100 to 102 of 1976. The data used here were published previously in Bruno et al. (2004)

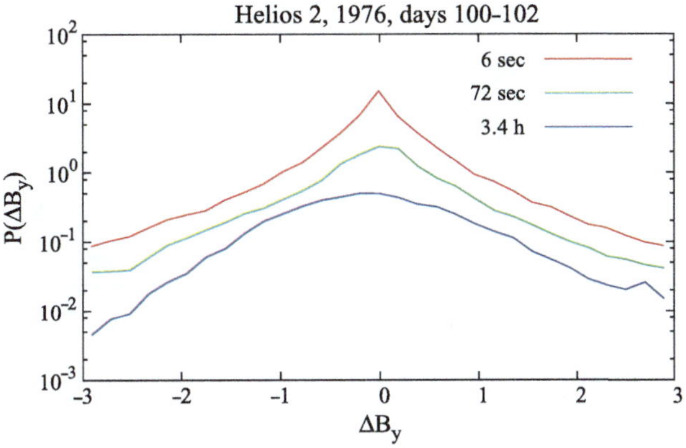

field is generally more intermittent than the velocity (Sorriso-Valvo et al. 1999, 2001). The possibility that this implies that magnetic structures are passively convected by the velocity field has been discussed, but no clear evidence was established, so that this is still an open question (Bershadskii and Sreenivasan 2004; Bruno et al. 2007).

The use of data from *Helios 2* spacecraft, which explored the inner heliosphere reaching about 0.3 AU, has allowed to study the radial evolution of intermittency, and its dependency on the wind type (fast or slow) (Bruno et al. 2003). The fast wind has revealed an important increase of intermittency as the wind blows away from the Sun, while the slow wind is less affected by the radial distance R. This suggests that some evolution mechanism must be active in the fast solar wind. This could be either due to the slower development of turbulence in the fast wind, with respect to the slow wind, or to the presence of superposed uncorrelated Alfvénic fluctuations, which could hide the structures responsible for intermittency in the fast wind closer to the Sun. These uncorrelated Alfvénic fluctuations, ubiquitous in the fast wind, are indeed observed to decay with R, as suggested for example by a parametric instability model (Malara et al. 2000, 2001; Bruno et al. 2003, 2004).

The ultimate responsible for emergence of intermittency are strong fluctuations of the fields with coupled phases over a finite range of scales. These are often referred to as coherent structures. Figure 5 shows an example of a coherent structure responsible for the non-Gaussian PDF tails in Fig. 4 at small scales: a shock wave with its normal quasi-perpendicular to the local mean field (Veltri and Mangeney 1999; Salem 2000; Veltri et al. 2005). This kind of structures may be responsible for the dissipation of turbulent energy in the collisionless solar wind.

A complication in the solar wind is that sharp structures, discontinuities, are ubiquitous. Discontinuities typically involve a rotation in the magnetic field direction, and sometimes variations in velocity, field magnitude and other plasma properties such as density and even temperature and composition (Owens et al. 2011). Parameters such as composition do not change much after the wind leaves the solar corona, so these might have been generated at its source. However, the vast majority of structures have no such signature: are these also part of the structure of the solar wind (Borovsky 2008), or are they generated dynamically by the turbulence (Greco et al. 2009, 2012)? These structures seem to be associated with enhanced temperature of the solar wind (Osman et al. 2011; Wu et al. 2013), so they might represent a source of energy dissipation via reconnection or enhanced damping. Discontinuities, as sharp jumps, also contribute to the intermittency of the solar wind turbulence. To what extent is the observed intermittency inherent to the plasma turbulence, therefore, as opposed to being an artifact of its generation in the corona? This is a currently unresolved issue and

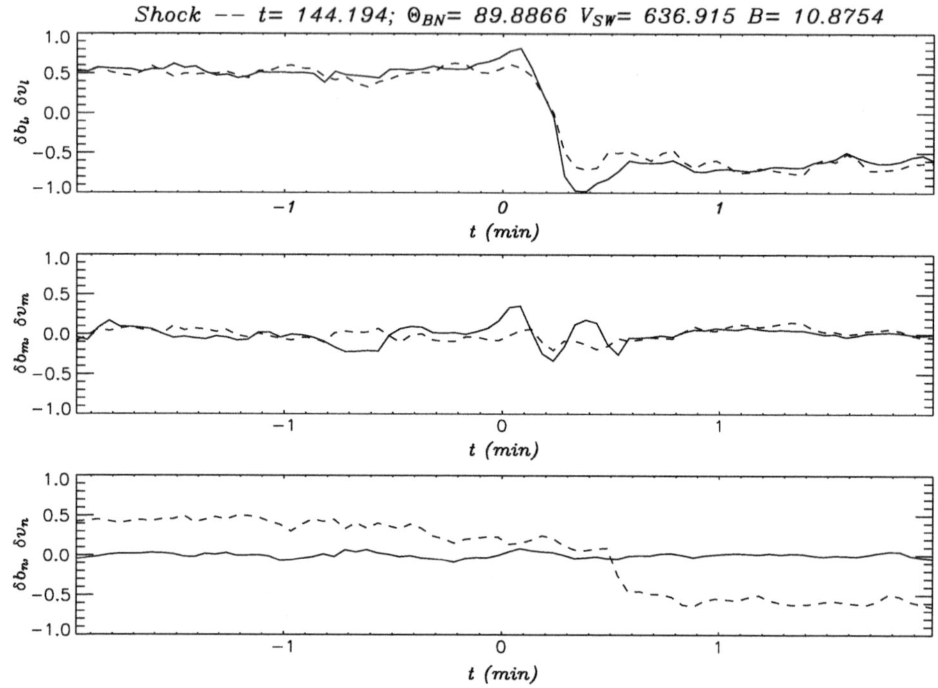

Fig. 5 Example of a coherent structure responsible for the non-Gaussian PDF tails in Fig. 4 at small scales: a quasi-perpendicular shock wave at a time scale of the order of $\tau = 30$ sec. Measurements of $\delta\mathbf{B}$ in the local minimum variance frame (*solid lines*) and velocity fluctuations $\delta\mathbf{v}$ in the same frame (*dashed lines*) as measured by *Wind* satellite in the fast solar wind (courtesy of C. Salem)

the topic of many recent works (Servidio et al. 2011, 2012; Zhdankin et al. 2012; Borovsky 2012a; Osman et al. 2012; Wu et al. 2013; Karimabadi et al. 2013).

2.3 Energy Transfer Rate

As we have mentioned in the introduction, any turbulent flow is characterized by power-law energy spectra, presence of intermittency and linear dependence between the third order structure function and scale. This last property is the only exact result for hydrodynamic turbulence, known as the K4/5 law, see Eq. (1). In plasmas, the incompressible MHD version of the K4/5 law has been obtained by Politano and Pouquet (1998) by using the Elsasser fields $\mathbf{Z}^{\pm}(t) = \mathbf{v}(t) \pm \mathbf{b}(t)/\sqrt{4\pi\rho}$ in place of velocity δv in Eq. (1) ($\mathbf{v}(t)$ and $\mathbf{b}(t)$ being the time dependent solar wind velocity and magnetic field).

The MHD equations can be conveniently written in terms of Elsasser variables $\mathbf{Z}^{\pm}$ as

$$\frac{\partial \mathbf{Z}^{\pm}}{\partial t} + \left(\mathbf{Z}^{\mp} \cdot \nabla\right)\mathbf{Z}^{\pm} = -\nabla P + \eta' \nabla^2 \mathbf{Z}^{\pm}, \tag{2}$$

where P is the total pressure (magnetic plus kinetic), and $\eta' = \eta = \nu$ is a dissipation coefficient[7]. Non-linear terms $(\mathbf{Z}^{\mp} \cdot \nabla)\mathbf{Z}^{\pm}$ in Eq. (2) are responsible for the transfer of energy between fluctuations at different scales, originating the turbulent cascade and the typical

[7]For simplicity, resistivity η is assumed to be equal to viscosity ν.

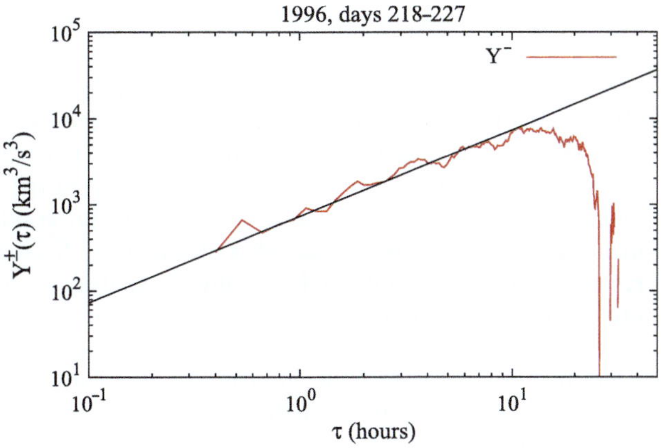

Fig. 6 The third order moment linear scaling law as evaluated in the 11 day time interval starting on day 218 of 1996, during the high latitude scan of *Ulysses* spacecraft. The heliocentric distance was 4.2 AU, the heliolatitude was 30°, and the mean wind speed of the sample was 735 km/s. The linear fit predicted by the law (3), is indicated. For this sample, the pseudo-energy transfer rate is estimated to be $\varepsilon^- = 212\ \mathrm{J\,kg^{-1}\,s^{-1}}$

Kolmogorov spectrum. The MHD version of the K4/5 law for ΔZ^+ is obtained by subtracting Eq. (2) for Z^- from the one for Z^+, evaluated at two generic points separated by the scale $\ell = V_{sw}\tau$ along the flow direction, and then by multiplying the result by ΔZ^+.

This provides an evolution equation for the pseudo-energy flux,[8] which includes terms accounting for anisotropy, inhomogeneity and dissipation. Under the hypotheses of isotropy, local homogeneity and vanishing dissipation (i.e. within the inertial range, far from the dissipation scale), the simple linear relation can be retrieved in the stationary state (Politano and Pouquet 1998):

$$Y^{\pm}(\tau) = \left\langle \left|\Delta \mathbf{Z}^{\pm}(\tau,t)\right|^2 \Delta Z_R^{\mp}(\tau,t)\right\rangle = \frac{4}{3}\varepsilon^{\pm}\ell, \tag{3}$$

where $Z_R^{\mp}$ is the radial component (i.e., along the mean solar wind flow $\mathbf{V}_{sw}$) of the Elsasser fields. For a detailed description of the derivation, see e.g. Danaila et al. (2001), Carbone et al. (2009a).

The turbulent cascade pseudo-energy fluxes $\varepsilon^{\pm}$ are defined as the trace of the dissipation rate tensors

$$\varepsilon_{ij}^{\pm} = \eta\langle(\partial_i Z_i^{\pm})(\partial_i Z_j^{\pm})\rangle.$$

$\varepsilon^{\pm}$ describe the energy transfer rate and dissipation rate between the Elsasser field structures on scales within the inertial range of MHD turbulence.

The relation (3) was first observed in numerical simulations of two dimensional MHD turbulence (Sorriso-Valvo et al. 2002; Pietarila Graham et al. 2006), and later in solar wind samples (MacBride et al. 2005; Sorriso-Valvo et al. 2007; MacBride et al. 2008), despite the observational difficulties (Podesta et al. 2009a) and the fact that solar wind turbulence is not isotropic (Sect. 2.1). An example of linear dependence between Y^- and the time scale τ from *Ulysses* high latitude data is shown in Fig. 6.

The observation of the third order moment scaling is particularly important, since it suggests the presence of a (direct or inverse) turbulent cascade[9] as the result of nonlinear interactions among fluctuations. It also suggests that solar wind turbulence is fully developed, as

[8] The pseudo-energy refers to the fact that the Elsasser fields, $\mathbf{Z}^+$ and $\mathbf{Z}^-$, are pseudo-vectors. The pseudo-energy associated to each Elsasser variable, $\varepsilon^{\pm}$, is not an invariant of the flow. An invariant of the flow is the total energy $(\varepsilon^+ + \varepsilon^-)/2$.

[9] The sign of the coefficient ε will give the direction of the cascade (i.e. the cascade is inverse for negative energy flux).

the dissipative effects have to be neglected in order to observe the linear scaling. It defines rigorously the extension of the inertial range, where a Kolmogorov like spectrum can be expected. In solar wind, the inertial range, as defined by the law of Politano and Pouquet (1998), Eq. (3), is found to be extremely variable, and can reach scales up to one day or even more (Sorriso-Valvo et al. 2007; Marino et al. 2012), much larger than usually assumed following typical estimates from the analysis of turbulent spectra. The variability of the inertial range extension, i.e. the range of scales where the linear relation (3) is observed, is in agreement with earlier multifractal analysis of solar wind fluctuations (Burlaga 1993). Moreover, recent results, obtained through conditioned analysis of solar wind fluctuations, have confirmed that, for high cross-helicity states, i.e. when $\langle \mathbf{v} \cdot \mathbf{b} \rangle / (\langle v^2 \rangle + \langle b^2 \rangle)$ is high, the inertial range observed in the spectrum extends to such larger scales (Wicks et al. 2013). It will be interesting as well to verify the influence of the solar wind expansion time τ_{exp} (in comparison with the non-linear time) on the extension of the inertial range (see our discussion in the introduction).

The third order moment law provides an experimental estimate of the mean energy transfer rates $\varepsilon^{\pm}$, a measurement which is not possible otherwise, as the solar wind dissipation mechanisms (and so the viscosity η) are unknown. Solar wind energy transfer rates have been shown to lie between $\sim 0.1\ \mathrm{kJ\,kg^{-1}\,s^{-1}}$ (in *Ulysses* high latitude fast wind data, far from the Earth) and up to $\sim 10\ \mathrm{kJ\,kg^{-1}\,s^{-1}}$ in slow ecliptic wind at 1 AU (Sorriso-Valvo et al. 2007; Marino et al. 2008, 2012; MacBride et al. 2008; Smith et al. 2009). The rate of occurrence of the linear scaling in the solar wind time series, and the corresponding energy transfer rate, have been related to several solar wind parameters. For example, the energy transfer rate has been shown to anti-correlate with the cross-helicity level (Smith et al. 2009; Stawarz et al. 2010; Marino et al. 2011, 2012; Podesta 2011), confirming that alignment between velocity and magnetic field reduces the turbulent cascade, as expected for Alfvénic turbulence (Dobrowolny et al. 1980; Boldyrev 2006). Relationships with heliocentric distance and solar activity have also been pointed out, with controversial results (Marino et al. 2011, 2012; Coburn et al. 2012).

The estimation of the turbulent energy transfer rate has also shown that the electromagnetic turbulence may explain the observed solar wind non-adiabatic profile of the total proton temperature (Vasquez et al. 2007; Marino et al. 2008; MacBride et al. 2008; Stawarz et al. 2009). However, this explanation does not take into account a possible ion temperature anisotropy, known to be important in the solar wind (see Sect. 3.2). Indeed, the weakly collisional protons exhibit important temperature anisotropies (and complicated departures from a Maxwellian shape, Marsch et al. 1982) and they have non double-adiabatic temperatures profiles. *Helios* observations indicate that protons need to be heated in the perpendicular direction from 0.3 to 1 AU, but in the parallel direction they need to be cooled at 0.3 AU. This cooling rate gradually transforms to a heating rate at 1 AU (Hellinger et al. 2011, 2013). It is not clear if the turbulent cascade may cool the protons in the parallel direction (and transform this cooling to heating by 1 AU).

The phenomenological inclusion of possible contributions of density fluctuations to the turbulent energy transfer rate resulted in enhanced energy flux, providing a more efficient mechanism for the transport of energy to small scales (Carbone et al. 2009b).

Anisotropic corrections to the third order law have also been explored using anisotropic models of solar wind turbulence (MacBride et al. 2008; Carbone et al. 2009a; Stawarz et al. 2009, 2010; MacBride et al. 2010; Osman et al. 2011).

It is important to keep in mind that the solar wind expansion, the large scale velocity shears and the stream-stream interactions importantly affect the local turbulent cascade (Stawarz et al. 2011; Marino et al. 2012). Their effect on the turbulent energy transfer rate needs to be further investigated (Wan et al. 2009; Hellinger et al. 2013).

3 Turbulence at Kinetic Scales

At 1 AU, the MHD scale cascade finishes in the vicinity of ion characteristic scales ~ 0.1–0.3 Hz in the spacecraft frame. Here the turbulent spectra of plasma parameters (magnetic and electric fields, density, velocity and temperature) change their shape, and steeper spectra are observed at larger wave-numbers or higher frequencies, e.g. (Leamon et al. 1998; Bale et al. 2005; Alexandrova et al. 2007; Chen et al. 2012a; Šafránková et al. 2013). There is a range of terminology used to describe this range, including "dissipation range", "dispersion range" and "scattering range". The possible physics taking place here includes dissipation of turbulent energy (Leamon et al. 1998, 1999, 2000; Smith et al. 2006; Schekochihin et al. 2009; Howes et al. 2011b), a further small scale turbulent cascade (Biskamp et al. 1996; Ghosh et al. 1996; Stawicki et al. 2001; Li et al. 2001; Galtier 2006; Alexandrova et al. 2007, 2008; Schekochihin et al. 2009; Howes et al. 2011b; Rudakov et al. 2011; Boldyrev and Perez 2012) or a combination of both.

The transition from the MHD scale cascade to the small scale range is sometimes called the *ion spectral break* due to the shape of the magnetic field spectrum and to the scales at which it occurs. The physical processes responsible for the break and the corresponding characteristic scale are under debate. If the MHD scale cascade was filled with parallel propagating Alfvén waves, the break point would be at the ion cyclotron frequency f_{ci}, where the parallel Alfvén waves undergo the cyclotron damping. The oblique kinetic Alfvén wave (KAW) turbulence is sensitive to the ion gyroradius ρ_i (Schekochihin et al. 2009; Boldyrev and Perez 2012) and the transition from MHD to Hall MHD occurs at the ion inertial length λ_i (Galtier 2006; Servidio et al. 2007; Matthaeus et al. 2008, 2010).

Recent *Cluster* measurements of magnetic fluctuations up to several hundred Hz in the solar wind (Alexandrova et al. 2009, 2012; Sahraoui et al. 2010) show the presence of another spectral change at electron scales. At scales smaller than electron scales, the plasma turbulence is expected to convert from electromagnetic to electrostatic (with the important scale being the Debye length, see, e.g., Henri et al. 2011), but this is beyond the scope of the present paper.

The energy partitioning at kinetic scales, the spectral shape and the properties of the small scale cascade are important for understanding the dissipation of electromagnetic turbulence in collisionless plasmas.

3.1 Turbulence Around Ion Scales

Figure 7 shows an example of the solar wind magnetic field spectrum covering the end of the MHD inertial range and ion scales. The data are measured at 1 AU by *Cluster*/FGM (open circles) and *Cluster*/STAFF-SC (filled circles), which is more sensitive than FGM at high frequencies. One may conclude that the transition from the inertial range to another power-law spectrum is around ion scales, such as the ion cyclotron frequency $f_{ci} = 0.1$ Hz, the ion inertial scale λ_i corresponding to $f_{\lambda_i} = V_{sw}/(2\pi\lambda_i) \simeq 0.7$ Hz and the ion Larmor radius ρ_i appearing at $f_{\rho_i} = V_{sw}/(2\pi\rho_i) \simeq 1$ Hz. However, which of these ion scales is responsible for the spectral break is not evident from Fig. 7.

Leamon et al. (2000) performed a statistical study of the spectral break values f_b at 1 AU for different ion beta conditions, $\beta_i = nk_B T_i/(B^2/2\mu_0) \in [0.03, 3]$[10], with μ_0 being the vacuum magnetic permeability. The best correlation is found with the ion inertial length while

[10] In this study, the authors used the statistical sample from Leamon et al. (1998), i.e., 33 turbulent spectra up to ~ 3 Hz measured by *Wind* spacecraft within the slow and fast streams, $V_{sw} \in [300, 700]$.

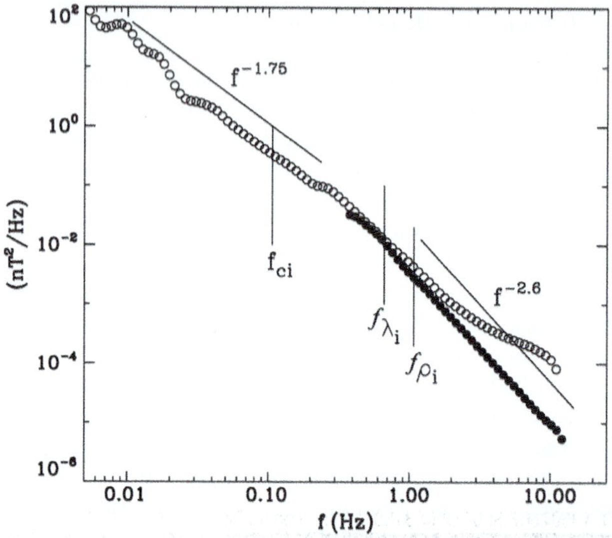

Fig. 7 Wavelet spectrum of magnetic fluctuations measured by *Cluster* in the solar wind up to 12.5 Hz for the time interval analyzed in Alexandrova et al. (2008). The *Cluster*/FGM spectrum is represented by open circles, *Cluster*/Search Coil (STAFF-SC) spectrum, by filled circles. The characteristic ion scales are marked by *vertical bars*

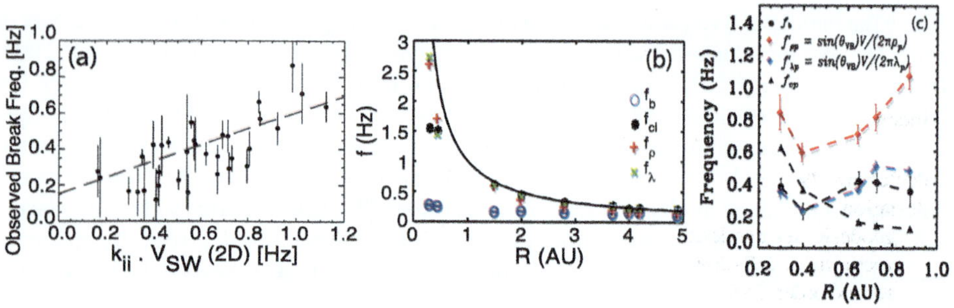

Fig. 8 (**a**) Observed ion break frequency f_b as a function of $f_{\lambda_i} = V_{sw} \sin\theta_{BV}/2\pi\lambda_i$, a correlation of 0.6 is observed (Leamon et al. 2000). (**b**) Radial evolution of f_b compared with the radial evolution of f_{ci}, $f_\rho = V_{sw}/2\pi\rho_i$ and $f_\lambda = V_{sw}/2\pi\lambda_i$: none of the ion scales follow the break (Perri et al. 2010). (**c**) Radial evolution of f_b (*black dots*) compared with f_{ci} (*black triangles*), $f_{\rho_p} = \sin\theta_{BV} V_{sw}/2\pi\rho_p$ (*red diamonds*) and $f_{\lambda_p} = \sin\theta_{BV} V_{sw}/2\pi\lambda_p$ (*blue diamonds*) (Bourouaine et al. 2012)

taking into account the 2D nature of the turbulent fluctuations, i.e. $k_\perp \gg k_\parallel$, see Fig. 8(a). A larger statistical sample of 960 spectra shows the dependence between f_b, and $f_{\lambda_i}\frac{B}{\delta B_b}$, where $\delta B_b/B$ is the relative amplitude of the fluctuations at the break scale (Markovskii et al. 2008). This result is still not explained. But, it is important to keep in mind that $\delta B_b/B$ is controlled by the ion instabilities in the solar wind when the ion pressure is sufficiently anisotropic (Bale et al. 2009), see Sect. 3.2 for more details.

A different approach has been used by Perri et al. (2010): the authors studied the radial evolution of the spectral break for distances $R \in [0.3, 5]$ AU. They showed that the ion break frequency is independent of the radial distance (see Fig. 8(b)). Bourouaine et al. (2012) explained this result by the quasi-bidimensional topology of the turbulent fluctuations, i.e. $k_\perp \gg k_\parallel$. When this wave vector anisotropy is taken into account, the Doppler shifted frequency $2\pi f = \mathbf{k} \cdot \mathbf{V}_{sw}$ can be approximated by $kV_{sw} \sin\theta_{BV}$. It appears that the ion inertial scale stays in the same range of frequencies as f_b, and a correlation of 0.7 is observed between f_b and $f_{\lambda_i} = V \sin\theta_{BV}/2\pi\lambda_i$, see Fig. 8(c).

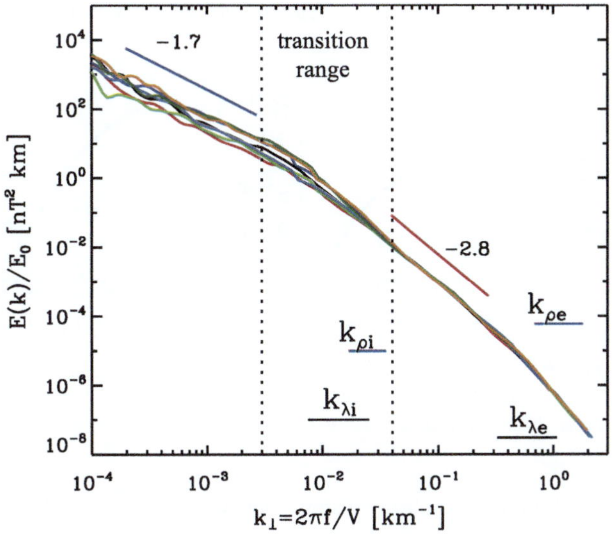

Fig. 9 7 solar wind spectra, analyzed in Alexandrova et al. (2009, 2010) under different plasma conditions as a function of the wave-vector $k_\perp$ perpendicular to the magnetic field. The spectra are superposed with a normalization factor E_0 at scales smaller than all ion scales: one observes divergence of the spectra in the transition range around the ion scales k_{ρ_i} and k_{λ_i}

As we have discussed above, the transition to kinetic Alfvén turbulence happens at the ion gyroradius ρ_i scale (Schekochihin et al. 2009; Boldyrev et al. 2012), while the dispersive Hall effect becomes important at the ion inertial length λ_i. Results of Leamon et al. (2000) and Bourouaine et al. (2012) indicate, therefore, that the Hall effect may be responsible for the ion spectral break. Note that Bourouaine et al. (2012) analyzed *Helios* data only within fast solar wind streams with $\beta_i < 1$, i.e. when $\lambda_i > \rho_i$.[11] It is quite natural that the largest characteristic scale (or the smallest characteristic wave number) affects the spectrum first (Spangler and Gwinn 1990). It will be interesting to verify these results for slow solar wind streams and high β_i regimes.

Just above the break frequency, $f > f_b$, the spectra are quite variable. Smith et al. (2006) show that within a narrow frequency range [0.4–0.8] Hz, the spectral index α varies between -4 and -2. This result was obtained using *ACE*/FGM measurements. However, one should be very careful while analyzing FGM data at frequencies higher than the ion break (i.e. at $f > 0.3$ Hz), where the digitalization noise becomes important (Lepping et al. 1995; Smith et al. 1998; Balogh et al. 2001). For example, in Fig. 7 the *Cluster*/FGM spectrum deviates from the STAFF spectrum at $f \geq 0.7$ Hz.[12]

Figure 9 shows several combined spectra, with *Cluster*/FGM data at low frequencies and *Cluster*/STAFF data at $f > f_b$. The spectra are shown as a function of the wave-vector $k_\perp$[13]. The spectra are superposed at $k_\perp > k_{\rho_i}, k_{\lambda_i}$, i.e. at scales smaller than all ion scales: while at these small scales all spectra follow the same law, around ion scales k_{ρ_i} and k_{λ_i} (named here a *transition range*) one observes a divergence of the spectra. The origin of this divergence is not completely clear. It is possible that ion damping (e.g. Denskat et al. 1983; Sahraoui et al. 2010), a competition between the convective and Hall terms (Kiyani et al.

[11] Ion plasma beta can be expressed in terms of ion scales: $\beta_i = 2\mu_0 n k_B T_i / B^2 = \rho_i^2/\lambda_i^2$.

[12] The digitalization noise at *Cluster*/FGM and at *ACE*/FGM is nearly the same, see Smith et al. (1998), Balogh et al. (2001).

[13] Cluster stays in the free solar wind not connected to the Earth's bow-shock, while the flow-to-field angle, θ_{BV}, is quasi-perpendicular. Therefore, only $k_\perp$ wave vectors are well resolved.

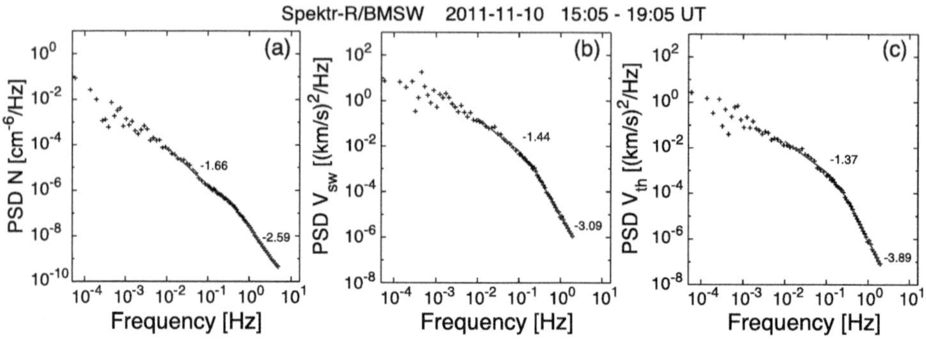

Fig. 10 Spectra of ion moments, (**a**) density, (**b**) velocity, (**c**) ion thermal speed, up to ~ 3 Hz as measured by *Spektr-R*/BMSW (Bright Monitor of Solar Wind) in the slow solar wind with $V_{sw} = 365$ km/s and $\beta_p \simeq 0.2$. Figure from Šafránková et al. (2013)

2013) or ion anisotropy instabilities (Gary et al. 2001; Matteini et al. 2007, 2011; Bale et al. 2009) may be responsible for the spectral variability within the transition range.

One of the important properties of the transition range is that the turbulent fluctuations become more compressible here (Leamon et al. 1998; Alexandrova et al. 2008; Hamilton et al. 2008; Turner et al. 2011; Salem et al. 2012; Kiyani et al. 2013). Let us define the level of compressibility of magnetic fluctuations as $\delta B_\|^2/\delta B_{tot}^2$, with δB_{tot}^2 being the total energy of the turbulent magnetic field fluctuations at the same scale as $\delta B_\|$ is estimated. If in the inertial range the level of compressibility is about 5 %, for $f > f_b$ it can reach 30 % and it depends on the plasma beta β_i (Alexandrova et al. 2008; Hamilton et al. 2008). The increase of the compressibility at kinetic scales has been attributed to the compressive nature of kinetic Alfvén or whistler turbulence (Gary and Smith 2009; Salem et al. 2012; TenBarge et al. 2012). On the other hand, it can be described by the compressible Hall MHD (Servidio et al. 2007). In particular, in the this framework, different levels of compressibility can also explain the spectral index variations in the transition range (Alexandrova et al. 2007, 2008).

The flattening of the electron density spectrum from $\sim f^{-5/3}$ to $\sim f^{-1}$, seen in Fig. 3, is observed within the same range of scales as the increase of the magnetic compressibility. The shape of this flattening is consistent with the transition between MHD scale Alfvénic turbulence and small scale KAW turbulence (Chandran et al. 2009; Chen et al. 2013a). More recently, Šafránková et al. (2013) measured the ion density spectrum within the transition range, finding similar results, as expected from the quasi-neutrality condition. In addition, they showed the ion velocity and temperature spectra in this range to be steeper with slopes around -3.4. An example of such spectra is shown in Fig. 10.

The transition range around ion scales is also characterized by magnetic fluctuations with quasi-perpendicular wave-vectors $k_\perp > k_\|$ and a plasma frame frequency close to zero (Sahraoui et al. 2010; Narita et al. 2011; Roberts et al. 2013). Sahraoui et al. (2010) interpret these observations as KAW turbulence, although Narita et al. (2011) found no clear dispersion relation. Magnetic fluctuations with nearly zero frequency and $k_\perp \gg k_\|$ can also be due to non-propagative coherent structures like current sheets (Veltri et al. 2005; Greco et al. 2010; Perri et al. 2012), shocks (Salem 2000; Veltri et al. 2005; Mangeney et al. 2001), current filaments (Rezeau et al. 1993), or Alfvén vortices propagating with a very slow phase speed $\sim 0.1V_A$ in the plasma frame (Petviashvili and Pokhotelov 1992; Alexandrova 2008). Such vortices are known to be present within the ion transition range

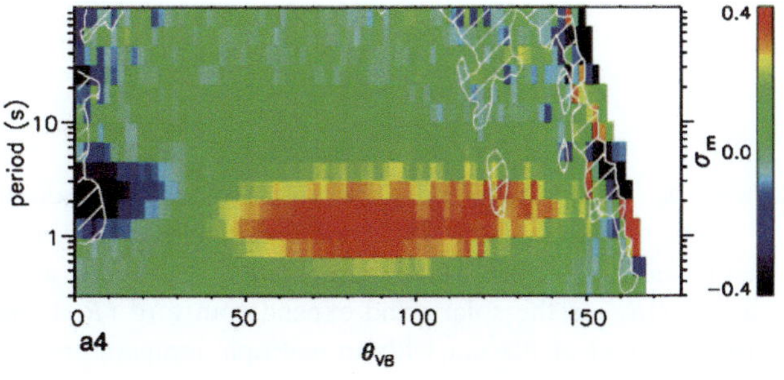

Fig. 11 Magnetic helicity σ_m for an outward magnetic sector as measured by *STEREO* spacecraft as a function of time scale τ (s) and angle to the magnetic field θ_{VB} (He et al. 2011b)

of the planetary magnetosheath turbulence, when ion beta is relatively low $\beta_i \leq 1$ (Alexandrova et al. 2006; Alexandrova and Saur 2008). Recent *Cluster* observations in the fast solar wind suggest that the ion transition range can be populated with KAWs and Alfvén vortices (Roberts et al. 2013).

As well as the spectrum of energy, the spectrum of magnetic helicity is also used to diagnose solar wind turbulence, and can tell us more details about the nature of the fluctuations (Matthaeus et al. 1982; Howes and Quataert 2010). Magnetic helicity is defined as $\langle \mathbf{A} \cdot \mathbf{B} \rangle$, where $\mathbf{B} = \nabla \times \mathbf{A}$, with $\mathbf{A}$ being the vector potential. It has been measured that at ion scales the magnetic helicity is anisotropic (He et al. 2011b). Figure 11 shows the reduced magnetic helicity[14] σ_m as a function of the time scale and of the local flow-to-field angle θ_{BV}. The authors found that, at time scales corresponding to the ion scales (1 to 10 s), there was a significant positive (negative) magnetic helicity signature for inward (outward) directed magnetic field in the parallel direction (i.e. for θ_{BV} close to 0 or to 180). This is consistent with left-hand parallel propagating Alfvén-ion-cyclotron waves. In the perpendicular direction, $\theta_{BV} \simeq 90°$, they found a magnetic helicity signature of the opposite sense: positive (negative) for outward (inward) field, consistent with the right-hand polarization, inherent to both whistler and kinetic Alfvén waves. Outside the range of frequencies (0.1–1) Hz, the magnetic helicity was generally zero. Podesta and Gary (2011) found the same result using *Ulysses* data and suggested the source of the parallel waves to be pressure anisotropy instabilities, which we will now discuss in more details.

3.2 Ion Scale Instabilities Driven by Solar Wind Expansion and Compression

The turbulent fluctuations, while cascading from the inertial range to the kinetic scales, will undergo strong kinetic effects in the vicinity of such ion scales as the ion skin depth or inertial scale λ_i, and near the thermal gyroradius ρ_i. At these small scales ion temperature anisotropy instabilities can occur (Gary et al. 2001; Marsch 2006; Matteini et al. 2007, 2011; Bale et al. 2009), and may remove energy from, or also inject it into, the turbulence.

As the solar wind expands into space, mass flux conservation leads to a density profile that falls roughly as $1/R^2$ (beyond the solar wind acceleration region); the magnetic field decays similarly, although the solar rotation and frozen flux condition ensure an azimuthal component to the field. If the solar wind plasma remains (MHD) fluid-like, then the double-adiabatic conditions (also called the Chew-Goldberger-Low or 'CGL') will apply (Chew et al. 1956) and will serve to modify adiabatically the plasma pressure components such that:

[14] I.e. the magnetic helicity measured along the satellite trajectory.

$$\frac{p_{\|}B^2}{\rho^3} = \text{const} \tag{4}$$

$$\frac{p_{\perp}}{\rho B} = \text{const}, \tag{5}$$

with $p_{\|,\perp}$ being the ion pressure along ($\|$) and perpendicular ($\perp$) to the mean field **B**.

Taken together, the CGL conditions suggest that an adiabatically transported fluid element should see its temperature ratio $T_{\perp}/T_{\|}$ fall as approximately $1/R^2$ between 10 and $100R_s$, as the solar wind expands outward (R_s being the radius of the Sun). Therefore a parcel of plasma with an isotropic temperature ($T_{\perp}/T_{\|} \sim 1$) at the edge of the solar wind acceleration region ($\sim 10R_s$) will arrive at 1 AU in a highly anisotropic state $T_{\|} \sim 100T_{\perp}$, if it remains adiabatic. Such a large temperature anisotropy has never been observed in the solar wind because the CGL conditions do not take into account wave-particle interactions or kinetic effects, which can control plasma via different types of instabilities.

Several early authors studied this possibility and looked for evidence of instability (Gary et al. 1976, 1996; Kasper 2002; Hellinger et al. 2006; Matteini et al. 2007; Bourouaine et al. 2010). Relatively recent results of Bale et al. (2009) using well-calibrated, statistical measurements from the *Wind* spacecraft have shown that the proton temperature anisotropy $T_{\perp}/T_{\|}$ is constrained by the $\beta_{\|}$[15]-dependent thresholds for the oblique firehose instability (for $T_{\perp}/T_{\|} < 1$) and the mirror-mode instability (for $T_{\perp}/T_{\|} > 1$) suggesting that the growth of ion-scale fluctuations acts to isotropize the plasma near the thresholds (Gary 1993). Indeed, a build-up of magnetic fluctuation power is observed near these thresholds (Bale et al. 2009) and the fluctuations seen near the mirror threshold and for $\beta_{\|} > 1$ are compressive, as would be expected from the growth of mirror waves (Hasegawa 1969). Figure 12 (left) shows time series data of magnetic and velocity fluctuations as the solar wind approaches the oblique firehose instability threshold: the top panel shows measurements of the ion temperature anisotropy (black dots) and the theoretical instability thresholds (Hellinger et al. 2006) as dotted lines. When the solar wind approaches the firehose threshold (black dotted line), enhanced fluctuation power is observed in the perpendicular components of the magnetic field and velocity, consistent with Alfvénic-like fluctuations excited by the firehose instability (Hellinger and Matsumoto 2000, 2001). Figure 12 (right) shows an example when the plasma conditions are close to both, mirror and firehose instability thresholds, and when both types of fluctuations, Alfvénic and compressive, are excited.

Figure 13 is reproduced from Bale et al. (2009) and shows statistically the effect seen in Fig. 12. One continuing puzzle here is the following: the instability thresholds, with the rate $\gamma \simeq 10^{-3}2\pi f_{ci}$, calculated by Hellinger et al. (2006) suggest that the ion cyclotron instability should be unstable at values of $T_{\perp}/T_{\|}$ lower than the mirror instability (at low $\beta_{\|}$), however there is no clear evidence in the data of an ion cyclotron limit. One reason for this may be that the mirror mode is non-propagating, and therefore more effective in pitch angle scattering. In any case, this is unresolved.

The clear existence of instability-limited anisotropies, and the measurement of the associated ion-scale fluctuations, bring to light a very important question: how much of the fluctuation power (magnetic, velocity, or other) measured near the ion scales in the solar wind is generated by instabilities, rather than driven by the turbulent cascade?

Figure 14 shows the probability distribution (see the yellow histogram) of parallel ion beta $\beta_{\|}$, using the *Wind* dataset described in Bale et al. (2009). The colored lines show the

[15] Parallel ion beta is defined with the parallel ion temperature, $\beta_{\|} = nk_BT_{\|}/(B^2/2\mu_0)$.

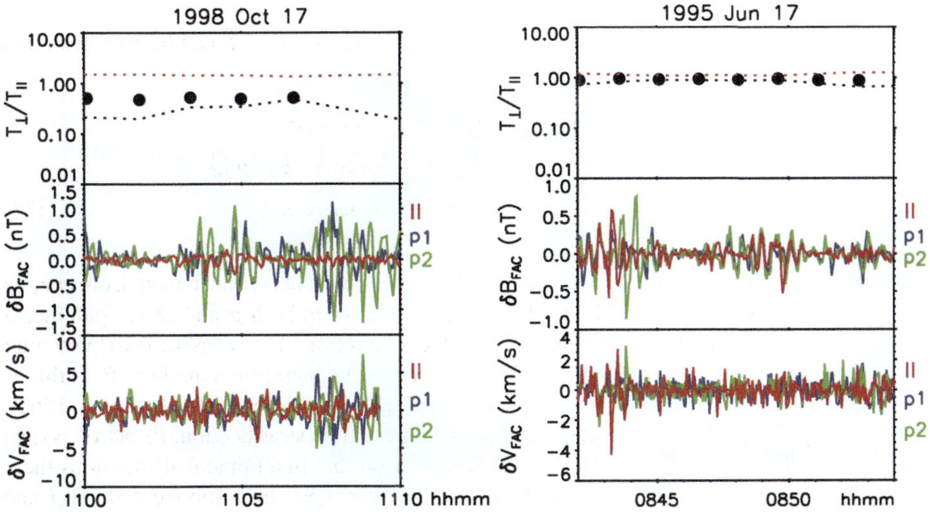

Fig. 12 (*Left*) time series data of measured proton temperature anisotropy (*dots*) and instability thresholds (*top panel*), of magnetic (*2nd panel*) and velocity (*3rd panel*) vector fluctuations in a field-aligned coordinate system (FAC), using 3 second measurements from the *Wind*/3DP instrument; *red lines* indicate fluctuations parallel to the mean field **B**, *p*1 (*violet*) and *p*2 (*green*) represent the two perpendicular components. As the measured proton anisotropy approaches the oblique firehose instability threshold (*black dotted line* in the *top panel*), Alfvénic-like fluctuations are excited and visible as perpendicular magnetic and velocity perturbations. (*Right*) the same format as left figure, but for the high ion beta regime, when the plasma conditions were close to both, mirror and firehose instability thresholds: both types of fluctuations, Alfvénic-like and compressive, are excited

cumulative distribution of "unstable" measurements, i.e. data points around and beyond the theoretical instability thresholds indicated in Fig. 13 by dotted lines. The black line gives the sum of all colored histograms. For solar wind intervals with $\beta_\parallel \geq\sim 3$, more than 20 % of the intervals would be unstable. However, the magnetic field fluctuation measurements, shown in Fig. 13, suggest that the power is enhanced well before the thresholds—hence the effect may be much larger.

It seems that the magnetic and velocity fluctuation power is injected near the ion scales by instabilities, whose energy source is solar wind expansion or compression, and that this effect is dependent on the plasma β. These quasi-linear ion instabilities co-exist with the non-linear turbulent cascade in the solar wind. Therefore, if the goal is to study cascade physics, care must be taken when studying ion scale fluctuations, to be certain that the plasma is very near to isotropic $T_\perp/T_\parallel \sim 1$ to avoid the quasi-linear ion instabilities. Interestingly, the bottom panel of Fig. 13, which shows the collisional age of protons,[16] demonstrates that the condition $T_\perp/T_\parallel \sim 1$ corresponds to a solar wind plasma that is collisionally well-processed ('old') and so remains 'fluid-like', rather than kinetic. The measurements of 'kinetic' turbulence must be qualified by considering the particle pressure anisotropies, and relative drifts between protons and α-particles and protons and electrons (Chen et al. 2013b; Perrone et al. 2013).

[16] The collisional age is defined as $\tau_{coll} = \nu_{pp} R/V_{sw}$, the Coulomb proton-proton collision frequency ν_{pp} multiplied by the transit time (or expansion time) from the Sun to 1 AU and is an estimate of the number of binary collisions in each plasma parcel during transit from the Sun to the spacecraft.

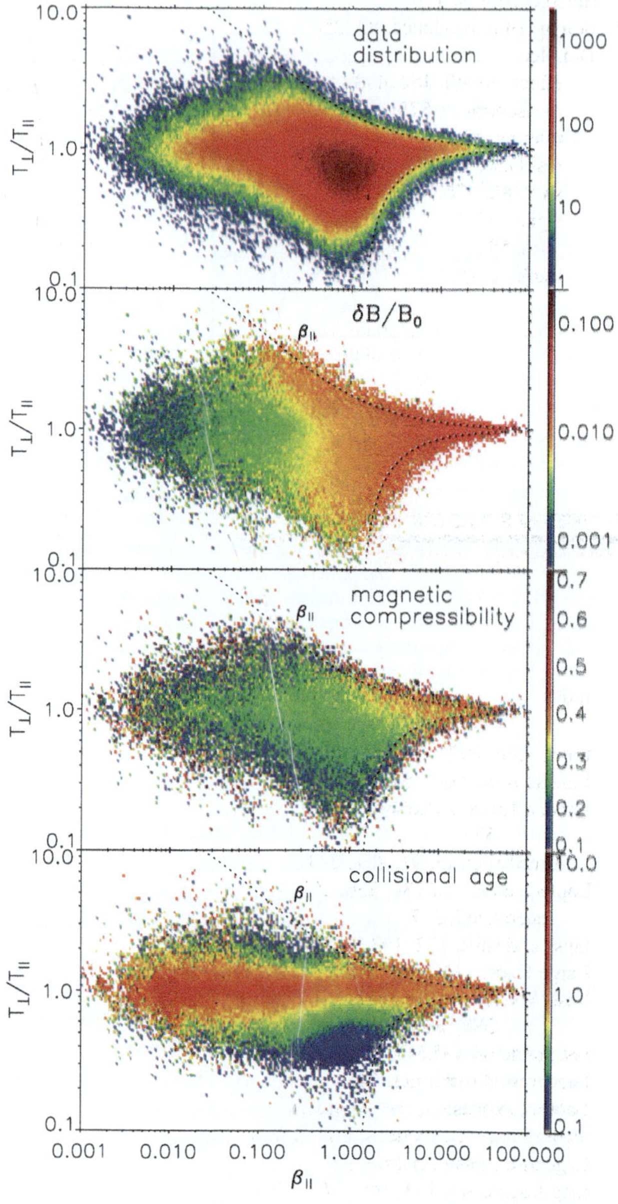

Fig. 13 Temperature anisotropy $T_\perp/T_{||}$ vs plasma parallel beta $\beta_{||}$ from Bale et al. (2009). The *upper panel* shows the constraint of plasma by the mirror (*upper dashed line*) and oblique firehose (*lower dashed line*) instabilities, as shown by Hellinger et al. (2006). The *second panel* shows a statistical enhancement of magnetic fluctuations $\delta B/B$ (calculated at $f = 0.3$ Hz, i.e. close to the ion spectral break) near the thresholds and at higher $\beta_{||}$. The *third panel* shows the distribution of the magnetic compressibility $\delta B_{||}/\delta B$ (at ion scales as well) and is consistent with mirror instability near that threshold. The *fourth panel* shows the collisional age of the ions (i.e. the number of collisions suffered by a thermal ion between the Sun and the spacecraft at 1 AU) in the same parameter plane

3.3 Small Scale Inertial Range Between Ion and Electron Scales, and Dissipation at Electron Scales

As far as the turbulent cascade crosses the ion scales and before reaching the electron scales (the satellite frequencies being $3 \leq f \leq 30$ Hz), magnetic spectra follow $\sim k_\perp^{-2.8}$ (Alexandrova et al. 2009; Chen et al. 2010a; Sahraoui et al. 2010), see Fig. 9. This spectral shape seems to be independent of the local plasma parameters, as far as the angle between the flow and the field θ_{BV} is quasi-perpendicular (Alexandrova et al. 2009, 2012).

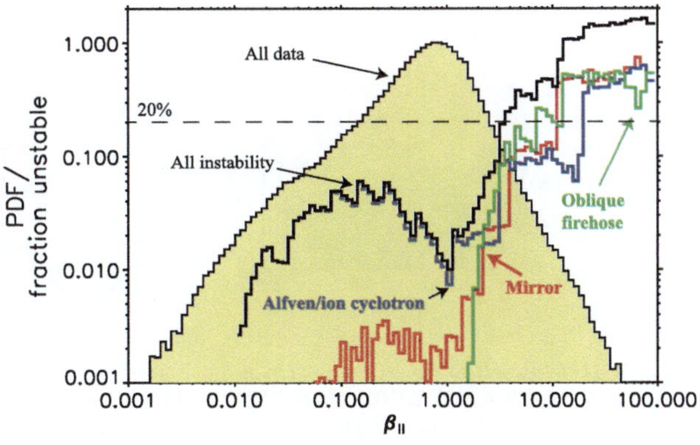

Fig. 14 Probability distribution of parallel ion beta $\beta_{\|}$ in the data set analyzed by Bale et al. (2009). The total distribution is shown in *yellow*; the most probable value of $\beta_{\|}$ in the solar wind is around 0.8. The *various colored lines* show the normalized histograms of occurrence of data at and beyond a certain threshold for different types of instabilities, as calculated by Hellinger et al. (2006), the *black line* gives the sum of all colored histograms: at high $\beta_{\|}$, more than 20 % of the solar wind is unstable

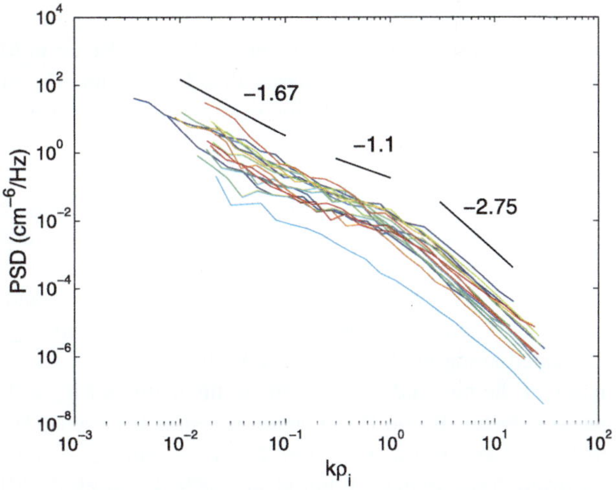

Fig. 15 17 electron density spectra normalized in scale to the ion gyroradius, showing a flattening at ion scales $\sim (k\rho_i)^{-1}$, as in Fig. 3, and a slope close to –2.75 between ion and electron scales Chen et al. (2013a) in agreement with the magnetic spectrum at these scales, see Fig. 9

The electron density spectrum between ion and electron scales was determined by Chen et al. (2012a, 2013a) using the high frequency measurements of spacecraft potential on *ARTEMIS*. Figure 15 shows 17 electron density spectra normalized to the ion gyroradius, measured for $\theta_{BV} > 45°$ in the solar wind. At large scales, the spectra are in agreement with previous observations (see Fig. 3). At small scales, for $k\rho_i \geq 3$ the electron density spectra follow the $\sim k^{-2.75}$ power-law, which is close to the typical value of –2.8 found in the magnetic field spectrum.

The observations of well defined power-laws in magnetic and density spectra between ion and electron scales suggest that at these scales there is a small scale inertial range (Alexandrova et al. 2007, 2008, 2009; Kiyani et al. 2009; Chen et al. 2010a, 2012a; Sahraoui et al. 2010) or an electron inertial range (Smith et al. 2012).

Kolmogorov arguments for Electron MHD lead to a $\sim k^{-7/3}$ magnetic energy spectrum (Biskamp et al. 1996, 1999; Cho and Lazarian 2004). More recent theories of strong KAW turbulence also predict a –7/3 spectrum for both density and magnetic field (Schekochihin et al. 2009). The fact that the observed spectra are typically steeper than this has been explained in several ways, including electron Landau damping (Howes et al. 2011b), compressibility effect (Alexandrova et al. 2007) and an intermittency correction resulting in a spectral index of $-8/3$ (Boldyrev and Perez 2012). The same spectral index of $-8/3$ can

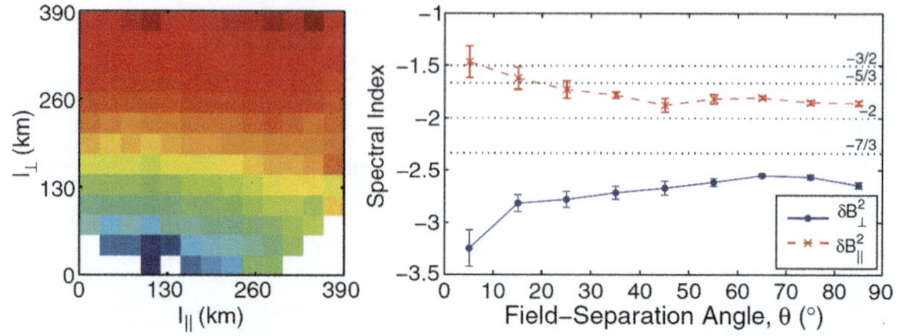

Fig. 16 (*Left*) Power in $\delta B_\perp$ (in color) as a function of parallel and perpendicular scale between ion and electron scales (Chen et al. 2010a). (*Right*) Spectral index as a function of angle θ_B (the angle between **B** and the separation vector between *Cluster* satellites) for the perpendicular $\delta B_\perp$ and parallel $\delta B_\parallel$ field components (Chen et al. 2010a)

be also obtained in quasi-bidimentional strong Electron MHD turbulence ($k_\perp \gg k_\parallel$) when parallel cascade is weak (Galtier et al. 2005). A model of Rudakov et al. (2011) of KAW turbulence with nonlinear scattering of waves by plasma particles gives spectral index between 2 and 3.

As we have mentioned, the magnetic and density spectra of Figs. 9 and 15 are measured for quasi-perpendicular θ_{BV}. Varying this angle, one may resolve turbulent fluctuations with different **k**, as discussed in Sect. 2.1. Chen et al. (2010a) used a multi-spacecraft technique to measure the wavevector anisotropy of the turbulence between ion and electron scales (up to ~ 10 Hz) using two-point structure functions. They found the turbulence to be anisotropic in the same sense as in the MHD scale cascade, with $k_\perp > k_\parallel$, corresponding to "eddies" elongated along the local mean field direction (Fig. 16, left). They also found the spectral index of the perpendicular magnetic fluctuations $\delta B_\perp$ to become steeper for small θ_B (the angle between **B** and the separation vector between *Cluster* satellites), i.e. for **k** parallel to **B** (Fig. 16, right), suggestive of strong whistler or kinetic Alfvén turbulence (Cho and Lazarian 2004; Schekochihin et al. 2009; Chen et al. 2010b; Boldyrev and Perez 2012). Note that two-point structure functions cannot resolve spectral indices steeper than −3, e.g. Abry et al. (1995, 2009), Chen et al. (2010a). So, it is possible that the parallel spectral index of $\delta B_\perp$ is steeper than what is shown in Fig. 16 (right).

Recently, Turner et al. (2011) studied anisotropy of the magnetic fluctuations up to ~ 20 Hz. The authors used the reference frame based on the mean magnetic field and velocity, which allow to check the axisymmetry and importance of the Doppler shift for $k_\perp$ fluctuations (Bieber et al. 1996). The authors found that the spectrum of magnetic fluctuations in the direction perpendicular to the velocity vector in the plane perpendicular to **B**, $\mathbf{V}_{sw\perp}$, is higher than the spectrum of δB along $\mathbf{V}_{sw\perp}$. This is consistent with a turbulence with $k_\perp \gg k_\parallel$, where the fluctuations with **k** along $\mathbf{V}_{sw\perp}$ are more affected by the Doppler shift than the fluctuations with **k** perpendicular to $\mathbf{V}_{sw\perp}$. These results are also in agreement with the magnetosheath observations between ion and electron scales (Alexandrova et al. 2008).

What happens at smaller scales? Several authors have suggested that the electromagnetic turbulent cascade in the solar wind dissipates at electron scales. These scales are usually called *electron dissipation range*, e.g. Smith et al. (2012).

Figure 17 is reproduced from Alexandrova et al. (2012). The upper panel shows a number of magnetic field spectra measured under different plasma conditions: 100 spectra from ion scales to a fraction of electron scales, and 7 spectra measured from the MHD range to a

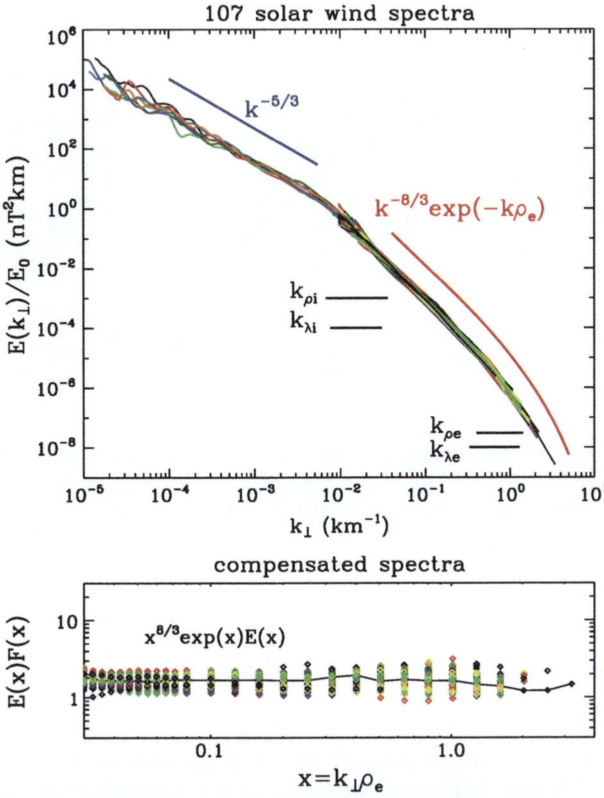

Fig. 17 100 magnetic field spectra in the kinetic range to a fraction of electron scales and 7 magnetic field spectra covering fluid and kinetic scales, with spectra compensated to $(k_\perp\rho_e)^{8/3}\exp(k_\perp\rho_e)$ in the *lower panel* (Alexandrova et al. 2012)

fraction of electron scales. At scales smaller than the ion scales ($k_\perp > k_{\rho i}, k_{\lambda i}$), all the spectra can be described by one algebraic function covering electron inertial and dissipation ranges,

$$E(k_\perp) = E_0 k_\perp^{-\alpha} \exp(-k_\perp \ell_d) \tag{6}$$

where $\alpha \simeq 8/3$ and where ℓ_d is found to be related to the electron Larmor radius ρ_e, with a correlation coefficient of 0.7. This law is independent of the solar wind properties, slow or fast, and of ion and electron plasma beta, indicating the universality of the turbulent cascade at electron scales. The compensated 100 spectra with the $k_\perp^{8/3}\exp(k_\perp\rho_e)$–function are shown in the bottom panel of Fig. 17: they are flat over about 2 decades confirming the choice of the model function

$$E(k_\perp) = E_0 k_\perp^{-8/3} \exp(-k_\perp \rho_e) \tag{7}$$

to describe solar wind spectrum at such small scales.

It is interesting that a similar curved spectrum is expected in the Interstellar Medium turbulence, but at ion scales (Spangler and Gwinn 1990; Haverkorn and Spangler 2013).

Another description of the spectrum within the electron inertial and dissipation ranges was proposed by Sahraoui et al. (2010). It consists of two power-laws separated by a break, see Fig. 18 (left). This double-power-law model can be formulated as

$$\tilde{E}(k_\perp) = A_1 k_\perp^{-\alpha_1}\big(1 - H(k_\perp - k_b)\big) + A_2 k_\perp^{-\alpha_2} H(k_\perp - k_b), \tag{8}$$

$H(k_\perp - k_b)$ being the Heaviside function, k_b the wave number of the break, $A_{1,2}$ the amplitudes of the two power-law functions with spectral indices $\alpha_{1,2}$ on both sides of k_b. This

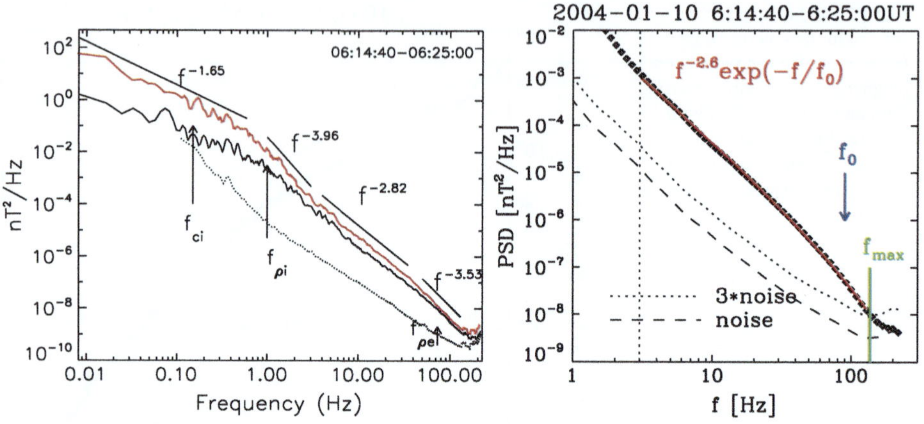

Fig. 18 (*Left*) Magnetic spectrum from Sahraoui et al. (2010), compared with $\sim f^{-2.8}$ for $4 \le f \le 35$ Hz and with $\sim f^{-3.5}$ for $50 \le f \le 120$ Hz, the break frequency is around 40 Hz. (*Right*) A zoom on the high frequency part of the spectrum on the left, fitted with $\sim f^{-2.6}\exp(-f/f_0)$, the exponential cut-off frequency $f_0 = 90$ Hz is close to the Doppler shifted ρ_e, $f_0 \simeq f_{\rho_e} = V_{sw}/2\pi\rho_e$. This last fitting function is equivalent to the model (7) for wave vectors

model has five free parameters. A statistical study of the solar wind magnetic spectra at high frequencies ($f > 3$ Hz) shows that α_1 does not vary a lot, $\alpha_1 = 2.86 \pm 0.08$ (Alexandrova et al. 2012). Then the amplitudes A_1 and A_2 are equal at the break point. Therefore we can fix two of the five parameters of model (8). This model has thus three free parameters, A_1, α_2 and k_b (in comparison with one free parameter, E_0, in Eq. (7)).

Figure 18 (left) shows the frequency spectrum from (Sahraoui et al. 2010), compared at high frequencies[17], $f > 3$ Hz, with the double power-law model (8) with $\alpha_1 \simeq 2.8$, $\alpha_2 \simeq 3.5$ and the spectral break at $f_b \simeq 40$ Hz. Figure 18 (right) shows the total power spectral density for the same dataset fitted with the exponential model (6), which can be written for frequency spectrum as $\sim f^{-\alpha}\exp(-f/f_0)$. The parameters of the fit are $\alpha \simeq 8/3$ and the exponential cut-off frequency $f_0 = 90$ Hz, which is close to the Doppler shifted electron gyro-radius ρ_e for this time interval. Therefore, the model (7) can be applied in this particular case as well.

In the statistical study by Alexandrova et al. (2012), the authors concluded that model function (7) describes all observed spectra, while the double-power-law model (8) cannot describe a large part of the observed spectra. Indeed the unique determination of the spectral break k_b with $A_1 = A_2$ at the break is not always possible because of the spectral curvature, and for low intensity spectra there are not enough data points to allow a good determination of α_2.

The equivalence between the electron gyro-radius ρ_e, in the solar wind turbulence, and the dissipation scale ℓ_d, in the usual fluid turbulence, can be seen also from Fig. 19 where the Universal Kolmogorov Function $E(k)\ell_d/\eta^2$ is shown as a function of $k\ell_d$ (Frisch 1995; Davidson 2004), for three different candidates for the dissipation scale ℓ_d, namely for ρ_i, λ_i and ρ_e; and for one time characteristic scale, namely the electron gyro-period f_{ce}^{-1}. For simplicity, the kinematic viscosity η is assumed to be constant, despite the varying plasma conditions. One can see that the ρ_i and λ_i normalizations are not efficient to collapse the spectra together. Normalization on λ_e gives the same result as for λ_i. At the same time,

[17] *Cluster*/Staff-SC measurements in the burst mode.

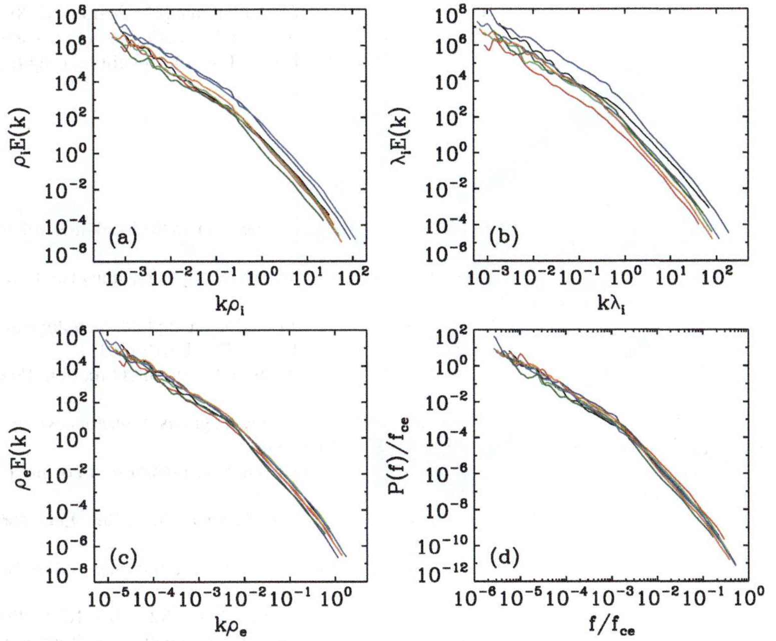

Fig. 19 Universal Kolmogorov function $\propto \ell_d E(k)$ for hypothesized dissipation scales ℓ_d as a function of (**a**) $k\rho_i$, (**b**) $k\lambda_i$, (**c**) $k\rho_e$ and (**d**) f/f_{ce}. Figure from Alexandrova et al. (2009), corrected for 3 STAFF-SA frequencies, as explained in Alexandrova et al. (2012)

the normalizations on ρ_e and f_{ce} bring the spectra close to each other, as expected while normalizing by ℓ_d. In addition to the spectral analysis presented in Fig. 17, the Universal Kolmogorov Function normalization gives an independent confirmation that the spatial scale which may play the role of the dissipation scale, in the weakly collisional solar wind, is the electron gyro-radius ρ_e.

It is important to mention, that the amplitude parameter E_0 of the exponential model (7) is found to be related to the solar wind plasma parameters (Alexandrova et al. 2011), see Fig. 20. The amplitude of the raw frequency spectra is found to be related to the ion thermal pressure as $\sim nk_BT_i$ (Fig. 20, upper line). This is similar to the amplitude of the inertial range spectrum, which is found to be correlated to the ion thermal speed $\sim V_{th}^2$ (Grappin et al. 1990). The amplitude of the k-spectra, as well as the amplitude of the normalized $k\rho_e$-spectra, appears to depend on the ion temperature anisotropy as $\sim (T_{i\perp}/T_{i\parallel})^{1.6\pm0.1}$ (Fig. 20, lower line). This last result suggests that the ion instabilities present around the ion break scale may indeed inject or remove energy from the cascade (see our discussion in Sect. 3.2). Therefore, the scales around the ion break (or 'transition range', see Fig. 9) may be seen, in part, as the energy injection scales for the small scale inertial range.

In usual fluid turbulence, the far dissipation range is described by $E(k) \sim k^3 \exp(-ck\ell_d)$ (with $c \simeq 7$) (Chen et al. 1993). The exponential tail is due to the resistive damping rate $\gamma \propto k^2$ valid in a collisional fluid. In the collisionless plasma of the solar wind there is no resistive damping, and thus the observation of the exponential spectrum within the electron dissipation range deserves an explanation.

Howes et al. (2011a) consider a model ("weakened cascade model") which includes the nonlinear transfer of energy from large to small scales in Fourier space and the damping

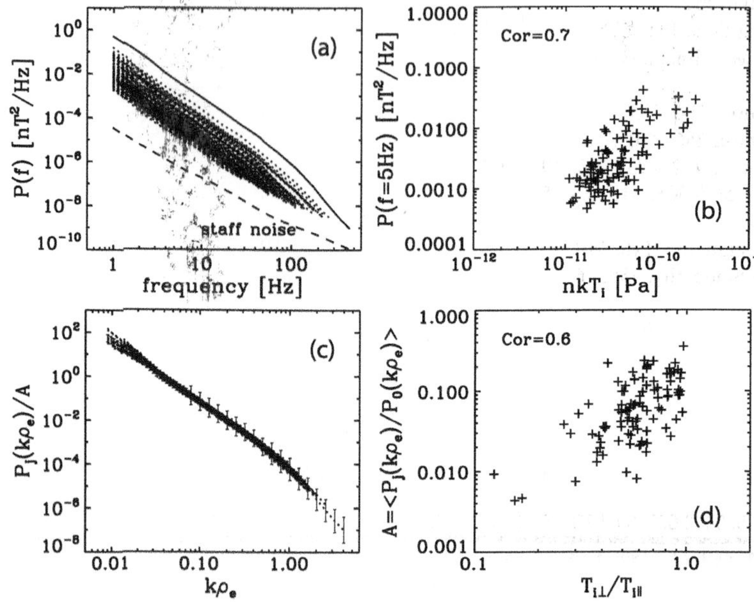

Fig. 20 (**a**) The 100 magnetic frequency spectra measured by *Cluster*/STAFF in the solar wind for $f > 1$ Hz, analyzed in Alexandrova et al. (2012); (**b**) intensity of the frequency spectra at a fixed frequency $f = 5$ Hz as a function of the ion thermal pressure nk_BT_i: dependence is $P \sim nk_BT_i$; (**c**) the same spectra as in panel (**a**) but shown as a function of $k\rho_e$ and superposed using an amplitude factor A (equivalent to E_0 in Fig. 17); (**d**) The amplitude A as a function of the ion temperature anisotropy $T_{i\perp}/T_{i\parallel}$: dependence $A \sim (T_{i\perp}/T_{i\parallel})^{1.6\pm0.1}$ is observed. Figure from Alexandrova et al. (2011)

of kinetic Alfvén waves. The spectral laws are respectively $E_k \propto k_\perp^{-5/3}$ at large scales and $E_k \propto k_\perp^{-7/3}$ between ion and electron scales. The damping becomes important at electron Larmor radius ρ_e scale. It is obtained by linearizing the Vlasov-Maxwell equations in the gyrokinetic limit ($k_\parallel \ll k_\perp$, with frequencies $f \ll f_{ci}$). For $k_\perp\rho_i \gg 1$ it has the form $\gamma \propto k_\parallel k_\perp^2$. Taking into account the assumption of critical balance $\tau_{nl} = \tau_A$ (i.e. $k_\perp v = k_\parallel V_A$) (Goldreich and Sridhar 1995), and the spectral index $-7/3$ (i.e. $v \sim k_\perp^{-2/3}$), one gets $k_\parallel \propto k_\perp^{1/3}$. Therefore, the damping term takes the form $\gamma \propto k_\perp^{2+1/3}$. The exponent of the damping rate is thus very close to the k^2 scaling of the Laplacian viscous term, which is known to lead in hydrodynamical turbulence to an exponential tail in the dissipation range. Indeed, when taking into account the damping term, Howes et al. (2011a) obtain numerically a final curved tail at scales smaller than electron scales. Superficially, this spectrum thus resembles the analytic form which we have found to be valid to describe the solar wind turbulence, Eq. (7).

As we have just seen, the model of Howes et al. (2011a) assumes the $k_\perp \gg k_\parallel$–anisotropy and very low frequencies $f \ll f_{ci}$. Present multi-satellite observations can not cover the electron inertial and dissipation ranges at scales smaller than the smallest satellite separation ~ 100 km. Only the one-satellite technique of Bieber et al. (1996) can be used. A first attempt to determine the distribution of wave-vectors **k** of the electromagnetic fluctuations within the electron inertial and dissipation ranges (for the observed frequencies [8, 500] Hz) was carried out in the magnetosheath by Mangeney et al. (2006). They show that the wavevectors **k** of the electromagnetic fluctuations are distributed within the plane nearly perpendicular to the mean field **B**, with an angle of $\sim \pm 5°$ around this plane. However, the authors have not found any agreement between the observed properties of magnetic fluctuations and KAW turbulence.

The nature of turbulence between ion and electron scales is still under debate. As with the MHD scale cascade, there are a number of observational and theoretical works, which identify turbulent fluctuations at small scales as having properties of linear wave modes (e.g., Denskat et al. 1983; Goldstein et al. 1994; Ghosh et al. 1996; Biskamp et al. 1996, 1999; Leamon et al. 1998; Cho and Lazarian 2004; Bale et al. 2005; Galtier 2006; Sahraoui et al. 2010, 2012; Howes et al. 2006, 2008, 2012b; Schekochihin et al. 2009; Gary and Smith 2009; Chandran et al. 2009; Chen et al. 2010b, 2013a; Salem et al. 2012; Klein et al. 2012; Boldyrev and Perez 2012). A recent analysis by Chen et al. (2013c) showed that the ratio of density to magnetic fluctuations in the range between ion and electron scales is very close to that expected for kinetic Alfvén waves, and not whistler waves, and concluded that the fluctuations in this range are predominantly strong kinetic Alfvén turbulence. The precise interplay between linear and non-linear physics is an important unsolved problem in plasma turbulence.

Solar wind observations and numerical simulations show that the fluctuations at kinetic scales have non-Gaussian distributions, indicating the presence of intermittency (Alexandrova et al. 2007, 2008; Kiyani et al. 2009; Wan et al. 2012; Wu et al. 2013). Recently, small scale coherent current sheets have been identified at scales close to electron scales (Perri et al. 2012). These features are consistent with strong, rather than weak (or wave) turbulence. The properties of the intermittency at small scales are not clear at the moment. There are two contradictory observations: (i) Alexandrova et al. (2008) show a scale dependent deviation from Gaussianity of the PDFs of the magnetic fluctuations δB_R (along the solar wind flow); (ii) Kiyani et al. (2009) show observations suggesting a scale-invariance within the small scales. Further work is needed to understand this discrepancy.

4 Discussion

In this paper we have discussed solar wind turbulence observations in a large range of scales: from the MHD scales to the electron characteristic scales.

At MHD scales, within the inertial range, the solar wind turbulence presents several general characteristics inherent to fully developed fluid turbulence: (i) energy spectra of different plasma parameters have well-defined power-laws; (ii) the probability distribution functions deviate from a Gaussian distribution, indicating stronger gradients at smaller scales (intermittency); (iii) the third order moments of turbulent fluctuations have the linear dependence on scale (the proportionality coefficient giving the energy transfer rate). The anisotropy of turbulence with respect to a mean magnetic field is shown to be important: the turbulence develops mostly in the plane perpendicular to **B**, i.e. with $k_\perp \gg k_\parallel$. The perpendicular magnetic spectrum follows $\sim k_\perp^{-5/3}$ scaling, while the parallel spectrum is steeper $\sim k_\parallel^{-2}$. The dominant fluctuations are Alfvénic in nature, i.e. $\delta B_\perp > \delta B_\parallel$, however, the velocity spectrum has a spectral slope of $-3/2$ and it does not follow the magnetic spectrum. There is a small fraction of the turbulent energy in compressible fluctuations. It is not clear whether they behave as a passive contaminant as in compressible neutral fluid turbulence or they are an active component of the turbulence in the solar wind. In other words, is it possible to describe these compressible fluctuations independently of the dominant Alfvénic cascade, or are they inherently coupled? This question is a matter of debate.

The MHD inertial range ends at ion characteristic scales. Here, different kinetic effects may take place and inject or remove energy from the turbulent cascade. In particular, the large scale energy reservoir related to the solar wind spherical expansion may be released into fluctuations, throughout instabilities, like mirror and oblique firehose instabilities, which becomes important for high ion betas ($\beta_i > 3$). Then these fluctuations may interact with particles and dissipate, or participate to the turbulent cascade at smaller scales. At lower beta, the plasma is stable and more or less isotropic: no additional energy is expected to arrive to the turbulent cascade. However, the exact energy partition between fluid and kinetic degrees of freedom at ion scales is still under debate. Around ion scales magnetic spectra are variable, and the compressibility increases. A spectral break seems to appear at the ion inertial scale, suggesting that dispersive effects (Hall effect) become significant.

Between ion and electron scales, a small scale turbulent cascade seems to be established. This cascade is characterized by a $k_\perp \gg k_\|$ anisotropy, as the MHD cascade. The $k_\perp$ magnetic and density spectra have a power-law shape with $\simeq -2.8$ spectral index. Fluctuations are more compressible than within the MHD inertial range and this compressibility seems to depend on the local plasma parameters, like the plasma β. Magnetic fluctuations are non-Gaussian, indicating the presence of the intermittency.

Approaching electron scales, the fluctuations are no more self-similar: the spectrum is no more a power-law, but an exponential cut-off is observed indicating an onset of dissipation. The dissipation range spectrum is observed to have a general shape. One algebraic function $\sim k_\perp^{-8/3} \exp(-k_\perp \rho_e)$ describes well the whole spectrum covering the small scale inertial range and the dissipation range.

The nature of the small scale cascade between ion and electron scales and the dissipation mechanism at electron scales are still under debate. The model of Howes et al. (2011a) can describe the observed exponential cut-off. The dissipation mechanism in this model is based on a quasi-linear description of the Landau damping of kinetic Alfvén waves onto electrons. Whether such description can apply on the solar wind observations is however under debate because of the presence of a significant degree of intermittency at kinetic scales.

To build a realistic model of the dissipation in the solar wind we need still to resolve an open question on the nature of the turbulent fluctuations: is it a mixture of linear waves or is it a strong turbulence with dissipation restricted to intermittent coherent structures? What is the topology of these structures—current sheets, shocks, solitons or coherent vortices?

Acknowledgements We all thank the International Space Science Institute (ISSI, Bern) for hospitality. S.D.B., C.H.K.C., T.H. and L.S.V. acknowledge the Marie Curie Project FP7 PIRSES-2010-269297—"Turboplasmas". OA thanks André Mangeney, Roland Grappin, Nicole Meyer, Robert Wicks and Petr Hellinger for discussions, Chadi Salem, Silvia Perri, William Matthaeus and Sofiane Bourouaine for providing figures, and Catherine Lacombe for reading this manuscript.

References

P. Abry, P. Gonçalves, P. Flandrin *Wavelets, spectrum analysis and 1/f processes. Wavelets and statistics*. Lecture Notes in Statistics (1995). http://perso.ens-lyon.fr/paulo.goncalves/pub/lns95.pdf. doi:10.1007/978-1-4612-2544-7_2

P. Abry, P. Gonçalves, J. Lévy Véhel, *Scaling, Fractals and Wavelets*. Digital Signal and Image Processing Series (ISTE/Wiley, London, 2009)

O. Alexandrova, Solar wind vs magnetosheath turbulence and Alfvén vortices. Nonlinear Process. Geophys. **15**, 95–108 (2008). doi:10.5194/npg-15-95-2008

O. Alexandrova, J. Saur, Alfvén vortices in Saturn's magnetosheath: Cassini observations. Geophys. Res. Lett. **35**, 15102 (2008). doi:10.1029/2008GL034411

O. Alexandrova, C. Lacombe, A. Mangeney, Spectra and anisotropy of magnetic fluctuations in the earth's magnetosheath: cluster observations. Ann. Geophys. **26**, 3585–3596 (2008). doi:10.5194/angeo-26-3585-2008
O. Alexandrova, A. Mangeney, M. Maksimovic, N. Cornilleau-Wehrlin, J.-M. Bosqued, M. André, Alfvén vortex filaments observed in magnetosheath downstream of a quasi-perpendicular bow shock. J. Geophys. Res. **111**(A10), 12208 (2006). doi:10.1029/2006JA011934
O. Alexandrova, V. Carbone, P. Veltri, L. Sorriso-Valvo, Solar wind cluster observations: turbulent spectrum and role of Hall effect. Planet. Space Sci. **55**, 2224–2227 (2007). doi:10.1016/j.pss.2007.05.022
O. Alexandrova, V. Carbone, P. Veltri, L. Sorriso-Valvo, Small-scale energy cascade of the solar wind turbulence. Astrophys. J. **674**, 1153–1157 (2008). doi:10.1086/524056
O. Alexandrova, J. Saur, C. Lacombe, A. Mangeney, J. Mitchell, S.J. Schwartz, P. Robert, Universality of solar-wind turbulent spectrum from MHD to electron scales. Phys. Rev. Lett. **103**(16), 165003 (2009). doi:10.1103/PhysRevLett.103.165003
O. Alexandrova, J. Saur, C. Lacombe, A. Mangeney, S.J. Schwartz, J. Mitchell, R. Grappin, P. Robert, Solar wind turbulent spectrum from MHD to electron scales, in *Twelfth International Solar Wind Conference*, vol. 1216 (2010), pp. 144–147. doi:10.1063/1.3395821
O. Alexandrova, C. Lacombe, A. Mangeney, R. Grappin Fluid-like dissipation of magnetic turbulence at electron scales in the solar wind. arXiv:1111.5649v1 (2011)
O. Alexandrova, C. Lacombe, A. Mangeney, R. Grappin, M. Maksimovic, Solar wind turbulent spectrum at plasma kinetic scales. Astrophys. J. **760**(2), 121 (2012). doi:10.1088/0004-637X/760/2/121
S.D. Bale, P.J. Kellogg, F.S. Mozer, T.S. Horbury, H. Reme, Measurement of the electric fluctuation spectrum of magnetohydrodynamic turbulence. Phys. Rev. Lett. **94**(21), 215002 (2005). doi:10.1103/PhysRevLett.94.215002
S.D. Bale, J.C. Kasper, G.G. Howes, E. Quataert, C. Salem, D. Sundkvist, Magnetic fluctuation power near proton temperature anisotropy instability thresholds in the solar wind. Phys. Rev. Lett. **103**, 211101 (2009). doi:10.1103/PhysRevLett.103.211101
A. Balogh, C.M. Carr, M.H. Acuña, M.W. Dunlop, T.J. Beek, P. Brown, K.-H. Fornaçon, E. Georgescu, K.-H. Glassmeier, J. Harris, G. Musmann, T. Oddy, K. Schwingenschuh, The cluster magnetic field investigation: overview of in-flight performance and initial results. Ann. Geophys. **19**, 1207–1217 (2001). doi:10.5194/angeo-19-1207-2001
A. Bershadskii, K.R. Sreenivasan, Intermittency and the passive nature of the magnitude of the magnetic field. Phys. Rev. Lett. **93**(6), 064501 (2004). doi:10.1103/PhysRevLett.93.064501
J.W. Bieber, W. Wanner, W.H. Matthaeus, Dominant two-dimensional solar wind turbulence with implications for cosmic ray transport. J. Geophys. Res. **101**, 2511–2522 (1996). doi:10.1029/95JA02588
D. Biskamp, *Nonlinear Magnetohydrodynamics* (Cambridge University Press, Cambridge, 1993)
D. Biskamp, E. Schwarz, J.F. Drake, Two-dimensional electron magnetohydrodynamic turbulence. Phys. Rev. Lett. **76**, 1264–1267 (1996). doi:10.1103/PhysRevLett.76.1264
D. Biskamp, E. Schwarz, A. Zeiler, A. Celani, J.F. Drake, Electron magnetohydrodynamic turbulence. Phys. Plasmas **6**, 751–758 (1999). doi:10.1063/1.873312
S. Boldyrev, Spectrum of magnetohydrodynamic turbulence. Phys. Rev. Lett. **96**(11), 115002 (2006). doi:10.1103/PhysRevLett.96.115002
S. Boldyrev, J.C. Perez, Spectrum of weak magnetohydrodynamic turbulence. Phys. Rev. Lett. **103**(22), 225001 (2009). doi:10.1103/PhysRevLett.103.225001
S. Boldyrev, J.C. Perez, Spectrum of kinetic-Alfvén turbulence. Astrophys. J. **758**, 44 (2012). doi:10.1088/2041-8205/758/2/L44
S. Boldyrev, J.C. Perez, J.E. Borovsky, J.J. Podesta, Spectral scaling laws in magnetohydrodynamic turbulence simulations and in the solar wind. Astrophys. J. **741**, 19 (2011). doi:10.1088/2041-8205/741/1/L19
S. Boldyrev, J.C. Perez, Y. Wang, Residual Energy in Weak and Strong MHD Turbulence, Numerical modeling of space plasma flows (astronum 2011), in *Proceedings of a 6th internation conference*, Velancia, Spain, 13–17 June, 2011, ed. by N.V. Pogorelov, J.A. Font, E. Audit, G.P. Zank, ASP Conference Series, vol. 459 (Astronomical Society of the Pacific, San Francisco, 2012), p. 3 Publication Date: 07/2012
J.E. Borovsky, Flux tube texture of the solar wind: strands of the magnetic carpet at 1 AU? J. Geophys. Res. **113**(A12), 8110 (2008). doi:10.1029/2007JA012684
J.E. Borovsky, Looking for evidence of mixing in the solar wind from 0.31 to 0.98 AU. J. Geophys. Res. **117**(A16), 6107 (2012a). doi:10.1029/2012JA017525
J.E. Borovsky, The velocity and magnetic field fluctuations of the solar wind at 1 AU: statistical analysis of Fourier spectra and correlations with plasma properties. J. Geophys. Res. **117**(A16), 5104 (2012b). doi:10.1029/2011JA017499
S. Bourouaine, E. Marsch, F.M. Neubauer, Correlations between the proton temperature anisotropy and transverse high-frequency waves in the solar wind. Geophys. Res. Lett. **37**, 14104 (2010). doi:10.1029/2010GL043697

S. Bourouaine, E. Marsch, F.M. Neubauer, Temperature anisotropy and differential streaming of solar wind ions. correlations with transverse fluctuations. Astron. Astrophys. **536**, 39 (2011). doi:10.1051/0004-6361/201117866

S. Bourouaine, O. Alexandrova, E. Marsch, M. Maksimovic, On spectral breaks in the power spectra of magnetic fluctuations in fast solar wind between 0.3 and 0.9 AU. Astrophys. J. **749**, 102 (2012). doi:10.1088/0004-637X/749/2/102

R. Bruno, V. Carbone, The solar wind as a turbulence laboratory. Living Rev. Sol. Phys. **2**, 4 (2005). doi:10.12942/lrsp-2005-4

R. Bruno, V. Carbone, P. Veltri, E. Pietropaolo, B. Bavassano, Identifying intermittency events in the solar wind. Planet. Space Sci. **49**(12), 1201–1210 (2001). Nonlinear Dynamics and Fractals in Space. http://www.sciencedirect.com/science/article/pii/S0032063301000617, doi:10.1016/S0032-0633(01)00061-7.

R. Bruno, V. Carbone, L. Sorriso-Valvo, B. Bavassano, Radial evolution of solar wind intermittency in the inner heliosphere. J. Geophys. Res. **108**, 1130 (2003). doi:10.1029/2002JA009615

R. Bruno, V. Carbone, L. Primavera, F. Malara, L. Sorriso-Valvo, B. Bavassano, P. Veltri, On the probability distribution function of small-scale interplanetary magnetic field fluctuations. Ann. Geophys. **22**, 3751–3769 (2004). doi:10.5194/angeo-22-3751-2004

R. Bruno, R. D'Amicis, B. Bavassano, V. Carbone, L. Sorriso-Valvo, Magnetically dominated structures as an important component of the solar wind turbulence. Ann. Geophys. **25**, 1913–1927 (2007). doi:10.5194/angeo-25-1913-2007

L.F. Burlaga, Intermittent turbulence in the solar wind. J. Geophys. Res. **96**, 5847–5851 (1991). doi:10.1029/91JA00087

L.F. Burlaga, Intermittent turbulence in large-scale velocity fluctuations at 1 AU near solar maximum. J. Geophys. Res. **98**(A10), 17467–17473 (1993). doi:10.1029/93JA01630.

V. Carbone, L. Sorriso-Valvo, R. Marino, On the turbulent energy cascade in anisotropic magnetohydrodynamic turbulence. Europhys. Lett. **88**, 25001 (2009a). doi:10.1209/0295-5075/88/25001

V. Carbone, R. Marino, L. Sorriso-Valvo, A. Noullez, R. Bruno, Scaling laws of turbulence and heating of fast solar wind: the role of density fluctuations. Phys. Rev. Lett. **103**(6), 061102 (2009b). doi:10.1103/PhysRevLett.103.061102

V. Carbone, P. Veltri, R. Bruno, Experimental evidence for differences in the extended self-similarity scaling laws between fluid and magnetohydrodynamic turbulent flows. Phys. Rev. Lett. **75**, 3110–3113 (1995). doi:10.1103/PhysRevLett.75.3110. http://link.aps.org/doi/10.1103/PhysRevLett.75.3110

L.M. Celnikier, C.C. Harvey, R. Jegou, P. Moricet, M. Kemp, A determination of the electron density fluctuation spectrum in the solar wind, using the ISEE propagation experiment. Astron. Astrophys. **126**, 293–298 (1983)

B.D.G. Chandran, E. Quataert, G.G. Howes, Q. Xia, P. Pongkitiwanichakul, Constraining low-frequency Alfvénic turbulence in the solar wind using density-fluctuation measurements. Astrophys. J. **707**, 1668–1675 (2009). doi:10.1088/0004-637X/707/2/1668

C.H.K. Chen, T.S. Horbury, A.A. Schekochihin, R.T. Wicks, O. Alexandrova, J. Mitchell, Anisotropy of solar wind turbulence between ion and electron scales. Phys. Rev. Lett. **104**, 255002 (2010a). doi:10.1103/PhysRevLett.104.255002

C.H.K. Chen, R.T. Wicks, T.S. Horbury, A.A. Schekochihin, Interpreting power anisotropy measurements in plasma turbulence. Astrophys. J. **711**, 79–83 (2010b). doi:10.1088/2041-8205/711/2/L79

C.H.K. Chen, A. Mallet, T.A. Yousef, A.A. Schekochihin, T.S. Horbury, Anisotropy of Alfvénic turbulence in the solar wind and numerical simulations. Mon. Not. R. Astron. Soc. **415**, 3219 (2011a). doi:10.1111/j.1365-2966.2011.18933.x

C.H.K. Chen, S.D. Bale, C. Salem, F.S. Mozer, Frame dependence of the electric field spectrum of solar wind turbulence. Astrophys. J. **737**, 41 (2011b). doi:10.1088/2041-8205/737/2/L41

C.H.K. Chen, C.S. Salem, J.W. Bonnell, F.S. Mozer, S.D. Bale, Density fluctuation spectrum of solar wind turbulence between ion and electron scales. Phys. Rev. Lett. **109**(3), 035001 (2012a). doi:10.1103/PhysRevLett.109.035001

C.H.K. Chen, A. Mallet, A.A. Schekochihin, T.S. Horbury, R.T. Wicks, S.D. Bale, Three-dimensional structure of solar wind turbulence. Astrophys. J. **758**, 120 (2012b). doi:10.1088/0004-637X/758/2/120

C.H.K. Chen, G.G. Howes, J.W. Bonnell, F.S. Mozer, K.G. Klein, S.D. Bale, Kinetic scale density fluctuations in the solar wind. Solar Wind 13 Proceedings **1539**, 143–146 (2013a). arXiv:1210.0127

C.H.K. Chen, S.D. Bale, C.S. Salem, B.A. Maruca, Residual energy spectrum of solar wind turbulence. Astrophys. J. **770**, 125 (2013b). doi:10.1088/0004-637X/770/2/125

C.H.K. Chen, S. Boldyrev, Q. Xia, J.C. Perez, The nature of subproton scale turbulence in the solar wind. Phys. Rev. Lett. **110**, 225002 (2013c). doi:10.1103/PhysRevLett.110.225002

S. Chen, G. Doolen, J.R. Herring, R.H. Kraichnan, S.A. Orszag, Z.S. She, Far-dissipation range of turbulence. Phys. Rev. Lett. **70**, 3051–3054 (1993). doi:10.1103/PhysRevLett.70.3051

G.F. Chew, M.L. Goldberger, F.E. Low, The Boltzmann equation and the one-fluid hydromagnetic equations in the absence of particle collisions. Proc. R. Soc. Lond. Ser. A, Math. Phys. Sci. **236**, 112–118 (1956). doi:10.1098/rspa.1956.0116

J. Cho, A. Lazarian, The anisotropy of electron magnetohydrodynamic turbulence. Astrophys. J. **615**, 41–44 (2004). doi:10.1086/425215

J. Cho, E.T. Vishniac, The anisotropy of magnetohydrodynamic Alfvénic turbulence. Astrophys. J. **539**, 273–282 (2000). doi:10.1086/309213

J.T. Coburn, C.W. Smith, B.J. Vasquez, J.E. Stawarz, M.A. Forman, The turbulent cascade and proton heating in the solar wind during solar minimum. Astrophys. J. **754**, 93 (2012). doi:10.1088/0004-637X/754/2/93

L. Danaila, F. Anselmet, T. Zhou, R.A. Antonia, Turbulent energy scale budget equations in a fully developed channel flow. J. Fluid Mech. **430**, 87–109 (2001). doi:10.1017/S0022112000002767.

P.A. Davidson, *Turbulence: an Introduction for Scientists and Engineers* (Oxford University Press, Oxford, 2004)

K.U. Denskat, H.J. Beinroth, F.M. Neubauer, Interplanetary magnetic field power spectra with frequencies from 2.4 × 10 to the −5th Hz to 470 Hz from HELIOS-observations during solar minimum conditions. J. Geophys. **54**, 60–67 (1983)

M. Dobrowolny, A. Mangeney, P. Veltri, Fully developed anisotropic hydromagnetic turbulence in interplanetary space. Phys. Rev. Lett. **45**, 144–147 (1980). doi:10.1103/PhysRevLett.45.144

T. Dudok de Wit, O. Alexandrova, I. Furno, L. Sorriso-Valvo, G. Zimbardo, Methods for characterising microphysical processes in plasmas. Space Sci. Rev. (2013). doi:10.1007/s11214-013-9974-9

U. Frisch, *Turbulence* (Cambridge University Press, Cambridge, 1995)

S. Galtier, Wave turbulence in incompressible Hall magnetohydrodynamics. J. Plasma Phys. **72**, 721–769 (2006). doi:10.1017/S0022377806004521

S. Galtier, A. Pouquet, A. Mangeney, On spectral scaling laws for incompressible anisotropic magnetohydrodynamic turbulence. Phys. Plasmas **12**(9), 092310 (2005). doi:10.1063/1.2052507

P.C. Gary, C.W. Smith, W.H. Matthaeus, N.F. Otani, Heating of the solar wind by pickup ion driven Alfvén ion cyclotron instability. Geophys. Res. Lett. **23**, 113–116 (1996). doi:10.1029/95GL03707

S.P. Gary, *Theory of Space Plasma Microinstabilities* (Cambridge University Press, Cambridge, 1993)

S.P. Gary, C.W. Smith, Short-wavelength turbulence in the solar wind: linear theory of whistler and kinetic Alfvén fluctuations. J. Geophys. Res. **114**, 12105 (2009). doi:10.1029/2009JA014525

S.P. Gary, M.D. Montgomery, W.C. Feldman, D.W. Forslund, Proton temperature anisotropy instabilities in the solar wind. J. Geophys. Res. **81**, 1241–1246 (1976). doi:10.1029/JA081i007p01241

S.P. Gary, R.M. Skoug, J.T. Steinberg, C.W. Smith, Proton temperature anisotropy constraint in the solar wind: ACE observations. Geophys. Res. Lett. **28**, 2759–2762 (2001). doi:10.1029/2001GL013165

S. Ghosh, E. Siregar, D.A. Roberts, M.L. Goldstein, Simulation of high-frequency solar wind power spectra using Hall magnetohydrodynamics. J. Geophys. Res. **101**, 2493–2504 (1996). doi:10.1029/95JA03201

P. Goldreich, S. Sridhar, Toward a theory of interstellar turbulence. II. Strong Alfvénic turbulence. Astrophys. J. **438**, 763–775 (1995). doi:10.1086/175121

P. Goldreich, S. Sridhar, Magnetohydrodynamic turbulence revisited. Astrophys. J. **485**, 680 (1997). doi:10.1086/304442

M.L. Goldstein, D.A. Roberts, C.A. Fitch, Properties of the fluctuating magnetic helicity in the inertial and dissipation ranges of solar wind turbulence. J. Geophys. Res. **99**, 11519–11538 (1994). doi:10.1029/94JA00789

H.L. Grant, R.W. Stewart, A. Moilliet, Turbulence spectra from a tidal channel. J. Fluid Mech. **12**, 241–268 (1962). doi:10.1017/S002211206200018X

R. Grappin, J. Leorat, A. Pouquet, Dependence of MHD turbulence spectra on the velocity field-magnetic field correlation. Astron. Astrophys. **126**, 51–58 (1983)

R. Grappin, A. Mangeney, E. Marsch, On the origin of solar wind MHD turbulence—HELIOS data revisited. J. Geophys. Res. **95**, 8197–8209 (1990). doi:10.1029/JA095iA06p08197

R. Grappin, M. Velli, A. Mangeney, "Alfvénic" versus "standard" turbulence in the solar wind. Ann. Geophys. **9**, 416–426 (1991)

A. Greco, W.H. Matthaeus, S. Servidio, P. Chuychai, P. Dmitruk, Statistical analysis of discontinuities in solar wind ACE data and comparison with intermittent MHD turbulence. Astrophys. J. **691**, 111–114 (2009). doi:10.1088/0004-637X/691/2/L111

A. Greco, S. Servidio, W.H. Matthaeus, P. Dmitruk, Intermittent structures and magnetic discontinuities on small scales in MHD simulations and solar wind. Planet. Space Sci. **58**, 1895–1899 (2010). doi:10.1016/j.pss.2010.08.019

A. Greco, W.H. Matthaeus, R. D'Amicis, S. Servidio, P. Dmitruk, Evidence for nonlinear development of magnetohydrodynamic scale intermittency in the inner heliosphere. Astrophys. J. **749**, 105 (2012). doi:10.1088/0004-637X/749/2/105

K. Hamilton, C.W. Smith, B.J. Vasquez, R.J. Leamon, Anisotropies and helicities in the solar wind inertial and dissipation ranges at 1 AU. J. Geophys. Res. **113**(A12), 1106 (2008). doi:10.1029/2007JA012559

A. Hasegawa, Drift mirror instability of the magnetosphere. Phys. Fluids **12**, 2642–2650 (1969). doi:10.1063/1.1692407

M. Haverkorn, S.R. Spangler, Plasma diagnostics of the interstellar medium with radio astronomy. Space Sci. Rev. (2013, submitted)

J.-S. He, E. Marsch, C.-Y. Tu, Q.-G. Zong, S. Yao, H. Tian, Two-dimensional correlation functions for density and magnetic field fluctuations in magnetosheath turbulence measured by the cluster spacecraft. J. Geophys. Res. **116**(A15), 06207 (2011a). doi:10.1029/2010JA015974

J. He, E. Marsch, C. Tu, S. Yao, H. Tian, Possible evidence of Alfvén-cyclotron waves in the angle distribution of magnetic helicity of solar wind turbulence. Astrophys. J. **731**, 85 (2011b). doi:10.1088/0004-637X/731/2/85

P. Hellinger, H. Matsumoto, New kinetic instability: oblique Alfvén fire hose. J. Geophys. Res. **105**, 10519–10526 (2000). doi:10.1029/1999JA000297

P. Hellinger, H. Matsumoto, Nonlinear competition between the whistler and Alfvén fire hoses. J. Geophys. Res. **106**, 13215–13218 (2001). doi:10.1029/2001JA900026

P. Hellinger, P. Trávníček, J.C. Kasper, A.J. Lazarus, Solar wind proton temperature anisotropy: linear theory and WIND/SWE observations. Geophys. Res. Lett. **33**, 09101 (2006). doi:10.1029/2006GL025925

P. Hellinger, L. Matteini, Š. Štverák, P.M. Trávníček, E. Marsch, Heating and cooling of protons in the fast solar wind between 0.3 and 1 AU: Helios revisited. J. Geophys. Res. **116**, 9105 (2011). doi:10.1029/2011JA016674

P. Hellinger, P.M. Trávníček, Š. Štverák, L. Matteini, M. Velli, Proton thermal energetics in the solar wind: Helios reloaded. J. Geophys. Res. **118** (2013). doi:10.1002/jgra.50107

P. Henri, F. Califano, C. Briand, A. Mangeney, Low-energy Langmuir cavitons: asymptotic limit of weak turbulence. Europhys. Lett. **96**, 55004 (2011). doi:10.1209/0295-5075/96/55004

J.C. Higdon, Density fluctuations in the interstellar medium: evidence for anisotropic magnetogasdynamic turbulence. I. Model and astrophysical sites. Astrophys. J. **285**, 109–123 (1984). doi:10.1086/162481

B. Hnat, S.C. Chapman, G. Rowlands, Intermittency, scaling, and the Fokker-Planck approach to fluctuations of the solar wind bulk plasma parameters as seen by the WIND spacecraft. Phys. Rev. E **67**(5), 056404 (2003). doi:10.1103/PhysRevE.67.056404

B. Hnat, S.C. Chapman, G. Rowlands, Compressibility in solar wind plasma turbulence. Phys. Rev. Lett. **94**(20), 204502 (2005). doi:10.1103/PhysRevLett.94.204502

T.S. Horbury, M. Forman, S. Oughton, Anisotropic scaling of magnetohydrodynamic turbulence. Phys. Rev. Lett. **101**(17), 175005 (2008). doi:10.1103/PhysRevLett.101.175005

T.S. Horbury, M.A. Forman, S. Oughton, Spacecraft observations of solar wind turbulence: an overview. Plasma Phys. Control. Fusion **47**, 703–717 (2005). doi:10.1088/0741-3335/47/12B/S52

G.G. Howes, E. Quataert, On the interpretation of magnetic helicity signatures in the dissipation range of solar wind turbulence. Astrophys. J. **709**, 49–52 (2010). doi:10.1088/2041-8205/709/1/L49

G.G. Howes, S.C. Cowley, W. Dorland, G.W. Hammett, E. Quataert, A.A. Schekochihin, Astrophysical gyrokinetics: basic equations and linear theory. Astrophys. J. **651**, 590–614 (2006). doi:10.1086/506172

G.G. Howes, S.C. Cowley, W. Dorland, G.W. Hammett, E. Quataert, A.A. Schekochihin, A model of turbulence in magnetized plasmas: implications for the dissipation range in the solar wind. J. Geophys. Res. **113**(A12), 5103 (2008). doi:10.1029/2007JA012665

G.G. Howes, J.M. TenBarge, W. Dorland, A weakened cascade model for turbulence in astrophysical plasmas. Phys. Plasmas **18**(10), 102305 (2011a). doi:10.1063/1.3646400

G.G. Howes, J.M. TenBarge, W. Dorland, E. Quataert, A.A. Schekochihin, R. Numata, T. Tatsuno, Gyrokinetic simulations of solar wind turbulence from ion to electron scales. Phys. Rev. Lett. **107**(3), 035004 (2011b). doi:10.1103/PhysRevLett.107.035004

G.G. Howes, S.D. Bale, K.G. Klein, C.H.K. Chen, C.S. Salem, J.M. TenBarge, The slow-mode nature of compressible wave power in solar wind turbulence. Astrophys. J. **753**, 19 (2012a). doi:10.1088/2041-8205/753/1/L19

G.G. Howes, S.D. Bale, K.G. Klein, C.H.K. Chen, C.S. Salem, J.M. TenBarge, The slow-mode nature of compressible wave power in solar wind turbulence. Astrophys. J. **753**, 19 (2012b). doi:10.1088/2041-8205/753/1/L19

P.S. Iroshnikov, Turbulence of a conducting fluid in a strong magnetic field. Astron. Zh. **40**, 742 (1963)

P.A. Isenberg, M.A. Lee, J.V. Hollweg, The kinetic shell model of coronal heating and acceleration by ion cyclotron waves. 1. Outward propagating waves. J. Geophys. Res. **106**, 5649–5660 (2001). doi:10.1029/2000JA000099

K. Issautier, A. Mangeney, O. Alexandrova, Spectrum of the electron density fluctuations: preliminary results from Ulysses observations. AIP Conf. Proc. **1216**, 148–151 (2010). doi:10.1063/1.3395822

D. Jankovicova, Z. Voros, J. Simkanin, The influence of solar wind turbulence on geomagnetic activity. Nonlinear Process. Geophys. **15**(1), 53–59 (2008). doi:10.5194/npg-15-53-2008

H. Karimabadi, V. Roytershteyn, M. Wan, W.H. Matthaeus, W. Daughton, P. Wu, M. Shay, B. Loring, J. Borovsky, E. Leonardis, S.C. Chapman, T.K.M. Nakamura, Coherent structures, intermittent turbulence, and dissipation in high-temperature plasmas. Phys. Plasmas **20**(1), 012303 (2013). doi:10.1063/1.4773205

J.C. Kasper Solar wind plasma: kinetic properties and micro-instabilities. Ph.D. thesis, Massachusetts Institute Of Technology (2002)

J.C. Kasper, A.J. Lazarus, S.P. Gary, Hot solar-wind helium: direct evidence for local heating by Alfvén-cyclotron dissipation. Phys. Rev. Lett. **101**(26), 261103 (2008). doi:10.1103/PhysRevLett.101.261103

J.C. Kasper, B.A. Maruca, M.L. Stevens, A. Zaslavsky, Sensitive test for ion-cyclotron resonant heating in the solar wind. Phys. Rev. Lett. **110**(9), 091102 (2013). doi:10.1103/PhysRevLett.110.091102

P.J. Kellogg, T.S. Horbury, Rapid density fluctuations in the solar wind. Ann. Geophys. **23**, 3765–3773 (2005). doi:10.5194/angeo-23-3765-2005

K.H. Kiyani, S.C. Chapman, Y.V. Khotyaintsev, M.W. Dunlop, F. Sahraoui, Global scale-invariant dissipation in collisionless plasma turbulence. Phys. Rev. Lett. **103**(7), 075006 (2009). doi:10.1103/PhysRevLett.103.075006

K.H. Kiyani, S.C. Chapman, F. Sahraoui, B. Hnat, O. Fauvarque, Y.V. Khotyaintsev, Enhanced magnetic compressibility and isotropic scale invariance at sub-ion Larmor scales in solar wind turbulence. Astrophys. J. **763**, 10 (2013). doi:10.1088/0004-637X/763/1/10

K.G. Klein, G.G. Howes, J.M. TenBarge, S.D. Bale, C.H.K. Chen, C.S. Salem, Using synthetic spacecraft data to interpret compressible fluctuations in solar wind turbulence. Astrophys. J. **755**, 159 (2012). doi:10.1088/0004-637X/755/2/159

A. Kolmogorov, The local structure of turbulence in incompressible viscous fluid for very large Reynolds' numbers. Dokl. Akad. Nauk SSSR **30**, 301–305 (1941a)

A.N. Kolmogorov, The local structure of turbulence in incompressible viscous fluid for very large Reynolds' numbers. Dokl. Akad. Nauk SSSR **30**, 299–303 (1941b)

R.H. Kraichnan, Inertial-range spectrum of hydromagnetic turbulence. Phys. Fluids **8**, 1385–1387 (1965)

R.J. Leamon, C.W. Smith, N.F. Ness, W.H. Matthaeus, H.K. Wong, Observational constraints on the dynamics of the interplanetary magnetic field dissipation range. J. Geophys. Res. **103**, 4775 (1998). doi:10.1029/97JA03394

R.J. Leamon, C.W. Smith, N.F. Ness, H.K. Wong, Dissipation range dynamics: kinetic Alfvén waves and the importance of β_e. J. Geophys. Res. **104**, 22331–22341 (1999). doi:10.1029/1999JA900158

R.J. Leamon, W.H. Matthaeus, C.W. Smith, G.P. Zank, D.J. Mullan, S. Oughton, MHD-driven kinetic dissipation in the solar wind and corona. Astrophys. J. **537**, 1054–1062 (2000). doi:10.1086/309059

R.P. Lepping, M.H. Acũna, L.F. Burlaga, W.M. Farrell, J.A. Slavin, K.H. Schatten, F. Mariani, N.F. Ness, F.M. Neubauer, Y.C. Whang, J.B. Byrnes, R.S. Kennon, P.V. Panetta, J. Scheifele, E.M. Worley, The wind magnetic field investigation. Space Sci. Rev. **71**, 207–229 (1995). doi:10.1007/BF00751330

M.P. Leubner, Z. Voros, A nonextensive entropy approach to solar wind intermittency. Astrophys. J. **618**(1), 547 (2005). http://stacks.iop.org/0004-637X/618/i=1/a=547. doi:10.1086/425893

H. Li, S.P. Gary, O. Stawicki, On the dissipation of magnetic fluctuations in the solar wind. Geophys. Res. Lett. **28**, 1347–1350 (2001). doi:10.1029/2000GL012501

Y. Lithwick, P. Goldreich, Compressible magnetohydrodynamic turbulence in interstellar plasmas. Astrophys. J. **562**, 279–296 (2001). doi:10.1086/323470

Q.Y. Luo, D.J. Wu, Observations of anisotropic scaling of solar wind turbulence. Astrophys. J. **714**, 138–141 (2010). doi:10.1088/2041-8205/714/1/L138

B.T. MacBride, M.A. Forman, C.W. Smith, Turbulence and third moment of fluctuations: Kolmogorov's 4/5 law and its MHD analogues in the solar wind, in *Solar Wind 11/SOHO 16, Connecting Sun and Heliosphere*, ed. by B. Fleck, T.H. Zurbuchen, H. Lacoste, ESA Special Publication, vol. 592 (2005), p. 613

B.T. MacBride, C.W. Smith, M.A. Forman, The turbulent cascade at 1 AU: energy transfer and the third-order scaling for MHD. Astrophys. J. **679**, 1644–1660 (2008). doi:10.1086/529575

B.T. MacBride, C.W. Smith, B.J. Vasquez, Inertial-range anisotropies in the solar wind from 0.3 to 1 AU: Helios 1 observations. J. Geophys. Res. **115**(A14), 7105 (2010). doi:10.1029/2009JA014939

F. Malara, L. Primavera, P. Veltri, Nonlinear evolution of parametric instability of a large-amplitude nonmonochromatic Alfvén wave. Phys. Plasmas **7**, 2866–2877 (2000). doi:10.1063/1.874136

F. Malara, L. Primavera, P. Veltri, Nonlinear evolution of the parametric instability: numerical predictions versus observations in the heliosphere. Nonlinear Process. Geophys. **8**, 159–166 (2001). doi:10.5194/npg-8-159-2001

A. Mangeney, Intermittency and regularity in the Alfvénic range of solar wind turbulence, in *American Institute of Physics Conference Series*, ed. by P.-L. Sulem, M. Mond, American Institute of Physics Conference Series, vol. 1439 (2012), pp. 26–41. doi:10.1063/1.3701349

A. Mangeney, R. Grappin, M. Velli, Magnetohydrodynamic turbulence in the solar wind, in *Advances in Solar System Magnetohydrodynamics*, ed. by E.R. Priest, A.W. Hood (1991), p. 327

A. Mangeney, C. Salem, P.L. Veltri, B. Cecconi, Intermittency in the solar wind turbulence and the Haar wavelet transform, in *Sheffield Space Plasma Meeting: Multipoint Measurements Versus Theory*, ed. by B. Warmbein, ESA Special Pub., vol. 492 (2001), p. 53

A. Mangeney, C. Lacombe, M. Maksimovic, A.A. Samsonov, N. Cornilleau-Wehrlin, C.C. Harvey, J.-M. Bosqued, P. Trávníček, Cluster observations in the magnetosheath. Part 1. Anisotropies of the wave vector distribution of the turbulence at electron scales. Ann. Geophys. **24**, 3507–3521 (2006). doi:10.5194/angeo-24-3507-2006

P.K. Manoharan, M. Kojima, H. Misawa, The spectrum of electron density fluctuations in the solar wind and its variations with solar wind speed. J. Geophys. Res. **99**, 23411 (1994). doi:10.1029/94JA01955

R. Marino, L. Sorriso-Valvo, V. Carbone, A. Noullez, R. Bruno, B. Bavassano, Heating the solar wind by a magnetohydrodynamic turbulent energy cascade. Astrophys. J. **677**, 71–74 (2008). doi:10.1086/587957

R. Marino, L. Sorriso-Valvo, V. Carbone, P. Veltri, A. Noullez, R. Bruno, The magnetohydrodynamic turbulent cascade in the ecliptic solar wind: study of Ulysses data. Planet. Space Sci. **59**, 592–597 (2011). doi:10.1016/j.pss.2010.06.005

R. Marino, L. Sorriso-Valvo, R. D'Amicis, V. Carbone, R. Bruno, P. Veltri, On the occurrence of the third-order scaling in high latitude solar wind. Astrophys. J. **750**, 41 (2012). doi:10.1088/0004-637X/750/1/41

S.A. Markovskii, B.J. Vasquez, C.W. Smith, Statistical analysis of the high-frequency spectral break of the solar wind turbulence at 1 AU. Astrophys. J. **675**, 1576–1583 (2008). doi:10.1086/527431

J. Maron, P. Goldreich, Simulations of incompressible magnetohydrodynamic turbulence. Astrophys. J. **554**, 1175–1196 (2001). doi:10.1086/321413

E. Marsch, Kinetic physics of the solar corona and solar wind. Living Rev. Sol. Phys. **3**, 1 (2006). doi:10.12942/lrsp-2006-1

E. Marsch, S. Bourouaine, Velocity-space diffusion of solar wind protons in oblique waves and weak turbulence. Ann. Geophys. **29**, 2089–2099 (2011). doi:10.5194/angeo-29-2089-2011

E. Marsch, A. Mangeney, Ideal MHD equations in terms of compressive Elsaesser variables. J. Geophys. Res. **92**, 7363–7367 (1987). doi:10.1029/JA092iA07p07363

E. Marsch, C.-Y. Tu, Spectral and spatial evolution of compressible turbulence in the inner solar wind. J. Geophys. Res. **95**, 11945–11956 (1990). doi:10.1029/JA095iA08p11945

E. Marsch, C.-Y. Tu, Evidence for pitch angle diffusion of solar wind protons in resonance with cyclotron waves. J. Geophys. Res. **106**, 8357–8362 (2001). doi:10.1029/2000JA000414

E. Marsch, R. Schwenn, H. Rosenbauer, K.-H. Muehlhaeuser, W. Pilipp, F.M. Neubauer, Solar wind protons—three-dimensional velocity distributions and derived plasma parameters measured between 0.3 and 1 AU. J. Geophys. Res. **87**, 52–72 (1982). doi:10.1029/JA087iA01p00052

L. Matteini, S. Landi, P. Hellinger, F. Pantellini, M. Maksimovic, M. Velli, B.E. Goldstein, E. Marsch, Evolution of the solar wind proton temperature anisotropy from 0.3 to 2.5 AU. Geophys. Res. Lett. **34**, 20105 (2007). doi:10.1029/2007GL030920

L. Matteini, P. Hellinger, S. Landi, P.M. Trávníček, M. Velli Ion kinetics in the solar wind: coupling global expansion to local microphysics. Space Sci. Rev., 128 (2011). doi:10.1007/s11214-011-9774-z

W.H. Matthaeus, M.L. Goldstein, Low-frequency 1/f noise in the interplanetary magnetic field. Phys. Rev. Lett. **57**, 495–498 (1986). doi:10.1103/PhysRevLett.57.495

W.H. Matthaeus, M. Velli, Who needs turbulence? A review of turbulence effects in the heliosphere and on the fundamental process of reconnection. Space Sci. Rev. **160**, 145–168 (2011). doi:10.1007/s11214-011-9793-9

W.H. Matthaeus, M.L. Goldstein, C. Smith, Evaluation of magnetic helicity in homogeneous turbulence. Phys. Rev. Lett. **48**, 1256–1259 (1982). doi:10.1103/PhysRevLett.48.1256

W.H. Matthaeus, M.L. Goldstein, D.A. Roberts, Evidence for the presence of quasi-two-dimensional nearly incompressible fluctuations in the solar wind. J. Geophys. Res. **95**, 20673–20683 (1990). doi:10.1029/JA095iA12p20673

W.H. Matthaeus, S. Servidio, P. Dmitruk, Comment on "Kinetic simulations of magnetized turbulence in astrophysical plasmas". Phys. Rev. Lett. **101**(14), 149501 (2008). doi:10.1103/PhysRevLett.101.149501

W.H. Matthaeus, S. Servidio, P. Dmitruk, Dispersive effects of Hall electric field in turbulence. AIP Conf. Proc. **1216**, 184–187 (2010). doi:10.1063/1.3395832

W.H. Matthaeus, S. Servidio, P. Dmitruk, V. Carbone, S. Oughton, M. Wan, K.T. Osman, Local anisotropy, higher order statistics, and turbulence spectra. Astrophys. J. **750**, 103 (2012). doi:10.1088/0004-637X/750/2/103

N. Meyer-Vernet, *Basics of the Solar Wind* (Cambridge University Press, Cambridge, 2007)
L.J. Milano, W.H. Matthaeus, P. Dmitruk, D.C. Montgomery, Local anisotropy in incompressible magnetohydrodynamic turbulence. Phys. Plasmas **8**, 2673–2681 (2001). doi:10.1063/1.1369658
W.-C. Müller, R. Grappin, Spectral energy dynamics in magnetohydrodynamic turbulence. Phys. Rev. Lett. **95**(11), 114502 (2005). doi:10.1103/PhysRevLett.95.114502
Y. Narita, S.P. Gary, S. Saito, K.-H. Glassmeier, U. Motschmann, Dispersion relation analysis of solar wind turbulence. Geophys. Res. Lett. **38**, 5101 (2011). doi:10.1029/2010GL046588
K.T. Osman, W.H. Matthaeus, A. Greco, S. Servidio, Evidence for inhomogeneous heating in the solar wind. Astrophys. J. **727**, 11 (2011). doi:10.1088/2041-8205/727/1/L11
K.T. Osman, W.H. Matthaeus, B. Hnat, S.C. Chapman, Kinetic signatures and intermittent turbulence in the solar wind plasma. Phys. Rev. Lett. **108**(26), 261103 (2012). doi:10.1103/PhysRevLett.108.261103
M.J. Owens, R.T. Wicks, T.S. Horbury, Magnetic discontinuities in the near-earth solar wind: evidence of in-transit turbulence or remnants of coronal structure? Sol. Phys. **269**(2), 411–420 (2011). doi:10.1007/s11207-010-9695-0
S. Perri, A. Balogh, Differences in solar wind cross-helicity and residual energy during the last two solar minima. Geophys. Res. Lett. **37**, 17102 (2010). doi:10.1029/2010GL044570
S. Perri, V. Carbone, P. Veltri, Where does fluid-like turbulence break down in the solar wind? Astrophys. J. **725**, 52–55 (2010). doi:10.1088/2041-8205/725/1/L52
S. Perri, M.L. Goldstein, J.C. Dorelli, F. Sahraoui, Detection of small-scale structures in the dissipation regime of solar-wind turbulence. Phys. Rev. Lett. **109**(19), 191101 (2012). doi:10.1103/PhysRevLett.109.191101
D. Perrone, F. Valentini, S. Servidio, S. Dalena, P. Veltri, Vlasov simulations of multi-ion plasma turbulence in the solar wind. Astrophys. J. **762**, 99 (2013). doi:10.1088/0004-637X/762/2/99
V.I. Petviashvili, O.A. Pokhotelov, *Solitary Waves in Plasmas and in the Atmosphere* (Gordon & Breach Science Pub, New York, 1992). ISBN2881247873
J. Pietarila Graham, D.D. Holm, P. Mininni, A. Pouquet, Inertial range scaling, Kármán-Howarth theorem, and intermittency for forced and decaying Lagrangian averaged magnetohydrodynamic equations in two dimensions. Phys. Fluids **18**(4), 045106 (2006). doi:10.1063/1.2194966
J.J. Podesta, Dependence of solar-wind power spectra on the direction of the local mean magnetic field. Astrophys. J. **698**, 986–999 (2009). doi:10.1088/0004-637X/698/2/986
J.J. Podesta, On the energy cascade rate of solar wind turbulence in high cross helicity flows. J. Geophys. Res. **116**(A15), 05101 (2011). doi:10.1029/2010JA016306
J.J. Podesta, S.P. Gary, Magnetic helicity spectrum of solar wind fluctuations as a function of the angle with respect to the local mean magnetic field. Astrophys. J. **734**, 15 (2011). doi:10.1088/0004-637X/734/1/15
J.J. Podesta, D.A. Roberts, M.L. Goldstein, Spectral exponents of kinetic and magnetic energy spectra in solar wind turbulence. Astrophys. J. **664**, 543–548 (2007). doi:10.1086/519211
J.J. Podesta, M.A. Forman, C.W. Smith, D.C. Elton, Y. Malécot, Y. Gagne, Accurate estimation of third-order moments from turbulence measurements. Nonlinear Process. Geophys. **16**, 99–110 (2009a). doi:10.5194/npg-16-99-2009
J.J. Podesta, B.D.G. Chandran, A. Bhattacharjee, D.A. Roberts, M.L. Goldstein, Scale-dependent angle of alignment between velocity and magnetic field fluctuations in solar wind turbulence. J. Geophys. Res. **114**(A13), 1107 (2009b). doi:10.1029/2008JA013504
H. Politano, A. Pouquet, Von Kármán-Howarth equation for magnetohydrodynamics and its consequences on third-order longitudinal structure and correlation functions. Phys. Rev. E **57**, 21 (1998). doi:10.1103/PhysRevE.57.R21
L. Rezeau, A. Roux, C.T. Russell, Characterization of small-scale structures at the magnetopause from ISEE measurements. J. Geophys. Res. **98**(17), 179–186 (1993). doi:10.1029/92JA01668
O.W. Roberts, X. Li, B. Li, Kinetic plasma turbulence in the fast solar wind measured by cluster. Astrophys. J. **769**, 58 (2013). doi:10.1088/0004-637X/769/1/58
L. Rudakov, M. Mithaiwala, G. Ganguli, C. Crabtree, Linear and nonlinear landau resonance of kinetic Alfvén waves: consequences for electron distribution and wave spectrum in the solar wind. Phys. Plasmas **18**(1), 012307 (2011). doi:10.1063/1.3532819
F. Sahraoui, M.L. Goldstein, G. Belmont, P. Canu, L. Rezeau, Three dimensional anisotropic k spectra of turbulence at subproton scales in the solar wind. Phys. Rev. Lett. **105**, 131101 (2010). doi:10.1103/PhysRevLett.105.131101
F. Sahraoui, G. Belmont, M.L. Goldstein, New Insight into Short-wavelength Solar Wind Fluctuations from Vlasov Theory. Astrophys. J **748**(2), 100 (2012)
C. Salem, Ondes, turbulence et phénoménes dissipatifs dans le vent solaire à partir des observations de la sonde wind. Ph.D. thesis, Univ. Paris VII (2000)
C. Salem, A. Mangeney, S.D. Bale, P. Veltri, Solar wind magnetohydrodynamics turbulence: anomalous scaling and role of intermittency. Astrophys. J. **702**, 537–553 (2009). doi:10.1088/0004-637X/702/1/537

C.S. Salem, G.G. Howes, D. Sundkvist, S.D. Bale, C.C. Chaston, C.H.K. Chen, F.S. Mozer, Identification of kinetic Alfvén wave turbulence in the solar wind. Astrophys. J. **745**, 9 (2012). doi:10.1088/2041-8205/745/1/L9

A.A. Schekochihin, S.C. Cowley, W. Dorland, G.W. Hammett, G.G. Howes, E. Quataert, T. Tatsuno, Astrophysical gyrokinetics: kinetic and fluid turbulent cascades in magnetized weakly collisional plasmas. Astrophys. J. Suppl. Ser. **182**, 310–377 (2009). doi:10.1088/0067-0049/182/1/310

S. Servidio, V. Carbone, L. Primavera, P. Veltri, K. Stasiewicz, Compressible turbulence in Hall magnetohydrodynamics. Planet. Space Sci. **55**, 2239–2243 (2007). doi:10.1016/j.pss.2007.05.023

S. Servidio, P. Dmitruk, A. Greco, M. Wan, S. Donato, P.A. Cassak, M.A. Shay, V. Carbone, W.H. Matthaeus, Magnetic reconnection as an element of turbulence. Nonlinear Process. Geophys. **18**, 675–695 (2011). doi:10.5194/npg-18-675-2011

S. Servidio, F. Valentini, F. Califano, P. Veltri, Local kinetic effects in two-dimensional plasma turbulence. Phys. Rev. Lett. **108**(4), 045001 (2012). doi:10.1103/PhysRevLett.108.045001

J.V. Shebalin, W.H. Matthaeus, D. Montgomery, Anisotropy in MHD turbulence due to a mean magnetic field. J. Plasma Phys. **29**, 525–547 (1983). doi:10.1017/S0022377800000933

C.W. Smith, J. L'Heureux, N.F. Ness, M.H. Acuña, L.F. Burlaga, J. Scheifele, The ACE magnetic fields experiment. Space Sci. Rev. **86**, 613–632 (1998). doi:10.1023/A:1005092216668

C.W. Smith, K. Hamilton, B.J. Vasquez, R.J. Leamon, Dependence of the dissipation range spectrum of interplanetary magnetic fluctuations on the rate of energy cascade. Astrophys. J. **645**, 85–88 (2006). doi:10.1086/506151

C.W. Smith, J.E. Stawarz, B.J. Vasquez, M.A. Forman, B.T. MacBride, Turbulent cascade at 1 AU in high cross-helicity flows. Phys. Rev. Lett. **103**(20), 201101 (2009). doi:10.1103/PhysRevLett.103.201101

C.W. Smith, B.J. Vasquez, J.V. Hollweg, Observational constraints on the role of cyclotron damping and kinetic Alfvén waves in the solar wind. Astrophys. J. **745**, 8 (2012). doi:10.1088/0004-637X/745/1/8

L. Sorriso-Valvo, E. Yordanova, V. Carbone, On the scaling properties of anisotropy of interplanetary magnetic turbulent fluctuations. Europhys. Lett. **90**(5), 59001 (2010). doi:10.1209/0295-5075/90/59001

L. Sorriso-Valvo, V. Carbone, P. Veltri, G. Consolini, R. Bruno, Intermittency in the solar wind turbulence through probability distribution functions of fluctuations. Geophys. Res. Lett. **26**, 1801–1804 (1999). doi:10.1029/1999GL900270

L. Sorriso-Valvo, V. Carbone, P. Giuliani, P. Veltri, R. Bruno, V. Antoni, E. Martines, Intermittency in plasma turbulence. Planet. Space Sci. **49**, 1193–1200 (2001). http://dx.doi.org/10.1016/S0032-0633(01)00060-5

L. Sorriso-Valvo, V. Carbone, A. Noullez, H. Politano, A. Pouquet, P. Veltri, Analysis of cancellation in two-dimensional magnetohydrodynamic turbulence. Phys. Plasmas **9**, 89–95 (2002). doi:10.1063/1.1420738

L. Sorriso-Valvo, R. Marino, V. Carbone, A. Noullez, F. Lepreti, P. Veltri, R. Bruno, B. Bavassano, E. Pietropaolo, Observation of inertial energy cascade in interplanetary space plasma. Phys. Rev. Lett. **99**(11), 115001 (2007). doi:10.1103/PhysRevLett.99.115001

S.R. Spangler, C.R. Gwinn, Evidence for an inner scale to the density turbulence in the interstellar medium. Astrophys. J. **353**, 29–32 (1990). doi:10.1086/185700

J.E. Stawarz, C.W. Smith, B.J. Vasquez, M.A. Forman, B.T. MacBride, The turbulent cascade and proton heating in the solar wind at 1 AU. Astrophys. J. **697**, 1119–1127 (2009). doi:10.1088/0004-637X/697/2/1119

J.E. Stawarz, C.W. Smith, B.J. Vasquez, M.A. Forman, B.T. MacBride, The turbulent cascade for high cross-helicity states at 1 AU. Astrophys. J. **713**, 920–934 (2010). doi:10.1088/0004-637X/713/2/920

J.E. Stawarz, B.J. Vasquez, C.W. Smith, M.A. Forman, J. Klewicki, Third moments and the role of anisotropy from velocity shear in the solar wind. Astrophys. J. **736**, 44 (2011). doi:10.1088/0004-637X/736/1/44

O. Stawicki, S.P. Gary, H. Li, Solar wind magnetic fluctuation spectra: dispersion versus damping. J. Geophys. Res. **106**, 8273–8282 (2001). doi:10.1029/2000JA000446

G.I. Taylor, The spectrum of turbulence. Proc. R. Soc. A **164**, 476–490 (1938)

J.M. TenBarge, J.J. Podesta, K.G. Klein, G.G. Howes, Interpreting magnetic variance anisotropy measurements in the solar wind. Astrophys. J. **753**, 107 (2012). doi:10.1088/0004-637X/753/2/107

C.-Y. Tu, E. Marsch, MHD structures, waves and turbulence in the solar wind: observations and theories. Space Sci. Rev. **73**, 1–2 (1995)

A.J. Turner, G. Gogoberidze, S.C. Chapman, B. Hnat, W.-C. Müller, Nonaxisymmetric anisotropy of solar wind turbulence. Phys. Rev. Lett. **107**(9), 095002 (2011). doi:10.1103/PhysRevLett.107.095002

J. Šafránková, Z. Němeček, L. Přech, G.N. Zastenker, Ion kinetic scale in the solar wind observed. Phys. Rev. Lett. **110**(2), 025004 (2013). doi:10.1103/PhysRevLett.110.025004

B.J. Vasquez, V.I. Abramenko, D.K. Haggerty, C.W. Smith, Numerous small magnetic field discontinuities of Bartels rotation 2286 and the potential role of Alfvénic turbulence. J. Geophys. Res. **112**(A11), 11102 (2007). doi:10.1029/2007JA012504

P. Veltri, MHD turbulence in the solar wind: self-similarity, intermittency and coherent structures. Plasma Phys. Control. Fusion **41**, 787–795 (1999). doi:10.1088/0741-3335/41/3A/071

P. Veltri, A. Mangeney, Scaling laws and intermittent structures in solar wind MHD turbulence, in *Solar Wind Nine*, ed. by S.R. Habbal, R. Esser, J.V. Hollweg, P.A. Isenberg. American Institute of Physics Conference Series, vol. 471 (1999), p. 543

P. Veltri, G. Nigro, F. Malara, V. Carbone, A. Mangeney, Intermittency in MHD turbulence and coronal nanoflares modelling. Nonlinear Process. Geophys. **12**, 245–255 (2005). doi:10.5194/npg-12-245-2005

A. Verdini, R. Grappin, R. Pinto, M. Velli, On the origin of the 1/f spectrum in the solar wind magnetic field. Astrophys. J. **750**, 33 (2012). doi:10.1088/2041-8205/750/2/L33

M. Wan, S. Servidio, S. Oughton, W.H. Matthaeus, The third-order law for increments in magnetohydrodynamic turbulence with constant shear. Phys. Plasmas **16** (2009). doi:10.1063/1.3240333

M. Wan, W.H. Matthaeus, H. Karimabadi, V. Roytershteyn, M. Shay, P. Wu, W. Daughton, B. Loring, S.C. Chapman, Intermittent dissipation at kinetic scales in collisionless plasma turbulence. Phys. Rev. Lett. **109**(19), 195001 (2012). doi:10.1103/PhysRevLett.109.195001

R.T. Wicks, T.S. Horbury, C.H.K. Chen, A.A. Schekochihin, Power and spectral index anisotropy of the entire inertial range of turbulence in the fast solar wind. Mon. Not. R. Astron. Soc. **407**, 31–35 (2010). doi:10.1111/j.1745-3933.2010.00898.x

R.T. Wicks, T.S. Horbury, C.H.K. Chen, A.A. Schekochihin, Anisotropy of imbalanced Alfvénic turbulence in fast solar wind. Phys. Rev. Lett. **106**, 045001 (2011). doi:10.1103/PhysRevLett.106.045001

R.T. Wicks, A. Mallet, T.S. Horbury, C.H.K. Chen, A.A. Schekochihin, J.J. Mitchell, Alignment and scaling of large-scale fluctuations in the solar wind. Phys. Rev. Lett. **110**(2), 025003 (2013). doi:10.1103/PhysRevLett.110.025003

P. Wu, S. Perri, K. Osman, M. Wan, W.H. Matthaeus, M.A. Shay, M.L. Goldstein, H. Karimabadi, S. Chapman, Intermittent heating in solar wind and kinetic simulations. Astrophys. J. **763**, 30 (2013). doi:10.1088/2041-8205/763/2/L30

A.M. Yaglom, O lokalnoi strukture polya temperatur v turbulentnom potoke. Dokl. Akad. Nauk SSSR **69**, 743–746 (1949)

S. Yao, J.-S. He, E. Marsch, C.-Y. Tu, A. Pedersen, H. Rème, J.G. Trotignon, Multi-scale anti-correlation between electron density and magnetic field strength in the solar wind. Astrophys. J. **728**, 146 (2011). doi:10.1088/0004-637X/728/2/146

V. Zhdankin, S. Boldyrev, J. Mason, J.C. Perez, Magnetic discontinuities in magnetohydrodynamic turbulence and in the solar wind. Phys. Rev. Lett. **108**(17), 175004 (2012). doi:10.1103/PhysRevLett.108.175004

DOI 10.1007/978-1-4899-7413-6_4
Reprinted from *Space Science Reviews* Journal, DOI 10.1007/s11214-013-9967-8

Solar Surface and Atmospheric Dynamics

The Photosphere

V. Martínez Pillet

Received: 11 November 2012 / Accepted: 30 January 2013 / Published online: 1 March 2013

Abstract Various aspects of the magnetism of the quiet sun are reviewed. The suggestion that a small scale dynamo acting at granular scales generates what we call the quiet sun fields is studied in some detail. Although dynamo action has been proved numerically, it is argued that current simulations are still far from achieving the complexity that might be present on the Sun. We based this statement not so much on the low magnetic Reynolds numbers used in the simulations but, above all, in the smallness of the kinetic Reynolds numbers employed by them. It is argued that the low magnetic Prandtl number at the solar surface may pose unexpected problems for the identification of the observed internetwork fields with dynamo action at granular scales. Some form of turbulent dynamo at bigger (and deeper) scales is favored. The comparison between the internetwork fields observed by Hinode and the magnetism inferred from Hanle measurements are converging towards a similar description. They are both described as randomly oriented, largely transverse fields in the several hecto-Gauss range. These similarities are ever making more natural to assume that they are the same. However, and because of the large voids of magnetic flux observed in the spatial distribution of the internetwork fields, it is argued that they are not likely to be generated by dynamo action in the intergranular lanes. It is concluded that if a dynamo is acting at granular scales, the end product might have not been observed yet at current spatial resolutions and sensitivities with the Zeeman effect. Thus an effort to increase these resolutions and polarimetric sensitivities must be made. New ground- and space-based telescopes are needed. The opportunity offered by the Solar Orbiter mission to observe the Quiet Sun dynamics at the poles is seen as one of the most important tests for confirming the existence, or otherwise, of a granularly driven surface dynamo.

Keywords Quiet Sun magnetism · Turbulent dynamo

V. Martínez Pillet (✉)
Instituto de Astrofísica de Canarias, 38200 La Laguna, Tenerife, Spain
e-mail: vmp@iac.es

1 Introduction

A consensus about the existence of a small-scale dynamo (SSD[1]) operating at the solar photosphere is being consolidated in today's solar physics (see, e.g., Vögler and Schüssler 2007; Abbett 2007; Pietarila Graham et al. 2010; Stein 2012, for a review). High resolution magnetograms from ground and (mostly) space-based telescopes observed in the internetwork are often used to indicate that such a surface dynamo exist (e.g. Danilovic et al. 2010a; Lites 2011). On theoretical grounds, simulations of various kinds have been used to suggest how universal the various ingredients of such a dynamo seems to be (Moll et al. 2011). They all indicate that turbulent shear stresses acting on the inertial range act as the main mechanism able to efficiently convert kinetic into magnetic energy. The conclusions from the various simulations of turbulence in a conducting fluid seems to be that it would have been a lot harder to explain the absence of a surface SSD than its presence. With an emphasis on the observational side, we review in this work the current status of this consensus and try to pinpoint which aspects are more solidly established and which are less settled.

Section 2 will be the only one that concentrates on simulations and the theoretical aspects related to the problem of the existence of SSDs on the Sun. It will address the most controversial argument questioning the existence of such a mechanism, namely the fact that the solar convective zone has a magnetic Prandtl number that is orders of magnitude smaller than one, while simulations work near the $\sim$1 regime most of the time. Low magnetic Prandtl numbers are known to severely discourage dynamo action since the early simulations of convectively driven turbulent dynamos (Nordlund et al. 1992; Schekochihin et al. 2004a, 2005). Recently, progress has been achieved, however, that indicates that an SSD is indeed possible in the low magnetic Prandtl regime (Iskakov et al. 2007; Schekochihin et al. 2007; Brandenburg 2011). But the situation is not conclusive and the papers addressing this issue often resort to the fact that a mixed polarity field is observed at the solar surface as the firmest indication that such a mechanism should exists. However, and in the absence of a clear proof that this observed (internetwork) fields originate from an SSD—and such a proof is not available yet—the only progress to settle this issue will come from further work in the simulation front.

The observational arguments that have been put forward to favor the presence of a solar surface dynamo are discussed later in Sect. 3. There are basically two such arguments. First, the evidence from the observed Hanle effect in lines such as the Sr I 4607 Å line and the careful modeling of these signals indicate the existence of a tangled field with a mean strength of $\langle B \rangle \sim 130$ G (Trujillo Bueno et al. 2004) some few hundred kilometers above the solar surface (see also Trujillo Bueno et al. 2006, and references therein). This number was originally derived under some model assumptions that made it uncertain to within a factor two. However, a recent study (Shchukina and Trujillo Bueno 2011) of the predicted Hanle signals from the MHD simulations described in Pietarila Graham et al. (2009b) has eliminated some of this model dependency and confirmed such large mean field strengths. As a mixed polarity tangled field at unresolved scales leaves basically no trace in the Zeeman profiles, these fields have always been a prime candidate to be considered as originated from a surface SSD. If the tangling occurs at scales near or above present resolutions, some signatures can be detected, though. It is unclear if these fields have been observed using the Zeeman effect (but see Lites et al. 2009; Bellot Rubio and Orozco Suárez 2012, and Sect. 3). The second observational argument in favor of an SSD comes from the Hinode

[1]Small scale here refers to generation of magnetic fields at scales smaller than the energy injection one, the granulation.

spectropolarimeter (SP) instrument (Tsuneta et al. 2008; Kosugi et al. 2007) and its unprecedented characterization of the internetwork fields using the Zeeman effect (Lites et al. 2008; Ishikawa and Tsuneta 2011; Orozco Suárez and Bellot Rubio 2012). While transverse fields were known to exist in the quiet sun, as originally found by the Advanced Stokes Polarimeter (ASP; Lites et al. 1996), it was totally unexpected that these fields have a predominant transverse character. This transverse nature seems to fit in a natural way with an origin related to a turbulent dynamo as shown by recent simulations (see Schüssler and Vögler 2008).

A possible outcome given this state of affairs could be as follows. The existence of an SSD acting at granular scales at the solar surface can eventually be confirmed from a set of improved SSDs simulations (along the lines described in Sect. 2). In them, a continuous distribution of fields is obtained that is able to explain the Hanle depolarization levels from those fields created at the smaller scales and the largely horizontal internetwork fields from those at larger scales. The separation between these two sets of fields does not have to be sharp and a range of spatial scales can contribute to both the Zeeman and Hanle results (or, perhaps, that the two observed processes are due to fields exactly at the same scales as it can be inferred from the recent results of Bellot Rubio and Orozco Suárez 2012). Note that, such a field distribution would solely depend on the existence of the always present turbulent convective motions near the surface and, thus, should be independent of latitude and of activity cycle phase. While this conclusion seems rather plausible given the current evidence, the aim of this work is to address some of the known problems that might prevent such an outcome. In particular, Sect. 4 describes some observations recently obtained with the IMaX/SUNRISE magnetograph (Martínez Pillet et al. 2011; Solanki et al. 2010) that show regions that display very little magnetic activity, either measured as residual signals in time-averaged deep magnetograms or as evidenced by a lack of flux emergence episodes in the form of small-scale loops (as discovered by Martínez González et al. 2012). It is unclear how these voids are compatible with a granularly driven SSD. That the situation is far from clear has been corroborated recently by the study of Stenflo (2012), based on SDO/HMI magnetograms, who proposes the existence of a basal flux of order 3 G that is suggested to be an upper limit to the efficiency of an SSD at the solar surface. According to this result all of the internetwork fields observed with Hinode/SP will not be generated through such a mechanism and only the Hanle depolarizing fields could be originated through it (if at all).

In spite of this somewhat confusing situation, it is important to stress that our understanding about the nature and the properties of the quiet sun fields has improved enormously in recent years. But it is clear that a number of important questions remains on the theoretical/modelling side and on the observational front. Section 5 finishes this work proposing a way forward to further advance in this understanding of the quiet sun magnetism. Not surprisingly, we promote an effort to increase the polarimetric sensitivity and the spatial and temporal resolutions of both, the Hanle and the Zeeman observations. Studying the statistics of the quiet sun fields at various latitudes will also prove crucial.

2 Small Scale Dynamo Action at Low P_m. Implications for the Solar Case

The seminal reference that triggered the present debate on the existence of a convectively driven turbulent dynamo at the solar surface was the work of Cattaneo (1999), although the debate is older (see, e.g., Petrovay and Szakaly 1993; Lin 1995). It is important to point out that the former work mentioned, both, the granular and supergranular scales as possible contributors to such non-helical dynamo. The simulations used closed upper and lower boundaries with vertical fields in both of them. The Reynolds and magnetic Reynolds numbers could be clearly defined for this simulation thanks to the fixed computational grid used to solve the MHD equations. They were $\mathrm{Re} = ul/\nu = 200$ and $\mathrm{Re}_m = ul/\eta = 1000$, with l the characteristic length of the energy injecting convective cells, u the velocity of these cells and ν and η the molecular viscosity and the magnetic diffusivity, respectively. These numbers are large enough to ensure the development of turbulence. But it is important to note that Re was five times smaller than Re_m (magnetic Prandtl number of $\mathrm{P}_m = \mathrm{Re}_m/\mathrm{Re} = \nu/\eta = 5 > 1$). Under these circumstances, the magnetic field sees a smooth mean flow efficiently acting on it. The numerical simulation resulted in dynamo action saturating at 20 % of the kinetic energy flow. The crucial ingredient was the chaotic nature of the driving flows. Figure 2 of this paper already showed that, at the surface, the strong fields were localized in the downflow lanes, while the cell interiors showed no (vertical) field signature. The situation was different in deeper layers where fluctuating fields were filling basically the whole volume. At the time of the publication, the dominant transverse nature of the internetwork was not known (see Sect. 3) and this aspect was not analyzed. For this reason, the profiles synthesized by Sánchez Almeida et al. (2003) using these simulations concentrated on the study of the asymmetries induced in the Stokes V profiles (circular polarization) and its comparison with those observed in the internetwork. A shortage of asymmetries indicated that the simulations still did not achieve as much complexity as present in the Sun. However, using this synthesis, and after including effects such as telescope diffraction, it was predicted that when improving the spatial resolution from 1 arcsec to 0.15 arcsec, one should detect four times more Stokes V polarization signals.

A number of assumptions made in this simulation (such as the Boussinesq approximation) have been relaxed in more recent works. The more realistic ones (in terms of their proximity to the physical conditions on the Sun) are those made with the MURaM code (Vögler and Schüssler 2007; Schüssler and Vögler 2008; Pietarila Graham et al. 2010). In particular, they have addressed the important question of the role played by the closed boundary conditions assumed in Cattaneo (1999). Stein et al. (2003) pointed out that, in the solar convective zone, fields are submerged efficiently to the bottom of the convective zone by strong and concentrated downflows showing little recirculation near the surface. This recirculation was artificially enhanced in the simulations of Cattaneo (1999) by the use of closed boundary conditions. Stein et al. (2003) concluded that diverging upflows sweep the fluid into downflows, often vortical, where stretching and twisting becomes effective (and balanced by diffusion) but all these fields are rather rapidly submerged down into the bulk of the convective zone. The energy added to the flux that visits the surface was a very small fraction of the global budget of magnetic energy and the effect cannot be considered a local dynamo. This criticism has, however, been superseded by the MUraM simulations which used an open boundary and allow for a non-zero pointing flux at the bottom boundary. The way in which this boundary condition was implemented in the simulations of Vögler and Schüssler (2007) was by imposing an artificially increased magnetic diffusivity there. This diffusivity ensured that horizontal fields moving downward in the simulation leave the box unimpeded while, at the same time, prevented horizontal flux from entering the domain. All flux leaving the bottom boundary in these simulations was created by dynamo action inside the box. Vögler and Schüssler (2007) concluded that while the downward pumping of flux outside of the domain does indeed reduce the growth rate of the dynamo, it does not shut it down. They predict that as long as a sufficiently high Re_m is used (above the critical magnetic Reynolds number, Re_m^C), dynamo action will be observed in the simulations. Specifically, they see exponential growth of the magnetic energy for Re_m of 2600, while they find a decrease for Re_m below 1300. Now, here we should caution that the exact values of Re_m and Re (and thus of P_m) are not as easily defined as in Cattaneo (1999) simulations. The need to resolve shocks with artificial viscosity schemes or the implementation of the bottom boundary diffusivity necessarily implies that these numbers are more difficult to ascertain. An estimate of the effective Reynolds and Prandtl numbers for the MURaM simulations was given by Pietarila Graham et al. (2010) using various moments of the velocity and magnetic spectra (or Taylor microscales). The various dynamo runs available from this code turn out to have $\mathrm{Re}_m \in [2100, 8300]$ (with the latter value using a grid resolution of 4 km) and $\mathrm{P}_m \in [0.8, 2]$. This set of simulations all displayed an SSD generating magnetic fields inside their volume. The conclusion was that current MURaM simulations (with $\mathrm{P}_m \sim 1$) will show dynamo action as long as $\mathrm{Re}_m > \mathrm{Re}_m^C \sim 2000$. The main mechanism identified for this generation was the stretching and twisting of field lines by fluid motions in the inertial range of the spectrum of velocity fluctuations. In particular, it was concluded that dynamo action is concentrated in the turbulent downflows as field line stretching against magnetic tension is very efficient there (e.g. Stein 2012, for a review). In order to clarify the nature of the observed dynamo action, Moll et al. (2011) have analyzed the underlying physical mechanism under various physical conditions and assumptions. They concluded that for the cases studied (incompressible MHD, Boussinesq convection and compressible solar convection), the field is amplified by similar inertial range shear stresses that are independent of the conditions at the injection scale. The inclusion of compressibility effects or the asymmetry between upflows and downflows generated by the strong stratification did not influence the result. They, thus, termed the dynamo mechanism as universal.

One concern remains about the existence of a possible SSD on the Sun, though. The problem has been known since the early studies of dynamo action in conducting fluids. It was originally formulated by Batchelor (1950) who studied how turbulent motions stretch the field lines and amplify the magnetic energy as long as this process remains unimpeded by ohmic diffusion. Field line stretching is produced by fluid motions in the inertial range whose dissipation scale is set by the viscosity of the fluid. High viscosity ν couples the field lines to the flow and allows it to bend them efficiently. Large magnetic diffusivity η decouples the plasma (and the flows) from the field lines and prevents the bending. Thus, it was always clear that the efficiency of the SSD was going to be controlled by the interplay of these two effects as measured by the ratio $\mathrm{P}_m = \nu/\eta$. Clearly, both conditions $\mathrm{Re}_m \gg 1$ (to favor field line stretching) and $\mathrm{P}_m \gg 1$ (to couple fluid motions and field lines) boost local dynamo action. Based on an analogy between vorticity and magnetic fields Batchelor (1950) even concluded that for $\mathrm{P}_m < 1$ no SSD was possible. This conclusion was later criticized by a number of authors as the analogy cannot include the different initial and boundary conditions seen by these two fields (see, e.g. Boldyrev and Cattaneo 2004). However, simulations in the early 90's (see, e.g., Nordlund et al. 1992), including compressibility and strong stratification, already resulted in dynamo action only for $\mathrm{P}_m \geq 1$ with efficient shutting down of the dynamo for $\mathrm{P}_m < 1$. Other simulations encountering the same problem are discussed in Boldyrev and Cattaneo (2004) and in Schekochihin et al. (2005). Thus, the question of the existence of an SSD at low Prandtl numbers has received some attention in recent years.

Before we briefly describe the results from the numerical studies about the existence of an SSD at low Prandtl numbers, it is important to remember what are the actual numbers that occur in the solar convective zone and get an idea of how far or how close are we from simulating these conditions. To this end, Fig. 1 shows Re, Re_m and P_m^{-1} as computed in the mixing-length based model of Spruit (1974). The velocities at the injection scale u are obtained as part of the model and the characteristic length is assumed here to be $l = z/2$, with z the depth inside the convective zone. The magnetic Reynolds number changes from 10^5 at the photosphere to 10^9 at the bottom of the convective zone, whereas the Reynolds number stays constant at a level of 10^{12}. This makes $\mathrm{P}_m \in [10^{-7}, 10^{-3}]$, with the smallest

Fig. 1 The Reynolds number, magnetic Reynolds number and (inverse) Prandtl number in the solar convective zone according to the mixing-length model of Spruit (1974). The X near the surface marks the tipical values achieved of both Re and Re_m in the simulations

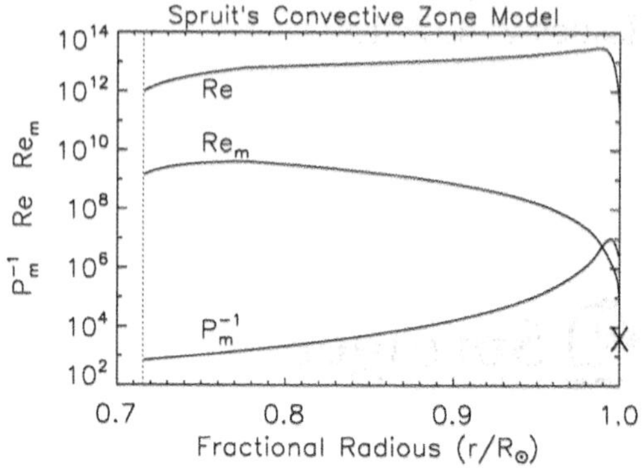

value reached at the photosphere. Thus, everywhere in the Sun, we have $\mathrm{Re} \gg \mathrm{Re}_m \gg 1$ and $\mathrm{P}_m \ll 1$ which is exactly the regime where the existence of an SSD becomes problematic. It is generally believed that for a sufficiently high Re_m there will always be an SSD at work. But the above mentioned simulations prompted a deeper study about the existence and nature of an SSD under solar conditions. As already mentioned, the simulations by Cattaneo (1999) had $\mathrm{P}_m = 5$ and those from the MURaM code always move close to the $\mathrm{P}_m \sim 1$ case (there is one with $\mathrm{P}_m = 0.8$ that is discussed below). Note that the regime where Re_m and Re are similar is actually very favorable for the numerical codes as similar grid sizes resolve the dissipative scales of both, magnetic and velocity fields.

What is the physical argument behind this difficulty to generate an SSD when $\mathrm{P}_m \ll 1$? Under such conditions the viscous (l_ν) and resistive (l_η) scales follow $l_\eta/l_\nu \sim \mathrm{P}_m^{-3/4} \gg 1$ (Schekochihin et al. 2004a, 2005) and the resistive scale l_η falls in the middle of the inertial range. Turbulent eddies of scales $l > l_\eta$ do the necessary field line bending and twisting for dynamo action at a rate of u_l/l (with u_l the typical flow velocity at this scale). If $\mathrm{P}_m > 1$ only these eddies occur and the field lines always see a spatially smooth flow. However, if $\mathrm{P}_m < 1$ one has now eddies below the resistive scale $l < l_\eta$. These eddies act on the field as a turbulent diffusion with diffusivity $u_l l$ and destroy magnetic energy. It is the predominance of this last process what can make dynamo action impossible as it was seen in the previously mentioned simulations. Using an incompressible spectral MHD code and the PENCIL code,[2] Schekochihin et al. (2005) studied what are the possible asymptotic limits when $\mathrm{Re} \gg \mathrm{Re}_m$ and the corresponding values of the Re_m^C for the existence of a dynamo. The above described effect always translates into a sharp increase in Re_m^C as $\mathrm{P}_m \to 0$ (see also Pietarila Graham et al. 2009a), but does not prevent the existence of a dynamo in this regime. The two asymptotic limits are, first, as $\mathrm{Re} \to \infty$, $\mathrm{Re}_m^C \to$ const, so that dynamo action is possible for higher Re_m and, second, $\mathrm{Re}_m^C \to \infty$ with $\mathrm{Re}_m^C/\mathrm{Re} \to \mathrm{P}_m^C = \mathrm{const}$, in which case no dynamo is possible (turbulent diffusion efficiently dissipates magnetic energy at small scales). Which exactly of the two asymptotic limits prevails has been under much debate in recent years. While Schekochihin et al. (2004a) and Schekochihin et al. (2005) (see their Fig. 2) favor the existence of a P_m^C (no dynamo) from simulations of incompressible magnetoconvection reaching values of P_m as small as 0.15, Boldyrev and Cattaneo (2004) provided analytical arguments favoring the existence of a Re_m^C.

While the debate in the mid last decade did not look promising for confirming the existence of an SSD at the solar surface, the situation has changed recently (even if not com-

[2]See http://www.nordita.dk/software/pencil-code.

pletely settled). Schekochihin et al. (2007) (see also Iskakov et al. 2007) performed simulations of incompressible MHD turbulence reaching values of $P_m = 0.1$ and with a sufficiently high Re_m that indicated a plateau region where a Re_m^C is observed (see their Fig. 1b). Admittedly, the number of such simulations proving the existence of this plateau is very small but the authors consider it enough numerical certainty. The results from Brandenburg (2011) using the PENCIL code (that includes compressible effects) and low values of P_m resulted in dynamo action being activated as well. Pietarila Graham et al. (2010) also find growth of magnetic energy in the one case they analyzed with $P_m = 0.8$. Thus, the most advanced existing numerical simulations of small-scale dynamo action in turbulent MHD currently favor the occurrence of such a process in the low P_m regime. However, a number of caveats remain:

- First, and most importantly, Re_m^C increases with decreasing P_m. The exact factor depends on the specificities of the simulations. Boldyrev and Cattaneo (2004) suggest a factor 7 increase in Re_m^C when shifting from the $P_m \gg 1$ to the $P_m \ll 1$ case. They use an analytical model of isotropic and homogeneous turbulence that includes the extra roughness of the velocity field for $P_m < 1$. Schekochihin et al. (2007) find from incompressible forced turbulence a factor 3 increase. As in the MURaM simulations one has $Re_m^C \sim 2000$ (Pietarila Graham et al. 2010), this means that as soon as we move into the low-P_m regime, we need magnetic Reynolds numbers above at least 6000 to be able to trigger dynamo action. These magnetic Reynolds numbers are not currently achieved by this code. In particular, none of the runs used by Danilovic et al. (2010a) would be able to actually sustain dynamo action. On top of that, the saturated field strength is known to decrease with decreasing P_m, thus the expected field strengths will be smaller than those computed for $P_m \sim 1$. The exact amount of this reduction is still a very controversial issue (see Schekochihin et al. 2004a, 2007; Brandenburg 2011) and its magnitude for the solar case unknown. But the net effect will be a reduction in the fields as compared to those computed for $P_m > 1$.
- To complete the demonstration of the existence of a dynamo driven by fluid motions in the inertial range at low P_m values, a growth rate of the dynamo scaling with $Re_m^{1/2}$ must be obtained from the simulations. Neither Schekochihin et al. (2007) nor Pietarila Graham et al. (2010) have reached that (see Fig. 3 of the latter work). Simulations with an increased resolution are needed to finally settle this issue. Schekochihin et al. (2007) concludes that, as long as this is not achieved, the mechanism that sustains the growth of the magnetic field fluctuations in the low-P_m regime will remain basically *unknown*.
- Schekochihin et al. (2007) and Iskakov et al. (2007) concluded that in the $P_m \ll 1$ regime, the magnetic energy spectra is fundamentally different from that found in the $P_m \gg 1$. The spatial distribution of the growing magnetic fields is qualitatively different too (see Fig. 2 in Schekochihin et al. 2007). This indicates that the use of simulations in the $P_m \sim 1$ range to compute the ensuing Stokes profiles and its comparison with those observed may not be justified.
- The often simulated case with $P_m \sim 1$ has the same spectral energy properties and field distribution as the $P_m \gg 1$ case (Schekochihin et al. 2004b). This is probably why the case with $P_m = 0.8$ simulated by Pietarila Graham et al. (2010) was so similar to the those in the range of $P_m \in [1, 2]$.

Let us finalize this section by stressing that if the situation looks confusing, it is because this has indeed been the case in this topic for some time (see Iskakov et al. 2007; Schekochihin et al. 2007, who speak about a frustrating outcome). One argument commonly given to promote the existence of an SSD is the observation of a mixed polarity field in the internetwork regions of the Sun (and many of the above mentioned works use this argument one way or another). Figure 2 of, both, Cattaneo (1999) and Vögler and Schüssler (2007) clearly indicate that this is a reasonable argument. The point we want to stress here is that the same applies to Fig. 10 from Stein and Nordlund (2006), which *does not* include an SSD. Pietarila Graham et al. (2009a) estimate for this simulation a $Re_m \sim 600$, which is known to be too small to develop dynamo action. However, their mixed polarity distribution located in the intergranular lines looks as 'solar' as in the other cases. In the work of Stein and Nordlund (2006) emphasis is made on diverging upflows bringing flux to the surface, expulsion to the intergranular lanes and sweeping of field lines into strong downflows that carry the flux into deeper layers. These simulations extent typically further down than those that concentrate in SSD generation and also include larger scales such as those associated with the mesogranulation. As shown in Fig. 1, the deeper we move into the Sun, the larger Re_m and P_m (although still smaller than one). Thus, a valid question is if it is not more natural to ask if a solar SSD exists at meso- and supergranular scales and, if so, whether they dominate over that might exists at granular ones. This point will be further discussed in Sect. 4.

3 Observed Signatures of Small Scale Dynamo Action at the Solar Surface

We now turn to the observational aspect of the discussion and ask the question: Have we seen the fields produced by a possible SSD operating at the solar surface? As it will become evident, there has been, as before, solid observational progress and areas with much confusion. Basically, two candidates exist that are often considered as by-products of an SSD, the internetwork fields and the, so-called, hidden fields that generate the Hanle depolarization signatures.

3.1 Zeeman Signals

There is no question that Hinode/SP data has produced a major quantitative and qualitative jump forward in our understanding of the internetwork fields. The publication by Lites et al. (2008) of a gigantic slit scanned map with consistent 10^{-3} polarimetric sensitivity and homogeneous spatial resolution of 0.3 arcsec changed our view of the quiet sun magnetism. In this map, a myriad of patches with predominant linear polarization signatures was discovered. The ASP already found the existence of episodic burst of largely transverse fields (the Horizontal Internetwork Features, HIF, Lites et al. 1996) but they were thought to be rather sporadic. The only previous indication of their existence and global character came from the SOLIS instrument as found by Harvey et al. (2007). But no prediction from the simulations or estimate of their magnitude was available. The existence of this ubiquitous horizontal field has now received full confirmation from the SUNRISE/IMaX data (Danilovic et al. 2010b) who could make the first study of their evolution (see the animation in Solanki et al. 2010) and establish a solid statistics of their lifetimes. Both instruments, Hinode/SP and SUNRISE/IMaX coincide in locating these HIF at the borders of the upflowing granules for a large fraction of their evolution. Lites et al. (2008) emphasized that the linear polarization signatures were not co-spatial with line-of-sight fields that were more frequently found in the intergranular lanes.

Before the publication of Lites et al. (2008) most of the discussion on the internetwork flux concentrated on obtaining its mean unsigned flux (see Solanki 2009, for a review on this topic previous to the impact of the Hinode measurements). The idea was that an intricately complex field with mixed polarities observed in the best available Stokes V magnetograms was the outstanding description of the internetwork fields. Increased resolution (or sensitivity) will result into ever increasing amounts of longitudinal signals observed as there was less and less cancellation due to instrumental effects. Either because of lack of reliable measurements of the field inclination or because of a habit to focus studies of solar magnetism exclusively in longitudinal magnetograms, no mention to its possible transverse character was traditionally made. As the field strengths were expected to be near or below equipartition values (less than 500 G, see Keller et al. 1994), these fields were not thought to be necessarily vertical either. Before the Hinode results, measurements of $\langle |B_L| \rangle$ (the spatially averaged unsigned longitudinal flux in the internetwork) were routinely being made and its variation with the spatial resolution closely followed (Sánchez Almeida et al. 2003). In a way, Hinode/SP results have made this emphasis obsolete. We now know that these fields are largely transverse and one should mainly care about $\langle |B_T| \rangle$ or simply about the spatially averaged $\langle |B| \rangle$. We should caution here that these magnitudes are obtained by observations of different Stokes parameters that have different sensitivities to the real magnetic field components on the Sun and to the fraction of the observed pixel that they occupy (the filling factor). Thus, the steps to compute $\langle |B| \rangle$ from the observed $\langle |B_L| \rangle$ and $\langle |B_T| \rangle$ are more problematic than what one might anticipate. Lites et al. (2008) estimated that the quiet sun map obtained by Hinode/SP had a $\langle |B_L^{app}| \rangle$ of 11 G (or Mx cm^{-2}) and a $\langle |B_T^{app}| \rangle$ of 55 G. These estimates were based on using integrals of the Stokes parameters that were calibrated against magnetic fluxes but with no account for the fraction of the pixel occupied by the fields. This is why they are named 'apparent' fluxes. An analysis performed by Orozco Suárez et al. (2007) of the same data, but this time using a Milne-Eddington (M-E) inversion code including a filling factor as a free parameter, resulted also in a predominantly transverse nature of the internetwork, albeit with a smaller ratio of transverse to longitudinal apparent fluxes.

It is no exaggeration to say that the ratio measured by Lites et al. (2008) $\langle |B_T^{app}| \rangle / \langle |B_L^{app}| \rangle \sim 5$ came as a surprise and was, thus, subjected to a deep scrutiny by the community.

Several factors can create a systematic bias in this ratio. Spatially averaging noise affects a positively defined quantity such as B_T in a way different than what it does to a signed quantity (B_L). As the Zeeman effect has a sensitivity different for each of these two components, the visibility of a given field strength is different depending on whether it is a field aligned with the LOS or perpendicular to it. In particular, fields close to the noise levels translate into different visibility thresholds. Last but not least, there is the already mentioned difference in how the filling factor couples to the real field strengths and inclinations for each one of the two components and depending on the specific method of analysis used. These effects and their impact into the factor 5 obtained by the first Hinode/SP measurements have been studied by various authors (Asensio Ramos 2009; Borrero and Kobel 2011; Stenflo 2011; Sánchez Almeida and Martínez González 2011; Steiner and Rezaei 2012, see the latter for a review). It all translates into understanding how exactly noise influences the final result given the method of analysis one follows and the various thresholds for inclusion of a given pixel or not. Depending on the specific case, different values for $\langle |B_T^{app}| \rangle$, $\langle |B_L^{app}| \rangle$, or inverted parameters B, inclination, azimuth and filling factor are obtained. Figure 2 shows the central portion of the same magnetogram used in Lites et al. (2008) with the top panels displaying the continuum intensity and B_L^{app} scaled to ± 20 G. Network bright points are visible (e.g., top-right of the figure) and the corresponding large Stokes V signals evident in the magnetogram. The bottom two images provided $|B_L^{app}|$ (left) and $|B_T^{app}|$ (right) both scaled within *the same range* [0,140] G. Almost all of the signals seen in the left image are identified with network regions as can be recognized by inspection with the two top panels. Thus, they will be excluded when computing $\langle |B_L^{app}| \rangle$ for internetwork regions. In contrast, all of the signals seen in the right panel correspond to internetwork and contribute to the $\langle |B_T^{app}| \rangle$ average. But note also in this last panel that regions with no apparent signals

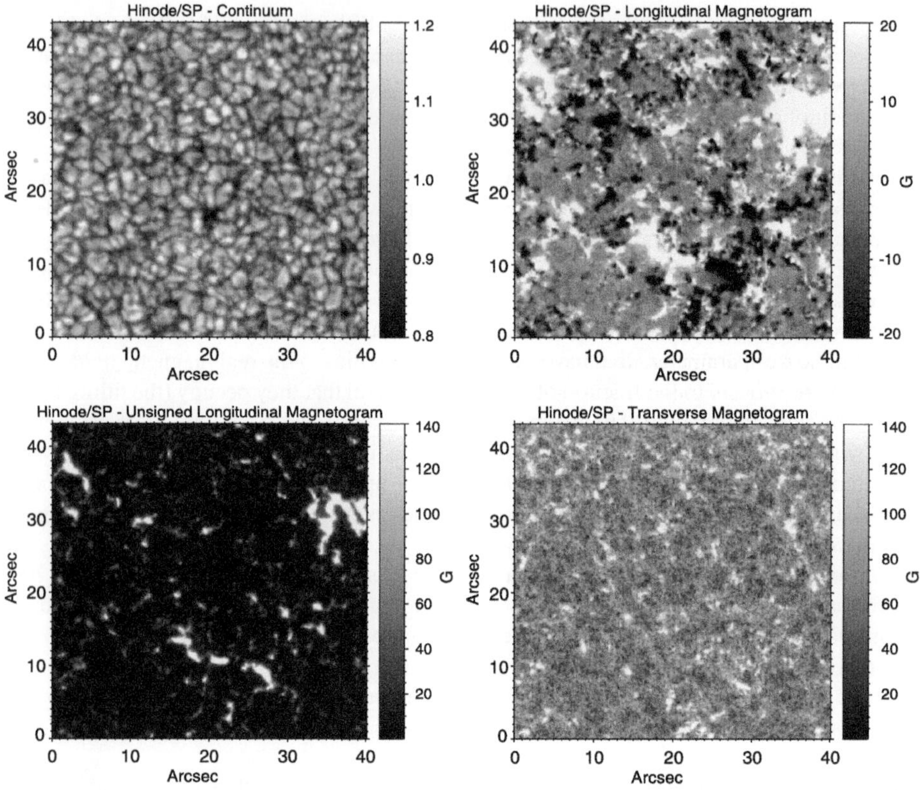

Fig. 2 Continuum intensity from the central portion of the Lites et al. (2008) Hinode/SP map (*top left*), signed magnetogram from the same portion (*top right*), unsigned magnetogram (*bottom left*) and transverse magnetogram (*bottom right*)

do not show up as dark black as in the unsigned longitudinal image, but they show a grey shade. This background is created by noise and it is larger than for the longitudinal magnetogram simply because of the effects described above. These regions should neither be just included when computing $\langle|B_T^{app}|\rangle$ nor completely excluded (as some transverse fields might exist there). It is in all of these details that reside some of the contradictory numbers that have been published. The most promising venue to clarify the situation is that of reducing noise. With the current instrumentation this can only be done by using longer integrations. This strategy has been recently pursued very successfully by Orozco Suárez and Bellot Rubio (2012) and Bellot Rubio and Orozco Suárez (2012). They use slit integrations of 6.1 minutes that reach a polarimetric sensitivity close to 10^{-4}, one order of magnitude better than commonly achieved. For visible lines, this improvement lowers down the detectability threshold for average transverse fields by a factor 2–3. Bellot Rubio and Orozco Suárez (2012) show that slit positions with almost 10 minute effective integration time that harbor linear polarization signals basically everywhere (60 % above 4.5σ).

The increased exposure times impose a penalty in the sense that the spatial resolution is decreased and evolutionary effects are intermingled in the final results.

This should not give the impression that these signals are always present on the Sun and detectable if sufficient sensitivity is available as they might have occurred in these pixels for only a fraction of the exposing time. Nevertheless, the data obtained with these long exposures is perfectly suited to reduced the noise induced bias present in the previous analysis of Hinode/SP. The M-E inversions performed by Orozco Suárez and Bellot Rubio (2012) and Bellot Rubio and Orozco Suárez (2012), consistently show that the internetwork fields have intrinsic field strength typically in the range of 100–200 G with basically no kG present, a strongly peaked field inclination distribution near 90°, with most of the pixels displaying inclinations in the range [45°, 135°], and an azimuth with no preferred orientation. Filling factors move in the range of [0.2, 0.4]. The authors inverted only those pixels that had a sufficiently large linear polarization signal (in either Stokes Q and/or U) to ensure a reliable result from the inversion. These results are largely free from most of the concerns expressed so far on the nature of internetwork fields. One criticism that remains to this analysis is the use of the M-E approximation and inversions able to reproduce the asymmetries are desirable. But the main conclusions from these recent analysis are likely to be confirmed by these more complex inversions, as those based in the M-E approximation are known to provide robust atmospheric means even in the presence of complex stratifications (see Westendorp Plaza et al. 1998). Thus internetwork fields do not have an isotropic distribution of inclination as it has been argued in a number of recent works (Asensio Ramos 2009; Stenflo 2011; Sánchez Almeida and Martínez González 2011) and their field strengths are typically on the few hG range.

Inversion codes allow an inference of the filling factor as a separate free parameter (admittedly, the most model dependent of all of them). Thus Orozco Suárez and Bellot Rubio (2012) were able to give real mean (not apparent) fluxes. For the real fluxes, they obtained, $\langle|B_T|\rangle \sim 198$ G, $\langle|B_L|\rangle \sim 64$ G and $\langle|B|\rangle \sim 220$ G (here the $\langle\rangle$ average means those pixels with large enough Stokes Q and U signals to allow a proper inversion, not the complete map). Their ratio, now, is $\langle|B_T|\rangle/\langle|B_L|\rangle \sim 3.1$. Note that because this ratio has eliminated filling factor effects and the inclusion of only those points that were inverted, it is not directly comparable to the number provided by Lites et al. (2008). However, it confirms the largely transverse nature of the internetwork fields as originally shown in that work. By eliminating filling factor effects, this new ratio allows for a cleaner comparison with numerical simulations.

Interestingly, although not anticipated, numerical simulations of magnetoconvection seem to have no problem in generating large amounts of horizontal fields. Soon after the publication of the results from Hinode, the SSDs simulations from Schüssler and Vögler (2008) (MURaM code) and those from Steiner et al. (2008) explained that these large amounts of transverse fields were present in their simulation boxes at different heights. The simulations from Steiner et al. (2008) were not dynamo simulations and used instead imposed fields in both vertical and horizontal directions. These two initial conditions generated a predominant horizontal field in the region where the Fe I line pair observed by Hinode forms. Thus, while SSDs are capable of generating a dominantly transverse field, it is not an exclusive property of them. In the work of Steiner et al. (2008), it was through the well known flux expulsion mechanism of vertical fields to the intergranular lanes that horizontal field lines were expelled above the granules in the overshooting region, generating the predominant transverse fields. The quantitative comparison with the observed fields was more complicated and both simulations fell short of the values observed. Danilovic et al. (2010a) used the MURaM simulations and performed spectral synthesis including instrument degradation and noise to compare the values predicted from the simulations with those observed by Lites et al. (2008). The result was that while the factor five in the ratio of apparent mean fluxes came naturally out of the SSD simulations, the absolute flux levels were close to those observed only if the SSD fields were multiplied artificially be a factor 2–3. After this artificial increase, the average values in the SSD simulation are closer to 100 G in the formation region of the Fe I lines which is nicely compatible with the peak field strength in Orozco Suárez and Bellot Rubio (2012). Similarly, the simulations presented by Steiner et al. (2008) resulted in average fields of the order of 20 G for the transverse component and suffer from the same problem as the SSD simulation. One could argue that the small Re_m numbers achieved in the SSD simulations and the field strength introduced in the simulations were too low and simply increasing them will explain the higher fluxes encountered by the observations. In any case, what was clearly established from all these studies was the fact that the natural state of a magnetic field component below equipartition strengths and closely coupled with the solar granulation is that of a predominant transverse field component as found by Hinode/SP. Another conclusion is that while SSDs are compatible with this result, the latter cannot be offered as a demonstration of their existence at the solar surface as a non-dynamo magnetoconvective simulation found the same results.

Another attempt to investigate whether the internetwork fields are generated by dynamo action has been presented recently by Lites (2011). In this work, the polarity imbalance of the internetwork field regions in 45 Hinode/SP maps is studied. Arguably, polarity balance is considered a necessary outcome of an SSD. However, and as stated by Lites (2011), this is not a sufficient condition as after a sufficiently large number of turnover times, the same polarity balance is to be expected in non-dynamo simulations. The difficulty here stems from the fact that to measure the internetwork polarity imbalance one needs to carefully isolate these fields from network ones. If the internetwork fields originate somehow by the shredding of nearby network fields, one expects the internetwork to have the same sign of the imbalance and a proportionality between the two. Interestingly enough, while no scaling with the unsigned flux was found, a suggestive correlation between the signed flux imbalance in the internetwork and that in the nearby network was measured (see Fig. 3 of the paper). However, being conclusive with this result is difficult as the isolation between internetwork and network fields is always problematic. This approach deserves further study, probably including some stray-light correction that deconvolves the wings of the spread function. This would allow to decontaminate the internetwork fields from the surrounding network contribution and allow a more reliable study of the resulting polarity imbalances.

It is clear that it will be very difficult to conclusively demonstrate that the hG, predominantly transverse internetwork fields originate from a granulation driven SSD. As mentioned in the above paragraph, we can only aim at disproving the SSD hypothesis rather than expect a firm confirmation of its presence. In Sect. 4, we present a recent result that, if consolidated, could be considered as one such refutation.

3.2 Hanle Signals

Scattering polarization in spectral lines and its modification via the Hanle effect allows to study a completely different parameter space of solar magnetism not accessible with the Zeeman effect (see the reviews in Trujillo Bueno et al. 2006; de Wijn et al. 2009; Stenflo 2011). A tangled field distribution within the resolution element is invisible through the Zeeman effect but can leave a clear imprint in the linear polarization profiles as long as the field strengths are below the Hanle saturation values (typically, a few hundred Gauss). While strong homogeneous vertical fields are hardly hidden in Stokes V through the Zeeman effect, weak disorganized transverse fields show up easily in Stokes Q and U of selected spectral lines thanks to the Hanle effect. This different sensitivity of Hanle effect has provided recently (Trujillo Bueno 2011) a seemingly alternative description of the Quiet Sun fields to that discussed above (and inferred from the Zeeman effect). Ever since the early studies (Stenflo 1982), the Hanle signals have been interpreted as being due to a tangled

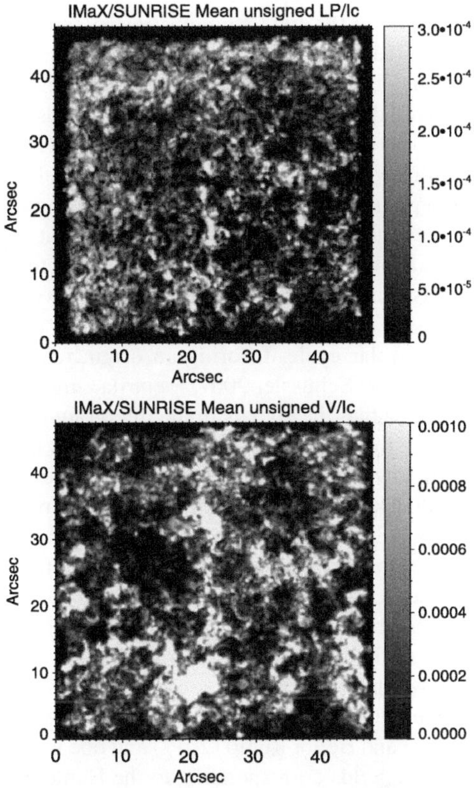

Fig. 3 SUNRISE/IMaX deep linear polarization magnetogram integrated over a 30 minute period (*top*). SUNRISE/IMaX deep circular polarization magnetogram integrated over the same period (*bottom*). Both are computed according to the definitions in Eq. (1)

mixed polarity field. For this reason, it has always been natural to associate these fields with the outcome of a turbulent dynamo (see, e.g. Vögler and Schüssler 2007; Pietarila Graham et al. 2010) This alternative description can be summarized as follows. The center-to-limb variation of the depolarization signals observed in the optically thick Sr I 4607 Å line suggests the presence of a tangled field with characteristics strengths of $B \sim \langle 130 \rangle$ G (as mentioned in the Introduction). This value is computed using realistic atmospheric models of the solar surface and complex 3D radiative transfer calculations (Trujillo Bueno et al. 2004; Shchukina and Trujillo Bueno 2011). In contrast, the analysis of the depolarization signals of a set of optically thin molecular lines suggests a mean field of the order of only $\langle 10 \rangle$ G (Trujillo Bueno et al. 2004; Kleint et al. 2011). These two distinct fields can be made compatible if one realizes that the molecular lines are entirely formed in the hot smooth upflowing granules and that these lines are blind to the fields present in the turbulent intergranular lanes (Trujillo Bueno et al. 2004). The Sr I line however sees the fields in both regions and, in order to give rise to average values of 130 G over the whole solar surface, Trujillo Bueno et al. (2004) concluded that one must have $\langle B \rangle \gtrsim 200$ G (very close to the Hanle saturation regime for this line) inside the lanes. This intergranular fields would generate most of the Sr I depolarization while having no effect in the linear polarization signals from the molecular lines. It is interesting to point out that when the Sr I depolarization levels are computed with the MURaM SSD simulations, Shchukina and Trujillo Bueno (2011) find that the depolarization levels are far too low. This is not surprising as these SSD simulations reach fields of the order of 20–30 G, very far from the typical 130 G that is needed. The only way in which an agreement with the observed depolarization values could be achieved was by multiplying everywhere the field strengths in the simulations by a factor 12. This factor is four times larger than that needed by Danilovic et al. (2010a) to match the observed apparent mean Zeeman fluxes of Hinode/SP with those in the MURaM simulations.

This description of the quiet sun fields as inferred from the Hanle signals is generally accepted and no clear alternative exist. However, it is clear that one would like to see it confirmed by an analysis that is less model dependent (see, Kleint et al. 2011; Stenflo 2012, who point out that the analysis made using molecular lines is differential in contrast to that made with the atomic lines). The fact that the Sr I line is so close to the saturation regime has also brought some criticism of the actual interpretation of the observed depolarization levels (Sánchez Almeida 2005). Indeed, very little variation of the Sr I polarization signals has been detected and this can be used to argue that the tangled hidden field is independent of the solar cycle, favoring an origin related to SSD action (Trujillo Bueno et al. 2004; Vögler and Schüssler 2007; Pietarila Graham et al. 2010). One other explanation, of course, could be that the signals in the intergranular lanes change with the activity cycle but we do not see the effect because they are always in the Hanle saturation regime.

In any case, the existence of this turbulent unresolved field is well established and one would like to understand if it bears any relation with the internetwork fields observed by Hinode/SP and described before. Note that when we say unresolved here, we refer to the Hanle observations used in the analysis that were obtained over large spatial scales (several arcseconds) and with exposure times of the order of one minute or so. Taken at face value, what the Hanle measurements need are small scale (arcsecond scale or below) non-vertical field patches (the Hanle effect is insensitive to vertical fields), with field strengths below equipartition with granulation ($\sim$400 G). It is evident that these are the properties of the internetwork fields described by Bellot Rubio and Orozco Suárez (2012) and Orozco Suárez and Bellot Rubio (2012) and one is tempted to conclude that the Hinode/SP fields are also the fields corresponding to the Hanle signals. Bellot Rubio and Orozco Suárez (2012) specifically mentions this possibility. One would be tempted to go further and state that it would be rather strange to have the Sun harboring two families of fields with so many things in common but that are totally unrelated. Thus, we also favor here the identification of the Hinode/SP internetwork fields with those that produce the Hanle depolarization signals (see also Lites et al. 2009). Perhaps, the only difficulty we encounter with this identification is the well known fact that the HIFs (the internetwork transverse fields) have a clear preference to be located at the borders of granules (Lites et al. 2008; Danilovic et al. 2010b) not in the intergranular lanes as required by the Sr I depolarization measurements. However, this could be a minor problem as a granular border might be sufficiently close to what is needed. An evident way to test this identification would be to invert a volume of the Hinode/SP internetwork observations with an inversion code that provides the complete atmospheric stratification such as the SIR code (Ruiz Cobo and del Toro Iniesta 1992). Then, perform the Hanle synthesis as in Trujillo Bueno et al. (2004) and Kleint et al. (2011) for the atomic and molecular lines and compare the result with the observations. In this exercise some assumptions about the upper layers might be needed to extend the retrieved atmospheres over the range of formation of Sr I line, but such an extension can be reasonably done.

In any event, it is clear that spatially resolved Hanle depolarization measurements (Stenflo 2012) that tell us where exactly the Sr I depolarization signals occur in the Sun are urgent.

4 Deep Magnetograms and 'Dead' Calm Areas: Implications

As commented in Sect. 1, dynamo action concentrates in the turbulent downflows as field line stretching and amplification is more efficient there. A dramatic visualization of this can be seen in Fig. 1 of Schüssler and Vögler (2008). Basically all of the intergranular lanes are seen to participate in this dynamo swing. One can then expect that the observable effects of an SSD driven by the granulation would be distributed homogeneously over spatial scales similar to that of the granulation after a period of time long compared with the lifetime of an intergranular lane (which is similar to that of the granules themselves, 10 minutes, cf. Title et al. 1989). In particular, the internetwork fields seen by Hinode/SP must be uniformly distributed over granulation scales. Thus, if we are able to somehow make a statistics of the location of these fields, the inferred spatial distribution must reflect the spatial scale of the granulation and have no voids larger than the typical size of one granule (or at least, the probability of occurrence of such voids will be small, see below). On the contrary, if one finds areas on the quiet sun where many intergranular lanes have existed but none of the expected effects of a presumed granular SSD are seen, one can conclude that no such SSD has been detected.

While snapshots from SSDs have been used to make spectral synthesis including the degradation effects of telescopes and detectors (Danilovic et al. 2010a; Shchukina and Trujillo Bueno 2011), what one needs is a complete time series that allows the study of the generation and disappearance of the dynamo fields at current spatial resolutions and polarimetric sensitivities. This study is unfortunately not available yet. From an observational perspective, we have seen how internetwork fields evolve at the solar surface with unprecedented detail. Both Hinode/SP and SUNRISE/IMaX have convincingly shown that internetwork fields are fed into the surface in the form of emerging small scale loops (rather than in the form of, for example, spontaneous appearances of newly created flux patches in the intergranular lanes). Hinode/SP with its superior spectral resolution and coverage has allowed to study a large number of such events (see Centeno et al. 2007; Martínez González and Bellot Rubio 2009; Ishikawa et al. 2010; Viticchié 2012; Gömöry et al. 2010, the latter for ground observations in the infrared) with great detail. Basically a horizontal patch is first detected in linear polarization that later displays two opposite polarity footpoints that move apart. The distance between the footpoints of these small scale loops is typically 1 Mm, the lifetime 10 minutes and the magnetic flux of around 10^{17} Mx (see the statistic in Martínez González and Bellot Rubio 2009). Danilovic et al. (2010b), using data from SUNRISE/IMaX, identified a large number (thousands) of HIFs that they associate with flux emergence in the form of loops. Such a large amount of occurrences emphasize this process as the main source for internetwork flux at present resolutions and sensitivities. A case in point is that described by Guglielmino et al. (2012) who analyzes in detail what is probably the largest quiet sun loop ever observed. In this case, a maximum footpoint separation of 4.5 Mm is achieved with a magnetic flux content of 6×10^{17} Mx and a duration of 25 minutes. This quiet sun bipole, that has one order of magnitude less flux than the smallest ephemeral region studied by Hagenaar et al. (2003), is arguably not generated by any process that occurs at granular scales. It is known that dynamo simulations generate similar loop-like structures (as the horizontal fields described in Sect. 2 are part of them). But a comparative study of the maximum footpoint separation, flux content, etc., is missing.

A recent study about loop emergence in the quiet sun has recently been published that is relevant for our discussion. Using the two time series of the first flight day obtained with the IMaX instrument, Martínez González et al. (2012) studied the emergence of 497 magnetic loops identified in them. They estimate an event rate of 0.25 loop h^{-1} $arcsec^{-2}$. If we associate a typical linear size of 2 arcsec for a granule and a lifetime of 10 minutes, this rate can be translated into 0.17 loops per granule (it takes 6 granules to get one quiet sun loop). Each one of these time series lasted for about 30 minutes, so granulation was efficiently created and destroyed over their timespan. Such a large number of detected loops allowed them to study the spatial distribution of these magnetic flux emergence processes. The result they obtained was that the spatial distribution of loop events was far from homogeneous at granular scales. They found what they termed "dead calm" areas where simply no loop was seen to emerge during the time series. One could argue that these calm areas can be created by chance and that their existence is simply a mere coincidence. However, the authors perform a statistical study of the likelihood of such voids given their size (70–100 arcsec^2) under the assumption of a spatially uniform loop emergence probability. They modeled the probability of finding one such large circular void with a resulting estimate of 3×10^{-4}. Two such dead calm areas were cleanly identified.

Let us show from another perspective how unlikely this result is. In voids of this size, one can fit around 20 typical granules at any given time. As the time series covered 3 granular lifetimes, 60 granules existed inside them which would have given rise to at least 10 loops at the above rate of creation, but none was found. And it occurred in two unrelated regions. How is it possible that if magnetic loop emergence is the observable outcome of SSD action, one encounters regions where this is not activated? While we find this result highly incompatible with the existence of a granularly driven SSD, we need a solid comparison with simulations including all of the possible observational biases to provide a firm answer.

We must note that a non-uniform distribution of quiet sun fields has been known for some time. They were identified by Lites et al. (2008, and references therein), who pointed out the existence of mesogranular scale voids in the Hinode/SP map. The preference of internetwork fields to be located at mesogranular scales has been clearly demonstrated using the same SUNRISE/IMaX data as that used to identify the voids (see Yelles Chaouche et al. 2011), but note that these voids were much larger than a mesogranule and have a scale closer to that of a small supergranule (see, e.g. Meunier et al. 2007, who give radius in the range of 8 to 30 arcsec).

One could argue that whereas no loop emerged in these dead calm areas, they were probably not devoid of some subtle form of internetwork field presence. However, inspection of the IMaX data with a scaling close to the noise levels readily shows that the locations of these voids clearly harbored less activity than the rest of the observed area. To prove this point more clearly, we produced two time averages of the IMaX time series that are shown in Fig. 3. The two quantities that are displayed correspond to deep magnetograms computed as:

$$\overline{LP} = \frac{1}{N}\sum_{i=1}^{N}\left(\frac{\sqrt{Q_i^2 + U_i^2}}{I_c} - \epsilon_{LP}\right), \qquad \overline{V} = \frac{1}{N}\sum_{i=1}^{N}\left(\frac{|V_i|}{I_c} - \epsilon_V\right) \tag{1}$$

where ϵ_V and ϵ_{LP} are quantities inferred from the data that allow convenient reduction of the noise when doing the time averages of the otherwise positively defined quantities. The deep magnetograms in Fig. 3 evidence the same two voids as detected by Martínez González et al. (2012). They are centred at coordinates [15, 25] and [35, 37]. The voids are visible in both the linear and the circular polarization deep magnetograms. Note that the scaling of both magnetograms saturates at 3×10^{-4} for $\overline{LP}$ and at 10^{-3} for $\overline{V}$. The calibration constants published by Martínez Pillet et al. (2011) would have translated these values into 45 and 5 G respectively. Measurements below these fluxes are at the limit of state of the art imaging magnetographs. These deep magnetograms show that they had reduced levels of magnetic activity and it is correct to refer to them as magnetically calm (maybe, not dead). The question, of course, is what was special about the many intergranular lanes that populated these regions that prevented them from displaying the magnetic activity levels seen elsewhere.

5 Where Do We Go from Here?: Higher Sensitivities and Higher Latitudes

The simulations described in Sect. 1 have reached $\mathrm{Re}_m \in [5000, 8000]$ (Pietarila Graham et al. 2009b). However, they fail to provide the magnetization levels needed to explain, both, the Hanle depolarization signals of the Sr I line and the Hinode/SP fluxes (see Danilovic et al. 2010a; Shchukina and Trujillo Bueno 2011; Orozco Suárez and Bellot Rubio 2012). This is often explained arguing that the simulations are still far from the magnetic Reynolds numbers of the Sun. But the values that are achieved by them are *only* about one order of magnitude below the expected values at the solar surface. In contrast, they reach, at best, similar Re values, which are 10^7 times smaller than what we encounter on the Sun (see Fig. 1, where the values achieved by the simulations are marked by an X near the surface). There are about 5 orders of magnitudes in the inertial range below the resistive scale, l_η, populated with cells that generate an enhanced turbulent diffusion that are not present in the simulations. As mentioned in Sect. 1, recent numerical simulations show that this might not be a problem to obtain dynamo action in the low P_m regime. But they also tell us that the value of the critical magnetic Reynolds number needed Re_m^C to sustain dynamo action increases sharply when $\mathrm{P}_m < 1$. A factor 3–7 increase in the value of Re_m^C is expected. As in the MURaM simulations this number is $\sim$2000, we expect this code to display a dynamo only when $\mathrm{Re}_m > 6000$ or larger as soon as they use Prandtl numbers in the right ballpark of the problem. Let us see the implications for some of the inferences that are made using the available numerical simulations. In the case of the MURaM runs, Danilovic et al. (2010a) found a plausible scaling of the saturated field strengths with $\sim\mathrm{Re}_m^{1/2}$. Their run G with $\mathrm{Re}_m \sim 5200$ has a mean field of 30 G (at $\log\tau \approx -1$). If we scale it to a solar value of $\mathrm{Re}_m \sim 10^5$ following this square root scaling, we obtain 130 G. This is the value needed to explain the Hanle depolarization measurements of the Sr I line (and the most probable field strength found in the internetwork by Orozco Suárez and Bellot Rubio 2012). This nice agreement was already pointed out by Shchukina and Trujillo Bueno (2011). However, for the reasons explained above, this is probably a mere coincidence. If this simulation would have been done with the same Re_m but with P_m of, say 0.1 (as in the simulations proving dynamo action at low Prandtl numbers), no dynamo action would have been found. The field strength to introduce in the above scaling would have been 0 G instead of 30 G.

In discussing the accepted view about the existence of dynamo action at the solar surface, we have gone a step further and suggested that the observed spatial distribution of quiet sun fields seems to be at odds with a granularly driven SSD. The argument used to make this claim was that in such a dynamo, all the magnetic byproducts must necessarily have a uniform spatial distribution at scales above that of a granule. We have found, however, that there are voids of magnetic activity (dead calm areas) in, both, the average apparent longitudinal and transverse fluxes of deep magnetograms and in the distribution of emerging loops. In fact, we have translated the loop emergence rate found by Martínez González et al. (2012) into a rate of 1 loop per 6 granules which can also be thought of as 1 emerging loop per mesogranule. It is clear that mesogranulation scales are very relevant for the quiet sun fields. SSD simulations including them are needed to see if they explain the presence of these voids in magnetic activity and emerging loop frequency. Given the fact that as we go deeper into the Sun, one reaches higher Re_m and P_m values, it is very tempting to suggest that SSDs acting in a range of convective scales larger than the granular ones are those that give rise to the presently observed internetwork fields. Note that in his work, Cattaneo (1999) already mentioned larger scales, such as those of the supergranulation, as possible places where to host dynamo activity. This might still be compatible with an SSD at granular scales that generates fields much weaker and that have not yet been observed by any of our currently available diagnostic techniques.

We have also suggested that the internetwork fields revealed by Hinode/SP and those that generate the Hanle depolarization of some atomic lines might have a lot more in common than previously thought (see however Lites et al. 2009; Bellot Rubio and Orozco Suárez 2012). The reason for this identification is based on the results from Orozco Suárez and Bellot Rubio (2012) and Bellot Rubio and Orozco Suárez (2012) who have shown that the quiet sun fields share the same field strengths, inclinations and azimuths than those needed by the fields detected with the Hanle effect. This is, of course, compatible with a continuous spectrum of fields in which the ranges corresponding to the two types of fields (the internetwork and the 'hidden' Hanle fields) simply overlap over a much larger fraction than thought so far. Of course, the internetwork fields will also include a fraction of vertical kG fields that contributes nothing to the Hanle depolarization and the weak granular fields that depolarize the light in molecular lines never make an imprint in the Fe I Zeeman lines observed by Hinode/SP at current sensitivities. But, of this continuous distribution of fields, the range of hG strengths with largely transverse orientations and spatially organized at granular scales, contributes simultaneously to, both, the Zeeman and the Hanle observations. If this result is confirmed (as, e.g., with the SIR inversions and the Hanle computations mentioned in Sect. 3) an important step to clarify the currently complex debate of the nature of the quiet sun magnetism would be achieved. We also want to stress that the evidence that the internetwork field component is composed of largely transverse fields renders the debate about the mean value of $\langle |B_L^{app}| \rangle$ (or of $\langle |B_L| \rangle$ for that matter) obsolete. Stokes V (longitudinal) magnetograms of the internetwork simply show a rather incomplete picture of these fields.

Note that many of the results commented above have benefited from high polarimetric sensitivities. Hanle measurements have always been very demanding in polarimetric accuracy. The deep magnetograms of SUNRISE/IMaX and the long integrations with Hinode/SP used by Orozco Suárez and Bellot Rubio (2012) both were at the 10^{-4} polarimetric sensitivity. This is not a coincidence. Much of the future progress will be achieved with sensitivities in this range. Those more regularly reached in present day observations, 10^{-3}, are due to instrumental limitations that have nothing to do with physical processes in the Sun. Polarimeters observing the solar photosphere with sensitivities of 10^{-4} and sub-arcsecond resolutions using the Zeeman and the Hanle effects will consolidate (or refute) many of the aspects commented here. The need for high spatial resolution observations of the Sr I depolarization cannot be emphasized enough (Stenflo 2012). These targets demand large apertures similar to those planned for future facilities such as the 4 m class ground-based telescopes (ATST, EST Keil et al. 2011; Collados et al. 2010) and the Japanese led Solar-C mission (1.5 m aperture). All these facilities will likely have to be used outside of the diffraction limit to pursue high sensitivity spectropolarimetry thanks to their large collecting areas Keller (1999).

Finally, observing regions of the Sun hardly reachable from the ecliptic will also help to clarify the nature of the quiet sun magnetism. The ESA-led Solar Orbiter mission (Müller et al. 2012) will carry on-board a magnetograph (the Polarimetric and Helioseismic Imager, PHI; Gandorfer et al. 2011) similar to that of SDO/HMI (Scherrer et al. 2012) and will observe the dynamics of the solar poles from an inclination of 35° with respect to the solar equator. Observing the poles is crucial in this discussion because they represent the regions at farther distances from the activity belts in the Sun. In the absence of an SSD mechanism working at the solar surface, the origin of the internetwork fields can only be explained as a result of the cascading down towards the smallest scales (see the discussion in Schüssler and Vögler 2008) of the global dynamo fields. This effect is inevitably present on the Sun, but whether it affects only to network fields or to internetwork ones can best be discerned by observing their latitudinal properties with good spatial resolution and sensitivity.

Acknowledgements This work has been partially funded by the Spanish MINECO through Project No. AYA200AYA2011-29833-C06. Comments on an original version of the manuscript by D. Orozco, K. Petrovay and an unknown referee are gratefully acknowledged. ISSI support to attend the meeting is also acknowledged.

References

W.P. Abbett, The magnetic connection between the convection zone and corona in the quiet Sun. Astrophys. J. **665**, 1469–1488 (2007)

A. Asensio Ramos, Evidence for quasi-isotropic magnetic fields from Hinode quiet-Sun observations. Astrophys. J. **701**, 1032–1043 (2009)

G.K. Batchelor, On the spontaneous magnetic field in a conducting liquid in turbulent motion. Proc. R. Soc. Lond. Ser. A **201**, 405–416 (1950)

L.R. Bellot Rubio, D. Orozco Suárez, Pervasive linear polarization signals in the quiet Sun. Astrophys. J. **757**, 19 (2012)

S. Boldyrev, F. Cattaneo, Magnetic-field generation in Kolmogorov turbulence. Phys. Rev. Lett. **92**(14), 144501 (2004)

J.M. Borrero, P. Kobel, Inferring the magnetic field vector in the quiet Sun. I. Photon noise and selection criteria. Astron. Astrophys. **527**, 29 (2011)

A. Brandenburg, Nonlinear small-scale dynamos at low magnetic Prandtl numbers. Astrophys. J. **741**, 92 (2011)

F. Cattaneo, On the origin of magnetic fields in the quiet photosphere. Astrophys. J. Lett. **515**, 39–42 (1999)

R. Centeno, H. Socas-Navarro, B. Lites, M. Kubo, Z. Frank, R. Shine, T. Tarbell, A. Title, K. Ichimoto, S. Tsuneta, Y. Katsukawa, Y. Suematsu, T. Shimizu, S. Nagata, Emergence of small-scale magnetic loops in the quiet-Sun internetwork. Astrophys. J. Lett. **666**, 137–140 (2007)

M. Collados, F. Bettonvil, L. Cavaller, I. Ermolli, B. Gelly, A. Pérez, H. Socas-Navarro, D. Soltau, R. Volkmer (EST Team), European Solar Telescope: progress status. Astron. Nachr. **331**, 615 (2010)

S. Danilovic, M. Schüssler, S.K. Solanki, Probing quiet Sun magnetism using MURaM simulations and Hinode/SP results: support for a local dynamo. Astron. Astrophys. **513**, 1 (2010a)

S. Danilovic, B. Beeck, A. Pietarila, M. Schüssler, S.K. Solanki, V. Martínez Pillet, J.A. Bonet, J.C. del Toro Iniesta, V. Domingo, P. Barthol, T. Berkefeld, A. Gandorfer, M. Knölker, W. Schmidt, A.M. Title, Transverse component of the magnetic field in the solar photosphere observed by SUNRISE. Astrophys. J. Lett. **723**, 149–153 (2010b)

A.G. de Wijn, J.O. Stenflo, S.K. Solanki, S. Tsuneta, Small-scale solar magnetic fields. Space Sci. Rev. **144**, 275–315 (2009)

A. Gandorfer, S.K. Solanki, J. Woch, V. Martínez Pillet, A. Álvarez Herrero, T. Appourchaux, The solar orbiter mission and its polarimetric and helioseismic imager (SO/PHI). J. Phys. Conf. Ser. **271**(1), 012086 (2011)

P. Gömöry, C. Beck, H. Balthasar, J. Rybák, A. Kučera, J. Koza, H. Wöhl, Magnetic loop emergence within a granule. Astron. Astrophys. **511**, 14 (2010)

S.L. Guglielmino, V. Martínez Pillet, J.A. Bonet, J.C. del Toro Iniesta, L.R. Bellot Rubio, S.K. Solanki, W. Schmidt, A. Gandorfer, P. Barthol, M. Knölker, The frontier between small-scale bipoles and ephemeral regions in the solar photosphere: emergence and decay of an intermediate-scale bipole observed with SUNRISE/IMaX. Astrophys. J. **745**, 160 (2012)

H.J. Hagenaar, C.J. Schrijver, A.M. Title, The properties of small magnetic regions on the solar surface and the implications for the solar dynamo(s). Astrophys. J. **584**, 1107–1119 (2003)

J.W. Harvey, D. Branston, C.J. Henney, C.U. Keller (SOLIS and GONG Teams), Seething horizontal magnetic fields in the quiet solar photosphere. Astrophys. J. Lett. **659**, 177–180 (2007)

R. Ishikawa, S. Tsuneta, The relationship between vertical and horizontal magnetic fields in the quiet Sun. Astrophys. J. **735**, 74 (2011)

R. Ishikawa, S. Tsuneta, J. Jurčák, Three-dimensional view of transient horizontal magnetic fields in the photosphere. Astrophys. J. **713**, 1310–1321 (2010)

A.B. Iskakov, A.A. Schekochihin, S.C. Cowley, J.C. McWilliams, M.R.E. Proctor, Numerical demonstration of fluctuation dynamo at low magnetic Prandtl numbers. Phys. Rev. Lett. **98**(20), 208501 (2007)

S.L. Keil, T.R. Rimmele, J. Wagner, D. Elmore (ATST Team), ATST: the largest polarimeter, in *Solar Polarization 6*, ed. by J.R. Kuhn, D.M. Harrington, H. Lin, S.V. Berdyugina, J. Trujillo-Bueno, S.L. Keil, T. Rimmele. Astronomical Society of the Pacific Conference Series, vol. 437 (2011), p. 319

C. Keller, The advanced solar telescope: I. Science goals, in *High Resolution Sol. Phys.: Theory, Observations, and Techniques*, ed. by T.R. Rimmele, K.S. Balasubramaniam, R.R. Radick. Astronomical Society of the Pacific Conference Series, vol. 183 (1999), p. 169

C.U. Keller, F.-L. Deubner, U. Egger, B. Fleck, H.P. Povel, On the strength of solar intra-network fields. Astron. Astrophys. **286**, 626–634 (1994)

L. Kleint, A.I. Shapiro, S.V. Berdyugina, M. Bianda, Solar turbulent magnetic fields: non-LTE modeling of the Hanle effect in the C_2 molecule. Astron. Astrophys. **536**, 47 (2011)

T. Kosugi, K. Matsuzaki, T. Sakao, T. Shimizu, Y. Sone, S. Tachikawa, T. Hashimoto, K. Minesugi, A. Ohnishi, T. Yamada, S. Tsuneta, H. Hara, K. Ichimoto, Y. Suematsu, M. Shimojo, T. Watanabe, S. Shimada, J.M. Davis, L.D. Hill, J.K. Owens, A.M. Title, J.L. Culhane, L.K. Harra, G.A. Doschek, L. Golub, The Hinode (Solar-B) mission: an overview. Sol. Phys. **243**, 3–17 (2007)

H. Lin, On the distribution of the solar magnetic fields. Astrophys. J. **446**, 421 (1995)

B.W. Lites, Hinode observations suggesting the presence of a local small-scale turbulent dynamo. Astrophys. J. **737**, 52 (2011)

B.W. Lites, et al., The horizontal magnetic flux of the quiet-Sun internetwork as observed with the Hinode spectro-polarimeter. Astrophys. J. **672**, 1237–1253 (2008)

B.W. Lites, K.D. Leka, A. Skumanich, V. Martinez Pillet, T. Shimizu, Small-scale horizontal magnetic fields in the solar photosphere. Astrophys. J. **460**, 1019 (1996)

B.W. Lites, M. Kubo, H. Socas-Navarro, T. Berger, Z. Frank, R. Shine, T. Tarbell, A.M. Title, K. Ichimoto, Y. Katsukawa, S. Tsuneta, Y. Suematsu, T. Shimizu, S. Nagata, Has Hinode revealed the missing turbulent flux of the quiet Sun?, in *Solar Polarization 5: In Honor of Jan Stenflo*, ed. by S.V. Berdyugina, K.N. Nagendra, R. Ramelli. Astronomical Society of the Pacific Conference Series, vol. 405 (2009), p. 173

M.J. Martínez González, L.R. Bellot Rubio, Emergence of small-scale magnetic loops through the quiet solar atmosphere. Astrophys. J. **700**, 1391–1403 (2009)

M.J. Martínez González, R. Manso Sainz, A. Asensio Ramos, E. Hijano, Dead calm areas in the very quiet Sun. Astrophys. J. **755**, 175 (2012)

V. Martínez Pillet, J.C. Del Toro Iniesta, A. Álvarez-Herrero, V. Domingo, J.A. Bonet, L. González Fernández, A. López Jiménez, C. Pastor, J.L. Gasent Blesa, P. Mellado, J. Piqueras, B. Aparicio, M. Balaguer, E. Ballesteros, T. Belenguer, L.R. Bellot Rubio, T. Berkefeld, M. Collados, W. Deutsch, A. Feller, F. Girela, B. Grauf, R.L. Heredero, M. Herranz, J.M. Jerónimo, H. Laguna, R. Meller, M. Menéndez, R. Morales, D. Orozco Suárez, G. Ramos, M. Reina, J.L. Ramos, P. Rodríguez, A. Sánchez, N. Uribe-Patarroyo, P. Barthol, A. Gandorfer, M. Knoelker, W. Schmidt, S.K. Solanki, S. Vargas Domínguez, The imaging magnetograph eXperiment (IMaX) for the sunrise Balloon-Borne solar observatory. Sol. Phys. **268**, 57–102 (2011)

N. Meunier, R. Tkaczuk, T. Roudier, M. Rieutord, Velocities and divergences as a function of supergranule size. Astron. Astrophys. **461**, 1141–1147 (2007)

R. Moll, J. Pietarila Graham, J. Pratt, R.H. Cameron, W.-C. Müller, M. Schüssler, Universality of the small-scale dynamo mechanism. Astrophys. J. **736**, 36 (2011)

D. Müller, R.G. Marsden, O.C. St. Cyr, H.R. Gilbert, Solar orbiter. Sol. Phys. **193** (2012)

A. Nordlund, A. Brandenburg, R.L. Jennings, M. Rieutord, J. Ruokolainen, R.F. Stein, I. Tuominen, Dynamo action in stratified convection with overshoot. Astrophys. J. **392**, 647–652 (1992)

D. Orozco Suárez, L.R. Bellot Rubio, Analysis of quiet-Sun internetwork magnetic fields based on linear polarization signals. Astrophys. J. **751**, 2 (2012)

D. Orozco Suárez, L.R. Bellot Rubio, J.C. del Toro Iniesta, S. Tsuneta, B.W. Lites, K. Ichimoto, Y. Katsukawa, S. Nagata, T. Shimizu, R.A. Shine, Y. Suematsu, T.D. Tarbell, A.M. Title, Quiet-Sun internetwork magnetic fields from the inversion of Hinode measurements. Astrophys. J. Lett. **670**, 61–64 (2007)

K. Petrovay, G. Szakaly, The origin of intranetwork fields: a small-scale solar dynamo. Astron. Astrophys. **274**, 543 (1993)

J. Pietarila Graham, R. Cameron, M. Schüssler, Turbulent small-scale dynamo action in solar surface simulations. Astrophys. J. **714**, 1606–1616 (2010)

J. Pietarila Graham, S. Danilovic, M. Schüssler, The small-scale solar surface dynamo (keynote), in *The Second Hinode Science Meeting: Beyond Discovery-Toward Understanding*, ed. by B. Lites, M. Cheung, T. Magara, J. Mariska, K. Reeves. Astronomical Society of the Pacific Conference Series, vol. 415 (2009a), p. 43

J. Pietarila Graham, S. Danilovic, M. Schüssler, Turbulent magnetic fields in the quiet Sun: implications of Hinode observations and small-scale dynamo simulations. Astrophys. J. **693**, 1728–1735 (2009b)

B. Ruiz Cobo, J.C. del Toro Iniesta, Inversion of Stokes profiles. Astrophys. J. **398**, 375–385 (1992)

J. Sánchez Almeida, On the Sr I λ4607 Å Hanle depolarization signals in the quiet Sun. Astron. Astrophys. **438**, 727–732 (2005)

J. Sánchez Almeida, M. Martínez González, The magnetic fields of the quiet Sun, in *Solar Polarization 6*, ed. by J.R. Kuhn, D.M. Harrington, H. Lin, S.V. Berdyugina, J. Trujillo-Bueno, S.L. Keil, T. Rimmele. Astronomical Society of the Pacific Conference Series, vol. 437 (2011), p. 451

J. Sánchez Almeida, T. Emonet, F. Cattaneo, Polarization of photospheric lines from turbulent dynamo simulations. Astrophys. J. **585**, 536–552 (2003)

A.A. Schekochihin, S.C. Cowley, J.L. Maron, J.C. McWilliams, Critical magnetic Prandtl number for small-scale dynamo. Phys. Rev. Lett. **92**(5), 054502 (2004a)

A.A. Schekochihin, S.C. Cowley, S.F. Taylor, J.L. Maron, J.C. McWilliams, Simulations of the small-scale turbulent dynamo. Astrophys. J. **612**, 276–307 (2004b)

A.A. Schekochihin, N.E.L. Haugen, A. Brandenburg, S.C. Cowley, J.L. Maron, J.C. McWilliams, The onset of a small-scale turbulent dynamo at low magnetic Prandtl numbers. Astrophys. J. Lett. **625**, 115–118 (2005)

A.A. Schekochihin, A.B. Iskakov, S.C. Cowley, J.C. McWilliams, M.R.E. Proctor, T.A. Yousef, Fluctuation dynamo and turbulent induction at low magnetic Prandtl numbers. New J. Phys. **9**, 300 (2007)

P.H. Scherrer, J. Schou, R.I. Bush, A.G. Kosovichev, R.S. Bogart, J.T. Hoeksema, Y. Liu, T.L. Duvall, J. Zhao, A.M. Title, C.J. Schrijver, T.D. Tarbell, S. Tomczyk, The Helioseismic and Magnetic Imager (HMI) investigation for the Solar Dynamics Observatory (SDO). Sol. Phys. **275**, 207–227 (2012)

M. Schüssler, A. Vögler, Strong horizontal photospheric magnetic field in a surface dynamo simulation. Astron. Astrophys. **481**, 5–8 (2008)

N. Shchukina, J. Trujillo Bueno, Determining the magnetization of the quiet Sun photosphere from the Hanle effect and surface dynamo simulations. Astrophys. J. Lett. **731**, 21 (2011)

S.K. Solanki, Photospheric magnetic field: quiet Sun, in *Solar Polarization 5: In Honor of Jan Stenflo*, ed. by S.V. Berdyugina, K.N. Nagendra, R. Ramelli. Astronomical Society of the Pacific Conference Series, vol. 405 (2009) p. 135

S.K. Solanki, P. Barthol, S. Danilovic, A. Feller, A. Gandorfer, J. Hirzberger, T.L. Riethmüller, M. Schüssler, J.A. Bonet, V. Martínez Pillet, J.C. del Toro Iniesta, V. Domingo, J. Palacios, M. Knölker, N. Bello González, T. Berkefeld, M. Franz, W. Schmidt, A.M. Title, SUNRISE: instrument, mission, data, and first results. Astrophys. J. Lett. **723**, 127–133 (2010)

H.C. Spruit, A model of the solar convection zone. Sol. Phys. **34**, 277–290 (1974)

R.F. Stein, Solar surface magneto-convection. Living Rev. Sol. Phys. **9**, 4 (2012)

R.F. Stein, Å. Nordlund, Solar small-scale magnetoconvection. Astrophys. J. **642**, 1246–1255 (2006)

R.F. Stein, D. Bercik, Å. Nordlund, Solar surface magneto-convection, in *Current Theoretical Models and Future High Resolution Solar Observations: Preparing for ATST*, ed. by A.A. Pevtsov, H. Uitenbroek. Astronomical Society of the Pacific Conference Series, vol. 286 (2003), p. 121

O. Steiner, R. Rezaei, Recent advances in the exploration of the small-scale structure of the quiet solar atmosphere: vortex flows, the horizontal magnetic field, and the Stokes-V line-ratio method, in *Fifth Hinode Science Meeting*, ed. by L. Golub, I. De Moortel, T. Shimizu. Astronomical Society of the Pacific Conference Series, vol. 456 (2012), p. 3

O. Steiner, R. Rezaei, W. Schaffenberger, S. Wedemeyer-Böhm, The horizontal internetwork magnetic field: numerical simulations in comparison to observations with Hinode. Astrophys. J. Lett. **680**, 85–88 (2008)

J.O. Stenflo, The Hanle effect and the diagnostics of turbulent magnetic fields in the solar atmosphere. Sol. Phys. **80**, 209–226 (1982)

J.O. Stenflo, Unsolved problems in solar polarization, in *Solar Polarization 6*, ed. by J.R. Kuhn, D.M. Harrington, H. Lin, S.V. Berdyugina, J. Trujillo-Bueno, S.L. Keil, T. Rimmele. Astronomical Society of the Pacific Conference Series, vol. 437 (2011), p. 3

J.O. Stenflo, Basal magnetic flux and the local solar dynamo. Astron. Astrophys. (2012). doi:10.1051/0004-6361/201219833

A.M. Title, T.D. Tarbell, K.P. Topka, S.H. Ferguson, R.A. Shine (SOUP Team), Statistical properties of solar granulation derived from the SOUP instrument on Spacelab 2. Astrophys. J. **336**, 475–494 (1989)

J. Trujillo Bueno, Modeling scattering polarization for probing solar magnetism, in *Solar Polarization 6*, ed. by J.R. Kuhn, D.M. Harrington, H. Lin, S.V. Berdyugina, J. Trujillo-Bueno, S.L. Keil, T. Rimmele. Astronomical Society of the Pacific Conference Series, vol. 437 (2011), p. 83

J. Trujillo Bueno, A. Asensio Ramos, N. Shchukina, The Hanle effect in atomic and molecular lines: a new look at the Sun's hidden magnetism, in *Astronomical Society of the Pacific Conference Series*, ed. by R. Casini, B.W. Lites. Astronomical Society of the Pacific Conference Series, vol. 358 (2006), p. 269

J. Trujillo Bueno, N. Shchukina, A. Asensio Ramos, A substantial amount of hidden magnetic energy in the quiet Sun. Nature **430**, 326–329 (2004)

S. Tsuneta, K. Ichimoto, Y. Katsukawa, S. Nagata, M. Otsubo, T. Shimizu, Y. Suematsu, M. Nakagiri, M. Noguchi, T. Tarbell, A. Title, R. Shine, W. Rosenberg, C. Hoffmann, B. Jurcevich, G. Kushner, M. Levay, B. Lites, D. Elmore, T. Matsushita, N. Kawaguchi, H. Saito, I. Mikami, L.D. Hill, J.K. Owens, The solar optical telescope for the Hinode mission: an overview. Sol. Phys. **249**, 167–196 (2008)

B. Viticchié, On the polarimetric signature of emerging magnetic loops in the quiet Sun. Astrophys. J. Lett. **747**, 36 (2012)

A. Vögler, M. Schüssler, A solar surface dynamo. Astron. Astrophys. **465**, 43–46 (2007)

C. Westendorp Plaza, J.C. del Toro Iniesta, B. Ruiz Cobo, V. Martinez Pillet, B.W. Lites, A. Skumanich, Optical tomography of a sunspot. I. Comparison between two inversion techniques. Astrophys. J. **494**, 453 (1998)

L. Yelles Chaouche, F. Moreno-Insertis, V. Martínez Pillet, T. Wiegelmann, J.A. Bonet, M. Knölker, L.R. Bellot Rubio, J.C. del Toro Iniesta, P. Barthol, A. Gandorfer, W. Schmidt, S.K. Solanki, Mesogranulation and the solar surface magnetic field distribution. Astrophys. J. **727**, 30 (2011)

DOI 10.1007/978-1-4899-7413-6_5
Reprinted from *Space Science Reviews* Journal, DOI 10.1007/s11214-013-0009-3

Astrophysical Hydromagnetic Turbulence

A. Brandenburg · A. Lazarian

Received: 1 March 2013 / Accepted: 21 July 2013 / Published online: 31 August 2013

Abstract Recent progress in astrophysical hydromagnetic turbulence is being reviewed. The physical ideas behind the now widely accepted Goldreich–Sridhar model and its extension to compressible magnetohydrodynamic turbulence are introduced. Implications for cosmic ray diffusion and acceleration is being discussed. Dynamo-generated magnetic fields with and without helicity are contrasted against each other. Certain turbulent transport processes are being modified and often suppressed by anisotropy and inhomogeneities of the turbulence, while others are being produced by such properties, which can lead to new large-scale instabilities of the turbulent medium. Applications of various such processes to astrophysical systems are being considered.

Keywords Magnetic fields · Turbulence · Sun: magnetic fields · ISM: magnetic fields

1 Introduction

Hydromagnetic or magnetohydrodynamic (MHD) turbulence plays an important role in many astrophysical settings. In a recent review by Brandenburg and Nordlund (2011b), properties of turbulence were discussed for the solar wind, stellar convection zones, the interstellar medium, accretion discs, galaxy clusters, and the early Universe. In an earlier review by Brandenburg and Subramanian (2005), a detailed account of dynamo theory with emphasis on helical dynamos was given. In that review, and also in Brandenburg et al. (2012c),

A. Brandenburg (✉)
Nordita, KTH Royal Institute of Technology, Stockholm University, Roslagstullsbacken 23, 10691 Stockholm, Sweden
e-mail: brandenb@nordita.org

A. Brandenburg
Department of Astronomy, Stockholm University, 10691 Stockholm, Sweden

A. Lazarian
Department of Astronomy, University of Wisconsin-Madison, 475 N. Charter St., Madison, WI 53706, USA
e-mail: lazarian@astro.wisc.edu

the small-scale dynamo was discussed in detail. Applications to galactic dynamos were discussed by Beck et al. (1996). Aspects of magnetic reconnection and particle acceleration in turbulent flows have recently been reviewed by Lazarian et al. (2012b). In the present review we begin with turbulence in the interstellar medium, discuss how turbulence is affected by magnetic fields and compressibility, address then applications to cosmic ray scattering and turn then to dynamo-generated magnetic fields as well as to anisotropic and inhomogeneous flows that are affected by stratification and rotation.

2 Turbulence in the Interstellar Medium

The ISM is turbulent on scales ranging from AUs to kpc (Armstrong et al. 1995; Elmegreen and Scalo 2004), with an embedded magnetic field that influences almost all of its properties. MHD turbulence is accepted to be of key importance for fundamental astrophysical processes, e.g. star formation, propagation and acceleration of cosmic rays. It is therefore not surprising that attempts to obtain spectra of interstellar turbulence have been numerous since the 1950s (Münch 1958). However, various directions of research achieved varying degrees of success. For instance, studies of turbulence statistics of ionized media accompanied by theoretical advancements in understanding scattering and scintillation of radio waves in ionized media (Goodman and Narayan 1985) were rather successful (cf. Spangler and Gwinn 1990). This work provided information about the statistics of the electron density on scales 10^8–10^{15} cm (Armstrong et al. 1995). These measurements have been recently combined with data from the Wisconsin Hα Mapper, which also measures electron density fluctuation, but on larger scales. The resulting extended spectrum presented in Chepurnov and Lazarian (2010) shows that the Kolmogorov $-5/3$ spectrum of electron density fluctuations extends to several more decades to larger scales; see Fig. 1.

In spite of their success, these sort of measurements provide only density statistics, which is a rather indirect measure of turbulence. Velocity statistics is a much more direct turbulence measure. Although it is clear that Doppler broadened lines are affected by turbulence, recovering the velocity statistics is extremely challenging without adequate theoretical insight. Indeed, both the z component of velocity and density contribute to fluctuations of the energy density $\rho_s(\boldsymbol{X}, V_z)$ in Position-Position-Velocity (PPV) space.

Traditionally, information on turbulence spectra is obtained using the measure of Doppler shifts termed Velocity Centroids, $\sim \int V_z \rho_s dV_z$, where the integration is taking place over the range of the velocities relevant to the object under study. In this situation it is easy to see that the Velocity Centroids are also proportional to $\int V_z \rho ds$, where ρ is the actual three-dimensional density and the integration is performed along the line of sight (Lazarian and Esquivel 2003).

Usually the Velocity Centroids are normalized by the intensity integrated over the line of sight (Stenholm 1990), and the work of Lazarian and Esquivel (2005) showed that this normalization does not change the statistical properties of the measure. However, the numerical and analytical analysis in Lazarian and Esquivel (2005) and Esquivel et al. (2007) showed that the Velocity Centroids fail for studying supersonic turbulence. This provides bad news for the studies of velocity statistics in molecular clouds and the diffuse cold ISM (Dickman and Kleiner 1985; Miesch et al. 1999; Miville-Deschênes et al. 2003) The studies for HII regions (O'dell and Castaneda 1987) are less strongly affected, as in most cases the turbulence there is subsonic.

There have been attempts to analyze PPV data cubes in other ways. For instance, Crovisier and Dickey (1983), Green (1993), and Stanimirovic et al. (1999) analyzed power

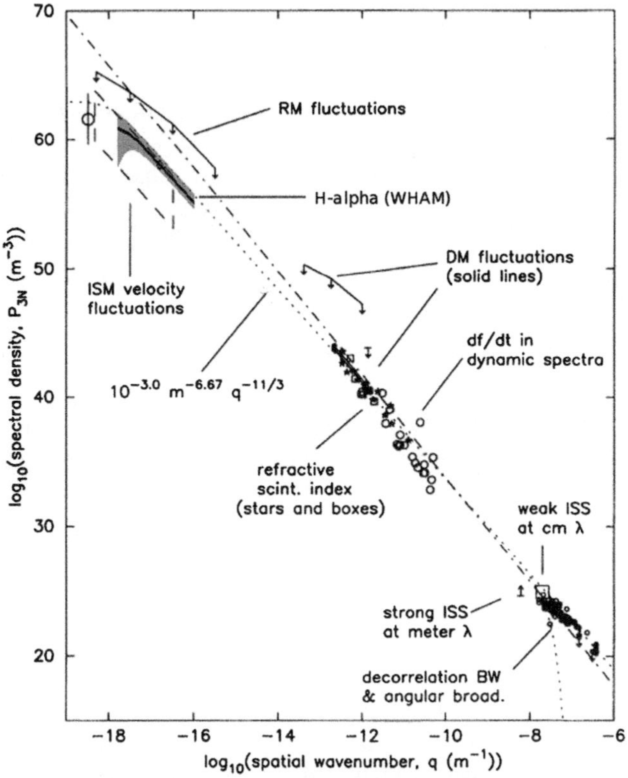

Fig. 1 Turbulence in the interstellar gas obtained from electron density fluctuations. The "Big Power Law in the Sky" of Armstrong et al. (1995) is here extended using data from the Wisconsin Hα Mapper (WHAM). The slope corresponds to that of Kolmogorov turbulence. Adapted from Chepurnov and Lazarian (2010)

spectra of velocity channels of HI data. The spatial spectrum of fluctuations of these velocity slices of PPV revealed power-law dependences, but the physical meaning of these dependences remained unclear. (Indeed, some of the authors erroneously identified the spectral index of intensity perturbations in slices of PPV data with the spectral index of the underlying turbulence spectrum. The nature of the variations of the spectral index in different studies was unclear.)

The analytical study of the statistical properties of the PPV energy density ρ_s has been initiated by Lazarian and Pogosyan (2000). There the observed statistics of ρ_s was related to the underlying 3D spectra of velocity and density in the astrophysical turbulent volume. Initially, the volume was considered transparent, but later the treatment was generalized to volumes with self-absorption and to studies of turbulence using absorption lines (Lazarian and Pogosyan 2004, 2006, 2008). The resulting theory of mapping of fluctuations in Position-Position-Position space with turbulent velocity into PPV space was successfully tested in a number of studies (Padoan et al. 2006, 2009; Chepurnov and Lazarian 2009; Burkhart et al. 2013). This theory lays the foundation for two separate techniques, Velocity Channel Analysis (VCA) and Velocity Correlation Spectrum (VCS) which were applied by a number of groups to different data sets including HI, C O^{13}, ^{18}C O; see more in Lazarian (2009). The results can be briefly summarized as follows: the tested supersonic media exhibit a velocity spectrum that is steeper than the spectrum of Kolmogorov turbulence and a density spectrum that is shallower. This result is, in fact, expected for supersonic MHD turbulence (Beresnyak et al. 2005; Kowal et al. 2007).

We emphasize that VCA and VCS are two related techniques based on solid analytical foundations. The theory of the VCA in Lazarian and Pogosyan (2000, 2004) and VCS in Lazarian and Pogosyan (2006, 2008) describe the non-linear mapping provided by velocity fluctuations from the turbulent volume to the Position-Position-Velocity (PPV) space. Therefore the technique provides the true spectrum of velocity and density fluctuations, irrespectively of the sources and sinks of turbulence. The energy injection associated with localized injection of turbulence, e.g. with the outflows should be detected as the changes in the spectral slope corresponding to the scales of energy injection.

3 The Picture of Alfvénic Turbulence

The picture of MHD turbulence has been developing over decades and pioneering works by Iroshnikov (1963) and Kraichnan (1965) are definitely to be mentioned. The Iroshnikov-Kraichnan model was the extension of Kolmogorov's *isotropic* turbulence model and it is the assumption of anisotropy that was a deficiency of this model. The notion of anisotropic turbulence was established later in important works, notably, by Shebalin et al. (1983) for incompressible turbulence and Higdon (1984) for the compressible turbulence. These papers provided the ground for the further advance.

Quantitative insight into MHD turbulence has been obtained in the seminal paper by Goldreich and Sridhar (1995), hereafter referred to as GS95. This paper quantified the properties of the anisotropic cascade and provided foundations for further theoretical development in the field. We may mention parenthetically that the original paper could not provide the perfect picture of MHD turbulence theory and a number of key aspects were clarified and corrected in subsequent studies. For instance, the original claim in GS95 and Goldreich and Sridhar (1997) about the role of 3-wave interactions were later corrected, and for weak MHD turbulence the point of view expressed in Ng and Bhattacharjee (1996) was adopted. Similarly, the notion of a *local* system of reference that is essential for understanding critical balance, which is a corner stone of our modern understanding of GS95 theory, was missing in the original paper. In fact, the closure relations that are used in GS95 to justify the model are written in the system of reference related to the mean field and therefore cannot be used as a proof. The importance of a local system of reference was understood only in subsequent theoretical and numerical studies by Lazarian and Vishniac (1999), henceforth LV99, Cho and Vishniac (2000), as well as Maron and Goldreich (2001).

3.1 Incompressible MHD Turbulence

While having a long history of ideas, the theory of MHD turbulence has become testable recently with the advent of numerical simulations (Biskamp 2003), which confirmed (see Cho and Lazarian 2005, and references therein) the prediction of magnetized Alfvénic eddies being elongated in the direction of the magnetic field (Shebalin et al. 1983; Higdon 1984) and provided results consistent with quantitative relations for the degree of eddy elongation obtained by GS95.

MHD turbulence theory is in many respects similar to the famous Kolmogorov (1941) theory of turbulence. In the latter theory, energy is injected at large scales, creating large eddies which do not dissipate energy through viscosity[1] but transfer energy to smaller eddies. The process continues until the cascade reaches the eddies that are small enough to dissipate energy over an eddy turnover time. In the absence of compressibility the hydrodynamic cascade of energy is $\sim v_l^2/\tau_{\mathrm{casc},l} = \mathrm{const}$, where v_l is the velocity at the scale l and the cascading time for the eddies of size l is $\tau_{\mathrm{casc},l} \approx l/v_l$. From this the well known relation $v_l \sim l^{1/3}$ follows.

In MHD turbulence, in the presence of dynamically important magnetic fields, eddies become anisotropic. At the same time, one can imagine eddies mixing magnetic field lines perpendicular to the direction of the magnetic field. For these eddies, the original Kolmogorov treatment is applicable resulting in perpendicular motions scaling as $v_l \sim l_\perp^{1/3}$, where $l_\perp$ denotes eddy scales measured perpendicular to the magnetic field. These mixing motions induce Alfvénic perturbations that determine the parallel size of the magnetized eddy. A cornerstone of the GS95 theory is *critical balance*, i.e. the equality of the eddy turnover time $l_\perp/v_l$ and the period of the corresponding Alfvén waves $\sim l_\parallel/V_\mathrm{A}$, where $l_\parallel$ is the parallel eddy scale and V_A is the Alfvén velocity. Making use of the earlier expression for v_l, one can easily obtain $l_\parallel \sim l_\perp^{2/3}$, which reflects the tendency of eddies to become more and more elongated as energy cascades to smaller scales.

It is important to stress that the scales $l_\perp$ and $l_\parallel$ are measured with respect to a system of reference related to the direction of the local magnetic field "seen" by the eddy. This notion was not present in the original formulation of the GS95 theory and was added to it later by Lazarian and Vishniac (1999), henceforth LV99, and Cho and Vishniac (2000). The local system of reference was also used in numerical studies in Cho and Vishniac (2000), Maron and Goldreich (2001), and Cho et al. (2002) that tested GS95 theory. In terms of mixing motions, it is rather obvious that the free Kolmogorov-type mixing is possible only with respect to the local magnetic field of the eddy rather than the mean magnetic field of the flow.

While the arguments above are far from being rigorous, they correctly reproduce the basic scalings of magnetized turbulence when the velocity is equal to V_A at the injection scale L. The most serious argument against this picture is the ability of eddies to perform mixing motions perpendicular to the magnetic field. This problem was addressed in LV99, where the self-consistency of the GS95 theory was related to fast reconnection of the magnetic field in turbulent fluids. A more rigorous discussion of a self-consistent treatment of MHD turbulence and magnetic reconnection is presented in Eyink et al. (2011).

The GS95 theory is formulated assuming isotropic injection of energy at scale L and the injection velocity equal to the Alfvén velocity in the fluid V_A, i.e. the Alfvén Mach number $M_\mathrm{A} \equiv (V_\mathrm{L}/V_\mathrm{A}) = 1$, where V_L is the injection velocity. Thus, it provides the description of trans-Alfvénic turbulence. This model was later extended for both sub-Alfvénic, i.e. $M_\mathrm{A} < 1$, and super-Alfvénic, i.e. $M_\mathrm{A} > 1$, cases (see LV99 and Lazarian 2006, respectively; see also Table 1). Indeed, if $M_\mathrm{A} > 1$, then, instead of the driving scale L one can use the scale

$$l_\mathrm{A} = L M_\mathrm{A}^{-3}, \tag{1}$$

which is the scale at which the turbulent velocity equals V_A. For $M_\mathrm{A} \gg 1$, magnetic fields are not dynamically important at the largest scales and the turbulence at those scales follows the isotropic Kolmogorov cascade $v_l \sim l^{1/3}$ over the range of scales $[L, l_\mathrm{A}]$. At the same

[1] The Reynolds number $Re \equiv L_\mathrm{f} V/\nu = (V/L_\mathrm{f})/(\nu/L_\mathrm{f}^2)$ characterizes the ratio of the eddy turnover rate $\tau_\mathrm{eddy}^{-1} = V/L_\mathrm{f}$ and the viscous dissipation rate $\tau_\mathrm{dis}^{-1} = \eta/L_\mathrm{f}^2$. Therefore large values of Re correspond to negligible viscous dissipation of large eddies over the cascading time τ_casc which is equal to τ_eddy in Kolmogorov turbulence.

Table 1 Regimes and ranges of MHD turbulence

Type of MHD turbulence	Injection velocity	Range of scales	Motion type	Ways of study
Weak	$V_L < V_A$	$[L, l_{trans}]$	Wave-like	Analytical
Strong sub-Alfvénic	$V_L < V_A$	$[l_{trans}, l_{min}]$	Eddy-like	Numerical
Strong super-Alfvénic	$V_L > V_A$	$[l_A, l_{min}]$	Eddy-like	Numerical

L and l_{min} are injection and dissipation scales
l_{trans} and l_A are given by (2) and (1), respectively

time, if $M_A < 1$, the turbulence obeys GS95 scaling (also called "strong" MHD turbulence) not from the scale L, but from a smaller scale

$$l_{trans} = LM_A^2, \tag{2}$$

while in the range $[L, l_{trans}]$ the turbulence is "weak".

The properties of weak and strong turbulence are rather different. Weak turbulence is wave-like turbulence with wave packets undergoing many collisions before transferring energy to small scales. Unlike strong turbulence, weak turbulence allows an exact analytical treatment (Galtier et al. 2000). By contrast, in strong turbulence intensive interactions between wave packets prevent the use of a perturbative approach. Numerical experiments have supported the GS95 ideas both for incompressible MHD turbulence (Cho and Vishniac 2000; Maron and Goldreich 2001; Cho et al. 2002; Beresnyak and Lazarian 2010, 2011) and for the Alfvénic component of compressible MHD turbulence (Cho and Lazarian 2002, 2003; Kowal and Lazarian 2010). [The compressible MHD turbulence simulations of Beresnyak et al. (2005) and Kowal et al. (2007) demonstrated that the density spectrum becomes more shallow and isotropic as the Mach number increases.]

While there are ongoing debates whether the original GS95 theory must be modified to better describe MHD turbulence, we believe that we do not have compelling evidence that GS95 is not adequate. The most popular one is the modification of the GS95 model by Boldyrev (2005, 2006), who, motivated by the spectral index of $-3/2$ observed in simulations of Maron and Goldreich (2001), proposed that the difference of the GS95 predictions and the numerical experiments arises from the dynamical alignment of velocity and magnetic fields. However, Beresnyak and Lazarian (2009, 2010) showed that present day numerical simulations may not have enough resolution to reveal the actual inertial range of MHD turbulence and the existing numerical simulations may be dominated by the bottleneck effect that distorts the actual slope of turbulence. Incidentally, the bottleneck effect already played a trick with the researchers when supersonic simulations suggested a $-5/3$ spectrum of supersonic turbulence (Boldyrev et al. 2002) which later was proven to be a bottleneck effect of shock wave turbulence with the expected -2 spectrum (Kritsuk et al. 2007). Such a spectrum has been confirmed with several different codes (Kritsuk et al. 2011). In addition, the $-5/3$ spectral index agrees well with the resolution studies by Beresnyak (2011, 2012). Thus, within the present review we will refer to GS95 when we shall talk about strong MHD turbulence.

The issue of the spectral slope is of both theoretical and practical importance. Although the differences between spectral slopes of $5/3$ and $3/2$ or even 2 do not look large, they correspond to very different physical pictures. The spectrum of $3/2$ corresponds to interactions decreasing with the scale of turbulent motions, $5/3$ corresponds to a strongly Kolmogorov-type picture of eddies, while 2 corresponds to a spectrum of shocks. The anisotropies predicted in these different pictures of turbulence are also different. They are in fact extremely important for cosmic ray propagation; see Yan and Lazarian (2004) and references therein. We also note that even a small difference in the slope can result in substantial differences in the energy at small scales due to the enormous extent of the astrophysical turbulent cascade. Finally, as GS95 has now the status of the accepted model of turbulence, it is essential to test all the predictions of this theory, including the predicted 5/3 spectral slope.

Usually, one considers balanced turbulence, i.e. the situation when the flows of energy in opposite directions are equal. In a more general case the turbulence is imbalanced, i.e. the flow of energy from one side dominates the flow from the opposite direction. The existing models of imbalanced turbulence are hotly debated at the moment and their predictions are being tested (Lithwick et al. 2007; Beresnyak and Lazarian 2008; Perez and Boldyrev 2009). Here we will just mention that in the case of astrophysical turbulence, compressibility may decrease the degree of imbalance, making the simple GS95 model applicable in spite of the presence of sources and sinks of energy.

3.2 Compressible MHD Turbulence

The statistical decomposition of 3D MHD turbulence into fundamental modes, i.e. Alfvén, slow and fast, was performed in Fourier space by Cho and Lazarian (2002, 2003), henceforth CL02 and CL03, respectively, and later using wavelets by Kowal and Lazarian (2010). The idea of the decomposition is presented in Fig. 2. The procedure was tested with the decomposition in real space in special cases when such a decomposition was possible, for instance, in the case of slow modes in a low plasma-β medium.

The most important result of this decomposition was establishing the relevance of Alfvénic turbulence scaling to a compressible medium. As we see in Fig. 3, the anisotropy of the Alfvénic component corresponds to the GS95 predictions. In general, the study of trans-Alfvénic turbulence with different Mach numbers in CL02 and CL03 revealed that GS95 scaling is valid for *Alfvén modes*:

$$\text{Alfvén:}\quad E^{\mathrm{A}}(k) \propto k^{-5/3}, \quad k_{\parallel} \propto k_{\perp}^{2/3}.$$

Slow modes also follow the GS95 model for both high β and mildly supersonic low β cases:

$$\text{Slow:}\quad E^{\mathrm{s}}(k) \propto k^{-5/3}, \quad k_{\parallel} \propto k_{\perp}^{2/3}.$$

For the highly supersonic low β case, the kinetic energy spectrum of slow modes tends to be steeper, which may be related to the formation of shocks.

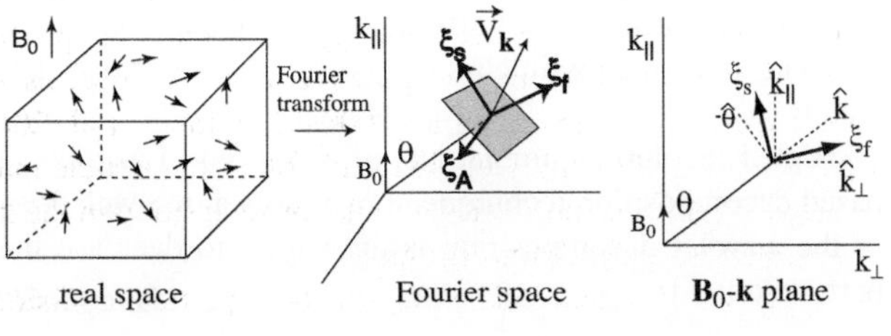

Fig. 2 Graphical representation of the mode separation method. We separate the Alfvén, slow and fast modes by the projection of the velocity Fourier component $\mathbf{v}_k$ on the bases $\hat{\xi}_A$, $\hat{\xi}_s$ and $\hat{\xi}_f$, respectively. Adapted from CL03

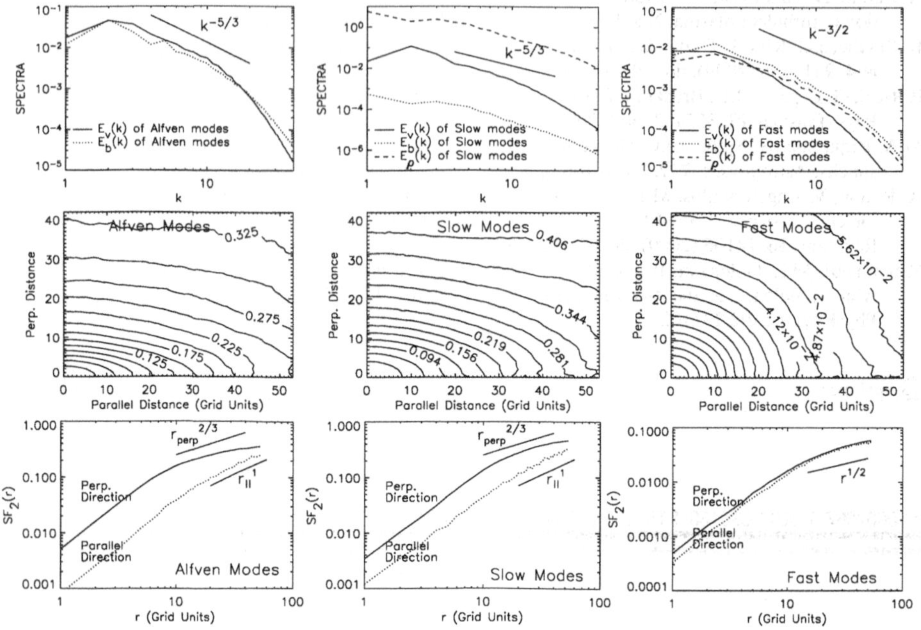

Fig. 3 Highly supersonic low β ($\beta \sim 0.02$ and $M_s \sim 7$). $V_A \equiv B_0/\sqrt{4\pi\rho} = 1$. a (sound speed) $= 0.1$. $\delta V \sim 0.7$. Alfvén modes follow the GS95 scalings. Slow modes follow the GS95 anisotropy. Fast modes are isotropic

Fast mode spectra are compatible with acoustic turbulence scaling relations:

$$\text{Fast:} \quad E^f(k) \propto k^{-3/2}, \quad \text{isotropic spectrum.}$$

The coupling between fast and Alfvén modes was shown to be weak and therefore the cascades of fast and Alfvén modes weakly affect each other (CL02). At the same time, Alfvén modes cascade to slow modes, which are otherwise passive in the cascade. This corresponds to the theoretical expectations discussed in GS95, Lithwick and Goldreich (2003), and CL03.

In terms of energy transfer from Alfvénic to compressible modes CL02 suggested the theory-motivated expression

$$\frac{\delta E_{\text{comp}}}{\delta E_{\text{Alf}}} \approx \frac{\delta V_A V_A}{V_A^2 + c_s^2}, \tag{3}$$

where δE_{comp} and δE_{Alf} are the energies of compressible and Alfvén modes, respectively. Equation (3) suggests that the drain of energy from Alfvénic cascade is marginal when the amplitudes of perturbations are weak, i.e. $(\delta V)_A \ll V_A$. Results of numerical calculations in CL02 are consistent with the expression above. The marginal transfer of energy between Alfvénic and compressible motions justifies considering the Alfvénic and fast cascades separately.

Higher resolution simulations in Kowal and Lazarian (2010) used a different wavelet-based decomposition technique. The results agree well with those in CL03. The advantage of the wavelet decomposition is the ability to decrease the error for the case of strongly perturbed fields.

4 Implications of MHD Turbulence for Diffusion Processes

4.1 Diffusion of Heat by MHD Turbulence

Transport processes are known to be affected by turbulence. A big issue related to MHD turbulence is the nature of turbulent eddies. If magnetic field lines are perfectly frozen into fluid, then one cannot talk about mixing motions at the scale of dynamically important magnetic fields. On the contrary, if magnetic reconnection is fast enough to resolve the knots of intersecting magnetic fluxes that naturally arise in GS95 turbulence, mixing motions perpendicular to the local magnetic field should be similar to those in hydrodynamical fluids. This problem was addressed in LV99, where it was shown that magnetic reconnection induced by turbulence makes the GS95 picture of a perpendicular cascade self-consistent. A more recent study by Eyink et al. (2011) revealed the deep connection between turbulence and magnetic reconnection. This provides a theoretical justification for discussing hydrodynamic-type turbulent advection of heat in the presence of dynamically important magnetic fields.[2]

In addition, in hot plasmas, the motion of electrons along wandering magnetic fields is important. The statistics of magnetic field wandering was described in LV99 for different regimes of turbulence and provides the necessary foundations for a quantitative description of the heat transfer process. This is the process that we start our discussion with.

Let us initially disregard the dynamics of fluid motions on diffusion, i.e. we consider diffusion induced by particles moving along wandering turbulent magnetic field lines, whose motions we disregard for the sake of simplicity. Magnetized turbulence with a dynamically important magnetic field is anisotropic with eddies elongated along the direction of local magnetic field (henceforth denoted by $\|$), i.e. $l_\perp < l_\|$, where $\perp$ denotes the direction perpendicular to the local magnetic field. Consider isotropic injection of energy at the outer scale L and dissipation at the scale $l_{\perp,\mathrm{min}}$. This scale corresponds to the minimal dimension of the turbulent eddies.

Initially, the problem of heat transport by electrons moving in turbulent magnetic fields was considered by Narayan and Medvedev (2001) for trans-Alfvénic turbulence. The treatment for both sub-Alfvénic and super-Alfvénic turbulence was presented in Lazarian (2006); henceforth L06.

It is easy to notice that the separations of magnetic field lines at scales below the damping scale of turbulence, i.e. for $r_0 < l_{\perp,\mathrm{min}}$, are mostly influenced by the motions at the smallest scale. This scale $l_{\perp,\mathrm{min}}$ results in Lyapunov-type growth $\sim r_0 \exp(l/l_{\|,\mathrm{min}})$. This growth is similar to that obtained in earlier models with a single scale of turbulent motions; see Rechester and Rosenbluth (1978), henceforth RR78, and Chandran and Cowley (1998). Indeed, as the largest shear that causes field line divergence is due to the marginally damped motions at the scale around $l_{\perp,\mathrm{min}}$ the effect of larger eddies can be neglected and we are dealing with the case of single-scale "turbulence" described by RR78.

The electron Larmor radius presents the minimal perpendicular scale of localization, while the other relevant scale is the Ohmic diffusion scale corresponding to the scale of damped motions. Thus, conservatively it is natural to associate r_0 with the size of the cloud

---

[2]The arguments in Eyink et al. (2011) should be distinguished from the arguments based on attempted renormalization of the effective magnetic Reynolds numbers in Blackman and Field (2008). Eyink et al. (2011) do not introduce artificial "turbulent diffusivities" but appeal to the established and tested concept of Richardson diffusion.

of electrons of the electron Larmor radius $r_{\mathrm{Lar,particle}}$. Applying the original RR78 theory, they found that the electrons should travel over a distance

$$L_{\mathrm{RR}} \sim l_{\|,\mathrm{min}} \ln(l_{\perp,\mathrm{min}}/r_{\mathrm{Lar},e}) \tag{4}$$

to get separated by $l_{\perp,\mathrm{min}}$.

Within the single-scale "turbulence model", which formally corresponds to $Lss = l_{\|,\mathrm{min}} = l_{\perp,\mathrm{min}}$, the distance L_{RR} is called Rechester–Rosenbluth distance. For the intracluster medium parameters, for which the problem was discussed originally, the logarithmic factor in (4) is of the order of 30. This causes a 30-fold decrease of the thermal conductivity for the single-scale models.[3]

The single-scale turbulence model is just a toy model to study the effects of turbulent motions. However, one can use this model to describe what is happening below the scale of the smallest eddies. Indeed, shear and, correspondingly, magnetic field line divergence are maximal for the marginally damped eddies at the dissipation scale. Thus, for scales less than the damping scale the action of the critically damped eddies is dominant and the results of (4) are applicable. The additional traveling distance of L_{RR} is of marginal importance for diffusion of heat over distances $\gg L_{\mathrm{RR}}$.

For the diffusion in super-Alfvénic turbulence the Alfvénic scale l_{A} given by (1) is important. It acts as the characteristic scale of magnetic fluctuations. Assuming that the mean free path of electrons is less than l_{A}, L06 obtained:

$$\kappa_e \equiv \Delta^2/\delta t \approx (1/3) l_{\mathrm{A}} v_e, \quad l_{\mathrm{A}} < \lambda, \tag{5}$$

where v_e is the electron velocity. In the opposite limit of effective scattering $\lambda < l_{\mathrm{A}}$, we have $\kappa \sim \lambda v_e$ with the coefficient of proportionality equal to $1/5$ according to Narayan and Medvedev (2001).

For sub-Alfvénic turbulence, the turbulence gets into the regime of strong GS95 type turbulence, which is described by (2). The diffusivity becomes anisotropic with the diffusion coefficient parallel to the mean field, $\kappa_{\|,\mathrm{particle}} \approx 1/3\kappa_{\mathrm{unmagn}}$ being larger than the coefficient for diffusion perpendicular to the magnetic field (L06):

$$\kappa_{\perp,e} = \kappa_{\|,e} M_{\mathrm{A}}^4, \quad M_{\mathrm{A}} < 1. \tag{6}$$

As discussed above, turbulent motions themselves can induce advective transport of heat. Appealing to the LV99 model of reconnection, one can conclude that turbulence with $M_{\mathrm{A}} \sim 1$ should be similar to hydrodynamic turbulence, i.e.

$$\kappa_{\mathrm{dynamic}} \approx C_{\mathrm{dyn}} L V_{\mathrm{L}}, \quad M_{\mathrm{A}} > 1, \tag{7}$$

where $C_{\mathrm{dyn}} \sim 0(1)$ is a constant, which for hydro turbulence is around $1/3$ (Lesieur 1990). If we deal with heat transport, for fully ionized non-degenerate plasmas we assume $C_{\mathrm{dyn}} \approx 2/3$ to account for the advective heat transport by both protons and electrons.

The advection of heat in the regime of sub-Alfvénic turbulence is reduced compared to the super-Alfvénic case and given by expression (L06):

$$\kappa_{\mathrm{dynamic}} \approx (\beta/3) L V_{\mathrm{L}} M_{\mathrm{A}}^3, \quad M_{\mathrm{A}} < 1, \tag{8}$$

where $\beta \approx 4$.

Figure 4 illustrates the existing ideas on processes of heat conduction in astrophysical plasmas. They range from heat insulation by unrealistically laminar magnetic field (a), to

[3] For the single-scale model, $L_{\mathrm{RR}} \sim 30L$ and the diffusion over distance Δ takes L_{RR}/Lss steps, i.e. $\Delta^2 \sim L_{\mathrm{RR}} L$, which decreases the corresponding diffusion coefficient $\kappa_{e,\mathrm{single}} \sim \Delta^2/\delta t$ by a factor 30.

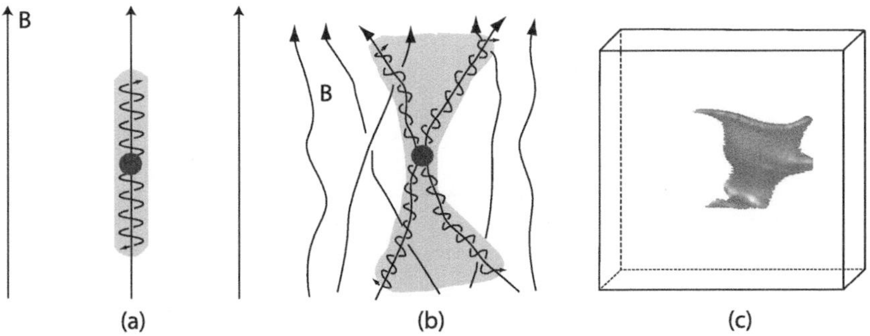

Fig. 4 Enßlin and Vogt (2006). (**a**) Textbook picture of electrons moving along magnetic field lines in thermal conduction process. (**b**) Actual motions of electrons in realistic turbulent plasmas where reconnection and spontaneous stochasticity of magnetic field are present. (**c**) Numerical simulations of heat advection in magnetized turbulence. From Cho and Lazarian (2004)

heat diffusion in turbulent magnetic field (b) and to heat advection by turbulent flows (c). field, to heat diffusion in turbulent magnetic field and to heat advection by turbulent flows. The relative efficiencies of the two latter processes depend on parameters of the turbulent plasma. The observational data for two clusters are also shown and it is clear that for the clusters of galaxies discussed, the turbulent advection of heat is the dominant process. The dominance of turbulent motions gets even more prominent if one takes into account that instabilities in the collisionless plasma of galaxies are likely to dramatically decrease the mean free path of electrons.

In thermal plasma, electrons are mostly responsible for thermal conductivity. The schematics of the parameter space for $\kappa_{\text{particle}} < \kappa_{\text{dynamic}}$ is shown in Fig. 5, where the Mach number M_s and the Alfvén Mach number M_A are the variables. For $M_A < 1$, the ratio of diffusivities arising from fluid and particle motions is $\kappa_{\text{dynamic}}/\kappa_{\text{particle}} \sim \beta\alpha M_s M_A(L/\lambda)$; see (6) and (8). The square root of the ratio of the electron to proton mass $\alpha = (m_e/m_p)^{1/2}$, which provides the separation line between the two regions in Fig. 2, is given by $\beta\alpha M_s \sim (\lambda/L)M_A$. For $1 < M_A < (L/\lambda)^{1/3}$ the mean free path is less than l_A which results in κ_{particle} being some fraction of κ_{unmagn}, while κ_{dynamic} is given by (7). Thus $\kappa_{\text{dynamic}}/\kappa_{\text{particle}} \sim \beta\alpha M_s(L/\lambda)$, i.e. the ratio does not depend on M_A (horizontal line in Fig. 5). When $M_A > (L/\lambda)^{1/3}$ the mean free path of electrons is constrained by l_A. In this case $\kappa_{\text{dynamic}}/\kappa_{\text{particle}} \sim \beta\alpha M_s M_A^3$; see (7) and (5). This results in the separation line $\beta\alpha M_s \sim M_A^{-3}$ in Fig. 5.

The application of the MHD approach to turbulent plasma has of course its limitations. For instance, in terms of magnetic reconnection, it is shown in Eyink et al. (2011) that the model of turbulent reconnection described in LV99 is applicable to current sheets if the broadening of the current sheet introduced through the wandering of magnetic field lines is larger than the Larmor radius of thermal ions. This makes the model not applicable to magnetosphere, where more sophisticated, e.g. based on PIC simulations, modeling is required.

4.2 Diffusion of Magnetic Fields in Turbulent Molecular Clouds

MHD turbulence induces not only mixing motions advecting heat, but it also induces the transport of magnetic field and matter in molecular clouds. This process, first discussed in

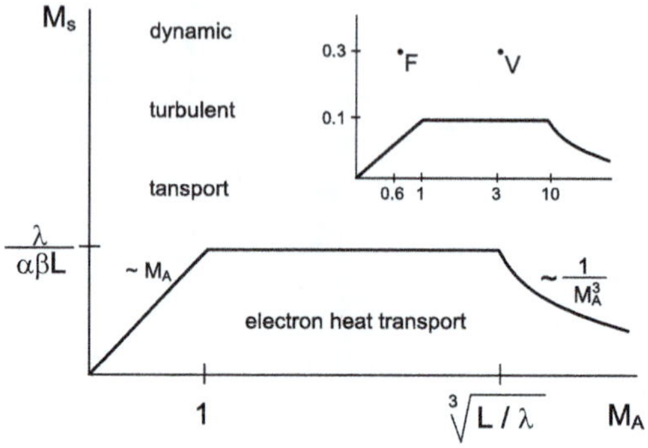

Fig. 5 Parameter space for particle diffusion or turbulent diffusion to dominate: application to heat transfer. Sonic Mach number M_S is plotted against the Alfvén Mach number M_A. The heat transport is dominated by the dynamics of turbulent eddies is above the curve (area denoted "dynamic turbulent transport") and by thermal conductivity of electrons is below the curve (area denoted "electron heat transport"). Here λ is the mean free path of the electron, L is the driving scale, and $\alpha = (m_e/m_p)^{1/2}$, $\beta \approx 4$. *Example of theory application*: The *panel in the right upper corner* of the figure illustrates heat transport for the parameters for a cool core Hydra cluster (point "F"), "V" corresponds to the illustrative model of a cluster core in Enßlin and Vogt (2006). Relevant parameters were used for L and λ. From L06

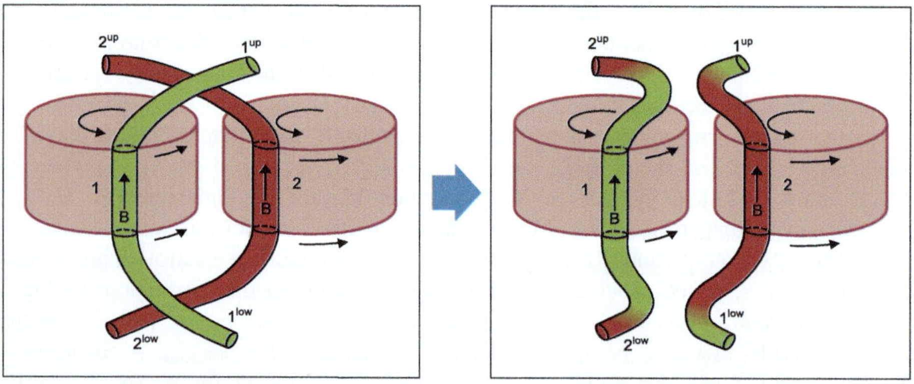

Fig. 6 Reconnection diffusion: exchange of flux with entrained matter. Illustration of the mixing of matter and magnetic fields due to reconnection as two flux tubes of different eddies interact. Only one scale of turbulent motions is shown. In real turbulent cascade such interactions proceed at every scale of turbulent motions. From Lazarian et al. (2012b)

Lazarian (2005) and Lazarian and Vishniac (2009), was later tested numerically in Santos-Lima et al. (2010, 2012) and showed high efficiency for removing magnetic fields from clouds and accretion disks. Lazarian et al. (2012a) showed that the process that they termed "reconnection diffusion" can explain why in observations by Crutcher et al. (2010) the envelopes had a lower mass to flux ratio than the cloud cores. In contrast, the usually considered ambipolar diffusion process predicts the opposite situation.

The elementary process of reconnection diffusion is illustrated in Fig. 6, where the densities of plasma along magnetic flux tubes belonging to different eddies are different. The process of fast turbulent reconnection (LV99) creates new flux tubes with columns of entrained dense and rarefied plasmas. The situation is similar to the earlier discussed case with plasma moving along magnetic fields and equalizing the pressure within the newly formed flux tubes. As a result, eddies with initially different plasma pressures exchange matter and equalize the plasma pressure. This process can be described as the diffusion of plasma perpendicular to the mean magnetic field. In reality, for turbulence with the extended inertial range, the shredding of the columns of plasmas with different density proceeds at all turbulence scales, making the speed of plasma motion irrelevant for the diffusion. For the case of strong turbulence, the diffusion of matter and magnetic field is given by (8). In the presence of the gravitational potential, the matter gets concentrated towards the center of the potential well. This was seen in the numerical simulations in Santos-Lima et al. (2010). The physical justification of the process is based on the nature of the GS95 cascade and the LV99 model of turbulent reconnection. The deep relation between the two is discussed in Eyink et al. (2011).

4.3 Cosmic Ray Scattering, Acceleration and Perpendicular Diffusion

MHD turbulence plays an important role in accelerating energetic particles. First of all, the second order Fermi acceleration can arise directly from the scattering of particles by turbulence (Melrose 1968). Properties of MHD turbulence that we discussed above are essential to understanding this process. If turbulence is injected at large scales, the anisotropy of Alfvénic modes at small scales makes them inefficient for scattering and acceleration of cosmic rays (Chandran 2000; Yan and Lazarian 2002).[4] In this situation, fast modes were identified in Yan and Lazarian (2002) as the major scattering and acceleration agent for cosmic rays and energetic particles in interstellar medium (see also Yan and Lazarian 2004, 2008). This conclusion was extended for solar environments in Petrosian et al. (2006) and intracluster medium in Brunetti and Lazarian (2007). Turbulent magnetic field in the pre-shock and post-shock environment are important for the first order Fermi acceleration associated with shocks (Schlickeiser 2002). In particular, magnetic field enhancement compared to its typical interstellar values is important in the pre-shock region for the acceleration of high energy particles. The turbulent dynamo can provide a way of generating magnetic field in the precursor of the shock. In Beresnyak et al. (2009) it was shown that the interactions of the density inhomogeneities pre-existing in the interstellar medium with the precursor generate strong magnetic fields in the shock precursor, which allows particle acceleration up to the energy of 10^{16} eV.

While discussing heat transport by thermal electrons streaming along turbulent magnetic fields, we have discussed the perpendicular diffusion that is also relevant for the turbulent transport of cosmic rays perpendicular to the mean magnetic field. The relation between the parallel and perpendicular diffusivities in this case is also given by (6); see Yan and Lazarian (2008). The important factor in this equation is $M_{\rm A}^4$. This dependence follows from the modern theory of MHD turbulence and it is very different from the dependence of $M_{\rm A}^2$ discussed in the literature (Jokipii 1974).

A stream of cosmic ray protons propagating parallel or antiparallel to a large-scale magnetic field can lead to important instabilities such as the Bell instability (Bell 2004). This is reviewed extensively in a companion paper by Bykov et al. (2013). The combined presence

[4] The resonant scattering is happening on the magnetic scales of the order of the cosmic ray gyroradius. If the Alfvénic eddies are strongly elongated, the particles interacts with many eddies within its radius and the scattering effect is dramatically reduced. Scattering efficiency and the acceleration efficiencies are closely related for the second order Fermi acceleration of cosmic rays by turbulence (see Schlickeiser 2002).

of a cosmic ray current and a parallel magnetic field gives rise to a pseudoscalar in the problem, and hence to an α effect which can lead to large-scale dynamo action (Rogachevskii et al. 2012). In the following, we discuss magnetic field amplification by dynamo action in more detail.

5 MHD Turbulence with Dynamo-Generated Magnetic Fields

In this section we discuss the case where the magnetic field is produced self-consistently by the action of turbulence through dynamo action. We discuss here mainly the results of numerical simulations.

5.1 Definitions and Conventions

In the following we characterize turbulent flows by the Reynolds number, which quantifies the ratio of advective to viscous accelerations, $\boldsymbol{u}\cdot\boldsymbol{\nabla}\boldsymbol{u}$ and $\nu\nabla^2\boldsymbol{u}$, respectively. Here, $\boldsymbol{u}$ is the velocity and ν is the kinematic viscosity. Throughout the remainder of this review, we define the Reynolds number as

$$Re = u_{\rm rms}/\nu k_{\rm f}, \tag{9}$$

where $u_{\rm rms} = \langle \boldsymbol{u}^2\rangle^{1/2}$ is the rms velocity within some appropriate volume and $k_{\rm f}$ is the wavenumber of the energy-carrying eddies, which is also known as the integral or correlation wavenumber. It can be defined through a weighted average of the inverse wavenumber over the kinetic energy spectrum, $E_{\rm K}(k,t)$, where $k=|\boldsymbol{k}|$ is the modulus of the wave vector $\boldsymbol{k}$, and t is time. The kinetic energy spectrum is normalized such that

$$\int_0^\infty E_{\rm K}(k,t)\,{\rm d}k = \frac{1}{2}\rho_0\langle\boldsymbol{u}^2\rangle, \tag{10}$$

where $\rho_0 = \langle\rho\rangle$ is the volume average of the gas density ρ. For incompressible and weakly compressible flows, it is customary to ignore fluctuations of ρ in the definition of $E_{\rm K}(k,t)$. In fact, there is no unique way of incorporating density. For supersonic turbulence, this is very much a current research topic in its own right. We refer here to the papers of Kritsuk et al. (2007), Galtier and Banerjee (2011), and Banerjee and Galtier (2013).

Returning to the case of incompressible or weakly compressible (subsonic) turbulence, a formal definition of $k_{\rm f}$ can be written as

$$k_{\rm f}^{-1} = \int k^{-1}E_{\rm K}(k,t)\,{\rm d}k \Big/ \int E_{\rm K}(k,t)\,{\rm d}k. \tag{11}$$

Note that $k_{\rm f}=k_{\rm f}(t)$ is in general time-dependent, which can be important in studies of decaying turbulence. An important example is helical MHD turbulence, because it drives an inverse cascade which manifests itself in a time-dependent decrease of $k_{\rm f}(t)$; see Tevzadze et al. (2012) and Kahniashvili et al. (2013) for recent examples. In most of the cases considered below we consider a time average of $k_{\rm f}$.

MHD turbulence is additionally characterized by the *magnetic* Reynolds number,

$$Re_M = u_{\rm rms}/\eta k_{\rm f}, \tag{12}$$

where η is the magnetic diffusivity. The ratio $Re_M/Re = \nu/\eta = Pr_M$ is the magnetic Prandtl number. Furthermore, a magnetic energy spectrum $E_{\rm M}(k,t)$ can be defined such that

$$\int_0^\infty E_{\rm M}(k,t)\,{\rm d}k = \frac{1}{2}\mu_0^{-1}\langle\boldsymbol{B}^2\rangle, \tag{13}$$

where $\boldsymbol{B}$ is the magnetic field and μ_0 is the vacuum permeability. Analogously to (11) we can then also define a magnetic correlation wavenumber $k_{\rm M}(t)$. The relative alignment between $\boldsymbol{u}$ and $\boldsymbol{B}$ is characterized by the so-called cross helicity, $\langle \boldsymbol{u} \cdot \boldsymbol{B} \rangle$, and its scale dependence is characterized by the cross helicity spectrum $E_{\rm C}(k,t)$ with the normalization $\int E_{\rm C}(k,t)\,{\rm d}k = \langle \boldsymbol{u} \cdot \boldsymbol{B} \rangle$. This quantity is a pseudoscalar and changes sign for a mirror-reflected image of the turbulence. Other important helicities are the kinetic helicity, $\langle \boldsymbol{w} \cdot \boldsymbol{u} \rangle$, with $\boldsymbol{w} = \nabla \times \boldsymbol{u}$ being the vorticity, the current helicity, $\langle \boldsymbol{J} \cdot \boldsymbol{B} \rangle$, with $\boldsymbol{J} = \nabla \times \boldsymbol{B}/\mu_0$ being the current density, and, in particular, the magnetic helicity, $\langle \boldsymbol{A} \cdot \boldsymbol{B} \rangle$, with $\boldsymbol{A}$ being the magnetic vector potential such that $\boldsymbol{B} = \nabla \times \boldsymbol{A}$.

In some cases we also discuss the evolution of a passive scalar, whose concentration is governed by a corresponding diffusivity κ. The relevant non-dimensional parameter is the Péclet number, $Pe = u_{\rm rms}/\kappa k_{\rm f}$.

5.2 Dynamo Instability and Spectrum

In the absence of an imposed magnetic field, the zero-field limit is unstable to dynamo action when the magnetic Reynolds number exceeds a critical value,

$$Re_M > Re_{M,\rm crit} \quad \text{(dynamo instability)}. \tag{14}$$

In practice, this means that the theory of Kolmogorov turbulence is not directly applicable to most astrophysical flows when the gas is ionized and therefore electrically conducting.

In this section we restrict ourselves to non-helical isotropic turbulence, i.e., $\langle \boldsymbol{w} \cdot \boldsymbol{u} \rangle \ll k_{\rm f}\langle \boldsymbol{u}^2 \rangle$. In that case, only random or turbulent magnetic fields can be expected. This possibility was already anticipated by Batchelor (1950), but the relevant theory was only developed later by Kazantsev (1968). He assumed that the velocity field was given by a smooth large-scale random flow and found that the resulting magnetic field has typical wavenumbers close to the resistive cutoff wavenumber, $k_\eta = \langle \mu_0^2 \boldsymbol{J}^2/\eta^2 \rangle^{1/4}$, and much larger than $k_{\rm f}$. In fact, his work predicted a $k^{3/2}$ spectrum for the magnetic field in the wavenumber range $k_{\rm f} < k < k_\eta$.

The first numerical solutions of such dynamos have been performed by Meneguzzi et al. (1981) at a resolution of just 64^3 collocation points. Those where the "golden years" of numerical turbulence research. For the first time, many of the ideas in turbulence could be put to the test and, although the numerical resolution was still poor, it was clear that it could only be a matter of time until all the newly emerging results will be confirmed at better resolution.

In the following years, small-scale dynamo action emerged in several direct numerical simulations (DNS). At first it appeared that kinetic helicity had only a minor effect in Cartesian simulations (Meneguzzi and Pouquet 1989; Kida et al. 1991; Nordlund et al. 1992). This was later understood to be an artefact of the lack of scale separation, i.e., $k_{\rm f}/k_1$ was not big enough (Haugen et al. 2004). Meanwhile, global convection simulations in spherical shells did produce large-scale magnetic fields (Gilman 1983; Glatzmaier 1985). Remarkably, although there was general awareness of the concepts of large-scale and small-scale dynamos, which was also clearly spelled out in an early review of Vainshtein and Zeldovich (1972), the theory of Kazantsev (1968) was still not yet widely cited in the West. This has changed by the late 1990s (e.g., Gruzinov et al. 1996; Subramanian 1998; Kulsrud 1999), and by the early 2000s many groups investigated the small-scale dynamo systematically (Cho and Vishniac 2000; Schekochihin et al. 2002, 2004a, 2004b; Haugen et al. 2003, 2004).

Although the resolution has improved significantly over the past two decades, some important aspects of small-scale dynamos was evident already early on. In particular,

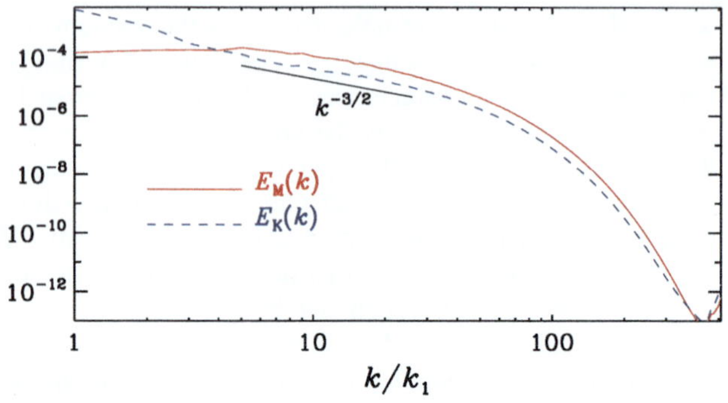

Fig. 7 Magnetic, kinetic and total energy spectra. $Re = Re_M = 960$ using 1024^3 meshpoints. Courtesy of Nils Erland Haugen (Haugen et al. 2003)

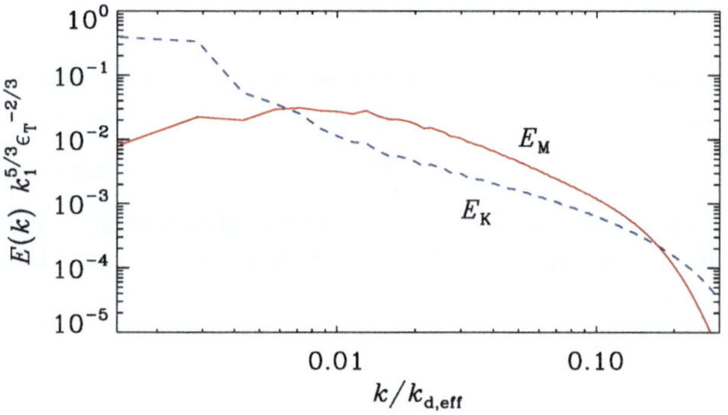

Fig. 8 Magnetic and kinetic energy spectra for runs with 512^3 meshpoints and hyperviscosity with hyperresistivity (*solid line*) and Smagorinsky viscosity with hyperresistivity (*red, dashed line*). Note the mutual approach of kinetic and magnetic energy spectra before entering the dissipative subrange. Adapted from Haugen and Brandenburg (2006)

Meneguzzi et al. (1981) and Kida et al. (1991) found that the magnetic energy spectrum reaches a maximum at a wavenumber k_M that is by a factor of ≈ 6 larger than k_f, which is where the kinetic energy has its maximum. This was an aspect that was later motivated by the work of Subramanian (1998), who proposed that k_M/k_f should be of the order of $Re_{M,crit}^{1/2}$. This result was indeed borne out by all the DNS obtained so far. In Fig. 7 we reproduce the result of Haugen et al. (2003) using 1024^3 meshpoints. For larger values of Pr_M, $Re_{M,crit}$ increases, so k_M also increases, making it harder to confirm the expected scaling in that regime. Indeed, Schekochihin et al. (2004b) propose that at large values of Pr_M the field shows folded structures. While Brandenburg and Subramanian (2005) confirmed the presence of folded structures in a simulation with $Pr_M = 50$, they found them rather the exception and showed other cases where the field was not folded. Recent simulations by Bhat and Subramanian (2013) confirmed that, after sufficiently many turnover times, k_M/k_f is of the order of $Re_{M,crit}^{1/2}$ even when $Pr_M = 50$.

Note that, at the position where the magnetic energy spectrum peaks, the magnetic field is in super-equipartition with the kinetic energy by a factor of 2–3. Initially, this was a somewhat surprising result in view of the work of GS95, according to which one might have expected equipartition. Subsequent work using large eddy simulations suggested that this super-equipartition would not persist deeper into the inertial range, provided Re and Re_M are large enough. Indeed, a trend toward equipartition can be seen in the compensated energy spectra of Haugen and Brandenburg (2006); see also Fig. 8, where a Smagorinsky subgrid scale model was used for the momentum equation and hyper-resistivity in the induction equation.

5.3 Small-Scale Dynamo Action at Small Values of Pr_M

The question of what happens in the case of $Pr_M \ll 1$ has always been on people's mind. Small values of Pr_M are characteristic of denser bodies such as stars, planets, and especially liquid metals. Only in recent years a clearer picture has emerged of what happens in the limit $Pr_M \to 0$. By comparing the onset of dynamo action, it became clear that $Re_{M,\text{crit}}$ grew larger and larger as one approached the value $Pr_M = 0.1$ (Schekochihin et al. 2005). Crucial insight was gained through a paper by Iskakov et al. (2007), who found that $Re_{M,\text{crit}}$ has a local maximum at $Pr_M = 0.1$, and that it decreases again as Pr_M is decreased further. Early work of Rogachevskii and Kleeorin (1997) did already predict an increased value of $Re_{M,\text{crit}}$ in the limit of small values of Pr_M, but not really a local maximum. Boldyrev and Cattaneo (2004) argue that the reason for an increased value of $Re_{M,\text{crit}}$ is connected with the "roughness" of the velocity field, as quantified by the scaling exponent ζ in velocity differences $\delta u_\ell \sim \ell^\zeta$ over spatial separations ℓ. In the diffusive subrange, $\zeta = 1$, so the velocity is smooth, but in the inertial range we have $\zeta \approx 0.4$, so velocity gradients diverge and the velocity field is therefore called "rough."

The connection with roughness also helped explaining the occurrence of a maximum in $Re_{M,\text{crit}}$ as Pr_M goes through 0.1. Indeed, the reason for this is that near $Pr_M = 0.1$ the resistive wavenumber is about 10 times smaller than the viscous one and thus right within the "bottleneck" where the spectrum is even shallower than in the rest of the inertial range, with a local scaling exponent $\zeta \to 0$, corresponding to turbulence that is in this regime rougher still, explaining thus the apparent divergence of $Re_{M,\text{crit}}$.

The physical reality of the bottleneck effect remains still a matter of debate, but the work of Falkovich (1994) suggests that it is related to the fact that near the viscous cut-off wavenumber the flow becomes harder to stir, and that triangle interactions between a wavenumber in the bottleneck range with wavenumbers in the dissipative subrange experience a difficulty in disposing of their energy. It is claimed in Beresnyak and Lazarian (2010) that the MHD turbulence while formally local, is more diffusive in terms of the interactions involved. This property termed "diffuse locality", may explain that the bottleneck effect in hydrodynamics is much more prominent that in MHD. Thus, one may suspect that even the highest resolution simulations would still not be showing the actual inertial range, but are influenced by an extended bottleneck effect Beresnyak and Lazarian (2009). This may be the reason why the numerically measured spectrum is a bit shallower than the GS95 prediction. A numerical study in Beresnyak (2011) seems to support this conclusion.

It has recently become possible to demonstrate that in the nonlinear regime, when the magnetic field affects the flow, the hydrodynamic bottleneck effect tends to be suppressed as the field strength becomes appreciable, so the divergence in the roughness disappears and there is a smooth dependence of the saturation field strength on the value of Pr_M; see Brandenburg (2011a) for details. In Fig. 9 we show the saturation energy of small-scale dynamos as a function of Pr_M using the data of Tables 1 and 2 of Brandenburg (2011a). It is clear that the position $Pr_M = 0.1$ is no longer special and that dynamo action is possible for small values of Pr_M as well. For $Re_M = 160$ the value of $B_{\text{rms}}/B_{\text{eq}}$ is still Re_M-dependent, but this may be an artefact of the dynamo being close to onset. For $Re_M = 220$ the dynamo is more clearly supercritical and, although there are only two data points, the results are now more clearly consistent with $B_{\text{rms}}/B_{\text{eq}}$ being independent of Re_M.

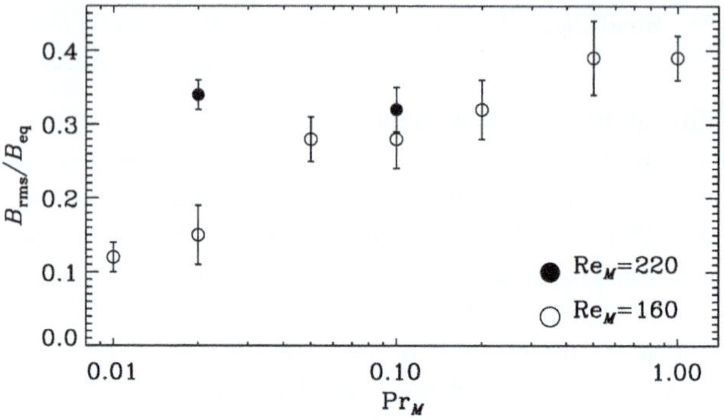

Fig. 9 Saturation field strengths for small-scale dynamos as a function of Pr_M for two values of Re_M. Note that for $Re_M = 160$ (*open symbols*) the dynamo is close to onset and the saturation field strength declines with decreasing values of Pr_M, while for $Re_M = 220$ (*filled symbols*) the field strength changes only weakly although only two data points are available

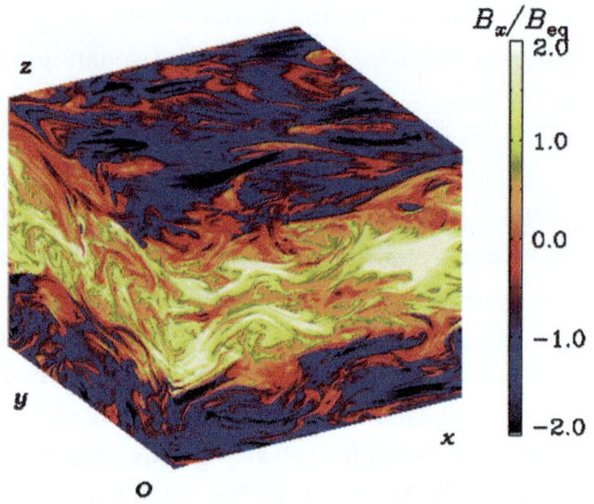

Fig. 10 Visualization of B_x on the periphery of the computational domain for a run with $Re_M = 600$ and a resolution of 512^3 mesh points. Note the clear anisotropy with structures elongated in the direction of the field (which lies in the xy plane). Adapted from Brandenburg et al. (2008a)

5.4 Helically Driven Turbulence

Eigenfunctions of the curl operator provide an ideal means of stirring the flow. In wavenumber space, these take the form (Haugen et al. 2004)

$$\boldsymbol{f}_k = \mathsf{R} \cdot \boldsymbol{f}_k^{(\text{nohel})} \quad \text{with } \mathsf{R}_{ij} = \frac{\delta_{ij} - \mathrm{i}\sigma\epsilon_{ijk}\hat{k}_k}{\sqrt{1+\sigma^2}}, \tag{15}$$

where σ is a measure of the helicity of the forcing and $\sigma = 1$ for positive maximum helicity of the forcing function. Furthermore,

$$\boldsymbol{f}_k^{(\text{nohel})} = (\boldsymbol{k} \times \hat{\boldsymbol{e}})/\sqrt{\boldsymbol{k}^2 - (\boldsymbol{k} \cdot \hat{\boldsymbol{e}})^2} \tag{16}$$

is a non-helical forcing function, where $\hat{\boldsymbol{e}}$ is an arbitrary unit vector not aligned with $\boldsymbol{k}$; note that $|\boldsymbol{f}_k|^2 = 1$ and $\boldsymbol{f}_k \cdot (\mathrm{i}\boldsymbol{k} \times \boldsymbol{f}_k)^* = 2\sigma k/(1+\sigma^2)$, so the relative helicity of the forcing function in real space is $2\sigma/(1+\sigma^2)$. When $\sigma = 0$, the forcing function is non-helical, and so is the resulting flow. This case is special, as was demonstrated on various occasions. Firstly, helical turbulence introduces an α effect which means that a weak *large-scale* magnetic field becomes destabilized and will be amplified. In Fig. 10 we show a visualization of one of the field components on the periphery of a Cartesian domain with periodic boundary conditions. Note the presence of both large-scale and small-scale components. Secondly, in

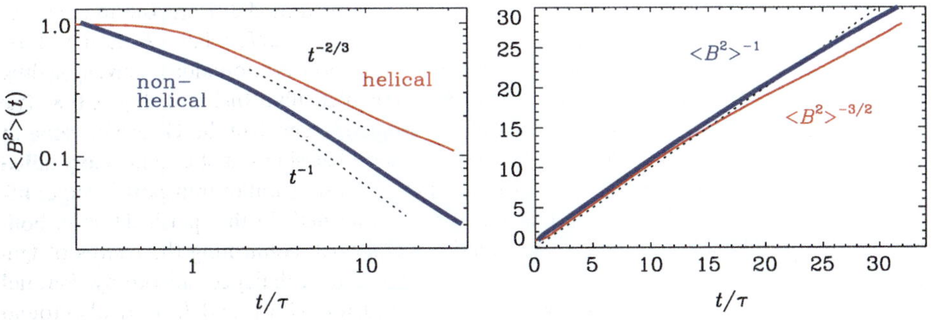

Fig. 11 Decay of magnetic energy with and without initial helicity (*left*) and the approximately linear evolution of $\langle \boldsymbol{B}\rangle^{-3/2}$ and $\langle \boldsymbol{B}\rangle^{-1}$ in the two cases (*right*)

the absence of forcing, a fully helical magnetic field decays more slowly than a non-helical one. Specifically, we have (Biskamp and Müller 1999; Biskamp 2003)

$$\langle \boldsymbol{B}^2\rangle(t) = \frac{\langle \boldsymbol{B}^2\rangle(0)}{(1+t/\tau)^{2/3}}, \tag{17}$$

where $\tau = \sqrt{\mu_0\rho_0}\langle \boldsymbol{A}\cdot\boldsymbol{B}\rangle/\langle \boldsymbol{B}^2\rangle^{3/2}$ is the typical decay time scale. In Fig. 11 we compare results of two simulations of Kahniashvili et al. (2013), one with an initial magnetic helicity and the other one without. Note the slower decay proportional to $t^{-2/3}$ in the helical case compared to the faster t^{-1} decay in the non-helical case. In both cases, time has been normalized by $\tau = \sqrt{\mu_0\rho_0}/k_{f0}B_{rms}$, where $k_{f0} = k_f(t=0) \approx 15k_1$. The rms velocity is about 20 % of the B_{rms} in the helical case and about 28 % in the non-helical case. The Reynolds number based on $k_f(t)$, which decreases with time either like $t^{-2/3}$ in the helical case, or like $t^{-1/2}$ in the non-helical case, increases from 50 to 100 during the coarse of both simulations. Even if the magnetic field is initially not fully helical, the relative helicity will increase, because magnetic energy decays faster than magnetic helicity; see Tevzadze et al. (2012). These considerations are important for primordial magnetic fields generated during cosmological phase transitions, because the inverse cascade allow the fields to reach appreciable length scales at the present time (Brandenburg et al. 1996; Banerjee and Jedamzik 2004; Kahniashvili et al. 2010).

The α effect is the reason behind the large-scale dynamo effect leading to the global magnetic field observed in many astrophysical bodies (Moffatt 1978; Parker 1979; Krause and Rädler 1980). The resulting magnetic field is helical and its helicity has the same sign as α. However, because of total magnetic helicity conservation, no net magnetic helicity can be produced. Therefore the α effect produces magnetic helicity of opposite signs at large and small length scales at the same time. In Fig. 12 we show magnetic and kinetic energy spectra compensated by $k^{1.5}$ together with compensated magnetic and kinetic helicity spectra, normalized by $k/2$ and $1/2k$, respectively. This normalization allows us to see whether or not the realizability conditions, $E_M(k) \ge 2kH_M(k)$ and $E_K(k) \ge 2H_K(k)/k$, are close to being saturated. Note also that $H_M(k)$ changes sign and becomes negative at $k/k_1 = 1$ (thin line), and is positive at all larger values of k/k_1 (thick line).

The case of homogeneous helical turbulence is a particularly interesting example, because accurate estimates can be made about the saturation field strength and the magnetic helicity balance, for example. However, such circumstances are not usually found in realis-

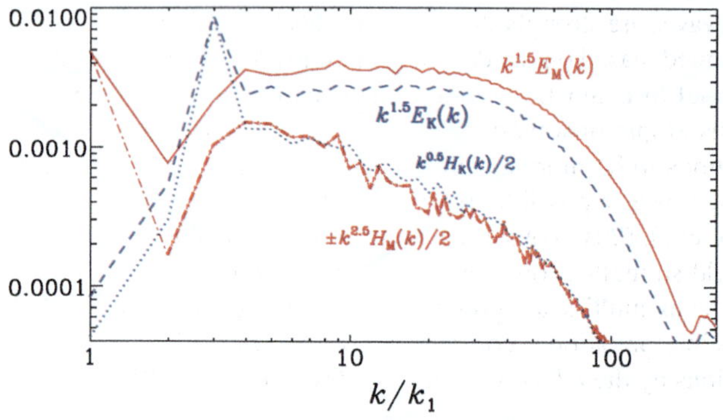

Fig. 12 Compensated time-averaged spectra of kinetic and magnetic energy (*dashed* and *solid lines*, respectively), as well as of kinetic and magnetic helicity (*dotted* and *dash-dotted lines*, respectively), for a run with $Re_M \approx 600$. Note that $H_M(k)$ changes sign and becomes negative $k/k_1 = 1$ (indicated by a *thin line*), and is positive at all larger values of k/k_1 (indicated by a *thicker line*). Adapted from Brandenburg et al. (2008a), where however $H_K(k)$ and $H_M(k)$ are compensated by $k^{1.2}$ and $k^{3.2}$, respectively

tic applications. The significance of homogeneity is that then the divergence of the magnetic helicity fluxes vanishes and does not affect the magnetic helicity evolution, so we have

$$\frac{\mathrm{d}}{\mathrm{d}t}\langle \boldsymbol{A}\cdot\boldsymbol{B}\rangle = -2\eta\mu_0\langle \boldsymbol{J}\cdot\boldsymbol{B}\rangle. \tag{18}$$

Furthermore, in a homogeneous system, $\langle \boldsymbol{A}\cdot\boldsymbol{B}\rangle$ is gauge-invariant, so in the steady state we have

$$\langle \boldsymbol{J}\cdot\boldsymbol{B}\rangle = 0 \quad \text{(steady state)}. \tag{19}$$

This is remarkable and applies even (and especially) in the case of helical forcing when large-scale fields can be generated by the α effect.

For the rest of this review, it will be crucial to distinguish between large-scale and small-scale magnetic fields. We do this by making use of the following decomposition:

$$\boldsymbol{U} = \overline{\boldsymbol{U}} + \boldsymbol{u}, \quad \boldsymbol{B} = \overline{\boldsymbol{B}} + \boldsymbol{b}. \tag{20}$$

In the previous sections of this review, there was no mean flow, so $\boldsymbol{U} = \boldsymbol{u}$, but from now on we shall denote the full velocity by a capital letter. Likewise, the vorticity of $\boldsymbol{U}$ is given by $\boldsymbol{W} = \nabla \times \boldsymbol{U}$.

In rotating astrophysical bodies, a commonly used average is the azimuthal one. However, in the present case of fully periodic Cartesian domains, the resulting large-scale fields can be described by planar averages, such as xy, yz, or xz averages. The resulting mean fields, $\overline{\boldsymbol{B}}$, depend then still on z, x, or y, in addition to t. Examples of such fields are those proportional to $(\sin kz, \cos kz, 0)$, $(0, \sin kx, \cos kx)$, and $(\cos ky, 0, \sin ky)$, respectively. All these examples obey

$$\nabla \times \overline{\boldsymbol{B}} = k\overline{\boldsymbol{B}} \tag{21}$$

and are thus eigenfunctions of the curl operator with eigenvalue k. In particular, it follows then that $\overline{\boldsymbol{J}}\cdot\overline{\boldsymbol{B}} = k\overline{\boldsymbol{B}}^2/\mu_0$ is uniform. This can only be compatible with (19), if there is a residual (small-scale or fluctuating) magnetic field, $\boldsymbol{b} = \boldsymbol{B} - \overline{\boldsymbol{B}}$, which obeys $\langle \boldsymbol{j}\cdot\boldsymbol{b}\rangle = -\langle \overline{\boldsymbol{J}}\cdot\overline{\boldsymbol{B}}\rangle$. Here, $\boldsymbol{j} = \nabla \times \boldsymbol{b}/\mu_0$ is the corresponding current density. Assuming $\langle \boldsymbol{j}\cdot\boldsymbol{b}\rangle = \epsilon_f k_f \langle \boldsymbol{b}^2\rangle/\mu_0$, we find that $\overline{\boldsymbol{B}}^2/\langle \boldsymbol{b}^2\rangle = \epsilon_f k_f/k$, which can exceed unity in cases of fully helical

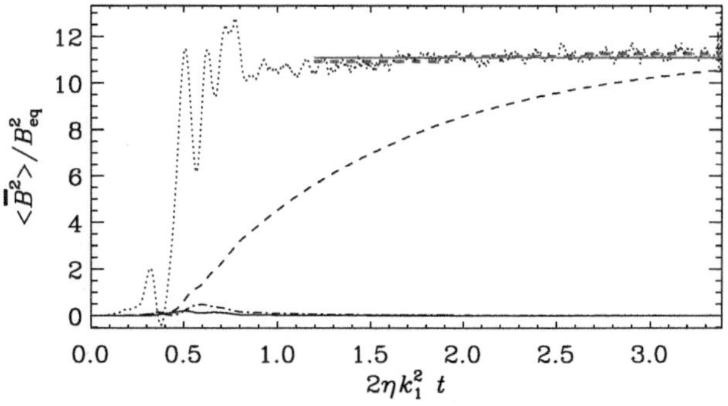

Fig. 13 Example showing the evolution of the normalized $\langle\overline{\boldsymbol{B}}^2\rangle$ (*dashed*) and that of $\langle\overline{\boldsymbol{B}}^2\rangle + \mathrm{d}\langle\overline{\boldsymbol{B}}^2\rangle/\mathrm{d}(2\eta k^2 t)$ (*dotted*), compared with its average in the interval $1.2 \le 2\eta k_1^2 t \le 3.5$ (*horizontal blue solid line*), as well as averages over 3 subintervals (*horizontal red dashed lines*). Here, $\overline{\boldsymbol{B}}$ is evaluated as an xz average, $\langle\boldsymbol{B}\rangle_{xz}$. For comparison we also show the other two averages, $\langle\boldsymbol{B}\rangle_{xy}$ (*solid*) and $\langle\boldsymbol{B}\rangle_{yz}$ (*dash-dotted*), but their values are very small. Adapted from Candelaresi and Brandenburg (2013)

forcing ($\epsilon_{\rm f} \to \pm 1$). We recall that the parameter $\epsilon_{\rm f}$ is related to the helicity parameter σ in the forcing function (15) via $\epsilon_{\rm f} = 2\sigma/(1+\sigma^2)$. Repeating this calculation for the late saturation phase of a dynamo, we have

$$\frac{\overline{\boldsymbol{B}}^2}{\langle \boldsymbol{b}^2\rangle} \approx \frac{\epsilon_{\rm f} k_{\rm f}}{k}\left[1 - e^{-2\eta k^2 (t - t_{\rm sat})}\right], \tag{22}$$

with a suitable integration constant $t_{\rm sat}$, having to do with just properties of the initial field strength. This equation describes the late ($t > t_{\rm sat}$), resistively dominated saturation phase of a helically driven dynamo of α^2 type. By differentiating this equation again, we can find that the final saturation field strength, $\overline{B}_{\rm sat} = \overline{B}_{\rm rms}(t \to \infty)$, obeys (Candelaresi and Brandenburg 2013)

$$\overline{\boldsymbol{B}}_{\rm sat}^2 \approx \overline{\boldsymbol{B}}^2 + \mathrm{d}\overline{\boldsymbol{B}}^2/\mathrm{d}(2\eta k^2 t). \tag{23}$$

This equation allows one to compute the value of $\overline{B}_{\rm sat}$ based on the measured rate at which $\overline{\boldsymbol{B}}^2$ increases. It is now routinely used to estimate $\overline{B}_{\rm sat}$ without actually reaching the final state; see Fig. 13 for an example.

5.5 Turbulent Mixing and Non-diffusive Transport

Turbulent flows are known to be capable of enhanced mixing. A prime example is the mixing of a passive scalar concentration $C(\boldsymbol{x}, t)$, whose evolution is governed by the equation

$$\frac{\partial C}{\partial t} = -\boldsymbol{\nabla}\cdot(\boldsymbol{U}C) + \kappa\nabla^2 C. \tag{24}$$

Loosely speaking, turbulent mixing can be modeled as an enhanced diffusivity in the corresponding evolution equation for the *mean* passive scalar concentration $\overline{C}(\boldsymbol{x}, t)$, which then takes the form

$$\frac{\partial \overline{C}}{\partial t} = -\boldsymbol{\nabla}\cdot(\overline{\boldsymbol{U}}\,\overline{C}) + \kappa_{\rm T}\nabla^2\overline{C}, \tag{25}$$

where $\kappa_{\rm T} = \kappa + \kappa_{\rm t}$ is the sum of molecular (or atomic) and turbulent diffusivities.

In a more precise formulation, κ_t becomes not only a tensor, κ_{ij}, but also an integral kernel that takes into account that on the right-hand side of (25) higher-order derivatives of $\overline{C}$ in space and time appear. In particular, there can in principle also be a term of the form $\nabla \cdot (\boldsymbol{\gamma}^C \overline{C})$ on the right-hand side which describes turbulent pumping or turbophoresis, and $\boldsymbol{\gamma}^C$ is a vector. This term acts like advection, but without any material motion. (In a kernel formulation, such a term could in principle be subsumed into the integral kernel.) However, under isotropic conditions, $\boldsymbol{\gamma}^C$ must vanish and the diffusivity tensor κ_{ij} becomes an isotropic tensor $\kappa_t \delta_{ij}$. Analogous equations can also be derived for the magnetic induction equation and the momentum equation. In both cases this can lead to physically new effects such as the mean-field (or large-scale) dynamo instability and the negative effective magnetic pressure instability (NEMPI), which will be discussed further below. The former exists in isotropic turbulence, while the latter requires inhomogeneity and sufficiently strong density stratification.

In the simulations presented in Sect. 5.4 we found the development of large-scale fields of Beltrami type. Such fields do indeed emerge as eigenfunctions of the related mean-field induction equation with constant coefficients,

$$\frac{\partial \overline{\boldsymbol{B}}}{\partial t} = \nabla \times (\overline{\boldsymbol{U}} \times \overline{\boldsymbol{B}} + \alpha \overline{\boldsymbol{B}} - \eta_T \mu_0 \overline{\boldsymbol{J}}). \tag{26}$$

Significant progress in this field has recently become possible through the numerical determination of the full set of turbulent transport coefficients. This method is known as the test-field method and involves the solution of additional evolution equations for the magnetic fluctuations arising from a given test field. One needs enough test fields to obtain all tensor components. By allowing the test fields to attain suitable variability in space and time, it is possible to determine then also the full integral kernel in spectral space.

The results obtained so far have shown that

$$\alpha \approx \alpha_0 \equiv -\epsilon_f u_{rms}/3 \tag{27}$$

and

$$\eta_t \approx \eta_{t0} \equiv u_{rms}/3k_f \tag{28}$$

for $Re_M \gtrsim 1$. For $Re_M \lesssim 1$, both coefficients increase linearly with increasing Re_M (Sur et al. 2008). In the nonlinear regime, there can also be velocity fluctuations generated through the presence of a mean field, but this requires what is known as magnetic background turbulence, i.e., magnetic fluctuations that would be present even without a mean magnetic field. This is in principle possible when there is small-scale dynamo action. This case can be treated with a correspondingly modified test-field method (Rheinhardt and Brandenburg 2010).

In the following we state several important results obtained by using the test-field method. We did already mention that for fully helical turbulence and large values of Re_M, α and η_t attain values of the order of $\pm u_{rms}/3$ and $u_{rms}/3k_f$, respectively. In turbulence, both coefficients possess a wavenumber dependence that is of the form of a Lorentzian proportional to $(1 + k^2/k_f^2)^{-1}$, corresponding to an exponential integral kernel proportional to $\exp[-(z - z')k_f]$; see Brandenburg et al. (2008b). When the mean field is non-steady, the memory effect can become important and this leads to a dependence of the form $(1 - i\omega\tau)^{-1}$, corresponding to a kernel proportional to $\exp(-|t - t'|/\tau)$ for $t' < t$, and 0 otherwise. Here, $\tau \approx (u_{rms}k_f)^{-1}$ is the correlation time.

In the limit $\ell = 1/k_f \to 0$ and $\tau \to 0$, the integral kernels become δ functions in space and time. However, this approximation breaks down at the bottom of the solar convection

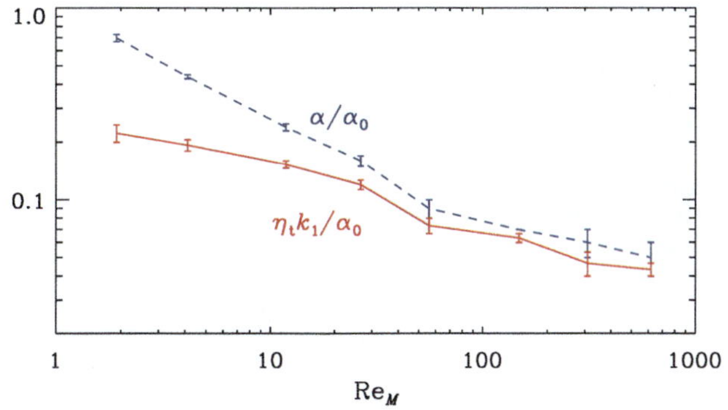

Fig. 14 Re_M-dependence of α and $\tilde{\eta}_t$. Both curves are normalized by α_0. Adapted from Brandenburg et al. (2008a)

zone, where the resulting mean magnetic field in dynamo models often shows structures on scales much smaller than ℓ (Chatterjee et al. 2011). Furthermore, nonlocality in time is violated when the mean magnetic field is either growing or decaying. Ignoring this can lead to discrepancies that are well detectable with the test-field method (Hubbard and Brandenburg 2009; Rädler et al. 2011). Finally, when the mean magnetic field depends strongly on both space and time, the integral kernel in spectral space becomes approximately proportional to $(1 - \mathrm{i}\omega\tau + \ell^2 k^2)^{-1}$. This form has the advantage that it can easily be treated in real space by solving an evolution equation in time with a positive diffusion term, i.e.,

$$\left(1 + \tau\frac{\partial}{\partial t} - \ell^2\nabla^2\right)\overline{\mathcal{E}}_i = \alpha_{ij}\overline{B}_j + \eta_{ijk}\overline{B}_{j,k}. \tag{29}$$

Here, α_{ij} and η_{ijk} are the usual α effect and turbulent diffusivity tensors for $\omega \to 0$ and $k \to 0$, and equal to $\alpha\delta_{ij}$ to $\eta_t\epsilon_{ijk}$ in the isotropic case. This equation has been studied in some detail by Rheinhardt and Brandenburg (2012). It is a special form of the telegraph equation, which has been studied in similar contexts (Brandenburg et al. 2004; Chamandy et al. 2013).

Using the quasi-kinematic test-field method, Brandenburg et al. (2008a) showed that in the case of a saturated dynamo, both α and η_t remain weakly Re_M-dependent; see Fig. 14. Note that no fully asymptotic regime has been obtained yet, so it remains unclear when or whether this will happen. It is clear, however, that α must approach $\eta_t k_1$ at large Re_M for the system to be in the stationary saturated state. However, in view of the astrophysical importance of turbulent dissipation, the remaining weak dependence of η_t on Re_M is expected to disappear eventually.

6 Inhomogeneous MHD Turbulence

6.1 Density Stratification

Stratification refers to nonuniformity that is usually caused by gravity. As a consequence, pressure increases in the direction of the gravity, and this causes similar changes in density and/or temperature. The turbulence intensity can itself also be stratified. This usually comes as a consequence of density stratification, but one can envisage circumstances in which the forcing is nonuniform. Such non-uniformity affects turbulent transport–not just diffusive but also non-diffusive transport, similar to the pumping velocity proportional to $\boldsymbol{\gamma}^C$, described in Sect. 5.5. Both effects are astrophysically important. Stratification usually leads

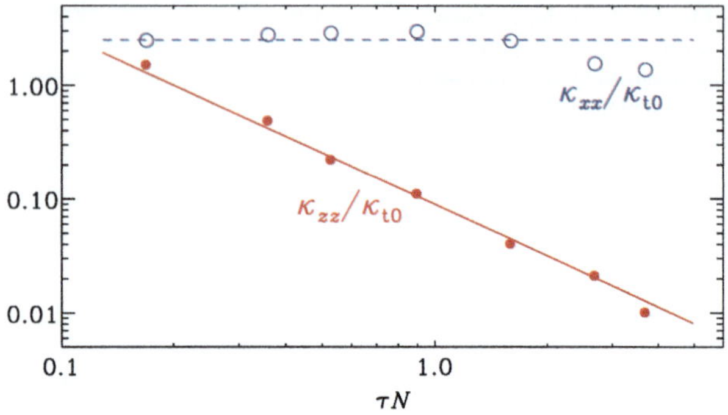

Fig. 15 Dependence of κ_{xx}/κ_{t0} (*open symbols*) and κ_{zz}/κ_{t0} (*filled symbols*) on the normalized buoyancy frequency. The *dashed line* shows that $\kappa_{xx}/\kappa_{t0} \approx 2.5$ while the *solid line* gives $\kappa_{zz}/\kappa_{t0} \approx 0.09(\tau N)^{-3/2}$. Adapted from Kitchatinov and Brandenburg (2012)

to a suppression of diffusive transport. An example is shown in Fig. 15, where we show the suppression of the vertical passive scalar diffusivity as a function of the stratification, which is here measured by the normalized Brunt–Väisälä frequency, N, with $N^2 = -\boldsymbol{g} \cdot \boldsymbol{\nabla} s/c_p$ and s being the specific entropy. For details of this, see the work of Kitchatinov and Brandenburg (2012).

Suppression of turbulent transport, for example, is critical for understanding the depletion of primordial elements (e.g., lithium) by mixing with deeper layers in the stably stratified lower overshoot layer of the convection zones of stars with outer convection zones. The suppression is here caused mainly by the stabilizing entropy gradient [reversing the gradient of s leads the negative values of N, corresponding the onset of convection with exponential growth proportional to $\exp(\mathrm{Im} N t)$]. In the following, we shall focus on another manifestation of stratification, namely the expansion of rising structures as they ascent into less dense surroundings. For that purpose, we make the assumption of an isothermal equation of state, which is a simplification that leads to a constant pressure scale height and suppresses also the stabilizing effect from the entropy gradient. For further discussion on this, see the papers by Brandenburg et al. (2012a) and Käpylä et al. (2012b) in the context of NEMPI; see Sect. 5.5.

6.2 Stratified Turbulence with a Vertical Field

In the presence of stratification and an imposed magnetic field along the direction of stratification, there is the possibility of producing another pseudoscalar called cross helicity. On theoretical grounds, one expects (Rüdiger et al. 2011)

$$\langle \boldsymbol{u} \cdot \boldsymbol{b} \rangle \propto \boldsymbol{g} \cdot \overline{\boldsymbol{B}}. \tag{30}$$

More specifically, it turns out that

$$\langle \boldsymbol{u} \cdot \boldsymbol{b} \rangle = -\eta_t \overline{B}/H_\rho, \tag{31}$$

where H_ρ is the density scale height. This does indeed turn out to be the case, as has been shown using simulations of forced isothermal turbulence in the presence of gravity. The result is shown in Fig. 16, where we plot $\langle \boldsymbol{u} \cdot \boldsymbol{b} \rangle$ as a function of Re_M.

6.3 Effects of Rotation

In the presence of stratification and/or rotation, MHD turbulence is subject to a range of new effects. These phenomena are associated with the vectors $\boldsymbol{g}$ (gravity) and $\boldsymbol{\Omega}$ (angular velocity), which introduce preferred directions to the flow. They do so in different ways, because

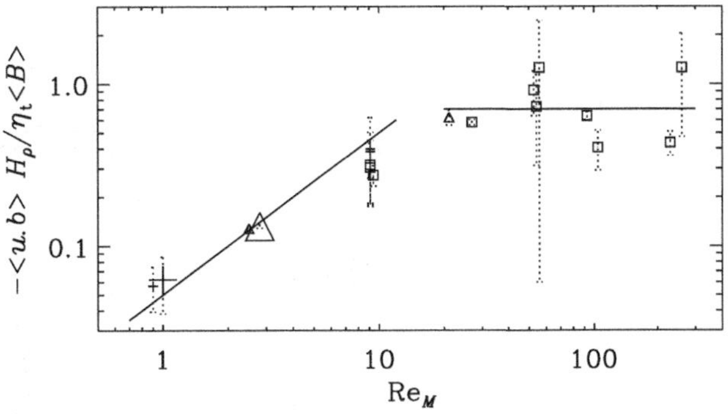

Fig. 16 Dependence of the normalized cross helicity on Rm for various field strength $B_z/B_{\rm eq} < 0.1$, $\mathrm{Pm} = 1$, $k_{\rm f}/k_1 = 2.2$, and $H_\rho k_1 = 2.5$. The straight line denotes the fit $\langle \boldsymbol{u} \cdot \boldsymbol{b}\rangle/\tau g\langle \boldsymbol{B}\rangle = 0.05$ Rm

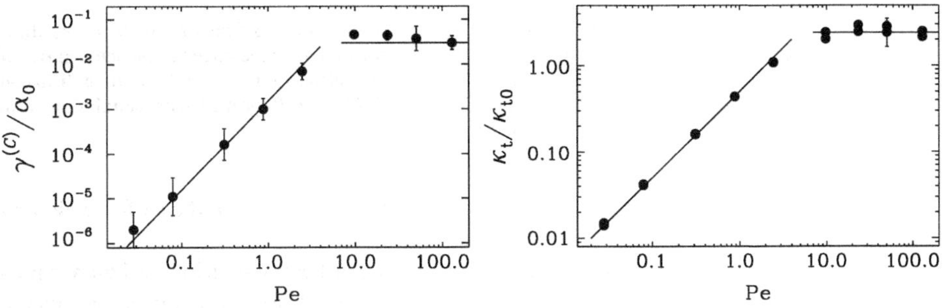

Fig. 17 Dependence of passive scalar pumping velocity, $\gamma^{(C)}$, and passive scalar diffusivity $\kappa_{\rm t}$ on Peclet number, *Pe*. The scale separation ratio is $k_{\rm f}/k_1 = 5$

$\boldsymbol{g}$ is a polar vector (identical to its mirror image) while $\boldsymbol{\Omega}$ is an axial vector (antiparallel to its mirror image). This means that turbulent transport effects characterized by some effective velocity must be proportional to another polar vector. This can then either be the vector $\boldsymbol{g}$ or, in forced turbulence simulations, where it is possible to produce helical turbulence, it can be the vector $\hat{\boldsymbol{\Omega}}$. In that case there is kinetic helicity, $\langle \boldsymbol{w} \cdot \boldsymbol{u}\rangle$, which is a pseudoscalar, so $\langle \boldsymbol{w} \cdot \boldsymbol{u}\rangle\boldsymbol{\Omega}$ would also be a polar vector, allowing pumping even in the homogeneous case if there is rotation and helicity.

In Fig. 17 we show such an example, where there is fully helical turbulence that is initially isotropic, but because of rotation it becomes anisotropic and there is now a polar vector that leads to turbulent pumping with the velocity

$$\gamma^{(C)} \approx 0.075\langle \boldsymbol{w} \cdot \boldsymbol{u}\rangle\boldsymbol{\Omega}/(u_{\rm rms}k_{\rm f})^2. \tag{32}$$

A similar result has previously been obtained by Pipin (2008) and Mitra et al. (2009) for shear flows, where the resulting mean vorticity vector acts as the relevant pseudovector. However, these situations are somewhat artificial, because helicity does not normally occur in the absence of additional stratification, so any pumping would still be indirectly associated with the stratification vector, although it can now attain a direction proportional to $\hat{\boldsymbol{\Omega}}$ or the mean vorticity.

Owing to the presence of stratification and rotation, the turbulence attains helicity and can then produce an α effect. This has been studied in great detail in the past using analytic methods and, more recently, the test-field method. In Fig. 18 we show an example from Brandenburg et al. (2012b), where the turbulence is governed by only one preferred direc-

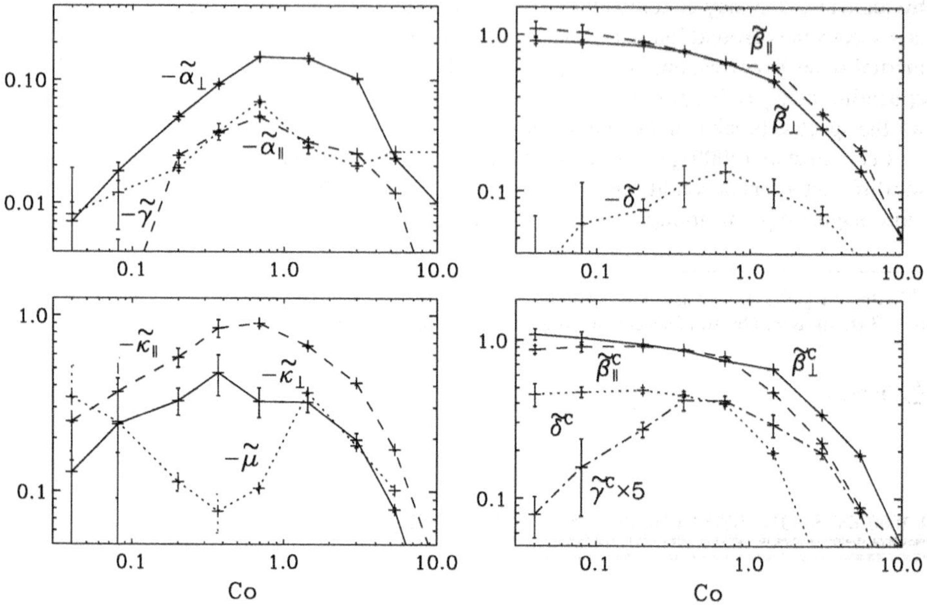

Fig. 18 Dependence of transport coefficients in a model with rotation and density stratification as a function of the Coriolis number, $Co = 2\Omega/u_{\rm rms}k_{\rm f}$. The other relevant parameters are $Re_M \approx 10$, $\mathrm{Gr} = g/c_{\rm s}^2 k_{\rm f} \approx 0.16$, $k_{\rm f}/k_1 = 5$, for $\nu = \eta = \kappa$

tion, and $\hat{\boldsymbol{\Omega}}$ and $\boldsymbol{g}$ are therefore assumed to be parallel. In that case, $\overline{\boldsymbol{\mathcal{E}}}$ can be represented in the form

$$\begin{aligned}\overline{\boldsymbol{\mathcal{E}}} = & -\alpha_\perp \overline{\boldsymbol{B}} - (\alpha_\parallel - \alpha_\perp)(\hat{\boldsymbol{e}} \cdot \overline{\boldsymbol{B}})\hat{\boldsymbol{e}} - \gamma \hat{\boldsymbol{e}} \times \overline{\boldsymbol{B}} \\ & - \beta_\perp \mu_0 \overline{\boldsymbol{J}} - (\beta_\parallel - \beta_\perp)(\hat{\boldsymbol{e}} \cdot \mu_0 \overline{\boldsymbol{J}})\hat{\boldsymbol{e}} - \delta \hat{\boldsymbol{e}} \times \mu_0 \overline{\boldsymbol{J}} \\ & - \kappa_\perp \overline{\boldsymbol{K}} - (\kappa_\parallel - \kappa_\perp)(\hat{\boldsymbol{e}} \cdot \overline{\boldsymbol{K}})\hat{\boldsymbol{e}} - \mu \hat{\boldsymbol{e}} \times \overline{\boldsymbol{K}} \end{aligned} \tag{33}$$

with nine coefficients $\alpha_\perp, \alpha_\parallel, \ldots, \mu$.

Clearly, because of stratification and rotation, the turbulence is no longer isotropic, so α will also no longer be isotropic. In the simplest case when both $\boldsymbol{g}$ and $\boldsymbol{\Omega}$ are parallel, α has components parallel and perpendicular to their direction. The α effect is of particular interest, because it can lead to large-scale magnetic field generation. Another effect that is known to lead to large-scale dynamo action is the Rädler or $\boldsymbol{\Omega} \times \boldsymbol{J}$ effect (Rädler 1969). Unlike the α effect, it exists already with just rotation and no stratification. Its astrophysical relevance is however still to be demonstrated. Note also that in all practical situations there must still be an additional source of energy, because $\boldsymbol{\Omega} \times \overline{\boldsymbol{J}}$ has no component along $\overline{\boldsymbol{J}}$ and does therefore not provide energy to the system.

6.4 Stratified Turbulence with an Imposed Magnetic Field

In the presence of an imposed magnetic field there is an important effect that deserves to be mentioned. In mean-field parameterizations of the Reynolds stress, there are terms that are quadratic in the mean magnetic field and contribute to a decrease of the Reynolds stress if there is a weak magnetic field. This suppression was discussed by Rüdiger (1974) and Rädler (1974) in connection with the understanding of the quenching of the α effect by a mean magnetic field. However, later it was understood that it also leads to a suppression of the turbulent pressure and that this suppression is stronger than the explicitly added magnetic pressure from the mean field, $\overline{\boldsymbol{B}}^2/2\mu_0$. This means that the contribution of the mean field to the total turbulent pressure,

$$p_{\rm tot} = p_{\rm gas} + p_{\rm turb} = p_{\rm gas} + p_{\rm turb}^{(0)} + \left[1 - q_{\rm p}\left(\overline{\boldsymbol{B}}^2/B_{\rm eq}^2\right)\right]\overline{\boldsymbol{B}}^2/2\mu_0, \tag{34}$$

which is embodied by the last term, $[1 - q_{\rm p}(\overline{\boldsymbol{B}}^2/B_{\rm eq}^2)]\overline{\boldsymbol{B}}^2/2\mu_0$, can be negative (Kleeorin et al. 1989, 1990, 1993, 1996; Kleeorin and Rogachevskii 1994; Rogachevskii and Kleeorin 2007). Here, $q_{\rm p}(\overline{\boldsymbol{B}}^2/B_{\rm eq}^2)$ is a non-dimensional quenching function describing the suppression of the total stress, which consists of Reynolds and Maxwell stress. It is only a function of $\overline{\boldsymbol{B}}^2/B_{\rm eq}^2$, so even for a uniform $\overline{\boldsymbol{B}}^2$ it can show spatial variation if $B_{\rm eq}^2$ changes, for example as a result of density stratification. This allows the full dependence of $q_{\rm p}$ on $\overline{\boldsymbol{B}}^2/B_{\rm eq}^2$ to be probed in a single simulation (Brandenburg et al. 2012a; Kemel et al. 2012). The effect of the Maxwell stress turns out the be weaker than that of the Reynolds stress and it has the opposite effect, as was demonstrated by numerical calculations (Brandenburg et al. 2010).

In a stratified layer with a sub-equipartition magnetic fields this negative effective magnetic pressure can lead to an instability producing spontaneously magnetic flux concentrations (Kleeorin et al. 1989, 1993; Rogachevskii and Kleeorin 2007). This has recently been confirmed with DNS (Brandenburg et al. 2011b; Kemel et al. 2012) and is being discussed in connection with explaining the spontaneous formation of active regions (Kemel et al. 2013) and sunspots (Brandenburg et al. 2013). In Fig. 19 we show horizontal and vertical cuts through a magnetic spot from the simulation of Brandenburg et al. (2013) in the presence of an imposed vertical field. In the horizontal cut, again, strong fields correspond to dark shades. The vertical cut is with a different color table where strong fields now correspond to light shades. It shows that the magnetic field (in units of the local equipartition field strength) decreases with height. Note also that the mean magnetic field fans out toward the bottom of the domain. Applying this finding to the origin of sunspots, it suggest that, contrary to common belief (cf. Brandenburg 2005), those structures may not be deeply anchored.

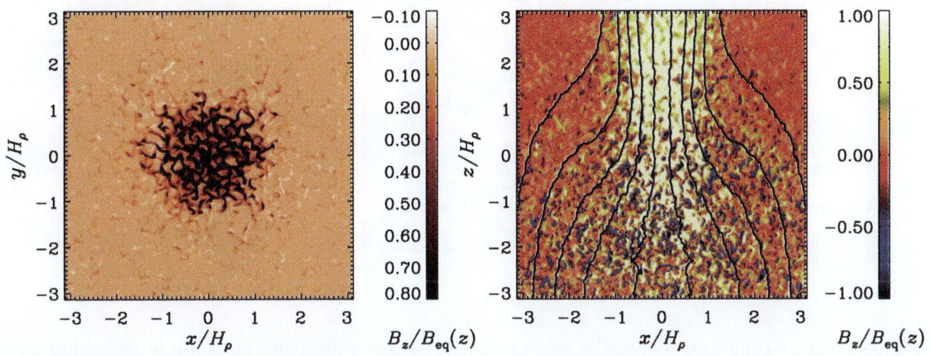

Fig. 19 Cuts of the vertical magnetic field in units of the equipartition field strength, $\overline{B}_z/B_{\rm eq}(z)$, through the horizontal plane at the top boundary (*left*) and the vertical plane through the middle of the spot (*right*). Field lines of the numerically averaged mean field are superimposed. Adapted from Brandenburg et al. (2013)

6.5 Solar Dynamo and Magnetic Helicity Fluxes

One of the main applications of mean-field theory has always been to explain the Sun's global magnetic field, its 11 year cycle, and the migration of magnetic field from mid to low latitudes, in addition, of course, eventually the formation of sunspots themselves. In the last few years, several groups have engaged in tackling the problem of the Sun's global magnetic field by performing numerical simulations of rotating turbulent convection in spherical shells using either spherical harmonics (Miesch and Toomre 2009; Brown et al. 2010, 2011), an implicit solver (Ghizaru et al. 2010; Racine et al. 2011), or finite differences in spherical wedges (Käpylä et al. 2010, 2012a) to overcome the timestep constraint at the poles. The results from all groups trying to model the Sun agree in that they show equipartition-strength magnetic fields in the bulk of the convection zone (rather than highly super-equipartition-strength magnetic fields just at the bottom of the convection zone), with magnetic activity concentrated toward low latitudes and, in some cases, cyclic reversals of the magnetic field direction, resembling the solar 22 year cycle.

A major breakthrough has been achieved through the recent finding of equatorward migration of magnetic activity belts in the course of the cycle (Käpylä et al. 2012a); see Fig. 20. These results are robust and have now been reproduced in extended simulations that include a simplified model of an outer corona (Warnecke et al. 2013). Interestingly, the convection simulations of all three groups produce cycles only at rotation speeds that exceed those of the present Sun by a factor of 3–5 (Brown et al. 2011). Both lower and higher rotation speeds give, for example, different directions of the dynamo wave (Käpylä et al. 2012a). Different rotation speeds correspond to different stellar ages (from 0.5 to 8 gigayears for rotation periods from 10 to 40 days), because magnetically active stars all have a wind and are subject to magnetic braking (Skumanich 1972). In addition, all simulations are subject to systematic "errors" in that they poorly represent the small scales and emulate in that way an effective turbulent viscosity and magnetic diffusivity that is larger than in reality; see the corresponding discussion in Sect. 4.3.2 of Brandenburg et al. (2012a) in another context. In future simulations, it will therefore be essential to explore the range of possibilities by including stellar age as an additional dimension of the parameter space.

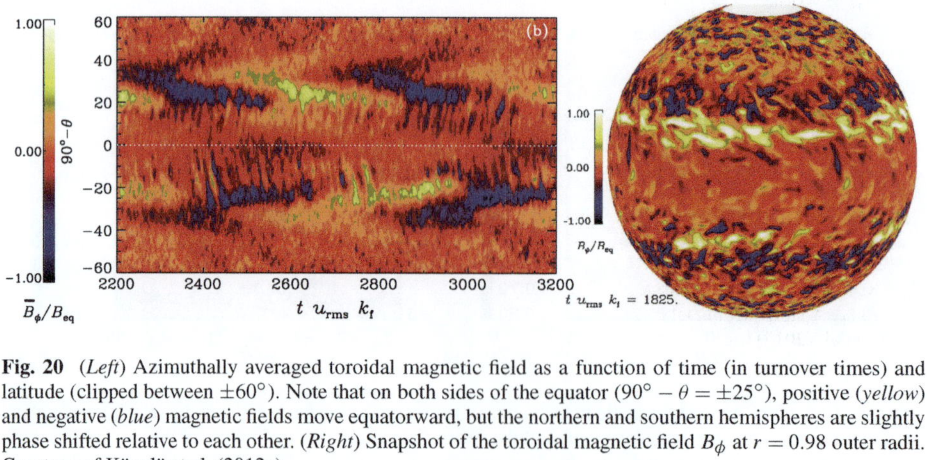

Fig. 20 (*Left*) Azimuthally averaged toroidal magnetic field as a function of time (in turnover times) and latitude (clipped between ±60°). Note that on both sides of the equator ($90° - \theta = \pm 25°$), positive (*yellow*) and negative (*blue*) magnetic fields move equatorward, but the northern and southern hemispheres are slightly phase shifted relative to each other. (*Right*) Snapshot of the toroidal magnetic field B_ϕ at $r = 0.98$ outer radii. Courtesy of Käpylä et al. (2012a)

In support of our statement that a poor representation of the small scales in DNS emulates artificially enhanced turbulent viscosity and turbulent magnetic diffusivity, let us recall that η_t and α are scale-dependent. As discussed before in Sect. 5.5, they decrease with increasing k in a Lorentzian fashion. The relative importance of Ω effect over the α effect depends on the ratio of C_Ω and a similar parameter $C_\alpha = \alpha/\eta_t k$ characterizing the strength of the α effect. Both C_Ω and the ratio C_Ω/C_α would be underestimated in a large eddy simulation in which $\eta_t(k)k$ and $\alpha(k)$ are too big, so one would need to compensate for this shortcoming by increasing Ω to recover cyclic dynamo action.

As alluded to in Sect. 5.4, magnetic helicity fluxes play a major role in the dynamo by alleviating the otherwise catastrophic quenching of the dynamo (Blackman and Brandenburg 2003). Recent work using a simple model with a galactic wind has shown, for the first time, that this may indeed be possible. We recall that the evolution equation for the mean magnetic helicity density of fluctuating magnetic fields, $\overline{h}_f = \overline{\boldsymbol{a} \cdot \boldsymbol{b}}$, is

$$\frac{\partial \overline{h}_f}{\partial t} = -2\overline{\boldsymbol{\mathcal{E}}} \cdot \overline{\boldsymbol{B}} - 2\eta\mu_0 \overline{\boldsymbol{j} \cdot \boldsymbol{b}} - \nabla \cdot \overline{\boldsymbol{\mathcal{F}}}_f, \tag{35}$$

where we allow two contributions to the flux of magnetic helicity from the fluctuating field $\overline{\boldsymbol{\mathcal{F}}}_f$: an advective flux due to the wind, $\overline{\boldsymbol{\mathcal{F}}}_f^{w} = \overline{h}_f \overline{\boldsymbol{U}}_w$, and a turbulent–diffusive flux due to turbulence, modeled by a Fickian diffusion term down the gradient of $\overline{h}_f$, i.e., $\overline{\boldsymbol{\mathcal{F}}}_f^{\text{diff}} = -\kappa_h \nabla \overline{h}_f$. Here, $\overline{\boldsymbol{\mathcal{E}}} = \overline{\boldsymbol{u} \times \boldsymbol{b}}$ is the electromotive force of the fluctuating field. The scaling of the terms on the right-hand side with Re_M has been considered before by Mitra et al. (2010) and Hubbard and Brandenburg (2010). They also drew attention to the fact that, even though $\overline{\boldsymbol{\mathcal{F}}}_f$ is gauge-invariant, the time average of $\nabla \cdot \overline{\boldsymbol{\mathcal{F}}}_f$ is not, *provided* the system is statistically stationary and $\partial \overline{h}_f/\partial t$ vanishes on average.

In Fig. 21 we show the basic result of Del Sordo et al. (2013). As it turns out, below $Re_M = 100$, the $2\eta\mu_0 \overline{\boldsymbol{j} \cdot \boldsymbol{b}}$ term dominates over $\nabla \cdot \overline{\boldsymbol{\mathcal{F}}}_f$, but because of the different scalings (slopes being -1 and $-1/2$, respectively), the $\nabla \cdot \overline{\boldsymbol{\mathcal{F}}}_f$ term is expected to becomes dominant for larger values of Re_M (about 3000). Unexpectedly, however, $\nabla \cdot \overline{\boldsymbol{\mathcal{F}}}_f$ becomes approximately constant already for $Re_M \gtrsim 100$ and $2\eta\mu_0 \overline{\boldsymbol{j} \cdot \boldsymbol{b}}$ shows now a shallower scaling (slope $-1/2$). This means that the two curves would still cross at a similar value. Our data suggest, however, that $\nabla \cdot \overline{\boldsymbol{\mathcal{F}}}_f$ may even rise slightly, so the crossing point is now closer to $Re_M = 1000$.

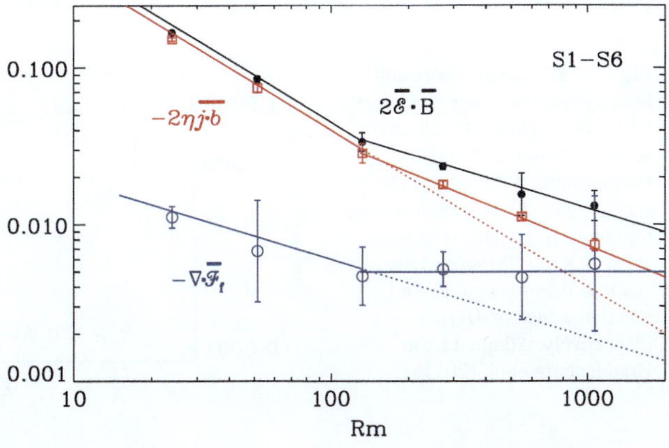

Fig. 21 Scaling properties of the vertical slopes of $2\overline{\boldsymbol{\mathcal{E}}} \cdot \overline{\boldsymbol{B}}$, $-2\eta\mu_0 \overline{\boldsymbol{j} \cdot \boldsymbol{b}}$, and $-\nabla \cdot \overline{\boldsymbol{\mathcal{F}}}_f$. The three quantities vary approximately linearly with z, so the three labels indicate their non-dimensional values at $k_1 z = 1$. The *dotted lines* show the extrapolated initial scaling for low Re_M. Adapted from Del Sordo et al. (2013)

7 Solar Wind Observations

Solar wind observations provide a good way of determining the energy spectrum of MHD turbulence. As already mentioned in Sect. 5.2, recent work by Boldyrev et al. (2011) provides an explanation of why the kinetic and magnetic energy spectra have slightly different spectral indices in that the magnetic energy spectrum is slightly steeper ($\propto k^{-1.6}$) than that of the kinetic energy ($\propto k^{-1.4}$). This was found previously by Podesta et al. (2007). Indeed, looking again at Fig. 8, it is clear that different slopes of kinetic and magnetic energy spectra is a consequence of the super-equipartition just below $k = k_{\rm f}$ and the subsequent trend toward equipartition for larger values of k.

Solar wind observations have long been able to provide estimates about the magnetic helicity spectrum (Matthaeus et al. 1982). We recall that, even though the magnetic helicity is gauge-dependent, its spectrum is not. Technically, this is because the computation of the spectrum involves an integration over all space. In practice, this is not possible, of course. However, by making use of statistical homogeneity and the Taylor hypothesis of the equivalence of spatial and temporal Fourier spectra, Matthaeus et al. (1982) were able to express the magnetic helicity spectrum as

$$H(k_R) = 4\,\mathrm{Im}\big(\hat{B}_T \hat{B}_N^\star\big)/k_R, \tag{36}$$

where $\hat{B}_T(k_R)$ and $\hat{B}_N(k_R)$ are the Fourier transforms of the two magnetic field components perpendicular to the radial direction away from the Sun, and (R, T, N) refers to the components of a locally Cartesian heliospheric coordinate system. Here, k_R is the wavenumber, which is related to the temporal frequency via $\omega = u_R k_R$, where $u_R \approx 800\ \mathrm{km\,s^{-1}}$ is the wind speed at high heliographic latitudes. Note that in (36), the expression for $H(k_R)$ is manifestly gauge-invariant.

Most spacecrafts have probed low heliographic latitudes, where the helicity is governed by fluctuations around zero. In recent years, however, it has been possible to estimate the magnetic helicity spectrum also at high heliographic latitudes using data from the Ulysses spacecraft that flew in a near-polar orbit. However, even at high heliographic latitudes the magnetic helicity is still strongly fluctuating and a clear sign of magnetic helicity can only be seen by averaging the spectra over broad, logarithmically spaced wavenumber bins; see Fig. 22. One can define the *relative* spectral magnetic helicity, $2\mu_0 E_{\rm M}(k)/k H_{\rm M}(k)$, which is a non-dimensional quantity between -1 and 1. It turns out that it is just a few percent. Nevertheless, the magnetic helicity is negative at large scales (small wavenumbers, $k < 30\ \mathrm{AU}^{-1}$ corresponding to frequencies below 0.03 mHz) and positive at smaller scales (large wavenumbers); see Brandenburg et al. (2011a). This agrees, at least qualitatively, with

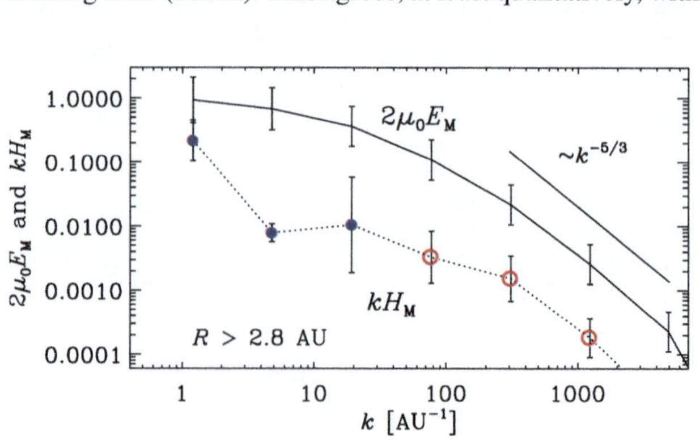

Fig. 22 Magnetic energy and helicity spectra, $2\mu_0 E_{\rm M}(k)$ and $k H_{\rm M}(k)$, respectively, for two separate distance intervals. Furthermore, both spectra are scaled by $4\pi R^2$ before averaging within each distance interval above 2.8 AU. *Filled* and *open symbols* denote negative and positive values of $H_{\rm M}(k)$, respectively. Adapted from Brandenburg et al. (2011a)

earlier results by Smith and Bieber (1993) that at low frequencies the magnetic helicity is negative in the northern hemisphere. At much higher frequencies (beyond 100 mHz), positive magnetic helicity in the northern hemisphere has now also been found by Podesta and Gary (2011).

When comparing with numerical simulations, it should be noted that virtually all observed spectra are based on one-dimensional measurements, while those of numerical simulations are based on the full three-dimensional velocity field. The two are related to each other via

$$E_{\mathrm{M}}^{\mathrm{1D}}(k_R) = \int_{k_R}^{\infty} E_{\mathrm{M}}^{\mathrm{3D}}(k)\,\mathrm{d}\ln k, \tag{37}$$

$$H_{\mathrm{M}}^{\mathrm{1D}}(k_R) = \int_{k_R}^{\infty} H_{\mathrm{M}}^{\mathrm{3D}}(k)\,\mathrm{d}\ln k. \tag{38}$$

This transformation is well known for the energy spectrum (cf., Tennekes and Lumley 1972; Dobler et al. 2003), and was recently generalized to the case with helicity (Brandenburg et al. 2011a). The resulting one- and three-dimensional spectra agree in the case of pure power laws, but near the dissipative cutoff wavenumbers there is a sharp departure from power law behavior. This is significant in view of the fact that energy spectra of three-dimensional simulations indicate the presence of a so-called bottleneck effect (Falkovich 1994). This corresponds to an uprise of the compensated energy spectrum, $k^{5/3}E_{\mathrm{K}}(k)$, near the dissipative cutoff wavenumber $k_\nu = \langle \boldsymbol{w}^2/\nu^2\rangle^{1/4}$. This bottleneck effect is much weaker or absent in one-dimensional spectra (Dobler et al. 2003; Beresnyak and Lazarian 2009). The bottleneck might therefore be a real effect. Although it happens at such small scales that is should not be astrophysically significant, it does play a role in three-dimensional simulations and can lead to effects whose astrophysical significance needs to be assessed carefully in view of the fact that the growth rate of small-scale dynamos depends on the shape of the spectrum at the resistive scale; see Sect. 5.3.

8 Concluding Comments

The last decades have been marked by important advances in our understanding of MHD turbulence. To a substantial degree this happened as numerical simulations became capable of producing high resolution MHD cubes. Therefore, MHD turbulence became a theory that can be tested. As a result of both analytical and numerical studies, as well as observational measurements of turbulence, the GS95 model of MHD turbulence has been established as the most promising model. While we believe that the model is not complete in detail (e.g. in terms of intermittency), it is able to describe the astrophysically important properties of turbulence, for instance, the scale dependence of local anisotropy important for cosmic ray propagation, and the magnetic field wandering important for heat transfer and magnetic reconnection.

The physical ideas of the GS95 model have been extended and applied to successfully describing compressible MHD turbulence. It has been shown that the low coupling of fast and Alfvén modes allows the independent treatment of the Alfvénic cascade, which is very important; see Sect. 3.2. Indeed, it allows one to use the GS95 scaling for describing Alfvénic modes in moderately compressible fluids, which is of major astrophysical significance.

However, there are many issues that require further studies. Those include the properties of highly compressible, highly supersonic MHD turbulence, scaling and properties of fast modes etc. Last but not the least, more work is required for the highly debated subject of imbalanced turbulence. The corresponding studies call for extensive numerical efforts to test the existing theories. We hope that many of these currently controversial issues will be solved in the near future. This has important applications for turbulent dynamos of all sorts. It is now clear that nonlinear turbulent dynamos work also at small magnetic Prandtl numbers, even though the excitation conditions for kinematic dynamos become prohibitively high at low magnetic Prandtl numbers of around 0.1. As discussed in Sect. 5.3, the reason for this has meanwhile been identified as the bottleneck effect in turbulence.

Large-scale dynamos are affected by similar subtleties. They are in particular subject to the possibility of catastrophic quenching, which means that dynamos and their underlying turbulent transport coefficients remain dependent on the magnetic Reynolds numbers. Astrophysical dynamos are believed to be independent of Re_M, but we now know that most dynamos in DNS are probably not yet in that regime, but there is not much doubt that such a regime exists that is independent of the magnetic Reynolds number. In practice, this is accomplished by magnetic helicity fluxes. Regarding solar and stellar dynamo theory, the reason for equatorward migration of magnetic activity belts is still not understood. This is an example where simulations might now lead the way toward explaining the observed solar behavior, but more progress is needed to fully understand the physics behind the behavior seen in simulations.

Acknowledgements We thank Andre Balogh for providing an inspiring atmosphere at the International Space Science Institute in Bern in 2012, which has led to new collaborations and scientific progress. Computing resources were provided by the Swedish National Allocations Committee at the Center for Parallel Computers at the Royal Institute of Technology in Stockholm and the High Performance Computing Center North in Umeå. This work was supported in part by the European Research Council under the AstroDyn Research Project No. 227952 and the Swedish Research Council under the project grants 621-2011-5076 and 2012-5797. A.L. acknowledges the support of the NSF grant AST-1212096, the NASA grant NNX09AH78G, the Vilas Associate Award as well as the support of the NSF Center for Magnetic Self-Organization. In addition, A.L. thanks the International Institute of Physics (Natal, Brazil) for its hospitality during the work on this review.

References

J.W. Armstrong, B.J. Rickett, S.R. Spangler, Electron density power spectrum in the local interstellar medium. Astrophys. J. **443**, 209–221 (1995)

R. Banerjee, K. Jedamzik, Evolution of cosmic magnetic fields: from the very early universe, to recombination, to the present. Phys. Rev. D **70**, 123003 (2004)

S. Banerjee, S. Galtier, Exact relation with two-point correlation functions and phenomenological approach for compressible magnetohydrodynamic turbulence. Phys. Rev. E **87**, 013019 (2013)

G.K. Batchelor, On the spontaneous magnetic field in a conducting liquid in turbulent motion. Proc. R. Soc. Lond. A **201**, 405–416 (1950)

R. Beck, A. Brandenburg, D. Moss, A. Shukurov, D. Sokoloff, Galactic magnetism: recent developments and perspectives. Annu. Rev. Astron. Astrophys. **34**, 155–206 (1996)

A.R. Bell, Turbulent amplification of magnetic field and diffusive shock acceleration of cosmic rays. Mon. Not. R. Astron. Soc. **353**, 550–558 (2004)

A. Beresnyak, Spectral slope and Kolmogorov constant of MHD turbulence. Phys. Rev. Lett. **106**, 075001 (2011)

A. Beresnyak, Basic properties of magnetohydrodynamic turbulence in the inertial range. Mon. Not. R. Astron. Soc. **422**, 3495–3502 (2012)

A. Beresnyak, A. Lazarian, Strong imbalanced turbulence. Astrophys. J. Lett. **682**, 1070–1075 (2008)

A. Beresnyak, A. Lazarian, Comparison of spectral slopes of magnetohydrodynamic and hydrodynamic turbulence and measurements of alignment effects. Astrophys. J. Lett. **702**, 1190–1198 (2009)

A. Beresnyak, A. Lazarian, Scaling laws and diffuse locality of balanced and imbalanced magnetohydrodynamic turbulence. Astrophys. J. Lett. **722**, L110–L113 (2010)

A. Beresnyak, A. Lazarian, J. Cho, Density scaling and anisotropy in supersonic magnetohydrodynamic turbulence. Astrophys. J. Lett. **624**, L93–L96 (2005)

A. Beresnyak, T.W. Jones, A. Lazarian, Turbulence-induced magnetic fields and structure of cosmic ray modified shocks. Astrophys. J. **707**, 1541–1549 (2009)
P. Bhat, K. Subramanian, Fluctuation dynamos and their Faraday rotation signatures. Mon. Not. R. Astron. Soc. **429**, 2469–2481 (2013)
D. Biskamp, *Magnetohydrodynamic Turbulence* (Cambridge University Press, Cambridge, 2003)
D. Biskamp, W.-C. Müller, Decay laws for three-dimensional magnetohydrodynamic turbulence. Phys. Rev. Lett. **83**, 2195–2198 (1999)
E.G. Blackman, A. Brandenburg, Doubly helical coronal ejections from dynamos and their role in sustaining the solar cycle. Astrophys. J. Lett. **584**, L99–L102 (2003)
E.G. Blackman, G.B. Field, Dimensionless measures of turbulent magnetohydrodynamic dissipation rates. Mon. Not. R. Astron. Soc. **386**, 1481–1486 (2008)
S. Boldyrev, On the spectrum of magnetohydrodynamic turbulence. Astrophys. J. Lett. **626**, L37–L40 (2005)
S. Boldyrev, Spectrum of magnetohydrodynamic turbulence. Phys. Rev. Lett. **96**, 115002 (2006)
S.A. Boldyrev, F. Cattaneo, Magnetic-field generation in Kolmogorov turbulence. Phys. Rev. Lett. **92**, 144501 (2004)
S. Boldyrev, Å. Nordlund, P. Padoan, Scaling relations of supersonic turbulence in star-forming molecular clouds. Astrophys. J. **573**, 678–684 (2002)
S. Boldyrev, J.C. Perez, J.E. Borovsky, J.J. Podesta, Spectral scaling laws in magnetohydrodynamic turbulence simulations and in the solar wind. Astrophys. J. **741**, L19 (2011)
A. Brandenburg, The case for a distributed solar dynamo shaped by near-surface shear. Astrophys. J. **625**, 539–547 (2005)
A. Brandenburg, Nonlinear small-scale dynamos at low magnetic Prandtl numbers. Astrophys. J. **741**, 92 (2011a)
A. Brandenburg, Å. Nordlund, Astrophysical turbulence modeling. Rep. Prog. Phys. **74**, 046901 (2011b)
A. Brandenburg, K. Subramanian, Astrophysical magnetic fields and nonlinear dynamo theory. Phys. Rep. **417**, 1–209 (2005)
A. Brandenburg, K. Enqvist, P. Olesen, Large-scale magnetic fields from hydromagnetic turbulence in the very early universe. Phys. Rev. D **54**, 1291–1300 (1996)
A. Brandenburg, P. Käpylä, A. Mohammed, Non-Fickian diffusion and tau-approximation from numerical turbulence. Phys. Fluids **16**, 1020–1027 (2004)
A. Brandenburg, K.-H. Rädler, M. Rheinhardt, K. Subramanian, Magnetic quenching of alpha and diffusivity tensors in helical turbulence. Astrophys. J. **676**, 740-L52 (2008a)
A. Brandenburg, K.-H. Rädler, M. Schrinner, Scale dependence of alpha effect and turbulent diffusivity. Astron. Astrophys. **482**, 739–746 (2008b)
A. Brandenburg, N. Kleeorin, I. Rogachevskii, Large-scale magnetic flux concentrations from turbulent stresses. Astron. Nachr. **331**, 5–13 (2010)
A. Brandenburg, N. Kleeorin, I. Rogachevskii, Self-assembly of shallow magnetic spots through strongly stratified turbulence. Astrophys. J. Lett. (2013, submitted). arXiv:1306.4915
A. Brandenburg, K. Subramanian, A. Balogh, M.L. Goldstein, Scale-dependence of magnetic helicity in the solar wind. Astrophys. J. **734**, 9 (2011a)
A. Brandenburg, K. Kemel, N. Kleeorin, D. Mitra, I. Rogachevskii, Detection of negative effective magnetic pressure instability in turbulence simulations. Astrophys. J. **740**, L50 (2011b)
A. Brandenburg, K. Kemel, N. Kleeorin, I. Rogachevskii, The negative effective magnetic pressure in stratified forced turbulence. Astrophys. J. **749**, 179 (2012a)
A. Brandenburg, K.-H. Rädler, K. Kemel, Mean-field transport in stratified and/or rotating turbulence. Astron. Astrophys. **539**, A35 (2012b)
A. Brandenburg, D. Sokoloff, K. Subramanian, Current status of turbulent dynamo theory: from large-scale to small-scale dynamos. Space Sci. Rev. **169**, 123–157 (2012c)
B.P. Brown, M.K. Browning, A.S. Brun, M.S. Miesch, J. Toomre, Persistent magnetic wreaths in a rapidly rotating sun. Astrophys. J. **711**, 424–438 (2010)
B.P. Brown, M.S. Miesch, M.K. Browning, A.S. Brun, J. Toomre, Magnetic cycles in a convective dynamo simulation of a young solar-type star. Astrophys. J. **731**, 69 (2011)
G. Brunetti, A. Lazarian, Compressible turbulence in galaxy clusters: physics and stochastic particle re-acceleration. Mon. Not. R. Astron. Soc. **378**, 245–275 (2007)
B. Burkhart, A. Lazarian, V. Ossenkopf, J. Stutzki, The turbulence power spectrum in optically thick interstellar clouds. Astrophys. J. **771**, 123 (2013)
A.M. Bykov, A. Brandenburg, M.A. Malkov, S.M. Osipov, Microphysics of cosmic ray driven plasma instabilities. Space Sci. Rev. (2013). doi:10.1007/s11214-013-9988-3
S. Candelaresi, A. Brandenburg, How much helicity is needed to drive large-scale dynamos? Phys. Rev. E **87**, 043104 (2013)

B.D.G. Chandran, Scattering of energetic particles by anisotropic magnetohydrodynamic turbulence with a Goldreich-Sridhar power spectrum. Phys. Rev. Lett. **85**, 4656–4659 (2000)

B.D.G. Chandran, S.C. Cowley, Thermal conduction in a tangled magnetic field. Phys. Rev. Lett. **80**, 3077–3080 (1998)

L. Chamandy, K. Subramanian, A. Shukurov, Galactic spiral patterns and dynamo action. I. A new twist on magnetic arms. Mon. Not. R. Astron. Soc. **428**, 3569–3589 (2013)

P. Chatterjee, G. Guerrero, A. Brandenburg, Magnetic helicity fluxes in interface and flux transport dynamos. Astron. Astrophys. **525**, A5 (2011)

A. Chepurnov, A. Lazarian, Turbulence spectra from Doppler-broadened spectral lines: tests of the velocity channel analysis and velocity coordinate spectrum techniques. Astrophys. J. **693**, 1074–1083 (2009)

A. Chepurnov, A. Lazarian, Extending the big power law in the sky with turbulence spectra from Wisconsin Hα mapper data. Astrophys. J. **710**, 853–858 (2010)

J. Cho, A. Lazarian, Compressible sub-Alfvénic MHD turbulence in low-β plasmas. Phys. Rev. Lett. **88**, 245001 (2002). CL02

J. Cho, A. Lazarian, Compressible magnetohydrodynamic turbulence: mode coupling, scaling relations, anisotropy, viscosity-damped regime and astrophysical implications. Mon. Not. R. Astron. Soc. **345**, 325–339 (2003). CL03

J. Cho, A. Lazarian, Thermal conduction in magnetized turbulent gas. J. Korean Astron. Soc. **37**, 557–562 (2004)

J. Cho, A. Lazarian, Grain alignment by radiation in dark clouds and cores. Astrophys. J. **631**, 361–370 (2005)

J. Cho, A. Lazarian, E.T. Vishniac, Simulations of magnetohydrodynamic turbulence in a strongly magnetized medium. Astrophys. J. **564**, 291–301 (2002)

J. Cho, E.T. Vishniac, The generation of magnetic fields through driven turbulence. Astrophys. J. **538**, 217–225 (2000)

J. Crovisier, J.M. Dickey, The spatial power spectrum of galactic neutral hydrogen from observations of the 21-cm emission line. Astron. Astrophys. **122**, 282–296 (1983)

R.M. Crutcher, B. Wandelt, C. Heiles, E. Falgarone, T.H. Troland, Magnetic fields in interstellar clouds from Zeeman observations: inference of total field strengths by Bayesian analysis. Astrophys. J. **725**, 466–479 (2010)

F. Del Sordo, G. Guerrero, A. Brandenburg, Turbulent dynamo with advective magnetic helicity flux. Mon. Not. R. Astron. Soc. **429**, 1686–1694 (2013)

R.L. Dickman, S.C. Kleiner, Large-scale structure of the Taurus molecular complex. Part 3. Methods for turbulence. Astrophys. J. **295**, 479–484 (1985)

W. Dobler, N.E.L. Haugen, T.A. Yousef, A. Brandenburg, Bottleneck effect in three-dimensional turbulence simulations. Phys. Rev. E **68**, 026304 (2003)

T.A. Enßlin, C. Vogt, Magnetic turbulence in cool cores of galaxy clusters. Astron. Astrophys. **453**, 447–458 (2006)

A. Esquivel, A. Lazarian, S. Horibe, J. Cho, V. Ossenkopf, J. Stutzki, Statistics of velocity centroids: effects of density-velocity correlations and non-Gaussianity. Mon. Not. R. Astron. Soc. **381**, 1733–1744 (2007)

G.L. Eyink, A. Lazarian, E.T. Vishniac, Fast magnetic reconnection and spontaneous stochasticity. Astrophys. J. **743**, 51 (2011)

B.G. Elmegreen, J. Scalo, Interstellar turbulence. I. Observations and processes. Annu. Rev. Astron. Astrophys. **42**, 211–273 (2004)

G. Falkovich, Bottleneck phenomenon in developed turbulence. Phys. Fluids **6**, 1411–1414 (1994)

S. Galtier, S. Banerjee, Exact relation for correlation functions in compressible isothermal turbulence. Phys. Rev. Lett. **107**, 134501 (2011)

S. Galtier, S.V. Nazarenko, A.C. Newell, A. Pouquet, A weak turbulence theory for incompressible magnetohydrodynamics. J. Plasma Phys. **63**, 447–488 (2000)

M. Ghizaru, P. Charbonneau, P.K. Smolarkiewicz, Magnetic cycles in global large-eddy simulations of solar convection. Astrophys. J. **715**, L133–L137 (2010)

P.A. Gilman, Dynamically consistent nonlinear dynamos driven by convection in a rotating spherical shell. II. Dynamos with cycles and strong feedbacks. Astrophys. J. Suppl. Ser. **53**, 243–268 (1983)

P. Goldreich, S. Sridhar, Toward a theory of interstellar turbulence. 2. Strong Alfvénic turbulence. Astrophys. J. **438**, 763–775 (1995). GS95

P. Goldreich, S. Sridhar, Magnetohydrodynamic turbulence revisited. Astrophys. J. **485**, 680–688 (1997)

G.A. Glatzmaier, Numerical simulations of stellar convective dynamos. II. Field propagation in the convection zone. Astrophys. J. **291**, 300–307 (1985)

J. Goodman, R. Narayan, Slow pulsar scintillation and the spectrum of interstellar electron density fluctuations. Mon. Not. R. Astron. Soc. **214**, 519–537 (1985)

D.A. Green, A power spectrum analysis of the angular scale of galactic neutral hydrogen emission towards $L = 140°$, $B = 0°$, Monthly Notices Roy. Mon. Not. R. Astron. Soc. **262**, 327–342 (1993)
A. Gruzinov, S. Cowley, R. Sudan, Small-scale-field dynamo. Phys. Rev. Lett. **77**, 4342–4345 (1996)
N.E.L. Haugen, A. Brandenburg, Hydrodynamic and hydromagnetic energy spectra from large eddy simulations. Phys. Fluids **18**, 075106 (2006)
N.E.L. Haugen, A. Brandenburg, W. Dobler, Is nonhelical hydromagnetic turbulence peaked at small scales? Astrophys. J. **597**, L141–L144 (2003)
N.E.L. Haugen, A. Brandenburg, W. Dobler, Simulations of nonhelical hydromagnetic turbulence. Phys. Rev. E **70**, 016308 (2004)
J.C. Higdon, Density fluctuations in the interstellar medium: evidence for anisotropic magnetogas turbulence. I. Model and astrophysical sites. Astrophys. J. **285**, 109–123 (1984)
A. Hubbard, A. Brandenburg, Memory effects in turbulent transport. Astrophys. J. **706**, 712–726 (2009)
A. Hubbard, A. Brandenburg, Magnetic helicity fluxes in an α^2 dynamo embedded in a halo. Geophys. Astrophys. Fluid Dyn. **104**, 577–590 (2010)
R.S. Iroshnikov, Turbulence of a conducting fluid in a strong magnetic field. Sov. Astron. **7**, 566–571 (1963)
A.B. Iskakov, A.A. Schekochihin, S.C. Cowley, J.C. McWilliams, M.R.E. Proctor, Numerical demonstration of fluctuation dynamo at low magnetic Prandtl numbers. Phys. Rev. Lett. **98**, 208501 (2007)
J.R. Jokipii, Pitch-angle scattering of charged particles in a random magnetic field. Astrophys. J. **194**, 465–469 (1974)
T. Kahniashvili, A. Brandenburg, A.G. Tevzadze, B. Ratra, Numerical simulations of the decay of primordial magnetic turbulence. Phys. Rev. D **81**, 123002 (2010)
T. Kahniashvili, A.G. Tevzadze, A. Brandenburg, A. Neronov, Evolution of primordial magnetic fields from phase transitions. Phys. Rev. D **87**, 083007 (2013)
P.J. Käpylä, M.J. Korpi, A. Brandenburg, D. Mitra, R. Tavakol, Convective dynamos in spherical wedge geometry. Astron. Nachr. **331**, 73–81 (2010)
P.J. Käpylä, M.J. Mantere, A. Brandenburg, Cyclic magnetic activity due to turbulent convection in spherical wedge geometry. Astrophys. J. Lett. **755**, L22 (2012a)
P.J. Käpylä, A. Brandenburg, N. Kleeorin, M.J. Mantere, I. Rogachevskii, Negative effective magnetic pressure in turbulent convection. Mon. Not. R. Astron. Soc. **422**, 2465–2473 (2012b)
A.P. Kazantsev, Enhancement of a magnetic field by a conducting fluid. Sov. Phys. JETP **26**, 1031–1034 (1968)
K. Kemel, A. Brandenburg, N. Kleeorin, D. Mitra, I. Rogachevskii, Spontaneous formation of magnetic flux concentrations in stratified turbulence. Sol. Phys. **280**, 321–333 (2012)
K. Kemel, A. Brandenburg, N. Kleeorin, D. Mitra, I. Rogachevskii, Active region formation through the negative effective magnetic pressure instability. Sol. Phys. (2013). doi:10.1007/s11207-012-0031-8
S. Kida, S. Yanase, J. Mizushima, Statistical properties of MHD turbulence and turbulent dynamo. Phys. Fluids A **3**, 457–465 (1991)
L.L. Kitchatinov, A. Brandenburg, Transport of angular momentum and chemical species by anisotropic mixing in stellar radiative interiors. Astron. Nachr. **333**, 230–236 (2012)
N. Kleeorin, I. Rogachevskii, Effective Ampère force in developed magnetohydrodynamic turbulence. Phys. Rev. E **50**, 2716–2730 (1994)
N.I. Kleeorin, I.V. Rogachevskii, A.A. Ruzmaikin, Negative magnetic pressure as a trigger of large-scale magnetic instability in the solar convective zone. Sov. Astron. Lett. **15**, 274–277 (1989)
N.I. Kleeorin, I.V. Rogachevskii, A.A. Ruzmaikin, Magnetic force reversal and instability in a plasma with advanced magnetohydrodynamic turbulence. Sov. Phys. JETP **70**, 878–883 (1990)
N. Kleeorin, M. Mond, I. Rogachevskii, Magnetohydrodynamic instabilities in developed small-scale turbulence. Phys. Fluids **5**, 4128–4134 (1993)
N. Kleeorin, M. Mond, I. Rogachevskii, Magnetohydrodynamic turbulence in the solar convective zone as a source of oscillations and sunspots formation. Astron. Astrophys. **307**, 293–309 (1996)
A.N. Kolmogorov, The local structure of turbulence in incompressible viscous fluid for very large Reynolds numbers. CR Acad. Sci. USSR **30**, 299–303 (1941)
G. Kowal, A. Lazarian, A. Beresnyak, Density fluctuations in MHD turbulence: spectra, intermittency, and topology. Astrophys. J. **658**, 423–445 (2007)
G. Kowal, A. Lazarian, Velocity field of compressible magnetohydrodynamic turbulence: wavelet decomposition and mode scalings. Astrophys. J. **720**, 742–756 (2010)
R.H. Kraichnan, Inertial-range spectrum of hydromagnetic turbulence. Phys. Fluids **8**, 1385–1387 (1965)
F. Krause, K.-H. Rädler, *Mean-Field Magnetohydrodynamics and Dynamo Theory* (Pergamon Press, Oxford, 1980)
A.G. Kritsuk, M.L. Norman, P. Padoan, R. Wagner, The statistics of supersonic isothermal turbulence. Astrophys. J. **665**, 416–431 (2007)

A.G. Kritsuk, Å. Nordlund, D. Collins, P. Padoan, M.L. Norman, T. Abel, R. Banerjee, C. Federrath, M. Flock, D. Lee, P.S. Li, W.-C. Müller, R. Teyssier, S.D. Ustyugov, C. Vogel, H. Xu, Comparing numerical methods for isothermal magnetized supersonic turbulence. Astrophys. J. **737**, 13 (2011)

R.M. Kulsrud, A critical review of galactic dynamos. Annu. Rev. Astron. Astrophys. **37**, 37–64 (1999)

A. Lazarian, Astrophysical implications of turbulent reconnection: from cosmic rays to star formation, in *Magnetic Fields in the Universe*, ed. by E. de Gouveia Dal Pino, G. Lugones, A. Lazarian. AIP, vol. 784, (2005), pp. 42–54

A. Lazarian, Enhancement and suppression of heat transfer by MHD turbulence. Astrophys. J. Lett. **645**, L25–L28 (2006). L06

A. Lazarian, Obtaining spectra of turbulent velocity from observations. Space Sci. Rev. **143**, 357–385 (2009)

A. Lazarian, A. Esquivel, Statistics of velocity from spectral data: modified velocity centroids. Astrophys. J. **592**, L37–L40 (2003)

A. Lazarian, A. Esquivel, Velocity centroids as tracers of the turbulent velocity statistics. Astrophys. J. **631**, 320–350 (2005)

A. Lazarian, D. Pogosyan, Velocity modification of HI power spectrum. Astrophys. J. **537**, 720–748 (2000). LP00

A. Lazarian, D. Pogosyan, Velocity modification of the power spectrum from an absorbing medium. Astrophys. J. **616**, 943–965 (2004)

A. Lazarian, D. Pogosyan, Statistics of fluctuations along velocity coordinate: effects of absorption. Astrophys. J. **652**, 1348 (2006)

A. Lazarian, D. Pogosyan, Studying velocity turbulence from Doppler-broadened absorption lines: statistics of optical depth fluctuations. Astrophys. J. **686**, 350–362 (2008)

A. Lazarian, E.T. Vishniac, Reconnection in a weakly stochastic field. Astrophys. J. **517**, 700–718 (1999). LV99

A. Lazarian, E.T. Vishniac, Model of reconnection of weakly stochastic magnetic field and its implications. Rev. Mex. Astron. Astrofís. Conf. Ser. **36**, 81–88 (2009)

A. Lazarian, A. Esquivel, R. Crutcher, Magnetization of cloud cores and envelopes and other observational consequences of reconnection diffusion. Astrophys. J. **757**, 154 (2012a)

A. Lazarian, L. Vlahos, G. Kowal, H. Yan, A. Beresnyak, E.M. de Gouveia Dal Pino, Turbulence, magnetic reconnection in turbulent fluids and energetic particle acceleration. Space Sci. Rev. **173**, 557–622 (2012b)

M. Lesieur, *Turbulence in Fluids* (Nijhoff, Dordrecht, 1990)

Y. Lithwick, P. Goldreich, Imbalanced weak magnetohydrodynamic turbulence. Astrophys. J. **582**, 1220–1240 (2003)

Y. Lithwick, P. Goldreich, S. Sridhar, Imbalanced strong MHD turbulence. Astrophys. J. **655**, 269–274 (2007)

J. Maron, P. Goldreich, Simulations of incompressible magnetohydrodynamic turbulence. Astrophys. J. **554**, 1175–1196 (2001)

W.H. Matthaeus, M.L. Goldstein, C. Smith, Evaluation of magnetic helicity in homogeneous turbulence. Phys. Rev. Lett. **48**, 1256–1259 (1982)

D.B. Melrose, The emission and absorption of waves by charged particles in magnetized plasmas. Astrophys. Space Sci. **2**, 171–235 (1968)

M. Meneguzzi, U. Frisch, A. Pouquet, Helical and nonhelical turbulent dynamos. Phys. Rev. Lett. **47**, 1060–1064 (1981)

M. Meneguzzi, A. Pouquet, Turbulent dynamos driven by convection. J. Fluid Mech. **205**, 297–312 (1989)

M.S. Miesch, J. Toomre, Turbulence, magnetism, and shear in stellar interiors. Ann. Rev. Fluid Dyn. **41**, 317–345 (2009)

M.S. Miesch, J. Scalo, J. Bally, Velocity field statistics in star-forming regions. I. Centroid velocity observations. Astrophys. J. **524**, 895–922 (1999)

D. Mitra, P.J. Käpylä, R. Tavakol, A. Brandenburg, Alpha effect and diffusivity in helical turbulence with shear. Astron. Astrophys. **495**, 1–8 (2009)

D. Mitra, S. Candelaresi, P. Chatterjee, R. Tavakol, A. Brandenburg, Equatorial magnetic helicity flux in simulations with different gauges. Astron. Nachr. **331**, 130–135 (2010)

M.-A. Miville-Deschênes, G. Joncas, E. Falgarone, F. Boulanger, High resolution 21 cm mapping of the Ursa Major Galactic cirrus: power spectra of the high-latitude H I gas. Astron. Astrophys. **411**, 109–121 (2003)

H.K. Moffatt, *Magnetic Field Generation in Electrically Conducting Fluids* (Cambridge Univ. Press, Cambridge, 1978)

G. Münch, Internal motions in the Orion nebula. Rev. Mod. Phys. **30**, 1035–1041 (1958)

R. Narayan, M.V. Medvedev, Thermal conduction in clusters of galaxies. Astrophys. J. **562**, L129–L132 (2001)

C.S. Ng, A. Bhattacharjee, Interaction of shear-Alfvén wave packets: implication for weak magnetohydrodynamic turbulence in astrophysical plasmas. Astrophys. J. **465**, 845–854 (1996)

Å. Nordlund, A. Brandenburg, R.L. Jennings, M. Rieutord, J. Ruokolainen, R.F. Stein, I. Tuominen, Dynamo action in stratified convection with overshoot. Astrophys. J. **392**, 647–652 (1992)

C.R. O'dell, H.O. Castaneda, Evidence for turbulence in H II regions. Astrophys. J. **317**, 686–692 (1987)

P. Padoan, M. Juvela, A. Kritsuk, M.L. Norman, The power spectrum of supersonic turbulence in Perseus. Astrophys. J. Lett. **653**, 125–128 (2006)

P. Padoan, M. Juvela, A. Kritsuk, M.L. Norman, The power spectrum of turbulence in NGC 1333: outflows or large-scale driving? Astrophys. J. Lett. **707**, L153–L157 (2009)

E.N. Parker, *Cosmical Magnetic Fields* (Clarendon, Oxford, 1979)

J.C. Perez, S. Boldyrev, Role of cross-helicity in magnetohydrodynamic turbulence. Phys. Rev. Lett. **102**, 025003 (2009)

V. Petrosian, H. Yan, A. Lazarian, Damping of magnetohydrodynamic turbulence in solar flares. Astrophys. J. **644**, 603–612 (2006)

V.V. Pipin, The mean electro-motive force and current helicity under the influence of rotation, magnetic field and shear. Geophys. Astrophys. Fluid Dyn. **102**, 21–49 (2008)

J.J. Podesta, D.A. Roberts, M.L. Goldstein, Spectral exponents of kinetic and magnetic energy spectra in solar wind turbulence. Astrophys. J. **664**, 543–548 (2007)

J.J. Podesta, S.P. Gary, Magnetic helicity spectrum of solar wind fluctuations as a function of the angle with respect to the local mean magnetic field. Astrophys. J. **734**, 15 (2011)

É. Racine, P. Charbonneau, M. Ghizaru, A. Bouchat, P.K. Smolarkiewicz, On the mode of dynamo action in a global large-eddy simulation of solar convection. Astrophys. J. **735**, 46 (2011)

K.-H. Rädler, On some electromagnetic phenomena in electrically conducting turbulently moving matter, especially in the presence of Coriolis forces. Geod. Geophys. Veröff., Reihe 2 **13**, 131–135 (1969)

K.-H. Rädler, On the influence of a large-scale magnetic field on turbulent motions in an electrically conducting medium. Astron. Nachr. **295**, 265–273 (1974)

K.-H. Rädler, A. Brandenburg, F. Del Sordo, M. Rheinhardt, Mean-field diffusivities in passive scalar and magnetic transport in irrotational flows. Phys. Rev. E **84**, 4 (2011)

A.B. Rechester, M.N. Rosenbluth, Electron heat transport in a Tokamak with destroyed magnetic surfaces. Phys. Rev. Lett. **40**, 38–41 (1978)

M. Rheinhardt, A. Brandenburg, Test-field method for mean-field coefficients with MHD background. Astron. Astrophys. **520**, A28 (2010)

M. Rheinhardt, A. Brandenburg, Modeling spatio-temporal nonlocality in mean-field dynamos. Astron. Nachr. **333**, 71–77 (2012)

I. Rogachevskii, N. Kleeorin, Intermittency and anomalous scaling for magnetic fluctuations. Phys. Rev. E **56**, 417–426 (1997)

I. Rogachevskii, N. Kleeorin, Magnetic fluctuations and formation of large-scale inhomogeneous magnetic structures in a turbulent convection. Phys. Rev. E **76**, 056307 (2007)

I. Rogachevskii, N. Kleeorin, A. Brandenburg, D. Eichler, Cosmic ray current-driven turbulence and mean-field dynamo effect. Astrophys. J. **753**, 6 (2012)

G. Rüdiger, On the Reynolds stresses in mean-field hydrodynamics III. two-dimensional turbulence and the problem of differential rotation. Astron. Nachr. **295**, 229–252 (1974)

G. Rüdiger, L.L. Kitchatinov, A. Brandenburg, Cross helicity and turbulent magnetic diffusivity in the solar convection zone. Sol. Phys. **269**, 3–12 (2011)

R. Santos-Lima, A. Lazarian, E.M. de Gouveia Dal Pino, J. Cho, Diffusion of magnetic field and removal of magnetic flux from clouds via turbulent reconnection. Astrophys. J. **714**, 442–461 (2010)

R. Santos-Lima, E.M. de Gouveia Dal Pino, A. Lazarian, The role of turbulent magnetic reconnection in the formation of rotationally supported protostellar disks. Astrophys. J. **747**, 21 (2012)

A.A. Schekochihin, J.L. Maron, S.C. Cowley, J.C. McWilliams, The small-scale structure of magnetohydrodynamic turbulence with large magnetic Prandtl numbers. Astrophys. J. **576**, 806–813 (2002)

A.A. Schekochihin, S.C. Cowley, J.L. Maron, J.C. McWilliams, Critical magnetic Prandtl number for small-scale dynamo. Phys. Rev. Lett. **92**, 054502 (2004a)

A.A. Schekochihin, S.C. Cowley, S.F. Taylor, J.L. Maron, J.C. McWilliams, Simulations of the small scale turbulent dynamo. Astrophys. J. **612**, 276–307 (2004b)

A.A. Schekochihin, N.E.L. Haugen, A. Brandenburg, S.C. Cowley, J.L. Maron, J.C. McWilliams, Onset of small scale dynamo at small magnetic Prandtl numbers. Astrophys. J. **625**, L115–L118 (2005)

R. Schlickeiser, *Cosmic Ray Astrophysics*. Astron. Astrophys. Lib. (Springer, Berlin, 2002)

J.V. Shebalin, W.H. Matthaeus, D. Montgomery, Anisotropy in MHD turbulence due to a mean magnetic field. J. Plasma Phys. **29**, 525–547 (1983)

A. Skumanich, Time scales for CA II emission decay, rotational braking, and lithium depletion. Astrophys. J. **171**, 565–567 (1972)

C.W. Smith, J.W. Bieber, Detection of steady magnetic helicity in low-frequency IMF turbulence, in *23rd International Cosmic Ray Conference*, vol. 3, ed. by D.A. Leahy, R.B. Hicks, D. Venkatesan (World Scientific, Singapore, 1993), pp. 493–496

S. Spangler, C. Gwinn, Evidence for an inner scale to the density turbulence in the interstellar medium. Astrophys. J. **353**, L29–L32 (1990)

S. Stanimirovic, L. Staveley-Smith, J.M. Dickey, R.J. Sault, S.L. Snowden, The large-scale HI structure of the Small Magellanic Cloud. Mon. Not. R. Astron. Soc. **302**, 417–436 (1999)

L.G. Stenholm, Molecular cloud fluctuations. II. Methods of analysis of cloud maps. Astron. Astrophys. **232**, 495–509 (1990)

K. Subramanian, Can the turbulent galactic dynamo generate large-scale magnetic fields? Mon. Not. R. Astron. Soc. **294**, 718–728 (1998)

S. Sur, A. Brandenburg, K. Subramanian, Kinematic alpha effect in isotropic turbulence simulations. Mon. Not. R. Astron. Soc. **385**, L15–L19 (2008)

H. Tennekes, J.L. Lumley, *First Course in Turbulence* (MIT Press, Cambridge, 1972)

A.G. Tevzadze, L. Kisslinger, A. Brandenburg, T. Kahniashvili, Magnetic fields from QCD phase transitions. Astrophys. J. **759**, 54 (2012)

H. Yan, A. Lazarian, Scattering of cosmic rays by magnetohydrodynamic turbulence. Phys. Rev. Lett. **89**, 281102 (2002)

H. Yan, A. Lazarian, Cosmic-ray scattering and streaming in compressible magnetohydrodynamic turbulence. Astrophys. J. **614**, 757–769 (2004)

H. Yan, A. Lazarian, Cosmic-ray propagation: nonlinear diffusion parallel and perpendicular to mean magnetic field. Astrophys. J. **673**, 942–953 (2008)

S.I. Vainshtein, Ya.B. Zeldovich, Origin of magnetic fields in astrophysics. Sov. Phys. Usp. **15**, 159–172 (1972)

J. Warnecke, P.J. Käpylä, M.J. Mantere, A. Brandenburg, Solar-like differential rotation in a convective dynamo with a coronal envelope. Astrophys. J. (2013, submitted). arXiv:1301.2248

DOI 10.1007/978-1-4899-7413-6_6
Reprinted from *Space Science Reviews* Journal, DOI 10.1007/s11214-013-9988-3

Microphysics of Cosmic Ray Driven Plasma Instabilities

A.M. Bykov · A. Brandenburg · M.A. Malkov · S.M. Osipov

Received: 1 February 2013 / Accepted: 26 April 2013 / Published online: 10 May 2013

Abstract Energetic nonthermal particles (cosmic rays, CRs) are accelerated in supernova remnants, relativistic jets and other astrophysical objects. The CR energy density is typically comparable with that of the thermal components and magnetic fields. In this review we discuss mechanisms of magnetic field amplification due to instabilities induced by CRs. We derive CR kinetic and magnetohydrodynamic equations that govern cosmic plasma systems comprising the thermal background plasma, comic rays and fluctuating magnetic fields to study CR-driven instabilities. Both resonant and non-resonant instabilities are reviewed, including the Bell short-wavelength instability, and the firehose instability. Special attention is paid to the longwavelength instabilities driven by the CR current and pressure gradient. The helicity production by the CR current-driven instabilities is discussed in connection with the dynamo mechanisms of cosmic magnetic field amplification.

Keywords Magnetic fields · Cosmic rays · Collisionless shocks · Supernova remnants

A.M. Bykov (✉) · S.M. Osipov
A.F. Ioffe Institute for Physics and Technology, 194021 Saint Petersburg, Russia
e-mail: byk@astro.ioffe.ru

S.M. Osipov
e-mail: osm2004@mail.ru

A.M. Bykov
St. Petersburg State Politechnical University, Saint Petersburg, Russia

A. Brandenburg
Nordita, Royal Institute of Technology and Stockholm University, Roslagstullsbacken 23, 10691 Stockholm, Sweden
e-mail: brandenb@nordita.org

A. Brandenburg
Department of Astronomy, Stockholm University, 10691 Stockholm, Sweden

M.A. Malkov
University of California, San Diego, La Jolla, California 92093, USA
e-mail: mmalkov@ucsd.edu

1 Introduction

Acceleration of cosmic rays (CRs) in the Galaxy by the first order Fermi mechanism is believed to be very efficient. Most of the theoretical studies of shock acceleration agree on its potential to convert, under *favorable conditions*, 50 % or more of shock mechanical energy into the CR energy. Observational estimates of the supernova remnant (SNR) shock power require, *on the average*, a 15–30 % conversion efficiency to maintain the observed CR energy against losses from the Galaxy (see, e.g., Berezinskii et al. 1990; Drury et al. 1989). However, this acceleration mechanism is fast enough only if it is self-sustained; accelerated particles must be scattered across the shock at an enhanced rate (to gain energy rapidly) by magnetic irregularities amplified by the particles themselves. Relying on the background magnetic irregularities (interstellar medium [ISM] turbulence) would result only in a very slow acceleration.

Fortunately, freshly accelerated CRs indeed comprise enough free energy to drive plasma instabilities thus bootstrapping their own acceleration (see, e.g., Zweibel 1979). While they are accumulated in a relatively thin layer near a shock front, their pressure gradient is built up. Furthermore, they stream through the inflowing plasma so that their pitch-angle distribution is anisotropic. They also provide an electric current and induce a return current in the upstream plasma.

Instabilities driven by the above sources of free energy may loosely be categorized as follows. First, an ion-cyclotron type, resonant instability (driven by the CR anisotropy) amplifies Alfven and magnetosonic waves, with no major changes to their dispersive properties and the macroscopic state of the medium near the shock. However, the amplified waves make the CR pressure and current to build-up rapidly through an enhanced CR scattering and energy gain. Second, there is a non-resonant firehose type instability driven by the CR pressure anisotropy. In contrast to the resonant instability, the firehose instability changes the Alfven wave dispersive properties by making the growing mode aperiodic. So does the current driven non-resonant instability. The renewed interest to this instability has been sparked by Bell (2004), who revealed its potential to strongly amplify the background magnetic field. Indeed, a formal analytic solution in which the instability driver is balanced by the nonlinearity indicates that the instability saturates only at very high amplitudes, $\delta B \gg B_0$ (see, e.g., Bell and Lucek 2001; Bell 2005; Marcowith et al. 2006; Caprioli et al. 2008; Vladimirov et al. 2009; Malkov et al. 2012). Finally, the CR pressure gradient in the shock precursor drives acoustic perturbations. All these instabilities should be treated on a unified basis, as they are driven by the anisotropic inhomogeneous CR plasma component near a shock front. An attempt of such treatment is presented below. However a complete nonlinear study of these phenomena is a formidable task, yet to be accomplished.

While the above instabilities, clearly associated with collisionless shocks, will be central to the present review, CRs are also known to drive instabilities crucial to their confinement regardless of the way they are accelerated. For example, a sufficiently dense CR cloud released into the ISM will drive Alfven waves which, in turn, will scatter the CRs, thus delaying their escape (see, e.g., Ptuskin et al. 2008; Ohira et al. 2011; Malkov et al. 2013; Yan et al. 2012). Moving further out to the CR confinement in the galaxy, the so-called Parker instability is known to be important, in addition to the Alfven wave self-generation by escaping CRs.

The diffusive shock acceleration (DSA) mechanism is based on repeated shock crossings with a $\sim u_s/c$ particle energy gain per cycle (see Krymskii 1977; Bell 1978; Blandford and Eichler 1987; Berezhko and Krymskiĭ 1988; Jones and Ellison 1991). While doing so, particles diffusively escape from the shock up to a distance $L_p \sim \kappa(p)/u_s$. Here κ is the momentum dependent diffusion coefficient and u_s is the shock velocity. One should expect then an extended ($\sim L_p$) shock precursor populated by accelerated protons and electrons so that synchrotron radiating electrons may make it visible. High-resolution X-ray observations have revealed thin X-ray synchrotron filaments and fast evolving clumps in synchrotron emitting supernova shells. The filaments are much thinner than L_p because the TeV regime electrons are confined in a narrow layer around the shock. Most likely they are limited by fast synchrotron cooling due to the X-ray emission in a highly amplified magnetic field (see for review Cassam-Chenaï et al. 2007; Reynolds 2008; Vink 2012; Helder et al. 2012). The synchrotron emission clumps with a year time scale variability observed with *Chandra* observatory by Uchiyama et al. (2007) can be associated with strong intermittency of the amplified magnetic fields (Bykov et al. 2008). Moreover, a quasi-regular set of strips of synchrotron emission resolved with *Chandra* in Tycho's SNR by Eriksen et al. (2011) potentially can be used to study a specific angular dependence and the spectral properties of nonlinear mechanisms of magnetic field amplification by CR-driven instabilities (Bykov et al. 2011).

According to the widely accepted view, the particle diffusion coefficient κ should be close to the Bohm value, $\kappa \sim c r_g(p)/3$, which requires strong magnetic fluctuations $\delta B_k \sim B_0$ at the resonant scale $k \sim 1/r_g(p)$. The high level of fluctuations is achieved through one of the instabilities driven by accelerated particles. A number of CR driven instabilities have been suggested to generate magnetic field fluctuations. The first one is the well known ion cyclotron resonant instability of a slightly anisotropic (in pitch angle) CR distribution (see, e.g., Sagdeev and Shafranov 1961; Zweibel 1979; Schlickeiser 2002; Amato 2011). The free energy source of this instability is potentially sufficient to generate magnetic field fluctuations needed to scatter CRs ahead of the shock (see, e.g., Bell 1978; McKenzie and Voelk 1982).

$$(\delta B/B_0)^2 \sim M_A P^{cr}/\rho u_s^2, \tag{1}$$

where $M_A \gg 1$ is the Alfvenic Mach number, P^{cr} is the CR pressure, ρ is the gas density and u_s is the shock velocity. However, the actual turbulence level was shown to remain moderate, $\delta B \sim B_0$ as this is a resonant kinetic instability that is usually suppressed by a quasilinear isotropisation or particle trapping effects easily (see, e.g., McKenzie and Voelk 1982; Achterberg and Blandford 1986; Zweibel 2003).

The second instability, is a nonresonant instability driven by the CR current. The advantage of this instability seems to be twofold. First, it cannot be stabilized by the quasilinear deformation of the CR distribution function since in the upstream plasma frame the driving CR current persists, once the CR cloud is at rest in the shock frame. Second, it generates a broad spectrum of waves, and the longest ones were claimed to be stabilized only at the level $\delta B \gg B_0$, due to the lack of efficient stabilization mechanism at such scales (see, e.g., Bell 2004). Within the context of the CR acceleration, this instability was studied by Achterberg (1983) (see also Shapiro et al. 1998), but the fast regime of the nonresonant instability was found by Bell and Lucek (2001) and Bell (2004), and therefore the instability is often referred to as Bell's instability. Bell (2004) pointed out that in the instability is driven by a fixed CR return current through the Ampere force $\mathbf{j}_{cr} \times \mathbf{B}$. It should be noted, however, that the dissipation of the return current due to the anomalous resistivity still needs to be addressed. The effect of a finite plasma temperature on the instability was studied by Zweibel and Everett (2010). Actually, as we will show below, both the resonant and the Bell instabilities are interconnected, they are driven by the CR drift relative the background plasma. Moreover, in the case of the modes propagating along the mean magnetic fields the two instabilities are simultaneously influencing the same modes. The dispersion relations of the modes are strongly influenced by the presence of the CR current are markedly different from the standard MHD modes. The dispersion relations of the modes strongly influenced by the presence of the CR current are markedly different from the standard MHD modes. The dispersion relation in the longwavelength regime (where the mode wavelengths are larger than the bulk CR gyroradii) can be also strongly modified by the ponderomotive forces induced by Bell's turbulence. The longwavelength instability has two regimes (Bykov et al. 2011b, 2012). The first regime is prominent in the intermediate range where the mode wavelength is above the CR gyroradii but below the CR mean free path. It is discussed in Sect. 4.4 and is associated with a dynamo type instability driven by the nonzero helicity, which is, in turn, produced by the short scale CR-driven turbulence. The intermediate wavenumber range is rather narrow in the case of the Bohm-type CR diffusion. The modes with wavelengths larger than the CR mean free path are subject of non-resonant long-wavelength instability caused by the ponderomotive force acting on the background plasma that is induced by Bell's turbulence. We discuss the long wavelength instability below in Sect. 4.5.

The third instability is an acoustic instability (also known as Drury's instability) driven by the pressure gradient of accelerated CRs upstream (Dorfi and Drury 1985; Drury and Falle 1986; Drury and Downes 2012; Schure et al. 2012). The pressure gradient is clearly a viable source of free energy for the instability. So, among the macroscopic quantities varying across a strong shock, the pressure jump is the most pronounced one in that it does not saturate with the Mach number, unlike the density or velocity jumps.

The acoustic instability has received somewhat less attention than the first two. Moreover, in many numerical studies of the CR shock acceleration, special care is taken to suppress it. The suppression is achieved by using the fact that a change of stability occurs at that point in the flow where $\partial \ln \kappa / \partial \ln \rho \simeq -1$ (for both stable and unstable wave propagation directions, of course, if such point exists at all). Here ρ is the gas density. Namely, one requires this condition to hold identically all across the shock precursor, i.e., where the CR pressure gradient $\nabla P^{cr} \neq 0$. Not only is this requirement difficult to justify physically, but, more importantly, an *artificial* suppression of the instability eliminates its *genuine* macroscopic and microscopic consequences, as briefly discussed below.

Among the macroscopic consequences an important one is the vorticity generation through the baroclinic effect (misalignment of the density and pressure gradients $\nabla \rho \times \nabla P \neq 0$, e.g. Ryu et al. 1993; Kulsrud et al. 1997). Here ∇P may be associated with a quasi-constant macroscopic CR-gas pressure gradient ∇P^{cr}, generally directed along the shock normal. Variations of $\nabla \rho$ are locally decoupled from P^{cr}, unlike in the situation in a gas with a conventional equation of state where $P = P(\rho)$ and where the baroclinic term vanishes. The vorticity generation obviously results (just through the frozen in condition) in magnetic field generation, so that the field can be amplified by the CR pressure gradient. More importantly, this process amplifies the *large scale field*, required for acceleration of *high energy particles*. Furthermore, the amplification takes place well ahead of the gaseous subshock. The both requirements are crucial for improving high energy particle confinement and making the shock precursor shorter, in agreement with the observations. Large scales should be present in the ambient plasma as a seed for their amplification by the acoustic instability and could be driven (or seeded) by wave packet modulations. Apart from that, they result from the coalescence of shocks generated by the instability, and from the scattering of Alfven waves in k-space by these shocks to larger scales (Malkov and Diamond 2006, 2009; Diamond and Malkov 2007). Note that the Bell instability is essentially a short scale instability (the maximum growth rate is at scales smaller than the gyro-radii of accelerated particles). At larger scales the magnetic field growth rate is dominated by the modified resonant and the longwavelength nonresonant instabilities (Bykov et al. 2011b). It should be noted that vorticity (and thus magnetic field) can be efficiently generated also at the subshock (see, e.g., McKenzie and Westphal 1970; Bykov 1982, 1988; Kevlahan 1997; Kulsrud et al. 1997; Giacalone and Jokipii 2007; Beresnyak et al. 2009; Fraschetti 2013). This would be too late for improving particle confinement and reducing the scale of the shock precursor. A more favorable for acceleration scenario is the above discussed field amplification in the CR shock precursor.

Now the question is which instability dominates the CR dynamics? Given the finite precursor crossing time, it is reasonable to choose the fastest growing mode and consider the development of a slower one under conditions created by the fast mode after its saturation. The Bell instability is likely to be efficient at the outskirt of the shock precursor where the CR current is dominated by the escaping CRs of the highest energies. The pressure gradient and the pitch angle anisotropy are strong enough to drive the acoustic and resonant instability in the shock precursor (see, e.g., Pelletier et al. 2006). Recall that the anisotropy is typically inversely proportional to the local turbulence level which is usually decrease with the distance from the shock

Within the main part of the shock precursor, both the CR-pressure gradient and CR current are strong, so that the nonresonant CR-driven instabilities are likely to be the strongest candidates to govern the shock structure. In fact, these instabilities are coupled, not only by the common energy source but also dynamically. But first, it is important to identify conditions under which one of the instabilities dominates.

2 Cosmic Plasmas with Cosmic Rays: the Governing Equations

In this section we discuss the governing equations for MHD-type flows of a cold background plasma interacting with cosmic rays. In most cases the cosmic ray particles are not subject to binary Coulomb or nuclear interactions with the background plasma particles. The interaction between the two components is due to both regular and fluctuating electromagnetic fields produced by the CRs. The momentum equation for the background plasma, including the Lorentz force associated with these fields is given by

$$\widetilde{\rho}\left(\frac{\partial \widetilde{\mathbf{u}}}{\partial t} + (\widetilde{\mathbf{u}}\nabla)\widetilde{\mathbf{u}}\right) = -\nabla \widetilde{p}_g + \frac{1}{c}\widetilde{\mathbf{j}} \times \widetilde{\mathbf{B}} + e(\widetilde{n}_p - \widetilde{n}_e)\widetilde{\mathbf{E}}, \tag{2}$$

where $\widetilde{\mathbf{B}}$ is the magnetic field induction, $\widetilde{\mathbf{E}}$—the electric field, $\widetilde{\mathbf{u}}$—the bulk plasma velocity, $\widetilde{p}_g$—the plasma pressure, $\widetilde{\mathbf{j}}$—the electric current carried by the background plasma. We assume quasi-neutrality for the whole system consisting of background plasma protons of number density $\widetilde{n}_p$, electrons of number density $\widetilde{n}_e$, and cosmic rays of number density $\widetilde{n}_{cr}$. For simplicity we consider cosmic-ray protons only such that $\widetilde{n}_p + \widetilde{n}_{cr} = \widetilde{n}_e$, and typically $\widetilde{n}_{cr} \ll \widetilde{n}_p$.

The magnetic field is assumed to be frozen into the background plasma

$$\widetilde{\mathbf{E}} = -\frac{1}{c}[\widetilde{\mathbf{u}} \times \widetilde{\mathbf{B}}]. \tag{3}$$

Both the background electric current $\widetilde{\mathbf{j}}$ and the electric current of accelerated particles $\widetilde{\mathbf{j}}^{cr}$ are the sources of magnetic fields in Maxwell's equations, where the Faraday displacement current was omitted for the slow MHD-type processes

$$\nabla \times \widetilde{\mathbf{B}} = \frac{4\pi}{c}\left(\widetilde{\mathbf{j}} + \widetilde{\mathbf{j}}^{cr}\right). \tag{4}$$

Then, for the quasi-neutral background plasmas, using Eqs. (2), (3) and (4), one can write the induction equation and the equation of motion of the background plasma in the form used by Bell (2004), Bykov et al. (2011b) and Schure and Bell (2011)

$$\frac{\partial \widetilde{\mathbf{B}}}{\partial t} = \nabla \times (\widetilde{\mathbf{u}} \times \widetilde{\mathbf{B}}), \tag{5}$$

$$\widetilde{\rho}\left(\frac{\partial \widetilde{\mathbf{u}}}{\partial t} + (\widetilde{\mathbf{u}}\nabla)\widetilde{\mathbf{u}}\right) = -\nabla \widetilde{p}_g + \frac{1}{4\pi}(\nabla \times \widetilde{\mathbf{B}}) \times \widetilde{\mathbf{B}} - \frac{1}{c}\left(\widetilde{\mathbf{j}}^{cr} - e\widetilde{n}_{cr}\widetilde{\mathbf{u}}\right) \times \widetilde{\mathbf{B}}. \tag{6}$$

The microscopic CR-dynamics can be described by a kinetic equation for the single-particle distribution function $\widetilde{f}$ that has the form

$$\frac{\partial \widetilde{f}}{\partial t} + \mathbf{v} \cdot \frac{\partial \widetilde{f}}{\partial \mathbf{r}} + e\widetilde{\mathbf{E}} \cdot \frac{\partial \widetilde{f}}{\partial \mathbf{p}} - \frac{ec}{\mathcal{E}}\widetilde{\mathbf{B}} \cdot \widehat{\mathcal{O}}\widetilde{f} = 0, \tag{7}$$

where the CR particle energy is $\mathcal{E}$, $\widehat{\mathcal{O}}$ is the momentum rotation operator (see, e.g., Toptygin 1983; Bykov et al. 2012). There are no Coulomb collisions in the kinetic equation (7), but the microscopic electromagnetic fields are fluctuating in a wide dynamical range due to collective plasma effects. The coarse grained distribution function of the CR particles $f = \langle \widetilde{f} \rangle$ obeys the equation that can be obtained by averaging the microscopic equation Eq. (7) over an ensemble of appropriate short-scale fluctuations

$$\frac{\partial f}{\partial t} + \mathbf{v} \cdot \frac{\partial f}{\partial \mathbf{r}} + e\mathbf{E} \cdot \frac{\partial f}{\partial \mathbf{p}} - \frac{ec}{\mathcal{E}}\mathbf{B} \cdot \widehat{\mathcal{O}} f = I[f, f']. \tag{8}$$

Here $\widetilde{f} = f + f'$, $\widetilde{\mathbf{B}} = \mathbf{B} + \mathbf{b}'$, $\widetilde{\mathbf{E}} = \mathbf{E} + \mathbf{E}'$, $\mathbf{B} = \langle \widetilde{\mathbf{B}} \rangle$, $\mathbf{E} = \langle \widetilde{\mathbf{E}} \rangle$—are the averaged fields, and therefore $\langle \mathbf{b}' \rangle = 0$, $\langle \mathbf{E}' \rangle = 0$. The ensemble of fluctuations can be of external origin or produced by the same population of charged particles we only assumed at this point that the collision operator

$$I[f, f'] = -e\left\langle \mathbf{E}' \cdot \frac{\partial f'}{\partial \mathbf{p}} \right\rangle + \frac{ec}{\mathcal{E}}\langle \mathbf{b}' \cdot \widehat{\mathcal{O}} f' \rangle, \tag{9}$$

is a functional of the averaged distribution function f and can be expressed through the statistical momenta of the fluctuating field. The collision operator describes the momentum and energy exchange between CRs and the background plasma and therefore it must be accounted for in the averaged governing equations for both the CRs and background plasma.

The momentum exchange rate is the first moment of Eq. (9)

$$\int \mathbf{p} I[f] d^3 p = -e\langle n'_{cr}\mathbf{E}' \rangle + \frac{1}{c}\langle \mathbf{j}'_{cr} \times \mathbf{b}' \rangle, \tag{10}$$

where n'_{cr}, $\mathbf{j}'_{cr}$—are the fluctuating parts of the CR number density and the CRs electric current defined by

$$n'_{cr} = e\int f' d^3 p, \tag{11}$$

$$\mathbf{j}'_{cr} = e\int \mathbf{v}(p) f' d^3 p, \tag{12}$$

where $\mathbf{v}(p)$—is the CR particle velocity, and $\langle f' \rangle = 0$.

Then, by averaging the last term in Eq. (6), one can get

$$\frac{1}{c}\langle(\widetilde{\mathbf{j}}_{cr} - en_{\mathrm{cr}}\widetilde{\mathbf{u}}) \times \widetilde{\mathbf{B}}\rangle = \frac{1}{c}(\mathbf{j}^{cr} - en_{cr}\mathbf{u}) \times \mathbf{B} - e\langle n'_{cr}\mathbf{E}'\rangle + \frac{1}{c}\langle \mathbf{j}'_{\mathrm{cr}} \times \mathbf{b}'\rangle, \tag{13}$$

where n_{cr}, $\mathbf{j}_{\mathrm{cr}}$—are the averaged CR number density and their electric current, $\widetilde{\mathbf{j}}_{cr} = \mathbf{j}_{\mathrm{cr}} + \mathbf{j}'_{\mathrm{cr}}$, $\widetilde{n}_{cr} = n_{\mathrm{cr}} + n'_{\mathrm{cr}}$. Note that Eq. (10) and the last two terms on the right hand side of Eq. (13) are coincident. Therefore, we conclude that the CR scattering due to the stochastic electromagnetic fields accounted for in the kinetic equation Eq. (8) by the collision operator must be simultaneously included into the equation of motion of the background plasma using Eq. (13).

The averaged induction equation Eq. (5) can be expressed as

$$\frac{\partial \mathbf{B}}{\partial t} = \nabla \times (\mathbf{u} \times \mathbf{B}), \tag{14}$$

and the averaged equation of motion Eq. (6) for the background plasma

$$\rho\left(\frac{\partial \mathbf{u}}{\partial t} + (\mathbf{u}\nabla)\mathbf{u}\right) = -\nabla p_g + \frac{1}{4\pi}(\nabla \times \mathbf{B}) \times \mathbf{B} - \frac{1}{c}(\mathbf{j}^{cr} - en_{\mathrm{cr}}\mathbf{u}) \times \mathbf{B} - \int \mathbf{p}I[f]d^3p, \tag{15}$$

where p_g—is the averaged pressure of background plasma. Note that Eqs. (14) and (15) is also valid for CRs consisting of electrons and positrons, with n_{cr} being the difference between the positron and the electron number densities, while $\mathbf{j}^{cr}$—the total electric current of the particles.

In a few cases, namely, for weakly fluctuating magnetic fields or, for strong magnetic fluctuations but at scales smaller than the CR gyroradii, some closure procedures exist to reduce the collision operator $I[f, f']$ to $I[f]$ (see, e.g., Toptygin 1983; Bykov et al. 2012). It is instructive, nevertheless, to derive the force density $\int \mathbf{p}I[f]d^3p$ for the most simple case of $I[f]$. The simplest form of the collision operator is the relaxation time approximation in the rest frame of the background plasma

$$I[f] = -\nu(f - f_{\mathrm{iso}}), \tag{16}$$

where f_{iso}—is the isotropic part of the momentum distribution f, and ν is the collision frequency due to CR particle-wave interactions (e.g., Bykov et al. 2011b). This approach usually implies that the scatterers have no mean (or drift) velocity relative to the rest frame of the background plasma. This is not always true, if the plasma instabilities that are producing the magnetic field fluctuations are highly anisotropic. However, it can be used to illustrate the importance of the momentum exchange between CRs and the background plasma.

Using the parameterisation $\nu = a\Omega$, where $\Omega = \frac{ecB_0}{\mathcal{E}}$, $\mathbf{B}_0$ is the mean magnetic field, and a—is the CR collisionality parameter, from Eq. (16), one can obtain

$$\int \mathbf{p}I[f]d^3p = -\frac{aB_0}{c}\mathbf{j}_{\mathrm{cr}}. \tag{17}$$

This is the force density in Eq. (15).

3 Instabilities Driven by Anisotropic CR Distributions: the Kinetic Approach

Consider incompressible modes propagating along the mean homogeneous magnetic field $\mathbf{B}_0$ in the rest frame of the background plasma. The linear dispersion relation can be obtained by the standard perturbation analysis of Eqs. (14), (15) and (8), assuming the small perturbations of magnetic field $\mathbf{b}$, plasma bulk velocity $\mathbf{u}$ and the CR distribution f to be $\propto \exp(ikx - i\omega t)$. The unperturbed anisotropic CR distribution, that is the source of the instability free energy, can be represented as

$$f_0^{cr} = \frac{n_{cr}N(p)}{4\pi}\left[1 + 3\beta\mu + \frac{\chi}{2}(3\mu^2 - 1)\right], \tag{18}$$

where $\mu = \cos\theta$, θ—is the CR particle pitch-angle, $n_{\rm cr}$—CR number density. The multipole moments of the CR angular distribution are parameterized by β (the dipole) and χ (the quadrupole). We assume below $\beta \le 1$ and $\chi \le 1$. The unperturbed state can be a steady state of a system with CRs where both the anisotropy and the spectral distribution $N(p)$ are determined by the energy source and sink as well as the magnetic field geometry through the kinetic equation Eq. (8) with some appropriate boundary conditions. The most interesting application of the formalism is related to diffusive shock acceleration model (see, e.g., Blandford and Eichler 1987; Malkov and Drury 2001; Bykov et al. 2012; Schure et al. 2012). In that case the normalized power-law CR spectrum is appropriate:

$$N(p) = \frac{(\alpha - 3)p_0^{(\alpha-3)}}{[1 - (\frac{p_0}{p_m})^{\alpha-3}]p^\alpha}, \qquad p_0 \le p \le p_m, \tag{19}$$

where α—is the spectral index, p_0 and p_m—are the minimal and maximal CR momenta, respectively. In the DSA applications it is convenient to express the dipole anisotropy parameters through the shock velocity $u_{\rm s}$ as $\beta = \frac{u_{\rm s}}{c}$.

Then dispersion equation has the form:

$$\omega^2 = v_a^2\left\{k^2 \mp k\left[(1 \pm ia)\left(k_0 A_0(x_0, x_m) + \frac{4\pi e n_{cr}\chi}{B_0}A_1(x_0, x_m)\right) - k_0\right]\right\}, \tag{20}$$

where $v_a = \frac{B_0}{\sqrt{4\pi\rho}}$, $k_0 = \frac{4\pi}{c}\frac{j_0^{cr}}{B_0}$, $j_0^{cr} = en_{cr}u_{\rm s}$, $x = \frac{kcp}{eB_0}$, $x_0 = \frac{kcp_0}{eB_0}$, $x_m = \frac{kcp_m}{eB_0}$,

$$A_{0,1}(x_0, x_m) = \int_{p_0}^{p_m} \sigma_{0,1}(p)N(p)p^2 dp \tag{21}$$

$$\sigma_0(p) = \frac{3}{4}\int_{-1}^{1}\frac{(1-\mu^2)}{1 \mp x\mu \pm ia}d\mu, \tag{22}$$

$$\sigma_1(p) = \frac{3}{4}\int_{-1}^{1}\frac{(1-\mu^2)\mu}{1 \mp x\mu \pm ia}d\mu, \tag{23}$$

where the $\pm$ signs correspond to the two possible circular polarizations defined by $\mathbf{b} = b(\mathbf{e}_y \pm i\mathbf{e}_z)$, with the x-axis along the mean field $\mathbf{B}_0$. The functions $A_{0,1}(x_0, x_m)$ are expressed in elementary functions in the Appendix. In the collisionless limit $a \to 0$ the contribution of the pole to the imaginary part of Eq. (22) describes the well known resonant instability (e.g., Zweibel 1979; Amato 2011), while the real part (the principal part of the integral) is responsible for the instability discovered by Bell (2004) (see also Achterberg 1983).

The kinetic approach we used here to derive the dispersion equation allows us to unify the instabilities due to both the dipole and quadrupole-type CR anisotropy. The finite mean free path of the CRs is characterized by the collisionality parameter a. The approach used above allows one to study the instabilities driven by the CR anisotropy for arbitrary relations between the mode wavelength, the CR mean free path and the CR gyroradii. It is instructive to demonstrate the transition between the collisionless case (i.e. $a = 0$), where the CR mean free part is much larger than the mode wavelength, and the opposite case with the collisionality parameter $a \to 1$ (Bohm's diffusion limit). In the collisionless limit (i.e. $a = 0$) the instabilities due to dipole type anisotropy ($\chi = 0$) were discussed by Bell (2004), Pelletier et al. (2006), and Amato and Blasi (2009). The firehose instability of a highly relativistic plasma without a dipole anisotropy was discussed by Noerdlinger and Yui (1968). Schure and Bell (2011) derived a dispersion equation for the mono-energetic particle distribution instead of the power-law distribution in Eq. (19) used here, and the dipole-type initial anisotropy (i.e. χ =0). The firehose instability of the anisotropic CR pressure with nonzero χ was studied by Bykov et al. (2011a).

4 Growth Rates of Incompressible Modes Propagating Along the Mean Magnetic Field

In Fig. 1 we illustrate the growth rates derived from Eq. (20) for a particular choice of parameters of the CR distribution functions typical for the upstream distribution of CRs accelerated by the diffusive acceleration at a shock of velocity $\frac{u_s}{c} = 0.01$, with $\alpha = 4$, and $\frac{p_m}{p_0} = 100$. The DSA spectrum may span many decades, but we choose the two-decade range of the particle spectrum to model the instability far upstream of the shock where the longwavelength fluctuation amplification is the most efficient. The CR distribution function and the CR current normalizations are fixed here by the dimensionless parameter $k_0 r_{g0} = 100$, where $r_{g0} = \frac{cp_0}{eB_0}$.To estimate the normalization of the CR distribution we assumed that about 10 % of the shock ram pressure is converted into the CR energy. For the CR spectrum of the index $\alpha = 4$ the fraction of CRs above the momentum p_0 is $\propto p_m/p_0$, while $r_{g0} \propto p_0$. Therefore the spatial dependence of the key governing parameter of the Bell instability $k_0 r_{g0}$ depends basically on the energy dependent CR anisotropy. The bulk of the CRs are confined in the accelerator and therefore would have anisotropy about u_s/c (apart from the particles at the very end of the CR spectrum escaping from the system).

4.1 Nonresonant Shortwavelength Instability

It is instructive to consider the short-scale CR-current driven modes produced by Bell's instability as an asymptotic case of the general Eq. (20), for different wavenumbers k in the collisionless case $a = 0$, following Bell (2004) and Bykov et al. (2011a). In Fig. 1, we illustrate the growth rate dependence on the collisionality parameter.

In the wavenumber range $k_0 r_{g0} > k r_{g0} > 1$, corresponding to the instability discovered by Bell (2004), the growth of the right hand polarized mode (panel a in Fig. 1) is much faster than the left hand mode (panel b in Fig. 1). This results in fast helicity production. In the collisionless limit the right hand mode has the growth rate

$$\gamma_b = v_a \sqrt{k_0 k - k^2}. \tag{24}$$

Equation (24) follows from Eq. (20), neglecting the response of the CR current on the magnetic fluctuations, i.e., $A_0(x_0, x_m) \to 0$ and $A_1(x_0, x_m) \to 0$. The weak CR-current response

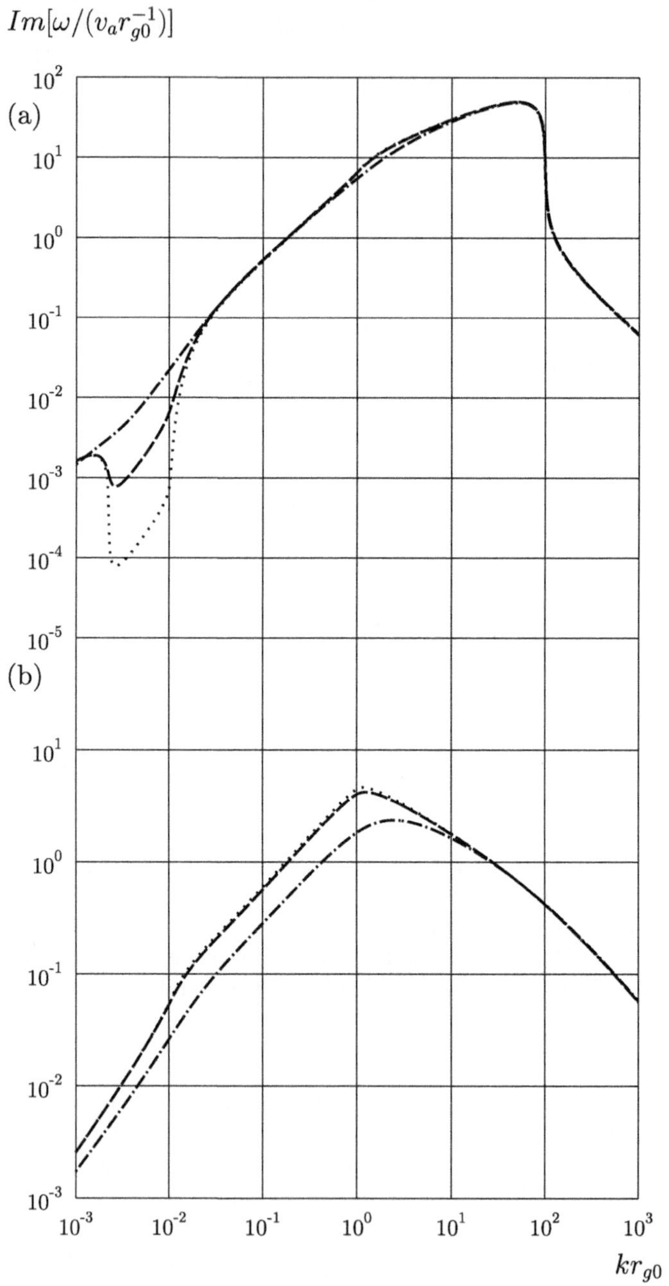

Fig. 1 The growth rates for the two circularly polarized modes. The right hand polarized mode (*panel a*) and the left hand mode (*panel b*) are derived from Eq. (20). We illustrate the growth rate dependence on the collisionality parameter a. *Dotted line* corresponds to $a = 0.01$, *dashed line*—$a = 0.1$, and *dot-dashed line*—$a = 1$. The quadrupole anisotropy is $\chi = 6(u_s/c)^2$. Note that in the *bottom panel* the *dashed* and *dotted lines* are very close

is the main cause of the Bell-type instability. Indeed, the CR current induces the compensatory reverse current in the background plasma and if the current is not responding to a magnetic field variation, then the magnetic fluctuation is growing due to the Ampere force. The CR current only weakly responds to the magnetic field fluctuations with wavenumbers $k_0 r_{g0} > k r_{g0} > 1$, and they grow. From Eq. (24) one may see that $\gamma_b \sim k^{\frac{1}{2}}$ for $k \ll k_0$.

4.2 The Resonant Instability

In the collisionless case for the wavenumber regimes $x_m > 1$, but $x_0 < 1$, the resonant contribution dominates the pole in the integrand in Eq. (22). Therefore, the resonant mode growth can be seen in Fig. 1 in the regime $0.01 < k r_{g0} < 1$, where both circular polarization modes are growing with the very close rates $\propto k$ for $\alpha = 4$ (compare panels a and b in Fig. 1). Collisions do not change the mode growth drastically for $a < 0.1$, but in the limit of strong collisions with $a = 1$ the left hand mode grows slower than the right hand polarized mode. This may also result in helicity production.

4.3 A Nonresonant Longwavelength Instability: the Firehose Mode

In the longwavelength regime where $x_m = \frac{k c p_m}{e B_0} \ll 1$, within the collisionless case, the dispersion relation in Eq. (20) can be approximated, following Bykov et al. (2011a), as

$$\omega^2 = v_a^2 k^2 \left\{ 1 \mp \frac{r_{g0}^2}{5} \left[k_0 x_m \pm \frac{4\pi e n_{cr} \chi}{B_0} \frac{\ln \frac{p_m}{p_0}}{(1 - \frac{p_0}{p_m})} \right] \right\}. \tag{25}$$

As it follows from Eq. (25), in the regime dominated by the dipole CR anisotropy ($\chi \to 0$) only the left-polarized mode is growing with the rate $\propto k^{\frac{3}{2}}$ (see Schure and Bell 2011). For a finite quadrupole-type CR anisotropy χ at small enough wavenumbers the modes of both circular polarizations are growing again with the very close rates $\propto k$ (see in Fig. 1). The instability due to the quadrupole-type CR anisotropy corresponds to the well known firehose instability in a plasma with anisotropic pressure. Indeed, the CR pressure anisotropy derived from the CR distribution Eq. (18) is

$$P_{\parallel}^{cr} - P_{\perp}^{cr} = \frac{3}{5} \chi P^{cr}, \tag{26}$$

where

$$P^{cr} = \frac{1}{3} n_{cr} \int_0^{\infty} v(p) N(p) p^3 dp. \tag{27}$$

The dispersion relation for the modes produced by only the quadrupole-type anisotropy of CR distribution can be obtained from Eq. (25) if one neglects the dipole-type contribution $x_m \to 0$. Then, it is reduced to the standard hydrodynamic dispersion relation of the firehose instability

$$\omega = \pm \sqrt{v_a^2 - \frac{P_{\parallel} - P_{\perp}}{\rho}} k, \tag{28}$$

where $P_{\parallel} - P_{\perp}$—is the pressure anisotropy along the mean magnetic field direction (see, e.g., Blandford and Eichler 1987; Treumann and Baumjohann 1997). The dispersion relation Eq. (28) is justified for the modes with the wavenumbers above the CR ion gyroradii. The dependence of the growth rates of the firehouse instability on the collisionality parameter can be seen in Fig. 1. It should be noted that the growth rates of the firehose modes of both polarizations in the regime $kr_{g0} < 1$ are declining functions of the collisionality parameter. Their growth rates would be equal in the case of lack of the CR current. Contrary, the growth rates of the current driven modes are different for the two polarizations. The growth rate of the right hand polarized CR-current driven mode is sensitive to the collisionality parameter (see Schure and Bell 2011).

4.4 A Nonresonant Long-Wavelength Instability: the Cosmic-Ray Current Driven Dynamo

Bell's instability results in the fast growth of short-scale modes with wavelengths shorter than the gyroradius of the cosmic-ray particles and in the presence of CR-current it may produce strong short-scale turbulence (e.g., Bell and Lucek 2001; Bell 2004; Zirakashvili and Ptuskin 2008; Zirakashvili et al. 2008; Reville et al. 2008; Vladimirov et al. 2009; Rogachevskii et al. 2012). Moreover, the shortscale turbulence is helical, and at the wavenumbers below $2k_0$ its kinetic energy density dominates over the magnetic energy density making a favorable condition for a pure α-dynamo effect (see Bykov et al. 2011b). The strong short-scale turbulence influences the background plasma dynamics on scales larger than the CR gyroradii. Bykov et al. (2011b) derived the mean field dynamic equations averaged over the ensemble of short-scale motions for plasma systems with CR-current. The averaged equation of motion can be presented as

$$\begin{aligned}\frac{\partial \mathbf{V}}{\partial t} + (\mathbf{V}\nabla)\mathbf{V} &= -\frac{1}{\rho}\nabla P_g - \langle(\mathbf{u}\nabla)\mathbf{u}\rangle + \frac{1}{4\pi\rho}\langle(\nabla \times \mathbf{b}) \times \mathbf{b}\rangle \\ &\quad + \frac{1}{4\pi\rho}\big((\nabla \times \mathbf{B}) \times \mathbf{B}\big) - \frac{1}{c\rho}\big(\big(\mathbf{j}^{\mathrm{cr}} - en_{\mathrm{cr}}\mathbf{V}\big) \times \mathbf{B}\big) \\ &\quad - \int \mathbf{p}I[f]d^3p, \end{aligned} \tag{29}$$

where $\mathbf{V}$ is the mean velocity of the plasma. The magnetic induction equation for the mean magnetic field $\mathbf{B}$ reads

$$\frac{\partial \mathbf{B}}{\partial t} = c\nabla \times \overline{\mathcal{E}} + \nabla \times (\mathbf{V} \times \mathbf{B}) + \nu_m \Delta \mathbf{B}. \tag{30}$$

Here $\overline{\mathcal{E}} = \langle \mathbf{u} \times \mathbf{b}\rangle$ is the average turbulent electromotive force and ν_m is the magnetic diffusivity. The averaged equations Eqs. (29) and (30) are designed to be applied to the dynamics of modes with scales larger than r_{g0}, i.e., CR particles are magnetized on these scales.

The ponderomotive forces $\langle(\mathbf{u}\nabla)\mathbf{u}\rangle$ and $\frac{1}{4\pi\rho}\langle(\nabla \times \mathbf{b}) \times \mathbf{b}\rangle$ in Eq. (29) describe the momentum exchange of the background plasma with the Bell mode turbulence. The averaged turbulent electromotive force $\overline{\mathcal{E}}$ results in the magnetic induction evolution. It is important that in the case under consideration the ponderomotive forces in Eq. (29) depend on the CR current through the Bell mode turbulence moments. To express the electromotive and ponderomotive forces through the CR current (Bykov et al. 2011b) followed the mean field closure procedure similar to the approach proposed by Blackman and Field (2002) in the dynamo theory (see for a review Brandenburg 2009a). The closure procedure is introduced by the parameter τ_{cor}. The correlation time τ_{cor} which is the relaxation time of triple correlations and is approximately equal to the turnover time of the Bell turbulence. The dependence of the electromotive force and the ponderomotive force on the CR current is determined by the kinetic coefficients α_t and κ_t, correspondingly. The kinetic coefficients are determined by the r.m.s. amplitude of Bell's turbulence $\langle b_B^2 \rangle$ and $\tau_{\rm cor}$. The short scale turbulence produced by the Bell mode instability is helical and therefore there is also a contribution to the electromotive force $\propto \alpha_t \overline{B}$ resulting in the α-dynamo effect. Then, the dispersion equation for the modes of wavelengths longer than r_{g0} in a plasma with anisotropic relativistic CRs can be derived from Eqs. (29) and (30) by the standard linear perturbation analysis:

$$\omega^2 - k^2 v_a^2 \mp \omega i k \frac{\alpha_t}{4\pi\rho}\left[\frac{1}{2}\left(k_0 A_0(x_0, x_m) + \frac{4\pi e n_{cr} \chi}{B_0} A_1(x_0, x_m)\right) + \frac{3}{2} k_0\right]$$
$$\pm k v_a^2 \left(1 + \frac{\kappa_t}{B_0}\right)\left[\left(k_0 A_0(x_0, x_m) + \frac{4\pi e n_{cr} \chi}{B_0} A_1(x_0, x_m)\right) - k_0\right]$$
$$+ i a k v_a^2 \left(k_0 A_0(x_0, x_m) + \frac{4\pi e n_{cr} \chi}{B_0} A_1(x_0, x_m)\right) = 0. \qquad (31)$$

The dispersion relation Eq. (31) was derived for the systems where the unperturbed CR-current is directed along the unperturbed magnetic field, and the short scale turbulence consists of Bell's modes. It is convenient to introduce two dimensionless parameters $N_B = \frac{\sqrt{\langle b_B^2 \rangle}}{B_0}$—Bell's turbulence r.m.s. amplitude, and the dimensionless mixing length ξ, instead of the correlation time $\tau_{\rm cor}$. The mixing length is defined here as $2\pi\xi/k_0 = \tau_{cor}\sqrt{\langle v^2 \rangle} \approx \tau_{cor}\sqrt{\xi \langle b_B^2 \rangle/(4\pi\rho)}$. Then $\alpha_t \approx \langle b_B^2 \rangle \tau_{\rm cor} \approx 8\pi^2 \sqrt{\xi} N_B v_a k_0^{-1} \rho$ and $\kappa_t = \pi N_B B_0$.

In Fig. 2 we illustrate the long wavelength mode growth derived from Eq. (31) for $\xi = 3$. The corresponding mixing length is close to the scale of the maximal growth rate of the short scale Bell's instability. The α-dynamo effect dominates the growth rate of a polarized mode shown in Fig. 2 (panel b) in the intermediate wavenumber regime $a < k r_{g0} < 1$. One should have in mind that in the case of Bohm's CR diffusion $a \sim 1$ and therefore the intermediate wavenumber regime is rather limited. It should be noted that the helicity of the unstable, long-wavelength mode studied above is opposite to that of the short-scale Bell mode. This provides, at least in principle, the possibility of balancing the global helicity of the system by combining short and long-wavelength modes. Care must be taken however, since numerical models indicate a high saturation amplitude of the Bell mode making a nonlinear analysis necessary to address the helicity balance issue. We will discuss some nonlinear simulations below in Sect. 5.

4.5 The Cosmic-Ray Current Driven Instability in the Hydrodynamic Regime

The nonresonant modes in a hydrodynamic regime, where the wavelength is longer than the mean free path, i.e., $k r_{g0} < a$, are unstable, as it follows from Eq. (31) (see, for details, Bykov et al. 2011b). Both circular polarizations in panels a and b in Fig. 2 grow with the same rate given by

$$\gamma \approx \sqrt{\frac{\pi N_B}{2}} \sqrt{k k_0 a} v_a. \qquad (32)$$

The transition from the intermediate wavenumber regime $a < k r_{g0} < 1$, dominated by the dynamo effect discussed in Sect. 4.4, where the mode growth rate can be approximated by

$$\gamma \approx 4\pi \sqrt{\xi} N_B v_a k, \qquad (33)$$

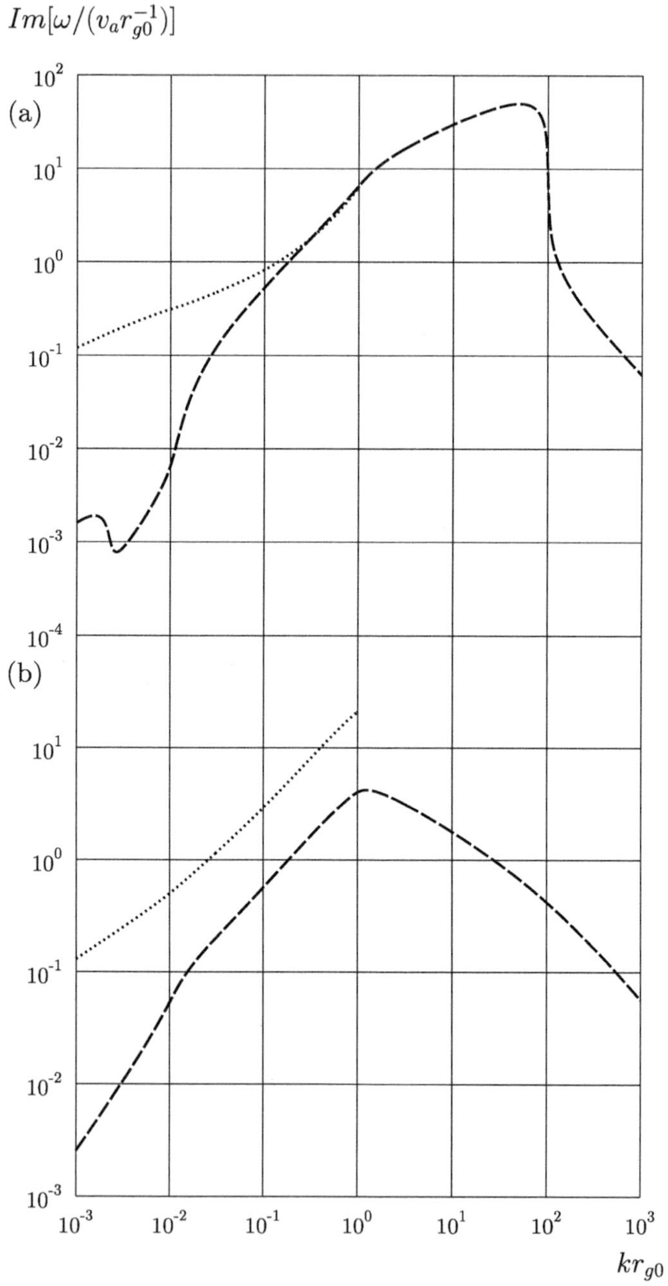

Fig. 2 The growth rates of the longwavelength modes of two circular polarizations. The right hand polarized mode (*panel a*) and the left hand mode (*panel b*) are propagating along the mean magnetic field as function of the wavenumber. The *dotted line curves* are derived from the dispersion equation Eq. (31) for the collisionality parameter $a = 0.1$, the dimensionless *r.m.s.* amplitude of Bell's turbulence $N_B = 1$, and the mixing parameter $\xi = 3$. The *dashed curves* given for comparison are the growth rates derived from Eq. (20) which are shown in Fig. 1

to the hydrodynamical regime with $kr_{g0} < a$ where the growth rate is $\propto k^{1/2}$ according to Eq. (32), is clearly seen in Fig. 2 (panel b). Note that for the mode polarization shown with the dotted line in the panel a of Fig. 2, no dynamo-type instability occurs, but the hydrodynamical regime instability is present. This mode grow fast in the short wavelength regime $kr_{g0} > 1$ due to Bell's instability.

The effect of the short-scale turbulence on the hydrodynamic regime instability enters Eq. (31) through the turbulent coefficient κ_t/B_0. The turbulent ponderomotive force is large enough in both the intermediate and hydrodynamical regimes, and the CR current response in the long-wavelength regime can no longer be neglected. The current cannot be treated as a fixed external parameter, as is normally done for the short-scale Bell instability, and therefore the MHD models of the Bell turbulence that assume a constant CR-current (see, e.g., Bell and Lucek 2001; Zirakashvili and Ptuskin 2008; Zirakashvili et al. 2008; Reville et al. 2008; Vladimirov et al. 2009; Rogachevskii et al. 2012) cannot be directly applied to the nonlinear models of the longwavelength instabilities discussed above. Particle-in-cell simulations with very limited dynamical range performed by Riquelme and Spitkovsky (2009, 2010) indicate the importance of the CR backreaction effect on the CR-driven instabilities. Therefore the nonlinear dynamics of the long-wave CR-driven turbulence in a wide dynamical range remains to be investigated. In the next section we illustrate the nonlinear evolution of the short scale turbulence driven by a fixed CR current, using high resolution MHD simulations.

5 Numerical Solutions of the Bell–Dynamo Instability

Significant insights have been possible through high-resolution direct numerical simulations (DNS) and large eddy simulations (LES) of the Bell instability and its subsequent saturation. In this section we describe some of the main results and, in particular, the connection with the dynamo instability. The simulations have been carried out in a Cartesian domain of size L^3, so the smallest wavenumber in that domain is $k_1 = 2\pi/L$. The system is characterized by the non-dimensional parameter

$$\mathcal{J} = \frac{4\pi}{c} \frac{j^{\rm cr}}{k_1 B_0}. \tag{34}$$

In the ideal case ($\nu_M = 0$), the Bell instability is excited when $\mathcal{J} > 1$ and the normalized wavenumber of the fastest growing mode is $k/k_1 = \mathcal{J}/2$. The normalized growth rate of this fastest growing mode is $\gamma_b/v_{A0}k_1 = \mathcal{J}/2$. In Fig. 3 we reproduce the results of numerical simulations of Bell (2004) for $\mathcal{J} = 2$ using 128^3 mesh points and Zirakashvili et al. (2008) for $\mathcal{J} = 16$ using 256^3 mesh points. These simulations confirmed the analytically expected linear growth rates. Interestingly, the saturation of the instability was never perfect. Instead, the magnetic field still continued to grow at a slow rate. Rogachevskii et al. (2012) have argued that this slow growth after the end of the exponential growth phase of the instability is the result of a mean-field α effect. The purpose of this section is to elaborate on this possibility.

We begin by discussing first the recent DNS of Rogachevskii et al. (2012) for $\mathcal{J} = 80$ and $\mathcal{J} = 800$ at a resolution of 512^3 mesh points and discuss also new results for $\mathcal{J} = 800$ at a resolution of 1024^3 mesh points. In all cases, explicit viscosity ν and magnetic diffusivity $\nu_{\rm M}$ are used, so the fastest growing modes in those cases have somewhat smaller wavenumbers than in the ideal case. This is quantified by the Lundquist number $\mathrm{Lu} = v_a/\nu_{\rm M}k_1$ and the ideal case corresponds then to $\mathrm{Lu} \to \infty$. For example, Rogachevskii et al. (2012) used

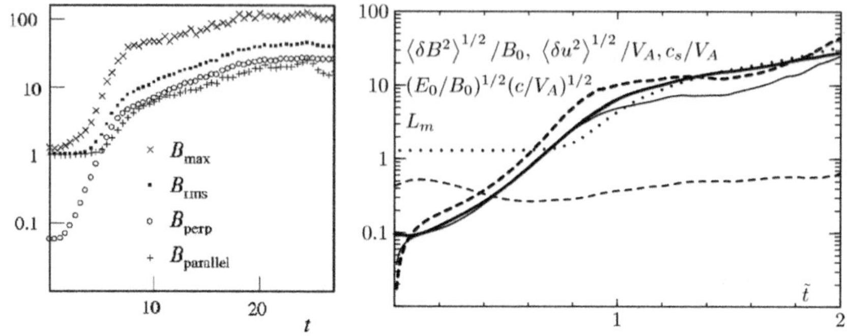

Fig. 3 Numerical solutions of the Bell instability for $\mathcal{J} = 2$ using 128^3 mesh points (left hand side, Bell 2004) and $\mathcal{J} = 16$ using 256^3 mesh points (right hand side, Zirakashvili et al. 2008). Note the continued growth of the magnetic field at the end of the linear growth phase at $t \approx 10$ on the *left* and $t \approx 1$ in the *right*. Courtesy of Tony Bell (*left panel*) and Vladimir Zirakashvili (*right panel*)

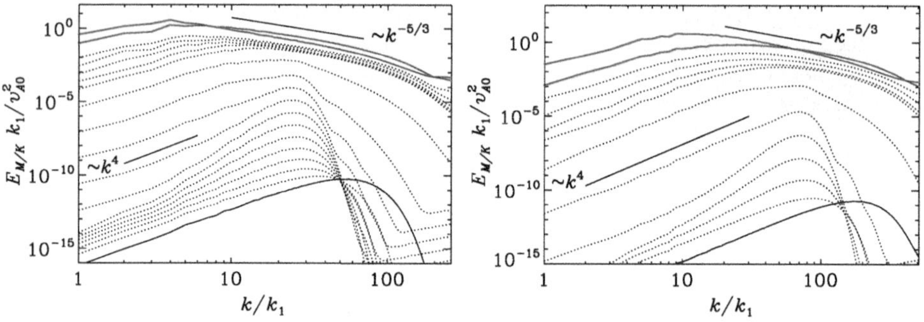

Fig. 4 Time evolution of $E_M(k, t)$ for $\mathcal{J} = 80$ (*left*) and $\mathcal{J} = 800$ (*right*) at resolutions 512^3 and 1024^3, respectively. The *solid lines* refer to the initial spectra proportional to k^4 for small values of k and the *red* and *blue lines* represent the last instant of E_M and E_K, respectively. The *straight lines* show the k^4 and $k^{-5/3}$ power laws

$\text{Lu} = 80$, in which case the fastest growing mode has $k_z/k_1 \approx 21$ for $\mathcal{J} = 80$ while for $\mathcal{J} = 800$ it has $k_z/k_1 \approx 63$. The DNS show that most of the power is at somewhat larger wavenumbers; see Fig. 4, where we show magnetic energy spectra for both cases.

In Fig. 5 we show the temporal evolution of spectral magnetic energy E_M and the spectral kinetic energy E_K at selected wavenumbers. These curves show an exponential growth at early times, followed by a slower growth at later times. At the wavenumbers of the Bell mode, the growth rate from linear theory is reproduced. At smaller wavenumbers, the growth is at first slower, and then it is even faster than the growth rate of the Bell mode. This is a consequence of mode coupling (Rogachevskii et al. 2012). Comparing with Fig. 4, we see that after some time a k^4 energy spectrum is established. Such an energy spectrum is also known as Batchelor spectrum and can be derived under the constraints of solenoidality and causality (Durrer and Caprini 2003). When the k^4 spectrum is established, the growth of spectral energy at small wavenumbers is no longer described by linear theory, but follows the growth of the Bell mode.

In Fig. 6 we show visualizations of B_x/B_0 on the periphery of the computational domain for $\mathcal{J} = 80$ using 512^3 mesh points and $\mathcal{J} = 800$ using 1024^3 mesh points at two different times. One clearly sees that at early times, the magnetic field shows a layered structure

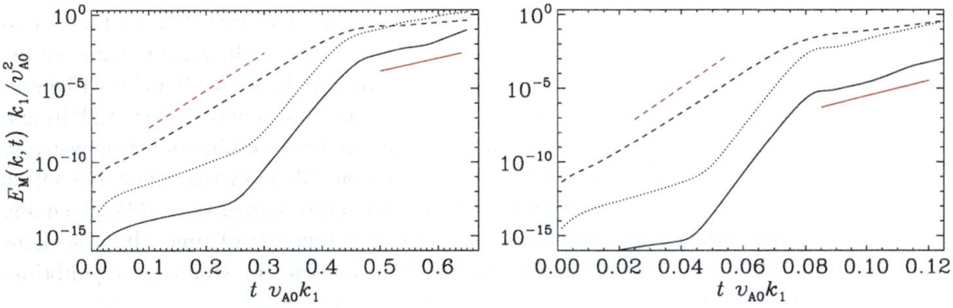

Fig. 5 Time evolution of $E_M k_1/v_{A0}^2$ for $\mathcal{J} = 80$ (*left*) at wavenumbers $k/k_1 = 1$ (*solid line*), 5 (*dotted*), and 21 (*dashed*) and $\mathcal{J} = 800$ (*right*) at wavenumbers $k/k_1 = 1$ (*solid line*), 10 (*dotted*), and 63 (*dashed*). The *short straight lines* show the growth of the energies for the Bell (*dashed*) and dynamo (*solid*) instabilities

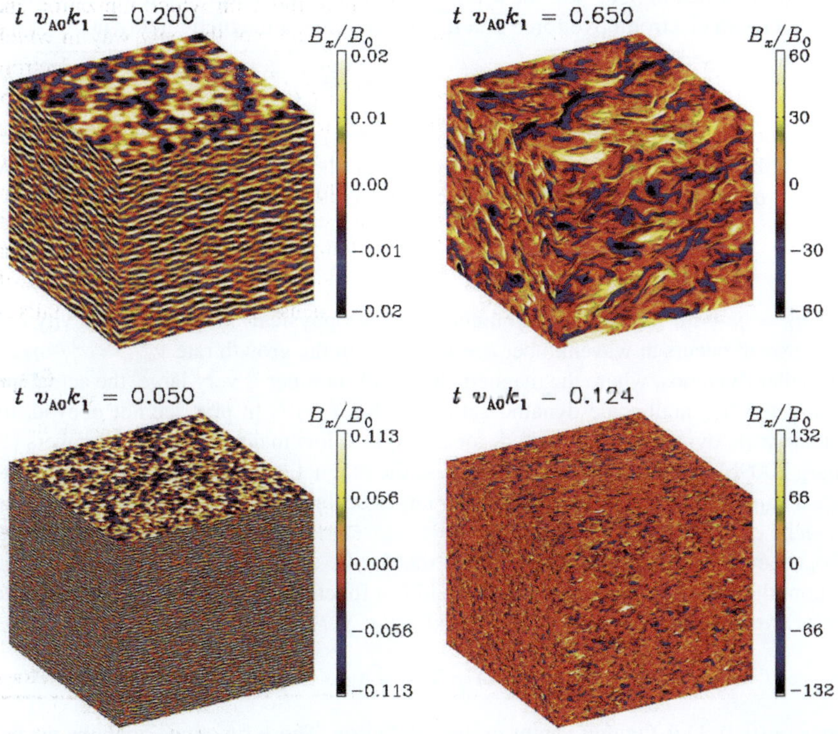

Fig. 6 Visualization of B_x/B_0 on the periphery of the computational domain $\mathcal{J} = 80$ using 512^3 mesh points (*upper row*) and $\mathcal{J} = 800$ using 1024^3 mesh points (*lower row*) with Lundquist number $\mathrm{Lu} = 80$ in both cases

with a high wavenumber in the z direction. At later times, the magnetic field breaks up and becomes turbulent. In both cases, larger scale structures develop, as one also sees from the energy spectra in Fig. 4.

It should be pointed out that, owing to the persistent growth of magnetic and kinetic energy, the Reynolds numbers grow eventually beyond the limit of what can be resolved at a given resolution. Unlike some of the earlier LES, where numerical effective viscosity and diffusivity keep the small scales resolved, in the DNS of Rogachevskii et al. (2012) this is not the case and the numerical code (in this case the PENCIL CODE[1]) eventually 'crashes'. The point when this happens can be delayed by using higher resolution. This is why we show here the results for $\mathcal{J}=800$ at a resolution of 1024^3 mesh points, where the simulation can be carried out for about 0.126 Alfvén times, compared to only 0.09 Alfvén times at a resolution of 512^3 mesh points used in Rogachevskii et al. (2012). Remeshing the 1024^3 run to 2048^3 mesh points, we were able to continue until 0.142 Alfvén times, after which we were unable to continue the run due to a disk problem.

The Bell instability is driven by the simultaneous presence of an external magnetic field $\mathbf{B}_0$ and an external current $\mathbf{j}_{\rm cr}$, giving therefore rise to a pseudo-scalar $\mathbf{j}_{\rm cr}\cdot\mathbf{B}_0$; here, $\mathbf{B}_0$ is an axial vector while $\mathbf{j}_{\rm cr}$ is a polar vector. In stellar magnetism, the presence of a pseudo-scalar is caused by rotation $\varOmega$ (an axial vector) and gravity $\mathbf{g}$ (a polar vector). This property is generally held responsible for the production of magnetic fields by what is known as the α effect. As explained in Sect. 4.4, the α effect denotes the presence of a tensorial connection between a mean electromotive force $\overline{\mathcal{E}}=\overline{\mathbf{u}\times\mathbf{b}}$ and a mean magnetic field via

$$\overline{\mathcal{E}}_i=\alpha_{ij}\overline{B}_j+\eta_{ijk}\overline{B}_{j,k}+\cdots, \tag{35}$$

where higher order derivatives (indicated by commas) of the mean magnetic field are also present. If the tensors α_{ij} and η_{ijk} were isotropic and the evolution characterized by just two quantities, $\alpha=\delta_{ij}\alpha_{ij}/3$ and $\eta_t=\epsilon_{ijk}\eta_{ijk}/6$, the growth of the mean magnetic field would occur at the rate

$$\gamma_{\rm dynamo}=\alpha k-\eta_T k^2, \tag{36}$$

where $\eta_T=\eta_t+\nu_M$ is the total (turbulent plus microphysical) magnetic diffusivity and the fastest growth occurs at wavenumber $k=\alpha/2\eta_T$ with the growth rate $\gamma_{\max}=\alpha^2/4\eta_T$.

In stellar dynamos, where the magnetic Reynolds number is very large, the actual growth is dominated by small-scale dynamo action, so Eq. (36) is in practice not obeyed, unless the small-scale dynamo is not excited, for example at low magnetic Prandtl numbers (Brandenburg 2009b). However, in the present case the magnetic energy spectra show that at late times, magnetic power moves gradually to larger scales. This is why we now ask whether this can be explained by the α effect.

Rogachevskii et al. (2012) have shown that in the case of $\mathbf{j}_{\rm cr}$ and $\mathbf{B}_0$ pointing in the z direction, the large-scale mean magnetic field is a function of x and y and can be written in terms of two scalar functions $\overline{A}_\parallel(x,y,t)$ and $\overline{B}_\parallel(x,y,t)$ with

$$\overline{\mathbf{B}}(x,y,t)=\nabla\times(\hat{\mathbf{z}}\overline{A}_\parallel)+\hat{\mathbf{z}}\overline{B}_\parallel, \tag{37}$$

where $\hat{\mathbf{z}}=(0,0,1)$ is the unit vector in the z direction. These functions obey the mean field equations

$$\partial\overline{A}_\parallel/\partial t=\alpha_A\overline{B}_\parallel+\eta_A\nabla^2\overline{A}_\parallel, \tag{38}$$

$$\partial\overline{B}_\parallel/\partial t=\alpha_B\overline{J}_\parallel+\eta_B\nabla^2\overline{B}_\parallel, \tag{39}$$

where $\overline{J}_\parallel=-\nabla^2\overline{A}_\parallel$ is the xy dependent part of the mean current density in the z direction. We consider a homogeneous system, so the coefficients α_A, α_B, η_A, and η_B are constant and

[1] http://www.pencil-code.googlecode.com.

we can seek solutions of a form proportional to $\exp(\lambda t + \mathbf{ik} \cdot \mathbf{x})$. In this case, the dynamo growth rate is still described by Eq. (36), provided we substitute

$$\alpha \to \alpha^{\text{eff}} = \left(\alpha_A \alpha_B + \epsilon_\eta^2 k^2\right)^{1/2} \quad \text{and} \quad \eta_t \to \eta_t^{\text{eff}} = (\eta_A + \eta_B)/2, \tag{40}$$

where $\epsilon_\eta = (\eta_A - \eta_B)/2$ quantifies the anisotropy of the turbulent diffusivity.

To determine these coefficients from the DNS, we use the so-called test-field method of Schrinner et al. (2005), which was originally used in spherical coordinates. The implementation in Cartesian coordinates is described in Brandenburg (2005) and especially in Brandenburg et al. (2012), where the mean magnetic field was allowed to depend on all three spatial coordinates, and not just on one, as was assumed in Brandenburg (2005). Under the assumption that the turbulence is governed by only one preferred direction, which is here the case, the number of coefficients reduces to 9, and homogeneity reduces this number further to 5, so in the present case we have

$$\overline{\mathcal{E}} = \alpha_\perp \overline{\mathbf{B}}_\perp + \alpha_\parallel \overline{\mathbf{B}}_\parallel - \beta_\perp \overline{\mathbf{J}}_\perp + \beta_\parallel \overline{\mathbf{J}}_\parallel - \mu \hat{\mathbf{z}} \times \overline{\mathbf{K}}_\perp, \tag{41}$$

where $\overline{\mathbf{J}} = \nabla \times \overline{\mathbf{B}}$ characterizes the antisymmetric part of the magnetic derivative tensor and $\overline{K}_i = (\overline{B}_{i,j} + \overline{B}_{j,i})\hat{z}_j/2$ the symmetric part. We have followed here the notation of Brandenburg et al. (2012), except that there the two α coefficients were defined with the opposite sign. Comparing with the coefficients used in Eqs. (38) and (39), we find that $\alpha_A = \alpha_\parallel$, $\alpha_B = \alpha_\perp$, $\eta_A = \beta_\parallel$, and $\eta_B = \beta_\perp - \mu/2$. In Fig. 7 we show the time dependence of the various parameter combinations. In the early kinematic phase ($t v_{A0} k_1 < 0.08$), the root mean square velocity, u_{rms}, as well as $\alpha_\parallel$ and $\alpha_\perp$ grow exponentially. At later times, $\alpha_\parallel$ continues to grow, while $\alpha_\perp$ remains small and approximately constant. The other turbulent transport coefficients also grow exponentially in the kinematic phase, and at later times $\beta_\parallel$, η_t, and ϵ_η continue to grow, while $\beta_\perp$ and μ remain small and can even become negative. The resulting effective dynamo number, which is proportional to the product $\alpha_\parallel \alpha_\perp$, reaches values well above the critical value of unity. The estimated and actual growth rates agree roughly and have a value of around 10 in units of $\eta_t k_1$.

6 Instabilities Driven by the Nearly Isotropic CR Distributions

In many astrophysical objects the CR mean free path due to the particle scattering by magnetic fluctuations carried by the background plasmas is below the characteristic scale sizes of the plasma flow. In that case the angular distribution of the CRs is nearly isotropic with a small anisotropic part (i.e. both $\beta \ll 1$ and $\chi \ll 1$ in Eq. (18)). Then one can use the diffusion approximation that assumes

$$f^{cr}(\mathbf{r}, \mathbf{p}) = \frac{1}{4\pi}\left[N^{cr}(\mathbf{r}, p) + \frac{3}{vp}\mathbf{p}\mathbf{J}^{cr}(\mathbf{r}, p)\right], \tag{42}$$

where the diffusive current of CRs is

$$J_\alpha^{cr} = -\kappa_{\alpha\beta}\nabla_\beta N^{cr} - \frac{p}{3}\frac{\partial N^{cr}}{\partial p}u_\alpha, \tag{43}$$

$\kappa_{\alpha\beta}$ is the momentum-dependent CR diffusion tensor. Then the kinetic equation Eq. (8) reduces to the advection-diffusion equation for the isotropic part of CR distribution

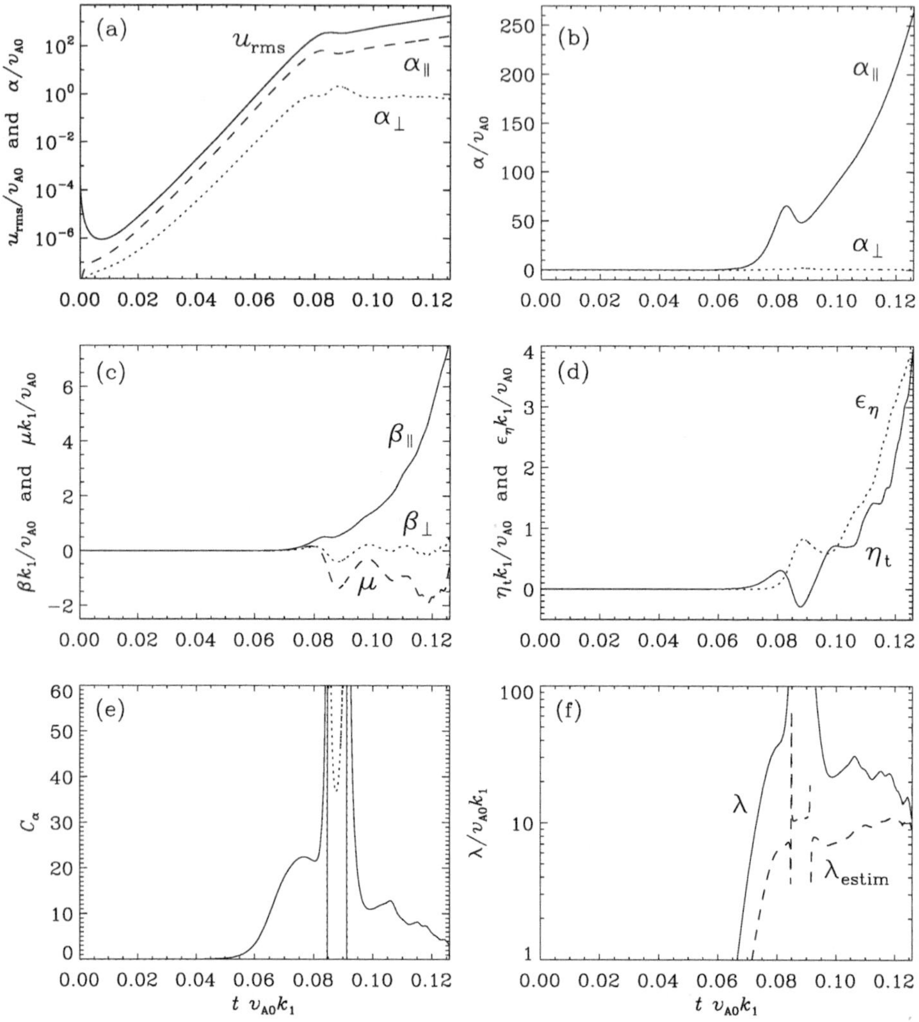

Fig. 7 Time evolution of the model parameters for $\mathcal{J} = 800$ and $\mathrm{Lu} = 80$ using 1024^3 mesh points. (**a**) Exponential growth and subsequent near-saturation of $u_{\rm rms}$, $\alpha_\parallel$, and $\alpha_\perp$ (all normalized by $v_{\rm A0}$) in linear-logarithmic representation. (**b**) Evolution of $\alpha_\parallel$ and $\alpha_\perp$ (normalized by $v_{\rm A0}$) in double linear representation, showing that $\alpha_\perp$ is much smaller than $\alpha_\parallel$. (**c**) Evolution of $\beta_\parallel$, $\beta_\perp$, and μ (normalized by $v_{\rm A0}/k_1$). (**d**) Evolution of $\eta_{\rm t}$ and ϵ_η (normalized by $v_{\rm A0}/k_1$). (**e**) Evolution of C_α (negative values are shown as *dotted lines*), and (**f**) growth of the fasted growing mode

$N^{cr}(\mathbf{r}, p, t)$

$$\frac{\partial N^{cr}}{\partial t} = \nabla_\alpha \kappa_{\alpha\beta} \nabla_\beta N^{cr} - (\mathbf{u}\nabla) N^{cr} + \frac{p}{3} \frac{\partial N^{cr}}{\partial p} \nabla \mathbf{u}, \tag{44}$$

where $\mathbf{u}(\mathbf{r}, t)$ is the bulk velocity of the background plasma (see, e.g., Toptygin 1983). It is assumed here for simplicity that the scatterers are carried with the plasma bulk velocity, though it is possible to account for the scatterers drift velocity (see, e.g., Skilling 1975). The advantage of this approach is that it is valid for collision operators $I[f]$ more general than just the simple relaxation time approximation given by Eq. (16). In the diffusion approximation the exact form of the collision operator determines the form of the diffusion tensor and its momentum dependence. Therefore, the results obtained within the diffusion approximation are valid for different collision operators.

To explore the effect of CRs on the background plasma one should calculate the first moment of the kinetic equation Eq. (8) for CRs that is the momentum exchange rate between the CRs and the background plasma:

$$\frac{\partial \mathcal{P}_\alpha}{\partial t} + \nabla_\alpha P^{cr} + \nabla_\beta \Pi'_{\alpha\beta} = \left[\frac{1}{c}\left(\mathbf{j}^{cr} - en_{\mathrm{cr}}\mathbf{u}\right) \times \mathbf{B} + \int \mathbf{p} I[f] d^3p\right]_\alpha, \tag{45}$$

where P^{cr} is the CR pressure, the CR momentum density

$$\mathcal{P}(\mathbf{r}, t) = \int \mathbf{p} f d^3p, \tag{46}$$

and the reduced CR momentum flux density $\Pi'_{\alpha\beta}$ is defined by

$$\Pi'_{\alpha\beta} = \int p_\alpha v_\beta f d^3p - P^{cr}\delta_{\alpha\beta}. \tag{47}$$

In the diffusion approximation for the steady state (e.g., in the shock rest frame) the first and the third terms in the left hand side of Eq. (45) are small and then Eq. (15) can be reduced to

$$\rho\left(\frac{\partial \mathbf{u}}{\partial t} + (\mathbf{u}\nabla)\mathbf{u}\right) = -\nabla\left(p_g + P^{cr}\right) + \frac{1}{4\pi}(\nabla \times \mathbf{B}) \times \mathbf{B}. \tag{48}$$

The equation can be applied to longwavelength perturbations. It should be supplied with the continuity equation:

$$\frac{\partial \rho}{\partial t} + \nabla(\rho \mathbf{u}) = 0, \tag{49}$$

the energy equations for the background plasma:

$$\frac{\partial p_g}{\partial t} + (\mathbf{u}\nabla)p_g + \gamma_g p_g \nabla \mathrm{u} = 0, \tag{50}$$

the MHD induction equation

$$\frac{\partial \mathbf{B}}{\partial t} = \nabla \times (\mathbf{u} \times \mathbf{B}), \quad \nabla \mathbf{B} = 0, \tag{51}$$

and the equation for CR-pressure variations

$$\frac{\partial P^{cr}}{\partial t} + (\mathbf{u}\nabla)P^{cr} + \gamma_{cr} P^{cr} \nabla \mathrm{u} = \nabla_\alpha \overline{\kappa}_{\alpha\beta} \nabla_\beta P^{cr}, \tag{52}$$

where $\overline{\kappa}_{\alpha\beta}$ is the CR diffusion tensor averaged over the CR distribution function, γ_g and γ_{cr}—are the adiabatic indexes of the plasma and CRs, respectively.

7 Acoustic Instability Driven by the CR Pressure Gradient

It was found by Drury (1984), Dorfi and Drury (1985), Drury and Falle (1986) and Drury and Downes (2012) that the force density in Eq. (48) associated with the CR pressure gradient that does not depend on the density of the background plasma results in a specific instability. The effect of magnetic field on the instability was studied by Berezhko (1986) and Chalov (1988b). The analytical study of the instability can be performed for the modes with the wavenumbers below the scale size of the CR pressure gradient $L \sim P^{cr}/|\nabla P^{cr}|$. In the generic case of the diffusive shock acceleration $L \sim (c/u_s) \times r_g/a$. Following Drury and Falle (1986) and Chalov (1988b) for the wavenumber range $kL > 1$, but $kr_g/a < 1$ the mode growth and damping can be derived from the continuity equation for the wave action.

The mode growth rate Γ in the simplified geometry where the CR pressure gradient is directed along the unperturbed magnetic field was derived using a standard linear analysis of Eqs. (48)–(52) by Chalov (1988a), who obtained the following expression

$$\Gamma = \frac{v_m^2 - v_a^2}{2v_m^2 - (v_s^2 + v_a^2)} \left\{ -\frac{\gamma_{cr} P_0^{cr}}{\rho_0} \frac{k^2}{\kappa_{0\parallel} k_\parallel^2 + \kappa_{0\perp} k_\perp^2} \frac{v_m^2 - v_a^2 \frac{k_\parallel^2}{k^2}}{v_m^2 - v_a^2} \right.$$
$$\left. \pm \frac{\nabla P_0^{cr}}{\rho_0 v_m} \frac{k_\parallel}{k} \left[1 + \frac{\varsigma \kappa_{0\parallel} k^2}{\kappa_{0\parallel} k_\parallel^2 + \kappa_{0\perp} k_\perp^2} \frac{v_m^2 - v_a^2 \frac{k_\parallel^2}{k^2}}{v_m^2 - v_a^2} \right] \right\}. \tag{53}$$

Here v_s is the sound speed of the background plasma, P_0^{cr} is the unperturbed CR pressure, ∇P_0^{cr} is the gradient of the unperturbed CR pressure, $k_\parallel$ and $k_\perp$ are the components of the mode wavevector parallel and transverse to the unperturbed magnetic field, respectively, and $\kappa_{0\parallel}$, $\kappa_{0\perp}$ are the components of the averaged CR diffusion tensor. It is assumed that the CR diffusion tensor components scale with the background plasma density as $\overline{\kappa}_{\parallel,\perp} \sim \rho^{\varsigma}$. The phase velocity of the mode is

$$v_m = \left[v_s^2 + v_a^2 \pm \frac{1}{2} \sqrt{\left(v_s^2 + v_a^2\right)^2 - 4 v_s^2 v_a^2 \frac{k_\parallel^2}{k^2}} \right]^{\frac{1}{2}}. \tag{54}$$

The first term in Eq. (53) is the wave damping rate due to the irreversible stochastic Fermi II CR acceleration effect (Achterberg 1979; Bykov and Toptyghin 1979; Ptuskin 1981), while the second and the third terms describe the growth/damping of the modes due to the acoustic instability studied by Drury and Falle (1986). A more general treatment with an arbitrary direction of the unperturbed magnetic field was performed by Chalov (1988b). He accounted for the response of the CR diffusion tensor to both the density and magnetic field variations and found that the latter does not change the character of the angular dependence of the growth rate significantly. A similar angular dependence of the long-wave mode growth rate due to the CR current driven instability (discussed above in Sect. 4.5) was found by Bykov et al. (2011b).

In the space plasma with the modest level of the magnetic field fluctuations the local CR diffusion is anisotropic. For magnetized CR particles ($a \ll 1$) the diffusion parallel to the mean magnetic field dominates over the CR diffusion transverse to the mean field, i.e., $\kappa_{0\parallel} \gg \kappa_{0\perp}$. The growth rate of the acoustic instability in the anisotropic system is maximal for the modes propagating nearly transverse to the mean magnetic field ($\vartheta \to \frac{\pi}{2}$). Here ϑ is the angle between the mode wavevector and the mean magnetic field.

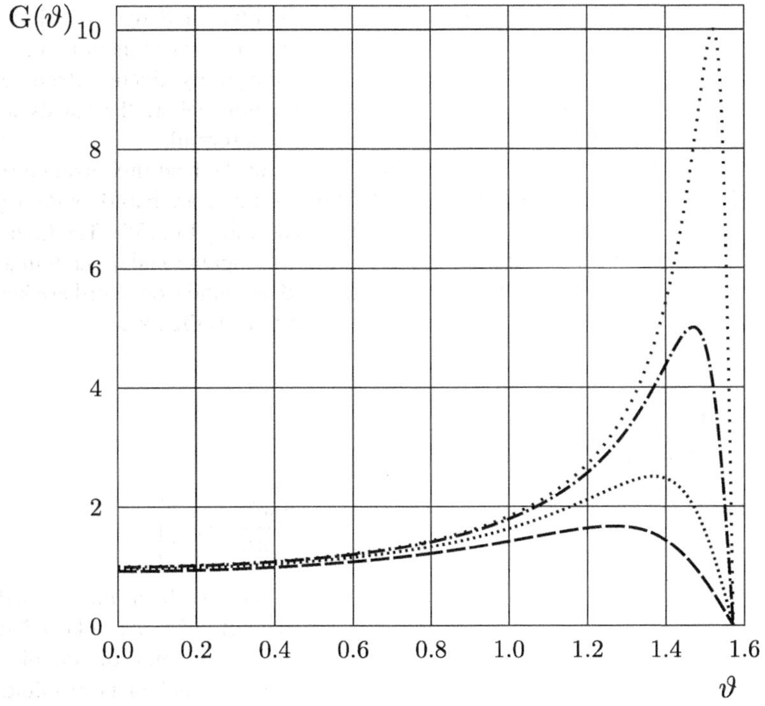

Fig. 8 The characteristic angular dependence of the growth rate of the acoustic instability Eq. (53) for $a = 0.3$ (the *dashed curve*), $a = 0.2$ (the *dotted curve*), $a = 0.1$ (the *dot-dashed curve*), $a = 0.05$ (the *rare dot curve*)

The angular dependence of the growth rate Eq. (53) can be approximated by

$$G_0(\vartheta) = \frac{\cos\vartheta}{\cos^2\vartheta + \frac{\kappa_{0\perp}}{\kappa_{0\parallel}}\sin^2\vartheta}, \tag{55}$$

where we used $\frac{k_\parallel}{k} = \cos\vartheta$, $\frac{k_\perp}{k} = \sin\vartheta$. The anisotropy of the CR diffusion is determined by the CR particle magnetization (e.g., Toptygin 1983), that is the inverse collisionality parameter a, and therefore, $\frac{\kappa_{0\perp}}{\kappa_{0\parallel}} \propto a^2$. The maximal growth rate is therefore achieved for the mode propagating at $\cos\vartheta_{\max} = a$, where $G_{\max}(\vartheta_{\max}) = \frac{1}{2a}$. The angular dependence of the growth rate of the acoustic instability is illustrated in Fig. 8 for various values of collisionality parameter a.

The linear perturbation analysis discussed above is based on the diffusion approximation of the CR dynamics in Eqs. (48)–(52) and, therefore, it is valid for the modes of the wavenumbers above the mean free path of the CRs. A numerical model of the acoustic instability in the nonlinear regime was performed recently by Drury and Downes (2012), who found a significant amplification of magnetic field. The authors assumed a fixed CR diffusion gradient with no response of the CR pressure to the fluctuations, that may affect the model results.

8 Self-Confinement of CRs Near Their Acceleration Sites

Apart from being a central issue for the acceleration in SNR shocks, the CR-driven instabilities are fast becoming an integral part of CR escape models. One common difficulty with the observational verification of the proton escape is that, in contrast to electrons, they likely remain invisible until they reach some dense material in SNR surroundings. Only there generate they enough π^0 mesons in collisions with other protons and the mesons in turn decay into gamma photons which may be detected. Not surprisingly, the escape of CRs from an SNR is a hot topic of today research in gamma-ray astronomy.

The backbone of the DSA is a self-confinement of accelerated particles by scattering off various magnetic perturbations that particles drive by themselves while streaming ahead of the shock. Most important of them were discussed at some length in this review. Logically, this process should also control the ensuing propagation of CRs, before their density drops below the wave instability threshold. Strictly speaking the CR release (escape) from the accelerator should be treated together with the acceleration, as it does not occur at once for all the particles. But this would be a combination of two difficult enough problems and most of the progress in CR escape was made by considering it separately from acceleration.

Remarkably, even within this limited approach, and under rather loose formulation of the problem, no consensus on the escape mechanism has been reached so far; the dividing lines seem to run across the following issues: (i) does the escape occur isotropically or along the local magnetic field? (ii) does the scattering by the background MHD turbulence control the CR propagation alone or self-excited waves need to be included? (iii) if so, is a quasilinear saturation of self-excited waves sufficient or nonlinear processes of wave damping are crucial to the particle propagation? (iv) if they are, which particular mechanism(s) should be employed?

Starting with (i–ii) we note that most of the early models, and some of the recent ones that target specific remnants, assume isotropic CR propagation from a point source impeded only by the background turbulence (one may call them test particle models, e.g. Aharonian and Atoyan 1996; Gabici et al. 2009; Ellison and Bykov 2011; Gabici 2011). It should be noted, however, that e.g., Rosner and Bodo (1996) and Nava and Gabici (2013) adopted a field aligned propagation while Drury (2011) included the finite radius of a SNR shock in the CR escape description. Given the topic of the present short review, however, we focus in this section on models that explicitly include the self-excited waves. Brief reviews of other aspects of CR propagation in the galaxy were given recently by Gabici (2011) and Ptuskin (2012).

The role of self-confinement effects in the CR escape, their subsequent propagation and how these phenomena are treated in different models, can be best demonstrated by writing the following equations that self-consistently describe the CR diffusion and wave generation

$$\frac{d}{dt}P_{\rm CR}(p)=\frac{\partial}{\partial z}\frac{\kappa_{\rm B}}{I}\frac{\partial P_{\rm CR}}{\partial z} \tag{56}$$

$$\frac{d}{dt}I=-v_{\rm a}\frac{\partial P_{\rm CR}}{\partial z}-\Gamma I. \tag{57}$$

Here v_a is the Alfvén velocity, κ_B is the CR diffusion coefficient in Bohm regime, $\kappa_B = cr_g/3$, and the time derivative is taken along the characteristics of unstable Alfvén waves, forward propagating along the field (z-direction):

$$\frac{d}{dt}=\frac{\partial}{\partial t}+v_{\rm a}\frac{\partial}{\partial z} \tag{58}$$

Eq. (56) above is essentially a well-known convection-diffusion equation, written for the dimensionless CR partial pressure P_{CR} instead of their distribution function $f(p,t)$. We normalized it to the magnetic energy density $\rho v_{\mathrm{a}}^2/2$:

$$P_{\mathrm{CR}} = \frac{4\pi}{3}\frac{2}{\rho v_{\mathrm{a}}^2} v p^4 f, \tag{59}$$

where v and p are the CR speed and momentum, and ρ—the plasma density. The total CR pressure is normalized to $d\ln p$, similarly to the wave energy density I:

$$\frac{\langle \delta B^2\rangle}{8\pi} = \frac{B_0^2}{8\pi}\int I(k)d\ln k = \frac{B_0^2}{8\pi}\int I(p)d\ln p$$

Eq. (57) is a wave kinetic equation in which the energy transferred to the waves equals the total work done by the particles, $(u+v_{\mathrm{a}})\nabla P_{\mathrm{CR}}$, less the work done on the fluid, $u\nabla P_{\mathrm{CR}}$ (Drury 1983) (we neglect the bulk flow velocity u, here and in Eq. (58) assuming that the active phase of acceleration ended by this time). The above interpretation of the wave generation indicates that it operates in a maximum efficiency regime. A formal quasilinear derivation of this equation assumes that the particle momentum p is related to the wave number k by the 'sharpened' resonance condition $kp = eB_0/c$ instead of the conventional cyclotron resonance condition $kp_{\parallel} = eB_0/c$ (Skilling 1975), (note that here $k = k_{\parallel}$). We assume that $\partial P_{\mathrm{CR}}/\partial z \le 0$ at all times, so that only the forward propagating waves are unstable. The latter inequality is ensured by the formulation of initial value problem symmetric with respect to $z=0$, so we consider the CR escape into the half-space $z>0$ with the boundary condition $\partial P_{\mathrm{CR}}/\partial z = 0$ at $z=0$.

Papers on CR self-confinement discussed below use equations that are largely similar to Eqs. (56)–(57) but different assumptions are made regarding geometry of particle escape from the source (see (i) above), the character and strength of wave damping Γ (iv), and the role of quasilinear wave saturation (iii). Fujita et al. (2011) and Yan et al. (2012) utilize the isotropic escape models (in this case $\partial/\partial z$ should be replaced by $\partial/\partial r$, etc.) while Ptuskin et al. (2008) and Malkov et al. (2013) assume that particles propagate predominantly along the local large-scale field. Note that Yan et al. (2012) considered the escape from an active accelerator (in Eq. (58), one should include the flow bulk velocity, $v_a \to v_a + u$ in this case) and, in addition, they introduce a step-wise increase in CR diffusivity at a certain particle momentum above which particles escape the accelerator. These assumptions make it difficult to compare their results with those of the remaining three papers. In these, Fujita et al. (2011) presented the results of numerical integration of Eqs. (56)–(57) (in a spherical symmetry) with neglected damping term Γ. The results indicate a considerable delay of diffusion from the source due to a self-confinement.

However, in the regions where magnetic perturbations are weak, i.e. $I \ll 1$, the field aligned CR transport is appropriate, as the perpendicular diffusion is suppressed, $\kappa_{\perp} \simeq I^2\kappa_{\parallel} \ll \kappa_{\parallel} \simeq \kappa_B/I$. Taking into account the condition $I_{\mathrm{ISM}} \ll 1$, such regime appears inevitable far away from the source and at late times when particles are spread over a large volume and the waves are driven only weakly. At earlier times and close to the region of the initial localization of CRs, an estimate $\kappa_{\perp} \sim \kappa_{\parallel} \sim \kappa_{\mathrm{B}}$ appears to be adequate. Both analytical models by Ptuskin et al. (2008) and Malkov et al. (2013), however, do not embrace the general case and rely on the assumption $\kappa_{\perp} \ll \kappa_{\parallel}$ thus considering a field-aligned escape. At the same time, they are different in further simplifications made, that lead to rather different results, both quantitatively and qualitatively.

Ptuskin et al. (2008) neglect dI/dt on the l.h.s of Eq. (57) thus balancing the driving term with the damping term on its r.h.s and assume a Kolmogorov dissipation for Γ,

$$\Gamma = kv_a\sqrt{I}/(2C_K)^{3/2} \tag{60}$$

with $C_K \approx 3.6$ and $k \simeq 1/r_g(p)$ being the resonant wave number. Therefore, only one equation (56) needs to be solved which lead to the following self-similar solution (in notations and normalization used in Eqs. (56)–(57))

$$P_{CR} = \frac{4 \cdot 3^{-3/2}}{t'^{3/2}\sqrt{\sigma + (kz)^4/t'^6}} \tag{61}$$

where the dimensionless time $t' = (\kappa_B k^2/2C_K)t$, $\sigma = \Gamma^8(1/4)/\pi^2 3^6 \eta^4$, and Γ is the gamma function. The single important parameter this solution depends on is the integrated (along the field line) CR partial pressure

$$\eta = 2k \int_0^\infty P_{CR} dz \tag{62}$$

Therefore, the CR density decays at the source as $\propto t^{-3/2}$ and the flat-topped, self-confined part of the CR distribution spreads as $z \propto t^{3/2}$, both pointing at the superdiffusive CR transport. The reason is clearly in a very strong wave damping due to the Kolmogorov dissipation. For the same reason this solution does not recover the test particle asymptotic result $P_{CR} \propto t^{-1/2} \exp(-z^2/4D_{\mathrm{ISM}}t)$, physically expected in $z, t \to \infty$ limit in the interstellar medium with the background diffusion coefficient D_{ISM}.

An alternative choice of damping mechanism is the Goldreich and Sridhar (1997) MHD spectrum, which seems to be more appropriate in $I \lesssim 1$ regime under not too strong MHD cascade (Farmer and Goldreich 2004; Beresnyak and Lazarian 2008; Yan et al. 2012). The damping rate in this case is

$$\Gamma = v_a\sqrt{\frac{k}{L}} \tag{63}$$

where L is the outer scale of turbulence which may be as large as $100pc$. Not only is this damping orders of magnitude (roughly a factor $\sqrt{r_g/L}$) lower than the Kolmogorov one but, as it does not depend on I and can be considered as coordinate independent, it allows for the following ('quasilinear') integral of the system of Eqs. (56) and (57):

$$P_{\mathrm{CR}}(z,t) = P_{\mathrm{CR0}}(z') - \frac{\kappa_{\mathrm{B}}}{v_{\mathrm{a}}}\frac{\partial}{\partial z}\ln\frac{I(z,t)}{I_0(z')} \tag{64}$$

Here $P_{\mathrm{CR0}}(z)$ and $I_0(z)$ are the initial distributions of the CR partial pressure and the wave energy density, respectively, and $z' = z - v_{\mathrm{a}}t$. Substituting P_{CR} in Eq. (57) and neglecting slow convection with v_a in Eq. (58), we arrive at the following diffusion equation for I

$$\frac{\partial I}{\partial t} = \frac{\partial}{\partial z}\frac{\kappa_B}{I}\frac{\partial I}{\partial z} - \Gamma I - v_a\frac{\partial P_{CR0}}{\partial z}.$$

The equation is supplemented with the boundary condition $I \to I_{ISM}$, for $|z| \to \infty$. Outside of the region where $P_{CR} \neq 0$, the last term on the r.h.s. may be neglected. The second term may be eliminated by replacing $I\exp(\Gamma t) \to I$, $\int_0^t \exp(\Gamma t)dt \to t$. However, if Γ is taken

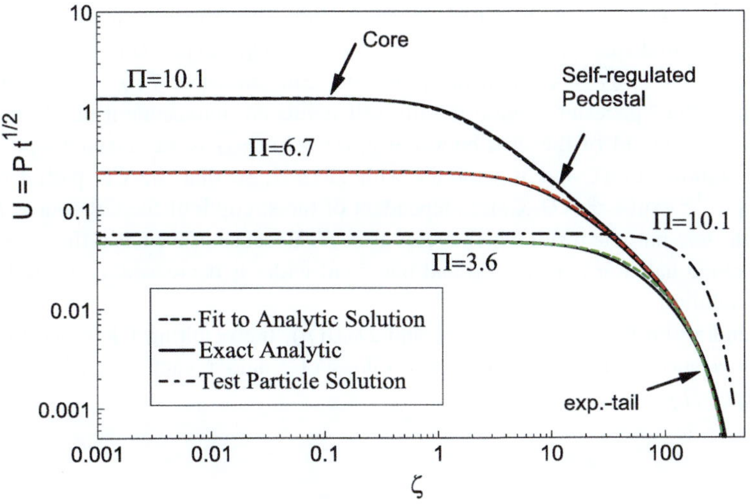

Fig. 9 Spatial distribution of CR partial pressure (as a function of $\zeta = z/\sqrt{a v_a t}$, multiplied by $v_a^{3/2}\sqrt{at}/\kappa_B$) shown for integrated values of this quantity $\Pi = 3.6$; 6.7; 10.1 and for the background diffusivity $D_{ISM} = 10^4$. Exact analytic solutions are shown with the *solid lines* while the interpolations given by Eq. (65) are shown with the *dashed lines*. For comparison, a formal test particle solution for $\Pi = 10.1$ is also shown with the *dot-dashed line*. Note the three characteristic zones of the CR confinement: the innermost flat top core, the scale invariant $(1/\zeta)$ pedestal, and the exponential decay zone

in the form of Eq. (63), it is fairly small due to the factor $\sqrt{r_g/L} \ll 1$. We may simply neglect it. The solution for I and $P_{CR}(z,t)$ may be found in an implicit form (see Malkov et al. 2013 for details). However, there exists a very accurate convenient interpolation formula that can be represented as follows

$$P_{\mathrm{CR}} = \frac{2\kappa_{\mathrm{B}}(p)}{v_{\mathrm{a}}^{3/2}\sqrt{L_c t}}\left[\zeta^{5/3} + (D_{\mathrm{NL}})^{5/6}\right]^{-3/5} e^{-\zeta^2/4D_{\mathrm{ISM}}} \tag{65}$$

where L_c is the size of the initial CR cloud, $\zeta = z/\sqrt{v_a L_c t}$, and $D_{\mathrm{NL}} = C(\Pi) D_{\mathrm{ISM}} \times \exp(-\Pi)$, with Π being a normalized integrated pressure

$$\Pi = \frac{v_a}{\kappa_B}\int_0^\infty P_{CR} dz$$

and D_{ISM} is the normalized background diffusivity

$$D_{\mathrm{ISM}} = \frac{\kappa_B}{v_a L_c} I_{\mathrm{ISM}}^{-1}$$

while $C \sim 1$, for $\Pi \gg 1$ and $C \sim \Pi^{-2}$, for $\Pi \ll 1$.

The representation of the solution given in Eq. (65) is convenient in that the function $\sqrt{t} P_{CR}(\zeta)$ does not depend on t, so that the solution can be shown for all t, z with only one curve, Fig. 9. To summarize these results, the self-regulated normalized ($\mathcal{P}_{CR} = v_a L_c P_{\mathrm{CR}}/\kappa_{\mathrm{B}}(p)$) CR partial pressure profile $\mathcal{P}_{CR}$ comprises the following three zones ($\Pi \gg 1$): (i) a quasi-plateau (core) at small $z/\sqrt{t} < \sqrt{D_{\mathrm{NL}}}$ of a height $\sim (D_{\mathrm{NL}} t)^{-1/2}$, which is elevated by a factor $\sim \Pi^{-1} \exp(\Pi/2) \gg 1$, compared to the test particle solution because of the strong quasi-linear suppression of the CR diffusion coefficient with respect to its background (test particle) value $D_{\rm ISM}$: $D_{\rm NL} \sim D_{\rm ISM} \exp(-\Pi)$ (ii) next to the core, where $\sqrt{D_{\rm NL}} < z/\sqrt{t} < \sqrt{D_{\rm ISM}}$, the profile is scale invariant, $\mathcal{P}_{CR} \approx 2/z$. The CR distribution in this "pedestal" region is fully self-regulated, independent of Π and $D_{\rm ISM}$ for $\Pi \gg 1$, (iii) the tail of the distribution at $z/\sqrt{t} > \sqrt{D_{\rm ISM}}$ is similar *in shape* to the test particle solution in 1D but it saturates with $\Pi \gg 1$, so that the CR partial pressure is $\propto (D_{\rm ISM} t)^{-1/2} \exp(-z^2/4D_{\rm ISM} t)$, independent of the strength of the CR source Π, in contrast to the test-particle result which scales as $\propto \Pi$. Because of the CR diffusivity reduction, the CR cloud half-life is increased and the cloud width is decreased, compared to the test particle solution.

Depending on the functions $\Pi(p)$ and $D_{\rm ISM}(p)$, the resulting CR spectrum generally develops a spectral break for the fixed values of z and t such that $z^2/t \sim D_{\rm NL}(p) \sim D_{\rm ISM} \exp(-\Pi)$.

9 Summary

Cosmic rays, being a highly non-equilibrium component, often comprise an energy density that is comparable to the ram pressure of energetic plasma flows and magnetic fields in astrophysical sources with high energy release such as supernova remnants, fast stellar winds, and astrophysical jets of different scales. CRs may also play a role in the global dynamics of interstellar gas in galaxies, in particular, they may support galactic winds. In the presence of gravitation, the buoyancy of CRs and magnetic field at galactic scales may result in the magnetic Parker instability (Parker 1966, 1967; Shu 1974; Ryu et al. 2003; Hanasz et al. 2009). The local CR diffusion is an important factor for the Parker instability to occur.

The microphysical instabilities discussed above lay the groundwork for detailed simulations of the global interstellar matter dynamics. In this review we addressed the recent progress in understanding of the CR-driven instabilities with special attention to non-relativistic shocks. We started with a quasi-linear analysis of the growth rates of the instabilities driven by anisotropic and inhomogeneous CR distributions. Time dependent nonlinear simulations are needed to draw conclusions about the saturation level and the spectra of magnetic fluctuations produced by the non-equilibrium CR distributions. We used numerical simulations to illustrate the nonlinear dynamics of magnetic fluctuations. The CR-driven instabilities are shown to be crucial for modeling particle acceleration sources and the CR escape from the sources into the interstellar matter.

Acknowledgements We thank Andre Balogh and the ISSI staff for providing an inspiring atmosphere at the International Space Science Institute Workshop in Bern in 2012, which has led to new collaborations and scientific progress. Computing resources to A.B. were provided by the Swedish National Allocations Committee at the Center for Parallel Computers at the Royal Institute of Technology in Stockholm and the High Performance Computing Center North in Umeå. A.B. was supported in part by the European Research Council under the AstroDyn Research Project No. 227952 and the Swedish Research Council under the project grant 621-2011-5076. A.M.B. and S.M.O. acknowledge support from the RAS Programs P21 and OFN 16, and from the Ministry of Education and Science of Russian Federation (Agreement No. 8409, 2012). Some of the simulations were performed at the Joint Supercomputing Centre (JSCC RAS) and the Supercomputing Centre at Ioffe Institute, St. Petersburg. M.M. acknowledges support by the Department of Energy, Grant No. DE-FG02-04ER54738.

Appendix

The dispersion equation (20) can be expressed in the elementary functions by evaluating Eqs. (22) and (23) as

$$
\begin{aligned}
\sigma_0(p) = {} & \frac{3}{2x^2} + \frac{3}{8x}\left(1 - \frac{1}{x^2} + \left(\frac{a}{x}\right)^2\right)\Psi_1 - \frac{3a}{2x^3}\Psi_2 \\
& \mp i\left\{\frac{3}{4x}\left(1 - \frac{1}{x^2} + \left(\frac{a}{x}\right)^2\right)\Psi_2 - \frac{3a}{2x^2} + \frac{3a}{4x^3}\Psi_1\right\},
\end{aligned} \tag{66}
$$

$$
\begin{aligned}
\sigma_1(p) = {} & \mp\frac{1}{x} \pm \frac{3}{2x^3} \mp \frac{3}{2x}\left(\frac{a}{x}\right)^2 \pm \frac{3}{8x^2}\left(1 - \frac{1}{x^2} + 3\left(\frac{a}{x}\right)^2\right)\Psi_1 \\
& \pm \frac{3a}{4x^2}\left(1 - \frac{3}{x^2} + \left(\frac{a}{x}\right)^2\right)\Psi_2 - \\
& - i\left\{\frac{3}{4x^2}\left(1 - \frac{1}{x^2} + 3\left(\frac{a}{x}\right)^2\right)\Psi_2 - \frac{3a}{x^3}\right. \\
& \left. - \frac{3a}{8x^2}\left(1 + \left(\frac{a}{x}\right)^2 - \frac{3}{x^2}\right)\Psi_1\right\},
\end{aligned} \tag{67}
$$

$$
\Psi_1(x) = \ln\left[\frac{(x+1)^2 + a^2}{(x-1)^2 + a^2}\right], \tag{68}
$$

$$
\Psi_2(x) = \mathrm{arctg}\left(\frac{x+1}{a}\right) + \mathrm{arctg}\left(\frac{x-1}{a}\right), \tag{69}
$$

References

A. Achterberg, The energy spectrum of electrons accelerated by weak magnetohydrodynamic turbulence. Astron. Astrophys. **76**, 276–286 (1979)

A. Achterberg, Modification of scattering waves and its importance for shock acceleration. Astron. Astrophys. **119**, 274–278 (1983)

A. Achterberg, R.D. Blandford, Transmission and damping of hydromagnetic waves behind a strong shock front—implications for cosmic ray acceleration. Mon. Not. R. Astron. Soc. **218**, 551–575 (1986)

F.A. Aharonian, A.M. Atoyan, On the emissivity of π^0^-decay gamma radiation in the vicinity of accelerators of galactic cosmic rays. Astron. Astrophys. **309**, 917–928 (1996)

E. Amato, The streaming instability: a review. Mem. Soc. Astron. Ital. **82**, 806 (2011)

E. Amato, P. Blasi, A kinetic approach to cosmic-ray-induced streaming instability at supernova shocks. Mon. Not. R. Astron. Soc. **392**, 1591–1600 (2009). arXiv:0806.1223

A.R. Bell, The acceleration of cosmic rays in shock fronts. I. Mon. Not. R. Astron. Soc. **182**, 147–156 (1978)

A.R. Bell, Turbulent amplification of magnetic field and diffusive shock acceleration of cosmic rays. Mon. Not. R. Astron. Soc. **353**, 550–558 (2004)

A.R. Bell, The interaction of cosmic rays and magnetized plasma. Mon. Not. R. Astron. Soc. **358**, 181–187 (2005)

A.R. Bell, S.G. Lucek, Cosmic ray acceleration to very high energy through the non-linear amplification by cosmic rays of the seed magnetic field. Mon. Not. R. Astron. Soc. **321**, 433–438 (2001)

A. Beresnyak, T.W. Jones, A. Lazarian, Turbulence-induced magnetic fields and structure of cosmic ray modified shocks. Astrophys. J. **707**, 1541–1549 (2009). arXiv:0908.2806

A. Beresnyak, A. Lazarian, Wave decay in magnetohydrodynamic turbulence. Astrophys. J. **678**, 961–967 (2008). arXiv:0805.0630

E.G. Berezhko, Instability in a shock propagating through gas with a cosmic-ray component. Sov. Astron. Lett. **12**, 352–354 (1986)
E.G. Berezhko, G.F. Krymskiĭ, REVIEWS OF TOPICAL PROBLEMS: Acceleration of cosmic rays by shock waves. Sov. Phys. Usp. **31**, 27–51 (1988)
V.S. Berezinskii, S.V. Bulanov, V.A. Dogiel, V.L. Ginzburg et al., *Astrophysics of Cosmic Rays* (North-Holland, Amsterdam, 1990)
E.G. Blackman, G.B. Field, New dynamical mean-field dynamo theory and closure approach. Phys. Rev. Lett. **89**(26), 265007–265010 (2002). arXiv:astro-ph/0207435
R. Blandford, D. Eichler, Particle acceleration at astrophysical shocks: a theory of cosmic ray origin. Phys. Rep. **154**, 1–75 (1987)
A. Brandenburg, Turbulence and its parameterization in accretion discs. Astron. Nachr. **326**, 787–797 (2005). arXiv:astro-ph/0510015
A. Brandenburg, Advances in theory and simulations of large-scale dynamos. Space Sci. Rev. **144**, 87–104 (2009a). arXiv:0901.0329
A. Brandenburg, Large-scale dynamos at low magnetic Prandtl numbers. Astrophys. J. **697**, 1206–1213 (2009b). arXiv:0808.0961
A. Brandenburg, K.H. Rädler, K. Kemel, Mean-field transport in stratified and/or rotating turbulence. Astron. Astrophys. **539**, A35 (2012). arXiv:1108.2264
A.M. Bykov, Interstellar turbulence and shock waves. Sov. Astron. Lett. **8**, 320 (1982)
A.M. Bykov, A model for the generation of interstellar turbulence. Sov. Astron. Lett. **14**, 60 (1988)
A.M. Bykov, D.C. Ellison, S.M. Osipov, G.G. Pavlov et al., X-ray stripes in Tycho's supernova remnant: synchrotron footprints of a nonlinear cosmic-ray-driven instability. Astrophys. J. **735**, L40 (2011). arXiv:1106.3441
A.M. Bykov, D.C. Ellison, M. Renaud, Magnetic fields in cosmic particle acceleration sources. Space Sci. Rev. **166**, 71–95 (2012). arXiv:1105.0130
A.M. Bykov, P.E. Gladilin, S.M. Osipov, Particle acceleration at supernova shocks in young stellar clusters. Mem. Soc. Astron. Ital. **82**, 800 (2011a). arXiv:1111.2587
A.M. Bykov, S.M. Osipov, D.C. Ellison, Cosmic ray current driven turbulence in shocks with efficient particle acceleration: the oblique, long-wavelength mode instability. Mon. Not. R. Astron. Soc. **410**, 39–52 (2011b). arXiv:1010.0408
A.M. Bykov, I.N. Toptyghin, Particle acceleration by astrophysical supersonic turbulence, in *16th International Cosmic Ray Conference*, vol. 2 Kyoto, Japan, (1979), p. 66
A.M. Bykov, Y.A. Uvarov, D.C. Ellison, Dots, clumps, and filaments: the intermittent images of synchrotron emission in random magnetic fields of young supernova remnants. Astrophys. J. **689**, L133–L136 (2008). arXiv:0811.2498
D. Caprioli, P. Blasi, E. Amato, M. Vietri, Dynamical effects of self-generated magnetic fields in cosmic-ray-modified shocks. Astrophys. J. **679**, L139–L142 (2008). arXiv:0804.2884
G. Cassam-Chenaï, J.P. Hughes, J. Ballet, A. Decourchelle, The blast wave of Tycho's supernova remnant. Astrophys. J. **665**, 315–340 (2007). arXiv:astro-ph/0703239
S.V. Chalov, Diffusive shock instability in plasma modified by cosmic-rays. Sov. Astron. Lett. **14**, 114 (1988a)
S.V. Chalov, Instability of the structure of strong oblique MHD cosmic-ray shocks. Astrophys. Space Sci. **148**, 175–187 (1988b)
P.H. Diamond, M.A. Malkov, Dynamics of mesoscale magnetic field in diffusive shock acceleration. Astrophys. J. **654**, 252–266 (2007). arXiv:astro-ph/0605374
E.A. Dorfi, L.O. Drury, A cosmic ray driven instability, in *International Cosmic Ray Conference*, vol. 3, ed. by F.C. Jones International Cosmic Ray Conference, (1985), pp. 121–123
L.O. Drury, An introduction to the theory of diffusive shock acceleration of energetic particles in tenuous plasmas. Rep. Prog. Phys. **46**, 973–1027 (1983)
L.O. Drury, Reaction effects in diffusive shock acceleration. Adv. Space Res. **4**, 185–191 (1984)
L.O. Drury, Escaping the accelerator: how, when and in what numbers do cosmic rays get out of supernova remnants? Mon. Not. R. Astron. Soc. **415**, 1807–1814 (2011). arXiv:1009.4799
L.O. Drury, T.P. Downes, Turbulent magnetic field amplification driven by cosmic ray pressure gradients. Mon. Not. R. Astron. Soc. **427**, 2308–2313 (2012). arXiv:1205.6823
L.O. Drury, S.A.E.G. Falle, On the stability of shocks modified by particle acceleration. Mon. Not. R. Astron. Soc. **223**, 353 (1986)
L.O. Drury, W.J. Markiewicz, H.J. Voelk, Simplified models for the evolution of supernova remnants including particle acceleration. Astron. Astrophys. **225**, 179–191 (1989)
R. Durrer, C. Caprini, Primordial magnetic fields and causality J. Cosmol. Astropart. Phys. **11**, 010 (2003). arXiv:astro-ph/0305059

D.C. Ellison, A.M. Bykov, Gamma-ray emission of accelerated particles escaping a supernova remnant in a molecular cloud. Astrophys. J. **731**, 87 (2011). arXiv:1102.3885

K.A. Eriksen, J.P. Hughes, C. Badenes, R. Fesen et al., Evidence for particle acceleration to the knee of the cosmic ray spectrum in Tycho's supernova remnant. Astrophys. J. **728**, L28 (2011). arXiv:1101.1454

A.J. Farmer, P. Goldreich, Wave damping by magnetohydrodynamic turbulence and its effect on cosmic-ray propagation in the interstellar medium. Astrophys. J. **604**, 671–674 (2004). arXiv:astro-ph/0311400

F. Fraschetti, Turbulent amplification of magnetic field driven by dynamo effect at rippled shocks. (2013). arXiv:1304.4956

Y. Fujita, F. Takahara, Y. Ohira, K. Iwasaki, Alfvén wave amplification and self-containment of cosmic rays escaping from a supernova remnant. Mon. Not. R. Astron. Soc. **415**, 3434–3438 (2011). arXiv:1105.0683

S. Gabici, Cosmic ray escape from supernova remnants. Mem. Soc. Astron. Ital. **82**, 760 (2011). arXiv:1108.4844

S. Gabici, F.A. Aharonian, S. Casanova, Broad-band non-thermal emission from molecular clouds illuminated by cosmic rays from nearby supernova remnants. Mon. Not. R. Astron. Soc. **396**, 1629–1639 (2009). arXiv:0901.4549

J. Giacalone, J.R. Jokipii, Magnetic field amplification by shocks in turbulent fluids. Astrophys. J. **663**, L41–L44 (2007)

P. Goldreich, S. Sridhar, Magnetohydrodynamic turbulence revisited. Astrophys. J. **485**, 680 (1997). arXiv:astro-ph/9612243

M. Hanasz, K. Otmianowska-Mazur, G. Kowal, H. Lesch, Cosmic-ray-driven dynamo in galactic disks. a parameter study. Astron. Astrophys. **498**, 335–346 (2009). arXiv:0812.3906

E.A. Helder, J. Vink, A.M. Bykov, Y. Ohira et al., Observational signatures of particle acceleration in supernova remnants. Space Sci. Rev. **173**, 369–431 (2012). arXiv:1206.1593

F.C. Jones, D.C. Ellison, The plasma physics of shock acceleration. Space Sci. Rev. **58**, 259–346 (1991)

N.K.R. Kevlahan, The vorticity jump across a shock in a non-uniform flow. J. Fluid Mech. **341**, 371–384 (1997)

G.F. Krymskii, A regular mechanism for the acceleration of charged particles on the front of a shock wave, in *Akademiia Nauk SSSR Doklady*, vol. 234, (1977), pp. 1306–1308

R.M. Kulsrud, R. Cen, J.P. Ostriker, D. Ryu, The protogalactic origin for cosmic magnetic fields. Astrophys. J. **480**, 481 (1997). arXiv:astro-ph/9607141

M.A. Malkov, P.H. Diamond, Nonlinear shock acceleration beyond the bohm limit. Astrophys. J. **642**, 244–259 (2006). arXiv:astro-ph/0509235

M.A. Malkov, P.H. Diamond, Nonlinear dynamics of acoustic instability in a cosmic ray shock precursor and its impact on particle acceleration. Astrophys. J. **692**, 1571–1581 (2009)

M.A. Malkov, P.H. Diamond, R.Z. Sagdeev, F.A. Aharonian, I.V. Moskalenko, Analytic Solution for Self-regulated Collective Escape of Cosmic Rays from their Acceleration Sites. Astrophys. J. **768**(1), 13 (2013). doi:10.1088/0004-637X/768/1/73

M.A. Malkov, L.O. Drury, Nonlinear theory of diffusive acceleration of particles by shock waves. Rep. Prog. Phys. **64**, 429–481 (2001)

M.A. Malkov, R.Z. Sagdeev, P.H. Diamond, Magnetic and density spikes in cosmic-ray shock precursors. Astrophys. J. **748**, L32 (2012). arXiv:1110.0257

A. Marcowith, M. Lemoine, G. Pelletier, Turbulence and particle acceleration in collisionless supernovae remnant shocks. II. cosmic-ray transport. Astron. Astrophys. **453**, 193–202 (2006). arXiv:astro-ph/0603462

J.F. McKenzie, H.J. Voelk, Non-linear theory of cosmic ray shocks including self-generated Alfvén waves. Astron. Astrophys. **116**, 191–200 (1982)

J.F. McKenzie, K.O. Westphal, Interaction of hydromagnetic waves with hydromagnetic shocks. Phys. Fluids **13**, 630–640 (1970)

L. Nava, S. Gabici, Anisotropic cosmic ray diffusion and gamma-ray production close to supernova remnants, with an application to W28. Mon. Not. R. Astron. Soc. **429**, 1643–1651 (2013). arXiv:1211.1668

P.D. Noerdlinger, A.K.M. Yui, Persistence of the firehouse instability in highly relativistic plasmas. Astrophys. J. **151**, 901 (1968)

Y. Ohira, K. Murase, R. Yamazaki, Gamma-rays from molecular clouds illuminated by cosmic rays escaping from interacting supernova remnants. Mon. Not. R. Astron. Soc. **410**, 1577–1582 (2011). arXiv:1007.4869

E.N. Parker, The dynamical state of the interstellar gas and field. Astrophys. J. **145**, 811 (1966)

E.N. Parker, The dynamical state of the interstellar gas and field. III. Turbulence and enhanced diffusion. Astrophys. J. **149**, 535 (1967)

G. Pelletier, M. Lemoine, A. Marcowith, Turbulence and particle acceleration in collisionless supernovae remnant shocks. I. Anisotropic spectra solutions. Astron. Astrophys. **453**, 181–191 (2006). arXiv:astro-ph/0603461

V. Ptuskin, Propagation of galactic cosmic rays. Astropart. Phys. **39**, 44–51 (2012)

V.S. Ptuskin, Influence of cosmic rays on propagation of long magneto hydrodynamic waves. Astrophys. Space Sci. **76**, 265–278 (1981)

V.S. Ptuskin, V.N. Zirakashvili, A.A. Plesser, Non-linear diffusion of cosmic rays. Adv. Space Res. **42**, 486–490 (2008)

B. Reville, S. O'Sullivan, P. Duffy, J.G. Kirk, The transport of cosmic rays in self-excited magnetic turbulence. Mon. Not. R. Astron. Soc. **386**, 509–515 (2008). arXiv:0802.0109

S.P. Reynolds, Supernova remnants at high energy. Annu. Rev. Astron. Astrophys. **46**, 89–126 (2008)

M.A. Riquelme, A. Spitkovsky, Nonlinear study of Bell's cosmic ray current-driven instability. Astrophys. J. **694**, 626–642 (2009). arXiv:0810.4565

M.A. Riquelme, A. Spitkovsky, Magnetic amplification by magnetized cosmic rays in supernova remnant shocks. Astrophys. J. **717**, 1054–1066 (2010). arXiv:0912.4990

I. Rogachevskii, N. Kleeorin, A. Brandenburg, D. Eichler, Cosmic-ray current-driven turbulence and mean-field dynamo effect. Astrophys. J. **753**, 6 (2012). arXiv:1204.4246

R. Rosner, G. Bodo, The origin of filaments in the interstellar medium. Astrophys. J. **470**, L49 (1996)

D. Ryu, H. Kang, T.W. Jones, The stability of cosmic-ray-dominated shocks—a secondary instability. Astrophys. J. **405**, 199–206 (1993)

D. Ryu, J. Kim, S.S. Hong, T.W. Jones, The effect of cosmic-ray diffusion on the Parker instability. Astrophys. J. **589**, 338–346 (2003). arXiv:astro-ph/0301625

R.Z. Sagdeev, V.D. Shafranov, On the instability of a plasma with anisotropic distribution of velocities in magnetic field. Sov. Phys. JETP **12**(1), 130–132 (1961)

R. Schlickeiser, *Cosmic Ray Astrophysics* (Springer, Berlin, 2002)

M. Schrinner, K.H. Rädler, D. Schmitt, M. Rheinhardt et al., Mean-field view on rotating magnetoconvection and a geodynamo model. Astron. Nachr. **326**, 245–249 (2005)

K.M. Schure, A.R. Bell, A long-wavelength instability involving the stress tensor. Mon. Not. R. Astron. Soc. **418**, 782–788 (2011). arXiv:1107.5817

K.M. Schure, A.R. Bell, L.O'C. Drury, A.M. Bykov, Diffusive shock acceleration and magnetic field amplification. Space Sci. Rev. **173**, 491–519 (2012). arXiv:1203.1637

V.D. Shapiro, K.B. Quest, M. Okolicsanyi, Non-resonant firehose instability: consequences for the theory of cosmic ray acceleration. Geophys. Res. Lett. **25**, 845–848 (1998)

F.H. Shu, The Parker instability in differentially-rotating disks. Astron. Astrophys. **33**, 55 (1974)

J. Skilling, Cosmic ray streaming. I. Effect of Alfvén waves on particles. Mon. Not. R. Astron. Soc. **172**, 557–566 (1975)

I.N. Toptygin, *Cosmic rays in interplanetary magnetic fields* (1983)

R.A. Treumann, W. Baumjohann, *Advanced Space Plasma Physics* (1997)

Y. Uchiyama, F.A. Aharonian, T. Tanaka et al., Extremely fast acceleration of cosmic rays in a supernova remnant. Nature **449**, 576–578 (2007)

J. Vink, Supernova remnants: the X-ray perspective. Astron. Astrophys. Rev. **20**, 49 (2012). arXiv:1112.0576

A.E. Vladimirov, A.M. Bykov, D.C. Ellison, Spectra of magnetic fluctuations and relativistic particles produced by a nonresonant wave instability in supernova remnant shocks. Astrophys. J. **703**, L29–L32 (2009). arXiv:0908.2602

H. Yan, A. Lazarian, R. Schlickeiser, Cosmic-ray streaming from supernova remnants and gamma-ray emission from nearby molecular clouds. Astrophys. J. **745**, 140 (2012). arXiv:1111.2410

V.N. Zirakashvili, V.S. Ptuskin, Diffusive shock acceleration with magnetic amplification by nonresonant streaming instability in supernova remnants. Astrophys. J. **678**, 939–949 (2008). arXiv:0801.4488

V.N. Zirakashvili, V.S. Ptuskin, H.J. Völk, Modeling Bell's nonresonant cosmic-ray instability. Astrophys. J. **678**, 255–261 (2008). arXiv:0801.4486

E.G. Zweibel, Energetic particle trapping by Alfvén wave instabilities, in *Particle Acceleration Mechanisms in Astrophysics*, ed. by J. Arons, C. McKee, C. Max. American Institute of Physics Conference Series, vol. 56, (1979), pp. 319–328

E.G. Zweibel, Cosmic-ray history and its implications for galactic magnetic fields. Astrophys. J. **587**, 625–637 (2003). arXiv:astro-ph/0212559

E.G. Zweibel, J.E. Everett, Environments for magnetic field amplification by cosmic rays. Astrophys. J. **709**, 1412–1419 (2010). arXiv:0912.3511

DOI 10.1007/978-1-4899-7413-6_7
Reprinted from *Space Science Reviews* Journal, DOI 10.1007/s11214-013-9966-9

Nonclassical Transport and Particle-Field Coupling: from Laboratory Plasmas to the Solar Wind

D. Perrone · R.O. Dendy · I. Furno · R. Sanchez · G. Zimbardo · A. Bovet · A. Fasoli · K. Gustafson · S. Perri · P. Ricci · F. Valentini

Received: 30 October 2012 / Accepted: 30 January 2013 / Published online: 8 March 2013

Abstract Understanding transport of thermal and suprathermal particles is a fundamental issue in laboratory, solar-terrestrial, and astrophysical plasmas. For laboratory fusion experiments, confinement of particles and energy is essential for sustaining the plasma long enough to reach burning conditions. For solar wind and magnetospheric plasmas, transport properties determine the spatial and temporal distribution of energetic particles, which can be harmful for spacecraft functioning, as well as the entry of solar wind plasma into the magnetosphere. For astrophysical plasmas, transport properties determine the efficiency of particle acceleration processes and affect observable radiative signatures. In all cases, transport depends on the interaction of thermal and suprathermal particles with the electric and magnetic fluctuations in the plasma. Understanding transport therefore requires us to understand these interactions, which encompass a wide range of scales, from magnetohydrodynamic to kinetic scales, with larger scale structures also having a role. The wealth of transport studies during recent decades has shown the existence of a variety of regimes that differ from the classical quasilinear regime. In this paper we give an overview of nonclassical plasma transport regimes, discussing theoretical approaches to superdiffusive and subdiffusive transport, wave–particle interactions at microscopic kinetic scales, the influence of coherent structures

D. Perrone (✉) · G. Zimbardo · S. Perri · F. Valentini
Department of Physics, University of Calabria, Ponte P. Bucci, Cubo 31C, 87036 Rende, Italy
e-mail: denise.perrone@fis.unical.it

R.O. Dendy
Culham Science Centre, Euratom/CCFE Fusion Association, Abingdon, Oxfordshire OX14 3DB, UK

R.O. Dendy
Centre for Fusion, Space and Astrophysics, Department of Physics, Warwick University, Coventry CV4 7AL, UK

I. Furno · A. Bovet · A. Fasoli · K. Gustafson · P. Ricci
Centre de Recherches en Physique des Plasmas (CRPP), Ecole Polytechnique Fédérale de Lausanne (EPFL), Association Euratom-Suisse, 1015 Lausanne, Switzerland

R. Sanchez
Department of Physics, Universidad Carlos III, Leganés 28911, Madrid, Spain

and of avalanching transport, and the results of numerical simulations and experimental data analyses. Applications to laboratory plasmas and space plasmas are discussed.

Keywords Transport · Wave–particle interaction · Laboratory plasmas · Space plasmas · Anomalous diffusion

1 Introduction

This paper addresses the problems of nonclassical, nondiffusive transport and of the coupling between particles and fields on an extended range of scales. We give an overview of these problems in laboratory plasmas and in space and astrophysical plasmas, and examine the inter-relationships, with the aim of assisting the exchange of novel ideas and techniques. In magnetized, nearly collisionless plasmas, the main contribution to transport comes from the influence of electric and magnetic turbulence on the particle motion, either in the form of drift velocities perpendicular to the background magnetic field, or because of resonant effects such as pitch-angle scattering, or because of the field-line random walk due to low frequency magnetic fluctuations. In addition, if turbulence influences transport, then transport influences the equilibrium plasma structure and hence dynamical processes such as particle acceleration that may depend on this structure. Plasma fluid or kinetic instabilities can modify the spectrum of electromagnetic fluctuations, for example by means of a nonlinear cascade, feeding further energy into turbulence. Thus, a complex coupling between plasma turbulence and transport, combining fluid and kinetic properties, arises. Due to the complexity of the problem, transport is sometimes considered from a test-particle perspective, as distinct from a fully self-consistent treatment. In the last two decades, a number of transport regimes different from the classical quasilinear regime have gained increasing attention. These include anomalous transport regimes including subdiffusion and superdiffusion (Zumofen et al. 1989), diffusive but non-quasilinear regimes such as percolation (e.g., Zimbardo et al. 2000a, 2012), and avalanching transport as well as the influence of coherent structures. Due to the broadness of these topics, we do not give a systematic coverage of the fields of transport and wave–particle interactions, but rather highlight a number of current approaches.

The topics addressed in this review all illustrate the fact that space and laboratory plasmas are complex systems in the technical sense (Dendy et al. 2007). That is, their phenomenology is governed by the interaction of multiple physical processes, each of which operates within a distinct range of length scales and time scales. These processes are nonlinearly coupled, and together span a very wide dynamic range. As an example, let us consider a plasma environment which is characterized by the presence of suprathermal particles, i.e. with energy larger than characteristic bulk plasma temperatures, possibly generated by turbulent acceleration, external sources, or, in the case of fusion devices, nuclear reactions. Understanding the nature of the transport of suprathermal particles is a key challenge for the description of a wide range of plasma systems, ranging from magnetically confined plasmas for fusion to space plasmas. In particular, this subject is receiving much attention in fusion research since, in future fusion reactors, additional heating will be needed to reach the burning regime, which will create strong suprathermal ion components, for example via the injection of energetic neutral beams (NBI). When burning conditions are reached, the majority of the heating will come from the fusion-generated, highly-energetic alpha particles. Good confinement of the slowing-down alpha particles and of the NBI-generated fast ions is therefore crucial. Despite its importance, the interaction between highly energetic ions and small-scale turbulence has not been extensively investigated to date. One of the reasons is that, in present fusion devices, suprathermal ions do not play a crucial role. The other more fundamental reason is that, as the suprathermal ions usually have a gyroradius that is larger than the turbulence scale, their interaction with turbulence is expected to be relatively weak. Gyro-averaging effects would then weaken or almost entirely suppress the interaction of alphas with turbulence, leading to subdiffusive transport. However, the question of how complete this suppression becomes in different regimes is still open. In addition, the results in Günter et al. (2005) indicate that a significant redistribution of suprathermal ions can be induced by turbulence. So far, no direct measurements of fluctuation induced fast-ion transport in toroidal devices have been performed. Thus there exists a strong need for experimental data with which to compare and validate the relevant theoretical models and numerical simulations. For these reasons, theoretical and experimental studies of suprathermal ion transport in turbulent plasmas are currently operating on basic plasma physics experiments, which allow a relatively simple experimental environment with easy access for diagnostics and well established plasma scenarios. In this review, among other examples, we will describe experimental and theoretical efforts to understand suprathermal particle transport in the TORPEX device (Fasoli et al. 2006, 2010).

In the solar wind, understanding transport is important for predicting the arrival of solar energetic particles (SEP) from the sun—one of the main concerns of space weather—and for understanding how, when, and where solar wind plasma enters the magnetosphere. The observation of non-thermal, non-isotropic particle distribution functions in the solar wind calls for a better understanding of wave–particle interactions and pitch-angle diffusion, possibly in a nonlinear regime. Further, the transport properties influence cosmic ray acceleration processes, notably diffusive shock acceleration. Both numerical and analytical studies show transport depends both on magnetic fluctuation level and on the turbulence anisotropy (Pommois et al. 1999; Zimbardo 2005). Here, too, the wave–particle interaction depends on the gyroradius, so that a number of different transport regimes are obtained.

The organization of this review is as follows: in Sect. 2, we discuss the theoretical framework appropriate to describe nondiffusive transport, emphasizing the presence of long-range correlation and non-Gaussian statistics. Concepts such as the continuous-time random walk (CTRW) and fractional diffusion equations are briefly introduced. In Sect. 3 we describe the study of wave–particle interactions by means of kinetic Vlasov simulations: such self-consistent studies gives information on small scale turbulence dissipation and on pitch angle scattering in the nonlinear regime. In Sect. 4, the effect of finite Larmor radius is investigated with regard to transport in laboratory and solar wind plasmas. We discuss results from Vlasov simulations and test-particle simulations, where both subdiffusive and superdiffusive regimes are found in many cases. Section 5 addresses the interaction between particle physics on the ion gyroscale, small-scale turbulence which may retain features of underlying linear wave physics, and larger-scale coherent nonlinear structures. In a typical laboratory plasma, this involves coupled physics unfolding over three orders of magnitude in length scale, and many more in time scale. Section 6 deals with avalanching transport, which typically arises (Dendy and Helander 1997) in physical systems where correlation exists on all length scales, leading to scale-free phenomenology and non-Gaussian fluctuation statistics and emitted signals (Dendy and Chapman 2006). In Sect. 7 we briefly discuss how the extension of diffusive shock acceleration (the standard model for the acceleration of cosmic rays) (Fisk and Lee 1980; Lee and Fisk 1982) to the case of superdiffusion leads to a new expression for the energy spectral index of accelerated particles. In Sect. 8 we give our conclusions.

2 Overview of Novel Transport Regimes

In nature, many systems exhibit complex and non-diffusive behavior. These systems usually share the characteristics of having many degrees of freedom that interact dominantly via nonlinear interactions, and in an environment that fosters self-organization, memory effects, long-range correlations, and non-diffusive transport. By non-diffusive transport, we mean transport processes in which fluxes are not trivially related to the local instantaneous values of the gradients of the transported fields; instead a more complicated relation exists between these quantities, which can include non-local kernels both in space and time. These types of transport behavior have been observed both in space and laboratory plasmas, and a substantial effort has been made to characterize these regimes and understand the physical mechanisms that cause them. Examples are provided by the dynamics of solar flares (Lu and Hamilton 1991), magnetic substorms in Earth's magnetosphere (Klimas et al. 2000), and radial turbulent transport in fusion devices and basic plasma experiments (Carreras et al. 1996; Newman et al. 1996; Fasoli et al. 2010) to name just a few.

2.1 Propagators of Non-diffusive Transport Equations

Many ways exist to characterize the complexity of transport. A relatively simple technique is to define the Green function (or *propagator*) of the effective equation characterizing the transport process. Say, for instance, that we consider the standard diffusive transport equation in one-dimension:

$$\frac{\partial n}{\partial t} = D\frac{\partial^2 n}{\partial x^2}. \tag{1}$$

The propagator of this equation is provided by the temporal evolution of the initial condition $n(x,0) = \delta(x - x_0)$. It is well known that this Green function is $n(x,t) = G((x - x_0)/\sqrt{2Dt})$, where $G(x)$ is the standard Gaussian distribution. Thus, the propagator is just a Gaussian function whose variance increases linearly with time. This behavior is usually called *diffusive transport*. However, in many cases it is possible to measure these propagators directly: for example, by following the evolution in time of a spatially localized initial concentration of a tracer population. This means that we do not need to assume any form for the transport equation in advance. In some cases, the inferred propagator is not a Gaussian with a variance that grows linearly in time. Sometimes the variance of the propagator, although corresponding to a Gaussian distribution, increases faster than linear (i.e., $\sigma^2 \propto t^{2H}$, $H > 1/2$). We call this behavior *superdiffusive transport*. In other cases, $\sigma^2 \propto t^{2H}$, $H < 1/2$, and we speak of *subdiffusion*.

Another frequent occurrence in practice is that the shape of the observed propagator is non-Gaussian. The central limit theorem, which lies at the heart of why Gaussian distributions are so frequently exhibited by physical processes, also provides alternative stable distributions in the case where some of its assumptions are relaxed. For instance, if one relaxes the constraint of having a finite variance, the central limit theorem yields the so-called *Lévy distributions* as the limiting distribution of the sum of stochastic processes (Samorodnitsky and Taqqu 1994). In the case of symmetry with respect to x, Lévy distributions can be simply expressed as the inverse Fourier transform of their characteristic function, $\exp(-C|k|^\alpha)$, where C is a scale parameter (e.g., Zaslavsky 2002). Clearly, for $\alpha = 2$ we recover a Gaussian, while for $\alpha < 2$ Lévy laws decay algebraically as $|x|^{-(1+\alpha)}$, with $\alpha \in (0, 2)$, instead of exponentially as Gaussians do. Lévy laws are usually a signature of the existence of transport via avalanching processes, non-local interactions, and long-range correlations, whose characteristic size is not set by the underlying physical process, but by the size of the system instead. This causes the divergence of the variance of the propagator. The empirical identification of propagators corresponding to this class of distributions with fat tails—that is an excess of large events compared to Gaussian—is usually a signature of non-diffusive, avalanche-driven transport.

The transport exponents α and H are very useful to characterize transport. Their diffusive values are $\alpha = 2$ and $H = 1/2$, but values in the range $\alpha \in (0, 2]$ and $H \in (0, 2]$ are possible. The physical meaning of these exponents is rather different. In particular, α can be related to the statistics of the underlying microscopic displacements that cause the overall motion, in a view reminiscent of the random-walk or Langevin descriptions (Sanchez et al. 2006). On the other hand, H is related to the correlation or memory existent in the transport process. For instance, for $H < 1$, the exponent $\beta = \alpha H$ is such that $\beta = 1$ implies a Markovian microscopic process. Here, β appears in the exponent of the waiting time power-law distribution, see below conversely $\beta < 1$ implies that the underlying microscopic transport process is non-Markovian, at least in a Lagrangian way (Sanchez et al. 2006).

Several propagators of effective transport equations based on fractional differential equations, see below, are known either analytically or from series expansions (Mainardi et al. 1996; del-Castillo-Negrete et al. 2004b; Sanchez et al. 2008). Comparison between these and empirically inferred propagators can help identify the best fractional transport equation for a particular system.

2.2 Lagrangian Statistics and Correlations

There are other ways to measure the transport exponents, in addition to constructing the propagators. For instance, when one has access to Lagrangian information concerning the transport process (say, via tracer trajectories), it can be shown that α can be obtained from the statistics of the Lagrangian displacements, whilst H can be obtained from the temporal correlation of their ordered time series (Mier et al. 2008). Also called the Hurst exponent in this context, H can be obtained by many different techniques such as the R/S rescaled range analysis (e.g., Feder 1988), or the detrended fluctuation analysis among other methods (Dendy and Chapman 2006).

It is also possible to relate these exponents to the correlation function of the Lagrangian velocities or displacements, $C_L(\tau) \equiv \langle v(t)v(t+\tau)\rangle$, where the angle bracket refers to an ensemble average over Lagrangian trajectories. A homogeneous system is usually assumed, so that the correlation function depends only on the time delay τ. In the diffusive case, this function is related to the effective diffusivity of the process via the Taylor-Green-Kubo (TGK) relation (e.g., Shalchi 2011),

$$D = \int_0^\infty C_L(\tau)d\tau. \tag{2}$$

It is also possible to infer from this relation when super- or sub-diffusion may occur depending on the velocities correlation function. Superdiffusion occurs if $C_L(\tau)$ is not integrable, which requires that

$$C_L(\tau) \sim \tau^{-2(1-H)} \tag{3}$$

for large τ (Mandelbrot and van Ness 1968). Here we have purposely expressed the power-law in terms of the Hurst exponent. Since $H > 1/2$ for superdiffusion, it follows that the integral in Eq. (2) diverges, corresponding to the presence of long range correlations. On the other hand, subdiffusion requires that $H < 1/2$, which ensures that the integral in the TGK relation is finite. However, a finite value would imply diffusive behavior, and it therefore follows that the integral of $C_L(\tau)$ must be exactly zero to yield subdiffusion. This implies that the long-time behavior of the autocorrelation function, although correctly given by the previous scaling in terms of the Hurst exponent, must be negative. This is a consequence of the negative correlation that is always responsible for subdiffusive behavior. Indeed, it can be shown (e.g., Feder 1988) that for $H \neq 1/2$ the probability of past and future Lagrangian displacements Δx_i is correlated, that is for $H > 1/2$ positive increments in the past imply positive increments in the future—this is called a persistent process. Conversely, for $H < 1/2$ a positive increment in the past implies a negative increment in the future—this is an antipersistent process.

In the case where the statistics of the Lagrangian velocities are Lévy-like (and characterized with some exponent α, as we mentioned before), the TGK relation breaks down: the integral diverges for any value of H due to the lack of a finite variance of the statistics of the velocities. In a spatiotemporally finite system, however, the correlation function is still meaningful, and many of the preceding statements remain valid, although $C_L(\tau)$ will non-trivially scale with the size of the system (Sanchez et al. 2006). It is however important to note, in that case, and for a finite system,

$$C_L(\tau) \sim \tau^{-(2-\alpha H)}. \tag{4}$$

2.3 Continuous-Time Random Walk Model

Yet another, particularly transparent and easily generalizable, microscale approach to describe the spreading of a population of particles is the continuous-time random walk (CTRW) (Montroll and Weiss 1965). This approach consists of following the trajectories of particles, the random walkers, through a sequence of steps. The CTRW is defined by $\phi(\Delta\mathbf{r}, \Delta t)$, which is the probability distribution function (PDF) for a random walker to make a step of length $\Delta\mathbf{r}$ that takes a time Δt. The time taken to make a step is also called the waiting time, because of models where the random walk is described as a consequence of instantaneous jumps separated by a time Δt. In the simplest (and initial) formulation, homogeneity in space and time is assumed, as well as separability of the jumping and waiting processes. This means that

$$\phi(\Delta\mathbf{r}, \Delta t) = p(\Delta\mathbf{r}) \cdot \psi(\Delta t), \tag{5}$$

where p is known as the step-size distribution, and ψ as the waiting time distribution. We note that it is not difficult to remove these constraints from the CTRW formulation (Shugard and Reiss 1976; van Milligen et al. 2004). As shown by Montroll and Weiss (1965), the probability distribution $\phi(\Delta x, \Delta t)$ allows one to compute the time behavior of the mean-square derivation by inserting the Fourier-Laplace transform $\widehat{\tilde{\phi}}$ into the Fourier-Laplace transform of the propagator $P(x, t)$, as given by the Montroll-Weiss equation (e.g., Ragot and Kirk 1997)

$$\widehat{\tilde{P}}(k, s) = \frac{1 - \tilde{\psi}(s)}{s} \frac{1}{1 - \widehat{\tilde{\phi}}(k, s)}. \tag{6}$$

Standard diffusive transport is recovered from the CTRW in the limit of long times and large distances, if p and ϕ have finite mean and variances. This limit is exact if p is a Gaussian law, and ψ an exponential. However, non-diffusive transport is exhibited by this simple

CTRW realization if p is chosen to follow a power-law distribution known as a Lévy law. Lévy laws are characterized by one exponent $0 < \alpha \leq 2$ that defines the asymptotic behavior of the distribution: $p(\Delta x) \sim |\Delta x|^{-(1+\alpha)}$, for large argument. Here, we have assumed one-dimensionality for simplicity. This asymptotic behavior guarantees a divergent variance for the distribution of step-sizes. Similarly, for the waiting-time distribution it is a requirement that $\psi(\Delta t) \sim (\Delta t)^{-(1+\beta)}$ for large argument, where $0 < \beta \leq 1$. Again, this guarantees a divergent mean waiting-time. A Gaussian p is recovered in the case $\alpha > 2$, and an exponential is recovered when $\beta \to 1$ (Samorodnitsky and Taqqu 1994).

The separable CTRW model is very useful for analyzing real transport situations. It can be implemented by collecting sufficient statistics of step-sizes and waiting-times from the motion of a population of particles or tracers. However is not always straightforward to come up with a reasonable definition of what a step-size and a waiting-time should be. Indeed, there are many cases in which particles are never at rest, for instance, when advected by turbulence. It is then a question of intuition to choose these quantities correctly (e.g., Zimbardo et al. 2000b). Assuming this is done appropriately, it is straightforward to show that the variance of the distribution of the tracers scales as:

$$\sigma^2 \sim t^{2H}, \quad H = \beta/\alpha. \tag{7}$$

The transport exponent H again tells us whether diffusive ($H = 1/2$), superdiffusive ($H > 1/2$) or subdiffusive ($H < 1/2$) transport takes place in the system.

It is also interesting to note that, in the long-time, large-distance limit, the CTRW can be shown to be well represented by the following transport equation under very general conditions (Sanchez et al. 2005),

$$\frac{\partial^\beta n}{\partial t^\beta} = D \frac{\partial^\alpha n}{\partial |x|^\alpha}, \tag{8}$$

where the fractional derivatives require introduction. Fractional derivatives are a class of integro-differential operators, which can be defined in a number of ways (e.g., del-Castillo-Negrete et al. 2004a; Sanchez et al. 2005). The Riemann-Liouville definition is

$$\frac{\partial^\mu \phi(x,t)}{\partial x^\mu} = {}_aD_x^\mu \phi(x,t) = \frac{1}{\Gamma(m-\mu)} \frac{\partial^m}{\partial x^m} \int_a^x \frac{\phi(x',t)}{(x-x')^{1+\mu-m}} dx' \tag{9}$$

and

$$\frac{\partial^\mu \phi(x,t)}{\partial (-x)^\mu} = {}^bD_x^\mu \phi(x,t) = \frac{(-1)^m}{\Gamma(m-\mu)} \frac{\partial^m}{\partial x^m} \int_x^b \frac{\phi(x',t)}{(x'-x)^{1+\mu-m}} dx' \tag{10}$$

where $m - 1 < \mu < m$, with integer m. The symmetric Riesz fractional derivative, used in the rhs of Eq. (8) above, is given by

$$\frac{\partial^\alpha}{\partial |x|^\alpha} = -\frac{1}{2\cos(\pi\alpha/2)} \left[{}_{-\infty}D_x^\alpha + {}^\infty D_x^\alpha \right]. \tag{11}$$

Fractional derivatives reduce to ordinary ones when μ is integer, while their integral form makes them an appropriate tool to study non-local phenomena, long range correlations, and scale-free transport. In fact, propagators for this family of equations (8) are known (Mainardi et al. 1996), and this connects the CTRW approach with the transport exponents α and β

described in Sect. 2.1. For instance, use of the Riesz fractional derivative in Eq. (8), for $\beta = 1$ leads to a Fourier transform of the form

$$\frac{\partial \widehat{n}(k)}{\partial t} = -D|k|^{\alpha}\widehat{n}(k) \tag{12}$$

whose solution is the characteristic function of a Lévy distribution.

A commonly used non-Gaussian CTRW based on a probabilistic description involving a Lévy random walk was introduced by Klafter et al. (1987). The difference from the CTRW just described is that here, in the random walk of single particles of a given constant energy, a large displacement Δx is associated with a correspondingly long time $\Delta t \equiv \Delta x/v$. Then, in one dimension, the probability ϕ of a random walker making a free path of length Δx (forward or backward) in a time Δt is given as

$$\phi(\Delta x, \Delta t) = A|\Delta x|^{-(1+\alpha)}\delta(\Delta x - v\Delta t), \quad |\Delta x| > \Delta x_0. \tag{13}$$

Here it is important to have coupling between free path length and free path duration, as expressed by the delta function, in order to ensure the conservation of the particle energy, i.e. constant velocity (note that Klafter et al. (1987) uses instead the notation $\mu \equiv 1 + \alpha$). The constant velocity requirement may however be relaxed when considering, for instance, the $\delta \mathbf{E} \times \mathbf{B}$ drift velocity induced by electric fluctuations if $\delta \mathbf{E}$ is a stochastic quantity. For $|\Delta x| < \Delta x_0$, a regular bell-shaped distribution of ϕ is assumed. For $\alpha < 2$ in Eq. (13) (i.e., a Lévy law), the mean square value of Δx, and hence the mean free path λ, is divergent:

$$\langle \Delta x^2 \rangle = \int \Delta x^2 \phi(\Delta x, \Delta t)\, d\Delta x\, d\Delta t \to \infty, \quad \alpha < 2. \tag{14}$$

This means that the central limit theorem, which leads to normal diffusion for finite $\langle \Delta x^2 \rangle$, requires instead a Lévy distribution for the propagator, and the normal diffusion coefficient is divergent. Therefore, this CTRW for $\alpha < 2$ represents another way to describe anomalous transport. By inserting Eq. (13) into Eq. (6), it can be shown that the transport regime depends on the index α of the free path probability distribution,

$$\sigma^2(t) \propto \begin{cases} t^2 & 0 < \alpha < 1, \\ t^{3-\alpha} & 1 < \alpha < 2, \\ t & \alpha > 2 \end{cases} \tag{15}$$

so that superdiffusion with $H = (3 - \alpha)/2$ is obtained for $1 < \alpha < 2$ (Klafter et al. 1987). For $\alpha < 1$, ballistic transport is obtained; for $\alpha > 2$, standard diffusion.

2.4 Examples

In the previous subsections we have discussed several possible nondiffusive behaviors, including both subdiffusion and superdiffusion, as well as several models useful to diagnose their presence in practical situations. In this subsection, we describe several examples in which these non-diffusive dynamics are present.

We start with subdiffusive transport which, as mentioned, requires the existence of antipersistent correlations. It is interesting to note that one natural and important case of antipersistent correlations is found for plasma transport perpendicular to the average magnetic field, when either collisions or pitch-angle scattering cause diffusive motion along the magnetic field, i.e., $\sigma_z^2 = 2D_{\parallel} t$. In the presence of magnetostatic fluctuations, the magnetic field lines may diffuse in the perpendicular direction as $\sigma_\perp^2 = 2D_m \Delta z$, and the resulting perpendicular particle transport is (Rechester and Rosenbluth 1978)

$$\sigma_\perp^2 = 2D_m \sqrt{2D_\parallel t}, \tag{16}$$

which clearly is subdiffusive in time. This transport regime is called double diffusion in laboratory plasmas (Krommes et al. 1983) and compound diffusion in astrophysical plasmas (Duffy et al. 1995; Kóta and Jokipii 2000; Zimbardo et al. 2009; Bitane et al. 2010; Hornsby et al. 2010), and corresponds to $H = 1/4$. Physically, the antipersistence is due to the fact that pitch angle scattering forces particles to trace back the magnetic field lines (in the magnetostatic case), so that a positive perpendicular displacement is followed by a negative one.

A further topical example of non-diffusive behavior is the transport of suprathermal ions in a simple magnetized torus (SMT), in which plasmas are confined by a toroidal magnetic field and a smaller vertical component. This magnetic geometry with open helicoidal field lines incorporates the main ingredients for drift and interchange instabilities, namely pressure gradients and magnetic field line curvature. By numerically integrating the trajectories of suprathermal ions in simulated SMT turbulent fields, and by exploring wide ranges of particle energy and turbulence amplitude, it has been recently shown (Gustafson et al. 2012a) that the ions have a complex motion, which in general cannot be considered diffusive. The simulations show that suprathermal ion dispersion in the direction perpendicular to the average magnetic field starts with a brief ballistic phase followed by a turbulence interaction phase, which shows the entire spectrum of suprathermal ion spreading: superdiffusive, diffusive, or subdiffusive, depending on particle energy and turbulence amplitude (Gustafson et al. 2012b).

Other interesting examples can be found in models of cosmic ray acceleration and transport. Recalling normal diffusion implies that the diffusion coefficient D can be estimated using the TGK relation to yield

$$D = \int_0^\infty \langle v_x(0) v_x(t) \rangle dt \approx v_x^2 \tau. \tag{17}$$

Assuming that the mean free path $\lambda = v\tau$ is given by the mean collision (or correlation) time τ times the velocity $v = (v_x^2 + v_y^2 + v_z^2)^{1/2}$, we can estimate the diffusion coefficient as $D \simeq \frac{1}{3} v^2 \tau = \frac{1}{3} \lambda v$. On the other hand, in the case of superdiffusion the above integral diverges. The divergence of D implies that λ also diverges, since the particle velocity is finite. In particular, for high energy cosmic rays $v \equiv c$, so that the constant velocity model of Eq. (13) is appropriate. This property has an immediate impact on models of cosmic ray acceleration and transport, since the mean free path is a fundamental parameter for particle propagation, the acceleration time, and the maximum reachable energy (e.g., Lagage and Cesarsky 1983a, 1983b; Bieber et al. 1994; Reames 1999). The divergence of D and of λ requires a non Gaussian approach to transport and acceleration in these problems. Some success has been achieved by applying the second CTRW construct introduced in Sect. 2.3, that incorporates a coupling between step-size and waiting-times that depends on the particle velocity (Klafter et al. 1987; Perri and Zimbardo 2012; Gustafson and Ricci 2012; Shlesinger et al. 1982).

3 Wave–Particle Interaction at the Kinetic Scales

Thus far we have considered the statistical description of nondiffusive transport. But what are the physical reasons for random particle motions at the microscopic level? In either weakly collisional or collisionless plasmas, the transport is influenced by electric and magnetic fluctuations and by wave–particle interactions, which cause pitch angle scattering and transverse particle drifts, and by low frequency magnetic fluctuations, which cause field lines to trace a random walk (Jokipii 1966). The strength of these effects varies with the particle species and energy, and with the properties of turbulence, so that a large number of different transport regimes can arise, as shown by independent numerical simulations (Zimbardo et al. 2000a; Pommois et al. 2007; Shalchi and Kourakis 2007; Gustafson et al. 2012b). Detailed study of wave–particle interactions is necessary to understand transport. The portfolio of tools for such study in the kinetic regime has recently been augmented by Vlasov simulations.

In a plasma system where the effects of particle collisions can be considered negligible (such as the solar wind or, in many cases, laboratory plasmas), resonant wave–particle interaction represents the main process by which particles and fields can exchange energy. Wave–particle resonance is central to many physical phenomena including wave damping, particle acceleration, growth of instabilities, generation of anisotropies and, in general, departure from the local thermodynamic equilibrium configuration. In the framework of kinetic theory, the resonant wave–particle interaction is described by the Vlasov equation. In the Vlasov description, complete statistical information about the plasma state is stored in the particle distribution function that represents the probability density in phase space. Analytical solutions of the combined Vlasov-Maxwell equations are known only in a few simplified linear cases, while the nonlinear regime must be investigated numerically.

An important tool to investigate the complexity of the kinetic plasma dynamics is given by the direct numerical simulations. One of the most adopted approaches is represented by the Lagrangian Particle in Cell (PIC) methods (Birdsall and Langdon 1985). Within the PIC approach, the equations of motion of a large number of macroparticles are numerically integrated under the effect of the self-consistent electromagnetic fields. At each time step, the macroscopic plasma variables (density, velocity and current) are obtained by collecting the particles in each grid point of a uniformly spaced grid and then used for the integration of the Maxwell equations for fields. The phase space particle distribution can be evaluated in the same statistically way. In recent years, the PIC codes have been extensively used for the description of the kinetic dynamics of turbulent space plasmas, particularly focusing on many interesting physical aspects, like wave–particle interaction (Araneda et al. 2008), particle heating (Araneda et al. 2009) and turbulence (Gary et al. 2008; Saito et al. 2008; Parashar et al. 2010, 2011; Camporeale and Burgess 2011).

Nowadays, thanks to the impressive increase of the computational technology, the Eulerian approach for the numerical solution of the Vlasov equation has become a valid alternative to the PIC methods. A Eulerian Vlasov code (Mangeney et al. 2002; Valentini et al. 2005, 2007) integrates numerically the Vlasov equation by time-advancing the particle distribution function on a uniform fixed grid in phase space under the effects of the self-consistent electric and magnetic fields. The particle density, the mean velocity and the current density needed for the solution of the Maxwell equations are evaluated at each time step as the velocity moments of the distribution function.

Since the Eulerian Vlasov algorithms advance in time the six-dimensional array that contains the particle distribution function, they are, in many cases, extremely demanding from the point of view of both the execution time and the memory and data storage requirements. However, while in general PIC codes are affected by an intrinsic statistical noise due to the limited number of particles that can be launched in a typical simulation, Eulerian schemes are essentially noise-free. This allows for a clean and precise description of the plasma dynamics even in situations where the PIC results are no longer reliable, like, for example, in the description of the high energy tails of the particle velocity distributions, in the analyses of low density processes or in the study of the tail at short spatial scales of the solar-wind turbulent cascade, where the energy level of the fluctuations is generally very low.

The numerical description of a Vlasov-Maxwell plasma system requires, in the most general case, to perform simulations in a six-dimensional phase space configuration, for both electrons and ions. Due to the large time scale separation between ion and electron dynamics, full electron-ion Vlasov numerical experiments are out of reach of the presently available computing resources. Nevertheless, significant analyses can be performed in phase space configurations with reduced dimensionality or by focusing on the kinetic dynamics of one particle species at a time. From this latter consideration, the so-called hybrid Vlasov-Maxwell algorithm (Valentini et al. 2007) has been recently developed. Within this hybrid model the Vlasov equation is numerically solved, through a Eulerian scheme, for the ion distribution function, while the electrons are treated as a fluid. Although the kinetic description in this hybrid model is restricted only to the ion species, still the computational cost of the hybrid Vlasov-Maxwell simulations is very high, especially when multi-dimensional problems are treated. For this reason, a massive parallelization procedure, based on the use of the Massage Passing Interface (MPI) protocol, has been performed so as to exploit the power of contemporary high performance computation.

In the last years, the Eulerian hybrid Vlasov-Maxwell code has been extensively employed in 1D-3V (one dimension in physical space and three in velocity space) phase space configuration, for the analysis of the kinetic effects on protons during the development of the solar-wind turbulent cascade towards kinetic scales, along the direction of the mean magnetic field (Valentini et al. 2008, 2010, 2011c; Valentini and Veltri 2009). The basic idea behind this numerical research thread is to shed light on the physical mechanisms that replace energy dissipation at short wavelengths in a system, like the solar wind, where collisional viscosity is absent. In this range of short spatial scales, kinetic effects are considered to be the best candidates in governing the system dynamics. Through the analysis of the numerical results of these hybrid Vlasov-Maxwell simulations, a novel branch of electrostatic fluctuations, propagating at a frequency which is a small fraction of the proton plasma frequency, has been identified. These waves, dubbed ion-bulk (IBk) waves (Valentini et al. 2011a, 2011b, 2011c), have acoustic type dispersion and phase velocity comparable to the proton thermal speed. The excitation of these fluctuations was also obtained in laboratory experiments with non-neutral plasmas (Anderegg et al. 2009). At variance with the well-known ion-acoustic waves, the IBk fluctuations can survive against Landau damping (Landau 1946) even for values of the electron to proton temperature ratio of order unity, typical values for the solar-wind environment (as well as for many other plasmas). For these reasons, the IBk fluctuations seem to represent a very efficient channel to carry the solar-wind energy coming from the large MHD scales towards small kinetic scales in the longitudinal component of the energy spectrum.

These hybrid Vlasov-Maxwell simulations have also shown that the resonant interaction of protons with these electrostatic fluctuations produces significant distortions in the longitudinal proton velocity distribution with the generation of a field-aligned beam of accelerated particles that stream at a typical speed close to the local Alfvén velocity. Figure 1(a) displays the longitudinal x–v_x (x being the direction of the ambient magnetic field) phase space contour plot of the proton distribution function. This contour plot shows the generation of a localized trapped particle region (delimited by the vertical white dashed lines) moving with mean velocity close to the phase speed of the IBk waves. The presence of this trapped particle population affects the velocity distribution of protons. This is shown in the plot (b) of Fig. 1, where the v_x–v_y level lines of the proton distribution function, integrated over v_z and averaged over x in the spatial region of trapped particles, are reported.

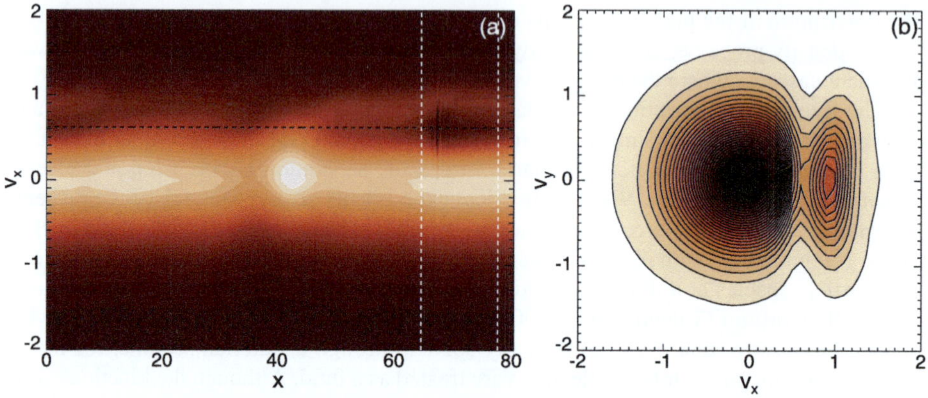

Fig. 1 (**a**): Longitudinal x–v_x phase space contour plot of the proton distribution function. This contour plot displays the generation of localized trapped region. (**b**): Level lines of proton distribution function in the velocity plane v_x–v_y. The generation of a well-defined field-aligned beam of protons is recovered (Valentini et al. 2008)

The generation of these double peaked proton velocity distributions has been recovered in many "in situ" spacecraft observations in the solar wind (Gurnett et al. 1979; Marsch et al. 1982a; Marsch 2006). These results show that non-Maxwellian distribution functions with multiple peaks or bumps can be spontaneously created by the wave–particle interactions in the kinetic regime. The generation of field aligned beams means that pitch-angle scattering is not very effective, as it would smooth out the field aligned beams. This shows that a population of particles can propagate along the magnetic field with little or no scattering, thus creating the long "persistent" displacements which are at the basis of the Lévy random walk and superdiffusive transport. Therefore the study of wave–particle interaction can give information on the microphysics of nondiffusive transport regimes.

More recently, the 2D-3V version of the hybrid Vlasov-Maxwell code has been also used to investigate the role of local kinetic effects in plasma turbulence, in the plane perpendicular to an ambient magnetic field (Servidio et al. 2012). In these simulations, during the evolution of turbulence, coherent structures and vortices appear in the bi-dimensional maps of the inplane magnetic field. Nearby the regions of high magnetic stress, magnetic reconnection events can occur locally as the results of the generation of small scales along the turbulent cascade. Figure 2 displays the contour map of the proton temperature anisotropy, defined as the ratio between perpendicular and parallel temperature, with respect to the direction of the local magnetic field. The black/white lines indicate the isosurface of the magnetic potential of the inplane magnetic field. It is clearly shown that, in the region of high magnetic stress, the particle velocity distributions depart from the typical configuration of thermodynamical equilibrium showing the generation of temperature anisotropy, both along and across the local magnetic field.

At present, an updated version of the hybrid Vlasov-Maxwell code has been implemented to take into account also the kinetic dynamics of alpha particles (Perrone et al. 2011) with the purpose of providing a more realistic description of the solar wind. In fact the interplanetary medium is constituted predominantly of protons, but a small amount of doubled ionized helium is also present. In recent works, through the statistical analysis of the solar wind data from the Helios spacecraft (Bourouaine et al. 2010, 2011a, 2011b), it has been pointed out that the dynamics of alpha particles can present important signatures of kinetic effects, like temperature anisotropy and heating, that in many situations are more evident than in the

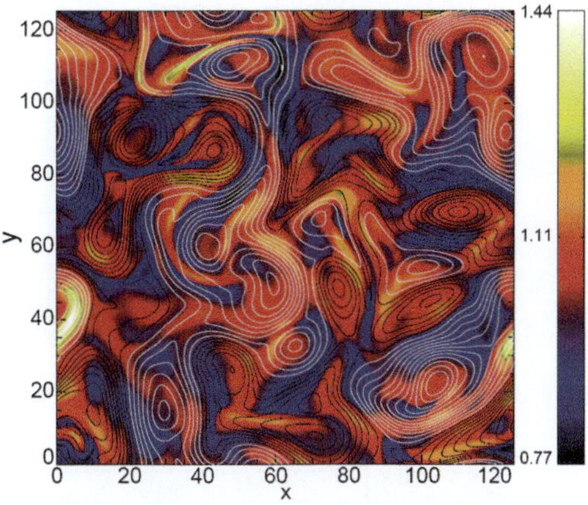

Fig. 2 Contour plot of the proton temperature anisotropy $T_{\perp}/T_{\parallel}$. The *black/white lines* indicate the isosurface of the magnetic potential of the inplane magnetic field; the different colors of the contour lines indicate the different direction of rotation of the vortices (Servidio et al. 2012)

case of protons. In 2011 the updated version of the hybrid Vlasov-Maxwell code (protons and alphas) has been used in 1D-3V phase space configuration (Perrone et al. 2011) to generalize the study of the role of kinetic effects in the longitudinal component of the solar-wind turbulent spectra, including the kinetic dynamics of the alpha particles.

4 Finite Larmor Radius Effects and Nonclassical Transport in Turbulence

There are essentially two approaches to capturing the effects of finite Larmor radius on particle transport in turbulent plasmas, whose relevance to a given physical situation depends on the type of turbulence considered. If the underlying magnetic field is not laminar, in the sense that the field lines deviate from each other or are braided together, it is essential to capture these features in statistical terms, and combine them (Duffy et al. 1995; Hornsby et al. 2010) with a model for particle orbits. For example, guiding centre drift combined with finite Larmor radius excursion may transfer particles between locally neighboring field lines which ultimately diverge substantially. Conversely large Larmor radius excursions may average out the effects of small scale (sub Larmor radius) variations in magnetic field.

If the underlying magnetic field can be considered locally laminar, the gradients of pressure and temperature within the plasma typically act as sources of free energy that can excite electrostatic turbulence. This is usually the dominant source of particle and energy transport in laboratory plasmas, and is expected to arise, beside electromagnetic turbulence, in space and astrophysical plasmas wherever gradients are steep; in foreshock regions, for example Kirk and Dendy (2001), Schmitz et al. (2002a, 2002b); Lee et al. (2005a, 2005b). The effects of the resultant electrostatic turbulence on particle transport, given finite Larmor radius, can be captured in two ways. First, one can implant test particles in a simulation, whose orbits respond to the ambient turbulent fields, but do not generate or act upon them. The ensemble statistics are then calculated. This is particularly appropriate to situations where the electrostatic turbulence can be modeled in quasi-fluid terms. For example, in Hasegawa-Mima turbulence, the effect of finite Larmor radius on the transport of ions, whose guiding centres drift with the $\mathbf{E} \times \mathbf{B}$ velocity, was first examined in Manfredi and Dendy (1996, 1997). These studies contribute to the understanding of the differential transport of fusion-born alpha particles (3.5 MeV initial energy; large Larmor radius) compared to thermal ions (10 keV; small Larmor radius) in magnetically confined plasmas. The approach to finite Larmor radius adopted in Manfredi and Dendy (1997) is equivalent to multiplying the kth

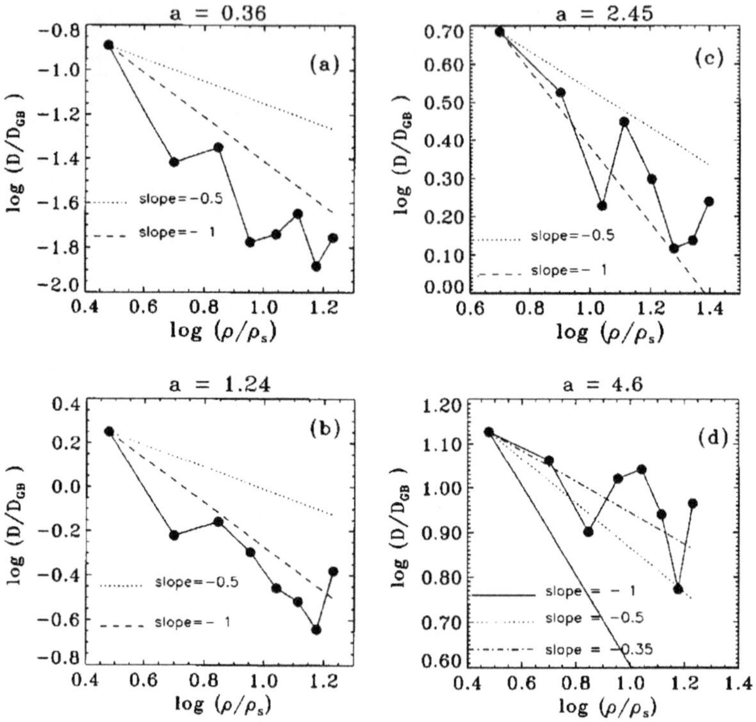

Fig. 3 Scaling of the diffusion coefficient against Larmor radius on a logarithmic scale (base 10) for four values of the normalized amplitude: (**a**) $a = 0.36$; (**b**) $a = 1.24$; (**c**) $a = 2.45$; and (**d**) $a = 4.6$. The slope -0.5 line (*dotted line*) corresponds to the large amplitude scaling, while the slope -1 line (*broken line*) corresponds to small amplitude scaling. In (**d**) the slope -0.35 is also indicated, which corresponds to a theoretical result for large amplitude. The different regimes of small and large amplitude are clearly visible. Reprinted with permission from Manfredi and Dendy (1997). Copyright (1997), American Institute of Physics

Fourier component, $E(k)$, of the electrostatic field, by an effective amplitude $J_0(k\rho)$, where ρ is the Larmor radius and J_0 is a Bessel function (Gustafson et al. 2008). The transport depends also on the normalized strength a of the field, which scales as the ratio of its root-mean-square amplitude and frequency; see Eq. (8) of Manfredi and Dendy (1997). Figure 3 (Fig. 14 of Manfredi and Dendy (1997)) shows the scaling of ion transport with Larmor radius in Hasegawa-Mima turbulence at different amplitudes.

Subsequent investigations (Annibaldi et al. 2000, 2002) draw attention to the non-Gaussian, strange kinetics statistical properties of the associated transport processes, arising from the interplay between finite Larmor radius ion orbits with coherent nonlinear structures in the turbulence. For example, Fig. 4 (corresponding to Figs. 13 to 15 of Annibaldi et al. (2002)) provides clear evidence of subdiffusion, superdiffusion, and ballistic motion, in addition to normal diffusion, depending on the character of the Hasegawa-Mima turbulence quantified by the plasma β and the magnitude of the Larmor radius. More sophisticated models for electrostatic turbulence have subsequently been studied from the same perspective: for example, the Hasegawa-Wakatani model, in which electrostatic potential and density are treated as independent coupled variables, has recently been extended to incorporate a gradient in the background magnetic field. The associated $\mathbf{E} \times \mathbf{B}$ transport of test particles is studied in Dewhurst et al. (2009), and finite Larmor radius is incorporated in Gustafson

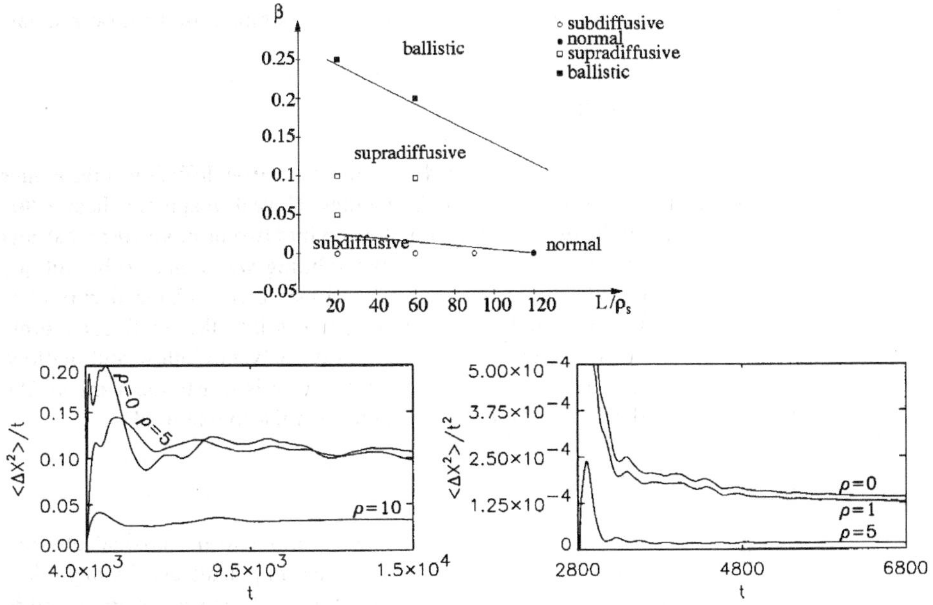

Fig. 4 *Top*: Dependence of ion transport regimes on plasma inhomogeneity parameter β and the ratio of system size L to Larmor radius ρ. *Bottom*: (*Left*) Dependence of transport on Larmor radius for $\beta = 0$ and $L = 120$, for $\rho = 0, 5$, and 10. Mean squared displacement divided by time reveals normal transport with diffusion coefficient decreasing with ρ. (*Right*) Dependence of transport on Larmor radius for $\beta = 0.2$ and $L = 60$, for $\rho = 0, 1$, and 5. Mean squared displacement divided by squared time reveals ballistic transport with coefficient decreasing with ρ. Reprinted with permission from Annibaldi et al. (2002). Copyright (2002), American Institute of Physics

et al. (2008) and Dewhurst et al. (2010). The latter includes combinations of small-scale turbulence with large scale coherent nonlinear structures, addressed below in Sect. 5. A key physics point emerging from these studies is the centrality of the Weiss field Q, which is proportional to the difference between squared stress and squared vorticity in the turbulence, in governing the interaction between finite Larmor radius particles and coherent nonlinear structures. This applies both to the spatial distribution of local concentrations of Q (positive or negative), see for example Figs. 1 and 3 of Annibaldi et al. (2002) and Fig. 2 of Dewhurst et al. (2010), and to the statistics of fluctuations in Q.

4.1 Test Particle Simulations of Transport in the Presence of Magnetic Turbulence

The transport of plasma particles in the presence of magnetic turbulence can be studied by test particle numerical simulations. Making reference to the solar wind, the background magnetic field is assumed to be constant (although plasma transport in nonhomogeneous systems like the magnetosphere is also relevant (see, e.g., Zimbardo et al. 2010)), while the turbulence properties are varied. In most numerical models, turbulence is represented by a discrete number of modes, with different models having different degrees of anisotropy, dimensionality (1D, 2D, or 3D), and spectral shape. In particular, turbulence anisotropy is described either by the so-called slab turbulence, in which the turbulence wave vectors are parallel to the background field $\mathbf{B}_0$, or by the so-called 2D turbulence, in which the turbulence wave vectors are distributed only in the plane perpendicular to $\mathbf{B}_0$. Other turbulence models, even nonaxisymmetric, have been developed.

Anomalous transport corresponding to perpendicular subdiffusion was reported by Qin et al. (2002a) in the case of slab turbulence, while parallel diffusion was found to be normal. The perpendicular subdiffusion is a result of particles tracing back the field lines after pitch angle diffusion, and is related to the so-called compound diffusion (Kóta and Jokipii 2000; Webb et al. 2006) discussed in Sect. 2.4. The latter is dependent on the rate of exponential separation of close field lines (e.g., Rechester and Rosenbluth 1978), which is smaller for slab anisotropy and is larger for quasi 2D anisotropy (Zimbardo et al. 2009; Bitane et al. 2010). If the exponential separation of field lines is fast enough, particles which have been pitch angle back scattered will not trace back the original field line, and normal diffusion results. This also means that the antipersistent correlations implied by compound diffusion are destroyed by the stochasticity of field lines. Indeed, for a composite model of slab plus 2D turbulence with 80 % of fluctuation energy in the 2D spectrum, Qin et al. (2002b) have recovered diffusion also perpendicular to $\mathbf{B}_0$, in agreement with the fact that the exponentiation of field lines is faster for 2D turbulence.

The influence of turbulence anisotropy on the structure of magnetic flux tubes has been studied by Zimbardo et al. (2004) with a fully 3D spectrum where quasi-slab turbulence is represented by a cigar-shaped distribution of wave vectors along $\mathbf{B}_0$, while the quasi-2D turbulence is represented by a pancake-shaped distribution of wave vectors in the plane perpendicular to $\mathbf{B}_0$. The axes of the constant amplitude ellipsoids in phase space are given by the inverse of the correlation lengths $l_\parallel$ and $l_\perp$ (Zimbardo et al. 2000a), with $l_\parallel/l_\perp \gg 1$ ($\ll 1$) corresponding to the quasi-2D (quasi-slab) anisotropy. Zimbardo et al. (2004) have shown that indeed the complexity of the magnetic flux tubes grows with $l_\parallel/l_\perp$ and with the Kubo number, see Fig. 5.

Zimbardo (2005) and Zimbardo et al. (2006) studied the effect of turbulence anisotropy on particle transport with the above 3D anisotropic numerical model. Here and in next sections, the anomalous transport exponent is indicated by $\gamma \equiv 2H$, where H is the Hurst exponent. It is found that for quasi-slab turbulence, transport is anomalous, corresponding to parallel superdiffusion with $\gamma \simeq 1.2$, and to perpendicular subdiffusion with $\gamma \simeq 0.8$. The flatness of the distribution of field line positions is much larger than the Gaussian value of 3, confirming that anomalous diffusion is related to non-Gaussian statistics and heavy tailed distributions. On the other hand, going to the isotropic case, parallel transport is superdiffusive with $\gamma \simeq 1.4$, while perpendicular transport is normal. In the quasi-2D case, both parallel and perpendicular transport are found to be normal, confirming the results of Qin et al. (2002b). These results emphasize the importance of turbulence anisotropy. Parallel superdiffusion with $\gamma \simeq 1.3$ and perpendicular subdiffusion with $\gamma \simeq 0.75$ were also found by Shalchi and Kourakis (2007) by injecting test particles in a composite turbulence model with 20 % slab turbulence and 80 % 2D turbulence, thus expanding the range of cases when superdiffusion can be found. On the other hand, the transport regime also depends on the particle energy, in particular through the ratio between the particle Larmor radius ρ in the background magnetic field and the turbulence correlation lengths. Pommois et al. (2007) find that for $\rho/l_\perp = 0.001$–0.01, parallel superdiffusion and perpendicular subdiffusion can be found, while normal diffusion is obtained for $\rho/l_\perp \sim 0.1$ (see Fig. 6). These results correspond to the fact that the turbulence energy density scales as $E(k) \sim k^{-5/3}$, so that higher energy particles, having larger Larmor radii, resonate with longer wavelength fluctuations which have larger amplitude; correspondingly, pitch angle scattering is faster, and the long parallel displacements which give rise to superdiffusion are no longer found. Therefore normal diffusion results. Interestingly, from Fig. 6 one can see that for $\rho/l_\perp \sim 1$ parallel superdiffusion is obtained again: this is because in the turbulence model considered by Pommois et al. (2007) most of the fluctuation energy is at scales smaller than $l_\perp$. In such a case, most fluctuations are average out by the large Larmor radius, and pitch angle scattering is reduced.

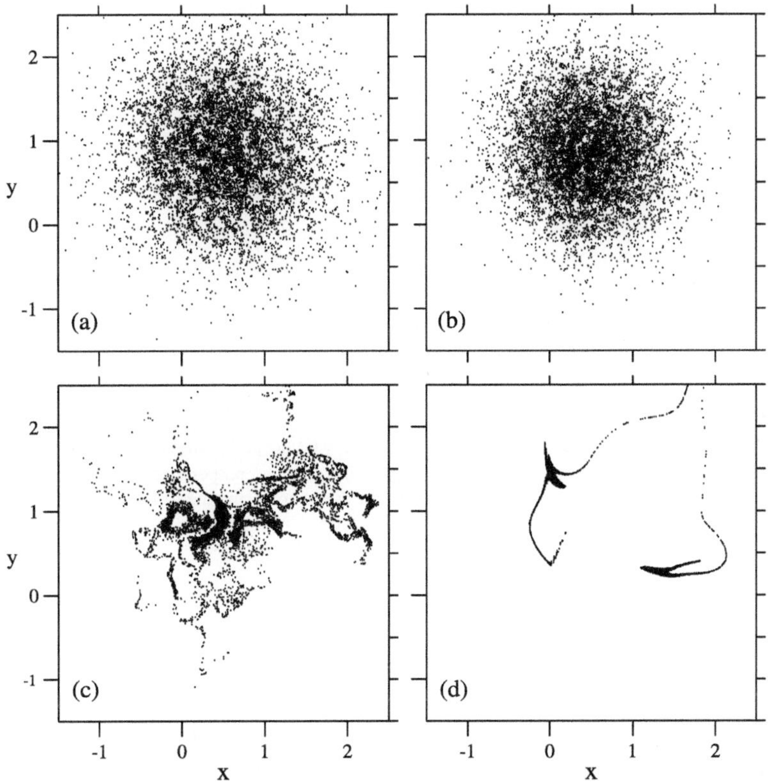

Fig. 5 Cross section of the magnetic flux tube structure of an initially circular flux tube, for different degrees of turbulence anisotropy: (**a**) quasi-2D anisotropy, (**b**) isotropic turbulence; (**c**) and (**d**) quasi-slab anisotropy, for two different ratios $l_\parallel/l_\perp$. Reprinted with permission from Zimbardo et al. (2004). Copyright 2004 by the American Geophysical Union

Recently, Tautz and Shalchi (2010) have found that perpendicular transport is clearly subdiffusive, with $\gamma \simeq 0.6$, in the case of slab turbulence. Normal perpendicular diffusion is almost recovered in the case of isotropic or two-component (slab + 2D) turbulence, but a weakly subdiffusive behavior remains, corresponding to $\gamma = 0.83$–0.94. According to Tautz and Shalchi (2010), perpendicular subdiffusion is the rule rather than the exception. An interesting recent study shows that perpendicular subdiffusion is obtained for magnetostatic slab turbulence, while the inclusion of wave time dependent electric and magnetic field leads to parallel superdiffusion as well as to particle energization (Tautz 2010; Gustafson et al. 2012a). Such an energization is due to stochastic (second order) Fermi acceleration. A similar process has also been investigated by Perri et al. (2007) and by Perri et al. (2011) in two-dimensional and three-dimensional numerical models where test particles interact with time dependent electromagnetic fluctuations. The stochastic interaction leads to superdiffusion in real space, diffusion in momentum space, and to a particle energization characterized by both energy gains and energy losses, which are typical of a second-order Fermi process. The consequences of occasional collisions for particle transport in magnetic turbulence were quantified using a computational Vlasov-Fokker-Planck model by Hornsby et al. (2010). This enables one to identify the connection to classic transport paradigms important for laboratory plasmas, such as quasi-linear and gyro-Bohm, and the extent to which

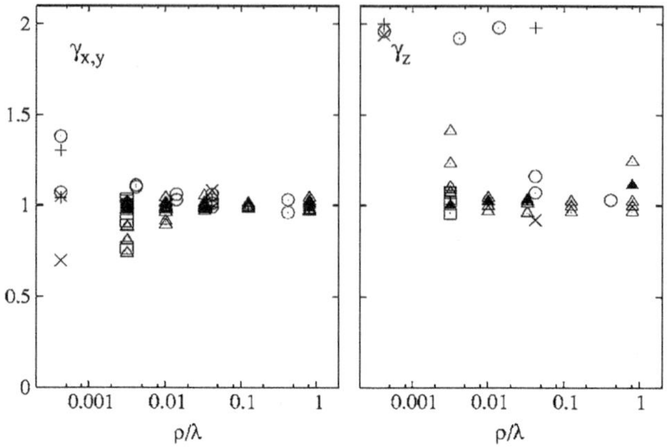

Fig. 6 Anomalous diffusion exponent γ versus the ratio of Larmor radius ρ over turbulence correlation length λ. *Left panel*: exponents γ_x and γ_y for transport in the plane perpendicular to the average magnetic field. *Right panel*: exponent γ_z for transport parallel to the average magnetic field. The different symbols denote different turbulence levels: *plus signs*, $\delta B/B_0 = 0.05$; *circles*, $\delta B/B_0 = 0.1$; *crosses*, $\delta B/B_0 = 0.2$; *triangles*, $\delta B/B_0 = 0.5$; *squares*, $\delta B/B_0 = 1.0$. Reprinted with permission from Pommois et al. (2007). Copyright (2007), American Institute of Physics

these provide a reliable guide depending on the strength of the turbulence. All these results agree with the main trend of the above quoted works, and we note that the various studies use substantially different numerical realizations of turbulence, so that the cases when anomalous transport is obtained emerge as being independent of the numerical scheme.

4.2 Vlasov Simulations

Let us now turn to cases where the turbulence is fundamentally kinetic in nature, so that the self-consistent evolution of particle dynamics, including finite Larmor radius, together with fields is central to the model. It is still possible to adopt a test particle approach in such cases, and to compute ensemble transport properties. This applies to Vlasov and particle-in-cell (PIC) kinetic simulations of turbulence. For example, Fig. 16 of Manfredi et al. (1996) shows diffusion coefficients inferred in this way from finite ion Larmor radius Vlasov gyrokinetic simulations of ion temperature gradient-driven turbulence. A systematic analysis of fields experienced by finite Larmor radius ions undergoing acceleration in PIC simulations of turbulence at solar system and astrophysical shocks is given in Lee et al. (2005a, 2005b).

The role of kinetic effects in turbulent plasmas has became a subject of increasing interest within space plasma physics. The interstellar medium is generally observed from spacecraft measurements (Bruno and Carbone 2005; Marsch 2006) to be in a fully turbulent regime. Along the turbulent cascade, nonlinear local couplings transfer energy from low to high frequencies. At low MHD frequencies, the energy spectra of the magnetic fluctuations display a slope in agreement with the Kolmogorov 5/3 law for fluid turbulence. When the energy is transferred towards increasingly high frequencies (increasingly small scales), kinetic effects come into play, leading to significant changes in the spectral properties (Alexandrova et al. 2009; Sahraoui et al. 2010). The first evident modification has been identified at length scales of the order of typical ion kinetic scales, namely the inertial length and Larmor radius (Bale et al. 2005). For values of the plasma parameter β_p (ratio between kinetic and magnetic pressures) of order unity, typical of the solar wind plasma away from the Sun, the inertial length and Larmor radius are comparable in size. Linked to these spectral changes, the signatures of finite Larmor radius effects have been also recovered from the analysis of particle velocity distributions. Significant distortions and non-Maxwellian features have been observed both for the proton and the alpha particle velocity distributions. For example, in recent work by Bourouaine et al. (2010, 2011a, 2011b) the generation of temperature anisotropy has been analyzed for the solar wind particle velocity distributions from the Helios spacecraft. It is found that more significant effects are visible for alpha particles than for protons (Chen et al. 2011). In this case the cyclotron resonance predicts that alpha particles stay in resonance with the waves having a lower frequency than those that the protons resonate with, and thus they can receive by absorption more wave energy for the turbulent spectrum than protons (Miller 1998).

The hybrid Vlasov-Maxwell code, briefly described in Sect. 3, has been employed in 2D-3V phase space configuration to identify the role of kinetic effects in the range of spatial scales close to the proton Larmor radius, in a plasma composed of kinetic ions (protons and alpha particles) and fluid electrons in conditions typical of the solar-wind environment. Here we discuss the numerical results for the kinetic dynamics of alpha particles and protons in decaying turbulence. The plasma dynamics is investigated in a doubly periodic x–y spatial domain perpendicular to a background magnetic field (Perrone et al. 2013). In the initial equilibrium the ion species have homogeneous densities and Maxwellian velocity distributions. The equilibrium configuration is perturbed by a 2D spectrum of fluctuations for the magnetic and proton velocity fields. Energy is injected with random phases and wave numbers in the range $0.1 < k\rho_p < 0.3$, where $k = 2\pi m/L$, with $2 \leq m \leq 6$. The rms of the initial magnetic perturbations is $\delta B/B_0 \simeq 0.3$. Neither density disturbances nor parallel variances are imposed at $t = 0$. The proton plasma beta is $\beta_p = 2{v_{th,p}}^2/V_A^2 = 2$ and the electron to proton temperature ratio is $T_e/T_p = 1$. For the alpha particles we set $Z_\alpha = 2$; $m_\alpha/m_p = 4$, $n_{0,\alpha}/n_{0,p} = 5\ \%$ and $T_\alpha/T_p = 1$. With this choice, the alpha particle thermal speed is $v_{th,\alpha} = v_{th,p}/2$.

The statistical analysis is performed at a given instant of time at which the level of the turbulent activity has attained its maximum value. During the evolution of turbulence, coherent structures and strongly sheared flows appear in the 2D pattern in physical space. In Fig. 7 [panel (a)], the shaded contours of the out-of-plane total current density j_z are represented together with the contour lines of the magnetic potential A_z of the inplane magnetic field. Different directions of rotation of the vortices in the contour lines of A_z are indicated by different colors (black and white). This figure shows that the current density becomes very intense in between the magnetic islands, and reconnection events occur at the X-points of A_z, indicated in the figure by red crosses.

In order to quantify the kinetic effects on alpha particles and protons, generated when the energy is transferred towards short scales along the turbulent cascade, the temperature anisotropy for each ion species has been computed as the ratio between perpendicular and parallel temperatures with respect to the local magnetic field: $A_i = T_\perp^{(i)}/T_\parallel^{(i)}$, where i stands for protons and alpha particles, respectively. The temperature of each ion species can be evaluated as the second order velocity moment of the particle distribution function, through a direct integration in velocity space:

$$T_i = \frac{m_i}{3n_i} \int (\mathbf{v} - \mathbf{u_i})^2 f_i d^3v \quad (i = p, \alpha) \tag{18}$$

where f_i is the ion distribution function, $n_i = \int f_i d^3v$ is the particle density and $\mathbf{u}_i = \int \mathbf{v} f_i d^3v/n_i$ is the ion-bulk velocity. Even though the initial conditions of the simulations are set up so as to have isotropic temperatures for both protons and alpha particles at $t = 0$

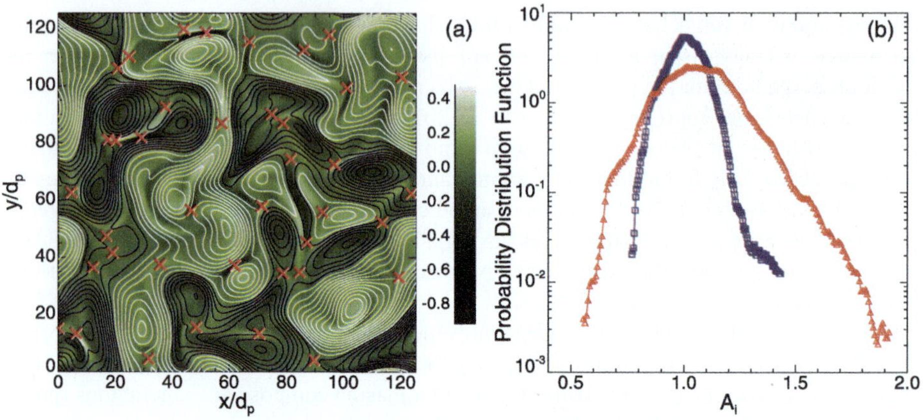

Fig. 7 Panel (**a**): Contour plot of the out-of-plane total current density (j_Z). The isolines of the magnetic potential A_Z are indicated by *black/white lines*. The positions of the X-points, where reconnection occurs, are indicated by *thick red crosses*. Panel (**b**): PDF of the temperature anisotropy A_i of protons (*blue-square line*) and alpha particles (*red-triangle line*) at the maximum of the turbulent activity $(t = 40)$

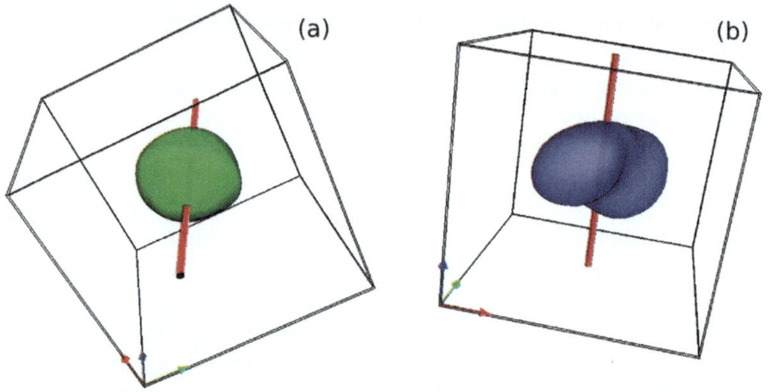

Fig. 8 Isosurfaces of the proton (**a**) and alpha particle (**b**) velocity distributions at two different spatial locations, where $A_p, A_\alpha > 1$. The direction of the local magnetic field is displayed as a *red tube*

$(A_i = 1)$, it is found that, during the system evolution, the anisotropy index A_i can depart significantly from unity for both ion species. To quantify this statement, the probability distribution function (PDF) of A_i has been evaluated at the time when the simulated turbulence is strongest. The results are shown in Fig. 7 [panel (b)] for protons (blue-square line) and alpha particles (red-triangle line). Both ion species develop temperature anisotropy with respect to the directions parallel and perpendicular to the local magnetic field. This anisotropy is significantly more evident for alpha particles than for protons, and this corresponds with recent solar-wind observations from the Helios spacecraft (Bourouaine et al. 2010, 2011a, 2011b).

The isosurfaces of the particle velocity distribution, evaluated at the spatial locations where the anisotropy index reaches its maximum value, are plotted in Fig. 8 for protons [panel (a)] and alpha particles [panel (b)]. In each panel, the direction of the local magnetic field is displayed as a red tube. From this figure, it is evident that, while the proton velocity distribution remains quite close to the Maxwellian spherical shape, the alpha-particles

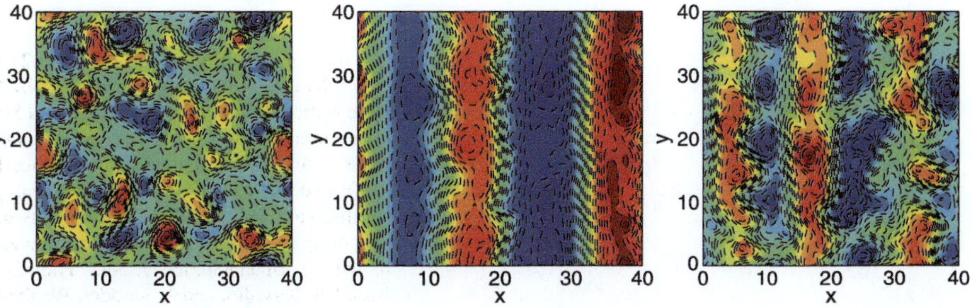

Fig. 9 Electrostatic potential in the plane perpendicular to the magnetic field obtained from a generalized Hasegawa-Wakatani model in three regimes: (*Left*) Small scale turbulence with damped zonal flows; (*Centre*) Self-generated zonal flows dominate; (*Right*) Energy of zonal flows and of small scale turbulence constrained to be equal. Reprinted with permission from Dewhurst et al. (2010). Copyright (2010), Institute of Physics

velocity distribution is evidently shaped by kinetic effects, displaying a certain elongation with the formation of a bubble structure in the direction perpendicular to the local magnetic field. These typical non-Maxwellian velocity distributions are common features of solar-wind plasmas (Marsch et al. 1982a, 1982b; Bourouaine et al. 2010, 2011a, 2011b).

The numerical results described here suggest that Eulerian Vlasov simulations can provide a noiseless multi-ion description of collisionless plasmas in physical conditions close to reality and represent an indispensable tool for the interpretation of the complex phenomenology recovered in the solar-wind observations.

5 Non-diffusive Transport Arising from the Combination of Small Scale Turbulence with Large Scale Coherent Nonlinear Structures

In well diagnosed laboratory plasmas, coexistence and interaction (Fujisawa 2011; Diamond et al. 2011) between small scale drift turbulence (Tynan et al. 2009) and large scale coherent nonlinear structures—zonal flows (Diamond et al. 2005), streamers (Yamada et al. 2008), and other objects that exhibit long range correlation (Inagaki et al. 2011)—is an established feature which is central to the phenomenology of the global system (Wagner 2007). Approaches to modelling this span the zero-dimensional Lotka-Volterra predator-prey paradigm (Malkov and Diamond 2009), nonlinear few-wave coupling (Manfredi et al. 2001), and large scale numerical simulations, which however are challenged by the need to incorporate a wide range of physically relevant lengthscales.

It is therefore interesting to include finite Larmor radius test particle dynamics in a plasma model which can incorporate the coexistence and interaction of small scale turbulence and coherent nonlinear structures. Figure 9 shows three regimes of electrostatic turbulence in the generalized Hasegawa-Wakatani model of Dewhurst et al. (2009), with small scale turbulence coexisting with zonal flows in the right-hand panel. The corresponding transport properties of ensembles of test particles that have finite Larmor radius ρ of different sizes, obtained in Dewhurst et al. (2010), are shown in Figs. 10 and 11. In Fig. 10, for convenience we quantify the mean squared spatial dispersion in terms of a diffusion coefficient, however Fig. 11 emphasizes that the transport is not necessarily diffusive. Furthermore, where it is diffusive, the rate of transport is substantially different between the cases where small scale turbulence is dominant and where it coexists with zonal flows. It is evident that the transport properties are sensitive, in particular, to the size of Larmor radius ρ

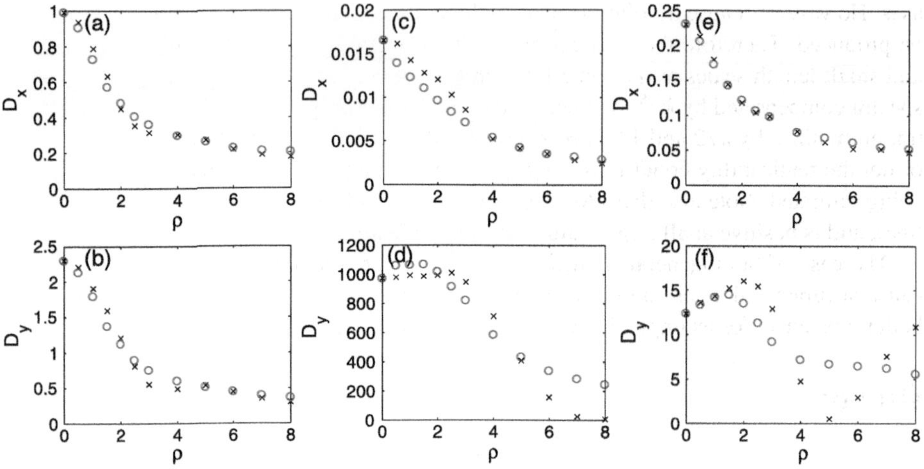

Fig. 10 Test particle diffusion coefficients, as functions of Larmor radius ρ, in the different turbulence regimes of Fig. 9. Directions of axes are perpendicular (x) and parallel (y) to the dominant orientation of zonal flows, where they exist. *Crosses* indicate results when all the test particles share the same Larmor radius ρ; *circles* indicate results when the Larmor radii are distributed around a most probable value ρ. Reprinted with permission from Dewhurst et al. (2010). Copyright (2010), Institute of Physics

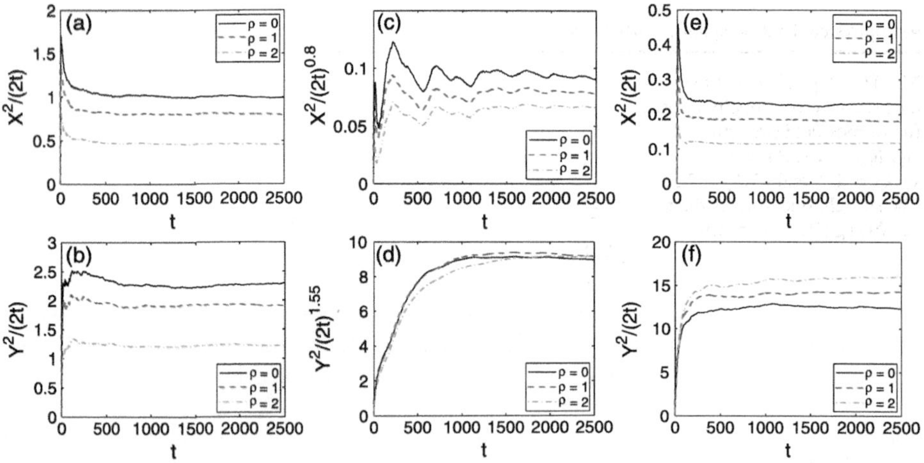

Fig. 11 Test particle dispersion in the x (*top*) and y (*bottom*) directions, as functions of time, for three different values of Larmor radius ρ. The three *vertical pairings* correspond to the three turbulence regimes of Fig. 9. (*Left*) normal diffusion; (*Centre*) subdiffusion in x and superdiffusion in y; (*Right*) normal diffusion at rates substantially lower in x and higher in y than in the (*Left*) case. Reprinted with permission from Dewhurst et al. (2010). Copyright (2010), Institute of Physics

relative to typical scale of the small scale turbulent vortices ($\rho = 3$) and the half-wavelength between zonal flows ($\rho = 8$) (Hauff and Jenko 2008).

Another useful test-bed is found in the injection of suprathermal ions into turbulence produced in an simple magnetized torus (SMT). Here, in Fig. 12, we present three simulated, representative examples of transport for an injected beam of mono-energetic ions (Gustafson and Ricci 2012): subdiffusion at $\mathcal{E} = 250$, diffusion at $\mathcal{E} = 25$, and superdiffusion at $\mathcal{E} = 5$,

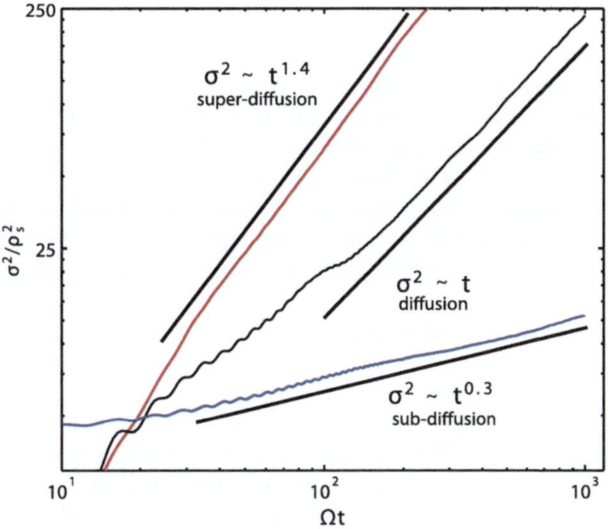

Fig. 12 Radial positional dispersion σ^2 for turbulent suprathermal ion transport for three different injection energies [$\mathcal{E} = 5$ (*red, topmost curve*), $\mathcal{E} = 25$ (*black, middle curve*) and $\mathcal{E} = 250$ (*blue, lower curve*)] showing representative examples of superdiffusion, diffusion, and subdiffusion, respectively. We express σ in terms of ρ_S, the ion sound radius for the SMT plasma, and t in terms of Ω, the suprathermal ion Larmor frequency. Reprinted with permission from Gustafson and Ricci (2012). Copyright (2012), American Institute of Physics

Table 1 Measured power law exponent for $\sigma^2(t)$, $p(\Delta r)$ and $\Delta r(\Delta t)$

$\gamma \pm 0.1$	$\mu \pm 0.1$	$\nu \pm 0.1$
1.4	3.0	0.7
1.0	3.5	0.9
0.3	3.5	0.15

where $\mathcal{E} \equiv mv_0^2/(2T_e)$ is the initial ion energy normalized to the plasma electron temperature, T_e. For these cases, the turbulence amplitude in the SMT is such that $\xi = 0.8$, where $\xi \equiv e\ \delta\phi/T_e$ and $\delta\phi$ is the standard deviation of the amplitude of the electrostatic potential. In Fig. 12, the mean square deviation of particle radial positions, $\sigma^2 \sim t^\gamma$, is shown as a function of time. The value of γ is determined by best-fitting the growth of σ^2 with a power-law curve during the turbulence interaction regime (described below). The values of γ for the three different cases are summarized in Table 1. Error in the fitted values is of order ± 0.1. The significant differences in the values of γ for different energies result from the interplay of turbulent diffusion, gyroaveraging, and geometrical effects. Essentially, the spreading is superdiffusive when both gyroaveraging and vertical $\nabla\mathbf{B}$ drifts are small, while the spreading becomes subdiffusive, $\gamma \rightarrow 0$, when vertical drift averaging becomes dominant over turbulent diffusion, as described in more detail in Gustafson et al. (2012b). Note that periodic oscillations in $\sigma^2(t) \equiv \langle \Delta r^2(t) \rangle$ are caused by the ion gyromotion.

The SMT study thus establishes that suprathermal ion spreading can be subdiffusive, diffusive, or superdiffusive depending on the ion energy and turbulence amplitude. In previous works, superdiffusion and subdiffusion have been used to model transport in plasmas, see e.g. Rechester and Rosenbluth (1978), del Castillo-Negrete (1998), Pommois et al. (2001), Ruffolo et al. (2003), Hauff et al. (2007), Sanchez et al. (2008), Burillo et al. (2009). The coexistence of the three regimes was also observed in Abdullaev (2000). The SMT simulations show that the diffusion approximation is valid only locally, or, stated otherwise, the effective local suprathermal ion diffusivities can show strong time dependence. For example, they can be significantly different from thermal ion diffusivities computed for similar turbulent plasmas in Ricci and Rogers (2009).

The assumption of coupled step sizes, Δr, and step durations, Δt in the CTRW framework, see Sect. 2.3, determines γ as a function of the two parameters ν and $\mu = \alpha + 1$ (Klafter et al. 1987). In particular μ can be found from the step-size PDF, $p(\Delta r) = \int \phi d\Delta t$, and ν is inferred from the relation between Δt and Δr. The SMT results from TORPEX simulations give superdiffusive and diffusive values of ν half the magnitude of those found by Zimbardo et al. (2000b). This is sensible, since that study of magnetic turbulence used a constant, unidirectional accelerating electric field, leading to larger Δr for a given Δt. The analytical predictions of the dependence of γ on ν and μ presented in Klafter et al. (1987) succeed in predicting the character of the SMT transport with gratifying accuracy (see Table 1). For example, the subdiffusive γ is predicted by Klafter et al. (1987) to be $\gamma = \nu\mu' - 1 \simeq 0.4$ with $\mu' = \mu - 1 + 1/\nu$, in close agreement with the observed $\gamma = 0.3$. Similarly, all three examples of suprathermal ion transport match predictions of the Klafter walk theory (Gustafson and Ricci 2012).

Having reviewed the theory and modeling of suprathermal ion transport in turbulent plasmas for the SMT configuration, we discuss here how it can be extended to the interpretation of TORPEX experimental data and show an exploratory comparison between experiments and simulations. TORPEX is an SMT (1 m major radius, 0.2 m minor radius) characterized by low plasma densities ($n_e \approx 10^{16}$–10^{17} m^{-3}) and temperatures ($T_e \sim 5$–20 eV). TORPEX is equipped with an extensive set of diagnostics allowing high-resolution measurements of plasma parameters and wave fields throughout the plasma cross-section. Plasmas of different gases can be produced and sustained by microwaves in the electron cyclotron frequency range, $f = 2.45$ GHz. A number of turbulence regimes have been characterized experimentally and validated numerically (Ricci and Rogers 2010; Ricci et al. 2009, 2011) for TORPEX. Here, a scenario with $B_t = 74$ mT and $B_v = 2$ mT is used, resulting in a SMT with vertical magnetic field line return distance $\Delta D \approx 17$ cm, which is dominated by an ideal interchange mode with wavenumbers $k_\parallel \approx 0$, and, $k_\perp \approx 37$ m^{-1}. This scenario, similar to those extensively studied in Furno et al. (2008a, 2008b), Müller et al. (2007), Theiler et al. (2008, 2009), Diallo et al. (2008), Podestà et al. (2008), Labit et al. (2011), Furno et al. (2011) using electrostatic probes, is characterized by the presence of a region on the low-field side where coherent structures are observed to propagate radially outward resulting in intermittent non-Gaussian transport of particles, heat, momentum and current.

Suprathermal Li^{6+} ions are injected using a miniaturized ion source, such that Li^{6+} ion currents up to 10 μA can be obtained (Plyushchev et al. 2006). The source is motorized and can be continuously moved over a toroidal distance of 50 cm between each discharge. Ion energy and current density profiles are measured using a miniaturized gridded energy analyzer (GEA), which consists of two identical GEAs facing opposite directions for background noise subtraction. Each detector has small dimensions (15 mm in diameter, 70 mm in length and in inlet diameter of 8 mm), and is able to measure fast ion currents as small as 0.1 μA. Synchronous detection is used to increase the signal-to-noise ratio by modulating the emitter bias voltage at a given frequency ($\sim$1 kHz). The GEA detector is installed on a two-dimensional moving system, which enables reconstruction of the ion current density profile with a spatial precision of 5 mm over almost the entire poloidal cross section at each toroidal position. In the series of experiments described here, Li^{6+} ions with energy of $\sim$70 eV $\gg T_e$ are injected horizontally in the coherent structure region. The time average electron density at the injection location is $\approx 5 \times 10^{15}$ m^{-3} and the standard deviation of the floating potential time series, indicating the level of fluctuations, is $\approx$1 V. Figure 13 shows an example of a fast ion current density profile at a toroidal distance of $\approx$ 54 cm from the source. The red cross indicates the position of the injection, showing the displacement of the beam spot due to the vertical drift. Measurements are made with and without plasma, in the

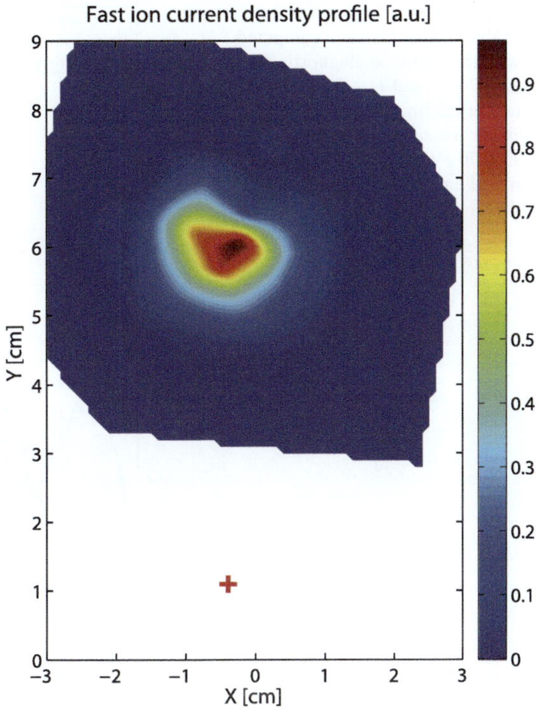

Fig. 13 Experimental current density profile of fast ions with energy $E = 70$ eV measured at $\sim$54 cm from the fast ion source. The position of the fast ion source is indicated by the *red cross*

presence of magnetic fields. The radial spatial variances of the fast ion current profiles as a function of the toroidal angle are shown in Fig. 14.

To interpret the experimental data, trajectories of tracer Li^{6+} ions are integrated in a simulated turbulent electrostatic field driven by the ideal interchange, which is calculated from 2D implementation of the drift reduced Braginskii fluid equations (Gustafson et al. 2012a). Turbulence simulations are performed with different values of the particle and heat sources in order to match experimental profiles. At the injection point, the fluctuation level of the floating potential is higher in the simulation than in the experiment. In order to match the potential fluctuations, the simulated plasma potential fluctuations, are rescaled. Tracer Li^{6+} ion trajectories are computed using source parameters based on measurements done without a magnetic field, and 10000 particles are launched with initial parameters modeled with Gaussian distributions. A synthetic diagnostic, mimicking the detector, computes 3D (Bovet et al. 2012; Gustafson et al. 2012a) profiles of the fast-ion current density for comparison with the experimental data. Figure 14 displays the radial variance of the beam profiles along the toroidal direction obtained with the synthetic diagnostic, from the simulations, and from experimental measurements. They show remarkable agreement. The oscillations of the variance of the beam due to the Larmor motion of the particles are clearly evident. The turbulent broadening of the beam is revealed by the radial variance of the beam which increases as a function of the distance from the source. Numerical simulations at later times indicate that, in these conditions, fast ions undergo subdiffusive transport with $\gamma \approx 0.78$. The same figure also shows the variance of the beam profiles of simulated data for Li^{6+} energy $\approx$50 eV, revealing a transition to a superdiffusive regime with $\gamma \approx 1.2$. Experimental measurements continue to investigate this new regime.

Fig. 14 Radial σ^2 as a function of the toroidal distance computed with the synthetic diagnostic for $E = 70$ eV and $E = 50$ eV from numerical simulations. Experimental data for the case $E = 70$ eV are also shown revealing good agreement with the simulations

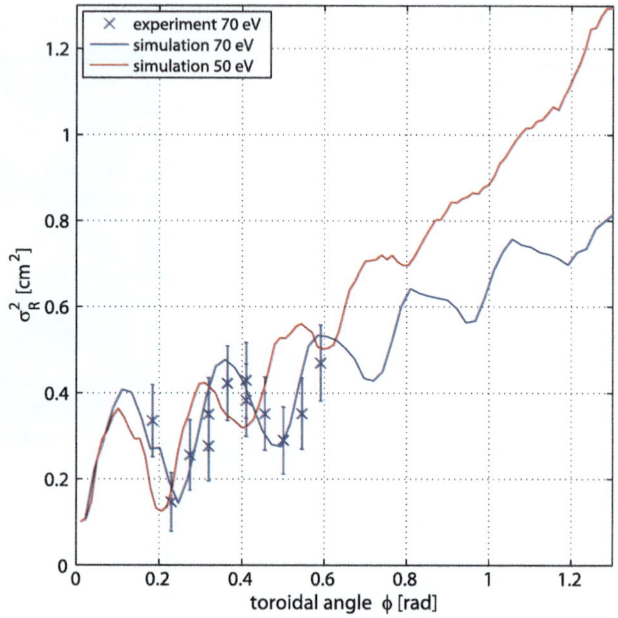

5.1 Shock Waves

Shock waves are distinctive large scale nonlinear structures which are observed both in laboratory plasmas and in the solar wind. In space and astrophysical plasmas, shock waves are considered to accelerate particles and cosmic rays, which then propagate both upstream and downstream of the shock. Clearly, such a propagation is influenced by the magnetic turbulence in the medium (Giacalone 2004, 2011). A relevant example where particle transport can be studied *in situ* is that of particles accelerated at shock waves in the solar wind: the energetic particles reach energies of order of 1–10 MeV, and are clearly distinguishable from the background plasma. This gives the possibility to distinguish between diffusive and non-diffusive transport by studying the energetic particle profiles upstream of the shock. Indeed, Perri and Zimbardo (2007, 2008a) have shown that while in the case of normal diffusion the flux of energetic particles upstream of the shock corresponds to an exponential decay, in the case of superdiffusion one has a power law decay. In particular, Perri and Zimbardo (2007, 2008a, 2009a) have shown that electron transport upstream of the shocks associated with corotating interaction regions (CIRs) detected by the Ulysses spacecraft in the solar wind at 4–5 AU is superdiffusive, with $\gamma \simeq 1.1$–1.7. Also, ion transport upstream of CIR shocks is found to be normal in most cases, although a slightly superdiffusive case with $\gamma \simeq 1.1$ is found at a CIR shock detected by Voyager 2 at 6.9 AU. The more decidedly superdiffusive behavior of electrons has been ascribed to the fact that electrons have smaller gyroradii then ions, and therefore the resonant interaction with turbulence happens at larger wavenumbers, where the wave power is less: this favors weak pitch angle scattering and hence superdiffusion (Perri and Zimbardo 2007, 2008a). On the other hand, analyzing the Voyager 2 data for low energy particles, Perri and Zimbardo (2009b) have shown that ion transport upstream of the solar wind termination shock at 84 AU is superdiffusive, too, with $\gamma \simeq 1.3$. Perri and Zimbardo (2009b) have interpreted this result as due to the decrease of the magnetic fluctuation amplitude, which causes weaker pitch angle scattering.

Recently Sugiyama and Shiota (2011), using ACE data, have found superdiffusive proton transport with $\gamma \simeq 1.3$ upstream of a shock at 1 AU driven by a coronal mass ejection. They also find that the level of magnetic fluctuations is not so small for the event considered (in contrast to what can be assumed for the heliopause termination shock at 84 AU), therefore pitch angle scattering should not be so weak. Sugiyama and Shiota (2011) propose that superdiffusion is due to the fact that the wave particle interaction falls into a nonlinear regime, where the quasilinear pitch angle diffusion coefficient no longer applies. In other words, quasilinear theory overestimates the pitch angle diffusion rate. Thus the experimental observation of superdiffusion gives information on the efficiency of wave particle interactions in the nonlinear regime, too, and this compares well with the results of Vlasov simulations reported in Sect. 3.

6 Avalanching Transport Conditioned by Critical Gradient Instablities

The exploitation of thermonuclear fusion plasmas as a mean of producing electricity requires one to confine the thermal energy a hot plasma of temperature $T \simeq$ a few tens of KeV and density $n \simeq 10^{20}\ \mathrm{m}^{-3}$ for a relatively long time $\tau \simeq$ a few seconds. As it is well known, this plasma must exceed the Lawson condition $n \cdot \tau > 3 \cdot 10^{20}\ \mathrm{s/m}^3$ to reach ignition. Magnetically confined plasmas, such as those in a tokamak, attempt to reach these conditions by trapping the plasma inside a set of closed, toroidal magnetic surfaces nested around a magnetic axis. Tokamaks and stellarators are realizations of this idea. However, the large radial gradients in pressure, temperature and density that these plasmas must sustain act as free-energy sources for many instabilities. These drive the turbulence which dominated the radial transport inside all these devices. The control of these losses is essential for the success of the fusion program. Traditionally, they have been quantified by using *effective transport coefficients*, obtained from direct measurement or from theoretical models with varying degree of complexity. However, there is much evidence to suggest that radial turbulent transport in these devices can behave in a very non-diffusive manner. This reduces confidence in the extrapolation of these coefficients to parameter regimes outside our current range of operation, which is necessary when designing new and larger devices.

Adoption of the sandpile paradigm (Bak et al. 1988; Dendy and Helander 1997, 1998; Helander et al. 1999; Dendy et al. 2007) in studies of global energy confinement and transport in laboratory and space plasmas is driven by both observational and theoretical considerations. The essence of the sandpile paradigm reflects everyday intuition, as follows. The system is fuelled, for example by the addition of grains at its centre. Local redistribution of sand arises when a critical gradient is exceeded. This may trigger redistribution at neighboring points, and so on progressively, giving rise to avalanching transport events, after which the gradient in affected regions is reset to a value below critical. The probability distribution of the frequency of occurrence of avalanches as a function of their magnitude may be scale-free, for example power-law, implying the existence of correlations on all scales. This portfolio of basic ingredients—energy fuelling, energy storage, and energy release through potentially large scale impulsive events that are conditioned by a critical gradient—maps well to the basic physics of macroscopic plasma systems. For example, a planetary magnetosphere can be viewed as a plasma and energy storage system, driven by the solar wind and subject to impulsive energy release events in the magnetotail and at the dayside magnetopause, which occur when critical gradients associated with the reconnection process are exceeded. Sandpile models have proven successful for magnetospheres (Chapman et al. 1998, 1999; Watkins et al. 1999, 2001) and for accretion disks (Dendy et al. 1998), as has a similar approach to reconnection in the solar corona (Hughes et al. 2003).

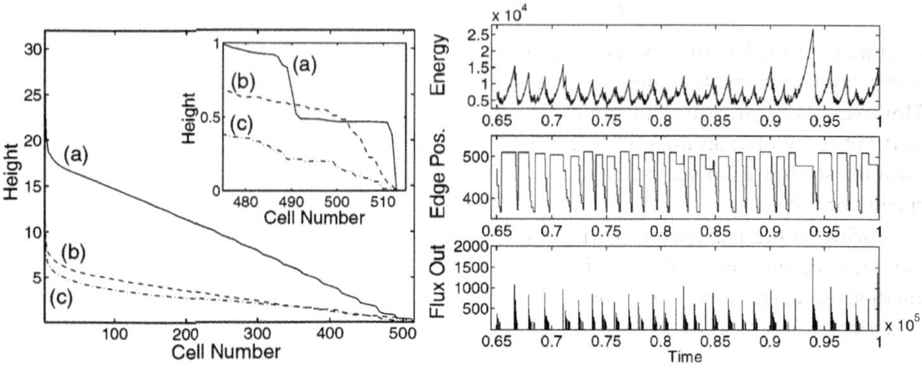

Fig. 15 Tokamak-like enhanced confinement, edge pedestals, and edge-localized pulses of energy loss in a sandpile model. (*Left*) Time averaged height profiles of a 512-cell sandpile for three different values of the model's sole control parameter; *inset* are edge profiles. (*Right*) Time sequence and magnitude of energy loss events for the corresponding regimes. Reproduced from Chapman et al. (2001a). Copyright (2001) by the American Physical Society

Empirical support for the application of the sandpile paradigm can arise from direct observation of avalanche events in experiments, and in numerical simulations thereof, together with the identification of event statistics that are power law or otherwise strongly non-Gaussian. Such observations are widespread in space and laboratory plasma systems (Dendy and Chapman 2006; Dendy et al. 2007). The primary benefit from the application of the sandpile paradigm arises from the fact that it can model global phenomenology which is otherwise prohibitively difficult to capture or explain. In particular, many real plasmas are complex systems in the technical sense. That is, their global phenomenology emerges from the interaction of multiple plasma physics processes, operating across a very broad range of lengthscales and timescales, coupled nonlinearly together. Examples of global plasma phenomenology, on which the sandpile paradigm sheds light, include: in fusion plasmas (Chapman et al. 2001a, 2001b, 2003; Graves et al. 2002), the existence of different classes of energy confinement regime (Wagner 2007), linked to impulsive energy loss events, see Fig. 15; and in space plasma, the measured distribution of ionospheric current flows that are ultimately driven by impulsive energy release events in the magnetotail (Chapman et al. 1998).

Carreras et al. (1996), Newman et al. (1996) were the first authors to propose a dynamical model to try to explain the extensive experimental evidence of non-diffusive behavior in tokamaks, notably in the so-called L-mode discharges. This regime of operation is characterized by very stiff profiles, whose shape is rather independent of the location of the heating sources, and in which superdiffusive propagation of hot and cold pulses is usually observed. They conjectured that the plasma profiles (of pressure, density and temperature) were probably lying in the neighborhood of their critical values for the onset of instabilities, pushed there by the constant external heating. In such a near-marginal state, it is not rare that modes excited at a certain position can propagate radially, in an avalanche-like manner, as the transport induced by the local turbulence suppresses it by flattening the local profile at the price of steepening the profiles in neighbor locations, that can in turn go unstable and propagate the avalanche. In this way, the radial transport in the tokamak in these plasmas acts very much like the sandpile used to proposed the concept of self-organized criticality in the 1980s (Bak et al. 1987).

Multiple simulations of plasma turbulence in near-marginal conditions have indeed exhibited this behavior since the seminal work by Carreras et al. (1996). There, the statistics of the avalanches observed were studied, and it was found that they exhibited strong power-laws, very reminiscent of those reported by Bak et al. (1987) in their sandpile studies. More recently, del-Castillo-Negrete et al. (2004b) showed, using the propagator method sketched in Sect. 2, that the radial propagators obtained for tracers in near-marginal cylindrical interchange turbulence clearly exhibited non-Gaussian shape ($\alpha \sim 0.75$) and superdiffusive behavior ($H \sim 0.66$). Consistent with these results, Mier et al. (2008) showed that the statistics of Lagrangian radial displacements in cylindrical dissipative-trapped-electron plasma turbulence also exhibited strong non-Gaussian statistics ($\alpha \sim 1$) and superdiffusive behavior ($H \sim 0.75$).

More recently, it has also been discovered that non-diffusive transport may occur in fusion plasmas whose profiles have gradients that are well above marginal values. This requires the existence of a strongly radially-sheared, poloidal flow. These poloidal flows are frequently self-induced by the turbulence via the Reynolds stresses, and contribute to limit the amplitude of the turbulent fluctuations by taking energy from them and reducing the radial size of turbulent eddies. Although this behavior had been known for some time, it had been assumed that the net effect of the presence of these flows on transport was to reduce the effective transport coefficients, thus improving plasma confinement. Sanchez et al. (2008) showed, however, that the presence of these radially-sheared poloidal flows causes a selection of the sign of the axial vorticity which is consistent with the sign of the shear of the flow. The dynamical consequence of this symmetry breaking is that transport is not only reduced, but its nature also changes into a subdiffusive character. That is, Hurst exponents obtained with the propagator method in this situation consistently yield values $H < 0.5$ in the radial direction. The values of H are smaller as the shear in the poloidal flow decreases. Simultaneously, transport along the poloidal direction becomes superdiffusive. The simultaneous observation of subdiffusion and superdiffusion in perpendicular directions is actually common, and has also been observed in other situations. In addition to radial subdiffusion, transport across these radially-sheared poloidal flows can also exhibit Lévy-like statistics. They can be linked to a predator-prey interaction between turbulent fluctuations and the local flow shear, which can translate into radial avalanches in spite of the fact that the overall transport remains subdiffusive (Sanchez et al. 2011).

7 Implications of Nonclassical Transport for Astrophysical Plasmas

As shown above, nondiffusive transport has direct consequences for the confinement of laboratory plasmas. In a similar way, the fact that energetic particles in space can propagate superdiffusively has a number of consequences. For instance, solar energetic electrons are usually considered to propagate either diffusively or "scatter-free" (Lin 1974), i.e., in a ballistic way. However, superdiffusive propagation is also a possibility, and this influences both the time of arrival of solar energetic particles, as well as their time profile (Lin 2005; Perri and Zimbardo 2008b). A relevant application is that of particle acceleration at shocks: indeed, the most popular acceleration theory is the so called diffusive shock acceleration (DSA), which assumes normal diffusion (Fisk and Lee 1980; Dosch and Shalchi 2010). One of the most important results of DSA is the spectral index s of the particle differential energy spectrum, which for relativistic energies reads as

$$s = \frac{r+2}{r-1} \tag{19}$$

where $r = V_1/V_2$ is the compression ratio at the shock, and V_1 and V_2 are the upstream and the downstream plasma speeds, respectively. This theory has been extended to the case of nondiffusive transport by Duffy et al. (1995) and by Kirk et al. (1996). They were motivated by the possibility of subdiffusive transport, corresponding to compound diffusion described by Eq. (16), in the case of a nearly perpendicular shock. In particular, Kirk et al. (1996) have shown that the energy spectral index of DSA is modified in the case of nondiffusive transport, and that the change depends essentially on the scaling properties of the propagator. For instance, in the case of normal diffusion one has the Gaussian propagator, and the scaling variable is $\xi = x/\sqrt{4Dt}$. Duffy et al. (1995) and Kirk et al. (1996) were concerned mainly with subdiffusion with $\gamma < 1$, so that $\xi = x/(kt)^{\gamma/2}$. In the case of superdiffusion described by a Lévy random walk, the propagator has power-law tails and the scaling variable is $\xi = x/(kt)^{1/(\mu-1)}$ (Zumofen and Klafter 1993), k being a constant with proper physical dimensions. A superdiffusive shock acceleration (SSA) scenario has been proposed by Perri and Zimbardo (2012), who have shown that in such a case the spectral index of the differential energy spectrum for relativistic particles is given by

$$s = \frac{6}{r-1}\frac{2-\gamma}{3-\gamma} + 1 \tag{20}$$

and for nonrelativistic particles is given by

$$s = \frac{3}{r-1}\frac{2-\gamma}{3-\gamma} + 1 . \tag{21}$$

This spectral index recovers that of DSA in the limit of normal diffusion, $\gamma = 1$. Notice that, while the spectral index in Eq. (19) depends only on the compression ratio of the shock, the one obtained in the framework of SSA also depends on the exponent of the mean square particle displacement. Therefore, spectral indices smaller than those of DSA can be obtained for $\gamma > 1$. In other words, the assumption of superdiffusive propagation allows one to explain the observation of very hard spectra for relativistic electrons at shell type supernova remnants (Perri and Zimbardo 2012), as well as the harder-than-DSA spectrum observed for MeV ions by the LECP onboard Voyager 2 at the heliospheric termination shock (Decker et al. 2008). Indeed, Decker et al. (2008) find a differential flux spectral index s_J of about 1.25 (the flux is proportional to the particle velocity, and for nonrelativistic particle this yield an extra -0.5 in the flux spectra index). Assuming the compression ratio $r = 2$, obtained by the Plasma Science Instrument, a spectral index $s_J = s - 0.5 = 2$ is obtained from DSA. To recover the observed spectral index, a compression ratio of 3 would be required by DSA. On the other hand, there is substantial uncertainty on the observed compression ratio: taking into account the solar wind speed decrease observed roughly forty days before the Voyager 2 termination shock crossing and using a two fluid model which considers the energetic particle momentum flux, too, Florinski et al. (2009) have obtained a compression ratio of about 2.4. Therefore it is possible to find a better agreement between the spectral index observed at the termination shock and that foreseen by shock acceleration assuming $r = 2.4$ and the spectral index in Eq. (21) obtained for nonrelativistic particles by SSA. Using the superdiffusion exponent $\gamma \sim 1.3 \pm 0.07$ reported by Perri and Zimbardo (2009b) at the termination shock, a differential flux spectral index of 1.38 ± 0.2 is obtained, in reasonable agreement with the Voyager 2 observations when the error bars are taken into account. Since different particle species can exhibit different transport properties (as shown in Perri and Zimbardo 2007, 2008a), this suggests that the energy spectra, too, can be characterized by different indices depending on the particle species considered. For instance,

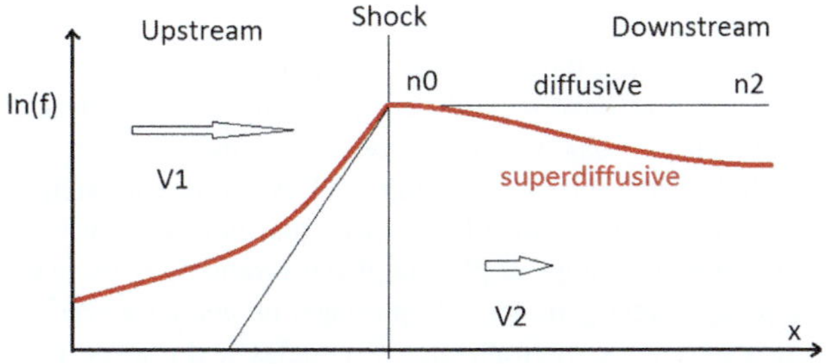

Fig. 16 Comparison between the particle distribution function f spatial profiles for particles accelerated at a shock, in the case of normal diffusion (*black line*) and in the case of superdiffusion (*red line*)

the recent PAMELA observations of Galactic cosmic rays show that hydrogen and helium ions have slightly different spectral indices (Adriani et al. 2011), which could be due to different transport regimes.

In addition, one has a change of perspective for the density profile of energetic particles accelerated at the shock, see Fig. 16. That is, while normal diffusion leads to an exponential decay upstream of the shock and to a constant density downstream, superdiffusion leads to a power-law decay upstream and to a far downstream density n_2 which is lower than the density at the shock n_0. The possible occurrence of this class of density profile should also be investigated in the observational data, both in space and in astrophysical shocks.

Another observable effect was considered by Ragot and Kirk (1997) for the propagation of relativistic electrons away from Coma cluster of galaxies. As the electrons are transported away from the source they cool, and the spectrum softens. In particular, the expected spatial dependence of the spectral index is sensitive to the kind of transport regime. Assuming that the effects of particle acceleration are negligible in the outer part of the Coma cluster, Ragot and Kirk (1997) find that superdiffusion can reproduce the observed rapid softening of the spectrum with distance.

8 Conclusions

In this review we have examined some attributes of nondiffusive transport and wave–particle interactions in both laboratory and space plasmas. We have considered the statistical description of anomalous diffusion, based on non-Gaussian random walks and long-range correlations, as well as the use of fractional transport equations. We have also illustrated the results of a number of numerical simulations, describing wave–particle interactions in the kinetic regime, and particle or plasma transport in the presence of electric and magnetic turbulence. These simulations show that nondiffusive transport can be found in many different plasmas. These include laboratory devices for plasma confinement, electron and ion transport in solar coronal loops, and energetic particle transport in the solar wind. Experimental indications of superdiffusion are found for the propagation of particles accelerated at interplanetary shocks, as in the case of co-rotating interaction regions and coronal mass ejections, and for the propagation of relativistic electrons in the Coma cluster. Avalanching transport in toroidal plasmas and superdiffusive acceleration at shocks has also been described.

We have discussed in some detail the suprathermal ion dynamics in the SMT configuration of the TORPEX device. Although this configuration is relatively simple, it contains all the basic elements for suprathermal ion dynamics in a number of physical systems. The framework established through study of the SMT scenario can be used for interpreting suprathermal ion dynamics in fusion devices with high-energy neutral beams and alpha particle production, cosmic ray propagation, and solar wind interaction with the magnetosphere.

Moreover, we expect that the results presented herein are relevant to suprathermal ion dynamics in other basic plasma physics configurations, where superdiffusive and subdiffusive ion transport have recently been measured (Zhou et al. 2010, 2011), and in magnetically confined plasmas, where it may be possible to model the interaction of energetic particles with Alfvénic turbulence using the Lévy walk approach.

The examples considered show that nonclassical transport is fundamental to the most important problems in plasma physics, ranging from plasma confinement in toroidal fusion devices to energetic particle propagation and acceleration in space plasmas. Therefore, understanding the microphysics of nonclassical transport is a high priority across a variety of plasmas.

Acknowledgements This work was part-funded by the RCUK Energy Programme under grant EP/I501045 and the European Communities under the contract of Association between EURATOM and CCFE. The views and opinions expressed herein do not necessarily reflect those of the European Commission. The TORPEX experiments and simulations were supported in part by the Swiss National Science Foundation. The authors wish to acknowledge the valuable support of the CRPP technical team. K.G. was supported by US-NSF IRFP grant OISE-0853498. The hybrid Vlasov-Maxwell numerical simulations discussed in the present work were performed on the Fermi supercomputer at Cineca (Bologna, Italy), within the project ASWTURB 2011 (HP10BO2REM), supported by the Italian Super-computing research allocation (ISCRA), and within the European project PRACE Pra04-771 (Partnership for advanced computing in Europe). D.P. is supported by the Italian Ministry for University and Research (MIUR) PRIN 2009 funds (grant number 20092YP7EY). G.Z. acknowledges support from the European Union FP7, Marie Curie project 269198 – Geoplasmas. S. P.'s research has been supported by "Borsa Post-doc POR Calabria FSE 2007/2013 Asse IV Capitale Umano - Obiettivo Operativo M.2".

References

S. Abdullaev, Structure of motion near saddle points and chaotic transport in Hamiltonian systems. Phys. Rev. E **62**, 3508 (2000)

O. Adriani et al. (PAMELA collaboration), PAMELA measurements of cosmic-ray proton and helium spectra. Science **332**, 69 (2011)

O. Alexandrova, J. Saur, C. Lacombe et al., Universality of the solar wind turbulent spectrum from MHD to electron scales. Phys. Rev. Lett. **103**, 165003 (2009)

F. Anderegg, C.F. Driscoll, D.H. Dubin et al., Electron acoustic waves in pure ion plasmas. Phys. Plasmas **16**, 055705 (2009)

S.V. Annibaldi, G. Manfredi, R.O. Dendy et al., Evidence for strange kinetics in Hasegawa-Mima turbulent transport. Plasma Phys. Control. Fusion **42**, L13 (2000)

S.V. Annibaldi, G. Manfredi, R.O. Dendy, Non-Gaussian transport in strong plasma turbulence. Phys. Plasmas **9**, 791 (2002)

J.A. Araneda, E. Marsch, A.F. Viñas, Proton core heating and beam formation via parametrically unstable Alfvén-cyclotron waves. Phys. Rev. Lett. **100**, 125003 (2008)

J.A. Araneda, Y. Maneva, E. Marsch, Preferential heating and acceleration of α particles by Alfvén-cyclotron waves. Phys. Rev. Lett. **102**, 175001 (2009)

P. Bak, C. Tang, K. Wiesenfeld, Self-organized criticality: an explanation of $1/f$ noise. Phys. Rev. Lett. **59**, 381 (1987)

P. Bak, C. Tang, K. Wiesenfeld, Self-organized criticality. Phys. Rev. A **38**, 364 (1988)

S.D. Bale, P.J. Kellogg, F.S. Mozer et al., Measurement of the electric fluctuation spectrum of magnetohydrodynamic turbulence. Phys. Rev. Lett. **94**, 215002 (2005)

J.W. Bieber, W.H. Matthaeus, C.W. Smith, Proton and electron mean free paths: the Palmer consensus revisited. Astrophys. J. **420**, 294 (1994)

C.K. Birdsall, A.B. Langdon, *Plasma Physics via Computer Simulation* (McGraw-Hill, New York, 1985)

R. Bitane, G. Zimbardo, P. Veltri, Electron transport in coronal loops: the influence of the exponential separation of magnetic field lines. Astrophys. J. **719**, 1912 (2010)

S. Bourouaine, E. Marsch, F.M. Neubauer, Correlations between the proton temperature anisotropy and transverse high frequency waves in the solar wind. Geophys. Res. Lett. **37**, L14104 (2010)

S. Bourouaine, E. Marsch, F.M. Neubauer, On the relative speed and temperature ratio of solar wind alpha particles and protons: collisions versus wave effects. Astrophys. J. **728**, L3 (2011a)

S. Bourouaine, E. Marsch, F.M. Neubauer, Temperature anisotropy and differential streaming of solar wind ions. Correlations with transverse fluctuations. Astron. Astrophys. **536**, A39 (2011b)
A. Bovet, A. Fasoli, I. Furno et al., Investigation of fast ion transport in TORPEX. Nucl. Fusion **52**, 094017 (2012)
R. Bruno, V. Carbone, The solar wind as a turbulence laboratory. Living Rev. Sol. Phys. **2**, 4 (2005)
G.S. Burillo, B.P. van Milligen, A. Thyagaraja, Analysis of the radial transport of tracers in a turbulence simulation. Phys. Plasmas **16**, 042319 (2009)
E. Camporeale, D. Burgess, The dissipation of solar wind turbulent fluctuations at electron scales. Astrophys. J. **730**, 114 (2011)
B.A. Carreras, V.E. Lynch, D.E. Newman, A model realization of self-organized criticality for plasma confinement. Phys. Plasmas **3**, 2903 (1996)
S.C. Chapman, N. Watkins, R.O. Dendy et al., A simple avalanche model as an analogue for magnetospheric activity. Geophys. Res. Lett. **25**, 2397 (1998)
S.C. Chapman, R.O. Dendy, G. Rowlands, A sandpile model with dual scaling regimes for laboratory, space and astrophysical plasmas. Phys. Plasmas **6**, 4169 (1999)
S.C. Chapman, R.O. Dendy, B. Hnat, A sandpile model with tokamak-like enhanced confinement phenomenology. Phys. Rev. Lett. **86**, 2814 (2001a)
S.C. Chapman, R.O. Dendy, B. Hnat, A simple avalanche model for astroplasma and laboratory confinement systems. Phys. Plasmas **8**, 1969 (2001b)
S.C. Chapman, R.O. Dendy, B. Hnat, Self organisation of edge and internal pedestals in a sandpile. Plasma Phys. Control. Fusion **45**, 301 (2003)
C.H.K. Chen, A. Mallet, T.A. Yousef et al., Anisotropy of Alfvénic turbulence in the solar wind and numerical simulations. Mon. Not. R. Astron. Soc. **415**, 3219 (2011)
R.-B. Decker, S.M. Krimigis, E.C. Roelof et al., Mediation of the solar wind termination shock by non-thermal ions. Nature **454**, 67 (2008)
D. del-Castillo-Negrete, Asymmetric transport and non-Gaussian statistics of passive scalars in vortices in shear. Phys. Fluids **10**, 576 (1998)
D. del-Castillo-Negrete, B.A. Carreras, V.E. Lynch, Fractional diffusion in plasma turbulence. Phys. Plasmas **11**, 3854 (2004a)
D. del-Castillo-Negrete, B.A. Carreras, V.E. Lynch, Nondiffusive transport in plasma turbulence: a fractional diffusion approach. Phys. Rev. Lett. **94**, 065003 (2004b)
R.O. Dendy, P. Helander, Sandpiles, silos and tokamak phenomenology: a brief review. Plasma Phys. Control. Fusion **39**, 1947 (1997)
R.O. Dendy, P. Helander, On the appearance and non-appearance of self-organised criticality in sandpiles. Phys. Rev. E **57**, 3641 (1998)
R.O. Dendy, P. Helander, M. Tagger, On the role of self-organised criticality in accretion systems. Astron. Astrophys. **337**, 962 (1998)
R.O. Dendy, S.C. Chapman, Characterisation and interpretation of strongly nonlinear phenomena in fusion, space, and astrophysical plasmas. Plasma Phys. Control. Fusion **48**, B313 (2006)
R.O. Dendy, S.C. Chapman, M. Paczuski, Fusion, space, and solar plasmas as complex systems. Plasma Phys. Control. Fusion **49**, A95 (2007)
J.M. Dewhurst, B. Hnat, R.O. Dendy, The effects of nonuniform magnetic field strength on density flux and test particle transport in drift wave turbulence. Phys. Plasmas **16**, 072306 (2009)
J.M. Dewhurst, B. Hnat, R.O. Dendy, Finite Larmor radius effects on test particle transport in drift wave-zonal flow turbulence. Plasma Phys. Control. Fusion **52**, 025004 (2010). doi:10.1088/0741-3335/52/2/025004
A. Diallo, A. Fasoli, I. Furno et al., Dynamics of plasma blobs in a shear flow. Phys. Rev. Lett. **101**, 115005 (2008)
P.H. Diamond, S.-I. Itoh, T.S. Hahm, Zonal flows in plasmas—a review. Plasma Phys. Control. Fusion **47**, R35 (2005)
P.H. Diamond, A. Hasegawa, K. Mima, Vorticity dynamics, drift wave turbulence, and zonal flows: a look back and a look ahead. Plasma Phys. Control. Fusion **53**, 124001 (2011)
A. Dosch, A. Shalchi, Diffusive shock acceleration at interplanetary perpendicular shock waves: influence of the large scale structure of turbulence on the maximum particle energy. Adv. Space Res. **46**, 1208 (2010)
P. Duffy, J.-G. Kirk, Y.-A. Gallant et al., Anomalous transport and particle acceleration at shocks. Astron. Astrophys. **302**, L21 (1995)
A. Fasoli, B. Labit, M. McGrath et al., Electrostatic turbulence and transport in a simple magnetized plasma. Phys. Plasmas **13**, 055902 (2006)
A. Fasoli, A. Burckel, L. Federspiel et al., Electrostatic instabilities, turbulence and fast ion interactions in the TORPEX device. Plasma Phys. Control. Fusion **52**, 124020 (2010)
J. Feder, *Fractals* (Plenum, New York, 1988)

L.A. Fisk, M.A. Lee, Shock acceleration of energetic particles in corotating interaction regions in the solar wind. Astrophys. J. **237**, 620 (1980)
V. Florinski, R.B. Decker, J.A. le Roux et al., An energetic-particle-mediated termination shock observed by Voyager 2. Geophys. Res. Lett. **36**, L12101 (2009)
A. Fujisawa, Experimental studies of mesoscale structure and its interactions with microscale waves in plasma turbulence. Plasma Phys. Control. Fusion **53**, 124015 (2011)
I. Furno, B. Labit, M. Podestà et al., Experimental observation of the blob-generation mechanism from interchange waves in a plasma. Phys. Rev. Lett. **100**, 055004 (2008a)
I. Furno, B. Labit, A. Fasoli et al., Mechanism for blob generation in the TORPEX toroidal plasma. Phys. Plasmas **15**, 055903 (2008b)
I. Furno, M. Spolaore, C. Theiler et al., Direct two-dimensional measurements of the field-aligned current associated with plasma blobs. Phys. Rev. Lett. **106**, 245001 (2011)
S.P. Gary, S. Saito, H. Li, Cascade of whistler turbulence: particle-in-cell simulations. Geophys. Res. Lett. **35**, L02104 (2008)
J. Giacalone, Large-scale hybrid simulations of particle acceleration at a parallel shock. Astrophys. J. **609**, 452 (2004)
J. Giacalone, Cosmic-ray transport and interaction with shocks. Space Sci. Rev. (2011). doi:10.1007/s11214-011-9763-2
J.P. Graves, R.O. Dendy, K.I. Hopcraft et al., The role of clustering effects in non-diffusive transport in tokamaks. Phys. Plasmas **9**, 1596 (2002)
S. Günter, C. Angioni, M. Apostoliceanu et al., Overview of ASDEX upgrade results—development of integrated operating scenarios for ITER. Nucl. Fusion **45**, S98 (2005)
D.A. Gurnett, E. Marsch, W. Pilipp et al., Ion-acoustic waves and related plasma observations in the solar wind. J. Geophys. Res. **84**, 2029 (1979)
K. Gustafson, D. Del-Castillo-Negrete, W. Dorland, Finite Larmor radius effects on nondiffusive tracer transport in a zonal flow. Phys. Plasmas **15**, 102309 (2008)
K. Gustafson, P. Ricci, A. Bovet et al., Suprathermal ion transport in simple magnetized torus configurations. Phys. Plasmas **19**, 062306 (2012a)
K. Gustafson, P. Ricci, I. Furno et al., Nondiffusive suprathermal ion transport in simple magnetized toroidal plasmas. Phys. Rev. Lett. **108**, 035006 (2012b)
K. Gustafson, P. Ricci, Lévy walk description of suprathermal ion transport. Phys. Plasmas **19**, 032304 (2012)
T. Hauff, F. Jenko, S. Eule, Intermediate non-Gaussian transport in plasma core turbulence. Phys. Plasmas **14**, 102316 (2007)
T. Hauff, F. Jenko, Mechanisms and scalings of energetic ion transport via tokamak microturbulence. Phys. Plasmas **15**, 2307 (2008)
P. Helander, S.C. Chapman, R.O. Dendy et al., Exactly solvable sandpile with fractal avalanching. Phys. Rev. **59**, 6356 (1999)
W.A. Hornsby, A.R. Bell, R.J. Kingham et al., A code to solve the Vlasov Fokker-Planck equation applied to particle transport in magnetic turbulence. Plasma Phys. Control. Fusion **52**, 075011 (2010)
D. Hughes, M. Paczuski, R.O. Dendy et al., Solar flares as cascades of reconnecting magnetic loops. Phys. Rev. Lett. **90**, 131101 (2003)
S. Inagaki, T. Tokuzawa, K. Itoh et al., Observations of long-distance radial correlation in toroidal plasma turbulence. Phys. Rev. Lett. **107**, 115001 (2011)
J.R. Jokipii, Cosmic-ray propagation. I. Charged particles in a random magnetic field. Astrophys. J. **146**, 480 (1966)
J.G. Kirk, P. Duffy, Y.A. Gallant, Stochastic particle acceleration at shocks in the presence of braided magnetic fields. Astron. Astrophys. **314**, 1010 (1996)
J.G. Kirk, R.O. Dendy, Shock acceleration of cosmic rays: a critical review. J. Phys. G **27**, 1589 (2001)
J. Klafter, A. Blumen, M.F. Shlesinger, Stochastic pathway to anomalous diffusion. Phys. Rev. A **35**, 3081 (1987)
A.J. Klimas, J.A. Valdivia, D. Vassiliadis et al., Self-organized criticality in the substorm phenomenon and its relation to localized reconnection in the magnetospheric plasma sheet. J. Geophys. Res. **105**(A8), 18765 (2000)
J. Kóta, J.R. Jokipii, Velocity correlation and the spatial diffusion coefficients of cosmic rays: compound diffusion. Astrophys. J. **531**, 1067 (2000)
J.A. Krommes, C. Oberman, R.B. Kleva, Plasma transport in stochastic magnetic fields. Part 3. Kinetics of test particle diffusion. J. Plasma Phys. **30**, 11 (1983)
B. Labit, C. Theiler, A. Fasoli et al., Blob-induced toroidal momentum transport in simple magnetized plasmas. Phys. Plasmas **18**, 032308 (2011)
P.O. Lagage, C.J. Cesarsky, Cosmic-ray shock acceleration in the presence of self-excited waves. Astron. Astrophys. **118**, 223 (1983a)

P.O. Lagage, C.J. Cesarsky, The maximum energy of cosmic rays accelerated by supernova shocks. Astron. Astrophys. **125**, 249 (1983b)
L.D. Landau, On the vibrations of the electronic plasma. J. Phys. (Moscow) **10**, 25 (1946)
M.A. Lee, L.A. Fisk, Shock acceleration of energetic particles in the heliosphere. Space Sci. Rev. **32**, 205 (1982)
R.E. Lee, S.C. Chapman, R.O. Dendy, Ion acceleration processes at reforming collisionless shocks. Phys. Plasmas **12**, 012901 (2005a)
R.E. Lee, S.C. Chapman, R.O. Dendy, Reforming perpendicular shocks in the presence of pickup protons: initial ion acceleration. Ann. Geophys. **23**, 643 (2005b)
R.P. Lin, Non-relativistic solar electrons. Space Sci. Rev. **16**, 189 (1974)
R.P. Lin, Relationship of solar flare accelerated particles to solar energetic particles (SEPs) observed in the interplanetary medium. Adv. Space Res. **35**, 1857 (2005)
E. Lu, R. Hamilton, Avalanches of the distribution of solar flares. Astrophys. J. **380**, L89 (1991)
F. Mainardi, Y. Luchko, G. Pagnini, The fundamental solution of the space-time fractional diffusion equation. Fract. Calc. Appl. Anal. **4**, 153–192 (1996)
M.A. Malkov, P.H. Diamond, Weak hysteresis in a simplified model of the L-H transition. Phys. Plasmas **16**, 012504 (2009)
B.B. Mandelbrot, J.W. van Ness, Fractional Brownian motions, fractional noises and applications. SIAM Rev. **10**, 422 (1968)
G. Manfredi, R.O. Dendy, Test-particle transport in strong electrostatic drift turbulence with finite Larmor radius effects. Phys. Rev. Lett. **76**, 4360 (1996)
G. Manfredi, M. Shoucri, R.O. Dendy et al., Vlasov gyrokinetic simulations of ion-temperature-gradient driven instabilities. Phys. Plasmas **3**, 202 (1996)
G. Manfredi, R.O. Dendy, Transport properties of energetic particles in a turbulent electrostatic field. Phys. Plasmas **4**, 628 (1997)
G. Manfredi, C.M. Roach, R.O. Dendy, Zonal flow and streamer generation in drift turbulence. Plasma Phys. Control. Fusion **43**, 825 (2001)
A. Mangeney, F. Califano, C. Cavazzoni et al., A numerical scheme for the integration of the Vlasov-Maxwell system of equations. J. Comput. Phys. **179**, 405 (2002)
E. Marsch, K.-H. Mühlhäuser, R. Schwenn et al., Solar wind protons: three-dimensional velocity distributions and derived plasma parameters measured between 0.3 and 1 AU. J. Geophys. Res. **87**, A1 (1982a)
E. Marsch, K.-H. Mühlhäuser, R. Schwenn et al., Solar wind helium ions: observations of the Helios solar probes between 0.3 and 1 AU. J. Geophys. Res. **35**, A1 (1982b)
E. Marsch, Kinetic physics of the solar corona and solar wind. Living Rev. Sol. Phys. **3**, 1 (2006)
J.A. Mier, R. Sanchez, L. Garcia et al., Characterization of non-diffusive transport in plasma turbulence via a novel Lagrangian method. Phys. Rev. Lett. **101**, 165001 (2008)
J.A. Miller, Particle acceleration in impulsive solar flares. Space Sci. Rev. **86**, 79 (1998)
E.W. Montroll, G.H. Weiss, Random walks on lattices. II. J. Math. Phys. **6**, 167 (1965)
S.H. Müller, A. Diallo, A. Fasoli et al., Plasma blobs in a basic toroidal experiment: origin, dynamics, and induced transport. Phys. Plasmas **14**, 110704 (2007)
D.E. Newman, B.A. Carreras, P.H. Diamond et al., The dynamics of marginality and self-organized criticality as a paradigm for turbulent transport. Phys. Plasmas **3**, 1858 (1996)
T.N. Parashar, S. Servidio, B. Breech et al., Kinetic driven turbulence: structure in space and time. Phys. Plasmas **17**, 102304 (2010)
T.N. Parashar, S. Servidio, M.A. Shay et al., Effect of driving frequency on excitation of turbulence in a kinetic plasma. Phys. Plasmas **18**, 092302 (2011)
S. Perri, G. Zimbardo, Evidence of superdiffusive transport of electrons accelerated at interplanetary shocks. Astrophys. J. **671**, L177 (2007)
S. Perri, F. Lepreti, V. Carbone et al., Position and velocity space diffusion of test particles in stochastic electromagnetic fields. Europhys. Lett. **78**, 40003 (2007)
S. Perri, G. Zimbardo, Superdiffusive transport of electrons accelerated at corotating interaction regions. J. Geophys. Res. **113**, A03107 (2008a). doi:10.1029/2007JA012695
S. Perri, G. Zimbardo, Observations of anomalous transport of energetic electrons in the heliosphere. Astrophys. Space Sci. Trans. **4**, 27 (2008b)
S. Perri, G. Zimbardo, Ion and electron superdiffusive transport in the interplanetary space. Adv. Space Res. **44**, 465 (2009a)
S. Perri, G. Zimbardo, Ion superdiffusion at the solar wind termination shock. Astrophys. J. **693**, L118 (2009b)
S. Perri, G. Zimbardo, A. Greco, On the energization of protons interacting with 3-D time-dependent electromagnetic fields in the Earth's magnetotail. J. Geophys. Res. **116**, A05221 (2011). doi:10.1029/2010JA016328

S. Perri, G. Zimbardo, Superdiffusive shock acceleration. Astrophys. J. **750**, 87 (2012)
D. Perrone, F. Valentini, P. Veltri, The role of alpha particles in the evolution of the solar-wind turbulence toward short spatial scales. Astrophys. J. **741**, 43 (2011)
D. Perrone, F. Valentini, S. Servidio, S. Dalena, P. Veltri, Vlasov simulations of multi-ion plasma turbulence in the solar wind. Astrophys. J. **762**, 99 (2013)
G. Plyushchev, A. Diallo, A. Fasoli et al., Fast ion source and detector for investigating the interaction of turbulence with suprathermal ions in a low temperature toroidal plasma. Rev. Sci. Instrum. **77**, 10F503 (2006)
M. Podestà, A. Fasoli, B. Labit et al., Cross-field transport by instabilities and blobs in a magnetized toroidal plasma. Phys. Rev. Lett. **101**, 045001 (2008)
P. Pommois, P. Veltri, G. Zimbardo, Anomalous and Gaussian transport regimes in anisotropic three-dimensional magnetic turbulence. Phys. Rev. E **59**, 2244 (1999)
P. Pommois, P. Veltri, G. Zimbardo, Field line diffusion in solar wind magnetic turbulence and energetic particle propagation across heliographic latitudes. J. Geophys. Res. **106**, 24965 (2001)
P. Pommois, G. Zimbardo, P. Veltri, Anomalous, non-Gaussian transport of charged particles in anisotropic magnetic turbulence. Phys. Plasmas **14**, 012311 (2007)
G. Qin, W.H. Matthaeus, J.W. Bieber, Subdiffusive transport of charged particles perpendicular to the large scale magnetic field. Geophys. Res. Lett. **29**, 1048 (2002a). doi:10.1029/2001GL014035
G. Qin, W.H. Matthaeus, J.W. Bieber, Perpendicular transport of charged particles in composite model turbulence: recovery of diffusion. Astrophys. J. **578**, L117 (2002b)
B.R. Ragot, J.G. Kirk, Anomalous transport of cosmic ray electrons. Astron. Astrophys. **327**, 432 (1997)
D.V. Reames, Particle acceleration at the Sun and in the heliosphere. Space Sci. Rev. **90**, 413 (1999)
A.B. Rechester, M.N. Rosenbluth, Electron heat transport in a Tokamak with destroyed magnetic surfaces. Phys. Rev. Lett. **40**, 38 (1978)
P. Ricci, C. Theiler, A. Fasoli et al., Langmuir probe-based observables for plasma-turbulence code validation and application to the TORPEX basic plasma physics experiment. Phys. Plasmas **16**, 055703 (2009)
P. Ricci, B.N. Rogers, Transport scaling in interchange-driven toroidal plasmas. Phys. Plasmas **16**, 062303 (2009)
P. Ricci, B.N. Rogers, Turbulence phase space in simple magnetized toroidal plasmas. Phys. Rev. Lett. **104**, 145001 (2010)
P. Ricci, C. Theiler, A. Fasoli et al., Methodology for turbulence code validation: quantification of simulation-experiment agreement and application to the TORPEX experiment. Phys. Plasmas **18**, 032109 (2011)
D. Ruffolo, W.H. Matthaeus, P. Chuychai, Trapping of solar energetic particles by the small-scale topology of solar wind turbulence. Astrophys. J. **597**, L169 (2003)
F. Sahraoui, M.L. Goldstein, G. Belmont et al., Three dimensional anisotropic *k* spectra of turbulence at subproton scales in the solar wind. Phys. Rev. Lett. **105**, 131101 (2010)
S. Saito, S.P. Gary, H. Li et al., Whistler turbulence: particle-in-cell simulations. Phys. Plasmas **15**, 102305 (2008)
G. Samorodnitsky, M.S. Taqqu, *Stable Non-Gaussian Distributions* (Chapman & Hall, New York, 1994)
R. Sanchez, B.A. Carreras, B. ph. van Milligen, Fluid limits of nonintegrable CTRWs in terms of fractional differential equations. Phys. Rev. E **71**, 011111 (2005)
R. Sanchez, B.A. Carreras, D.E. Newman et al., Renormalization of tracer turbulence leading to fractional differential equations. Phys. Rev. E **74**, 016305 (2006)
R. Sanchez, D.E. Newman, J.-N. Leboeuf et al., Nature of transport across sheared zonal flows in electrostatic ion-temperature-gradient gyrokinetic plasma turbulence. Phys. Rev. Lett. **101**, 205002 (2008)
R. Sanchez, D.E. Newman, J.-N. Leboeuf et al., Nature of turbulent transport across sheared zonal flows: insights from gyrokinetic simulations. Plasma Phys. Control. Fusion **53**, 074018 (2011)
H. Schmitz, S.C. Chapman, R.O. Dendy, Electron pre-acceleration mechanisms in the foot region of high Alfvénic Mach number shocks. Astron. Astrophys. **579**, 327 (2002b)
H. Schmitz, S.C. Chapman, R.O. Dendy, The influence of electron temperature and magnetic field on cosmic ray injection at high Mach number shocks. Astron. Astrophys. **570**, 637 (2002a)
S. Servidio, F. Valentini, F. Califano et al., Local kinetic effects in two-dimensional plasma turbulence. Phys. Rev. Lett. **108**, 045001 (2012)
A. Shalchi, I. Kourakis, A new theory for perpendicular transport of cosmic rays. Astron. Astrophys. **470**, 405 (2007)
A. Shalchi, Applicability of the Taylor-Green-Kubo formula in particle diffusion theory. Phys. Rev. E **83**, 046402 (2011)
M.F. Shlesinger, J. Klafter, Y.M. Wong, Random walks with infinite spatial and temporal moments. J. Stat. Phys. **27**, 499 (1982)
W.J. Shugard, H. Reiss, Transient nucleation in H_2O–H_2SO_4 mixtures: a stochastic approach. J. Chem. Phys. **65**, 2827 (1976)

T. Sugiyama, D. Shiota, Sign for super-diffusive transport of energetic ions associated with a coronal-mass-ejection-driven interplanetary shock. Astrophys. J. **731**, L34 (2011)

R.C. Tautz, Simulation results on the influence of magneto-hydrodynamic waves on cosmic ray particles. Plasma Phys. Control. Fusion **52**, 045016 (2010)

R.C. Tautz, A. Shalchi, On the diffusivity of cosmic ray transport. J. Geophys. Res. **115**, A03104 (2010). doi:10.1029/2009JA014944

C. Theiler, A. Diallo, A. Fasoli et al., The role of the density gradient on intermittent cross-field transport events in a simple magnetized toroidal plasma. Phys. Plasmas **15**, 042303 (2008)

C. Theiler, I. Furno, P. Ricci et al., Cross-field motion of plasma blobs in an open magnetic field line configuration. Phys. Rev. Lett. **103**, 065001 (2009)

G.R. Tynan, A. Fujisawa, G. McKee, A review of experimental drift turbulence studies. Plasma Phys. Control. Fusion **51**, 113001 (2009)

F. Valentini, P. Veltri, A. Mangeney, A numerical scheme for the integration of the Vlasov–Poisson system of equations, in the magnetized case. J. Comput. Phys. **210**, 730 (2005)

F. Valentini, P. Trávníček, F. Califano et al., A hybrid-Vlasov model based on the current advance method for the simulation of collisionless magnetized plasma. J. Comput. Phys. **225**, 753 (2007)

F. Valentini, P. Veltri, F. Califano et al., Cross-scale effects in solar-wind turbulence. Phys. Rev. Lett. **101**, 025006 (2008)

F. Valentini, P. Veltri, Electrostatic short-scale termination of solar-wind turbulence. Phys. Rev. Lett. **102**, 225001 (2009)

F. Valentini, F. Califano, P. Veltri, Two-dimensional kinetic turbulence in the solar wind. Phys. Rev. Lett. **104**, 205002 (2010)

F. Valentini, F. Califano, D. Perrone et al., New ion-wave path in the energy cascade. Phys. Rev. Lett. **106**, 165002 (2011a)

F. Valentini, F. Califano, D. Perrone et al., Excitation of nonlinear electrostatic waves with phase velocity close to the ion-thermal speed. Plasma Phys. Control. Fusion **53**, 105017 (2011b)

F. Valentini, D. Perrone, P. Veltri, Short-wavelength electrostatic fluctuations in the solar wind. Astrophys. J. **739**, 54 (2011c)

B.Ph. van Milligen, R. Sanchez, B.A. Carreras, Probabilistic finite-size transport models for fusion: anomalous transport and scaling laws. Phys. Plasmas **11**, 2272 (2004)

F. Wagner, A quarter-century of H-mode studies. Plasma Phys. Control. Fusion **49**, B1 (2007)

N.W. Watkins, S.C. Chapman, R.O. Dendy et al., Robustness of collective behaviour in strongly driven avalanche models: magnetospheric implications. Geophys. Res. Lett. **26**, 2617 (1999)

N.W. Watkins, M.P. Freeman, S.C. Chapman et al., Testing the SOC hypothesis for the magnetosphere. J. Atmos. Sol.-Terr. Phys. **63**, 1435 (2001)

G.M. Webb, G.P. Zank, E.Kh. Kaghashvili et al., Compound and perpendicular diffusion of cosmic rays and random walk of the field lines. I. Parallel particle transport models. Astrophys. J. **651**, 211 (2006)

T. Yamada, S.-I. Itoh, T. Maruta et al., Anatomy of plasma turbulence. Nat. Phys. **4**, 721 (2008)

G.M. Zaslavsky, Chaos, fractional kinetics, and anomalous transport. Phys. Rep. **371**, 461 (2002)

S. Zhou, W.W. Heidbrink, H. Boehmer et al., Turbulent transport of fast ions in the large plasma device. Phys. Plasmas **17**, 092103 (2010)

S. Zhou, W.W. Heidbrink, H. Boehmer et al., Dependence of fast-ion transport on the nature of the turbulence in the large plasma device. Phys. Plasmas **18**, 082104 (2011)

G. Zimbardo, P. Veltri, P. Pommois, Anomalous, quasilinear, and percolative regimes for magnetic-field-line transport in axially symmetric turbulence. Phys. Rev. E **61**, 1940 (2000a)

G. Zimbardo, A. Greco, P. Veltri, Superballistic transport in tearing driven magnetic turbulence. Phys. Plasmas **7**, 1071 (2000b)

G. Zimbardo, P. Pommois, P. Veltri, Magnetic flux tube evolution in solar wind anisotropic magnetic turbulence. J. Geophys. Res. **109**, A02113 (2004). doi:10.1029/2003JA010162

G. Zimbardo, Anomalous particle diffusion and Lévy random walk of magnetic field lines in three-dimensional solar wind turbulence. Plasma Phys. Control. Fusion **47**, B755 (2005)

G. Zimbardo, P. Pommois, P. Veltri, Superdiffusive and subdiffusive transport of energetic particles in solar wind anisotropic magnetic turbulence. Astrophys. J. **639**, L91 (2006)

G. Zimbardo, R. Bitane, P. Pommois et al., Kolmogorov entropy of magnetic field lines in the percolation regime. Plasma Phys. Control. Fusion **51**, 015005 (2009)

G. Zimbardo, A. Greco, L. Sorriso-Valvo et al., Magnetic turbulence in the geospace environment. Space Sci. Rev. **156**, 89 (2010)

G. Zimbardo, S. Perri, P. Pommois et al., Anomalous particle transport in the heliosphere. Adv. Space Res. **49**, 1633 (2012)

G. Zumofen, A. Blumen, J. Klafter et al., Lévy-walks for turbulence: a numerical study. J. Stat. Phys. **54**, 1519 (1989)

G. Zumofen, J. Klafter, Scale-invariant motion in intermittent chaotic systems. Phys. Rev. E **47**, 851 (1993)

DOI 10.1007/978-1-4899-7413-6_8
Reprinted from *Space Science Reviews* Journal, DOI 10.1007/s11214-013-9970-0

Collisional and Radiative Processes in Optically Thin Plasmas

Stephen J. Bradshaw · John Raymond

Received: 18 October 2012 / Accepted: 20 February 2013 / Published online: 15 March 2013

Abstract Most of our knowledge of the physical processes in distant plasmas is obtained through measurement of the radiation they produce. Here we provide an overview of the main collisional and radiative processes and examples of diagnostics relevant to the microphysical processes in the plasma. Many analyses assume a time-steady plasma with ion populations in equilibrium with the local temperature and Maxwellian distributions of particle velocities, but these assumptions are easily violated in many cases. We consider these departures from equilibrium and possible diagnostics in detail.

Keywords Microphysical processes

1 Introduction

Radiation is often the dominant cooling mechanism for optically thin astrophysical plasmas, which means that it determines the energy budget. It also provides most of the diagnostics for plasma parameters such as density, temperature and composition. It is therefore necessary to understand the dominant collisional and radiative processes in the plasma in order to answer astrophysical questions about the heating or energy dissipation in the plasma. In most cases, the radiation arises from collisions between electrons and ions, but interactions of electrons with a magnetic field or radiation field can also be important.

The subsections of the introduction briefly summarize the processes that dominate in most astrophysical settings, including the wavelength ranges where they are observed and their identifying signatures. In this section we emphasize radiative signatures relevant to microphysical plasma processes, such as differences between electron and ion temperatures, turbulence, and non-Maxwellian velocity distributions.

S.J. Bradshaw (✉)
Department of Physics and Astronomy, Rice University, 6100 Main St., Houston, TX 77005, USA
e-mail: stephen.bradshaw@rice.edu

J. Raymond
Center for Astrophysics, 60 Garden St., Cambridge, MA 02138, USA

The most detailed diagnostics for the physical parameters of plasmas and the microphysical processes taking place are generally based upon atomic and molecular lines and continua. In Sect. 2 we turn to a discussion of radiative processes and the diagnostics that are available. In Sect. 3 we present the theory of line formation in the coronal approximation and describe the dominant collisional and radiative processes. Section 4 comprises a discussion of the factors that influence the charge state, including the key ionization and recombination processes, the charge state in temperature equilibrium and the circumstances under which the charge state can become decoupled from the local temperature. The microphysics that arise when the electron distributions exhibit strong departures from Maxwellian are introduced in Sect. 5; we review the kinetic equations that describe the evolution of the distribution function and the different formalisms that have been adopted for handling collisions. In addition we address the consequences for the heat flux in terms of saturation and de-localization, and for the excitation and ionization rate coefficients which affect the ionization state and, in turn, the radiative losses. The optically-thin radiative loss function itself is the subject of Sect. 6 together with its dependence on the ionization state and the electron distribution. In Sect. 7 we return to a detailed review of the observational signatures and diagnostics that provide evidence for the importance of non-equilibrium ionization and non-Maxwellian electron distributions in the solar atmosphere. Finally, we present a summary of our review and look to the future in Sect. 8.

1.1 Bremsstrahlung

Bremsstrahlung is continuum radiation produced by an electron when it is accelerated in the electric field of an ion. The spectral shape is $P_\nu \sim \exp(-h\nu/kT)$. The mechanism is well understood from basic electromagnetic theory (Rybicki and Lightman 1979), but relativistic corrections are needed for very high temperatures and photon energies (Nozawa et al. 2009). Bremsstrahlung dominates the X-ray continua of many astrophysical sources, though the continua due to radiative recombination and 2-photon processes should not be ignored (Raymond and Smith 1977), and there could be a contribution from synchrotron emission in young SNRs.

Bremsstrahlung emission in the X-rays generally arises from thermal plasmas, but bremsstrahlung is also seen from beams of non-thermal electrons in solar flares (Kontar et al. 2011). Bremsstrahlung emission is also referred to as free-free emission, particularly when observed at longer wavelengths. For example, free-free emission is observed from planetary nebulae and H II regions in the radio, and it is especially valuable as a measure of the ionizing flux from the central star, because it is unaffected by reddening.

The signature of bremsstrahlung emission is a smooth continuum with an exponential cutoff at $h\nu \sim kT$. For normal astrophysical abundances it will be accompanied (and energetically dominated) by spectral line emission unless the temperature is so high that the abundant elements are ionized to their bare nuclei.

1.2 Synchrotron and Cyclotron Emission

The emission from electrons gyrating in a magnetic field can be accurately predicted from electromagnetic theory (Rybicki and Lightman 1979). Relativistic electrons dominate the radio and X-ray synchrotron emission from SNRs, the Galactic Halo, AGN and jets from X-ray binaries. Non-relativistic cyclotron emission can be important in the solar corona and in accreting magnetic white dwarfs.

Synchrotron emission dominates the radio emission of supernova remnants, and in the fast shocks in young SNRs it produces narrow filaments of X-ray emission. The sharpness of the X-ray filaments is used to derive limits on the diffusion coefficient for energetic particles in the acceleration region (Long et al. 2003) and show that the magnetic field is amplified well beyond the values expected for compression in the shock (Vink and Laming 2003; Bamba et al. 2005).

Cyclotron and synchrotron emission are highly polarized, but turbulence randomizes the field directions and Faraday rotation can change the polarization direction and depolarize the emission from an extended region. Bykov et al. (2009) demonstrate how turbulence will affect the X-ray polarization on small scales, and Dickel et al. (1991) have shown that the radio polarization indicates radial, rather than tangential magnetic fields near the edge of Tycho's SNR. Polarization maps in the radio provide a unique method for observing the turbulent structure of the galactic magnetic field (Haverkorn and Heesen 2012).

Synchrotron emission dominates the radio and X-ray spectra of pulsar wind nebulae (PWNe), jets from AGN and gamma-ray bursts. It is straightforward to determine the power law slope of the emitting electrons from the slope of the spectrum. The ambiguity between magnetic field strength and the number of emitting electrons can sometimes be resolved based on spectral breaks due to optical depth or synchrotron cooling.

The emission and absorption occur between quantized Landau levels in the solar corona at radio wavelengths (Dulk et al. 1979), in magnetic cataclysmic variables in the optical, and in accreting neutron stars in the X-ray. The emission at harmonics of the cyclotron frequency can be used to determine the magnetic field strength. The lowest harmonics often are optically thick and the higher ones optically thin. At the transition, the radiation can be strongly polarized. For example, Brosius and White (2006) used radio measurements above the solar limb to obtain the magnetic field strength above a sunspot.

The signatures of synchrotron emission from relativistic electrons are a power law spectrum and a substantial polarization fraction. Gyro emission from non-relativistic thermal electrons typically shows a spectral peak corresponding to a modest harmonic of the cyclotron frequency, with substantial polarization.

1.3 Compton and Inverse Compton Heating and Cooling

The interaction between a photon and an electron can transfer energy either way. As for bremsstrahlung and synchrotron emission, the physical process is well understood (Rybicki and Lightman 1979). Hot plasma above an accretion disk will experience Compton heating by hard X-rays from the central source and Compton cooling by softer photons from the disk. Energetic electrons can interact with synchrotron photons produced by the same electron population (synchrotron self Compton emission).

In supernova remnants, the energetic electrons can produce TeV gamma rays by inverse Compton interaction with the cosmic microwave background (CMB) or with locally enhanced IR or optical radiation. It is currently debated whether the gamma ray emission observed from several SNRs arises from inverse Compton emission by energetic electrons or from decay of pions produced by interactions between cosmic ray protons and dense ambient plasma. Consideration of the lower energy gamma rays observed by FERMI can help to resolve the ambiguity. Inverse Compton gamma ray observations provide at least a lower limit to the maximum energy of the accelerated electrons, and they provide the number of energetic electrons. The latter, in combination with the synchrotron X-ray emission also determines the magnetic field strength.

The signature of inverse Compton emission is a high energy continuum. It is most important when the radiation field is strong and the plasma density and magnetic field are relatively low.

1.4 Plasma Emission

Plasma emission is one of the most direct links between microphysical processes in a plasma and Earth-based observations. It is most commonly seen in the form of type II and type III radio bursts from the Sun. In either case, the radio emission is produced by a several step process. A beam of electrons produced in a flare or at a shock front penetrates into thermal plasma, giving an unstable bump-on-tail velocity distribution. That distribution produces Langmuir waves at the plasma frequency as it flattens into a stable distribution. The Langmuir waves can produce backscattered Langmuir waves and ion acoustic waves, and subsequent interactions between the beam-driven waves and these secondary waves produce radio emission at the plasma frequency of $9n_e^{1/2}$ kHz and twice the plasma frequency (Pick and Vilmer 2008). Schmidt and Cairns (2012) present an analytical formalism for the type II radiation from a shock. The emission tends to be strongest at nearly perpendicular shocks (Cairns 2011).

Though plasma emission has been reported from the coronae of a few active stars, it is seldom observed from astrophysical sources. Most objects bright enough to observe are very dense and optically thick. Type II emission is almost certainly produced by shock waves in supernova remnants, but radiation at the kHz frequencies given by the density of the ISM does not reach Earth.

Since the emission is at the plasma frequency or first harmonic, the measured frequency directly gives the density in the emitting region. The drift rate of the frequency gives the shock speed for a type II burst if the density structure is known. However, shock speeds inferred from type II drift rates do not agree well with shock speeds measured by coronagraphs (Mancuso 2007), either because an inappropriate density structure is assumed or because different parts of the CME shock emit as the CME evolves due, for instance, to selection of a particular angle between the field and the shock where emission is efficient.

The signatures of plasma emission from the solar corona are enormous brightness temperatures and narrow bands of emission near the plasma frequency.

1.5 Dust Emission

In many cases a plasma is optically thin to radiation from dust, even though the emission from individual dust grains is optically thick at some wavelengths. Grains in relatively hot plasmas that are heated to temperatures of order 10–100 K emit at sub-millimeter and infrared wavelengths. The spectrum is a blackbody modified by the opacity of the grain, so it may contain features such as the silicate bump at 9.7 microns that can reveal the nature of the grain material.

Behind the fast shock wave of a supernova remnant, dust is heated to temperatures around 100 K, mainly by collisions with electrons, even as it is gradually eroded by sputtering due to collisions with ions. Infrared emission by dust can be the main radiative energy loss from shock waves faster than about 300 km/s (Arendt et al. 1992). The spectrum and the intensity falloff behind the shock can be used to infer the post-shock density and the destruction rate of the dust (Williams et al. 2006, 2008, 2011; Sankrit et al. 2010).

Dust also absorbs and scatters light at optical, UV and X-ray wavelengths. The wavelength dependence of the absorption, in particular the 2200 Å feature, in combination with the IR emission spectrum, is used to infer the size distribution and composition of the dust (Draine 2003). If the dust column density is fairly high, a detectable halo of X-rays appears around a bright X-ray point source (Smith et al. 2002), from which one can derive the grain size distribution and the location of the grains along the line of sight.

The signature of dust emission is a blackbody-like spectrum at IR or sub-millimeter wavelengths, sometimes with discrete features due to silicates, polycyclic aromatic hydrocarbons (PAHS) or other features. The brightness is proportional to the amount of dust and grain size distribution. It is also sensitive to temperature, which in turn is sensitive to either the radiation field that heats the grains or the density and temperature of the i gas in which they are immersed.

1.6 Ion-Ion Collisions

While most of the radiation detectable at Earth is produced by electrons, energetic collisions between ions produce observable gamma rays. These are most clearly seen during solar flares, when energetic ions strike the dense gas of the chromosphere to produce broad and narrow nuclear de-excitation lines, positrons that subsequently annihilate to produce 0.511 MeV photons, and neutron capture lines (Vilmer et al. 2011). Gamma ray spectra from RHESSI and other instruments can be used to infer the composition of both the chromosphere and the energetic ions, the spectral shape of the accelerated particles and their energy content at MeV energies.

At higher energies, cosmic rays can collide with nuclei in the ambient gas to produce pions, which can decay into gamma rays. Though it is often difficult to tell whether TeV gamma rays are produced by pion decay or inverse Compton interaction between ambient photons and energetic electrons, observations of supernova remnants with the ground-based arrays H.E.S.S, MAGIC and VERITAS, and with the FERMI satellite, offer constraints on the acceleration of hadrons in strong shock waves. The nature of the gamma ray emission from many SNRs is still under debate, but the gamma rays from some old SNRs interacting with dense clouds can be attributed to pion decay (e.g., Uchiyama et al. 2012 FERMI observations of W44).

2 Atomic and Molecular Spectral Line Diagnostics

Atomic spectral lines can appear in emission or absorption. Emission lines usually arise following excitation by electron impact or recombination into an excited level, though they can also be produced by ion impact (Laming et al. 1996) or photoabsorption (Noci et al. 1987). Cooling by emission of atomic or molecular lines often dominates the energy budget of the plasma, and the intensities of the spectral lines provide powerful diagnostics for the physical parameters of the plasma. This Section provides an overview of atomic emission line diagnostics, and Sect. 3 provides a rigorous discussion of the line formation process.

Intensity ratios of lines within a single ion can be used to infer the electron temperature and density of the gas. Electron temperature diagnostics generally hinge on the Boltzmann factor, $\exp^{-\Delta E/k_B T}$, where ΔE is the energy difference between the two upper levels (Fig. 1 left diagram). Such a ratio works best for $\Delta E \sim k_B T$, so that optical line ratios are effective for T around 10^4 K, where $\Delta E \sim k_B T \sim 1$ eV. UV line ratios are effective around 10^5 K and X-ray line ratios above 10^6 K. Often the desirable spectral lines lie at much different wavelengths, so that it is hard to obtain a ratio with a single instrument, but the technique has been applied to solar spectra (David et al. 1998).

The density can be inferred from ratios involving a metastable level. The population of that level will be small at low densities. It approaches a constant value given by the statistical weight and Boltzmann factor above a critical density $n_{crit} = A_{21}/q_{21}$, where A_{21} is the Einstein A value and q_{21} is the de-excitation rate coefficient. The ratio of a line which

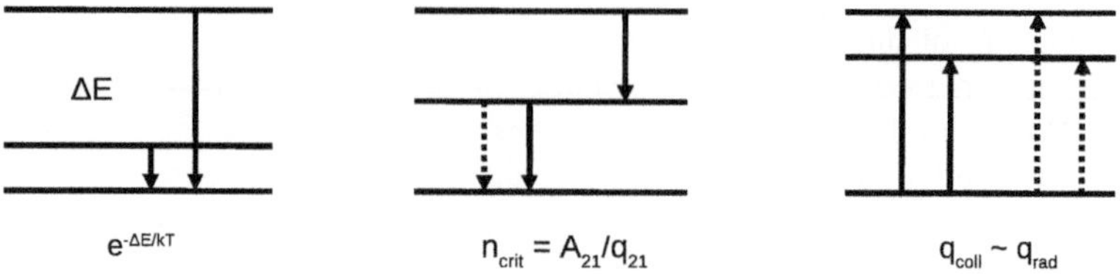

Fig. 1 Temperature diagnostics are generally based on the ratio of Boltzmann factors, $\exp^{-\Delta E/k_B T}$, in the excitation rates of two spectral lines (*left diagram*). Density diagnostics can be based on the competition between radiative decay and collisional de-excitation when the Einstein A value is comparable to the density times the collisional rate coefficient (*middle panel*) or on the relative contributions of collisional excitation and radiative excitation (*right panel*)

involves the metastable level to a line which does not will be sensitive to density (Fig. 1 center diagram). Because the A values increase rapidly with transition energy and q values decline, n_{crit} increases rapidly from values around 10^2 to 10^4 cm^{-3} for optical forbidden lines to 10^8 to 10^{10} cm^{-3} for UV lines and 10^{11} to 10^{15} cm^{-3} for X-ray lines.

A less commonly used density diagnostic takes advantage of the fact that some lines formed in the solar corona include both collisionally excited and radiatively excited components. The ratio of those components is proportional to the density and perhaps plasma velocity and line width (Noci et al. 1987). It is interesting to note that for an ion X, the ratios indicated in the center and right panels of Fig. 1 give $\langle n_X n_e^2/(n_{crit}+n_e)\rangle / \langle n_X n_e/(n_{crit}+n_e)\rangle$ and $\langle n_X n_e\rangle / \langle n_X W\rangle$, respectively. Here W is the dilution factor of the radiation (Sect. 3.4). Thus different density estimates are differently weighted averages that do not necessarily agree. In principle, comparison of differently weighted averages could yield unique information about the distributions of electron density and density of the diagnostic ion within the observed volume, but that requires very good accuracy for both diagnostics (Lee et al. 2008).

Ratios of emission or absorption lines of different elements can also be used to derive the relative elemental abundances. In practice that is often tricky because in many cases only 1 or 2 ions of each element can be observed, so the ionization state of each element must be accurately known. This usually requires a model that involves ionization and recombination rates, each having perhaps a 20 % uncertainty, and it often involves an assumption of ionization equilibrium (Sect. 4.2) that may not be justified (Sects. 4.3 and 7.1). These difficulties are somewhat mitigated if one can use ions such as He-like and H-like ions that dominate the ionization distribution over broad temperature ranges.

The profiles of optically thin emission or absorption lines provide a direct measurement of the velocity distribution of atoms, molecules or ions along the line of sight. Therefore, they provide good diagnostics for the ion kinetic temperatures, turbulence and in principle non-Maxwellian velocity distributions (Sect. 5), though there may be ambiguities among the different interpretations.

Line profiles directly give ion temperatures when bulk motions do not dominate. In low density regions of the solar corona, the line widths of oxygen ions exceed those of hydrogen, indicating that the kinetic temperature of O is more than 16 times that of H (Kohl et al. 1997; Cranmer et al. 2008; Frazin et al. 2003).

Collisionless shock waves are another good example of the application of line profile diagnostics. Neutral hydrogen that passes through a strong shock does not feel the collisionless shock itself, but finds itself immersed in the hot post-shock flow. Diagnostics based on Balmer line profiles from these shocks are discussed in Bykov et al. (2013). Most observed line profiles can be fit with a Gaussian or a sum of Gaussians, so they are consistent with

Maxwellian distributions. The broad Hα profile of a bright knot produced by a 2000 km/s shock in Tycho's supernova remnant is not Maxwellian, suggesting either a power-law tail or a pickup-ion contribution, though an interpretation as a sum of Maxwellian contributions cannot be excluded (Raymond et al. 2010).

Line profiles can be directly used to determine the level of turbulent velocity fluctuations if thermal and bulk velocities do not dominate. Comparison of lines from elements of different masses can help to resolve the ambiguity between thermal and turbulent line widths. Line widths have been used to estimate the level of turbulence in reconnection current sheets during solar eruptions (Bemporad 2008). Another application has been study of turbulence in interstellar gas using the Velocity Coordinate Spectrum method to combine line profiles and their spatial variations (Chepurnov et al. 2010). These statistical methods, along with methods based on polarization (Burkhart et al. 2012) can reveal the turbulence spectrum and whether the turbulence is subsonic.

3 Optically-Thin Emission Lines

3.1 Line Formation

High temperature ($> 10^6$ K) and low density ($< 10^{13}$ K) astrophysical plasmas are optically-thin to visible, EUV and X-ray radiation. Photons at these wavelengths are generally able to propagate through these environments unhindered by opacity effects, such as absorption and re-emission, and scattering, and therefore retain a record of the plasma conditions at the site of emission. Most of the radiation in the region of 10^6 K is due to the emission of photons by electron transitions in ions, giving rise to spectral lines. The radiated power per unit volume, commonly referred to as the *emissivity*, depends on: (a) the number of ions that are present; and (b) the fraction of those ions in the excited state that corresponds to the transition. For a given transition (in the notation of Mason and Monsignori Fossi 1994):

$$P(\lambda_{j,i}) = N_j\left(X^{+m}\right)A_{j,i}\Delta E_{j,i}\ \left[\text{erg cm}^{-3}\,\text{s}^{-1}\right], \tag{1}$$

where $N_j(X^{+m})$ [cm^{-3}] is the number density of ions of charge $+m$ in excited state j, $A_{j,i}$ [s^{-1}] is the Einstein coefficient and $\Delta E_{j,i}$ is the energy of the emitted photon. The quantity $N_j(X^{+m})$ can be rewritten as a series of ratios that can be measured observationally or experimentally, or calculated theoretically (Mason and Monsignori Fossi 1994). The total energy flux due to the transition, at a distance R from the emitting volume of plasma, can be found by integrating the emissivity over the volume and dividing by the surface area of the sphere with radius R:

$$I(\lambda_{j,i}) = \frac{1}{4\pi R^2}\int_V P(\lambda_{j,i})dV\ \left[\text{erg cm}^{-2}\,\text{s}^{-1}\,\text{sr}^{-1}\right]. \tag{2}$$

3.2 The Coronal Model

A convenient approximation for optically-thin plasmas, such as the solar corona, allows a decoupling of the processes that determine the excitation state from those that determine the charge state. This can be justified by noting that changes in the energy level populations of the emitting ions occur far more frequently than changes in the charge state. The processes that determine the excitation state are discussed in this Section and those that determine the charge state are discussed in Sect. 4. In optically-thin plasmas energy levels become populated by electron collisional excitation from the ground-state (g) of each ion, and they become depopulated by spontaneous radiative decay. It is assumed that timescales of photon absorption and electron collisional de-excitation are far longer. This is called the *coronal model* approximation and in statistical equilibrium the number of collisional transitions from the ground-state g to the excited state j must be equal to the number of spontaneous radiative decays back to the ground-state.

$$N_g(X^{+m})N_e C^e_{g,j} = N_j(X^{+m})A_{j,g}\ \left[\mathrm{cm^{-3}\,s^{-1}}\right]. \tag{3}$$

$C^e_{g,j}$ [cm^3 s^{-1}] is the electron collisional excitation rate coefficient between the ground-state and level j. If collisions are relatively infrequent then $A_{j,g} \gg N_e C^e_{g,j}$ and it follows that $N_g(X^{+m}) \gg N_j(X^{+m})$. There are many more ions in the ground-state than in excited states. In a typical transition at EUV wavelengths $A_{j,g} = 10^{10}$ [s^{-1}] and $N_e C^e_{g,j} = 1$ [s^{-1}] and so for every collisional excitation there is an almost immediate radiative decay to satisfy the requirements of statistical equilibrium. We note that Eq. (3) pertains to 2-level atoms, but radiative cascades from higher levels, following excitation or recombination, may dominate under particular circumstances, such as transitions from Fe XVII 3s levels (Beiersdorfer et al. 2004).

The statistical equilibrium relationship given in Eq. (3) and the fact that $\frac{N_g(X^{+m})}{N(X^{+m})} \approx 1$ leads to an expression for the emissivity in terms of the collisional excitation rate:

$$P(\lambda_{j,g}) = \frac{N(X^{+m})}{N(X)}\frac{N(X)}{N(H)}\frac{N(H)}{N_e} C^e_{g,j}\Delta E_{j,g} N_e^2. \tag{4}$$

The spectral line intensity is proportional to N_e^2 as expected.

3.3 Collisional Processes

The rate at which collisional transitions occur depends on the interaction cross-section presented to incident particles by the target and on the flux of incident particles. The flux of incident particles can be written:

$$F = nvf(E)dE\ \left[\text{particles cm}^{-2}\,\text{s}^{-1}\right], \tag{5}$$

where n is the number density of particles, v is the incident particle velocity, E the kinetic energy of the incident particles and $f(E)$ the particle distribution function. Since particle-particle interactions are mostly via collisions then it is common to assume that the distribution function is a collisionally relaxed Maxwellian of the form:

$$f(E) = 2\sqrt{\frac{E}{\pi}}\left(\frac{1}{k_B T}\right)^{\frac{3}{2}} \exp\left(-\frac{E}{k_B T}\right)\ \left[\text{particles erg}^{-1}\right]. \tag{6}$$

The electron collisional excitation rate coefficient is found by integrating the electron distribution function over the interaction cross-section.

$$C^e_{i,j} = \int_{\Delta E}^{\infty} Q_{i,j} v f(E) dE\ \left[\mathrm{cm^3\,s^{-1}}\right]. \tag{7}$$

ΔE is the energy difference between level i and j, and this is the lower limit to the integral because an incident particle must have at least this much energy in order to excite the transition. $Q_{i,j}$ [cm^2] is the interaction cross-section. In simplified form

$$C^e_{i,j} = \frac{8.63 \times 10^{-6}}{\omega_i \sqrt{T}} \Upsilon_{i,j} \exp\left(-\frac{\Delta E}{k_B T}\right), \tag{8}$$

where ω_i is the statistical weight of level i, which is the number of different spin and angular momentum states that have energy E_i (the number of degenerate states in energy E_i), and $\Upsilon_{i,j}(T)$ is the thermally averaged collision strength (Mason and Monsignori Fossi 1994). $\omega_i = 2n_q^2$ for hydrogen (where n_q is the principle quantum number).

3.4 Radiative Processes

Spontaneous radiative decay of electrons from excited states is the dominant depopulation mechanism in optically-thin plasmas. The generalised radiative decay coefficient is:

$$R_{j,i} = A_{j,i}\left(1 + \frac{W}{\exp(\frac{\Delta E}{k_B T}) - 1}\right) [\mathrm{s}^{-1}]. \tag{9}$$

The first term of Eq. (9) takes account of spontaneous emission. The second term accounts for the stimulated component of the emission in the presence of a background continuum radiation field, described by a Planck function. In the case of radiative decay in the solar corona the background radiation field would have a temperature of 5800 K, characteristic of the photosphere. W is a dilution factor that describes the decay of the radiation field with radial distance, where:

$$W = \frac{1}{2}\left[1 - \left(1 - \frac{R_0^2}{r^2}\right)\right]. \tag{10}$$

In the case of the Sun, R_0 would be the solar radius and r the distance from the centre of the Sun to the height in the atmosphere at which W must be calculated. As $r \to \infty$ (e.g. sufficiently far above the surface that $r \gg R_0$) then $W \to 0$ and the stimulated component of the emission can be neglected so that $R_{j,i} = A_{j,i}$. The stimulated component of the solar radiative flux is also negligible at far UV and shorter wavelengths; however, photoexcitation of UV lines such as the Lyman series and O VI is very important beyond about 1.3 solar radii.

4 The Charge State of a Plasma

4.1 Ionization and Recombination

The charge state of the ions in a plasma is governed by the rate at which electrons are freed from their bound states and the rate at which free electrons are captured into bound states. Bound-free transitions are called *ionization* and free-bound transitions are called *recombination*. Collisional excitation and radiative decay occur on timescales far shorter than ionization and recombination timescales, and so these processes can be de-coupled from the excitation and decay processes. Ionization (recombination) can then be considered to take place from (to) the ground-state of the ion, though it is worth noting that at transition region densities (e.g. $n \approx 10^{10}$ cm^{-3}) ionization and recombination from metastable levels can become important (Vernazza and Raymond 1979). In optically-thin plasma, such as solar and stellar coronae, the important ionization processes are: collisional ionization; and excitation-autoionization. The important recombination processes are: radiative recombination; and dielectronic recombination.

Collisional Ionization as in the case of collisional excitation, the dominant process of ionization is by electron collisions (photo-ionization is negligible at the energies of interest). Where collisional excitation is generally due to electrons in the bulk of the distribution (e.g. a Maxwellian), ionization arises from electrons in the high-energy tail. Since the number density of electrons in the tail is relatively low then collisional ionization is relatively infrequent compared with collisional excitation. The process of collisional ionization can be written (again employing the standard notation of Mason and Monsignori Fossi 1994):

$$X_i^{+m} + e^- \rightarrow X_{i'}^{+m+1} + 2e^-. \tag{11}$$

The ion in the state i loses an electron and a new ion is created in the state i'. The incident electron must have sufficient energy to free the bound electron and retain enough to remain unbound.

Radiative Recombination similarly to radiative decay, an important recombination process is the capture of an energetic free electron into a lower energy, bound state, leading to the emission of a photon. The radiative recombination process can be written:

$$X_{i'}^{+m+1} + e^- \rightarrow X_i^{+m} + \Delta E. \tag{12}$$

Dielectronic Recombination the dominant recombination mechanism at high temperatures, as shown by Burgess (1964). The dielectronic recombination process can be written:

$$X_{i'}^{+m+1} + e^- \rightarrow \left(X_{i''}^{+m}\right)^{**} \rightarrow X_i^{+m} + \Delta E. \tag{13}$$

Equation (13) shows that an ion with $m+1$ missing electrons may capture a free electron into a particular outer energy level while simultaneously exciting an inner electron to a higher energy level *instead* of emitting a photon. The $()^{**}$ notation indicates a doubly excited state. The excited inner electron may then decay to its original level (or another, if low-lying fine structure states are available), with the emission of a photon, leaving the ion in a singly excited state because the captured electron remains in an outer energy level. At this point the recombination is complete. Dielectronic recombination is the dominant recombination mechanism for most ions at high temperatures, especially those with $\Delta n_q = 0$ transitions from the ground state. Dielectronic recombination can also be somewhat density dependent, because the emission of a photon often leaves the recombined ion in a highly excited state that can be ionized before it decays to the ground state.

Excitation-Autoionization if the two excited electrons in the second stage of Eq. (13) together have more energy than is needed to remove a single electron from the ground state, then the ion is energetically able to autoionize. This means that it can decay to the ground state with the ejection of one of the excited electrons:

$$\left(X_{i''}^{+m}\right)^{**} \rightarrow X_i^{+m} + e^-. \tag{14}$$

Note that the process described by Eq. (14) is the inverse process to the first stage of dielectronic recombination in Eq. (13). The doubly excited ion has two choices: (1) emit a photon; or (2) autoionize (if the total energy of the excited electrons exceeds the threshold for ionization).

Charge Transfer charge transfer between ionized species and neutral hydrogen is not usually important in the solar corona, but it can modify the ionization state in astrophysical plasmas, especially cool plasma photoionized by a hard radiation field. The cross-section for resonant charge transfer is very large and this sometimes makes up for a low neutral fraction.

4.2 The Charge State in Equilibrium

Ionization and recombination rate coefficients depend strongly on temperature and, to a somewhat lesser extent, on density. At higher temperatures the free electrons have a greater average kinetic energy and so are able to collisionally release even the strongly bound, inner electrons of the target ions. At lower temperatures the free electrons are less energetic and can be captured even into the low ionization energy, outer bound states of the ions. It is useful to observe that ions are typically found at a temperature such that the ionization potential is $\approx 5k_BT$ in equilibrium. A full set of ionization and recombination rate coefficients (e.g. Arnaud and Rothenflug 1985; Arnaud and Raymond 1992; Mazzotta et al. 1998; Bryans et al. 2009; Dere 2007) for a given element allows the distribution among the charge states for the ions of that element to be calculated as a function of temperature. We note here that published rate coefficients tend to be calculated assuming that the free electrons have relaxed into a Maxwellian electron distribution. We consider the consequences of the breakdown of this assumption and the calculation and the consequences of departures from an underlying Maxwellian in Sect. 5.

One may ask what proportion of helium atoms are neutral, singly ionized and doubly ionized at a particular temperature. This is the charge or ionization state of the element. At a temperature of 10^6 K helium is fully ionized and so the population fractions are: He I (neutral) $= 0.0$; He II (singly-ionized) $= 0.0$; and He III (doubly-ionized) $= 1.0$. At 10^5 K (adopting the ionization rates of Dere et al. (2009) and the recombination rates of Mazzotta et al. (1998) the ionization state of helium is: He I $= 0.0$; He II $= 0.131$; and He III $= 0.869$. 13 % of helium is singly ionized and 87 % of helium is fully ionized at 10^5 K. The population fractions for all the ions of a particular element must sum to 1.0 in order to conserve the particle number.

The population fraction for each ion peaks at the temperature at which the ionization and recombination rates are equal. More ionizations would act to deplete the ion population in favour of a higher charge state, and more recombinations would deplete the population in favour of a lower charge state. The ionization states for helium given above are only reached when the ionization state is in *equilibrium* with the electron temperature of the plasma. Strictly speaking, as $t \rightarrow \infty$ at $T = 10^5$ K then He I $\rightarrow 0.0$, He II $\rightarrow 0.131$ and He III $\rightarrow 0.869$. The reason for this is that collisional processes are not instantaneous. It takes a certain period of time for ionization and recombination events to arrange the ions into the charge states that correspond to the current electron temperature. As long as the ionization and recombination timescales are much shorter than the timescale on which the temperature changes then the ionization state can be considered in equilibrium with the temperature, and therefore depends only on the temperature. The break-down of this condition will be discussed in Sect. 4.3.

One consequence of de-coupling ionization and recombination from the processes of excitation and radiative decay is that one may assume ionization (recombination) occurs from (to) the ground state of the ion, and so the rate of change of the population fraction of a particular ion i of element X can be written:

$$\frac{dX_i}{dt} = n(I_{i-1}X_{i-1} + R_iX_{i+1} - I_iX_i - R_{i-1}X_i). \tag{15}$$

In the notation of Eq. (15) element X might be helium and then $X_{i=0}$ would be neutral helium (He I), and so forth. n [cm^{-3}] is the electron number density, and I_i and R_i are the temperature dependent *total* ionization and recombination rate coefficients, respectively, with units [$\text{cm}^3\ \text{s}^{-1}$]. In equilibrium $\frac{d}{dt} = 0$ so that:

$$I_{i-1}X_{i-1} + R_i X_{i+1} = I_i X_i + R_{i-1} X_i. \tag{16}$$

The LHS of Eq. (16) comprises the processes that lead to the creation of ion X_i (ionization from lower charge states and recombination from higher charge states). The RHS comprises the processes that lead to the destruction of X_i (ionization to higher charge states and recombination to lower charge states). In equilibrium the principle of detailed balance implies that the rate of ionization to X_i is equal to the rate of recombination from X_i, and the rate of ionization from X_i is equal to the rate of recombination to X_i. This can be expressed in the form of two de-coupled equations:

$$I_{i-1}X_{i-1} = R_{i-1}X_i; \tag{17}$$

$$R_i X_{i+1} = I_i X_i. \tag{18}$$

The ionization state can then be fully specified subject to the final constraint:

$$\sum_{i=0}^{Z} X_i = 1.0, \tag{19}$$

where Z is the atomic number of the element X. Making use of Eqs. (17) and (18) it can be seen that:

$$X_{i-1} = \frac{R_{i-1}}{I_{i-1}} X_i \quad \text{and} \quad X_{i+1} = \frac{I_i}{R_i} X_i. \tag{20}$$

Given a set of ionization and recombination rate coefficients the ionization state can be calculated by choosing a suitable value for X_i. The most abundant ion i of element X is the one for which $I_i(T) \approx R_i(T)$ at the temperature of interest. The population fraction of this ion can then be assigned some arbitrary quantity $X_i = X_i'$ usually chosen to avoid computational overflow errors since the population fractions can vary over many orders of magnitude (this is not so much of an issue in the case of double-precision arithmetic). It is then straightforward to calculate $[X_{i-1}', X_{i-2}', \ldots, X_0']$ and $[X_{i+1}', X_{i+2}', \ldots, X_Z']$ recursively from Eq. (20) and find the true population fractions by normalising the values of X_i' to 1.0:

$$X_i = \frac{X_i'}{\sum_{i=0}^{Z} X_i'}. \tag{21}$$

4.3 Non-equilibrium Charge States

In circumstances where the electron temperature has been held fixed for a long time *or* the temperature is changing slowly, then the ionization state of the plasma is in equilibrium and depends on the temperature only. A slowly changing temperature in the present context means that it changes more slowly than the timescales on which the processes that change the ionization state of the plasma operate. If the temperature change is sufficiently slow then collisions have ample time to arrange the charge states of the element such that they are in equilibrium with the temperature.

Table 1 Population fractions, and ionization and recombination rate coefficients for the series of ions Fe XIV, XV, XVI. These data are based on the ionization/recombination rate coefficients from/to a given ion provided by Mazzotta et al. (1998) and Dere (2007). The rate coefficients are in units of [$\mathrm{cm^3\,s^{-1}}$]

Charge state	Population fraction	Ionization rate		Recombination rate	
	10^6 K	10^6 K	2.5×10^6 K	10^6 K	2.5×10^6 K
Fe XIV	4.60×10^{-4}	4.13×10^{-12}	1.13×10^{-10}	1.35×10^{-10}	5.16×10^{-11}
Fe XV	1.41×10^{-5}	1.04×10^{-12}	6.07×10^{-11}	1.04×10^{-10}	5.09×10^{-11}
Fe XVI	1.40×10^{-7}	3.78×10^{-13}	3.60×10^{-11}	9.66×10^{-12}	2.39×10^{-11}

Consider now a plasma that is heated by some mechanism from 10^6 K to 10^7 K in just 1 second, but it takes several minutes for collisions to change the ionization state. In this scenario a plasma of electron temperature 10^7 K is created with an ionization state (and consequently an emission spectrum) that is characteristic of a 10^6 K plasma in equilibrium. The time derivative in Eq. (15) cannot now be neglected (however, the bulk velocity will be neglected from the total derivative in the following treatment) and a non-equilibrium ionization state arises. In order to determine whether non-equilibrium ionization is important in a particular physical scenario of interest, the equilibration timescale of the ionization state at the new temperature can be estimated from Eq. (15). If it is significantly greater than the timescale of the temperature change itself then non-equilibrium ionization will be important.

Taking a somewhat less extreme example, suppose that a plasma is heated from 10^6 K to 2.5×10^6 K essentially instantaneously. The equilibrium population of Fe XV reaches its maximum at 2.5×10^6 K and so how long does it take to equilibrate in this scenario? Based on the data provided in Table 1 we can write:

$$\begin{aligned}\text{Rate of loss of Fe X} &= n\big[-I_i(T+\Delta T)X_i(T) - R_{i-1}(T+\Delta T)X_i(T)\big]\\ &= n \times 1.41 \times 10^{-5} \times \big(-6.07 \times 10^{-11} - 5.16 \times 10^{-11}\big)\\ &= -n \times 1.58 \times 10^{-15}\ \big[\mathrm{s^{-1}}\big];\end{aligned} \tag{22}$$

$$\begin{aligned}\text{Rate of gain of Fe X} &= n\big[I_{i-1}(T+\Delta T)X_{i-1}(T) + R_i(T+\Delta T)X_{i+1}(T)\big]\\ &= n \times \big(1.13 \times 10^{-10} \times 4.60 \times 10^{-4} + 5.09 \times 10^{-11} \times 1.40 \times 10^{-7}\big)\\ &= n \times 5.20 \times 10^{-14}\ \big[\mathrm{s^{-1}}\big];\end{aligned} \tag{23}$$

$$\begin{aligned}\text{Net rate of change of Fe X} &= n\big(-1.58 \times 10^{-15} + 5.20 \times 10^{-14}\big)\\ &= n \times 5.04 \times 10^{-14}\ \big[\mathrm{s^{-1}}\big].\end{aligned} \tag{24}$$

For an electron density characteristic of the solar corona $n = 10^9\ \mathrm{cm^{-3}}$ then the equilibration timescale is given by:

$$\tau = \frac{1}{10^9 \times 5.04 \times 10^{-14}} \approx 20{,}000\ [\mathrm{s}]. \tag{25}$$

If the plasma temperature is changed effectively instantaneously from 10^6 K to 2.5×10^6 K and then held constant at the new temperature, then the population of Fe XV will approach

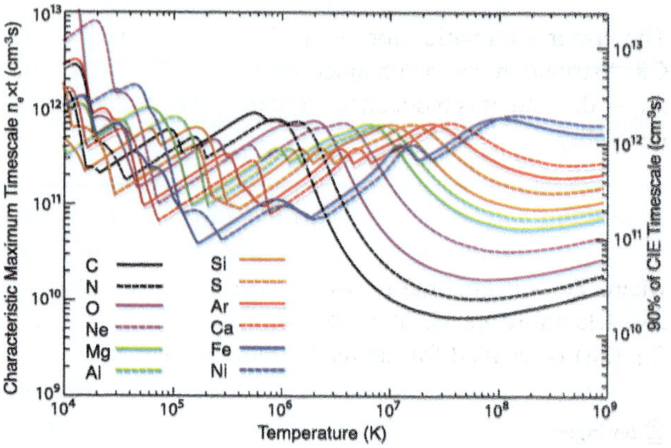

Fig. 2 Smith and Hughes (2010): the *left axis* measures the density-weighted timescale [cm^{-3} s] for several abundant elements to achieve one e-folding toward ionization equilibrium in a constant temperature plasma; the *right axis* measures the density-weighted timescale for the elements to reach within 10 % of their equilibrium population

equilibrium on an e-folding timescale of 20,000 s. In consequence, heating on timescale much shorter than 20,000 s will give rise to a non-equilibrium ionization state; for example, the population of Fe XV is guaranteed to be out of equilibrium if heating in the solar corona is impulsive (of duration shorter than the characteristic cooling timescale). Heating on timescales significantly longer than 20,000 s allows the ionization state to evolve in equilibrium with the electron temperature. We note that even coronal densities of order 10^{12} to 10^{13} cm^{-3} may not be sufficient to maintain the ion population close to equilibrium during particularly explosive heating such as occurs during solar flares. The estimate of the timescale provided by Eq. (25) should be regarded as an absolute upper-limit. The intermediate population fractions of Fe XV and its neighbouring charge states as the system equilibrates are not accounted for in the approximation. In essence, the rate of change of the population fraction is proportional to the magnitude of the population itself and it will therefore equilibrate more rapidly as it grows. Figure 2 is from Smith and Hughes (2010) and shows the characteristic equilibration e-folding time-scales for a number of astrophysically abundant elements. Bradshaw (2009) presents a freely available numerical code that solves the time-dependent ionization equations for all elements up to nickel ($Z = 28$), given any tabulated electron temperature and density evolution as a function of time (the time-steps need not be uniform). We describe specific examples of scenarios in which non-equilibrium ionization might arise in Sect. 7.

5 Non-Maxwellian Electron Distributions

Temperature changes on timescales much shorter than those on which ionization and recombination can change the charge state of the plasma are not the only way in which the ionization state may be different than expected for a given temperature. If the electron distribution function is driven away from Maxwellian with the addition of a significant population to the high-energy tail of the distribution, then ions of greater charge can be created at some temperature that is lower than the temperature at which they arise in equilibrium. Non-Maxwellian distributions can arise in several ways; for example, in an astrophysical context they may be expected to occur in circumstances where a region of very hot plasma is separated from a much cooler region by a steep temperature gradient, as is the situation in the solar atmosphere. Collisionless electrons may then stream from the hot, less-dense plasma down the temperature gradient into the cooler, denser plasma driving the tails of the electron distributions in these regions away from Maxwellian. Non-Maxwellian distributions can also be induced when a beam of particles is accelerated by some mechanism, such as magnetic reconnection, and interacts with the background plasma. A non-Maxwellian distribution can be created in laboratory plasma by laser-heating.

Departures of the electron population from Maxwellian distributions have implications for several properties of the plasma, among them the excitation and ionization states (and, consequently the spectral emission) of its component ions and the transport of energy by thermal conduction. It is therefore highly desirable to take account of these effects in numerical modeling studies, but this is an extremely difficult task to achieve self-consistently. There are two general approaches: (1) carry out particle-in-cell type calculations where the distribution functions can be obtained directly; (2) carry out calculations based on the fluid equations derived by taking successive moments of the underlying distribution function. The first approach is discussed elsewhere in this volume. The difficulty of the second approach is that solutions to the fluid equations can quickly become inconsistent with the assumptions on which their derivation is based. For example, in the case of steep temperature gradients the mean-free-path of even thermal electrons can approach (and exceed!) the temperature scale length and then the plasma cannot be considered collisional on the characteristic scale length of the fluid. However, the validity of the fluid equations is contingent on the collisionality of the system on the relevant spatial scale. One advantage of the second approach over the first is that plasma systems can be modelled across a much larger range of spatial scales. In particle-type codes one is generally confined to studying phenomena on a particular scale, such as the width of a conduction front or a shock, or the scale of the diffusion region in reconnection. In fluid codes, the solution can range across many spatial scales from a few meters to hundreds of thousands of kilometers (e.g. in the case of the Sun's atmosphere). The challenge is to develop a method by which physical phenomena on particle scales can be self-consistently included, when needed, in a code that operates predominantly on fluid scales. This requires the distribution function to be calculated from a suitable kinetic equation in tandem with the time-advancement of the system of fluid equations in order that corrections can be made to the fluid variables.

5.1 Kinetic Equations

The key to calculating the distribution function in a collisional or weakly-collisional plasma is the manner in which the collision term of the Boltzmann equation is treated. One of two approaches is usually adopted. The first is to handle collisions via a phenomenological term based on the expectation that the plasma particles will ultimately relax into a collisional/Maxwellian distribution on some timescale that depends on the degree of collisionality of the plasma. This was first suggested by Bhatnagar et al. (1954) and kinetic equations of this form are commonly referred to as BGK equations (based on the initials of the authors of that paper).

$$\left(\frac{\partial f_s}{\partial t}\right)_{\text{collisions}} = \nu_{ss}(F_s - f_s) + \nu_{ss'}(F_{ss'} - f_s), \tag{26}$$

where s, s' denote the particle species (e.g. electrons and protons), ν_{ss}, $\nu_{ss'}$ are the species and inter-species collision frequencies, and F denotes a Maxwellian distribution determined by the local properties of the plasma (e.g. temperature, density and bulk flow). Morse (1963) studied the energy and momentum exchange between non-equipartition gases in the cases of Maxwell, Coulomb and hard sphere interactions, and Morse (1964) showed how to choose free parameters for the cross-collision terms in BGK-type models to conserve density, momentum and energy. This work was limited by the underlying assumption that $n_e/\tau_{ei} = n_i/\tau_{ie}$, which for a fully-ionized hydrogen plasma ($n_e = n_i$) implies that electrons and ions are equally affected by their mutual collisions ($\tau_{ei} = \tau_{ie}$) when in reality they relax on a timescale longer by the square root of the mass ratio ($\tau_{ie} = \sqrt{m_i/m_e}\tau_{ei}$). Greene (1973) then developed a simple improvement for BGK-type models of electron-ion collisions to produce the correct relation between the time scales of ion-electron momentum exchange and ion thermalisation. This work showed how to obtain the correct timescale ordering and how to choose the correct parameters for the Maxwellians in the cross-collision terms to conserve density, momentum and energy.

The second approach to handling collisions is to assume that changes in the velocities of charged particles are due to the cumulative effect of long-range encounters via inverse square forces (e.g. Landau 1936). The collision integral can then be written (e.g. Ljepojevic and Burgess 1990):

$$\left(\frac{\partial f_s}{\partial t}\right)_{\text{collisions}} = -\sum_i \frac{\partial}{\partial v_s^i}\left(f_s\langle\Delta v_s^i\rangle\right) + \frac{1}{2}\sum_{i,j}\frac{\partial^2}{\partial v_s^i \partial v_s^j}\left(f_s\langle\Delta v_s^i \Delta v_s^j\rangle\right), \tag{27}$$

where

$$\langle\Delta v_s^i\rangle = \sum_{s'}\int f_{s'}(\mathbf{v}'_{s'})\int_{\theta_{min}}^{\theta_{max}} \sigma_{ss'}(g,\Omega) g \Delta v_s^i d^2\Omega d^3\mathbf{v}'_{s'} \tag{28}$$

and

$$\langle\Delta v_s^i \Delta v_s^j\rangle = \sum_{s'}\int f_{s'}(\mathbf{v}'_{s'})\int_{\theta_{min}}^{\theta_{max}} \sigma_{ss'}(g,\Omega) g \Delta v_s^i \Delta v_s^j d^2\Omega d^3\mathbf{v}'_{s'}. \tag{29}$$

The terms of Eqs. (27) to (29) are described in detail in Sect. 2(a) of Ljepojevic and Burgess (1990). Equations of the form of (27) are commonly referred to as Fokker-Planck (FP) equations. Cohen et al. (1950) adopted a method of approximating the distribution function by representing it as a Maxwellian plus a small perturbation to calculate the electrical conductivity of a gas. Their approach is valid in the presence of weak spatial gradients and weak electromagnetic fields. The distribution function then takes the form $f_s = f_0 + f_1$ where $f_0 = F_s$ and

$$f_1 = F_s D(v_s)\mu, \tag{30}$$

where μ is the cosine of the pitch angle. D is a function of the particle speed found by substituting $f_s = f_0 + f_1$ for f_s in the Boltzmann equation, linearising the collisional integral in f_1, and solving the integro-differential equation. Cohen et al. (1950) neglected electron-electron interactions from the collision operator but Spitzer and Härm (1953), in what is now considered the 'classical' treatment, followed the same approach and included electron-electron interactions in their collision operator. They also extended the solutions to completely ionized gases with different mean nuclear charges and calculate the electrical and thermal conductivities of the gas. The solution to the integro-differential equation in the classical treatment has the form (for electrons)

$$f_e = F_e\left(1 - \lambda_0\left[\frac{ZD_E}{A}\left(\frac{eE}{k_B T_e} + \frac{1}{P_e}\frac{\partial P_e}{\partial s}\right) - 2\frac{ZD_T}{B}\frac{1}{T_e}\frac{\partial T_e}{\partial s}\right]\mu\right). \tag{31}$$

The quantities ZD_E/A and ZD_T/B are tabulated in Spitzer and Härm (1953) as functions of the electron speed normalised to the thermal speed and λ_0 is the mean-free-path of thermal electrons. Above a particular speed v_{crit} the calculated value of f_1 becomes comparable to f_0 and the assumptions made to derive Eq. (31) are invalidated. The approximation of Spitzer and Härm (1953) is only valid in the low-velocity regime $v < v_{crit}$ and the upper limit of the regime depends on the strength of the electric field E, and the temperature and pressure gradients.

Rosenbluth et al. (1957) derived the Fokker-Planck equation for arbitrary distribution functions in the case where two-body interactions are governed by a force that obeys the inverse square law. The coefficients Δv and $\Delta v \Delta v$ in the Fokker-Planck operator were written in terms of two fundamental integrals/potentials that depend on the distribution function of the background particles (including those of the same species). Expanding the distribution function as a set of Legendre functions of the pitch angle, the Fokker-Planck equation is cast into the form of an infinite set of one-dimensional, coupled non-linear integro-differential equations. Approximating the distribution function by a finite series, the Fokker-Planck equations can be solved numerically. Keeping one term of the series corresponds to the approximate solution of Chandrasekhar (1943) and keeping two terms yields the solution of Cohen et al. (1950). Rosenbluth et al. (1957) showed that

$$\left(\frac{\partial f_s}{\partial t}\right)_{\text{collisions}} = -\Gamma \frac{\partial}{\partial v_s^i}\left(f \frac{\partial \mathbf{H}}{\partial v_s^i}\right) + \frac{1}{2}\frac{\partial^2}{\partial v_s^i \partial v_s^j}\left(f \frac{\partial^2}{\partial v_s^i \partial v_s^j}\mathbf{G}\right), \tag{32}$$

with the Rosenbluth potentials

$$\mathbf{H}(\mathbf{v_s}) = \sum_{s'} \frac{m_s + m_{s'}}{m_{s'}} \int d^3 v'_{s'} f_{s'}(\mathbf{v}'_{s'}) \left| v_s - v'_{s'} \right|^{-1}$$

$$\mathbf{G}(\mathbf{v_s}) = \sum_{s'} \int d^3 v'_{s'} f_{s'}(\mathbf{v}'_{s'}) \left| v_s - v'_{s'} \right| \tag{33}$$

$$\text{where } \Gamma = \frac{4\pi (Z_s e)^2 (Z_s e)^2 \ln \Lambda_{ss}}{m_s^2}.$$

Ljepojevic and Burgess (1990) presented a step-by-step description of a method for calculating the distribution function in the presence of strong gradients. In their method the low-velocity part of the distribution is given by solution of Spitzer and Härm (1953). The high-velocity tail of the distribution function is given as a solution to the high-velocity form of the Fokker-Planck equation which is derived from Eqs. (32) and (33) by neglecting the interaction between the high-velocity particles themselves and considering only their interaction with the low-velocity (near Maxwellian) part of the distribution function. One may then derive a linearised form of the Boltzmann equation with the Fokker-Planck collision operator that applies to high-velocity particles. For electrons:

$$\left(\frac{\partial f_e}{\partial t}\right)_{\text{collisions}} = \frac{1}{v_e^2}\frac{\partial}{\partial v_e}\left[v_e^2 \nu(v_e)\left(\frac{k_B T_e}{m_e}\frac{\partial f_e}{\partial v_e} + v_e f_e\right)\right] - \nu(v_e)\frac{\partial}{\partial \mu}\left[(1-\mu^2)\frac{\partial f_e}{\partial \mu}\right]. \tag{34}$$

The full derivation of Eq. (34) is given by Ljepojevic and Burgess (1990) on pages 73 to 88 of their article. They also describe in detail a numerical treatment for its solution following non-dimensionalisation and transformation into a form more convenient for numerical work. The solutions in the low-velocity and high-velocity regime are combined, subject to suitable matching conditions (e.g. smoothness), in a region of the velocity space where both methods are approximately valid; two thermal speeds was found to be the optimal value.

Given the significant complexities that are involved in working with the Fokker-Planck equation it is tempting to revert to BGK-type approximations of the collision operator. However, one must be careful. Livi and Marsch (1986) compared the collisional relaxation of a double-beam and a bi-Maxwellian distribution function for a Fokker-Planck and a BGK collision operator. They found that moments of the distribution function up to and including temperature (the 2nd moment) were in good agreement between the two schemes when the frictional energy-loss rate was used as the effective collision frequency in the BGK operator, but that the heat flux (the 3rd moment) exhibited differences due to its sensitivity to the shape/skew of the distribution function, which enters the Fokker-Planck operator via the second derivative of the distribution function w.r.t. velocity. Ljepojevic and MacNeice (1988) calculated contributions to the heat flux in a solar flare atmosphere from the tail of the distribution function using the high-velocity form of the Fokker-Planck equation (Ljepojevic and Burgess 1990) and compared the results with a BGK-type calculation. They found that the BGK technique can estimate contributions from the high-energy tail to the heat flux to within order of magnitude.

As computers became more powerful, detailed numerical treatments of the Fokker-Planck equation became feasible. Shoub (1983) provides a detailed discussion of the break-down of the Spitzer and Härm (1953) calculation of the electron distribution function and describes an approach to deriving and then solving numerically the high-velocity form of the Landau-Fokker-Planck equation. Implications of the break-down of the local Maxwellian approximation are discussed for: energy balance in the upper chromosphere and low TR; the He resonance line spectrum; the Schmahl-Orrall observation of continuum absorption by neutral H, and the origin of the 20,000 K temperature plateau. However, Shoub (1983) was unable to say anything quantitative about the heat flux since the kinetic equation was only solved to six thermal speeds. Had Shoub (1983) applied the same transformation following non-dimensionalisation as employed by Ljepojevic and Burgess (1990) then it would have been possible to significantly extend the calculation in velocity space. Ljepojevic (1990) used the approach described in Ljepojevic and Burgess (1990) to show that distribution functions are near Maxwellian in the commonly used FAL (Fontenla et al. 1993) models of the photosphere to mid-TR and so the models are valid in their given form. MacNeice et al. (1991) applied the same approach to the transition region of a flaring loop and found a substantial enhancement in the tail populations throughout that region of the atmosphere. We discuss the results of some attempts to apply these calculations of distribution functions to fluid models in order to take account of the consequences of non-Maxwellian distributions in the following Section.

5.2 Heat Flux/Transport

The fluid equations are derived by taking successive moments of the Boltzmann equation when it is written in terms of distribution functions that exhibit only small deviations from a fully-relaxed Maxwellian distribution. Since the statistical treatment of a particle ensemble in terms of a fluid is valid only in this collisional limit, then only small deviations can be tolerated. In general, departures from Maxwellian are treated as a perturbation and the distribution is expanded in terms of some parameter that should remain small, such as the ratio of the electron mean-free-path to the temperature scale length (the Knudsen number, Kn), in order to derive non-linear terms of the fluid equations such as the heat flux.

The transport of heat by thermal conduction is the dominant transport process in hot but tenuous astrophysical plasmas. It determines the temperature and thus the density structure, via the temperature-dependent scale length, in the solar atmosphere (for example) and so it is important to handle it as accurately as possible. The most commonly used form for the heat flux is that given by Spitzer and Härm (1953), valid in the limit of weak gradients and weak electric fields:

$$F_c = -\kappa \nabla T, \tag{35}$$

where the conductivity $\kappa = \kappa_0 T^{5/2}$ (for a fully ionized hydrogen plasma) and the constant is the quantity calculated by Spitzer ($\kappa_0 \approx 10^{-6}$). Despite its strong non-linearity the form represented by Eq. (35) is convenient to implement in fluid-based numerical codes, but cannot guarantee an accurate representation of the heat flux if it is used indiscriminately. Experimental and numerical results have shown that its range of applicability is actually quite limited. Equation (35) indicates that the heat flux can increase indefinitely provided that the temperature gradient continues to steepen, but eventually a physical limit must be reached when there are no more particles remaining to support the implied heat flux. This is the free-streaming limit, essentially the maximum heat flux that the plasma can sustain, and may be estimated by assuming that the majority of the particles (e.g. electrons) stream down the temperature gradient at the thermal speed (Bradshaw and Cargill 2006) (more sophisticated numerical treatments indicate the free-streaming limit is about 1/6 of this value). At the very least, then, a limiter should be deployed in any numerical model that uses Eq. (35), in order to constrain the heat flux to physically justifiable values.

There have been a number of efforts to derive systems of fluid equations that take account of stronger departures from Maxwellian to be implemented in numerical models. Campbell (1984) found a solution to the Boltzmann equation that extends the Chapman-Enskog approximation to large temperature gradients and electric fields, to calculate electron transport in a fully ionized gas. The collision term was written in the form of a collisional relaxation with a velocity-dependent relaxation time defined in terms of the scattering length. The distribution function was assumed to be separable with the angular dependence represented by a slowly varying function. Calculating the moments of this distribution function led to correction factors to the classical (Spitzer and Härm 1953) transport coefficients as a function of the temperature gradient scale-length and an inherently flux-limited heat flow. Killie et al. (2004) derived a complete set of fluid equations for fully ionized gases that improve the treatment of Coulomb collisions by taking into account the shape of the distribution function to better calculate the heat flux and the thermal force. They chose an analytical velocity distribution function with a Maxwellian core plus a high-velocity correction term proportional to v^3, and obtained transport equations by inserting their choice of distribution function into the Boltzmann equation with a Fokker-Planck collision operator. Chiuderi et al. (2011) derived a set of two-fluid equations applicable to weakly collisional plasmas by using a relaxation approach to the collision operator and selecting 'mixed' Maxwellian distributions for the two interacting species that conserve momentum and energy. The collisional term in their treatment depends on an 'average' or 'representative' collisional timescale that is velocity-independent.

Gray and Kilkenny (1980) described the results of experiments in which the ratio of the electron mean-free-path to the temperature scale-length was found to be about 0.5 and enhanced low-frequency turbulence was observed. They used a numerical simulation of the experimental set-up to show that ratios of 0.5 implied a heat flux limited to less than 5 % of the free-streaming limit in the hot part of the plasma. They also found T_e/T_i in the same region sufficient to excite heat-flux driven ion acoustic turbulence, thus explaining the low frequency turbulence observed in the experiment. The observed level of turbulence in the experiment was enough to account for the predicted low thermal conductivity in the numerical model, which was due to electron scattering from interactions with the ion acoustic turbulence. Bell et al. (1981) and Matte and Virmont (1982) studied electron heat transport down steep temperature gradients in laser-induced plasmas by numerically solving the Fokker-Planck equation. The heat flux was found to be substantially smaller than that predicted by the classical theory or the free-streaming value when the mean-free-path reached a fraction of only one-hundredth of the temperature scale length ($Kn = 10^{-2}$). Shoub (1983) found significant deviations from Maxwellian in the tail of the distribution for $Kn = 10^{-3}$, but was unable to provide a quantitative estimate of the heat flux. Owocki and Canfield (1986) used a BGK-type method to calculate the electron distribution function in the solar transition region to study the effect of a high-energy tail on the heat transport and collisional excitation and ionization rates. For the case studied they found that non-classical transport does not significantly alter the excitation or ionization state of ions with emission lines that form predominantly in the lower transition region (with excitation energies in the range 10 eV, because electrons with these low energies thermalise quickly), but the non-classical heat flux in this region does depend sensitively on the temperature gradient in the upper transition region.

In the case of pronounced departures from Maxwellian distributions it is clear that correction factors and localised approaches to calculating the distribution function, and hence the heat flux, are not sufficient. For example, contributions to local quantities from non-local sources may lead to strong departures from local Maxwellian distributions. Such kinetic behaviour is inherently incompatible with the fluid approximation in which it is assumed that the properties of the plasma can be determined entirely locally (e.g. the heat flux as a function of temperature and the temperature gradient). Since it is generally not feasible to solve a kinetic equation (certainly not a time-dependent form) in tandem with the fluid equations to correct for the consequences of kinetic behaviour, then the challenge is to find an alternative; e.g. a computationally tractable approach that can be implemented in an otherwise fluid-based treatment, and that permits one to account for purely kinetic effects (e.g. non-local influences) on quantities such as the heat flux in regions where the Knudsen number grows large. One such approach is to adopt a delocalisation formula.

Luciani et al. (1983) found a delocalisation formula for the heat flux, using a set of solutions to the Fokker-Planck equation. Delocalisation formulae are based on delocalisation kernels that operate on calculations of the heat flux made using Eq. (35). The kernel essentially acts to 'smear' the classical heat flux out over the computational domain in a manner that mimics the spatial profile of the heat flux that would be found from a full Fokker-Planck calculation. The formula presented by Luciani et al. (1983) has the form:

$$F_c(s) = \int w(s, s') F_{SH}(s') ds' \tag{36}$$

and the delocalisation kernel is

$$w(s, s') = \frac{1}{2\lambda(s')} \exp\left[-\left| \int_{s'}^{s} \frac{n(s'')}{\lambda(s')n(s')} ds'' \right| \right]. \tag{37}$$

The quantity λ is an effective range for the electrons, related to the mean-free-path. In the limit of shallow temperature and density gradients the kernel w behaves like a δ-function, where $\int w(s, s')ds' = 1$ and Eq. (36) reduces to $F_c = F_{SH}$. Despite the double integration, the delocalisation formula is straightforward to efficiently implement in a fluid code to replace the heat flux in the form of Eq. (35). Luciani et al. (1985) found an analytical justification for the delocalisation formula and Bendib et al. (1988) developed an improvement

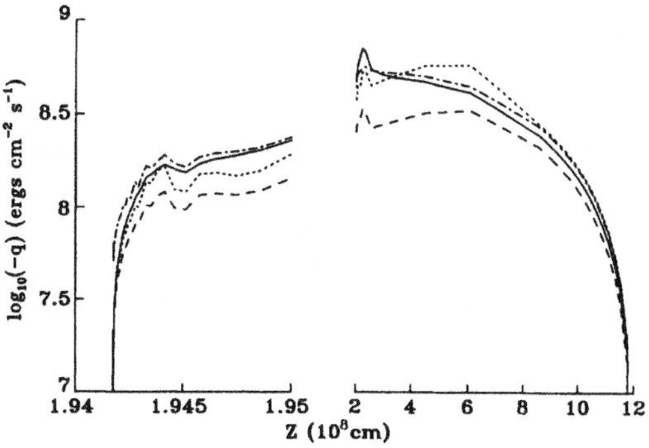

Fig. 3 The heat flux in a solar active region loop calculated by Ljepojevic and MacNeice (1989), Ljepojevic and Burgess (1990) (*solid line*), Spitzer and Härm (1953) (*dotted line*), Campbell (1984) (*dashed line*), and Luciani et al. (1983, 1985) (*dot-dashed line*). Image credit: Ljepojevic and MacNeice (1989)

that takes the presence of an electric field into account. More recently, Alouani-Bibi et al. (2004) studied non-local electron heat transport using a number of different approximations to the Rosenbluth potentials in the Fokker-Planck equation to find delocalisation kernels for non-local heat flux formulae to be used in fluid codes.

A number of authors have compared the different approaches to calculating the heat flux and have implemented them in numerical models in order to apply them to particular problems in which non-Maxwellian electron distributions are expected to arise. Smith (1986) discussed classical (Spitzer and Härm 1953), locally limited (Campbell 1984) and non-local (Luciani et al. 1983, 1985) algorithms for the heat flux and their application to heat transport in the case of the steep temperature gradients (thin conduction fronts) that arise during the impulsive phase of solar flares. Karpen and DeVore (1987) investigated how these different formulations for the heat flux affect the physical characteristics of the corona, transition region and chromosphere in numerical models of solar flares. Both sets of authors found that the heat flux in the hot part of the plasma obtained with the non-local treatment was smaller than the locally limited and classical values, whereas the heat flux in the colder parts of the plasma (e.g. in the transition region and chromosphere) was significantly enhanced compared with the locally limited and classical values. In consequence, both flux limiting and delocalisation play an important role in the evolution of the plasma. In the case of flares this leads to a 'bottling up' of energy in the corona, allowing it to reach much higher temperatures, and the earlier onset of weaker chromospheric evaporation.

Ljepojevic and MacNeice (1989) calculated the heat flux in a solar active region coronal loop from distribution functions obtained using the sophisticated model described in Ljepojevic and Burgess (1990), and compared it with the heat flux given by the classical treatment of Spitzer and Härm (1953), the correction coefficients to the classical treatment given by Campbell (1984) and the heat flux given by the delocalisation formula of Luciani et al. (1983, 1985) (Fig. 3). They concluded that the classical treatment failed completely in the lower corona, predicting a strong heat flux flowing down the temperature gradient when the kinetic equation yielded heat flux flowing up the temperature gradient, and the possibility that the role of the heat flux could be misinterpreted in the energy balance of the corona. Landi and Pantellini (2001) also found that the heat flux can flow up the temperature gradient in the case of supra-thermal tails characterised by κ distributions with $\kappa < 5$. Encouragingly, Ljepojevic and MacNeice (1989) did find relatively good agreement between the delocalisation formula and the more sophisticated kinetic calculation. West et al. (2008) implemented the delocalisation formula in the HYDRAD (e.g. Bradshaw and Mason 2003; Bradshaw et al. 2012) code to investigate the lifetime of hot, nanoflare-heated plasma in the solar corona. The aim of this work was to determine whether the bottling up of energy in the corona due to the severe heat flux limiting that arises in the limit of large Knudsen numbers provided sufficient time for the ionization state to equilibrate following rapid heating. Alouani-Bibi and Matte (2002) developed a non-local model of electron heat flow in laser-heated plasmas, taking into account super-Gaussian deformation of the electron distribution function. Alouani-Bibi and Matte (2003) derived an analytical description of electron-ion energy exchange by Coulomb collisions in the presence of super-Gaussian electron distributions, and found the ratio T_i/T_e at which the collisional electron-ion energy exchange cancels increases from 1 in a Maxwellian plasma to 1.98 in a super-Gaussian plasma.

5.3 Excitation, Ionization and Radiation

The specific nature of the local distribution function can have an important effect on the rate of excitation and ionization via collisions. Excitation is generally a consequence of interactions between ions and electrons in the bulk of the distribution, but ionization is particularly sensitive to the tail population. The stronger heat fluxes at the base of steep temperature gradients found in a number of the studies described in Sect. 5.2 imply enhanced tail populations of streaming electrons, which can feasibly alter the ionization state such that it can no longer be considered a strong function of the local temperature (and, to a lesser extent, the local density) alone. Collisional excitation rate coefficients can be calculated by substituting a suitable distribution function into Eq. (7) and the ionization rate can be calculated by inserting the appropriate ionization cross-section (usually pertaining to the ground-state) in place of $Q_{i,j}$. Investigations of the effect of non-Maxwellian distributions on the ionization state have proceeded along two general lines: (a) calculate the distribution function by solving some simplified form of the Boltzmann equation (e.g. BGK, Fokker-Planck); or (b) choose an analytical form for the distribution function with the properties of a Maxwellian at low-velocities/energies, but which permits an enhanced tail population where the degree of enhancement can be controlled by a single parameter. A popular generalisation of the Maxwellian distribution that fulfills these requirements is the κ-distribution:

$$f_\kappa(E) = A_\kappa \left(\frac{m}{2\pi k_B T}\right)^{3/2} \frac{\sqrt{E}}{(1 + \frac{E}{(\kappa-1.5)k_B T})^{\kappa+1}}; \tag{38}$$

$$A_\kappa = \frac{\Gamma(\kappa+1)}{\Gamma(\kappa-0.5)(\kappa-1.5)^{3/2}}. \tag{39}$$

The κ-distribution has the form of a Maxwellian in the limit $\kappa \to \infty$. The most probable energy of a particle in the distribution is $E_p = (\kappa - 1.5)k_B T/\kappa$ and the mean energy of the distribution is $\langle E \rangle = 3k_B T/2$ (i.e. independent of κ and the same as the Maxwellian at the same temperature). Yoon et al. (2006) and Rhee et al. (2006) demonstrated that κ-distributions can be induced by spontaneous scattering (absent in collisional treatments) when electron beams are accelerated by weakly turbulent processes.

Owocki and Scudder (1982) used κ-distributions to study the ionization state of gases with non-Maxwellian electron distributions, finding changes from the ionization temperature assuming an underlying Maxwellian distribution of up to a factor of 2. The importance of the high-velocity tail to the ionization state depends on the ratio of the ionization potential to the mean thermal energy of the electrons. Owocki and Scudder (1982) also found that the high ionization energy required for the O VIII $\leftrightarrow$ O IX transition means that oxygen ionization at solar coronal temperatures is more sensitive to the tail of the distribution than elements of lower ionization energy (such as iron) found within that temperature range.

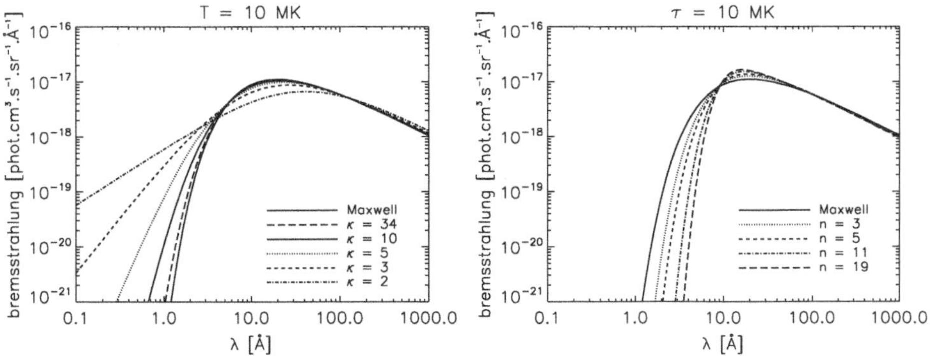

Fig. 4 The X-ray and UV continuum at 10 MK calculated from κ and n distributions representing different strengths of departure from Maxwellian. Image credit: Dudík et al. (2012)

Dzifčáková and Kulinová (2003) calculated excitation, ionization and recombination rates for κ-distributions for a range of values of the parameter κ. They found changes in the level populations and the relative ion abundances. A synthetic spectrum was calculated which showed that some C III, C IV and O IV lines are sensitive to the shape of the distribution function and their intensities enhanced by a factor 2–6 in the presence of strongly non-thermal distributions. Dzifčáková (2006) investigated the influence of κ-distributions in the solar corona on Fe VIII–XV excitation and ionization, and on the line intensities associated with those transitions. They concluded that it ought to be possible to diagnose the value of κ that would best characterise the electron distribution from ratios of Fe IX 171 Å, Fe XII 195 Å and Fe XV 284 Å lines provided that the plasma density is known.

Dzifčáková and Mason (2008) calculated non-Maxwellian electron excitation rates for ions of astrophysical interest. They demonstrated a method for extracting collision strengths from the Maxwell-averaged collision strengths (Upsilons) that are provided by the CHIANTI atomic database and then integrated these over the specific non-Maxwellian distribution in order to calculate the corresponding excitation rate. κ-distributions, employing a range of values of κ, were used to calculate synthetic spectra for Fe XV and XVI in the 50–80 Å range for comparison with solar observations. For consistency in the generation of the synthetic spectra, Dzifčáková and Mason (2008) used the equilibrium ionization states of Fe derived for a range of κ by Dzifčáková (2002). However, they found no conclusive evidence for non-Maxwellian distributions in the particular flare dataset that was compared with the synthetic spectra.

Dudík et al. (2011) calculated the bound-bound and free-free radiative losses arising from plasmas with non-Maxwellian electron distribution functions using κ- and n-distributions (Eqs. (5) and (6) of Dudík et al. 2011). It was found that changes in the radiative loss function due to non-Maxwellian distributions are greater than errors in the atomic data and errors due to missing contributions from free-bound continuum. While radiative loss functions for κ distributions are generally weaker than for Maxwellians, the opposite is true for n-distributions. They also found that the contribution from bremsstrahlung changes by only a few percent, except in the extreme case of $\kappa = 2$. Following on from this earlier work, Dudík et al. (2012) calculated the X-ray, UV and radio continuum arising from non-Maxwellian distributions using κ- and n-distributions (Fig. 4). They found that at flare temperatures and hard X-ray energies both the bremsstrahlung and the free-bound spectra are dependent on the assumed distribution, and concluded that the low energy part of κ distributions can be determined from observations of the continuum.

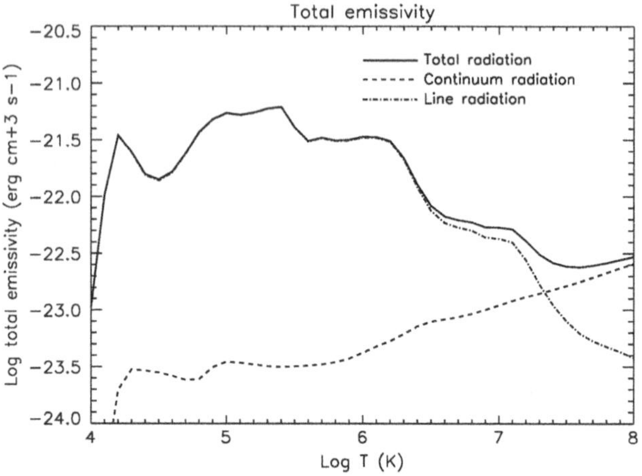

Fig. 5 The optically-thin radiative loss function. Image credit: Landi and Landini (1999)

6 The Radiative Loss Function

The power per unit volume of plasma emitted by a single spectral line is given by Eq. (4). The total power per unit volume is then the sum total of the power emitted by the many thousands of spectral lines which belong to the chemical elements that comprise the plasma. In the case of optically-thin astrophysical plasmas this quantity can be written in a conveniently compact form:

$$E_R = N_e N_H \Lambda(T_e) \left[\mathrm{erg\,cm^{-3}\,s^{-1}}\right], \tag{40}$$

where N_H is the number density of hydrogen atoms. In a fully ionized hydrogen plasma $N_e = N_H$. E_R is generally referred to as the *radiative volumetric loss rate*. $\Lambda(T_e)$ is called the *optically-thin radiative loss function* (or the total emissivity of the plasma, as shown in Fig. 5) and it encapsulates a vast amount of atomic data. The radiative loss function depends upon the element abundances, the ionization state of the elements, and transition energies and probabilities. These must be determined for hundreds of ions and in many cases thousands of spectral lines per ion, in order that accurate radiative loss functions can be calculated. As atomic data is improved and updated then so must the radiative loss function. The most convenient way to keep abreast of developments is to use a comprehensive and regularly updated atomic database, such as Chianti (Dere et al. 1997; Landi et al. 1997, https://www.chiantidatabase.org), which contains carefully assessed atomic data and the appropriate functionality for calculating spectra and Λ. A number of assumptions concerning the properties of the plasma, such as the nature electron distribution function and the time-dependence of the ionization state, are implicit in the most commonly used calculations of Λ for astrophysical applications. The limits of these assumptions have been demonstrated in Sects. 4 and 5 and so the emissivity of individual spectral lines and the radiative loss function must be recalculated, whenever these limits are reached, to take proper account of the microphysical processes that can arise in astrophysical plasmas.

Following Eq. (4) the *emissivity* of a single transition between energy levels j and g in a particular ion is given by

$$\epsilon_{j,g} = \frac{N(X^{+m})}{N(X)} \frac{N(X)}{N(H)} \frac{N(H)}{N_e} C^e_{g,j} \Delta E_{j,g} \left[\mathrm{erg\,cm^3\,s^{-1}}\right]. \tag{41}$$

The *ion emissivity* is then obtained by summing over all of the transitions that may occur within the ion:

$$\Lambda_{X_i} = \sum_{\lambda} \epsilon_{j,g}, \tag{42}$$

so that Λ_{X_i} is the radiative loss function for the particular ionization state i of element X. The radiative loss function for the element can be found by summing over the number of charge states:

$$\Lambda_X = \sum_i \Lambda_{X_i} \tag{43}$$

and the total radiative loss function is the sum over the number of elements of interest:

$$\Lambda = \sum_X \Lambda_X. \tag{44}$$

The radiative loss function is strongly dependent on the electron temperature T_e in equilibrium, but it is clear from Eq. (41) how important the ionization state (the first factor on the right-hand side) and the collisional excitation rate $C^e_{g,j}$ are to accurately calculating it. When the ionization state exhibits strong departures from equilibrium then the temperature dependence of the radiative loss function can be lifted, the rate of energy loss by radiation, and the intensities of individual spectral lines, may not be characteristic of the actual electron temperature. Furthermore, the ionization state (via the ionization and recombination rate coefficients) and the collisional excitation rate depend on the underlying electron distribution which is generally assumed to be Maxwellian, but circumstances can easily arise astrophysical plasmas when this assumption is certainly not valid (Sect. 5).

7 Signatures and Diagnostics of Non-equilibrium Processes

Analytical analyses can identify the conditions under which non-equilibrium processes become important to understanding the properties and behaviour of astrophysical plasmas, and numerical models demonstrate that such conditions are commonplace in the optically-thin astrophysical plasma systems that are the focus of a great deal of current research interest. In this Section we consider potential signatures of non-equilibrium processes and the evidence for their manifestation in observational datasets.

7.1 Non-equilibrium Ionization

Griem (1964) discussed the potential for departures from equilibrium of the ionization state in cases where the dynamical evolution of the plasma occurs on timescales that are shorter than those of ionization and recombination. He cited the particular example of transporting the ion population across a strong temperature gradient, as might be the case in the solar transition region. Joslyn et al. (1979a, 1979b) investigated steady flows across a range of temperature gradients and found that ionization equilibrium in the transition region is an acceptable assumption for iron at flow speeds no greater than 20 km/s, but that carbon and oxygen ion populations can be driven away from equilibrium at flow speeds of only 1 km/s. Raymond and Dupree (1978) and Dupree et al. (1979) carried out a similar study related to steady flows in the transition region and also found significant departures of the ion populations from equilibrium. Borini and Noci (1982) investigated the ionization state in coronal loops in the temperature range 0.2 MK to 2 MK and showed that considerable deviations from equilibrium ionization could arise in average to low intensity loops characterised by high-speed flows. They reported a pronounced effect for cooler loops, which despite exhibiting lower speed flows were found to have steeper temperature and density gradients than hot loops.

Noci et al. (1989) calculate the number density of carbon ions for a selection of coronal loops models in the case of steady-state, sub-sonic flows (siphon flows) and found departures from equilibrium of the ionization state for flows of only a few km/s at the loop apex and for a factor of 10 slower at the base of the transition region. Spadaro et al. (1990a) calculated the spectral line profiles of carbon ions formed in the transition region that are commonly used in spectroscopic diagnostic studies. They used the number densities of the carbon ions found by Noci et al. (1989) and found predominantly blue-shifted emission lines, which could not be reconciled with observations that show both up- and down-flows in the transition region. The absence of red-shifted emission was attributed to the assumption of spatially uniform heating. Spadaro et al. (1990b) focused on the corona and calculated the emissivities of carbon and oxygen, both in and out of equilibrium, and found substantial differences between the resulting radiative loss functions. In the case of up-flows (down-flows) the radiative losses were generally enhanced (suppressed). One may understand this by considering an ion of relatively low charge state transported into a region of temperature significantly higher than the formation temperature of the ion in equilibrium; the ion will tend to emit more strongly since a greater proportion of the electrons in the bulk of the distribution will have sufficient energy to excite its emission lines.

Spadaro and Ventura (1994a, 1994b) studied the effect of non-equilibrium ion populations on the line intensities of O VI and H I ions that originate in solar wind source regions. They calculated the intensity and line profiles for equilibrium and non-equilibrium ionization balance based on a steady flow model, finding significant deviations from equilibrium beyond 3–4 solar radii for O VI and beyond 5 solar radii for the Lyman α emission from H I. These results are significant for estimates of the solar wind speed that rely on the Doppler-dimming technique, which estimates the speed from variations in the line intensities compared with their expected values in the absence of a steady outflow.

Spadaro et al. (1994) investigated the signatures that may be observable when non-equilibrium ion populations are present and should be considered when carrying out spectroscopic diagnostics using line ratios. Since non-equilibrium ion populations are displaced from their temperatures of peak abundance in equilibrium, the temperature-dependent Boltzmann factors that appear in the expression for the excitation rate coefficients for the spectral lines are changed considerably, which results in changes to the energy level populations, the line intensities and, consequently, the values of the line ratios. The values of the temperature sensitive line intensities arising from the non-equilibrium C IV and O IV–VI populations calculated by Noci et al. (1989) (carbon) and Spadaro et al. (1990b) (oxygen) were compared with the same line intensities computed in equilibrium. In the presence of a non-equilibrium ion population the line intensities were found to be reduced for both up-flows and down-flows across the transition region temperature gradient. The C IV population was found to be the most sensitive to non-equilibrium ionization, with decreases in the line ratio by an order of magnitude in the case of down-flows. In response to discrepancies identified by Keenan et al. (1992) between C IV line intensities observed during highly dynamic events and theoretical predictions of the same line intensities, Spadaro et al. (1995) used a siphon-flow model and non-equilibrium ion populations to recalculate the predicted line intensities. However, they found only a marginal improvements in the agreement between the observed and predicted intensities when non-equilibrium ionization was accounted for, and concluded that the observed intensities could not be reconciled with a sub-sonic, siphon-flow model. Esser et al. (1998) examined the effect on the ionization state of the solar wind when the acceleration process occurs at much lower heights in the solar atmosphere than previously considered, based on flow speeds estimated from chromospheric, transition region and coronal emission lines. These observations yielded flow speeds for O VI ions that are a factor of 3–4 greater than indicated by earlier work, which imply the ion populations may depart from equilibrium as they are transported at speed across the steep temperature gradients found in the lower atmosphere. In this case, the use of charge state ratios to estimate equilibrium temperatures is unlikely to be valid. Esser et al. (1998) found outflow models with speeds in the region of 130–230 km/s to predict charge state ratios consistent with those observed. Edgar and Esser (2000) considered the effect of non-equilibrium ionization on the ratio of Ne VI to Mg VI lines in the solar transition region, which is used as a diagnostic of the first ionization potential (FIP) effect. In the presence of a strong heating or cooling effect the populations of ions of low FIP are enhanced relative to those of higher FIP. They calculated the non-equilibrium populations of these ions for simple flows across the transition region and showed that their spectral line ratios depend on non-equilibrium effects, as well as on the temperature and density.

These investigations into the consequences of non-equilibrium ionization assumed steady-state conditions where only flows may affect the ionization state in the presence of a steep temperature gradient. In general, this is due to the assumption of some form of constant heating that maintains the plasma in a steady-state condition, but this need not be so. There exist mechanisms by which energy can be impulsively released into the plasma on short timescales (e.g. a collisionless shock or magnetic reconnection) leading to temperature changes on timescales that are short compared with the ionization time. Local temperature enhancements can give rise to localised pressure gradients which may in turn drive flows. Consequently, a detailed understanding of the consequences of non-equilibrium ionization requires a treatment of both local, temporal changes in the plasma properties and the fast transport of ions by flows. A shock is perhaps the simplest case, since it drives the plasma from one nearly steady state to another, and if the shock is strong the ionization state can be far from equilibrium. Ma et al. (2011) used the compression, density and heating determined from optical and radio observations of a CME-driven shock to compute the time-dependent ionization in the post-shock flow, and they found that it matched the observed rise times of emission in the AIA bands.

Hansteen (1993) presented a numerical model that solved the time-dependent ion population equations in tandem with the hydrodynamic equations, taking account of departures from ionization equilibrium on the radiative losses for ions formed below 0.3 MK. The model was used to study the dynamic response of a coronal loop to energy released impulsively near the apex. It was found that the line shifts predicted for C IV, O IV and O VI by the model were consistent with the persistent red-shifts observed in transition region lines (e.g. Brekke et al. 1997). The amplitude of the predicted line shift was shown to depend on the ionization timescale of the emitting ion. It was also found that the radiative losses could change by a factor of 2 due to the influence of flows and waves on the ion population. Teriaca et al. (1999a) noted the presence of blue-shifts at temperatures characteristic of the transition region in the quiet Sun and in active regions and Teriaca et al. (1999b) suggested that impulsive heating localised in the transition region at the temperature of peak O VI abundance in equilibrium (0.3 MK) might account for the presence of red-shifts and blue-shifts. The heating (whether located at the loop apex or in the transition region) generates compression waves and by including the partial reflection of the downward propagating wave from the chromosphere, and allowing for non-equilibrium ionization, reasonable agreement was found between the observed Doppler-shifts and those predicted by the numerical model (red-shifts in the cooler C IV lines and blue-shifts in the warmer O VI lines). Doyle et al. (2002) found that the higher the temperature at which a heating event occurs then the greater the delay in the response from the mid-transition region lines in terms of changes in the Doppler-shift.

Bradshaw and Mason (2003) studied the response of the plasma and the ionization state to a small-scale, impulsive energy release at the apex of a coronal loop, characteristic of nanoflare heating and solved the ion population equations for the 15 most abundant elements of the solar atmosphere (including C, O, Ne, Mg, Si and Fe). The ionization state was used to calculate the radiative loss function in the energy equation, thereby coupling the energy balance with the ionization state. They concluded that broad/narrow-band imaging instruments can miss small-scale heating events entirely due to the weak sensitivity of the non-equilibrium emissivity to the changing temperature compared to the emissivity for equilibrium ionization, which fell by a factor of up to 5. The non-equilibrium emission remained relatively steady throughout the heating event, despite a factor 2 change in the temperature on a timescale of 30 s. In order to diagnose non-equilibrium ionization they proposed searching for signatures in line ratios of ion pairs that are populous in the temperature range of interest but have different characteristic lifetimes (e.g. C IV and O VI in the transition region, or different ions of Fe at coronal temperature). Bradshaw et al. (2004) investigated non-equilibrium ionization in a small compact flare, using the same numerical model as Bradshaw and Mason (2003), and localised the energy release in the corona to drive the flare evolution by thermal conduction. During the impulsive phase they found the emissivities of He I, He II and C IV in the transition region to be strongly enhanced above their expected equilibrium values, which was then followed by a significant reduction leading to an increase in the amount of chromospheric plasma ablated into the corona (less energy radiated in the transition region leaves more energy available to drive ablation). During the initial energy release the charge state of the coronal ions was seen to evolve substantially out-of-equilibrium with the increasing temperature and line ratio measurements would yield plasma temperatures that are much greater than the formation temperature of the emitting ion. During the gradual phase the emissivity at transition region temperatures was suppressed relative to equilibrium with reduced downflow velocities, since the enthalpy flux did not have to work as hard to power the transition region, and commensurately increased radiative cooling time-scales. The flare emission as it would be detected by TRACE in its 171 Å and 195 Å wavelength bands was computed and it was found that the filter ratio technique can give reasonably good estimates of the plasma temperature in quiescence. However, when the populations of Fe VIII, Fe IX, Fe X and Fe XII exhibited non-equilibrium effects the temperatures derived from filter ratio measurements were unreliable.

Bradshaw and Cargill (2006) and Reale and Orlando (2008) considered strong or 'explosive' heating, on short timescales to high temperatures, in an initially rarefied coronal loop atmosphere, to determine the consequences for the evolution of the ionization state. They found extremely strong departures from ionization equilibrium and concluded that for sufficiently short heating events the charge states characteristic of the highest temperatures reached (10–30 MK) could never be created before the onset of fast cooling by thermal conduction and coronal filling by chromospheric ablation (Fig. 6). In consequence, the emission measure peaks at temperatures significantly lower than the peak temperature of the plasma and forward modeling emission in the wavelength range of Hinode-EIS showed that no 'hot' (e.g. $>$ 10 MK) component of the plasma would be detected. Heating models that assume ionization equilibrium predict such a hot component, but no observational evidence has yet been found and so non-equilibrium ionization presents one possibility to reconcile observations with current theory. In the future, observations of the solar X-ray continuum could

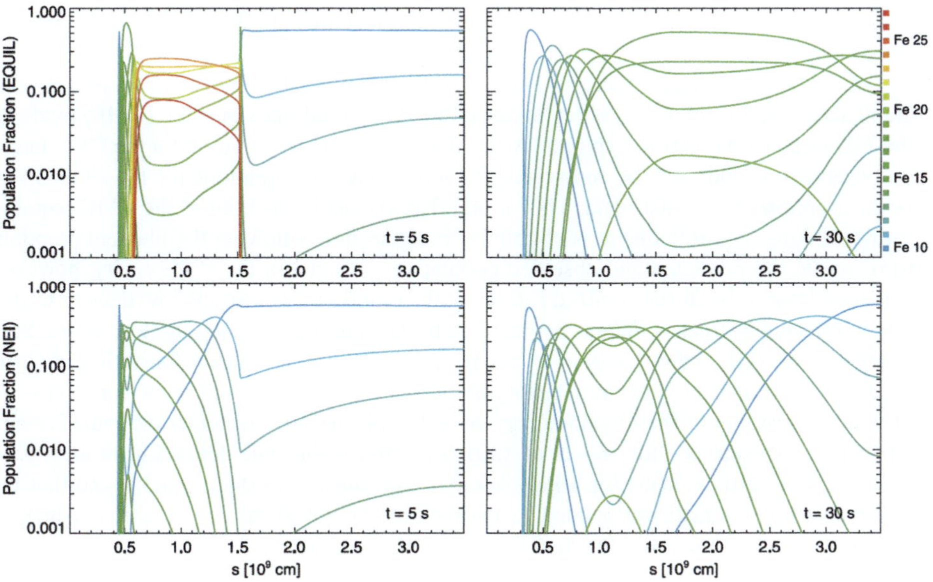

Fig. 6 Ion population fractions along a heated coronal loop calculated in equilibrium and non-equilibrium. The highly charged states associated with the hottest plasmas in equilibrium are never reached in the non-equilibrium calculation before the onset of cooling. Image credit: Reale and Orlando (2008)

be used to confirm, or otherwise, the presence of a hot component to the emission. The X-ray continuum is due mostly to bremsstrahlung from H and He, which is not sensitive to non-equilibrium ionization effects. There is a significant contribution from radiative recombination in over ionized plasmas and radiative recombination continua (RRC) are seen in a few X-ray binaries (Cyg X-3: Paerels et al. 2000) and supernova remnants (IC 443: Yamaguchi et al. 2009). In under-ionized plasmas, such as may be created in the case of rapid heating in the solar corona, the RRC is weak compared to the bremsstrahlung. In any case the RRC scales as $\exp(h\nu/k_B T)$ and so provided the edges of the most abundant elements are avoided, the continuum shape can be used to diagnose the temperature. Bradshaw and Klimchuk (2011) conducted a more extensive survey of the parameter space of energy release magnitudes and timescales, and carried out forward modeling to predict the emission that would be detected in the passbands of the recently launched SDO-AIA, in order to perform a more detailed evaluation of the potential for non-equilibrium ionization to explain the high temperature part of the emission measure. The study led to several conclusions: (1) Deviations from equilibrium were found to be greatest for short-duration nanoflares at low initial coronal densities. (2) Hot emission lines were the most affected and could be suppressed to the point of invisibility. (3) For many of the heating scenarios considered the emission detected in several of the SDO-AIA channels (131, 193, and 211 Å) was predicted to be dominated by warm, overdense, cooling plasma. (4) It was found to be difficult to avoid creating coronal loops that emit strongly at 1.5 MK and in the range 2–6 MK, which are the most commonly observed kind, for a broad range of nanoflare scenarios; the mere abundance of such loops does not help to constrain the heating parameter space. (5) The Fe XV (284.16 Å) emission predicted by most of the models was about 10 times brighter than the predicted Ca XVII (192.82 Å) emission, consistent with observations. Bradshaw and Klimchuk (2011) concluded that small-scale, impulsive heating that induces non-equilibrium ionization leads to predictions for observable quantities that are entirely consistent with what is actually observed.

On larger spatial scales Rakowski et al. (2007) examined the ionization state of several elements derived from in-situ observations of a halo coronal mass ejection (CME) made by the Advanced Composition Explorer (ACE). They assumed an evolution for the CME based on observations and models, and solved the ion population equations for the elements to be compared with the ACE data. They found that plasma in the core of the CME required further heating, possibly due to post-eruptive reconnection following the filament eruption, to reconcile the predicted and observed populations. Plasma in the CME cavity, however, was found not to be further ionized following the eruption, because the low density in that region effectively freezes the ion populations in the state they existed in close to the Sun. Murphy et al. (2011) and Landi et al. (2010) used the time-dependent ionization in CME ejecta to constrain the temperature history of the expanding plasma and show that an amount of heat comparable to the kinetic energy must be injected to counteract the radiative and adiabatic expansion cooling. Ko et al. (2010) examined the time-dependent ionization in post-eruption current sheets for a Petschek-type reconnection exhaust, and found that the observable line intensities depended strongly on the height of the reconnection X-line, as well as the density and magnetic field in the surrounding plasma.

7.2 Non-Maxwellian Distributions

The heat flux is sensitive to the underlying distribution of the particles that carry it and since properties of the distribution may also manifest in the emission spectra, then one may consider predicting observable signatures of the heat flux based on the results of numerical experiments and then searching for them in observational datasets. Karpen et al. (1989) used the results of their earlier flare calculations (Karpen and DeVore 1987) to investigate the effect of different heat flux formulations on the X-ray resonance lines of Ca XIX and Mg XI and included non-equilibrium ionization in their computation of the spectral lines. By comparing the results of spectroscopic diagnostics carried out with the predicted and observed emission lines, they found the properties of the flare plasma to be most consistent with the non-local formulation of the heat flux. Jiang et al. (2006) found thermal conductivity suppressed relative to the classical value in a loop-top source during the late decay phase of a flare, strongly indicative of the onset of flux limiting. The measured cooling timescale was longer than that predicted by classical thermal conduction, but shorter than for radiative cooling. However, they were unable to definitively determine whether plasma wave turbulence was providing additional heating and/or suppressing conduction by scattering electrons.

Esser and Edgar (2000, 2002) address the issue that electron temperatures observed at the solar wind acceleration site in the inner corona are too low to give rise to the ion populations observed in-situ in the solar wind, by considering non-Maxwellian electron distributions. They show that reconciling the low electron temperatures and the relatively highly charged ions requires a number of conditions to be satisfied in the inner corona. (1) The electron distribution function must be near-Maxwellian at the coronal base. (2) A departure from Maxwellian must then occur rapidly as a function of height, reaching essentially interplanetary properties within a few solar radii. (3) Ions of different elements must have different speeds to separate their freezing-in distances enough that they encounter different distributions.They also show that the required distributions are very sensitive to the electron temperature, density, and ion flow speed profiles in the coronal region where the ions form.

A number of studies have adopted forms of non-Maxwellian electron distributions to explain differences between the predicted and observed properties of emission lines. Dufton et al. (1984) found that discrepancies between the observed and theoretically predicted ratios of lines from Si III could plausibly be explained by non-Maxwellian electron distributions. However, Anderson et al. (1996) showed that the $\kappa = 2.5$ distribution used in velocity filtration models of coronal heating overpredicts the intensities of lines normally formed near 10^5 K by a factor of 100. Pinfield et al. (1999) presented evidence that observations made by SoHO-SUMER showing enhancements in the predicted Si III 1313 Å line intensity by a factor of 5 in active regions, and by a factor of 2 in the quiet Sun and coronal holes, could also be explained by non-Maxwellians. Ralchenko et al. (2007) showed that the excess brightness of some hotter lines (low lying transitions in ions formed at temperatures greater than 2 MK) in the quiet corona may be accounted for by a two-component Maxwellian, where a high-energy component adding 5 % electrons in the temperature range 300–400 eV is needed to account for the excess brightness.

Dzifčáková et al. (2008) carried out thermal and non-thermal diagnostics of a solar flare observed with RESIK and RHESSI. They found that in comparison with a synthetic isothermal or multithermal spectra, a non-thermal synthetic spectrum fitted the observed Si XII dielectronic satellite lines much more closely (with error less than 10 %), and concluded that evidence for significant deviations of the free electron distribution from Maxwellian during the impulsive phase of a solar flare can be diagnosed using X-ray spectral observations. Dzifčáková et al. (2011) explained features of the RESIK X-ray flare spectra using a Maxwellian or n-distribution for the bulk and a power-law tail, finding that the power-law tail has only a small effect on the satellite-to-allowed Si XIId/Si XIII ratio, which is sensitive to the shape of the bulk distribution and allows the parameter n to be diagnosed. Kulinová et al. (2011) carried out diagnostics of non-thermal distributions in solar flares observed with RESIK and RHESSI. They used two independent diagnostic methods, both indicating the flare plasma affected by the electron beam can have a non-thermal component in the 2–(-5) keV range, which was found to be well-described by the n-distribution. Spectral line analysis also revealed that the n-distribution does not occupy the same spatial location as the thermal component detected by RHESSI at 10 keV. Karlický et al. (2012) investigated the physical meaning of n-distributions in solar flares. The electron component of the return current in a beam-plasma system was shown to have the form of a moving Maxwellian and this was found to be very similar to the high-energy part of an n-distribution.

Dudík et al. (2009) calculated TRACE EUV filter responses to emission arising from non-Maxwellian distributions and showed that for κ-distributions the resulting responses to emission are more broadly dependent on temperature, and their maxima are flatter than for the Maxwellian electron distribution. Dzifčáková and Kulinová (2010) computed a set of synthetic spectra for various κ-distributions with varying electron densities and mean energies in the spectral range corresponding to the Hinode/EIS and Coronas-F/SPIRIT detectors. Strong EUV lines of Fe in various degrees of ionization were used to analyze the sensitivity of the line ratios to the shape of the distribution function, electron density, and temperature. It was found that EUV coronal Fe lines are generally not very suitable for diagnosing the non-thermal distributions due to their high sensitivity to electron density, but pairs of Fe XVII lines were reasonably good candidates for non-thermal diagnostics. Finally, Dzifčáková and Kulinová (2011) was able to explain the observed intensity of the Si III spectrum in coronal holes, the quiet Sun and active region transition regions by adopting an underlying κ-distribution for the electrons.

8 Summary and Future Directions

We have reviewed a number of the microphysical processes occurring in optically-thin, astrophysical plasma environments, such as supernova remnants and the solar corona, that specifically influence their internal physics such as energy transport and atomic processes and, in consequence, their emission spectra. In particular, we have described the theory of spectral line formation in the coronal approximation and how it is affected by the decoupling of the ion population from the local temperature (non-equilibrium ionization), that arises when collisional processes are unable to keep pace with heating or cooling, and by the formation of non-Maxwellian particle distribution functions, which are also related to the collisionality of the plasma. Calculations of the charge state of the plasma both in and out of equilibrium, and the most computationally tractable formulations of the kinetic equations that are solved to find the underlying particle distribution function have been presented. A selection of methods have been discussed by which the heat flux can be calculated in a hot plasma, when even near-thermal electrons have long mean-free-paths in relation to the characteristic spatial scales of the system, inducing strongly non-Maxwellian distributions, without recourse to solving a full kinetic equation. In addition, the ways in which non-Maxwellian distributions alter the rates of excitation and ionization have been considered. The results and findings from practical applications of these calculations have been shown throughout and the physics of these processes has been connected with the total radiative energy loss from the system. Finally, evidence for observational signatures of non-equilibrium ionization and non-Maxwellian particle distributions has been presented in association with discussions of the diagnostics that have been used to reveal their influence.

In the future, as astrophysical plasmas are probed with ever greater spatial, temporal and spectral resolution, we expect the microphysical processes that we have discussed here to become increasingly important to developing a full understanding of the physics that drives and governs these systems. The forthcoming Interface Region Imaging Spectrograph (IRIS: De Pontieu 2009) will provide detailed spectroscopic observations of exactly the region of the Sun's atmosphere where non-equilibrium ionization and non-Maxwellian particle distributions may play a large role in forming the spectral emission lines and, consequently, in determining what information about those regions can be extracted from the spectra by diagnostic studies. The upper-chromosphere and lower-transition region are highly dynamic environments where non-equilibrium ionization has been predicted to be a significant factor in emission from ions such as C IV and Si IV (Li-like and Na-like, respectively) that are undergoing heating (Judge et al. 2012); their enhanced emission would lead to over-estimates of the density if equilibrium ionization were assumed when interpreting the observations. IRIS may be able to shed light on the strength of departures from equilibrium in this regime.

Furthermore, streaming particles that enhance the tails of the particle distributions in the interface region may play a role in producing emission from ions of higher charge state than would be predicted from the local temperature alone. A larger tail population provides more electrons with sufficient energy to ionize the ambient plasma to a greater degree. The source of the streaming particles may be a hot ($\geq$ 10 MK) component of the coronal emission due to in-situ heating in a high-altitude region where the energy per particle is large, leading to high temperatures and mean-free-paths of lengths on the order of the spatial scales of the magnetic structures, even for near-thermal electrons. Observations of the corona at the highest spatial resolution so far achieved (75 km: Cirtain et al. 2013) indicate the presence of entwined bundles of magnetic flux that may be reminiscent of the long-theorized braiding of magnetic field lines leading to reconnection and heating (e.g. Parker 1983). The hot component may be extremely difficult, if not impossible, to observe directly if the plasma is initially tenuous (Bradshaw and Cargill 2006;

Reale and Orlando 2008), but if reconnection does lead to high temperatures and streaming particles, or direct particle acceleration, in the corona, then suitably sophisticated numerical models may be able to predict detectable signatures from lower altitude emission as indirect evidence that can be searched for in real observations by instruments such as IRIS.

Finally, the initial state of what ultimately becomes coronal plasma after heating occurs is another question that is worthy of attention. For example, does heating occur during active region emergence or following the draining of the material dredged up from below the surface as it rises? If the plasma carried to high altitudes cools below $\approx 20{,}000$ K then it becomes significantly partially ionized and, no longer supported by the Lorentz force due to the emerging field, the neutral atoms rain back onto the surface. In this scenario one might expect the active region plasma to be in an initially tenuous state and the energy per particle relatively high in the case of direct heating, giving rise to very high temperatures where the magnetic field strength and free energy are greatest (in the core of the active region). This may also be the case if the heating is intermittent and the corona is allowed to drain substantially between heating events. Evidence is beginning to accumulate to suggest that the frequency of heating in active regions increases with its age (Ugarte-Urra and Warren 2012); young active regions are heated by low-frequency events (e.g. Mulu-Moore et al. 2011; Bradshaw et al. 2012) and older active regions are heated with greater frequency (e.g. Warren et al. 2010; Reep et al. 2013). The physics of heat flux saturation and non-local thermal conduction must come into play when considering the energy transport and the overall energy balance of a hot but tenuous atmosphere and the treatments that extend the classical heat flux, described in Sect. 5.2, must be revisited. In the case of steady heating, where the atmosphere is near hydrostatic, the flux saturation regime is not reached but non-local thermal conduction may still be important in the high-temperature cores of active regions.

Acknowledgement The authors would like to thank Dr. Helen Mason for comments on Sects. 3 and 4 and Professor Peter Cargill for his comments and suggestions for improving the manuscript as a whole. We also thank the anonymous referee for their valuable input.

References

F. Alouani-Bibi, J.-P. Matte, Phys. Rev. E **66**, 066414 (2002)
F. Alouani-Bibi, J.-P. Matte, Phys. Plasmas **10**, 4 (2003)
F. Alouani-Bibi, M.M. Shoucri, J.-P. Matte, Comput. Phys. Commun. **164**, 60 (2004)
S.W. Anderson, J.C. Raymond, A. van Ballegooijen, Astrophys. J. **457**, 939 (1996)
R.G. Arendt, E. Dwek, D. Leisawitz, Astrophys. J. **400**, 562 (1992)
M. Arnaud, J. Raymond, Astrophys. J. **398**, 394 (1992)
M. Arnaud, R. Rothenflug, Astron. Astrophys. Suppl. Ser. **60**, 425 (1985)
A. Bamba, R. Yamazaki, T. Yoshida, T. Terasawa, K. Koyama, Astrophys. J. **621**, 793 (2005)
P. Beiersdorfer, M. Bitter, S. von Goeler, K.W. Hill, Astrophys. J. **610**, 616 (2004)
A.R. Bell, R.G. Evans, D.J. Nicholas, Phys. Rev. Lett. **46**, 243 (1981)
A. Bemporad, Astrophys. J. **689**, 572 (2008)
A. Bendib, J.F. Luciani, J.-P. Matte, Phys. Fluids **31**, 711 (1988)
P.L. Bhatnagar, E.P. Gross, M. Krook, Phys. Rev. **94**, 511 (1954)
G. Borini, G. Noci, Sol. Phys. **77**, 153 (1982)
S.J. Bradshaw, Astron. Astrophys. **502**, 409 (2009)
S.J. Bradshaw, P.J. Cargill, Astron. Astrophys. **458**, 987 (2006)
S.J. Bradshaw, J.A. Klimchuk, Astrophys. J. Suppl. Ser. **194**, 26 (2011)
S.J. Bradshaw, J.A. Klimchuk, J.W. Reep, Astrophys. J. **758**, 53 (2012)
S.J. Bradshaw, H.E. Mason, Astron. Astrophys. **407**, 1127 (2003)
S.J. Bradshaw, G. Del Zanna, H.E. Mason, Astron. Astrophys. **425**, 287 (2004)
P. Brekke, D.M. Hassler, K. Wilhelm, Sol. Phys. **175**, 349 (1997)
J.W. Brosius, S.M. White, Astrophys. J. Lett. **641**, L69 (2006)

P. Bryans, E. Landi, D.W. Savin, Astrophys. J. **691**, 1540 (2009)
A. Burgess, Astrophys. J. **139**, 776 (1964)
B. Burkhart, A. Lazarian, B.M. Gaensler, Astrophys. J. **749**, 145 (2012)
A.M. Bykov, Yu.A. Uvarov, J.B.G.M. Bloemen, J.W. den Herder, J.S. Kaastra, Mon. Not. R. Astron. Soc. **399**, 1119 (2009)
A.M. Bykov, M.A. Malkov, J.C. Raymond, Space Sci. Rev. (2013, this issue)
I.H. Cairns, in *The Sun, the Solar Wind, and the Heliosphere*, ed. by M.P. Miralles, J. Almeida (Springer, New York, 2011), p. 267
P.M. Campbell, Phys. Rev. A **30**, 365 (1984)
S. Chandrasekhar, Rev. Mod. Phys. **15**, 1 (1943)
A. Chepurnov, A. Lazarian, S. Stanimirović, C. Heiles, J.E.G. Peek, Astrophys. J. **714**, 1398 (2010)
C. Chiuderi, P. Pietrini, G. Torricelli-Ciamponi, J. Geophys. Res. **116**, A04107 (2011)
J.W. Cirtain, L. Golub, A.R. Winebarger, B. De Pontieu, K. Kobayashi, R.L. Moore, R.W. Walsh, K.E. Korreck, M. Weber, P. McCauley, A. Title, S. Kuzin, C.E. DeForest, Nature **493**, 501 (2013)
R.S. Cohen, L. Spitzer, P. Routly, Phys. Rev. **80**, 230 (1950)
S.R. Cranmer, A.V. Panasyuk, J.L. Kohl, Astrophys. J. **678**, 1480 (2008)
C. David, A.H. Gabriel, F. Bely-Dubau, A. Fludra, P. Lemaire, K. Wilhelm, Astron. Astrophys. **336**, L90 (1998)
B. De Pontieu, Fall AGU. Abstract #SH33B-1499, 2009
K.P. Dere, Astron. Astrophys. **466**, 771 (2007)
K.P. Dere, E. Landi, H.E. Mason, B.C. Monsignori Fossi, P.R. Young, Astron. Astrophys. Suppl. Ser. **125**, 149 (1997)
K.P. Dere, E. Landi, P.R. Young, G. Del Zanna, M. Landini, H.E. Mason, Astron. Astrophys. **498**, 915 (2009)
J.R. Dickel, W.J.M. van Breugel, R.G. Strom, Astron. J. **101**, 2151 (1991)
J.G. Doyle, M.S. Madjarska, I. Roussev, L. Teriaca, J. Giannikakis, Astron. Astrophys. **396**, 255 (2002)
B.T. Draine, Annu. Rev. Astron. Astrophys. **41**, 241 (2003)
J. Dudík, A. Kulinová, E. Dzifčáková, M. Karlický, Astron. Astrophys. **505**, 1255 (2009)
J. Dudík, E. Dzifčáková, M. Karlický, A. Kulinová, Astron. Astrophys. **529**, 103 (2011)
J. Dudík, J. Kašparová, E. Dzifčáková, M. Karlický, Š. Mackovjak, Astron. Astrophys. **539**, 107 (2012)
P.L. Dufton, A.E. Kingston, F.P. Keenan, Astrophys. J. **280**, L35 (1984)
G.A. Dulk, D.B. Melrose, S.M. White, Astrophys. J. **234**, 1137 (1979)
A.K. Dupree, R.T. Moore, P.R. Shapiro, Astrophys. J. **229**, L101 (1979)
E. Dzifčáková, Sol. Phys. **208**, 91 (2002)
E. Dzifčáková, Sol. Phys. **234**, 243 (2006)
E. Dzifčáková, A. Kulinová, Sol. Phys. **218**, 41 (2003)
E. Dzifčáková, A. Kulinová, Sol. Phys. **263**, 25 (2010)
E. Dzifčáková, A. Kulinová, Astron. Astrophys. **531**, 122 (2011)
E. Dzifčáková, H.E. Mason, Sol. Phys. **247**, 301 (2008)
E. Dzifčáková, A. Kulinová, C. Chifor, H.E. Mason, G. Del Zanna, J. Sylwester, B. Sylwester, Astron. Astrophys. **488**, 311 (2008)
E. Dzifčáková, M. Homola, J. Dudík, Astron. Astrophys. **531**, 111 (2011)
R.J. Edgar, R. Esser, Astrophys. J. **538**, 167 (2000)
R. Esser, R.J. Edgar, Astrophys. J. **532**, 71 (2000)
R. Esser, R.J. Edgar, Adv. Space Res. **30**, 481 (2002)
R. Esser, R.J. Edgar, N.S. Brickhouse, Astrophys. J. **498**, 448 (1998)
J.M. Fontenla, E.H. Avrett, R. Loeser, Astrophys. J. **406**, 319 (1993)
R.A. Frazin, S.R. Cranmer, J.L. Kohl, Astrophys. J. **597**, 1145 (2003)
D.R. Gray, J.D. Kilkenny, Plasma Phys. **22**, 81 (1980)
J.M. Greene, Phys. Fluids **16**, 2022 (1973)
H.R. Griem, *Plasma Spectroscopy* (McGraw-Hill, New York, 1964)
V.H. Hansteen, Astrophys. J. **402**, 741 (1993)
M. Haverkorn, V. Heesen, Space Sci. Rev. **166**, 133 (2012)
Y.W. Jiang, L. Siming, W. Liu, V. Petrosian, Astrophys. J. **638**, 1140 (2006)
J. Joslyn, R.H. Munro, T.E. Holzer, Sol. Phys. **64**, 57 (1979a)
J. Joslyn, R.H. Munro, T.E. Holzer, Astrophys. J. Suppl. Ser. **40**, 793 (1979b)
P.G. Judge, B. De Pontieu, S.W. McIntosh, K. Olluri, Astrophys. J. **746**, 158 (2012)
M. Karlický, E. Dzifčáková, J. Dudík, Astron. Astrophys. **537**, 36 (2012)
J.T. Karpen, C.R. DeVore, Astrophys. J. **320**, 904 (1987)
J.T. Karpen, C.C. Cheng, G.A. Doschek, C.R. DeVore, Astrophys. J. **338**, 1184 (1989)
F.P. Keenan, E.S. Conlon, L.K. Harra, V.M. Burke, K.G. Widing, Astrophys. J. **385**, 381 (1992)
M.A. Killie, Å.M. Janse, Ø. Lie-Svendsen, E. Leer, Astrophys. J. **604**, 842 (2004)

Y.-K. Ko, J.C. Raymond, B. Vršnak, E. Vujić, Astrophys. J. **722**, 625 (2010)
J.L. Kohl et al., Sol. Phys. **175**, 613 (1997)
E.P. Kontar et al., Space Sci. Rev. **159**, 301 (2011)
A. Kulinová, J. Kašparová, E. Dzifčáková, J. Sylwester, B. Sylwester, M. Karlický, Astron. Astrophys. **533**, 81 (2011)
J.M. Laming, J.C. Raymond, B.M. McLaughlin, W.P. Blair, Astrophys. J. **472**, 267 (1996)
L.D. Landau, Phys. Z. Sowjetunion **10**, 154 (1936) (Transl. in Collected Papers of L. D. Landau (ed. D. ter Haar), pp. 163–170. Oxford: Pergamon Press (1965))
E. Landi, M. Landini, Astron. Astrophys. **347**, 401 (1999)
S. Landi, F.G.E. Pantellini, Astron. Astrophys. **372**, 686 (2001)
E. Landi, G. Del Zanna, P.R. Young, K.P. Dere, H.E. Mason, Astrophys. J. **744**, 99 (1997)
E. Landi, J.C. Raymond, M.P. Miralles, H. Hara, Astrophys. J. **751**, 21 (2010)
K.-S. Lee, Y.-J. Moon, K.-S. Kim, J.-Y. Lee, K.-S. Cho, G.S. Choe, Astron. Astrophys. **486**, 1009 (2008)
S. Livi, E. Marsch, Phys. Rev. A **34**, 533 (1986)
N.N. Ljepojevic, J. Quant. Spectrosc. Radiat. Transf. **44**, 203 (1990)
N.N. Ljepojevic, A. Burgess, Proc. R. Soc. Lond. A **428**, 71 (1990)
N.N. Ljepojevic, P. MacNeice, Sol. Phys. **117**, 123 (1988)
N.N. Ljepojevic, P. MacNeice, Phys. Rev. A **40**, 981 (1989)
K.S. Long, S.P. Reynolds, J.C. Raymond, P.F. Winkler, K.K. Dyer, R. Petre, Astrophys. J. **586**, 1162 (2003)
J.F. Luciani, P. Mora, J. Virmont, Phys. Rev. Lett. **51**, 1664 (1983)
J.F. Luciani, P. Mora, R. Pellat, Phys. Fluids **28**, 835 (1985)
S. Ma et al., Astrophys. J. **738**, 160 (2011)
P. MacNeice, J. Fontenla, N.N. Ljepojevic, Astrophys. J. **369**, 544 (1991)
S. Mancuso, Astron. Astrophys. **463**, 1137 (2007)
H.E. Mason, B.C. Monsignori Fossi, Astron. Astrophys. Rev. **6**, 123 (1994)
J.-P. Matte, J. Virmont, Phys. Rev. Lett. **49**, 1936 (1982)
P. Mazzotta, G. Mazzitteli, S. Colafrancesco, N. Vitorio, Astron. Astrophys. Suppl. Ser. **133**, 403 (1998)
T.F. Morse, Phys. Fluids **6**, 1420 (1963)
T.F. Morse, Phys. Fluids **7**, 2012 (1964)
F.M. Mulu-Moore, A.R. Winebarger, H.P. Warren, Astrophys. J. **742**, 6 (2011)
N.A. Murphy, J.C. Raymond, K.E. Korreck, Astrophys. J. **735**, 17 (2011)
G. Noci, J.L. Kohl, G.L. Withbroe, Astrophys. J. **315**, 706 (1987)
G. Noci, D. Spadaro, R.A. Zappala, S.K. Antiochos, Astrophys. J. **338**, 1131 (1989)
S. Nozawa, K. Takahashi, Y. Kohyama, N. Itoh, Astron. Astrophys. **499**, 661 (2009)
S.P. Owocki, R.C. Canfield, Astrophys. J. **300**, 420 (1986)
S.P. Owocki, J.D. Scudder, in *SAO 2nd Cambridge Workshop on Cool Stars, Stellar Systems, and the Sun*, vol. 1 (1982), p. 107
F. Paerels, J. Cottam, M. Sako, D.A. Liedahl, A.C. Brinkman, R.L.J. van der Meer, J.S. Kaastra, P. Predehl, Astrophys. J. **533**, L135 (2000)
E.N. Parker, Astrophys. J. **264**, 642 (1983)
M. Pick, N. Vilmer, Astron. Astrophys. Rev. **16**, 1 (2008)
D.J. Pinfield, F.P. Keenan, M. Mathioudakis, K.J.H. Phillips, W. Curdt, K. Wilhelm, Astrophys. J. **527**, 1000 (1999)
C.E. Rakowski, M.J. Laming, S.T. Lepri, Astrophys. J. **667**, 602 (2007)
Y. Ralchenko, U. Feldman, G.A. Doschek, Astrophys. J. **659**, 1682 (2007)
J.C. Raymond, B.W. Smith, Astrophys. J. Suppl. Ser. **35**, 419 (1977)
J.C. Raymond, A.K. Dupree, Astrophys. J. **222**, 379 (1978)
J.C. Raymond, P.F. Winkler, W.P. Blair, J.-J. Lee, S. Park, Astrophys. J. **712**, 901 (2010)
F. Reale, S. Orlando, Astrophys. J. **684**, 715 (2008)
J.W. Reep, S.J. Bradshaw, J.A. Klimchuk, Astrophys. J. **764**, 193 (2013)
T. Rhee, C.-M. Ryu, P.H. Yoon, J. Geophys. Res. **111**, 09107 (2006)
M.N. Rosenbluth, W.M. MacDonald, D.L. Judd, Phys. Rev. **107**, 1 (1957)
G.B. Rybicki, A.P. Lightman, *Radiation Processes in Astrophysics* (Wiley-Interscience, New York, 1979)
R. Sankrit et al., Astrophys. J. **712**, 1092 (2010)
J.M. Schmidt, I.H. Cairns, J. Geophys. Res. **117**, A04106 (2012)
E.C. Shoub, Astrophys. J. **266**, 339 (1983)
D.F. Smith, Astrophys. J. **302**, 836 (1986)
R.K. Smith, J.P. Hughes, Astrophys. J. **718**, 583 (2010)
R.K. Smith, R.J. Edgar, R.A. Shafer, Astrophys. J. **581**, 562 (2002)
D. Spadaro, R. Ventura, Astron. Astrophys. **289**, 279 (1994a)
D. Spadaro, R. Ventura, Astron. Astrophys. **281**, 245 (1994b)

D. Spadaro, G. Noci, R.A. Zappala, S.K. Antiochos, Astrophys. J. **355**, 342 (1990a)
D. Spadaro, R.A. Zappala, S.K. Antiochos, G. Lanzafame, G. Noci, Astrophys. J. **362**, 370 (1990b)
D. Spadaro, P. Leto, S.K. Antiochos, Astrophys. J. **427**, 453 (1994)
D. Spadaro, S. Orlando, G. Peres, P. Leto, Astron. Astrophys. **302**, 285 (1995)
L. Spitzer, R. Härm, Phys. Rev. **89**, 977 (1953)
L. Teriaca, D. Banerjee, J.G. Doyle, Astron. Astrophys. **349**, 636 (1999a)
L. Teriaca, J.G. Doyle, R. Erdélyi, L.M. Sarro, Astron. Astrophys. **352**, L99 (1999b)
Y. Uchiyama et al., Astrophys. J. Lett. **749**, L35 (2012)
I. Ugarte-Urra, H.P. Warren, Astrophys. J. **761**, 21 (2012)
J.E. Vernazza, J. Raymond, Astrophys. J. **228**, L89 (1979)
N. Vilmer, A.L. MacKinnon, G.J. Hurford, Space Sci. Rev. **159**, 167 (2011)
J. Vink, J.M. Laming, Astrophys. J. **584**, 758 (2003)
H.P. Warren, A.R. Winebarger, D.H. Brooks, Astrophys. J. **711**, 228 (2010)
M.J. West, S.J. Bradshaw, P.J. Cargill, Sol. Phys. **252**, 89 (2008)
B.J. Williams et al., Astrophys. J. Lett. **652**, L33 (2006)
B.J. Williams et al., Astrophys. J. **687**, 1054 (2008)
B.J. Williams et al., Astrophys. J. **729**, 65 (2011)
H. Yamaguchi, M. Ozawa, K. Koyama, K. Masai, J.S. Hiraga, M. Ozaki, D. Yonetoku, Astrophys. J. **705**, L6 (2009)
P.H. Yoon, T. Rhee, C.-M. Ryu, J. Geophys. Res. **111**, 09106 (2006)

DOI 10.1007/978-1-4899-7413-6_9
Reprinted from *Space Science Reviews* Journal, DOI 10.1007/s11214-013-0021-7

Recent Evolution in the Theory of Magnetic Reconnection and Its Connection with Turbulence

Homa Karimabadi · Vadim Roytershteyn · William Daughton · Yi-Hsin Liu

Received: 14 February 2013 / Accepted: 29 August 2013 / Published online: 28 September 2013

Abstract The concept of reconnection is found in many fields of physics with the closest analogue to magnetic reconnection being the reconnection of vortex tubes in hydrodynamics. In plasmas, magnetic reconnection plays an important role in release of energy associated with the magnetic shear into particle energy. Although most studies to date have focused on 2D reconnection, the availability of 3D petascale kinetic simulations have brought the complexity of 3D reconnection to the forefront in collisionless reconnection studies. Here we briefly review the latest advances in 2D and compare and contrast the results with recent 3D studies that address role of anomalous transport in reconnection, effects of turbulence on the rate and structure, among others. Another outcome of recent research is the realization of a deeper link between turbulence and reconnection where the common denominator is the generic formation of electron scale sheets which dissipate the energy through reconnection. Finally, we close the review by listing some of the major outstanding problems in reconnection physics.

Keywords Magnetic · Reconnection · Turbulence

H. Karimabadi (✉)
University of California at San Diego, La Jolla, CA 92093-0407, USA
e-mail: homakar@gmail.com

H. Karimabadi · V. Roytershteyn
SciberQuest, Inc., Del Mar, CA 92014, USA

V. Roytershteyn
e-mail: vroytersh@gmail.com

W. Daughton · Y.-H. Liu
Los Alamos National Laboratory, MS F699, Los Alamos, NM, USA

W. Daughton
e-mail: daughton@lanl.gov

Y.-H. Liu
e-mail: yhliu10@gmail.com

1 Introduction

A commonly occurring structure in magnetized plasmas is a current sheet: a quasi-equilibrium structure supporting a finite magnetic shear. Magnetic reconnection is an important process that enables release of energy associated with the magnetic shear into other forms of energy. Magnetic reconnection (henceforth called "reconnection") is surprisingly robust, operating in a wide variety of conditions, from fully collisional to collisionless, from laminar to fully turbulent medium, and from nonrelativistic to relativistic regimes (Priest and Forbes 2000; Biskamp 2005; Zweibel and Yamada 2009; Uzdensky 2011). This diversity has led to at least four distinct lines of research that have historically evolved with minimal cross-pollination: (a) reconnection of single, laminar current sheet with applications to laboratory plasmas (Yamada et al. 2010; Roytershteyn et al. 2010, 2013a), planetary magnetospheres (Fuselier and Lewis 2011; Hesse et al. 2011; Mozer et al. 2011a, 2011b; Paschmann et al. 2013; Egedal et al. 2013; Scudder et al. 2012), solar wind (Gosling 2010, 2012), and the Sun (Daughton and Roytershteyn 2012; Cassak and Shay 2012); (b) reconnection of current sheet in the presence of imposed magnetohydrodynamic (MHD) turbulence with applications to astrophysical plasmas (Lazarian and Vishniac 1999; Lapenta and Lazarian 2012; Lazarian et al. 2012), (c) reconnection in filamentary current sheets generated as part of the turbulence cascade with applications to solar wind (Matthaeus and Lamkin 1985; Matthaeus and Velli 2011; Servidio et al. 2011; Karimabadi et al. 2013); and (d) theoretical studies on the general definition and characterization of reconnection in 3D with applications so far mostly to laboratory and the Sun (Hesse and Schindler 1988; Priest and Forbes 2000; Pontin 2011; Yeates and Hornig 2011; Boozer 2012). Many scientists have made important contributions to the field of reconnection and it is impossible to reference all the papers. As a compromise we have cited above mostly review articles with each containing a comprehensive list of references in each specialized area.

There have been several recent reviews, each emphasizing a particular aspect of reconnection and/or regime. Our focus here is to provide an overview of recent theoretical advances in collisionless reconnection. One of the key discoveries has been the generic formation of electron layers that can affect the structure of the reconnection layer, lead to time-dependent reconnection, and contribute to self-generation of turbulence in 3D. We start by a brief overview of two-dimensional (2D) reconnection models and the discovery of two new regimes of reconnection as a function of guide field. The second part of this review concerns recent extension of the 2D studies to 3D made possible by the recent advent of petascale computing. One of the outcomes of these studies has been the realization that reconnection and turbulence are intimately related. We discuss this connection and give suggestions for future work.

2 2D Reconnection

Most theoretical models have focused on 2D steady state reconnection in the presence of a single X-line. In this limit, resistive MHD simulations have established a direct link between the length of the diffusion region and the type of reconnection possible. Within the MHD model, the length of the diffusion region is determined by the localization scale of the resistivity. A continuum of solutions are obtained (Priest and Forbes 2000) ranging from fast reconnection when resistivity is localized (Petschek 1964) to inefficient reconnection (Sweet 1958; Parker 1957) when resistivity is uniform. It is instructive to start with the Sweet-Parker

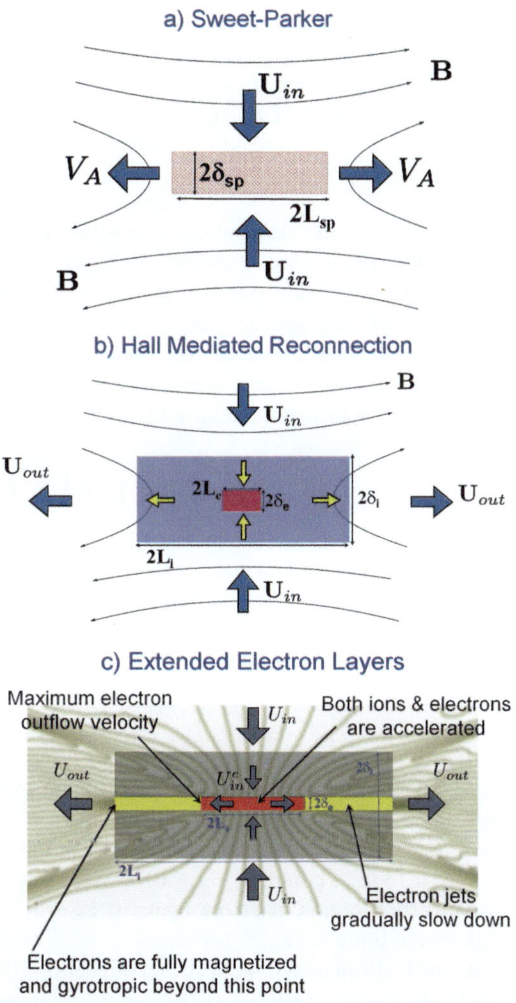

Fig. 1 Evolution in our understanding of the diffusion region in reconnection in 2D anti-parallel geometry. (**a**) Sweet-Parker model. (**b**) Hall-mediated reconnection model where the electron diffusion region was thought to remain on electron scales in length. (**c**) Current model of reconnection where the electron layers have been observed to extend to macroscopic scales in length. Also shown are electron streamlines. Adapted from Karimabadi et al. (2007)

model (SP) (Fig. 1a) which is a steady state solution of 2D, incompressible, resistive MHD. Through the use of conservation laws and neglecting the resistive heating within the layer, one obtains the following expressions for the reconnection rate ($R \equiv U_{in}/V_A$):

$$R = \frac{U_{in}}{V_A} = \frac{\delta_{sp}}{L_{sp}} = \frac{1}{\sqrt{S}}. \tag{1}$$

Here U_{in} is the inflow velocity, $V_A = B_o/\sqrt{4\pi m_i n}$ is the Alfven speed, B_o is the reconnecting component of the magnetic field upstream of the layer, n is the plasma density in the current sheet, δ_{sp} and L_{sp} are the half-width and the half-length of the layer, and $S = 4\pi V_A L_{sp}/\eta c^2$ is the Lundquist number. The fact that $U_{out} = V_A$ in the SP solution has been used in the derivation of the above equations.

In contrast to MHD, in a kinetic plasma electrons and ions decouple from the magnetic field on different scales (Sonnerup 1979; Mandt et al. 1994) and the diffusion region consists of an inner electron region and an outer ion region as illustrated in (Fig. 1b). An estimate of the relative role of each diffusion region in controlling reconnection can be obtained as follows. The maximum possible reconnection rate through the diffusion region can be estimated by assuming a limiting outflow velocity and then imposing mass conservation as in the Sweet-Parker analysis. Applying it to the electron diffusion region with half-thickness of δ_e and half-length of L_e yields:

$$R_e = \frac{\delta_e}{L_e}\frac{U_{e_{out}}}{V_{Ae}}. \tag{2}$$

Here $U_{e_{out}}$ is the electron outflow velocity and V_{Ae} is the electron Alfven speed. Since $U_{e_{out}} \leq V_{Ae}$ (neglecting pressure gradients which could permit faster or slower outflow speeds), the maximum possible rate is set by the aspect ratio δ_e/L_e. In general the outflow speed of each species can be lower than the Alfven speed for that species. However, in some cases such as the SP, the outflow speed is the Alfven speed and the rate is given exactly by the aspect ratio. Using the same analysis, one can derive the maximum rate R_i through the ion diffusion. Taking the ratio, we find:

$$\frac{R_i}{R_e} = \left(\frac{B_e}{B_i}\right)^2 \left(\frac{n_i}{n_e}\right)^{1/2} \left(\frac{m_i}{m_e}\right)^{1/2}. \tag{3}$$

Here B_e and n_e are the values of the magnetic field and density just upstream of the electron diffusion region and B_i and n_i are those just upstream of the ion diffusion region. For a given ion reconnection rate R_i and assuming $U_{e_{out}} = V_{Ae}$ and $U_{i_{out}} = V_{Ai}$, one may use these expressions to estimate the aspect ratio of the electron diffusion required for electrons to start becoming the bottleneck:

$$\frac{L_e}{\delta_e} = \frac{1}{R_i}\left(\frac{B_e}{B_i}\right)^2 \left(\frac{m_i}{m_e}\right)^{1/2}. \tag{4}$$

The width of the electron diffusion region in the collisionless regime is fairly well understood and is thought to be on the order of electron skin depth $\delta_e \sim d_e$. Simulations show $B_i \sim 2B_e$, although there exists variations in this ratio in different regimes. Taking $m_i/m_e = 1836$, we find $L_e/\delta_e \sim 107$. Assuming $R_i \sim 0.1$ and $\delta_e \sim d_e$, this means the electron diffusion region has to elongate to $L_e \sim 107 d_e \sim 2.5 d_i$ for electrons to form a Sweet-Parker type bottleneck.

In the Hall-mediated model of reconnection (Birn et al. 2001; Hesse et al. 2001; Pritchett 2001; Shay et al. 2001) (Fig. 1b), it was assumed that the length of the electron diffusion region L_e would remain microscopic. In such a scenario, it follows from Eq. (4) that the reconnection rate is controlled by the ion diffusion region. This, together with empirical results suggesting that the Hall effect may be localizing the diffusion region, provided an explanation of fast reconnection. However, subsequent larger-scale fully kinetic simulations (Daughton et al. 2006; Fujimoto 2006; Klimas et al. 2008) found that the electron diffusion region does not remain on electron scales as had been predicted but elongates to ion scales, becoming unstable to secondary island formation. The structure of the electron layer was found to consist of two parts as seen in Fig. 1c (Karimabadi et al. 2007; Shay et al. 2007). The inner region is characterized by the locale where electrons reach a peak outflow speed near the electron Alfvén velocity and was found to extend to several ion inertial lengths. Ions also approach 80 % of their peak velocity in this inner region but remain sub-Alfvénic. The electron frozen-in condition is violated over a wider outer region characterized by highly collimated electron jets that are gradually decelerated and thermalized. From Eq. (4), the extension of the electron inner layer to ion scales makes it possible for electrons to form a bottleneck and control the reconnection rate. Evidence for this effect has been reported (Daughton et al. 2006).

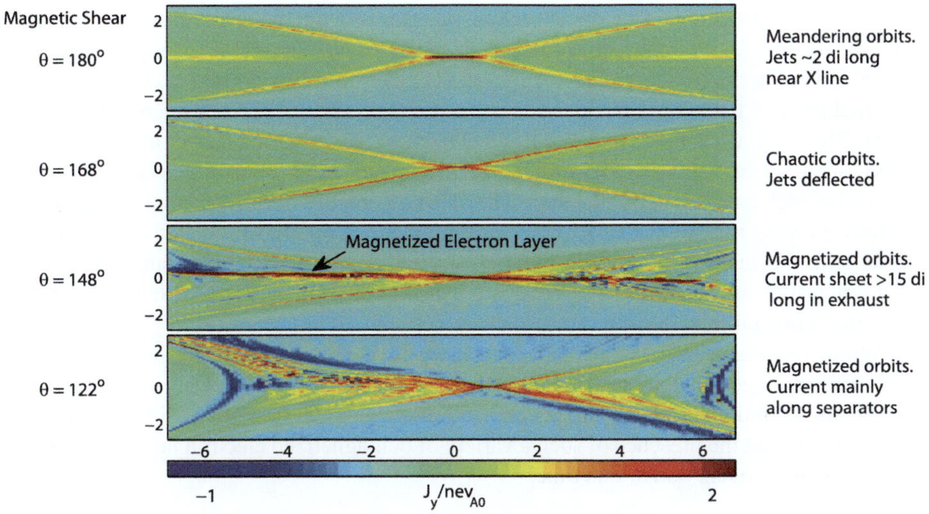

Fig. 2 Plot of the out-of-plane current from 2D fully kinetic simulations at realistic mass ratio, illustrating the four regimes as the magnetic shear varies. A new regime at intermediate guide field contains a long magnetized electron current sheet embedded in the exhaust. Adapted from Le et al. (2013)

These findings have required re-thinking of the physics of fast reconnection in collisionless plasmas, which remains an open question. It is not currently understood the mechanism(s) that controls the length of the ion diffusion region to maintain an aspect ratio of $\sim$0.1 for a wide range of parameters. There are suggestions that the length may depend on several factors including the external boundary conditions (Chacón et al. 2007; Simakov and Chacón 2008; Zocco et al. 2009; Sullivan et al. 2009).

The above studies were primarily for the anti-parallel case. The addition of a guide field gives rise to additional complexity in the structure of the electron layers that are formed (Drake et al. 2006). More careful studies of the effect of the guide field on the structure of the electron layers revealed two new regimes, in addition to the well known anti-parallel and strong guide field regimes. Figure 2 shows the out-of-plane current in the four regimes for a scan of magnetic shear which is a measure of the relative strength of the guide field to the reconnecting field. In the anti-parallel case (magnetic shear of 180°), an unmagnetized electron layer is formed in the center of the reconnection layer. For weak guide fields, the centrally located elongated electron layer disappears as the guide field prevents electrons from meandering about the magnetic field reversal (Goldman et al. 2011). For moderate guide fields, electron pressure anisotropy develops in the exhaust, supporting the longest electron layers ever seen in the simulations (Le et al. 2013). These extended electron layers are not really the 'proper' electron diffusion region since they can be highly non-ideal, but there is no normal influx into them, and they do not control the rate. The electron layer in this case extends to tens of ion inertial lengths and provides an explanation for the very long $\sim 60 d_i$ electron current sheet inferred from Cluster spacecraft data in the Earth's magnetosheath during reconnection (Phan et al. 2007). Here d_i is the ion inertial length. For strong guide fields, the electron pressure cannot compete with the tension of the guide field to approach the firehose condition. The central electron layer does not form and the current tends to peak only near the pair of diagonally opposed separator field lines.

There is growing observational support for these new results such as detection of a very long demagnetized electron layer ($\sim$60 ion inertial length) in the magnetosheath (Phan et al.

2007), and identification of secondary magnetic islands in the magnetotail and at the magnetopause (see the review by Paschmann et al. 2013). However, the above theoretical uncertainties also impact the observational search for the electron diffusion region. Different observational proxies for finding the electron diffusion layers have been considered (Paschmann 2008; Mozer et al. 2011b; Hesse et al. 2011). More recently, Scudder et al. (2012) developed a new set of proxies for identification of demagnetized electron layers and used it to search for such events in the Polar data. They found arguably the best example of an electron diffusion layer to date. It possesses the extremely rare properties predicted by simulations: flows with thermal electron Mach number of order 1, electron pressure anisotropy of 10, and supporting signatures of non-perturbative drift expansion parameters that correlate well with those in a tailored asymmetric guide geometry simulation.

While the generation of elongated electron layers is now widely accepted, their implications for reconnection continue to be hotly debated. The role of the electron layers in these newly discovered regimes and the generalization of the definition of the diffusion region to take into account the presence of such a variety of demagnetized electron layers (e.g., Fig. 2) remain poorly understood at this time.

3 3D Reconnection

Reconnection in three-dimensional (3D) magnetic fields gives rise to a richer variety of configurations, effects, and complexities that do not arise in 2D. For example, in 3D, the number of sites of current sheet formation and reconnection is greatly increased. Some of the proposed mechanisms for generation of current sheets in 3D include generic footpoint motion (Parker 1972), presence of strong gradients in the magnetic field (Longcope and Strauss 1994; Priest and Demoulin 1995; Boozer 2012), and braided magnetic field structures (Pontin et al. 2011). Another complexity of 3D reconnection is the possibility of field line exponentiation or chaos in field line mapping. It was recognized early on that the magnetic field lines can be stochastic, where adjacent field lines random walk away from each other (Jokipii and Parker 1968; Rechester and Rosenbluth 1978). This can arise due to the presence of pre-existing turbulence but in three dimensional magnetic fields (finite gradients in all three directions), even slow spatial variations can give rise to exponential divergence of magnetic field lines. The implications of the latter for magnetic reconnection has been emphasized by Boozer (2012).

Theories and simulations of reconnection in 3D magnetic fields have been mainly within the MHD limit. We refer the reader to a number of recent articles and reviews on this topic (Priest and Forbes 2000; Pontin 2011; Yeates and Hornig 2011; Boozer 2012). Instead of describing 3D reconnection in all of its generality, we describe here selected examples of the recent 3D kinetic simulations of reconnection. These studies were focused on direct extension of the results and models described in Sect. 2 to 3D. As such, the simulations had rather simple quasi-2D initial and boundary conditions corresponding to isolated current sheets, but allowed complex 3D dynamics to arise self-consistently as the reconnection proceeds. Crucially, the results described below were obtained using fully kinetic simulations, i.e. they use a description of plasma that is rigorously valid in the collisionless regimes.

3.1 Is There Evidence for Anomalous Transport?

When binary Coulomb collisions are considered to be the only origin of the resistivity in plasma, Sweet-Parker reconnection model fails spectacularly in describing reconnection rates in weakly collisional or collisionless systems such as the solar corona or the Earth's magnetosphere. Since the early days of reconnection research, this fact has motivated the search for a mechanism that would significantly enhance effective resistivity (Huba et al. 1977) or viscosity (Biskamp 2005). Such enhanced transport coefficients are traditionally referred to as anomalous and are usually presumed to originate from microscopic turbulence associated with instabilities triggered within reconnecting current sheets. More recently, it has become apparent that collisionless reconnection can proceed with high enough rates to explain the relevant observations even in the absence of microscopic turbulence (see Sect. 2). Consequently, the focus of investigations concerning the role of microscopic instabilities on reconnection has shifted somewhat from explaining the rates to a more generic question of when microscopic instabilities are triggered within reconnection layers and whether they significantly modify the details of reconnection process.

Lower-hybrid drift instability (LHDI) is a current-driven instability that has long been considered a potential candidate for generation of anomalous resistivity. Fluctuations in the lower-hybrid frequency range are routinely observed in the vicinity of reconnection layers in both space (Bale et al. 2002; Eastwood et al. 2009; Mozer et al. 2011a) and laboratory experiments (Carter et al. 2002; Ji et al. 2004; Fox et al. 2010). However it remained unclear until recently if such fluctuations played a fundamental role in the reconnection process. A series of recent papers (Mozer et al. 2011a; Roytershteyn et al. 2012; Pritchett et al. 2012) resolved that question and demonstrated that LHDI fluctuations are unlikely to play a significant role in reconnection under parameter typical of the Earth's magnetosphere or the laboratory experiments such as Magnetic Reconnection Experiment (MRX) (Roytershteyn et al. 2013a).

For example, Roytershteyn et al. (2012) considered the influence of LHDI on reconnection in asymmetric antiparallel geometry. This configuration is characterized by large density gradients across reconnection layers that are favorable for the excitation of LHDI. In contrast to many previous investigations, they considered the long-time dynamics of self-consistently generated current sheets. It was found that LHDI has the strongest influence on the reconnection process in regimes with large asymmetry, low values of $T_i/T_e \ll 1$, and low plasma β on both sides of the current sheet. In such regimes, LHDI may directly influence the reconnection mechanism in the vicinity of the X-line, with the sum of fluctuation-induced terms accounting for $\sim$60 % of the average electric field close to the X line (Roytershteyn et al. 2012). Another way that LHDI can influence reconnection is by broadening the separatrix current layer, decreasing the growth rate of tearing instabilities that generate secondary flux ropes along the separatrix. Figure 3 compares the thickness of the reconnection in 2D where there is no LHDI and in 3D where LHDI is present. LHDI is found to increase the width of the layer by a factor of 2–4. Since for most applications such as the magnetosphere $T_i/T_e \geq 1$, LHDI is not expected to play a significant role as a source of anomalous resistivity.

In addition LHDI, a variety of other instabilities may be triggered in reconnecting current sheets under appropriate conditions. For example, Che et al. (2011) considered 3D reconnection in the strong guide field regime and for force-free current sheets. They reported a fast growing electron shear instability, resulting in strong turbulent viscosity and associated broadening of the layer. However, the short time duration and limited spatial volume of their simulations left open the question of whether these results would persist in large systems. In order to distinguish between transient effects and to allow coupling to 3D flux rope dynamics, Liu et al. (2013) re-examined this problem using 3D simulations with similar parameters as in Che et al. (2011) but with $\sim 300\times$ larger volume and duration over $\sim 12\times$ longer. In contrast to Che et al. (2011), Liu et al. (2013) found that the dominant instability is collisionless tearing, with no evidence of turbulent broadening or turbulent viscosity in the electron

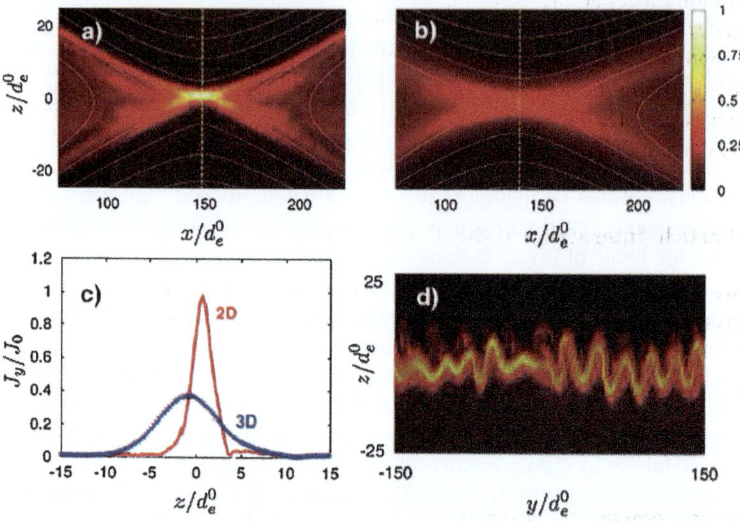

Fig. 3 Panels (**a**) and (**b**) show current density in 2D and 3D simulations, respectively. Here J_y is the out--of-the-plane component of the current and is normalized to its peak value in the 2D case; (**c**) profiles of J_y across the layer at x positions marked by *dashed lines* in panels (**a**) and (**b**); panel (**d**) shows a profile of J_y in a y–z plane at $x/d_e^0 = 140$ in the 3D simulation. Adapted from Roytershteyn et al. (2012)

layers. They also found that the parallel electric fields are supported predominantly by the electron pressure tensor and electron inertia as in 2D, while turbulent dissipation remains small.

3.2 Bifurcation of the Diffusion Region

Traditional ideas of reconnection have been based on a single electron diffusion at the center of the reconnection layer. However, recent 3D simulations of weak magnetic shear $\leq 80°$ regime found that the electron diffusion region is broken up into two or more current sheets (Liu et al. 2013) as shown in Fig. 4. The cause of the splitting was found to be oblique tearing modes which produce flux ropes while simultaneously drive enhanced current at multiple resonance surfaces. Surprisingly, this physics persists into the nonlinear regime leading to multiple electron layers embedded within a larger Alfvenic inflow and outflow. Despite the rich 3D dynamics, the energy conversion time scale is nearly the same as 2D. And the mechanism for breaking the frozen-in condition is similar to that in 2D.

3.3 Connection to Turbulence

Turbulence is often considered to be the greatest unsolved problem in classical physics. Compared to ordinary fluids, the presence of a magnetic field in plasma and its high electrical conductivity cause additional complexity in the properties of turbulence and result in an intimate link to magnetic reconnection. There is indeed growing evidence that formation of reconnecting current sheets may be a generic feature of strong turbulence in magnetized plasmas (Matthaeus and Velli 2011). This has long been observed in MHD simulations of turbulence, where the structure and dynamics of such current sheets depend on the imposed value of resistivity and viscosity (possibly numerical). More recent kinetic studies

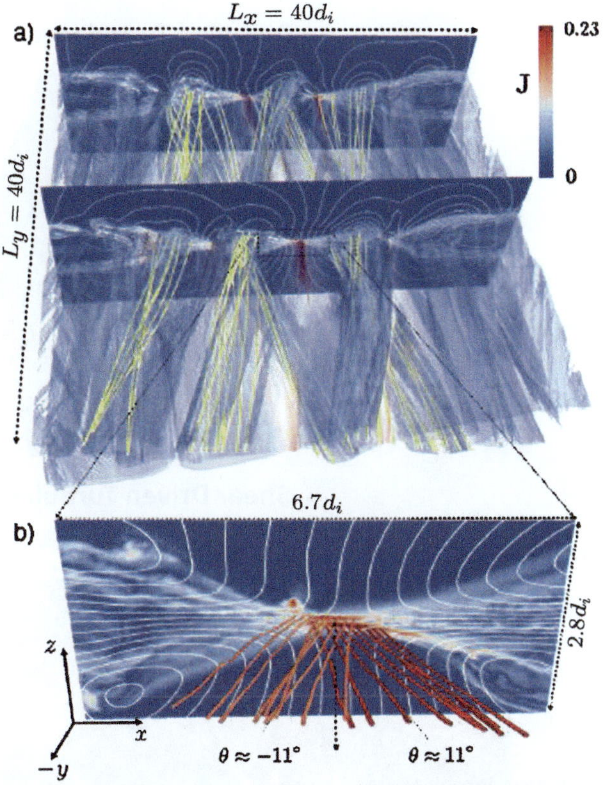

Fig. 4 (**a**) Three-dimensional structure of reconnection for the large simulation with a guide field of 2.5. Shown is an isosurface of particle density, colored by the current density along with sample magnetic field lines (*yellow*). Cutting planes also show the current along with streamlines of the in-plane ion flow velocity (*white*). (**b**) Closeup of the electron diffusion region along with sample streamlines of the current density (*red*). From Liu et al. (2013)

of this problem, such as those illustrated in Fig. 5a–b, show formation of current sheets on ion and electron scales (Servidio et al. 2011; Wan et al. 2012; Karimabadi et al. 2013; Wu et al. 2013).

To illustrate the link between reconnection and turbulence, we show in Fig. 5 plots of the current density from fully kinetic simulations of three very different configurations: decaying turbulence (Roytershteyn et al. 2013b), shear driven turbulence (Karimabadi et al. 2013), and strong guide field reconnection (Liu et al. 2013), where a 2D cut of the simulation in Fig. 4 is shown. The common feature among all these examples is the formation of long current sheets with widths on the order of electron skin depth which continually form and break up due to reconnection.

In the first two examples in Fig. 5, current sheets are formed as part of the turbulence cascade whereas in the example in Fig. 5c the system starts with a large scale laminar current sheet and turbulence is self-generated due to reconnection. Note a close similarity in the current structure and the resulting spectrum with the shear driven case (see Fig. 6). This is partly due to the formation of velocity shears in the exhaust of the guide field reconnection run. In contrast to mechanisms for generation of stochasticity that rely on three-dimensional magnetic field structures, plasma turbulence and wandering of field lines can also be self-generated starting from a laminar, 1D current sheet (gradient only in one direction) as shown in Fig. 5c. Galeev et al. (1986) noted that in the presence of a guide field tearing instability can occur at multiple resonance surfaces lying on different planes. Using analytical theory they concluded that a broad spectrum of tearing modes would be excited, each generating a magnetic island. As the magnetic islands get larger they overlap, leading to stochastic

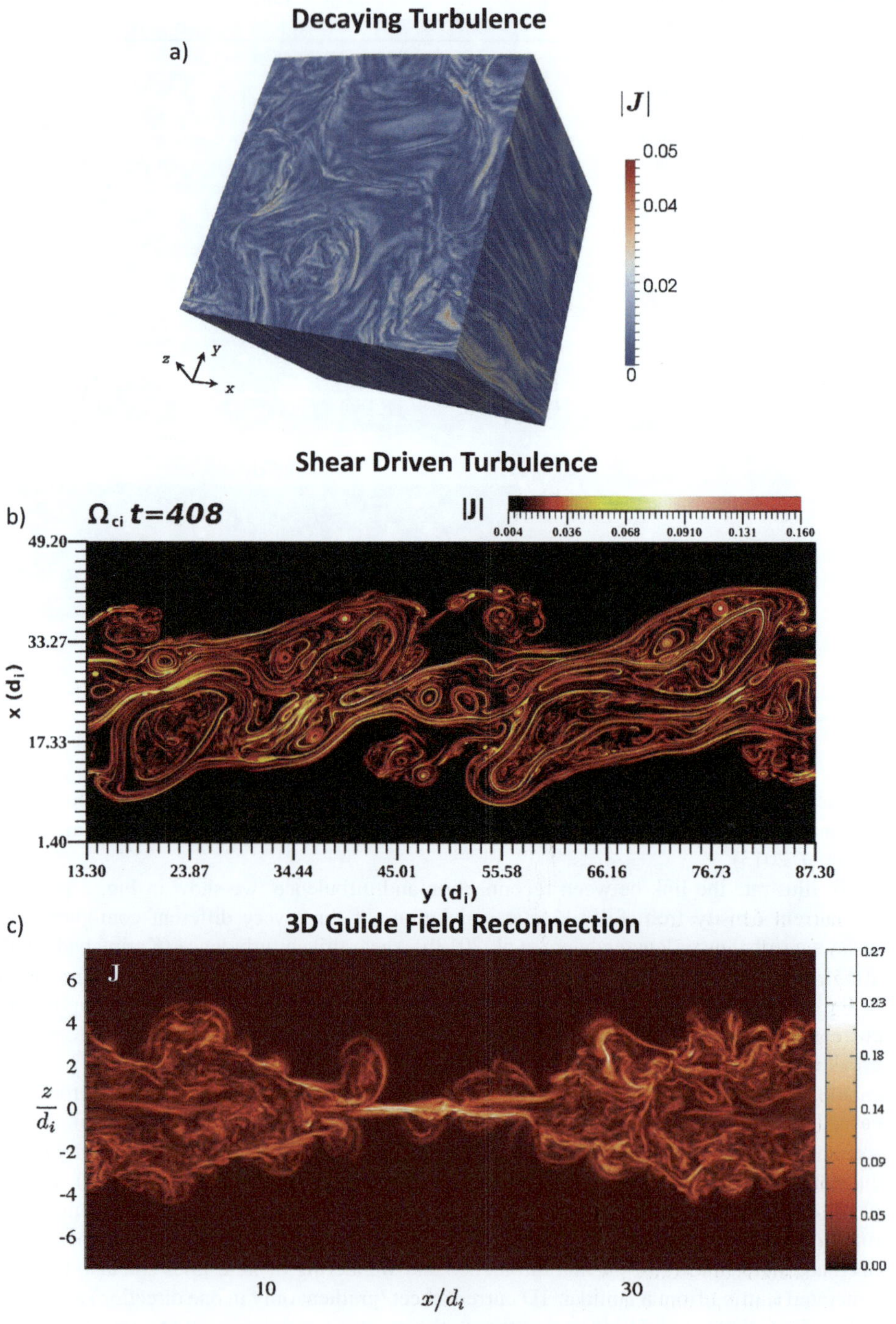

Fig. 5 Intensity plot of the current density from simulations of (**a**) decaying turbulence (Roytershteyn et al. 2013b), (**b**) shear driven turbulence (Karimabadi et al. 2013), and (**c**) guide field reconnection of $B_g = 4B_o$ (Liu et al. 2013)

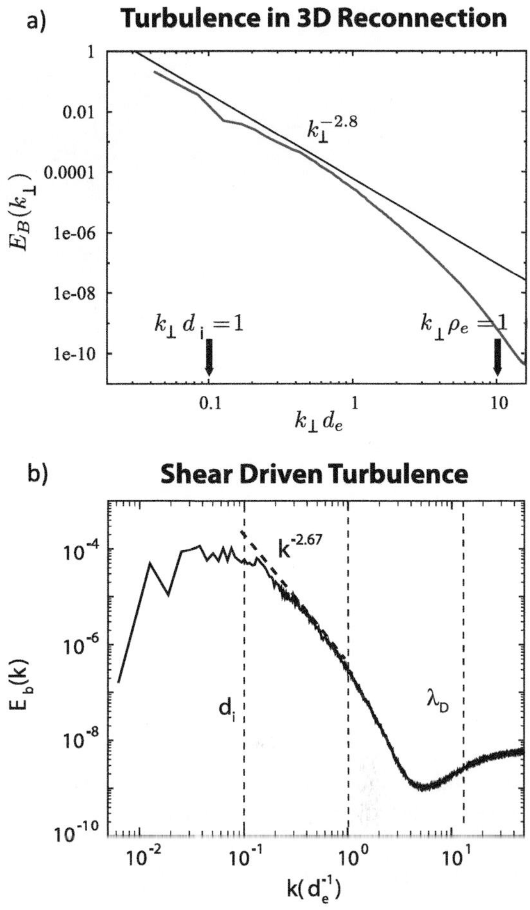

Fig. 6 Spectrum of magnetic field for (**a**) simulation shown in Fig. 5c and (**b**) a 2D shear driven turbulence (Karimabadi et al. 2013)

magnetic field lines and turbulence evolution. They referred to this mechanism as percolation model. Similar ideas for generating turbulence have been studied in fusion plasmas using resistive magnetohydrodynamics (MHD) (Carreras et al. 1980) and two-fluid models (Borgogno et al. 2005). More recently, this mechanism was explored within the collisionless regime using 3D fully kinetic simulations (Daughton et al. 2011). It was found that the 3D evolution is dominated by the formation and interaction of helical magnetic structures known as flux ropes. In contrast to previous theories, the majority of flux ropes are produced by secondary instabilities within the electron layers. New flux ropes spontaneously appear within these layers leading to a turbulent evolution where electron physics plays a central role. Another source of turbulence in these simulations is velocity shear in the exhaust (Leonardis et al. 2013).

Another scenario where reconnection and turbulence are interwined is where a large scale current sheet is embedded in pre-existing turbulence. An example is the magnetopause current layer which is embedded in the turbulent magnetosheath. However, there has been no studies of the effects of pre-existing turbulence of reconnection in the collisionless regime. In MHD, several analytical models have been developed (see review by Karimabadi and Lazarian 2013) that invoke turbulent broadening of the current layer to obtain fast reconnection rates. Since the reconnection rate is given by the aspect ratio in these models, the

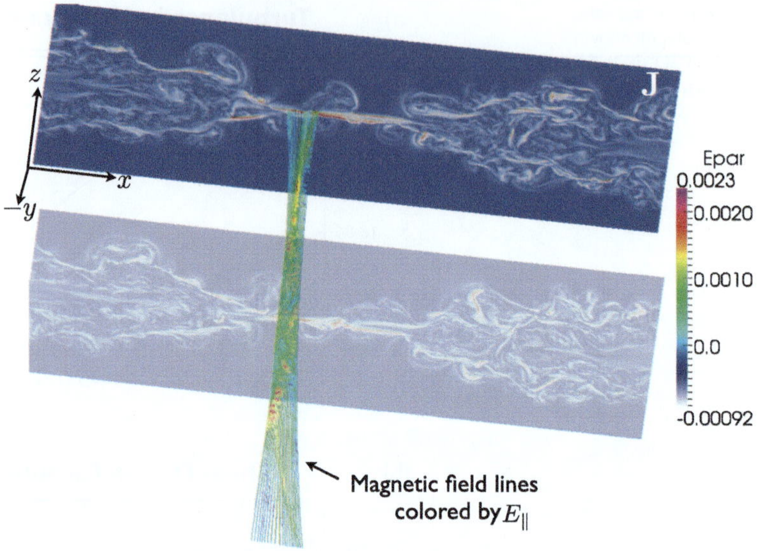

Fig. 7 Calculation of the reconnection rate in 3D. Two cuts of the current density is shown. Several seeds are placed in the central electron sheet and $E_{\|}$ is integrated along these magnetic field lines once through the system. The magnetic field lines are colored by $E_{\|}$. See Liu et al. (2013) for details of the simulation

degree of turbulent broadening has to be large enough to yield widths as large as a fraction of the length of the current sheet. Given lack of evidence for such turbulent broadening in the collisionless regime, the applicability of such models to the collisionless regime appears questionable and requires further study.

3.4 Does Turbulence Increase the Rate?

In steady state, reconnection rate is proportional to the aspect ratio. There are two ways to keep the aspect ratio within a fraction of unity: (a) broaden the thickness while keeping the length fixed, or (b) keep the thickness fixed and decrease the length. In resistive MHD, turbulence is required for fast reconnection, either in form of plasmoid formation which shortens the length of the layer or broadening of the sheet due to pre-existing turbulence. However, as we showed, there is no evidence so far that turbulence leads to significant broadening in the collisionless regime. So the question arises as to whether turbulence can increase the reconnection rate in the collisionless regime?

Recently Liu et al. (2013) calculated the reconnection rate from a 3D simulation of force free current sheet with a strong guide field (Fig. 4), using the procedure proposed by Hesse and Schindler (1988). Applying this approach is complicated by the fact that the magnetic field lines are chaotic and it is difficult to identify a transition back into an ideal region. Instead they seeded 20 points along the central electron sheet and integrated $E_{\|}$ along these magnetic field lines once through the system as shown in Fig. 7. Despite time averaging the fields over an interval $\sim 1/\omega_{ci}$, the parallel electric field has large fluctuations due to short-wavelength fluctuations, and the integrated value along neighboring field lines can vary by 50 %. Nevertheless, the average for these 20 field lines passing through the diffusion region current sheet is $\langle E_{\|}\rangle \sim 0.023 V_A B_o/c$ which is very close to the corresponding 2D simulation near the X-point $E_{\|} \sim 0.025 V_A B_o/c$. The fact that reconnection is already fast

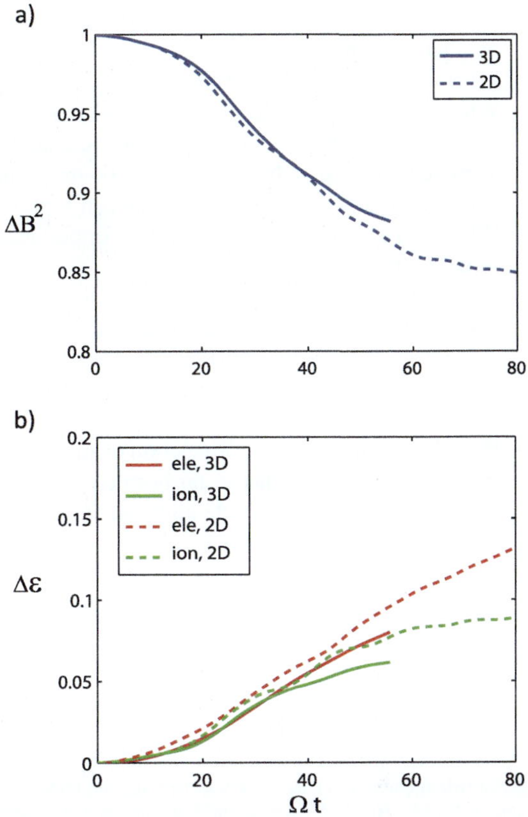

Fig. 8 Comparison of the change in the magnetic, ion and electron kinetic energies in 2D versus 3D simulations of force free current sheets

in the collisionless regime may be one contributing factor in lack of increase in the rate due to turbulence.

In addition to reconnection rate, it is also instructive to compare the change of magnetic and kinetic energy in 2D and 3D. This is shown in Fig. 8 for a similar run to that in Fig. 4 but for $B_g = 0.5B_o$. No appreciable differences in the change in the energy is observed in this case. Despite similar energetic in 2D and 3D, the partitioning of energy may be different in 3D. One interesting open question is whether the presence of turbulence can lead to a more efficient generation of energetic tails.

4 Summary and Outlook

These are exciting times in reconnection physics. Recent advances are revealing strong relationships between properties of reconnection in the different regimes and this has fostered greater collaboration across all areas of reconnection research. Although most of the studies are still limited to 2D, there is increasing shift towards 3D due to the advent of petascale computers. This is allowing tests of fundamental theories and concepts in 3D reconnection. There remain many important open questions, such as:

- What is the role of the different types of electron layers that are formed as a function of guide field in the macroscopic features of reconnection such as its rate, steadiness and energy conversion?

- What is the proper definition of the diffusion layer in systems beyond the idealized anti-parallel 2D configurations?
- What is the physics that localizes the diffusion region?
- What is the proper fluid closure model that would capture the basic elements of fully kinetic reconnection? For large magnetic islands, reconnection ensues when the islands approach each other but after a few bounces the reconnection can shut off. There are currently no fluid closure models that can capture this evolution.
- What is the relationship and relative importance of exponentiation, null, non-null and separator reconnection in 3D?
- What is the significance of magnetic nulls in a turbulent collisionless plasma such as the magnetosheath where there are many small scale fluctuations due to waves (e.g., Alfven ion cyclotron and mirror) that can create a very large number of nulls?
- What are the conditions for efficient generation of energetic ions and electrons in reconnection?
- How does properties of reconnection change as a function of plasma beta? Only a limited range on plasma beta has so far been considered. But the range of plasma beta can vary by many orders of magnitude in different environments where reconnection is operative.
- What is the effect of pre-existing turbulence on reconnection? This remains an open question even in the collisional regime. For example, in the turbulent reconnection model of Lazarian and Vishniac (1999) the width of the current sheet is controlled by the scale and strength of the driving turbulence. The resulting width is much wider than the Sweet-Parker thickness for most parameter regimes. It is not clear how this would affect the plasmoid instability and it may even suppress it since the plasmoid instability is based on the Sweet-Parker thickness.

Acknowledgements This work was partially supported by NSF through EAGER 1105084 and 1202152 and by NASA's NNH11CC65C grant. Simulations were performed on Kraken provided by the NSF at NICS, on Pleiades provided by NASA's HEC Program, and resources of the National Center for Computational Sciences at Oak Ridge National Laboratory (Jaguar/Lens), which is supported by DOE under Contract No. DE-AC05-00OR22725. Some of the research and simulations were part of the Blue Waters sustained-petascale computing project, which is supported by the NSF (OCI 07-25070) and the state of Illinois. Visualization and analysis were performed on Nautilus and Longhorn systems using ParaView and visualization software developed by the NICS RDAV group. We especially thank Burlen Loring for help with the visualization and preparation of the figures. We acknowledge useful conversations with A. Bhattacharjee, J. Birn, Paul Cassak, L. Chacon, M. Hesse, A. Lazarian, N. F. Loureiro, W. Matthaeus, S. Servidio, and J. Scudder. We thank A. Boozer for comments on an earlier draft, A. Le for discussions and preparation of one of the figures, and for the referee for many useful comments and suggestions that led to a significant improvement in the manuscript.

References

S. D. Bale, F. S. Mozer, T. D. Phan, Observation of lower hybrid drift instability in the diffusion region at a reconnecting magnetopause. Geophys. Rev. Lett. **29**, (2002)

J. Birn, J. Drake, M. Shay, B. Rogers, R. Denton, M. Hesse, M. Kuznetsova, Z. Ma, A. Bhattacharjee, A. Otto, P. Pritchett, Geospace environmental modeling (GEM) magnetic reconnection challenge. J. Geophys. Res. **106**, 3715 (2001)

D. Biskamp, *Magnetic Reconnection in Plasmas* (Cambridge University Press, Cambridge, 2005)

A.H. Boozer, Magnetic reconnection in space. Phys. Plasmas **19**, 092902 (2012)

D. Borgogno, D. Grasso, F.P.F. Califano, F. Pegoraro, D. Farina, Aspects of threedimensional magnetic reconnection. Phys. Plasmas **12**, 032309 (2005)

B. Carreras, H. Hicks, J. Holmes, B. Waddell, Nonlinear coupling of tearing modes with self-consistent resistivity evolution in tokamaks. Phys. Fluids **23**, 1811 (1980)

T. A. Carter, H. Ji, F. Trintchouk, M. Yamada, R.M.. Kulsrud, Measurement of lower-hybrid drift turbulence in a reconnecting current sheet. Phys. Rev. Lett. **88**, (2002)

P.A. Cassak, M.A. Shay, Magnetic reconnection for coronal conditions: reconnection rates, secondary islands and onset. Space Sci. Rev. **172**, 283 (2012)

L. Chacón, A.N. Simakov, A. Zocco, Steady-state properties of driven magnetic reconnection in 2D electron magnetohydrodynamics. Phys. Rev. Lett. **99**, 235001 (2007)

H. Che, J.F. Drake, M. Swisdak, A current filamentation mechanism for breaking magnetic field lines during reconnection. Nature **474**, 184 (2011)

W. Daughton, V. Roytershteyn, Emerging parameter space map of magnetic reconnection in collisional and kinetic regimes. Space Sci. Rev. **172**, 271 (2012)

W. Daughton, J. Scudder, H. Karimabadi, Fully kinetic simulations of undriven magnetic reconnection with open boundary conditions. Phys. Plasmas **13**, 072101 (2006)

W. Daughton, V. Roytershteyn, H. Karimabadi, L. Yin, B. Albright, B. Bergen, K. Bowers, Role of electron physics in the development of turbulent magnetic reconnection in collisionless plasmas. Nat. Phys. (2011). doi:10.1038/NPHYS1965

J.F. Drake, M. Swisdak, K.M. Schoeffler, B.N. Rogers, S. Kobayashi, Formation of secondary islands during magnetic reconnection. Geophys. Res. Lett. **33**, L13105 (2006)

J. P. Eastwood, T. D. Phan, S. D. Bale, A. Tjulin, Observations of turbulence generated by magnetic reconnection. Phys. Rev. Lett. **102** (2009)

J. Egedal, A. Le, W. Daughton, A review of pressure anisotropy caused by electron trapping in collisionless plasma, and its implications for magnetic reconnection. Phys. Plasmas **20**, 061201 (2013). doi:10.1063/1.4811092

W. Fox, M. Porkolab, J. Egedal, N. Katz, A. Le, Laboratory observations of electron energization and associated lower-hybrid and Trivelpiece-Gould wave turbulence during magnetic reconnection. Phys. Plasmas **17** (2010)

K. Fujimoto, Time evolution of the electron diffusion region and the reconnection rate in fully kinetic and large system. Phys. Plasmas **13**, 072904 (2006)

S. Fuselier, W. Lewis, Properties of near-Earth magnetic reconnection from in-situ observations. Space Sci. Rev. **160**, 55704 (2011)

A.A. Galeev, M.M. Kuznetsova, L.M. Zelenyi, Magnetopause stability threshold for patchy reconnection. Space Sci. Rev. **44**, 1 (1986)

M.V. Goldman, G. Lapenta, D.L. Newman, S. Markidis, H. Che, Jet deflection by very weak guide fields during magnetic reconnection. Phys. Rev. Lett. **107**, 135001 (2011)

J.T. Gosling, Magnetic reconnection in the solar wind: an update, in *Twelfth International Solar Wind Conference*. AIP Conference Proceedings, vol. 1216, ed. by M. Maksimovic, K. Issautier, N. Meyer-Vernet, M. Moncuquet, R. Pantellini, Melville, NY, USA (2010), p. 188

J.T. Gosling, Magnetic reconnection in the solar wind. Space Sci. Rev. **172**, 187 (2012)

M. Hesse, K. Schindler, A theoretical foundation of general magnetic reconnection. J. Geophys. Res. **93**, 5558 (1988)

M. Hesse, J. Birn, M. Kuznetsova, Collisionless magnetic reconnection: electron processes and transport modeling. J. Geophys. Res. **106**, 3721 (2001)

M. Hesse, T. Neukirch, K. Schindler, M. Kuznetsova, S. Zenitani, The diffusion region in collisionless magnetic reconnection. Space Sci. Rev. **160**, 3 (2011)

J.D. Huba, N.T. Gladd,K. Papadopoulos, The lower-hybrid-drift instability as a source of anomalous resistivity for magnetic field line reconnection. Geophys. Res. Lett. **4**, 125 (1977)

H. Ji, S. Terry, M. Yamada, R.M.. Kulsrud, A. Kuritsyn, Y. Ren, Electromagnetic fluctuations during fast reconnection in a laboratory plasma. Phys. Rev. Lett. **92**, 115001 (2004)

J.P. Jokipii, E.N. Parker, Random walk of magnetic lines of force in astrophysics. Phys. Rev. Lett. **21**, 44 (1968)

H. Karimabadi, A. Lazarian, Magnetic reconnection in the presence of externally driven and self-generated turbulence. Phys. Plasmas (2013 accepted)

H. Karimabadi, W. Daughton, J. Scudder, Multi-scale structure of the electron the electron diffusion region. Geophys. Res. Lett. **34**, 13104 (2007)

H. Karimabadi, V. Roytershteyn, M. Wan, W.H. Matthaeus, W. Daughton, P. Wu, M.A. Shay, B. Loring, J. Borovsky, E. Leonardis, S.C. Chapman, T.K.M. Nakamura, Coherent structures, intermittent turbulence, and dissipation in high-temperature plasmas. Phys. Plasmas **20**, 012303 (2013)

A. Klimas, M. Hesse, S. Zenitani, Particle-in-cell simulations of collisionless reconnection with open outflow boundaries. Phys. Plasmas **15**, 082102 (2008)

G. Lapenta, A. Lazarian, Achieving fast reconnection in resistive MHD models via turbulent means. Nonlinear Process. Geophys. **19**, 251 (2012)

A. Lazarian, E.T. Vishniac, Reconnection in a weakly stochastic field. Astrophys. J. **517**, 700 (1999)

A. Lazarian, L. Vlahos, G. Kowal, H. Yan, A. Beresnyak, E.M. de Gouveia Dal Pino, Turbulence, magnetic reconnection in turbulent fluids and energetic particle acceleration. Space Sci. Rev. **173**, 557 (2012)

A. Le, J. Egedal, W. Daughton, H. Karimabadi, O. Ohia, V.S. Lukin, Regimes of the electron diffusion region in magnetic reconnection. Phys. Rev. Lett. **110**, 135004 (2013)

E. Leonardis, S.C. Chapman, W. Daughton, V. Roytershteyn, H. Karimabadi, Identification of intermittent multifractal turbulence in fully kinetic simulations of magnetic reconnection. Phys. Rev. Lett. **110**, 205002 (2013)

Y.-H. Liu, W. Daughton, H. Karimabadi, H. Li,V. Roytershteyn, Bifurcated structure of the electron diffusion region in three-dimensional magnetic reconnection. Phys. Rev. Lett. **110**, 265004 (2013)

D.W. Longcope, H.R. Strauss, The form of ideal current layers in line-tied magnetic fields. Astrophys. J. **437**, 851 (1994)

M.E. Mandt, R.E. Denton, J.F. Drake, Transition to whistler mediated magnetic reconnection. Geophys. Res. Lett. **21**, 73 (1994)

W. Matthaeus, S. Lamkin, Rapid reconnection caused by finite amplitude fluctuations. Phys. Fluids **28**, 303 (1985)

W. Matthaeus, M. Velli, Who needs turbulence? A review of turbulence effects in the heliosphere and on the fundamental process of reconnection. Space Sci. Rev. **160**, 1 (2011)

F. S. Mozer, M. Wilber, J. F. Drake, Wave associated anomalous drag during magnetic field reconnection. Phys. Plasmas **18**, 102902 (2011a)

F.S. Mozer, D. Sundkvist, J.P. McFadden, P.L. Pritchett, I. Roth, Satellite observations of plasma physics near the magnetic field reconnection X line. J. Geophys. Res. **116**, A12224 (2011b)

E.N. Parker, Sweet's mechansm for merging magnetic fields in conducting fluids. J. Geophys. Res. **62**, 509 (1957)

E.N. Parker, Topological dissipation and the small-scale fields in turbulent gases. Astrophys. J. **174**, 499 (1972)

G. Paschmann, Recent in-situ observations of magnetic reconnection in near-Earth space. Geophys. Res. Lett. **35**, L19109 (2008)

G. Paschmann, M. Øieroset,T. Phan, In-situ observations of reconnection in space. Space Sci. Rev. (2013). doi:10.1007/s11214-012-9957-2

H. Petschek, Magnetic field annihilation, in *AAS-NASA Symposium on the Physics of Solar Flares*, Washington, DC, ed. by W. Hess (1964), pp. 425–439

T.D. Phan, J.F. Drake, M.A. Shay, F.S. Mozer, J.P. Eastwood, Evidence for an elongated (>60 ion skin depths) electron diffusion region during fast magnetic reconnection. Phys. Rev. Lett. **99**, 255002 (2007)

D.I. Pontin, Three-dimensional magnetic reconnection regimes: a review. Adv. Space Res. **47**, 1508 (2011)

D.I. Pontin, A.L. Wilmot-Smith, G. Hornig, K. Galsgaard, Dynamics of braided coronal loops, II: cascade to multiple small-scale reconnection events. Astron. Astrophys. **525**, A57 (2011)

E.R. Priest, P. Demoulin, Three-dimensional magnetic reconnection without null points, I: basic theory of magnetic flipping. J. Geophys. Res. **100**, 23443 (1995)

E. Priest, T. Forbes, *Magnetic Reconnection* (Cambridge University Press, Cambridge, 2000)

P. Pritchett, Geospace environmental modeling magnetic reconnection challenge: simulations with a full particle electromagnetic code. J. Geophys. Res. **106**, 3783 (2001)

P. L. Pritchett, F. S. Mozer, M. Wilber, Intense perpendicular electric fields associated with three-dimensional magnetic reconnection at the subsolar magnetopause. J. Geophys. Res. **117**, A06212 (2012)

A.B. Rechester, M.N. Rosenbluth, Electron heat transport in a Tokamak with destroyed magnetic surfaces. Phys. Rev. Lett. **40**, 38 (1978)

V. Roytershteyn, W. Daughton, L. Yin, B. Albright, K. Bowers, S. Dorfman, Y. Ren, H. Ji, M. Yamada, H. Karimabadi, Driven magnetic reconnection near the dreicer limit. Phys. Plasmas **17**, 055706 (2010)

V. Roytershteyn, W. Daughton, H. Karimabadi, F.S. Mozer, Influence of the lower-hybrid drift instability on magnetic reconnection in asymmetric configurations. Phys. Rev. Lett. **108**, 185001 (2012)

V. Roytershteyn, S. Dorfman, W. Daughton, H. Ji, M. Yamada, H. Karimabadi, Electromagnetic instability of thin reconnection layers: comparison of 3D simulations with MRX observations. Phys. Plasmas **20**, 061212 (2013a)

V. Roytershteyn, H. Karimabadi, et al., Coherent structures and intermittent dissipation in 3D fully kinetic simulation of collisionless plasma turbulence Phys. Rev. Lett. (2013b in preparation)

J.D. Scudder, R.D. Holdaway, W. Daughton, H. Karimabadi, V. Roytershteyn, C.T. Russell, J.Y. Lopez, First resolved observations of the demagnetized electron-diffusion region of an astrophysical magnetic-reconnection site. Phys. Rev. Lett. **108**, 225005 (2012)

S. Servidio, P. Dmitruk, A. Greco, M. Wan, S. Donato, P.A. Cassak, M.A. Shay, V. Carbone, W. Matthaus, Magnetic reconnection as an element of turbulence. Nonlinear Process. Geophys. **18**, 675 (2011)

M. Shay, J. Drake, B. Rogers, R. Denton, Alfvénic collisionless magnetic reconnection and the Hall term. J. Geophys. Res. **106**, 3759 (2001)

M. Shay, J. Drake, M. Swisdak, Two-scale structure of the electron dissipation region during collisionless magnetic reconnection. Phys. Rev. Lett. **99**, 155002 (2007)

A.N. Simakov, L. Chacón, Quantitative, comprehensive, analytical model for magnetic reconnection in hall magnetohydrodynamics. Phys. Rev. Lett. **101**, 105003 (2008)

B.U.Ö. Sonnerup, Magnetic field reconnection, in *Solar System Plasma Processes*, vol. 1320, ed. by L.J. Lanzerotti, C.F. Kennel, E.N. Parker (North-Holland, Washington, 1979), p. 45

B.P. Sullivan, A. Bhattacharjee, Y.-M. Huang, Extension of the electron dissipation region in collisionless Hall magnetohydrodynamics reconnection. Phys. Plasmas **16**, 102111 (2009)

P.A. Sweet, The neutral point theory of solar flares, in *Electromagnetic Phenomena in Cosmical Physics*. IAU Symp., vol. 1320, ed. by B. Lehnert (Cambridge University Press, New York, 1958), p. 123

D.A. Uzdensky, Magnetic reconnection in extreme astrophysical environments. Space Sci. Rev. **160**, 45 (2011)

M. Wan, W.H. Matthaeus, H. Karimabadi, V. Roytershteyn, M.A. Shay, P. Wu, W. Daughton, B. Loring, S.C. Chapman, Intermittent dissipation at kinetic scales in collisionless plasma turbulence. Phys. Rev. Lett. **109**, 195001 (2012)

P. Wu, S. Perri, K. Osman, M. Wan, W.H. Matthaues, M.A. Shay, M.L. Goldstein, H. Karimabadi, S.C. Chapman, Intemittent heating is solar wind and kinetic simulations. Astrophys. J. Lett. **763**, L30 (2013)

M. Yamada, R. Kulsrud, H. Ji, Magnetic reconnection. Rev. Mod. Phys. **82**, 603 (2010)

A.R. Yeates, G. Hornig, A generalized flux function for three-dimensional magnetic reconnection. Phys. Plasmas **19**, 102118 (2011)

A. Zocco, L. Chacón, A.N. Simakov, Current sheet bifurcation and collapse in electron magnetohydrodynamics. Phys. Plasmas **16**, 110703 (2009)

E.G. Zweibel, M. Yamada, Magnetic reconnection in astrophysical and laboratory plasmas. Annu. Rev. Astron. Astrophys. **47**, 291 (2009)

DOI 10.1007/978-1-4899-7413-6_10
Reprinted from *Space Science Reviews* Journal, DOI 10.1007/s11214-013-0022-6

Notes on Magnetohydrodynamics of Magnetic Reconnection in Turbulent Media

Philippa Browning · Alex Lazarian

Received: 18 March 2013 / Accepted: 12 September 2013 / Published online: 1 October 2013

Abstract Astrophysical fluids have very large Reynolds numbers and therefore turbulence is their natural state. Magnetic reconnection is an important process in many astrophysical plasmas, which allows restructuring of magnetic fields and conversion of stored magnetic energy into heat and kinetic energy. Turbulence is known to dramatically change different transport processes and therefore it is not unexpected that turbulence can alter the dynamics of magnetic field lines within the reconnection process. We shall review the interaction between turbulence and reconnection at different scales, showing how a state of turbulent reconnection is natural in astrophysical plasmas, with implications for a range of phenomena across astrophysics. We consider the process of magnetic reconnection that is fast in magnetohydrodynamic (MHD) limit and discuss how turbulence—both externally driven and generated in the reconnecting system—can make reconnection independent on the microphysical properties of plasmas. We will also show how relaxation theory can be used to calculate the energy dissipated in turbulent reconnecting fields. As well as heating the plasma, the energy dissipated by turbulent reconnection may cause acceleration of non-thermal particles, which is briefly discussed here.

Keywords Magnetic reconnection · Turbulence · Magnetohydrodynamics · Cosmic plasma · Fast reconnection · Solar flares · Dynamos

1 Introduction

Magnetic fields are observed in many astrophysical objects and it is accepted that these fields play an essential role in the dynamics (e.g. Crutcher 1999; Beck 2002; Vallée 1997, 1998). Magnetic fields are vital for magneto-rotational instability, transport and acceleration of cosmic rays and other energetic particles, accretion of matter, and activity in stellar atmospheres

P. Browning (✉)
Jodrell Bank Centre for Astrophysics, University of Manchester, Manchester, UK
e-mail: p.browning@manchester.ac.uk

A. Lazarian
Dept of Astronomy, University of Wisconsin, Madison, WI, USA

(e.g. Balbus and Hawley 1998; Parker 1992; Schlickeiser and Lerche 1985; Melrose 2009; Elmegreen and Scalo 2004; Kotera and Olinto 2011). Crucially, magnetic fields provide a means to store energy as well as to transport it.

For laminar fluids in the limit of zero resistivity, the topology of the field lines is a constant of motion and the magnetic flux threading any fluid element is constant. This is the basis of the textbook notion that in the limit of very small resistivity, which is typical for astrophysical objects, the magnetic flux is "frozen in" and magnetic field lines resist passing through one another or changing their topology (Moffat 1978).[1]

At the same time, changes of magnetic field topology are necessary for the generation of magnetic fields by dynamos, and for many other phenomena. Magnetic reconnection is a process that is responsible for topology changes and the annihilation of magnetic field on faster time-scales than the (usually) extremely slow, process of global resistive diffusion. It is now well-established that magnetic reconnection is a ubiquitous process across the universe, with important consequences for restructuring of magnetic fields and efficient release of stored magnetic energy, such as in solar and stellar flares, coronal heating and planetary magnetospheres; for reviews, see Biskamp (1996, 2000), Priest and Forbes (2000), Yamada (2007), Zweibel and Yamada (2009), Yamada et al. (2011). Furthermore, reconnection is common in laboratory plasmas, both in magnetically-confined fusion devices (e.g. Ono et al. 2012; Stanier et al. 2013) and in experiments specifically devised to study reconnection (e.g. Yamada et al. 1990; Brown 1999; Yamada 2007). Whilst reconnection theory originated mainly through 2D analytical models (Parker 1957; Sweet 1958; Furth et al. 1963; Petschek 1964; Hahm and Kulsrud 1985), recent research on astrophysical reconnection has increasingly emphasised the role of 3D geometries and complex topologies. One strand of such complexity is a close association between magnetic reconnection and turbulence, which is the focus of this paper.

Turbulence is an almost universal process in astrophysical plasmas; as well as being widely observed in laboratory plasmas, and associated there with the important and yet ill-understood phenomenon of anomalous transport. For instance, in the interstellar medium, supernovae explosions are thought to be the source of ubiquitously observed turbulent motions. It is generally accepted that the "Big Power Law in the Sky" (see Fig. 1) indicates the presence of turbulence on scales from tens of parsecs to thousands of kilometres (Armstrong et al. 1995; Chepurnov and Lazarian 2010). Among other sources, evidence for this comes from studies of atomic hydrogen spectra in molecular clouds and galaxies (Lazarian and Pogosyan 1999; Stanimirović and Lazarian 2001; Padoan et al. 2006, 2009; Chepurnov et al. 2010), see also the review by Lazarian (2009) and references therein; as well as recent studies of emission lines and Faraday rotation (see Burkhart et al. 2010; Gaensler et al. 2011). Much work over recent years has been devoted to understanding fluid turbulence in the presence of a magnetic field (e.g. Zank and Matthaeus 1993; Lazarian and Cho 2005).

In fact, the phenomena of reconnection and turbulence in magnetised plasmas, traditionally viewed as separate problems, should be studied together. This is inevitably a vast subject, and in this review we touch only on some aspects of this, focussing only on magnetohydrodynamic (MHD) models. Recently, there has also been much emphasis on the interaction between turbulence and reconnection in collisionless reconnection (e.g. Daughton et al. 2011; Karimabadi et al. 2013) and Hall-MHD (e.g. Dmitruk and Matthaeus 2006), but here we consider only fluid models. The many interesting recent developments in kinetic models of turbulence and reconnection are discussed elsewhere in this volume. The interaction between turbulence and reconnection within MHD are also reviewed by Lazarian et al.

[1] As we discuss later in Sect. 3.4, the very notion of frozen-in magnetic fields requires serious revisions.

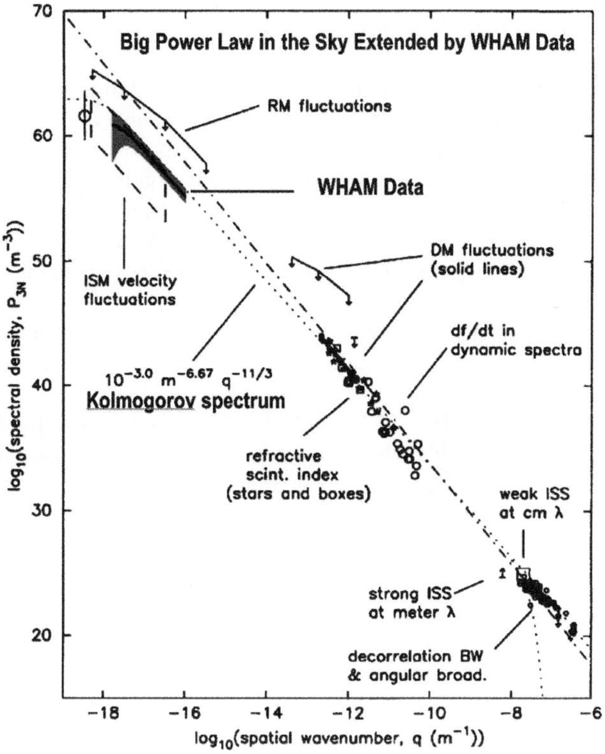

Fig. 1 Turbulence in the interstellar gas as revealed by electron density fluctuations. "Big Power Law in the Sky" in extended using WHAM data (Armstrong et al. 1995). The slope corresponds to that of Kolmogorov turbulence (see discussion in Armstrong et al. (1995)). Modified from Chepurnov and Lazarian (2010)

(2012b), focused more on a review of MHD turbulence (which is not covered here) and on cosmic ray acceleration. A complementary review is provided by Servidio et al. (2011), considering the role of reconnecting current sheets within 2D turbulence, in both MHD and Hall-MHD frameworks.

Reconnection and turbulence may interact in a number of ways. Firstly, a realisation that flows are turbulent on fluid scales can substantially modify the dynamics of reconnection and lead to *fast reconnection*, as described in the following section and Sect. 3. Secondly, turbulence on kinetic scales is also ubiquitous, and these microphysical processes interact strongly with the global (fluid scale) dynamics (e.g. Karimabadi et al. 2013). Whilst discussion of kinetic processes is beyond the scope of this paper, we discuss synergies between the MHD and plasma-based approaches to reconnection in Sect. 4, and some effects of microphysics on fluid processes are mentioned in Sect. 5. Thirdly, within a single reconnecting current sheet, there is natural tendency for fragmentation into a series of plasmoids, leading again to a turbulent scenario, as discussed briefly in Sect. 5. Fourthly, the dynamics of a magnetised plasma involving instabilities and loss of equilibrium naturally lead to fine structure and turbulence, as is discussed in Sect. 6.1—thus reconnection tends to occur in a multiplicity of localised current sheets (or similar structures), with turbulence naturally arising. Reconnection in such complex fields is usually 3D rather than 2D, see Sect. 6.2. Finally, an important consequence of turbulent reconnection in cosmic plasmas is the efficient dissipation of stored magnetic energy—leading both to plasma heating and the energisation of charged particles. An approach to the former, based on the idea of relaxation to a minimum energy state, is outlined in Sect. 7, whilst the latter is reviewed in Sect. 8.

2 "Classical" Reconnection Models and the Quest for Fast Reconnection

The first analytic model for magnetic reconnection was proposed independently by Parker (1957) and Sweet (1958). *Sweet-Parker reconnection* has the virtue that it relies on a robust and straightforward geometry. Two regions with uniform magnetic fields are separated by thin current sheet. The speed of reconnection is given roughly by the resistivity divided by the sheet thickness, i.e. $V_{rec} \approx \eta/\Delta$, where η is the resistivity. This suggests that the reconnection can be very fast for small Δ and by decreasing Δ, one can make reconnection as fast as may be required. This is not true, however. Indeed, the conducting plasma in the current sheet is constrained to move along the local field lines, and is ejected from the edge of the current sheet at the Alfvén speed, V_A. Since the width of the current sheet, Δ, limits the flux of expelled fluid, Δ should be made as large as possible to enable faster reconnection. This results, for steady-state reconnection, in a compromise, with the overall reconnection speed reduced from the Alfvén speed by the square root of the Lundquist number, $S \equiv LV_A/\eta$, where L is the length of the current sheet. In most astrophysical contexts, S is very large and the Sweet-Parker reconnection speed, $V_{SP} \approx V_A S^{-1/2}$, is very small. Fast reconnection requires that the reconnection rate be independent of η or depend on S logarithmically.

Another early paradigm for reconnection, which treats the transient development of reconnection rather than assuming a steady-state, is the *tearing instability* (Furth et al. 1963). A magnetic field with a reversal—or more generally, a sheared field—may be linearly-unstable to small perturbations, forming a chain of growing magnetic islands at the reversal surface (or resonant surface, where $\mathbf{k} \cdot \mathbf{B} = 0$). However, the growth rate scales with the geometric mean of the Alfven and diffusive times, which again is too slow to explain phenomena such as solar flares.

A further analytical model for reconnection is *forced reconnection* (Hahm and Kulsrud 1985), in which an external disturbance at the boundary triggers reconnection and energy-release, in a field which may be tearing-mode stable. In the original model, a sinusoidal disturbance is applied at the boundary of a field-reversal in a slab, leading initially to the formation of a discontinuous current sheet which subsequently relaxes through reconnection into a lower-energy state with a chain of magnetic islands. This model can be generalised to a sheared force-free field (Vekstein and Jain 1998), with implications for solar coronal heating. However, again the time-scale for energy-release is slow.

The realization that Sweet-Parker reconnection—and related time-dependent models such a linear tearing mode—are inadequate to explain magnetic reconnection for solar flares was immediately apparent, and this gave rise to much research on models of fast reconnection—see reviews Biskamp (2000), Priest and Forbes (2000)—that used different outflow conditions, avoiding the contradictory requirements on Δ of the Sweet-Parker model. The first proposal was to replace the long current sheet with an X-point configuration, so that the "sheet" thickness and length are comparable. The magnetic field lines diverge from the "point" of reconnection, forming an X-type structure over the scale of the system. This is the basis for the ingenious Petschek's model of fast reconnection (Petschek 1964). The stability and the conditions for emergence of such structures were an issue for extensive research over the years that followed.

It is easy to see that self-consistent X-point reconnection requires that the outflow prevents a general collapse into a narrow current sheet. Otherwise we would expect that the same bulk forces that brought the magnetic field lines together would lead to Sweet-Parker reconnection, which corresponds to the collapse of the X-point to an extended Y-sheet. Petschek (1964) proposed that slow-mode shocks on either side of the X-point would serve this purpose. Moreover, these shocks are responsible for converting most of the magnetic energy into kinetic energy. The X-point in this model has an overall size which depends on resistivity, but since the magnetic field decreases logarithmically when approaching the current sheet (due to the assumption of the current-free magnetic field in the inflow region), the resulting reconnection speed depends on $\ln S$ and can be an appreciable fraction of V_A. Numerical simulations with uniform resistivity (Biskamp 1996) have showed that in the MHD limit, the shocks fade away and the contact region evolves into Sweet-Parker reconnection. The suggested way to make the Petschek configuration stable was by introducing a local non-uniform resistivity (Parker 1973; Ugai and Tsuda 1977; Scholer 1989; Ugai 1992; Yan et al. 1992; Forbes 2001; Shibata and Magara 2011), which remained the favourite way of accounting for fast reconnection for many years.

Plasma effects, such as anomalous resistivity (resistivity that depends on the current), formally present the best bet for stabilising the reconnecting X-point and attaining Petschek reconnection: see Sect. 5. However, there were several important issues that remained unresolved. First, it is not clear that this kind of fast reconnection persists on scales greater than the ion inertial scale (see Bhattacharjee et al. 2003). Several numerical studies (Wang et al. 2001; Smith et al. 2004; Fitzpatrick 2004) have found large-scale reconnection speeds which depend on resistivity: hence, these are not true "fast reconnection". Second, in many circumstances the magnetic field geometry does not allow the formation of X-point reconnection. For example, a saddle-shaped current sheet—the generic configuration of fluxes pressing against each other when one flux tube partially engulfs the other as they pulled in opposite directions—cannot be spontaneously replaced by an X-point. The energy required to do so is comparable to the magnetic energy liberated by reconnection, and must be available beforehand. Finally, the requirement that reconnection occurs only in collisionless plasmas is very restrictive. Many astrophysical fluids are collisional. For example, while reconnection in stellar coronae can be collisionless (at least, on the "dissipation" length-scale), stellar chromospheres are collisional. Magnetic reconnection does happen in both regions.[2] Different phases of the interstellar medium are also collisional and, while magnetic reconnection is more difficult to observe in those environments, there is an indirect way to infer fast magnetic reconnection there (Lazarian et al. 2012b).

As an example, we mention that Yamada (2007) estimated that the scale of the reconnecting current sheet should not exceed about 40 times the electron mean-free-path. This condition is not satisfied in many environments, e.g. the InterStellar Medium (ISM). The conclusion that stellar interiors and atmospheres, accretion disks, and the ISM does not allow fast reconnection is drastic and unpalatable. On the other hand, an intriguing possibility is that in some environments, such as the solar corona, a local collisionless state may be maintained, allowing fast reconnection (Uzdensky 2007).

Is there a way to make magnetic reconnection independent of resistivity within the MHD framework? Turbulence may be the primary suspect, as most of astrophysical environments are observed to be turbulent, as discussed in Sect. 1. We now discuss how turbulence may resolve the fast reconnection problem.

3 The Role of Turbulence in Fast Reconnection

Turbulence is known to accelerate diffusion processes, making them independent of the microphysical parameters. This reasoning was behind the ill-founded concept of turbulent

[2] In some of these environments, collisions with neutral gas may dominate over Coulomb collisions, strongly affecting the reconnection process, so the distinction between "collisionless" and "collisional" may be an over-simplification. We do not discuss this interesting issue further here.

diffusivity within models of the kinematic dynamo (Moffat 1978). There, in direct analogy with the turbulent diffusivity of a passive scalar, it was assumed that magnetic fields can be mixed by turbulent motions up to the resistive scale. Naturally, in such circumstances, the magnetic reconnection problem becomes trivial.

The problem with the "magnetic turbulent diffusion" idea is that realistic astrophysical magnetic fields are important and therefore the small-scale magnetic mixing of oppositely-directed magnetic fields is not applicable. In fact, the dynamical importance of magnetic fields and the inability of the turbulence to bend them appreciably on the resistive scales must be taken into account from the very early stages of the action of astrophysical dynamo.

Thus, in addressing the role of turbulence in astrophysical reconnection, one should consider dynamically important magnetic fields. Lazarian and Vishniac (1999) [hereafter LV99] proposed a 3D model for fast reconnection which depends on the presence of turbulence, and magnetic field wandering.

We must note that the idea that turbulence and fieldline wandering can enhance the reconnection rate has a long history. For instance, Speiser (1970) considered the effects of turbulence on microscopic resistivity—further discussed in Sect. 5. The "Tangled Discharge Model" was devised to explain relaxation in a Reversed Field Pinch and other fusion devices through the existence of stochastic fields subject to turbulent reconnection (Rusbridge 1977, 1991). Similarly, Jacobson and Moses (1984) proposed that the current diffusivity should be modified to include the diffusion of electrons across the large scale magnetic field due to the small-scale field line stochasticity. These models involving field line wandering are also closely linked to the idea of relaxation to a minimum energy state, see Sect. 7 below, and provide a very useful picture of the interactions between turbulence and reconnection; but they do not solve the fast reconnection problem. Further models considering the effects of 2D turbulence on reconnection are discussed in Sect. 3.3 below. We now proceed to show how 3D fluid turbulence may provide a solution to the fast reconnection problem.

3.1 The LV99 Model

LV99's model (see Fig. 2) uses the properties of turbulence to predict broad outflows from extended current sheets. The diffusivity of magnetic field line trajectories in a turbulent plasma implies that flows can follow local magnetic field lines without being confined to the current sheet. When the turbulent diffusivity is less than the ohmic resistivity, this model reduces to the Sweet-Parker reconnection model.

Let us consider the differences between the Sweet-Parker model of laminar reconnection (Sect. 2) and LV99 which accounts for turbulence. The latter can be seen as a generalization of the Sweet-Parker model (see Fig. 2) in the sense that the two regions of differing magnetic directions are pressed up against one another over a broad contact region. This is a generic configuration, which should arise naturally whenever a magnetic field has a non-trivial configuration, whose energy could be lowered through reconnection. The outflow of plasma and reconnected flux will fluctuate as the turbulence evolves and the field line connections change, but the long term average will reflect the turbulent diffusion of the field lines. Consequently, the essential difference between the Sweet-Parker model and the LV99 model is that in the former, the outflow is limited by microphysical Ohmic diffusivity; while in LV99, the large-scale magnetic field wandering determines the thickness of outflow. The latter depends only on the turbulence properties and is independent of resistivity. This ensures that the LV99 reconnection is fast.

For extremely weak turbulence, when the value of Δ arising from magnetic field wandering becomes smaller than the width of the Sweet-Parker layer $LS^{-1/2}$, the two models

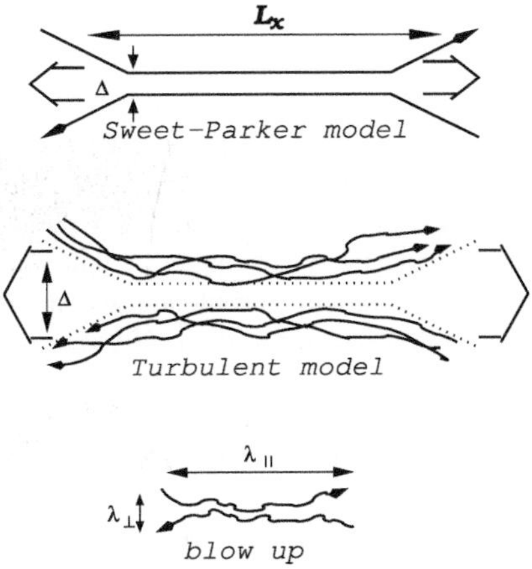

Fig. 2 *Upper plot*: The Sweet-Parker reconnection model. The outflow is confined to a thin layer of width Δ, which is set by Ohmic diffusivity. The length of the current sheet is a macroscopic scale $L \gg \Delta$. Magnetic field lines are assumed to be laminar. *Middle plot*: Reconnection in the presence of stochastic magnetic field lines. The stochasticity introduced by turbulence is weak and the mean field provides a clear direction. The outflow width is set by the diffusion of the magnetic field lines, which is a macroscopic process, independent of resistivity. *Lower plot*: An individual small-scale reconnection region. The reconnection over small patches of magnetic field determines the local reconnection rate. The global reconnection rate is substantially larger as many independent patches reconnect simultaneously. Conservatively, the LV99 model assumes that the small scale events happen at a slow Sweet-Parker rate. From Lazarian et al. (2004)

are indistinguishable and the reconnection proceeds at the Sweet-Parker rate. Similarly, we expect that when the turbulence-induced Δ is much larger than the Larmor radius of ions, then the plasma effects should not be important.

With the Goldreich and Sridhar (1995) model of turbulence, LV99 obtained:

$$V_{rec} = V_A \left(\frac{l}{L}\right)^{1/2} \left(\frac{v_l}{V_A}\right)^2, \tag{1}$$

where l and v_l are the energy injection scale and velocity at the injection scale, respectively. This expression assumes that energy is injected isotropically at a scale l smaller than the length of the current sheet L, which for sub-Alfvénic turbulence leads to the generation of weakly-interacting waves at that scale. The waves transfer energy to modes with larger values of $k_\perp$ until at a scale $l_{trans} = LM_A^2$, where M_A is the Alfven Mach number, strong turbulence sets in Lazarian (2006). It is important to note that the strongly turbulent eddies have a characteristic velocity of $v_{turb} \approx V_A(v_l/V_A)^2$. In other words, the reconnection speed is the large-eddy, strong-turbulent velocity multiplied factors which depend on whether the current sheet length is smaller or larger than the large eddies (whose length is approximately the injection scale). In this sense, the reconnection speed is insensitive to the exact mechanism for turbulent power injection.

It is important to note three features of Eq. (1). First, and most important, it is independent of resistivity. This is, by definition, fast reconnection. Second, we usually expect the reconnection speed to be close to the turbulent eddy speed; the geometric ratios that enter the expression, i.e. the injection scale l divided by the length of the reconnection layer L, are typically of order unity. Reconnection will occur on dynamical time scales. Finally, we note that in particular situations when the turbulence is extremely weak the reconnection speed can be much slower than the Alfvén speed. Strong magnetic field prior to reconnection presents such a case, which ensures that magnetic flux of opposite polarities can be accumulated prior to reconnection in solar flares (see LV99, Lazarian and Vishniac 2009).

More recently, Eq. (1) was derived using ideas based on the well-known concept of Richardson diffusion (Eyink et al. 2011). From the theoretical perspective this new derivation avoids rather complex considerations of the cascade of reconnection events that were presented in LV99. Eyink et al. (2011) also show that the LV99 model is closely connected with the idea of "spontaneous stochasticity" of magnetic fields in turbulent fluids.

The deep connection between magnetic turbulence and magnetic reconnection is thus evident. LV99 showed that the Goldreich and Sridhar (1995) model, which envisages mixing motions perpendicular to magnetic field lines, becomes self-consistent in the presence of magnetic reconnection given by Eq. (1). Eyink et al. (2011) showed that the established Lagrangian properties of 3D MHD turbulence require the LV99 reconnection with given by Eq. (1).

We note that a theoretical model of turbulent reconnection was suggested recently by Gao et al. (2012), based similarly to LV99 on the Goldreich and Sridhar (1995) theory of Alfvenic turbulence. Unlike LV99 theory, these authors apply a mean-field approach to the problem of magnetic reconnection and utilize the concept of hyper-resistivity, which has some difficulties (e.g. Eyink et al. 2011). It was further argued in Eyink et al. (2011) that any mean-field proof of fast reconnection is not tenable unless the reconnection rates obtained are strictly independent of the length-scales and time-scales of averaging.

3.2 Numerical Testing of LV99 Model

The first test of the LV99 model using three-dimensional (3D) simulations was performed in Kowal et al. (2009). LV99 is a model formulated in the MHD regime. Thus a set of MHD equations was solved in Kowal et al. (2009, 2012b). The boundary conditions and the manner of energy injection were varied between different simulations. In order to avoid the complications of strong compressibility, high-beta simulations were used.

For numerical simulations it is easier to control not v_l, but the energy injection power P. For sub-Alfenic injection the power in the turbulent cascade is $P \sim v_{turb}^2(V_A/l)$ or $v_l^4/(lV_A)$. The amount of energy injected during one Alfvén time unit t_A, which is constant in our models, is $t_A P \sim (L/V_A)v_l^4/(lV_A)$. Therefore $v_l^2 \sim (l/L)^{1/2}(Pt_A)^{1/2}V_A$. Substituting v_l^2 in Eq. (1) results in

$$V_{rec} \sim \left(\frac{l}{L}\right)(t_A P)^{1/2} \propto l\, P^{1/2}, \tag{2}$$

which is the prediction that was tested in the numerical studies. In what follows we refer to the injection power and scale using P_{inj} and l_{inj}, respectively.

The setup of the reconnection simulations is illustrated in Fig. 3, which is a 2D cut, indicating the location of the diffusion region. The top and bottom of the computational domain contain equal and opposite field components in the $\hat{x}$ direction, as well as a sheared component B_z (see the left panel of Fig. 3). Magnetic field lines enter through the top and bottom and are bent by the inflow V_{in} as they move into the diffusion region. The diffusion region has a length Δ in the $\hat{x}$ direction and a thickness δ in the $\hat{y}$ direction (see the left panel of Fig. 3).

The numerical box is periodic in the $\hat{z}$ direction and the diffusion region extends through the entire domain. The projection of the magnetic topology on the xz plane shows that the fieldlines in the upper region (solid lines in the right panel of Fig. 3) and in the lower region (dashed lines) are offset by an angle α determined by the strength of the sheared component B_{0z}. Once the incoming magnetic lines enter the diffusion region, they are reconnected

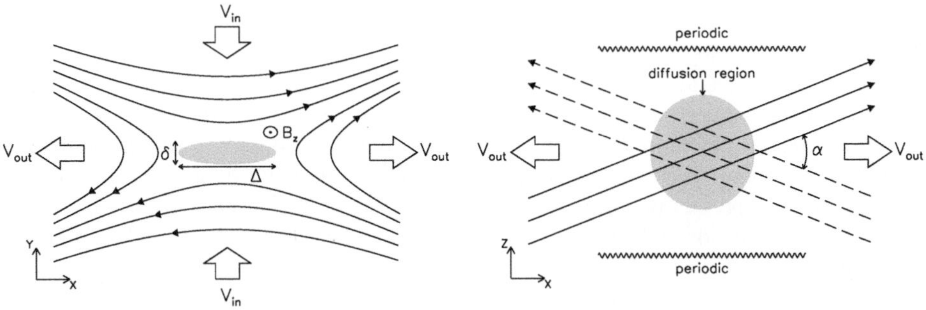

Fig. 3 A schematic of magnetic field configuration projected on the xy (*left*) and xz (*right*) planes. *Left:* xy projection of the magnetic field lines. The *gray area* describes the diffusion region where the incoming field lines reconnect. The longitudinal and transverse scales of the diffusion region are given by Δ and δ, respectively. We use outflow and inflow boundary conditions in the $\hat{x}$ and $\hat{y}$ directions, respectively. *Right:* xz projection of the magnetic field lines as seen from the top. *Solid* and *dashed lines* show the incoming field lines from the upper and lower parts of the domain, respectively. We see that the oppositely-directed field lines are not anti-parallel but are offset by an angle α determined by the strength of the sheared component B_z. The $\hat{z}$ boundary conditions can be open or periodic, depending on the model (from Kowal et al. 2009)

and the product of this process is ejected along the x direction with a speed V_{out} (the left panel of Fig. 3).

In Kowal et al. (2009), the turbulence was driven using a method described by Alvelius (1999), in which the driving term was implemented in the spectral space with discrete Fourier components concentrated around a wave vector k_{inj} corresponding to the injection scale $l_{inj} = 1/k_{inj}$. In Kowal et al. (2012a), a new method of turbulence driving was employed. Individual eddies with random locations of their centers and random orientations, either to velocity or magnetic field, at random moments in time were introduced. This guarantees the randomness of the forcing with the new method. This new method drives turbulence directly in real space, in contrast with the previous approach; therefore, it can be applied locally. The turbulence is driven in a subvolume of the domain, whose size is determined by two scales: the radius r_d on the xz plane around the center of the domain and the height h_d describing the thickness of the driving region from the midplane.

All models are evolved without turbulence for several dynamical times in order to allow the system to achieve stationary laminar reconnection. Then, at a given time t_b we start driving turbulence by increasing its amplitude to the desired level, until t_e. In this way we let the system adjust to a new state. From time t_e, the turbulence is driven with the full power P_{inj}.

The reconnection rate was measured using the method introduced in Kowal et al. (2009) and described by the formula

$$V_{\rm rec} = \frac{1}{2|B_{x,\infty}|L_z}\left[\oint \mathrm{sign}(B_x)\boldsymbol{E}\cdot d\boldsymbol{l} - \partial_t \int |B_x| dA\right] \tag{3}$$

where B_x is the strength of reconnecting magnetic component, $\boldsymbol{E}$ is the electric field, dA is an area element of an xz plane across which we perform the integration, $d\boldsymbol{l}$ is the line element separating two regions of the YZ plane defined by the sign of B_x, $|B_{x,\infty}|$ is the asymptotic absolute value of B_x, and L_z is the width of the box. The measures obtained in this way were in good agreement with the measures obtained by calculating the inflow velocity of the plasma and magnetic field at the boundaries.

In Fig. 4, examples of xy-cuts (upper row) and xz-cuts (lower row) through the box are shown. The driving is applied in real space and a large number of individual eddies is injected in the magnetic field with random locations and random orientations.

Figure 5 shows averaged values of the reconnection speed V_{rec} in models with turbulent power P_{inj} varying by more than one order-of-magnitude, from 0.1 to 2.0, for both models with Fourier driving (black symbols) and with real-space driving (blue and red symbols). Filled symbols represent the averaged reconnection rate in the presence of turbulence. The dotted line corresponds to the reconnection rate without turbulence. The error bars show the time variance of V_{rec}.

The reconnection rates for models with different types of driving confirm the theoretical dependence of V_{rec} on the injected power, which scales as $\sim P_{inj}^{1/2}$. There is no significant difference between models in which turbulence was driven in velocity and in magnetic field. This is in good agreement with the LV99 expectations that the reconnection rate should not depend on the nature of the turbulent driving.

These recent numerical studies confirmed the main predictions of LV99.[3] This motivated applications of the model to explain astrophysical phenomena from star formation to cosmic ray acceleration and gamma-ray bursts (Lazarian 2005, 2009; de Gouveia Dal Pino and Lazarian 2005; Zhang and Yan 2011; Santos-Lima et al. 2010; Lazarian and Yan 2012; Lazarian et al. 2012a). The most dramatic consequence of LV99 theory is that the magnetic flux is not frozen in turbulent fluids (Vishniac and Lazarian 1999). This prediction was strongly supported by the study in Eyink et al. (2011), where LV99 theory was related to the concept of the Richardson diffusion of magnetic field. The latter was recently demonstrated numerically in Eyink et al. (2013), providing an independent test of LV99. Future work should consider the dependence on the guide-field, which is important for the solar corona and other astrophysical applications.

3.3 2D and 3D Turbulent Reconnection

Studies of 2D reconnection were performed many years ago by Matthaeus and Lamkin (1985, 1986). The authors studied 2D magnetic reconnection in the presence of external turbulence, including the effects of multiple X-points as reconnection sites. An enhancement of the reconnection rate was reported, but this claim was not supported by the calculation of a long term average reconnection rate. Interestingly enough, a more recent study along the approach in Matthaeus and Lamkin (1985) was performed by Watson et al. (2007), where the effects of small-scale turbulence on 2D reconnection were carefully studied. However, no significant effects of the turbulence on reconnection were reported. A more optimistic conclusion is reached in Servidio et al. (2010) where reconnection on small scales with dimensionless inflow speed 0.1–0.3 was reported. However, this was happening on small scales, where the Sweet-Parker rates for the numerical set-up are comparable and therefore the implications for the large-scale reconnection were unclear. In fact, the theoretical model in Matthaeus and Lamkin (1985) does not predict the global reconnection rate.

We note that these studies did not include the effect of magnetic field wandering, which is at the core of the LV99. In view of the differences between 2D and 3D turbulence (see the corresponding discussion in Eyink et al. (2011)) it is not clear how to relate the 2D results with 3D reconnection. The aforementioned papers on 2D turbulent reconnection did

[3]After LV99 was published, Kim and Diamond (2001) produced a study arguing that turbulence will not change reconnection rates. The theoretical arguments in this study have been criticized by Lazarian et al. (2004) and Eyink et al. (2011).

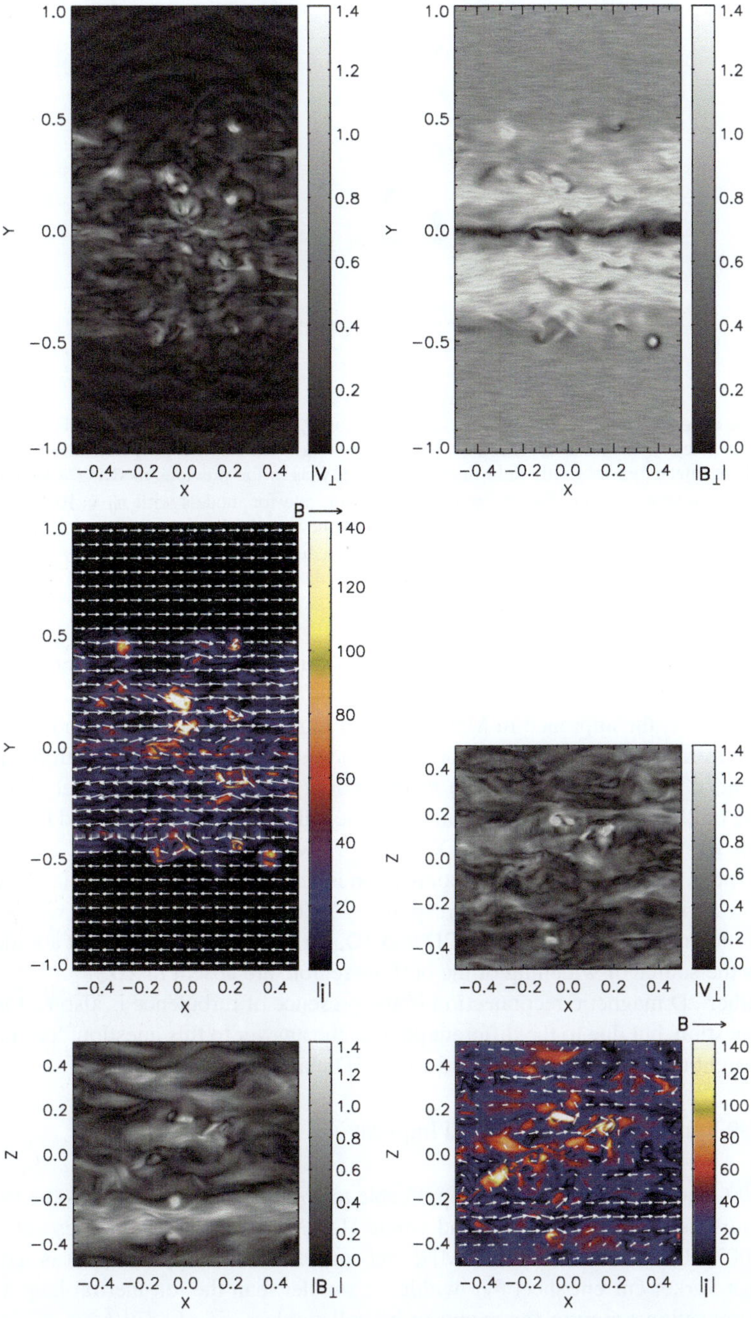

Fig. 4 Topology and strength of the velocity field (*left panel*) and magnetic field (*middle panel*) in the presence of fully developed turbulence for an example model with the new driving method at time $t = 10$. In the *right panel* we show the distribution of the absolute value of current density $|\mathbf{J}|$ overlapped with the magnetic vectors. The images show the xy-cut (*upper row*) and xz-cut (*lower row*) of the domain at the midplane of the computational box. Turbulence is injected with power $P_{inj} = 1$ at scale $k_{inj} = 8$ directly in the magnetic field. From Kowal et al. (2012a)

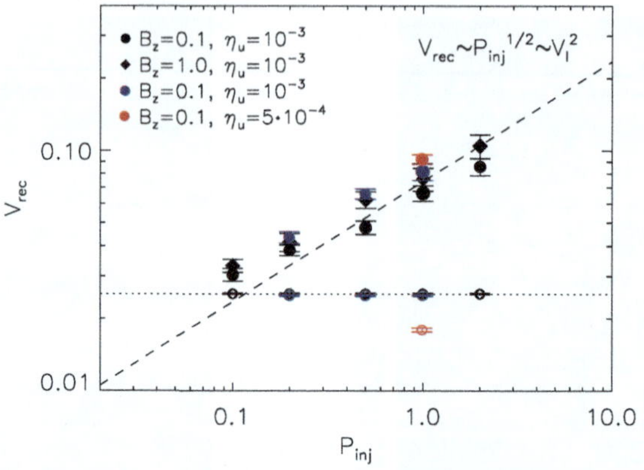

Fig. 5 The dependence of the reconnection speed V_{rec} on P_{inj} including the new models, in dimensionless units as described in Kowal et al. (2012a). *Blue symbols* show models with the new driving, in which the eddies where injected in magnetic field instead of velocity, as in the previous models (*black symbols*). The *dotted line* corresponds to the Sweet-Parker reconnection rate for models with $\eta_u = 10^{-3}$. A unique red symbol shows the reconnection rates from a model with the new driving in velocity performed with higher resolution ($512 \times 1024 \times 512$) and resistivity coefficient reduced to $\eta_u = 5 \cdot 10^{-4}$. *Error bars* represent the time variance of V_{rec}. The size of symbols corresponds to the error of V_{rec}. From Kowal et al. (2012a)

not make predictions of how the reconnection should change with the level and properties of turbulence.[4]

Furthermore, the approach in Matthaeus and Lamkin (1985, 1986) is radically different from that in LV99. The former authors associate the increase of reconnection with the development of X-point, Petsheck type reconnection: thus in, e.g. Servidio et al. (2010), the modeling of 2D turbulent reconnection under the influence of Matthaeus and Lamkin (1985) approach was focused on detecting X-points. However, X-points are not a part of LV99 model. There, it is shown that the reconnection is fast even if the small-scale events happen at the low Sweet-Parker rate and correspond to Y-point reconnection. It is well known that MHD turbulence is very different in 2D and 3D, e.g. the Alfven modes that are identified in LV99 as the source of widening of the outflow region, are absent in 2D.

Whether 2D magnetic reconnection in the presence of turbulence is also fast is an interesting question, but due to the different physics the answer to this question does not provide much insight to understanding the real world 3D magnetic reconnection

3.4 LV99 Range of Applicability and Implications

The LV99 model of reconnection is applicable to collisional media, such as the ISM, which is both turbulent and magnetized, and where Hall-MHD reconnection does not work (Yamada 2007). For instance, for Hall-MHD reconnection to be applicable, it is required that the Sweet-Parker current sheet δ_{SP} width ibe smaller than the ion inertial length d_i. Thus, the "reconnection criterion for media to be collisionless" is $(L/d_i)^{1/2}/(\omega_c \tau_e) < 1$, which

[4]By itself, the measurements of current densities in the aforementioned papers do not constrain the reconnection speed much. Indeed, one can get transient enhancements of current densities within slow Sweet-Parker reconnection by decreasing Δ. One cannot reconnect much flux this way, however, as the decrease of Δ would choke off the outflow and related magnetic reconnection.

presents a stringent constraint on the possible rate of collisions. As a result magnetic reconnection is mediated by the Hall-MHD only if the extent of the current sheet L does not exceed 10^{12} cm. These are scales too small compared to the scales at which magnetic fields in the ISM interact.

At the same time, the LV99 model works for collisionless media and it shows that the microphysics of collisionless reconnection events does not change the resulting reconnection rates. This point was further analyzed in Eyink et al. (2011) who concluded that for most of astrophysical collisionless plasmas, the LV99 model should be applicable, provided that the plasma is turbulent. The most stringent criterion for the application of LV99 theory coincides with the applicability of the MHD approximation—see Sect. 4 below.

Solar flares inspired much of the earlier research on reconnection (Pneuman 1981; Priest and Forbes 2002). Stochastic reconnection provides an explanation for solar flares that does not involve plasma microphysics. Indeed, an important prediction of the LV99 model is related to the *reconnection instability* that arises in the situation when the initial structure of the flux prior to reconnection is laminar. This allows magnetic flux to accumulate. Eventually, tearing and other instabilities enhance the reconnection rate and provide 3D turbulence. This turbulence excites faster reconnection, creating positive feedback which results in a flare (Lazarian and Vishniac 2009). This "reconnection instability" can explain the bursty character of reconnection in solar flares and also gamma ray bursts (Lazarian et al. 2003; Zhang and Yan 2011). Furthermore, the reconnection instability can be triggered by turbulence from adjacent reconnection sites, as observed in Sych et al. (2009).

We should mention that observations of solar flares are consistent with LV99 predictions. For instance, observations of the thick reconnection current outflow regions observed in solar flares (Ciaravella and Raymond 2008) were predicted within LV99 model—at a time when the competing plasma Hall models were predicting X-point localized reconnection. However, as plasma models have evolved to include tearing and formation of magnetic islands (Drake et al. 2010)—see Sect. 5—one has to be more quantitative in comparing observations with the predictions from the competing theories. The corresponding comparison was done in Eyink et al. (2013). There it was shown that the differences between the LV99 predictions and the measured thickness of the reconnection layers in Ciaravella and Raymond (2008) arises from the isotropic manner of turbulence driving assumed for the sake of simplicity in LV99 theory. If a more relevant anisotropic driving arising from magnetic reconnection is accounted for, a good quantitative agreement between the measured thickness of the reconnection regions and LV99 predictions can be obtained.

In the process of testing LV99 theory, one should keep in mind that, unlike Sweet-Parker reconnection, the turbulent scenario exhibits outflow layers which consists of a multitude of fractal current sheets. The thickness of an individual current sheet may be very narrow and these may be dominated by anomalous plasma effects. Thus additional care should be applied while testing the theory with in situ measurements.

The concept of flux-freezing violation induces numerous consequences for different fields of astrophysics. For instance, star formation theory was formulated in the assumption of magnetic flux being well frozen into plasmas with flux-to-mass ratio being changed due to the diffusion of neutrals. Magnetic flux diffusion induced by turbulent reconnection leads to the removal of magnetic flux from the star formation regions. This possibility was discussed in Lazarian (2005) and was numerically confirmed in Santos-Lima et al. (2010). Further theoretical studies of a new way of removing of magnetic flux from clouds and cores are presented in Lazarian et al. (2012a).

We note that the turbulence that we consider may be either pre-existing or generated within the reconnection layer itself. In most astrophysical situations the former situation is more common. Therefore, the issue that one has to answer dealing with reconnection in astrophysical systems is how the pre-existing turbulence is accounted for in the reconnection processes. Apparently, the laminar approximation is not a good assumption for most of astrophysical high-Reynolds-number fluids.

4 Synergy of MHD and Plasma-Based Approaches to Reconnection

While our review is focused on fast magnetic reconnection that takes place in MHD regime, in this section we sketch the relation with some different directions of reconnection research. First of all, if the original models of magnetic reconnection, i.e. Sweet-Parker and Petschek reconnection, are distinctly regular and laminar, the models that now are considered most promising, whether turbulent, as we discuss in this review, or plasmoid-type (Karimabadi et al. 2013), include the distinct effects of stochasticity. This is a remarkable shift of paradigm. Indeed, when LV99 model was introduced, its main competitor was X-point Hall reconnection.

As we mentioned earlier, the LV99 theory based on MHD is not applicable to plasmas unless the expected turbulent broadening of the outflow region is *substantially* larger than the ion Larmor radius. Indeed, below the latter scale, no MHD description of the turbulent magnetic field wandering is applicable. Only when the reconnection region is $\sim \alpha\rho_i$ then one can talk about MHD turbulence broadening of the outflow region. The exact value of α is difficult to define theoretically. A possible guess is that it should be sufficiently large, e.g. in the range of 10^2 or 10^3 for the magnetic wandering not to be strongly affected by plasma effects. For thinner current sheets we are in the regime where we expect to be the domain of plasmoid reconnection. This is definitely the regime of magnetic reconnection in magnetosphere (see the discussion in Eyink et al. (2011)) and possibly over parts of the solar wind. The two examples present important cases of space plasmas that are intensively studied through in situ observations.

The criterion of the turbulent broadening depends on the level of turbulence; in the case when the initial magnetic configuration is laminar or only slightly turbulent, plasma instabilities, e.g. tearing mode, are expected to dominate the initial dynamics of magnetic reconnection. However, both PIC simulations (e.g. Daughton et al. 2011; Karimabadi et al. 2013) and MHD simulations (e.g. Hood et al. 2009; Beresnyak 2013) (see also Sect. 6 below) testify that turbulence is generated in the process of reconnection. This turbulence should modify the reconnection. One possible effect of turbulence is the suppression of instabilities, e.g. the tearing instability, and the transition to pure turbulent reconnection (see also Rapazzo et al. 2013) . If this is true, then plasma effects can trigger turbulent magnetic reconnection. However, the identification of true turbulence in 3D MHD simulations is difficult, due to the limited range of spatial scales. Further work is needed to identify whether, for example, the fine-scale structures and complex flows identified by Hood et al. (2009) correspond to a turbulent cascade of energy to small-scales.

5 Some Other Approaches to Turbulent Reconnection in MHD

We now consider some different aspects of the interaction between turbulence and reconnection within the MHD framework. An important approach towards understanding the destabilisation of an initially laminar reconnection region can be traced back to the work of Shibata and Tanuma (2001). This paper proposes a *fractal reconnection* scenario, in which an initial current sheet thins and becomes subject to tearing instability (if the sheet length is longer than the critical wavelength for onset of tearing); the same process repeats on increasingly smaller scales, leading to a fractal distribution of magnetic islands on scales down to a microphysical limit such as the ion Larmor radius. At the same time, islands or plasmoids ejected during the reconnection process may coalesce with others to form larger plasmoids. Such a process is indeed observed in MHD simulations (Tanuma et al. 2001). The instabilities, like tearing instability, open up the reconnection layer enabling a wide outflow.

Recently, Barta et al. (2010, 2011) used a 2.5D numerical MHD simulation with Adaptive Mesh Refinement to demonstrate the formation of plasmoids on a very wide range of scales. Further recent studies of fragmentation and coalescence of plasmoids include Loureiro et al. (2009, 2012), Bhattacharjee et al. (2009). These instabilities can lead to fast reconnection (Uzdensky et al. 2010). Similar merging and coalescence of plasmoids is also widely observed in PIC simulations (e.g. Drake et al. 2010; Karlicky et al. 2012; Markidis et al. 2012; Huang and Bhattarcharjee 2013).

Another idea which suggests a complex pattern of reconnections is "recursive reconnection" (Parnell et al. 2008). This numerical model, comprising two opposite-polarity sources interacting with an overlying field, illustrates the role of magnetic topology described further in Sect. 6. Reconnection can cause both opening and closing of magnetic flux, as closed field interacts with open field: notably, this can happen recursively, with field lines opening and closing repeatedly. This both enhances the global reconnection rate and leads to more distributed heating.

Finally, we mention the interaction between MHD models of reconnection and kinetic-scale turbulence—see also Karimabadi et al. (2013). In many astrophysical plasmas, there is a complex interplay between large-scale fluid phenomena and microscales which are kinetic. Indeed reconnection on MHD scales may itself generate kinetic instabilities (see Brown et al. 2013, this issue), whilst kinetic turbulence may influence the global reconnection dynamics.

Numerical simulations of collisionless plasma, such as Shay et al. (1998, 2004) and much subsequent work, have been encouraging in showing that collisionless dissipation on small scales may play an important role in achieving fast reconnection; for example, in the solar corona. Furthermore, "anomalous resistivity", driven by kinetic turbulence within current sheets, in many cases almost certainly plays a far more important role in reconnection which is described by MHD on the global scale, than classical Spitzer resistivity. Most MHD simulations of reconnection actually rely on some kind of anomalous resistivity, although the resistivity model is not usually physically-motivated: see Gordovskyy et al. (2013b) for a recent attempt to include a more realistic resistivity model, based on ion-acoustic turbulence, in a MHD simulation. Predictions of anomalous resistivity from current-driven instabilities leading to microturbulence, such as by Petkaki et al. (2006) and Buechner and Elkina (2006), have important consequences for MHD reconnection. However, a full understanding of the coupling between global fluid models and kinetic physics remains a subject for future investigation.

6 Complex and 3D Fields and the Formation of Fine Structure

Using the solar corona as an example, we will now show that a state of topological complexity with many localised current sheets—a state of turbulent reconnection—is natural in astrophysical plasmas. A wealth of observations from a series of space-borne telescopes—such as Solar Dynamic Observatory, and, most recently, Hi-C (Cirtain et al. 2013)—indicate that the corona is highly dynamic and is full of fine structure (on scales at least down to present resolution limits).

6.1 Current Sheet Formation in the Solar Corona

The strong solar coronal magnetic field (with low $\beta = 2\mu_0 p/B^2$) is rooted in the dense photosphere, where turbulent velocities move the footpoints of the magnetic field. The existence of current sheets in the solar coronal fields has been predicted to arise in the following ways: (i) Complex footpoint motions in a simple initial field geometry, or "field line braiding". (ii) Complex initial fields with simple motions—including footpoint displacements in fields with X-points, separators, separatrices etc., emergence of new magnetic flux and the effects of discrete photospheric flux tubes. (iii) Simple initial fields with simple motions e.g. forced magnetic reconnection and kink instability due to twisting motions. We now discuss each of these three ideas.

The idea of current sheet formation due to *braiding* of the coronal flux tubes originates with Parker (1972). It was proposed that as the footpoints of coronal flux tubes are slowly moved by complex photospheric motions, the flux tubes develop a braided pattern for which no smooth force-free equilibrium can be found. In consequence, the field, in the ideal MHD limit, develops discontinuities (infinitely-thin current sheets). The coronal field thus is in a state of turbulent reconnection, heating the coronal plasma through a process dubbed "topological dissipation". Parker proposed a simple theoretical paradigm for this process, consisting of an initially uniform, straight field embedded in a perfectly-conducting plasma, between two conducting planes (representing the photosphere), subject to slow motions of the photospheric footpoints which lack symmetry. The subsequent equilibrium states of the field are constrained by the footpoint connectivity.

The "Parker" model has led to a much subsequent work over 40 years; such as, recently Low (2006), Rapazzo et al. (2007, 2013), Berger and Asgari-Targhi (2009). It is broadly established that fine structure, including localised, very strong currents, indeed arises. However, there is still no consensus on the inevitability of true discontinuities (infinite currents) in the ideal MHD limit, although Van Ballegoijen (1985) provided a methodology to find a force-free equilibrium for any continuous footpoint motions, suggesting there are no discontinuities. For example, Ng et al. (2012) use Reduced-MHD (RMHD) simulations with random footpoint motions to show that the energy dissipation rate is independent of the resistivity, and that reconnection is much faster than Sweet-Parker. Wilmot-Smith et al. (2010) and Pontin et al. (2011) create a field representing a simple braid of three flux tubes, which is close to force-free equilibrium, using a Lagrangian relaxation scheme. They then use a 3D MHD simulation to show that this develops many small-scale current sheets and subsequently dissipates free magnetic energy by reconnection. It should be noted, however, that these simulations consider loops with very weak transverse magnetic field components and hence with very little free magnetic energy (i.e. the fields are always very close to a current-free state); hence, their relevance to heating coronal Active Regions is questionable. The same difficulty applies to the many works which use the RMHD approach, in which a weak transverse field component is assumed.

Recent high resolution solar observations from Hi-C (Cirtain et al. 2013) indicate the presence of untwisting coronal loops, which are suggested to be consistent with dissipation of braided fields. However, as noted by Cargill (2013), these observations are equally consistent with kink-unstable twisted loops (see Sect. 7 below). Indeed, both braiding and twisted loop models results in a network of localised thin current layers, with turbulent reconnection.

It is very easy to form current sheets by footpoint motions of fields with complex topology (i.e. with regions of different fieldline connectivity). For example, shearing the footpoints of a 2D field configuration with a separatrix and multiple flux domains naturally forms a current sheet along the separatrix (e.g. Vekstein et al. 1992), since neighbouring field lines on different sides of the separatrix may have very different footpoint displacements, leading to a discontinuity in the field. Similarly, the fact that the photospheric footpoints of the coronal field are actually concentrated into discrete isolated flux tubes naturally leads to the formation of discontinuities (in the ideal limit) within the coronal magnetic field. Thus, motions within flux sources, or motions of the flux sources themselves, inevitably generates coronal reconnection (Browning et al. 1986; Lothian and Browning 1995). This scenario has been dubbed "flux tube tectonics", and explored in some depth (Priest et al. 2002). Furthermore, models of new flux emerging into the corona from the solar interior involve the interaction of two flux systems, again resulting in current sheet formation and considerable topological structure, as demonstrated in 3D numerical simulations (e.g. Archontis et al. 2005; Parnell et al. 2010).

In fact, even the simplest field topologies and footpoint velocity profiles can generate fine-scale structure in the corona. One example has already been mentioned in Sect. 2—namely, forced reconnection. Here, a simple sinusoidal disturbance of a sheared force-free field in a slab can generate a current sheet, leading to reconnection and a chain of magnetic islands (Hahm and Kulsrud 1985; Vekstein and Jain 1998; Jain et al. 2005). If the external disturbance consists of multiple modes in different directions, then a series of current sheets will be formed at the various resonant surfaces, leading to overlapping islands with stochastic field lines (Onofri et al. 2004). Another example is the nonlinear kink instability (see Sect. 7): even smooth rotational motions of the footpoints of simple cylindrical flux tubes can lead to the formation of a fragmented current structure with turbulent reconnection.

An important consequence of the ubiquitous presence of small-scale structure in the solar atmosphere is that the resulting state of multiple reconnections may heat the coronal plasma. Whilst it is generally accepted that large solar flares are a result of release of stored magnetic energy by reconnection in a large-scale current sheet, a promising scenario for maintaining the overall high temperature of the corona is the combined effect of many very small flare-like events known as "nanoflares" (Parker 1988). We return to the coronal heating problem in Sect. 7.

In summary, in the solar corona the topology is complex, and hence multiple reconnection sites are likely to exist (Maclean et al. 2009), with reconnection usually being 3D. The coronal field is predicted to be in a constant state of turbulent reconnection. This theoretical picture is supported by observations. For example, observations of Type III radio bursts, which are produced by beams of non-thermal electrons in the corona, indicate that energy release is highly fragmented both in space and time (Chen et al. 2013). A similar picture is likely to pertain in other astrophysical plasmas.

6.2 3D Reconnection Models

Whilst classical reconnection models are two-dimensional, as outlined in Sect. 2, most astrophysical phenomena are not well-described by 2D models, and much recent research shows that the reconnection in 3D differs in quite fundamental ways. Even the definition of reconnection is less clear. Schindler et al. (1988) suggest that reconnection can be identified through the existence of localised parallel electric fields. Reconnection in 3D can occur at 3D magnetic nulls—but does not require the presence of nulls; in contrast to 2D reconnection which requires an X-point (a 2D null) or a field reversal (although 2.5D reconnection, i.e. including a guide-field, also does not require nulls). Furthermore, the topological properties of 3D reconnection can be quite different. Whereas in 2D reconnection, there is a simple pair-wise reconnection of flux tubes, with two incoming flux tubes "breaking and re-joining" to form two flux tubes in the outflow, the situation is more complex in 3D (Priest et al. 2003; Cargill et al. 2010). In general, due to counter-rotating flows within the diffusion region, it is not possible to uniquely match pairs of inflow and outflow fieldlines (Priest et al. 2003; Pontin et al. 2005). The properties of 3D reconnection are reviewed by Pontin (2011).

The topology of the magnetic field may be summarised through knowledge of the "magnetic skeleton" (Bungey et al. 1996). This comprises: *field sources* on the photospheric boundary (which are usually, more or less, discrete); *magnetic null points* (usually 3D); a set of flux domains bounded by *separator surfaces* (the 3D analogue of 2D separatrices); and *separator lines*, which are the intersection of separators. Such topological features form the likely sites of magnetic reconnection, as well as *Quasi Separatrix Surfaces* (Titov et al. 2002; Demoulin 2006) which are layers of strong divergence in field line connectivity. Reconnection may occur at 3D null points (Priest and Titov 1996; Craig and Fabling 1996) but also in the absence of nulls, with separator lines playing an important role (Longcope and Cowley 1996; Parnell et al. 2010).

Understanding the nature of reconnection in 3D—and the inevitable interactions with turbulence—is an important subject of future research.

7 Relaxation in Turbulent Reconnecting Plasmas

An important consequence of turbulent reconnection in astrophysical plasmas is the dissipation of stored magnetic energy. A large-scale energy release may cause a solar flare, whereas the combination of many smaller energy releases may be the source of heating required to maintain coronal plasma at temperatures of over a million degrees Kelvin. In order to predict the energy release, it is not necessary to calculate the detailed dynamics, if it is assumed that the field relaxes towards the state of lowest possible magnetic energy. The appropriate constraint for such relaxation, in the presence of multiple localised reconnections, is that the global magnetic helicity

$$K = \int_V \mathbf{A} \cdot \mathbf{B} dV, \tag{4}$$

is conserved, where $\mathbf{A}$ is the vector potential (Taylor 1974, 1986). This follows since reconnection transfers helicity between flux tubes, without creating or destroying helicity. Also, it can be shown that, if dissipation is confined to narrow current layers, with width much less than the global length scale, then fractional helicity dissipation is much less than energy dissipation (Berger 1984; Browning 1988). Note that in a volume in which magnetic fieldlines cross the bounding surface (such as the corona, in which field lines cross the photosphere), then the relative helicity must be used (Berger and Field 1984; Finn and Antonsen 1985). The minimum energy state with conserved helicity is a constant-α or linear force-free field:

$$\nabla \times \mathbf{B} = \alpha \mathbf{B}, \tag{5}$$

where α, which is the ratio of parallel current to magnetic field, is spatially constant. Relaxation theory was developed by Taylor (1974) to explain the reversal of toroidal field in Reverse Field Pinch devices, and has been successfully applied also to other laboratory devices such as spheromaks (Taylor 1986; Jarboe 1994).

Heyvaerts and Priest (1984) first applied this idea to the solar corona, proposing that the coronal field is continually stressed by slow photospheric footpoint motions, causing it evolve through a series of nonlinear force-free fields satisfying (5) but with spatially-varying α. The field then relaxes to a minimum energy state, conserving helicity whilst releasing free magnetic energy as heat. This idea was developed by many others (e.g. Browning and Priest 1986; Browning et al. 1986; Dixon et al. 1988; Vekstein et al. 1993; Wolfson

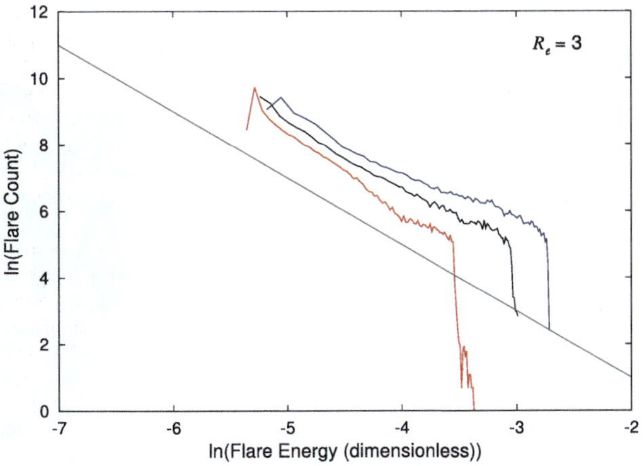

Fig. 6 The distribution of event frequencies against the magnitude of the energy release for cylindrical twisted loops with different aspect ratios (L/R, where L and R are, respectively, the loop length of radius). This shows a power law distribution as expected for nanoflares. From Bareford et al. (2011)

et al. 1994; Kusano et al. 1994; Lothian and Browning 2000), and generalised to different boundary conditions (Browning 1988; Dixon et al. 1989)—but it remained unclear how far free magnetic energy could build up before the field relaxed. Thus, the energy release depended on an unknown "relaxation time". This problem was resolved by Browning and Van der Linden (2003), who suggested that relaxation could be triggered by the onset of ideal instability, such as kink instability in a twisted loop. In the case of a twisted cylindrical loop in which the current profile arising from photospheric motions was parameterised by a family with piecewise-constant α, the energy release was shown to depend on the current profile at the point of instability onset (Browning and Van der Linden 2003).

Subsequently, Bareford et al. (2010, 2011) developed a relaxation-based model of coronal heating due to random photospheric footpoint driving. The field in a cylindrical loop evolves randomly through a series of equilibria until the threshold for ideal kink instability is reached: at this point, the energy dissipated during a helicity-conserving relaxation to a constant-α field is calculated. The process is repeated, with a cycle of stressing the field followed by relaxation: thus, a distribution of heating events or "nanoflares" is built up. In the case of localised twisting motions, in which the loop thus carries zero-net current, a power-law distribution of event size vs. occurrence frequency can be generated (Bareford et al. 2011); see Fig. 6. The average rate of energy dissipation, due to the repeated driving and relaxation, is sufficient for heating a coronal Active Region. This provides an *a priori* prediction of the distribution of nanoflare energies, whose combined effect heats the corona, as first postulated by Parker (1988).

A strength of these models is that they require significant free energy to be stored before the onset of kink instability and heating, thus providing sufficient heating for Active Regions requirements; see the discussion in Bareford et al. (2010), showing agreement with the requirements on transverse field set by Parker (1988). This contrasts with RMHD models and similar models which use a small transverse field component. It is interesting to investigate the behaviour of repeated heating events (e.g. Jain et al. 2005). Preliminary 3D MHD simulations (Gordovskyy, private communication) suggest that a repeated series of heating events may indeed be produced if the footpoint driving is maintained. On the other hand, Rapazzo et al. (2013) suggest that, if the driving motions are turbulent, the fields may become sufficiently incoherent that kink instability cannot arise.

Observational evidence to support Taylor's hypothesis—of helicity-conserving relaxation to a minimum energy state—is found in a wide range of laboratory experiments (e.g.

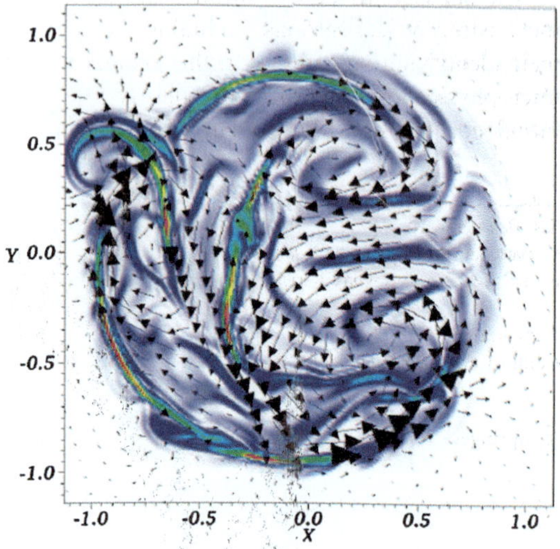

Fig. 7 The distribution of currents (colour scale) and velocities (*arrows*) at the loop midplane during the later stages of relaxation in a kink-unstable twisted loop. From Hood et al. (2009)

see Taylor (1986) and references therein). Furthermore, relaxation towards a constant-α state is observed in solar flares (Nandy et al. 2003). Further verification of Taylor relaxation in coronal loops, as well as understanding of *how* the relaxation takes place, is provided by 3D MHD simulations of twisted cylindrical coronal loops (Browning et al. 2008; Hood et al. 2009; Bareford et al. 2013). These simulations consider an initially force-free cylindrical twisted loop which is linearly unstable to the ideal kink mode. In the nonlinear phase of the kink instability, a helical current ribbon forms, leading to fast reconnection and dissipation of magnetic energy, with the field subsequently reaching a new equilibrium with lower energy.

Browning et al. (2008) show that helicity dissipation is much less than energy dissipation, and that the final relaxed state is close to a constant-α field, consistent with Taylor relaxation. The relaxation mainly occurs as the initial helical current sheet breaks up and fragments (Hood et al. 2009), leading to distributed reconnection throughout the loop volume, with a turbulent velocity profile, in which outflows from one current sheet drive reconnection in its neighbour (see Fig. 7). The initially monolithic current sheet stretches and bifurcates, with the smaller sheets then subsequently splitting repeatedly, in a manner somewhat analogous to the 2D "fractal reconnection" models described in Sect. 5. The α profile becomes more uniform due to the multiple reconnections, as suggested long ago in the Tangled Discharge Model (Rusbridge 1977, 1991); again, we see an association between reconnection and fieldline wandering (see Sect. 3). The loop is heated throughout its volume.

The Taylor hypothesis only predicts the state of lowest possible energy, which may not actually be attained if there are other constraints on the dynamics (Bhattacharjee et al. 1980), or simply if there is not enough free energy to drive the relaxation. In the case of a loop twisted by localised photospheric motions, which has zero net-current, the extent of the disruption within the nonlinear phase of the instability is confined to a region with about 1.5 times the original loop radius; this leads to a partial relaxation in which the magnetic field outside this is undisturbed (Hood et al. 2009; Bareford et al. 2013).

Yeates et al. (2010), based on simulations of an initially braided coronal loop, propose that additional topological constraints, associated with the mapping of field lines from one end of the loop to the other, may prevent full Taylor relaxation. However, their simulations consider initial fields with little much free energy, and their final state in energetic terms is actually very close to the constant-α state—which is a potential field in this case. So Taylor theory still provides a good estimate of the energy release, despite the incomplete relaxation.

Nevertheless, it is clear that full relaxation can never really occur in an astrophysical plasma which is necessarily unbounded—unlike finite-volume laboratory plasmas. So relaxation must necessarily be limited spatially by the extent of the region of turbulent reconnection: a new theory of partial relaxation for unstable cylindrical loops which accounts for this has been recently proposed by Bareford et al. (2013). Furthermore, relaxation is similarly restricted by the onset of such a turbulent reconnecting state: it is quite possible for a field to remain in a state with free energy—which is not the minimum energy state—until relaxation is somehow triggered. As discussed above, an example of such a trigger is the onset of ideal MHD instability, but there are likely to be many other possible mechanisms.

A full understanding of relaxation in astrophysical plasmas requires much further research.

8 Turbulent Reconnection and Particle Energisation

The origin of high-energy particles in solar flares is a long-standing problem in astrophysics, as well the acceleration of cosmic rays. Proposed mechanisms include both turbulence or acceleration by direct electric fields in a reconnecting current sheet (see reviews Miller et al. (1997), Zharkova et al. (2011)). However, turbulence models tend to rely on an arbitrary turbulent field which is not related to the reconnection which is the primary source of energy-release in a flare, whilst reconnection models with a localised coronal current sheet suffer with difficulties, such as bringing in a large number of particles into a small current sheet volume. Many aspects of flare particle acceleration can be explained by combining turbulence with reconnection. This topic has been extensively reviewed elsewhere (e.g. Vlahos et al. 2009; Cargill et al. 2012), as well as some aspects being discussed by Lazarian et al. (2012b); we mention here some aspects of this topic which relate directly to our earlier discussions of turbulent reconnection.

Firstly, the turbulent reconnection LV99 paradigm, described in Sect. 3, provides effective particle acceleration through First-order Fermi acceleration—in which particles bounce between converging magnetic mirrors. Figure 8 illustrates the First-order Fermi acceleration that takes place within a turbulent reconnection region (de Gouveia Dal Pino and Lazarian 2005; Lazarian 2006).[5] The acceleration happens as particles bounce back and forth within shrinking magnetic loops (see Fig. 8). Recently, the acceleration of cosmic rays in reconnection has been invoked to explain results on the anomalous cosmic rays obtained by Voyager spacecrafts (Lazarian and Opher 2009; Drake et al. 2010), the local anisotropy of cosmic rays (Lazarian et al. 2012b) and the acceleration of cosmic rays in clusters of galaxies (Brunetti and Lazarian 2011). Naturally, this process of acceleration widespread and not limited to these examples. Numerical studies of the particle acceleration have shown differences in the acceleration process between 2D and 3D and confirmed the first-order Fermi nature of the acceleration in turbulent reconnection layers (Kowal et al. 2011, 2012b).

The breakup of long current sheets into a chain of magnetic islands (Sect. 5) also has consequences for particle acceleration (Kliem 1994; Li and Lin 2012). Gordovskyy et al.

[5]The predicted spectrum without taking the backreaction of the accelerated particles is $N(E)dE \sim E^{-5/2}dE$. Considerations in Drake et al. (2006) suggest that the spectrum of the particles can get shallower if the backreaction is taken into account.

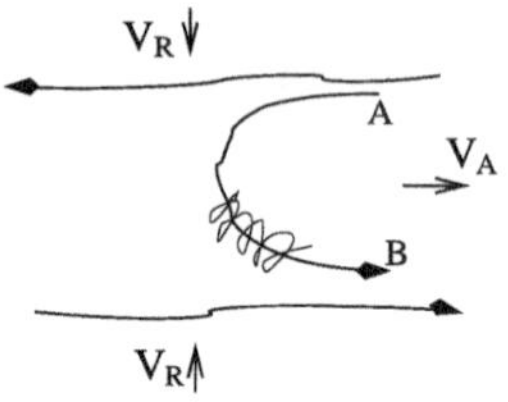

Fig. 8 First order-Fermi acceleration as cosmic rays bounce within a 3D loop of reconnected flux that shrinks due to magnetic reconnection. From Lazarian (2005)

(2010a, 2010b) showed that two populations of accelerated particles arise in this case: a particle population trapped in the growing magnetic islands gains substantial energy, whereas particles remaining on open fieldlines are predominantly thermal.

Particles can be accelerated in fields with turbulent current sheets, such as those arising from braiding footpoint motions (Turkmani et al. 2006) (see Sect. 6.1). Also, the fragmented current sheets in unstable twisted loops (see Sect. 7) provide an effective means for distributed acceleration of charged particles in flares, with acceleration produced by the parallel electric fields within the current sheets. Using a relativistic guiding-centre test particle code coupled to 3D MHD situations, Gordovskyy and Browning (2011) and Gordovskyy and Browning (2012) show that particles gain energy through a series of almost discontinuous jumps, as they randomly encounter current sheets, in a manner reminiscent of earlier cellular automaton models (Vlahos et al. 2004)—see Fig. 9.

As the loop first becomes kink-unstable, particles are accelerated in the helical current ribbon which is quite radially localised, and leads to a narrow distribution of energetic particles. However, as described in Sect. 7, the current sheet breaks up into a distributed structure. Thus, particles are accelerated throughout the loop, and the loop quickly fills with energetic particles. The loop expands somewhat as it reconnects with the surrounding field, and hence the spatial extent of the energetic particles grows in time. The particle energy spectrum develops a non-thermal tail.

The particle spectrum is somewhat sensitive to the assumed resistivity profile. Assuming a rather high value of a uniform background resistivity (Turkmani et al. 2006) tends to over-estimate the number of accelerated particles (Gordovskyy and Browning 2012). If a localised current-dependent resistivity is used, the fraction of energetic particles can be relatively small, typically 5–10 %, validating the use of test particle modelling in this case. The increased numerical resolution of the more recent simulations may also explain the decrease in the fraction of energetic particles compared with earlier work, since the current sheets are better resolved; see further discussion in Cargill et al. (2012).

In cylindrical loop models, energetic particles are inevitably quickly lost through the ends of the loop. This may be mitigated by the effects of fieldlines convergence at the footpoints (Gordovskyy et al. 2013a); but in fact, few energetic particles are mirrored as they tend to have very small pitch-angles, a consequence of direct electric field acceleration. The confinement of particles within the loop is substantially modified by the occurrence of collisions within the dense chromosphere at the loop footpoints: this causes pitch- angle scattering, allowing more particles to be reflected and confined with the loop (Gordovskyy et al. 2013a). The time-evolving energy spectrum in a loop with field line convergence, and incorporating the effects of collisions with the stratified background plasma, is shown in Fig. 10.

Recently, a more self-consistent model of particle acceleration and transport in a twisted coronal loop has been developed: this considers a curved loop, twisted by localised footpoint motions, within a gravitationally-stratified atmosphere. The effects of transport and acceleration are incorporated, including collisions as well as acceleration by direct electric fields in a fragmented current sheet, with an anomalous resistivity dependent on the particle

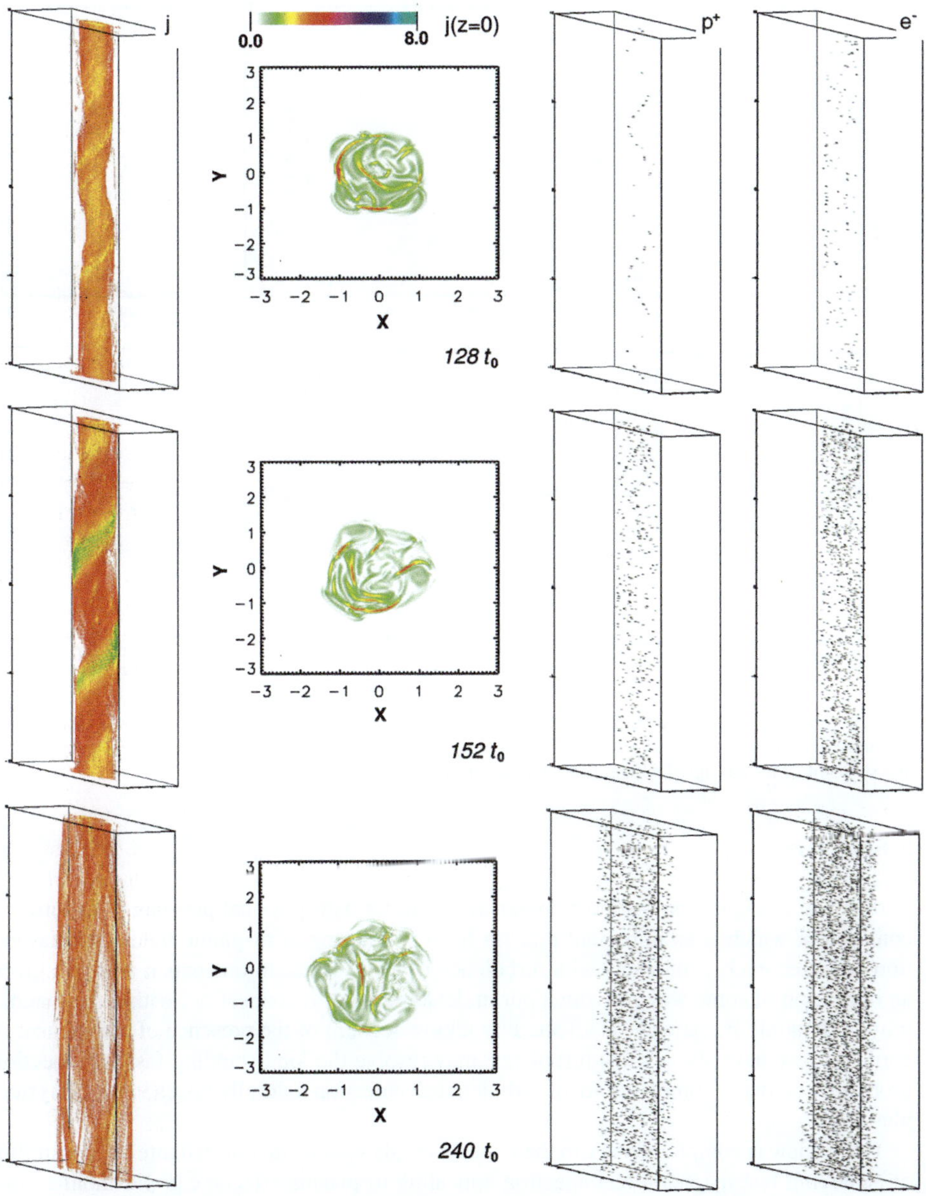

Fig. 9 Particle acceleration in a kink-unstable twisted loop due to fragmented current sheets. Showing (from *left* to *right*), current isosurfaces, current at the midplane and spatial distributions of protons and electrons, at successive times. From Gordovskyy and Browning (2012)

drift velocity (Gordovskyy et al. 2013b). Particles are mainly accelerated both at the loop top and near the footpoints. Synthesised Hard X-ray emission is calculated, with footpoint and looptop sources which evolve through the flare event.

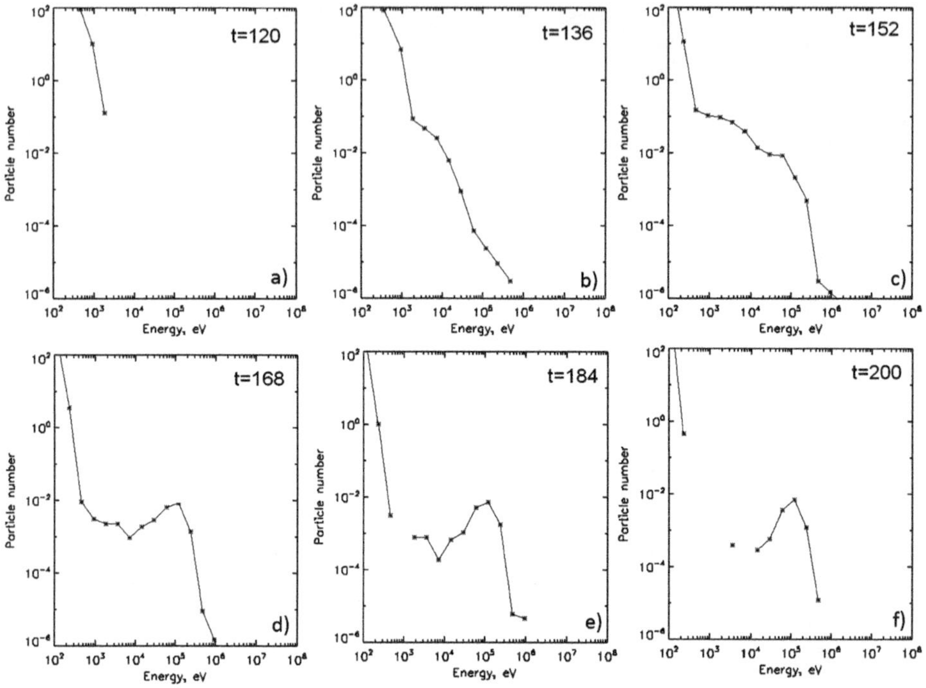

Fig. 10 Electron energy spectra in a kink-unstable twisted loop with field line convergence towards the chromosphere, including the effects of collisions with the dense chromospheric plasma on the test particles, at a series of times through the flare. From Gordovskyy et al. (2013a)

9 Summary

Turbulence is a common—almost universal—state for astrophysical plasmas. Magnetic reconnection, which is an important process for restructuring of magnetic fields and dissipation of stored energy, interacts with turbulence in many ways. In this paper, we have given an indication of some ways in which our understanding of reconnection, within the magnetohydrodynamic framework, is affected by taking account of the presence of turbulence. In particular, we have shown how turbulence may resolve the long-standing fast reconnection problem, and that a complex state of turbulent reconnection naturally arises in astrophysical plasmas.

Our review is complementary to the reviews in this volume that investigate the possibility of achieving fast magnetic reconnection appealing to plasma effects, e.g. plasmoid reconnection. The relation between these two approaches requires further studies; in particular, at sufficiently small scales, the MHD description of the turbulent field wandering is not applicable and thus the model of turbulent reconnection we described is not applicable either. At the same time, turbulence arising from the reconnection process that is reported in both PIC and MHD simulations initiated with laminar magnetic fields is indicative that turbulent reconnection may take over from reconnection dominated by plasma effects or tearing. Further investigation of these situations is necessary.

The work described here, from many different perspectives, support the notion that turbulence is intrinsic element of fast reconnection in most astrophysical environments and that attempts to treat instabilities in reconnection systems, e.g. tearing instability, without accounting for the turbulence that these instabilities generate, are of limited relevance. At the same time, turbulent reconnection is necessary for making modern models of MHD turbulence self-consistent, allowing resolution of the magnetic knots produced by magnetic eddies, mixing matter and magnetic fields as a part of the self-similar turbulent cascade.

Furthermore, using the solar corona as an example, we have shown that a state of turbulent reconnection, with many small-scale current sheets interacting through complex flow fields, naturally arises. Even in the case of very simple laminar driving motions (such as rotation), the magnetic field naturally develops into a state filled with reconnecting current sheets and turbulent flows. The energy release in such complex fields can be determined using the idea that the plasma undergoes helicity-conserving relaxation to a minimum energy state. This has important consequences for understanding the heating the solar corona. As well as heating the plasma, the energy dissipated by magnetic reconnection may be transferred to non-thermal ions and electrons: a turbulent reconnecting plasma is an effective particle accelerator.

Acknowledgements AL and PKB are grateful to the International Space Science Institute for hosting the Workshop at which this paper was formulated. PKB acknowledges financial support from the UK STFC. AL is supported by the Center for Magnetic Self-Organization in Laboratory and Astrophysical Plasmas, NSF grant AST 1212096, NASA grant NNX09AH78G as well as the Vilas Associate Award. The hospitality of the International Institute of Physics (Natal, Brazil) is acknowledged.

References

K. Alvelius, Random forcing of three-dimensional homogeneous turbulence. Phys. Fluids **11**, 1880–1889 (1999)

V. Archontis, F. Moreno-Insertis, K. Galsgaard, A.W. Hood, The three dimensional interaction between emerging magnetic flux and the large-scale coronal field; reconnection, current sheets and jets. Astrophys. J. **635**, 1299 (2005)

J.W. Armstrong, B.J. Rickett, S.R. Spangler, Electron density power spectrum in the local interstellar medium. Astrophys. J. **443**, 209–221 (1995)

S.A. Balbus, J.F. Hawley, Instability, turbulence, and enhanced transport in accretion disks. Rev. Mod. Phys. **70**, 1–53 (1998)

M.R. Bareford, P.K. Browning, R.A.M. Van der Linden, A nanoflare distribution generated by repeated relaxations triggered by kink instability. Astron. Astrophys. **521**, A70 (2010)

M.R. Bareford, P.K. Browning, R.A.M. Van der Linden, The flare-energy distributions generated by ensembles of kink-unstable zero-net current coronal loops. Sol. Phys. **273**, 93–115 (2011)

M.R. Bareford, A.W. Hood, P.K. Browning, Coronal heating by the partial relaxation of twisted loops. Astron. Astrophys. **550**, A40 (2013)

M. Barta, J. Buechner, M. Karlicky, Multi-scale MHD approach to the current sheet filamentation in solar coronal reconnection. Adv. Space Res. **45**, 10–17 (2010)

M. Barta, J. Buechner, M. Karlicky, J. Skala, Spontaneous current sheet fragmentation and cascading reconnection in solar flares. Astrophys. J. **737**, 24 (2011)

R. Beck, Magnetic fields in spiral arms and bars, in *Disks of Galaxies: Kinematics, Dynamics and Perturbations*, ed. by E. Athanassoula, A. Bosma, R. Mujica. ASP Conference Proceedings, vol. 275 (Astron. Soc. Pac., San Francisco, 2002), pp. 331–342. ISBN:1-58381-117-6

A. Beresnyak, On the rate of spontaneous magnetic reconnection (2013). arXiv:1301.7424

M.A. Berger, Rigorous new limits on magnetic helicity dissipation in the solar corona. Geophys. Astrophys. Fluid Dyn. **30**, 79 (1984)

M.A. Berger, M. Asgari-Targhi, Self-organised braiding and the structure of solar coronal loops. Astrophys. J. **705**, 347–355 (2009)

M.A. Berger, G.B. Field, The topological properties of magnetic helicity. J. Fluid Mech. **147**, 133 (1984)

A. Bhattacharjee, R.L. Dewar, D.A. Monticello, Energy principle with global invariants for toroidal plasma with global invariants. Phys. Rev. Lett. **45**, 347 (1980)

A. Bhattacharjee, Z.W. Ma, X. Wang, Recent developments in collisionless reconnection theory: applications to laboratory and astrophysical plasmas. Lect. Notes Phys. **614**, 351–375 (2003)

A. Bhattacharjee, Y.-M. Huang, H. Yang, B. Rogers, Fast reconnection in high-Lundquist-number plasmas due to the plasmoid instability. Phys. Plasmas **16**, 112102 (2009)
D. Biskamp, Magnetic reconnection in plasmas. Astrophys. Space Sci. **242**, 165–207 (1996)
D. Biskamp, *Magnetic Reconnection in Plasmas* (Cambridge University Press, Cambridge, 2000)
M. Brown, Experimental studies of magnetic reconnection. Phys. Plasmas **6**, 1717 (1999)
M. Brown et al., Microphysics of cosmic plasmas: hierarchies of plasma instabilities from MHD to kinetic. Space Sci. Rev. (2013). doi:10.1007/s11214-013-0005-7
P.K. Browning, Helicity injection and relaxation in a solar-coronal magnetic loop with a free surface. J. Plasma Phys. **40**, 263 (1988)
P.K. Browning, E.R. Priest, Heating of coronal arcades by magnetic tearing turbulence, using the Taylor-Heyvaerts hypothesis. Astron. Astrophys. **159**, 129 (1986)
P.K. Browning, R.A.M. Van der Linden, Solar corona heating by relaxation events. Astron. Astrophys. **400**, 355 (2003)
P.K. Browning, T. Sakurai, E.R. Priest, Coronal heating in closely-packed flux tubes, a Taylor-Heyvaerts relaxation theory. Astron. Astrophys. **158**, 217 (1986)
P.K. Browning, C. Gerrard, A.W. Hood, R. Kevis, R.A.M. Van der Linden, Heating the corona by nanoflares; simulations of energy-release triggered by a kink instability. Astron. Astrophys. **485**, 837 (2008)
G. Brunetti, A. Lazarian, Acceleration of primary and secondary particles in galaxy clusters by compressible MHD turbulence: from radio haloes to gamma-rays. Mon. Not. R. Astron. Soc. **410**, 127 (2011)
J. Buechner, N. Elkina, Anomalous resistivity of current-driven isothermal plasmas due to phase-space structuring. Phys. Plasmas **13**, 082304 (2006)
T.N. Bungey, V.S. Titov, E.R. Priest, Basic topological elements of coronal magnetic fields. Astron. Astrophys. **208**, 33 (1996)
B. Burkhart, S. Stanimirović, A. Lazarian, G. Kowal, Characterizing magnetohydrodynamic turbulence in the small Magellanic cloud. Astrophys. J. **708**, 1204–1220 (2010)
P.J. Cargill, Towards ever smaller length scales. Nature **493**, 485 (2013)
P. Cargill, C. Parnell, P.K. Browning, I. de Moortel, A.W. Hood, Magnetic reconnection, from proposal to paradigm. Astron. Geophys. **51**, 3.31 (2010)
P.J. Cargill, L. Vlahos, G. Baumann, J.F. Drake, A. Nordlund, Current fragmentation and particle acceleration in solar flares. Space Sci. Rev. **173**, 223–245 (2012)
B. Chen, T.S. Bastian, S.M. White et al., Tracing electron beams in the solar corona with radio dynamic imaging spectroscopy. Astrophys. J. **763**, L21 (2013)
A. Chepurnov, A. Lazarian, Extending the big power law in the sky with turbulence spectra from Wisconsin Hα mapper data. Astrophys. J. **710**, 853–858 (2010)
A. Chepurnov, A. Lazarian, S. Stanimirović, C. Heiles, J.E.G. Peek, Velocity spectrum for H I at high latitudes. Astrophys. J. **714**, 1398–1406 (2010)
A. Ciaravella, J.C. Raymond, The current sheet associated with the 2003 November 4 coronal mass ejection: density, temperature, thickness, and line width. Astrophys. J. **686**, 1372–1382 (2008)
J.W. Cirtain, L. Golub, A.R. Winebarger et al., Energy release in the solar corona from spatially resolved magnetic braids. Nature **493**, 501 (2013)
I.J.D. Craig, R.B. Fabling, Exact solutions for steady state, spine, and fan magnetic reconnection. Astrophys. J. **462**, 969 (1996)
R.M. Crutcher, Magnetic fields in molecular clouds: observations confront theory. Astrophys. J. **520**, 706–713 (1999)
W. Daughton, V. Roytershteyn, H. Karimabadi, Y. Yin, B.J. Albright, B. Bergen, K.J. Bowers, Role of electron physics in the development of turbulent magnetic reconnection in collisionless plasmas. Nat. Phys. **7**, 539 (2011)
E.M. de Gouveia dal Pino, A. Lazarian, Production of the large scale superluminal ejections of the microquasar GRS 1915+105 by violent magnetic reconnection. Astron. Astrophys. **441**, 845–853 (2005)
P. Demoulin, Extending the concept of separatrices to QSLs for magnetic reconnection. Space Sci. Rev. **37**, 1269 (2006). Reconnection in the Sun and at magnetospheres, eds. J. Buechner and X. Deng
A.M. Dixon, P.K. Browning, E.R. Priest, Coronal heating by relaxation in a sunspot magnetic field. Geophys. Astrophys. Fluid Dyn. **40**, 293 (1988)
A.M. Dixon, M.A. Berger, P.K. Browning, E.R. Priest, A generalisation of the Woltjer minimum energy principle. Astron. Astrophys. **225**, 156–166 (1989)
P. Dmitruk, W.H. Matthaeus, Structure of the electromagnetic field in three-dimensional Hall MHD turbulence. Phys. Plasmas **13**, 042307 (2006)
J.F. Drake, M. Swisdak, H. Che, M.A. Shay, Electron acceleration from contracting magnetic islands during reconnection. Nature **443**, 553–556 (2006)
J.F. Drake, M. Opher, M. Swisdak, J.N. Chamoun, A magnetic reconnection mechanism for the generation of anomalous cosmic rays. Astrophys. J. **709**, 963–974 (2010)

B.G. Elmegreen, J. Scalo, Interstellar turbulence I: observations and processes. Annu. Rev. Astron. Astrophys. **42**, 211–273 (2004)
G.L. Eyink, A. Lazarian, E.T. Vishniac, Fast magnetic reconnection and spontaneous stochasticity. Astrophys. J. **743**, 51 (2011)
E.L. Eyink, E. Vishniac, C. Lalescu et al., Flux freezing breakdown in high conductivity MHD turbulence. Nature **497**, 466 (2013)
J.M. Finn, T.M. Antonsen, Magnetic helicity, what is it and what is it good for? Commun. Plasma Phys. Control. Fusion **9**, 111 (1985)
R. Fitzpatrick, Scaling of forced magnetic reconnection in the Hall-magnetohydrodynamic Taylor problem. Phys. Plasmas **11**, 937–946 (2004)
T.G. Forbes, The nature of Petschek-type reconnection. Earth Planets Space **53**, 423 (2001)
H.P. Furth, J. Killeen, M.N. Rosenbluth, Finite resistivity instabilities of a sheet pinch. Phys. Fluids **6**, 459 (1963)
B.M. Gaensler, M. Haverkorn, B. Burkhart, K.J. Newton-McGee, R.D. Ekers, A. Lazarian, N.M. McClure-Griffiths, T. Robishaw, J.M. Dickey, A.J. Green, Low-Mach-number turbulence in interstellar gas revealed by radio polarization gradients. Nature **478**, 214–217 (2011)
P. Goldreich, S. Sridhar, Toward a theory of interstellar turbulence. 2: strong Alfvenic turbulence. Astrophys. J. **438**, 763–775 (1995). (GS95)
M. Gordovskyy, P.K. Browning, Particle acceleration by magnetic reconnection in a twisted coronal loop. Astrophys. J. **729**, 101 (2011)
M. Gordovskyy, P.K. Browning, High energy particles in confined solar flares: acceleration in reconnecting unstable twisted coronal loops. Sol. Phys. **277**, 299 (2012)
M. Gordovskyy, P.K. Browning, G.E. Vekstein, Particle acceleration in a transient magnetic reconnection event. Astron. Astrophys. **51**, A21 (2010a)
M. Gordovskyy, P.K. Browning, G.E. Vekstein, Particle acceleration in a fragmenting periodic reconnecting current sheets in solar flares. Astrophys. J. **720**, 1603 (2010b)
M. Gordovskyy, P.K. Browning, N. Bian, E. Kontar, Effects of collisions and magnetic convergence on high energy particles in solar coronal loops. Sol. Phys. **284**, 489–498 (2013a)
M. Gordovskyy, P.K. Browning, N. Bian, E. Kontar, Acceleration and transport of energetic flare particles in an unstable twisted loop in a gravitationally-stratified atmosphere. Astron. Astrophys. (2013b, in press)
Z.B. Gao, P.H. Diamond, X.G. Wong, Magnetic reconnection, helicity-dynamics and hyperdiffusion. Astrophys. J. **757**, 173 (2012)
T.S. Hahm, R.M. Kulsrud, Forced magnetic reconnection. Phys. Fluids **28**, 2412–2418 (1985)
J. Heyvaerts, E.R. Priest, Coronal heating by reconnection in DC current systems; a theory based on Taylor hypothesis. Astron. Astrophys. **137**, 63 (1984)
A.W. Hood, P.K. Browning, R.A.M. Van der Linden, Coronal heating by magnetic reconnection in loops with zero-net current. Astron. Astrophys. **506**, 913 (2009)
Y.-M. Huang, A. Bhattarcharjee, Plasmoid instability in high-Lundquist-number magnetic reconnection. Phys. Plasmas **20**, 055702 (2013)
A.R. Jacobson, R.W. Moses, Nonlocal dc electrical conductivity of a Lorentz plasma in a stochastic magnetic field. Phys. Rev. A **29**, 3335–3342 (1984)
R. Jain, P.K. Browning, K. Kusano, Solar coronal heating by forced magnetic reconnection; multiple reconnection events. Phys. Plasmas **12**, 012904 (2005)
T.R. Jarboe, Review of spheromak research. Plasma Phys. Control. Fusion **36**, 945 (1994)
H. Karimabadi, V. Roytershteyn, M. Wan, W.H. Matthaeus, W. Daughton et al., Coherent structures, intermittent turbulence, and dissipation in high-temperature plasmas. Phys. Plasmas **20**, 012303 (2013)
M. Karlicky, M. Barta, D. Nickeler, Fragmentation during merging of plasmoids in the magnetic field reconnection. Astron. Astrophys. **541**, A86 (2012)
E.-j. Kim, P.H. Diamond, On turbulent reconnection. Astrophys. J. **556**, 1052–1065 (2001)
B. Kliem, Particle orbits, trapping and acceleration in a filamentary current sheet model. Astrophys. J. **90**, 719 (1994)
K. Kotera, A.V. Olinto, The astrophysics of ultrahigh-energy cosmic rays. Annu. Rev. Astron. Astrophys. **49**, 119–153 (2011)
G. Kowal, A. Lazarian, E.T. Vishniac, K. Otmianowska-Mazur, Numerical tests of fast reconnection in weakly stochastic magnetic fields. Astrophys. J. **700**, 63–85 (2009)
G. Kowal, E.M. de Gouveia Dal Pino, A. Lazarian, Magnetohydrodynamic simulations of reconnection and particle acceleration: three-dimensional effects. Astrophys. J. **735**, 102 (2011)
G. Kowal, A. Lazarian, E.T. Vishniac, K. Otmianowska-Mazur, Reconnection studies under different types of turbulence driving. Nonlinear Process. Geophys. **19**, 297–314 (2012a)
G. Kowal, E.M. de Gouveia Dal Pino, A. Lazarian, Acceleration in turbulence and weakly stochastic reconnection. Phys. Rev. Lett. **108**, 241102 (2012b)

K. Kusano, Y. Suzuki, H. Kubo, T. Miyoshi, K. Nishikawa, 3D simulation study of the MHD relaxation process in the solar corona. Astrophys. J. **433**, 361 (1994)

A. Lazarian, Astrophysical implications of turbulent reconnection: from cosmic rays to star formation, in *Magnetic Fields in the Universe: From Laboratory and Stars to Primordial Structures*, vol. 784 (AIP, New York, 2005), p. 42

A. Lazarian, Theoretical approaches to particle propagation and acceleration in turbulent intergalactic medium. Astron. Nachr. **327**, 609 (2006)

A. Lazarian, Obtaining spectra of turbulent velocity from observations. Space Sci. Rev. **143**, 357–385 (2009)

A. Lazarian, J. Cho, Scaling, intermittency and decay of MHD turbulence. Phys. Scr. T **116**, 32 (2005)

A. Lazarian, M. Opher, A model of acceleration of anomalous cosmic rays by reconnection in the heliosheath. Astrophys. J. **703**, 8 (2009)

A. Lazarian, D. Pogosyan, Velocity modification of HI spectrum and clouds in velocity space. Bull. Am. Astron. Soc. **31**, 1449 (1999)

A. Lazarian, E.T. Vishniac, Reconnection in a weakly stochastic field. Astrophys. J. **517**, 700–718 (1999). (LV99)

A. Lazarian, E.T. Vishniac, Model of reconnection of weakly stochastic magnetic field and its implications. Rev. Mex. Astron. Astrofís., Ser. Conf. **36**, 81–88 (2009)

A. Lazarian, H. Yan, Magnetic reconnection in turbulent plasmas and gamma ray bursts. AIP Conf. Ser. **1505**, 101 (2012)

A. Lazarian, V. Petrosian, H. Yan, J. Cho, Physics of gamma-ray bursts: turbulence, energy transfer and reconnection (2003). arXiv:astro-ph/0301181

A. Lazarian, E.T. Vishniac, J. Cho, Magnetic field structure and stochastic reconnection in a partially ionized gas. Astrophys. J. **603**, 180–197 (2004)

A. Lazarian, A. Esquivel, R. Crutcher, Magnetisation of cloud cores and envelopes and other consequences of reconnection diffusion. Astrophys. J. **757**, 154 (2012a)

A. Lazarian, L. Vlahos, G. Kowal, H. Lan, A. Beresnyak, E. del Pino, Turbulence, magnetic reconnection in turbulent fluids and energetic particle acceleration. Space Sci. Rev. **173**, 107 (2012b)

Y. Li, J. Lin, Acceleration of electrons and protons in reconnecting current sheets including single or multiple X-points. Sol. Phys. **279**, 91 (2012)

D.W. Longcope, S.C. Cowley, Current sheet formation along three-dimensional magnetic separators. Phys. Plasmas **3**, 2885 (1996)

R.M. Lothian, P.K. Browning, Energy dissipation and helicity in coronal loops of variable cross-section. Sol. Phys. **161**, 289 (1995)

R.M. Lothian, P.K. Browning, Energy dissipation and helicity in coronal loops of variable cross-section. Sol. Phys. **194**, 205–227 (2000)

N.F. Loureiro, D.A. Uzdensky, A.A. Schekochihin, S.C. Cowley, T.A. Yousef, Turbulent magnetic reconnection in two dimensions. Mon. Not. R. Astron. Soc. **399**, L146–L150 (2009)

N.F. Loureiro, T. Samataney, A.A. Schekochihin, D.A. Uzdensky, Magnetic reconnection and stochastic plasmoid chains in high-Lundquist-number plasmas. Phys. Plasmas **19**, 042303 (2012)

B.C. Low, Spontaneous current sheets in an ideal hydromagnetic fluid. Astrophys. J. **649**, 1064 (2006)

R.S. Maclean, J. Buechner, E.R. Priest, Relationship between the topological skeleton, current concentrations, and 3D magnetic reconnection sites in the solar atmosphere. Astron. Astrophys. **501**, 321–333 (2009)

S. Markidis, P. Henri, G. Lapenta, A. Divin, M. Goldman, D. Newman, E. Laure, Kinetic simulations of plasmoid chain dynamics. Phys. Plasmas **20**, 082105 (2012)

W.H. Matthaeus, S.L. Lamkin, Rapid magnetic reconnection caused by finite amplitude fluctuations. Phys. Fluids **28**, 303–307 (1985)

W.H. Matthaeus, S.L. Lamkin, Turbulent magnetic reconnection. Phys. Fluids **29**, 2513–2534 (1986)

D.B. Melrose, Acceleration mechanisms (2009). arXiv:0902.1803

J.A. Miller, P.J. Cargill, A.G. Emslie et al., Critical issues for understanding particle acceleration in impulsive solar flares. J. Geophys. Res. **103**, 14631–14659 (1997)

H.K. Moffat, *Magnetic Field Generation in Electrically Conducting Fluids* (Cambridge University Press, London, 1978)

D. Nandy, M. Hahn, R.C. Canfield, D.W. Longcope, Detection of a Taylor-like plasma relaxation process in the Sun. Astrophys. J. **597**, L73–L76 (2003)

C.S. Ng, S. Lin, A. Bhattarcharjee, High Lundquist number scaling in three-dimensional simulations of Parker's model of coronal heating. Astrophys. J. **747**, 109 (2012)

Y. Ono, H. Tanabe, T. Yamada et al., Ion and electron heating characteristics of magnetic reconnection in tokamak plasma merging experiments. Plasma Phys. Control. Fusion **54**, 12409 (2012)

M. Onofri, L. Primavera, F. Malara, P. Veltri, Three-dimensional simulations of magnetic reconnection in slab geometry. Phys. Plasmas **11**, 4837 (2004)

P. Padoan, M. Juvela, A. Kritsuk, M.L. Norman, The power spectrum of supersonic turbulence in Perseus. Astrophys. J. Lett. **653**, L125–L128 (2006)
P. Padoan, M. Juvela, A. Kritsuk, M.L. Norman, The power spectrum of turbulence in NGC 1333: outflows or large-scale driving? Astrophys. J. Lett. **707**, L153–L157 (2009)
E.N. Parker, Sweet's mechanism for merging magnetic fields in conducting fluids. J. Geophys. Res. **62**, 509–520 (1957)
E.N. Parker, Topological dissipation and small-scale magnetic fields in turbulent gases. Astrophys. J. **174**, 499 (1972)
E.N. Parker, Comments on the reconnexion rate of magnetic fields. J. Plasma Phys. **9**, 49 (1973)
E.N. Parker, Nanoflares and the solar X-ray corona. Astrophys. J. **330**, 474 (1988)
E.N. Parker, Fast dynamos, cosmic rays, and the galactic magnetic field. Astrophys. J. **401**, 137–145 (1992)
C. Parnell, A.J. Haynes, K. Galsgaard, Recursive reconnection and magnetic skeletons. Astrophys. J. **675**, 1656 (2008)
C. Parnell, R.C. Maclean, A. Haynes, The detection of numerous magnetic separators in a 3D simulation of emerging magnetic flux. Astrophys. J. Lett. **725**, L214 (2010)
P. Petkaki, M.P. Freeman, T. Kirk, C.E.J. Watt, R.B. Horne, Anomalous resistivity and the nonlinear evolution of the ion-acoustic instability. J. Geophys. Res. **11**, A01205 (2006)
H.E. Petschek, Magnetic field annihilation. NASA Spec. Publ. **50**, 425–439 (1964)
G.W. Pneuman, Two-ribbon flares—post-flare loops, in *Solar Flare Magnetohydrodynamics* (1981), pp. 379–428
D.I. Pontin, Three-dimensional magnetic reconnection regimes; a review. Adv. Space Res. **47**, 158 (2011)
D.I. Pontin, G. Hornig, E.R. Priest, Kinematic reconnection at a magnetic null point: fan-aligned current. Geophys. Astrophys. Fluid Dyn. **99**, 77 (2005)
D.I. Pontin, A.L. Wilmot-Smith, G. Hornig, K. Galsgaard, Dynamics of braided coronal loops. Cascade to multiple small-scale reconnection events. Astron. Astrophys. **525**, A57 (2011)
E.R. Priest, T. Forbes, *Magnetic Reconnection* (Cambridge University Press, Cambridge, 2000)
E.R. Priest, T. Forbes, The magnetic nature of solar flares. Astron. Astrophys. Rev. **10**, 313–377 (2002)
E.R. Priest, V.S. Titov, Magnetic reconnection at three-dimensional null points. Philos. Trans. R. Soc. Lond. A **354**, 2951–2992 (1996)
E.R. Priest, J. Heyvaerts, A.M. Title, A flux-tube tectonics model for solar coronal heating driven by the magnetic carpet. Astrophys. J. **576**, 533 (2002)
E.R. Priest, D. Pontin, G. Hornig, On the nature of three-dimensional magnetic reconnection. J. Geophys. Res. **108**(A7), 1285 (2003)
A.F. Rapazzo, M. Velli, G. Einaudi, R.B. Dahlburg, Coronal heating, weak MHD turbulence and scaling laws. Astrophys. J. **657**, L47 (2007)
A.F. Rapazzo, M. Velli, G. Einaudi, Fieldlines twisting in a noisy corona: implications for energy storage and release, and initiation of solar eruptions. Astrophys. J. **771**, 76 (2013)
M.G. Rusbridge, Model of field reversal in diffuse pinch. Plasma Phys. Control. Fusion **19**, 499 (1977)
M.G. Rusbridge, The relationship between the tangled discharge and dynamo models of magnetic relaxation. Plasma Phys. Control. Fusion **33**, 1381 (1991)
R. Santos-Lima, A. Lazarian, E.M. de Gouveia Dal Pino, J. Cho, Diffusion of magnetic field and removal of magnetic flux from clouds via turbulent reconnection. Astrophys. J. **714**, 442–461 (2010)
R. Schlickeiser, I. Lerche, Cosmic gas dynamics. I—Basic equations and the dynamics of hot interstellar matter. Astron. Astrophys. **151**, 151–156 (1985)
M. Scholer, Undriven magnetic reconnection in an isolated current sheet. J. Geophys. Res. **94**, 8805–8812 (1989)
S. Servidio, W.H. Matthaeus, M.A. Shay, P. Dmitruk, P.A. Cassak, M. Wan, Statistics of magnetic reconnection in two-dimensional magnetohydrodynamic turbulence. Phys. Plasmas **17**, 032315 (2010)
M.A. Shay, J.F. Drake, R.E. Denton, D. Biskamp, Structure of the dissipation region during collisionless magnetic reconnection. J. Geophys. Res. **103**, 9165–9176 (1998)
M.A. Shay, J.F. Drake, M. Swisdak, B.N. Rogers, The scaling of embedded collisionless reconnection. Phys. Plasmas **11**, 2199–2213 (2004)
K. Shibata, T. Magara, Solar flares: magnetohydrodynamic processes. Living Rev. Sol. Phys. **8**, 6 (2011)
K. Shibata, S. Tanuma, Plasmoid-induced-reconnection and fractal reconnection. Earth Planets Space **53**, 473–482 (2001)
K. Schindler, M. Hesse, J. Birn, General magnetic reconnection, parallel electric fields and helicity. J. Geophys. Res. **93**, 5547 (1988)
S. Servidio, P. Dmitruk, A. Greco, M. Wan, S. Donato, P.A. Cassak, M. Shay, V. Carbone, W.H. Matthaeus, Magnetic reconnection as an element of turbulence. Nonlinear Process. Geophys. **18**, 675–695 (2011)
D. Smith, S. Ghosh, P. Dmitruk, W.H. Matthaeus, Hall and turbulence effects on magnetic reconnection. Geophys. Res. Lett. **310**, L02805 (2004)

T.W. Speiser, Conductivity without collisions or noise. Planet. Space Sci. **18**, 613 (1970)
A. Stanier, P.K. Browning, M. Gordovskyy, K.G. McClements, M.P. Gryazenevich, V.S. Lukin, Two-fluid simulations of driven reconnection in the Mega Ampere Spherical Tokamak. Phys. Plasmas (2013, submitted). arXiv:1308.2855
S. Stanimirović, A. Lazarian, Velocity and density spectra of the small Magellanic cloud. Astrophys. J. Lett. **551**, L53–L56 (2001)
P.A. Sweet, The neutral point theory of solar flares, in *Electromagnetic Phenomena in Cosmical Physics*, ed. by B. Lehnert. Conf. Proc. IAU Symposium, vol. 6 (Cambridge University Press, Cambridge, 1958), pp. 123–134
R. Sych, V.M. Nakariakov, M. Karlicky, S. Anfinogentov, Relationship between wave processes in sunspots and quasi-periodic pulsations in active region flares. Astron. Astrophys. **505**, 791–799 (2009)
S. Tanuma, T. Yokoyama, T. Kudoh, K. Shibata, 2D MHD numerical simulations of magnetic reconnection generated by a supernova shock in the interstellar medium: generation of X-ray gas in a galaxy. Astrophys. J. **551**, 312–332 (2001)
J.B. Taylor, Relaxation of toroidal plasma and generation of reverse magnetic fields. Phys. Rev. Lett. **33**, 1139–1142 (1974)
J.B. Taylor, Relaxation and magnetic reconnection in plasmas. Rev. Mod. Phys. **58**, 741–763 (1986)
V.S. Titov, G. Hornig, P. Demoulin, Theory of magnetic connectivity in the solar corona. J. Geophys. Res. (2002)
R. Turkmani, P.J. Cargill, K. Galsgaard, L. Vlahos, H. Isliker, Particle acceleration in stochastic current sheets in stressed coronal active regions. Astron. Astrophys. **449**, 749 (2006)
M. Ugai, Computer studies on development of the fast reconnection mechanism for different resistivity models. Phys. Fluids B **4**, 2953–2963 (1992)
M. Ugai, T. Tsuda, Magnetic field-line reconnexion by localized enhancement of resistivity. I—Evolution in a compressible MHD fluid. J. Plasma Phys. **17**, 337–356 (1977)
D.A. Uzdensky, The fast collisionless reconnection condition and the self-organization of solar coronal heating. Astrophys. J. **671**, 2139 (2007)
D.A. Uzdensky, N.F. Loureiro, A.A. Schekochhin, Fast magnetic reconnection in the plasmoid-dominated regime. Phys. Rev. Lett. **105**, 23 (2010)
J.P. Vallée, Observations of the magnetic fields inside and outside the Milky Way, starting with globules (~ 1 parsec), filaments, clouds, superbubbles, spiral arms, galaxies, superclusters, and ending with the cosmological universe's background surface (at ~ 8 teraparsecs). Fundam. Cosm. Phys. **19**, 1–89 (1997)
J.P. Vallée, Observations of the magnetic fields inside and outside the solar system: from meteorites ($\sim$ 10 attoparsecs). Asteroids, planets, stars, pulsars, masers to protostellar cloudlets (< 1 parsec). Fundam. Cosm. Phys. **19**, 319–422 (1998)
A.A. Van Ballegoijen, Electric currents in the solar corona and the existence of magnetostatic equilibrium. Astrophys. J. **298**, 421 (1985)
G.E. Vekstein, R. Jain, Energy release and plasma heating by forced magnetic reconnection. Phys. Plasmas **5**, 1506 (1998)
G.E. Vekstein, E.R. Priest, C.D.C. Steele, MHD equilibria and cusp formation at an X-type neutral line by footpoint shearing. Astrophys. J. **384**, 333 (1992)
G.E. Vekstein, E.R. Priest, C.D.C. Steele, On the problem of magnetic coronal heating by turbulent relaxation. Astrophys. J. **417**, 781 (1993)
E. Vishniac, A. Lazarian, Reconnection in the interstellar medium. Astrophys. J. **591**, 193 (1999)
L. Vlahos, H. Isliker, F. Lepreti, Particle acceleration in an evolving network of unstable current sheets. Astrophys. J. **608**, 540 (2004)
L. Vlahos, S. Krucker, P.J. Cargill, The solar flare, a strongly turbulent particle accelerator, in *Turbulence in Space Plasmas*, ed. by L. Vlahos, P.J. Cargill (Springer, Berlin, 2009)
X. Wang, A. Bhattacharjee, Z.W. Ma, Scaling of collisionless forced reconnection. Phys. Rev. Lett. **87**, 265003 (2001)
P.G. Watson, S. Oughton, I.J.D. Craig, The impact of small-scale turbulence on laminar magnetic reconnection. Phys. Plasmas **14**, 032301 (2007)
A.L. Wilmot-Smith, D. Pontin, G. Hornig, Dynamics of braided coronal loops. I. Onset of magnetic reconnection. Astron. Astrophys. **516**, A5 (2010)
R. Wolfson, G.E. Vekstein, E.R. Priest, Nonlinear evolution of the coronal magnetic field under reconnective relaxation. Astrophys. J. **428**, 345 (1994)
M. Yamada, Progress in understanding magnetic reconnection in laboratory and space astrophysical plasmas. Phys. Plasmas **14**, 058102 (2007)
M. Yamada, Y. Ono, M. Hayakawa, K. Katsurai, F.W. Perkins, Magnetic reconnection in plasma toroids with cohelicity and counterhelicity. Phys. Rev. Lett. **55**, 721–725 (1990)
M. Yamada, R. Kulsrud, H. Ji, Magnetic reconnection. Rev. Mod. Phys. **86**, 603–664 (2011)

M. Yan, L.C. Lee, E.R. Priest, Fast magnetic reconnection with small shock angles. J. Geophys. Res. **97**, 8277–8293 (1992)

A.R. Yeates, G. Hornig, A.L. Wilmot-Smith, Topological constraints on magnetic relaxation. Phys. Rev. Lett. **105**, 085002 (2010)

G. Zank, W.H. Matthaeus, Nearly incompressible fluids II: magnetohydrodynamics, turbulence and waves. Phys. Fluids A **5**, 257 (1993)

B. Zhang, H. Yan, The internal-collision-induced magnetic reconnection and turbulence (ICMART) model of gamma-ray bursts. Astrophys. J. **726**, 90 (2011)

V. Zharkova et al., Recent advances in understanding particle acceleration processes in solar flares. Space Sci. Rev. **159**, 357–420 (2011)

E.G. Zweibel, M. Yamada, Magnetic reconnection in astrophysical and laboratory plasmas. Annu. Rev. Astron. Astrophys. **47**, 291–332 (2009)

DOI 10.1007/978-1-4899-7413-6_11
Reprinted from *Space Science Reviews* Journal, DOI 10.1007/s11214-013-0005-7

Microphysics of Cosmic Plasmas: Hierarchies of Plasma Instabilities from MHD to Kinetic

M.R. Brown · P.K. Browning · M.E. Dieckmann · I. Furno · T.P. Intrator

Received: 5 March 2013 / Accepted: 18 June 2013 / Published online: 31 August 2013

Abstract In this article, we discuss the idea of a hierarchy of instabilities that can rapidly couple the disparate scales of a turbulent plasma system. First, at the largest scale of the system, L, current carrying flux ropes can undergo a kink instability. Second, a kink instability in adjacent flux ropes can rapidly bring together bundles of magnetic flux and drive reconnection, introducing a new scale of the current sheet width, ℓ, perhaps several ion inertial lengths (δ_i) across. Finally, intense current sheets driven by reconnection electric fields can destabilize kinetic waves such as ion cyclotron waves as long as the drift speed of the electrons is large compared to the ion thermal speed, $v_D \gg v_i$. Instabilities such as these can couple MHD scales to kinetic scales, as small as the proton Larmor radius, ρ_i.

Keywords MHD · Turbulence · Reconnection · Relaxation · Instability

1 Introduction

Cosmic plasmas are complex, exhibiting processes at all scales from meters to megaparsecs. There is a tendency to study these processes in isolation but natural plasma systems surely

M.R. Brown (✉)
Dept. of Physics and Astronomy, Swarthmore Coll., Swarthmore, PA 19081, USA
e-mail: doc@swarthmore.edu

P.K. Browning
Univ. of Manchester, Jodrell Bank Ctr. Astrophys., Manchester M13 9PL, UK

M.E. Dieckmann
Dept. of Science and Technology (ITN), Linkoping University, Campus Norrkoping, 60174 Norrkoping, Sweden

I. Furno
Ecole Polytech Fed Lausanne, CRPP, Ctr Rech Phys Plasmas, Lausanne, Switzerland

T.P. Intrator
Los Alamos Natl. Lab, Los Alamos, NM 87545, USA

find ways to couple scales together. Indeed a fully turbulent fluctuation spectrum is the result of a cascade of couplings from large scales to small.

We discuss here the idea of a hierarchy of instabilities that can rapidly couple the disparate scales of a turbulent plasma system. First, at the largest scale of the system, L, current carrying flux ropes can undergo a kink instability. Second, a kink instability in adjacent flux ropes can rapidly bring together bundles of magnetic flux and drive reconnection, introducing a new scale of the current sheet width, ℓ, perhaps several ion inertial lengths (δ_i) across. Finally, intense current sheets driven by reconnection electric fields can destabilize kinetic waves such as ion cyclotron waves as long as the drift speed of the electrons is large compared to the ion thermal speed, $v_D \gg v_i$. Instabilities such as these can couple MHD scales to kinetic scales, as small as the proton Larmor radius, ρ_i.

The paradigm of the turbulent cascade is that energy is added to the system at the largest scale, L. Nonlinearities break up the flow and introduce fluctuations at perhaps $L/2$ in an eddy-turnover time, L/v, where v is a typical flow speed. This is a slow process. Indeed, filling a turbulent spectrum down to the dissipation scale presents theoretical challenges. We suggest in this chapter that there are processes that can directly and rapidly couple the very largest scales to the very smallest.

In Sect. 2, we discuss some theoretical background, beginning with MHD processes at the largest scales (kink instability, reconnection, and relaxation) and kinetic instabilities at the smallest scales. In Sect. 3, some observations and experiments on hierarchies of instabilities including the RSX and SSX experiments are presented. In Sect. 4, results of a PIC code are presented, demonstrating kinetic instabilities driven by large scale field-aligned current, and in Sect. 5 we present a summary.

2 Theoretical Discussion

2.1 MHD Framework: Reconnection and Relaxation

Magnetic reconnection is a localised breakdown of the frozen-in field condition in a highly-conducting plasma, with resulting change in magnetic field line connectivity. The process is characterised by a separation of length-scales between the global scale and the local scale of dissipation. Reconnection results in dissipation of magnetic energy—much more rapidly than simple Ohmic diffusion—and is an important process in many space, astrophysical and laboratory plasmas. We briefly summarize some key elements of reconnection theory here: full reviews include Priest and Forbes (2000), Biskamp (2005), Birn and Priest (2007), Zweibel and Yamada (2009).

In its simplest 2D form, reconnection involves the merging of anti-parallel magnetic fields with flow across magnetic seperatrices. Classical 2D models are: *spontaneous reconnection*, involving linear instability, such as tearing mode (Furth et al. 1963), and *steady-state reconnection*, notably the Sweet–Parker and Petschek models; the latter incorporates standing slow shocks in the inflow region and allows fast reconnection. *Forced reconnection* is triggered by boundary disturbances, and may occur in tearing-stable configurations. In the basic paradigm (Hahm and Kulsrud 1985), the boundary of a 2D field reversal is perturbed by a slow, transient, sinusoidal displacement; initially a current sheet forms, which subsequently reconnects into a chain of magnetic islands. This has been generalised to sheared force-free fields by Vekstein and Jain (1998), who calculate how the energy release during the reconnection, a transition from the higher-energy state with an infinitely-thin current sheet to the lower-energy state with islands, depends on the equilibrium shear.

More recently, much work has been devoted to understanding 3D reconnection, which differs in many fundamental ways from 2D (see review by Pontin 2011); the symmetries required for 2D models are inevitably absent in naturally-occurring plasmas such as in the solar corona, and 3D models are essential. Even the definition of reconnection in 3D is less clear, but a broadly-accepted definition is the existence of a non-zero integrated parallel electric field in a localised region (Schindler et al. 1988). In 3D reconnection, field lines do not generally "cut and join" in pairs (as in 2D), and field line connectivity changes continuously throughout the dissipation region (Priest and Pontin 2009; Pontin 2011). In fully 3D reconnection—or in 2D reconnection with a guide field—reconnection can occur both at magnetic null points and in the absence of nulls. The latter includes Quasi Separatrix Layers, in which there are strong gradients in magnetic connectivity (Demoulin et al. 1996) and separator lines (Longcope 2001; Parnell et al. 2010), which are the intersection of two separatrix surfaces bounding flux domains. Reconnection at 3D nulls has been studied both through analytical models (Priest and Titov 1996; Craig et al. 1997; Craig and Watson 2000) as well as numerical simulations (Pontin et al. 2007; Priest and Pontin 2009; Santos et al. 2011).

Relaxation theory (Taylor 1974, 1986), in which the field relaxes to a state of minimum magnetic energy with conserved magnetic helicity, is a very useful tool for predicting the final state of a plasma subjected to multiple magnetic reconnections. Whilst first developed to explain the toroidal field reversal in Reverse Field Pinch devices, it has been successfully applied also to tokamaks, spherical tokamaks and spheromaks, as well as to the solar corona and other astrophysical plasmas. The minimum energy state is a constant-α or linear force-free field

$$\nabla \times \mathbf{B} = \alpha \mathbf{B} \tag{1}$$

2.2 From Kink Instabilities to Reconnection in the Solar Corona

2.2.1 Models of Nonlinear Kink Instability

A very long-standing problem in solar physics is to explain the high temperature of the solar corona—which is over a million degrees Kelvin (Browning 1991; Klimchuk 2006)—compared with the surface temperature of around 6000 K. Another problem is to understand the process of energy release in solar flares (Benz 2008). These two phenomena may both be manifestations of the physical process of magnetic reconnection, whereby magnetic energy is stored in the corona and converted into thermal and kinetic energy on rapid time-scales by reconnection. Indeed, it was proposed by Parker (1988) that the corona could be heated by the combined effect of a large number of small flare-like events known as "nanoflares". Much effort has thus been devoted to modelling reconnection in the solar corona, and determining how the conditions for reconnection, such as current sheets, may arise (reviewed by Birn and Priest 2007). It is expected that coronal heating occurs due to frequently-occurring reconnection events, releasing free magnetic energy injected by slow motions of the photospheric footpoints of the coronal field.

One approach to coronal heating uses relaxation theory. The coronal field evolves through a sequence of (in general) nonlinear force-free states $\mathbf{j} \times \mathbf{B} = \mathbf{0}$ or $\nabla \times \mathbf{B} = \alpha \mathbf{B}$ (where $\alpha = \frac{\mu_0 j_{//}}{B}$, releasing energy as it relaxes to a constant-α state (Heyvaerts and Priest 1984)). The energy dissipation rate depends on how much energy is stored before a relaxation event occurs: this was undetermined within the Heyvaerts and Priest (1984) framework. This was resolved (Browning and van der Linden 2003) by proposing that relaxation could be triggered by the onset of the ideal kink instability in a twisted loop.

The coronal field contains numerous magnetic loops, which in general comprise twisted magnetic fields: the magnetic field may emerge already twisted from below the solar surface (Moreno-Insertis and Emonet 1996) or photospheric motions with vorticity may create twist by rotating the field footpoints. Such twisted loops may be subject to the ideal kink instability if sufficiently strongly twisted, even taking account of the stabilizing effects of line-tying of the footpoints at the dense photosphere (Hood and Priest 1979). Numerical simulations of the ideal kink instability demonstrate that current sheets form in the nonlinear phase (Baty and Heyvaerts 1996; Velli et al. 1997; Lionello et al. 1998; Linton et al. 1998; Baty 2000), initially taking the form of a helical current ribbon at a quasi-resonant surface, leading to magnetic reconnection.

Browning et al. (2008) and Hood et al. (2009) used 3D MHD simulations of a kink-unstable twisted flux tube in order to investigate the energy release process and the relaxation towards a minimum energy state. These simulations used the Lagrangian Remap code LARE3D (Arber et al. 2001) to solve the 3D MHD equations, without conduction or radiation, on a staggered Cartesian grid. The initial equilibrium was a cylindrical flux tube with a piecewise-constant profile of $\alpha(r)$, allowing a two parameter family of nonlinear force-free fields (Browning and van der Linden 2003).

First, the linear stability threshold for the ideal kink mode, taking account of line-tying, was determined using the CILTS code, for the family of equilibria; field profiles just within the unstable region were selected as initial conditions for the nonlinear 3D MHD simulations. Browning et al. (2008) consider various initial field profiles from points around the linear stability threshold curve; these field profiles all carry a net current, so that the surrounding envelope of potential field has a non-zero azimuthal field component. The initial growing disturbance matches very well the eigenfunction of the most unstable mode from the linear theory, and the amplitude initially grows exponentially, with a growth rate corresponding to the linear prediction. As the evolution reaches its nonlinear phase, a strong helical current sheet is formed and magnetic reconnection initiates, reducing the magnetic energy whilst the kinetic energy passed through a series of peaks. Eventually, the magnetic energy decline levels out, and the kinetic energy reduces significantly (although a full stationary state is never attained). The final "relaxed" state magnetic field components are shown to be well-matched by a constant-α field—despite the fact that the current profile remains spiky and it is not clear that α is constant. Furthermore, the energy release is well predicted by relaxation theory.

Hood et al. (2009) noted that it is more realistic to consider field profiles in which the net current in the loop vanishes, corresponding to footpoint motions localised within the loop cross-section. In this case, the eigenfunction of the unstable kink mode is strongly localised, and the resulting nonlinear disturbance does not extend far beyond the initial loop radius. It is shown that the initial helical current sheet stretches and splits into two, and then breaks up further, resulting in a turbulent relaxation phase with reconnection occurring in a fragmented current sheet distributed throughout the loop. This distributed reconnection mixes the α profile very effectively, causing the field to relax towards the constant-α state, and also results in heating throughout the loop. In this model, in which radiation and thermal conduction are neglected, plasma can be locally heated to temperatures of up to 10^8 K. Conduction and radiation reduce the peak temperature and create a more uniform temperature profile (Botha et al. 2011). Simulations of a wider family of fields with zero net-current show that the kink instability generates both reconnection within the loop and reconnection with the surrounding axial field, causing some expansion of the loop (Bareford et al. 2013). The final state of the field can be well-described by a new partial relaxation model, in which the field relaxes within a limited volume.

It is not viable to perform 3D numerical simulations for a large set of initial field profiles; if we are interested mainly in the energy released by reconnection, then it is appropriate to use relaxation theory. Bareford et al. (2010, 2011) have modelled the energy dissipation resulting from driving of a coronal loop by slow, random photospheric footpoint motions, for families of field profiles both with and without net-current. The coronal field responds to the driving by evolving through a sequence of non-linear force-free equilibria: when the threshold for the linear ideal kink instability is reached, the field becomes disrupted. The energy release is determined by assuming a helicity-conserving relaxation to a linear force-free state. A Monte–Carlo approach is used to generate a large number of relaxation events—the energy-release depending on the equilibrium profile when instability is reached. For the case of localised footpoint twisting, a power-law frequency distribution of heating events can result (Bareford et al. 2011), representing the distribution of nanoflares. The average heating rate determined by this process is shown to be compatible with the requirements for Active Region heating.

2.2.2 Observations of Kink Unstable Loops in the Solar Corona

There is very extensive evidence that the solar corona is—in general—in a non-potential state and thus elementary flux tubes are likely to be twisted. Numerical simulations of flux tubes emerging into the corona from the solar interior, such as described by Archontis et al. (2005), require the flux tube to be twisted so that it retains its identity as it rises. Observations also suggest that newly-emerging flux tubes carry currents (Leka et al. 1996), whilst coronal field lines may also be twisted by footpoint motions. The values of α required for kink instability are, for a loop of radius 1000 km, of the order $3 \times 10^{-6}\ \mathrm{m}^{-1}$.

Direct observations of kink instability are difficult, since the coronal magnetic field is not directly measured. Furthermore, the simulations suggest that the temperature distribution following a nonlinear kink instability event is quite symmetric and would display no obvious helical character. Nevertheless, some observations are indicative of kink instability associated with magnetic reconnection and energy release in the solar corona. Liu and Alexander (2009) analyze Hard X-ray emission in two kinking filament events, and show that the emission is consistent with simulations of curved kink-unstable flux tubes (Török et al. 2004; Török and Kliem 2005). Srivastava et al. (2010) observe a small flare using SOHO, Hinode and TRACE, demonstrating the presence of a highly-twisted magnetic flux tube with a total twist of about 12π (which would almost certainly be linearly unstable). The flare emission is interpreted as arising in current sheets generated by the kink. Imaging microwave observations of a partially-occulted flare (Karlický and Kliem 2010) also provide evidence of reconnection in a kinking loop.

2.3 From Kink Instabilities to Reconnection in Laboratory Plasmas

2.3.1 Spheromaks

Spheromaks are magnetically-confined plasmas contained (ideally) within a simply-connected vessel. Spheromaks may be formed and sustained using coaxial helicity injection, in which a plasma ring is ejected from a Marshall gun into a conducting chamber; the plasma may be sustained against resistive decay by maintaining an electric field from the gun which directly drives the open magnetic flux connected to the electrodes. The presence of instabilities with toroidal mode number $n = 1$ during the driven phase of spheromak operation was first noted by Knox et al. (1986). Duck et al. (1997) mapped the column of open magnetic flux in the SPHEX spheromak, demonstrating that this formed a rotating helical structure, which is proposed to arise from the nonlinear saturation of an ideal current-driven kink instability. Furthermore, this nonlinear kink instability plays a crucial role in the relaxation process—distributing helicity from the directly-driven open flux to the main plasma torus (closed flux). A Poynting flux of energy from the column to the torus was previously directly measured (Browning et al. 1992), which can now be seen to be a direct consequence of the helical structure of the direct column. Furthermore, direct measurements of the dynamo electric field, showed that the $n = 1$ fluctuations were associated with an "anti-dynamo" in the central column, whereas a turbulent dynamo was found in the torus, somehow triggered by the helical disturbance due to the rotating column (Al-Karkhy et al. 1993). The current drive or dynamo process involves small-scale instabilities, including magnetic reconnection. The close association of the $n = 1$ mode (the saturated kink instability) with current drive is further demonstrated by observations on SPHEX in conditions when the $n = 1$ mode is intermittent (Willett et al. 1999): when the mode is present (absent), the plasma current in the torus grows (decays). Subsequent work on SSPX supports the picture of the crucial role of kink instability of the central column in current drive and relaxation in spheromaks (Hooper et al. 1999; Wang et al. 2002; Stallard et al. 2003; Cohen et al. 2009).

A theoretical framework for these observations of the role of the kink-unstable central column in current drive is provided by Brennan et al. (1999, 2002, 2009). The linear stability of an idealized 1D model of the spheromak equilibrium is analyzed, taking into account both line-tying of the field lines at the gun and the compression of the column by the surrounding torus. As expected, the column is kink unstable if it carries sufficiently large current—but this can be stabilized by plasma current in the surrounding closed-flux torus. The current-drive mechanism is thus self-limiting, switching off once the torus current becomes too large. This is consistent with experimental measurements of intermittent $n = 1$ mode (the observable signature of a kink-unstable central column), which demonstrate that the magnetic field naturally exists close to a state of marginal kink instability. An example of the nonlinear phase of the kink instability from a 3D MHD simulation is presented in Fig. 1.

Kink instability leading to reconnection is also observed in a line-tied screw pinch plasma (Bergerson et al. 2006), in a similar scenario to the spheromak central column.

2.3.2 Edge Localized Modes in Tokamaks

Tokamaks are inherently stable and large-scale kink modes are not present. However, current-driven instabilities, dubbed "peeling modes" may be found in the edge region of the plasma, and these may be responsible for the onset of repetitive disturbances known as Edge Localized Modes (ELMs). ELMs tend to occur in H mode operation, and may cause serious losses of heat and particles from the plasma, with potentially serious consequences for future fusion devices. Gimblett et al. (2006a, 2006b) have proposed a model of ELMS in which the disturbance is triggered by a peeling mode, leading to a localized relaxation in the edge region, generating a region of flattened current profile (a local Taylor state) that is bounded by two current sheets. This allows the size of the ELM (the width of the relaxed region) to be calculated as a function of the initial current profile. The resulting dependence of ELM size of edge q is rather complex, displaying a series of peaks related to rational resonances in the edge q.

Recent measurements of ELM mitigation on JET, using resonant magnetic perturbations, show an intriguing "multi-resonance" effect in the dependence of the ELM frequency on edge q (Liang et al. 2010, 2011). This can be explained in terms of the relaxation-based

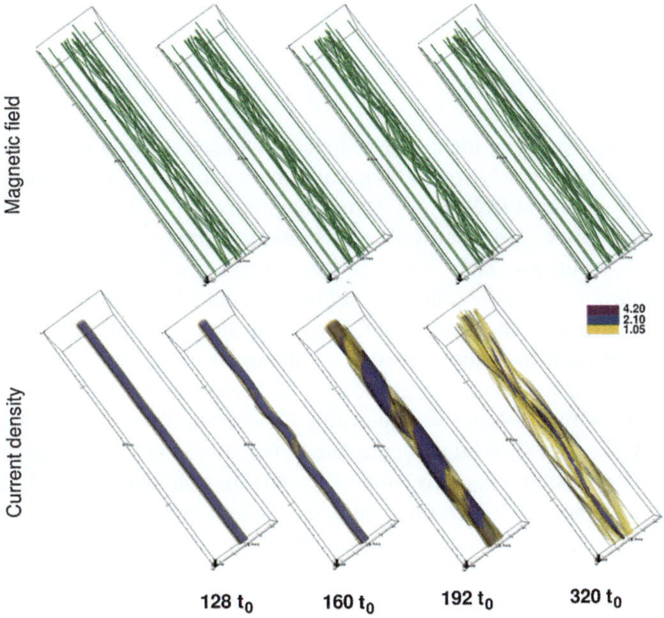

Fig. 1 The fieldlines and current isosurfaces during the nonlinear phase of the kink instability from 3D MHD simulations of an unstable twisted loop, at successive times

model of ELM size, assuming that the ELM repetition period is proportional to the width of the relaxed region created by the ELM.

2.4 Coupling to Kinetic Scales

In a famous 1962 paper, Drummond and Rosenbluth (1962) showed that in a current-carrying plasma system with electrons drifting at speed v_D with respect to ions, electrostatic ion cyclotron waves with $\omega \cong \Omega_{ci}$ are destabilized if $v_D \geq 5v_i$ (roughly) where v_i is the ion thermal speed. The waves propagate across the magnetic field with $k_\perp \rho_i \cong 1$, where $k_\perp$ is the wave number orthogonal to the magnetic field and ρ_i is the ion's thermal gyroradius. The theory is appropriate for $\beta \ll 1$ and $T_e \cong T_i$. Section 4 presents results of a PIC simulation of this scenario. Weibel (1970) modified the theory to include counter-streaming ions in a stationary background of electrons. Again, ion cyclotron waves propagating at nearly right angles to the magnetic field are destabilized. This instability is independent of the mass ratio M_i/m_e.

The instability predicted by Weibel (1970) was verified experimentally in the mid-1970's by Hendel et al. (1976) in a Q-machine. The experimental setup featured a long magnetized target plasma and a coaxial ion beam. The beam density and speed could be tuned relative to the target density and target ion thermal speed, v_i. The target plasma had $T_e \approx T_i \cong 0.3$ eV. The researchers observed waves just below the ion cyclotron frequency with $k_\perp \rho_i \approx 1$ as expected. Instability was observed with a normalized ion beam speed v_{beam}/v_i as low as 2.5. For $v_{beam}/v_i \gg 1$ the instability frequency approached Ω_{ci}. Preliminary evidence of a Drummond-like instability is observed in SSX and discussed in Sect. 3.

3 Observations and Experiments

3.1 Prior Experimental Results

The kink instability has been identified and studied since the early days of magnetic confinement fusion. A particularly compelling experimental identification of the kink instability was performed by Hsu and Bellan (2003) in a magnetized current-carrying plasma column. They found that at low values of current-to-flux ratio (essentially the α parameter introduced in Sect. 2), the column was stable with a helical magnetic field. At higher levels of current-to-flux, a kink developed in the column resulting in a detached plasma. They identified the threshold as the Kruskal–Shafranov limit, and implicated the kink in the dynamo process which converts toroidal to poloidal flux in a spheromak. These are closely related to the SPHEX experimental results discussed above (Al-Karkhy et al. 1993) in which a helical disturbance in a rotating column is responsible for current drive and dynamo in a spheromak. The SPHEX group showed that correlated fluctuations of velocity and magnetic field ($\langle \mathbf{v} \times \mathbf{b} \rangle$) were responsible for conversion of toroidal to poloidal flux. In later experiments (Duck et al. 1997), the SPHEX group identified a "dough-hook" shaped rotating $n = 1$ kink instability of the central open flux tube as the mechanism responsible for current-drive in the spheromak.

In a dramatic series of experiments, Moser and Bellan (2012) tracked the dynamics of a kinking plasma column and identified a growing Rayleigh-Taylor instability at small scales driven by initial kink instability at the largest scale. The drive for the Rayleigh-Taylor instability comes from the effective gravity of the accelerating kink ($g_{eff} = 4 \times 10^{10}$ m/s^2). The kink instability is at the scale of the system, while the Rayleigh-Taylor instability is at a non-MHD micro-scale identified as the ion inertial length (c/ω_{pi}), an order of magnitude smaller. This experiment clearly shows the rapid coupling of scales from MHD to kinetic.

3.2 RSX Experiments

In our plasma universe there is a universal tendency to develop filaments of electric current and tubes of magnetic field. For example *flux ropes* are both twisted along their own axis and writhe or gyrate helically, and are ubiquitous structures on the sun, the rest of the heliosphere, astrophysical objects and laboratory plasmas. Flux ropes form basic building blocks for magnetic dynamics, are analogues of macroscopic magnetic field lines, and are irreducibly three dimensional (3D). Moreover, much of the interesting and important physics occurs beyond the purview of MHD, i.e. at smaller spatial scales or with non ideal two fluid and/or kinetic features.

Large scale kink instabilities of flux ropes were shown to grow on the Reconnection Scaling Experiment (RSX) (Furno et al. 2003). The new feature was that the kink onset threshold depended on boundary conditions (Furno et al. 2006; Sun et al. 2008) that could be adjusted between line tied and free. These two boundary conditions correspond for example to CME eruption flux ropes that are anchored ("line tied") at one end to solar coronal holes with the other end "free" to drift in the solar system.

This experiment showed for the first time how a plasma instability (in this case the kink) initiated magnetic reconnection (Furno et al. 2005). The subsequent saturation or stagnation corresponded to a dynamic balance between magnetic $J \times B$ forces and flux rope inertia in the mass flows (Intrator et al. 2009). The dynamics of the plasma, fields and the supporting plasma current system had features that were intrinsically 3D, and restoring forces due to magnetic field line bending competed with kink driven flows and plasma compression (Sun

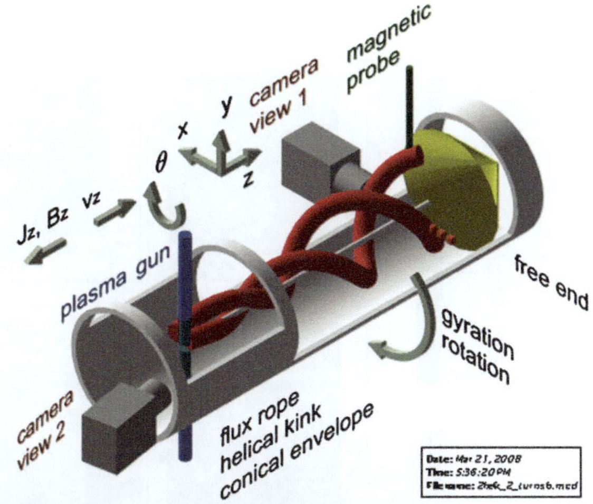

Fig. 2 RSX experiment schematic, showing coordinate system, geometry, background axial magnetic field Bz (*single arrows*). A 19 gun array plasma source (*double arrows*) creates axial flux rope current sheets. Two kinking flux sheets terminate at the external conical anode (*triple arrow*) which allows adjustable axial boundary conditions. Magnetic probes are inserted through 3D probe positioners. Radial excursion from the reference z axis is exaggerated ×4 for clarity

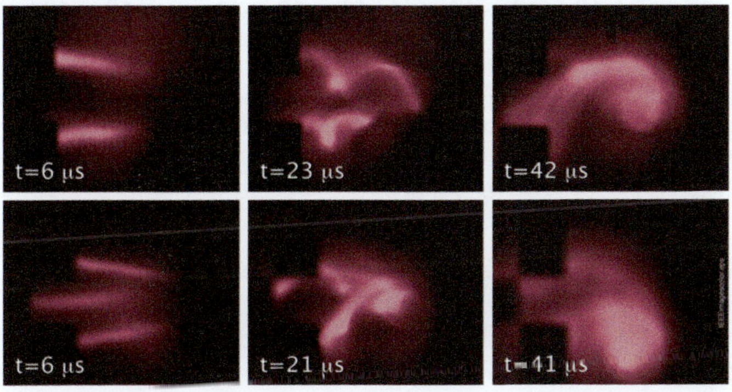

Fig. 3 Fast gated camera frames from an end on perspective view of two (*top row*) and three (*bottom row*) RSX flux ropes that kink, twist and writhe, collide, reconnect and/or bounce (Hemsing et al. 2005)

et al. 2010). It turned out that for this reason, flux ropes bounced off each other much of the time instead of undergoing merging reconnection.

Magnetic structures in RSX are typically strongly inhomogeneous on local spatial scales, down to the electron inertial length, as we will show in Fig. 3. This could lead to local violations of MHD, strong shear, and the existence of plasmoids or magnetic islands. Unsteady intermittent structures like these are critical ingredients for the development of turbulence (Daughton et al. 2011).

RSX is an experimental facility that can create one, two or multiple flux ropes, and study then onset and evolution of these intrinsically 3D objects. The geometry is linear, with axial boundary conditions that can be line tied, free or in between. The Reconnection Scaling Experiment (RSX) (see Fig. 2) has the capability to study the dynamics of flux ropes in 3D, without symmetries that are found in most other (toroidal) experiments.

RSX creates flux ropes that collide, magnetically reconnect, and form turbulent knots, islands, kinks, and disruptions. We use experimental probes inside the flux ropes to investigate macroscopic and unsteady wandering magnetic field lines, and evolution of islands. Structure is measured down to the dissipation scale length which is the electron inertial length $d_i = c/\omega_{pi}$, and typically 1–3 mm in size. The RSX schematic in Fig. 2 includes a view

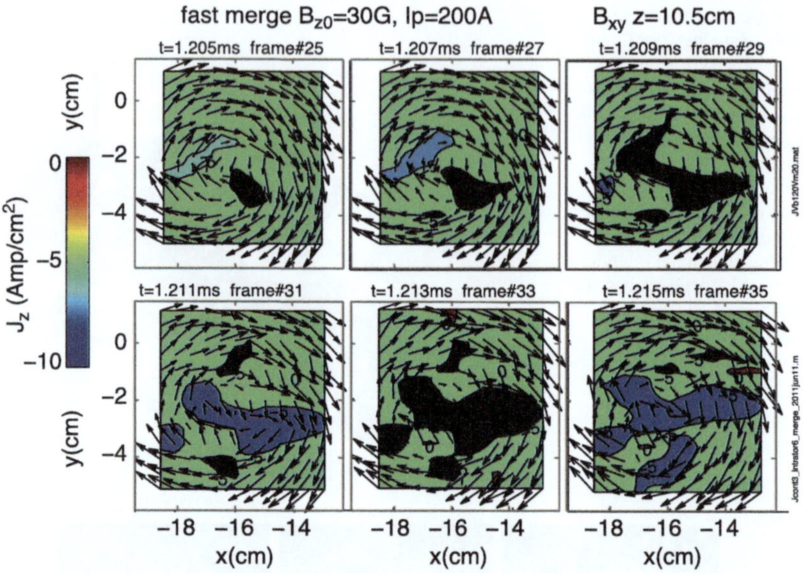

Fig. 4 Detailed RSX fast reconnection data from the data set shown in for $J_z(x, y)$ contours in the x–y cut-plane, showing magnetic field arrows for the case of very fast flux rope merging. Two flux ropes shown with $J_z \approx -10$ Amp/cm^2 (*blue contours*) attract each other, merge, and then break up later in time. $B_{z0} = 30$ G, $I_p = 200$ A, $V_{bias} = 120$ V, $z = 48$ cm from the gun nozzle, *arrows* denote $\vec{B}$

of two plasma guns located at $z = 0$ which are radially inserted into the center of the RSX cylindrical vacuum vessel (2 m length × 0.2 m radius). These forms two cylindrical plasma columns embedded in an externally imposed axial guide magnetic field $B_{z0} \approx 10$–1000 G. A screw pinch is formed by driving a current in the plasma, and the profiles of plasma pressure, magnetic fields, and current density match well with simple analytic models for screw pinch equilibria.

Figure 4 shows examples of probe data for magnetics on sub mm spatial scales, built up over many repeatable shots. RSX data in Fig. 4 show that reconnecting flux ropes can be "shredded" due to 3D effects, where flux ropes can merge only later to be torn apart by (out of plane B_z) field line tension forces. A reconnection current sheet is typically considered to have a width that is the Sweet Parker L_{SP} length scale. In the Sweet-Parker picture of reconnection, the ratio to the thickness of the current sheet δ_{SP} is $L_{SP}/\delta_{SP} = S^{1/2}$, where S is the Lundquist number. Wide thin current sheets have been theoretically (Loureiro et al. 2007; Samtaney et al. 2009; Bhattacharjee et al. 2009) and computationally (Lapenta and Bettarini 2011) shown to be virulently unstable to the creation of islands, plasmoids or flux ropes in 3D. An instability threshold in current sheet size $L_{SP}/\rho_i > 10^2$ was predicted by many, including Daughton et al. (2011). The system size in Fig. 4 is far smaller than this threshold, yet shredding of the current sheet appears to be taking place. It may be that island creation occurs down to kinetic spatial scales, and that these scales are still difficult to handle correctly in theory and simulations.

3.3 SSX Experiments

The Swarthmore Spheromak Experiment (SSX) (Brown 1999) is a flexible facility used to study plasma merging and magnetic reconnection with a variety of boundary shapes.

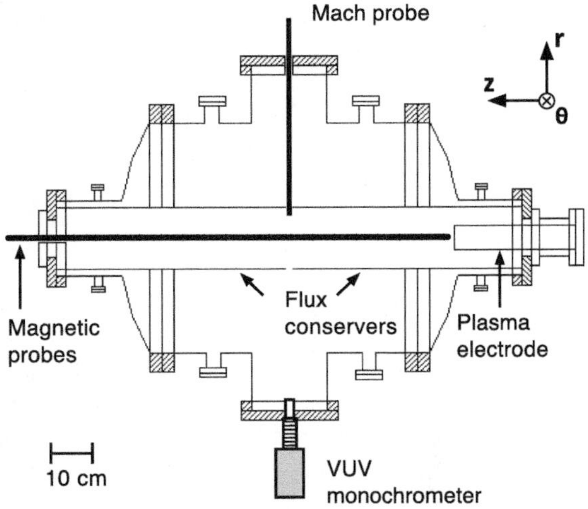

Fig. 5 Schematic of the SSX plasma wind tunnel. For the experiments discussed here, a second plasma source was used and additional magnetic probes were inserted at the midplane

The SSX device features a $L \cong 1$ m long, high vacuum chamber in which we generate $n \geq 10^{20}\ \mathrm{m}^{-3}$, $T \geq 20$ eV, $B \cong 0.1$ T hydrogen plasmas. Plasma plumes are generated by pulsed magnetized plasma guns at either end of the device. Plasmas are accelerated to high velocity ($\cong$50 km/s) by the discharge current in the guns ($\leq$100 kA) and injected into a highly evacuated target volume called a flux conserver. The flux conserver is usually cylindrical in shape and bounded by a thick, highly conducting copper shell. In a typical experiment, plasma plumes are injected at both ends of a flux conserver and dynamical merging and relaxation ensue.

For this study, we have implemented plasma sources in a high aspect ratio "wind tunnel" configuration (see Fig. 5). The wind tunnel has dimensions $R = 0.08$ m and $L = 1.0$ m. The plasma gun can inject a magnetized plasma plume of either right-handed (RH) or left-handed (LH) magnetic helicity from either end of the machine. The magnetic helicity of the plume also determines the helical pitch of magnetic field lines in the final relaxed state in the wind tunnel. From line-averaged measurements of n_e, T_e, T_i, and B, we measure a plasma beta in the wind tunnel of about $\beta \sim 0.5$ (Gray et al. 2013).

Plasma merging studies in the SSX wind tunnel show evidence of large-scale MHD activity (reconnection) driving a kinetic instability. The experiment features merging at $\pm$50 km/s of two opposite helicity plasma plumes. The peak value of $|B| = 0.25$ T, corresponding to a gyro-frequency of $f_{ci} = 3.8$ MHz. With a background ion temperature of about 20 eV, the proton gyro-radius is about $\rho_i = 0.2$ cm. A high resolution probe array monitored reconnection activity at the midplane at a bandwidth a factor of 10 above f_{ci} and at a spatial resolution about ρ_i. The vector magnetic field $\mathbf{B}$ was measured at 16 locations along a radius at the midplane (B_r, B_θ, B_z). Digitization at 14 bits enables measurement of magnetic fields from 1 gauss to nearly 1 Tesla in the same discharge. Ion heating and flow was monitored using ion Doppler spectroscopy (IDS) at the midplane where the main interaction occurs.

A typical reconnection event in the SSX wind tunnel occurs 10–30 μs after initial merging. There is a pile-up of magnetic flux and a spontaneous reconnection event occurs. The event results in a field reversal of $\pm$0.15 T in a few cm so the current density is $J \geq 10^7\ \mathrm{A/m}^2$. Since $J = nev$ and $n = 10^{21}\ \mathrm{m}^{-3}$, typical electron drift speeds in the current sheet are $v_D \geq 100$ km/s while proton thermal speed is $v_i \cong 40$ km/s. This dynamic merg-

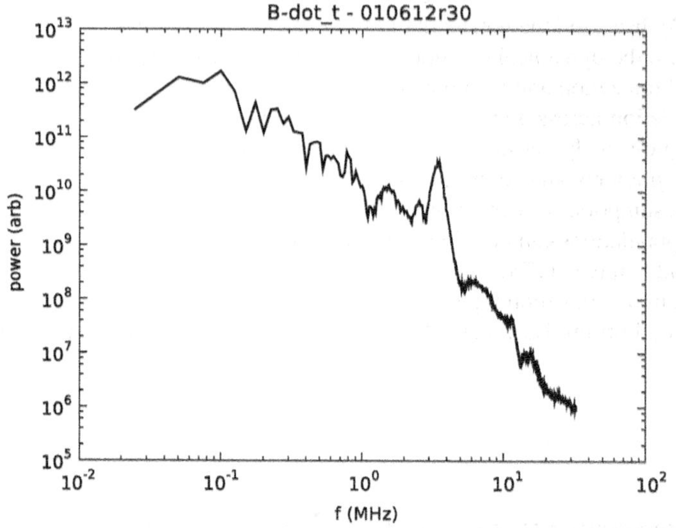

Fig. 6 Magnetic field fluctuation spectrum of the entire discharge (B_θ) from a high-resolution magnetic probe array at the midplane. The spectrum is close to power-law with $f^{-5/3}$ below the proton gyro-frequency

ing and reconnection of twisted plasma plumes at ± 50 km/s drives broadband fluctuations and reconnection-driven turbulence.

Immediately after the reconnection event, and localized to regions typically about 1 ρ_i, we observe bursts of wave activity at 3.5 MHz, near the ion cyclotron frequency. This activity is evident in preliminary analysis of a magnetic field fluctuation spectrum of the entire discharge (see Fig. 6). The power spectrum is otherwise close to a power-law with $f^{-5/3}$ below the proton gyro-frequency, and steeper above the proton gyro-frequency. Correlated in time with the wave bursts, we observe a population of ions heated to $T_i = 40$ eV and flowing at 30 km/s (see Fig. 7). The waves are strongly damped and drive intense ion heating and flows. This measurement is consistent with ion cyclotron waves destabilized by an intense current sheet with large v_D, but more studies need to be done.

4 Kinetic Instabilities Driven by the Field-Aligned Current: PIC Simulations

4.1 Background

The change of the magnetic field topology during reconnection induces field-aligned electric fields, which accelerate the electrons. The resulting relative drift between the ions and electrons along the magnetic field **B** gives rise to an unstable plasma configuration. Electrostatic instabilities driven by a drifting plasma component typically arise from a resonance between the beam dispersion relation $\omega/k = |\mathbf{V}|$ and a wave mode supported by the plasma. The nature of the instability, by which the plasma relaxes, depends on how the drift speed **V** of ions relative to electrons compares to the phase speed of the plasma waves and to the electron thermal speed v_{te}.

If **V** exceeds significantly v_{te}, the Buneman instability develops (Buneman 1959). It is an electrostatic instability and the waves grow fastest if their wave vector is parallel to **V** and, thus, to **B**. This case is illustrated in Fig. 8(a). The Buneman instability can also drive

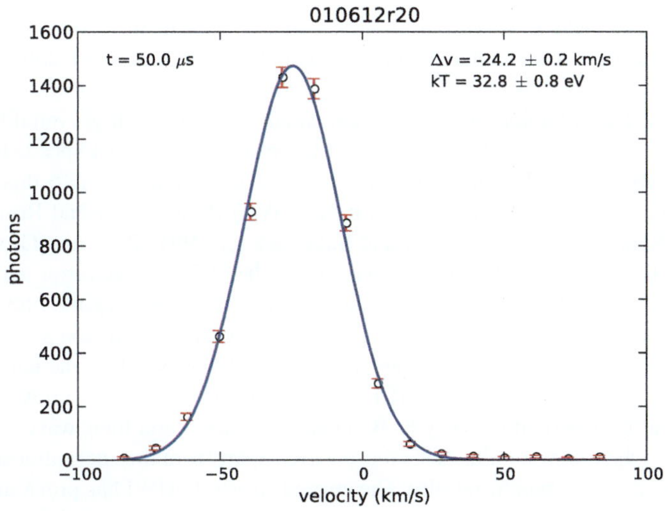

Fig. 7 Burst of plasma flow with $T_i = 40$ eV and flowing at 30 km/s immediately after reconnection event. The background proton temperature is 20 eV

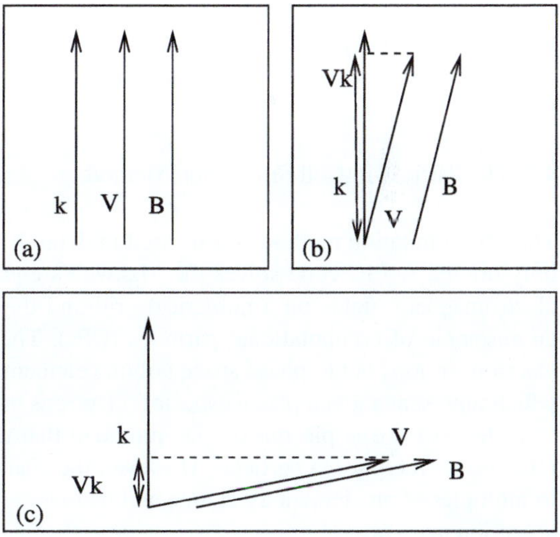

Fig. 8 Cases (**a**–**c**) denote the possible configurations if the drift speed **V** between ions and electrons is parallel to the magnetic field **B**. The *arrows* denoted as **k** are the wave vectors of the unstable electrostatic waves. The beam velocity vector **V** is always parallel to the magnetic field **B**. The projection V_k of **V** onto **k** fulfills $V_k \leq |\mathbf{V}|$. (**a**) and (**b**) result in a Buneman instability if $|\mathbf{V}|$ and V_k are significantly larger than the electron thermal speed v_{te}, respectively. The projected speed $V_k \ll v_{te}$ in (**c**) and only the ICHWI can be driven

oblique modes with a wave vector **k**, if the drift speed component $\mathbf{V}_k$ along **k** exceeds v_{te}. This case is illustrated in Fig. 8(b). The Buneman instability saturates by the formation of a chain of electron phase space holes, which were first found in simulations of counter-streaming electron beams (Roberts and Berk 1967). Electron phase space holes are kinetic-scale bipolar electric field structures, in which a fraction of the electrons is trapped by the wave potential (Bernstein et al. 1957; Schamel 1986; Dieckmann et al. 2012). A chain of electron phase space holes is unstable to the coalescence instability (Roberts and Berk 1967) and, as a result, only solitary phase space holes survive. Solitary phase space holes are stable in one dimension (Bernstein et al. 1957) and they can be observed over limited time intervals also in three dimensions (Sarri et al. 2010). Their life-time is extended by the guiding magnetic field **B** (Ergun et al. 1998) that is present in reconnection studies and they are thus encountered in related numerical simulations (Drake et al. 2003; Lapenta et al. 2010).

As discussed above a field-aligned current can drive an instability, even if the drift speed between ions and electrons does not exceed the electron's thermal speed. It can destabilize electrostatic ion cyclotron harmonic (ICH) waves and we refer to this instability as the ion cyclotron harmonic wave instability (ICHWI) (Drummond and Rosenbluth 1962; Kindel and Kennel 1971; Rasmussen and Schrittwieser 1991; Basu and Grossbard 2011). ICH waves, like their high-frequency counterpart the electron cyclotron harmonic (ECH) waves (Bernstein 1958), are practically undamped in the linear regime only if they move almost perpendicularly to **B**. The field-aligned driving current in the ion's rest frame, which is provided by the drifting electrons, can couple to ICH waves only if the latter have a component of **k** along **V** (Fig. 8(c)). The ICHWI does thus drive waves with wave vectors that are oriented almost perpendicularly to **B**. Their orientation and their wavelength, which is comparable to the ion gyroradius, sets these waves apart from the short almost field-aligned waves driven the Buneman instability. The current-driven ICHWI has previously been modelled by a hybrid scheme, which approximates electrons by an inertia-less fluid and uses a kinetic scheme for the ions (Seyler and Providakes 1987). A spatial modulation of the ion density constituted the final state of the ICHWI. Later on this instability has been examined with particle-in-cell (PIC) simulations in periodic and open boundary systems (Ishiguro et al. 1997a, 1997b). We present here results from PIC simulations (see Sect. 4.2) of the Buneman instability and the ICHWI, which elucidate their growth and saturation mechanisms. The simulations are restricted to one spatial dimension and we consider the cases in Figs. 8(a,c).

4.2 The Particle-in-Cell Simulation Method

The PIC simulation method is discussed in detail by Dawson (1983) and we outline it here only briefly. A PIC code solves the Maxwell's equations for the macroscopic (collective) electromagnetic fields on a numerical grid and the relativistic Lorentz force equation for an ensemble of computational particles (CPs). The CP's do not correspond to individual electrons or ions but to phase space volume elements. The ensemble properties of the CP's, which represent a given plasma species (electrons or ions) in a PIC simulation, approximate well those of a true plasma species provided that the CP's have the same charge-to-mass ratio as the true plasma particles. However, the charge q_j of a CP does not have to be equal to multiples of an elementary charge and its mass m_j does not have to be the electron mass m_e or ion mass m_i.

The PIC code algorithm works as follows. The microscopic current density $\mathbf{j}_j \propto q_j \mathbf{v}_j$ of the j'th CP that moves at the speed $\mathbf{v}_j$ is distributed over the closest grid points. The exact distribution method depends on the interpolation order. The assignment of the entire microcurrent to the next grid point (NGP) is possible but, typically, higher order polynomial schemes are used due to their superior numerical properties. Common are second order or cubic interpolation schemes. By summing up the current contributions of all CP's, we obtain the macroscopic plasma current density $\mathbf{J}(\mathbf{x})$. The equations

$$\nabla \times \mathbf{E} = -\frac{\partial \mathbf{B}}{\partial t} \tag{2}$$

$$\nabla \times \mathbf{B} = \mu_0 \mathbf{J} + \mu_0 \epsilon_0 \frac{\partial \mathbf{E}}{\partial t} \tag{3}$$

are used to update in time the macroscopic electric **E** and magnetic **B** fields. In what follows we always assume that **B** is constant; although the EPOCH PIC code we use is electromagnetic, the electrostatic instabilities leave **B** unchanged. The equations $\nabla \cdot \mathbf{B} = 0$ and $\nabla \cdot \mathbf{E} = \rho/\epsilon_0$ are typically solved as constraints or through correction steps. The position of each CP is updated in time by $\mathbf{x}_{i,new} = \mathbf{x}_{i,old} + \mathbf{v}_{i,old}\Delta_t$, where Δ_t is the time step. A numerical approximation of the relativistic Lorentz equation

$$\frac{d\mathbf{p}_i}{dt} = q_i\big(\mathbf{E}(\mathbf{x}_i) + \mathbf{v}_i \times \mathbf{B}(\mathbf{x}_i)\big). \tag{4}$$

updates the momentum of each CP. The electric $\mathbf{E}(\mathbf{x}_i)$ and magnetic $\mathbf{B}(\mathbf{x}_i)$ fields are interpolated from the grid to the particle position using the same interpolation scheme as that between the microscopic and macroscopic currents. We resolve in what follows all three components of the particle momentum and one spatial (x) dimension and use the periodic boundary conditions.

4.3 Buneman Instability

The Buneman instability develops as the result of a drift between the ions and electrons that exceeds the electron thermal speed. We introduce two plasma species in the simulations. Electrons with the charge-to-mass ratio $-e/m_e$ and equally dense protons with the correct ratio e/m_p. The electrons are at rest in the rest frame of the simulation and the protons move at the mean speed **V** in the positive x-direction. We apply a spatially uniform background current to compensate the net protonic current. The plasma is thus initially charge and current neutral. No initial electromagnetic fields are introduced; a $\mathbf{B} \neq 0$ has no effect on the Buneman instability if only unstable waves with $\mathbf{k} \parallel \mathbf{B}$ are considered.

The Buneman instability is described by the simple dispersion relation $1 - \omega_p^2/\omega^2 + \omega_{pi}^2/(\omega - kV)^2 = 0$ in the cold plasma limit with the proton plasma frequency ω_{pi}. The proton beam speed **V** and the wavenumber **k** are both taken to be parallel to x and they thus reduce to scalars. The most unstable Buneman modes have the phase speed $\omega_u/k_u \approx V$ and they are co-moving with the proton beam. The frequency $\omega_u \approx \omega_p$ in the electron rest frame and $k_u \approx \omega_p/V$ or $k_u\lambda_D \approx v_{te}/V$. The exponential growth rate of the cold Buneman instability $\omega_i \approx (3m_e/16m_i)^{1/3}$ is $\approx 0.05\omega_p$. We consider three cases. We start with the case $V \gg v_{te}$ to illustrate the growth and saturation of the cold Buneman instability. We then reduce the proton beam speed in two steps to get an idea how thermal effects affect the saturation state. We resolve in all cases a simulation box with length $L_1 = 1700\lambda_D$ by 2400 simulation cells and we represent each species by 1600 CPs per cell. We examine the cases $V/v_{te} = 15$, 5 and 2.5.

The time-evolution of the electrostatic field of the Buneman instability driven by a spatially uniform proton beam in one dimension is visualized best with a k-t diagram. We perform a Fourier transform over space of the electric field $E_x(x,t)$ along the simulation direction and compute the power spectrum by multiplying the result $E_x(k,t)$ with its complex conjugate. We display the power spectrum $|E_x(k,t)|^2$ as a function of the time t. The power spectra of all three case studies are normalized to that with $V/v_{te} = 15$ to illustrate its change with the beam speed.

Case 1, $V/v_{te} = 15$ A strong wave appears at $t\omega_p \approx 70$ at a wavenumber $k\lambda_D \approx 0.07$, which equals the estimate $k_u\lambda_D \approx v_{te}/V$ from cold plasma theory. Its harmonics nk_u (Yoon et al. 2003) are observed during $145 < t\omega_p < 180$. The spectrum shows that wave power is transferred to lower k after $t\omega_p \approx 200$, which is a consequence of the coalescence (Roberts

Fig. 9 Panel (**a**) shows the time evolution of the power spectrum $\log_{10}|E_x(k,t)|^2$, normalized to its peak value. Panel (**b**) shows the electron phase space distribution in a sub-interval of the box at the time $t\omega_p \approx 180$

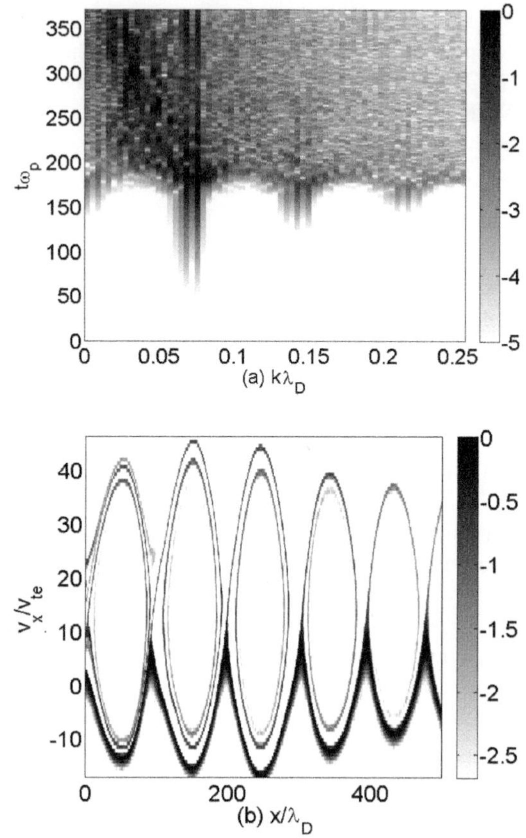

and Berk 1967) of the electron phase space holes visible in Fig. 9(b). The elliptical structures correspond to the electrons, which are trapped in the wave potential. Multiple beams of trapped electrons are observed, because the electrons have been picked up by a wave with a growing potential. The center of each electron phase space hole corresponds to a stable equilibrium point. These equilibrium points move at the phase speed that is just below the $V = 15v_{te}$ of the proton beam (Buneman 1959). The broadband electrostatic turbulence visible after $t\omega_p \approx 180$ in Fig. 9(a) is a consequence of the spatially varying shapes and sizes of the electron phase space holes in Fig. 9(b).

Case 2, $V/v_{te} = 5$ A reduction of the proton beam speed by a factor 3 leaves the instability qualitatively unchanged, which is demonstrated by Fig. 10. The wave is growing here at $k\lambda_D \approx 0.2$, again in excellent agreement with the estimate $k_u\lambda_D \approx v_{te}/V$ from cold theory. The peak power is reduced by more than an order of magnitude compared to the first case; the lower phase speed of the wave in the electron reference frame means that it starts to interact nonlinearly with the electrons at a lower wave amplitude. The order-of-magnitude decrease is in line with analytic predictions (Ishihara et al. 1980). The electron phase space distribution shows again a chain of electron phase space holes when the instability saturates and a gradual transfer of power to lower wavenumbers after $t\omega_p \approx 200$.

Case 3, $V/v_{te} = 2.5$ A further decrease of the proton beam speed changes also qualitatively the growth and saturation of the instability. Although we observe growing waves with

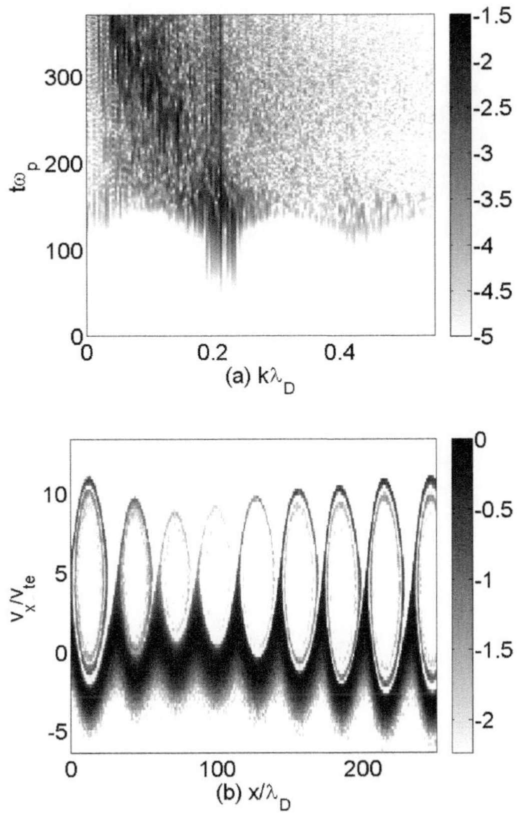

Fig. 10 Panel (**a**) shows the time evolution of the power spectrum $\log_{10} |E_x(k,t)|^2$, normalized to the peak value of Case 1. Panel (**b**) shows the electron phase space distribution in a sub-interval of the box at the time $t\omega_p \approx 140$

the expected $k_u\lambda_D \approx 0.4$, the band of unstable waves in Fig. 11(a) is wider and the system is practically stationary for $t\omega_p > 400$. A shift in time of the wave spectrum to lower values of k prior to this time does, however, demonstrate that at least initially the electron phase space holes merge also in this simulation. The electron phase space holes are now filled with electrons; the wave has been growing in a velocity interval that has been populated with electrons, while the phase speed of the wave in the first two cases was much higher than the peak speed of the electrons. The latter is limited to about $|v| \leq 3$ in the rest frame of the electrons due to the representation of the Maxwellian velocity distribution by a large but finite number of CPs in our PIC simulations. The larger dynamical range of the phase space density accessible to a Vlasov code pushes the boundary between the cold Buneman instability (Case 1) and the kinetic Buneman instability (Case 3) to higher beam speeds (Dieckmann et al. 2004a).

A further decrease of the drift speed between protons and electrons results in a drastic decrease of the growth rate of the Buneman instability, because the waves experience Landau damping by the hot electrons. The long growth times implies that this instability can no longer develop during fast reconnection processes.

4.4 Ion-Cyclotron Instability

A relative speed between the proton beam and the electrons below $V \approx v_{te}$ implies that the dispersion relation $\omega/k = V$ of the proton beam in the electron frame of reference can

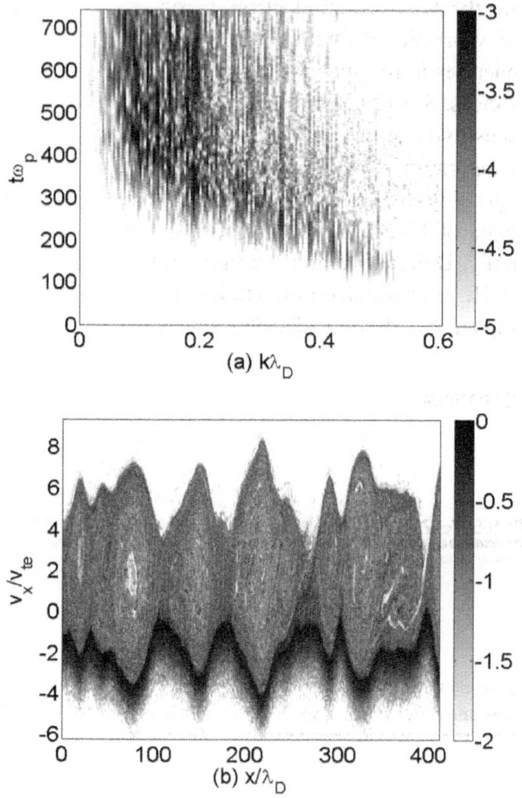

Fig. 11 Panel (**a**) shows the time evolution of the power spectrum $\log_{10}|E_x(k,t)|^2$, normalized to the peak value of Case 1. Panel (**b**) shows the electron phase space distribution in a sub-interval of the box at the time $t\omega_p \approx 140$

no longer couple to the undamped Langmuir modes with their phase speed $\omega_p/k > v_{te}$. Any other electrostatic instability would require the presence of undamped waves with a lower phase speed, which is not the case for $\mathbf{V} \parallel \mathbf{k}$ in the absence of a second ion beam (Forslund and Shonk 1970). Since we also demand $\mathbf{V} \parallel \mathbf{B}$, instabilities would have to involve waves with $\mathbf{k} \not\parallel \mathbf{B}$. The projection V_k of the beam velocity $\mathbf{V}$ onto an oblique wave vector $\mathbf{k}$ (see Fig. 8(c)) implies $V_k \ll |\mathbf{V}|$ and any unstable wave would have to have a phase speed $\omega/k \ll v_{te}$. The rich spectrum of electrostatic waves with a $\mathbf{k}$ that is quasi-perpendicular to $\mathbf{B}$ impedes their simple description in form of solutions of the linear dispersion relation.

A PIC simulation produces statistical noise with a broad range of frequencies and wavenumbers. The noise in the electrostatic field component along a 1D simulation box is electrostatic and is associated with statistical fluctuations of the particle numbers and, thus, of the charge density. The noise is strongest, if its frequency ω and wave number k correspond to a plasma eigenmode. A PIC simulation demonstrating this relation has been performed by Dieckmann et al. (2004b). We can determine with the help of the noise the electrostatic wave spectrum, which is supported by the plasma. We perform for this purpose a 1D simulation and take $\mathbf{B}$ to be perpendicular to the simulation direction. We consider one electron species and one species of ions with a mass $m_i = 25m_e$. Both species are at rest in the simulation box and have the temperature 250 eV. We resolve the length $L_2 = 5100\lambda_D$ of the simulation box by 4800 simulation cells and set $\omega_{ce} \equiv e|\mathbf{B}|/m_e = \omega_p/2$. L thus equals 2550 thermal gyro-radii v_{te}/ω_{ce} of the electrons or 510 gyroradii v_{ti}/ω_{ci} of the ions that move with the ion thermal speed v_{ti}. The simulation time $t\omega_p = 4500$ or $t\omega_{ci} = 90$. We

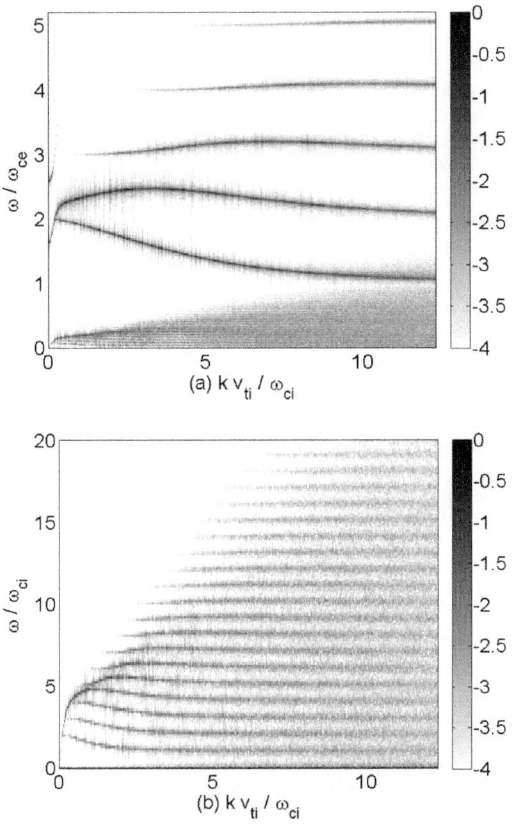

Fig. 12 The 10-logarithmic power spectrum of the noise as a function of ω/ω_{ce} and kv_{ti}/ω_{ci}: Panel (**a**) shows the high-frequency branches of the electron cyclotron harmonic (ECH) waves with $\omega \geq \omega_{ce}$. The low frequency branches of the ion cycloton harmonic (ICH) waves have $\omega < \omega_{ce}$. Panel (**b**) enlarges the low-frequency interval and gives frequencies in units of the ion cyclotron frequency ω_{ci}. All wave numbers are normalized to the ion thermal gyroradius

sample the electric field component $E_x(x,t)$ along the simulation direction and Fourier transform it over space and time to get $E_x(k,\omega)$.

The full power spectrum $|E_x^2(k,\omega)|$ is displayed in Fig. 12. The modes with $\omega/\omega_{ce} > 1$ are the ECH modes. The ECH mode in the branch $2 < \omega/\omega_{ce} < 3$ contains the upper-hybrid frequency and it couples to the slow extraordinary mode at low k. The wave modes with $\omega/\omega_{ce} < 1$ are the ICH modes. Figure 12(b) enlarges their frequency interval. Clearly, ICH branches exist with frequencies exceeding ω_{ci}. The ICH branches couple at low wave numbers to the lower-hybrid mode. The slow extraordinary mode and the lower-hybrid mode are fluid modes and their dispersion relation is only weakly affected by a change in temperature, while the wave numbers of the kinetic ECH/ICH waves scale linearly with the inverse thermal speed. The ECH / ICH modes will get damped by gyroresonance with the electrons, when we go to the required $\mathbf{k} \not\perp \mathbf{B}$.

Now we examine the ICHWI. We select a magnetic **B** field direction, which is oriented in the x–y plane and forms an angle $\theta = 0.2$ radians with respect to the y-direction. The strength of the magnetic field gives $\omega_p = 2\omega_{ce}$. We use an ion-to-electron mass ratio $m_i/m_e = 200$. The temperature of electrons and ions is 250 eV. The length of the 1D simulation box $L_3 = 82.5v_{ti}/\omega_{ci}$ along the x-direction is resolved by 2400 grid cells and we represent ions and electrons by 8000 CPs per cell each. The ions are at rest and the electrons flow along **B** at the speed $v_D = v_{te}$. The projection of the drift speed along the direction of the wavevectors, which is resolved by the 1D simulation, is thus $v_k = 0.2v_D$ and $\theta V_D/c_s = 2.2$ with $c_s^2 = \gamma kT_e/m_i$ and $\gamma = 5/3$. The latter implies that we are in the

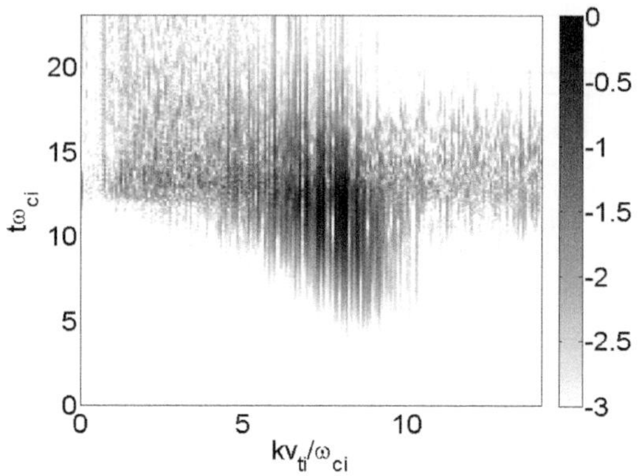

Fig. 13 The time evolution of the power spectrum $\log_{10}|E_x(k,t)|^2$ of the electrostatic field in the simulation box. The power spectrum is normalized to its peak value

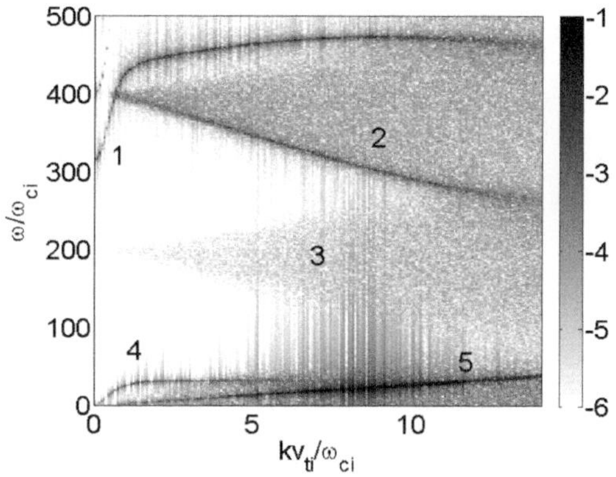

Fig. 14 The 10-logarithmic power spectrum $|E_x(k,\omega)|^3$ of the initial electrostatic field, normalized to its peak value. Mode 1 is the slow extraordinary mode, which couples to the upper hybrid frequency branch of the ECH modes at $kv_{ti}/\omega_{ci} \approx 1$. Mode 2 is an ECH mode. The electrostatic noise 3 around ω_{ce} is generated by gyroresonance of the waves with the electrons that move along the oblique **B**. Mode 4 is the oblique ion acoustic mode and 5 is the electron beam mode

kinetic regime of the ICHWI (Seyler and Providakes 1987). The high electron drift speed and the reduced ion mass accelerate the growth of the ICHWI (Drummond and Rosenbluth 1962). We compensate for the electronic net current and the plasma is initially charge- and current-neutral.

We compute the k-t power spectrum of the electrostatic field $|E_x(k,t)|^2$ in the simulation direction by Fourier transforming $E_x(x,t)$ over x and by taking the square of the amplitude modulus $E_x(k,t)$. Figure 13 shows its evolution during the simulation time $T\omega_{ci} = 23$. Strong electrostatic waves start to grow at $t\omega_{ci} = 5$ at around $5 < kv_{ti}/\omega_{ci} < 10$ and the electrostatic waves thus have a wavelength close to a thermal ion Larmor radius, which is what we expect for the ICHWI (Drummond and Rosenbluth 1962). The wave power grows rapidly by about three orders of magnitude within $t\omega_{ci} \approx 3$ and saturates; the exponential growth rate is thus not small compared to ω_{ci}.

The dispersion relation provided by the electrostatic fluctuations helps us to identify the mode, which is destabilized by the electron current. We compute the power spectrum $|E_x(k,\omega)|^2$ by Fourier transforming $E_x(x,t)$ over the full length of the simulation box and over the time interval $0 < t\omega_{ci} < 5$, when the electrostatic field of the ICHWI is weak. This allows us to detect the weak noise moving on the undamped solutions of the dispersion relation, together with the signal caused by the instability. The latter will dominate the spectrum at later times. Figure 14 shows the result. We find two ECH modes in the dispersion relation, the one in the upper-hybrid frequency branch and the one (2) in the lowest ECH branch. As expected both modes do not participate in the instability. We find a maximum of the power

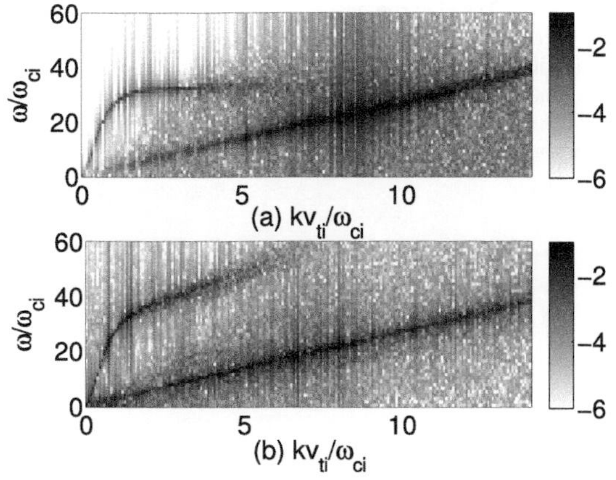

Fig. 15 Panel (**a**) shows the power spectrum $|E_x(k,\omega)|^2$ sampled during the time $0 < t\omega_{ci} < 5$, while (**b**) shows the power spectrum sampled during the time $18 < t\omega_{ci} < 23$. Both power spectra are normalized to their respective maxima and the *gray scale* is 10-logarithmic

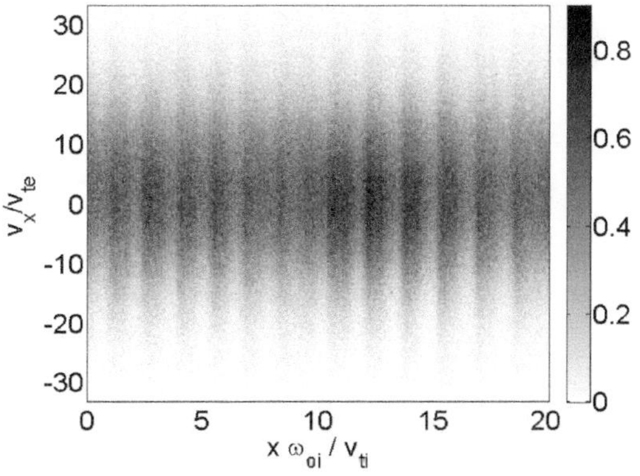

Fig. 16 The electron phase space density in the $x - v_x$ plane and in a subinterval of the simulation box at the time $t\omega_{ci} = 10$. The *gray scale* is linear

on the dispersion relation $\omega/k = v_k$ of the electron beam in the interval $5 < kv_{ti}/\omega_{ci} < 10$. This part of the dispersion relation is thus driving the ICHWI. The unstable band of frequencies on the electron beam mode is well below that of the oblique ion acoustic mode, which is visible in the dispersion relation at lower k and goes over into the fast ion acoustic wave at $kv_{ti}/\omega_{ci} \approx 1$ (Seyler and Wahlund 1996). Only the ICH modes exist in the interval in ω, k, in which the beam is destabilized, and the simulation further suggests that the electron beam couples to a high harmonic of the ICH modes. A high harmonic has a real part of the frequency $\omega_r \gg \omega_{ci}$, which can support the large observed growth rate of the ICHWI.

It is illustrative to compare in Fig. 15 the low-frequency part of the dispersion relation displayed by Fig. 14 with its counterpart sampled in the time interval $18 < t\omega_{ci} < 23$, after the ICHWI has saturated and after the strong waves have collapsed. The dispersion relations of the electron beam and of the fast ion acoustic wave at $kv_{ti}/\omega_{ci} < 1$ remain practically unchanged. However, the slope of the dispersion relation of the oblique ion acoustic mode has increased drastically during the simulation. The phase speed of the ion acoustic modes is proportional to the ion sound speed in the plasma and its increase must arise from an increase in the electron temperature.

The simulation data supports this conclusion. The distribution of the electron phase space density in the $x - v_x$ plane in Fig. 16 shows a distribution that is modulated along the x-direction. The density is modulated on a length scale comparable to an ion thermal gyroradius, which is a consequence of the strong electrostatic waves driven by the ICHWI (see

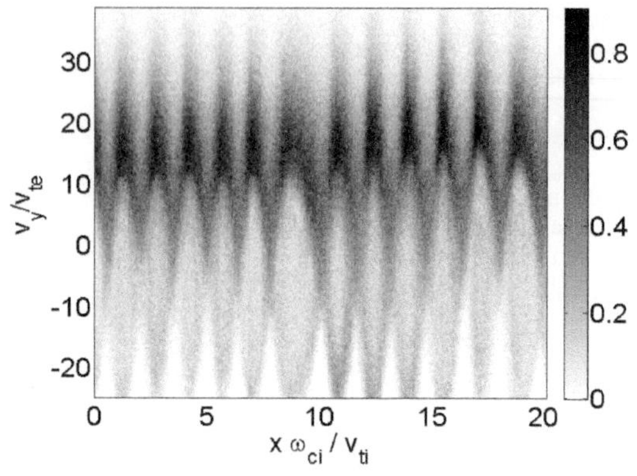

Fig. 17 The electron phase space density in the $x - v_y$ plane and in a subinterval of the simulation box at the time $t\omega_{ci} = 10$. The *gray scale* is linear

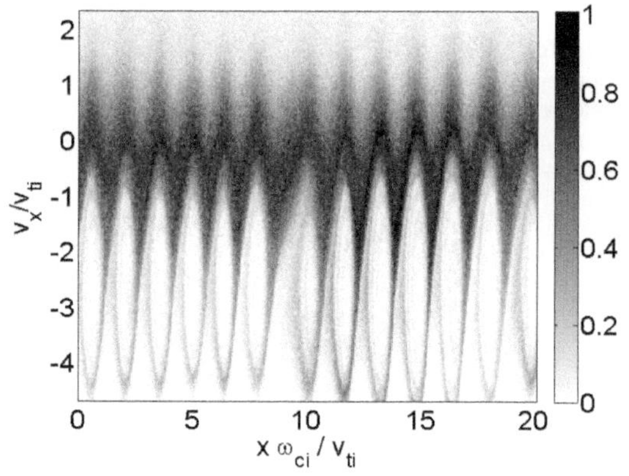

Fig. 18 The ion phase space distribution in the $x - v_x$ plane sampled at the time $t\omega_{ci} = 10$. The *gray scale* is linear

Fig. 13). The electron density is large over the velocity interval $|v_x/v_{te}| < 15$. Initially most electrons were located in an interval with a width $\sim 2v_{te}$ and they have thus been heated up. Figure 17 demonstrates that the electrons have also been heated up along v_y. We find oscillations of the electron mean speed along v_y on spatial scales of an ion thermal gyroradius. The strong electron heating along v_x and v_y implies that the ion acoustic speed $c_s \propto T_e^{1/2}$ has increased by a factor of a few. Here T_e is the electron temperature. It is indeed the electron heating that changes the dispersion relation of the ion acoustic mode in Fig. 15.

The strong electrostatic waves have a low phase speed and they should interact strongly also with the ions. The ion phase space distribution sampled at the time $t\omega_{ci} = 10$ and shown in Fig. 18 confirms this. We find large phase space vortices centred at a speed $\approx -2.5v_{ti}$ and a spatial width $\sim v_{ti}/\omega_{ci}$. These are ion phase space holes (Schamel and Bujarbarua 1980). The phase speed ω/k of the electrostatic wave that produced them corresponds to the central speed of these structures. The phase speed is thus $\sim -2.5v_{ti}$, since the vortices move in the negative x-direction. Combining $\omega/k \approx -2.5v_{ti}$ and $k \approx 6\omega_{ci}/v_{ti}$ (see Fig. 15) we find that the frequency of the wave that is driving them is $|\omega/\omega_{ci}| \approx 20$.

This chain of ion phase space holes is unstable and we find a diffuse ion phase space distribution at the final time $t\omega_{ci} = 23$ in Fig. 19. The bulk ions have been heated by the ICHWI, since their initial thermal spread of a few v_{ti} has increased to about $10v_{ti}$. Figure 13 shows that the life-time of this chain of ion phase space holes, which is tied to strong electrostatic fields, is comparable to about $T_L\omega_{ci} \approx 10$. Ion phase space holes are at least in unmagnetized plasma stabilized by an electron temperature that exceeds the ion temperature

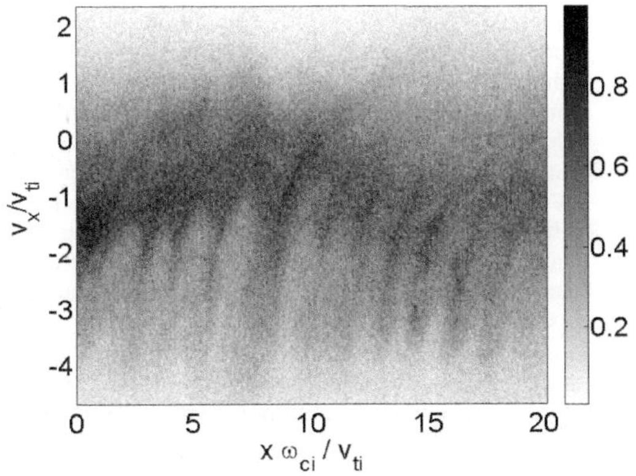

Fig. 19 The ion phase space distribution in the $x - v_x$ plane sampled at the time $t\omega_{ci} = 23$. The *gray scale* is linear

(Schamel and Bujarbarua 1980), so we anticipate that the life time of these ion phase space holes can in some laboratory settings be much longer than for our simulation setup.

Our comparison of the Buneman and ICHWI instabilities shows some clear differences in the initial conditions that drive them and in the final state. These differences help us distinguishing them in reconnection experiments. Buneman instabilities are driven by fast beams of electrons. They do thus require strong field-aligned electric fields, which may not always be generated by reconnection processes. They saturate by the formation of electron phase space holes. Such structures move fast with about the electron thermal speed and they correspond to a localized positive electrostatic potential, which results in the electron phase space vortices. Electron phase space holes typically move along a guiding magnetic field, which stabilizes them. The ion cyclotron harmonic wave instability (ICHWI), on the other hand, requires much lower drift speeds of the electrons than the Buneman instability and can thus be triggered more easily in reconnection experiments. It grows much slower and saturates by forming ion phase space holes that move across the magnetic field. These structures move at speeds comparable to the ion thermal speed and correspond to a localized negative potential.

5 Summary

We have assembled some evidence from experiments, simulation, and theory supporting the idea that large-scale plasma motions can rapidly couple to micro-scales. The paradigm of a hierarchy of closely coupled instabilities is at play in a wide variety of cosmic and laboratory plasma systems. Indeed, the explanation of rapid events such as the heating of a solar flare require a mechanism coupling the largest scales to the smallest. The RSX experiments (and those of Moser and Bellan) show that a large-scale instability such as the ideal kink can initiate physics at smaller scales via reconnection or a secondary instability (such as the Rayleigh-Taylor). Intense current sheets with large electron drift compared to ion thermal speed can initiate kinetic instabilities such as those observed in the SSX experiment and in PIC simulations.

References

A. Al-Karkhy, P.K. Browning, G. Cunningham, S.J. Gee, M.G. Rusbridge, Observations of the magnetohydrodynamic dynamo effect in a spheromak plasma. Phys. Rev. Lett. **70**, 1814–1817 (1993)

T.D. Arber, A.W. Longbottom, C.L. Gerrard, A.M. Milne, A staggered grid, Lagrangian–Eulerian remap code for 3-D MHD simulations. J. Comput. Phys. **171**, 151–181 (2001)
V. Archontis, F. Moreno-Insertis, K. Galsgaard, A.W. Hood, The three-dimensional interaction between emerging magnetic flux and a large-scale coronal field: reconnection, current sheets, and jets. Astrophys. J. **635**, 1299–1318 (2005)
M.R. Bareford, P.K. Browning, R.A.M. van der Linden, A nanoflare distribution generated by repeated relaxations triggered by kink instability. Astron. Astrophys. **521**, 70 (2010)
M.R. Bareford, P.K. Browning, R.A.M. van der Linden, The flare-energy distributions generated by kink-unstable ensembles of zero-net-current coronal loops. Sol. Phys. **273**, 93–115 (2011)
M.R. Bareford, A.W. Hood, P.K. Browning, Coronal heating by the partial relaxation of twisted loops. Astron. Astrophys. **550**, 40 (2013)
B. Basu, N.J. Grossbard, Ion-cyclotron instability in current-carrying Lorentzian (kappa) and Maxwellian plasmas with anisotropic temperatures: a comparative study. Phys. Plasmas **18**, 092106 (2011)
H. Baty, Magnetic topology during the reconnection process in a kinked coronal loop. Astron. Astrophys. **360**, 345–350 (2000)
H. Baty, J. Heyvaerts, Electric current concentration and kink instability in line-tied coronal loops. Astron. Astrophys. **308**, 935–950 (1996)
A.O. Benz, Flare observations. Living Rev. Sol. Phys. **5** (2008)
W.F. Bergerson, C.B. Forest, G. Fiksel, D.A. Hannum, R. Kendrick, J.S. Sarff, S. Stambler, Onset and saturation of the kink instability in a current-carrying line-tied plasma. Phys. Rev. Lett. **96**, 015004 (2006)
I.B. Bernstein, Waves in a plasma in a magnetic field. Phys. Rev. **109**, 10–21 (1958)
I.B. Bernstein, J.M. Greene, M.D. Kruskal, Exact nonlinear plasma oscillations. Phys. Rev. Lett. **108**, 546–550 (1957)
A. Bhattacharjee, Y.-M. Huang, H. Yang, B. Rogers, Fast reconnection in high-lundquist-number plasmas due to the plasmoid instability. Phys. Plasmas **16**, 112102 (2009)
J. Birn, E.R. Priest, *Reconnection of Magnetic Fields* (Cambridge University Press, Cambridge, 2007)
D. Biskamp, *Magnetic Reconnection in Plasmas* (Cambridge University Press, Cambridge, 2005)
G.J.J. Botha, T.D. Arber, A.W. Hood, Thermal conduction effects on the kink instability in coronal loops. Astron. Astrophys. **525**, 96 (2011)
D. Brennan, P.K. Browning, R.A.M. van der Linden, A.W. Hood, S. Woodruff, Stability studies and the origin of the $n = 1$ mode in the SPHEX spheromak experiment. Phys. Plasmas **6**, 4248–4259 (1999)
D.P. Brennan, P.K. Browning, R.A.M. van der Linden, A two-dimensional magnetohydrodynamic stability model for helicity-injected devices with open flux. Phys. Plasmas **9**, 3526–3535 (2002)
D.P. Brennan, P.K. Browning, J. Gates, R.A.M. van der Linden, Helicity-injected current drive and open flux instabilities in spherical tokamaks. Plasma Phys. Control. Fusion **51**, 045004 (2009)
M.R. Brown, Experimental studies of magnetic reconnection. Phys. Plasmas **6**, 1717–1724 (1999)
P.K. Browning, Mechanisms of solar coronal heating. Plasma Phys. Control. Fusion **33**, 539–571 (1991)
P.K. Browning, R.A.M. van der Linden, Solar coronal heating by relaxation events. Astron. Astrophys. **400**, 355–367 (2003)
P.K. Browning, G. Cunningham, R. Duck, S.J. Gee, K.J. Gibson, D.A. Kitson, R. Martin, M.G. Rusbridge, Injection and sustainment of plasma in a preexisting toroidal field using a coaxial helicity source. Phys. Rev. Lett. **68**, 1722–1725 (1992)
P.K. Browning, C. Gerrard, A.W. Hood, R. Kevis, R.A.M. van der Linden, Heating the corona by nanoflares: simulations of energy release triggered by a kink instability. Astron. Astrophys. **485**, 837–848 (2008)
O. Buneman, Instability, turbulence and conductivity in current-carrying plasma. Phys. Rev. Lett. **1**, 8–9 (1959)
B.I. Cohen, C.A. Romero-Talamás, D.D. Ryutov, E.B. Hooper, L.L. Lodestro, H.S. McLean, T.L. Stewart, R.D. Wood, The role of the $n = 1$ column mode in spheromak formation. Phys. Plasmas **16**, 042501 (2009)
I.J.D. Craig, P.G. Watson, Flare-like energy release by flux pile-up reconnection. Sol. Phys. **191**, 359–379 (2000)
I.J.D. Craig, R.B. Fabling, P.G. Watson, The power output of spine and fan magnetic reconnection solutions. Astrophys. J. **485**, 383 (1997)
W. Daughton, V. Roytershteyn, H. Karimabadi, L. Yin, B.J. Albright, B. Bergen, K.J. Bowers, Role of electron physics in the development of turbulent magnetic reconnection in collisionless plasmas. Nat. Phys. **7**, 539–542 (2011)
J.M. Dawson, Particle simulation of plasmas. Rev. Mod. Phys. **55**, 403–447 (1983)
P. Demoulin, J.C. Henoux, E.R. Priest, C.H. Mandrini, Quasi-separatrix layers in solar flares. I. Method. Astron. Astrophys. **308**, 643–655 (1996)
M.E. Dieckmann, B. Eliasson, A. Stathopoulos, A. Ynnerman, Connecting shock velocities to electron-injection mechanisms. Phys. Rev. Lett. **92**, 065006 (2004a)

M.E. Dieckmann, A. Ynnerman, S.C. Chapman, G. Rowlands, N. Andersson, Simulating thermal noise. Phys. Scr. **69**, 456 (2004b)
M.E. Dieckmann, A. Bret, G. Sarri, E. Perez Alvaro, I. Kourakis, M. Borghesi, Particle simulation study of electron heating by counter-streaming ion beams ahead of supernova remnant shocks. Plasma Phys. Control. Fusion **54**, 085015 (2012)
J.F. Drake, M. Swisdak, C. Cattell, M.A. Shay, B.N. Rogers, A. Zeiler, Formation of electron holes and particle energization during magnetic reconnection. Science **299**, 873–877 (2003)
W.E. Drummond, M.N. Rosenbluth, Anomalous diffusion arising from microinstabilities in a plasma. Phys. Fluids **5**, 1507–1513 (1962)
R.C. Duck, P.K. Browning, G. Cunningham, S.J. Gee, A. al-Karkhy, R. Martin, M.G. Rusbridge, Structure of the $n = 1$ mode responsible for relaxation and current drive during sustainment of the SPHEX spheromak. Plasma Phys. Control. Fusion **39**, 715–736 (1997)
R.E. Ergun, C.W. Carlson, J.P. McFadden, F.S. Mozer, L. Muschietti, I. Roth, R.J. Strangeway, Debye-scale plasma structures associated with magnetic-field-aligned electric fields. Phys. Rev. Lett. **81**, 826–829 (1998)
D.W. Forslund, C.R. Shonk, Numerical simulation of electrostatic counterstreaming instabilities in ion beams. Phys. Rev. Lett. **25**, 281–284 (1970)
I. Furno, T. Intrator, E. Torbert, C. Carey, M.D. Cash, J.K. Campbell, W.J. Fienup, C.A. Werley, G.A. Wurden, G. Fiksel, Reconnection scaling experiment: a new device for three-dimensional magnetic reconnection studies. Rev. Sci. Instrum. **74**, 2324–2331 (2003)
I. Furno, T.P. Intrator, E.W. Hemsing, S.C. Hsu, S. Abbate, Coalescence of two magnetic flux ropes via collisional magnetic reconnection. Phys. Plasmas **12**, 055702 (2005)
I. Furno, T.P. Intrator, D.D. Ryutov, S. Abbate, T. Madziwa-Nussinov, A. Light, L. Dorf, G. Lapenta, Current-driven rotating-kink mode in a plasma column with a non-line-tied free end. Phys. Rev. Lett. **97**, 015002 (2006)
H.P. Furth, J. Killeen, M.N. Rosenbluth, Finite-resistivity instabilities of a sheet pinch. Phys. Fluids **6**, 459–484 (1963)
C.G. Gimblett, R.J. Hastie, P. Helander, Model for current-driven edge-localized modes. Phys. Rev. Lett. **96**, 035006 (2006a)
C.G. Gimblett, R.J. Hastie, P. Helander, The role of edge current-driven modes in ELM activity. Plasma Phys. Control. Fusion **48**, 1531–1550 (2006b)
T. Gray, M.R. Brown, D. Dandurand, Observation of a relaxed plasma state in a quasi-infinite cylinder. Phys. Rev. Lett. **110**, (2013)
T.S. Hahm, R.M. Kulsrud, Forced magnetic reconnection. Phys. Fluids **28**, 2412–2418 (1985)
E.W. Hemsing, I. Furno, T.P. Intrator, Fast camera images of flux ropes during plasma relaxation. IEEE Trans. Plasma Sci. **33**, 448–449 (2005)
H.W. Hendel, M. Yamada, S.W. Seiler, H. Ikezi, Ion-beam-driven resonant ion-cyclotron instability. Phys. Rev. Lett. **36**, 319–322 (1976)
J. Heyvaerts, E.R. Priest, Coronal heating by reconnection in DC current systems—a theory based on Taylor's hypothesis. Astron. Astrophys. **137**, 63–78 (1984)
A.W. Hood, E.R. Priest, Kink instability of solar coronal loops as the cause of solar flares. Sol. Phys. **64**, 303–321 (1979)
A.W. Hood, P.K. Browning, R.A.M. van der Linden, Coronal heating by magnetic reconnection in loops with zero net current. Astron. Astrophys. **506**, 913–925 (2009)
E.B. Hooper, L.D. Pearlstein, R.H. Bulmer, MHD equilibria in a spheromak sustained by coaxial helicity injection. Nucl. Fusion **39**, 863–871 (1999)
S.C. Hsu, P.M. Bellan, Experimental identification of the kink instability as a poloidal flux amplification mechanism for coaxial gun spheromak formation. Phys. Rev. Lett. **90**, 215002 (2003)
T.P. Intrator, X. Sun, L. Dorf, G. Lapenta, I. Furno, Experimental onset threshold and magnetic pressure pileup for 3D reconnection. Nat. Phys. **5**, 521–526 (2009)
S. Ishiguro, T. Sato, H. Takamaru, Complexity Simulation Group, V-shaped dc potential structure caused by current-driven electrostatic ion-cyclotron instability. Phys. Rev. Lett. **78**, 4761–4764 (1997b)
S. Ishiguro, T. Sato, H. Takamaru, K. Watanabe, Complexity Simulation Group, Formation of wave-front pattern accompanied by current-driven electrostatic ion-cyclotron instabilities. Phys. Plasmas **4**, 2886–2892 (1997a). doi:10.1063/1.872613
O. Ishihara, A. Hirose, A.B. Langdon, Nonlinear saturation of the Buneman instability. Phys. Rev. Lett. **44**, 1404–1407 (1980)
M. Karlický, B. Kliem, Reconnection of a kinking flux rope triggering the ejection of a microwave and hard X-ray source I. Observations and interpretation. Sol. Phys. **266**, 71–89 (2010)
J.M. Kindel, C.F. Kennel, Topside current instabilities. J. Geophys. Res. **76**, 3055 (1971)
J.A. Klimchuk, On solving the coronal heating problem. Sol. Phys. **234**, 41–77 (2006)

S.O. Knox, C.W. Barnes, G.J. Marklin, T.R. Jarboe, I. Henins, H.W. Hoida, B.L. Wright, Observations of spheromak equilibria which differ from the minimum-energy state and have internal kink distortions. Phys. Rev. Lett. **56**, 842–845 (1986)
G. Lapenta, L. Bettarini, Spontaneous transition to a fast 3D turbulent reconnection regime. Europhys. Lett. **93**, 65001 (2011)
G. Lapenta, S. Markidis, A. Divin, M. Goldman, D. Newman, Scales of guide field reconnection at the hydrogen mass ratio. Phys. Plasmas **17**, 082106 (2010)
K.D. Leka, R.C. Canfield, A.N. McClymont, L. van Driel-Gesztelyi, Evidence for current-carrying emerging flux. Astrophys. J. **462**, 547 (1996)
Y. Liang, C.G. Gimblett, P.K. Browning, P. Devoy, H.R. Koslowski, S. Jachmich, Y. Sun, C. Wiegmann, Multiresonance effect in type-I edge-localized mode control with low n fields on JET. Phys. Rev. Lett. **105**, 065001 (2010)
Y. Liang, C.G. Gimblett, P.K. Browning, P. Devoy, A. Alfier, G. Calabrò, C. Giroud, D. Harting, H.R. Koslowski, S. Jachmich, J. Pearson, Y. Sun, C. Wiegmann, T. Zhang, JET-EFDA contributors, Observations of multi-resonance effect in ELM control with magnetic perturbation fields on the JET tokamak. Nucl. Fusion **51**, 073001 (2011)
M.G. Linton, R.B. Dahlburg, G.H. Fisher, D.W. Longcope, Nonlinear evolution of kink-unstable magnetic flux tubes and solar delta-spot active regions. Astrophys. J. **507**, 404–416 (1998)
R. Lionello, M. Velli, G. Einaudi, Z. Mikic, Nonlinear magnetohydrodynamic evolution of line-tied coronal loops. Astrophys. J. **494**, 840 (1998)
R. Liu, D. Alexander, Hard X-ray emission in kinking filaments. Astrophys. J. **697**, 999–1009 (2009)
D.W. Longcope, Separator current sheets: generic features in minimum-energy magnetic fields subject to flux constraints. Phys. Plasmas **8**, 5277–5290 (2001)
N.F. Loureiro, A.A. Schekochihin, S.C. Cowley, Instability of current sheets and formation of plasmoid chains. Phys. Plasmas **14**, 100703 (2007)
F. Moreno-Insertis, T. Emonet, The rise of twisted magnetic tubes in a stratified medium. Astrophys. J. **472**, 53 (1996)
A.L. Moser, P.M. Bellan, Magnetic reconnection from a multiscale instability cascade. Nature **482**, 379–381 (2012)
E.N. Parker, Nanoflares and the solar X-ray corona. Astrophys. J. **330**, 474–479 (1988)
C.E. Parnell, A.L. Haynes, K. Galsgaard, Structure of magnetic separators and separator reconnection. J. Geophys. Res. **115**, 2102 (2010)
D.I. Pontin, Three-dimensional magnetic reconnection regimes: a review. Adv. Space Res. **47**, 1508–1522 (2011)
D.I. Pontin, A. Bhattacharjee, K. Galsgaard, Current sheet formation and nonideal behavior at three-dimensional magnetic null points. Phys. Plasmas **14**, 052106 (2007)
E. Priest, T. Forbes, *Magnetic Reconnection* (Cambridge University Press, Cambridge, 2000)
E.R. Priest, D.I. Pontin, Three-dimensional null point reconnection regimes. Phys. Plasmas **16**, 122101 (2009)
E.R. Priest, V.S. Titov, Magnetic reconnection at three-dimensional null points. Proc. R. Soc. Lond. Ser. A, Math. Phys. Sci. **354**, 2951–2992 (1996)
J.J. Rasmussen, R.W. Schrittwieser, On the current-driven ion-cyclotron instability—a review. IEEE Trans. Plasma Sci. **19**, 457–501 (1991)
K.V. Roberts, H.L. Berk, Nonlinear evolution of a two-stream instability. Phys. Rev. Lett. **19**, 297–300 (1967)
R. Samtaney, N.F. Loureiro, D.A. Uzdensky, A.A. Schekochihin, S.C. Cowley, Formation of plasmoid chains in magnetic reconnection. Phys. Rev. Lett. **103**, 105004 (2009)
J.C. Santos, J. Büchner, A. Otto, Development of electric currents in a magnetic field configuration containing a magnetic null point. Astron. Astrophys. **525**, 3 (2011)
G. Sarri, M.E. Dieckmann, C.R.D. Brown, C.A. Cecchetti, D.J. Hoarty, S.F. James, R. Jung, I. Kourakis, H. Schamel, O. Willi, M. Borghesi, Observation and characterization of laser-driven phase space electron holes. Phys. Plasmas **17**, 010701 (2010)
H. Schamel, Electron holes, ion holes and double-layers phase-space structures in theory and experiment. Phys. Rep. **140**, 161–191 (1986)
H. Schamel, S. Bujarbarua, Solitary plasma hole via ion-vortex distribution. Phys. Fluids **23**, 2498 (1980)
K. Schindler, M. Hesse, J. Birn, General magnetic reconnection, parallel electric fields, and helicity. J. Geophys. Res. **93**, 5547–5557 (1988)
C.E. Seyler, J. Providakes, Particle and fluid simulations of resistive current-driven electrostatic ion cyclotron waves. Phys. Fluids **30**, 3113–3128 (1987)
C.E. Seyler, J.-E. Wahlund, Theory of nearly perpendicular plasma waves and comparison to Freja satellite observations. J. Geophys. Res. **101**, 21795–21813 (1996)

A.K. Srivastava, T.V. Zaqarashvili, P. Kumar, M.L. Khodachenko, Observation of kink instability during small B5.0 solar flare on 2007 June 4. Astrophys. J. **715**, 292–299 (2010)

B.W. Stallard, E.B. Hooper, S. Woodruff, R.H. Bulmer, D.N. Hill, H.S. McLean, R.D. Wood, Magnetic helicity balance in the sustained spheromak plasma experiment. Phys. Plasmas **10**, 2912–2924 (2003)

X. Sun, T. Intrator, L. Dorf, I. Furno, G. Lapenta, Transition of mhd kink stability properties between line-tied and non line tied boundary conditions. Phys. Rev. Lett. **100**, 205004 (2008)

X. Sun, T.P. Intrator, L. Dorf, J. Sears, I. Furno, G. Lapenta, Flux rope dynamics: experimental study of bouncing and merging. Phys. Rev. Lett. **105**, 255001 (2010)

J.B. Taylor, Relaxation of toroidal plasma and generation of reverse magnetic fields. Phys. Rev. Lett. **33**, 1139–1141 (1974)

J.B. Taylor, Relaxation and magnetic reconnection in plasmas. Rev. Mod. Phys. **58**, 741–763 (1986)

T. Török, B. Kliem, Confined and ejective eruptions of kink-unstable flux ropes. Astrophys. J. **630**, 97–100 (2005)

T. Török, B. Kliem, V.S. Titov, Ideal kink instability of a magnetic loop equilibrium. Astron. Astrophys. **413**, 27–30 (2004)

G.E. Vekstein, R. Jain, Energy release and plasma heating by forced magnetic reconnection. Phys. Plasmas **5**, 1506–1513 (1998)

M. Velli, R. Lionello, G. Einaudi, Kink modes and current sheets in coronal loops. Sol. Phys. **172**, 257–266 (1997)

Z. Wang, C.W. Barnes, G.A. Wurden, D.N. Hill, E.B. Hooper, H.S. McLean, R.D. Wood, S. Woodruff, Large-amplitude electron density and H fluctuations in the sustained spheromak physics experiment. Nucl. Fusion **42**, 643–652 (2002)

E.S. Weibel, Ion cyclotron instability. Phys. Fluids **13**, 3003–3006 (1970)

D.M. Willett, P.K. Browning, S. Woodruff, K.J. Gibson, The internal magnetic structure and current drive in the SPHEX spheromak. Plasma Phys. Control. Fusion **41**, 595–612 (1999)

P.H. Yoon, R. Gaelzer, T. Umeda, Y. Omura, H. Matsumoto, Harmonic Langmuir waves. I. Nonlinear dispersion relation. Phys. Plasmas **10**, 364–372 (2003)

E.G. Zweibel, M. Yamada, Magnetic reconnection in astrophysical and laboratory plasmas. Astron. Astrophys. Rev. **47**, 291–332 (2009)

DOI 10.1007/978-1-4899-7413-6_12
Reprinted from *Space Science Reviews* Journal, DOI 10.1007/s11214-012-9957-2

In-Situ Observations of Reconnection in Space

Götz Paschmann · Marit Øieroset · Tai Phan

Received: 3 October 2012 / Accepted: 4 December 2012 / Published online: 2 February 2013

Abstract This paper gives an overview of the insights into the magnetic reconnection process obtained by in-situ measurements across current sheets found in planetary magnetospheres and the solar wind. Emphasis is placed on results that might be of interest to the study of reconnection in regions where no in-situ observations are available. These results include the role of symmetric versus asymmetric boundary conditions, the identification of the onset conditions, the reconnection rates, and the spatial and temporal scales. Special attention is paid to observations in the so-called diffusion region surrounding the reconnection sites, where ions and eventually also electrons become demagnetized and reconnection is initiated.

Keywords Magnetic reconnection · Current sheets · Diffusion region · Magnetosphere · Solar wind

1 Introduction

This paper describes the insights on magnetic reconnection that have been obtained from in-situ observations in planetary magnetospheres and in the solar wind, and their comparison with theory and simulations. Emphasis is placed on results general enough to possibly be of interest to the study of reconnection in regions where no in-situ observation are available.

The reconnection scenario can be split into two regimes: the small region surrounding the reconnection site, commonly referred to as the diffusion (or dissipation) region, where non-MHD effects dominate and allow reconnection to occur, and the large-scale reconnection layer away from the diffusion region, for which a fluid description is appropriate, but where kinetic effects can also be important.

G. Paschmann (✉)
Max-Planck-Institut für extraterrestrische Physik, Garching, Germany
e-mail: goetz.paschmann@mpe.mpg.de

M. Øieroset · T. Phan
Space Sciences Laboratory, University of California, Berkeley, USA

We discuss the two regimes in reverse order. Section 2 deals with the large-scale aspects that have emerged from observations obtained in the reconnection layer downstream from the actual reconnection site. Section 3 then focusses on observations of the microphysics in and near the diffusion region. Section 4 summarizes the results and lists some open questions.

2 Large-Scale Aspects

To set the scene, Fig. 1 shows a 2D cut through Earth's magnetosphere. Reconnection is shown to occur at two sites, the subsolar magnetopause and the magnetotail current sheet. At the subsolar magnetopause, a southward directed interplanetary magnetic field (IMF) can connect with the northward directed magnetospheric magnetic field. Once reconnected, field lines are pulled over the polar caps, enter the magnetotail, and are reconnected once more across the tail current sheet, setting up a global circulation of plasma and magnetic field (Dungey 1961). Early observations demonstrating that magnetospheric activity responds to the southward turnings of the IMF (Fairfield and Cahill 1966), as well as measurements of the convection over the polar caps (Heppner 1972), provided strong support for the reconnection concept, long before any in-situ observations at the magnetopause or the magnetotail current sheets were available. In addition to the magnetopause and magnetotail, reconnection has also been observed to occur across current sheets in the solar wind, and in the magnetosheath, the region downstream of the bow shock.

2.1 Plasma Conditions

Boundary conditions and configurations at these various reconnection sites differ greatly. The magnetopause is usually characterized by highly asymmetric conditions on the two sides of the current layer, the density being lower and the magnetic field strength higher on the Earthward side. The magnetic shear across the magnetopause varies, reflecting the variable orientation of the interplanetary magnetic field carried by the solar wind. Thus the reconnection configuration at the magnetopause is generally characterized by substantial guide fields. In Earth's magnetotail, reconnection occurs across the current sheet that separates the oppositely directed magnetic fields in the northern and southern tail lobes. As a result, the boundary conditions are nearly symmetric and there is essentially no guide field,

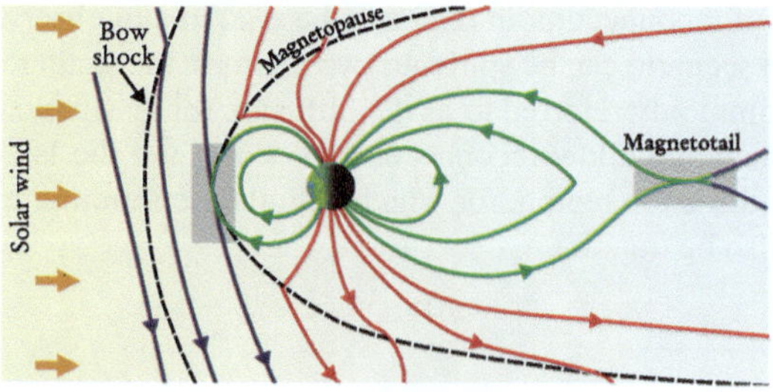

Fig. 1 2D cut through Earth's magnetosphere and upstream solar wind, showing reconnection at the subsolar magnetopause and in the magnetotail. Bow shock and magnetopause are indicated by *curved dashed lines*, with the magnetosheath in between. Interplanetary field lines are in *dark blue*, closed magnetospheric field lines in *green*, and reconnected field *lines in red* (From Day 2001)

at least not in the near-Earth magnetotail. Solar wind and magnetosheath current sheets, finally, are characterized by nearly symmetric plasma conditions, but the magnetic shear is variable.

Not only the boundary conditions at the various reconnection sites differ, but so do the key plasma parameters. The plasma β, the ratio between plasma pressure, Nk_BT, and magnetic field pressures, $B^2/2\mu_0$, is of order 1 in the solar wind, in the range 1–10 in the magnetosheath outside the magnetopause, but only of order 0.01 in the magnetotail lobes surrounding the magnetotail current sheet. The Alfvén speed, which sets the scale for the reconnection outflow velocities, is near 50 km s^{-1} in the solar wind, a few hundred km s^{-1} at the magnetopause, and 1000–2000 km s^{-1} in the tail lobes. The ion inertial length, λ_i, which is the characteristic length scale, is about 100 km in the solar wind, 50 km in the magnetosheath outside the magnetopause, and 1000 km around the magnetotail current sheet.

2.2 Reconnection Signatures

2.2.1 Normal Magnetic Field and Flow; Reconnection Electric Field

As reconnection implies the presence of a significant component of the magnetic field normal to the current layer, B_n, and a proportional inflow velocity, v_n, it would seem that the most direct way to prove the occurrence of reconnection would be to measure B_n and/or v_n. However, those quantities usually are small compared to the tangential components, and their determination is dependent on precise knowledge of the current sheet orientation, which makes their reliable determination difficult. Similarly, the reconnection electric field tangential to the current layer, E_t, is not only small, but has to be determined in the frame comoving with the current sheet. To get a rough estimate of B_n, one usually relies on minimum variance analysis of the magnetic field (Sonnerup and Cahill 1967), where B_n is identified with the minimum eigenvalue, but this technique has a number of pitfalls that must be considered before the resulting B_n's can be trusted (e.g., Sonnerup and Scheible 1998).

2.2.2 Accelerated Plasma Flows: The Walén Relation

Given the difficulties with the determination of B_n, v_n, and E_t, the clearest evidence for reconnection are detections of the accelerated plasma flows. As plasma flows across a current layer with $B_n \neq 0$, the components of the plasma velocity, $\boldsymbol{v}$, tangential to that layer change in response to the $\boldsymbol{j} \times \boldsymbol{B}_n$ force. For an ideal rotational discontinuity (RD), the change in the velocity is (Hudson 1971):

$$\Delta \boldsymbol{v} = \pm \Delta \boldsymbol{v}_{\mathrm{A}}, \tag{1}$$

where the symbol Δ refers to changes relative to some reference state (upstream of the magnetopause for example), and $\boldsymbol{v}_{\mathrm{A}}$ is the local Alfvén velocity, corrected for the effect of pressure anisotropy,

$$\boldsymbol{v}_{\mathrm{A}} = \boldsymbol{B}\sqrt{(1-\alpha)/\mu_0\rho}, \tag{2}$$

with $\alpha = (p_\parallel - p_\perp)\mu_0/B^2$. The positive (negative) sign on the r.h.s. of Eq. (1) applies if the normal components of the magnetic field and plasma velocity, B_n and v_n, have the same (opposite) signs. Equation (1) is the Walén relation expressed in the spacecraft frame. A more convenient frame is often the Hoffmann-Teller frame, as first demonstrated by Sonnerup et al. (1987). In the Hoffmann-Teller frame, the flow velocity, $\boldsymbol{v}'$, is aligned with the magnetic field, and the Walén-relation reduces to $\boldsymbol{v}' = \pm \boldsymbol{v}_{\mathrm{A}}$.

In spite of the fact that the flow acceleration $\Delta \boldsymbol{v}$ is due to the $\boldsymbol{j} \times \boldsymbol{B}_n$ force, the Walén relation does not contain B_n. This is because for an RD, as for any planar Alfvén wave, $\boldsymbol{v}_n$ is proportional to B_n. Thus the smaller B_n, the smaller the inflow velocity, and thus the less mass to accelerate. This independence means that a successful Walén test does not say anything about the reconnection rate, which in turn is proportional to B_n.

2.2.3 Kinetic Effects

Strong evidence for reconnection has also been obtained from the kinetic effects, which manifest themselves in the ion and electron velocity distribution functions (Cowley 1982, 1995). Among those are a D-shape of the 2D distributions of the transmitted ions, and interpenetrating ion beams, an example of which is shown in Fig. 9.

2.3 Earth's Magnetopause

In ideal MHD, the magnetopause is an impenetrable boundary forever separating the solar wind from the magnetosphere. Among the many possible plasma transfer mechanisms, it is primarily magnetic reconnection that allows solar wind plasma to penetrate the magnetopause and enter Earth's magnetosphere (Sibeck et al. 1999).

2.3.1 Structure, Flows, and Kinetic Effects

The left part of Fig. 2 shows the magnetopause reconnection configuration assuming antiparallel magnetic fields on the two sides of the current layer, i.e., 180° magnetic shear. Reconnection occurs in a small 'diffusion' region near the X that is considered a black box in all of Sect. 2. Outside the diffusion region, the MP current layer is akin to a rotational discontinuity. Plasma flowing across the RD gets accelerated by the $\boldsymbol{j} \times \boldsymbol{B}_n$ force, which can be visualized as the slingshot effect from the sharply bent magnetic field lines. The outflow forms a boundary layer on the Earthward side of the RD, with a plasma density dropping rapidly with increasing distance from the RD. Because of the large vertical extent of the RD, most of the plasma in the boundary layer entered way downstream from the X, never encountering the diffusion region at the X itself.

The right part of Fig. 2 shows the first in-situ observation of the predicted plasma jetting (Paschmann et al. 1979), made possible when high time-resolution 3D plasma measurements became available. The figure shows time-series of key plasma parameters for an outbound orbit by the ISEE spacecraft that included a complete magnetopause crossing, followed by partial re-crossings caused by magnetopause motion reversals. The magnetic shear across the magnetopause was 100° for this case. As expected for a crossing near the subsolar point, flow speeds were near-zero in the adjacent magnetosheath, but increased to almost 500 km s^{-1} upon crossing of the current layer, in agreement with the predictions from the Walén-relation.

More accelerated flows were subsequently detected both on the dayside and flank magnetopause (e.g., Sonnerup et al. 1981; Gosling et al. 1986). By now, the detection of the accelerated outflows and their comparison with the predictions from the Walén-relation have become the prime in-situ evidence for reconnection, as shown by the many references to Walén-relation tests in the review by Paschmann (2008). Further support for the magnetopause reconnection configuration illustrated in Fig. 2 has been provided by simultaneous observations by two spacecraft of the oppositely directed plasma outflows away from the X-line (Phan et al. 2000).

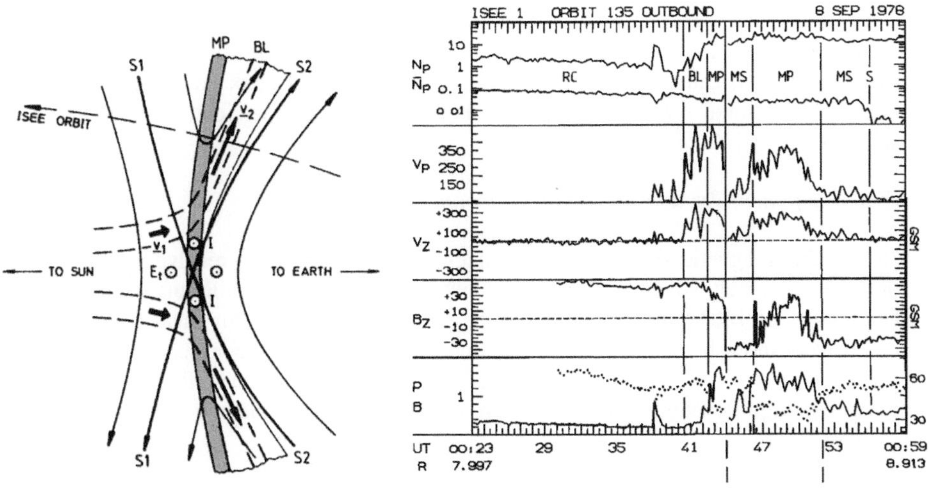

Fig. 2 *Left*: The magnetopause reconnection configuration for assumed antiparallel magnetic fields. The magnetopause (MP) is shown as a *grey-shaded* current layer, with a boundary layer (BL) on the Earthward side. Reconnection occurs at the X-line at the center, and the field lines emanating from the X form the separatrices, labeled S1 and S2. The *dashed lines* are stream lines, and the *solid arrows* indicate the plasma inflow and outflow velocities. E_t is the reconnection electric field, which is aligned with the current I. The magnetic field normal component is directed inward (outward) north (south) of the X-line. *Right*: First observation of plasma jetting at Earth's magnetopause. The figure shows the measurements along the spacecraft trajectory shown on the left. From top to bottom, it shows the plasma density (in cm^{-3}), the magnitude and z-component of the flow velocity (both in $\mathrm{km\,s}^{-1}$), followed by the z-component of the magnetic field (in nT), and the pressures of the magnetic field (*dotted line*) and plasma (*solid line*), both in nPa. The magnetopause is recognized by the magnetic field rotation from positive B_z in the magnetosphere to negative B_z in the magnetosheath (From Paschmann et al. 1979)

Fig. 3 Comparison of the measured changes in flow velocity, $\Delta \boldsymbol{V}_{\mathrm{obs}}$, with those predicted by the Walén-relation, $\Delta \boldsymbol{V}_{\mathrm{th}}$, for 11 high-speed flows observed at the dayside magnetopause. The velocities have been normalized to a common scale and orientation, such that the predicted velocity change is of unit length and oriented horizontally (Figure from Sonnerup et al. 1981)

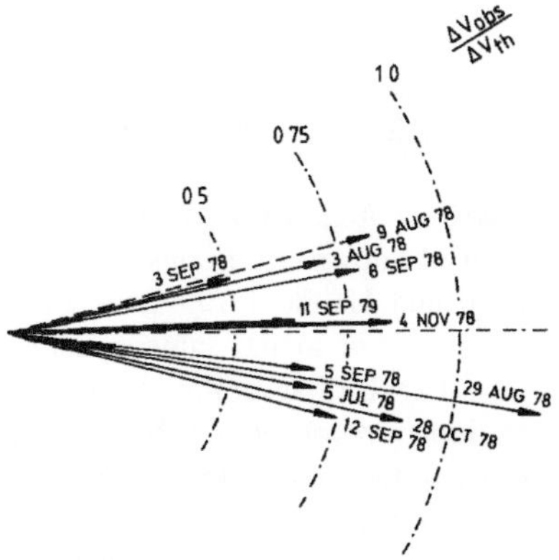

It should be noted that the accelerated flows do not always match the Walén-relation perfectly. While the observed flow direction agrees with the prediction usually quite well, the flow magnitude is often less than the predicted value. This is illustrated in Fig. 3, which shows that the measured velocity directions are close to the prediction, but that their magnitude can be as low as only 50 % of the prediction (Sonnerup et al. 1981). These discrepancies are likely due to deviations of the real situation from the locally 1D, time-stationary conditions assumed in the Walén-relation, or to obstructions in the flow.

In addition to the plasma flow jets, a number of kinetic effects are observed at the magnetopause, among them a D-shape of the 2D distributions of the transmitted ions (e.g., Fuselier et al. 1991) that is caused by a cut-off at the Hoffmann-Teller velocity, interpenetrating ion beams (Gosling et al. 1990), and edge effects from velocity dispersion (Gosling et al. 1990; Vaivads et al. 2010).

2.3.2 Component Versus Anti-parallel Reconnection

The variable magnetic shear across the magnetopause allowed to resolve a long-standing controversy, namely whether or not reconnection requires anti-parallel magnetic fields, i.e, shear-angles of 180°. Observations near the reconnection X-line for shear angles of ∼90° conclusively show that magnetopause reconnection can occur in the presence of strong guide fields (e.g., Retinò et al. 2005; Pu et al. 2007; Trenchi et al. 2008). Reconnection in the presence of significant guide fields is often referred to as component reconnection.

2.3.3 X-Line Location

The special geometry of the Earth's magnetic field, together with the variability of the IMF direction and the draping of the field lines over the magnetopause surface, imply that the magnetic shear angle depends on time and location, with values ranging from 0° to 180° along the magnetopause surface at any given time. Although of mainly geophysical interest, the question is therefore, where the reconnection X-lines will occur. In general it is located in the subsolar region for southerly directed IMF, while for northward orientations it can occur along the polar magnetopause, tailward of the polar cusps. For east-west directed IMF, there is a tendency for reconnection to occur along a tilted X-line across the dayside magnetopause, where the magnetic shear maximizes (Trattner et al. 2012).

2.3.4 Spatial and Temporal Scales

Another controversy that magnetospheric observations have resolved is whether reconnection is intrinsically transient or whether it can be quasi-stationary. In-situ and remote-sensing observations of magnetopause reconnection have provided evidence for both modes.

Quasi-stationary Versus Transient Reconnection That reconnection can be quasi-stationary is well-known from observations of its ionospheric signatures (e.g., Heppner 1972; Rich and Hairston 1994; Greenwald et al. 1995). In-situ measurements at the magnetopause lead to the same conclusion. Gosling et al. (1982) had observed reconnection jets at each of several crossing occurring within a 5-hour interval, which suggests quasi-stationary reconnection. More recently, fortuitous circumstances left the Cluster spacecraft close to the magnetopause for long times, with the measurements indicating the presence of reconnection jets at all crossing instances, for up to several hours (Phan et al. 2004; Retinò et al. 2005). Figure 4 shows excellent agreement between the measured flow velocities and those predicted by the Walén-relation for the many magnetopause crossings occurring within a half-hour interval. Remote-sensing optical observations of bright auroral features caused by ion beams from a magnetopause reconnection site that lasted for almost four hours have provided additional evidence (Frey et al. 2003).

On the other hand, magnetopause reconnection frequently is transient, causing what is commonly referred to as flux-transfer-events (FTEs). FTEs are recognized as bipolar pulses of the normal component of the magnetic field. Russell and Elphic (1979) envisioned FTEs as a pair of elbow-shaped flux tubes, having diameters of order one Earth

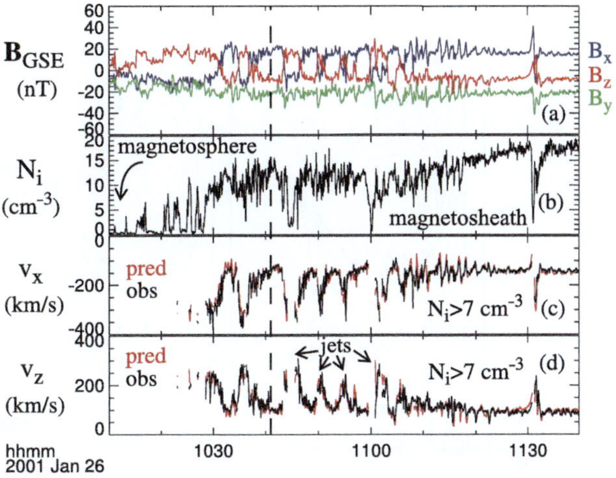

Fig. 4 Multiple magnetopause crossings by the Cluster 1 spacecraft: (**a**) magnetic field components in GSE; (**b**) ion number density; (**c–d**) x and z components of the predicted (*red*) and observed (*black*) ion bulk velocity, showing excellent agreement. The magnetopause crossings are recognized by the magnetic field rotation (Adapted from Phan et al. 2004)

radius (R_E), interconnecting the interplanetary with the terrestrial magnetic field, and moving away from the reconnection site, as shown in Fig. 5(a). In the models of Scholer (1988a) and Southwood et al. (1988), FTEs are caused by transient single X-line reconnection, leading to a pair of bulges moving away from the reconnection site at essentially the Alfvén speed (Fig. 5(b)). Ionospheric signatures of FTEs tend to support this picture of pulsed, but longitudinally extended reconnection (e.g., Lockwood et al. 1990b; Fear et al. 2012). Alternatively, elongated structures could be formed by multiple X-line reconnection (Lee and Fu 1985), produced by the tearing mode in the magnetopause current layer, leading to magnetic flux ropes connected to the Earth on one end and to the solar wind on the other, as illustrated in Fig. 5(c).

In either case the resulting structures are flux ropes with twisted (helical) fields (Song and Lysak 1989; Scholer 1995). The twisted magnetic fields will exert tension which will enhance the core magnetic field to preserve pressure balance (Paschmann et al. 1982). As

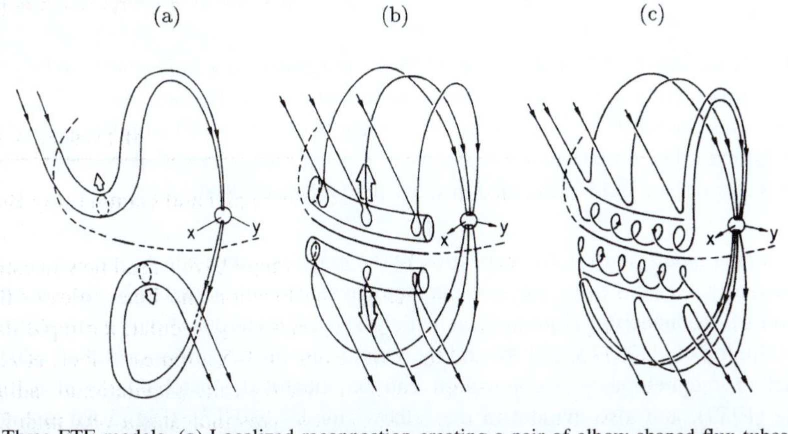

Fig. 5 Three FTE models. (**a**) Localized reconnection creating a pair of elbow-shaped flux tubes, as proposed by Russell and Elphic (1979); (**b**) Single X-line bursty reconnection, leading to a pair of bulges with a substantial longitudinal extent (Scholer 1988a; Southwood et al. 1988); (**c**) Multiple X-line reconnection (Lee and Fu 1985). From Scholer (2003), after Lockwood et al. (1990a)

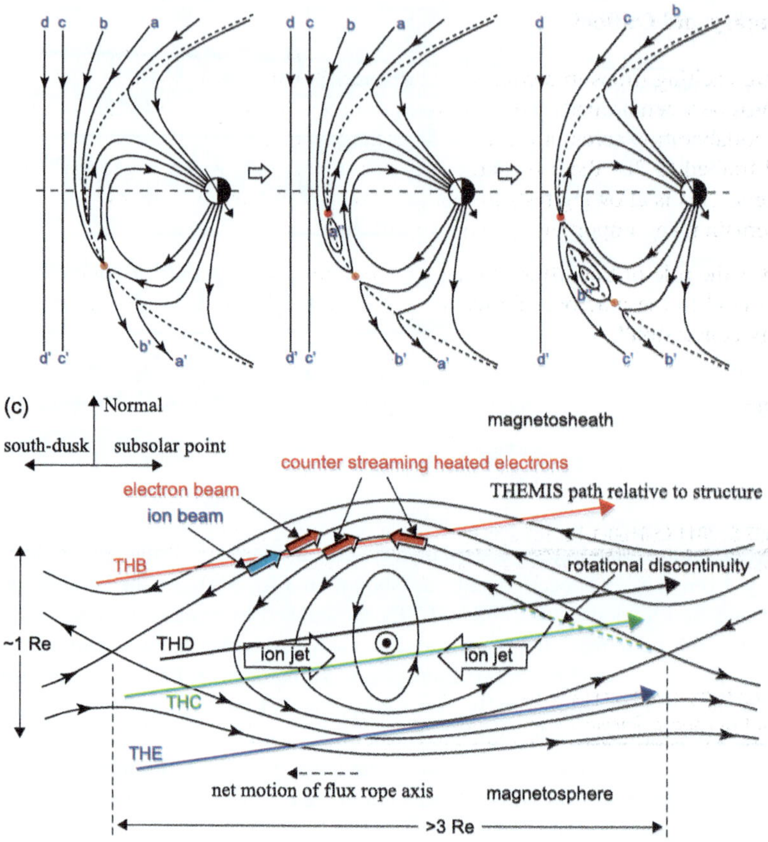

Fig. 6 *Top*: 2D views of the field line evolution in global MHD simulations showing the formation of a flux rope by reconnection at two sequentially activated X-lines marked in *red*; *bottom*: structure of the flux rope inferred from ion and electron beams observed along the trajectories of four THEMIS spacecraft (Adapted from Hasegawa et al. 2010)

argued by Sonnerup (1987) and Scholer (1988b), enhanced core fields can also be produced by a sweeping up of magnetopause flux by the reconnected field lines. As far as simulations are concerned, core-field enhancements are evident in the 3D MHD simulations by Ma et al. (1994). Global 3D simulations have also been able to model so-called crater FTEs, which possess a strong core field embedded in weak fields (Sibeck et al. 2008).

Large- Versus Small-Scale Reconnection at the magnetopause can be large-scale, extending essentially over the entire dayside magnetopause, as demonstrated by the simultaneous detection of reconnection signatures at widely spaced locations, both in-situ (Phan et al. 2006b; Dunlop et al. 2011a, 2011b) and by remote-sensing (e.g., Pinnock et al. 2003).

Whether magnetopause reconnection can be localized ('patchy'), as envisioned by Nishida (1989), and also evident in the 'elbow' model of FTEs (Fig. 5(a)) has not been fully resolved. Only a few ionospheric observations have been reported that suggest that FTEs can indeed be small-scale (e.g., Oksavik et al. 2004). It should be pointed out though that transient magnetotail reconnection can be localized, as discussed in Sect. 2.4.2.

Flux Ropes The recent multi-point measurements from the Cluster and THEMIS missions have revealed in considerable detail the structure of flux ropes. Figure 6 shows a magnetopause case, reported by Hasegawa et al. (2010), where a 2D flux-rope is formed by reconnection at two sequentially activated closely-spaced X-lines, as suggested by global MHD simulations (Raeder 2006). Hasegawa et al. also show maps of the transverse magnetic field within the flux rope, as obtained from Grad-Shafranov reconstruction, a technique that had been applied to FTEs already earlier (Sonnerup et al. 2004). Another flux rope, again flanked by two active X lines producing colliding plasma jets near the center, but with properties suggestive of 3D effects and suprathermal electron energization, has been reported by Øieroset et al. (2011). In a comprehensive study of almost 4000 FTEs observed by the THEMIS spacecraft (Zhang et al. 2012a), 41 flux ropes that were flanked by two X-lines could be identified, suggesting that multiple X-lines are rare or shortlived.

2.4 Earth's Magnetotail

As already mentioned in Sect. 2.1, reconnection in the magnetotail is characterized by nearly symmetric plasma conditions and a magnetic shear that is near 180° in the near-Earth magnetotail, but can deviate from 180° in the distant magnetotail. There are two preferred locations for magnetotail reconnection. First, there is the large-scale, nearly continuous reconnection in the distant tail, which on average must balance subsolar reconnection, because otherwise magnetic flux would be added to the magnetotail forever. The distant X-line lies typically around 140 R_E (Nishida et al. 1997), but can also be as close as 60 R_E (Øieroset et al. 2000). Second, there is near-Earth reconnection that is transient and associated with magnetospheric substorms (e.g., Angelopoulos et al. 1994, 2008; Nagai et al. 1998). The near-Earth X-lines form at distances between 10 and 30 R_E (Nishida and Nagayama 1973). There is a dependence on solar wind energy input, with the position near 15 R_E at high input, but beyond 20 R_E at low input (Nagai 2006). An example of the substorm-related bursty high-speed reconnection flows, observed during the passage of an X-line at a distance of 22 R_E, is shown in Fig. 7 (Angelopoulos et al. 2008).

2.4.1 Structure and Flows

For symmetric conditions, the outflow region is expected to be bounded by slow shocks and such shocks have indeed been reported (Feldman et al. 1985; Saito et al. 1998; Eriksson et al. 2004).

Initially, near-Earth magnetotail reconnection involves field lines embedded within the hot plasma sheet surrounding the current sheet, where the Alfvén-speed is typically several hundred km s^{-1}. If reconnection proceeds long enough, it will eventually reconnect the much more tenuous lobe flux tubes where the Alfvén-speed is typically 1000–2000 km s^{-1}. It is this fast reconnection phase that is generally believed to be the phase associated with magnetospheric substorm (e.g., Baker et al. 2002).

2.4.2 Spatial and Temporal Scales

In contrast to dayside magnetopause reconnection which at times can be quasi-steady and extended in space, reconnection in the near-Earth magnetotail is generally highly intermittent (Baumjohann et al. 1990; Angelopoulos et al. 1992) and patchy, producing narrow flow burst channels. The spatial scale of these flow channels is a few R_E, corresponding to a few tens of ion inertial lengths (Angelopoulos et al. 1997; Nakamura et al. 2004). Braking of

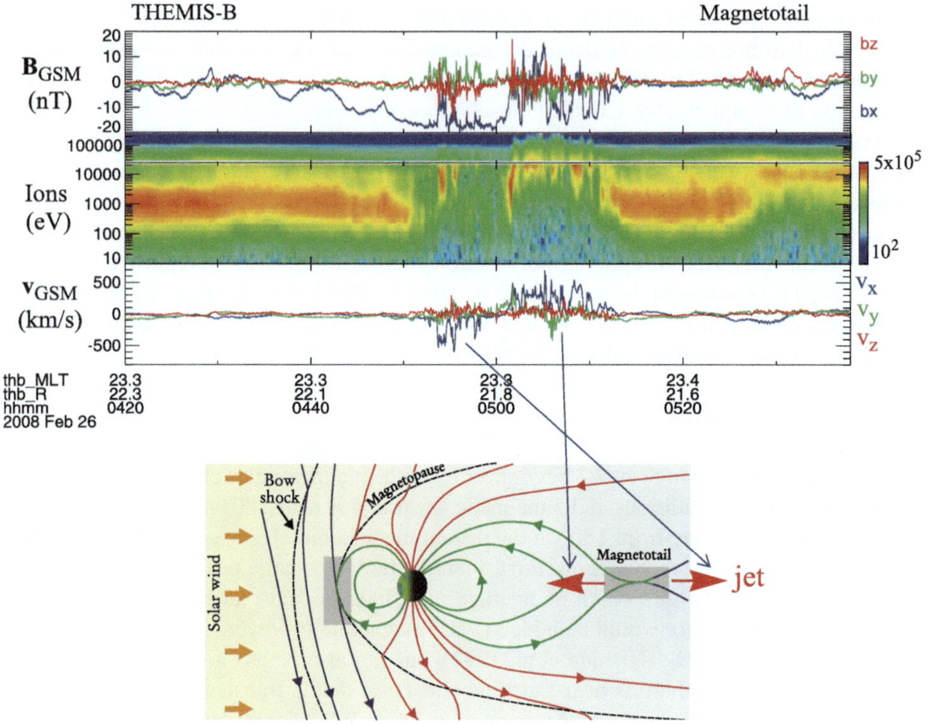

Fig. 7 Crossings of the magnetotail current sheet by one of the THEMIS spacecraft, showing bursty high-speed jets that switch direction from tailward to Earthward, in association with the passage of the reconnection X-line. *Top*: time-series of the three components of the measured magnetic field; energy-time spectrogram of the ions from 10 to 10^5 eV; the three components of the plasma bulk velocity. *Bottom*: schematic that illustrates the underlying configuration. Adapted from Angelopoulos et al. (2008)

these jets in the magnetic flux pile-up regions Earthward of the reconnection site is a strong candidate for electron acceleration (see Sect. 2.4.3).

Another transient reconnection feature in the magnetotail are magnetic flux ropes or plasmoids. On the largest scales, there are the plasmoids that are formed between the near-Earth and distant X-lines, subsequently propagating down-tail (Hones 1979), which can be remotely sensed because they compress the lobe magnetic field (Slavin et al. 1984). On the smallest scales, there are flux ropes that are formed by reconnection at multiple X-lines (e.g., Slavin et al. 2005). Recent multi-point measurements with the Cluster and THEMIS spacecraft have allowed to probe their internal structure (Eastwood et al. 2005; Imber et al. 2011; Beyene et al. 2011; Kiehas et al. 2012).

2.4.3 Electron Acceleration

The potential for accelerating electrons to high energies is one of the most appealing features of magnetic reconnection. In-situ measurements offer a unique opportunity to check the reality of this expectation.

At the magnetopause the identification of acceleration and heating is hindered by the energetic particles of magnetospheric origin that can enter the outflow region and mix with the incident solar wind plasma. There is as yet no evidence to suggest that reconnection in the solar wind ever produces substantial particle acceleration (Gosling 2011). However, observations in the magnetotail have clearly shown the association of energetic electrons up to several hundred keV with reconnection (Terasawa and Nishida 1976; Baker and Stone 1976).

The initial acceleration appears to occur in the diffusion region itself, as first reported by Øieroset et al. (2002) and later confirmed by Retinò et al. (2008). The smallness of the diffusion region limits its effectiveness for the overall acceleration. In the case presented by Øieroset et al., the energy density in the energetic electrons was less that 1 % of the energy density in the ion jets. But once these accelerated electrons have escaped the diffusion region and enter the outflow region, other processes can take over.

Hoshino et al. (2001) were the first to consider the magnetic flux pileup region, created when the fast reconnection outflow jets collide with the pre-existing plasmas at rest, as the site for the second step in the electron acceleration. Strong observational evidence for this scenario has been provided by the Cluster and THEMIS multi-point measurements in the outflow region close to the diffusion region (Imada et al. 2007; Ashour-Abdalla et al. 2011; Khotyaintsev et al. 2011; Vaivads et al. 2011). Regarding the actual mechanism, there is observational evidence for betatron and/or Fermi acceleration (Hada et al. 1981; Ashour-Abdalla et al. 2011; Khotyaintsev et al. 2011; Fu et al. 2011). Cluster multi-point measurements have also provided evidence for electron acceleration in association with magnetic islands (Chen et al. 2008b) and for island (or flux-rope) coalescence (Retinò et al. 2008).

The finding that much of the electron acceleration seems to occur not within the diffusion region itself, but through jet braking in the outflow is reminiscent of scenarios envisioned for acceleration in the solar corona (e.g., Tsuneta 1995; Shibata et al. 1995).

It should be noted that particle acceleration in cosmic plasmas was the subject of the first ISSI workshop in this series. In the volume resulting from that workshop, the chapter by Birn et al. (2012) provides a detailed discussion of both in-situ observations as well as simulation studies of ion and electron acceleration in Earth's magnetotail.

2.5 Solar Wind

Current sheets associated with directional discontinuities in the magnetic field are ubiquitous in the solar wind. That they can undergo reconnection and exhibit the associated plasma jetting has been discovered only fairly recently (Gosling et al. 2005). Study of these reconnection events offers valuable new insights because the boundary conditions are more stable and more symmetric than at the magnetopause, while the magnetic shear is as variable. There is also the advantage that the solar wind rapidly advects any embedded structure past an observing spacecraft. One thus obtains nearly instantaneous snapshots of the configuration, in contrast to the magnetopause and magnetotail current sheets, which are constantly reversing their motion. In the meantime, it has been established (see Gosling 2011) that such reconnection events are quite common and apparent in essentially all solar wind data sets, covering distances between 0.3 and 5 AU.

2.5.1 Structure, Flows, and Kinetic Effects

The overall structure is determined by plasma inflow across the RDs on both sides, with a wedge-shaped outflow (exhaust) region in between, as depicted on the left in Fig. 8. Because of the two back-to-back RDs, the entire structure appears as a bifurcated current sheet. On the right in Fig. 8, the observations for a pass through such a structure are presented. Because the inflow velocity on the two sides points in opposite directions, while the normal magnetic

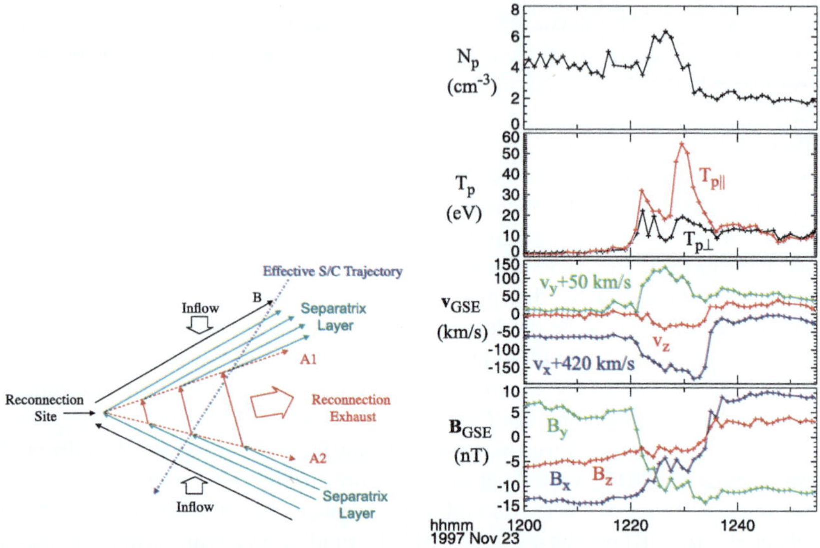

Fig. 8 *Left*: Schematic showing a slightly asymmetric solar wind reconnection configuration, consisting of an outflow region (exhaust), bounded on either side by thin current sheets labelled A1 and A2 that are akin to standing Alfvén waves, i.e., RDs. For simplicity, only one of the outflows emanating from the reconnection site is shown. *Right*: Observations by the ACE spacecraft during a crossing of a structure as depicted on the *left*. From top to bottom, the figure shows the plasma density, the parallel and perpendicular proton temperatures, the three components of the bulk velocity, and the corresponding magnetic field components, both in GSE coordinates (Adapted from Gosling et al. 2005)

field must be continuous and thus points in the same direction on both sides, the velocity and magnetic field variations at the boundaries are correlated on one side and anti-correlated on the other, as evident from Fig. 8. The outflow speed is much lower than at the magnetopause because the Alfvén velocity in the solar wind is much smaller. Nevertheless the observed flows closely meets the predictions from the Walén-relation. Multi-spacecraft observations of oppositely directed outflows, implying that an X-line was located between the observing spacecraft, have been reported by Davis et al. (2006) and Gosling et al. (2007c).

Under symmetric conditions, a slow shock (SS) is expected to occur inside the RDs on both sides of the exhaust. Although the enhanced proton density and temperature and decreased magnetic field strength observed in the central portion of the exhaust are qualitatively consistent with this expectation, the transitions across the boundaries of the exhaust in Fig. 8 are rather thick (a few hundred ion inertial lengths.) Sharper exhaust boundaries, reminiscent of slow shocks, have been seen in other solar wind exhausts (Phan et al. 2006a), although it is often found that these boundaries consist of merged RDs and slow-shocks (Teh et al. 2009; Sasunov et al. 2012). Sasunov et al. did, however, also report an event with a well-separated RD and SS pair. A remarkable result of the solar wind reconnection observations is the persistence of well-defined exhaust boundaries even at very large distances downstream from the X-line, of order of thousands or tens of thousands of ion inertial lengths.

In Sect. 2.2.3, interpenetrating beams were mentioned as one of the possible kinetic effects. In ideal MHD, the plasmas inflowing from the two sides do not mix, but are kept separated by a contact discontinuity. In practice this does not occur, as illustrated in Fig. 9, which shows the accelerated beams from the two sides interpenetrating in the middle of the

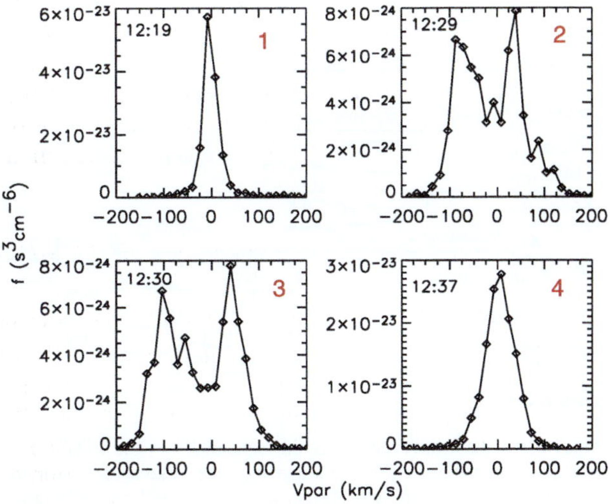

Fig. 9 Selected samples of the proton distribution function in the solar wind frame obtained before (*upper left panel*), during (*upper right* and *lower left panels*), and following (*lower right panel*) the reconnection exhaust crossing from Fig. 8 (From Gosling et al. 2005)

event (Gosling et al. 2005). The two beams have velocities separated by roughly twice the Alfvén speed, as expected from reconnection.

2.5.2 *Prevalence of Low Shear*

Solar wind reconnection events exhibit a prevalence of local magnetic shear angles <90°, with the smallest reported angle being only 11° (Gosling 2011), confirming the conclusion from the magnetopause observations that even large guide fields do not prevent reconnection.

2.5.3 *Spatial and Temporal Scales*

The presence of extended current sheets with stable boundary conditions in the solar wind allows studies of the large-scale properties of reconnection. Multi-spacecraft observations have indicated that solar wind reconnection events can have large spatial and temporal scales. The X-line can extend to several million kilometers (or tens of thousands of ion inertial lengths) and be observed over periods of several hours (or thousands of Alfvén transit times). An example is shown in Fig. 10 (Phan et al. 2006a). In one instance, the observing spacecraft remained in the exhaust as long as 3 hours (Gosling et al. 2007a). Even more extreme events, observed by many widely-spaced spacecraft, including STEREO-A and STEREO-B, have been reported by Gosling et al. (2007b) and Lavraud et al. (2009). The latter study suggests that reconnection rates might not have been constant over the duration of the event.

2.6 Magnetosheath

In the magnetosheath downstream from the quasi-parallel bow shock, the shocked solar wind plasma is highly turbulent. Figure 11 shows (on the right) the rapid fluctuations in magnetic field magnitude and direction, implying the occurrence of thin current sheets, which are necessary for reconnection to occur. In the lower left of the figure, the suggested formation of such current sheets between magnetic islands is illustrated. Analysis of high-resolution

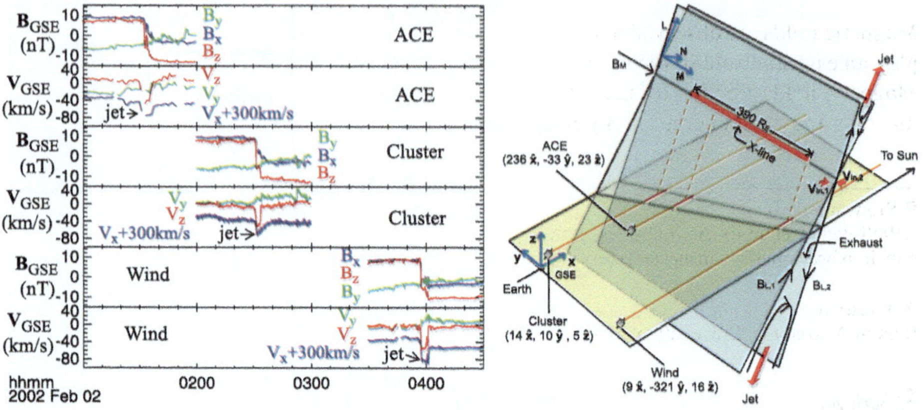

Fig. 10 A solar wind event observed sequentially by the widely spaced ACE, Cluster, and Wind spacecraft, implying that reconnection was occurring over large spatial and temporal scales. *Left*: The *top two panels* show the magnetic field and plasma flow velocities from ACE, and the subsequent panels show the same quantities for one of the Cluster spacecraft and for Wind. *Right*: Schematic of the encounters by ACE, Cluster and Wind with the exhaust region and its boundaries (*blue planes*) emanating from an extended (390 R_E) reconnection X-line The *yellow plane* is the ecliptic. Spacecraft positions (in R_E) are given in geocentric solar ecliptic coordinates (Adapted from Phan et al. 2006a)

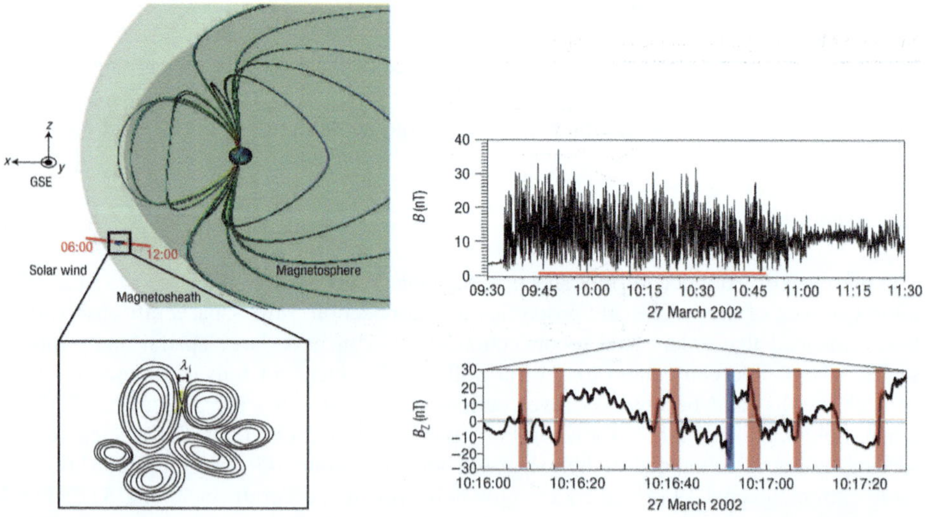

Fig. 11 *Left*: Cluster spacecraft crossing of a quasi-parallel bow shock and schematic illustration of current sheet formation between magnetic islands in the downstream magnetosheath. *Right*: time-series of magnitude and main component of the measured magnetic field, showing evidence of many current sheets (Adapted from Retinò et al. 2007)

multi-point measurements from Cluster (Retinò et al. 2007) for one such current layer crossing have revealed the microphysical (Hall-current) signatures that are evidence for magnetic reconnection (see Sect. 3.1). If reconnection occurs in a large percentage of these turbulent current sheets, the process could play an important role in the dissipation of turbulent energy (Sundkvist et al. 2007).

Reconnection in the magnetosheath has also been observed on larger scales. Thick (non-reconnecting) solar wind current sheets are compressed when they convect across the bow shock and further compressed as they pile up against the magnetopause. The thinning of the current sheet can trigger reconnection. This scenario was revealed by hybrid and MHD simulations (Lin and Xie 1997; Maynard et al. 2002; Omidi et al. 2009) and confirmed by multi spacecraft observations (Phan et al. 2007b). The interaction of current sheets with the bow shock and the magnetopause may have applications beyond the near-Earth space. It has been suggested that similar interactions could also occur across the termination shock and/or at the heliopause of our solar system producing anomalous cosmic rays (Drake et al. 2010), or in striped wind compression across termination shocks in pulsar wind nebulae (Lyubarsky 2003).

2.7 Other Planets

Reconnection signatures have been reported for the magnetospheres of five other planets, mostly based on in-situ magnetic field measurements.

Mercury, with its small intrinsic magnetic field and its closeness to the Sun, has a small magnetosphere, although its basic structure resembles that of Earth's. MESSENGER observations have revealed reconnection signatures at the magnetopause and in the magnetotail. In one instance, minimum variance analysis of the magnetic field implied that the magnetopause appeared as a rotational discontinuity, with a finite B_n that translated into a reconnection rate of 0.13 (Slavin et al. 2009).

Venus has no intrinsic magnetic field, but a magnetotail is formed by the draped IMF. There formation of a plasmoid due to magnetic reconnection has been observed recently by the Venus Express spacecraft (Zhang et al. 2012b).

At Mars, the Hall magnetic fields that are a signature of the reconnection diffusion region have been detected at tail and flank current sheets (Eastwood et al. 2008; Halekas et al. 2009), indicating that reconnection is occurring, as discussed in Sect. 3.1. Flux ropes have been observed there as well (Eastwood et al. 2012).

Jupiter has the largest planetary magnetosphere, its dynamics being dominated by planetary rotation rather than by the solar wind. Yet there is significant solar wind interaction, as manifested by a long magnetotail. In the magnetotail, magnetic signatures of localized and transient reconnection have been observed (Russell et al. 1998), and some auroral features have also been interpreted as resulting from magnetotail reconnection (Radioti et al. 2011). In a direct transfer of insights obtained for Earth's magnetosphere, magnetopause reconnection rates at Jupiter have been estimated using empirical relationships between solar wind parameters and reconnection voltages obtained for Earth, scaled to Jupiter conditions (Nichols et al. 2006).

Like Jupiter, Saturn has a large rapidly rotating magnetosphere, but reconnection might still play a role in its dynamics. In one magnetopause crossing, a non-zero normal component of the magnetic field has been reported from minimum variance analysis of the magnetic field, implying a reconnection rate of 0.10 (McAndrews et al. 2008). Interestingly, in another direct transfer of knowledge gained at Earth (see Sect. 2.8.2), Masters et al. (2012) have suggested that the higher plasma β in the magnetosheath of Saturn should prevent magnetopause reconnection except for large shear angles. The case reported by McAndrews et al. had indeed a fairly large magnetic shear (149°).

2.8 General Characteristics

In this section, we will discuss some general characteristics of reconnection that the in-situ observations have revealed.

2.8.1 *Occurrence Frequency*

Not all crossings of the magnetopause, magnetotail, or solar wind current sheets show evidence of reconnection. In fact, the majority of current sheet crossings in the solar wind and in the magnetotail show no reconnection signatures, while at the dayside magnetopause the occurrence rate of reconnection signatures is about 50 % (Paschmann et al. 1986). This raises the question what conditions must be fulfilled for reconnection to occur. The next section discusses such conditions.

2.8.2 *Onset Conditions*

Current Sheet Thickness For reconnection to occur, the current sheet must be sufficiently thin. In collisionless plasmas, it appears that the current sheet thickness has to be one ion inertial length or smaller, in order to initiate reconnection (e.g., Cassak et al. 2006).

According to the general understanding of magnetotail dynamics, the tail current sheet is usually too thick for reconnection to start. The bursty nature of near-Earth magnetotail reconnection suggests that sufficient thinning occurs only sporadically. Regardless of the cause for thinning, ISEE and Cluster multi-spacecraft observations have provided evidence for sub-ion-inertial-length current sheets just prior to magnetotail reconnection and associated substorm onset (e.g., Sanny et al. 1994; Runov et al. 2008).

The dayside magnetopause current sheet is usually thin due to the constant compression of the solar wind against the dayside magnetosphere. However, as mentioned earlier, half of the magnetopause crossings display no reconnection signatures even when the magnetic shear is high. This indicates that a thin current sheet is a necessary but not sufficient condition for reconnection.

Magnetic Shear and Plasma β While it is clear that current sheets must be sufficiently thin for reconnection to occur, the plasma β, i.e, the ratio of plasma to magnetic pressure, has long been suspected to be an important additional constraint. Early magnetopause observations (Paschmann et al. 1986) had suggested that β in the inflow (magnetosheath) region may be a controlling factor, with reconnection more likely to occur for small values of β. It now appears, based on recent solar wind and magnetopause observations, that it is not β alone that controls reconnection, but a combination of β (more precisely the difference between the β values on the two sides) and the magnetic shear (Phan et al. 2010, 2013, in press). The left part of Fig. 12 shows that for low $\Delta\beta$ reconnection occurred at current sheets with both low and high magnetic shear angle, whereas for large $\Delta\beta$ reconnection occurred only for high magnetic shear angles. The plot on the right shows that β itself does not organize the data as well as $\Delta\beta$. These observations are in quantitative agreement with a theoretical prediction (Swisdak et al. 2003, 2010) that reconnection is suppressed in high β plasmas at low magnetic shear angle due to diamagnetic drift of the reconnection X-line caused by plasma pressure gradients across the current sheets.

2.8.3 *Reconnection Rates*

Reconnection rates are notoriously difficult to obtain from in-situ observations because they require knowledge of the normal component of the magnetic field, B_n, or the plasma flow, v_n, which are both small and depend on precise knowledge of the current sheet orientation. Similar restrictions apply to the determination of the reconnection electric field, E_t. For the magnetopause, Fuselier and Lewis (2011) have compiled some of the reported values in

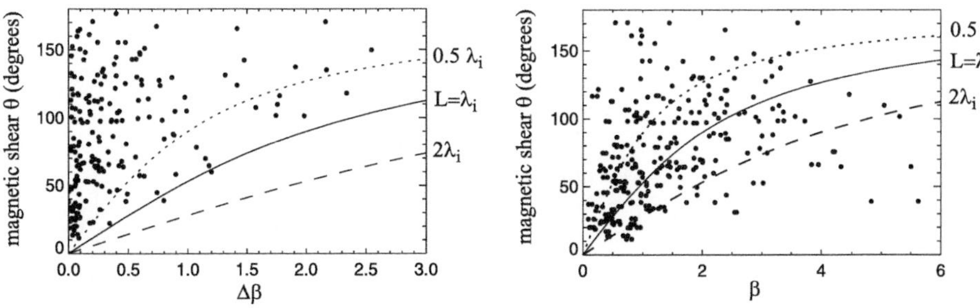

Fig. 12 *Left*: Plot of magnetic shear angle versus the difference of the plasma-β on the two sides of a set of 197 solar wind reconnection exhausts. The *three curves* are theoretical predictions from Swisdak et al. (2010) for different choices of the scale size, L (in units of the ion inertial length, λ_i), of the density gradient at a reconnection diffusion region. Reconnection should be suppressed below these curves. *Right*: Similar plot, but for β (Adapted from Phan et al. 2010)

the literature, ranging between <0.01 and 0.2. For one extended magnetopause reconnection event, Fuselier et al. (2010) have reported an average value of 0.08. Rosenqvist et al. (2008) have determined reconnection rates for the multiple magnetopause crossings shown in Fig. 4, and obtained values between 0.01 and 0.3, with an average of 0.14. For individual solar wind and magnetosheath events, values of ~0.03 and ~0.07, respectively, were inferred (Phan et al. 2006a, 2007b). The large variations in the reported values do not necessarily imply intrinsic variability of the reconnection rates, but may simply represent the (hard to quantify) uncertainties in the determinations. Note that simulations of fast reconnection give typical rates of 0.1–0.2 (Shay et al. 1998).

3 Microphysics

Magnetic reconnection is a cross-scale phenomenon. While the process is initiated in a small *diffusion region*, where ion and electrons are demagnetized, the consequences of reconnection are large-scale, as discussed in Sect. 2.

The manner in which the particles demagnetize is closely related to the relative importance of the different scale sizes: the resistive scale, the ion scale, and the electron scale. In the diffusion region the plasma frozen-in condition must be violated, implying that some of the terms on the r.h.s. of the generalized Ohm's law must be non-zero. Ohm's law can be written as (e.g., Rossi and Olbert 1970):

$$\boldsymbol{E} + \boldsymbol{v} \times \boldsymbol{B} = \eta \boldsymbol{j} + + \frac{1}{ne} \boldsymbol{j} \times \boldsymbol{B} - \frac{1}{ne} \nabla \cdot \boldsymbol{P}_e + \frac{m_e}{ne^2} \left[\frac{\partial \boldsymbol{j}}{\partial t} + \nabla \cdot (\boldsymbol{j}\boldsymbol{v} + \boldsymbol{v}\boldsymbol{j}) \right] \tag{3}$$

where $\boldsymbol{E}$ and $\boldsymbol{B}$ denote electric and magnetic fields, $\boldsymbol{v}$ is the velocity of the particle species, and $\boldsymbol{j}$ is the current density. Each of the terms on the right hand side of Eq. (3) can break the frozen-in condition and are associated with different scale sizes. Here $\eta \boldsymbol{j}$ is the resistive term, $\frac{1}{ne} \boldsymbol{j} \times \boldsymbol{B}$ is the Hall, or ion, term, $\frac{1}{ne} \nabla \cdot \boldsymbol{P}_e$ is the divergence of the electron pressure term, and $\frac{m_e}{ne^2}[\frac{\partial \boldsymbol{j}}{\partial t} + \nabla \cdot (\boldsymbol{j}\boldsymbol{v} + \boldsymbol{v}\boldsymbol{j})]$ is the electron inertia term. Thus the first term becomes important on the resistive scale, the second term is associated with the ion scale, and the last two terms become important on the electron scale.

If the resistive scale is larger than the ion scale, both electrons and ions demagnetize on the same scale, the resistive scale. However, if the resistivity is so small that the resistive

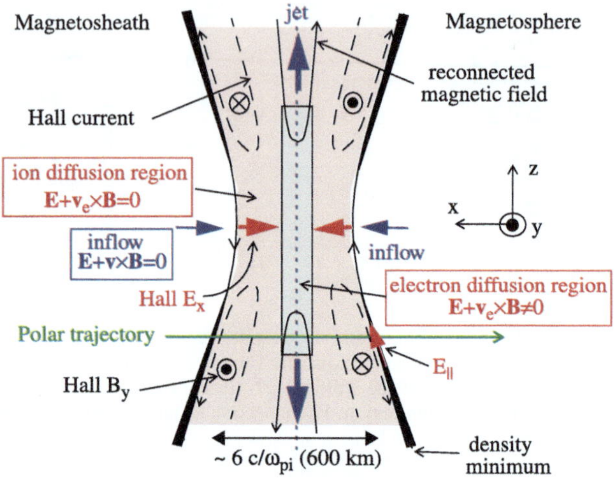

Fig. 13 The geometry of the diffusion region. Ions are decoupled from the electrons and magnetic field in the ion diffusion region, creating Hall magnetic and electric field patterns. Electrons are demagnetized in the much smaller electron diffusion region. The *green arrow* indicates the trajectory of the Polar spacecraft while collecting the data shown in Fig. 14(A) (From Mozer et al. 2002)

term can be ignored, the ions will demagnetize at the ion scale, due to the $\frac{1}{ne}\boldsymbol{j} \times \boldsymbol{B}$ (Hall) term, while the electrons will stay magnetized longer, until they reach the much smaller electron scale where they become demagnetized due to one of the electron terms. The ion and electron separation in the diffusion region leads to a system of Hall currents, which in turn induce the quadrupolar Hall magnetic fields, as shown in Fig. 13. A Hall electric field develops in the ion diffusion region and is directed towards the current sheet on both sides of the current sheet (Fig. 13). The $V_{e,y} = E_z \times B_x/B^2$ drift and its pulling of the magnetic field in the negative y direction results in the quadrupolar Hall B_y (Mandt et al. 1994; Shay et al. 1998; Pritchett 2005).

The quadrupolar Hall currents and magnetic fields are symmetric in magnitude on the two sides of the current sheet only when reconnection is symmetric, i.e. when the plasma on the two sides of the current sheet are similar. When reconnection is asymmetric, i.e. the reconnecting plasmas are different, the Hall currents and magnetic field patterns are also asymmetric, with larger magnitude Hall fields on the low-density side of the current sheet (Pritchett 2008). Reconnection is usually asymmetric at the Earth's magnetopause, and symmetric in the Earth's magnetotail. If a guide field is present, the Hall magnetic field will be superposed onto this guide field and therefore appear to be asymmetric even if reconnection is symmetric (Karimabadi et al. 1999; Pritchett and Coroniti 2004).

Until recently, much of our knowledge of processes in the diffusion region was derived solely from theoretical modeling. The small size of the diffusion region made it very challenging to observe in situ with a spacecraft. The width of the ion diffusion region is of the order of the ion inertial length, λ_i, which is only $\sim$50 km at the Earth's dayside magnetopause and $\sim$1000 km in the magnetotail. The length of the diffusion region is predicted to be 10 times larger. The electron diffusion region width is the electron inertial length, λ_e, which is 43 times smaller, only $\sim$1–2 km at the magnetopause and $\sim$25 km in the magnetotail. In this section we discuss recent findings from spacecraft encounters with both the ion-scale and the electron-scale diffusion region, as well as observations from additional reconnection-related regions where electron-scale physics has been found to be important. Finally, we discuss briefly the upcoming NASA Magnetospheric Multi-Scale (MMS) mission, which is dedicated to studies of electron-scale physics in reconnection.

3.1 The Ion Diffusion Region

In collisionless reconnection, the ion diffusion region is the region where ions are demagnetized while electrons are not, resulting in a Hall current and induced Hall magnetic fields, as discussed above and illustrated in Fig. 13. Confirming the presence of the Hall effect with in-situ spacecraft observations is significant since it implies that the resistive scale is smaller than the ion inertial length, and the ions and electrons are demagnetized at different scales. While the quadrupolar Hall currents and Hall magnetic field in the diffusion region were first predicted theoretically (Sonnerup 1979), the first in situ spacecraft observations of these phenomena were made approximately two decades later.

Hall Currents The first observational confirmation of the predicted ion and electron decoupling in the diffusion region were made along the separatrix layers. The separatrices are the surfaces defined by all magnetic field lines crossing the reconnection X-line, thus representing the most recently reconnected field lines. In Fig. 2, the field lines labeled S1 and S2 are cuts through the separatrix surfaces. Fujimoto et al. (1997) and Nagai et al. (2001) reported observations by the Geotail spacecraft along magnetotail separatrices of low-energy electrons streaming towards the X-line, while higher energy electrons were streaming away from the X-line. The directions of the low energy electrons were consistent with them being the Hall current carriers.

Hall Magnetic Fields In addition to observing the Hall current carrier along the separatrices, Nagai et al. (2001) reported associated out-of-plane magnetic fields consistent with the predicted directions of the Hall magnetic field. Subsequently, Hall magnetic fields were detected inside the diffusion region itself by the Wind spacecraft in Earth's distant magnetotail (Øieroset et al. 2001). The diffusion region was identified from the plasma jet reversal, which coincided with a reversal in the normal component of the magnetic field, indicating that the spacecraft crossed a reconnection diffusion region, going from the earthward to the tailward jet. Coinciding with these reversals, the out-of-plane magnetic field component also reversed sign and the observed polarities were consistent with the predicted polarities of the quadrupolar Hall magnetic field (see Fig. 14(C)).

Mozer et al. (2002) reported a fortuitous diffusion region encounter by the Polar spacecraft at the dayside magnetopause. We note that this event was a rare case when reconnection was nearly symmetric at the Earth's magnetopause, thus the Hall magnetic field pattern would still be quadrupolar, similar to the magnetotail Hall pattern. However, in contrast to the Wind crossing along the outflow direction of the tail diffusion region, the Polar crossing was normal to the current sheet, going from the magnetosheath to the magnetosphere south of an X-line, as indicated by the green horizontal line in Fig. 13, the out-of-plane magnetic field reversed sign from positive to negative, consistent with the predicted directions of the Hall magnetic fields (Fig. 14(A)).

A Cluster multi-spacecraft encounter with a diffusion region at the dayside magnetopause was reported by Vaivads et al. (2004), also showing out-of-plane magnetic fields consistent with the Hall magnetic field directions (Fig. 14(B)). In this multi-spacecraft encounter, two spacecraft observed the Hall magnetic field simultaneously in two of the quadrants, hence establishing even more firmly that the observed out-of-plane magnetic fields were spatial, not temporal structures. From the four-point timing analysis, the spatial scale shown along the bottom of Fig. 14(B) was determined, giving a current layer thickness of a few ion inertial lengths.

Confirming the presence of the Hall magnetic field in all four quadrants in one single event is challenging, even with multi-spacecraft observations. To overcome this difficulty,

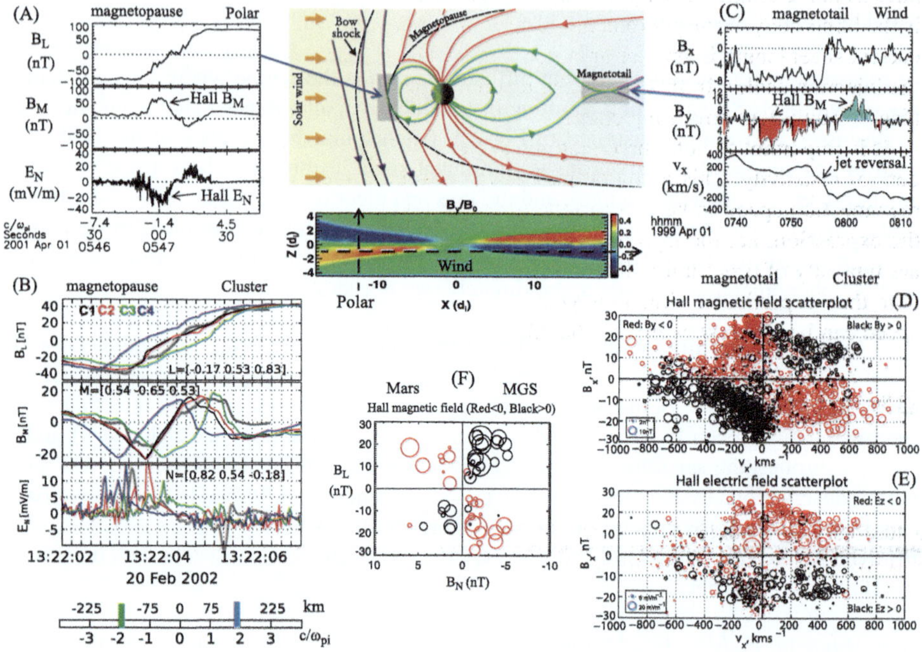

Fig. 14 Hall magnetic and electric fields, as observed at the magnetopause and magnetotail reconnection sites shown in the cartoon at the *top center*. (**A**) Magnetopause crossing by the Polar spacecraft, showing the Hall and reconnecting magnetic field components and the Hall electric field (adapted from Mozer et al. 2002). (**B**) Magnetopause crossing by the four Cluster spacecraft, showing the reconnecting magnetic field component, the out-of-plane magnetic field component, and the electric field normal to the magnetopause (adapted from Vaivads et al. 2004). (**C**) Crossing of the magnetotail current sheet by the Wind spacecraft, showing he bipolar Hall magnetic field and the reversal in the Earthward directed flow velocity (adapted from Øieroset et al. 2001). (**D**) Hall magnetic field B_y versus reconnecting magnetic field B_x and reconnection jet velocity V_x. Black corresponds to $B_y > 0$ and red to $B_y < 0$. (**E**) Hall electric field versus B_x and V_x. Black corresponds to $E_z > 0$ and red to $E_z < 0$. The size of the symbols indicates the magnitude of the data points (adapted from Eastwood et al. 2010b). (**F**) Hall magnetic fields observed by the Mars Global Surveyor (MGS) spacecraft, in a format similar to part (**D**), except that B_n is used to distinguish the two sides of the X (MGS data courtesy of J. Halekas). *Center*: simulation of the diffusion region, with the Hall field directions in *red* (courtesy M. Shay)

Eastwood et al. (2010b) performed a statistical study, using Cluster multi-point observations of 18 diffusion region encounters in the Earth's magnetotail, which, taken together, covered all four quadrants multiple times. The observed out-of-plane magnetic fields in these events were indeed consistent with the predicted Hall magnetic field in all four quadrants, as shown in Fig. 14(D).

The events included in Eastwood et al. (2010b) did not have any significant guide field. By contrast, Eastwood et al. (2010a) studied a separate diffusion region encounter by Cluster where a moderate guide field (20 % of the reconnecting field) was present and showed that the Hall magnetic and electric fields were asymmetric and shunted away from the current sheet, consistent with simulations.

In addition to the studies already mentioned, there has been several other reports of in-situ spacecraft encounters with the diffusion region, both in the Earth's magnetotail (e.g., Runov et al. 2003; Wygant et al. 2005; Borg et al. 2005; Nakamura et al. 2006; Asano et al. 2008), at the Earth's magnetopause (e.g., Pu et al. 2005; Zhang et al. 2008), in the magnetosheath (Phan et al. 2007a), and even at other planets (e.g., Eastwood et al. 2008; Halekas and Brain 2010). Figure 14(F) shows the amplitude of the out-of-plane magnetic field for 28 current sheet crossings at Mars. The observed polarities of the out-of-plane magnetic fields surrounding diffusion regions at Mars are consistent with the predictions for the Hall quadrupolar magnetic field, indicating that ion-electron decoupling also occur in reconnection at Mars (Halekas et al. 2009).

Hall Electric Fields Electric fields with directions consistent with the predicted direction of the Hall electric field (Shay et al. 1998; Pritchett 2005) has been reported in connection with observations of the Hall magnetic field, both at the Earth's magnetopause (Mozer et al. 2002; Vaivads et al. 2004) and in the magnetotail (Wygant et al. 2005; Borg et al. 2005; Eastwood et al. 2010b). Examples are shown in Fig. 14(A) and (B). In their statistical study Eastwood et al. (2010b) recorded the normal electric field and showed that its direction was consistent with the predicted direction of the Hall electric field for all 18 events (Fig. 14(E)).

3.2 The Inner Electron Diffusion Region

Within the ion diffusion region there is a thin layer where electrons demagnetize (Fig. 13). This inner electron diffusion region is located in close vicinity of the reconnection X-line.

According to theory, the electron diffusion region is characterized by a large out-of-plane current centered at the electron jet reversal (Shay and Drake 1998; Hesse et al. 1999). The two lower panels in the left part of Fig. 15 show horizontal cuts through a simulation of the reconnection region, including the electron diffusion region shown in the top panel. As shown in the middle panel, there is a large difference between electron and ion outflow speeds. The bottom panel shows the electron and ion velocities in the out-of-plane direction. While very little variation is seen in the out-of-plane ion velocity, there is a large into-the-plane electron jet located right at the jet reversal. This into-the-plane electron jet indicates the presence of a strong out-of-plane current right in the center where the ion and electron jets reverse sign.

Because the inner electron diffusion region is 43 times smaller than the ion diffusion region, there have been few reports of encounters with this region until recently. Chen et al. (2008a), with guidance from kinetic simulations, reported the encounter of the inner electron diffusion region by the Cluster spacecraft. More recently, Nagai et al. (2011) reported a fortuitous encounter with the electron diffusion region in the Earth's magnetotail, when the Geotail spacecraft traversed from the tailward to the earthward side of a reconnection X-line (Fig. 15, right). Panel (c) shows the electron and ion velocities in the outflow direction as the dotted and solid line, respectively. A large difference between electron and ion speeds is seen, similar to that in the simulation (Fig. 15, left). Furthermore, right at the jet reversal there was a strong into-the-plane electron jet, a key characteristic of the inner electron diffusion region.

The observed strong into-the-plane electron jet indicates that the spacecraft encountered the inner electron diffusion region. However, only two data points were collected in the region of strong current. While this indicated that the electron diffusion region was indeed encountered, it is not sufficient to study the detailed electron physics in this region. Investigating the electron-scale processes in the electron diffusion region is the goal of the upcoming MMS mission, which will perform very high-resolution plasma and field measurements (see Sect. 3.5).

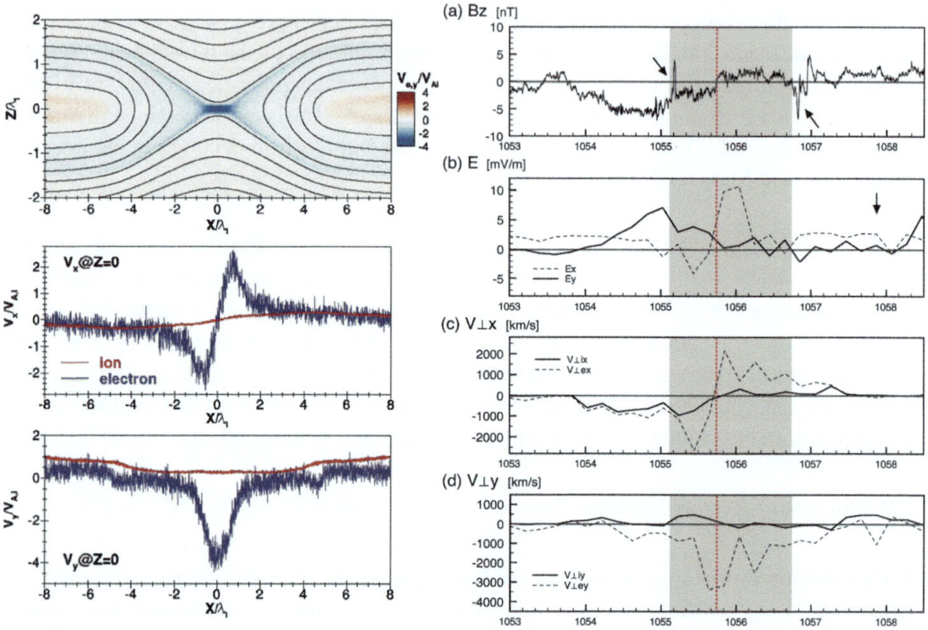

Fig. 15 *Left*: Simulations of a reconnection region. The *top panel* shows the diffusion region, with the electron part at the center. The *middle panel* shows the electron and ion velocities in the outflow direction, the *lower panel* the velocities in the out-of-plane direction, electrons in *blue* and ions in *red* (Figure courtesy of I. Shinohara). *Right*: Geotail observations on 15 May 2003. (**a**) Normal magnetic field B_Z, (**b**) electric field E_x, E_y, and (**c**, **d**) plasma perpendicular velocities. The presumed X-line crossing is indicated by a *red vertical line*. From Zenitani et al. (2012), after Nagai et al. (2011)

3.3 Elongated Electron Jet Layer

Full particle simulations in large simulation domains revealed the existence of a long (tens of ion inertial length) super-Alfvénic electron jet connected to the inner electron diffusion region (Daughton et al. 2006; Karimabadi et al. 2007; Shay et al. 2007). Cluster detected such an electron jet (extending at least 60 ion inertial lengths) in a magnetosheath reconnection event under nearly symmetric boundary conditions (Phan et al. 2007a). Simulations showed that in the case of asymmetric reconnection with a guide field, the extended electron jet exists over a shorter length than for symmetric reconnection, and the jet exists on one side of the X-line only (Mozer and Pritchett 2009). In contrast to the inner electron diffusion region, the extended electron jet is not accompanied by dissipation (Hesse et al. 2008).

3.4 Electron Physics Along the Separatrices

Connected to the diffusion region are the separatrices, referred to earlier. Ion and electron decoupling have been observed in thin (electron scale) layers, associated with the separatrices, at distances far away from the traditional diffusion region surrounding the X-line (Mozer et al. 2003; André et al. 2004; Khotyaintsev et al. 2006). These layers of electron-ion decoupling along the separatrices are characterized by large density fluctuations and large electric fields both perpendicular and parallel to the magnetic field.

Solitary waves, which are bipolar electric field pulses traveling parallel to the magnetic field, have been observed on the magnetospheric side of the magnetopause current sheet

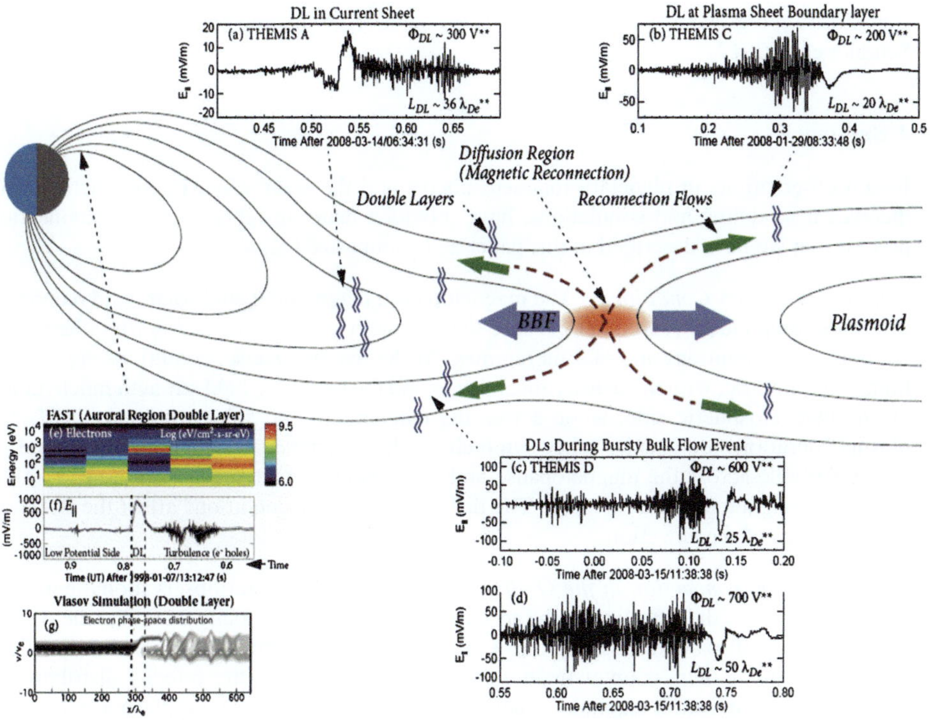

Fig. 16 FAST and THEMIS observations of double layers and electron holes in the magnetosphere. The occurrence regions are indicated in the cartoon. The data plots show time series of E_{parallel}, demonstrating that double layers and solitary waves have been observed in the magnetotail current sheet (*top*, *left*), in the plasma sheet boundary layer (*top*, *right*), in the auroral region (*bottom*, *left*) and in bursty bulk flows (*bottom*, *right*). THEMIS data plots adapted from Ergun et al. (2009). Figure courtesy of R. Ergun and L. Andersson

(Cattell et al. 2002), as well as in the Earth's magnetotail where they have been observed near the outer edge of the plasma sheet and in the current sheet itself (Cattell et al. 2005; Ergun et al. 2009; Andersson et al. 2009) (see Fig. 16). These waves, also referred to as electron holes, could play an important role in dissipation and diffusion if they occur in large numbers because they can affect the electron distribution function (Cattell et al. 2002).

3.5 NASA's Magnetospheric Multi-scale Mission

The ion diffusion region has now been encountered and explored by a variety of spacecraft, but the electron diffusion region is largely unexplored. This is primarily because current spacecraft instrumentation cannot resolve the electron diffusion region in detail. The upcoming Magnetospheric Multi-Scale (MMS) mission, scheduled for launch in 2014, consists of four spacecraft carrying identical plasma and field instruments with orders of magnitude higher time resolution plasma measurements than before (e.g., Burch and Drake 2009). The spacecraft will fly in a tetrahedral (pyramid) formation with close (as small as 10 km) separation, allowing them to determine three dimensional structures of the reconnection sites they encounter, both at the magnetopause and in the magnetotail. Achieving the MMS mission goal rests on the ability to unambiguously identify the inner electron diffusion region in the data. Studies of recent spacecraft encounters with the inner electron diffusion region have yielded methods for identifying this region (Zenitani et al. 2012; Scudder et al. 2012).

4 Summary

In-situ observations in planetary magnetospheres and the solar wind, in conjunction with theoretical modeling and simulations, have provided many insights into the reconnection process that we have described in this paper and summarize below.

Different Boundary Conditions The magnetopause, magnetotail, and solar wind reconnection sites complement each other because they allow to study reconnection for quite different boundary conditions and plasma regimes. At the magnetopause, conditions are usually highly asymmetric, with the density much lower and the magnetic field strength much higher on the magnetospheric side, while across the magnetotail and solar wind current sheets, plasma conditions are usually fairly symmetric. The magnetic shear, on the other hand, is highly variable across the magnetopause and solar wind current sheets, while nearly 180° across the magnetotail current sheet. The different boundary conditions affect the structure and dynamics of reconnection.

Signatures of Reconnection in the Outflow Region The dominant observational evidence for reconnection is the detection of the accelerated plasma bulk velocity at or near the Alfvén speed in the outflow (exhaust) region. A number of kinetic effects complement the fluid signatures, among them counterstreaming beams and cut-offs in the velocity distributions. Other basic reconnection signatures such as the normal magnetic field, plasma inflow and reconnection electric field are much more difficult to determine.

Nature of the Outflow Region Boundaries For asymmetric conditions, the dominant boundary feature is predicted to be a rotational discontinuity (RD). The RDs have been identified for the magnetopause through the outflow velocities meeting the Walén-relation. For symmetric conditions, the structure is expected to include slow shocks. In a few instances, these have been identified for magnetotail and solar wind reconnection. A remarkable result is the persistence of sharp and well-defined boundaries to very large distances downstream from the X-line.

Detection of the Reconnection Diffusion Region While most of the spacecraft encounters with the reconnection layer occur in the outflow (exhaust) region downstream of the diffusion region, the past decade has witnessed numerous reports of ion diffusion region encounters by spacecraft in the magnetotail, at the magnetopause and in the magnetosheath, as well as glimpses of the much smaller electron diffusion region. Advances in the theoretical understanding of the diffusion region, coupled with the spacecraft detections of the diffusion region, form the basis for the upcoming Magnetospheric Multi-Scale mission which will investigate the magnetic field breaking processes in the electron diffusion region using ultra-high-resolution plasma and field measurements on four spacecraft.

Reconnection Rates Reconnection rates are notoriously difficult to measure reliably because they are proportional to the normal components of the magnetic field or plasma flow, which are small and dependent on precise knowledge of the current sheet orientation. Nevertheless, some values have been reported, ranging between <0.01 and 0.2, where it is uncertain whether this range reflects true variability or simply the large uncertainties. On the other hand, the detections of Hall magnetic and electric fields in the diffusion region suggest that reconnection operates in the 'fast' regime in near-Earth space.

Spatial and Temporal Scales Reconnection in the near-Earth magnetotail associated with geomagnetic substorms is highly bursty, storing and explosively releasing magnetic energy similar to solar flares. In the solar wind and at the magnetopause there is evidence that reconnection can operate in a quasi-steady manner along extremely long X-lines. Thus in-situ observations have clarified that reconnection is not intrinsically transient, but can operate in a quasi-stationary, continuous fashion. It can operate over large spatial scales, but can also be patchy.

Flux Ropes Intermittent or continuous reconnection with modulated reconnection rate could produce flux ropes that are commonly observed at the magnetopause (referred to as Flux Transfer Events) and in the magnetotail (called plasmoids or magnetic islands). Their role in energizing particles has been suggested but has yet to be firmly established.

Reconnection Onset Conditions and Occurrence Frequency A large number of current sheet encountered by spacecraft in the magnetosphere and solar wind display no local signatures of reconnection. The question is what conditions need to be met for reconnection to occur. There is clear evidence that current sheets must be sufficiently thin (an ion inertial length or less). In addition to the thin current sheet requirement, a combination of plasma β (more precisely the difference, $\Delta\beta$, across the current sheet) and magnetic shear has a controlling effect. For small $\Delta\beta$, reconnection can happen even for low magnetic shear (or large guide field), while for large $\Delta\beta$, reconnection requires large shear. With these (and possibly additional) strict conditions it is not surprising that many current sheets do not undergo reconnection.

Anti-parallel Versus Component Reconnection A long-standing controversy that the in-situ observations have resolved is whether or not reconnection requires purely anti-parallel magnetic fields. The observations clearly show that it can happen in the presence of substantial guide fields.

Turbulent Reconnection Downstream of the quasi-parallel bow shock, the shocked solar wind (magnetosheath) plasma is highly turbulent and filled with thin current sheets, and some initial evidence for reconnection across such current sheets has been reported. However, the plasma measurements onboard current spacecraft do not have sufficient temporal resolution to firmly establish the occurrence rate of reconnection in these thin current sheets. The upcoming magnetospheric Multi-Scale mission, with its ultra-high-resolution plasma measurements, should be able to determine what fraction of the current sheets undergoes reconnection and whether the process plays a significant role in dissipating turbulent energy.

Electron Acceleration Multi-point measurements in and near the reconnection diffusion region have provided strong evidence for electron acceleration to hundreds of keV in a two- or multi-step process, beginning within the diffusion region itself, but becoming more pronounced when the electrons enter the outflow region, either in the flux pileup region created by flow jet braking, or within magnetic islands.

Reconnection at Other Planets In-situ magnetic field measurements have provided evidence for reconnection at five other planets (Mercury, Venus, Mars, Jupiter, and Saturn). Because the solar wind plasma-β value decreases systematically with increasing distance from the Sun, reconnection may be more frequent at inner planets and consequently may play a more important role in their interaction with the solar wind.

Acknowledgements At UC Berkeley, this research was funded in part by NSF grant AGS-1103303 and NASA grant NNX08AO83G.

References

L. Andersson, R.E. Ergun, J. Tao, A. Roux, O. Lecontel, V. Angelopoulos, J. Bonnell, J.P. McFadden, D.E. Larson, S. Eriksson, T. Johansson, C.M. Cully, D.N. Newman, M.V. Goldman, K.-H. Glassmeier, W. Baumjohann, New features of electron phase space holes observed by the THEMIS mission. Phys. Rev. Lett. **102**(22), 225004 (2009). doi:10.1103/PhysRevLett.102.225004

M. André, A. Vaivads, S.C. Buchert, A.N. Fazakerley, A. Lahiff, Thin electron-scale layers at the magnetopause. Geophys. Res. Lett. **31**, 3803 (2004). doi:10.1029/2003GL018137

V. Angelopoulos, W. Baumjohann, C.F. Kennel, F.V. Coronti, M.G. Kivelson, R. Pellat, R.J. Walker, H. Luehr, G. Paschmann, Bursty bulk flows in the inner central plasma sheet. J. Geophys. Res. **97**, 4027–4039 (1992). doi:10.1029/91JA02701

V. Angelopoulos, C.F. Kennel, F.V. Coroniti, R. Pellat, M.G. Kivelson, R.J. Walker, C.T. Russell, W. Baumjohann, W.C. Feldman, J.T. Gosling, Statistical characteristics of bursty bulk flow events. J. Geophys. Res. **99**, 21257 (1994). doi:10.1029/94JA01263

V. Angelopoulos, T.D. Phan, D.E. Larson, F.S. Mozer, R.P. Lin, K. Tsuruda, H. Hayakawa, T. Mukai, S. Kokubun, T. Yamamoto, D.J. Williams, R.W. McEntire, R.P. Lepping, G.K. Parks, M. Brittnacher, G. Germany, J. Spann, H.J. Singer, K. Yumoto, Magnetotail flow bursts: association to global magnetospheric circulation, relationship to ionospheric activity and direct evidence for localization. Geophys. Res. Lett. **24**, 2271–2274 (1997). doi:10.1029/97GL02355

V. Angelopoulos, J.P. McFadden, D. Larson, C.W. Carlson, S.B. Mende, H. Frey, T. Phan, D.G. Sibeck, K.-H. Glassmeier, U. Auster, E. Donovan, I.R. Mann, I.J. Rae, C.T. Russell, A. Runov, X.-Z. Zhou, L. Kepko, Tail reconnection triggering substorm onset. Science **321**, 931 (2008). doi:10.1126/science.1160495

Y. Asano, R. Nakamura, I. Shinohara, M. Fujimoto, T. Takada, W. Baumjohann, C.J. Owen, A.N. Fazakerley, A. Runov, T. Nagai, E.A. Lucek, H. Rème, Electron flat-top distributions around the magnetic reconnection region. J. Geophys. Res. **113**, 1207 (2008). doi:10.1029/2007JA012461

M. Ashour-Abdalla, M. El-Alaoui, M.L. Goldstein, M. Zhou, D. Schriver, R. Richard, R. Walker, M.G. Kivelson, K.-J. Hwang, Observations and simulations of non-local acceleration of electrons in magnetotail magnetic reconnection events. Nat. Phys. **7**, 360–365 (2011). doi:10.1038/nphys1903

D.N. Baker, E.C. Stone, Energetic electron anisotropies in the magnetotail - identification of open and closed field lines. Geophys. Res. Lett. **3**, 557–560 (1976). doi:10.1029/GL003i009p00557

D.N. Baker, W.K. Peterson, S. Eriksson, X. Li, J.B. Blake, J.L. Burch, P.W. Daly, M.W. Dunlop, A. Korth, E. Donovan, R. Friedel, T.A. Fritz, H.U. Frey, S.B. Mende, J. Roeder, H.J. Singer, Timing of magnetic reconnection initiation during a global magnetospheric substorm onset. Geophys. Res. Lett. **29**(24), 240000-1 (2002). doi:10.1029/2002GL015539

W. Baumjohann, G. Paschmann, H. Luehr, Characteristics of high-speed ion flows in the plasma sheet. J. Geophys. Res. **95**, 3801–3809 (1990). doi:10.1029/JA095iA04p03801

S. Beyene, C.J. Owen, A.P. Walsh, C. Forsyth, A.N. Fazakerley, S. Kiehas, I. Dandouras, E. Lucek, Cluster observations of a transient signature in the magnetotail: implications for the mode of reconnection. Ann. Geophys. **29**, 2131–2146 (2011). doi:10.5194/angeo-29-2131-2011

J. Birn, A.V. Artemyev, D.N. Baker, M. Echim, M. Hoshino, L.M. Zelenyi, Particle acceleration in the magnetotail and Aurora. Space Sci. Rev. (2012). doi:10.1007/s11214-012-9874-4

A.L. Borg, M. Øieroset, T.D. Phan, F.S. Mozer, A. Pedersen, C. Mouikis, J.P. McFadden, C. Twitty, A. Balogh, H. Rème, Cluster encounter of a magnetic reconnection diffusion region in the near-Earth magnetotail on September 19, 2003. Geophys. Res. Lett. **32**, 19105 (2005). doi:10.1029/2005GL023794

J.L. Burch, J.F. Drake, Reconnecting magnetic fields. Am. Sci. **97**, 392 (2009)

P.A. Cassak, J.F. Drake, M.A. Shay, A model for spontaneous onset of fast magnetic reconnection. Astrophys. J. Lett. **644**, 145–148 (2006). doi:10.1086/505690

C. Cattell, J. Crumley, J. Dombeck, J.R. Wygant, F.S. Mozer, Polar observations of solitary waves at the Earth's magnetopause. Geophys. Res. Lett. **29**(5), 050000-1 (2002). doi:10.1029/2001GL014046

C. Cattell, J. Dombeck, J. Wygant, J.F. Drake, M. Swisdak, M.L. Goldstein, W. Keith, A. Fazakerley, M. André, E. Lucek, A. Balogh, Cluster observations of electron holes in association with magnetotail reconnection and comparison to simulations. J. Geophys. Res. **110**, 1211 (2005). doi:10.1029/2004JA010519

L.J. Chen, N. Bessho, B. Lefebvre, H. Vaith, A. Fazakerley, A. Bhattacharjee, P.A. Puhl-Quinn, A. Runov, Y. Khotyaintsev, A. Vaivads, E. Georgescu, R. Torbert, Evidence of an extended electron current sheet and its neighboring magnetic island during magnetotail reconnection. J. Geophys. Res. **113**, 12213 (2008a). doi:10.1029/2008JA013385

L.-J. Chen, A. Bhattacharjee, P.A. Puhl-Quinn, H. Yang, N. Bessho, S. Imada, S. Mühlbachler, P.W. Daly, B. Lefebvre, Y. Khotyaintsev, A. Vaivads, A. Fazakerley, E. Georgescu, Observation of energetic electrons within magnetic islands. Nat. Phys. **4**, 19–23 (2008b). doi:10.1038/nphys777

S.W.H. Cowley, The causes of convection in the earth's magnetosphere—a review of developments during the IMS. Rev. Geophys. **20**, 531–565 (1982). doi:10.1029/RG020i003p00531

S.W.H. Cowley, Theoretical perspectives of the magnetopause: a tutorial review, in *Physics of the Magnetopause*, ed. by P. Song, B.U.O. Sonnerup, M.F. Thomsen. AGU Monograph, vol. 90 (American Geophysical Union, Washington, 1995), p. 29

W. Daughton, J. Scudder, H. Karimabadi, Fully kinetic simulations of undriven magnetic reconnection with open boundary conditions. Phys. Plasmas **13**(7), 072101 (2006). doi:10.1063/1.2218817

M.S. Davis, T.D. Phan, J.T. Gosling, R.M. Skoug, Detection of oppositely directed reconnection jets in a solar wind current sheet **33**, 19102 (2006). doi:10.1029/2006GL026735

C. Day, Spacecraft probes the site of magnetic reconnection in Earth's magnetotail. Phys. Today **54**(10), 100000-17 (2001). doi:10.1063/1.1420541

J.F. Drake, M. Opher, M. Swisdak, J.N. Chamoun, A magnetic reconnection mechanism for the generation of anomalous cosmic rays. Astrophys. J. **709**, 963–974 (2010). doi:10.1088/0004-637X/709/2/963

J.W. Dungey, Interplanetary magnetic field and the Auroral zones. Phys. Rev. Lett. **6**, 47–48 (1961). doi:10.1103/PhysRevLett.6.47

M.W. Dunlop, Q.-H. Zhang, Y.V. Bogdanova, M. Lockwood, Z. Pu, H. Hasegawa, J. Wang, M.G.G.T. Taylor, J. Berchem, B. Lavraud, J. Eastwood, M. Volwerk, C. Shen, J.-K. Shi, D. Constantinescu, H. Frey, A.N. Fazakerley, D. Sibeck, P. Escoubet, J.A. Wild, Z.-X. Liu, Extended magnetic reconnection across the dayside magnetopause. Phys. Rev. Lett. **107**(2), 025004 (2011a). doi:10.1103/PhysRevLett.107.025004

M.W. Dunlop, Q.-H. Zhang, Y.V. Bogdanova, K.J. Trattner, Z. Pu, H. Hasegawa, J. Berchem, M.G.G.T. Taylor, M. Volwerk, J.P. Eastwood, B. Lavraud, C. Shen, J.-K. Shi, J. Wang, D. Constantinescu, A.N. Fazakerley, H. Frey, D. Sibeck, P. Escoubet, J.A. Wild, Z.X. Liu, C. Carr, Magnetopause reconnection across wide local time. Ann. Geophys. **29**, 1683–1697 (2011b). doi:10.5194/angeo-29-1683-2011

J.P. Eastwood, D.G. Sibeck, J.A. Slavin, M.L. Goldstein, B. Lavraud, M. Sitnov, S. Imber, A. Balogh, E.A. Lucek, I. Dandouras, Observations of multiple X-line structure in the Earth's magnetotail current sheet: a cluster case study. Geophys. Res. Lett. **32**, 11105 (2005). doi:10.1029/2005GL022509

J.P. Eastwood, D.A. Brain, J.S. Halekas, J.F. Drake, T.D. Phan, M. Øieroset, D.L. Mitchell, R.P. Lin, M. Acuña, Evidence for collisionless magnetic reconnection at Mars. Geophys. Res. Lett. **35**, 2106 (2008). doi:10.1029/2007GL032289

J.P. Eastwood, M.A. Shay, T.D. Phan, M. Øieroset, Asymmetry of the ion diffusion region hall electric and magnetic fields during guide field reconnection: observations and comparison with simulations. Phys. Rev. Lett. **104**(20), 205001 (2010a). doi:10.1103/PhysRevLett.104.205001

J.P. Eastwood, T.D. Phan, M. Øieroset, M.A. Shay, Average properties of the magnetic reconnection ion diffusion region in the Earth's magnetotail: the 2001–2005 cluster observations and comparison with simulations. J. Geophys. Res. **115**, 8215 (2010b). doi:10.1029/2009JA014962

J.P. Eastwood, J.J.H. Videira, D.A. Brain, J.S. Halekas, A chain of magnetic flux ropes in the magnetotail of Mars. Geophys. Res. Lett. **39**, 3104 (2012). doi:10.1029/2011GL050444

R.E. Ergun, L. Andersson, J. Tao, V. Angelopoulos, J. Bonnell, J.P. McFadden, D.E. Larson, S. Eriksson, T. Johansson, C.M. Cully, D.N. Newman, M.V. Goldman, A. Roux, O. Lecontel, K.-H. Glassmeier, W. Baumjohann, Observations of double layers in Earth's plasma sheet. Phys. Rev. Lett. **102**(15), 155002 (2009). doi:10.1103/PhysRevLett.102.155002

S. Eriksson, M. Øieroset, D.N. Baker, C. Mouikis, A. Vaivads, M.W. Dunlop, H. Rème, R.E. Ergun, A. Balogh, Walén and slow-mode shock analyses in the near-Earth magnetotail in connection with a substorm onset on 27 August 2001. J. Gephys. Res. **109**(A18) (2004). doi:10.1029/2004JA010534

D.H. Fairfield, L.J. Cahill Jr., Transition region magnetic field and polar magnetic disturbances. J. Geophys. Res. **71**, 155–169 (1966). doi:10.1029/JZ071i001p00155

R.C. Fear, S.E. Milan, K. Oksavik, Determining the axial direction of high-shear flux transfer events: implications for models of FTE structure. J. Geophys. Res. **117**, 9220 (2012). doi:10.1029/2012JA017831

W.C. Feldman, D.N. Baker, S.J. Bame, J. Birn, J.T. Gosling, E.W. Hones Jr., S.J. Schwartz, Slow-mode shocks—a semipermanent feature of the distant geomagnetic tail. J. Geophys. Res. **90**, 233–240 (1985). doi:10.1029/JA090iA01p00233

H.U. Frey, T.D. Phan, S.A. Fuselier, S.B. Mende, Continuous magnetic reconnection at Earth's magnetopause. Nature **426**, 533–537 (2003)

H.S. Fu, Y.V. Khotyaintsev, M. André, A. Vaivads, Fermi and betatron acceleration of suprathermal electrons behind dipolarization fronts. Geophys. Res. Lett. **38**, 16104 (2011). doi:10.1029/2011GL048528

M. Fujimoto, M.S. Nakamura, I. Shinohara, T. Nagai, T. Mukai, Y. Saito, T. Yamamoto, S. Kokubun, Observations of earthward streaming electrons at the trailing boundary of a plasmoid. Geophys. Res. Lett. **24**, 2893 (1997). doi:10.1029/97GL02821

S.A. Fuselier, W.S. Lewis, Properties of near-Earth magnetic reconnection from in-situ observations. Space Sci. Rev. **160**, 95–121 (2011). doi:10.1007/s11214-011-9820-x

S.A. Fuselier, D.M. Klumpar, E.G. Shelley, Ion reflection and transmission during reconnection at the Earth's subsolar magnetopause. Geophys. Res. Lett. **18**, 139–142 (1991). doi:10.1029/90GL02676

S.A. Fuselier, S.M. Petrinec, K.J. Trattner, Antiparallel magnetic reconnection rates at the Earth's magnetopause. J. Geophys. Res. **115**, 10207 (2010). doi:10.1029/2010JA015302

J.T. Gosling, Magnetic reconnection in the solar wind. Space Sci. Rev. (2011). doi:10.1007/s11214-011-9747-2

J.T. Gosling, J.R. Asbridge, S.J. Bame, W.C. Feldman, G. Paschmann, N. Sckopke, C.T. Russell, Evidence for quasi-stationary reconnection at the dayside magnetopause. J. Geophys. Res. **87**, 2147–2158 (1982). doi:10.1029/JA087iA04p02147

J.T. Gosling, M.F. Thomsen, S.J. Bame, C.T. Russell, Accelerated plasma flows at the near-tail magnetopause. J. Geophys. Res. **91**, 3029–3041 (1986). doi:10.1029/JA091iA03p03029

J.T. Gosling, M.F. Thomsen, S.J. Bame, T.G. Onsager, C.T. Russell, The electron edge of the low latitude boundary layer during accelerated flow events. Geophys. Res. Lett. **17**, 1833–1836 (1990). doi:10.1029/GL017i011p01833

J.T. Gosling, R.M. Skoug, D.J. McComas, C.W. Smith, Direct evidence for magnetic reconnection in the solar wind near 1 AU. J. Geophys. Res. **110**, 1107 (2005). doi:10.1029/2004JA010809

J.T. Gosling, S. Eriksson, T.D. Phan, D.E. Larson, R.M. Skoug, D.J. McComas, Direct evidence for prolonged magnetic reconnection at a continuous X-line within the heliospheric current sheet. Geophys. Res. Lett. **34**, 6102 (2007a). doi:10.1029/2006GL029033

J.T. Gosling, S. Eriksson, L.M. Blush, T.D. Phan, J.G. Luhmann, D.J. McComas, R.M. Skoug, M.H. Acuna, C.T. Russell, K.D. Simunac, Five spacecraft observations of oppositely directed exhaust jets from a magnetic reconnection X-line extending 4.26×10^6 km in the solar wind at 1 AU. Geophys. Res. Lett. **34**, 20108 (2007b). doi:10.1029/2007GL031492

J.T. Gosling, S. Eriksson, D.J. McComas, T.D. Phan, R.M. Skoug, Multiple magnetic reconnection sites associated with a coronal mass ejection in the solar wind. J. Geophys. Res. **112**, 8106 (2007c). doi:10.1029/2007JA012418

R.A. Greenwald, K.B. Baker, J.R. Dudeney, M. Pinnock, T.B. Jones, E.C. Thomas, J.-P. Villain, J.-C. Cerisier, C. Senior, C. Hanuise, R.D. Hunsucker, G. Sofko, J. Koehler, E. Nielsen, R. Pellinen, A.D.M. Walker, N. Sato, H. Yamagishi, Darn/Superdarn: a global view of the dynamics of high-latitude convection. Space Sci. Rev. **71**, 761–796 (1995)

T. Hada, A. Nishida, T. Terasawa, E.W. Hones Jr., Bi-directional electron pitch angle anisotropy in the plasma sheet. J. Geophys. Res. **86**, 11211–11224 (1981). doi:10.1029/JA086iA13p11211

J.S. Halekas, D.A. Brain, Global distribution, structure, and solar wind control of low altitude current sheets at Mars. Icarus **206**, 64–73 (2010). doi:10.1016/j.icarus.2008.12.032

J.S. Halekas, J.P. Eastwood, D.A. Brain, T.D. Phan, M. Øieroset, R.P. Lin, In situ observations of reconnection hall magnetic fields at Mars: evidence for ion diffusion region encounters. J. Geophys. Res. **114**, 11204 (2009). doi:10.1029/2009JA014544

H. Hasegawa, J. Wang, M.W. Dunlop, Z.Y. Pu, Q.-H. Zhang, B. Lavraud, M.G.G.T. Taylor, O.D. Constantinescu, J. Berchem, V. Angelopoulos, J.P. McFadden, H.U. Frey, E.V. Panov, M. Volwerk, Y.V. Bogdanova, Evidence for a flux transfer event generated by multiple X-line reconnection at the magnetopause. Geophys. Res. Lett. **37**, 16101 (2010). doi:10.1029/2010GL044219

J.P. Heppner, Polar-cap electric field distributions related to the interplanetary magnetic field direction. J. Geophys. Res. **77**, 4877 (1972). doi:10.1029/JA077i025p04877

M. Hesse, S. Zenitani, A. Klimas, The structure of the electron outflow jet in collisionless magnetic reconnection. Phys. Plasmas **15**(11), 112102 (2008). doi:10.1063/1.3006341

M. Hesse, K. Schindler, J. Birn, M. Kuznetsova, The diffusion region in collisionless magnetic reconnection. Phys. Plasmas **6**, 1781–1795 (1999). doi:10.1063/1.873436

E.W. Hones Jr., Transient phenomena in the magnetotail and their relation to substorms. Space Sci. Rev. **23**, 393–410 (1979). doi:10.1007/BF00172247

M. Hoshino, T. Mukai, T. Terasawa, I. Shinohara, Suprathermal electron acceleration in magnetic reconnection. J. Geophys. Res. **106**, 25979–25998 (2001). doi:10.1029/2001JA900052

P.D. Hudson, Rotational discontinuities in an anisotropic plasma. Planet. Space Sci. **19**, 1693–1699 (1971). doi:10.1016/0032-0633(71)90129-2

S. Imada, R. Nakamura, P.W. Daly, M. Hoshino, W. Baumjohann, S. Mühlbachler, A. Balogh, H. Rème, Energetic electron acceleration in the downstream reconnection outflow region. J. Geophys. Res. **112**, 3202 (2007). doi:10.1029/2006JA011847

S.M. Imber, J.A. Slavin, H.U. Auster, V. Angelopoulos, A THEMIS survey of flux ropes and traveling compression regions: location of the near-Earth reconnection site during solar minimum. J. Geophys. Res. **116**, 2201 (2011). doi:10.1029/2010JA016026

H. Karimabadi, W. Daughton, J. Scudder, Multi-scale structure of the electron diffusion region. Geophys. Res. Lett. **34**, 13104 (2007). doi:10.1029/2007GL030306

H. Karimabadi, D. Krauss-Varban, N. Omidi, H.X. Vu, Magnetic structure of the reconnection layer and core field generation in plasmoids. J. Geophys. Res. **104**, 12313–12326 (1999). doi:10.1029/1999JA900089

Y.V. Khotyaintsev, A. Vaivads, A. Retinò, M. André, C.J. Owen, H. Nilsson, Formation of inner structure of a reconnection separatrix region. Phys. Rev. Lett. **97**(20), 205003 (2006). doi:10.1103/PhysRevLett.97.205003

Y.V. Khotyaintsev, C.M. Cully, A. Vaivads, M. André, C.J. Owen, Plasma jet braking: energy dissipation and nonadiabatic electrons. Phys. Rev. Lett. **106**(16), 165001 (2011). doi:10.1103/PhysRevLett.106.165001

S.A. Kiehas, V. Angelopoulos, A. Runov, M.B. Moldwin, C. Möstl, On the formation of tilted flux ropes in the Earth's magnetotail observed with ARTEMIS. J. Geophys. Res. **117**, 5231 (2012). doi:10.1029/2011JA017377

B. Lavraud, J.T. Gosling, A.P. Rouillard, A. Fedorov, A. Opitz, J.-A. Sauvaud, C. Foullon, I. Dandouras, V. Génot, C. Jacquey, P. Louarn, C. Mazelle, E. Penou, T.D. Phan, D.E. Larson, J.G. Luhmann, P. Schroeder, R.M. Skoug, J.T. Steinberg, C.T. Russell, Observation of a complex solar wind reconnection exhaust from spacecraft separated by over 1800 R_E. Sol. Phys. **256**, 379–392 (2009). doi:10.1007/s11207-009-9341-x

L.C. Lee, Z.F. Fu, A theory of magnetic flux transfer at the earth's magnetopause. Geophys. Res. Lett. **12**, 105–108 (1985). doi:10.1029/GL012i002p00105

Y. Lin, H. Xie, Formation of reconnection layer at the dayside magnetopause. Geophys. Res. Lett. **24**, 3145–3148 (1997). doi:10.1029/97GL03278

M. Lockwood, S.W.H. Cowley, P.E. Sandholt, Transient reconnection: search for ionospheric signatures. EOS Trans. **71**, 709–720 (1990a). doi:10.1029/EO071i020p00709-02

M. Lockwood, S.W.H. Cowley, P.E. Sandholt, R.P. Lepping, The ionospheric signatures of flux transfer events and solar wind dynamic pressure changes. J. Geophys. Res. **95**, 17113–17135 (1990b). doi:10.1029/JA095iA10p17113

Y.E. Lyubarsky, The termination shock in a striped pulsar wind. Mon. Not. R. Astron. Soc. **345**, 153–160 (2003). doi:10.1046/j.1365-8711.2003.06927.x

Z.W. Ma, A. Otto, L.C. Lee, Core magnetic field enhancement in single X line, multiple X line and patchy reconnection. J. Geophys. Res. **99**, 6125–6136 (1994). doi:10.1029/93JA03480

M.E. Mandt, R.E. Denton, J.F. Drake, Transition to whistler mediated magnetic reconnection. Geophys. Res. Lett. **21**, 73–76 (1994). doi:10.1029/93GL03382

A. Masters, J.P. Eastwood, M. Swisdak, M.F. Thomsen, C.T. Russell, N. Sergis, F.J. Crary, M.K. Dougherty, A.J. Coates, S.M. Krimigis, The importance of plasma β conditions for magnetic reconnection at Saturn's magnetopause. Geophys. Res. Lett. **39**, 8103 (2012). doi:10.1029/2012GL051372

N.C. Maynard, B.U.Ö. Sonnerup, G.L. Siscoe, D.R. Weimer, K.D. Siebert, G.M. Erickson, W.W. White, J.A. Schoendorf, D.M. Ober, G.R. Wilson, M.A. Heinemann, Predictions of magnetosheath merging between IMF field lines of opposite polarity. J. Geophys. Res. **107**, 1456 (2002). doi:10.1029/2002JA009289

H.J. McAndrews, C.J. Owen, M.F. Thomsen, B. Lavraud, A.J. Coates, M.K. Dougherty, D.T. Young, Evidence for reconnection at Saturn's magnetopause. J. Geophys. Res. **113**, 4210 (2008). doi:10.1029/2007JA012581

F.S. Mozer, P.L. Pritchett, Regions associated with electron physics in asymmetric magnetic field reconnection. Geophys. Res. Lett. **36**, 7102 (2009). doi:10.1029/2009GL037463

F.S. Mozer, S.D. Bale, T.D. Phan, Evidence of diffusion regions at a subsolar magnetopause crossing. Phys. Rev. Lett. **89**(1), 015002 (2002). doi:10.1103/PhysRevLett.89.015002

F.S. Mozer, S.D. Bale, T.D. Phan, J.A. Osborne, Observations of electron diffusion regions at the subsolar magnetopause. Phys. Rev. Lett. **91**(24), 245002 (2003). doi:10.1103/PhysRevLett.91.245002

T. Nagai, Location of magnetic reconnection in the magnetotail. Space Sci. Rev. **122**, 39–54 (2006). doi:10.1007/s11214-006-6216-4

T. Nagai, M. Fujimoto, Y. Saito, S. Machida, T. Terasawa, R. Nakamura, T. Yamamoto, T. Mukai, A. Nishida, S. Kokubun, Structure and dynamics of magnetic reconnection for substorm onsets with geotail observations. J. Geophys. Res. **103**, 4419–4440 (1998). doi:10.1029/97JA02190

T. Nagai, I. Shinohara, M. Fujimoto, M. Hoshino, Y. Saito, S. Machida, T. Mukai, Geotail observations of the hall current system: evidence of magnetic reconnection in the magnetotail. J. Geophys. Res. **106**, 25929–25950 (2001). doi:10.1029/2001JA900038

T. Nagai, I. Shinohara, M. Fujimoto, A. Matsuoka, Y. Saito, T. Mukai, Construction of magnetic reconnection in the near-Earth magnetotail with geotail. J. Geophys. Res. **116**, 4222 (2011). doi:10.1029/2010JA016283

R. Nakamura, W. Baumjohann, C. Mouikis, L.M. Kistler, A. Runov, M. Volwerk, Y. Asano, Z. Vörös, T.L. Zhang, B. Klecker, H. Rème, A. Balogh, Spatial scale of high-speed flows in the plasma sheet observed by cluster. Geophys. Res. Lett. **31**, 9804 (2004). doi:10.1029/2004GL019558

R. Nakamura, W. Baumjohann, Y. Asano, A. Runov, A. Balogh, C.J. Owen, A.N. Fazakerley, M. Fujimoto, B. Klecker, H. Rème, Dynamics of thin current sheets associated with magnetotail reconnection. J. Geophys. Res. **111**, 11206 (2006). doi:10.1029/2006JA011706

J.D. Nichols, S.W.H. Cowley, D.J. McComas, Magnetopause reconnection rate estimates for Jupiter's magnetosphere based on interplanetary measurements at ~5 AU. Ann. Geophys. **24**, 393–406 (2006). doi:10.5194/angeo-24-393-2006

A. Nishida, Can random reconnection on the magnetopause produce the low latitude boundary layer? Geophys. Res. Lett. **16**, 227–230 (1989). doi:10.1029/GL016i003p00227

A. Nishida, N. Nagayama, Synoptic survey for the neutral line in the magnetotail during the substorm expansion phase. J. Geophys. Res. **78**, 3782 (1973). doi:10.1029/JA078i019p03782

A. Nishida, T. Yamamoto, T. Mukai, The GEOTAIL mission: principal characteristics and scientific results. Adv. Space Res. **20**, 539–548 (1997). doi:10.1016/S0273-1177(97)00437-7

M. Øieroset, T.D. Phan, R.P. Lin, B.U.Ö. Sonnerup, Walén and variance analyses of high-speed flows observed by wind in the midtail plasma sheet: evidence for reconnection. J. Geophys. Res. **105**, 25247–25264 (2000). doi:10.1029/2000JA900075

M. Øieroset, T.D. Phan, M. Fujimoto, R.P. Lin, R.P. Lepping, In situ detection of collisionless reconnection in the Earth's magnetotail. Nature **412**, 414–417 (2001). doi:10.1038/35086520

M. Øieroset, R.P. Lin, T.D. Phan, D.E. Larson, S.D. Bale, Evidence for electron acceleration up to ~300 keV in the magnetic reconnection diffusion region of Earth's magnetotail. Phys. Rev. Lett. **89**(19), 195001 (2002). doi:10.1103/PhysRevLett.89.195001

M. Øieroset, T.D. Phan, J.P. Eastwood, M. Fujimoto, W. Daughton, M.A. Shay, V. Angelopoulos, F.S. Mozer, J.P. McFadden, D.E. Larson, K.-H. Glassmeier, Direct evidence for a three-dimensional magnetic flux rope flanked by two active magnetic reconnection X lines at Earth's magnetopause. Phys. Rev. Lett. **107**(16), 165007 (2011). doi:10.1103/PhysRevLett.107.165007

K. Oksavik, J. Moen, H.C. Carlson, High-resolution observations of the small-scale flow pattern associated with a poleward moving auroral form in the cusp. Geophys. Res. Lett. **31**, 11807 (2004). doi:10.1029/2004GL019838

N. Omidi, T. Phan, D.G. Sibeck, Hybrid simulations of magnetic reconnection initiated in the magnetosheath. J. Geophys. Res. **114**, 2222 (2009). doi:10.1029/2008JA013647

G. Paschmann, Recent in-situ observations in near-Earth space. Geophys. Res. Lett. **35**, 19109 (2008). doi:10.1029/2008GL035297

G. Paschmann, I. Papamastorakis, N. Sckopke, G. Haerendel, B.U.O. Sonnerup, S.J. Bame, J.R. Asbridge, J.T. Gosling, C.T. Russel, R.C. Elphic, Plasma acceleration at the earth's magnetopause—evidence for reconnection. Nature **282**, 243–246 (1979). doi:10.1038/282243a0

G. Paschmann, G. Haerendel, I. Papamastorakis, N. Sckopke, S.J. Bame, J.T. Gosling, C.T. Russell, Plasma and magnetic field characteristics of magnetic flux transfer events. J. Geophys. Res. **87**, 2159–2168 (1982). doi:10.1029/JA087iA04p02159

G. Paschmann, W. Baumjohann, N. Sckopke, I. Papamastorakis, C.W. Carlson, The magnetopause for large magnetic shear—AMPTE/IRM observations. J. Geophys. Res. **91**, 11099–11115 (1986). doi:10.1029/JA091iA10p11099

T.D. Phan, L.M. Kistler, B. Klecker, G. Haerendel, G. Paschmann, B.U.Ö. Sonnerup, W. Baumjohann, M.B. Bavassano-Cattaneo, C.W. Carlson, A.M. DiLellis, K.-H. Fornacon, L.A. Frank, M. Fujimoto, E. Georgescu, S. Kokubun, E. Moebius, T. Mukai, M. Øieroset, W.R. Paterson, H. Reme, Extended magnetic reconnection at the Earth's magnetopause from detection of bi-directional jets. Nature **404**, 848–850 (2000)

T.D. Phan, J.T. Gosling, M.S. Davis, R.M. Skoug, M. Øieroset, R.P. Lin, R.P. Lepping, D.J. McComas, C.W. Smith, H. Reme, A. Balogh, A magnetic reconnection X-line extending more than 390 Earth radii in the solar wind. Nature **439**, 175–178 (2006a). doi:10.1038/nature04393

T.D. Phan, H. Hasegawa, M. Fujimoto, M. Oieroset, T. Mukai, R.P. Lin, W. Paterson, Simultaneous geotail and wind observations of reconnection at the subsolar and tail flank magnetopause. Geophys. Res. Lett. **33**, 9104 (2006b). doi:10.1029/2006GL025756

T.D. Phan, J.F. Drake, M.A. Shay, F.S. Mozer, J.P. Eastwood, Evidence for an elongated electron diffusion region during fast magnetic reconnection. Phys. Rev. Lett. **99**(25), 255002 (2007a). doi:10.1103/PhysRevLett.99.255002

T.D. Phan, G. Paschmann, C. Twitty, F.S. Mozer, J.T. Gosling, J.P. Eastwood, M. Øieroset, H. Rème, E.A. Lucek, Evidence for magnetic reconnection initiated in the magnetosheath. Geophys. Res. Lett. **34**, 14104 (2007b). doi:10.1029/2007GL030343

T.D. Phan, J.T. Gosling, G. Paschmann, C. Pasma, J.F. Drake, M. Øieroset, D. Larson, R.P. Lin, M.S. Davis, The dependence of magnetic reconnection on plasma β and magnetic shear: evidence from solar wind observations. Astrophys. J. **719**, 199–203 (2010). doi:10.1088/2041-8205/719/2/L199

T.D. Phan, G. Paschmann, J.T. Gosling, M. Oieroset, M. Fujimoto, J.F. Drake, V. Angelopoulos, The dependence of magnetic reconnection on plasma β and magnetic shear: evidence from magnetopause observations. Geophys. Res. Lett. (2013). doi:10.1029/2012GL054528

T. Phan, M. Dunlop, G. Paschmann, B. Klecker, J. Bosqued, H. Rème, A. Balogh, C. Twitty, F. Mozer, C. Carlson, C. Mouikis, L. Kistler, Cluster observations of continuous reconnection at the magnetopause under steady interplanetary magnetic field conditions. Ann. Geophys. **22**, 2355–2367 (2004). doi:10.5194/angeo-22-2355-2004

M. Pinnock, G. Chisham, I.J. Coleman, M.P. Freeman, M. Hairston, J.-P. Villain, The location and rate of dayside reconnection during an interval of southward interplanetary magnetic field. Ann. Geophys. **21**, 1467–1482 (2003). doi:10.5194/angeo-21-1467-2003

P.L. Pritchett, Onset and saturation of guide-field magnetic reconnection. Phys. Plasmas **12**(6), 062301 (2005). doi:10.1063/1.1914309

P.L. Pritchett, Collisionless magnetic reconnection in an asymmetric current sheet. J. Geophys. Res. **113**, 6210 (2008). doi:10.1029/2007JA012930

P.L. Pritchett, F.V. Coroniti, Three-dimensional collisionless magnetic reconnection in the presence of a guide field. J. Geophys. Res. **109**, 1220 (2004). doi:10.1029/2003JA009999

Z.Y. Pu, C.J. Xiao, X.G. Zhang, Z.Y. Huang, S.Y. Fu, Z.X. Liu, M.W. Dunlop, Q.G. Zong, C.M. Carr, H. Réme, I. Dandouras, A. Fazakerley, T. Phan, T.L. Zhang, H. Zhang, X.G. Wang, Double Star TC-1 observations of component reconnection at the dayside magnetopause: a preliminary study. Ann. Geophys. **23**, 2889–2895 (2005). doi:10.5194/angeo-23-2889-2005

Z.Y. Pu, X.G. Zhang, X.G. Wang, J. Wang, X.-Z. Zhou, M.W. Dunlop, L. Xie, C.J. Xiao, Q.G. Zong, S.Y. Fu, Z.X. Liu, C. Carr, Z.W. Ma, C. Shen, E. Lucek, H. Rème, P. Escoubet, Global view of dayside magnetic reconnection with the dusk-dawn IMF orientation: a statistical study for double star and cluster data. Geophys. Res. Lett. **34**, 20101 (2007). doi:10.1029/2007GL030336

A. Radioti, D. Grodent, J.-C. Gérard, M.F. Vogt, M. Lystrup, B. Bonfond, Nightside reconnection at Jupiter: Auroral and magnetic field observations from 26 July 1998. J. Geophys. Res. **116**, 3221 (2011). doi:10.1029/2010JA016200

J. Raeder, Flux transfer events: 1. generation mechanism for strong southward IMF. Ann. Geophys. **24**, 381–392 (2006). doi:10.5194/angeo-24-381-2006

A. Retinò, M.B. Bavassano Cattaneo, M.F. Marcucci, A. Vaivads, M. André, Y. Khotyaintsev, T. Phan, G. Pallocchia, H. Rème, E. Möbius, B. Klecker, C.W. Carlson, M. McCarthy, A. Korth, R. Lundin, A. Balogh, Cluster multispacecraft observations at the high-latitude duskside magnetopause: implications for continuous and component magnetic reconnection. Ann. Geophys. **23**, 461–473 (2005). doi:10.5194/angeo-23-461-2005

A. Retinò, D. Sundkvist, A. Vaivads, F. Mozer, M. André, C.J. Owen, In situ evidence of magnetic reconnection in turbulent plasma. Nat. Phys. **3**, 236–238 (2007). doi:10.1038/nphys574

A. Retinò, R. Nakamura, A. Vaivads, Y. Khotyaintsev, T. Hayakawa, K. Tanaka, S. Kasahara, M. Fujimoto, I. Shinohara, J.P. Eastwood, M. André, W. Baumjohann, P.W. Daly, E.A. Kronberg, N. Cornilleau-Wehrlin, Cluster observations of energetic electrons and electromagnetic fields within a reconnecting thin current sheet in the Earth's magnetotail. J. Geophys. Res. **113**, 12215 (2008). doi:10.1029/2008JA013511

F.J. Rich, M. Hairston, Large-scale convection patterns observed by DMSP. J. Geophys. Res. **99**, 3827–3844 (1994). doi:10.1029/93JA03296

L. Rosenqvist, A. Vaivads, A. Retinò, T. Phan, H.J. Opgenoorth, I. Dandouras, S. Buchert, Modulated reconnection rate and energy conversion at the magnetopause under steady IMF conditions. Geophys. Res. Lett. **35**, 8104 (2008). doi:10.1029/2007GL032868

B. Rossi, S. Olbert, *Introduction to the Physics of Space* (McGraw-Hill, New York, 1970)

A. Runov, R. Nakamura, W. Baumjohann, R.A. Treumann, T.L. Zhang, M. Volwerk, Z. Vörös, A. Balogh, K.-H. Glaßmeier, B. Klecker, H. Rème, L. Kistler, Current sheet structure near magnetic X-line observed by cluster. Geophys. Res. Lett. **30**(11), 110000-1 (2003). doi:10.1029/2002GL016730

A. Runov, W. Baumjohann, R. Nakamura, V.A. Sergeev, O. Amm, H. Frey, I. Alexeev, A.N. Fazakerley, C.J. Owen, E. Lucek, M. André, A. Vaivads, I. Dandouras, B. Klecker, Observations of an active thin current sheet. J. Geophys. Res. **113**, 7 (2008). doi:10.1029/2007JA012685

C.T. Russell, R.C. Elphic, ISEE observations of flux transfer events at the dayside magnetopause. Geophys. Res. Lett. **6**, 33–36 (1979). doi:10.1029/GL006i001p00033

C.T. Russell, K.K. Khurana, D.E. Huddleston, M.G. Kivelson, Localized reconnection in the near Jovian magnetotail. Science **280**, 1061 (1998). doi:10.1126/science.280.5366.1061

Y. Saito, T. Mukai, T. Terasawa, Kinetic structure of the slow-mode shocks in the Earth's magnetotail, in *New Perspectives on the Earth's Magnetotail*, ed. by A. Nishida, D.N. Baker, S.W.H. Cowley (1998), p. 103

J. Sanny, R.L. McPherron, C.T. Russell, D.N. Baker, T.I. Pulkkinen, A. Nishida, Growth-phase thinning of the near-Earth current sheet during the CDAW 6 substorm. J. Geophys. Res. **99**, 5805–5816 (1994). doi:10.1029/93JA03235

Y.L. Sasunov, V.S. Semenov, M.F. Heyn, I.V. Kubyshkin, H.K. Biernat, Kelvin-Helmholtz stability of reconnection exhausts in the solar wind. Geophys. Res. Lett. **39**, 6104 (2012). doi:10.1029/2012GL051273
M. Scholer, Magnetic flux transfer at the magnetopause based on single X line bursty reconnection. Geophys. Res. Lett. **15**, 291–294 (1988a). doi:10.1029/GL015i004p00291
M. Scholer, Strong core magnetic fields in magnetopause flux transfer events. Geophys. Res. Lett. **15**, 748–751 (1988b). doi:10.1029/GL015i008p00748
M. Scholer, Models of flux transfer events, in *Physics of the Magnetopause*, ed. by P. Song, B.U.O. Sonnerup, M.F. Thomsen (American Geophysical Union, Washington, 1995), p. 235
M. Scholer, Magnetic reconnection on the Sun and in the Earth's magnetosphere, in *Energy Conversion and Particle Acceleration in the Solar Corona*, ed. by L. Klein. Lecture Notes in Physics, vol. 612 (Springer, Berlin, 2003), pp. 9–27
J.D. Scudder, R.D. Holdaway, W.S. Daughton, H. Karimabadi, V. Roytershteyn, C.T. Russell, J.Y. Lopez, First resolved observations of the demagnetized electron-diffusion region of an astrophysical magnetic-reconnection site. Phys. Rev. Lett. **108**(22), 225005 (2012). doi:10.1103/PhysRevLett.108.225005
M.A. Shay, J.F. Drake, The role of electron dissipation on the rate of collisionless magnetic reconnection. Geophys. Res. Lett. **25**, 3759–3762 (1998). doi:10.1029/1998GL900036
M.A. Shay, J.F. Drake, M. Swisdak, Two-scale structure of the electron dissipation region during collisionless magnetic reconnection. Phys. Rev. Lett. **99**(15), 155002 (2007). doi:10.1103/PhysRevLett.99.155002
M.A. Shay, J.F. Drake, R.E. Denton, D. Biskamp, Structure of the dissipation region during collisionless magnetic reconnection. J. Geophys. Res. **103**, 9165–9176 (1998). doi:10.1029/97JA03528
K. Shibata, S. Masuda, M. Shimojo, H. Hara, T. Yokoyama, S. Tsuneta, T. Kosugi, Y. Ogawara, Hot-plasma ejections associated with compact-loop solar flares. Astrophys. J. Lett. **451**, 83 (1995). doi:10.1086/309688
D.G. Sibeck, G. Paschmann, R.A. Treumann, S.A. Fuselier, W. Lennartsson, M. Lockwood, R. Lundin, K.W. Ogilvie, T.G. Onsager, T.-D. Phan, M. Roth, M. Scholer, N. Sckopke, K. Stasiewicz, M. Yamauchi, Chapter 5-plasma transfer processes at the magnetopause. Space Sci. Rev. **88**, 207–283 (1999). doi:10.1023/A:1005255801425
D.G. Sibeck, M. Kuznetsova, V. Angelopoulos, K.-H. Glaßmeier, J.P. McFadden, Crater FTEs: simulation results and THEMIS observations. Geophys. Res. Lett. **35**, 17 (2008). doi:10.1029/2008GL033568
J.A. Slavin, E.J. Smith, B.T. Tsurutani, D.G. Sibeck, H.J. Singer, D.N. Baker, J.T. Gosling, E.W. Hones, F.L. Scarf, Substorm associated traveling compression regions in the distant tail—ISEE-3 geotail observations. Geophys. Res. Lett. **11**, 657–660 (1984). doi:10.1029/GL011i007p00657
J.A. Slavin, E.I. Tanskanen, M. Hesse, C.J. Owen, M.W. Dunlop, S. Imber, E.A. Lucek, A. Balogh, K.-H. Glassmeier, Cluster observations of traveling compression regions in the near-tail. J. Geophys. Res. **110**, 6207 (2005). doi:10.1029/2004JA010878
J.A. Slavin, M.H. Acuña, B.J. Anderson, D.N. Baker, M. Benna, S.A. Boardsen, G. Gloeckler, R.E. Gold, G.C. Ho, H. Korth, S.M. Krimigis, R.L. McNutt, J.M. Raines, M. Sarantos, D. Schriver, S.C. Solomon, P. Trávníček, T.H. Zurbuchen, MESSENGER observations of magnetic reconnection in Mercury's magnetosphere. Science **324**, 606 (2009). doi:10.1126/science.1172011
Y. Song, R.L. Lysak, Evaluation of twist helicity of flux transfer event flux tubes. J. Geophys. Res. **94**, 5273–5281 (1989). doi:10.1029/JA094iA05p05273
B.U.Ö. Sonnerup, Magnetic field reconnection, in *Solar System Plasma Processes*, ed. by L.J. Lanzerotti, C.F. Kennel, E.N. Parker, vol. 3 (North-Holland, Washington, 1979), p. 45
B.U.Ö. Sonnerup, On the stress balance in flux transfer events. J. Geophys. Res. **92**, 8613–8620 (1987). doi:10.1029/JA092iA08p08613
B.U.O. Sonnerup, L.J. Cahill Jr., Magnetopause structure and attitude from explorer 12 observations. J. Geophys. Res. **72**, 171 (1967). doi:10.1029/JZ072i001p00171
B.U.Ö. Sonnerup, M. Scheible, Minimum and maximum variance analysis, in *Analysis Methods for Multi-Spacecraft Data*, ed. by G. Paschmann, P.W. Daly. ISSI Scientific Reports (ESA Publ. Div., Noordwijk, 1998), pp. 185–220
B.U.Ö. Sonnerup, H. Hasegawa, G. Paschmann, Anatomy of a flux transfer event seen by cluster. Geophys. Res. Lett. **31**, 11803 (2004). doi:10.1029/2004GL020134
B.U.Ö. Sonnerup, G. Paschmann, I. Papamastorakis, N. Sckopke, G. Haerendel, S.J. Bame, J.R. Asbridge, J.T. Gosling, C.T. Russell, Evidence for magnetic field reconnection at the Earth's magnetopause. J. Geophys. Res. **86**, 10049–10067 (1981). doi:10.1029/JA086iA12p10049
B.U.Ö. Sonnerup, I. Papamastorakis, G. Paschmann, H. Luehr, Magnetopause properties from AMPTE/IRM observations of the convection electric field—method development. J. Geophys. Res. **92**, 12137–12159 (1987). doi:10.1029/JA092iA11p12137
D.J. Southwood, C.J. Farrugia, M.A. Saunders, What are flux transfer events? Planet. Space Sci. **36**, 503–508 (1988). doi:10.1016/0032-0633(88)90109-2

D. Sundkvist, A. Retinò, A. Vaivads, S.D. Bale, Dissipation in turbulent plasma due to reconnection in thin current sheets. Phys. Rev. Lett. **99**(2), 025004 (2007). doi:10.1103/PhysRevLett.99.025004

M. Swisdak, B.N. Rogers, J.F. Drake, M.A. Shay, Diamagnetic suppression of component magnetic reconnection at the magnetopause. J. Geophys. Res. **108**, 1218 (2003). doi:10.1029/2002JA009726

M. Swisdak, M. Opher, J.F. Drake, F. Alouani Bibi, The vector direction of the interstellar magnetic field outside the heliosphere. Astrophys. J. **710**, 1769–1775 (2010). doi:10.1088/0004-637X/710/2/1769

W.-L. Teh, B.U.Ö. Sonnerup, Q. Hu, C.J. Farrugia, Reconstruction of a large-scale reconnection exhaust structure in the solar wind. Ann. Geophys. **27**, 807–822 (2009). doi:10.5194/angeo-27-807-2009

T. Terasawa, A. Nishida, Simultaneous observations of relativistic electron bursts and neutral-line signatures in the magnetotail. Planet. Space Sci. **24**, 855–866 (1976). doi:10.1016/0032-0633(76)90076-3

K.J. Trattner, S.M. Petrinec, S.A. Fuselier, T.D. Phan, The location of reconnection at the magnetopause: testing the maximum magnetic shear model with THEMIS observations. J. Geophys. Res. **117**, 1201 (2012). doi:10.1029/2011JA016959

L. Trenchi, M.F. Marcucci, G. Pallocchia, G. Consolini, M.B. Bavassano Cattaneo, A.M. Di Lellis, H. Rème, L. Kistler, C.M. Carr, J.B. Cao, Occurrence of reconnection jets at the dayside magnetopause: double star observations. J. Geophys. Res. **113**, 7 (2008). doi:10.1029/2007JA012774

S. Tsuneta, Particle acceleration and magnetic reconnection in solar flares. Publ. Astron. Soc. Jpn. **47**, 691–697 (1995)

A. Vaivads, Y. Khotyaintsev, M. André, A. Retinò, S.C. Buchert, B.N. Rogers, P. Décréau, G. Paschmann, T.D. Phan, Structure of the magnetic reconnection diffusion region from four-spacecraft observations. Phys. Rev. Lett. **93**(10), 105001 (2004). doi:10.1103/PhysRevLett.93.105001

A. Vaivads, A. Retinò, Y.V. Khotyaintsev, M. André, The Alfvén edge in asymmetric reconnection. Ann. Geophys. **28**, 1327–1331 (2010). doi:10.5194/angeo-28-1327-2010

A. Vaivads, A. Retinò, Y.V. Khotyaintsev, M. André, Suprathermal electron acceleration during reconnection onset in the magnetotail. Ann. Geophys. **29**, 1917–1925 (2011). doi:10.5194/angeo-29-1917-2011

J.R. Wygant, C.A. Cattell, R. Lysak, Y. Song, J. Dombeck, J. McFadden, F.S. Mozer, C.W. Carlson, G. Parks, E.A. Lucek, A. Balogh, M. Andre, H. Reme, M. Hesse, C. Mouikis, Cluster observations of an intense normal component of the electric field at a thin reconnecting current sheet in the tail and its role in the shock-like acceleration of the ion fluid into the separatrix region. J. Geophys. Res. **110**, 9206 (2005). doi:10.1029/2004JA010708

S. Zenitani, I. Shinohara, T. Nagai, Evidence for the dissipation region in magnetotail reconnection. Geophys. Res. Lett. **39**, 11102 (2012). doi:10.1029/2012GL051938

H. Zhang, Q.-G. Zong, T.A. Fritz, S.Y. Fu, S. Schaefer, K.H. Glassmeier, P.W. Daly, H. Rème, A. Balogh, Cluster observations of collisionless hall reconnection at high-latitude magnetopause. J. Geophys. Res. **113**, 3204 (2008). doi:10.1029/2007JA012769

H. Zhang, M.G. Kivelson, V. Angelopoulos, K.K. Khurana, Z.Y. Pu, R.J. Walker, R.L. McPherron, T.-S. Hsu, Q.G. Zong, T. Phan, Generation and properties of in vivo flux transfer events. J. Geophys. Res. **117**, 5224 (2012a). doi:10.1029/2011JA017166

T.L. Zhang, Q.M. Lu, W. Baumjohann, C.T. Russell, A. Fedorov, S. Barabash, A.J. Coates, A.M. Du, J.B. Cao, R. Nakamura, W.L. Teh, R.S. Wang, X.K. Dou, S. Wang, K.H. Glassmeier, H.U. Auster, M. Balikhin, Magnetic reconnection in the near Venusian magnetotail. Science **336**, 567 (2012b). doi:10.1126/science.1217013

DOI 10.1007/978-1-4899-7413-6_13
Reprinted from *Space Science Reviews* Journal, DOI 10.1007/s11214-012-9954-5

Kinetic Structure of Current Sheets in the Earth Magnetotail

Anton Artemyev · Lev Zelenyi

Received: 3 October 2012 / Accepted: 28 November 2012 / Published online: 14 December 2012

Abstract In this paper we present a short review of kinetic models of the thin current sheet and corresponding Cluster observations in the Earth magnetotail. We concentrate mainly on manifestations of non-fluid ion kinetic effects. We discuss the different approaches to model description of the proton component and show that current sheets observed by Cluster contain population of particles with transient trajectories, which, in fact, are main carriers of the transverse current in the Earth magnetotail. We describe the influence of the electron temperature anisotropy on the current sheet structure. We demonstrate that the decoupling of proton and electron motions in thin current sheets results in appearance of the earthward electrostatic field, which redistributes currents due to the cross-field drift. This effect can describe small or negative proton currents often observed by Cluster.

Keywords Earth magnetotail · Current sheets · Kinetic models

1 Introduction

The current sheet (CS) of the Earth magnetotail is a critical element of all substorm models (see, e.g., Baker et al. 1996; Lui et al. 2008; Angelopoulos et al. 2008; Sergeev et al. 2012, and references therein). Moreover, every model of particle acceleration due to magnetic reconnection involve CSs as a source of magnetic free energy (Priest and Forbes 2000). Besides many laboratory experimental investigations of the CS formation (see, e.g., Frank 2010; Yamada et al. 2010; Frank et al. 2011), the Earth magnetotail CS is the most accessible for direct studying by spacecraft. From the plasma-physical point of view, CSs represent a selfconsistent high-β plasma configuration which might have remarkably small thickness.

Starting from the first papers devoted to the Earth magnetotail (Ness 1965; Speiser and Ness 1967) the CS was supposed to be relatively thick with transverse spatial scale around one-two Earth radii. The density j_y of the current flowing in the neutral plane (where main component of the magnetic field, B_x, changes sign) was estimated to be about few nA/m^2

A. Artemyev (✉) · L. Zelenyi
Space Research Institute, RAS, 84/32 Profsoyuznaya Str, 117997 Moscow, Russia
e-mail: artemyev@iki.rssi.ru

(McComas et al. 1986; Sergeev et al. 1993). However, this data were collected by single or double spacecraft missions and no statistical data were available before Cluster mission.

The first Cluster multispacecraft data have shown that the conception of the thick CS (where current density and plasma density profiles coincide) is not really supported by observations. Thickness of the CS appeared to be about 1000–3000 km (Runov et al. 2006) or even thinner (Nakamura et al. 2006). Such CS was called thin current sheet (TCS). Amplitude of the current density in TCSs is, on average, about 10 nA/m^2 (Artemyev et al. 2011a; Davey et al. 2012) and can reach 100 nA/m^2 (Nakamura et al. 2008). TCS can be considered as a transient increase of the current density in the magnetotail. Thus statistics without additional selection of TCSs gives only $\sim$4 nA/m^2 for the average cross-tail current density (Kaufmann et al. 2001; Rong et al. 2011). Intense TCSs are embedded into a thick plasma sheet with spatial scale about few Earth radii. The embedding manifests itself in two effects: (1) amplitude of the magnetic field at the TCS boundary, B_0, is smaller than the lobe magnetic field amplitude B_{ext} (Sergeev et al. 1993; Runov et al. 2006; Artemyev et al. 2008a) and for the abundant statistics of Cluster TCS crossings $B_{ext}/B_0 \sim$ 3–2 (see, e.g., Artemyev et al. 2011a; Petrukovich et al. 2011); (2) current density decreases away from the TCS central region much faster than plasma density does (Asano et al. 2005; Runov et al. 2006; Artemyev et al. 2010). Besides the embedding, TCSs possess another unexpected property: the ion (proton) current density in TCS often could be substantially smaller than the electron one (Asano et al. 2004; Artemyev et al. 2009). Possible explanations of these TCS properties are suggested in this review.

Earlier analytical models of CSs can be separated into two classes: kinetic equilibria and fluid models. The first class includes the most known model of the Harris CS (Harris 1962) and several less investigated solutions of the Vlasov equations (Morozov and Solov'ev 1961; Grad 1961; Nicholson 1963; Alpers 1969; Kan 1973; Channell 1976). The second class consists of equilibria for MHD systems (Syrovatskii 1971; Bird and Beard 1972; Birn et al. 1975; Cowley 1978; Syrovatskii 1981). There are several recent models which can be considered as modifications of simple solutions presented by the two earlier classes (Schindler and Birn 2002; Mottez 2003; Yoon and Lui 2004). Advantages and shortcomings of all these theoretical models in context of modern spacecraft observations are also discussed below.

Additionally one should mention a separate class of TCS models based on the idea of transient ions in the magnetotail. Due to the presence of the TCS natural boundaries where magnetic field is approximately constant, a population of ions (protons) with open trajectories can be found (so called Speiser trajectories, see Speiser 1965). The current carried by ions moving along these transient orbits supports very intense TCS (Eastwood 1972, 1974) with very small spatial scale of order of the ion (proton) Larmor radius (Francfort and Pellat 1976; Sitnov et al. 2000; Zelenyi et al. 2000). Development of analytical models of the TCS with Speiser trajectories is based on the conception of the quasiadiabatic ion motion (Büchner and Zelenyi 1989) and was proposed by Kropotkin and Domrin (1996) and Kropotkin et al. (1997). The first analytical model for a realistic energy distribution of transient particles was presented by Sitnov et al. (2000), Zelenyi et al. (2000). Results of this model coincide with ones provided by the numerical simulations of TCS configurations with the substantial population of transient ions (Hamilton and Eastwood 1982; Burkhart et al. 1992a; Pritchett and Coroniti 1992). There are several generalization of this TCS model represented by two groups of equilibrium solutions (Sitnov et al. 2003, 2006; Zhou et al. 2009) and (Zelenyi et al. 2004, 2006, 2011; Petrukovich et al. 2011). Moreover, it is worth noting that effect of Speiser trajectories was partially included into the fluid model of the TCS by Steinhauer et al. (2008).

2 Current Sheet Models

Any kinetic model of a collisionless CS can be considered as a solution of the stationary Vlasov-Maxwell system

$$\begin{cases} \frac{\partial f_\alpha}{\partial \mathbf{r}}\mathbf{v} + \frac{q_\alpha}{m_\alpha}\frac{\partial f_\alpha}{\partial \mathbf{v}}(\mathbf{E} + \frac{1}{c}[\mathbf{v}\times\mathbf{B}]) = 0 \\ \operatorname{rot}\mathbf{B} = (4\pi/c)\sum_\alpha q_\alpha \int \mathbf{v} f_\alpha d\mathbf{v}, \quad \operatorname{div}\mathbf{B} = 0 \\ \operatorname{div}\mathbf{E} = 4\pi \sum_\alpha q_\alpha \int f_\alpha d\mathbf{v}, \quad \operatorname{rot}\mathbf{E} = 0 \end{cases} \tag{1}$$

Here f_α is a velocity distribution of particles with mass m_α and charge q_α ($\alpha = e$ for electrons and $\alpha = i$ for ions or protons). The generalization of this system for nonstationary CSs ($\partial/\partial t \neq 0$) can be found in Mahajan (1989). Spatial scale of the magnetic field inhomogeneity in the Earth magnetotail is much larger than the corresponding Debye length $r_D = \sqrt{\sum_\alpha 4\pi q_\alpha^2 n_\alpha / T_\alpha}$ where T_α is a particle temperature. Therefore, we can consider quasineutrality condition as the equation for the scalar potential φ where $\mathbf{E} = -\nabla\varphi$

$$\sum_\alpha q_\alpha \int f_\alpha d\mathbf{v} = 0 \tag{2}$$

Although, B_y component of the magnetic field often exists in the Earth magnetotail (Petrukovich 2011), we restrict our analysis by CS models with the classical geometry when only B_x and B_z components are taken into account (hereafter we use GSM coordinate system where y-axis is directed from dawn to dusk, x-axis is directed from the Earth to the Sun, while z-axis is directed from south to north). Generalized kinetic models with $B_y \neq 0$ for the magnetotail CS can be found in Artemyev (2011), Malova et al. (2012), Mingalev et al. (2012), while models of the magnetopause CS with B_x and B_y are discussed in review by Roth et al. (1996) and in recent papers (see Harrison and Neukirch 2009; Panov et al. 2011 and references therein).

Two components of the magnetic field B_z and B_x correspond to the single component of the vector potential $A_y(x, z)$ where $\operatorname{rot}(A_y\mathbf{e}_y) = \mathbf{B}$. Thus two equations for the magnetic field can be reduced to

$$\Delta A_y = -(4\pi/c)\sum_\alpha q_\alpha \int v_y f_\alpha d\mathbf{v} \tag{3}$$

2.1 Vlasov approach

Stationarity and homogeneity of the system along y (dawn-dusk) direction result in conservation of the particle total energy $H_\alpha = (1/2m_\alpha)(p_x^2 + p_z^2) + (1/2m_\alpha)(p_y - (q_\alpha/c)A_y)^2 + q_\alpha\varphi$ and the generalized momentum $p_y = m_\alpha v_y + (q_\alpha/c)A_y$. Any function $f_\alpha = f_\alpha(H_\alpha, p_y)$ of these integrals represents some solution of the stationary Vlasov equation. Therefore, one can introduce certain distribution $f_\alpha(H_\alpha, p_y)$ and reduce problem (1) to the integro-differential equation for A_y, while φ can be found as a function of A_y from the quasineutrality condition (2). This is the essence of the so called Vlasov approach (Schindler and Birn 2002; Schindler et al. 2012).

Choice of the velocity distribution as a shifted Maxwellian function $f_\alpha \sim \exp(-(H_\alpha - v_{D\alpha} p_y)/T_\alpha)$ with a constant drift velocity $v_{D\alpha}$ gives for 1D system (without the B_z component) the popular Harris solution (Harris 1962). For 2D systems (with $B_z(x, z)$, $B_x(x, z)$)

a generalization of the Harris solution was obtained by Kan (1973). Review of exact 2D solutions of Eq. (3) can be found in Yoon and Lui (2005), Vasko et al. (2012), while approximated solutions are discussed by Lui (2004). There are several specific functions $f_\alpha(H_\alpha, p_y)$ giving bifurcated CS solutions (Camporeale and Lapenta 2005; Génot et al. 2005; Israelevich et al. 2007) and solutions possessing TCS manifestations (Schindler and Birn 2002; Birn et al. 2004; Liu et al. 2010).

All stationary models that use distribution functions in a form $f_\alpha(H_\alpha, p_y)$ have several common properties. (1) Scalar potential is a function of a single component of the vector potential $\varphi = \varphi(A_y)$ and the corresponding electric field is orthogonal to the magnetic field $\mathbf{E} \cdot \mathbf{B} = -\nabla\varphi \cdot \mathbf{B} = -(\partial\varphi/\partial A_y)\nabla A_y \cdot [\nabla \times A_y \mathbf{e}_y] = 0$ everywhere (Schindler et al. 2012). (2) Anisotropy of the particle temperature occurs simultaneously with the nongyrotropy. Distributions f_α in the velocity space depend on $v_x^2 + v_z^2$ and on v_y. Therefore, one can develop an isotropic CS model or a CS equilibrium having temperature anisotropy $T_{xx} = T_{zz} \neq T_{yy}$. Here T_{yy} can be considered to be the perpendicular component of the temperature, while the parallel component as well as the second perpendicular component represent certain combinations of T_{xx} and $T_{zz} = T_{xx}$ (Mottez 2004). As a result, two perpendicular components are not equal to each other for any anisotropic CS with $T_{yy} \neq T_{xx}$. (3) The pressure balance along the magnetotail is supported by the gradient of the total plasma pressure only. As a result, a transverse CS spatial scale (thickness, L_z) and a longitudinal spatial scale, L_x, cannot be considered as two independent free parameters: the scaling relation $2L_x B_z/L_z B_0 \approx 1$ should be satisfied (Burkhart and Chen 1993).

2.2 Speiser trajectories in TCS

Vlasov models discussed in the previous section do not take into account effects corresponding to the transient Speiser trajectories. To include these effects one needs to introduce an additional invariant of motion, $I_z = \oint p_z dz$. Here we do not discuss the peculiarities of the Speiser ion motion and the accuracy of the conservation of I_z (detailed introduction to these problems can be found in Sonnerup 1971; Büchner and Zelenyi 1986; Büchner and Zelenyi 1989; Vainchtein et al. 2005).

TCS models with Speiser ions are based on the ion velocity distribution $f_i(H_i, I_z)$. There are several features of the TCS with $f_i(H_i, I_z)$: (1) the current density profile is substantially thinner (L_z is about ion Larmor radius) in comparison with the plasma density profile (Sitnov et al. 2000; Zelenyi et al. 2000). (2) The ion velocity distribution in the neutral plane $B_x = 0$ has a very characteristic structure. In the velocity space (v_x, v_y) this distribution looks like half of a ring. Such form corresponds to a peculiarity of ion orbits (Hamilton and Eastwood 1982; Burkhart et al. 1992a; Zelenyi et al. 2011): before leaving the $z = 0$ plane (where $B_x = 0$), ions make a half of a Larmor circle in the course of their rotation around B_z. (3) The fast field-aligned motion of Speiser ions produces the inertial force which could support the pressure balance without any pressure gradient along the magnetotail (Rich et al. 1972; Hill 1975; Burkhart et al. 1992a). Thus transient particles make it possible to construct a practically 1D TCS with $B_z \neq 0$ (Burkhart and Chen 1993), i.e. the scale L_x can become much larger than $L_z(B_0/B_z)$.

Current density carried by Speiser ions in the neutral plane has an amplitude $\sim q_i n_{sp} v_D$ where v_D is the bulk velocity of particles and n_{sp} is the density of Speiser particles. For 40 TCSs observed by Cluster, v_D was estimated to be $v_D \sim 1$–$5v_T$ where v_T is the ion thermal velocity, i.e. parameter $\varepsilon = v_T/v_D \approx 0.3$ (Artemyev et al. 2010). This current can be comparable with observed values of the cross-tail TCS current only if the density of Speiser population n_{sp} is much smaller than the total density of ions (Artemyev

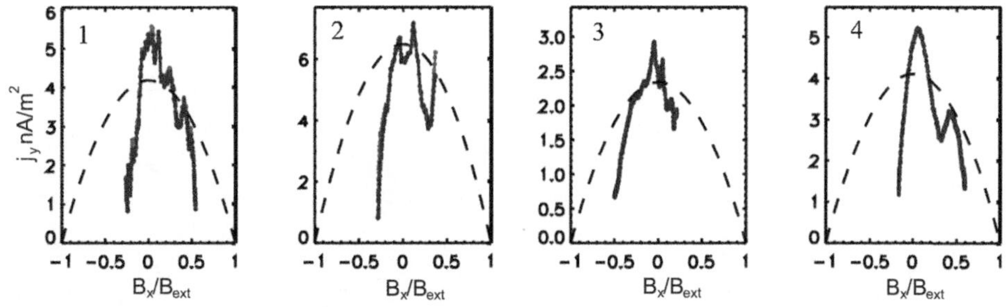

Fig. 1 Comparison of the Harris model (*dotted curves*) with the spacecraft observations. Figure is adopted from Runov et al. (2006)

et al. 2010). Therefore, TCSs with $f_i(H_i, I_z)$ should be embedded into the dense almost isotropic background population. This embedding can be performed in two ways: (1) one can consider the ion distribution $f_i(H_i, \eta I_z, p_y)$ and regulate impact of the Speiser population by parameter $\eta \geq 0$ (Sitnov et al. 2006); (2) the ion distribution can be considered as a sum $f_i = \eta f_i(H_i, I_z) + f_{i,bg}(H_i, p_y)$ where $f_{i,bg}(H_i, p_y)$ defines background population and parameter $\eta \geq 0$ regulates the contribution of Speiser ions (Zhou et al. 2009; Petrukovich et al. 2011).

3 Spacecraft Observations of the TCS: Comparison with Models

Multipoint magnetic field measurements make it possible to restore the current density j_y by the so-called curlometer technique (Paschmann and Schwartz 2000). The first Cluster statistical observations of the Earth magnetotail CS demonstrated the effect of embedding. The most obvious demonstration was given by Asano et al. (2005) and Runov et al. (2006). In both papers authors compared spacecraft observations with the Harris CS model where the current density profile coincides with the plasma density profile. The Harris model predicts a specific relation between the current density and the magnetic field $j_y/j_{\max} = 1 - (B_x/B_{ext})^2$. Asano et al. (2005) have shown that $j_y/j_{\max}$ calculated for $B_x/B_{ext} \approx 0.45$ is much smaller than the expected value $1 - (B_x/B_{ext})^2 \approx 0.8$ for the majority of observed TCSs. Therefore, j_y decreases with the increase of B_x faster than it can be obtained from the Harris model. Runov et al. (2006) have compared profiles $j_y(B_x)$ with the Harris parabolic dependence $j_y \sim 1 - (B_x/B_{ext})^2$. Examples of this comparison are presented in Fig. 1. Profiles of $j_y(B_x)$ are substantially narrower than the Harris parabolic shapes, i.e. Cluster spacecraft observe TCSs with a strong current j_y embedded into the thick CS with weak current density.

To describe this embedded TCS structure one needs to separate profiles of the plasma density (attributed to the background thick CS) and the current density of the TCS. Such separation is intrinsic for the TCS models with Speiser particles (Zelenyi et al. 2004; Sitnov et al. 2006) and these models are capable to match the observed profiles much better than the Harris configuration, see Fig. 2 and the comparison presented in Sitnov et al. (2006), Artemyev et al. (2008a).

Although the comparison of the theoretical and experimental current density profiles strongly supports the TCS models with the population of Speiser ions, results of such comparison cannot be considered as a decisive argument. As it was shown by Artemyev et al. (2009), Vlasov models (Birn et al. 2004; Yoon and Lui 2004) can also describe $j_y(B_x)$ profiles reasonably well if TCSs produced by these models would be embedded in the certain

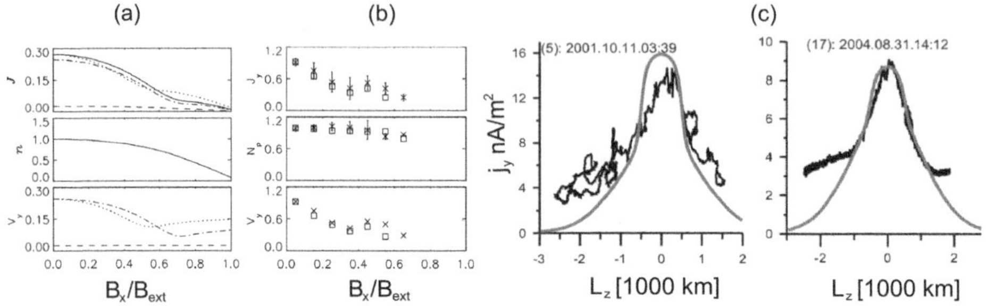

Fig. 2 Comparison of the spacecraft observations with the TCS models. Panel (**a**) shows the normalized current density, plasma density n and velocity $V_y = j_y/en$ profiles from the TCS model (Sitnov et al. 2006). Panel (**b**) shows the Cluster observations of the same parameters as panel (**a**) for the particular TCS: *various symbols* denotes measurements for $B_x < 0$ and $B_x > 0$. Both panels are adopted from Sitnov et al. (2006). Panel (**c**) shows the comparison of the TCS model (Zelenyi et al. 2004) (*grey curve*) with the Cluster observations (*black curves*). Figure is adopted from Artemyev et al. (2008a)

background current sheet. Therefore, the presence of the Speiser population and its role in TCS formation can be established only from investigation of ion velocity distributions (Ball et al. 2005; Zhou et al. 2009; Artemyev et al. 2010).

Because density of the Speiser population in the TCS is relatively small (often it does not exceed 10 % of the total proton density, see Artemyev et al. 2010) there are only two methods to distinguish the Speiser velocity distribution from the background plasma. The first approach works for the TCS observed in the vicinity of the reconnection region when the background plasma density is small. In this case the velocity distribution contains only two components: a very cold core (temperature < 10 MK (1 keV)) and a hot wing forming by the Speiser population (Ball et al. 2005; Zhou et al. 2009). The example of such velocity distribution is presented in Fig. 3(a): one can easily see the asymmetric flank of the velocity distribution with $v_y > 0$. The second approach can be applied to the TCS observed under quiet conditions with the substantial ion (proton) current density. Due to weak (if any) variations of the plasma density across the TCS, one can calculate the distribution function of particles residing in the central region as the difference of f_i observed in $B_x \approx 0$ and in the TCS boundaries where $|B_x| \approx B_0$. This differences Δf_i contains two parts: the volume of phase density $\Delta f_i > 0$ is assumed to be filled by Speiser particles, while the phase volume $\Delta f_i < 0$ corresponds to some population residing mostly at the TCS boundaries (Artemyev et al. 2010). The example of such velocity distribution Δf_i is shown in Fig. 3(b). One can notice the similarity between the flank of the distribution shown in Fig. 3(a) and the positive difference $\Delta f_i > 0$ shown in Fig. 3(b).

Both experimental velocity distributions can be compared with the numerical (Hamilton and Eastwood 1982; Burkhart et al. 1992a) and analytical (Zelenyi et al. 2010) distributions of Speiser particles (see Figs. 3(c), (d)). The common half-ring form is characteristic for all the distributions. Therefore, we have another indirect evidence of the important role of Speiser particles in the TCS formation.

One additional feature of the TCS with the finite Speiser population is the impact of the inertia of ion motion on the longitudinal pressure balance (Rich et al. 1972; Hill 1975; Cowley and Pellat 1979). This effect allows to maintain practically one-dimensionality of TCS with $B_z \neq 0$ and $L_x/L_z \gg B_0/B_z$ (Burkhart et al. 1992a; Ashour-Abdalla et al. 1994; Mingalev et al. 2009). Therefore, the presence of finite population of Speiser ions in the observed TCS should result in a large value of the parameter $\lambda = 2L_x B_z/B_0 L_z$. However, the longitudinal scale L_x of the magnetic field inhomogeneity cannot be estimated directly from

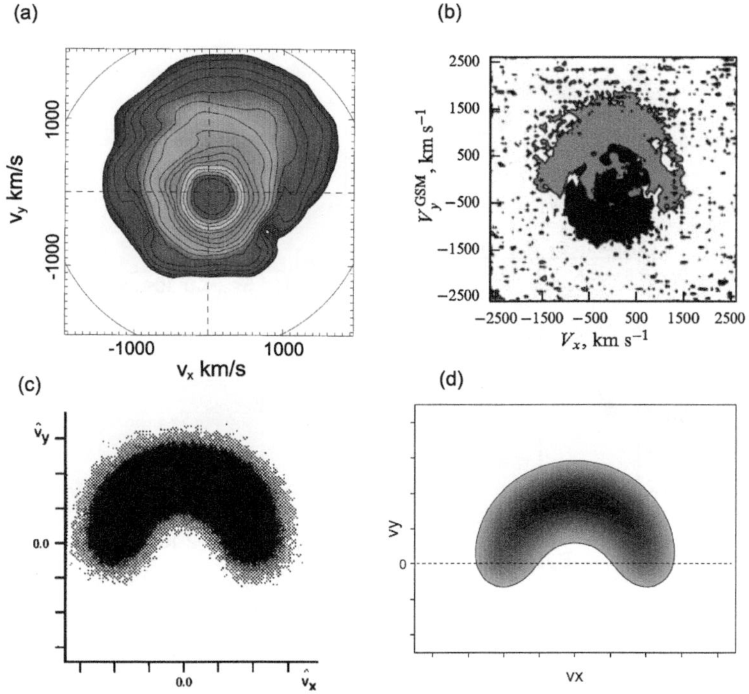

Fig. 3 The velocity distribution of ions in (v_x, v_y) plane. Panel (**a**) shows the velocity distribution observed by THEMIS spacecraft P1, figure is adopted from Zhou et al. (2009). Panel (**b**) shows the difference of the velocity distributions observed by Cluster spacecraft C4 in the TCS central region and in the boundary. *Grey color* indicates the positive difference: this volume of phase density is filled by particles presented only in the central region. Figure is adopted from Artemyev et al. (2010). Panel (**c**) shows the velocity distribution of Speiser particles in the neutral plane $B_x = 0$. This distribution is obtained in the numerical modeling. Figure is adopted from Burkhart et al. (1992a). Panel (**d**) shows the velocity distribution in the neutral plane obtained in the TCS model (Zelenyi et al. 2004). Figure is adopted from Zelenyi et al. (2010)

the closely packed Cluster observations ($L_x \sim 5$–$10 R_{\rm E}$ is much larger than the characteristic spacecraft separation ~100–10000 km).

To estimate L_x we use measurements of the electron temperature profiles across TCSs: T_e could be considered as a function of the magnetic field B_x and profiles $T_e(B_x)$ could be explicitly measured. These profiles can be considered as projections of the longitudinal profile $T_e(x)$ (i.e. $T_e(B_z)$) along field lines (Artemyev et al. 2011b). Therefore, using the simple model of the electron heating in the course of the earthward convection due to $\sim cE_y/B_z$ drift one can recover the profile $B_z(x)$ from the profile $T_e(x)$, while $T_e(x)$ can be obtained from the reverse projection of $T_e(B_x)$ along field lines. This indirect technique allows to estimate L_x for 62 TCS Cluster crossings (Artemyev et al. 2011b, 2012a).

The examples of the $T_e(B_x)$ profiles are shown in Fig. 4(a). Decrease of the electron temperature T_e with increase of $|B_x|$ corresponds to the tailward decrease of $T_e(x)$: electrons from the deep tail come to higher $|z|$ (i.e. larger $|B_x|$) moving along field lines. We estimate L_x by using the profiles $T_e(B_x)$, see statistical distribution in Fig. 4(b). The average value of L_x is about $10R_{\rm E}$. The current sheet thickness as well as B_z and B_0 can be obtained for each TCS crossing by using direct spacecraft measurements. Thus we can estimate the parameter $\lambda = 2L_xB_z/B_0L_z$ (see Fig. 4(b)). One can see, that the average value of λ is about ~ 7. Such high values of λ could be achieved only if significant part of the longitudinal stress balance

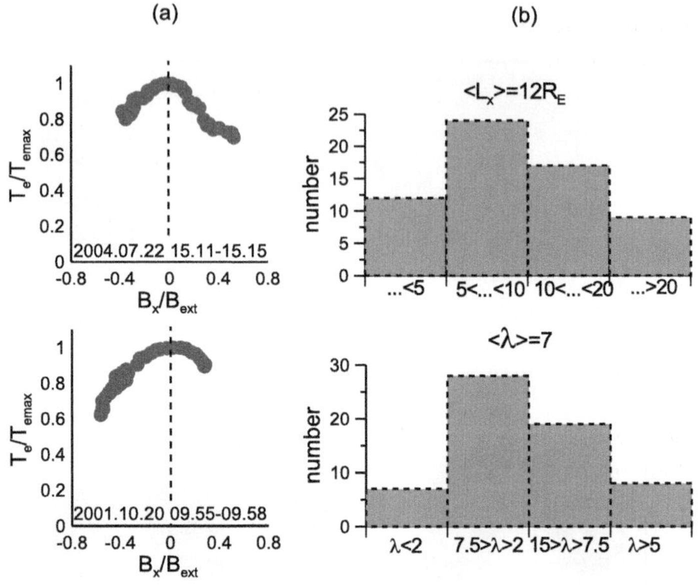

Fig. 4 Panel (**a**) shows two examples of profiles of electron temperature in TCS. Panel (**b**) shows distribution of estimates L_X (measured in R_E) and $\lambda = 2L_X B_z/B_0 L_z$. All figures are adopted from Artemyev et al. (2011b)

will be supported by the Speiser population. Therefore, one can conclude that Speiser ions represent the principal ingredient of the TCS plasma population.

4 Adiabatic Electrons in the TCS

The TCS thickness L_z is much larger than electron Larmor radius and the normal component B_z always has a finite value (far from the immediate vicinity of X-line). As a result, electrons, as a rule, are magnetized in the TCS and can be described as adiabatic particles with a conserved magnetic moment μ. Therefore, one can use anisotropic MHD approach to describe of the electron dynamics. In a stationary system after neglecting the electron mass the stress balance of the electron component can be written as Shkarofsky et al. (1966)

$$\nabla \hat{p}_e = -en_i \mathbf{E} + c^{-1}[\mathbf{j}_e \times \mathbf{B}] \tag{4}$$

where we assumed the quasineutrality $n_e \approx n_i$ (n_α is the density of particles of sort α and $q_e = -e$) and introduced pressure tensor $\hat{p}_e$ as

$$\nabla \hat{p}_e = \nabla_\perp p_{\perp e} + \nabla_\parallel p_{\parallel e} + \Lambda\left((\mathbf{B}\nabla)\mathbf{B} - 2\frac{\mathbf{B}}{B}(\mathbf{B}\nabla)B \right)$$

Here $\Lambda = (p_{\parallel e} - p_{\perp e})/B^2$ and $B = |\mathbf{B}|$. Projection of Eq. (4) on the transverse direction (perpendicular to $\mathbf{B}$) gives the transverse electron current density

$$j_{\perp e} = -cen_i \frac{[\mathbf{E} \times \mathbf{B}]}{B^2} - c\frac{[\nabla_\perp p_{\perp e} \times \mathbf{B}]}{B^2} + c\Lambda \frac{[\mathbf{B} \times (\mathbf{B}\nabla)\mathbf{B}]}{B^2} \tag{5}$$

The last term in the expression for $j_{\perp e}$ is the curvature drift current existing only for configurations with a finite pressure anisotropy. For the 1D TCS ($B_z \approx$ const) in the vicinity of the neutral plane ($B_x \sim 0$ and $B \sim B_z$) this term can be written as

$$c\Lambda \frac{[\mathbf{B} \times (\mathbf{B}\nabla)\mathbf{B}]}{B^2} \approx c\left(\frac{p_{\parallel e}}{p_{\perp e}} - 1\right)\frac{p_{\perp e}}{B_z^2}\frac{\partial B_x}{\partial z} \approx j_y\left(\frac{p_{\parallel e}}{p_{\perp e}} - 1\right)\frac{p_{\perp e}}{4\pi B_z^2} \tag{6}$$

where $j_y = (c/4\pi)\partial B_x/\partial z$ is the total density of current supporting the existence of the TCS. For magnetospheric conditions we estimate $p_{\perp e}/4\pi B_z^2 \sim 10$–$100$ ($p_{\perp e} \sim 0.01$–0.1 nPa and $B_z \sim 0.5$–5 nT). As a result, even 1 %–10 % of the electron temperature anisotropy results in the extremely high electron current density $j_{\perp e} \sim j_y$ (Zelenyi et al. 2011). Therefore, the effect of the electron temperature anisotropy, often observed in the magnetotail (Stiles et al. 1978; Artemyev et al. 2011a, 2012a), leads to a very strong (although very localized) transverse current. Thus the electron temperature anisotropy should be taken into account by any realistic TCS model.

There are two procedures for the development of TCS models that include effects of the electron anisotropy. For models obtained from the Vlasov approach one can introduce the velocity distribution of electrons as a function of the energy and the generalized momentum $f_e = f_e(H_e, p_y)$, where $H_e = (v_x^2 + v_z^2)m_e/2 + (p_y + (e/c)A_y)^2/2m_e - e\varphi$ and $p_y = m_e v_y - (e/c)A_y$. Therefore, the distribution f_e can be written as $f_e = f_e(v_x^2 + v_z^2, v_y)$. For such distribution the electron temperature tensor has three components T_{xx}, $T_{zz} = T_{xx}$, and T_{yy} (see CS models with $T_{yy} \neq T_{xx}$ proposed by Nicholson 1963; Mottez 2003). The dawn-dusk direction, y, is perpendicular to the magnetic field of the TCS, $\mathbf{B} = B_x\mathbf{e}_x + B_z\mathbf{e}_z$. Thus, T_{yy} can be considered to be the perpendicular component of the electron temperature $T_{\perp e}^{(1)}$. However, the second perpendicular component of the temperature tensor, $T_{\perp e}^{(2)} = T_{xx}(B_z/B)^2 + T_{zz}(B_x/B)^2 = T_{xx}$, is not equal to $T_{yy} = T_{\perp e}^{(1)}$, but is equal to $T_{\parallel e} = T_{xx}(B_x/B)^2 + T_{zz}(B_z/B)^2 = T_{xx}$. Therefore, we obtain the pressure nongyrotropy for CS models developed on the basis of the Vlasov approach (Mottez 2004). Moreover, as was shown by Schindler and Hesse (2010) even initially gyrotropic CSs with $f_e(H_e, p_y)$ could violate the gyrotropy during the slow evolution.

If we consider an additional invariant of motion μ (assumption about μ conservation already takes into account conservation of the generalized momentum p_y, see Cary and Brizard 2009), then the velocity distribution of electrons can be written as $f_e = f_e(H_e, \mu)$, where $H_e = m_e v_\parallel^2/2 + \mu B - e\varphi$ and $\mu = m_e v_\perp^2/2B$. Dependence of f_e on $v_\parallel$ and $v_\perp$ results in the temperature tensor with the two different components $T_{\parallel e} \neq T_{\perp e}^{(1)} = T_{\perp e}^{(2)}$.

To test both approaches we calculate the electron temperature tensor for 62 TCS crossings by the Cluster mission (see details in Artemyev et al. 2012a). Comparison of components of the temperature tensor T_{yy}, T_{zz} and T_{xx} with components $T_{\parallel e}$ and $T_{\perp e}$ is shown in Fig. 5. As one can see, $T_{zz} \approx T_{yy} \approx T_{\perp e}$, $T_{xx} \approx T_{\parallel e}$, and $T_{\parallel e}/T_{\perp e} \approx 1.1$. Therefore, the electron nongyrotropy is practically absent in the observed TCS. Moreover, the average electron gyroradius is much smaller than the vertical spatial scale of TCS, L_z. As a result, T_{xx} can be approximated by $T_{\parallel e}$ almost everywhere.

Estimates of the curvature currents and observed ratios of components of the temperature tensor demonstrate that one should use the distribution function $f_e(H_e, \mu)$ that depends on the magnetic moment for the description of the TCS electron component. Therefore, for electrons it is important to introduce the additional adiabatic invariant, as it was done for the ion component with I_z.

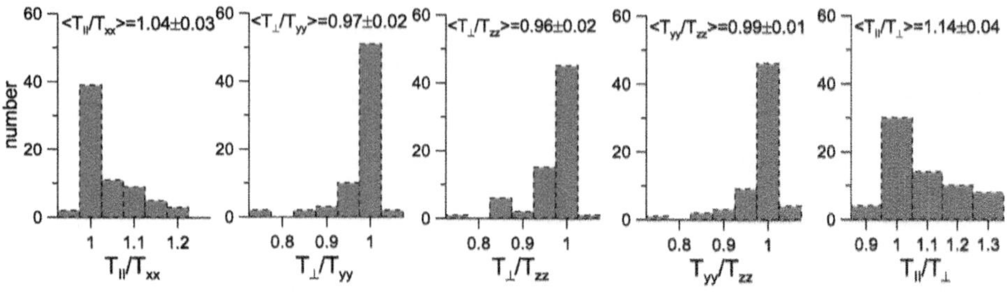

Fig. 5 Distribution of ratios of electron temperature tensor components for TCSs. Figure is adopted from Artemyev et al. (2012a)

5 Electric Field Due to the Decoupling of Electron and Ion Motions

Separation of ion and electron motions in the TCS results in appearance of the electrostatic field which is described by the scalar potential φ. Spatial distribution of φ can be obtained from the quasineutrality condition (2). Here, again, two approaches can be used. According to the Vlasov approach one should consider a velocity distribution as a function of the energy $H_\alpha = m_\alpha(v_x^2 + v_z^2)/2 + (p_y - (q_\alpha/c)A_y)^2/2m_\alpha + q_\alpha\varphi$ and the generalized momentum p_y. As a result, the particle density (as well as other system parameters) depends on A_y and φ only. The quasineutrality condition $n_e(A_y, \varphi) = n_i(A_y, \varphi)$ gives the dependence $\varphi = \varphi(A_y)$ (Yoon and Lui 2004; Birn et al. 2004; Liu et al. 2010; Schindler et al. 2012). Corresponding components of the electrostatic field are

$$\begin{cases} E_x = -\partial\varphi/\partial x = (\partial\varphi/\partial A_y)B_z \\ E_z = -\partial\varphi/\partial z = -(\partial\varphi/\partial A_y)B_x \end{cases} \tag{7}$$

and $\mathbf{E} \cdot \mathbf{B} = 0$. This electrostatic field can redistribute the current density between ions and electrons by means of the cross-field drift (Schindler and Birn 2002; Birn et al. 2004)

$$v_{\mathbf{E}\times\mathbf{B}} = c\frac{\mathbf{E}\times\mathbf{B}}{B^2} = c\frac{E_z B_x - E_x B_z}{B_x^2 + B_z^2} = -c\frac{\partial\varphi}{\partial A_y}$$

For various TCS models one can obtain different dependencies $\varphi = \varphi(A_y)$ and different drift velocities $v_{\mathbf{E}\times\mathbf{B}}$ (Yoon and Lui 2004; Birn et al. 2004; Schindler 2006). However, all TCS models of this type do not contain parallel electrostatic fields.

The situation is different if one takes into account effects of the anisotropy and consider electrons in the framework of the MHD theory. In this case projection of the stress balance (4) on the parallel direction $\mathbf{B}/B$ gives (Shkarofsky et al. 1966)

$$en_e E_\parallel = -\nabla_\parallel p_{\parallel e} + \Lambda\frac{\mathbf{B}}{B}(\mathbf{B}\nabla)B$$

where one can substitute $E_\parallel = -\nabla_\parallel\varphi$. For simplified systems with the uniform isotropic temperature one immediately obtains the Boltzmann distribution of electrons along field lines: $en_e E_\parallel = -T_e\nabla_\parallel n_e$ and $e\varphi/T_e = \ln n_e$.

For the TCS observed in the magnetotail one can assume the uniform plasma density $n_e = \text{const}$ (Artemyev et al. 2010) and the uniform temperature anisotropy $\alpha_e = T_{\parallel e}/T_{\perp e} > 1$ (Artemyev et al. 2012a) across the TCS. Then one can write

$$e\nabla_\parallel\varphi = T_{\perp e}\nabla_\parallel \ln\left(T_{\perp e}^{\alpha_e}/B^{\alpha_e - 1}\right) \tag{8}$$

At the boundary of the TCS ions as well as electrons become magnetized and decoupling of ion and electron motions vanishes, i.e. $\varphi_{\text{boundary}} = 0$. From the TCS boundary to the TCS central region magnetic field B decreases, while the electron temperature $T_{\perp e}$ increases (Artemyev et al. 2011b). As a result, we obtain a negative gradient $\nabla_{\|}\varphi < 0$, i.e. φ decreases towards the TCS boundaries. For 1D TCS we have $\nabla_{\|} = (B_z/B)(\partial/\partial z)$ and $\varphi_{B_x=0} > \varphi_{\text{boundary}}$. Thus the electrostatic potential acquires its maximum in the central region of the 1D TCS. Correspondingly E_z field is directed from the TCS central region towards the TCS boundaries and is equal to zero in the neutral plane (Zelenyi et al. 2011). Drift of particles due to the field E_z is $v_{\mathbf{E}\times\mathbf{B}} = cE_zB_x/B^2 > 0$. This drift decreases the electron current density in the periphery of the TCS (where B_x field has substantially large value) and is absent in the vicinity of the central region. Thus obtained E_z has the opposite direction compared to the models developed using the Vlasov approach (see Schindler and Birn 2002; Birn et al. 2004; Baumjohann et al. 2007). This is effect of the parallel electric field $E_{\|}$, which determines E_z in the 1D TCS as $E_z = E_{\|}(B/B_z)$. The parallel electric field is absent in Vlasov models where presence of a finite $B_z \neq 0$ results in the inhomogeneity along x with $E_x = -E_z(B_z/B_x)$.

It is interesting to note that Eq. (8) gives the following expression for E_z in the isotropic TCS ($\alpha_e = 1$): $E_z = -\partial\varphi/\partial z = -(1/e)\partial T_e/\partial z$. Substituting E_z into Eq. (5) we obtain $j_{\perp e} = -cen_e(E_zB_x/B^2) - cn_e(B_x/B^2)\partial T_e/\partial z = 0$. Thus, the distribution of electrons along field lines in isotropic systems with $B_z \neq 0$ reduces the drift electron current. Only anisotropic electrons can support the cross-tail current density in the 1D TCS (see, e.g., Cowley and Pellat 1979).

One can also consider effects of the weak inhomogeneity of the magnetic field component B_z along the x-direction ($\partial B_z/\partial x > 0$). In this case magnetic field lines are not equidistant and each line has its own length measured from the TCS central region up to the boundary where $\varphi = 0$. Therefore, there is a certain gradient of the scalar potential along the neutral plane $\partial\varphi/\partial x \neq 0$. Corresponding electrostatic field E_x should appear. We derive the expression for $E_x = E_x(z)$ for the simplified geometry of the magnetic field with $B_x(z)$ and $B_z(x)$. The main effect responsible for E_x formation is the dependence of the length of field lines on x. The length of a fragment of field lines is $\Delta s \approx \Delta z(B/B_z)$. The potential accumulated along this fragment is $\Delta\varphi \approx \Delta s(\partial\varphi/\partial s) \approx \Delta s(\partial\varphi/\partial z)(\partial z/\partial s) \approx -\Delta sE_zB_z/B$. Once we consider $\partial B_z/\partial x \neq 0$ one can write

$$\frac{\partial\Delta\varphi}{\partial x} \approx \frac{\partial\Delta s}{\partial x}\frac{\partial\varphi}{\partial s} = -\Delta z\frac{B_x^2}{B_zB}\frac{\partial\ln B_z}{\partial x}\frac{\partial\varphi}{\partial s}$$

After the integration over z one can obtain

$$\frac{\partial\varphi}{\partial x} \approx \frac{\partial\ln B_z}{\partial x}\int_{L_z}^{z} E_z(z')\frac{B_x^2(z')}{B_x^2(z') + B_z^2}dz'$$

where we keep $\partial\ln B_z/\partial x$ outside the integral due to the simplifying assumption that $\partial\ln B_z/\partial x \approx \text{const}$. Final expression for the electric field E_x can be written as Zelenyi et al. (2010):

$$E_x(z) \approx \frac{\partial\ln B_z}{\partial x}\int_{z}^{L_z}\frac{E_z(z')B_x^2(z')}{B_x^2(z') + B_z^2}dz' \tag{9}$$

The electrostatic field $E_x(z)$ has a maximum in the central region of the TCS and is equal to zero at the TCS boundaries where $|z| = L_z$. Of course, Eq. (9) gives only an estimate of the E_x field. To obtain distribution of the scalar potential $\varphi(x,z)$ in the 2D TCS in a more rigorous manner one needs to define the distribution of electron pressure and magnetic field components in the (x, z) plane. This requires the solution of the 2D system for the TCS with Speiser trajectories and the anisotropic electron pressure (such models still have not been developed, unfortunately).

Equation (9) defines simplified spatial distribution of the electrostatic field across the TCS. Using E_z from the 1D TCS model derived from Eq. (8) we obtain $E_x > 0$ in the central region of the TCS (Zelenyi et al. 2010). The amplitude of E_x can be estimated as $E_x \approx E_z L_z / L_x \approx \delta\varphi / L_x$ where $L_x \approx (\partial \ln B_z / \partial x)^{-1}$ and $\delta\varphi \approx T_e / e$ is the drop of the scalar potential between the TCS boundary and the central region. Then we obtain $E_x \approx T_e / e L_x$. The corresponding cross-field drift velocity in the vicinity of the neutral plane $B_x \approx 0$ is $v_{\mathbf{E}\times\mathbf{B}} = -cE_x / B_z \approx -cT_e / eB_z L_x < 0$. This drift enhances the electron current density in the central region and reduces the ion contribution to the total cross-tail current. The corresponding current density amplitude can be estimated as $j_{\mathbf{E}\times\mathbf{B}} = en_e |v_{\mathbf{E}\times\mathbf{B}}| \approx cn_p T_e / B_z L_x = 2cn_p T_e / B_0 L_z \lambda = ev_T n_p (\rho_p / \lambda L_z)(T_e / T_p)$ where $v_T = \sqrt{2T_p / m_p}$, $\rho_p = v_T m_p c / eB_0$ and T_p is the proton temperature. It is very important that the contributions from the $v_{\mathbf{E}\times\mathbf{B}}$ drift to the electron and ion currents exactly compensate each other. The presence of E_x does not change the total cross-tail currents but instead dramatically redistributes the partial contributions of ions and electrons. The total proton current density (with $v_{\mathbf{E}\times\mathbf{B}}$ drift taken into account) can be written as

$$j_p = j_{sp} + j_{bg,p} - ev_T n_p \left(\frac{T_e}{T_p} \frac{\rho_p}{\lambda L_z} \right)$$

where $j_{bg,p} \ll j_p$ is the current density of background protons, while $j_{sp} = ev_D n_{sp} \approx \varepsilon^{-1} ev_T n_{sp} \approx 3ev_T n_{sp}$ is the current density of Speiser protons with density n_{sp} (here we also take into account that the bulk velocity of Speiser protons in the neutral plane v_D is about three times the thermal velocity: $\varepsilon = v_T / v_D \approx 0.3$ for TCSs, see Artemyev et al. 2010). Thus, we obtain

$$j_p = ev_T n_{sp} \left(1 - \frac{n_p}{n_{sp}} \frac{T_e}{T_p} \frac{3\rho_p}{\lambda L_z} \right)$$

For $T_p / T_e \approx 3$–5, $L_z \approx \rho_p$, $n_{sp} \approx 0.1$–$0.2 n_p$, and $\lambda \approx 7$ (see discussion above) we obtain $j_p \approx 0$ or even $j_p < 0$. The negative total proton current density are often observed in the Earth magnetotail (Runov et al. 2006; Artemyev et al. 2009). Examples of such observations can be found in Fig. 6 where three profiles of the TCS current density from (Runov et al. 2006) are shown. One can see that j_p profiles do not follow to curlometer profiles j_{curl} and the local proton current density is negative.

The negative proton currents are, as a rule, compensated by the enhanced electron currents. Thus the total current $j_p + j_e$ fits the curlometer data (Artemyev et al. 2009; Zelenyi et al. 2010; Artemyev et al. 2011a). Average values of $j_p + j_e$ for the central region of the TCS $|B_x| < 5$ nT can be compared to the average value of j_{curl} to examine the reliability of the plasma data and the idea of the current redistribution due to the cross-field drifts (i.e. that the total current $j_p + j_e$ should be equal to j_{curl} even for the negative proton contribution). Artemyev et al. (2011a) used 62 TCS crossings by the Cluster mission to compare average $j_p + j_e$ and j_{curl} and results are shown in Fig. 7(b). One can see that the ratio $(j_p + j_e)/j_{curl}$ for TCS crossings with various y coordinate is quite close to 1 (the distribution of a number of crossings over y also shown in Fig. 7(a)).

Comparison between Eq. (9) for E_x obtained in the anisotropic MHD and Eq. (7) describing electrostatic fields in the system with $\varphi = \varphi(A_y)$ clarifies the main effect of the

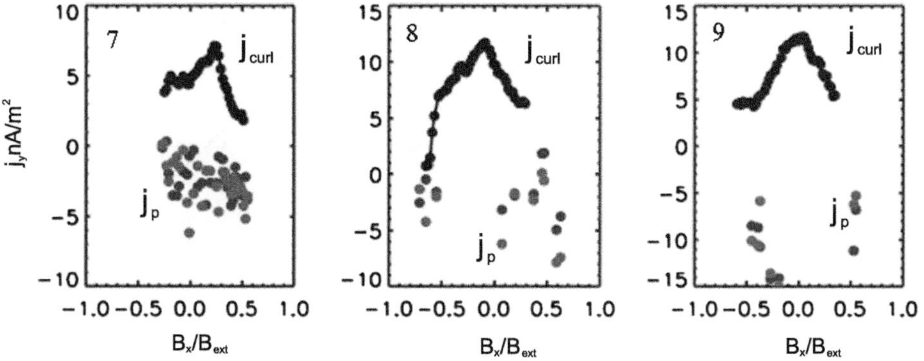

Fig. 6 Comparison proton current density j_p and curlometer current density j_{curl} for three TCSs. Figure is adopted from Runov et al. (2006)

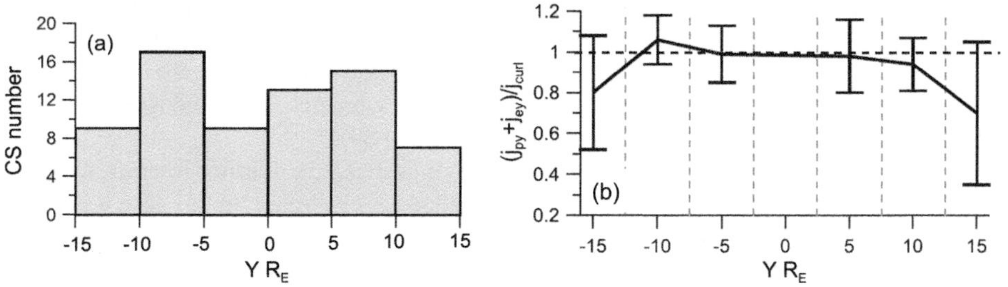

Fig. 7 Panel (**a**) shows distribution of TCS crossings over y coordinate. Panel (**b**) shows comparison of total particle current density and curlometer data. Figure is adopted from Artemyev et al. (2011a)

electron temperature anisotropy. Nonzero gradients of plasma parameters (pressure, temperature) along field lines leads to a value of finite parallel electric field $E_{\|} \neq 0$. In this case $E_x = E_{\|}(B/B_x) - (E_z/B_x)B_z$ can be positive even if $E_z/B_x > 0$, while Vlasov models require $E_x = -(E_z/B_x)B_z$. This is an important difference between models constructed based on two approaches. The presence of a parallel gradient of plasma parameters violates the conservation of the plasma density along field lines. This is the principal question for all models describing the magnetotail structure and dynamics. Observations of the negative proton current density indirectly supports the idea of the anisotropic electron distribution with $E_{\|} \neq 0$.

6 Discussion

In this paper we have shown that any realistic TCS models should include Speiser ions (described with the help of invariant I_z) and anisotropic electrons. Therefore, one needs to consider more narrow class of the three-parameter distributions $f_\alpha(H_\alpha, p_y, K_\alpha)$ instead of the very wide class of two-parameter distribution functions $f_\alpha = f_\alpha(H_\alpha, p_y)$. Here the additional adiabatic invariant is $K_\alpha = I_z$ for ions and $K_\alpha = \mu$ for electrons. Actually, the introduction of an additional invariant (I_z or μ) corresponds to the assumption of the existence of an additional intrinsic symmetry in the system. Conservation of the momentum

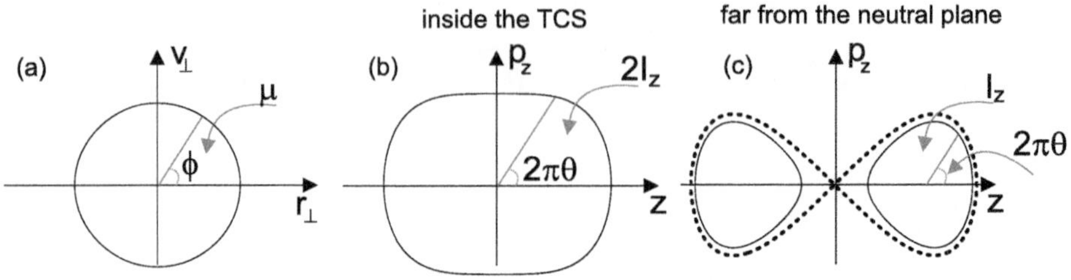

Fig. 8 Trajectories in the phase plane are shown by *thin curves* for the system with $\kappa \gg 1$ (panel (**a**)), for the system with $\kappa \ll 1$ *inside* the TCS (panel (**b**)) and far from the neutral plane (panel (**c**)). Areas surrounded by trajectories are equal to corresponding adiabatic (quasiadiabatic) invariants. For all trajectories periodic phases conjugate to the adiabatic (quasiadiabatic) invariants are shown

p_y yields a Hamiltonian H_α with two degrees of freedom and four corresponding independent variables: $H_\alpha = H_\alpha(p_x, x, p_z, z)$ (two pairs of conjugate variables (p_z, z) and (p_x, x)). This stationary system cannot be integrated: the existence of only one integral of motion, $H_\alpha = \text{const}$, is not sufficient to integrate a system with two degrees of freedom. However, for a particular relation between the scale of the magnetic field inhomogeneity and the particle Larmor radius ρ_α, this system has the additional symmetry. This ratio is defined by the parameter $\kappa = B_z/B_0\sqrt{L_z/\rho_\alpha}$ (Büchner and Zelenyi 1989).

For $\kappa \gg 1$, the magnetic field is strong enough to make the Larmor rotation the fastest motion in the system (this condition is often satisfied for electrons). In this case it is reasonable to introduce new pairs of conjugate variables $(p_x, x, p_z, z) \to (p_\parallel, s, \mu, \phi)$ where s is a coordinate along field lines and ϕ is the gyrophase ($\dot{\mu} = -\partial H/\partial\phi$ and $\dot{\phi} = \partial H/\partial\mu \sim \kappa^2 \gg 1$). Averaging over ϕ corresponds to the introduction of an additional symmetry associated to the rotation around magnetic field (the gyrotropy of the system). For the averaged Hamiltonian $\langle H\rangle_\phi$ we have $\partial\langle H\rangle_\phi/\partial\phi = 0$. Therefore, we obtain the conservation of the magnetic moment μ and the corresponding reduction of the number of independent variables and the dimension of the system. The system with $\langle H\rangle_\phi$ becomes one-dimensional and can be integrated.

Similar assumptions are valid for the opposite asymptotical regime with $\kappa \ll 1$ (typical for ions). In this case the additional symmetry of the system corresponds to the averaging of the Hamiltonian over the angle θ. To define θ we make the change of variables by the introduction of $I_z = \oint p_z dz = I_z(H, p_x, x)$: $(p_z, z) \to (I_z, \theta)$ where θ is the variable conjugate to I_z, i.e. $\dot{I}_z = -\partial H/\partial\theta$ and $2\pi\dot{\theta} = \partial H/\partial I_z$. For fixed κx, on the (z, p_z) phase plane particles move along some closed trajectories. The area surrounded by each of these trajectories is equal to I_z, while a phase of the periodic motion is θ (for details, see Büchner and Zelenyi 1986). In Fig. 8 we illustrate the definition of θ (we also show the system with $\kappa \gg 1$ for comparison). The topology of velocity distributions in the system with $\kappa \ll 1$ (contour lines $I_z = \text{const}$) was discussed by Ashour-Abdalla et al. (1991). The variables $(\kappa x, p_x)$ change much slower than θ because $(\kappa\dot{x}, \dot{p}_x) = \pm\kappa\partial H/\partial(p_x, \kappa x) \sim \kappa \ll 1$ and $2\pi\dot{\theta} = \partial H/\partial I_z \sim 1$. As a result, we can average over the "fast" periodic evolution of θ to obtain the Hamiltonian $\langle H\rangle_\theta$ with $\dot{I}_z = -\partial\langle H\rangle_\theta/\partial\theta = 0$ (Landau and Lifshitz 1988). The averaged system with $\langle H\rangle_\theta$ has one degree of freedom and two independent variables (x, p_x) and, thus, is integrable. Note that in the case $B_z \to 0$ (i.e. $\kappa \to 0$) I_z is an exact integral of motion (Sonnerup 1971).

Formally, the adiabatic invariant μ and the quasiadiabatic invariant I_z represent the same invariant (action) for two quantitatively different systems. The magnetic moment μ is conserved in the systems where the spatial scale of the magnetic field inhomogeneity is much larger than the typical Larmor radius. On the other hand, I_z is approximately constant if the spatial scale of the magnetic field inhomogeneity is much smaller than typical Larmor radius. Between these two regimes lies the region of chaotic motion, when $\kappa \sim 1$ and neither I_z nor μ are conserved even approximately, and particle trajectories cannot be integrated analytically (Büchner and Zelenyi 1989).

For typical conditions of the Earth magnetotail far from the X-line, the assumption of the conservation of μ provides an adequate description of the electron population, while I_z conservation describes motion of unmagnetized ions (protons). The introduction of I_z for electron description (Sitnov et al. 2006) means the presence of very hot quasiadiabatic electron population with Larmor radius of the order of Larmor radius of thermal protons. Such a situation can be realized for example in the vicinity of X-line where the magnetic field is weak and electrons could follow Speiser type unclosed orbits and electron-scale current sheets could emerge (see numerical modeling by Divin et al. 2010).

We hasten to mention that in our paper only stationary (or quasistationary) TCSs were discussed. However, the conditions of the TCS stability are also important for understanding the dynamics of the Earth magnetotail. Here we briefly discuss main results obtained for the TCS dynamics and new challenges of the TCS theory corresponding to the near-Earth TCS with the relatively strong gradient $\partial/\partial x$.

6.1 TCS Oscillations

Details of the investigations of the Vlasov equilibria stability can be found in several quite comprehensive reviews (Lui 2004; Schindler 2006). Here we briefly discuss the main stability properties of the TCS containing the Speiser population. There are two general branches of the TCS unstable modes: various lower-frequency drift perturbations are believed to be responsible for the TCS oscillations (so called flapping motions, see, e.g., Runov et al. 2005; Sergeev et al. 2006; Petrukovich et al. 2006, and references therein), while the development of tearing modes results in the magnetic reconnections (Coppi et al. 1966). Drift modes were investigated for the bifurcated TCS where ion distributions $f_i = f_i(H_i, p_y, \eta I_z)$ correspond to the generalized Harris equilibrium (for $\eta = 0$ distribution function f_i is identical to the classical Harris one) by Sitnov et al. (2004a). This configuration becomes unstable relative to large-scale kink perturbations. Shear of the proton bulk velocity along the current direction provides the energy source for the growth of this mode. Low frequency (below the local ion gyrofrequency) and high frequency (around the frequency of lower-hybrid resonance) oscillations of the TCS model (Sitnov et al. 2000; Zelenyi et al. 2004) were studied by Artemyev et al. (2008b) and Sitnov et al. (2004b), respectively. In particular, Artemyev et al. (2008b) have shown that the phase velocity of drift modes in the TCS can become very low due to the weakness of plasma density gradient in the central region of the TCS. Such slow drift modes can describe observed quasiperiodical oscillations of the TCS (Zelenyi et al. 2009). Moreover, the background plasma of the sheet where TCS is embedded can modify the TCS drift modes due to the presence of the additional large spatial scale of the density variations (Artemyev et al. 2012b).

6.2 Tearing Mode

The TCS thinning and following magnetic reconnection at the distances $x \in [-15, -30] R_{\mathrm{E}}$ (Baumjohann et al. 1999; Petrukovich et al. 2007; Angelopoulos et al. 2008; Petrukovich et al. 2009) are traditionally associated with the tearing instability (Schindler 1974; Galeev and Zelenyi 1976; Galeev 1979). While for the modification of the Harris current sheet

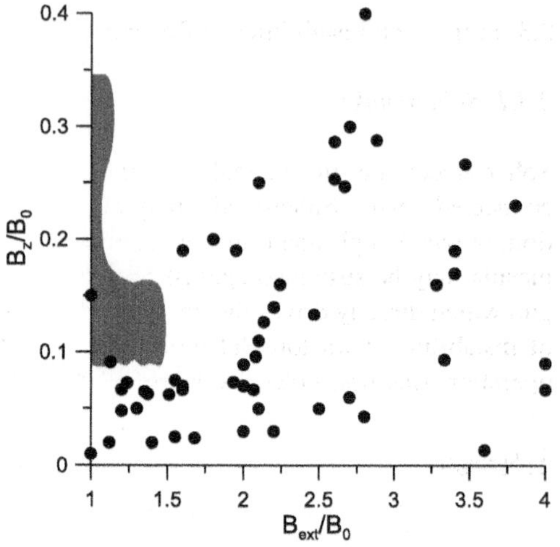

Fig. 9 Region of the tearing instability with the positive growth rate is shown in *grey color*. The *black dots* show positions of the stable (undestroyed) TCSs observed by the Cluster mission. Figure is adopted from Zelenyi et al. (2010)

model with $B_z \neq 0$ the tearing was found to have a negative growth rate (Pellat et al. 1991), the situation can be different for the TCS with a finite population of Speiser particles. The TCS can be unstable relative to the tearing mode inside a certain parameter range (Zelenyi et al. 2008) due to the combination of realistic physical effects missed in any modifications of the Harris equilibrium. Important destabilization effects are related to the stretching (when the pressure balance is partially supported by the inertia of the proton motion, i.e. parameter λ is growing and becomes $\lambda > 1$), embedding of the TCS in the wider plasma sheet (Burkhart et al. 1992b), and the strong electron curvature currents corresponding to the temperature anisotropy. Domain of the instability in parameter space can be compared with spacecraft observations, see Fig. 9. It was shown for Cluster observations that TCSs found under quiet conditions are located outside of the instability region (Fig. 9), while evolution of TCSs during the growth phase brings their state toward the instability domain (Zelenyi et al. 2010).

6.3 Convection and Near-Earth TCS

Although, some part of the longitudinal pressure balance in the TCS can be supported by the inertia of the Speiser ion motion (i.e. $\lambda > 1$), the other part should correspond to the gradient $\partial/\partial x$ and to the plasma convection with inhomogeneous bulk velocity $v_x(x) \neq 0$. The plasma convection $v_x(x)$ occurs due to the electrostatic field E_y (Angelopoulos et al. 1993) and represent the classical cross-field drift. Therefore, for a fixed x coordinate, the TCS cannot be considered as a totally self-consistent 1D structure, but its properties are controlled by the downtail plasma characteristics. The slow convection of ions from the downtail region can be described by the quasiadiabatic theory assuming conservation of I_z and one additional invariant $I_x = \oint p_x dx$ (Vainchtein et al. 2005). Combination of these two invariants allows to determine properties of trajectories in the system with weak gradients $\partial B_z/\partial x$ (Zelenyi et al. 1990). Conservation of I_x is a consequence of small values of B_z and weak gradients of B_z, i.e. L_x being substantially larger than proton Larmor radius in the neutral plane $\sim (B_0/B_z)\rho_p \sim 1\text{–}2R_E$. Additional condition is that the velocity of the convection should be small enough: distance passed by ions along x direction during one Larmor period in B_z should be substantially smaller than L_x. In fact TCSs formed at various x are coupled by the earthward convection (Zelenyi et al. 1990). This model seems to be adequate for $x < -15R_E$ where gradients $\partial B_z/\partial x$ are traditionally considered to be weak. However, an important problem of this model is the relation between the timescale of the earthward convection and the lifetime of the TCS (time interval of existing of a relatively stable and quasistationary TCS structure), i.e. the question of the convection regime in the magnetotail (Baumjohann 2002).

For the TCS observed in the near Earth region $x > -15R_{\mathrm{E}}$, the approximation of weak gradients $\partial B_z/\partial x$ becomes unrealistic. Moreover, this region is characterized by frequent transient increases of B_z corresponding to depolarization fronts (Nakamura et al. 2009; Runov et al. 2009; Saito et al. 2010). Therefore, structure of the TCS in this region can sometimes include local enhancement of B_z, so called plasma-bubbles (see Sergeev et al. 1992; Sitnov et al. 2007, and references therein) which drastically change the TCS stability criteria even for simplified Vlasov models (Sitnov and Schindler 2010) and single fluid models (Erkaev et al. 2007; Korovinskiy et al. 2011). Strong (and sometimes reversed) gradients $\partial B_z/\partial x$ lead to substantial modification of the current sheet physics and new instability modes can occur, e.g. ballooning mode with substantial role of the electron dynamics (Pritchett and Coroniti 2010; Pritchett and Coroniti 2011). This mode seems to be responsible for the TCS sausage-like oscillations (Panov et al. 2012a) and the subsequent magnetic reconnections (Panov et al. 2012b).

7 Conclusions

More than ten years of the magnetotail Cluster observation demonstrated that protons at Speiser trajectories play the principal role in the TCS formation and dynamics for $x < -15R_{\mathrm{E}}$. Thus, the new class of TCS models based on the conservation of I_z invariant for proton (ion) description as well as effects of the electron temperature anisotropy should be considered as the most adequate for investigation of the TCS structure and dynamics in this region. However, estimates of the longitudinal structure of the TCS at $x > -15R_{\mathrm{E}}$ and a couple of years of the THEMIS mission provide the new challenge for the application of the TCS theory to the near-Earth region. The kinetic description of current sheets embedded into strong dipole field with substantial role of the gradient $\partial B_z/\partial x$ is necessary. These new models could be considered as generalizations of the early model proposed by Whipple et al. (1991).

Acknowledgements Authors are grateful to A. Petrukovich for fruitful discussions, and to D. Vainshtein for valuable comments and suggestions. The work was supported by RFBR (Nos. 10-02-9311, 12-02-91158), by grants of Leading Schools NIII-623.2012.2, and by the Program (OFN-15) of the Division of Physical Sciences of the Russian Academy of Sciences. The authors thank the International Space Science Institute (ISSI) and the organizing committee for support and the opportunity to participate in the ISSI Workshop on Microphysics of cosmic plasmas.

References

W. Alpers, Steady state charge neutral models of the magnetopause. Astrophys. Space Sci. **5**, 425–437 (1969)

V. Angelopoulos et al., Characteristics of ion flow in the quiet state of the inner plasma sheet. Geophys. Res. Lett. **20**, 1711–1714 (1993). doi:10.1029/93GL00847

V. Angelopoulos et al., Tail reconnection triggering substorm onset. Science **321**, 931–935 (2008). doi:10.1126/science.1160495

A.V. Artemyev, A model of one-dimensional current sheet with parallel currents and normal component of magnetic field. Phys. Plasmas **18**(2), 022104 (2011). doi:10.1063/1.3552141

A.V. Artemyev et al., Comparison of multi-point measurements of current sheet structure and analytical models. Ann. Geophys. **26**, 2749–2758 (2008a)
A.V. Artemyev et al., Effect of the normal component of the magnetic field on the kink instability of the Earth's magnetospheric current sheet. Plasma Phys. Rep. **34**, 771–779 (2008b). doi:10.1134/S1063780X08090110
A.V. Artemyev et al., Thin embedded current sheets: cluster observations of ion kinetic structure and analytical models. Ann. Geophys. **27**, 4075–4087 (2009)
A.V. Artemyev et al., Proton velocity distribution in thin current sheets: cluster observations and theory of transient trajectories. J. Geophys. Res. **115**, 12255 (2010). doi:10.1029/2010JA015702
A.V. Artemyev et al., Cluster statistics of thin current sheets in the earth magnetotail: specifics of the dawn flank, proton temperature profiles and electrostatic effects. J. Geophys. Res. **116**, 0923 (2011a). doi:10.1029/2011JA016801
A.V. Artemyev et al., Hot electrons as tracers of large-scale structure of magnetotail current sheets. Geophys. Res. Lett. **38**, 14102 (2011b). doi:10.1029/2011GL047979
A.V. Artemyev et al., Adiabatic electron heating in the magnetotail current sheet: cluster observations and analytical models. J. Geophys. Res. **117**, 06219 (2012a). doi:10.1029/2012JA017513
A.V. Artemyev et al., Drift modes of a quasi-two-dimensional current sheet. Plasma Phys. Rep. **38**, 207–218 (2012b). doi:10.1134/S1063780X1202002X
Y. Asano et al., Statistical study of thin current sheet evolution around substorm onset. J. Geophys. Res. **109**, 5213 (2004). doi:10.1029/2004JA010413
Y. Asano et al., How typical are atypical current sheets? Geophys. Res. Lett. **32**, 3108 (2005). doi:10.1029/2004GL021834
M. Ashour-Abdalla, J. Büchner, L.M. Zelenyi, The quasi-adiabatic ion distribution in the central plasma sheet and its boundary layer. J. Geophys. Res. **96**, 1601–1609 (1991). doi:10.1029/90JA01921
M. Ashour-Abdalla et al., Consequences of magnetotail ion dynamics. J. Geophys. Res. **99**, 14891–14916 (1994). doi:10.1029/94JA00141
D.N. Baker et al., Neutral line model of substorms: past results and present view. J. Geophys. Res. **101**, 12975–13010 (1996). doi:10.1029/95JA03753
B.M. Ball et al., Nonadiabatic orbit features in ion distribution functions of fast flow magnetotail configurations. J. Geophys. Res. **110**, 4208 (2005). doi:10.1029/2004JA010676
W. Baumjohann, Modes of convection in the magnetotail. Phys. Plasmas **9**, 3665–3667 (2002)
W. Baumjohann et al., Substorm dipolarization and recovery. J. Geophys. Res. **104**, 24995–25000 (1999). doi:10.1029/1999JA900282
W. Baumjohann et al., Dynamics of thin current sheets: cluster observations. Ann. Geophys. **25**, 1365–1389 (2007)
M.K. Bird, D.B. Beard, The self-consistent geomagnetic tail under static conditions. Planet. Space Sci. **20**, 2057–2072 (1972). doi:10.1016/0032-0633(72)90062-1
J. Birn, K. Schindler, M. Hesse, Thin electron current sheets and their relation to auroral potentials. J. Geophys. Res. **109**, 2217 (2004). doi:10.1029/2003JA010303
J. Birn, R. Sommer, K. Schindler, Open and closed magnetospheric tail configurations and their stability. Astrophys. Space Sci. **35**, 389–402 (1975). doi:10.1007/BF00637005
J. Büchner, L.M. Zelenyi, Deterministic chaos in the dynamics of charged particles near a magnetic field reversal. Phys. Lett. A **118**, 395–399 (1986)
J. Büchner, L.M. Zelenyi, Regular and chaotic charged particle motion in magnetotaillike field reversals. I—Basic theory of trapped motion. J. Geophys. Res. **94**, 11821–11842 (1989). doi:10.1029/JA094iA09p11821
G.R. Burkhart, J. Chen, Particle motion in x-dependent Harris-like magnetotail models. J. Geophys. Res. **98**, 89–97 (1993). doi:10.1029/92JA01528
G.R. Burkhart et al., A particle model for magnetotail neutral sheet equilibria. J. Geophys. Res. **97**, 13799–13815 (1992a). doi:10.1029/92JA00495
G.R. Burkhart et al., Ion tearing in a magnetotail configuration with an embedded thin current sheet. J. Geophys. Res. **97**, 16749–16756 (1992b). doi:10.1029/92JA01523
E. Camporeale, G. Lapenta, Model of bifurcated current sheets in the Earth's magnetotail: equilibrium and stability. J. Geophys. Res. **110**, 7206 (2005). doi:10.1029/2004JA010779
J.R. Cary, A.J. Brizard, Hamiltonian theory of guiding-center motion. Rev. Mod. Phys. **81**, 693–738 (2009). doi:10.1103/RevModPhys.81.693
P.J. Channell, Exact Vlasov-Maxwell equilibria with sheared magnetic fields. Phys. Fluids **19**, 1541–1545 (1976). doi:10.1063/1.861357
B. Coppi, G. Laval, R. Pellat, Dynamics of the geomagnetic tail. Phys. Rev. Lett. **16**, 1207–1210 (1966)
S.W.H. Cowley, The effect of pressure anisotropy on the equilibrium structure of magnetic current sheets. Planet. Space Sci. **26**, 1037–1061 (1978)

S.W.H. Cowley, R. Pellat, A note on adiabatic solutions of the one-dimensional current sheet problem. Planet. Space Sci. **27**, 265–271 (1979)
E.A. Davey et al., The orientation and current density of the magnetotail current sheet: a statistical study of the effect of geomagnetic conditions. J. Geophys. Res. **117**, 7217 (2012). doi:10.1029/2012JA017715
A. Divin et al., Model of electron pressure anisotropy in the electron diffusion region of collisionless magnetic reconnection. Phys. Plasmas **17**(12), 122102 (2010). doi:10.1063/1.3521576
J.W. Eastwood, Consistency of fields and particle motion in the 'Speiser' model of the current sheet. Planet. Space Sci. **20**, 1555–1568 (1972)
J.W. Eastwood, The warm current sheet model, and its implications on the temporal behavior of the geomagnetic tail. Planet. Space Sci. **22**, 1641–1668 (1974)
N.V. Erkaev, V.S. Semenov, H.K. Biernat, Magnetic double-gradient instability and flapping waves in a current sheet. Phys. Rev. Lett. **99**(23), 235003 (2007). doi:10.1103/PhysRevLett.99.235003
P. Francfort, R. Pellat, Magnetic merging in collisionless plasmas. Geophys. Res. Lett. **3**, 433–436 (1976). doi:10.1029/GL003i008p00433
A.G. Frank, Dynamics of current sheets underlying flare-type events in magnetized plasmas. Phys. Usp. **53**, 941–947 (2010). doi:10.3367/UFNe.0180.201009h.0982
A.G. Frank, N.P. Kyrie, S.N. Satunin, Plasma dynamics in laboratory-produced current sheets. Phys. Plasmas **18**(11), 111209 (2011). doi:10.1063/1.3647576
A.A. Galeev, Reconnection in the magnetotail. Space Sci. Rev. **23**, 411–425 (1979)
A.A. Galeev, L.M. Zelenyi, Tearing instability in plasma configurations. Sov. Phys. JETP **43**, 1113 (1976)
V. Génot et al., Bifurcated current sheet: model and cluster observations. Planet. Space Sci. **53**, 229–235 (2005). doi:10.1016/j.pss.2004.09.048
H. Grad, Boundary layer between a plasma and a magnetic field. Phys. Fluids **4**, 1366–1375 (1961)
J.E.M. Hamilton, J.W. Eastwood, The effect of a normal magnetic field component on current sheet stability. Planet. Space Sci. **30**, 293–305 (1982)
E.G. Harris, On a plasma sheet separating regions of oppositely directed magnetic field. Nuovo Cimento **23**, 115–123 (1962)
M.G. Harrison, T. Neukirch, One-dimensional Vlasov-Maxwell equilibrium for the force-free Harris sheet. Phys. Rev. Lett. **102**(13), 135003 (2009). doi:10.1103/PhysRevLett.102.135003
T.W. Hill, Magnetic merging in a collisionless plasma. J. Geophys. Res. **80**, 4689–4699 (1975)
P.L. Israelevich, A.I. Ershkovich, R. Oran, Bifurcation of the tail current sheet in Jovian magnetosphere. Planet. Space Sci. **55**, 2261–2266 (2007). doi:10.1016/j.pss.2007.05.006
J.R. Kan, On the structure of the magnetotail current sheet. J. Geophys. Res. **78**, 3773–3781 (1973). doi:10.1029/JA078i019p03773
R.L. Kaufmann et al., Plasma sheet thickness and electric currents. J. Geophys. Res. **106**, 6179–6194 (2001). doi:10.1029/2000JA000284
D.B. Korovinskiy et al., Kink-like mode of a double gradient instability in a compressible plasma current sheet. Adv. Space Res. **48**, 1531–1536 (2011)
A.P. Kropotkin, V.I. Domrin, Theory of a thin one-dimensional current sheet in collisionless space plasma. J. Geophys. Res. **101**, 19893–19902 (1996). doi:10.1029/96JA01140
A.P. Kropotkin, H.V. Malova, M.I. Sitnov, Self-consistent structure of a thin anisotropic current sheet. J. Geophys. Res. **102**, 22099–22106 (1997). doi:10.1029/97JA01316
L.D. Landau, E.M. Lifshitz, *Mechanics*, vol. 1 (Pergamon, Oxford, 1988)
W.W. Liu, J. Liang, E.F. Donovan, Electrostatic field and ion temperature drop in thin current sheets: a theory. J. Geophys. Res. **115**, 3211 (2010). doi:10.1029/2009JA014359
A.T.Y. Lui, Potential plasma instabilities for substorm expansion onsets. Space Sci. Rev. **113**, 127–206 (2004). doi:10.1023/B:SPAC.0000042942.00362.4e
A.T.Y. Lui et al., Near-Earth substorm features from multiple satellite observations. J. Geophys. Res. **113**, 7 (2008). doi:10.1029/2007JA012738
S.M. Mahajan, Exact and almost exact solutions to the Vlasov-Maxwell system. Phys. Fluids, B Plasma Phys. **1**, 43–54 (1989). doi:10.1063/1.859103
H.V. Malova et al., Thin current sheets in the presence of a guiding magnetic field in Earth's magnetosphere. J. Geophys. Res. **117**, 4212 (2012). doi:10.1029/2011JA017359
D.J. McComas et al., The near-earth cross-tail current sheet—detailed ISEE 1 and 2 case studies. J. Geophys. Res. **91**, 4287–4301 (1986). doi:10.1029/JA091iA04p04287
O.V. Mingalev et al., Asymmetric configurations of a thin current sheet with a constant normal magnetic field component. Plasma Phys. Rep. **35**, 76–83 (2009). doi:10.1134/S1063780X09010097
O.V. Mingalev et al., Kinetic models of current sheets with a sheared magnetic field. Plasma Phys. Rep. **38**, 300–314 (2012). doi:10.1134/S1063780X12030063
A.I. Morozov, L.S. Solov'ev, A kinetic examination of some equilibrium plasma configurations. Sov. Phys. JETP **40**, 1316–1324 (1961)

F. Mottez, Exact nonlinear analytic Vlasov-Maxwell tangential equilibria with arbitrary density and temperature profiles. Phys. Plasmas **10**, 2501–2508 (2003)

F. Mottez, The pressure tensor in tangential equilibria. Ann. Geophys. **22**, 3033–3037 (2004). doi:10.5194/angeo-22-3033-2004

R. Nakamura et al., Cluster observations of an ion-scale current sheet in the magnetotail under the presence of a guide field. J. Geophys. Res. **113**, 7 (2008). doi:10.1029/2007JA012760

R. Nakamura et al., Evolution of dipolarization in the near-Earth current sheet induced by earthward rapid flux transport. Ann. Geophys. **27**, 1743–1754 (2009)

R. Nakamura et al., Thin current sheets in the magnetotail observed by cluster. Space Sci. Rev. **122**, 29–38 (2006). doi:10.1007/s11214-006-6219-1

N.F. Ness, The Earth's magnetic tail. J. Geophys. Res. **70**, 2989–3005 (1965)

R.B. Nicholson, Solution of the Vlasov equations for a plasma in an externally uniform magnetic field. Phys. Fluids **6**, 1581–1586 (1963)

E.V. Panov et al., Kinetic ballooning/interchange instability in a bent plasma sheet. J. Geophys. Res. **117**, A06228 (2012b)

E.V. Panov et al., Two types of tangential magnetopause current sheets: cluster observations and theory. J. Geophys. Res. **116**, 12204 (2011). doi:10.1029/2011JA016860

E.V. Panov et al., Observations of kinetic ballooning/interchange instability signatures in the magnetotail. Geophys. Res. Lett. **39**, 8110 (2012a). doi:10.1029/2012GL051668

G. Paschmann, S.J. Schwartz, *ISSI Book on Analysis Methods for Multi-spacecraft Data*. ESA Special Publication, vol. 449 (2000)

R. Pellat, F.V. Coroniti, P.L. Pritchett, Does ion tearing exist? Geophys. Res. Lett. **18**, 143–146 (1991). doi:10.1029/91GL00123

A.A. Petrukovich, Origins of plasma sheet B_y. J. Geophys. Res. **116**, 7217 (2011). doi:10.1029/2010JA016386

A.A. Petrukovich et al., Oscillatory magnetic flux tube slippage in the plasma sheet. Ann. Geophys. **24**, 1695–1704 (2006)

A.A. Petrukovich et al., Thinning and stretching of the plasma sheet. J. Geophys. Res. **112**, 10213 (2007). doi:10.1029/2007JA012349

A.A. Petrukovich et al., Tailward and Earthward flow onsets observed by cluster in a thin current sheet. J. Geophys. Res. **114**, 9203 (2009). doi:10.1029/2009JA014064

A.A. Petrukovich et al., Embedded current sheets in the Earth magnetotail. J. Geophys. Res. **116**, 25 (2011). doi:10.1029/2010JA015749

E. Priest, T. Forbes, *Magnetic Reconnection* (2000)

P.L. Pritchett, F.V. Coroniti, Formation and stability of the self-consistent one-dimensional tail current sheet. J. Geophys. Res. **97**, 16773–16787 (1992). doi:10.1029/92JA01550

P.L. Pritchett, F.V. Coroniti, A kinetic ballooning/interchange instability in the magnetotail. J. Geophys. Res. **115**, 06301 (2010). doi:10.1029/2009JA014752

P.L. Pritchett, F.V. Coroniti, Plasma sheet disruption by interchange-generated flow intrusions. Geophys. Res. Lett. **381**, 10102 (2011). doi:10.1029/2011GL047527

F.J. Rich, V.M. Vasyliunas, R.A. Wolf, On the balance of stresses in the plasma sheet. J. Geophys. Res. **77**, 4670–4676 (1972). doi:10.1029/JA077i025p04670

Z.J. Rong et al., Statistical survey on the magnetic structure in magnetotail current sheets. J. Geophys. Res. **116**, 9218 (2011). doi:10.1029/2011JA016489

M. Roth, J. de Keyser, M.M. Kuznetsova, Vlasov theory of the equilibrium structure of tangential discontinuities in space plasmas. Space Sci. Rev. **76**, 251–317 (1996)

A. Runov et al., THEMIS observations of an Earthward-propagating dipolarization front. Geophys. Res. Lett. **36**, 14106 (2009). doi:10.1029/2009GL038980

A. Runov et al., Electric current and magnetic field geometry in flapping magnetotail current sheets. Ann. Geophys. **23**, 1391–1403 (2005)

A. Runov et al., Local structure of the magnetotail current sheet: 2001 cluster observations. Ann. Geophys. **24**, 247–262 (2006)

M.H. Saito et al., Spatial profile of magnetic field in the near-Earth plasma sheet prior to dipolarization by THEMIS: feature of minimum B. Geophys. Res. Lett. **37**, 08106 (2010). doi:10.1029/2010GL042813

K. Schindler, A theory of the substorm mechanism. J. Geophys. Res. **79**, 2803–2810 (1974)

K. Schindler, *Physics of Space Plasma Activity* (Cambridge University Press, Cambridge, 2006)

K. Schindler, J. Birn, Models of two-dimensional embedded thin current sheets from Vlasov theory. J. Geophys. Res. **107**, 1193 (2002). doi:10.1029/2001JA000304

K. Schindler, M. Hesse, Conditions for the formation of nongyrotropic current sheets in slowly evolving plasmas. Phys. Plasmas **17**, 082103 (2010). doi:10.1063/1.3464198

K. Schindler, J. Birn, M. Hesse, Kinetic model of electric potentials in localized collisionless plasma structures under steady quasi-gyrotropic conditions. Phys. Plasmas **19**, 082904 (2012). doi:10.1063/1.4747162
V.A. Sergeev et al., A two satellite study of nightside flux transfer events in the plasma sheet. Planet. Space Sci. **40**, 1551–1572 (1992). doi:10.1016/0032-0633(92)90052-P
V.A. Sergeev, V. Angelopoulos, R. Nakamura, Recent advances in understanding substorm dynamics. Geophys. Res. Lett. **39**, 5101 (2012). doi:10.1029/2012GL050859
V.A. Sergeev et al., Structure of the tail plasma/current sheet at $\sim 11R_E$ and its changes in the course of a substorm. J. Geophys. Res. **98**, 17345–17366 (1993). doi:10.1029/93JA01151
V.A. Sergeev et al., Survey of large-amplitude flapping motions in the midtail current sheet. Ann. Geophys. **24**, 2015–2024 (2006)
I.P. Shkarofsky, T.W. Johnston, M.P. Bachnynski, *the Particle Kinetic of Plasmas* (Addison-Wesley, Reading, 1966)
M.I. Sitnov, K. Schindler, Tearing stability of a multiscale magnetotail current sheet. Geophys. Res. Lett. **37**, 8102 (2010). doi:10.1029/2010GL042961
M.I. Sitnov, P.N. Guzdar, M. Swisdak, A model of the bifurcated current sheet. Geophys. Res. Lett. **30**, 45 (2003). doi:10.1029/2003GL017218
M.I. Sitnov, P.N. Guzdar, M. Swisdak, Atypical current sheets and plasma bubbles: a self-consistent kinetic model. Geophys. Res. Lett. **341**, 15101 (2007). doi:10.1029/2007GL029693
M.I. Sitnov et al., Thin current sheet embedded within a thicker plasma sheet: self-consistent kinetic theory. J. Geophys. Res. **105**, 13029–13044 (2000). doi:10.1029/1999JA000431
M.I. Sitnov et al., A model of the bifurcated current sheet: 2. Flapping motions. Geophys. Res. Lett. **31**, 9805 (2004a). doi:10.1029/2004GL019473
M.I. Sitnov et al., Current-driven instabilities in forced current sheets. J. Geophys. Res. **109**, 3205 (2004b). doi:10.1029/2003JA010123
M.I. Sitnov et al., Structure and dynamics of a new class of thin current sheets. J. Geophys. Res. **111**, 8204 (2006). doi:10.1029/2005JA011517
B.U.Ö. Sonnerup, Adiabatic particle orbits in a magnetic null sheet. J. Geophys. Res. **76**, 8211–8222 (1971). doi:10.1029/JA076i034p08211
T.W. Speiser, Particle trajectories in model current sheets, 1, analytical solutions. J. Geophys. Res. **70**, 4219–4226 (1965). doi:10.1029/JZ070i017p04219
T.W. Speiser, N.F. Ness, The neutral sheet in the geomagnetic tail: its motion, equivalent currents, and field line connection through it. J. Geophys. Res. **72**, 131 (1967)
L.C. Steinhauer, M.P. McCarthy, E.C. Whipple, Multifluid model of a one-dimensional steady state magnetotail current sheet. J. Geophys. Res. **113**, 4207 (2008). doi:10.1029/2007JA012578
G.S. Stiles et al., Plasma sheet pressure anisotropies. J. Geophys. Res. **83**, 3166–3172 (1978)
S.I. Syrovatskii, Formation of current sheets in a plasma with a frozen-in strong magnetic field. Sov. JETP **33**, 933 (1971)
S.I. Syrovatskii, Pinch sheets and reconnection in astrophysics. Annu. Rev. Astron. Astrophys. **19**, 163–229 (1981). doi:10.1146/annurev.aa.19.090181.001115
D.L. Vainchtein et al., Quasiadiabatic description of nonlinear particle dynamics in typical magnetotail configurations. Nonlinear Process. Geophys. **12**, 101–115 (2005)
I.Y. Vasko et al., Kinetic models of 2D plane and axially symmetric current sheets. Phys. Plasma (2012, submitted)
E. Whipple, R. Puetter, M. Rosenberg, A two-dimensional, time-dependent, near-Earth magnetotail. Adv. Space Res. **11**, 133–142 (1991). doi:10.1016/0273-1177(91)90024-E
M. Yamada, R. Kulsrud, H. Ji, Magnetic reconnection. Rev. Mod. Phys. **82**, 603–664 (2010). doi:10.1103/RevModPhys.82.603
P.H. Yoon, A.T.Y. Lui, Model of ion- or electron-dominated current sheet. J. Geophys. Res. **109**, 11213 (2004). doi:10.1029/2004JA010555
P.H. Yoon, A.T.Y. Lui, A class of exact two-dimensional kinetic current sheet equilibria. J. Geophys. Res. **110**, 1202 (2005). doi:10.1029/2003JA010308
L.M. Zelenyi, A.V. Artemyev, A.A. Petrukovich, Earthward electric field in the magnetotail: cluster observations and theoretical estimates. Geophys. Res. Lett. **37**, 6105 (2010). doi:10.1029/2009GL042099
L.M. Zelenyi, D.V. Zogin, J. Büchner, Quasiadiabatic dynamics of charged particles in the tail of the magnetosphere. Cosm. Res. **28**, 369–380 (1990)
L.M. Zelenyi et al., Thin and superthin ion current sheets. Quasi-adiabatic and nonadiabatic models. Nonlinear Process. Geophys. **7**, 127–139 (2000)
L.M. Zelenyi et al., Nonlinear equilibrium structure of thin currents sheets: influence of electron pressure anisotropy. Nonlinear Process. Geophys. **11**, 579–587 (2004)

L.M. Zelenyi et al., "Matreshka" model of multilayered current sheet. Geophys. Res. Lett. **33**, 5105 (2006). doi:10.1029/2005GL025117

L.M. Zelenyi et al., Marginal stability of thin current sheets in the Earth's magnetotail. J. Atmos. Sol.-Terr. Phys. **70**, 325–333 (2008). doi:10.1016/j.jastp.2007.08.019

L.M. Zelenyi et al., Low frequency eigenmodes of thin anisotropic current sheets and cluster observations. Ann. Geophys. **27**, 861–868 (2009)

L.M. Zelenyi et al., Metastability of current sheets. Phys. Usp. **53**, 933–941 (2010). doi:10.3367/UFNe.0180.201009g.0973

L.M. Zelenyi et al., Thin current sheets in collisionless plasma: equilibrium structure, plasma instabilities, and particle acceleration. Plasma Phys. Rep. **37**, 118–160 (2011). doi:10.1134/S1063780X1102005X

X. Zhou et al., Thin current sheet in the substorm late growth phase: modeling of THEMIS observations. J. Geophys. Res. **114**, 3223 (2009). doi:10.1029/2008JA013777

DOI 10.1007/978-1-4899-7413-6_14
Reprinted from *Space Science Reviews* Journal, DOI 10.1007/s11214-013-9959-8

Mechanisms of Spontaneous Reconnection: From Magnetospheric to Fusion Plasma

Lev Zelenyi · Anton Artemyev

Received: 3 October 2012 / Accepted: 16 January 2013 / Published online: 1 March 2013

Abstract Very often space plasma is treated as collisionless. We check the validity of this paradigm considering various regimes of tearing mode (spontaneous reconnection) including effects of particle collisions and shear of magnetic field. We briefly describe Pitaevskii's effect of effective modification of collision frequency due to the finite particle Larmor radius in the presence of magnetic field. This effect results in a significant increase of the role of collisionality, especially in a weakly magnetized systems. Another popular paradigm is related with application of MHD description to collisionless or weakly collisional systems. We show, that for current sheets observed in the Earth magnetotail and magnetopause as well as for current sheets formed in Solar corona and in laboratory devices most appropriate is the kinetic semi-collisional tearing regime. Role of "collisions" could play usual Coulomb pair collisions of electrons and ions (e.g. in Solar corona) or effective collisions (scattering) of electrons with the microturbulence wave modes. Transition to real MHD modes requires either very large collisions frequencies and/or very large amplitudes of the magnetic field shear. The largest domain in the parameter space is occupied by the kinetic regimes of tearing mode growth where dissipation is provided either by Landau damping or by real (or effective) collisions.

Keywords Tearing instability · Magnetic reconnection

1 Introduction

Starting from the original paper by Giovanelli (1947) reconnection of magnetic field lines is considered as a main mechanism of magnetic energy dissipation. First MHD models describing quasi-stationary magnetic reconnection already included main elements of this process: diffusion region (Sweet 1958), slow shock waves (Petschek 1964) and thin

L. Zelenyi · A. Artemyev (✉)
Space Research Institute, RAS, Moscow, Russian Federation
e-mail: Ante0226@yandex.ru

L. Zelenyi
e-mail: lzelenyi@iki.rssi.ru

current sheet (Syrovatskii 1966). There are several comprehensive books (Parker 1994; Priest and Forbes 2000; Biskamp 2000; Birn and Priest 2007) devoted to various aspects of magnetic reconnection and related charged particle acceleration. The most of theories can be attributed to the one of two possible approaches: kinetic collisionless approach and fluid resistive approach.

Magnetic reconnection plays an important role in various plasma systems starting from rarefied collisionless plasma of interplanetary medium and planetary magnetospheres and going to a weakly collisional plasma of Solar corona and then to collision dominated plasma of laboratory devices (see review by Yamada et al. 2010). Initialization of the magnetic reconnection corresponds to the instability of current sheet separating magnetic fields with opposite polarities. Therefore, the problem of relationship between collision and collisionless reconnection regimes can be reformulated as a problem of current sheet instabilities in presence of collisions with arbitrary frequency ν_{eff} and of an magnetic shear (also of an arbitrary intensity). It should be noted that as was shown by Coppi et al. (1966b), heuristically even the case with $\nu_{eff} = 0$ could be reduced to collisional if one will take into account that Landau damping (providing necessary dissipation for the case with $\nu_{eff} \to 0$) could be roughly considered as supporting effective scattering of electrons with frequency $\nu_{eff} \sim v_{T_e}/\lambda$, where λ is the wavelength of the mode and v_{T_e} is electron thermal velocity.

First paper devoted to instability of current sheet relative to the tearing mode was written by Furth (1962). In this paper the stability of neutral current sheet with magnetic field reversal was considered relative to a periodical fluctuation of a normal component of magnetic field. The further development of theory of the tearing instability includes effects of electric field perturbation and effect of magnetic field shear (Laval et al. 1966). Investigation of the tearing mode based on energy variation principle was developed for current sheets (Schindler and Soop 1968) and generalized in the recent monograph by Schindler (2006).

Application of the tearing instability to collisionless plasma of the Earth magnetotail was done in pioneering work by Coppi et al. (1966b). Further investigations have shown that the principal role for this instability is played by the finite normal component of magnetic field, which magnetizes electrons and destroy corresponding Landau resonant damping (Schindler 1974; Galeev and Zelenyi 1976). Magnetized electrons provide the effect of tearing stabilization due to combination of the frozen-in condition and condition of quasi-neutrality. This effect could be so strong that the spontaneous reconnection mode will be stable for the entire parameter range (Pellat et al. 1991).

Stabilization of the magnetotail current sheet contradicts to numerous observations of magnetic reconnection (see, e.g., Angelopoulos et al. 2008). This problem can be solved by choice of proper initial equilibrium, which describes magnetotail current sheet with a number of additional realistic effects. For example it was shown that embedded thin current sheets often observed in the downtail are unstable relative to the tearing mode (Zelenyi et al. 2008, 2010). Alternative idea corresponds to current sheet with the reversed longitudinal gradient of the normal component of magnetic field. Such current sheets could become unstable relative to the tearing mode (Sitnov and Schindler 2010) or to a more exotic instabilities also resulting in magnetic reconnection (Pritchett and Coroniti 2011).

Although, stability problem for the magnetotail current sheet is of primary importance, in this review we consider mainly the stability of current sheet without normal component of magnetic field, but in presence of a shear magnetic field component having an arbitrary intensity. Therefore our consideration deals with current sheet of planetary magnetopause. Collisions (which is the primary goal of our paper) weakly influence the properties of equilibrium current sheet solutions, but once the system unstable—collisions strongly control the rate of instability growth. We will combine in our analysis effects of usual Coulomb collisions and effects of particle scattering at microturbulent fluctuations ("effective" collisions). These effects correspond to plasma microturbulence and can be important for current sheets in Solar corona and laboratory experiments.

2 Initial Equilibria

Harris current sheet (Harris 1962) can be considered as the simplest kinetic equilibrium model describing the basic properties of space current sheets (in absence of the normal component of magnetic field). This model corresponds to the velocity distribution f_α for particles with mass m_α and charge q_α introduced as shifted Maxwellian distribution:

$$f_\alpha = C_{0\alpha} \exp\bigl(-(H_\alpha - v_{D_\alpha} P_y)/T_\alpha\bigr) = C_\alpha N(z) \exp\bigl(-\bigl(v_x^2 + v_z^2 + (v_y - v_{D_\alpha})^2\bigr)/v_{T_\alpha}\bigr)$$

where $H_\alpha = m_\alpha(v_x^2 + v_y^2 + v_z^2)/2$ is particle energy, $P_y = m_\alpha v_y + (q_\alpha/c)A_y$ is generalized momentum, v_{D_α} is a constant particle drift velocity, $T_\alpha = m_\alpha v_{T_\alpha}^2/2$ is particle temperature, $C_\alpha = n_0(2\pi v_{T_\alpha})^{-3/2}$ is the constant of normalization with particle density in the central region of current sheet, n_0. Distribution of particle density is defined by function $N(z) = \cosh^{-2}(z/L)$, where L is the current sheet thickness. Here α denotes type of particles: $\alpha = i$ for ions and $\alpha = e$ for electrons.

Density of cross-tail current supported by distribution function f_α is $j_y(z) = (4\pi \times B_0/Lc)\cosh^{-2}(z/L)$ and resulting magnetic field acquires the simple form $B_x = B_0 \tanh(z/L)$. Therefore, in the central region of current sheet $z \sim 0$ magnetic field B_x changes sign. This region is filled by particles crossing $z = 0$ and oscillating in nonlinear potential. Corresponding equation of particle motion across current sheet has the form $\ddot{z} \approx -z(const - z^2)$ (Sonnerup 1971). Unmagnetized particles are trapped inside the region $|z| < R_\alpha$, where $R_\alpha = \sqrt{L\rho_\alpha}$ with $\rho_\alpha = v_{T_\alpha}/\Omega_\alpha$ and $\Omega_\alpha = |q_\alpha|B_0/m_\alpha c$ (Dobrowolny 1968). Reflecting from magnetic "walls" $z = \pm R_\alpha$, these particles move along current sheet plane and can therefore interact with unstable waves accordingly to the Landau mechanism (see review by Galeev 1979, and references therein). Several recent investigation of tearing and drift instabilities were devoted to the precise calculations of the impact of these resonant particles (see, e.g., Lapenta and Brackbill 1997; Daughton 1999; Daughton and Karimabadi 2005; Karimabadi et al. 2005).

It can be noticed, that the velocity distribution of particles in the Harris current sheet remains the same even if the guide component of magnetic field $B_y = const$ is applied to the sheet. To take into account inhomogeneous $B_y = B_y(z)$ one needs to consider $j_x(z)$ current and corresponding modification of the velocity distribution (see review by Roth et al. 1996, and references therein). However, for simplified geometry with $B_y = const$ we can restrict our analysis by the modified Harris equilibrium distribution.

We consider below normal component $B_z \neq 0$ only for illustrating effect of modification of collision frequency (so called Pitaevskii effect). This effect could be very important for weakly magnetized and weakly collisional plasmas. One should note, however, that presence of $B_z \neq 0$ brings principal topological change to the system configuration and results in appearance of very different plasma equilibria (see models of 1D current sheet with $B_z \neq 0$ in Kropotkin et al. 1997; Sitnov et al. 2000; Zelenyi et al. 2000). Modification of the current sheet and corresponding accompanying effects are described in review by Zelenyi et al. (2011).

3 Dissipative Effects

Besides the Landau kinetic mechanism of dissipation providing the growth of certain wave modes in the current sheets, there are more standard effects of direct energy dissipation due to effective (turbulent conductivity) or Coulomb particle collisions. For magnetospheric plasma system certain role in generation of effective conductivity can be played by whistler waves (see, e.g., Deng and Matsumoto 2001), by Alfven-whistler mode (Huang and et al. 2012), and by lower-hybrid waves (see Huba et al. 1977; Fujimoto et al. 2011). Independently of dispersion of waves forming turbulence, their interaction with particles can be approximately considered as effective collisions with certain collision frequency ν_{eff}. For example, collision frequency provided by weak lower-hybrid turbulence can be described by expression

$$\nu_{eff} = \omega_{LH}\left(\frac{1}{2}\rho_i |d \ln N/dz|\right)^3 (T_i/T_e)^2$$

where ω_{LH} is lower-hybrid frequency.

To take into account this effect in kinetic model of current sheet instability we use collision integral in Bhatnagar-Gross-Krook (BGK) form (Bhatnagar et al. 1954), which was originally designed for Coulomb collisions. So in the consideration below ν_{eff} could have the meaning either of effective or Coulomb collision frequency. We consider only effect of collisions on instability and neglect by their influence on the initial equilibrium. Collision integral for perturbation of the velocity distribution $f_{1\alpha}$ can be written as

$$\mathrm{St} f_{1\alpha} = -\sum_{\sigma} \nu_{\alpha\sigma}\left(f_{1\alpha} - \frac{f_{0\alpha}}{\int f_{0\alpha} d\mathbf{v}}\left(\int f_{1\alpha} d\mathbf{v} + \frac{m_\alpha}{T_\alpha}\mathbf{v}\int \mathbf{v} f_{1\alpha} d\mathbf{v}\right)\right)$$

where $\nu_{\alpha\sigma}$ is frequency of collisions of α-type and σ-type particles: $\nu_{ie} = \mu\nu_{ei}$ with $\mu = m_e/m_i$. Linearized Vlasov equation gives following expression for $f_{1\alpha}$ (Zelenyi and Taktakishvili 1981):

$$\begin{aligned} f_{1\alpha} = {} & \frac{q_\alpha}{cT_\alpha} f_{0\alpha}\left(A_{1y} v_{D_\alpha} - c\varphi_1 + i\int_{-\infty}^{0}\left(\left(\omega v_y(\tau) - v_{D_\alpha}\nu_\alpha\right)A_{1y} - \omega_\alpha c\varphi_1\right)\varepsilon_\alpha(\tau)d\tau\right) \\ & + \sum_{\sigma} \nu_{\alpha\sigma} f_{0\alpha}\int_{-\infty}^{0}\left(n_{1\alpha} + \mathbf{v}(\tau)\int \mathbf{v} f_{1\alpha} d\mathbf{v}\right) n^{-1}\varepsilon_\alpha(\tau)d\tau \end{aligned} \tag{1}$$

where $n = n_0 N(z)$, $n_{1\alpha} = \int f_{1\alpha} d\mathbf{v}$, $\omega_\alpha = \omega + i\nu_\alpha$, $\nu_\alpha = \sum_\sigma \nu_{\alpha\sigma}$, and

$$\varepsilon_\alpha = \exp\left(-i\omega_\alpha\tau + ik\left(x(\tau) - x\right)\right)$$

Wavenumber of perturbation is k. Condition of quasineutrality $n_{1e} = n_{1i}$ gives the perturbation of the scalar potential φ_1 as a function of the perturbation of the vector potential A_{1y}. Substituting (1) into Maxwell equation we obtain

$$\begin{aligned} & \frac{d^2 A_{1y}}{dz^2} - \left(k^2 + V_0(z) + V^{<}(z)\right)A_{1y} = 0 \\ & V_0(z) + V^{<}(z) = -\frac{4\pi q_\alpha}{c}\sum_{\alpha}\int v_y f_{1\alpha} d\mathbf{v} \end{aligned} \tag{2}$$

where we separate adiabatic impact $V_0 = -2L^{-2}\cosh^{-2}(z/L)$ and resonant impact $V^<$. Term $V^<$ can be taken as zero for $|z| > R_\alpha$. The corresponding dispersion relation could be obtained by matching solutions of (2) in the internal ($|z| < R_\alpha$) and external ($|z| > R_\alpha$) regions. This technique is straightforward and described in details in may early publications (Dobrowolny 1968; Galeev 1979; Zelenyi and Taktakishvili 1981)

$$\frac{L}{2}\int_{-\infty}^{+\infty}\left(V_e^<(z) + V_i^<(z)\right)dx = \frac{1-(kL)^2}{kL} \tag{3}$$

Dispersion relation (3) can be rewritten as

$$\frac{\omega_{pe}^2 L^2}{c^2}\frac{\bar{\gamma}}{\bar{\nu}_e}\left(\sqrt{\frac{\rho_e}{L}}\frac{\mathrm{A}_e(1-\mathrm{A}_i)+\mathrm{A}_i(1-\mathrm{A}_e)}{1-\mathrm{A}_e\mathrm{A}_i} + \sqrt{\frac{\rho_i}{L}}\mathrm{A}_i\right) = \frac{1-(kL)^2}{kL} \tag{4}$$

where we neglect electron current density in the region $R_e < |z| < R_i$ in comparison with ion current and introduce the following notations: ω_{pe} is plasma frequency, $\bar{\gamma} = \gamma/kv_{T_e}$, $\bar{\nu}_e = \nu_e/kv_{T_e}$, and

$$\mathrm{A}_\alpha = -i\nu_{\alpha\sigma}\frac{Z_{0\alpha}}{kv_{T_\alpha}}\left(1 + i\nu_{\alpha\alpha}\frac{Z_{0\alpha}}{kv_{T_\alpha}}\right)^{-1}, \quad \alpha \neq \sigma \tag{5}$$

Plasma integral $Z_{n\alpha}$ (Kramp function) has a form

$$Z_{n\alpha}(\omega_\alpha/kv_{T_\alpha}) = \frac{1}{\sqrt{\pi}}\int_{-\infty}^{+\infty}\frac{x^n e^{-x^2}dx}{x-(\omega_\alpha/kv_{T_\alpha}) - i\vartheta\,\mathrm{sign}k}, \quad \vartheta \to 0$$

There are several cases when analytical solutions of (4) could be obtained.

For very weak collisions $\bar{\nu}_e \sim \nu_i/kv_{T_i} \ll 1$ we have $\omega_e \ll kv_{T_i} \ll kv_{T_e}$. Expansion of $Z_{0\alpha}$ function around zero value of $\omega_\alpha/kv_{T_\alpha}$ gives the solution of (4):

$$\bar{\gamma} = \frac{2}{\sqrt{\pi}}\left(\frac{\rho_e}{L}\right)^{3/2}\frac{1-(kL)^2}{kL} = \bar{\gamma}_{0e} \tag{6}$$

This is classical growth rate of the electron tearing mode in absence of collisions and shear (see, e.g., Galeev and Zelenyi 1976).

For very strong collisions $\bar{\nu}_e \sim \nu_i/kv_{T_i} \gg 1$ we can expand $Z_{0\alpha}$ with $\omega_\alpha/kv_{T_\alpha} \gg 1$ and obtain equation for $\bar{\gamma}$

$$\bar{\gamma} = \bar{\gamma}_{0e}\bar{\nu}_e\left(1 + \frac{\Delta_i}{R_e}\left(1 + \bar{\nu}_e^{-2}\mu^{-1/2} + \bar{\gamma}\mu^{-1}\bar{\nu}_e^{-1}\right)^{-1}\right)^{-1} \tag{7}$$

where $\Delta_i = R_i$ if the mean free path $\lambda_{ei} = v_{T_e}/\nu_{ei}$ is larger than R_i, while $\Delta_i \sim \bar{\nu}_e\mu^{-1/2}\rho_e > R_i$, if the mean free path λ_{ei} is smaller than R_i. Here, therefore, we have three solutions of (7). Together with (6) we have

$$\bar{\gamma} = \bar{\gamma}_{0e}\begin{cases} 1, & \bar{\nu}_e \ll 1 \\ \bar{\nu}_e, & 1 \ll \bar{\nu}_e \ll \mu^{-1/8}, \quad \rho_e/L < \mu^{3/4} \\ \mu^{-1/4}/\bar{\nu}_e, & \mu^{-1/8} \ll \bar{\nu}_e \ll \mu^{-1/4}, \quad \rho_e/L < \mu \\ (\mu L/\rho_e)^{1/2}, & \mu^{1/4}(L/\rho_e)^{1/2} < \bar{\nu}_e \end{cases} \tag{8}$$

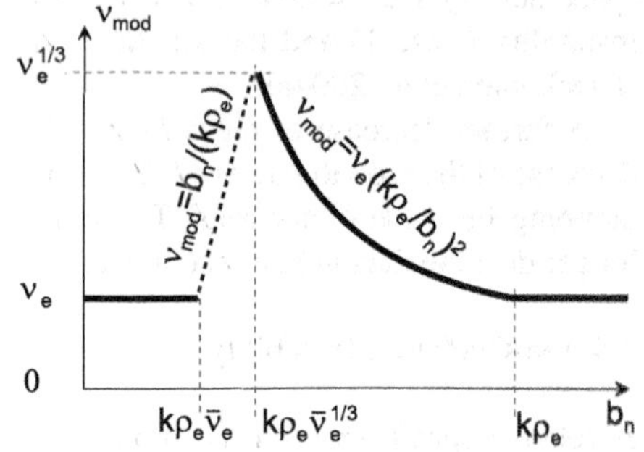

Fig. 1 Modified collision frequency as function of $b_n = B_z/B_0$. Figure is adopted from (Zelenyi and Taktakishvili 1981)

Equation (8) demonstrates how Coulomb (of effective) dissipation replaces the weak and sensitive Landau damping as a driving mechanism of spontaneous reconnection. This effect is especially important if Landau damping is switched off by the influence of a weak normal component of magnetic field and tearing mode becomes linearly stable (Schindler 1974; Galeev and Zelenyi 1976; Pellat et al. 1991; Quest et al. 1996). In this case the presence of collisions provides relatively slow, but persistent growth of reconnecting modes. Resistive mode (similar to the one described by (8) for $B_z = 0$ case, $\bar{\gamma} \sim \bar{\nu}_e$) emerges in this case, even for the modes stable in collisionless regime. These dissipative modes are clearly seen in numerical simulations (Lipatov and Zelenyi 1982), where numerical dissipation due to "shot noise" effects is unfortunately unavoidable.

4 Pitaevskii Effect

It is necessary to note, that the presence of finite normal component of magnetic field in the vicinity of the neutral plane $B_x \approx 0$ results in modification of the collision frequency (so called Pitaevskii effect, see Pitaevskii 1963). The nature of this effect is related with the kinetic character of collision process, which is missed both in τ-approximation and BGK-approach. Strictly speaking, collisions should be described by Landau collisional operator (Pitaevskii and Lifshitz 1981), where all details of distribution function become important (especially gradients of distribution function in a phase space). For short wavelength modes $k\rho_{en} > 1$ ($\rho_{en} = v_{T_e}/\Omega_{en}$ is electron Larmor radius in the vicinity of current sheet neutral plane with electron gyrofrequency in B_z field, Ω_{en}) perturbed distribution $f_{1\alpha} \sim \exp(ikx - i\omega t) \sim \exp(ik\rho_{en})\sin(\Omega_{en}t)$ becomes very inhomogeneous in the phase space and collisions act much more effectively to smoothen it. Pitaevskii took this effect into account and have shown that it could in a first approximation be reduced to the corresponding increase of collision frequency:

$$\bar{\nu}_e \to \bar{\nu}_{mod} = \bar{\nu}_e \times \begin{cases} k^2\rho_{en}^2, & k^2\rho_{en}^2 > 1 \\ 1, & k^2\rho_{en}^2 < 1 \end{cases}$$

Here we take into account that for current sheet geometry $\rho_{en} = \rho_e/b_n$, where $b_n = B_z/B_0$ is the dimensional value of the normal component of the magnetic field. The corresponding modification of ν_e can be found in Fig. 1 and have non-monotonous form.

Modified collision frequency ν_{mod} equals to ν_e for unmagnetized electrons, when $\nu_e < \Omega_{en}$. For strongly magnetized electrons $k\rho_{en} < 1$ (i.e. when the wavelength of perturbations becomes larger than electron gyroradius in B_z field) modified frequency also equals to ν_e. Effect of Pitaevskii starts working in the region with $k\rho_{en} > 1$. In this region modified collision frequency increases with decrease of b_n as $\nu_{mod} \sim b_n^{-2}$ until ν_{mod} is smaller than Ω_{en}. When $\nu_{mod} = \Omega_{en}$ (i.e. $b_n = k\rho_e \bar{\nu}_e^{1/3}$), effect of Pitaevskii stops working. We decided to remind about this effect, because it could be important even for a weakly collisional plasma in a weak magnetic field characteristic for interplanetary (interstellar) space and planetary magnetotails. Simple estimates of Reinolds numbers without taking into account the kinetics of collision process could significantly underestimate their role.

5 Effects of Magnetic Field Shear and Collisional Dissipation

Very often plasma configurations in space and laboratory have the additional component of magnetic field along the current supporting configuration. This component (toroidal in fusion devices) could have small (magnetotail) or large (magnetopause) values, so we will consider the general case when it could have an arbitrary value. In presence of a finite magnetic field B_y motion of particles in the neutral plane can become magnetized by this component. The critical value of B_y for such "magnetization" is defined as: $B_y^* = B_0\rho_\alpha/R_\alpha = B_0\sqrt{\rho_\alpha/L}$. If $B_y < B_y^*$ particles can be considered as unmagnetized, because Larmor radius in B_y is larger than the thickness of the central region of current sheet $|z| < R_\alpha$. For such weak B_y component its influence on system properties could be neglected. For $B_y > B_y^*$ all particles are magnetized (Galeev and Zelenyi 1978; Karimabadi et al. 2005). For these two regimes mechanisms of tearing mode growth are principally different (Drake and Lee 1977; Zelenyi and Taktakishvili 1987). When electrons get magnetized a finite dissipation due to Landau resonance interaction is replaced by the dissipation produced by electron inertia ($m_e \neq 0$).

Here we introduce dimensionless parameter $b_y = B_y/B_0$ and consider regimes of the tearing mode for various values of b_y. In contrast to the system with $B_y = 0$, tearing mode in the current sheet with magnetic field shear is very sensitive to any perturbation of the scalar potential φ (Coppi 1965; Galeev et al. 1986; Daughton and Karimabadi 2005). The spatial domain can be separated into two regions: (1) central region in the vicinity of so called singular surface which is the layer with $k_\parallel = k_x B_x(z)/|\mathbf{B}| = 0$. In this region perturbations of the electrostatic field $-\nabla_\parallel\varphi = -k_\parallel\varphi$ are small and can not compensate perturbation of the inductive field $-c^{-1}\partial A_\parallel/\partial t$, where $A_\parallel = A_{1y}(z)B_y/|\mathbf{B}|$. As a result, a finite electric field $E_\parallel = -\nabla_\parallel\varphi - c^{-1}\partial A_\parallel/\partial t$ exists in the vicinity of singular surface and frozen-in condition breaks down. (2) Outer region where inductive and potential parts of $E_\parallel$ compensate each other ($E_\parallel = 0$) and single fluid-approximation can be used. In the vicinity of the layer with $k_\parallel = 0$ equations for perturbed vector and scalar potentials for the general case with both shear ($b_y \neq 0$) and collisional effects ($\nu \neq 0$) taken into account can be written as (Zelenyi and Taktakishvili 1987)

$$\frac{d^2\varphi}{dz^2} = G(z), \qquad \frac{d^2 A_\parallel}{dz^2} = \frac{\rho_i^2}{2R_i^2 b_y^2}\frac{\omega}{k_\parallel c}G(z) \tag{9}$$

where ω is frequency of perturbation and

$$G(z) = \left(\varphi(z) - \frac{\omega A_\parallel}{k_\parallel c}\right)\frac{2R_i^2}{b_y^2\rho_i^2}\sum_\alpha \frac{Z_{1\alpha}(1+X_\sigma)}{R_\alpha^2 D_\alpha}, \qquad \alpha \neq \sigma$$

$$X_\sigma = Z_{1\alpha}\frac{2i\nu_{\alpha\sigma}\omega}{(k_\parallel v_{T_\alpha})^2 D_\alpha}, \qquad \alpha \neq \sigma$$

$$D_\alpha = 1 + \frac{i\nu_\alpha Z_{0\alpha}}{k_\| v_{T_\alpha}} + \frac{2i\nu_{\alpha\sigma}\omega Z_{1\alpha}}{(k_a v_{T_\alpha})^2}$$

Argument of $Z_{n\alpha}$ function is $\omega/k_\| v_{T_\alpha}$. To derive system of (9) we took into account several assumptions: (1) we neglect particle drift v_{D_α} in the vicinity of the layer $k_\| = 0$; (2) we assume that $d^2/dz^2 \gg k^2$ and $d^2/dz^2 \gg V_0(z)$; (3) in the vicinity of the layer $k_\| = 0$ we assume $A_\| \approx const \neq 0$, but we keep terms $d^2 A_\|/dz^2$; (4) we assume that $\rho_i/b_y \gg R_i$, R_e ($B_y < B_y^*$), i.e. $b_y \ll \sqrt{\rho_i/L}$ (but $b_y \gg \sqrt{\rho_e/L}$ for an external solution).

One can get good physical insight to the problem considering the Doppler-shift of perturbation for collisionless regime with $\gamma > \nu_e$: $\omega' = \omega - \omega_D$, where $\omega_D = k_\| v_\| \approx k_\| v_{T_e}$. While for collisional case $\gamma < \nu_e$ particles motion resembles the diffusion along magnetic field lines. In this case Doppler-shift can be written as $\omega_D \approx k_\|^2 v_{T_e}^2/\nu_e$ (Drake and Lee 1977). If the value of Doppler-shift is much smaller than time scale of electric field variation ($\omega_D \ll \omega$), particle can be accelerated by $E_\|$ in the vicinity of the layer $k_\| = 0$. Condition $\omega_D \ll \omega$ defines the width of the singular region:

$$\Delta_s = \begin{cases} \Delta_s^0 = \bar{\gamma} b_y L, & \gamma \gg \nu_e \\ \Delta_s^c = \sqrt{\bar{\gamma}\bar{\nu}_e} b_y L, & \gamma \ll \nu_e \end{cases}$$

System (9) determines the dispersion relation valid for $\Delta_s \ll L$:

$$\frac{1-(kL)^2}{kL} = \frac{L}{A_\|}\int_{-\infty}^{+\infty} \frac{d^2 A_\|}{dz^2} dz \tag{10}$$

This dispersion relation determines the growth rates for all regimes of the tearing mode depending on b_y and ν_e. The important role is played by relation between scales Δ_s and δ_φ, where δ_φ defines the scale of φ variation, i.e. perturbations of vector potential $A_\|$ are not compensated by perturbations of the scalar potential in the domain $|z| < \delta_\varphi$. For systems with $\Delta_s < \delta_\varphi$ kinetic regime of tearing mode is provided by resonant collisionless or collisional interaction with particles in the region $|z| < \Delta_s$ (see left panel of the scheme in Fig. 2). Electrostatic effects become important already outside the region of strong interaction of waves with electrons $|z| < \Delta_s$. For $|z| > \Delta_s$ Doppler shift $\sim k_\|(z) v_\| \sim k_\|(z) v_{T_e}$ strongly reduces the resulting value of the perturbed current $d^2 A_\|/dz^2 \sim j_\|$.

For the opposite case $\Delta_s > \delta_\phi$ (see right panel of the scheme in Fig. 2) electrostatic effects control the evolution of the system because the width of interaction region depends on the width of the domain, where the frozen in condition $E_\| = i\omega A_\|/c - ik_\|\varphi = 0$ is violated. For $|z| > \delta_\varphi$ $E_\| \to 0$ and interaction for the cases with $\Delta_s > \delta_\phi$ occurs in MHD regime, when the dissipation could be provided either by collisional ($\sim\nu_e$) or inertial ($\sim m_e$) resistivities.

Below we consider two different regimes of the tearing mode: (1) MHD regime, when inertia or resistivity produce perturbation of current density with spatial scale exceeding ion Larmor radius $\rho_{yi} = \rho_i/b_y$ ($\delta_\varphi \sim \rho_{yi} < \Delta_s$); (2) kinetic regime, when spatial scales of current perturbation are smaller than ion Larmor radius ($\delta_\varphi \sim \rho_{yi} > \Delta_s$).

5.1 Collisionless Systems

If collisional frequency is small ($\nu_e \ll \gamma$) one can neglect the real part of the frequency of perturbation ($\mathrm{Re}\omega = 0$, $\mathrm{Im}\omega = \gamma$) and consider only electron input to the growth of perturbations. In this case system (9) takes a form

$$\frac{d^2\varphi}{dz^2} = \frac{2T_i}{\rho_{yi}^2 T_e}\left(\varphi - \frac{\omega}{k_\| c} A_\|\right) Z_{1e}, \qquad \frac{d^2 A_\|}{dz^2} = \frac{T_i}{R_i^2 T_e}\frac{\omega}{k_\| c}\left(\varphi - \frac{\omega}{k_\| c} A_\|\right) Z_{1e} \tag{11}$$

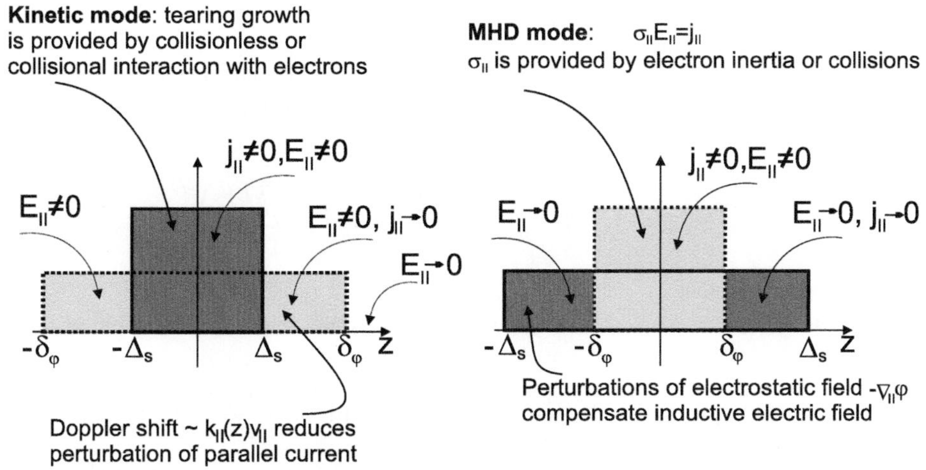

Fig. 2 Schematic view of relation between spatial scales Δ_s and δ_ϕ in kinetic and MHD regimes

For kinetic regime ($\varphi \sim 0$) we could substitute expression for $d^2A_\parallel/dz^2$ (11) into dispersion relation (10) with additional condition $Z_{1e} \approx 1$ valid for $|z| \gg \Delta_s^0$ to obtain growth rate (see Laval et al. 1966)

$$\bar{\gamma}_{01} = \frac{1}{\sqrt{\pi}}\left(1+\frac{T_i}{T_e}\right)\frac{1-(kL)^2}{kL}\left(\frac{\rho_e}{L}\right)^2\frac{1}{b_y}$$

First equation of system (11) gives estimate of the spatial scale $\delta_\varphi \approx \rho_{yi}\sqrt{T_i/2T_e}$. Then limit of this regime is defined by equation $\delta_\varphi \approx \Delta_s$: $b_y < (L/\rho_e)\sqrt{1/2\mu}$.

For MHD regime $\delta_\varphi < \Delta_s$ we can use expansion $Z_{1e} \approx (1/2)(z/\Delta_s)^2$ in the region $|z| < \Delta_s^0$. Substitution of expression for $d^2A_\parallel/dz^2$ (11) into dispersion relation (10) gives (Zelenyi and Taktakishvili 1987)

$$\bar{\gamma}_{02} = \sqrt{\mu}\left(\frac{\rho_e}{L}\right)^3\left(4\left(1+\frac{T_i}{T_e}\right)\frac{T_i}{T_e I}\frac{1-(kL)^2}{kL}\right)^2 \tag{12}$$

with $I = 2\pi\Gamma(3/4)/\Gamma(1/4)$. It is worth to notice, that these two regimes match at demagnetization point b_y^*, i.e. the ratio $\bar{\gamma}_{01}/\bar{\gamma}_{01}$ is some constant around unity, when $b_y = b_y^* = (L/\rho_e)\sqrt{1/2\mu}$ (Galeev and Zelenyi 1977). For the first time this instability (inertial MHD tearing mode) was found in the early paper by Coppi (1965). One can see that the such mode could exist only in the very exotic case $b_y > \sqrt{m_i/m_e}$ ($L \sim \rho_i$) or equivalently for extremely small plasma beta $\beta < m_e/m_i$.

5.2 Collisional Systems

For the case with strong collisions $\nu_\alpha/k_\parallel v_{T_\alpha} \gg 1$ system (9) can be rewritten as

$$\begin{aligned}\rho_{yi}^2\frac{d^2\varphi}{dz^2} &= \frac{2T_i}{T_e}\left(\varphi-\frac{\omega}{k_\parallel c}A_\parallel\right)\frac{k_\parallel^2 v_{T_e}^2}{2\gamma\nu_e + k_\parallel^2 v_{T_e}^2}\\ R_e^2\frac{d^2A_\parallel}{dz^2} &= \frac{\omega}{k_\parallel c}\left(\varphi-\frac{\omega}{k_\parallel c}A_\parallel\right)\frac{k_\parallel^2 v_{T_e}^2}{2\gamma\nu_e + k_\parallel^2 v_{T_e}^2}\end{aligned} \tag{13}$$

For kinetic regime ($\delta_\varphi > \Delta_s$ and $\varphi \to 0$) the solution of the first equation of system (13) in the region $|z| > \Delta_s^0$ is the same as the solution of the first equation of collisionless system (11). As a result, we obtain estimates of $\delta_\varphi \approx \rho_{yi}\sqrt{T_i/2T_e}$. Second equation of system (13) after substitution into (10) with $\varphi = 0$ gives the growth rate for so called semi-collision mode (Drake and Lee 1977):

$$\bar{\gamma}_{sc} = \bar{\nu}_e^{1/3}\left(\frac{\rho_e}{L}\right)^{4/3}\frac{1}{\pi^{1/3}b_y^{2/3}}\left(\left(1+\frac{T_i}{T_e}\right)\frac{1-(kL)^2}{kL}\right)^{2/3}$$

Growth rate $\bar{\gamma}_{sc}$ matches $\bar{\gamma}_{01}$ at $\bar{\nu}_e = (\rho_e/L)^2/b_y$.

For MHD regime ($\delta_\varphi < \Delta_s$) we can obtain the solution of the first equation of system (13):

$$\varphi(z) = \frac{\omega}{k_\| c}A_\| \frac{z^2}{4\delta_\varphi \Delta_s^c}\int_0^{\pi/2}\sqrt{\sin\theta}\exp\left(-\frac{z^2\cos\theta}{4\delta_\varphi \Delta_s^c}\right)d\theta$$

Substituting $\varphi(z)$ into the second equation of system (13) we obtain the expression for $A_\|$, which can be substituted into (10). As a result, we obtain growth rate of the well-known resistive Furth-Killeen-Rosenbluth mode (Furth et al. 1963):

$$\bar{\gamma}_{FKR} = \bar{\nu}_e^{3/5}\left(\frac{\rho_e}{L}\right)^{6/5}\mu^{1/5}\left(\frac{1-(kL)^2}{kL}\frac{2(T_i+T_e)}{T_e I}\right)^{4/5}$$

with $I = 2\pi\Gamma(3/4)/\Gamma(1/4)$. Growth rates γ_{sc} and γ_{FKR} match at $\Delta_s^c \sim \rho_{yi}$.

5.3 Role of B_y

Collisionless growth rate for resonant tearing mode in neutral current sheet with $B_y = 0$ was estimates as $\bar{\gamma}_{0e} \sim (\rho_e/L)^{3/2}$ (Coppi et al. 1966b). Therefore, estimates of collisionless growth rate for inertial mode $\bar{\gamma}_{01} \sim (\rho_e/L)^2/b_y$ with $B_y \neq 0$ becomes equal to $\bar{\gamma}_{0e}$ for $b_y = (\rho_e/L)^{1/2} \ll 1$. This value of b_y corresponds to magnetization of electrons ($B_y = B_y^*$), i.e. $\rho_e/b_y = R_e$ for $b_y = (\rho_e/L)^{1/2}$. For system with $b_y > (\rho_e/L)^{1/2}$ growth rate is determined by electron inertial resistivity and described by expression $\gamma_{01} \sim (\rho_e/L)^2/b_y$ until frequency of collisions is small enough $\bar{\nu}_e < \bar{\gamma}_{01}$. When $\bar{\nu}_e = \bar{\gamma}_{01}$ and $b_y > (\rho_e/L)^{1/2}$ (i.e. $\bar{\nu}_e > (\rho_e/L)^{3/2}$) the semi-collisional regime, where inertial resistivity is replaced by the collisional one (Drake and Lee 1977) establishes with $\bar{\gamma}_{sc} \sim \bar{\nu}_e^{1/3}/b_y^{2/3}$. As we mention above for very large $b_y > (L/\rho_e)\sqrt{1/2\mu}$ collisionless kinetic inertial mode transforms to MHD inertial mode and growth rates also match quite well at $B_y = B_y^*$. Therefore, we have dependence of the growth rate on b_y, where growth rate for intermediate regime $b_y \sim (\rho_e/L)^{1/2}$ can be obtained only by numerical solution of the corresponding dispersion equation (see Zelenyi and Taktakishvili 1987). So, we see that the interplay of different mictroscales of spontaneous reconnection process (scales of resonant or collisional electron interaction Δ_s, scale of the violation of the frozen in condition, δ_ϕ) determines the real modes and mechanisms of its operating.

6 Discussion and Conclusions

The general character of the growth rates for spontaneous reconnection modes as function of ν_e and b_y is shown in Fig. 3, where we also indicate parameter regions for various current sheets in different conditions existing in space and laboratory plasmas. Magnetotail current sheet is typically characterized by small value of magnetic shear $b_y \ll 1$ (Petrukovich 2011). The level of high-frequency wave activity responsible for effective collisions is also weak in this region (Coroniti 1985; Eastwood et al. 2009; Fujimoto et al. 2011). Therefore "MT" domain is situated at a lower left corner of Fig. 3 in $\bar{\nu}_e \ll 1$, $b_y \ll 1$ region. However, this domain could be extended by including current sheets observed under active conditions. For example, in the vicinity of the reconnection regions, where secondary X-lines can be formed due to the tearing instability of current sheet located in the outflow region, component B_y could be relatively strong ($B_y \sim B_0$, see, e.g., Nakamura and et al. 2008; Wang and et al. 2012). Moreover, in case of strong B_y electric field fluctuations related to flows of accelerated particles are often observed in outflow region supporting increase of effective collisions ν_e (Huang and et al. 2012) in agreement with theoretical estimates (Yoon and Lui 2006). However, for weak values of B_y the intensity of these effective collisions is low to be responsible for reconnection (Eastwood et al. 2009), and corresponding MT domain can be expanded to semicollisional region only for sufficiently large B_y. Here we also can mention alternative source of effective conductivity corresponding to stochastic ion and electron motion in the current sheets (Horton and Tajima 1990; Numata and Yoshida 2002). For magnetopause current sheet shear of magnetic field is often strong enough to provide $b_y > 1$ (Berchem and Russell 1982; Panov et al. 2008). Therefore, electrons and ions are magnetized in the vicinity of the neutral plane by B_y. In this case "MP" domain corresponds to the inertial mode with $\gamma = \gamma_{01}$. However, similar to "MT" domain effective collisions due to lower-hybrid and/or ion-cyclotron turbulence (Labelle and Treumann 1988; Panov et al. 2006) could expand "MP" domain up to semi-collisional regime of the tearing mode.

Current sheets detected in Solar corona ("SC" domain) correspond to strong (but finite) shear $m_i/m_e > b_y > 1$ and weak, but finite, collisions (Priest and Forbes 2000; Uzdensky 2003; Birn and Priest 2007). Development of the semi-collision tearing mode in these current sheets results in spontaneous magnetic reconnection (initiating the onset of Solar flares) and the subsequent electron acceleration. Here effective collisions due to high-frequency turbulence could also contribute to the growth rate and help to destabilize current sheet (Büchner 2007). We emphasize that the process of spontaneous reconnection in Solar corona and upper Solar atmosphere is mostly kinetic. Neither the value of shear component b_y, nor the degree of collisionality are strong enough to support this process to be accomplished in MHD regime.

Laboratory devices with relatively cool plasma, where magnetic field configurations with current sheet are produced, are located in Fig. 3 in the domain with strong electron collisions (Frank 2010; Yamada et al. 2010; Frank et al. 2011) enhanced by effective collisions (Ji et al. 2004). Magnetic reconnection in laboratory current sheets due to growth of collisional Furth-Killeen-Rosenbluth tearing mode (Furth et al. 1963) are often observed and described in details (see, e.g., Frank 2010). Moreover, laboratory devices can operate with relatively strong magnetic shear $B_y \sim B_0$ induced initially and growing with development of current sheet (Frank et al. 2005).

Finally tokamaks ("TK" domain) with high-temperature plasma are characterized by strong toroidal field (Wesson 2004; Steinhauer 2011) (strong shear $b_y \gg 1$ in our notations) and moderately strong collisions. Although the degree of collisionality could be enhanced by turbulence (see Budaev et al. 2011), tokamak domain most probably is located as semicollisional regime at Fig. 3 especially for future devices for real hot fusion plasma confinement. For tokamaks tearing instability plays important, although undesirable, role of destruction of magnetic surface (see review Boozer 2012b, and references therein).

Although, the analysis of reconnection mechanisms presented above is more appropriate for description of reconnection in space plasma, here we would like to discuss briefly the comparison with reconnection processes occurring in laboratory and tokamak experiments. The theory of collisional reconnection in tokamaks is originated from paper by Kadomtsev (1975) (see also review Kadomtsev 1987), where the model of resistive reconnection was developed. This model predicts reconnection time $\sim \nu_e^{-1/2}$ and is unable to explain the powerful sawtooth instability related to fast reconnection (von Goeler et al. 1974). Physics of this fast tokamak reconnection is essentially similar to models with large B_y presented in this review. It is believed that $m = 1$ modes in tokamaks results in formation of magnetic islands in the vicinity of a singular layer $\mathbf{k} \cdot \mathbf{B} = 0$ (Rutherford 1973; Rosenbluth et al. 1973; Zakharov 1980). Here the principal role is played by the electron inertia (Wesson 1990), because plasma beta is small enough ($\beta < m_i/m_e$) due to the large shear component of the magnetic field. Weak level of collisions in tokamak plasma leads to the dominance of this inertial mode, which can describe fast sawtooth reconnection (Porcelli 1991). Inertial $m = 1$ mode in tokamaks in principal corresponds to the current sheet thickness of the order of electron inertial length $d \sim m_e^{1/2}$ and develops with the growth rate $\gamma_{m=1} \sim \sqrt{\mu}(\rho_i/d) \sim \sqrt{\mu}$. The same estimates can be obtained for collisionless inertial tearing regime $\gamma_{02} \sim \sqrt{\mu}$ for $L \sim d \sim \sqrt{m_e}$ (see (12)).

Important role of additional effective collisions (or anomalous diffusion) for inertial $m = 1$ mode was considered by Drake and Kleva (1991). Authors have shown that stabilization of $m = 1$ mode due to diamagnetic drifts (effect similar to Doppler-shift effect for collisionless mode, see discussion above and in Drake and Lee (1977)) results in significal reduction of the growth rate. At the same time effective diffusion due to drift instability can provide the increase of reconnection rate. The stabilization effect of diamagnetic drifts for $m = 1$ mode was confirmed by experimental observation (Levinton et al. 1994) and numerical modeling (Zakharov et al. 1993). Other possible candidate for the increase of the tearing growth rate in tokamaks reconnection with large shear magnetic component is the gradient of electron pressure along field lines (Aydemir 1992; Grasso et al. 1999). Presence of the finite electron compressibility results in appearance of nonvanishing parallel electric field in the vicinity of the singular layer. In this case the structure of reconnection region resembles the one shown in Fig. 2 (left panel), where $j_\parallel \neq 0$ domain is embedded into $E_\parallel \neq 0$ domain (Kleva et al. 1995).

General model of two-fluid magnetic reconnection in tokamaks with two limits ($\beta < m_e/m_i$ and $\beta \sim 1$) can be found in Biskamp et al. (1997). In case of small plasma beta electron inertia plays the most important role, while large-β regime corresponds to separation of electron and ion motions and Hall reconnection. The comprehensive review by Porcelli et al. (2002) can be used to obtain more detailed information about inertial and Hall modes of magnetic reconnection in tokamaks, while papers by Park et al. (2006b, 2006a); Igochine et al. (2007) contain comparison of theoretical predictions and experimented observations.

Substantial difference between tearing modes developed in space and laboratory plasmas is provided by the difference of boundary conditions. The traditional approach to growth rate calculations consists in matching of solutions of perturbed Vlasov-Maxwell equations at the boundary separating inner region around the singular layer $\mathbf{k} \cdot \mathbf{B} = 0$ and outer region, where resonant wave-particle interaction or inertial effects can be neglected. Therefore, to determine solutions in the outer region one needs to introduce the certain external boundary conditions. The most appropriate approach for space systems consists in consideration of infinitely distant boundaries with corresponding solutions quickly decreasing with distance from the singular layer (an example of alternative approach can

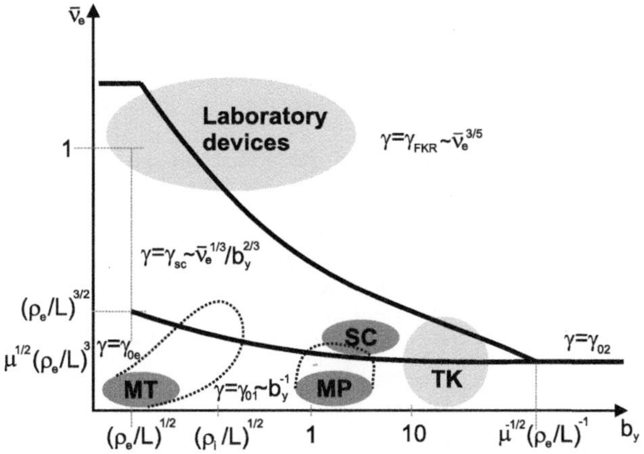

Fig. 3 Regimes of tearing mode: MT denotes magnetotail, MP denotes magnetopause, SC denotes solar corona, and TK denotes tokamaks. Figure is adopted from Zelenyi and Taktakishvili (1981)

be found in Zelenyi and Kuznetsova 1984). Situation is different for tokamaks configurations, where outer boundaries are accessible and have well defined physical properties like infinite conductivity (Coppi et al. 1966a; Wesson 1966). In this case, the set of external solutions is fully controlled by system geometry (see, e.g., Mikhailovskii 1978; Pegoraro and Schep 1986, and references therein). Additionally, characteristic cylindrical-like geometry of the tokamak system corresponds to appearance of local singularities of solutions (Newcomb 1960). Such singularities are absent in simplified plane geometry typical for space systems. These two problems are not encountered in major of space-plasma systems. Therefore, further comparison between spacecraft observations (and corresponding theories) with tokamak and laboratory reconnections requires accurate consideration of the geometry issue (see discussion in Boozer 2012a).

In conclusion, we can mention that for the major part of observed current sheets the semi-collisions regime of spontaneous reconnection seems to play the most important role. This regime principally cannot be described in a frame of MHD approach, until shear of magnetic field becomes unrealistically strong. On the other hand, unlikely that pure kinetic mode with unmagnetized electrons could be realized in realistic systems due to electron magnetization by even very weak magnetic fields. As a result, it is principal that the regime of current sheet destruction in the course of magnetic reconnection should be described in a frame of kinetic models with careful taking into account effects of collisions, which also exist in many seemingly collisionless configurations as effective collisions due to scattering of electrons at microturbulence fluctuations. In addition it should be kept in mind that weak magnetization ($k\rho_\alpha > 1$) of particle trajectories could substantially enhance collision frequencies formally defined in simplified τ- or BGK descriptions due to kinetic properties of the exact collisional operator. This effect known as Pitaevskii one could occur in a wide parameter range.

Acknowledgements The work was supported by RFBR (Nos. 10-02-93114, 12-02-91158), by grants for a Leading Scientific Schools NIII-623.2012.2, and by Program 22 of the Presidium of the Russian Academy of Sciences. The author thanks the International Space Science Institute (ISSI) and the organizing committee for support and the opportunity to participate in the ISSI Workshop on Microphysics of cosmic plasmas.

References

V. Angelopoulos, J.P. McFadden, D. Larson, et al., Tail reconnection triggering substorm onset. Science **321**, 931–935 (2008). doi:10.1126/science.1160495

A.Y. Aydemir, Nonlinear studies of $m = 1$ modes in high-temperature plasmas. Phys. Fluids B **4**, 3469–3472 (1992). doi:10.1063/1.860355

J. Berchem, C.T. Russell, Magnetic field rotation through the magnetopause—ISEE 1 and 2 observations. J. Geophys. Res. **87**, 8139–8148 (1982). doi:10.1029/JA087iA10p08139

P.L. Bhatnagar, E.P. Gross, M. Krook, A model for collision processes in gases. I. Small amplitude processes in charged and neutral one-component systems. Phys. Rev. **94**, 511–525 (1954). doi:10.1103/PhysRev.94.511

J. Birn, E.R. Priest, *Reconnection of Magnetic Fields: Magnetohydrodynamics and Collisionless Theory and Observations* (2007)

D. Biskamp, *Magnetic Reconnection in Plasmas* (2000)

D. Biskamp, E. Schwarz, J.F. Drake, Two-fluid theory of collisionless magnetic reconnection. Phys. Plasmas **4**, 1002–1009 (1997). doi:10.1063/1.872211

A.H. Boozer, Magnetic reconnection in space. Phys. Plasmas **19**(9), 092902 (2012a). doi:10.1063/1.4754715

A.H. Boozer, Theory of Tokamak disruptions. Phys. Plasmas **19**(5), 058101 (2012b). doi:10.1063/1.3703327

J. Büchner, Astrophysical reconnection and collisionless dissipation. Plasma Phys. Control. Fusion **49**, 325–339 (2007). doi:10.1088/0741-3335/49/12B/S30

V.P. Budaev, S.P. Savin, L.M. Zelenyi, Investigation of intermittency and generalized self-similarity of turbulent boundary layers in laboratory and magnetospheric plasmas: towards a quantitative definition of plasma transport features. Phys. Usp. **54**, 875–918 (2011). doi:10.3367/UFNe.0181.201109a.0905

B. Coppi, Current-driven instabilities in configurations with sheared magnetic fields. Phys. Fluids **8**, 2273–2280 (1965). doi:10.1063/1.1761195

B. Coppi, J.M. Greene, J.L. Johnson, Resistive instabilities in a diffuse linear pinch. Nucl. Fusion **6**, 101–117 (1966a)

B. Coppi, G. Laval, R. Pellat, Dynamics of the geomagnetic tail. Phys. Rev. Lett. **16**, 1207–1210 (1966b). doi:10.1103/PhysRevLett.16.1207

F.V. Coroniti, Space plasma turbulent dissipation—reality or myth? Space Sci. Rev. **42**, 399–410 (1985). doi:10.1007/BF00214995

W. Daughton, The unstable eigenmodes of a neutral sheet. Phys. Plasmas **6**, 1329–1343 (1999). doi:10.1063/1.873374

W. Daughton, H. Karimabadi, Kinetic theory of collisionless tearing at the magnetopause. J. Geophys. Res. (Space Phys.) **110**, 3217 (2005). doi:10.1029/2004JA010751

X.H. Deng, H. Matsumoto, Rapid magnetic reconnection in the Earth's magnetosphere mediated by whistler waves. Nature **410**, 557–560 (2001). doi:10.1038/35069018

M. Dobrowolny, Instability of a neutral sheet. Nuovo Cim., B **55**, 427–442 (1968). doi:10.1007/BF02711653

J.F. Drake, R.G. Kleva, Collisionless reconnection and the sawtooth crash. Phys. Rev. Lett. **66**, 1458–1461 (1991). doi:10.1103/PhysRevLett.66.1458

J.F. Drake, Y.C. Lee, Kinetic theory of tearing instabilities. Phys. Fluids **20**, 1341–1353 (1977). doi:10.1063/1.862017

J.P. Eastwood, T.D. Phan, S.D. Bale, A. Tjulin, Observations of turbulence generated by magnetic reconnection. Phys. Rev. Lett. **102**(3), 035001 (2009). doi:10.1103/PhysRevLett.102.035001

A.G. Frank, Dynamics of current sheets underlying flare-type events in magnetized plasmas. Phys. Usp. **53**, 941–947 (2010). doi:10.3367/UFNe.0180.201009h.0982

A.G. Frank, N.P. Kyrie, S.N. Satunin, Plasma dynamics in laboratory-produced current sheets. Phys. Plasmas **18**(11), 111209 (2011). doi:10.1063/1.3647576

A.G. Frank, S.Y. Bogdanov, V.S. Markov, G.V. Ostrovskaya, G.V. Dreiden, Experimental study of plasma compression into the sheet in three-dimensional magnetic fields with singular X lines. Phys. Plasmas **12**(5), 052316 (2005). doi:10.1063/1.1896376

M. Fujimoto, I. Shinohara, H. Kojima, Reconnection and waves: a review with a perspective. Space Sci. Rev. **160**, 123–143 (2011). doi:10.1007/s11214-011-9807-7

H.P. Furth, The 'mirror instability' for finite particle gyro-radius. Nucl. Fusion **1**, 169–174 (1962)

H.P. Furth, J. Killeen, M.N. Rosenbluth, Finite-resistivity instabilities of a sheet pinch. Phys. Fluids **6**, 459–485 (1963)

A.A. Galeev, Reconnection in the magnetotail. Space Sci. Rev. **23**, 411–425 (1979). doi:10.1007/BF00172248

A.A. Galeev, L.M. Zelenyi, Tearing instability in plasma configurations. Sov. Phys. JETP **43**, 1113 (1976)

A.A. Galeev, L.M. Zelenyi, Model of magnetic-field reconnection in a plane layer of collisionless plasma. JETP Lett. **25**, 380 (1977)

A.A. Galeev, L.M. Zelenyi, Magnetic reconnection in a space plasma, in *Theoretical and Computational Plasma Physics* (1978), pp. 93–116
A.A. Galeev, M.M. Kuznetsova, L.M. Zelenyi, Magnetopause stability threshold for patchy reconnection. Space Sci. Rev. **44**, 1–41 (1986). doi:10.1007/BF00227227
R.G. Giovanelli, Magnetic and electric phenomena in the Sun's atmosphere associated with sunspots. Mon. Not. R. Astron. Soc. **107**, 338 (1947)
D. Grasso, F. Pegoraro, F. Porcelli, F. Califano, Hamiltonian magnetic reconnection. Plasma Phys. Control. Fusion **41**, 1497–1515 (1999). doi:10.1088/0741-3335/41/12/306
E.G. Harris, On a plasma sheet separating regions of oppositely directed magnetic field. Nuovo Cimento **23**, 115–123 (1962)
W. Horton, T. Tajima, Decay of correlations and the collisionless conductivity in the geomagnetic tail. Geophys. Res. Lett. **17**, 123–126 (1990). doi:10.1029/GL017i002p00123
S.Y. Huang, et al., Observations of turbulence within reconnection jet in the presence of guide field. Geophys. Res. Lett. **39**, 11104 (2012). doi:10.1029/2012GL052210
J.D. Huba, N.T. Gladd, K. Papadopoulos, The lower-hybrid-drift instability as a source of anomalous resistivity for magnetic field line reconnection. Geophys. Res. Lett. **4**, 125–126 (1977). doi:10.1029/GL004i003p00125
V. Igochine, O. Dumbrajs, H. Zohm, A. Flaws, ASDEX Upgrade Team, Stochastic sawtooth reconnection in ASDEX upgrade. Nucl. Fusion **47**, 23–32 (2007). doi:10.1088/0029-5515/47/1/004
H. Ji, S. Terry, M. Yamada, R. Kulsrud, A. Kuritsyn, Y. Ren, Electromagnetic fluctuations during fast reconnection in a laboratory plasma. Phys. Rev. Lett. **92**(11), 115001 (2004). doi:10.1103/PhysRevLett.92.115001
B.B. Kadomtsev, Disruptive instability in tokamaks. Sov. J. Plasma Phys. **1**, 710–715 (1975)
B.B. Kadomtsev, Review article: magnetic field line reconnection. Rep. Prog. Phys. **50**, 115–143 (1987). doi:10.1088/0034-4885/50/2/001
H. Karimabadi, W. Daughton, K.B. Quest, Physics of saturation of collisionless tearing mode as a function of guide field. J. Geophys. Res. **110**, 3214 (2005). doi:10.1029/2004JA010749
R.G. Kleva, J.F. Drake, F.L. Waelbroeck, Fast reconnection in high temperature plasmas. Phys. Plasmas **2**, 23–34 (1995). doi:10.1063/1.871095
A.P. Kropotkin, H.V. Malova, M.I. Sitnov, Self-consistent structure of a thin anisotropic current sheet. J. Geophys. Res. **102**, 22099–22106 (1997). doi:10.1029/97JA01316
J. Labelle, R.A. Treumann, Plasma waves at the dayside magnetopause. Space Sci. Rev. **47**, 175–202 (1988). doi:10.1007/BF00223240
G. Lapenta, J.U. Brackbill, A kinetic theory for the drift-kink instability. J. Geophys. Res. **102**, 27099–27108 (1997). doi:10.1029/97JA02140
G. Laval, R. Pellat, M. Vuillemin, Instabilités Électromagnétiques des Plasmas Sans Collisions (cn-21/71), in *Plasma Physics and Controlled Nuclear Fusion Research, vol. II* (1966), pp. 259–277
F.M. Levinton, L. Zakharov, S.H. Batha, J. Manickam, M.C. Zarnstorff, Stabilization and onset of sawteeth in TFTR. Phys. Rev. Lett. **72**, 2895–2898 (1994). doi:10.1103/PhysRevLett.72.2895
A.S. Lipatov, L.M. Zelenyi, The study of magnetic islands dynamics. Plasma Phys. **24**, 1082–1089 (1982)
A.B. Mikhailovskii, Review of instability theory for high pressure Tokamak plasma, Technical report, 1978
R. Nakamura, et al., Cluster observations of an ion-scale current sheet in the magnetotail under the presence of a guide field. J. Geophys. Res. **113**, 7 (2008). doi:10.1029/2007JA012760
W.A. Newcomb, Hydromagnetic stability of a diffuse linear pinch. Ann. Phys. **10**, 232–267 (1960). doi:10.1016/0003-4916(60)90023-3
R. Numata, Z. Yoshida, Chaos-induced resistivity in collisionless magnetic reconnection. Phys. Rev. Lett. **88**(4), 045003 (2002). doi:10.1103/PhysRevLett.88.045003
E.V. Panov, et al., CLUSTER observation of collisionless transport at the magnetopause. Geophys. Res. Lett. **33**, 15109 (2006). doi:10.1029/2006GL026556
E.V. Panov, J. Büchner, M. Fränz, et al., High-latitude earth's magnetopause outside the cusp: cluster observations. J. Geophys. Res. **113**, 1220 (2008). doi:10.1029/2006JA012123
H.K. Park, A.J.H. Donné, N.C. Luhmann Jr., I.G.J. Classen, C.W. Domier, E. Mazzucato, T. Munsat, M.J. van de Pol, Z. Xia, Comparison study of 2D images of temperature fluctuations during sawtooth oscillation with theoretical models. Phys. Rev. Lett. **96**(19), 195004 (2006a). doi:10.1103/PhysRevLett.96.195004
H.K. Park, N.C. Luhmann Jr., A.J.H. Donné, I.G.J. Classen, C.W. Domier, E. Mazzucato, T. Munsat, M.J. van de Pol, Z. Xia, Observation of high-field-side crash and heat transfer during sawtooth oscillation in magnetically confined plasmas. Phys. Rev. Lett. **96**(19), 195003 (2006b). doi:10.1103/PhysRevLett.96.195003
E.N. Parker, Spontaneous current sheets in magnetic fields: with applications to stellar x-rays, in *Spontaneous Current Sheets in Magnetic Fields: With Applications to Stellar X-rays. International Series in Astronomy and Astrophysics*, vol. 1 (Oxford University Press, New York, 1994)

F. Pegoraro, T.J. Schep, Theory of resistive modes in the ballooning representation. Plasma Phys. Control. Fusion **28**, 647–667 (1986). doi:10.1088/0741-3335/28/4/003
R. Pellat, F.V. Coroniti, P.L. Pritchett, Does ion tearing exist? Geophys. Res. Lett. **18**, 143–146 (1991). doi:10.1029/91GL00123
A.A. Petrukovich, Origins of plasma sheet B_y. J. Geophys. Res. **116**, 7217 (2011). doi:10.1029/2010JA016386
H.E. Petschek, Magnetic field annihilation. NASA Spec. Publ. **50**, 425 (1964)
L.P. Pitaevskii, Effect of collisions on perturbation of body rotating in plasma. J. Exp. Theor. Phys. **44**, 969–979 (1963) (in Russian)
L.P. Pitaevskii, E.M. Lifshitz, *Physical Kinetics*, Course of Theoretical Physics, vol. 10 (Pergamon Press, New York, 1981)
F. Porcelli, Collisionless $m = 1$ tearing mode. Phys. Rev. Lett. **66**, 425–428 (1991). doi:10.1103/PhysRevLett.66.425
F. Porcelli, D. Borgogno, F. Califano, D. Grasso, M. Ottaviani, F. Pegoraro, Recent advances in collisionless magnetic reconnection. Plasma Phys. Control. Fusion **44**, 389 (2002)
E. Priest, T. Forbes, *Magnetic Reconnection* (2000)
P.L. Pritchett, F.V. Coroniti, Plasma sheet disruption by interchange-generated flow intrusions. Geophys. Res. Lett. **381**, 10102 (2011). doi:10.1029/2011GL047527
K.B. Quest, H. Karimabadi, M. Brittnacher, Consequences of particle conservation along a flux surface for magnetotail tearing. J. Geophys. Res. **101**, 179–184 (1996). doi:10.1029/95JA02986
M.N. Rosenbluth, R.Y. Dagazian, P.H. Rutherford, Nonlinear properties of the internal $m = 1$ kink instability in the cylindrical tokamak. Phys. Fluids **16**, 1894–1902 (1973). doi:10.1063/1.1694231
M. Roth, J. de Keyser, M.M. Kuznetsova, Vlasov theory of the equilibrium structure of tangential discontinuities in space plasmas. Space Sci. Rev. **76**, 251–317 (1996). doi:10.1007/BF00197842
P.H. Rutherford, Nonlinear growth of the tearing mode. Phys. Fluids **16**, 1903–1908 (1973). doi:10.1063/1.1694232
K. Schindler, A theory of the substorm mechanism. J. Geophys. Res. **79**, 2803–2810 (1974). doi:10.1029/JA079i019p02803
K. Schindler, *Physics of Space Plasma Activity* (Cambridge University Press, Cambridge, 2006). doi:10.2277/0521858976
K. Schindler, M. Soop, Stability of plasma sheaths. Phys. Fluids **11**, 1192–1195 (1968). doi:10.1063/1.1692083
M.I. Sitnov, K. Schindler, Tearing stability of a multiscale magnetotail current sheet. Geophys. Res. Lett. **37**, 8102 (2010). doi:10.1029/2010GL042961
M.I. Sitnov, L.M. Zelenyi, H.V. Malova, A.S. Sharma, Thin current sheet embedded within a thicker plasma sheet: self-consistent kinetic theory. J. Geophys. Res. **105**, 13029–13044 (2000). doi:10.1029/1999JA000431
B.U.Ö. Sonnerup, Adiabatic particle orbits in a magnetic null sheet. J. Geophys. Res. **76**, 8211–8222 (1971). doi:10.1029/JA076i034p08211
L.C. Steinhauer, Review of field-reversed configurations. Phys. Plasmas **18**(7), 070501 (2011). doi:10.1063/1.3613680
P.A. Sweet, The neutral point theory of solar flares, in *Electromagnetic Phenomena in Cosmical Physics*, ed. by B. Lehnert. IAU Symposium, vol. 6 (1958), p. 123
S.I. Syrovatskii, Dynamic dissipation of a magnetic field and particle acceleration. Sov. Astron. **10**, 270 (1966)
D.A. Uzdensky, Petschek-like reconnection with current-driven anomalous resistivity and its application to solar flares. Astrophys. J. **587**, 450–457 (2003). doi:10.1086/368075
S. von Goeler, W. Stodiek, N. Sauthoff, Studies of internal disruptions and $m = 1$ oscillations in tokamak discharges with soft-X-ray techniques. Phys. Rev. Lett. **33**, 1201–1203 (1974). doi:10.1103/PhysRevLett.33.1201
R. Wang, et al., Asymmetry in the current sheet and secondary magnetic flux ropes during guide field magnetic reconnection. J. Geophys. Res. **117**, 7223 (2012). doi:10.1029/2011JA017384
J. Wesson, Finite resistive instabilities of a sheet pinch. Nucl. Fusion **6**, 130–134 (1966)
J.A. Wesson, Sawtooth reconnection. Nucl. Fusion **30**, 2545–2549 (1990)
J. Wesson, *Tokamaks*, 3rd edn. (Oxford University Press, London, 2004). doi:10.1017/S0022377804003058
M. Yamada, R. Kulsrud, H. Ji, Magnetic reconnection. Rev. Mod. Phys. **82**, 603–664 (2010). doi:10.1103/RevModPhys.82.603
P.H. Yoon, A.T.Y. Lui, Quasi-linear theory of anomalous resistivity. J. Geophys. Res. **111**, 2203 (2006). doi:10.1029/2005JA011482
L. Zakharov, B. Rogers, S. Migliuolo, The theory of the early nonlinear stage of $m = 1$ reconnection in tokamaks. Phys. Fluids B **5**, 2498–2505 (1993). doi:10.1063/1.860735

L.E. Zakharov, On the nature of disruptive instability in a tokamak. JETP Lett. **31**, 714 (1980)

L.M. Zelenyi, M.M. Kuznetsova, Large-scale instabilities of the plasma sheet driven by particle fluxes at the boundary of the magnetosphere. Sov. J. Plasma Phys. **10**, 326–334 (1984)

L.M. Zelenyi, A.L. Taktakishvili, The influence of dissipative processes on the development of the tearing mode in current sheets. Sov. J. Plasma Phys. **7**, 1064–1075 (1981)

L.M. Zelenyi, A.L. Taktakishvili, Spontaneous magnetic reconnection mechanisms in plasma. Astrophys. Space Sci. **134**, 185–196 (1987). doi:10.1007/BF00636466

L.M. Zelenyi, M.I. Sitnov, H.V. Malova, A.S. Sharma, Thin and superthin ion current sheets. Quasi-adiabatic and nonadiabatic models. Nonlinear Process. Geophys. **7**, 127–139 (2000)

L.M. Zelenyi, A.V. Artemyev, H.V. Malova, V.Y. Popov, Marginal stability of thin current sheets in the Earth's magnetotail. J. Atmos. Sol.-Terr. Phys. **70**, 325–333 (2008). doi:10.1016/j.jastp.2007.08.019

L.M. Zelenyi, A.V. Artemyev, K.V. Malova, A.A. Petrukovich, R. Nakamura, Metastability of current sheets. Phys. Usp. **53**, 933–941 (2010). doi:10.3367/UFNe.0180.201009g.0973

L.M. Zelenyi, H.V. Malova, A.V. Artemyev, V.Y. Popov, A.A. Petrukovich, Thin current sheets in collisionless plasma: equilibrium structure, plasma instabilities, and particle acceleration. Plasma Phys. Rep. **37**, 118–160 (2011). doi:10.1134/S1063780X1102005X

DOI 10.1007/978-1-4899-7413-6_15
Reprinted from *Space Science Reviews* Journal, DOI 10.1007/s11214-013-9989-2

Topics in Microphysics of Relativistic Plasmas

Maxim Lyutikov · A. Lazarian

Received: 29 January 2013 / Accepted: 20 March 2013 / Published online: 10 July 2013

Abstract Astrophysical plasmas can have parameters vastly different from the more studied laboratory and space plasmas. In particular, the magnetic fields can be the dominant component of the plasma, with energy-density exceeding the particle rest-mass energy density. Magnetic fields then determine the plasma dynamical evolution, energy dissipation and acceleration of non-thermal particles. Recent data coming from astrophysical high energy missions, like magnetar bursts and Crab nebula flares, point to the importance of magnetic reconnection in these objects.

In this review we outline a broad spectrum of problems related to the astrophysical relevant processes in magnetically dominated relativistic plasmas. We discuss the problems of large scale dynamics of relativistic plasmas, relativistic reconnection and particle acceleration at reconnecting layers, turbulent cascade in force-fee plasmas. A number of astrophysical applications are also discussed.

Keywords Plasma astrophysics · High energy astrophysics

M. Lyutikov (✉)
Department of Physics, Purdue University, 525 Northwestern Avenue, West Lafayette, IN 47907-2036, USA
e-mail: lyutikov@purdue.edu

M. Lyutikov
INAF, Osservatorio Astrofisico di Arcetri Largo, Enrico Fermi 5, 50125 Firenze, Italy

M. Lyutikov
The Canadian Institute for Theoretical Astrophysics, University of Toronto, 60 St. George Street, Toronto, Ontario M5S 3H8, Canada

A. Lazarian
Department of Astronomy, University of Wisconsin-Madison, 475 Charter St., Madison, WI 53706, USA

1 Introduction

In many astrophysical settings the magnetic field controls the overall dynamics of plasma while the dissipation of magnetic energy may power the high energy emission. The relevant astrophysical settings include magnetars (strongly magnetized neutron stars possessing super-strong magnetic fields), pulsars and pulsar wind nebulae, jets of Active Galactic Nuclei and Gamma-Ray Bursters. All these objects are efficient emitters of X-rays and γ-rays and in the past two decades they have been subjects of intensive observational studies using a number of very successful high energy satellites. These objects seem to share one important property—their plasma is magnetically dominated, that is, the energy density of this plasma is dominated not by the rest mass-energy of matter but by the mass-energy of magnetic field. This is dramatically different from the laboratory plasmas, the magnetospheres of planets, and the interplanetary plasma.

Recently, these topics came to the front of astrophysical and plasma physical research, driven by a series of highly successful high energy mission like, *Swift*, *Fermi*, AGILE satellites and coming on-line of the very high energy γ-ray telescopes like HESS and VERITAS. *A number of observations point to the importance of magnetic dissipation in astrophysical high energy sources*, as we describe below. This signifies a shift of paradigm (from the fluid-dominated point of view) and requires a targeted study of plasma microphysics in a new regime. In this review we outline the related basic plasma physical problems and possible astrophysical applications.

1.1 Crab Nebula Flares: A New Type of Astrophysical Events

The constancy of the high energy Crab nebula emission has been surprisingly shown to be false by multiple day- to week-long flares, presenting a challenge to standard pulsar wind models (Kennel and Coroniti 1984). During these events, the Crab Nebula gamma-ray flux above 100 MeV exceeded its average value by a factor of several or higher (Abdo et al. 2011; Tavani et al. 2011; Buehler et al. 2012), while in other energy bands nothing unusual was observed (e.g. Abdo et al. 2011; Tavani et al. 2011), and references therein). Additionally, sub-flare variability timescales of $\sim$10 hours has been observed (Buehler et al. 2012). The prevailing conclusion from the observations of flares is that flares are associated with the nebular (and not the neutron star) and are mostly likely due to the highest energy synchrotron emitting electrons. Thus, the flares reflect the instantaneous injection/emission properties of the nebular and are not expected to produce a noticeable change in the inverse Compton (IC) component above $\sim$1 GeV. One of the most surprising property of the flares is their short time-scale variability, with typical duration two orders of magnitude smaller than the dynamical time-scale of the nebular.

These events question the dominant paradigm of shock acceleration in pulsar wind nebular (Lyutikov and Ouyed 2010a; Clausen-Brown and Lyutikov 2012). The key argument in favor of the reconnection origin of the flare is its SED: the peak frequency is above the classical synchrotron limit (de Jager et al. 1996; Lyutikov and Ouyed 2010a). This limit comes from assuming the electric field accelerating the emitting particles, E, is less than the emission region magnetic field, or $E = \eta B$, where $0 < \eta < 1$:

$$\mathcal{E}_{\rm ph}^{\rm max} = \frac{27}{16\pi} \eta \frac{mhc^3}{e^2} = 236\eta \ {\rm MeV} \tag{1}$$

Instead, these events offer tantalizing evidence in favor of relativistic reconnection (Uzdensky et al. 2011). Clausen-Brown and Lyutikov (2012) suggested that the flare arises due to intermitted reconnection in the downstream region, which produces relativistically moving blobs of plasma. We associate the duration of the flare with stochastically changing properties of plasma within the nebula. Second, the flares are apparently isolated, intermittent high flux events. Such intermittent behavior is often associated with power-law distributions of various kinds generated by astrophysical systems such as magnetically-driven Solar flares (Aschwanden 2005, and references therein).

Relativistic reconnection is a natural flaring mechanism in PWNe. The flare can be due to a highly localized emission region, or blob, so that the flare observables determine the intrinsic properties of the emission region. The natural flaring mechanism in this category is relativistic magnetic reconnection, which has been invoked by Crab Nebula flare models (Uzdensky et al. 2011) and fast flaring models in gamma-ray bursts (GRBs) and active galactic nuclei (AGN, Lyutikov and Ouyed 2006a; Giannios et al. 2009).

In reconnection, the magnetic energy of a localized region, a current sheet, is converted to random particle energy, (possibly) bulk relativistic motion, and radiation (for studies on reconnection in highly magnetized relativistic plasmas, see Lyutikov and Uzdensky 2003; Lyubarsky 2003; Uzdensky et al. 2011; McKinney and Uzdensky 2012). Reconnection in PWNe has already been studied in the past as a possible resolution of the well known σ-problem (Lyubarsky 2003; Komissarov 2012). In a similar vein, Lyutikov and Ouyed (2010a) proposes a model in which, reconnection occurs primarily along the rotation axis and equatorial region well beyond the light cylinder, thus qualitatively reproducing the jet/equatorial wisp morphology of the nebula.

We have developed (Clausen-Brown and Lyutikov 2012) a statistical model of the emission from Doppler boosted reconnection mini-jets, looking for analytical expressions for the moments of the resulting nebula light curve (e.g. time average, variance, skewness). The light curve has a flat power spectrum that transitions at short timescales to a decreasing power-law of index 2, Fig. 2. The flux distribution from mini-jets follows a decreasing power-law of index $\sim$1, implying the average flux from flares is dominated by bright rare events. The predictions for the flares' statistics can be tested against forthcoming observations. We find the observed flare spectral energy distributions (SEDs) have several notable features: A hard power-law index of $p \lesssim 1$ for accelerated particles that is expected in various reconnection models, including some evidence of a pile-up near the radiation reaction limit. Also, the photon energy at which the SED peaks is higher than that implied by the synchrotron radiation reaction limit, indicating the flare emission regions' Doppler factors are $\gtrsim$ few. Magnetic reconnection can be an important, if not dominant, mechanism of particle acceleration within the nebula.

If magnetic reconnection is what causes the flares, the reconnection process itself may leave a particular imprint on the SED in the form of a hard electron distribution. If much of the synchrotron emission occurs from particles near the synchrotron limit, then the emitting particles will display a SED that is close the single-particle synchrotron SED, see Fig. 1. The best flare SED observations to date are 11 SEDs taken during the April 2011 by Fermi/LAT team (Buehler et al. 2012). They were fitted with an empirical function of the form $\epsilon F_\epsilon \propto \epsilon^a \exp(\epsilon/\epsilon_c)$, where ϵ represents photon energy, and different values for the normalization and ϵ_c were used for each SED. The parameter a was assumed to be constant for all of the SEDs, and its best fit value is $a = 0.73 \pm 0.12$. The SED taken during the most luminous part of the flare, with a ϵF_ϵ maximum of $\sim 4 \times 10^{-9}$ $\mathrm{ergs}^{-1}\,\mathrm{s}^{-1}$ at a peak photon energy of $\epsilon_{\mathrm{peak}} = 375 \pm 26$ MeV, probably constrained the best fit value of a the most.

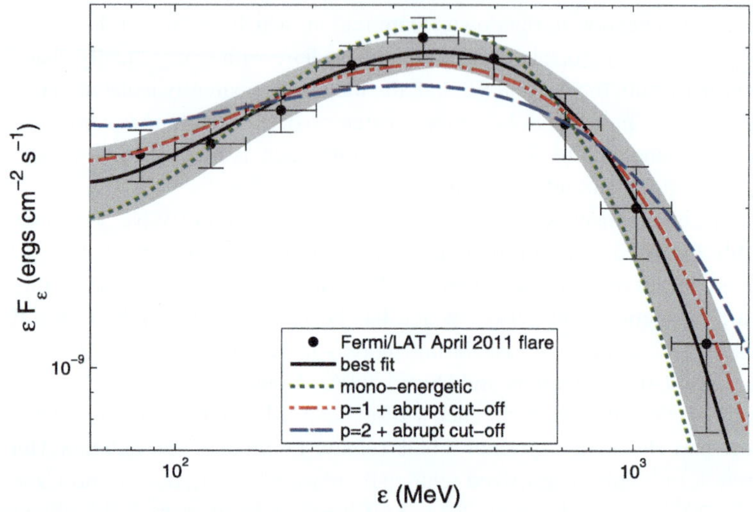

Fig. 1 The Fermi/LAT data from the most energetic part of the April 2011 Crab flare (Buehler et al. 2012) with the corresponding best fit curve and SEDs from three different electron energy distributions: a $p = 1$ power-law with an abrupt cut-off, the same for $p = 2$, and a mono-energetic electron distribution (Clausen-Brown and Lyutikov 2012). The *shaded area* represents the one-σ error region. The data favor steep injection spectrum with a pile-up, consistent with acceleration in reconnection regions

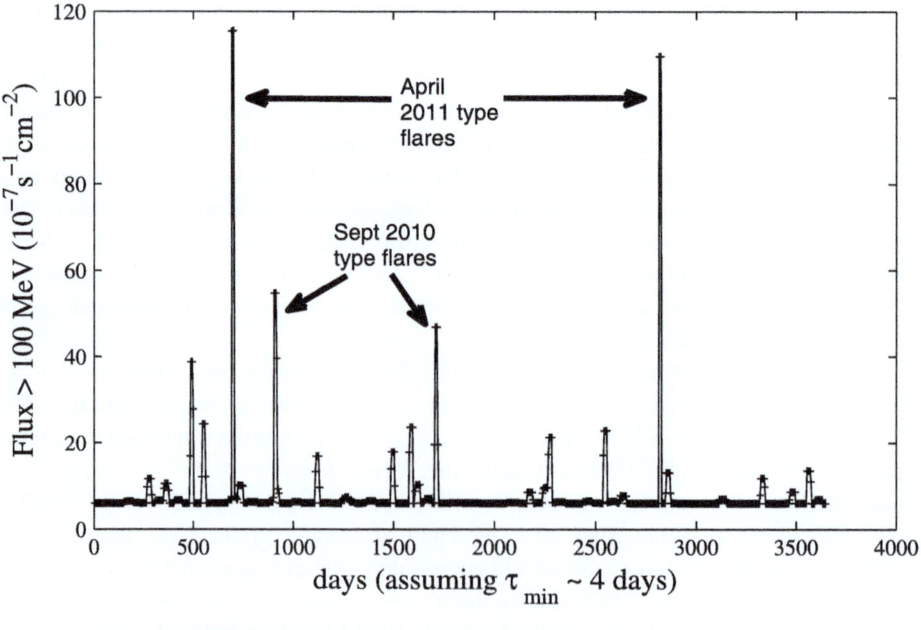

Fig. 2 Ten year simulated Crab nebula light curve with the reconnection model of Crab flares. The "April 2011 type flares" represent flares with increases of $\sim$30 over the nebular average as found in Buehler et al. (2012)

1.2 Magnetar Giant Flares

In phenomena possibly related to Crab flares, two closely related classes of young neutron stars—Anomalous X-ray Pulsars (AXPs) and the Soft Gamma-ray Repeaters (SGRs)—both show X-ray flares and, once localized, quiescent X-ray emission (Kouveliotou et al. 1998; Gavriil et al. 2002; for recent review see Woods and Thompson (2004)). Their high energy emission is powered by the dissipation of their super-strong magnetic fields, $B > 10^{15}$ G (Thompson and Duncan 1995). Two models of GFs are proposed. First, a GF may result from a *sudden untwisting of the internal magnetic field* (and twisting-up of the external magnetic field, Thompson and Duncan 1995, 2001). Alternatively, a *slow* untwisting of the internal magnetic field may lead to a gradual twisting of magnetospheric field lines, on time scales much longer than the GF, until it reaches a dynamical stability threshold due to increasing energy associated with the current-carrying magnetic field. Then follows a *sudden relaxation* of the twist outside the star and an associated dissipation and a change of magnetic topology lead to a GF, in analogy with solar flares and Coronal Mass Ejections (CMEs) (Lyutikov 2006).

The observed sharp rise of γ-ray flux during GF, on a time scale similar to the Alfvén crossing time of the inner magnetosphere, which takes $\sim$0.25 ms (Palmer et al. 2005). This unambiguously points to the magnetospheric origin of GFs, presumably during reconnection event in the magnetically-dominated magnetosphere (Lyutikov 2006).

The processes that cause magnetar X-ray flares (and possibly the persistent emission) may be similar to those operating in the solar corona. The energy of magnetar flares is accumulated inside the neutron star at the moment of its formation in the form of interior electric currents. These currents are then slowly pushed into the magnetosphere, gated by slow, plastic deformations of the neutron star crust. This leads to gradual twisting of the magnetospheric field lines, on time scales much longer than the magnetar's GF, and creates active magnetospheric regions similar to the Sun's spots. Initially, when the electric current (and possibly the magnetic flux) is pushed from the interior of the star into the magnetosphere, the latter slowly adjusts to the changing boundary conditions. As more and more current is pushed into the magnetosphere, it eventually reaches a point of dynamical instability. The loss of stability leads to a rapid restructuring of magnetic configuration, on the Alfvén crossing time scale, to the formation of narrow current sheets, and to the onset of magnetic dissipation. As a result, a large amount of magnetic energy is converted into the kinetic and bulk motion and radiation.

Observationally, a number of data point to the reconnection origin of the magnetar flares. Other predictions of the model, confirmed by observations are (i) the post-flare magnetosphere has a simpler structure, as the pre-flare network of currents has been largely dissipated; (ii) its spectrum is softer, since the hardness of the spectrum is a measure of the current strength in the bulk of magnetosphere (Thompson et al. 2002; Lyutikov and Gavriil 2005), with softer spectra corresponding to a smaller external current; and (iii) the spin-down rate, which depends on the amount of electric current flowing through the open field lines, decreases. The observations of two recent GFs, in SGR 1900+14 and SGR 1806-20, fully agree with these predictions. In both cases the persistent flux increased by a factor of two, its spectrum hardened (the power law index decreased from 2.2 to 1.5), and the spin-down rate increased in the months leading to the flare (Mereghetti et al. 2005). In the post-flare period, the pulsed fraction and the spin-down rate have significantly decreased and the spectrum softened (Rea et al. 2005). All these effects imply an increase of the external current before, and a decrease after the flare (Lyutikov 2006).

1.3 Reconnection in Magnetized Jets of Active Galactic Nuclei and Gamma Ray Bursts

Recent observations of AGNs in GeV and TeV energy range have raised new questions regarding the parameters of the central engine, and the location and kinematics of the high energy γ-ray, as well as X-ray and radio emission zones. In particular, the rapid flares reported for Mrk 501 and PKS 2155-304, on timescales of 3–5 minutes (Albert et al. 2007; Aharonian et al. 2007) imply an emitting size smaller than the gravitational radius $t_{lc} \sim$ hours of the supermassive black holes of these blazars. This indicates a very high Doppler factors δ, exceeding $\delta = 100$. A similar estimate also comes from the requirement that the TeV photons escape the production region.

While highly relativistic motion may appear to be a cure-all, the bulk Lorentz factor Γ can be directly constrained by VLBI observations of bright blobs moving with apparent speeds on the sky, β_{app}, that appear to be superluminal. This type of motion occurs when the emitting region is moving relativistically and close to the line of sight (Rees 1966). The apparent motion can exceed c due to propagation effects. If a blob is moving along with the bulk flow of a jet and its velocity vector makes an angle, θ_{ob}, with the line of sight, then its apparent motion transverse to the line of sight will be: $\beta_{app} = \frac{\beta_\Gamma \sin\theta_{ob}}{1-\beta_\Gamma \cos\theta_{ob}}$. The maximum β_{app} can reach is $\beta_\Gamma \Gamma$ when $\theta_{ob} \cong 1/\Gamma$. Thus, if the blob motion corresponds to the underlying bulk motion of the jet, measuring β_{app} can constrain the possible bulk Lorentz factor, Γ.

Following the suggestion by Lyutikov and Ouyed (2006a), a number of authors (Giannios et al. 2009; Ghisellini and Tavecchio 2008; Lazar et al. 2009; Kumar and Narayan 2009) proposed that fast time scale variability both in AGNs and GRBs is produced by "mini-jets", compact emitting regions that move relativistically *within* a jet of bulk $\Gamma \sim 10$ (in case of AGNs; for GRBs $\Gamma \sim 100$). Thus, the emission is beamed in the bulk outflow frame, e.g. due to relativistic motion of (using pulsar physics parlance) "fundamental emitters".

In GRBs, claims of high polarization (Coburn and Boggs 2003; Willis et al. 2005) offer direct measurements of the possibly dominant large-scale magnetic field. If confirmed, these observations argue in favor of magnetic reconnection as the main particle acceleration mechanism.

1.4 Reconnection in the Double Pulsar System PSR J0737-3039

Detection by McLaughlin (2004) of drifting sub-pulses of pulsar B in the Double Pulsar system PSR J0737−3039 with the frequency related to A period, presents an excellent opportunity to use Pulsar B as a probe of Pulsar A wind properties at ~1000 light cylinder radii of A, many orders of magnitude closer that have been possible so far. In particular, the fact that the observed modulation is at the frequency of A, and not a double frequency, already can be used as an indication that a large fraction of A wind is carried by relativistic MHD waves so that *directions* of electric and/or magnetic fields are important, not only total pressure of the wind.

A possible explanation is that the modulation of B by A is due to reconnection between magnetic fields in the wind and in the B magnetosphere. When magnetospheric field lines connect to the wind's magnetic field, they are "dragged" by the wind. Half a spin period of pulsar A later, when the wind's magnetic field changes polarity, the magnetospheric magnetic field disconnects from the wind and relaxes back to the position given by the impenetrable conductive boundary conditions. As the radio emission is produced along the local direction of a magnetic field, this periodic "dragging" and relaxation.

2 Dynamic Force-Free Plasma

In astrophysical settings the local microscopic plasma time scale (e.g. plasma frequency) is much shorter than the global dynamical time scales and there is plenty of charges available to screen the component of electric field along the magnetic field. In addition, astrophysical plasma is usually collisionless, making it an extremely good conductor. In addition, there is plenty of charges available to screen the component of electric field along the magnetic field $\mathbf{E} \cdot \mathbf{B} = 0$. Even in the extreme cases, when plasma may not be able to short out parallel electric field due to lack of available changes (charge separated flows) various radiative process (e.g. emission of curvature photon or through inverse Compton scattering) may lead to what is known as vacuum breakdown: abundant production of electron-positron pairs. The newly born pairs will create a charge density that would shut-off the accelerating electric field. The typical potential difference $\Delta V_{\rm vac} \sim kT/e$ needed to break down the vacuum in a GRB is typically in the MV-GV range. This is often orders of magnitude smaller than typically available EMF ($\sim 10^{15}$–10^{16} eV for pulsars, $\sim 10^{18}$–10^{20} eV for AGNs and GRBs).

The properties of plasma in the magnetospheres of pulsars and magnetars, pulsar winds, AGN and GRB jets are very different from those of more conventional Solar and laboratory plasmas. The principal difference is that it is relativistically strongly magnetized. In order to describe the level of magnetization it is convenient to use the so-called magnetization parameter $\sigma = 2(u_B/u_p)$, where $u_B = B^2/8\pi$ is the magnetic energy density and $u_p = \rho c^2$ is the rest mass-energy density. In traditional plasmas this parameter is very small. On the contrary, in some astrophysical settings it is likely to be very large. For example, in magnetars

$$\frac{\omega_B R_{NS}}{c}\left(\frac{m_e}{m_p}\right) \sim 10^{13} \leq \sigma \leq \frac{\omega_B}{\Omega}\left(\frac{m_e}{m_p}\right) \sim 10^{16} \qquad (2)$$

(the upper limit corresponds to the Goldreich-Julian density of electron-ion plasma whereas the lower limit corresponds to the poloidal current producing the toroidal magnetic fields of the same order as the poloidal one. Here $\omega_B = eB/m_e c$ is the cyclotron frequency, B is the magnetic field at the neutron star surface, R_{NS} is the neutron star radius, and Ω is its rotational frequency.

The parameter regime of highly magnetized plasma, $\sigma \gg 1$, implies that (i) the inertia of this plasma is dominated by the magnetic field and not by the particle rest mass, $B^2/8\pi \gg \rho c^2$, (ii) the propagation speed of Alfvén waves approaches the speed of light, (iii) the conduction current flows mostly along the magnetic field lines, (iv) the displacement current $(c/4\pi)\partial_t \mathbf{E}$ may be of the same order as the conduction current, $\mathbf{j}$, (v) the electric charge density, ρ_e, may be of the order of j/c. These are very different from the properties of laboratory plasmas, plasmas of planetary magnetospheres, and the interplanetary plasma, the cases where plenty of experimental data and theoretical results exist.

The large expected value of σ (or small $1/\sigma$) may be used as an expansion parameter in the equations of relativistic magnetohydrodynamics. The zero order equations describe the so-called relativistic force-free approximation. One may see this limit as the model where massless charged particles support currents and charge densities such that the total Lorentz force vanishes all the time (this also insures the ideal condition $\mathbf{E} \cdot \mathbf{B} = 0$.) This allows one to related the current to electro-magnetic fields (Gruzinov 1999)

$$\mathbf{J} = \frac{c}{4\pi} \frac{(\mathbf{E} \times \mathbf{B})\nabla \cdot \mathbf{E} + (\mathbf{B} \cdot \nabla \times \mathbf{B} - \mathbf{E} \cdot \nabla \times \mathbf{E})\mathbf{B}}{B^2} \qquad (3)$$

This may be considered as the Ohm's law for relativistic force-free electro-dynamics. (Note that this implies that the invariant $\mathbf{E} \cdot \mathbf{B} = 0$ and that electromagnetic energy is conserved, $\mathbf{E} \cdot \mathbf{J} = 0$.)

Under stationary and axisymmetric conditions, these equations guarantee that the angular velocity Ω is conserved along field lines. They also require a space charge density $\rho = \nabla \cdot \mathbf{E}/(4\pi)$ of magnitude $\sim \Omega B/c$ to develop. (Formally this, like the equation $\nabla \cdot \mathbf{B} = 0$, is just an initial condition.)

In the non-relativistic plasma the notion of force-free fields is often related to the stationary configuration attained asymptotically by the system (subject to some boundary conditions and some constraints, e.g. conservation of helicity). This equilibrium is attained on time scales of the order of the Alfvén crossing times. In strongly magnetized relativistic plasma the Alfvén speed may become of the order of the speed of light c, so that crossing times becomes of the order of the light travel time. But if plasma is moving relativistically its state is changing on the same time scale. This leads to a notion of dynamical force-free fields.

The force-free condition can be re-expressed by setting the divergence of the electromagnetic stress tensor to zero. This form has the merit that it brings out the analogy with fluid mechanics. Electromagnetic stress pushes and pulls electromagnetic energy which moves with an electromagnetic velocity $\mathbf{E} \times \mathbf{B}/B^2$, perpendicular to the electric and magnetic fields. This is the velocity of the frames (only defined up to an arbitrary Lorenz boost along the magnetic field direction) in which the electric field vanishes, (provided that the first electromagnetic invariant $B^2 - E^2 > 0$).

The limit $\sigma \to \infty$ is somewhat reminiscent of subsonic hydrodynamics as both the fast speed and the Alfvén speed approach the speed of light. For example, in case of slow processes, taking place on time scales much longer than light travel time, the Maxwell's equations may be written as continuity and momentum conservation (Komissarov et al. 2007):

$$\partial_t \rho + \nabla(2\rho \mathbf{V}) = 0, \qquad \partial_t \rho \mathbf{V} + \nabla\left(-\frac{\mathbf{B} \otimes \mathbf{B}}{4\pi} + \mathbf{I}\frac{B^2}{8\pi}\right) = 0 \tag{4}$$

where $\rho = B^2/8\pi c^2$ is the effective mass density of the electromagnetic field and $\mathbf{V} = \mathbf{E} \times \mathbf{B}/B^2$ is the drift speed of charged particles. This closed system of equations is very similar to non-relativistic MHD. This observation provides an interesting insight into the dynamics of two very different dynamical systems.

2.1 Time-Dependent Hyperbolic Grad-Shafranov Equations

Estimating possible electromagnetic signature of merging and collapsing neutron stars is most desirable for the gravitation waves searchers by LIGO and for identifying possible progenitors of short Gamma Ray Bursts. Collapse of a neutron star into black hole may proceed either through the accretion induced collapse (AIC) or during binary neutron star mergers. We expect at late stages both processes proceed along a somewhat similar path: in case of the merger, the two collapsing neutron stars form a transient supermassive neutron star which then collapses into the black hole. Both an accreting neutron star (in case of an AIC) and the transient supermassive neutron star are expected to be magnetized. In addition, in case of merging neutron stars the strong shearing of the matter may increase magnetic field well above the initial values.

Conventionally, in estimating the possible electromagnetic signatures it was first assumed that a fraction R_{NS}/R_G of the initial external magnetic energy (also built-up by the collapse and compression of the magnetic field) is radiated away on time scale of the order of the collapse time (Eardley and Press 1975). The main limitation of these models is that the external medium was treated as a vacuum. Electrodynamically, vacuum is a highly resistive mediums, with the impedance of the order of $4\pi/c = 477\ \Omega$. As a result, nothing prevents magnetic fields from becoming disconnected from the star and escaping to infinity. We expect that the magnetic field dynamics would be drastically different if the external magnetosphere were treated as a highly conducing medium.

Lyutikov (2011) demonstrated that the time evolution of the axisymmetric force-free magnetic fields can be expressed in terms of the hyperbolic Grad-Shafranov equation, under the assumption that the fields remain axially-symmetric. Qualitatively, there two separate types of non-stationarity: (i) due to the variations of the current $I(t)$ for a given shape of the flux function; (ii) due to the variations of the shape of the flux function for a given current I. Using these equations it is possible to find exact non-linear time-dependent Michel-type (split-monopole) structure of magnetospheres, e.g., driven by spinning and collapsing neutron star in Schwarzschild geometry:

$$
\begin{aligned}
&B_r = \left(\frac{R_s}{r}\right)^2 B_s, \qquad B_\phi = -\frac{R_s^2 \Omega \sin\theta}{r} B_s, \qquad E_\theta = B_\phi \\
&j_r = -2\left(\frac{R_s}{r}\right)^2 \cos\theta\, \Omega\, B_s \\
&P = (1-\cos\theta) B_s R_s^2 \\
&\Phi = -P\Omega \\
&I = -\frac{P(P-2B_s R_s^2)\Omega}{2B_s R_s^2} = \frac{1}{2} B_s R_s^2 \Omega \sin^2\theta
\end{aligned}
\tag{5}
$$

where P is the flux function, and Φ is the electric potential and $\Omega = \Omega(r-t)$ is *an arbitrary function*. Thus, we found exact solutions for time-dependent non-linear relativistic force-free configurations. Though the configuration is non-stationary (there is a time-dependent propagating wave), the form of the flux surfaces remains constant.

2.2 Limitations of Force-Free Approach

The generic limitation of the force-free formulation of MHD is that the evolution of the electromagnetic field leads, under certain conditions, to the formation of regions with $E > B$ (e.g. Uzdensky 2003), since there is no mathematical limitation on $B^2 - E^2$ changing a sign under a strict force-free conditions. In practice, the particles in these regions are subject to rapid acceleration through $\vec{E} \times \vec{B}$ drift, following by a formation of pair plasma via various radiative effects and reduction of the electric field. Thus, regions with $E > B$ are necessarily resistive. This breaks the ideal assumption and leads to the slippage of magnetic field lines with respect to plasma. In addition, evolution of the magnetized plasma often leads to formation of resistive current sheets, with the similar effect on magnetic field.

As a simple example demonstrating that the dynamical system described by the relativistic force-free limit has a natural tendency to violate the physical requirement of a negative first electromagnetic invariant, $E^2 - B^2 < 0$, consider a resistive decay of a line current. Suppose at time $t = 0$—there is a line current I_0 that decays for times $t > 0$ according to

$I = I_0(1 - t/\tau)$. This launches an outgoing rarefaction wave in which the EM field are given by

$$\begin{aligned} B_\phi &= \frac{I_0}{2\pi r}\left(1 - \frac{\sqrt{t^2 - r^2}}{\tau}\right) \\ E_z &= -\frac{I_0}{2\pi\tau} \ln \frac{t - \sqrt{t^2 - r^2}}{r} \end{aligned} \tag{6}$$

(for $r < t$). The resulting radial inward velocity, $v_r = E_z/B_\phi$ at some moment becomes larger than the speed of light. This example illustrates an important point: in force-free approximation plasma tends to develop dissipative regions where $E \to B$. This regions can develop non-locally, far from the cause that initiated the plasma motion (resistive decay at $r = 0$).

Another generic limitation of the force-free approach is related to the structure of the current sheet. Since in the force-free limit the plasma pressures are neglected, nothing prevents formation of very thing and thus highly dissipative current sheets. As a result, numerical models then can produce exceptionally high dissipation rates (e.g. Gruzinov 2012; Lehner et al. 2011). If small kinetic pressure in taken into account (e.g. Lyutikov and McKinney 2011; Tchekhovskoy and Spitkovsky 2012), the reconnection rate drops considerably. Another approach that allows (partial) stabilization of the current sheet is the resistive force-free plasma dynamics, which we discuss next.

3 Dissipation in Highly Magnetized Plasmas

3.1 Tearing Mode in Force-Free Plasma

One of the most important resistive instabilities in a conventional plasma is the so-called tearing instability. This is one of the principle unstable resistive modes, which plays a key role in various TOKAMAK discharges like the sawtooth oscillations and the major disruptions (e.g. Kadomtsev 1975), and leads to the unsteady reconnection of Solar flares (e.g. Shivamoggi 1985; Aschwanden 2002) and Earth's magnetotail (e.g. Galeev et al. 1978). The most important property of the tearing instability is the growth time that is much shorter than the resistive time.

In addition to being magnetically-dominated, microscopic plasma processes, like particle collisions or plasma turbulence, may contribute to resistivity and thus make plasma non-ideal. Resistivity will result in the decay of currents supporting the magnetic field; this, in turn, will influence the plasma dynamics. Introduction of resistivity into force-free formulation is not entirely self-consistent. The reason is that in force-free plasma the velocity along the field is not defined. Since plasma resistivity must be defined in the plasma rest-frame this creates a principal ambiguity.

The force-free tearing mode has been considered by Lyutikov (2003) (see also Gruzinov 2007; Li et al. 2012, for a somewhat different formulation of resistive force-free plasma). Lyutikov (2003) found that similar to the non-relativistic case, the resistive force-free current layers are unstable toward the formation of small-scale dissipative current sheets. He has also found that the growth rate of tearing instability, $\tau = \sqrt{\tau_d \tau_a}$, is intermediate between the short Alfvén time scale τ_a (which equals to the light crossing time in the force-free regime) and the long resistive time scale $\tau_d = l^2/\eta$, where l is the width of the current layer. This is exactly the same expression as in the non-relativistic case, which is rather surprising given

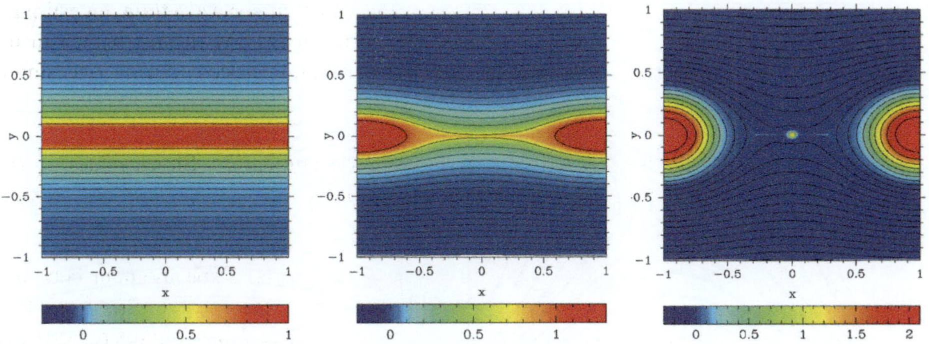

Fig. 3 Numerical simulations of the tearing instability in force-free plasma. The color images show the distribution of B_z and the contours show the magnetic field lines. In this simulations the perturbation has the wavelength corresponding to the maximum growth rate, $\tau \simeq \sqrt{\tau_d \tau_c}$, where τ_d is the resistive time and τ_c is the light crossing time of the current layer. The Lundquist number $L_u = \tau_d / \tau_c = 10^3$

the fact that the dynamic equations of force-free plasma are very different from the equations of non-relativistic MHD.

Numerical modeling of the tearing instability in strongly magnetized plasma (Komissarov et al. 2007) (see Fig. 3) fully confirm the analytical estimates.

In high Reynolds number plasma it is expected that tearing developing into turbulence. This has been observed in both PIC and MHD recent simulations (see below).

3.2 X-point Collapse in Force-Free Plasma

Consider a vicinity of an X-point. The non-current-carrying configuration has null lines intersecting at 90 degrees. Following the work on the collapse of a non-relativistic X-point (Dungey 1953; Imshennik and Syrovatskivi 1967; Priest and Forbes 2000), let us assume that the initial configuration is squeezed by a factor λ, so that the initial configuration has a vector potential $A_z \propto x^2 - y^2/\lambda^2$. In addition, we assume that there is an axial constant magnetic field B_z.

We are looking for time evolution of a vector potential of the type

$$A_z = -\left(\frac{x^2}{a(t)^2} - \frac{y^2}{b(t)^2}\right)\frac{B_0}{2L}, \tag{7}$$

where parameters B_0 and L charachterize the overall scaling of the magnetic field and the spacial scale of the problem.

The initial condition for the squeezed X-point and the ideal condition then require

$$\begin{aligned} b(t) &= \lambda / a(t) \\ \Phi &= xy\frac{B_z}{c}\partial_t \ln a \end{aligned} \tag{8}$$

(Since $\Delta\Phi = 0$ there is no induced charge density.) Parameter a characterizes the "squeezeness" of the configuration; $a = 1$ is the current-free case.

Faraday's law is then an identity, while the induction equation in the limit $x, y \to 0$ gives

$$\partial_t^2 \ln a = \mathcal{A}\left(\frac{a^4 - \lambda^2}{\lambda^4}\right), \quad \mathcal{A} = \frac{c^2}{L^2}\frac{B_0^2}{B_z^2} \tag{9}$$

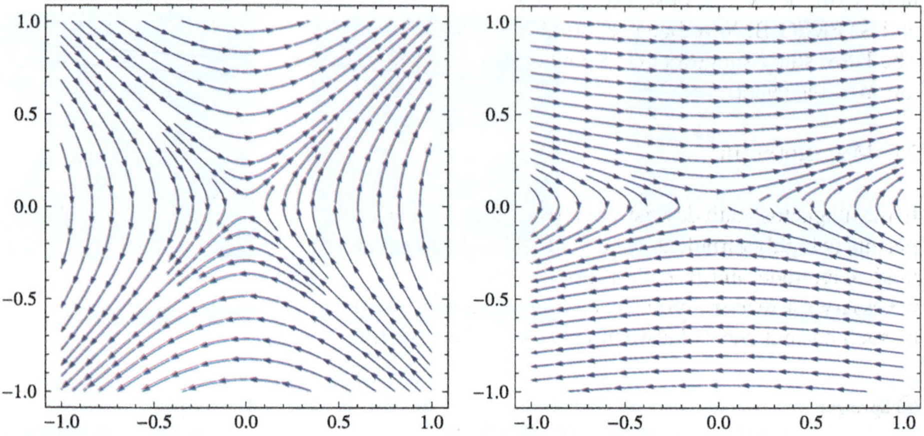

Fig. 4 Structure of the magnetic field in the x–y plane during X-point collapse in force-free plasma. The initial configuration on the left is slightly "squeezed". On dynamical time scale the X-point collapses to form a current sheet, right figure. The structure of the electric field in the x–y plane does not change during the collapse and qualitatively resembles the $t = 0$ configuration of the magnetic field

Solutions of the equations (9) show that $a(t)$ has a finite time singularity for $\lambda < 1$: in finite time a becomes infinite. At the moment when one of the parameters a or b becomes zero, the current sheet forms, see Fig. 4.

For small times $t \to 0$ (when $|a-1| \ll 1$), with initial conditions $a(0) = 1$, $a'(0) = 0$, and assuming that the initial "squeezing" is small, $\lambda = 1 - \epsilon$, $\epsilon \ll 1$, the solution is

$$a = 1 + \epsilon \sinh^2\left(\frac{ct}{L}\frac{B_0}{B_z}\right) \tag{10}$$

Thus, the typical collapse time is

$$\tau \sim \frac{L}{c}\frac{B_z}{B_0} \tag{11}$$

is of the order of the Alfvén (light) crossing time of the initial configuration. At these early times the electric field grows exponentially

At early times the particle drift follows a trajectory in the x–y plane $y \propto 1/x$. During the final collapse, in the limit $a \to \infty$, the particle distribution is further squeezed towards the neutral layer, $y \propto 1/(a^2\sqrt{\ln x})$ (though in this limit the drift approximation becomes inapplicable.) This shows that the X-point is stretched in one direction and is compressed in the other direction (note that $\nabla \mathbf{v} = 0$—collapse is incompressible at the initial stage).

Thus, in an ideal relativistic force-free plasma the X-point undergoes a finite time collapse. At the same time, particles are squeezed by the electromagnetic drift towards the neutral layer. The assumption of the force-free plasma will be broken down when the inflow velocity would become of the order of the Alfvén velocity. Then, the maximum electric field is $E \sim \beta_A B_0$ which is of the order of B_0, magnetic field in the bulk, for $\sigma \geq 1$. The collapse occurs faster for small axial field $B_z \leq B_0$.

3.3 Stationary Relativistic Reconnection

Magnetic reconnection is widely recognized as a very important phenomenon in many laboratory and astrophysical plasmas (Biskamp 2000, Priest and Forbes 2000). It has been studied very extensively over the last 40 years, and a significant progress has been made. However, historically, the main applications of the reconnection theory were confined to Solar physics, Earth's magnetosphere and the fusion projects. In all these cases, the magnetic energy density is much smaller than the particle rest mass-energy density, and the characteristic speeds are much less than the speed of light. Therefore it is not surprising that most of the progress has been made in the non-relativistic regime.

The analytical studies of dissipative effects in resistive magnetically dominated plasmas, though limited in their generality, provide an important first step towards the full understanding of the plasma dynamics under these extreme conditions and will serve as a guide for numerical investigations.

The first step is the generalization of non-relativistic models to the new regime Lyutikov and Uzdensky (2003), Lovelace et al. (2005) According to Lyutikov and Uzdensky (2003), the relativistic theory of the simplest model of magnetic reconnection—the Sweet–Parker model—involves two very large parameters: the Lundquist number, L_u, and the magnetization parameter σ. The simplest Sweet–Parker model of relativistic reconnection cannot be built self-consistently. The reason is that the convention non-relativistic model operates only with conservation laws, and not with the dynamical structure. In the relativistic case the conservation of energy then predicts acceleration to high Lorentz factors (Lyutikov and Uzdensky 2003), but this would lead to large pressure drop and would violate a force-balance across the current sheet (Lovelace et al. 2005).

Overall properties of stationary relativistic reconnection (like inflow and outflow velocities, Sweet-Parker versus Petschek models) remains an open question (for a recently review see Hoshino and Lyubarsky 2012). The shift in our understanding of non-relativistic reconnection layers, in particular as we discuss in more detail later, the importance of turbulence (Lazarian and Vishniac 1999) that is present in most astrophysical systems due to numerous instabilities that prey on high Reynolds number velocity fluids.

Interestingly enough, reconnection itself creates turbulence, which can be the cause of "reconnection instability" described in Lazarian and Vishniac (1999, 2009), which develops when the initial level of turbulence in the system is low or even the magnetic fields are originally laminar. As the outflow gets turbulent, the level of turbulence in the system and the reconnection rate increases inducing the positive feedback. This process may result in bursty reconnection of the time seen in solar flares, where the initial state of turbulence is low. This process can be also a driver of other dramatic energy bursts, e.g. gamma ray bursts (Lazarian et al. 2003, Zhang and Yan 2011; Lazarian and Yan 2012).

It is important that the initially laminar reconnection layer is subject to the tearing mode instability[1] (Loureiro et al. 2007), which both drive reconnection and ensure the turbulent state of the 3D reconnection layer (see Karimabadi 2013; Lazarian and Karimabadi 2013; Beresnyak 2013). All this puts in doubt many stationary and laminar reconnection models. Indeed, as described above, in astrophysics the relativistic reconnection is invoked for highly non-stationary processes and a laminar state of astrophysical fluid is more of an exception rather than a rule. Thus, one might expect that in the relativistic regime the current sheet is being fragmented and broadened by turbulence or/and subject to tearing.

3.3.1 Reconnection in the Presence of Turbulence

Properties of fluids are know to be strongly affected by turbulence. For instance, diffusion in turbulent fluids does not depend on molecular diffusivity. Thus, it is important to understand

[1] The estimates of the tearing mode in highly relativistic plasmas (see Sect. 3.1) indicate that the mode growth rate (surprisingly) follows the non-relativistic scaling.

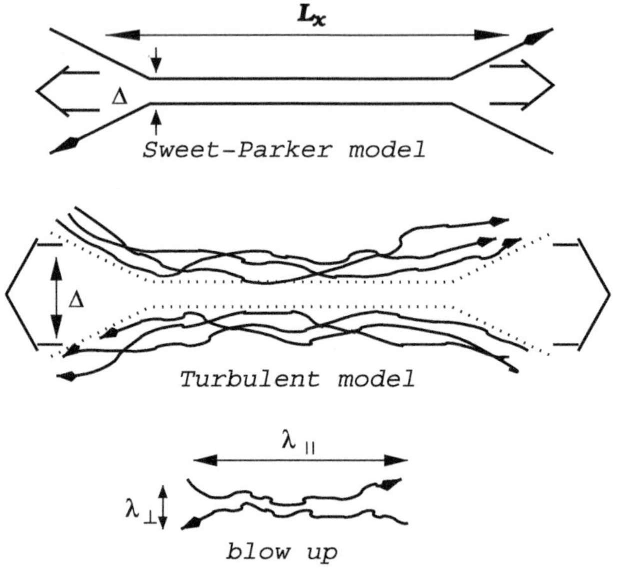

Fig. 5 *Upper plot*: Sweet-Parker model of reconnection. The outflow is limited by a thin slot Δ, which is determined by Ohmic diffusivity. The other scale is an astrophysical scale $L_x \gg \Delta$. *Middle plot*: Reconnection of weakly stochastic magnetic field according to LV99. The Goldreich-Sridhar (1995) model of MHD turbulence is used to account for the stochasticity of magnetic field lines. The outflow in the LV99 theory is limited by the diffusion of magnetic field lines, which depends on field line stochasticity. *Lower plot*: An individual small-scale reconnection region. The reconnection over small patches of magnetic field determines the local reconnection rate. The global reconnection rate is substantially larger as many independent patches come together. From Lazarian et al. (2004)

what can be the role of turbulence for the diffusion of magnetic field and reconnection that this transport can entail within relativistic plasmas. Below we provide arguments suggesting that turbulence can make magnetic reconnection fast.

In terms of non-relativistic fluids, the predictive model of turbulent reconnection was presented in Lazarian and Vishniac (1999) [henceforth LV99]. LV99 considered reconnection in the presence of sub-Alfvénic turbulence in magnetized plasmas. They identified stochastic wandering of magnetic field-lines as the most critical property of MHD turbulence which permits fast reconnection. As illustrated in Fig. 5, this line-wandering widens the outflow region and alleviates the controlling constraint of mass conservation.[2]

One can argue that the LV99 model that was shown to make reconnection fast carries over to relativistic fluids the same way as other, e.g. Sweet-Parker model does. In fact, it is clear from Fig. 5 that LV99 generalizes the Sweet-Parker model for the case of turbulent magnetic fields. The limitation that makes the Sweet-Parker reconnection slow both in relativistic and non-relativistic cases stems from the fact that the thickness of the outflow region is limited

[2]The LV99 model is radically different from its predecessors which also appealed to the effects of turbulence. For instance, unlike Speiser (1970) and Jacobson and Moses (1984) the model does not appeal to changes of microscopic properties of plasma. The nearest progenitor to LV99 was the work of Matthaeus and Lamkin (1985, 1986), who studied the problem numerically in 2D MHD and who suggested that magnetic reconnection may be fast due to a number of turbulence effects, e.g. multiple X points and turbulent EMF. However, these papers did not address the important role of magnetic field-line wandering, and did not obtain a quantitative prediction for the reconnection rate, as did LV99.

by effects related to plasma conductivity. These effects are related to microscopic scales and make $\Delta \ll L$ for the case of laminar fluids. In the case of LV99 reconnection the outflow is determined by macroscopic field wandering and therefore the Δ can get comparable with L. As the mass conservation dictates that the reconnection velocity is

$$V_{rec} \approx V_A \Delta / L \tag{12}$$

LV99 reconnection gets fast and it depends only on the intensity and injection scale of turbulence. In the relativistic limit V_A approaches the velocity of light and therefore one can expect Eq. (12) to hold with the change of V_A to c. At the same time, the exact scaling of Δ depends on the properties of relativistic turbulence. In any case, Δ if it determined by turbulence does not depend on microscopic plasma resistivity and therefore the reconnection should be fast[3] even if the scaling of turbulence in relativistic case differs from non-relativistic one.

On the basis of the recent studies of relativistic turbulence one may argue that LV99 model may be even directly applicable to the relativistic case. Indeed, the existing studies of the scaling and anisotropies of MHD turbulence (Cho 2005 and Cho and Lazarian 2012) testify that the relation between the parallel and perpendicular scales of eddies as well as the spectrum of turbulence are the same for relativistic and non-relativistic turbulence. As Δ in LV99 theory is determined by those properties of turbulence one can argue that the LV99 expressions for the reconnection rates can be directly relevant to relativistic reconnection.[4] This can be used at least as an educated guess for the discussion that we present further.

Irrespectively of the exact correspondence of the properties of relativistic and non-relativistic MHD turbulence one can argue that in the presence of turbulence Δ should increase and therefore the rate of magnetic reconnection should grow. This provides a prediction of flares of reconnection, explaining bursty energy release that is observed in solar flares as well as in many high energy phenomena, as it discussed in LV99. Indeed, if magnetic fluxes in contact are initially laminar or very weakly turbulent, Δ and therefore the reconnection rate may be slow initially. However this situation is unstable in the sense that if the outflow of plasma from the reconnection region gets turbulent, this will increase the turbulence of the ambient magnetic field and increase Δ. With larger Δ the outflow has larger Reynolds number and thus will get more turbulent. This results in "reconnection instability" (see more in Lazarian and Vishniac 2009). Similarly, one can argue that another process predicted in LV99 and reported in the observations by Sych et al. (2009), i.e. the initiation of reconnection by magnetic reconnection in adjacent regions should also be present in the relativistic case.

3.3.2 Acceleration at Relativistic Reconnection

Studies of the First order Fermi acceleration has been performed so far for non-relativistic reconnection (see de Gouveia dal Pino and Lazarian 2005; Lazarian 2005; Lazarian and Opher 2009; Drake et al. 2006, 2010; Lazarian and Desiati 2010; Lazarian et al. 2011, Kowal et al. 2012). We briefly summarize those below. The first order acceleration of particles entrained on the contracting magnetic loop can be understood from the Liouville theorem.

[3]Magnetic reconnection is universally defined as fast when it does not depend on resistivity.

[4]One may question the application of the field wondering concept to relativistic case. However, LV99 expression for Δ (and therefore also reconnection rates) was re-derived in ELV11 without appealing to this concept.

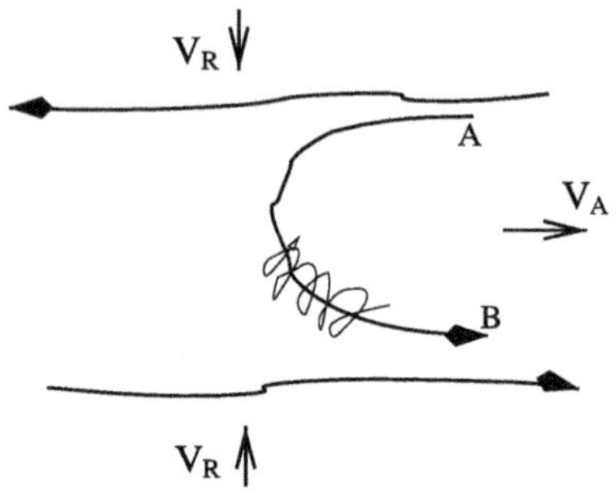

Fig. 6 Cosmic rays spiral about a reconnected magnetic field line and bounce back at points A and B. The reconnected regions move towards each other with the reconnection velocity V_R. From Lazarian (2005)

In the process of the magnetic tubes contraction a regular increase of the particle's energies is expected. The requirement for the process to proceed efficiently is to keep the accelerated particles within the contracting magnetic loop. This introduces limitations on the particle diffusivity perpendicular to the magnetic field direction. The subtlety of the point above is related to the fact that while in the first-order Fermi acceleration in shocks magnetic compression is important, the acceleration via the LV99 reconnection process is applicable even to incompressible fluids. Thus, unlike shocks, it is not the entire volume that shrinks for the acceleration, but only the volume of the magnetic flux tube. Thus high perpendicular diffusion of particles may decouple them from the magnetic field. Indeed, it is easy to see that while the particles within a magnetic flux rope depicted in Fig. 6 bounce back and forth between the converging mirrors and get accelerated, if these particles leave the flux rope fast, they may start bouncing between the magnetic fields of different flux ropes which may sometimes decrease their energy. Thus it is important that the particle diffusion both in the parallel and perpendicular directions to the magnetic field stay different. The particle anisotropy which arises from particles preferentially getting acceleration in terms of the parallel momentum may also be important.

Similarly, the first order Fermi acceleration can happen in terms of the perpendicular momentum. This is illustrated in Fig. 7. There the particle with a large Larmour radius is bouncing back and forth between converging mirrors of reconnecting magnetic field systematically getting an increase of the perpendicular component of its momentum. Both processes take place in reconnection layers.

3.4 Force-Free Turbulent Cascade

One of the most efficient ways of magnetic dissipation is through the turbulent cascade where the energy is transported from the large input scales to the small dissipation scales (Zakharov et al. 1992). The non-relativistic theory of turbulent cascade in magnetized plasma is in the stage of active development, both theoretical (Iroshnikov 1963; Kraichnan 1965; Montgomery and Turner 1981; Shebalin et al. 1983; Sridhar and Goldreich 1994; Goldreich and Sridhar 1997; Ng and Bhattacharjee 1996; Lazarian and Vishniac 1999; Galtier et al. 2000) and numerical (Maron and Goldreich 2001; Cho et al. 2002; Cho and Lazarian 2002, 2003; Kowal and Lazarian 2010). The consensus that is emerging from these studies is that in the non-relativistic case the Alfvén wave cascade is well decoupled from the fast and slow modes due to the fact that Alfvén waves are incompressible (slow mode follows Alfvén waves). Unlike the hydrodynamic turbulence, the Alfvén wave cascade is anisotropic with energy cascading mostly perpendicularly to the magnetic field.

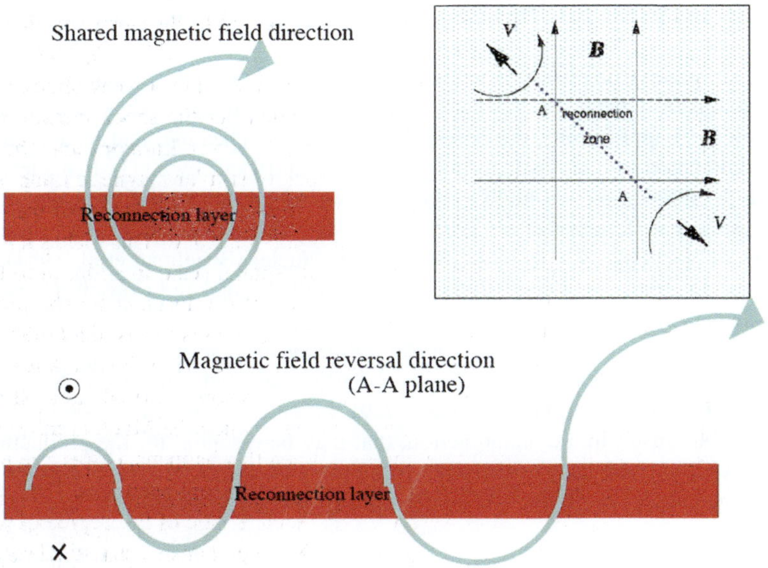

Fig. 7 Particles with a large Larmor radius gyrate about the magnetic field shared by two reconnecting fluxes (the latter is frequently referred to as "guide field". As the particle interacts with converging magnetized flow corresponding to the reconnecting components of magnetic field, the particle gets energy gain during every gyration. From Lazarian et al. (2012)

What is the structure of cascade in strongly magnetized plasma? First, we expect that like the traditional hydromagnetic turbulence the electromagnetic turbulence is local in phase space, so that the most important interaction is between waves with similar wave lengths. This is because the longer wavelength perturbations can be excluded via the relevant Lorentz transformation.

Previously, its properties have been addressed in Thompson and Blaes (1998), Troischt and Thompson (2004). Thompson and Blaes (1998) assumed that the relativistic Alfvén turbulence is similar to the non-relativistic one thus the energy cascade remains anisotropic. However, this is unlikely to true for the electromagnetic turbulence where the fast waves play as an important role as the Alfvén waves. Indeed, there exists strong three-wave coupling between the non-zero frequency Alfvén waves and fast waves: $A + F \to A$, $A + F \to F$ and $A + A \to F$. In the non-relativistic case, the three wave coupling between non-zero frequency Alfvén waves and fast waves does not exist because the resonance condition cannot be satisfied, and the coupling between three fast waves does not exists because of the vanishing coupling coefficient.

The three wave interaction coefficients are complicated, with different dependences on the angles of interacting waves. Generically, the coupling of Alfvén and fast waves is strong, so that the two cascades are well coupled. Two important questions need to be answered: what are the angular dependence of the cascades and what are the wave number dependence. It is feasible that both cascades are anisotropic in such a way that the stationary kinetic equation for both Alfvén and fast waves are satisfied. We consider this unlikely. Alternative possibility, which we favor, is that the interaction of two cascades may isotropise them.

4 Step Toward Relativistic Turbulence: Vortical Flows of Relativistic Fluid

Dynamics of relativistic plasma is a basic problem in fluid mechanics that has a wide range of applications from the physics of early Universe to heavy nuclei collisions to astrophysics. In cosmological applications, the post-inflation stage of reheating, that lead to matter creation, is dominated by relativistic turbulence (Micha and Tkachev 2003). In nuclear physics, the head-on collision of two highly relativistic nuclei creates a relativistically hot quark–gluon plasma that (Rischke et al. 1995). On a very different scale, a wide variety of astrophysical objects like jets from Active Galactic Nuclei (Begelman et al. 1984), Gamma Ray Bursts (Lyutikov and Ouyed 2006b) pulsar winds Kennel and Coroniti (1984) contain relativistic plasma.

Both, the quark–gluon plasma of nuclear collisions and astrophysical plasma are nearly ideal, with very small viscose contribution. Nearly ideal fluids are subject to the development of turbulence Landau and Lifshitz (1959). In the astrophysical set-up the relativistic turbulence may result in a dynamo action, that may be essential for the production of the high energy emission (Zhang et al. 2009). In addition, in case of relativistic supersonic flows the turbulence is necessarily for acceleration of cosmic rays at shocks Blandford and Eichler (1987).

Despite these important applications, the theory of relativistic turbulence is not developed, with only a few works addressing its statistical and dynamics properties (e.g. Dettmann and Frankel 1996; Goodman and MacFadyen 2008). Since the relativistic turbulence is generically compressive, we expect (following the analogy with the non-relativistic compressive turbulence) that the turbulence can be represented as a collection of interacting vortical and compressible modes. The compressible modes in the relativistic plasma are well known (e.g. Landau and Lifshitz 1959). On the other hand, the vortical modes in the relativistic fluid has not been considered so far.

One of the key features of relativistic vortices is that the flow compressibility must be taken into account. This make a majority of work on vortices, which often use a non-compressible approximation, (Landau and Lifshitz 1959; Lamb 1975), not applicable to relativistic vortices. On the other hand, relativistic vortices resemble in many ways the vortices in the compressible fluids (Green 1995). The centrifugal forces "pull away" gas from the axis and can lead to the development of a cavitated core.

First we consider a structure of simple rectilinear relativistic vortex. A simple rectilinear vortex with zero distributed vorticity has the following four-velocity structure, Fig. 8:

$$u_\phi = \frac{1}{\sqrt{(r/r_0)^2 - 1}} \tag{13}$$

For polytropic equation of state (EoS) with index Γ the density becomes zero at

$$r_i = \frac{1 + \Gamma\kappa_\infty}{\sqrt{\Gamma\kappa_\infty(2 + \Gamma\kappa_\infty)}} \frac{\Gamma_z}{2\pi\tilde{w}_\infty} \tag{14}$$

where κ_∞ and $\tilde{w}_\infty$ are vorticity and proper enthalpy measured at infinity. Thus, a simple relativistic vortex is cavitated—it has an empty vacuum core.

Vortical flows must have cores with special properties that are different from the bulk flow. The structure of the core of the both relativistic and non-relativistic vortices depends both on the parameters of the system and the history how a vortex has been created. Formally, for realistic isentropic EoS the rectilinear relativistic vortex develops a cavitated core. The development of the cavitated core occurs in the non-relativistic compressible fluid as

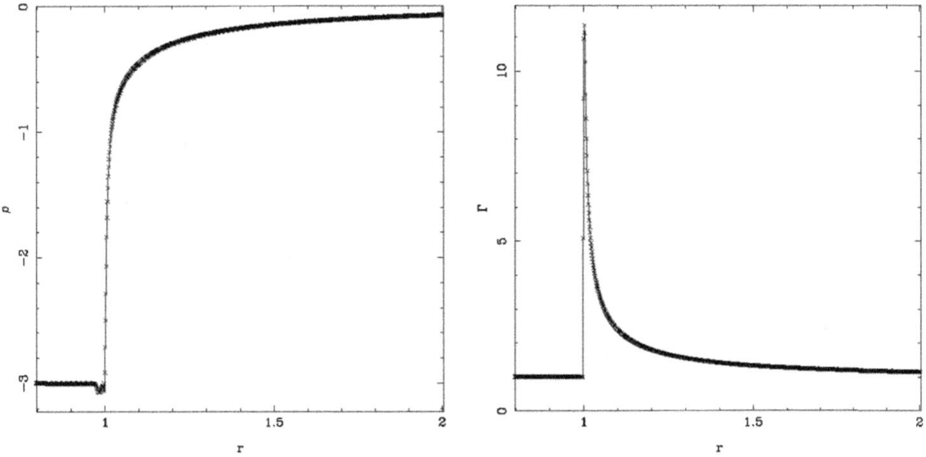

Fig. 8 Structure of relativistic vortex: proper density and Lorentz factor. One clearly sees a cavitated core. The inside of the vortex had to be filled with dynamically unimportant low density gas since the scheme does not allow full cavitation

well. For a typical laboratory fluid vortex with highly subsonic flow $M \ll 1$ and not exceptionally high Reynolds numbers Re, the viscose core forms on time of the order of the dynamical time. On the other hand, in astrophysics we expect $M \sim 1$, while the Reynolds number Re is huge. Thus, in astrophysical conditions the formation time of a viscose core is very long, so that, generically, an isolated rectilinear astrophysical vortex will be cavitated.

5 The Subtle Role of Global Electric Fields

Astrophysical plasmas, typically, do not tolerate large scale electric fields: charge non-neutrality in highly conducting medium is quickly suppressed by drawing-in of the opposite charges. As a result, the dynamical role of large-scale electric fields in the overall dynamics and particle acceleration is often under-appreciated. Magnetic fields, on the other hand, may both suppress charge neutralization and can themselves give rise to global inductive electric fields. Below we discuss two cases where global inductive electric fields and charge non-neutral plasma may be important in otherwise conventional plasmas.

5.1 Inductive Acceleration of UHERCs

Relativistic outflows carrying large scale magnetic fields have large inductive potential and may accelerate protons to ultra high energies. Lyutikov and Ouyed (2007) discussed a scheme of Ultra-High Energy Cosmic Ray (UHECR) acceleration due to drifts in magnetized, cylindrically collimated, sheared jets of powerful active galaxies.

Lyutikov and Ouyed (2007) model of UHECR acceleration relies on the observation that in a transversely sheared flow one sign of charges is located at a maximum of electric potential, as we describe in this section. Consider sheared flow carrying magnetic field. At each point there is electric field $\mathbf{E} = -\mathbf{v} \times \mathbf{B}/c$, so that the electric potential is determined by

$$\Delta\Phi = \frac{1}{c}\nabla \cdot (\mathbf{v} \times \mathbf{B}) = \frac{1}{c}\big(\mathbf{B} \cdot (\nabla \times \mathbf{v}) - \mathbf{v} \cdot (\nabla \times \mathbf{B})\big) \tag{15}$$

If system is stationary and current-free, (in a local rest frame, the second term in Eq. (15) vanishes at the position of a particle and is generally sub-dominant to the first term in the near vicinity) then $\nabla \times \mathbf{B} = 0$ and we find

$$\Delta\Phi = \frac{1}{c}(\mathbf{B} \cdot \nabla \times \mathbf{v}) \tag{16}$$

We have arrived at an important result: *depending on the sign of the quantity* $(\mathbf{B} \cdot \nabla \times \mathbf{v})$ *(which is a scalar) charges of one sign are near potential minimum, while those with the opposite sign are near potential maximum.* Since electric field is perpendicular both to velocity and magnetic field, locally, the electric potential is a function of only one coordinate along this direction. For $(\mathbf{B} \cdot \nabla \times \mathbf{v}) < 0$ ions are near potential maximum. A positively charged particle carried by such a plasma is in an unstable equilibrium if $\mathbf{B} \cdot \nabla \times \mathbf{v} < 0$, so that kinetic drift along the velocity shear would lead to fast, *regular* energy gain.

The procedure outlined above to calculate electric potential is beyond the limits of applicability of *non-relativistic* MHD, which assumes quasi-neutrality and thus neglects the dynamical effects associated with the potential (16). Thus, even in the low frequency regime with non-relativistic velocities, a conventional realm of MHD theory, one should use at least two fluid approach and also must retain both charge density as well as displacement current in Maxwell equations.

Under ideal fluid approximation particles cannot move across magnetic field lines, so that they cannot "sample" the electric potential (16). On the other hand, kinetic effects, like drift motions, may lead to regular radial displacement along the shear and thus along electric field. In this case one sign of charge will be gaining energy, while another sign will be losing energy. This is independent on whether the drift is along the shear or counter to the shear direction and thus is independent on the sign of the magnetic field gradient that induces the shear.

When the Larmor radius becomes comparable to shear scale the particle motion becomes unstable even for homogeneous flow. Particle trajectories can be found in quadratures in the general case Lyutikov and Ouyed (2007). In the non-relativistic limit equations of motion can be integrated exactly,

$$y = r_L \cos Z\omega_B\sqrt{1+\zeta/Z\omega_B}t, \qquad z = -\frac{r_L}{\sqrt{1+\zeta/Z\omega_B}} \sin Z\omega_B\sqrt{1+\zeta/\omega_B}t \tag{17}$$

where $\zeta = V'$. This clearly shows that for strong negative shear, $\zeta < -\omega_B$, particle trajectory is unstable and its energy growth exponentially. For positive shear, $\zeta > 0$, particle motion is stable.

5.2 Relativistic Effects at Cosmic Ray-modified Perpendicular Shocks

Acceleration of cosmic rays is one of the main problems of high energy astrophysics. Shock acceleration is the leading model (Blandford and Eichler 1987). Particle acceleration at quasi-parallel shocks (when the magnetic field in the upstream medium is nearly aligned with the shock normal) and quasi-perpendicular shocks (when the magnetic field in the upstream medium is nearly orthogonal to the shock normal) proceeds substantially differently. Most astrophysical shocks are quasi-perpendicular, yet theoretically acceleration at this type of shocks is less understood than in the case of quasi-parallel shocks. It is recognized that the feedback of accelerated cosmic rays may considerably modify the parallel shock structure (Axford et al. 1982).

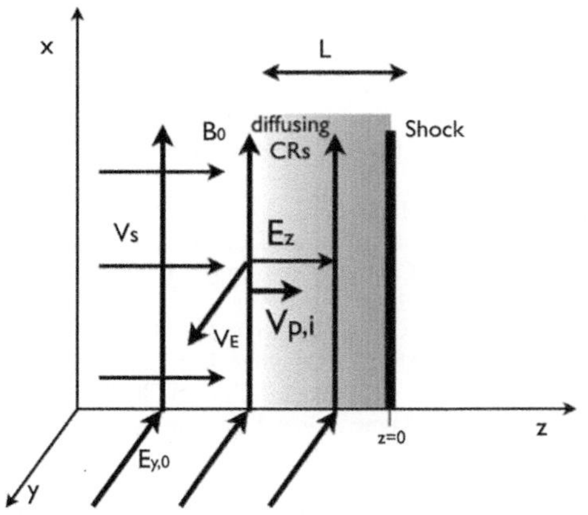

Fig. 9 Plasma flows in the frame of the shock. Far upstream, the incoming plasma moves with velocity v_S, magnetic field is along x direction and inductive electric field along $-y$ direction. At the shock, cosmic rays are accelerated and diffuse ahead of the shock a typical distance L, creating an electric field along the shock normal. Electric drift of cosmic rays in this induced electric field and the initial magnetic field produces z-dependent electric drift v_E in y direction. Acceleration of plasma in y direction in turn produces polarization drift of ions in z direction

Kinetic diffusion of cosmic rays ahead of perpendicular shocks induces large scale charge non-neutrality, which is typically neglected in the non-relativistic fluid approach. Cosmic rays diffusing ahead of the shock offset the charge balance in the incoming plasma, which becomes non-neutral, with electric field directed along the shock normal. The incoming upstream plasma will *partially* compensate this charge density by a combination of electric and polarization drifts. This creates a current along the shock normal, perpendicular to the magnetic field, see Fig. 9. Current-driven instabilities, in particular of the modified Buneman type, generate plasma turbulence with wave vectors preferentially perpendicular to the initial magnetic field, generating the field line wandering required for acceleration of cosmic rays in the first place. Thus, similar to parallel shocks, assumption of turbulence and cosmic ray acceleration leads to turbulence generation by cosmic rays themselves (Lyutikov and Ouyed 2010b).

References

Abdo et al., Science **331**, 739 (2011)
F. Aharonian et al., Astrophys. J. **664**, L71 (2007)
J. Albert et al., Astrophys. J. **669**, 862 (2007)
M.J. Aschwanden, Space Sci. Rev. **101**, 1 (2002)
M.J. Aschwanden, *Physics of the Solar Corona. An Introduction with Problems and Solutions*, 2nd edn. (2005)
W.I. Axford, E. Leer, J.F. McKenzie, Astron. Astrophys. **111**, 317 (1982)
M.C. Begelman, R.D. Blandford, M.J. Rees, Rev. Mod. Phys. **56**, 255 (1984)
A. Beresnyak, arXiv:1301.7424 (2013)
D. Biskamp, *Magnetic Reconnection in Plasmas* (Cambridge University Press, Cambridge, 2000)
R. Blandford, D. Eichler, Phys. Rep. **154**, 1 (1987)
Buehler et al., Astrophys. J. **749**, 26 (2012)
J. Cho, Astrophys. J. **621**, 324 (2005)
J. Cho, A. Lazarian, Phys. Rev. Lett. **88**, 245001 (2002)
J. Cho, A. Lazarian, Mon. Not. R. Astron. Soc. **345**, 325 (2003)
J. Cho, A. Lazarian (2012 in preparation)

J. Cho, A. Lazarian, E.T. Vishniac, Astrophys. J. **564**, 291 (2002)
E. Clausen-Brown, M. Lyutikov, Mon. Not. R. Astron. Soc. **426**, 1374 (2012)
W. Coburn, S.E. Boggs, Nature **423**, 415 (2003)
E.M. de Gouveia dal Pino, A. Lazarian, Astron. Astrophys. **441**, 845 (2005)
O.C. de Jager, P.F. Michelson, H.I. Nel, P.L. Nolan, P. Sreekumar, D.J. Thompson, Astrophys. J. **457**, 253 (1996)
C.P. Dettmann, N.E. Frankel, Phys. Rev. E **53**, 5502 (1996)
J.F. Drake, M. Swisdak, H. Che, M.A. Shay, Nature **443**, 553 (2006)
J.F. Drake, M. Opher, M. Swisdak, J.N. Chamoun, Astrophys. J. **709**, 963 (2010)
J.W. Dungey, Mon. Not. R. Astron. Soc. **113**, 180 (1953)
D.M. Eardley, W.H. Press, Annu. Rev. Astron. Astrophys. **13**, 381 (1975)
A.A. Galeev, F.V. Coroniti, M. Ashour-Abdalla, Geophys. Res. Lett. **5**, 707 (1978)
S. Galtier, S.V. Nazarenko, A.C. Newell, A. Pouquet, J. Plasma Phys. **63**, 447 (2000)
F.P. Gavriil, V.M. Kaspi, P.M. Woods, Magnetar-like X-ray bursts from an anomalous X-ray pulsar. Nature **419**, 142–144 (2002)
G. Ghisellini, F. Tavecchio, Mon. Not. R. Astron. Soc. **386**, L28 (2008)
D. Giannios, D.A. Uzdensky, M.C. Begelman, Mon. Not. R. Astron. Soc. **395**, L29 (2009)
P. Goldreich, S. Sridhar, Toward a theory of interstellar turbulence, 2: strong alfvenic turbulence. Astrophys. J. **438**, 763–775 (1995)
P. Goldreich, S. Sridhar, Astrophys. J. **485**, 680 (1997)
J. Goodman, A. MacFadyen, J. Fluid Mech. **604**, 325 (2008)
S.I. Green, *Fluid Vortices* (Springer, Berlin, 1995)
A. Gruzinov, ArXiv Astrophysics e-prints (1999)
A. Gruzinov, ArXiv e-prints (2007)
A. Gruzinov, ArXiv e-prints (2012)
M. Hoshino, Y. Lyubarsky, Space Sci. Rev. **78** (2012)
V.S. Imshennik, S.I. Syrovatskivi, Sov. Phys. JETP **25**, 656 (1967)
P. Iroshnikov, Sov. Astron. **7**, 566 (1963)
A.R. Jacobson, R.W. Moses, Phys. Rev. A **29**, 3335 (1984)
B.B. Kadomtsev, Sov. J. Plasma Phys. **1**, 710 (1975)
H. Karimabadi, V. Roytershteyn, M. Wan, W.H. Matthaeus, W. Daughton, P. Wu, M. Shay, B. Loring, J. Borovsky, E. Leonardis, S.C. Chapman, T.K.M. Nakamura, Coherent structures, intermittent turbulence, and dissipation in hightemperature plasmas. Phys. Plasmas **20**, 012303 (2013)
C.F. Kennel, F.V. Coroniti, Astrophys. J. **283**, 694 (1984)
S.S. Komissarov, arXiv:1207.3192 (2012)
S.S. Komissarov, M. Barkov, M. Lyutikov, Mon. Not. R. Astron. Soc. **374**, 415 (2007)
C. Kouveliotou, S. Dieters, T. Strohmayer, J. van Paradijs, G.J. Fishman, C.A. Meegan, K. Hurley, J. Kommers, I. Smith, D. Frail, T. Murakami, An X-ray pulsar with a superstrong magnetic field in soft γ-ray repeater SGR1806-20. Nature **393**, 235–237 (1998)
G. Kowal, A. Lazarian, Astrophys. J. **720**, 742 (2010)
G. Kowal, A. Lazarian, E.T. Vishniac, K. Otmianowska-Mazur, Nonlinear Process. Geophys. **19**, 297 (2012)
R. Kraichnan, Phys. Fluids **8**, 1385 (1965)
P. Kumar, R. Narayan, Mon. Not. R. Astron. Soc. **395**, 472 (2009)
H. Lamb, *Hydrodynamics*, 6th edn. (Cambridge University Press, Cambridge, 1975)
L.D. Landau, E.M. Lifshitz, *Fluid Mechanics* (Pergamon, Oxford, 1959)
A. Lazar, E. Nakar, T. Piran, Astrophys. J. **695**, L10 (2009)
A. Lazarian, *Magnetic Fields in the Universe: From Laboratory and Stars to Primordial Structures* (2005), p. 784
A. Lazarian, P. Desiati, Astrophys. J. **722**, 188 (2010)
A. Lazarian, H. Karimabadi, Phys. Plasmas (2013 in preparation)
A. Lazarian, M. Opher, Astrophys. J. **703**, 8 (2009)
A. Lazarian, E.T. Vishniac, Astrophys. J. **517**, 700 (1999)
A. Lazarian, E.T. Vishniac, Model of reconnection of weakly stochastic magnetic field and its implications. Rev. Mex. Astron. Astrofis. Conf. Ser., **36**, 81 (2009)
A. Lazarian, H. Yan, AIP Conf. Ser. **1505**, 101 (2012)
A. Lazarian, V. Petrosian, H. Yan, J. Cho, arXiv:astro-ph/0301181 (2003)
A. Lazarian, E. Vishniac, J. Cho, Astrophys. J. **603**, 180 (2004)
A. Lazarian, G. Kowal, E. Vishniac, E. de Gouveia Dal Pino, Planet. Space Sci. **59**, 537 (2011)
A. Lazarian, L. Vlahos, G. Kowal, H. Yan, A. Beresnyak, E. de Gouveia Dal Pino, Space Sci. Rev. (2012). doi:10.1007/S11214-9934-7
L. Lehner, C. Palenzuela, S.L. Liebling, C. Thompson, C. Hanna, ArXiv e-prints (2011)

J. Li, A. Spitkovsky, A. Tchekhovskoy, Astrophys. J. **746**, 60 (2012)
N.F. Loureiro, A.A. Schekochihin, S.C. Cowley, Phys. Plasmas **14**, 100703 (2007)
R.V.E. Lovelace, H. Li, A.V. Koldoba, G.V. Ustyugova, M.M. Romanova, Mon. Not. R. Astron. Soc. **358**, 113 (2005)
Y.E. Lyubarsky, Mon. Not. R. Astron. Soc. **345**, 153 (2003)
M. Lyutikov, Mon. Not. R. Astron. Soc. **346**, 540 (2003)
M. Lyutikov, Mon. Not. R. Astron. Soc. **367**, 1594 (2006)
M. Lyutikov, Phys. Rev. D **83**, 124035 (2011)
M. Lyutikov, F. Gavriil, Mon. Not. R. Astron. Soc. **368**, 690 (2005)
M. Lyutikov, J.C. McKinney, Phys. Rev. D **84**, 084019 (2011)
M. Lyutikov, R. Ouyed, Mon. Not. R. Astron. Soc. **369**, L5 (2006a)
M. Lyutikov, R. Ouyed, New J. Phys. **8**, 119 (2006b)
M. Lyutikov, R. Ouyed, Astropart. Phys. **27**, 473 (2007)
M. Lyutikov, R. Ouyed, Mon. Not. R. Astron. Soc. **405**, 1809 (2010a)
M. Lyutikov, R. Ouyed, Mon. Not. R. Astron. Soc. **407**, 1721 (2010b)
M. Lyutikov, D. Uzdensky, Astrophys. J. **589**, 893 (2003)
J. Maron, P. Goldreich, Astrophys. J. **554**, 1175 (2001)
W.H. Matthaeus, S.L. Lamkin, Phys. Fluids **28**, 303 (1985)
W.H. Matthaeus, S.L. Lamkin, Phys. Fluids **29**, 2513 (1986)
J.C. McKinney, D.A. Uzdensky, Mon. Not. R. Astron. Soc. **419**, 573 (2012)
M.A. McLaughlin, Astrophys. J. **613**, 57 (2004)
S. Mereghetti et al., Astrophys. J. **628**, 938 (2005)
R. Micha, I.I. Tkachev, Phys. Rev. Lett. **90**, 121301 (2003)
D. Montgomery, L. Turner, Phys. Fluids **24**, 825 (1981)
C.S. Ng, A. Bhattacharjee, Astrophys. J. **465**, 845 (1996)
D.M. Palmer et al., Nature **434**, 1107 (2005)
E. Priest, T. Forbes, *Magnetic Reconnection: MHD Theory and Applications* (Cambridge University Press, New York, 2000)
N. Rea, A. Tiengo, S. Mereghetti, G.L. Israel, S. Zane, R. Turolla, L. Stella, Astrophys. J. **627**, 133 (2005)
M.J. Rees, Nature **211**, 468 (1966)
D.H. Rischke, S. Bernard, J.A. Maruhn, Nucl. Phys. A **595**, 346 (1995)
J.V. Shebalin, W.H. Matthaeus, D. Montgomery, J. Plasma Phys. **29**, 525 (1983)
B.K. Shivamoggi, Astrophys. Space Sci. **114**, 15 (1985)
T.W. Speiser, Planet. Space Sci. **18**, 613 (1970)
S. Sridhar, P. Goldreich, Astrophys. J. **432**, 612 (1994)
R. Sych, V.M. Nakariakov, M. Karlicky, S. Anfinogentov, Astron. Astrophys. **505**, 791 (2009)
Tavani et al., Science **331**, 736 (2011)
A. Tchekhovskoy, A. Spitkovsky, ArXiv e-prints (2012)
C. Thompson, O. Blaes, Phys. Rev. D **57**, 3219 (1998)
C. Thompson, R.C. Duncan, Mon. Not. R. Astron. Soc. **275**, 255 (1995)
C. Thompson, R.C. Duncan, Astrophys. J. **561**, 980 (2001)
C. Thompson, M. Lyutikov, S.R. Kulkarni, Astrophys. J. **574**, 332 (2002)
P. Troischt, C. Thompson, Phys. Rev. D **70**, 124030 (2004)
D.A. Uzdensky, Astrophys. J. **598**, 446 (2003)
D. Uzdensky, B. Cerutti, M. Begelman, The Astrophysical Journal Letters **737** (2011)
D.A. Uzdensky, B. Cerutti, M.C. Begelman, Astrophys. J. **737**, L40+ (2011)
D.R. Willis, E.J. Barlow, A.J. Bird, D.J. Clark, A.J. Dean, M.L. McConnell, L. Moran, S.E. Shaw, V. Sguera, Astron. Astrophys. **439**, 245 (2005)
P.M. Woods, C. Thompson, arXiv:astro-ph/0406133 (2004)
V.E. Zakharov, V.S. L'Vov, G. Falkovich, *Kolmogorov Spectra of Turbulence I: Wave Turbulence*. Springer Series in Nonlinear Dynamics (Springer, Berlin, 1992)
B. Zhang, H. Yan, Astrophys. J. **726**, 90 (2011)
W. Zhang, A. MacFadyen, P. Wang, Three-dimensional relativistic magnetohydrodynamic simulations of the Kelvin-Helmholtz instability: magnetic field amplification by a turbulent dynamo. Astrophys. J. Lett. **692**, L40–L44 (2009)

DOI 10.1007/978-1-4899-7413-6_16
Reprinted from *Space Science Reviews* Journal, DOI 10.1007/s11214-013-0014-6

Plasma Diagnostics of the Interstellar Medium with Radio Astronomy

Marijke Haverkorn · Steven R. Spangler

Received: 14 March 2013 / Accepted: 7 August 2013 / Published online: 20 September 2013

Abstract We discuss the degree to which radio propagation measurements diagnose conditions in the ionized gas of the interstellar medium (ISM). The "signal generators" of the radio waves of interest are extragalactic radio sources (quasars and radio galaxies), as well as Galactic sources, primarily pulsars. The polarized synchrotron radiation of the Galactic non-thermal radiation also serves to probe the ISM, including space between the emitting regions and the solar system. Radio propagation measurements provide unique information on turbulence in the ISM as well as the mean plasma properties such as density and magnetic field strength. Radio propagation observations can provide input to the major contemporary questions on the nature of ISM turbulence, such as its dissipation mechanisms and the processes responsible for generating the turbulence on large spatial scales. Measurements of the large scale Galactic magnetic field via Faraday rotation provide unique observational input to theories of the generation of the Galactic field.

Keywords Interstellar matter-Milky Way · Plasmas-astrophysical · Turbulence-space plasma

1 Introduction

The purpose of this article is to discuss how radio propagation measurements provide diagnostics of the interstellar medium (ISM). By radio propagation measurements, we mean

M. Haverkorn
Department of Astrophysics/IMAPP, Radboud University Nijmegen, P.O. Box 9010, 6500 GL Nijmegen, The Netherlands
e-mail: m.haverkorn@astro.ru.nl

M. Haverkorn
Leiden Observatory, Leiden University, P.O. Box 9513, 2300 RA Leiden, The Netherlands

S.R. Spangler (✉)
Department of Physics and Astronomy, University of Iowa, Iowa City, IA 52242, USA
e-mail: steven-spangler@uiowa.edu

those in which a radio astronomical observable (such as the interferometric visibility, or the polarization position angle) has been modified by a medium between the source of the radio waves and the radio telescope. In this paper, we will be interested in plasma media. These measurements provide rather direct information on the ionized gas density (strictly speaking, the electron density), the interstellar magnetic field, and (sometimes) indirect information on flow velocities in the interstellar medium.

In addition to information on the mean plasma properties of the interstellar medium such as $\langle n_e \rangle$ and $\langle \mathbf{B} \rangle$, these propagation observations yield information on turbulent fluctuations in the interstellar plasma. In fact, it can be argued that the information on interstellar plasma turbulence is the most unique contribution of this type of observation to studies of the ISM.

This paper is intended, in part, to serve a tutorial and review function. However, there have been numerous reviews in the past on the probing of the interstellar medium by radio propagation measurements, and the implications of those measurements for the astrophysics of the ISM (see, in particular, Uscinski 1977; Rickett 1977, 1990). There is no point in repeating the material already published in those papers. In the present article, we will make detailed reference to those papers to make a number of important points. At the same time, we will stress remaining, open questions about the interstellar plasma and its turbulence. In some cases, those questions have been actively discussed for many years. We will also clearly point out and discuss those topics in which radio propagation measurements provide crucial data for some of the issues of greatest importance in contemporary astrophysics.

An underlying theme of this paper will be the conceptual unity of plasma processes that occur in the interstellar medium, the solar corona, the interplanetary medium, and finally, experiments in plasma physics laboratories. In the last decade or two, plasma physics laboratory experiments have succeeded in illuminating processes which also occur in astrophysical plasmas. These experiments deal with processes which are at the basis of plasma astrophysics. A partial list of the experiments which are contributing a new dimension to plasma astrophysics are measurement of Faraday rotation in laboratory plasmas, and its use in diagnosing the basic properties and processes in those plasmas (Brower et al. 2002; Ding et al. 2003), observation of the nonlinear interaction of Alfvén waves (Carter et al. 2006), and a number of experimental efforts to investigate the nature of magnetic field reconnection, a core process in astrophysics (e.g. Brown et al. 1989; Zweibel and Yamada 2009; Yamada et al. 2010). The unity of plasma physics and plasma astrophysics is exemplified by the interesting fact that the same radio propagation techniques, with the same radio telescopes, are, or can be, used to study the plasma physics of the interstellar medium, the corona, and the solar wind.

1.1 The Fundamentals: (1) Phases of the ISM

As has been noted for decades, the interstellar medium exists in a number of "phases" of different temperature, density, and ionization state. These different properties mean that fundamental plasma parameters such as the ion gyroradius, ion cyclotron frequency, plasma β, and Debye length differ from one phase to another. The different phases and their plasma parameters were discussed in Spangler (2001). Since this paper will discuss a number of parts of the ISM, we list in Table 1 the phases of the interstellar medium, together with their physical properties. Column 1 gives the astronomical name for the phase, column 2 gives the number density, and column 3 the temperature. Column 4 gives the plasma β (discussion below), and column 5 gives the volume filling factor of each phase. The numbers in this last column are taken directly from Table 1.1 of Tielens (2005), with the exception of the value for the Very Local Interstellar Medium (VLISM), which is taken from Frisch et al. (2011).

Table 1 Main phases of the interstellar medium

Astronomical name	Density (cm^{-3})	Temperature (K)	β	Filling factor
Molecular Cloud	$200-\geq 10^5$	≤ 100	9×10^{-8}	0.050 %
Cold Neutral Medium (CNM)	10–100	~ 100	...	1 %
HII regions	5–10	8000	15–30	3×10^{-3} %
Warm Neutral Medium (WNM)	0.1–0.5	8000	...	30 %
Warm Ionized Medium (WIM)	0.1–0.5	8000	0.29	25 %
Very Local Medium (VLISM)	0.11	6700	0.27	6–19 %
Coronal (HIM)	5×10^{-3}	10^6	1.8	50 %

The physical parameters listed in Table 1 represent averages over large volumes, and in some cases are quite uncertain. The main point of this table is to illustrate the great variety of physical conditions in the ISM. One of the best diagnosed phases in Table 1 is that of the VLISM, consisting of a group of clouds within about 15 parsecs of the Sun. Their properties are known well from high resolution spectroscopy as well as studies of the interaction of the heliosphere with these clouds. A discussion of the properties of the VLISM, as well as the means for deducing these characteristics, is given in Frisch et al. (2011) and Redfield (2009).

The plasma β is an important parameter in specifying the nature of any plasma. It is usually defined as Krall and Trivelpiece (1973)

$$\beta \equiv \frac{p}{B^2/8\pi} \tag{1}$$

where p is the gas pressure and B is the magnitude of the magnetic field. An alternative, and sometimes more meaningful definition is in terms of two fundamental wave speeds in a plasma, the ion acoustic speed c_s and the Alfvén speed V_A (Spangler et al. 1997)

$$\beta \equiv \frac{c_s^2}{V_A^2} \tag{2}$$

where $c_s = \sqrt{\frac{\gamma k_B(T_e+T_i)}{m_i}}$ (Nicholson 1983) and $V_A = \frac{B}{\sqrt{4\pi\rho}}$. In the above definitions, γ is the ratio of specific heats (taken to be 5/3 for the calculations below), k_B is Boltzmann's constant, T_e and T_i are the electron and ion temperatures, respectively, m_i is the mass of the ion which constitutes the gas (taken here to be hydrogen), and ρ is the mass density in the plasma. The two definitions of β in Eqs. (1) and (2) are nearly identical, differing only by a factor of γ, so they are the same for an isothermal equation of state.

The reason for defining β in terms of wave speeds rather than pressures is that the definition of Eq. (2) better suits the problem at hand, which is an understanding of turbulence in the plasmas of the interstellar medium. The ratio of wave speeds in Eq. (2) is critical in determining wave damping and instability properties, as well as other wave characteristics such as compressibility. To the extent that turbulence in astrophysical medium may be modeled in terms of wave properties, this definition of β is more appropriate.

We employ a number of simplifications in calculating c_s, V_A, and β. We assume $T_e = T_i$. Although this is not the case in the solar corona and solar wind, it is known to be the case for the Warm Ionized Medium (WIM, Haffner et al. 2009), and is probably the case in the clouds of the VLISM (Spangler et al. 2011b). We also assume a pure hydrogen plasma, with the important exception of the molecular clouds (see below). This choice avoids the sometimes complicated question of the degree of ionization of helium. Finally, a value of $B = 4 \times 10^{-6}$ G is chosen, with the exception of the molecular clouds.

An important restriction in our calculations is that β is calculated for the "ionized fluid", i.e. the gas that consists of electrons and ions. This restriction is not meaningful for fully-ionized media like the solar corona and the WIM, but is an important point for partially-ionized media like the VLISM and molecular clouds. This distinction most directly affects the Alfvén speed via the choice of ρ. In Table 1 we choose ρ to be the mass density of the ionized fluid, not the total density that includes the neutral gas. Our choice is justified for plasma waves and fluctuations with size scales much smaller than the ion-neutral collisional scale. For much larger scales corresponding to outer scales of turbulence in partially-ionized plasmas, neutral gas participates in the dynamics of the ionized fluid. The effective Alfvén speed is then lower, and the plasma β higher.

We have omitted values of β for the Warm Neutral Medium (WNM) and Cold Neutral Medium (CNM). At an excessively superficial level, one might think that the plasma β is not a meaningful parameter for these neutral gases. In reality, these phases will be ionized at some level. In fact, an interesting recent contribution to the discussion of phases of the ISM has been the advocacy of Heiles for a "fifth phase", which is partially ionized (Heiles 2011). It seems likely that the "fifth phase" of Heiles is the same as the WNM, with perhaps an elevated degree of ionization due to the proximity to an HII region. However, at the present time, the nature and characteristics of this partially ionized medium are not sufficiently specified to add to Table 1.

Given the comments in the previous paragraph, it might seem odd to include a full set of entries for the molecular cloud phase, which contains cold, predominantly neutral, and molecular as opposed to atomic gas. At the outset, it must be recognized that there is an enormous range of gas properties within the category of molecular clouds, from diffuse molecular gas, to dense cores, to protostars. All derived parameters such as c_s, V_A, β, and filling factor also have an enormous range. For this reason, we have chosen one restricted but well-discussed case in considering the plasma β.

At a very basic level, it is obvious that molecular cloud gas is partially ionized because one of the most important observational diagnostics is line radiation from molecular ions such as HCO^+, H_3^+, and N_2H^+. In an insufficiently appreciated paper, Smith (1992) uses data from millimeter wavelength observations of dense clouds to determine properties such as electron density and temperature, and then proceeds to show that these properties satisfy the classic criteria for the plasma state, such as a large number of electrons per Debye sphere. In Table 1, our value of β for molecular clouds is calculated for the parameters in Smith (1992).

An important result from Table 1 is the wide range of β in the different phases of the ISM. It is one of the reasons why the nature of turbulence in these different media may differ as well. The extremely low value in molecular clouds warrants immediate comment. The value for β is so low because, relative to other ISM plasmas, molecular clouds have very low temperatures, high magnetic field strength (for those clouds with H_2 densities greater than about 200 cm^{-3} (Crutcher et al. 2010)), and low ion densities due to low ionization fractions. The value for β here suggests that small scale turbulent fluctuations or plasma waves that do exist in the ionized fluid of molecular clouds will have the properties of waves in low or zero β plasmas. However, for turbulent fluctuations and waves on much larger scales that involve motion of the neutral fluid as well, both the Alfvén speed and β will be much lower.

Of the media listed in Table 1, the Warm Ionized Medium (WIM) is perhaps the one of greatest interest in plasma astrophysics. This situation is due to the substantial body of observational data on this medium; it is probably the best-diagnosed astrophysical plasma beyond the solar wind. A major contribution to our understanding of the properties of the WIM has been the long term program of observing the medium with imaging Fabry-Perot interferometers operating in the Hα line and other important spectral lines. This program was conceived and directed by R.J. Reynolds of the University of Wisconsin; as a consequence the WIM is often referred to as the "Reynolds Layer". In the past decade, these Hα observations have been greatly expanded by the Wisconsin H-Alpha Mapper (WHAM) instrument, under the direction of L.M. Haffner. An excellent review of the scientific results emergent from WHAM and a relevant bibliography is given in Haffner et al. (2009). Observations complementary to those of WHAM are provided by radio propagation measurements, primarily pulsar observations but also in some cases of extragalactic radio sources. Assembly of data on radio wave scattering of pulsars and extragalactic radio sources has led to the inference of the power spectrum of density fluctuations in the WIM (Armstrong et al. 1995). This topic is discussed further in Sect. 1.2 below.

As has long been noted, the pressures of the less dense phases of the ISM are, very roughly, comparable at a value of 1.0×10^{-13}–1.0×10^{-12} dynes/cm^2 (Ferrière 1998; Tielens 2005). By this standard, molecular clouds and HII regions are overpressured. In the case of molecular clouds, the gravitational potential contributes to confinement of the gas. HII regions are overpressured, expanding entities. The other phases, Cold Neutral Medium (CNM), Warm Neutral Medium (WNM), Warm Ionized Medium (WIM), Very Local Interstellar Medium (VLISM) and Coronal Phase or Hot Ionized Medium (HIM) have roughly comparable pressures and may, in fact, be in pressure equilibrium. In addition, the magnetic pressure of the interstellar medium is comparable to the aforementioned gas pressures, with a value of $\simeq 6.4 \times 10^{-13}$ dynes/cm^2 for $B_{ISM} = 4 \times 10^{-6}$ G. Finally, the pressure corresponding to the energy density of the Galactic cosmic rays is also similar, $\simeq 1.0 \times 10^{-12}$ dynes/cm^2, suggesting equilibration between the forms in which the ISM can store energy. This whole situation is summarized in the textbook by Tielens (2005), where a value of $\simeq 0.5 \times 10^{-12}$ dynes/cm^2 is quoted for the gas phases of the ISM (CNM, WNM, WIM, and HIM), and a pressure of $\sim 1.0 \times 10^{-12}$ dynes/cm^2 is assigned to both the magnetic and cosmic ray pressure.

The pressures and other properties of the various phases of the ISM, as well as the pressures of the interstellar magnetic field and cosmic rays were considered in detail by Ferrière (1998). Ferrière (1998) also estimated how these pressures change with Galactocentric radius and altitude above the Galactic plane. For the Galactic location of the Sun, and in the Galactic plane, Ferrière (1998) estimates (see Fig. 3 of that paper) a pressure of 6.0×10^{-13} dynes/cm^2 for the gas phase, and $\simeq 1.0 \times 10^{-12}$ dynes/cm^2 for both the magnetic and cosmic ray pressures.

The rough similarity between thermal gas pressure and magnetic pressure that seems to characterize the local Galactic ISM might not be universal. Beck (2007) discusses observations and analysis of the galaxy NGC 6946, characterized by a high star formation rate, and reports results on the variation of all pressures with galactocentric distance. Beck (2007) finds that the magnetic pressure is considerably larger than the value quoted above for the Milky Way, and that the ionized gas pressure is approximately an order of magnitude less than magnetic pressure. The interstellar plasma of NGC 6946 appears to be a low-β plasma, and the processes of energy equilibration have not progressed to completion.

The above considerations are relevant to the scope of this paper. Our view of the interstellar medium considers it as a dynamic plasma. Magnetohydrodynamics (MHD) describes the dynamics as an interaction of the gas and the magnetic field via pressure terms as discussed above and magnetic tension forces. In addition, plasmas interact with energetic particles such as the cosmic rays through resonant interactions with plasma turbulence.

Among the important, recent developments in this field has been continued progress in specifying the strength of the interstellar magnetic field $\mathbf{B}_{ISM}$, and its dependence on gas density (Crutcher et al. 2010). A plot of magnetic field strength (largely deduced through Zeeman effect measurements) versus gas density shows considerable scatter, and a trend towards larger values only for densities greater than about 200 cm^{-3}. Crutcher et al. (2010) also infer a median magnitude of the interstellar magnetic field in the low density phases of the ISM of 6 ± 1 μG, slightly higher than the value used in the calculations above. This is in agreement with equipartition estimates of the total magnetic field strength from synchrotron emission, and a factor of about three higher than the regular magnetic field component. At the present, there is no observational evidence for a change in the magnitude of $\mathbf{B}_{ISM}$ between different phases of the low density ISM.

1.2 The Fundamentals: (2) Radio Wave Propagation Through the ISM

This paper will concentrate on two radio propagation effects, acting on small (10^2–10^4 km) scales and large (pc) scales, respectively. These are angular broadening of a compact or pointlike source due to density turbulence in the interstellar medium, and Faraday rotation of linearly polarized radio waves from a radio source embedded in the ISM, or outside the Galaxy. We also briefly allude to other radio scintillation phenomena caused by small scale turbulence. In the latter topic, we include the signature of ISM turbulence in gradients of the polarization vector of synchrotron radiation.

1.2.1 Angular Broadening of Compact Sources

The basic physics content of radio wave propagation through the ISM is to be found in the expression for the refractive index of radio waves in a plasma. This is discussed in the proceedings of previous meetings of the International Space Science Institute, i.e. Spangler (2001, 2009). As discussed there, the refractive index depends on the plasma density and magnetic field. For radio wave propagation in the ISM, the magnetic field dependence is determined by the component of the magnetic field in the direction of wave propagation. The modification of the radio refractive index by the magnetic field is much smaller than the modification due to the plasma density. This is responsible for the well-known feature that radio propagation measurements primarily diagnose the plasma density of the ISM, with only a higher order contribution due to the Galactic magnetic field.

Turbulent fluctuations in the plasma density and magnetic field cause stochastic spatial and temporal fluctuations in the refractive index in the ISM. As a result, propagation through such a medium induces all manner of fluctuations in the received radio wave field (Uscinski 1977). The theory of how fluctuations in the refractive index generate corresponding fluctuations in properties of the wave electric field (various n-point correlations of the electric field) is generally attributed to Tatarski (1961). An excellent illustration of the effects of wave propagation through a random medium is given by dynamic spectra of pulsars, an example of which is shown in Fig. 1.[1] The spectrum of a pulsar is a measured as a function of time, and the set of spectra combined as shown in Fig. 1. In the absence of the turbulent interstellar medium, the flux density of the pulsar would be constant over the frequency range shown. The gray scale indicates the brightness of the pulsar, with dark shaded regions being bright. The variation in brightness is due to scattered radio waves alternatively constructively and destructively interfering at different frequencies and times. A discussion of pulsar

[1]We thank James Cordes of Cornell University for providing this graph.

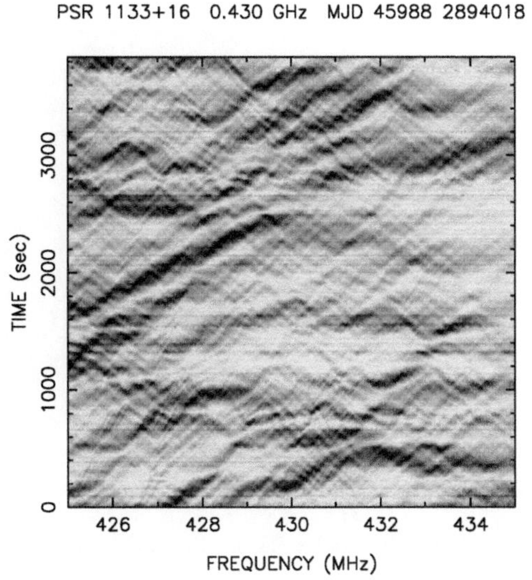

Fig. 1 A dynamic spectrum of the pulsar PSR1133+16. Wave propagation through the stochastic interstellar medium produces constructive and destructive interference at different spatial locations (here portrayed as time) and frequencies. Observations and figure provided by James Cordes, Cornell University

dynamic spectra and the information they contain is given in Cordes (1986). The specific observations shown in Fig. 1 are discussed in Lazio et al. (2004).

A major goal of the theory of wave propagation in a random medium is to relate, via an integral transform, the radio astronomical measurement as a function of the independent variable, to the density power spectral density as a function of wavenumber. Examples of major contributions to this literature are Uscinski (1977), Lee and Jokipii (1975), and Rickett (1977, 1990). An illustration of the various types of observable stochastic propagation phenomena is given in Fig. 1 of Spangler (2009).

1.2.2 Depolarization and Faraday Rotation of Synchrotron Radiation

Interstellar radio propagation measurements also provide information on the basic plasma state of the ISM, such as the plasma density, the vector magnetic field, as well as how these fields vary with position in the Galaxy.

Since variations in the magnetic field vector along the line of sight and/or within the angular size of a telescope beam will partially depolarize linearly polarized synchrotron emission, the observed degree of polarization traces the ratio of large-scale (regular) magnetic field strength to total magnetic field strength (Beck 2001). However, due to small-scale variations in this ratio caused by local structure (supernova remnants, variable turbulence parameters), this method is mostly utilized on kpc-scales in external galaxies.

Parsec-size scales in the magnetized ISM are typically probed using Faraday rotation. The Faraday rotation measure *RM* is directly proportional to the path integral along the line of sight (los) of the electron density n_e and line-of-sight component of the magnetic field $B_{\|}$:

$$\left(\frac{RM}{\text{rad m}^{-2}}\right) = 0.81 \int_{los} \left(\frac{n_e}{\text{cm}^{-3}}\right)\left(\frac{B_{\|}}{\mu\text{G}}\right)\left(\frac{dl}{\text{pc}}\right) \tag{3}$$

Measurements of *RM* therefore provide nearly unique information on the magnetic field in the tenuous, ionized component of the ISM. Equation (3) illustrates the fact that the measured *RM* is sensitive to the distribution of n_e and $B_{\|}$ along the line of

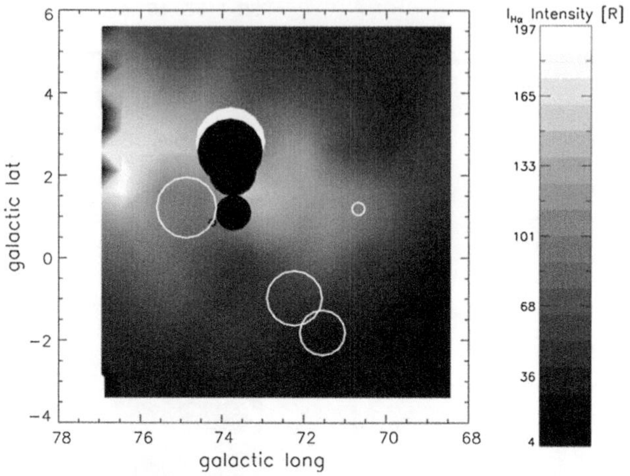

Fig. 2 Highly variable Faraday Rotation Measures in the Galactic plane in Cygnus, in the vicinity of the Cygnus OB1 star formation region. The sizes of the *plotted circles* are proportional to the magnitude of *RM*, specifically proportional to the logarithm of the absolute magnitude of *RM*. *Filled circles* correspond to positive *RM*, *open circles* are negative *RM*. The largest value of (absolute magnitude) of *RM* is 732 rad/m^2, and the smallest is 7.5 rad/m^2. The *gray scale* represents the intensity of Hα emission in this region; the Hα data may show the upper half of the plasma shell associated with the Cygnus OB1 association. Further discussion of these data is given in Whiting et al. (2009). Figure taken from Whiting et al. (2009), reproduced by permission of the American Astronomical Society

sight. If **B** rotates through a large angle such that $B_{\parallel}$ changes substantially or even reverses sign along the line of sight, the value of B inferred by the RM is much less than the true value that would be measured in-situ. This will also be the case if n_e and $B_{\parallel}$ are anticorrelated in the medium being probed, as discussed by Beck et al. (2003).

Traditionally, Faraday rotation is measured from the rotation of linear polarization angle θ as a function of wavelength $\theta = \theta_0 + RM\lambda^2$, where θ_0 is the intrinsic polarization angle at emission of the synchrotron radiation. An illustration of these ideas is given in Fig. 2, which shows the Faraday rotation measure *RM* along several lines of sight to extragalactic radio sources in the Galactic plane in Cygnus. The Faraday rotation of the synchrotron radiation emitted by these sources is dominated by the Milky Way. The large differences in magnitude, and even sign, of *RM* between closely-spaced lines of sight in Fig. 2 are an indicator of the role of young, luminous stars in this region, as they produce structure in the ISM via stellar winds and supernovae, and ionize the gas.

However, if synchrotron emission and the Faraday-rotating medium are mixed or alternating along the line of sight, the simple linearity of polarization angle change with λ^2 is no longer valid. This may be the case in the majority of Faraday rotation measurements. Faraday rotation measurements of the diffuse synchrotron emission in galaxies is the most obvious example, but extragalactic sources may also have several intrinsic *RM* components. In this case, every *RM* component i along the line of sight—now called Faraday depth ϕ_i to indicate that it only probes Faraday rotation along a part of the line of sight—adds its own polarization angle rotation of $\phi_i\lambda^2$, resulting in a non-linear polarization angle change with λ^2. However, this opens the possibility of a Fourier transform, with λ^2 as one of the conjugate variables, in order to disentangle the various ϕ_i components in the total observed signal. This method is called Rotation Measure synthesis (Burn 1966;

Brentjens and de Bruyn 2005). Rotation Measure synthesis takes as its basic observable field a complex polarization function formed from the Stokes parameters Q and U, $P \equiv Q + \imath U$. The observable P is a function of wavelength, or wavelength squared.

This Fourier transform relation between the observed polarization function as a function of wavelength squared $P(\lambda^2)$ and the Faraday dispersion function (or Faraday spectrum) is expressed as

$$P(\lambda^2) = \int_{-\infty}^{+\infty} F(\phi) e^{2i\phi\lambda^2} d\phi \tag{4}$$

$$F(\phi) = \int_{-\infty}^{+\infty} P(\lambda^2) e^{-2i\phi\lambda^2} d\lambda^2 \tag{5}$$

However, since integration over wavelengths from $-\infty$ to $+\infty$ is not possible by definition, in practice these equations include a window function $W(\lambda^2)$ which is non-zero where there is wavelength coverage and zero elsewhere:

$$P_{obs}(\lambda^2) = W(\lambda^2) P(\lambda^2) = W(\lambda^2) \int_{-\infty}^{+\infty} F(\phi) e^{2i\phi\lambda^2} d\phi \tag{6}$$

$$F_{obs}(\phi) = F(\phi) * R(\phi) = K \int_{-\infty}^{+\infty} P_{obs}(\lambda^2) e^{-2i\phi\lambda^2} d\lambda^2, \tag{7}$$

where

$$K = \left(\int_{-\infty}^{+\infty} W(\lambda^2) d\lambda^2 \right)^{-1} \tag{8}$$

This introduces the Rotation Measure Spread Function (RMSF) $R(\phi)$, which describes sidelobes in the Faraday depth signal due to imperfect wavelength coverage, very analogous to the dirty beam (or point spread function) in radio interferometry due to imperfect coverage of the (u, v) plane.

A nice illustration of this effect is given in Fig. 3, which is taken from Schnitzeler et al. (2009). The figure shows polarized synchrotron intensity and polarization angle as a function of wavelength squared of an extragalactic radio source. Variations in polarized intensity and non-linearity of polarization angle with λ^2 suggest that synchrotron emission and Faraday rotation are partially mixed. Indeed, the Faraday spectrum in the figure shows two Faraday depth peaks belonging to two different synchrotron emitting regions with different amounts of Faraday rotation along the line of sight. The two Faraday depth components might be due to a magneto-ionized medium in or around the extragalactic point source. Alternatively, both Faraday depth components could be due to the Galactic interstellar medium, with one of the emission components also being Galactic. In the latter case, the polarized flux of the Galactic component would have to fortuitously be comparable to the extragalactic source. Without additional information it is not possible to distinguish between these two possibilities.

2 Different Astrophysical Media, Common Phenomena

It is a remarkable fact that a number of astrophysical plasmas, i.e. distinct and very different media, have roughly similar radio propagation effects. The Faraday rotation measure (*RM*) due to the Earth's ionosphere is typically in the range 0.5–3 rad/m^2. This is of the order of the *RM* due to the solar corona for lines of sight that pass at a heliocentric distance $\sim 10\ R_\odot$,

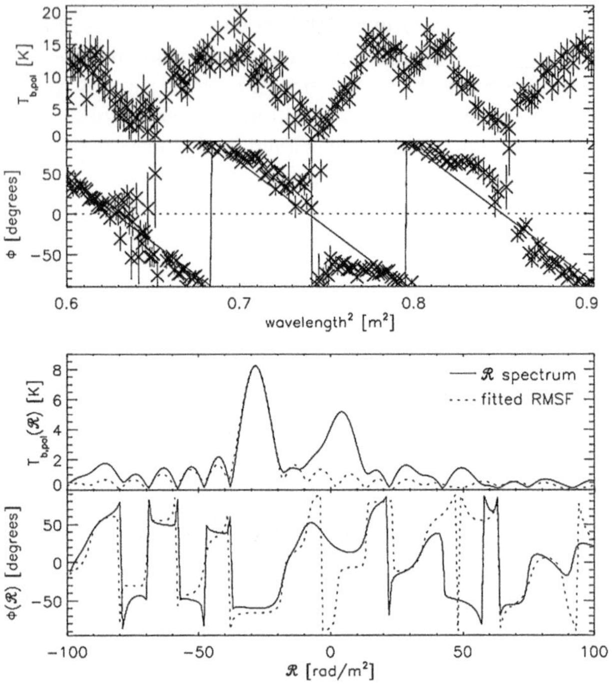

Fig. 3 Polarized intensity (*top*) and polarization angle (*2nd from top*) as a function of wavelength squared for synchrotron emission from an extragalactic radio source observed by Schnitzeler et al. (2009). The Faraday spectrum (*3rd from top*) as a function of Faraday depth ϕ shows two peaks, indicating synchrotron-emitting source components at two separate Faraday depths along this line of sight. Credit: Schnitzeler et al. A & A 494, 611, 2009, reproduced with permission © ESO

and a factor of a few smaller than the standard deviation in *RM* of the interstellar medium at high Galactic latitudes (e.g. Mao et al. 2010).

Another example is provided by Very Long Baseline Interferometry. Very Long Baseline Interferometers operating at frequencies of 1–5 GHz measure similar effects, and of similar magnitude, when observing extragalactic radio sources through the inner solar wind at heliocentric distances of 10–30 $R_\odot$ or through the Galactic plane (Spangler and Sakurai 1995; Spangler et al. 2002; Spangler and Cordes 1988, 1998; Fey et al. 1991).

The fact that similar observational effects are measured for very different media means that scientists who study the ISM should be in dialog with heliospheric scientists who are studying related scientific questions. Finally, in the case of Faraday rotation, there is also a connection with laboratory plasma research. Faraday rotation is used as a diagnostic of fusion plasmas (Brower et al. 2002; Ding et al. 2003). This offers ISM astronomers the possibility of laboratory "ground truth" for some of our diagnostics.

3 Well-Established Results from Radio Propagation Studies

There are several results from radio propagation studies that are so well established, and confirmed by different investigators, that they now have the status of basic properties of the ISM that should be explained by theories.

3.1 "The Big Power Law in the Sky"

One of the best-known results from radio propagation studies is that the spatial power spectrum of density fluctuations follows a Kolmogorov spectrum over at least 5, and perhaps 10 decades (Armstrong et al. 1995). It should be emphasized that the result of Armstrong et al. (1995) pertains to the Warm Ionized Medium (WIM) component of the ISM. Analogous results have been reported for diagnostics of the neutral gas (e.g. Chepurnov et al. 2010). The neutral gas resides in the WNM phase, which is spatially distinct from the WIM. As noted above, (Sect. 1.1) the degree of ionization in the WNM, while low, is not zero, and it may be considered as a weakly-ionized plasma. As such, the dynamics of the plasma component are relevant, since they are communicated to the neutrals via collisions. Whether the fluctuations studied by Chepurnov et al. (2010) are driven by plasma dynamics due to the minority ion species remains unknown.

It is intriguing that the solar wind density power spectrum is the same as that observed for the ISM (at least for the slow solar wind), although the inertial subrange is smaller, being perhaps 3 orders of magnitude (Bruno and Carbone 2005). In the case of the ISM, it is perhaps not sufficiently appreciated that there must be a corresponding power law spectrum of magnetic field fluctuations to explain the highly diffusive transport properties of cosmic rays with a very wide range of energies. This point has been made and emphasized by Jokipii (1977, 1988).

3.2 Spatial Variation in the Intensity of Interstellar Turbulence

A measure of the intensity of turbulence is the parameter C_N^2, which is the normalization constant of the density power spectrum. That is, if the spatial power spectrum of density fluctuations is $P_n(q)$ where q is the spatial wavenumber,

$$P_n(q) = C_N^2 q^{-\alpha} \tag{9}$$

and α is the spectral index of the power spectrum. This form of the power spectrum is the simplest, in which the power spectral density depends only on the magnitude of $\mathbf{q}$, and is thus isotropic. It is known that magnetohydrodynamic (MHD) turbulence is in fact anisotropic, with the large scale magnetic field determining the preferred direction (see comments in Sect. 3.4 below). In this case, the power spectral density $P_n(\mathbf{q})$ depends separately on $q_\perp$ and $q_\parallel$, the components of $\mathbf{q}$ perpendicular and parallel, respectively, to the large scale field (this is discussed in Spangler 1999, with a guide to the relevant literature). For most of this paper, we adopt Eq. (9) as a convenient approximation.

The parameter C_N^2 is directly related to the variance of the density fluctuations. It has been long realized that C_N^2 varies drastically from one part of the interstellar medium to another (Rickett 1977; Cordes et al. 1985). In some cases, it is clear that lines of sight with large C_N^2 traverse HII regions, or other regions with higher than normal plasma density. However, there appear to other lines of sight where no such obvious region of enhanced density exists. It remains unclear whether some of this variation could be due to true turbulent intermittency (Spangler and Cordes 1998; Spangler 1999).

3.3 The Galactic-Scale Magnetic Field

Large-scale Faraday rotation surveys of the sky show substantial organization of the *RM* in different parts of the sky, in the sense that the magnitude and sign of *RM* are correlated over significant parts of the sky. This is interpreted as evidence of a Galactic-scale magnetic field. Most likely, the large-scale magnetic field in the Milky Way generally follows the spiral arms, as is ubiquitously seen in synchrotron observations of external galaxies (Beck 2001). However, there is some evidence for local deviations from the spiral structure (Brown et al.

2007; Rae and Brown 2010; Van Eck et al. 2011), similar to M51 (Patrikeev et al. 2006). One reversal in the large-scale magnetic field direction just inside the Solar circle has been known for decades (e.g. Thomson and Nelson 1980), although it remains unclear why these large-scale reversals are not observed in external spirals. There is still much controversy about the number and location of any other large-scale field reversals in the Galactic disk. For an extensive review, see Haverkorn (2013).

3.4 Anisotropy of Turbulence

A major result from the theory of magnetohydrodynamic (MHD) turbulence, confirmed by observations of turbulence in the solar corona and solar wind, is that turbulence is anisotropic, in the sense that turbulent irregularities are stretched out along the large scale magnetic field. This result was obtained by Strauss (1976) for the case of irregularities in fusion plasmas, but the arguments presented by Strauss are also valid in the case of astrophysical plasmas. This has been broadly appreciated in the astrophysical, as well as plasma physics community since the work of Goldreich and Sridhar (1995), which advocated a view of MHD turbulence as comprised of counterpropagating Alfvén waves. The interaction of these Alfvén waves consequently develops a perpendicular cascade of turbulent energy. This anisotropy is also observed to be present in interstellar turbulence (see Brisken et al. 2010; Rickett 2011, for recent discussions of the more pronounced cases).

3.5 The Dissipation Range of Turbulence

Important progress has been made in the past decade in our understanding of the dissipation of plasma turbulence. This progress has been possible through improved measurements of solar wind turbulence, as well as novel theoretical developments (Howes et al. 2008; Alexandrova et al. 2009, 2012; Howes et al. 2011a, 2011b; Sahraoui et al. 2012). These investigations have identified spatial scales on which dissipation occurs, and advanced suggestions for the responsible mechanisms. It is now clear that a break in the power spectrum of magnetic field fluctuations in the solar wind occurs on scales comparable to, and smaller than the ion inertial length l_i,

$$l_i \equiv \frac{V_A}{\Omega_i} \tag{10}$$

where V_A is the Alfvén speed and Ω_i is the ion (proton) cyclotron frequency (Howes et al. 2011a; Alexandrova et al. 2009, 2012). There remains active discussion in the community as to whether the dissipation is due to Landau damping of highly oblique Alfvén waves (the assumption being that turbulent fluctuations on these scales have damping properties similar to linear plasma wave modes (Howes et al. 2008, 2011a, 2011b), dissipation of other, higher frequency modes propagating at large angles with respect to the mean magnetic fields (Sahraoui et al. 2012), or damping by other modes on electron inertial scales (Alexandrova et al. 2009, 2012). In keeping with the philosophy of this paper, we assume that these results are of great importance of our understanding of the ISM as well.

As will be discussed in Sect. 4, there is good evidence for the beginning of the dissipation range in interstellar turbulence at the ion inertial length, although there is also evidence for differences between solar wind and interstellar turbulence (see Sect. 4.5 below).

3.6 The Outer Scale of Interstellar Turbulence

The outer scale to magnetized turbulence in the Warm Ionized Medium (WIM) phase of the ISM is measured from structure functions of RM to be a few parsecs in the spiral arms, but up to ∼100 pc in the interarm regions (Haverkorn et al. 2006, 2008). This is in agreement with estimates of the outer scale of turbulence averaged over large parts in the sky (mostly towards the Galactic halo) of order 100 pc by (e.g. Lazaryan and Shutenkov 1990; Ohno and Shibata 1993; Chepurnov and Lazarian 2010). These scales are similar to the final sizes of supernova remnants, suggesting (combined with energy arguments, see MacLow 2004) that these are the dominant sources of turbulence in the WIM. The smaller outer scale found in spiral arms may be due to the abundance of HII regions (Minter and Spangler 1996; Haverkorn et al. 2004a). Small outer scales of a few parsecs have also been found in the highly polarized Fan region (Iacobelli et al. 2013) or from anisotropies in TeV cosmic ray distributions (Malkov et al. 2010).

4 Turbulent Microscales in the Interstellar Medium

4.1 Definition of Turbulent Microscales

In this section, we discuss the ways we can measure turbulence on very small scales in the interstellar medium. By very small scales, we mean those in the dissipation range. Knowledge of this part of the turbulent cascade is very important because it contains information on the way in which energy is taken from the large scales and transferred to other forms, presumably heat energy of the interstellar gas.

Our interest in this section will be particularly focused on two phases of the ISM, the WIM and HII regions surrounding young stars. The WIM is of interest because it appears to be the best-diagnosed phase of the ISM, as discussed in Sect. 1.1 above.

4.2 Turbulent Microscales in the Solar Wind

Once again, we use the solar wind, with its extensive and often sophisticated, in-situ measurements and substantial body of theoretical results as a model for interstellar turbulence. Spacecraft instruments provide in-situ measurements of virtually all plasma parameters of interest, and provide the best data set for discussions of MHD turbulence. Measurements of solar wind turbulence at a heliocentric distance of 1 A.U. (the bulk of spacecraft measurements) show a power-law power spectrum of magnetic field fluctuations with a single spectral index that extends from an outer scale with a size of one to a few solar radii, down to an inner scale of a few thousand kilometers (e.g. Bruno and Carbone 2005). A number of investigations have shown that this scale corresponds to the ion inertial length l_i defined in Eq. (10).

Radio propagation observations through the corona and inner solar wind (Coles and Harmon 1989; Harmon and Coles 2005; Spangler and Sakurai 1995) also show strong evidence for an enhancement in the power spectral density of plasma density fluctuations on the scale of the ion inertial length. This bulge on the approximate scale of the ion inertial length can also be seen in power spectra from in-situ measurements of plasma density in the solar wind at 1 A.U. (Chen et al. 2012). These observations, from direct, in-situ measurements as well as radio propagation observations, are interpreted as evidence that the fluctuations in the dissipation range have properties of obliquely-propagating, kinetic Alfvén waves, since such waves become more compressive at the ion inertial scale (Harmon 1989). In fact, the kinetic Alfvén nature of the fluctuations on the dissipation scale is the basis of the model of turbulence in the dissipation range advanced by Howes and colleagues (Howes et al. 2011a). The situation for the solar corona and solar wind seems quite consistent as regards both measurements of plasma density fluctuations and theoretical understanding of the entire turbulent cascade. The prominence of this directly-detected bulge appears to be less pronounced than that retrieved from radio propagation measurements in the corona and inner solar wind. This may indicate that the kinetic Alfvén wave component decays with increasing heliocentric distance in the solar wind. This would hardly be surprising, since many properties of the solar wind change with heliocentric distance (e.g. Bruno and Carbone 2005).

4.3 How We Measure Microscales in Interstellar Turbulence

The ion inertial length in the WIM phase of the interstellar medium is of order one hundred to a few hundred kilometers (Spangler and Gwinn 1990). At first, it seems amazing that any kind of astronomical measurement, made on a medium with an extent of kiloparsecs, could diagnose fluctuations on such a scale. Radio propagation measurements make this possible.

The subsequent discussion in this section will concentrate on one of the scintillation phenomena mentioned in Sect. 1.2, angular broadening. A point source of radio waves viewed through a turbulent medium will appear as a blurred, fuzzy object. Essentially the same phenomenon is encountered at optical wavelengths in the form of seeing disks of stars. A radio source that is blurred by interstellar turbulence will have a measured brightness distribution $I(x, y)$ which is more extended than the intrinsic image of the source. Here I is the intensity of the radiation, which is a function of two angular coordinates on the sky, x and y, customarily Right Ascension and Declination. This brightness distribution contains information on the intensity and spatial power spectrum of the density fluctuations. The brightness distribution is related to the observable quantity which is directly measured by the interferometer. The way a radio interferometer makes an image of a radio source is to measure the *complex visibility function* $V(u, v)$, which is directly related to the correlation between the radio wave electric field at two antennas of an interferometer (Thompson et al. 1986). The complex visibility function has units of Janskys (radiative flux), and is a function of the arguments u and v, the east-west and north-south components of the interferometer baseline, normalized by the wavelength of observation. A two dimensional Fourier transform relates the complex visibility function $V(u, v)$ and $I(x, y)$ (Thompson et al. 1986).

Observations have shown that the turbulent irregularities in the WIM and HII regions around OB associations, like those in the corona and solar wind, are anisotropic in the sense that is theoretically expected. A summary of the observational evidence as of 1999 is given in Spangler (1999). For the present purposes, we will employ the simplifying approximation of isotropic irregularities. The arguments given here can be generalized to the case of anisotropic scattering (Spangler 1988). In the case of isotropic scattering,

$$V(u, v) = V\left(\sqrt{u^2 + v^2}\right) = V(r) \tag{11}$$

where r is the (dimensional) interferometer baseline length, projected on the plane of the sky. In this case, the complex visibility function of a point source viewed through a turbulent medium is a real function, (Cordes et al. 1985; Spangler and Gwinn 1990)

$$V(r) = S_0 e^{-\frac{1}{2}D_\phi(r)} \tag{12}$$

where S_0 is the total flux density of the source, and $D_\phi(r)$ is the *phase structure function*, which contains information on the intensity and spatial power spectrum of the density fluctuations. For the case of scattering by a homogeneous, turbulent slab of thickness L, the phase structure function is (Cordes et al. 1985; Spangler and Gwinn 1990)

$$D_\phi(r) = 8\pi^2 r_e^2 \lambda^2 L \int_0^\infty dq q\left[1 - J_0(qr)\right] P_n(q) \tag{13}$$

The functions and variables in Eq. (13) are as follows. The classical electron radius is given by r_e, λ is the wavelength of observation, L is the thickness, or extent along the line of sight of the turbulent plasma, q is the magnitude of the turbulent wavenumber, $J_0(x)$ is a Bessel function of the first kind of order 0, and $P_n(q)$ is the spatial power spectrum of the density fluctuations. One expects the power spectra of all plasma parameters to be modified at wavenumbers corresponding to the reciprocal of the ion inertial length, ion gyroradius, or similar plasma microscale on which dissipation begins to become important. Spangler and Gwinn (1990) adopted the following simple model in which Eq. (9) is modified by having the power spectrum truncated on wavenumbers larger than a dissipation wavenumber q_0,

$$P_n(q) = C_N^2 q^{-\alpha} e^{-q/q_0} \tag{14}$$

Although Eq. (14) is highly simplified, and was adopted by Spangler and Gwinn (1990) for analytic convenience, the power spectrum of magnetic field fluctuations in the solar wind at 1 A.U. is truncated by an exponential function (Alexandrova et al. 2009, 2012). An important difference between the result of Alexandrova (2009, 2012) and the analysis presented below is that Alexandrova and coworkers found exponential truncation of the power spectrum on electron rather than ion scales (i.e. electron inertial length, or electron gyroradius).

Substitution of Eq. (14) into Eq. (13), and change of variables from $q \to y \equiv qr$ gives the following expression

$$D_\phi(r) = 8\pi^2 r_e^2 \lambda^2 \left(C_N^2 L\right) r^{\alpha-2} \int_0^\infty dy\left[1 - J_0(y)\right] y^{-(\alpha-1)} e^{-y/Q} \tag{15}$$

where $Q \equiv \frac{r}{l_d}$, with l_d being the dissipation scale, $l_d \simeq \frac{1}{q_0}$. The quantity $C_N^2 L$ is termed the *scattering measure*, and roughly determines the magnitude of angular broadening. Turbulent plasmas that have large C_N^2, large L, or both, will produce heavy angular broadening.

Equation (15) yields important insight on remote sensing diagnosis of interstellar (and heliospheric) turbulence. Let us start with the case $Q \to \infty$, which corresponds to an infinitely small dissipation scale. In this case, the spectrum is power law for all wavenumbers larger than that corresponding to the outer scale. In this case, the integral in Eq. (15) is a number which depends only on the index α. For the Kolmogorov spectrum ($\alpha = 11/3$) the value of the integral is 1.117, and the structure function $D_\phi(r) \propto r^{5/3}$.

Equation (15) also illustrates one of the most intriguing aspects of radio wave propagation, and demonstrates why radio astronomical measurements can contribute much to a discussion of plasma microscales in the ISM. Since Eq. (15) is an integral over wavenumber (this is explicit in Eq. (13)), the integrand shows which wavenumbers dominate the measurement. The integrand in Eq. (15), which we note by the function $I(y, Q)$, is shown in Fig. 4 for the case $I(y, \infty)$ and $\alpha = 11/3$ (Kolmogorov spectrum).

The function $I(y, \infty)$ is monotonically decreasing with increasing y. At first, this would seem to indicate that the lowest wavenumbers in the spectrum dominate the measurement. However, that is not true for the case of a Kolmogorov spectrum. The measurement (D_ϕ) is determined by an integral over y. This is because for $y \leq 2$, each progressively higher decade in y makes a larger contribution to the integral. The integral is dominated by contributions with $y \sim 1$–10 where there is an inflection in the function $I(y, \infty)$. When an

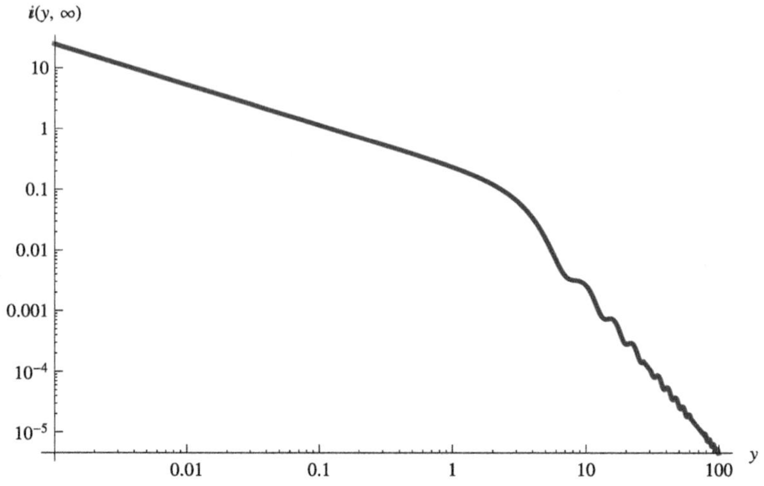

Fig. 4 Plot of the function $I(y, Q)$ for a Kolmogorov spectrum ($\alpha = 11/3$) and $Q = \infty$. The integrand tells what wavenumbers in the turbulent spectrum dominate the measurement on a given interferometer baseline. This occurs for $y \equiv qr \geq 1$, which corresponds to irregularities with sizes of order the interferometer baseline

interferometer measures a broadened radio source, it is responding to irregularities with a wide range of wavenumbers. However, the dominant contribution to the visibility measurement is from irregularities with sizes comparable to the baseline length. This baseline length ranges from tens of kilometers in the case of the Very Large Array, to a few thousand kilometers in the case of Very Long Baseline Interferometers. Obviously, for this statement to be relevant, a measurement with a given interferometer must be affected or even dominated by propagation effects. This point was made in the context of interstellar scattering in Spangler (1988).

Equation (15) also shows how the presence of turbulent dissipation is manifest in radio propagation measurements. When Q is finite, corresponding to a finite dissipation scale, the $e^{-y/Q}$ term will depress the value of the integral. The value of $D_\phi(r)$ at short baselines, where dissipation is pronounced, is less than a value extrapolated from larger values of r according to an $r^{5/3}$ relation. In the dissipation range, $D_\phi(r)$ has a steeper dependence than $r^{5/3}$. To illustrate these points, Fig. 5 shows two structure functions, one without an inner scale and the other with an outer scale of 300 km.

4.4 Observational Results on Turbulent Dissipation Scales in the Interstellar Medium

These issues were discussed in Spangler and Gwinn (1990), who showed that observers who interpret their angular broadening data in terms of a spectral index α would report values which depend on the baselines used in the measurement. Angular broadening measurements on short baselines would yield a value of $\alpha \simeq 4.0$, whereas measurements on long baselines in the inertial subrange would yield $\alpha \simeq 3.67$. This may be seen by reference to Fig. 5. The inferred spectral index is determined by the slope of $D_\phi(r)$ versus r on a log-log plot, such as Fig. 5. Spangler and Gwinn (1990) assembled the data on angular broadening measurements that were available at that time, and showed that a dependence of the inferred value of α on the interferometer baselines used did seem to be present in the data. The result was shown in Fig. 1 of Spangler and Gwinn (1990). From these data they found that there could be a

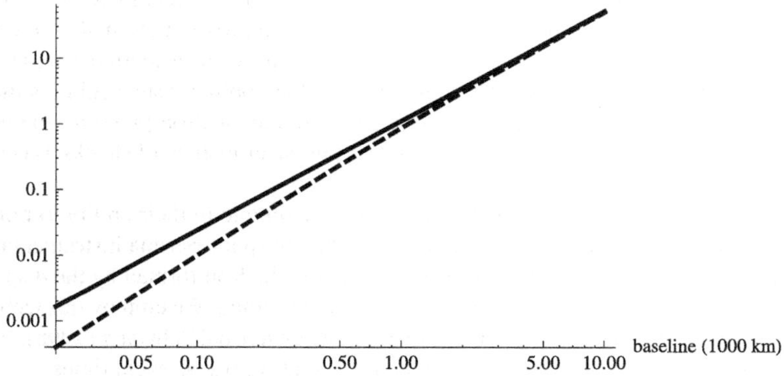

Fig. 5 Theoretical phase structure functions due to turbulence in the interstellar medium. The *plotted lines* correspond to theoretical structure functions $D_\phi(r)$ given by Eq. (12). Both adopt a Kolmogorov spectrum of density irregularities. The *solid curve* is for a Kolmogorov spectrum with no inner scale ($l_d \rightarrow 0$). The *dashed line* corresponds to a spectrum with an inner scale $l_d = 300$ km. Sensitive, carefully calibrated interferometer measurements can distinguish between these two cases

break in the interstellar density power spectrum with an inner scale of 50–200 km. More importantly, they pointed out that a scale in this range was actually expected, if the inner scale corresponds to the ion inertial length l_i defined in Eq. (10), as is the case for scattering in the corona and solar wind (Sect. 4.2).

A more direct determination of an inner scale, made from comparison of measured $D_\phi(r)$ values with the theoretical expression in Eqs. (10) and (12), was made by Molnar et al. (1995). Molnar et al. (1995) made and analyzed angular broadening measurements similar to those of Spangler (1988), but of the radio source Cygnus X-1. It is viewed through the HII region associated with the Cygnus OB2 association. Molnar et al. (1995) essentially made a fit of Eqs. (12) and (15) (but also including the effect of anisotropy of scattering) to their data, and found a satisfactory model to be a Kolmogorov underlying spectrum and an inner scale of 300 kilometers.

An additional, and particularly compelling result has been the recent report by Rickett et al. (2009). They studied the form of the broadening profile of pulses from the pulsar PSRJ1644-4559. Late in the pulse, radiation is being received from highly scattered rays that are probing very small scale irregularities. Rickett et al. (2009) make the important point that the amount of radiation received late in the pulse is only consistent with a Kolmogorov spectrum that breaks at an inner scale as expressed by Eq. (14), or something similar. A power spectrum that remained Kolmogorov to infinitely high spatial wavenumber would cause more pulse power to be observed late in the pulse than is actually seen. Rickett et al. (2009) use their data to extract a value for the inner scale of 70–100 km.

These three independent investigations using radio propagation data have therefore concluded that there is a spectral break in the density power spectrum of the interstellar medium, and that this inner scale is consistent with the ion inertial length. It should be emphasized that the lines of sight analyzed by Spangler and Gwinn (1990), Molnar et al. (1995), and Rickett et al. (2009) all traversed HII regions. There is, as yet, no observational data that can determine the inner scale to the turbulence in the WIM. We do not know if such an inner scale would be at the ion inertial length.

4.5 A Break or a Bulge?

One interesting, and at the present preliminary feature emergent from these investigations regards the transition to the dissipation range in interstellar turbulence. In HII region plasmas, the dissipation range appears to consist of a smooth steepening, without the bulge in the density power spectrum on the ion inertial length, as exists in the corona and solar wind (Sect. 4.2). Given the admittedly limited present information, it appears that the interstellar spectrum of density fluctuations has no compressive bulge at the inner scale. If confirmed by subsequent investigations, it could point to an important distinction between turbulence in the interstellar medium and that in the solar corona and solar wind. The results from Rickett et al. (2009) seem particularly compelling, because a bulge in the interstellar density power spectrum on the ion inertial scale would produce more pulse power at late arrival times than is actually seen (this point is clearly illustrated in Fig. 7 of Rickett et al. 2009).

If this bulge is missing in the interstellar density spectrum, what does it signify? Does it imply that kinetic Alfvén waves are not present in the turbulent field, or that the small scale irregularities in the interstellar medium do not evolve in a manner similar to kinetic Alfvén waves? In that case, what is the nature of the fluctuations over such a large inertial subrange in the ISM? As mentioned in Sect. 4.2, the solar wind results may provide guidance; the results of Chen et al. (2012) indicate that the prominence of kinetic Alfvén waves decreases with increasing heliocentric distance. Interstellar turbulence is comparatively much older in terms of the number of eddy turnover times, so it is certainly plausible that the kinetic Alfvén wave component of ISM turbulence has dissipated.

Another possible resolution is also suggested by studies of the solar wind. As shown in Table 1, HII regions have large values of β, $\beta \gg 1$. Chandran et al. (2009), in a discussion of solar wind density fluctuations, showed that the compressibility of kinetic Alfvén waves decreases with increasing β (see Fig. 3 of Chandran et al. 2009). Kinetic Alfvén waves may well be present in HII regions, but are relatively incompressive and make a small contribution to the density fluctuations in these plasmas.

These ruminations need more extensive and more convincing observational demonstration. Fortunately, the instruments and observational techniques are operational and available. The instruments currently available for angular broadening measurements are greatly improved over those used in the measurements cited above (Spangler and Cordes 1988, 1998; Spangler and Gwinn 1990; Molnar et al. 1995). Those investigations used Very Long Baseline Interferometers with much smaller bandwidths and correlator capability than are now available with the Very Long Baseline Array (VLBA) of the National Radio Astronomy Observatory (NRAO). In addition, the LOFAR low frequency radio telescope in Europe is now operational and has the capability of making novel angular broadening measurements. Finally, the work of Rickett et al. (2009) also demonstrates the advances that have been made in pulsar measurements of the ISM, utilizing new, state-of-the-art pulsar processors on large single dish telescopes such as the Parkes antenna or the Green Bank Telescope of NRAO. Future investigations with these powerful new instruments could illuminate the interesting question as to whether turbulent dissipation processes are the same in the solar wind and the plasma components of the interstellar medium.

5 Turbulent Mesoscales in the Interstellar Medium

5.1 Rotation Measure Synthesis Results

The interpretation of three-dimensional Faraday depth cubes (i.e. polarization intensity maps in spatial coordinates where Faraday depth is the third dimension) is anything but straightforward.

In addition to the artefacts introduced by a non-Gaussian rotation measure spread function, as explained in Sect. 1.2.2, a number of other effects contribute to the difficulty of translating rotation measure cubes into physical properties of the interstellar medium.

Firstly, in analogy to aperture synthesis, a limited range in wavelength squared causes a limited sensitivity to large-scale Faraday depth structures. In contrast with aperture synthesis, in RM synthesis this can lead to a situation where the maximum detectable scale is smaller than the Faraday depth resolution. Therefore, only Faraday-thin components[2] and sharp gradients in Faraday depth, such as the edges of Faraday-thick components, will show up in a Faraday spectrum. The dependence on wavelength range of the Faraday depth resolution $\delta\phi$, maximum detectable scale $\Delta\phi_{max}$ and maximum detectable Faraday depth ϕ_{max} are given by Brentjens and de Bruyn (2005) as

$$\delta\phi \approx \frac{2\sqrt{3}}{\Delta\lambda^2} \tag{16}$$

$$\Delta\phi_{max} \approx \frac{\pi}{\lambda^2_{min}} \tag{17}$$

$$|\phi_{max}| \approx \frac{\sqrt{3}}{\delta\lambda^2} \tag{18}$$

Secondly, different Faraday depth features in a Faraday spectrum only contain information about the amount of their Faraday depth, but not necessarily about their distance. If along a line of sight Faraday depth increases monotonically, i.e. if no magnetic field reversals exist along the line of sight, then the distance order of Faraday depth components is known. However, in the general ISM, with many multi-scale magnetic field reversals, distance to Faraday components is usually unknown. Only in exceptional cases, if one has complementary rotation measures from a number of pulsars with known distances along similar lines of sight, or if the Faraday depth component has a counterpart with a known distance in another tracer, is it possible to estimate the distance to a Faraday component.

Taking these caveats into account, a number of studies on RM synthesis of diffuse Galactic synchrotron emission have been done, which show mostly consistent results. Brentjens (2011) examines a ~4° × 7° field around the Perseus galaxy cluster, which mostly displays Galactic synchrotron emission, at a broad frequency range around 350 MHz. Synchrotron emitting components are detected at multiple Faraday depths between −50 and +100 rad m^{-2}, are Faraday thin and spatially thin ($\lesssim$40 pc), and are well separated in Faraday depth space, suggesting that they are flanked by Faraday-rotating-only parts of the ISM.

The same effect is noticed by Iacobelli et al. (2013), who study the "ring stucture" in the Fan region (see e.g. Haverkorn et al. 2003; Bernardi et al. 2009) in RM synthesis around 150 MHz. They also identify separate Faraday depth components, viz. the ring structure and a foreground component which they associate with the Local Bubble. Similarly, Pizzo (2010) notice in their Galactic foreground studies in the direction of the galaxy cluster Abell 2255 at multiple frequency bands from 150 MHz to 1200 MHz three distinct ranges of Faraday depth with widely different morphologies.

These early RM synthesis studies of Galactic diffuse synchrotron emission consistently conclude that the synchrotron emission is detected in discrete, often Faraday thin, structures

[2]Faraday-thin component is defined as a gaseous medium observed at a wavelength where the change of polarization angle through Faraday rotation is small. Brentjens and de Bruyn (2005) define Faraday-thin as $\phi\lambda^2 \ll 1$ and Faraday-thick as $\phi\lambda^2 \gg 1$. A Faraday-thin component displays negligible internal Faraday depolarization.

with widely different morphologies, interspersed with Faraday-rotating-only components. It is tempting to interpret these observations as actual small-scale variability in synchrotron emission in the ISM, or discrete regions of excess emission. However, two other effects are at play as well. Synchrotron emission dominates in locations where $\mathbf{B} = \mathbf{B}_{\perp}$, while Faraday rotation only depends on $B_{\parallel}$. This may cause the observed apparent anti-correlation between synchrotron emission and Faraday rotation. Secondly, the insensitivity of the technique to large Faraday-thick (emitting and Faraday rotating) structures, which may mimic Faraday-thin emission components at the edges of the Faraday depth range, will play a role.

Wavelet analysis can be successfully applied to Faraday depth cubes to recognize magnetic features such as turbulence or large-scale magnetic field reversals in nearby spiral galaxies or the intracluster medium in galaxy clusters (Beck et al. 2012). Low frequency data ($\sim$100 MHz) are needed to provide the necessary Faraday depth resolution, while broad frequency coverage (up to several GHz) is crucial to make broad Faraday structures detectable. In practice this requires combination of broad-band data from various telescopes, such as the Global Magneto-Ionic Medium Survey (GMIMS Wolleben et al. 2009).

5.2 Polarization Gradients

Linearly polarized intensity maps of diffuse synchrotron emission consistently show narrow one-dimensional structures of complete depolarization named *depolarization canals* (Haverkorn et al. 2000). Some of these depolarization canals are observational artefacts due to missing short spacings in radio interferometric observations, while other canals point to locations of sharp jumps in rotation measure, i.e. sudden changes in electron density and/or parallel magnetic field in the ISM (Shukurov and Berkhuijsen 2003; Haverkorn et al. 2004b). In addition, not all of these sudden changes in ISM conditions are visible as depolarization canals.

The method of gradients in linear polarization was devised to obtain a complete census of these locations of sudden change of conditions in the ISM (Gaensler et al. 2011). The vectorial polarization gradient is calculated from the Stokes parameters (Q, U) as

$$|\nabla \mathbf{P}| = \left\{ \left(\frac{\partial Q}{\partial x} \right)^2 + \left(\frac{\partial Q}{\partial y} \right)^2 + \left(\frac{\partial U}{\partial x} \right)^2 + \left(\frac{\partial U}{\partial y} \right)^2 \right\}^{1/2} \tag{19}$$

This gives a random-looking pattern of mostly one-dimensional locations of high polarization gradient, of which depolarization canals are a subset (see Fig. 6).

These polarization gradient filaments can be characterized by the moments of the polarization gradient distribution. Simulations of magnetohydrodynamic (MHD) turbulence show that the third and fourth order moments (skewness γ and kurtosis β) increase monotonically with Mach number and depend on sonic Mach number. Comparison of the observed values $\gamma = 0.3$ and $\beta = 0.9$ with simulated values for varying sonic Mach numbers indicates that the magnetic turbulence in the ISM (at least in the field given in Fig. 6) is mildly subsonic to transonic (Gaensler et al. 2011). This is in agreement with estimates of the sonic Mach number in the warm ionized medium from emission measure distributions (Hill et al. 2008).

Simulations also show that the filaments in polarization gradients can be caused by either interacting shocks or random fluctuations in MHD turbulence (Burkhart et al. 2012). These authors also introduce the genus method to characterize the polarization gradient maps. For subsonic turbulence as in the ISM, where magnetic field fluctuations dominate the polarization gradient topology, the topology is 'clumpy', as opposed to supersonic turbulence which shows a "Swiss cheese" topology.

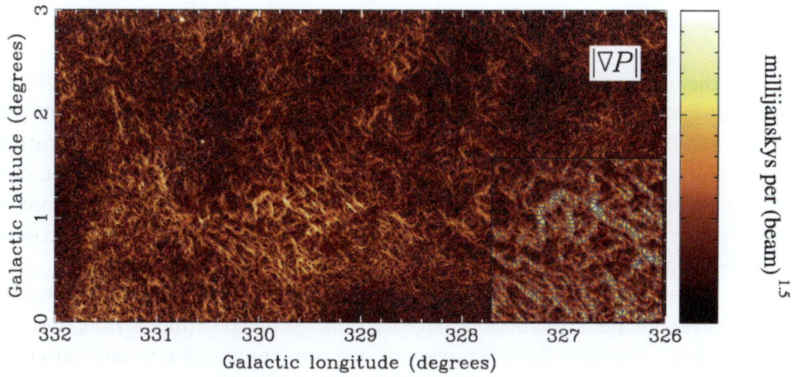

Fig. 6 The map of gradient of linear polarization $|\nabla \mathbf{P}|$ as defined in Eq. (19) for a field around the Galactic plane. The *inset panel* shows a $0.9^\circ \times 0.9^\circ$ close-up of the brightest structure in $|\nabla \mathbf{P}|$, where the direction of the gradient is plotted only for strong (>5 mJy beam$^{1.5}$) gradient amplitudes. Figure reproduced from Gaensler et al. (2011)

6 Mysteries of Interstellar Turbulence

In the paper to this point, we have reviewed the remarkable amount of information revealed by radioastronomical studies on turbulence in the ISM. However, there remain a number of phenomena and effects that are not understood. In this section, we present what may be considered an agenda for future ISM radio propagation studies, that might clarify these issues. We discuss topics in which additional measurements may contribute in a major way to advances in our understanding of interstellar turbulence, or cases in which emerging observational results appear difficult to understand, given our current vision of the interstellar medium and the turbulence in it. These "mysteries" often involve input from the theory of plasma turbulence, and frequently rely on the latest results in that field. In other cases, they represent known observational results of long standing that have eluded explanation.

6.1 The Existence of a Cascade in Interstellar Turbulence

Is there really a cascade in interstellar turbulence from the outer scale of 4 parsecs (or 100 pc) to the dissipation scale of order 100–500 km? Even in the case of well-studied solar wind turbulence this issue is not entirely resolved. Observations of solar wind turbulence indicate that it is comprised of Alfvén waves propagating in both directions with respect to the large scale interplanetary magnetic field, i.e. towards and away from the Sun[3]. Within the context of the most commonly discussed theories of solar wind turbulence, these counterpropagating waves are necessary for the existence of nonlinearities that produce the turbulent cascade. In the case of the interstellar medium, observations are not adequate to demonstrate that counterpropagating Alfvèn waves are present, and it is not clear how such a discrimination could be done. In fact, it is not even clear that interstellar turbulence can be described as an ensemble of Alfvén waves.

[3]A more precise and technical statement would be that both the positive and negative "Elsasser Variables" are present in solar wind turbulence.

6.2 Do We Understand the Flat RM Structure Functions?

The form of the rotation measure structure function for a turbulent, Faraday rotating medium with fluctuations in density δn_e and magnetic field δB_i (B_i being a component of the magnetic field) was derived in Minter and Spangler (1996). The expression presented there assumed vanishing correlation between δn_e and all magnetic field components, but did not exclude the possibility of a correlation between the density and the magnitude of the field. The expression of Minter and Spangler (1996) agrees with that of Simonetti and Cordes (1988) in the limiting case of density fluctuations in a uniform magnetic field.

Rotation Measure structure functions should have a logarithmic slope of $5/3$ if both density and magnetic field fluctuations have a Kolmogorov spectrum, and an observationally indistinguishable value of $3/2$ if a Kraichnan spectrum applies. This results holds if the two lines of sight are separated by a distance which is in the inertial subrange of the turbulence. Such a slope is rarely measured; a number of independent investigations have found that the RM structure functions on angular lags of several tenths of a degree to several degrees have logarithmic slopes of $\sim 1/2$, or even flatter.

The interpretation of this result has been that the angular lags probed correspond to large scales in the interstellar medium, of order the outer scale or larger (Minter and Spangler 1996; Haverkorn et al. 2006, 2008). Minter and Spangler (1996) suggest that the aforementioned data are probing 2D turbulence that exists in sheets, and that the fully 3D component of the turbulence is on spatial scales less than the thickness of these sheets, with corresponding angular scales of order a few tenths of a degree or less. These analyses are, in fact, the basis of the claim that the outer scale is of order 1–5 parsecs in extent.

Nonetheless, it would be comforting to actually measure, in a clear and unambiguous fashion, the transition from a $\sim 1/2$ logarithmic slope on angular lags $\geq 1^\circ$ to $\sim 5/3$ on angular lags $\leq 0.2^\circ$. This would securely establish the value of the outer scale in the WIM turbulence. Without such measurements, we will continue to be tormented by the specter of an ISM in which we *are* measuring the inertial range of the turbulence (assuming this to be a meaningful concept), and that we lack an explanation for its logarithmic slope. Such a flat slope for an inertial subrange of interstellar turbulence would constitute a major paradox for our understanding of interstellar turbulence. A flat spectrum (power law index $\sim\frac{1}{2}$ instead of $\simeq\frac{5}{3}$) would predict density fluctuations on the scales responsible for radio wave scattering (see Sect. 4.3) that are far too large to be compatible with the observed magnitude of radio scintillations of pulsars and extragalactic radio sources.

6.3 What is the Significance of the "Pulsar Arcs"

One of the most intriguing developments of the last decade in the study of interstellar turbulence has been the discovery of "pulsar arcs" (Walker et al. 2004; Cordes et al. 2006). This phenomenon seems most easily explicable if the turbulence responsible for interstellar turbulence is confined to one, or at most, a few thin sheets. A recent overview of the observational properties of the arcs and their interpretation is given in Rickett (2011).

The existence of the arcs is thus linked to one of the most intriguing "mysteries" of ISM turbulence, i.e. whether that turbulence is widely distributed through the Galaxy, or confined to spatially restricted and widely separated regions of intense turbulence. The resolution of this matter has obvious import for our understanding of the mechanisms which generate the turbulence.

As noted above, the existence of the arcs for many pulsars seems to suggest that the turbulence exists in thin sheets. However, the research to definitely prove this has not yet been done. Cordes et al. (2006) point to the desirability of calculations that would investigate the properties, including the existence of arcs caused by extended turbulent media.

Other types of radio scattering measurements can also address the question of the distribution of the turbulence. A particularly promising approach was investigated by Gwinn et al. (1993) who compared angular broadening and pulse broadening measurements for a sample of 10 pulsars. As discussed in Gwinn et al. (1993), the angular width and the temporal width of pulse broadening have different dependences on the distribution of turbulent plasma along the line of sight. In principle, a comparison between pulse broadening and angular broadening can distinguish between uniformly distributed turbulence and that concentrated in a thin screen. Gwinn et al. (1993) concluded that their data were consistent with a uniform distribution of turbulence, except for pulsars such as the Crab Nebula and Vela Nebula pulsars, for which the turbulence is partially contained in a screen that is naturally associated with a supernova remnant.

Another investigation into this matter (Bhat et al. 2004) used only pulse broadening measurements. These authors utilized the fact that the shape of the pulse broadening function is different for screens and uniform, extended media. The goal of the analysis of Bhat et al. (2004) was to determine if screen or extended media better fit the observations of a sample of 98 pulsars. They found that some pulsars in their sample were better fit by extended media, and others by thin screens. This study would therefore indicate that there is no general rule regarding the distribution of turbulence in interstellar space.

A final relevant study is that of Linsky et al. (2008), who convincingly associated the turbulence responsible for intraday flux variations of two quasars with a region of interaction between two of the clouds in the Very Local Interstellar Medium (VLISM). In this specific case, the radio wave scattering is dominated by turbulence in a relatively thin region of interaction between two independent media. However, it is not clear if these conclusions for the VLISM are applicable to the general ISM.

In summary, at the present time observations are ambiguous as to whether the small scale turbulence which is responsible for radio wave scintillations, and which is a focus of attention in this paper, uniformly fills the ISM, is confined to thin layers on presumptive interfaces in the ISM, or is a combination of these two limiting cases. Different observational studies reach different conclusions. Future research could improve our state of knowledge. Work in the theory of scintillations could indicate if thin turbulent screens are necessary for the existence of pulsar arcs, or if these features can also arise in extended, turbulent media. Equally promising would be a new investigation along the lines of Gwinn et al. (1993), utilizing angular broadening measurements made with the Very Long Baseline Array (VLBA), an instrument which has greater sensitivity and accuracy than the instrument used in Gwinn et al. (1993), and pulse broadening analyses as in Bhat et al. (2004).

6.4 What Generates Interstellar Turbulence?

Plasma turbulence, as revealed by the density fluctuations responsible for radio scintillations, appears to be very widely distributed in the interstellar medium. There are also indicators of turbulence in the neutral gas, such as spectral line widths enhanced over their thermal values. Evidence exists for turbulence in most, or all of the phases of the ISM listed in Table 1, and this turbulence exists on a wide range of spatial scales. A current review of interstellar turbulence in general is Elmegreen and Scalo (2004). The question then arises as to the mechanism responsible for its generation. A general consensus holds that the free energy source is expanding supernova remnants, stellar superbubbles, and expanding HII regions, as well as magnetorotational, shear, or other instabilities associated with Galactic rotation (Norman and Ferrara 1996; Elmegreen and Scalo 2004; MacLow and Klessen 2004; Hill et al. 2012). The small scale fluctuations that we detect in radio scintillations might be generated by baroclinic effects[4] at the expanding interfaces between supernova remnants, stellar bubbles, and the ISM. There is also the most obvious possibility, in which these small scale irregularities arise as a consequence of a cascade from large injection scales. This suggestion encounters the difficulty that known dissipation mechanisms are not restricted to small spatial scales (see Sect. 6.6 below).

A number of studies have estimated the volumetric power input to interstellar turbulence from a variety of astronomical sources; these are summarized in Elmegreen and Scalo (2004). The relative contributions of these sources are estimated by Norman and Ferrara (1996). Elmegreen and Scalo (2004) quote a global turbulent input power density of 3×10^{-26} ergs/sec/cm^3, attributing this theoretical estimate to MacLow and Klessen (2004). Interesting, and perhaps fortuitously, this turbulent power density is very close in magnitude to both the estimated volumetric heating rate in the WIM from the dissipation of turbulence due to ion-neutral collisions, and the cooling rate of the WIM (Minter and Spangler 1997; Spangler 2003). These sums are then consistent, though not uniquely so, with a picture in which turbulent energy is input by supernova and stellar associations on the scale of parsecs or tens of parsecs, cascades down to scales comparable to and smaller than the ion-neutral collisional scale, where it is dissipated and then radiated away by the glow of the WIM. Elmegreen and Scalo (2004) estimate that the power input from Galactic rotation is significantly smaller than the numbers above. Elmegreen and Scalo (2004) also make the important point that turbulent input power densities seems to be significantly higher in the denser parts of the ISM, suggesting distinct and segregated turbulent generation mechanisms in different phases of the ISM.

A final point about generation of turbulence is that there may be a problem with the distribution or diffusion of turbulence throughout the ISM. The preceding discussion assumes that all of the processes involved in turbulent power input, i.e. generation at the outer scale, cascade through wavenumber, and dissipation of small spatial scales, are spatially co-located. However, supernova remnants and stellar bubbles occupy a very small fraction of the ISM, and turbulent damping limits the extent to which turbulence can propagate from the generation site to locations throughout the WIM (Spangler 2007; Spangler et al. 2011a). Observations of scintillations of pulsars and extragalactic radio sources, on the other hand, indicate that turbulence is widely distributed through at least the WIM phase of the ISM.

However, it must be admitted that the possible difficulty raised in Spangler (2007) and Spangler et al. (2011a) has not generated "weeping and the gnashing of teeth" in the interested community. The response from that community has been that the role of supernova remnants is to excite a global Galactic system of flows, which then generate turbulence throughout the Galaxy via velocity shear. An evaluation of this matter will depend on a better understanding of whether ISM turbulence is produced by a limited number of point sources, or by processes such as shear that occur throughout the Galaxy.

6.5 Removal of Fast Magnetosonic Waves from Interstellar Turbulence

Fast Magnetosonic waves are one of the three MHD wave modes, so one would expect them to comprise part of interstellar turbulence. Cho and Lazarian (2002, 2003) and Klein et al.

[4]Baroclinic effects involve the generation of fluid vorticity by misaligned gradients of pressure and density.

(2012) have formally investigated the partition of MHD turbulence into fluctuations possessing the properties of these modes, as well as the generation of slow mode and fast mode-like fluctuations from predominantly Alfvénic turbulence. However, it has been argued on observational grounds that Fast Mode waves can only constitute a negligibly small fraction of the energy in interstellar turbulence. The argument is based on the rapid damping of Fast Mode waves on thermal ions for conditions appropriate to the Warm Ionized Medium. If a sizeable fraction of the energy in interstellar turbulence is in the form of Fast Mode waves, then the large power input to the interstellar medium would exceed the cooling capacity of the WIM gas (Spangler 1991, 2003). Interestingly enough, these waves also seem to be absent from the solar corona and the solar wind at 1 A.U. Harmon and Coles (2005) make a convincing argument that a substantial contribution of Fast Mode waves to the coronal turbulence budget is incompatible with spaced-receiver propagation measurements. Klein et al. (2012) also argue that Fast Modes waves, or fluctuations possessing Fast Mode properties, constitute an insignificant portion of solar wind turbulence at 1 A.U. The analysis of Klein et al. (2012) is based on calculations of simulated turbulence, consisting of a superposition of Fast Mode waves, Slow Mode waves, and Alfvén waves. These simulated realizations of turbulence are compared with actual measurements of solar wind turbulence, especially the density-magnetic field correlation function. Klein et al. (2012) find that the realizations that resemble the true, observed turbulence are those with an insignificant fraction of Fast Mode waves. It should be mentioned before leaving this topic that the absence of Fast Mode waves in heliospheric plasmas is a characteristic of plasmas far from shocks or other sources of unstable particle distributions. Shocks produce ion streaming instabilities which in turn generate beautiful, large amplitude Fast Magnetosonic waves, the best known examples of which are the waves upstream of the Earth's bow shock (for an entry point to a large literature Hoppe et al. 1981). However, it seems to be the case that these Fast Mode waves are confined to relatively thin layers that bound strong shocks in the solar wind. To conclude this subsection, whether such a minor role for the Fast Mode in astrophysical turbulence is due to enhanced damping, or the turbulence generation mechanisms remains to be determined by future research.

6.6 Do We Understand the Lack of a Spectral Break at the Ion-Neutral Collisional Scale?

As discussed in Sect. 3.1, there is observational evidence for a power law spectrum of interstellar turbulence from scales of at least a few parsecs, down to scales as small as 100 km. The existence of a power law spectrum seems to indicate, on general grounds of dimensional analysis, that there are no fundamental scales between the outer scale on which stirring is done, and the inner scale where dissipation occurs. This assumption is in stark contrast to the situation for the WIM phase of the interstellar medium, in which a fundamentally-defined scale, the collisional scale $l_c \equiv \frac{V_A}{\nu_{in}}$, where V_A is the Alfvén speed and ν_{in} is the ion-neutral collisional scale, lies between the inner and outer scales ($l_c \simeq 10^{15}$–10^{16} cm). In the WIM the neutral atoms are helium, which is partially or fully neutral. This oddity was discussed by Armstrong et al. (1995).

Cho and Lazarian (2003, this paper also contains references to earlier results by these authors) and Oishi and MacLow (2006) present numerical simulations showing that irregularities exist on scales smaller than the ion-neutral collisional scale, and conclude that ion-neutral collisional effects do not truncate the turbulent cascade. A different conclusion appears to be reached by Shaikh and Zank (2008). Shaikh and Zank (2008) find that while the turbulent cascade continues at wavenumbers larger than that corresponding to the ion-neutral collisional scale, the fundamental physics of the nonlinear interaction is modified.

These authors claim that the magnetic field and velocity power spectra of the ionized fluid are significantly steepened in comparison with the same spectra of a fully-ionized plasma. This result would seem to be discordant with the observed Kolmogorov spectrum of density fluctuations down to much smaller spatial scales. Further discussion of this interesting question should include recognition of the intrinsic anisotropy of MHD turbulence (see Sect. 3.4). For Alfvénic turbulence, the ion-neutral collisional interaction is determined by the parallel wavenumber $k_{\|}$, whereas the irregularities responsible for radio wave scintillation almost certainly have $\frac{k_{\perp}}{k_{\|}} \gg 1$. Whether the anisotropy of turbulence, which will be large at these scales, is enough to resolve this question remains to be determined.

6.7 The Outer Scale of Interstellar Turbulence and Cosmic Ray Propagation

In Sect. 3.6 above, we noted that several independent investigations find that the outer scale of turbulence in the WIM must be of order a few parsecs. There is an associated curiosity that was raised in Spangler (2001), but does not appear to have been discussed since. If there is a break in the turbulence corresponding to an outer scale of 4 parsecs, then there should be an associated change in the transport properties of cosmic rays which are resonant with such irregularities, i.e. those with energies of 10^{15}–10^{16} eV. In fact, the famous "knee" in the cosmic ray spectrum occurs here, but other mechanisms are normally invoked for its existence. Should the change in turbulence properties at scales that resonate with such cosmic rays be added to the mechanisms considered? In considering this matter, it should be recognized that the outer scale of turbulence in the Galactic halo is probably large, of order 100 pc (see Sect. 3.6 above). The larger fluctuations in the Galactic halo could resonate with higher energy cosmic rays than the smaller fluctuations in the WIM of the spiral arms. The fluctuations in the halo may dominate the Galactic transport of cosmic rays.

6.8 Can Observations Provide a Connection to Theories of Kinetic Processes in Turbulence?

One of the most intriguing recent developments in the study of plasma turbulence has been elucidation of the role of kinetic processes in turbulence (i.e. those described by the Vlasov equation rather than MHD), and the observational support for these ideas in spacecraft measurements of solar wind turbulence (Sect. 3.5 above). Can we find similar evidence of kinetic processes in the interstellar medium? Do the same kinetic processes which appear crucial in the solar wind, such as Landau damping of kinetic Alfvén waves, play an important role in the interstellar medium? In kinetic damping processes, energy will flow to either ions or electrons, depending on which is resonant with the fluctuations being damped. However, observations of the WIM plasma, the best diagnosed astrophysical plasma, show temperature equilibration between electrons and different ion species (Haffner et al. 2009). A similar situation occurs in the clouds of the Very Local Interstellar Medium, where the same temperature characterizes neutral atoms as well as several ions with different masses (and therefore cyclotron frequencies, Spangler et al. 2011b).

6.9 Why Is the Spectrum of Plasma Turbulence the Same in HII Regions, the Warm Ionized Medium, and the Solar Wind?

The Armstrong et al. (1995) result of a Kolmogorov density fluctuation spectrum over several decades pertains to the WIM component of the interstellar medium, having been established from observations of relatively nearby pulsars and extragalactic radio sources whose lines of sight are at high galactic latitudes. Radio wave propagation measurements made on heavily-scattered lines of sight that pass through HII regions (e.g. Spangler 1988; Molnar et al. 1995; Rickett et al. 2009) are also consistent with a Kolmogorov density spectrum. Finally, the plasma of the solar corona and inner solar wind also has a Kolmogorov spectrum, particularly in the slow solar wind. In the solar wind, the spectral of magnetic field and flow velocity are observed to possess inertial subranges with spectra that are close to Kolmogorov. Whether the spectra are both exactly Kolmogorov, both slightly flatter than Kolmogorov, or different for the two fields remains a point of contention. Boldyrev et al. (2011) claim that there is a slight difference in power law indices of the magnetic and velocity spectra, but this has been disputed by Beresnyak and Lazarian (2010). Regardless of the resolution of these important matters dealing with the physical nature of plasma turbulence, it remains empirically the case that density, magnetic field, and velocity have "Kolmogorov-like" spectra in the solar wind, and plasmas in the WIM and HII regions have similar density spectra. This result is, perhaps, somewhat unexpected since the mechanisms for generation of the turbulence at the outer scale are presumably quite different in these different media, as might be the plasma β that determines the dissipation mechanisms. Is the similarity of the turbulence in these quite different media a consequence of the universality of turbulence?

7 Summary and Conclusions

Radio propagation observations yield a surprising amount of quantitative information about the plasma state of the interstellar medium, particularly for turbulence in the WIM phase of the ISM and HII regions. The measurements emphasized in this paper have been Faraday rotation of linearly polarized signals that have propagated through the ISM during the passage from extragalactic radio sources to the Earth, frequency-dependent polarization characteristics of the Galactic synchrotron radiation, and scintillations of Galactic and extragalactic radio sources due to small scale density fluctuations in the ISM. From this information, we can deduce the amplitude and spectral properties of interstellar turbulence. We have information on the outer and inner scales of this turbulence; the corresponding inertial subrange extends over roughly 10 decades. In some respects, interstellar turbulence resembles the extensively studied turbulence in the solar wind. However, there appear to be significant differences as well. In spite of impressive progress in this field, there are several (at least) poorly-understood aspects of interstellar turbulence that warrant the term "mysteries". Several of these aspects are discussed in Sect. 6 above. Some of these could be addressed in a significant way with new observations on new or substantially upgraded radio telescopes such as the VLBA and LOFAR.

Acknowledgements This work was supported at the University of Iowa by grants AST09-07911 and ATM09-56901 from the National Science Foundation of the United States. M.H. acknowledges the support of research programme 639.042.915, which is partly financed by the Netherlands Organisation for Scientific Research (NWO). The authors acknowledge the work of Jacob J. Buffo of the University of Iowa in the analysis of model structure functions, contained in Figs. 4 and 5. We thank James Cordes of Cornell University for providing the beautiful pulsar dynamic spectrum in Fig. 1 as an illustration of one of the phenomena of radio wave propagation in a random medium. We also appreciate the advice and recommendations of Dr. Cordes on the question of the spatial uniformity of interstellar turbulence. Finally, we thank Olga Alexandrova, Rainer Beck, and Alex Lazarian for interesting, helpful, and collegial readings of this paper.

References

O. Alexandrova et al., Phys. Rev. Lett. **103**, 165003 (2009)
O. Alexandrova, C. Lacombe, A. Mangeney, R. Grappin, M. Maksimovic, Astrophys. J. **760**, 121 (2012)
J.W. Armstrong, B.J. Rickett, S.R. Spangler, Astrophys. J. **443**, 209 (1995)
R. Beck, Space Sci. Rev. **99**, 243 (2001)
R. Beck, Astron. Astrophys. **470**, 539 (2007)
R. Beck, A. Shukurov, D. Sokoloff, R. Wielebinski, Astron. Astrophys. **411**, 99 (2003)
R. Beck, P. Frick, R. Stepanov, D. Sokoloff, Astron. Astrophys. **543**, A113 (2012)
A. Beresnyak, A. Lazarian, Astrophys. J. **722**, L110 (2010)
G. Bernardi et al., Astron. Astrophys. **500**, 965 (2009)
N.D.R. Bhat, J.M. Cordes, F. Camilo, D.J. Nice, D.R. Lorimer, Astrophys. J. **605**, 759 (2004)
S. Boldyrev, J.C. Perez, J.E. Borovsky, J.J. Podesta, Astrophys. J. **741**, L19 (2011)
M.A. Brentjens, Astron. Astrophys. **526**, A9 (2011)
M.A. Brentjens, A.G. de Bruyn, Astron. Astrophys. **441**, 1217 (2005)
W.F. Brisken et al., Astrophys. J. **708**, 232 (2010)
D.L. Brower et al., Phys. Rev. Lett. **88**, 185005 (2002)
M.R. Brown, C.D. Cothran, M. Landreman, D. Schlossberg, W.H. Matthaeus, Astrophys. J. **577**, L63 (1989)
J.C. Brown et al., Astrophys. J. **663**, 258 (2007)
R. Bruno, V. Carbone, Living Rev. Sol. Phys. **2**, 4 (2005)
B. Burkhart, A. Lazarian, B.M. Gaensler, Astrophys. J. **749**, 145 (2012)
B.J. Burn, Mon. Not. R. Astron. Soc. **133**, 67 (1966)
T.A. Carter, B. Brugman, P. Pribyl, W. Lynbarger, Phys. Rev. Lett. **96**, 155001 (2006)
B.D.G. Chandran, E. Quataert, G.G. Howes, Q. Xia, P. Pongkitiwanichakul, Astrophys. J. **707**, 1668 (2009)
C.H.K. Chen, C.S. Salem, J.W. Bonnell, F.S. Mozer, S.D. Bale, Phys. Rev. Lett. **109**, 035001 (2012)
A. Chepurnov, A. Lazarian, Astrophys. J. **710**, 853 (2010)
A. Chepurnov, A. Lazarian, S. Stanimirovic, C. Heiles, J.E.G. Peek, Astrophys. J. **714**, 1398 (2010)
J. Cho, A. Lazarian, Phys. Rev. Lett. **88**, 245001 (2002)
J. Cho, A. Lazarian, Mon. Not. R. Astron. Soc. **345**, 325 (2003)
W.A. Coles, J.K. Harmon, Astrophys. J. **337**, 1023 (1989)
J.M. Cordes, Astrophys. J. **311**, 183 (1986)
J.M. Cordes, J.W. Weisberg, V. Boriakoff, Astrophys. J. **288**, 221 (1985)
J.M. Cordes, B.J. Rickett, D.R. Stinebring, W.A. Coles, Astrophys. J. **637**, 346 (2006)
R.M. Crutcher, B. Wandelt, C. Heiles, E. Falgarone, T. Troland, Astrophys. J. **725**, 466 (2010)
W.X. Ding et al., Phys. Rev. Lett. **90**, 035002 (2003)
B.G. Elmegreen, J. Scalo, Annu. Rev. Astron. Astrophys. **42**, 211 (2004)
K. Ferrière, Astrophys. J. **497**, 759 (1998)
A.L. Fey, S.R. Spangler, J.M. Cordes, Astrophys. J. **372**, 132 (1991)
P.C. Frisch, S. Redfield, J.D. Slavin, Annu. Rev. Astron. Astrophys. **49**, 237 (2011)
B.M. Gaensler, M. Haverkorn, B. Burkhart et al., Nature **478**, 214 (2011)
P. Goldreich, S. Sridhar, Astrophys. J. **438**, 763 (1995)
C.R. Gwinn, N. Bartel, J.M. Cordes, Astrophys. J. **410**, 673 (1993)
L.M. Haffner et al., Rev. Mod. Phys. **81**, 969 (2009)
J.K. Harmon, J. Geophys. Res. **94**, 15399 (1989)
J.K. Harmon, W.A. Coles, J. Geophys. Res. **110**, A03101 (2005)
M. Haverkorn, in *Magnetic Fields in Diffuse Media*, ed. by E. Gouveia dal Pino, A. Lazarian (2013)
M. Haverkorn, P. Katgert, A.G. de Bruyn, Astron. Astrophys. **356**, L13 (2000)
M. Haverkorn, P. Katgert, A.G. de Bruyn, Astron. Astrophys. **404**, 233 (2003)
M. Haverkorn, B.M. Gaensler, N.M. McClure-Griffiths, J.M. Dickey, A.J. Green, Astrophys. J. **609**, 776 (2004a)
M. Haverkorn, P. Katgert, A.G. de Bruyn, Astron. Astrophys. **427**, 549 (2004b)
M. Haverkorn, B.M. Gaensler, J.C. Brown, N.S. Bizunok, N.M. McClure-Griffiths, J.M. Dickey, A.J. Green, Astrophys. J. **637**, L33 (2006)
M. Haverkorn, J.C. Brown, B.M. Gaensler, N.M. McClure-Griffiths, Astrophys. J. **680**, 362 (2008)
C. Heiles, Bull. Am. Astron. Soc. **43**, 21713208 (2011)
A.S. Hill, R.A. Benjamin, G. Kowal et al., Astrophys. J. **686**, 363 (2008)
A.S. Hill et al., Astrophys. J. **750**, 104 (2012)
M.M. Hoppe, C.T. Russell, L.A. Frank, T.E. Eastman, E.W. Greenstadt, J. Geophys. Res. **86**, 4471 (1981)
G.G. Howes et al., J. Geophys. Res. **113**, A05103 (2008)
G.G. Howes, J.M. TenBarge, W. Dorland, Phys. Plasmas **18**, 102305 (2011a)
G.G. Howes et al., Phys. Rev. Lett. **107**, 035004 (2011b)

M. Iacobelli, M. Haverkorn, P. Katgert, Astron. Astrophys. **549**, A56 (2013)
J.R. Jokipii, in *Proceedings of the 15th International Cosmic Ray Conference*, vol. 1 (1977), p. 15
J.R. Jokipii, Radio wave scattering in the interstellar medium. AIP Conf. Proc. **174**, 48 (1988)
K.G. Klein et al., Astrophys. J. **755**, 159 (2012)
N.A. Krall, A.W. Trivelpiece, *Principles of Plasma Physics* (McGraw-Hill, New York, 1973)
A.L. Lazaryan, V.P. Shutenkov, Sov. Astron. Lett. **16**, 297L (1990)
T.J.W. Lazio, J.M. Cordes, A.G. de Bruyn, J.P. Marquart, New Astron. Rev. **48**, 1439 (2004)
L.C. Lee, J.R. Jokipii, Astrophys. J. **196**, 695 (1975)
J.L. Linsky, B.J. Rickett, S. Redfield, Astrophys. J. **675**, 413 (2008)
M.M. MacLow, Astrophys. Space Sci. **289**, 323 (2004)
M.M. MacLow, R.S. Klessen, Rev. Mod. Phys. **76**, 125 (2004)
M.A. Malkov, P.H. Diamond, L. Drury, R.Z. Sagdeev, Astrophys. J. **721**, 705 (2010)
S.A. Mao, B.M. Gaensler, M. Haverkorn et al., Astrophys. J. **714**, 1170 (2010)
A.H. Minter, S.R. Spangler, Astrophys. J. **458**, 194 (1996)
A.H. Minter, S.R. Spangler, Astrophys. J. **485**, 182 (1997)
L.M. Molnar, R.L. Mutel, M.J. Reid, K.J. Johnston, Astrophys. J. **438**, 708 (1995)
D.R. Nicholson, *Introduction to Plasma Theory* (Wiley, New York, 1983)
C.A. Norman, A. Ferrara, Astrophys. J. **467**, 280 (1996)
H. Ohno, S. Shibata, Mon. Not. R. Astron. Soc. **262**, 953 (1993)
J.S. Oishi, M.M. MacLow, Astrophys. J. **638**, 2810 (2006)
I. Patrikeev, A. Fletcher, R. Stepanov et al., Astron. Astrophys. **458**, 441 (2006)
R. Pizzo, PhD thesis, Groningen University, 2010
K.M. Rae, J.C. Brown, in *Astronomical Society of the Pacific Conference Series*, vol. 438 (2010), p. 229
S. Redfield, Space Sci. Rev. **143**, 323 (2009)
B.J. Rickett, Annu. Rev. Astron. Astrophys. **15**, 479 (1977)
B.J. Rickett, Annu. Rev. Astron. Astrophys. **28**, 561 (1990)
B.J. Rickett, AIP Conf. Proc. **1366**, 107 (2011)
B.J. Rickett, S. Johnston, T. Tomlinson, J. Reynolds, Mon. Not. R. Astron. Soc. **395**, 1391 (2009)
F. Sahraoui, G. Belmont, M.L. Goldstein, Astrophys. J. **748**, 100 (2012)
D.H.F.M. Schnitzeler, P. Katgert, A.G. de Bruyn, Astron. Astrophys. **494**, 611 (2009)
D. Shaikh, G.P. Zank, Astrophys. J. **688**, 683 (2008)
A. Shukurov, E.M. Berkhuijsen, Mon. Not. R. Astron. Soc. **342**, 496 (2003)
J.H. Simonetti, J.M. Cordes, AIP Conf. Proc. **174**, 135 (1988)
D. Smith, Plasma Phys. Control. Fusion **34**, 1817 (1992)
S.R. Spangler, AIP Conf. Proc. **174**, 32 (1988)
S.R. Spangler, Astrophys. J. **376**, 540 (1991)
S.R. Spangler, Astrophys. J. **522**, 879 (1999)
S.R. Spangler, Space Sci. Rev. **99**, 261 (2001)
S.R. Spangler, Astron. Astrophys. **407**, 563 (2003)
S.R. Spangler, Astron. Soc. Pac. Conf. Ser. **265**, 307 (2007)
S.R. Spangler, Space Sci. Rev. **143**, 277 (2009)
S.R. Spangler, J.M. Cordes, Astrophys. J. **332**, 346 (1988)
S.R. Spangler, J.M. Cordes, Astrophys. J. **505**, 766 (1998)
S.R. Spangler, C.W. Gwinn, Astrophys. J. **353**, L29 (1990)
S.R. Spangler, T. Sakurai, Astrophys. J. **445**, 999 (1995)
S.R. Spangler, J.A. Leckband, I.H. Cairns, Phys. Plasmas **4**, 846 (1997)
S.R. Spangler et al., Astron. Astrophys. **384**, 654 (2002)
S.R. Spangler, A.H. Savage, S. Redfield, AIP Conf. Proc. **1366**, 97 (2011a)
S.R. Spangler, A.H. Savage, S. Redfield, Astrophys. J. **742**, 30 (2011b)
H.R. Strauss, Phys. Fluids **19**, 134 (1976)
V.I. Tatarski, *Wave Propagation in a Turbulent Medium* (McGraw-Hill, New York, 1961)
A.R. Thompson, J.M. Moran, G.W. Swenson Jr., *Interferometry and Synthesis in Radio Astronomy* (Wiley, New York, 1986)
R.C. Thomson, A.H. Nelson, Mon. Not. R. Astron. Soc. **191**, 863 (1980)
A.G.G.M. Tielens, *The Physics and Chemistry of the Interstellar Medium* (Cambridge University Press, Cambridge, 2005)
B.J. Uscinski, *The Elements of Wave Propagation in Random Media* (McGraw-Hill, New York, 1977)
C.L. Van Eck et al., Astrophys. J. **728**, 97 (2011)
M.A. Walker, D.B. Melrose, D.R. Stinebring, C.M. Zhang, Mon. Not. R. Astron. Soc. **354**, 43 (2004)
C.A. Whiting, S.R. Spangler, L.D. Ingleby, L.M. Haffner, Astrophys. J. **694**, 1452 (2009)
M. Wolleben et al., in *IAU Symposium*, vol. 259 (2009), p. 89
M. Yamada, R. Kulsrud, H. Ji, Rev. Mod. Phys. **82**, 603 (2010)
E.G. Zweibel, M. Yamada, Annu. Rev. Astron. Astrophys. **47**, 291 (2009)

DOI 10.1007/978-1-4899-7413-6_17
Reprinted from *Space Science Reviews* Journal, DOI 10.1007/s11214-013-9969-6

Microphysics of Quasi-parallel Shocks in Collisionless Plasmas

D. Burgess · M. Scholer

Received: 17 October 2012 / Accepted: 13 February 2013 / Published online: 1 March 2013

Abstract Shocks in collisionless plasmas require dissipation mechanisms which couple fields and particles at scales much less than the conventional collisional mean free path. For quasi-parallel geometries, where the upstream magnetic field makes a small angle to the shock normal direction, wave-particle coupling produces a broad transition zone with large amplitude, nonlinear magnetic pulsations playing an important role. At high Mach numbers, ion reflection and acceleration are dominant processes which control the structure and dissipation at the shock. Accelerated particles produce a precursor, or foreshock, characterized by low frequency magnetic waves which are convected by the plasma flow into the shock transition zone. The interplay between energetic particles, waves, ion reflection and acceleration leads to a complicated interdependent system. This review discusses the spacecraft observations which have motivated the current view of the high Mach number quasi-parallel shock, and the theories and simulation studies which have led to a better understanding of the microphysics on which the quasi-parallel shock depends.

Keywords Space plasma · Collisionless shock · Particle acceleration

1 Introduction

Shocks in collisionless plasmas require dissipation mechanisms which couple fields and particles at scales much less than the conventional collisional mean free path. They cannot be understood without a study of the processes which operate and the resulting scale lengths and structure—in other words, the microphysics of the shock. Collisionless shocks are not only of interest in themselves, but also for the key role they play in universal mechanisms in cosmic plasmas such as particle acceleration and flow-obstacle interactions. The structure

D. Burgess (✉)
Astronomy Unit, Queen Mary University of London, London E1 4NS, UK
e-mail: D.Burgess@qmul.ac.uk

M. Scholer
Max-Planck-Institut für extraterrestrische Physik, 85740 Garching, Germany

and dissipation mechanisms affect the injection of particles into acceleration processes and thus are important for the overall acceleration efficiency. They are also vital for understanding how the shock behaviour and resulting particle acceleration may depend on the plasma environment and shock parameters. A companion review describes particle acceleration at the terrestrial bow shock (Burgess et al. 2012), and the mechanisms described therein can mostly be generalized to other regimes.

In a collisionless plasma a shock is primarily characterized by the shock strength or Mach number M, and magnetic geometry. The latter can be defined by the angle θ_{Bn} between the upstream magnetic field and the shock normal: if $\theta_{Bn} > 45°$ the shock is described as quasi-perpendicular, and otherwise quasi-parallel. For sufficiently high Mach number fluid models with only resistive dissipation are inconsistent, i.e., no stable solution can be found. For such shocks it is found in observations and simulations that they are dominated by ion dissipation deriving from reflection (usually near specular reflection) of some fraction of the incident distribution at the shock ramp. Ion reflection at the quasi-perpendicular shock changes the structure of the shock, both on average and in terms of the fluctuations generated at the shock. Such shocks are usually termed super-critical, where the critical Mach number is that above which no resistive fluid solutions can be found. We will see shortly that the sub/super-critical distinction is less applicable to the quasi-parallel shock due to the lack of a single, monotonic shock ramp corresponding to that found in low Mach number fluid solutions. However, as discussed in this review, at high Mach number ion reflection appears to play a crucial role for dissipation processes also within the quasi-parallel shock.

At the quasi-perpendicular shock the reflected ions gain perpendicular energy and their subsequent gyration and motion downstream produces strong ion heating. The upstream magnetic geometry and the particle gyration ensures that they remain within a gyro-scale of the shock ramp, and the average profiles for the magnetic field and other plasma quantities have some similarity to the laminar solutions of fluid models. There may be considerable contributions from fluctuations and internal structure at shorter scales, but the overall scale is set by the particle gyration of reflected ions. However, as θ_{Bn} is decreased the magnetic geometry dictates that the particle parallel velocity obtained via reflection may, if large enough, allow the particle to move away upstream from where it was reflected. Similarly, considering an energetic ion, or heated ion downstream but just at the shock ramp, it may have enough parallel velocity to escape upstream of the shock. In either case, it is clear that energetic, shock associated upstream particles are an intrinsic feature of the high Mach number quasi-parallel shock, and this is borne out by observations at the terrestrial bow shock and other heliospheric shocks. Upstream suprathermal particles can drive waves via plasma instabilities, and which in turn controls the diffusion properties of the energetic particles in the upstream region. In this way the foreshock (that region of space ahead of the shock filled with waves and particles by virtue of the shock itself) plays a central role in diffusive shock acceleration theory. The presence of energetic particles in the foreshock complicates the question of the dominant scale lengths at the quasi-parallel shock. We have seen that the gyro-scale is a characteristic scale at the quasi-perpendicular shock. But at the quasi-parallel shock the characteristic length scales depend on the coupling between waves and energetic particles, which in turn depends on the mechanisms for extracting energetic particles from the thermal distribution.

Early ideas, such as the work of Golden et al. (1973) and others, addressed the problem of how to link the upstream state (cold, high velocity) with that downstream (heated, slowed) by postulating an extended transition which was modelled as a gradual overlap between upstream and downstream states. The underlying assumption was that the transition had some degree of monotonicity in terms of average properties. This upstream/downstream overlap is analogous to the idea that there was an "evaporation" of the heated downstream ions into an upstream precursor region, where they would produce a distribution function with parallel anisotropy unstable to the firehose instability (Parker 1961). The subsequent waves and relaxation would lead to the heating of ions as they passed through the precursor. Although conceptually attractive, these models, which relied on the idea of turbulent coupling, did not take into account many of the features of turbulence which are now believed to be important, such as the role of nonlinear interactions, intermittency, localized structures, etc. Parker's 1961 quote, however, seems prescient: "The relative importance of the precursor ... can be determined eventually by experiment or by sufficiently detailed machine calculations."

In terms of observations, these models, although important for introducing the idea of an extended coupling zone between upstream and downstream states, were dealt a blow by the early observations of the quasi-parallel terrestrial bow shock. Spacecraft observations will be discussed from a modern point of view in the next section. But it is instructive to consider the summary of Greenstadt et al. (1977) based on single spacecraft data, using instruments of only modest capability. They noted that the high Mach number quasi-parallel shock was characterized by "(1) Irregular large amplitude magnetic pulsations, sometimes in bursts, often separated by intervals of smaller amplitude upstream-like waves; (2) Thickness of $<2R_E$; (3) Large amplitude quasi-period transverse magnetic wave components; (4) Solar wind of nominally unreduced but significantly deflected streaming velocity; (5) Solar wind of elevated temperature, enhanced density, and distinct distribution with skewed high energy tails and irregular low-energy envelopes; ... (8) Interpulsation regions of upstream magnetic magnitude and wave structure but noisy, deflected, and partially thermalized plasma flow; (9) No direct evidence that the macrostructure was governed by fire hose instability as a dissipation mechanism." All these points survive in our current picture of the high Mach number quasi-parallel "pulsation" shock.

2 The Quasi-Parallel Pulsation Shock: Observations

Unlike traversals of the quasi-perpendicular bow shock, the terrestrial quasi-parallel shock often resembles a "hash" of multiple short-lived apparent transitions from upstream to downstream states, for example as in Fig. 1. Closer inspection shows a more complex structure, with variations of field, plasma speed and distribution function type. The quasi-parallel shock is associated with the ULF foreshock sited ahead of the shock and populated with energetic particles and ULF waves (see reviews in Burgess et al. 2005, 2012; Eastwood et al. 2005). One puzzling feature is that the amplitudes of structures seen in the quasi-parallel shock zone often exceed the downstream value that would be predicted from the shock jump (Rankine-Hugoniot) relations. This points to a lack of time-steady behaviour at the shock. A key breakthrough in the study of the terrestrial quasi-parallel bow shock was the dual spacecraft ISEE observations of Thomsen et al. (1990a) which showed that some of the magnetic pulsations within the shock transition had a "convective" signature. Typically the spacecraft velocity relative to the average bow shock position is small, of order a few km/s. Observations of the quasi-perpendicular shock occur when the bow shock position shifts and the shock moves across the spacecraft, either outwards or inwards. In contrast, when data from at least two spacecraft are combined, a convective signature in the relative timing of a feature in the time series indicates that the magnetic field structure is convecting with, or in the direction of the plasma. Magnetic pulsations in the quasi-parallel shock have a peak magnetic field amplitude which is at least, and often exceeds the average downstream

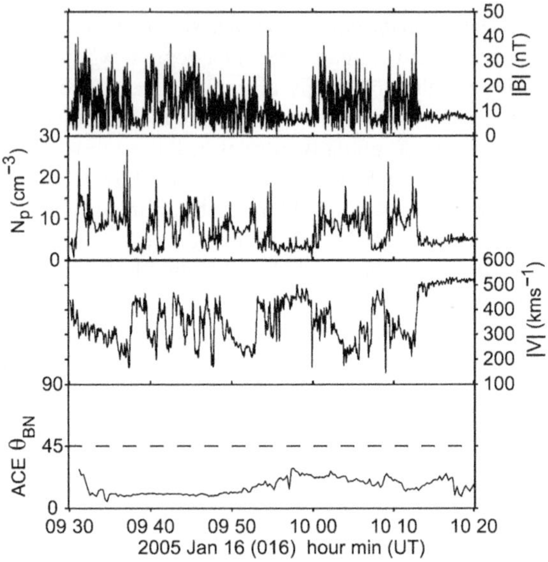

Fig. 1 Data from Cluster for a quasi-parallel shock crossing. Shown are magnetic field magnitude $|B|$ in nT, proton number density N_p in cm^{-3}, and plasma flow speed $|V|$ in $km\,s^{-1}$, all from Cluster 1. The *bottom panel* shows the angle θ_{Bn} estimated from upstream ACE data. From Lucek et al. (2008)

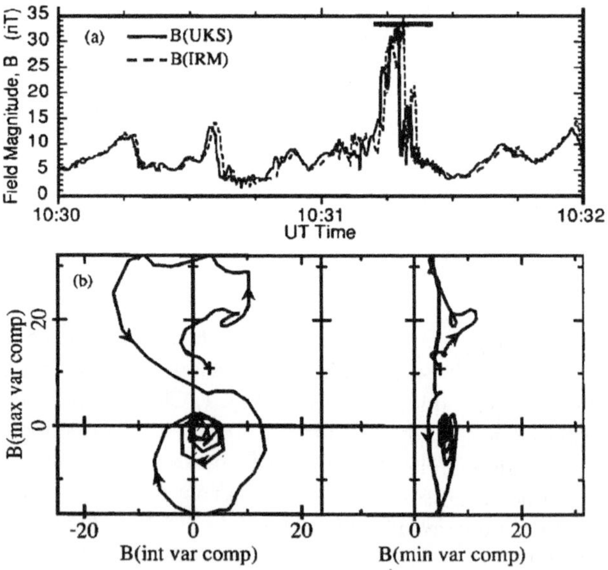

Fig. 2 Magnetic pulsation in data from AMPTE UKS and IRM, with (*below*) hodogram in minimum variance frame for the interval marked with a bar. From Schwartz et al. (1992)

value. The presence of a convective signature for some events therefore shows that these large values are achieved in magnetic field structures convecting in the flow, not merely by oscillations of a shock surface, as at the quasi-perpendicular shock.

Schwartz et al. (1992) distinguished between "isolated" pulsation events, with typical upstream solar wind on both sides, and "embedded" events which had a more complex form, with magnetosheath or partially heated solar wind on one or both sides. The isolated events have a well-defined monolithic appearance, lasting 5–20 s, a large peak magnetic field amplitude at least twice, but sometimes up to 5–6 times the solar wind value. Figure 2 shows a magnetic pulsation in data from AMPTE IRM and UKS. The relative timings are consistent with convection in the direction of, but with a speed less than the solar wind. The hodogram shows that the initial part is LH polarized (in the spacecraft frame), the latter part RH polarized and in addition there is a LH polarized wave train consistent with a convected whistler. Different combinations of LH and RH polarization can be found at different pulsations, although the intrinsically RH polarized wave train, when seen, corresponds to a standing structure on the steepened upstream edge. For a convected pulsation this is seen as a standing structure on the trailing edge as observed in time. Similar structures are seen at the steepened edges of "shocklets" seen in the ULF wave foreshock, albeit at smaller amplitudes. Timing analysis indicates that the pulsations are typically propagating super-Alfvénically in the plasma rest frame, but slower than the solar wind speed, so that they are convected with the solar wind. The pulsation propagation speed increases with the peak amplitude of the magnetic field strength. Mann et al. (1994) found speeds between 2–5 v_A, which implies that the trailing (in time, i.e., upstream in space) edges of the larger pulsations are shock-like.

The complex magnetic structure of the quasi-parallel pulsation shock is reflected in the observations of ion particle distribution functions. It should be remembered that throughout the quasi-parallel shock and connected foreshock there is a population of energetic "diffuse" ions. However, at lower supra-thermal energies the ion distributions are highly variable with clumps of particles in velocity space. Gosling et al. (1989) reported evidence of (1) cold, coherent ion beams with velocities roughly consistent with specular reflection at a sharp gradient, and (2) downstream ion distributions which indicated the presence of a cold, deflected core, and a hotter low density shell in velocity space. They pointed out that similar features were seen at the quasi-perpendicular shock, implying that ion specular reflection could contribute, in the same manner, to ion dissipation at the quasi-parallel shock. However, this cannot be the complete picture since the observation of such distributions is intermittent, and the idea of a single shock surface acting to specularly reflect ions has to be reconciled with the behaviour of convecting pulsations which would disrupt that surface. Thomsen et al. (1990b) studied the ion distributions (in 2-D velocity space) downstream and presented evidence of two types of distribution: one with a cooler, denser core plus hot shell, and the other being a less dense, broader and more "Maxwellian" in appearance. These distributions would be observed in alternation close to the nominal shock transition. Thomsen et al. (1990b) argued that these two types of distribution were not evolutionary because the distribution far downstream had properties intermediate between the two initial states, and that examples of both types of distribution could be found at crossings of the shock ramp. Many of these results are refinements of the early Heos and Ogo observations presented in Greenstadt et al. (1977). Onsager et al. (1990), again using the ISEE 2-D ion experiment, performed a survey of coherent ion reflection in quasi-parallel shocks and found that such beams were detected frequently throughout the shock transition, and were seen near the shock ramp (identified in the electron density) or near other shock-like features (presumably pulsations). On the other hand, there are some shock-like features (pulsations) which are not seen associated with coherent ion beams. When cold coherent beams were observed they were almost always consistent with specular reflection after only a small fraction of a gyro-period after reflection. The observations of Onsager et al. (1990) and Thomsen et al. (1990b) indicate that ion specular reflection is important at some shock crossings, but not all. Or, more plausibly, time variability is crucial in interpreting any particular crossing. Another issue to be considered is the relatively low time resolution of particle instruments compared to magnetic field variations near pulsations, so that time-aliasing of observations is a real possibility.

The Onsager et al. (1990) observation that the specularly reflected ions are only seen as a coherent clump in velocity space shortly after reflection indicates that the ions are scattered,

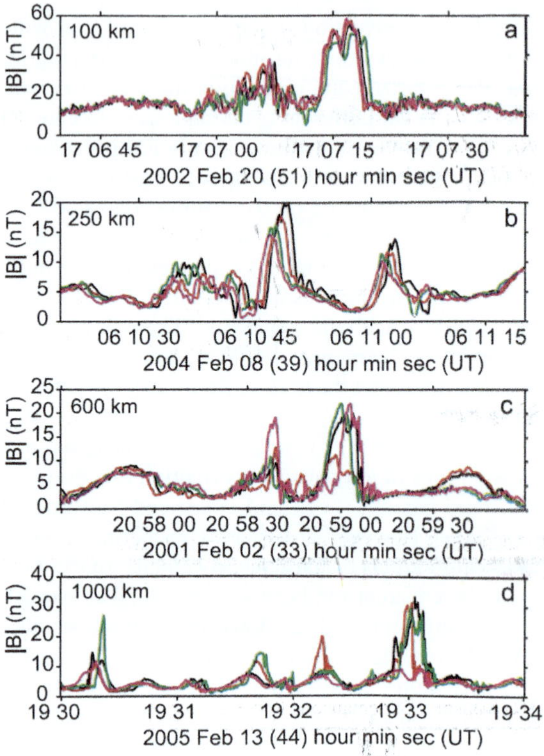

Fig. 3 Four examples of Cluster four spacecraft data for magnetic pulsations within the quasi-parallel shock transition. In each plot the *different colours* of the time series represent different spacecraft in the Cluster constellation. The approximate tetrahedron scale in km is marked for each case. From Lucek et al. (2008)

or coupled in some way to the rest of the plasma, relatively rapidly. It might be thought that the pulsations would be the source of that coupling, but another possibility is indicated by the observation that the dense, short-lived ion beams are associated with bursts of broad-band magnetic turbulence with frequencies up to at least 3 Hz, and possibly higher (Wilkinson et al. 1993). The magnetic turbulence was observed to be decorrelated between spacecraft at a separation of 150 km, making it difficult to fully analyse its properties. It was also found, with data from a 3D ion instrument, that cold ion beams could sometimes be seen which were inconsistent with specular reflection at the nominal shock orientation.

From the earliest studies it was acknowledged that single spacecraft observations cannot be used to distinguish the inherent propagation properties of fluctuations and pulsations, given the overall motion of the average bow shock. Multi-spacecraft observations give the possibility of establishing orientations of plasma structures, but only if there is sufficient time stability so that cross-correlation gives meaningful delay times between the spatially dispersed spacecraft. If there is considerable temporal evolution of a structure in the time that it convects past the spacecraft constellation, then it becomes impossible to determine orientations, and only some minimal information about scale lengths can be extracted. Cluster observations taken when the four spacecraft constellation had different separation scales are shown in Fig. 3 (Lucek et al. 2008). Different colours of the time series represent different spacecraft in the Cluster constellation. At a separation of 100 km (of order the ion inertial length) the time difference is less than one second, so that differences are more likely to be due to spatial gradients rather than temporal evolution. At separations of 250 km there is evidence of temporal evolution, and growth of the pulsation as it convects. Since the time difference is only of the order of a few seconds the growth is rapid, much less than

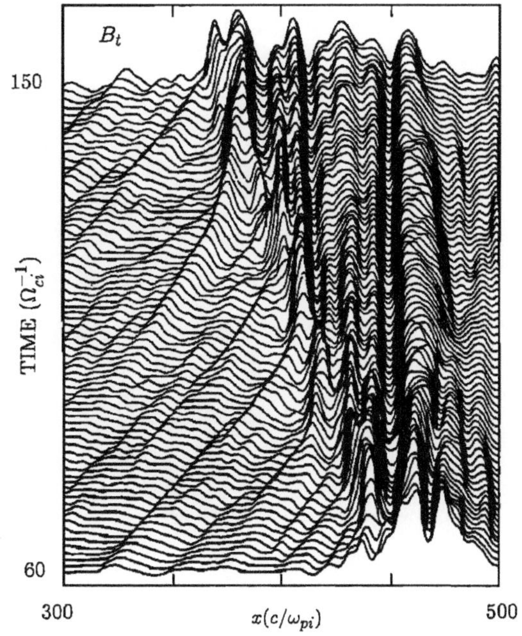

Fig. 4 Results from a hybrid simulation of a $\theta_{Bn} = 20°$, $M_A = 3.5$ shock. Shown are profiles of the magnetic field tangential to the shock stacked in time. From Scholer and Terasawa (1990)

the ion gyroperiod in the upstream magnetic field. For a separation of 600 km or 1000 km there are now considerable differences between the spacecraft profiles, and it is only possible to deduce the rather weak conclusion that temporal evolution and/or spatial gradients are important at those scales.

3 Simulations, Mechanisms and Microphysics

Hybrid simulations of quasi-parallel shocks above a Mach number of $M_A \sim 2$ have revealed that such shocks have a quasi-periodic cyclic behaviour in their structure, in that the abrupt transition of the magnetic field at the shock relaxes to a more gradual transition and is replaced by a new shock front ahead of the previous nominal shock position (Burgess 1989). Figure 4 (Scholer and Terasawa 1990) shows the time evolution of the transverse magnetic field component for a $\theta_{Bn} = 20°$, $M_A = 3.5$ shock as obtained from a 1-D hybrid simulation. The shock is launched off the right hand rigid wall of the simulation box, so that the shock propagates leftwards. From the left hand side upstream waves can be seen to run into the shock. These upstream waves are compressed closer to the shock; simultaneously the shock profiles becomes more gradual. The arriving wave steepens up at the upstream edge which becomes the newly reformed shock. Some of the reformation cycles are particularly prominent while others seem to be more like mini-cycles. On average reformation takes place every $20\Omega_c^{-1}$, corresponding to ~20 sec at Earth's bow shock.

The temporal development of the local shock normal angle and the occurrence of periodic bursts of backstreaming ions connected with the reformation process can be seen from Fig. 5 (Burgess 1995). These results are from a quasi-parallel 1-D hybrid shock simulation with $\theta_{Bn} = 30°$, $M_A = 6.5$, showing time evolution in the form of a grey scale plots of the total magnetic field B, the backstreaming ion density, and local instantaneous shock normal angle θ_{Bn}. The frame for the plots corresponds to the average shock rest frame. In all plots,

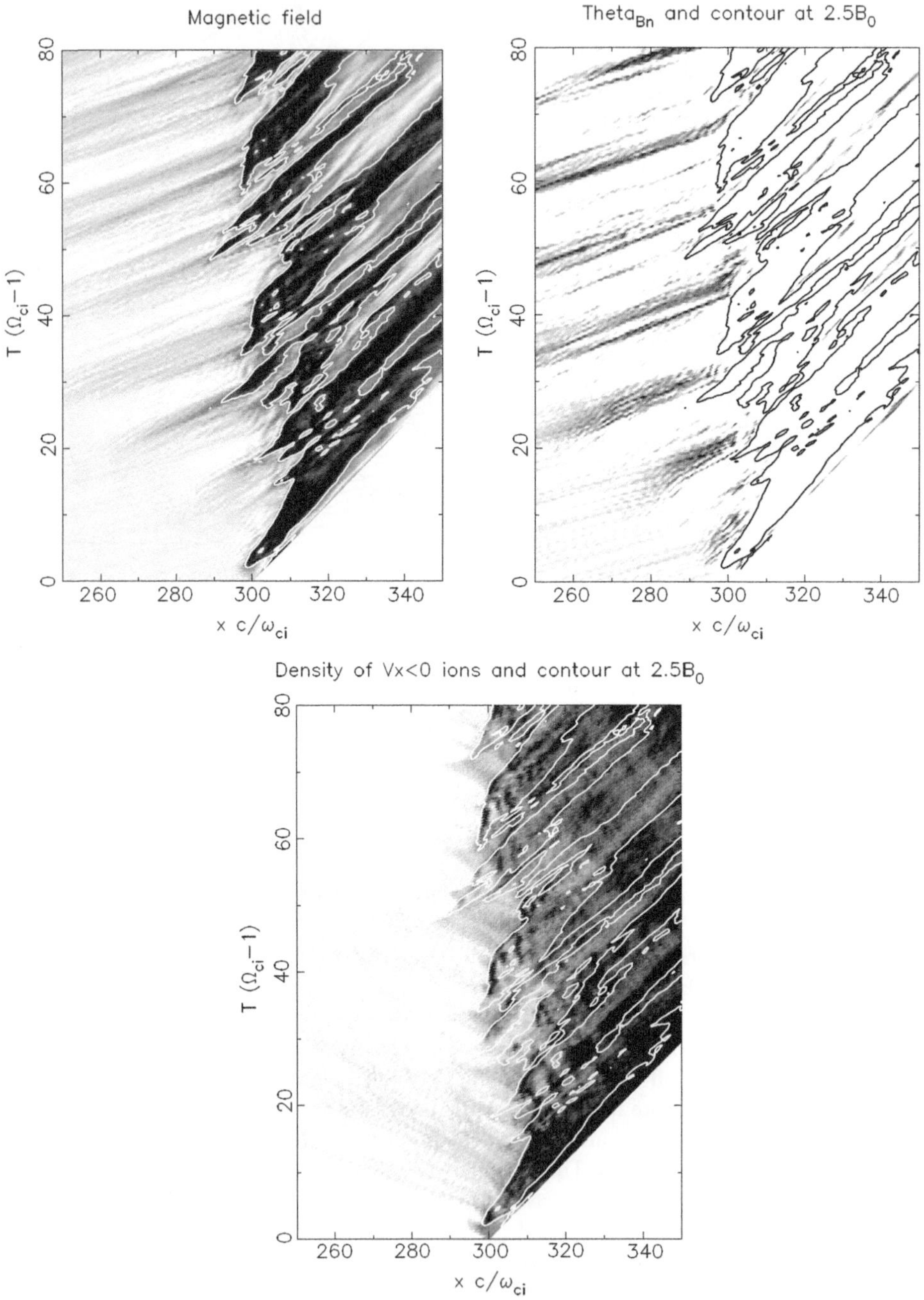

Fig. 5 Results from a quasi-parallel ($\theta_{Bn} = 30°$, $M_A = 6.5$) hybrid shock simulation showing time evolution in the shock frame of magnetic field (*white-black*: $0.9B_0 \rightarrow 3.5B_0$, with contour at $2.5B_0$), local instantaneous value of shock normal angle θ_{Bn} (*white-black*: $30° \rightarrow 5°$), and backstreaming ion density (*white-black*: $0.02 \rightarrow 2.0$). From Burgess (1995)

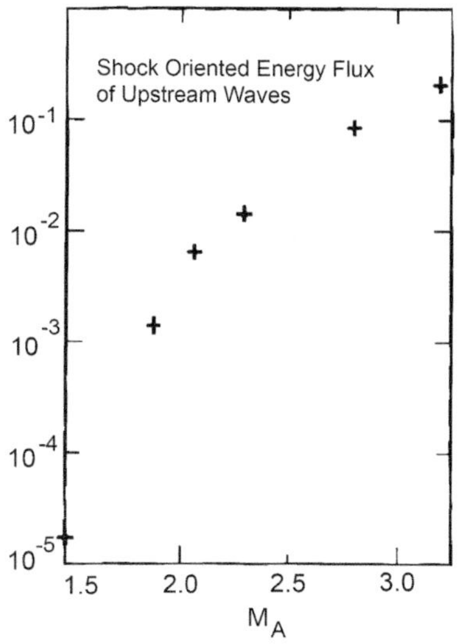

Fig. 6 Magnetic energy flux of upstream waves with downstream directed group velocity for 5 shocks with different Mach number ($\theta_{Bn} = 30°$). Adapted from Krauss-Varban and Omidi (1991)

a contour indicates where the total field is 2.5 times above its nominal upstream value, and thus the location of the main field jump. A major reformation cycle can be identified between $T = 42\Omega^{-1}$ and $T = 54\Omega^{-1}$, where the initial "shock" (i.e., location of an abrupt transition) moves backwards, and is replaced by another "shock" front ahead of the nominal shock position. This representation also reveals that there are mini-reformation cycles (e.g., at $T = 62\Omega^{-1}$), which some stack plot representations fail to show. From the local value of the shock normal angle, where the grey scale covers the range $\theta_{Bn} = 30°$ (white) to $\theta_{Bn} = 5°$ (black), it is clear from correlating these figures that the change from sharp shock transition to irregular occurs when a region of low instantaneous θ_{Bn} is convected into the field jump. The large-amplitude waves, as they arrive at the shock, force the shock to adjust to the newly changing upstream conditions. This results in a retreating shock ramp (in the average shock frame), and a decay of the sharp profile to a flatter, more extended and irregular transition. This in turn causes a switch-off of the production of backstreaming ions. The arrival of the upstream wave also correlates with the production of backstreaming ions ahead of the main jump in magnetic field amplitude. The correlation between low instantaneous θ_{Bn}, bursty production of backstreaming ions and the transition to an irregular magnetic structure is very good, even down to the mini-reformation cycles.

An apparent prerequisite for reformation to occur is that, in the shock frame, the group velocity of the upstream waves is directed downstream, i.e., the wave energy is transported back into the shock. Krauss-Varban and Omidi (1991) concluded, from hybrid simulations of quasi-parallel shocks with different Mach numbers, that the shock becomes unsteady when the energy flux of waves with downstream directed group velocity exceeds 10 % of the flux given by the upstream background magnetic field and the Alfven velocity Bv_A. Figure 6 shows the dramatic increase in the magnetic energy flux of upstream waves with downstream directed group velocity as a function of Mach number.

However, in order to drastically disturb the shock by changing the upstream condition and to steepen up to a new shock the waves have to grow in amplitude as they are convected into the shock. This is only possible by interaction with existing diffuse ions or with more specularly reflected ions closer to the shock. In a computer experiment a quasi-parallel shock has been followed for some time and a certain reformation cycle was recorded. Figure 7 (Scholer and Burgess 1992) shows in the upper panel a small part of the simulation domain,

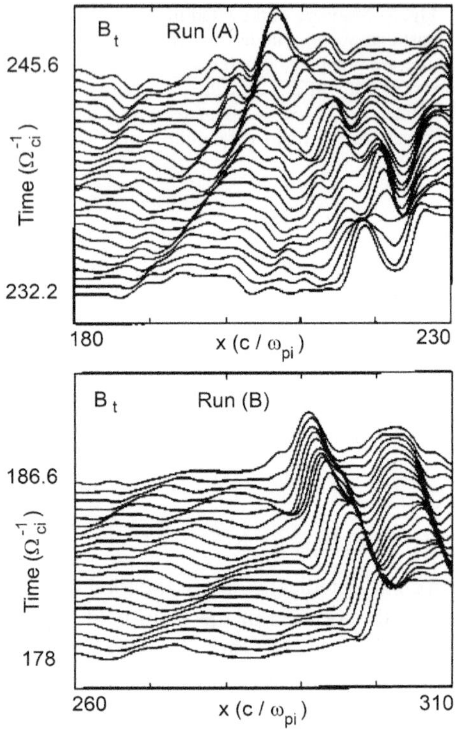

Fig. 7 Time development of the tangential magnetic field for a $\theta_{Bn} = 20°$, $M_A = 4.6$ shock. Results are shown in the simulation frame so that the shock travels to the left. Shown is part of the simulation box and the time evolution near the shock. In the run shown in the *bottom panel* all backstreaming ions have been eliminated within $80c/\omega_{pi}$ upstream of the shock after $t\Omega_c t = 150$. From Scholer and Burgess (1992)

and the approaching wave crest can be followed in time as the wave is convected toward the shock until the upstream edge becomes the reformed shock. The simulation run was repeated and at $\Omega_c t = 150$ (after the upstream wave train was established) all backstreaming ions in a region of 80 c/ω_{pi} upstream from the shock were removed. In particular no subsequent new backstreaming ions were allowed for. In the lower panel it can be seen that while the wave is convected into the shock it does not steepen up to a new shock front.

The mechanisms for the emergence of large amplitude magnetic field pulsations are worth considering in more detail. The consistent observation of large amplitude pulsations near the Earth's quasi-parallel bow shock and their occurrence in simulations have led to a picture of the quasi-parallel shock as being a patchwork of large amplitude pulsations with "inter-pulsation" plasma sandwiched between the pulsations (Schwartz and Burgess 1991). Figure 8 is a schematic of the relation between large amplitude pulsations (shaded ellipses), magnetic field (dashed lines), and bulk flow (double arrows). The pulsations are decelerated in the shock rest frame, which means that their velocity with respect to the plasma rest frame increases, they are deflected and merge as they convect toward what becomes the downstream state.

In 1-D simulations of quasi-parallel shocks with θ_{Bn} exceeding ~20° the low frequency upstream waves steepen up into large amplitude pulsations in a region very close to the shock. An incoming wave interacts with the shock, producing an associated increased density of diffuse and/or nearly specularly reflected ions. Figure 9 (Scholer 1993) shows for a shock with $\theta_{Bn} = 30°$, $M_A = 4.6$ the ion phase space density, the log of the backstreaming ion density (the solar wind density is normalized to 1), and the two magnetic field components B_z, B_y versus x within a region of $100c/\omega_{pi}$ around the shock (at $\sim 420c/\omega_{pi}$. As can be seen from the phase space plot there are no specularly reflected ions upstream of the

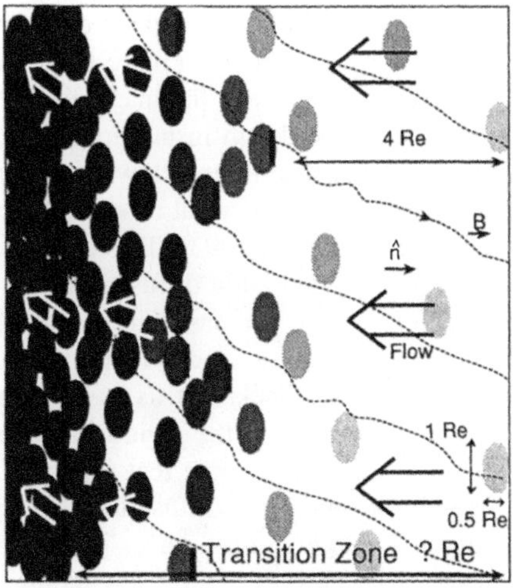

Fig. 8 Schematic of the relation between large amplitude pulsations (*shaded ellipses*), magnetic field (*dashed lines*), and bulk flow (*double arrows*). From Schwartz and Burgess (1991)

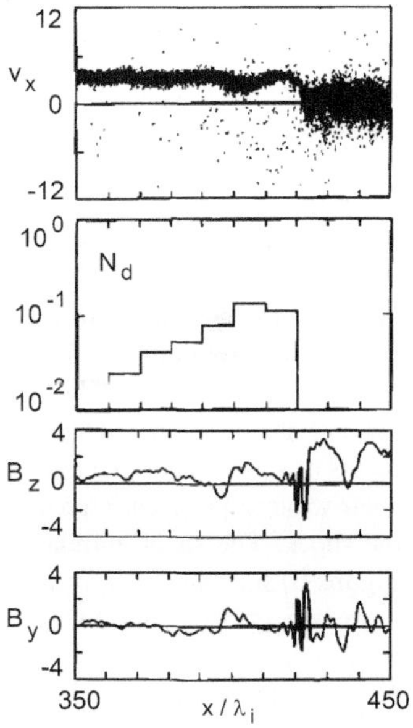

Fig. 9 Ion phase space, the number of backstreaming ions normalized to the far upstream ion density, B_z and B_y versus x normalized to the ion inertial length at one particular time during a 1-D hybrid simulation of a $\theta_{Bn} = 30°$, $M_A = 4.6$ shock. From Scholer (1993)

shock during this time; however, close to the shock the density of diffuse ions drastically increases and exceeds a value of 13 % of the far upstream ion density. This demonstrates that the growth of the shocklet with the upstream steepened edge near $x \sim 400c/\omega_{pi}$ is apparently due to interaction with the increasing number density of diffuse ions. Although this result may depend to some extent on how ions are counted as "diffuse." Attached to the shock ramp is a whistler wave which is due to dispersion during steepening of the shock ramp. In the steepened portion of the wave the upstream ions' bulk velocity decreases, indicating that steepening is in part also due to energy loss of the solar wind. As the large positive B_y of the shocklet reaches the shock ramp a cloud of ions is specularly reflected. However, the specularly reflected ions are trapped within the approaching pulsation which results in a further increase of the pulsation amplitude.

One dimensional simulations constrain the wave vectors to the aligned with the shock normal, but simulations in more than one spatial dimension result in upstream waves with wave vectors mainly in the magnetic field direction, in agreement with the expectations of linear theory for beam instabilities. In 2-D hybrid quasi-parallel shock simulations it has been seen that field-aligned low frequency waves evolve to have their wave vector direction closer to the shock normal direction once they have steepened up into shocklets or pulsations (Scholer et al. 1993). This can be explained by refraction of the waves/pulsations in the region of increasing diffuse ion density: as can be seen from Fig. 4 the velocity of the steepening pulses decreases in the shock frame as the pulse is convected toward the shock, i.e., the upstream directed velocity in the upstream plasma rest frame increases as the pulse approaches the shock. Let us assume that to zeroth order the pulses obey Snell's law as they are amplified in the region of increasing diffuse ion density close to the shock transition. Let us also assume that an upstream boundary parallel to the shock front separates a region of rather low diffuse ion density from a region of high diffuse ion density where the phase velocity of the waves/pulsations in the upstream rest frame is almost equal to the shock Mach number. Snell's law requires conservation of the shock frame wave frequency and tangential wave length at the boundary. This is only possible when the wavelength in the region of decreasing phase velocity in the shock frame ion also decreases. In order to keep the wavelength parallel to the boundary constant, the waves have to be refracted away from the magnetic field direction towards the boundary normal direction. Thus as the phase velocity in the shock frame decreases the wave fronts become more aligned with the shock front. Some observational evidence for this effect is presented in Lucek et al. (2008).

Simulations have also given evidence that pulsations, and subsequent associated reformation, can also be induced by specularly reflected ions. Hybrid shock simulations have been performed where the diffuse upstream ions have been removed from the system a distance $50c/\omega_{pi}$ upstream of the shock position in order to suppress the long wavelength upstream waves (Thomas et al. 1990). These artificial shocks nevertheless exhibit an unsteady behavior and are reforming. One mechanism for the unsteadiness is the periodically occurring bursts of specularly reflected ions which propagate upstream and interact with the incoming solar wind to produce large amplitude pulsations which are almost standing with respect to the shock. The shock retreats and these pulsations become the new reformed shock ramp. Figure 10 shows results from a $\theta_{Bn} = 30°$, $M_A = 6$ shock simulation where upstream waves have largely been suppressed. Shown is the magnetic field magnitude, the B_y component, the ion density, and ion phase space versus x for two different times which are $6\Omega_c^{-1}$ apart. Note that the range in x has been shifted by $5.5c/\omega_{pi}$ to the left between the left hand plot and the right hand plot. At $t = t_0$ a cold beam of specularly reflected ions (negative v_x) can be seen to emerge from the shock. This beam has strongly interacted $6\Omega_c^{-1}$ later with the incident ion beam: at the upstream edge the two have coupled and peaks in ion density n_i and magnetic field B build up at the upstream edge. The upstream edge has essentially taken over the role of the reformed shock ramp. The new downstream region is not uniform, i.e., the cool and hot parts of the incident and reflected ion beam plasmas have not yet completely phase mixed. This situation is rather similar to the two state temperature often found downstream of the ramp of Earth's bow shock (Thomsen et al. 1990b). This is suggestive that at a collisionless shock reformation can actually be induced by specularly reflected ions without any interaction with upstream waves. However, it has to be kept in mind that the simulations which show this involve the removal of foreshock ions, and therefore the generation of associated waves, which are an intrinsic part of the quasi-parallel shock.

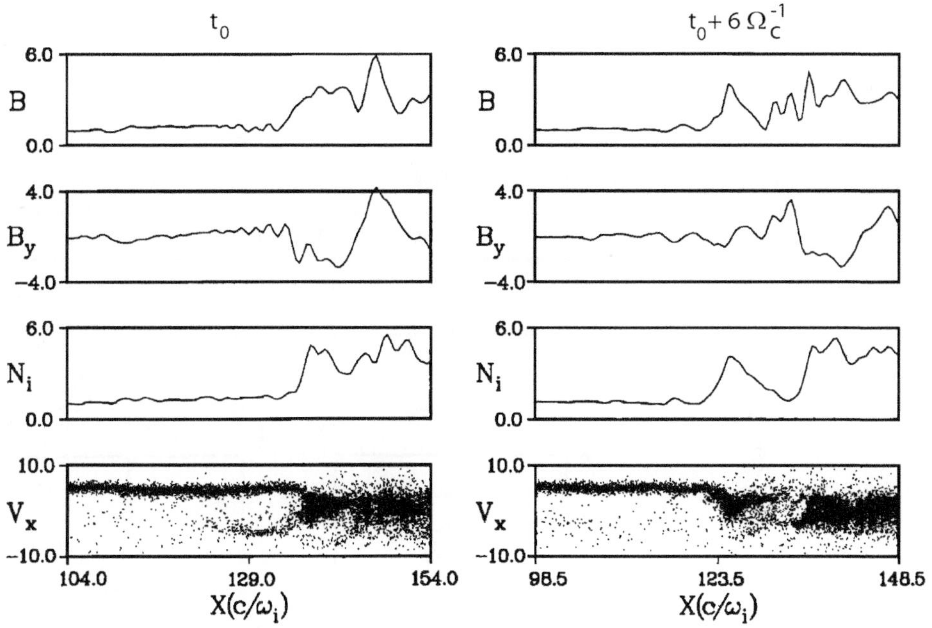

Fig. 10 Profiles of $|B|$, B_y, n_i, $v_x - x$ phase space versus x for a limited range of x around a simulated $\theta_{Bn} = 30^\circ$, $M_A = 6$ shock where upstream diffuse ions have been removed in order to largely suppress low frequency upstream waves. From Thomas et al. (1990)

Let us now focus in more detail on the incident ion-reflected ion beam interaction. The interaction of a spatially limited beam of cold ions with an incident cold ion beam has been the topic of considerable research. In the parallel limit, i.e., magnetic field parallel to the incident beam and a spatially limited beam flowing in the opposite direction, the interaction involves the excitation of ion/ion beam instabilities. For the high reflected beam densities of over 10–20 % it is expected that the nonresonant (firehose type) right hand beam instability has the largest growth rate. However, as shown by Onsager et al. (1991a) two important facts modify the interaction compared to the homogeneous plasma case: Firstly, the time a growing wave spends in contact with the finite length beam is important for its final amplitude. The large group velocity (relative to the beam) of the nonresonant mode therefore limits their growth. The resonant mode has a smaller growth rate; however the smaller group velocity allows the waves to interact longer with the reflected ions, so that they can grow to larger amplitudes. Secondly, the beam progressively spreads in space due to velocity dispersion. This leads to a cold, high velocity, low density beam, ahead of the main beam population, in which the resonant beam instability has a larger growth rate compared to the nonresonant instability. The coupling of incident beam and reflected beam leads to a transfer of the beam velocity difference into thermal energy. For a beam density ~40 % and a beam length of the order of $100c/\omega_{pi}$, it turns out from the simulations that 50 % of the relative velocity can be transferred into thermal energy within 5 ion gyro-times. However, the discussion of the interaction of a finite length beam with an incident beam in the limit of field-parallel propagation is largely academic. The magnetic field upstream of a shock is either inclined relative to the incident (and specular reflected) bulk flow due to a finite θ_{Bn} and/or the angle of the magnetic field relative to the velocity of the reflected ions is considerably modified due to the surrounding wave field.

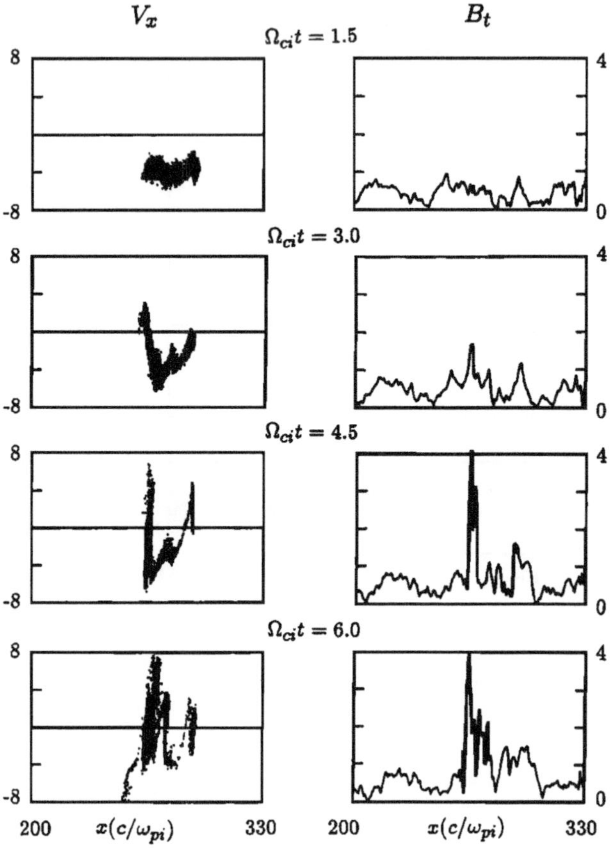

Fig. 11 $v_x - x$ phase space plots of (*left*) a finite length beam placed into an upstream low frequency wave field and (*right*) tangential magnetic field B_t at various times after the beam has been placed into the upstream region. From Scholer and Burgess (1992)

When treating the reflected particles as test particles in a uniform upstream flow, the distance they reach upstream can be found from their motion in the upstream magnetic field and $\mathbf{V}_u \times \mathbf{B}$ motional electric field (Onsager et al. 1991b). For $\theta_{Bn} > 30°$ the distance d normal to the shock where the normal component of velocity is zero is given by

$$d = M_A\big[\Omega_c\tau\big(2\cos^2\theta_{Bn} - 1\big) + 2\sin^2\theta_{Bn}\sin\Omega_c\tau\big]c/\omega_{pi}, \tag{1}$$

where the distance d is reached at the time τ given by

$$\Omega_c\tau = \arccos\left[\frac{1 - 2\cos^2\theta_{Bn}}{2\sin^2\theta_{Bn}}\right]. \tag{2}$$

For $M_A = 5$ and $\theta_{Bn} = 30°$ specularly reflected test particles reach zero normal velocity by $\Omega_c\tau = \pi$ and at $d \approx 8c/\omega_{pi}$. It has been suggested that the reflected ions will then accumulate and may considerably affect the upstream density, flow velocity, and magnetic field. An example of the ensuing interaction is shown in Fig. 11 (Scholer and Burgess 1992). A 1-D hybrid simulation of a $\theta_{Bn} = 20°$, $M_A = 4.6$ shock was run up to some time, at which point the region upstream of the shock was extracted and placed into a new system with open boundaries together with a finite length beam representing cold specularly reflected ions. The beam length is assumed to be $30c/\omega_{pi}$ and the beam density is 40 % of the upstream ion density. The upper two panels of Fig. 11 show the $v_x - x$ phase space of the beam ions and the magnitude of the magnetic field tangential to the shock B_t shortly after the beam has been placed into the simulation. Due to the large local θ_{Bn} in the upstream waves evident in the top right hand panel the beam ions immediately begin to interact with the incident ion population. At the upstream edge the beam ions are decelerated and deflected parallel to the shock plane. This leads to a local increase in beam density and in turn to a compression of the tangential magnetic field. The stronger field inhomogeneity results in a positive feedback loop generating the large magnetic field perturbation. At the same time the incident beam is decelerated at the upstream edge of the pulsation; the energy for the large amplitude pulsation comes in part from energy transfer of the bulk velocity of the incident ions. The positive feedback process between reflected ion deflection and magnetic field steepening produces the reformation seen in Fig. 11. Although this process may also work without upstream low frequency waves, it is however facilitated as a low frequency wave approaches the shock, thereby producing an increased tangential magnetic field component, and thus a larger local θ_{Bn}.

With a few exceptions, such as the work of Pantellini et al. (1992), the majority of early simulation studies of the quasi-parallel shock used the hybrid simulation technique, since it was not possible to carry out full particle (PIC) simulations with an extended upstream region; to follow the shock over ion gyration time scales; and to use a reasonably large ion to electron mass ratio. Pantellini et al. (1992) used an implicit PIC code, with, because of computational constraints, an unrealistic value of $m_i/m_e = 100$. With ever increasing computer capability, large-scale PIC simulations have become possible. Such PIC simulations are desirable in order to study the influence of whistler waves, either phase standing with the shock or attached to the upstream edge of steepening shocklets. These waves, with wavelengths less than an ion inertial length, cannot be properly resolved by hybrid simulations and their damping rate cannot be modelled correctly. Some early hybrid simulations have actually suggested that the dispersive whistler waves provide the required dissipation at the quasi-parallel shock (Kan and Swift 1983).

Results from a 1-D PIC simulation of a $\theta_{Bn} = 30°$, $M_A = 4.7$ shock with a mass ratio of $m_i/m_e = 100$ and $\omega_{pe}/\Omega_{ce} = \sqrt{10}$ are shown in Fig. 12 (Scholer et al. 2003). The total system size for this simulation is $200c/\omega_{pi}$. The left hand side of Fig. 12 shows from top to bottom B_z, the shock normal potential Φ, the ion bulk speed in the normal direction, and the ion density n_i versus x. Various magnetic field structures are numbered with 1 through 4. The shock transition was shortly before this time at the upstream edge of pulsation 1 and is now at the upstream edge of pulsation 2 as indicated by the sudden drop in bulk speed and by the increase of the ion density. Pulsation 3 has a steepened upstream edge and becomes later the reformed shock. Phase standing whistler waves are attached to the shock ramp (upstream edge of pulsation 2) and a smaller amplitude whistler train is attached to the steepening edge of pulsation 3. The left hand side of Fig. 12 shows the temporal development of pulsation 3 (stacked profiles of the magnetic field B_z) component. Arriving pulsation 3 runs into the whistler train attached to the ramp and the waves are damped, at the same time pulsation 3 steepens and emits a new whistler wave train at the upstream edge. The damping leads to a local heating of the ions in the trailing part of the pulsation. This suggests that dispersive whistlers also play an active role in shock dissipation; however this is due to the interaction with a newly arriving pulsation. During the emergence of a new shock ramp specularly reflected ions are produced; these ions are subsequently scattered in the large amplitude phase standing whistler. This leads to a deviation of the phase space position of specular reflected ions from the nominal one based on θ_{Bn}. But again nothing final can be said about wave amplitudes simply from results of 1-D simulations. Another issue to consider is the role of the artificially low m_i/m_e ratio, and further work is required to investigate to what extent the simulated structure changes as m_i/m_e approaches its real value.

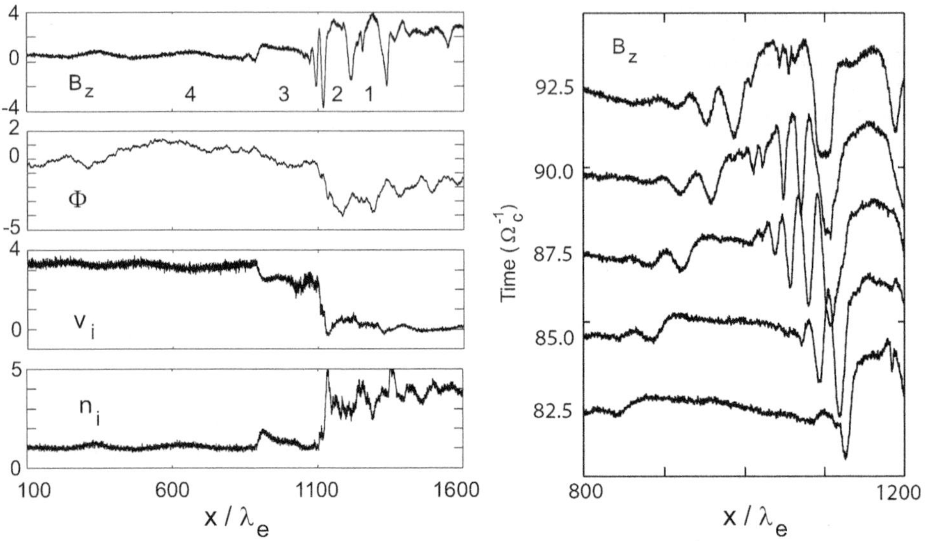

Fig. 12 (*Left*) PIC simulation results for a $\theta_{Bn} = 30°$, $M_A = 4.7$ shock. Magnetic field B_z component, shock normal potential Φ, ion bulk velocity and ion density versus x at one particular time. (*Right*) Time sequence of profiles of B_y; second profile from bottom corresponds to time of overview shown to the left. From Scholer et al. (2003)

The mechanism discussed for shock reformation so far involves the occurrence of large amplitude pulsations which become an integral part of the shock thermalization process. This mechanism involves the production of backstreaming particles, either suprathermal diffuse ions or specular reflected ions. In hybrid simulations of collisionless shocks an experiment can be made where all backstreaming ions are eliminated as soon as they are generated. It has been shown that in such an experiment the shock still reforms. This is due to an instability in a small shock interface region where the incident ion beam and part of the downstream hot ion distribution overlap. Since this instability occurs in a very limited region at the shock interface it has been termed the interface instability (Winske et al. 1990). Linear theory fails to describe such an instability since the region where this instability would be excited is possibly smaller than the wavelength of the (linearly unstable) waves. It is thought that the waves at the interface can rapidly grow to large amplitudes and mimic reformation. Figure 13 shows results from a numerical experiment where initially two beams fill the separate halves of space. The "upstream" beam has a velocity of $M_A = 5$, the downstream beam has twice the temperature of the upstream beam and 2.4 the upstream density and is at rest in the simulation system. The magnetic field is inclined by 5° with the x (simulation) direction. The upper panel of Fig. 13 shows the B_z magnetic field component stacked in time, the bottom panel exhibits the phase angle Φ of the magnetic field defined by $\tan\Phi = B_z/B_y$. One sees that large amplitude waves are generated at the interface (at $x = 200c/\omega_{pi}$, which propagate downstream and are subsequently replaced by new waves. From the increase of phase angle with x (lower panel) it can be concluded that these waves have right hand (positive) helicity.

More direct proof of an instability occurring at the shock interface can be seen from a 1-D hybrid simulation of a higher Mach number, more parallel shock with $M_A = 6$ and $\theta_{Bn} = 10°$ in Krauss-Varban (1995) (Fig. 14). The magnetic field displays a similar behaviour as the shock shown earlier in Fig. 4 (Scholer and Terasawa 1990), in that upstream fast magne-

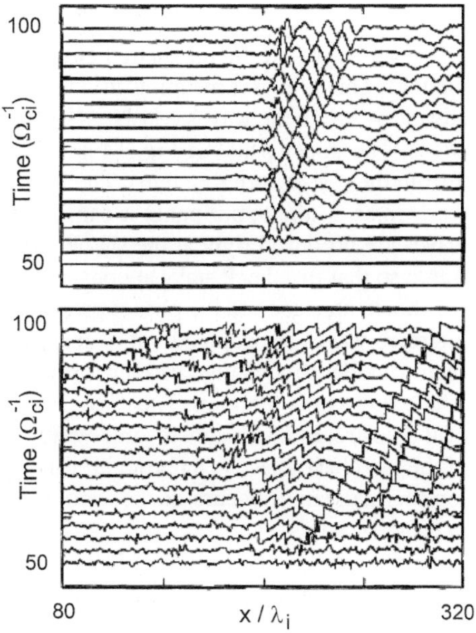

Fig. 13 Time sequence of (*top*) B_z and (*bottom*) phase angle profiles over portion of a 1-D hybrid simulation of a two-stream interaction (upstream parameters $\theta_{Bn} = 5°$, $\beta_i = 1$; downstream parameters $n_2/n_1 = 2.4$, $\beta_i = 2$). From Winske et al. (1990)

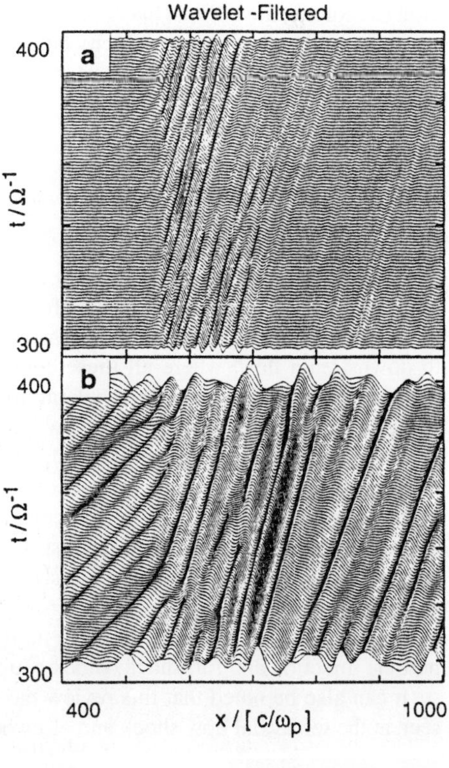

Fig. 14 Stack of wave-filtered B_y in the shock frame. (**a**) Wavelet components 4 and 5 showing the monochromatic waves immediately behind the shock; (**b**) wavelet components 1–3 demonstrating the upstream waves as they mode convert through the shock interface. From Krauss-Varban (1995)

tosonic waves steepen and at times disturb the shock transition. Instead of displaying a stack plot (in time) of the B_y magnetic field component the data has been filtered using a wavelet method. This allows a meaningful interpretation of localized structures and is of particular advantage when the waves consist of only a few cycles and/or have signatures of steepened edges. The upper panel of Fig. 14 shows contributions from high wavelet numbers (4 and 5 from the ${}_{10}\psi$-wavelet system). The lower panel shows the remainder after subtracting out the former. In the upper panel it can be seen that small wavelength rather monochromatic waves are produced right at the shock interface. (Note that the data are plotted in the shock frame so that the shock stays at constant x.) These small wavelength waves are produced at the shock interface and the phases propagate downstream. After a distance $\sim 150c/\omega_{pi}$ these interface waves are damped out. Conversely, the larger wavelength waves upstream can be seen to undergo a change of their properties at the shock interface and to continue in the whole downstream region. The change at the shock interface is due to the mode conversion of the upstream fast magnetosonic wave into a downstream Alfvén wave. Close inspection of the ion phase space at a particular time at various positions through the shock reveals that ion thermalization occurs in the whole transition region defined by the small wavelength interface waves.

In the medium and higher Mach number range up to $M_A \sim 8$ the interface waves are right hand polarized and their wavelength dependence on θ_{Bn} and M_A agrees with excitation by the ion/ion right hand resonant instability for the local plasma properties at the shock interface (Winske et al. 1990). These monochromatic waves are damped downstream over a distance of 100–200 ion inertial lengths (Krauss-Varban 1995). Beyond this distance downstream from the shock the waves produced upstream and then mode-converted at the shock are the main wave modes. At Mach numbers above $M_A \sim 8$ the resonant as well as the ion/ion right hand nonresonant, firehose type instability gets excited in the interface region (Scholer et al. 1997). This is due to the rapid increase of the growth rate of the nonresonant instability with relative velocity between beam and background plasma. These resonant and nonresonant waves then have very large amplitudes and are the dominant component of the far downstream turbulence.

4 Summary

The quasi-parallel shock operates by a complex interplay between a number of different components: particle injection/reflection; nonlinear wave-particle coupling including coherent structures; energetic particles and foreshock waves. The key to understanding the quasi-parallel shock depends on the microphysics of the extended transition zone from upstream to downstream flow, where all these components interact. A companion review (Burgess et al. 2012) describes particle acceleration mechanisms, and kinetic processes for particle injection, and consequently this review has concentrated on the magnetic structure and behaviour of waves and particles at the shock. However, it is worth emphasizing the point that it may be difficult to separate the processes of diffusive acceleration from the microphysics of the shock. Theories of diffusive acceleration often make assumptions abut wave-particle scattering that have to be tested against the microphysics of the shock transition. And, as it has been shown, the waves of the foreshock are convected into the shock transition, thereby affecting (or even controlling) the microphysics. This poses a puzzle: namely, how does the overall, global system arrange the fine tuning between energetic ion acceleration efficiency, particle injection and foreshock/shock coupling?

It can also be noted that this review has concentrated on the high Mach number shocks seen at the terrestrial bow shock and elsewhere in the heliosphere, but it is possible that new models of wave-particle-flow coupling within the quasi-parallel shock will be required for the very high Mach number regime as found in some astrophysical systems (e.g., Gargaté and Spitkovsky 2012).

We have discussed the quasi-parallel shock mostly in terms of computer simulations, but from the discussion of underlying concepts there arises another puzzle about the absolute scale of the shock. As a shock simulation continues in time energetic particles spread further and further away from the "nominal" shock position, extending the foreshock. The question arises whether there is any intrinsic scale which limits this process. Some studies (e.g., Giacalone et al. 1993) have used a "free-escape boundary" where upstream energetic particles are removed at some fixed distance, to model a shock with spatial or, equivalently, temporal limitations. Alternatively, Sugiyama (2011) has simulated in 1-D an extremely large system with no free-escape boundary, and found that the characteristic energy in the upstream (i.e., the energy of turn-over from a power-law behaviour) increases with time. How the influence of a very large foreshock impacts the microphysics of the shock transition zone (where most of the thermalization occurs) is still an open question, as is the behaviour when full three-dimensionality is taken into account.

Finally, we consider some the current challenges which have to be addressed in order to achieve progress in understanding the physics of the quasi-parallel shock. Although many of the original concepts of the pulsation shock were rooted in observations, a major part of the overall view of the shock dissipation process is now based on simulations. It has become important to return to modern observations to validate the results from simulations. Of course, many of the observational problems associated with the quasi-parallel shock remain, such as the difficulty of separating temporal and spatial variations. However, problems such as the ion injection mechanism and electron heating are relatively under-explored and would benefit from further observational work.

In terms of physical understanding, possible the greatest challenge of the quasi-parallel shock is its multi-scale nature. The pulsation shock layer is embedded in turbulence between the upstream foreshock and the downstream flow region. How important is the character of the ULF foreshock for the qualitative (and quantitative) behaviour of the pulsation shock layer? At smaller scales, there is yet again competition between scales: At quasi-parallel geometries wave dispersion plays a crucial role, but whistler mode dispersion, as is observed, implies electron scale damping which is not properly modelled in the hybrid simulations which dominate this field. To what extent does pulsation evolution depend on processes which act at scales smaller than can properly be included in the hybrid simulations? On the other hand, does pulsation growth have to be modelled using the large scales which can only (at the present time) be realised by hybrid simulations? These questions may be addressed by larger and more detailed PIC and multi-dimensional simulations. It would also be interesting to extend the simulations by using realistic mass ratio in order to analyse fine scale structure in the shock transition region. However, as one is forced to consider larger and larger scales associated with diffusive particle acceleration and foreshock waves, eventually the type and configuration of the shock driver or obstacle has to be taken into account. Because of the relatively long time scales for diffusive acceleration, and the long length scales for the foreshock, it becomes possible that to understand a particular quasi-parallel shock it might have to be necessary to take into account the totality of its environment, including sources of upstream turbulence, sources of already accelerated energetic particles, the three-dimensional configuration of the interaction, and so on. Within this complex interaction one would then hope to find those properties of the quasi-parallel shock that are universal and can be applied to other astrophysical systems where we do not have the riches of observations available from heliospheric shocks.

Acknowledgements D. Burgess acknowledges support of STFC grant ST/J001546/1.

References

D. Burgess, Cyclical behavior at quasi-parallel collisionless shocks. Geophys. Res. Lett. **16**, 345–349 (1989)

D. Burgess, Foreshock-shock interaction at collisionless quasi-parallel shocks. Adv. Space Res. **15**, 159–169 (1995). doi:10.1016/0273-1177(94)00098-L

D. Burgess, E. Möbius, M. Scholer, Ion acceleration at the Earth's bow shock. Space Sci. Rev. **173**(1–4), 5–47 (2012). doi:10.1007/s11214-012-9901-5

D. Burgess, E.A. Lucek, M. Scholer, S.D. Bale, M.A. Balikhin, A. Balogh, T.S. Horbury, V.V. Krasnoselskikh, H. Kucharek, B. Lembège, E. Möbius, S.J. Schwartz, M.F. Thomsen, S.N. Walker, Quasi-parallel shock structure and processes. Space Sci. Rev. **118**, 205–222 (2005). doi:10.1007/s11214-005-3832-3

J.P. Eastwood, E.A. Lucek, C. Mazelle, K. Meziane, Y. Narita, J. Pickett, R.A. Treumann, The foreshock. Space Sci. Rev. **118**, 41–94 (2005). doi:10.1007/s11214-005-3824-3

L. Gargaté, A. Spitkovsky, Ion acceleration in non-relativistic astrophysical shocks. Astrophys. J. **744**, 67 (2012). doi:10.1088/0004-637X/744/1/67

J. Giacalone, D. Burgess, S.J. Schwartz, D.C. Ellison, Ion injection and acceleration at parallel shocks—Comparisons of self-consistent plasma simulations with existing theories. Astrophys. J. **402**, 550–559 (1993). doi:10.1086/172157

K.I. Golden, L.M. Linson, S.A. Mani, Ion streaming instabilities with application to collisionless shock wave structure. Phys. Fluids **16**, 2319–2325 (1973). doi:10.1063/1.1694299

J.T. Gosling, M.F. Thomsen, S.J. Bame, C.T. Russell, Ion reflection and downstream thermalization at the quasi-parallel bow shock. J. Geophys. Res. **94**, 10027–10037 (1989). doi:10.1029/JA094iA08p10027

E.W. Greenstadt, F.L. Scarf, C.T. Russell, R.E. Holzer, V. Formisano, P.C. Hedgecock, M. Neugebauer, Structure of a quasi-parallel, quasi-laminar bow shock. J. Geophys. Res. **82**, 651–666 (1977). doi:10.1029/JA082i004p00651

J.R. Kan, D.W. Swift, Structure of the quasi-parallel bow shock—Results of numerical simulations. J. Geophys. Res. **88**, 6919–6925 (1983). doi:10.1029/JA088iA09p06919

D. Krauss-Varban, Waves associated with quasi-parallel shocks: Generation, mode conversion and implications. Adv. Space Res. **15**, 271–284 (1995). doi:10.1016/0273-1177(94)00107-C

D. Krauss-Varban, N. Omidi, Structure of medium Mach number quasi-parallel shocks—Upstream and downstream waves. J. Geophys. Res. **961**, 17715 (1991). doi:10.1029/91JA01545

E.A. Lucek, T.S. Horbury, I. Dandouras, H. Rème, Cluster observations of the Earth's quasi-parallel bow shock. J. Geophys. Res. (Space Phys.) **113**, 7 (2008). doi:10.1029/2007JA012756

G. Mann, H. Luehr, W. Baumjohann, Statistical analysis of short large-amplitude magnetic field structures in the vicinity of the quasi-parallel bow shock. J. Geophys. Res. **99**, 13315 (1994). doi:10.1029/94JA00440

T.G. Onsager, D. Winske, M.F. Thomsen, Interaction of a finite-length ion beam with a background plasma—Reflected ions at the quasi-parallel bow shock. J. Geophys. Res. **96**, 1775–1788 (1991a). doi:10.1029/90JA02008

T.G. Onsager, D. Winske, M.F. Thomsen, Ion injection simulations of quasi-parallel shock re-formation. J. Geophys. Res. **962**, 21183 (1991b). doi:10.1029/91JA01986

T.G. Onsager, M.F. Thomsen, J.T. Gosling, S.J. Bame, C.T. Russell, Survey of coherent ion reflection at the quasi-parallel bow shock. J. Geophys. Res. **95**, 2261–2271 (1990). doi:10.1029/JA095iA03p02261

F.G.E. Pantellini, A. Heron, J.C. Adam, A. Mangeney, The role of the whistler precursor during the cyclic reformation of a quasi-parallel shock. J. Geophys. Res. **97**, 1303–1311 (1992). doi:10.1029/91JA02653

E.N. Parker, A quasi-linear model of plasma shock structure in a longitudinal magnetic field. J. Nucl. Energy **2**, 146–153 (1961). doi:10.1088/0368-3281/2/1/323

M. Scholer, Upstream waves, shocklets, short large-amplitude magnetic structures and the cyclic behavior of oblique quasi-parallel collisionless shocks. J. Geophys. Res. **98**, 47–57 (1993). doi:10.1029/92JA01875

M. Scholer, D. Burgess, The role of upstream waves in supercritical quasi-parallel shock re-formation. J. Geophys. Res. **97**, 8319–8326 (1992). doi:10.1029/92JA00312

M. Scholer, T. Terasawa, Ion reflection and dissipation at quasi-parallel collisionless shocks. Geophys. Res. Lett. **17**, 119–122 (1990). doi:10.1029/GL017i002p00119

M. Scholer, M. Fujimoto, H. Kucharek, Two-dimensional simulations of supercritical quasi-parallel shocks: upstream waves, downstream waves, and shock re-formation. J. Geophys. Res. **98**, 18971 (1993). doi:10.1029/93JA01647

M. Scholer, H. Kucharek, V. Jayanti, Waves and turbulence in high Mach number nearly parallel collisionless shocks. J. Geophys. Res. **102**, 9821–9834 (1997). doi:10.1029/97JA00345

M. Scholer, H. Kucharek, I. Shinohara, Short large-amplitude magnetic structures and whistler wave precursors in a full-particle quasi-parallel shock simulation. J. Geophys. Res. (Space Phys.) **108**, 1273 (2003). doi:10.1029/2002JA009820

S.J. Schwartz, D. Burgess, Quasi-parallel shocks—A patchwork of three-dimensional structures. Geophys. Res. Lett. **18**, 373–376 (1991). doi:10.1029/91GL00138

S.J. Schwartz, D. Burgess, W.P. Wilkinson, R.L. Kessel, M. Dunlop, H. Luehr, Observations of short large-amplitude magnetic structures at a quasi-parallel shock. J. Geophys. Res. **97**, 4209–4227 (1992). doi:10.1029/91JA02581

T. Sugiyama, Time sequence of energetic particle spectra in quasiparallel shocks in large simulation systems. Phys. Plasmas **18**(2), 022302 (2011). doi:10.1063/1.3552026

V.A. Thomas, D. Winske, N. Omidi, Re-forming supercritical quasi-parallel shocks. I—One- and two-dimensional simulations. J. Geophys. Res. **951**, 18809–18819 (1990). doi:10.1029/JA095iA11p18809

M.F. Thomsen, J.T. Gosling, S.J. Bame, C.T. Russell, Magnetic pulsations at the quasi-parallel shock. J. Geophys. Res. **95**, 957–966 (1990a). doi:10.1029/JA095iA02p00957

M.F. Thomsen, J.T. Gosling, S.J. Bame, T.G. Onsager, C.T. Russell, Two-state ion heating at quasi-parallel shocks. J. Geophys. Res. **95**, 6363–6374 (1990b). doi:10.1029/JA095iA05p06363

W.P. Wilkinson, A.K. Pardaens, S.J. Schwartz, D. Burgess, H. Luehr, R.L. Kessel, M. Dunlop, C.J. Farrugia, Nonthermal ions and associated magnetic field behavior at a quasi-parallel earth's bow shock. J. Geophys. Res. **98**, 3889–3905 (1993). doi:10.1029/92JA01669

D. Winske, V.A. Thomas, N. Omidi, K.B. Quest, Re-forming supercritical quasi-parallel shocks. II—Mechanism for wave generation and front re-formation. J. Geophys. Res. **95**, 18821–18832 (1990). doi:10.1029/JA095iA11p18821

DOI 10.1007/978-1-4899-7413-6_18
Reprinted from *Space Science Reviews* Journal, DOI 10.1007/s11214-013-9972-y

The Dynamic Quasiperpendicular Shock: Cluster Discoveries

V. Krasnoselskikh · M. Balikhin · S.N. Walker · S. Schwartz · D. Sundkvist · V. Lobzin · M. Gedalin · S.D. Bale · F. Mozer · J. Soucek · Y. Hobara · H. Comisel

Received: 19 October 2012 / Accepted: 25 February 2013 / Published online: 28 March 2013

Abstract The physics of collisionless shocks is a very broad topic which has been studied for more than five decades. However, there are a number of important issues which remain unresolved. The energy repartition amongst particle populations in quasiperpendicular shocks is a multi-scale process related to the spatial and temporal structure of the electromagnetic fields within the shock layer. The most important processes take place in the close vicinity of the major magnetic transition or ramp region. The distribution of electromagnetic fields in this region determines the characteristics of ion reflection and thus defines the con-

V. Krasnoselskikh (✉)
LPC2E, CNRS-University of Orleans, Orleans, France
e-mail: vkrasnos@cnrs-orleans.fr

M. Balikhin · S.N. Walker
ACSE, University of Sheffield, Sheffield S1 3JD, UK

S. Schwartz
Blackett Laboratory, Imperial College London, London SW7 2AZ, UK

D. Sundkvist · S.D. Bale · F. Mozer
Space Sciences Laboratory, University of California, Berkeley, CA, USA

V. Lobzin
School of Physics, University of Sydney, Sydney, NSW, Australia

M. Gedalin
Department of Physics, Ben-Gurion University, Beer-Sheva, Israel

J. Soucek
Institute of Atmospheric Physics, Academy of Sciences of the Czech Republic, Prague, Czech Republic

Y. Hobara
Research Center of Space Physics and Radio Engineering, University of Electro-Communications, Tokyo, Japan

H. Comisel
Institute for Space Sciences, Bucharest, Romania

ditions for ion heating and energy dissipation for supercritical shocks and also the region where an important part of electron heating takes place. In other words, the ramp region determines the main characteristics of energy repartition. All these processes are crucially dependent upon the characteristic spatial scales of the ramp and foot region provided that the shock is stationary. The process of shock formation consists of the steepening of a large amplitude nonlinear wave. At some point in its evolution the steepening is arrested by processes occurring within the shock transition. From the earliest studies of collisionless shocks these processes were identified as nonlinearity, dissipation, and dispersion. Their relative role determines the scales of electric and magnetic fields, and so control the characteristics of processes such as ion reflection, electron heating and particle acceleration. The determination of the scales of the electric and magnetic field is one of the key issues in the physics of collisionless shocks. Moreover, it is well known that under certain conditions shocks manifest a nonstationary dynamic behaviour called reformation. It was suggested that the transition from stationary to nonstationary quasiperiodic dynamics is related to gradients, e.g. scales of the ramp region and its associated whistler waves that form a precursor wave train. This implies that the ramp region should be considered as the source of these waves. All these questions have been studied making use observations from the Cluster satellites. The Cluster project continues to provide a unique viewpoint from which to study the scales of shocks. During its lifetime the inter-satellite distance between the Cluster satellites has varied from 100 km to 10000 km allowing scientists to use the data best adapted for the given scientific objective.

The purpose of this review is to address a subset of unresolved problems in collisionless shock physics from experimental point of view making use multi-point observations onboard Cluster satellites. The problems we address are determination of scales of fields and of a scale of electron heating, identification of energy source of precursor wave train, an estimate of the role of anomalous resistivity in energy dissipation process by means of measuring short scale wave fields, and direct observation of reformation process during one single shock front crossing.

Keywords Collisionless shocks · Waves in plasmas · Nonstationarity · Shock scales · Plasma heating and acceleration · Wave-particle interactions

1 Introduction

Collisionless shocks (CS) are ubiquitous in the universe. They play an important role in the interaction of the solar wind with the planets (Russell 1977, 1985; Greenstadt and Fredricks 1979; Ness et al. 1974, 1981), they also are supposed to have vital role in fundamental astrophysical problems such as cosmic ray acceleration (Krymskii 1977; Axford et al. 1977; Bell 1978a, 1978b; Blandford and Ostriker 1978). CS's are of crucial importance for understanding physical processes in the vicinity of such astrophysical objects as supernova remnants (Koyama et al. 1995; Bamba et al. 2003), plasma jets (Piran 2005), binary systems and ordinary stars. In spite of this great variety of CS in the Universe only those shocks in the Solar system can be probed using in-situ observations. Moreover, comprehensive in-situ data exist only for interplanetary shocks and planetary bow shocks, however, it is worth noting that some astrophysical shocks are similar to those in the solar system. As was noted by Kennel et al. (1985) 'The density, temperature and magnetic field in the hot interstellar medium are similar to those in the solar wind, and the Mach numbers of supernova shocks at the phase when they accelerate the most cosmic rays are similar to

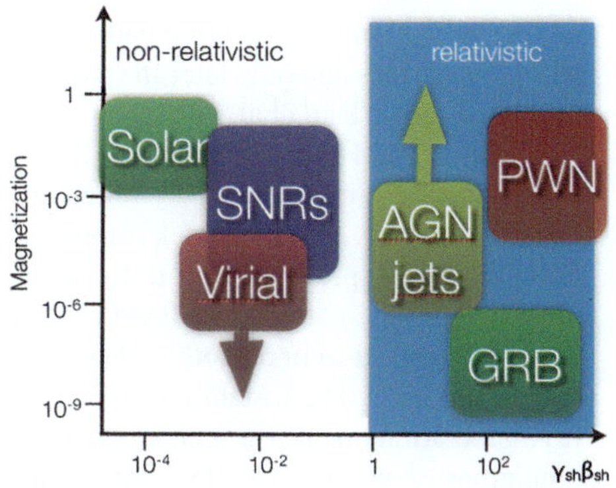

Fig. 1 (Courtesy of A. Spitkovsky) Parametric range of observations of collisionless shocks associated with different astrophysical objects

those of solar wind shocks'. Astrophysical shocks associated with different objects exhibit large differences in the parameters that characterise them. Figure 1 shows the variation of astrophysical shocks as a function of magnetisation (Y-axis), determined as $1/M_A$ where M_A is the Alfvén Mach number (the ratio of the upstream flow velocity to the characteristic velocity of propagation of magnetic perturbations in a plasma or Alfvén velocity) and the characteristic plasma pressure to magnetic pressure ratio (X-axis) where γ_{sh} is the ratio of the upstream flow velocity to the velocity of light and β_{sh} the ratio of total plasma particle thermal pressure to the magnetic field pressure in the reference frame of the upstream flow. Collisionless shocks associated with different astrophysical objects such as Supernovae Remnants (SNR), Active Galactic Nuclei jets (AGN), Pulsar Wind Nebulae (PWN), and Gamma Ray Bursts (GRB) are indicated.

As can be seen from this figure, different 'families' of shocks occupy different regions of parameter space. One can also see that the parameters of SNR shocks are quite similar to those of solar system shocks. It allows one to suggest that the studies of solar system shocks, and in particular Earth's bow shock, represents an interest for wider scientific community than only for geophysics. The majority of astrophysical and Solar system shocks are developed in magnetised plasmas.

Collisional shocks have been studied for many decades, beginning with the earliest observations of Mach (Mach and Arbes 1886; Mach and Salcher 1887). A shock occurs when an obstacle finds itself immersed in a supersonic gas flow. Before reaching the surface of the obstacle the flow should be decelerated to velocities lower than the velocity of sound so that it may flow around the body. This process of flow deceleration and the redistribution of its directed energy occurs over distances of the order of the collisional mean free path of gas particles. The energy taken from the flow during deceleration is mainly transformed into the thermal energy of the gas as it is heated. As a result the sound velocity in the gas increases and, after the shock transition, becomes larger than the remaining directed velocity of the flow so that the motion downstream of the shock is subsonic. Thus the shock represents the transition from supersonic directed motion to subsonic in the reference frame of the obstacle immersed into the flow.

The notion of the collisionless shock was introduced by several authors in the late 50's (Adlam and Allen 1958; Davis et al. 1958; Sagdeyev 1960). The modern form of the description was presented in an almost complete form in the famous review paper by Sagdeev (1966). The first problem to be overcome is related to the the existence of a shock. For collisional shocks (as mentioned above) the shock thickness is related to the mean free path of the gas particles. However, in space plasmas the mean free path can be as large as 5 AU (i.e. similar to the distance of Jupiter from the Sun)! So, how can a shock exist whose width is much smaller than the mean free path? Historically, a very similar problem first appeared in laboratory devices and only later in space plasmas. However the crucial issue in both cases is exactly the same (Paul et al. 1965; Kurtmullaev et al. 1965, 1967; Ascoli-Bartoli et al. 1966; Goldenbaum and Hintz 1965).

The solution proposed initially relied on the processes of anomalous dissipation, namely, anomalous resistivity. Thin shock transition contains quite strong variations of the magnetic field components perpendicular to its normal. This implies that there is a very intense current inside the shock transition layer. The current carriers, populations of charged particles, move with respect to one another. The plasma state supporting these intense currents is, in general, unstable. The instabilities in the plasma result in the generation of intense waves. Wave generation opens a new channel of impulse and energy exchange between the different populations of plasma particles. For instance, current carrying electrons can emit/generate the waves, and these waves can be absorbed by ions. This generation-absorption exchange using waves as a transmission media between plasma components leads to an exchange of energy and momentum between them. Typically, the characteristic scale of energy exchange between different particle populations can be much smaller than the mean free path of particles. As a result the characteristic scale of the dissipation process can be determined by this anomalous dissipation. Thus the principal difference between collisional and colisionless shocks is the change of the dissipation scale that is determined by additional processes involved, but the nature of the transition and its characteristics remain very similar. In both cases the shock redistributes the directed bulk plasma flow energy to plasma thermal energy. However, the dissipation rate and characteristic scales of collisionless shocks are determined by the anomalous process of energy dissipation. The notion of anomalous resistivity was already well known and widely used in plasma physics. The theory of anomalous resistivity based on current instabilities and the generation of ion-sound waves was directly applied to the theory of collisionless shocks by Galeev (1976). Later this idea was further developed in series of papers by Papadopoulos (1985a, 1985b), who noticed that in the case of currents perpendicular to the magnetic field the ion-sound instability is less efficient than the instability of lower-hybrid waves that propagate almost perpendicular to the magnetic field and the theory of anomalous resistivity in this case should account for these rather than for ion-sound waves.

From the earliest experimental studies of shocks in space and laboratory plasmas it was found that the characteristics of the shocks observed can be quite different even in the range of parameters that correspond to solar system shocks and those in laboratory plasmas. There were observations of quite small scales for the ramp with much longer precursor wave train, there were shocks consisting of a long transition region with large amplitude structures in the magnetic field filling a very large area in space. These early observations gave rise to attempts to classify shocks.

The first systematic classification was proposed by Formisano (1985). He noticed that there are three basic parameters of the upstream flow that are important for the classification. These are the angle between the magnetic field and shock front normal, θ_{Bn}, the plasma beta β, i.e. the ratio of total particle thermal pressure to the magnetic field pressure $\beta = 8\pi nT/B^2$, where T is the total plasma temperature, n is the plasma density, and B the magnitude of the magnetic field; and the magnetosonic Mach number $M_{Ms} = (V_{up}/V_{Ms})$, where V_{up} is the normal component of the velocity of the shock, V_{Ms} is the velocity of the magnetosonic wave propagating in the same direction as the shock. Later this classification was slightly modified and is used in the form proposed by Kennel et al. (1985). This paper divides shocks into two broad classes that are related to the ion dynamics, namely, quasiparallel and quasiperpendicular. The characteristic feature of the first group is that the ions that are reflected from the shock front can freely propagate to the upstream region. These shocks correspond to the range of angles between the magnetic field and the normal vector to the shock front $\theta_{Bn} < 45°$. In the second group, quasiperpendicular, part of the ion population is reflected. After reflection they turn back to the upstream region and can gain extra energy due to the acceleration by inductive electric field tangential to the shock surface and perpendicular to the magnetic field. Then they can cross the shock front (Woods 1969, 1971; Sckopke et al. 1983). This process can occur when $\theta_{Bn} > 45°$.

Low Mach number shocks can dissipate the necessary energy entirely through some anomalous resistivity within the current-carrying shock layer. The right-hand fast magnetosonic/whistler waves have phase and group velocities that increase with decreasing wavelength beyond the fluid regime. Thus, steepened fast mode shocks are expected to radiate short wavelength waves, and hence energy, into the unshocked oncoming flow. The shortest wavelength capable of standing in the flow then forms a "precursor wavetrain" that has been observed at these sub-critical shocks (Mellott 1985) and as we shall show later this process occurs in supercritical shocks also. However, above a critical Mach number, anomalous resistivity within the layer carrying the limited shock current is unable to convert the required amount of energy from directed bulk flow into thermal energy. At quasi-perpendicular shocks, where the magnetic field in the unshocked region makes an angle $\theta_{Bn} > 45°$ with the shock propagation direction (the shock "normal" $\hat{\mathbf{n}}$), a fraction of the incident ions are reflected by the steep shock ramp as described above. They gyrate around the magnetic field and gain energy due to acceleration by the transverse motional electric field ($-\mathbf{V} \times \mathbf{B}$). Returning to the shock layer they have sufficient energy to pass through into the downstream shocked region (Woods 1969, 1971; Sckopke et al. 1983). The separation of ions onto two groups, crossing front directly and after the reflection, results in the dispersal of particles in the velocity space. Group of reflected particles is separated from the bulk ion population due to an increase in peculiar velocity relative to the bulk motion. This process corresponds to the kinetic "heating" required by the shock jump conditions and it ensures the major part of energy dissipation necessary for directed energy transfer to thermal energy of plasma ion population. The process of ion thermalization takes place on rather large scale downstream of the shock front. The spatial length of the transition to ion thermal equilibrium can be treated in a similar fashion to that of the shock front thickness in collisional shocks. Detailed measurements of ion distributions onboard ISEE mission resulted in establishing all major characteristics of this process (Sckopke et al. 1983). This result is probably one of the most important obtained in this outstanding program. In theory this critical Mach number corresponds to the multi-fluid hydrodynamic limit in which the resistivity and viscosity cannot provide sufficient dissipation (Coroniti 1970). The reflection occurs on sufficiently smaller scales than thermalization due to a combination of magnetic forces and an electrostatic cross-shock potential. The main potential, which corresponds to the frame-invariant $\mathbf{E} \cdot \mathbf{B}$ electric field, is known as the deHoffmann-Teller potential (de Hoffmann and Teller 1950; Goodrich and Scudder 1984). It results directly from the leading electron pressure gradient term in the Generalised Ohm's Law (Scudder et al. 1986b). However in more detailed two-fluid descriptions, the quasiperpendicular shock has fine structure that depends upon the characteristics of the nonlinear shock profile (Galeev et al. 1988, 1989; Gedalin 1997; Krasnoselskikh et al. 2002). In this paper we shall present the results of the studies of quasiperpendicular shocks only. Quasiparallel shocks will be discussed in a paper by Burgess and Scholer (this review).

The basic ideas of shock formation can be understood by considering the propagation and evolution of large amplitude wave. In gas dynamics the wave corresponds to a sound wave whose evolution in terms of gas dynamics leads to the formation of discontinuities. In reality, however, narrow transition regions are formed in which the dynamics is dominated by dissipative processes. In plasmas, the characteristics of the main shock transition are determined not only by an interplay between nonlinearity and dissipation, but also by another important physical effect, wave dispersion. It is well known that a subcritical shock has a nonlinear whistler wave train upstream of its front (Sagdeev 1966; Mellott 1985). The presence of whistler/fast magnetosonic precursor wave trains in supercritical shocks as well, was experimentally established in Krasnoselskikh et al. (1991), Balikhin et al. (1997b), Oka et al. (2006). These whistler waves have rather large amplitudes and their role in energy transformation and redistribution between different particle populations and in the formation of the structure of the shock front is still an open question. The major transition of such a dispersive shock, the ramp, may behave in a similar fashion to either the largest peak of the whistler precursor wave packet (Karpman et al. 1973; Sagdeev 1966; Kennel et al. 1985; Galeev et al. 1988, 1989; Krasnoselskikh et al. 2002) or the dissipative shock region in which the major dissipation due to current driven instabilities occurs. The nonlinear steepening process can be described as the transfer of energy to smaller scales. The steepening can be terminated either by collisionless dissipation, as described above, or by wave dispersion. Typically the dissipative scale L_d exceeds the dispersive scale L_{disp}, the former is reached first and further steepening can be prevented by the dissipation that takes away energy. When steepening is balanced by dissipation, a dissipative subcritical shock forms. Most subcritical collisionless shocks observed in situ are supposed to be dissipative even though dispersive processes play a role in forming a dispersive precursor wave train. Such a shock is characterized by a monotonic transition in the magnetic field (magnetic ramp) of width L_d. The dissipative length is determined by the most important anomalous dissipative process. Its major features are the generation of intense short-scale waves and their dissipation. This form of the evolution of a nonlinear wave takes place at low Mach numbers. However, if the nonlinearity is strong enough (as determined by the velocity and density of the incoming flow), dissipation is not capable of stopping the steepening, and the gradients continue to grow, then energy transfer to smaller scales continues and the characteristic scale of the transition can become smaller. The next process that comes into play to counterbalance the steepening is dispersion. Dispersion becomes important when the gradients become comparable with the dispersive scale L_{disp}. In this case the shock front structure becomes multi-scale. The steepening is prevented by short-scale dispersive waves which are able to propagate away from the evolving shock front. These waves effectively remove some part of the energy and, most importantly, restrain further growth of the gradients. For perpendicular shocks the phase velocity of the dispersive waves decreases with decreasing scale and a wave train is formed downstream of the magnetic ramp. For shocks with a more oblique geometry (quasiperpendicular shocks) the phase velocity increases with decreasing spatial scale and an upstream wave train is formed. The upstream wave precursor is approximately phase standing in the upstream flow. Its amplitude decreases with the distance from the shock ramp due to dissipation processes as was discussed in the early theoretical papers describing subcritical shocks (Sagdeev 1965a, 1966).

The transition to reflection shock takes place when downstream bulk velocity reaches the downstream ion-sound speed. Supercritical reflection shocks have a more complex structure in comparison to subcritical shocks. In quasi-perpendicular shocks the upstream magnetic field does not allow reflected ions to travel far upstream before turning them back to the shock front. The upstream region in which the beam of reflected ions perform part of their Larmor orbit before being turned back to the shock is called a foot. The foot region is characterised by a 15–20 % increase in the magnitude of the magnetic field. The consideration of a Larmor orbit of a reflected ion gives relatively accurate estimate of the spatial size of the foot $L_f = 0.68 R_{Li} \sin\theta_{Bn}$ where R_{Li} is the gyroradius of ions moving with the velocity equal to normal component of the velocity of upstream flow (Woods 1969; Livesey et al. 1984). The coefficient 0.68 corresponds the case of specular reflection. For non-specular reflection this relation is slightly modified (Gedalin 1996). Downstream of the quasi-perpendicular shock's main transition the joint gyration of the bulk plasma ions and beam of reflected ions leads to an overshoot-undershoot structure. Again, the size of this overshoot/undershoot can be estimated in a straightforward manner in terms of the ion gyroradius. However, the main transition layer lies between the foot and the overshoot. This is the region where the most dramatic changes in the plasma parameters occur. In a supercritical, quasi-perpendicular shock this layer is characterised by the steepest increase of the magnetic field referred to as the ramp. The change of the electrostatic potential, reflection of ions, and electron thermalisation take place within the ramp and its spatial scale determines the major physical processes within the shock and the mechanisms for the interaction of the shock front with the incoming electrons and ions. For instance, several theoretical models suggest that in the ramp of high Mach number shocks very small scale electric fields can be present (Krasnoselskikh 1985; Galeev et al. 1988, 1989). There are several critical issues regarding supercritical quasiperpendicular shock physics for which alternative explanations for the observational features of the shock front have been proposed. Theoretical considerations (Galeev et al. 1988, 1989; Krasnoselskikh et al. 2002), that treat the supercritical shock front as being similar to a nonlinear dispersive wave, predict that the ramp scales (gradients) should decrease with increasing Mach number, eventually reaching characteristic values as small as several electron inertial lengths $L_e = c/\omega_{ep}$. Moreover, after some critical Mach number corresponding to nonlinear whistler critical Mach number whose value is approximately 1.4 times the linear whistler critical Mach number, the shock should become nonstationary. These critical Mach numbers determine the characteristic flow velocities when they become larger than the maximum velocity of a linear or nonlinear whistler wave propagating upstream along the shock front normal. Many computer simulations (Scholer and Matsukiyo 2004; Matsukiyo and Scholer 2006) come to the conclusion that the thickness of the ramp is determined by dissipative processes due to either the modified Buneman instability (MBI) or the modified two stream instability (MTSI). Both theoretical studies and computer simulations have associated pitfalls. Theoretical models can not accurately take into consideration the presence of reflected ions whilst simulations are carried out with an unrealistic ratio of the plasma frequency to gyrofrequency that strongly changes the ratio of electric to magnetic wave fields and often with an unrealistic ion to electron mass ratio. Both theoretical models and simulations predict the transition to nonstationary dynamics. However, they strongly differ in the determination of the scales of the electric and magnetic fields in the ramp region, in the energy sources for the upstream whistler waves that form the precursor wave train, and in the characteristics of the shock dynamics when it becomes nonstationary. For these reasons experimental studies of these questions are crucial for our understanding of the physical processes in quasiperpendicular collisionless shocks. Our Review aims to report the studies of all these questions making use mainly of Cluster measurements (adding some other data where it is necessary, in particular THEMIS data in studies of magnetic field scales of shocks).

The first critical issue we shall address is magnetic ramp width and spatial scale. The main motivation for the study of the magnetic ramp width L_r is that it is this scale that determines the nature of the shock, i.e., the dominant physical processes that counteract nonlinear steepening. The shock width can be determined either by the solitary structure of nonlinear whistler slightly modified by the presence of reflected ions or by characteristic anomalous resistivity scale associated with one of instabilities mentioned above. For instance if it is indeed the case that the ramp width increases with increasing Mach number as concluded by Bale et al. (2003), then the evolution of nonlinear whistler waves must be excluded from the processes that are involved in the formation of supercritical shocks. The characteristics of the major transition within the shock in which the flow deceleration and the magnetic field and electrostatic potential variations take place are determined by the interplay between nonlinearity, dispersion and dissipation. The presence of a population of reflected ions makes it difficult to construct a reliable theoretical model based on an analytical or semi-analytical description. However, the establishment of the scales of this transition allows one to determine the characteristics of the dominant physical processes in play. The ramp thickness is also crucial for a redistribution of energy between electrons and ions. An important characteristic involved in this process is the gradient scale of the transition. Two reasons cause the need in introducing this ramp gradient spatial scale. The first is the interaction between the incoming electrons and the electromagnetic field at the shock front. As was shown by Balikhin et al. (1993, 1998), Balikhin and Gedalin (1994), and Gedalin et al. (1995a, 1995b), an important effect of this interaction is the possible violation of adiabaticity even in the case when the width of the magnetic ramp considerably exceeds the formally calculated electron gyroradius. Two very different scenarios of electron heating can occur depending upon if the conditions for adiabaticity are satisfied or violated. This effect is crucially dependent upon the ramp spatial scale. The change between adiabatic/non-adiabatic regimes is related to the ability of the nonuniform electric field within the ramp to rectify the electron motion and increase their effective gyration radius (Balikhin et al. 1993, 1998; Balikhin and Gedalin 1994; Gedalin et al. 1995a, 1995b). The parameter that determines the transition from adiabatic to nonadiabatic motion of the electrons is the inhomogeneity of the magnetic field. The characteristic spatial scale of such a layer may be defined as the product of the change in magnetic field ΔB normalised to the upstream field B_0 and the spatial distance L_{br} over which this change occurs i.e. $l_{gr} = L_{Br}(B_0/\Delta B)$. To illustrate the effect of this parameter, one can consider two cases. For a weak shock for which $(B_d/B_0) \approx 1.2$, here B_d is the magnetic field magnitude after the shock transition (downstream), and whose ramp width is of the order 5–6 R_{Le} (electron Larmor radii) the electron motion will be adiabatic. However, in a stronger shock of similar magnetic ramp width for which the maximum magnetic field observed in the overshoot exceeds that of the upstream field by a factor 5–6 the electron behaviour becomes non-adiabatic. This makes it necessary to carry out the statistical study of the magnetic ramp spatial scale in addition to the ramp width (size). The ramp width and its gradient scale are also important for the problem of stability of the ramp region of the shock front. According to Krasnoselskikh (1985), Galeev et al. (1988, 1989) and Krasnoselskikh et al. (2002), the nonlinear whistler wave structure becomes unstable when the characteristic gradient exceeds some critical value. It was suggested by Krasnoselskikh et al. (2002) that it takes place when the Mach number becomes equal to nonlinear critical whistler Mach number. When this happens, dispersive processes can no longer counterbalance the nonlinearity and the shock front overturns. Thus, the characteristic gradient scale provides a rather universal characteristic of the degree of steepness of the shock front. Thus its determination completed a comprehensive statistical study of the magnetic ramp spatial gradient scale in addition to the ramp width (size). Many papers have been devoted to the magnetic field structure of collisionless shocks (e.g. Russell et al. 1983; Krasnoselskikh et al. 1991; Farris et al. 1991; Newbury and Russell 1996; Hobara et al. 2010; Mazelle et al. 2010). In particular, the spatial scales of its various regions have been comprehensively investigated (Balikhin et al. 1995; Farris et al. 1993; Hobara et al. 2010; Mazelle et al. 2010). ISEE and AMPTE measurements of the magnetic field profiles of the shock front structure led to evaluation of the scale sizes of the foot and overshoot regions that were supposed to be of the order of ion inertial length c/ω_{pi} and $3c/\omega_{pi}$ respectively, here ω_{pi} is the ion plasma frequency. The ramp scale has been estimated to be less than an ion inertial length with reports of one or two very particular shocks whose ramp scale was sufficiently smaller, of the order $0.1c/\omega_{pi}$ (Newbury and Russell 1996; Walker et al. 1999b). We report here the statistical studies of scales based on papers by Hobara et al. (2010) and Mazelle et al. (2010). Another critical issue we shall address hereafter is the electric field distribution inside the ramp region. The energy transfer to smaller scales due to steepening can achieve the scales comparable to electron inertial scale where the whistler waves become quasi-electrostatic. In nonlinear-dispersive scenario of the shock front description the field can have multiple short scale electric field spikes. The experimental study can answer the question whether they exist or not and to determine their characteristics. Studies of the magnetic field profile across the terrestrial bow shock significantly outnumber those based on electric field measurements. Despite the fundamental effect that the electric field has on the plasma dynamics across collisionless shocks, the complexity of the interpretation of electric field data has impeded studies of the electric field structure within the shock front. It is worth noting that only a handful of studies are dedicated to the electric field structure within the shock front (Heppner et al. 1978; Formisano 1982; Scudder et al. 1986a; Wygant et al. 1987; Balikhin et al. 2002, 2005; Bale and Mozer 2007; Walker et al. 2004; Hobara et al. 2008; Dimmock et al. 2012; Bale et al. 2008). There have been very few reports regarding the scale lengths of features observed in the electric field at quasi-perpendicular shocks. The scale size over which the potential varies at the front of a quasi-perpendicular bow shock is an issue that should be resolved in order to gain a full understanding of the physical processes that are occurring in the front. Several different points of view have been published on the relationship between the scale size of the magnetic ramp and that over which the change in potential is observed. Some studies (Eselevich et al. 1971; Balikhin et al. 1993, 2002; Formisano and Torbert 1982; Formisano 1982, 1985; Krasnoselskikh 1985; Leroy et al. 1982; Liewer et al. 1991; Scholer et al. 2003) have proposed that the spatial scale of electrostatic potential is of the same order or smaller than that of the magnetic ramp under certain conditions. Such shocks have been observed in numerous experimental and numerical studies of quasi-perpendicular supercritical shocks. On the other hand Scudder (1995) claimed that the potential scale length is larger than that of the magnetic scale length. Actual measurements of the electric field variations within the bow shock are very sparse. The main reason for this is due to the difficulties encountered when making electric field measurements. Only a small number of space-based measurements of the electric field during the shock front crossing have been reported. Heppner et al. (1978) reported observations of a short lived spikes in the electric field making use of ISEE measurements. However, being short duration, these features were not observed at every shock crossing. Subsequent investigations by Wygant et al. (1987) have shown the existence of spike-like features in the electric field both at the shock ramp and in the region just upstream. From the study of spin averaged ISEE-1 data, Formisano (1982) determined that the increase in the observed electric field intensity began just upstream of the magnetic ramp and lasted longer than the ramp crossing itself. Whilst the electric-field intensity in the regions upstream and downstream of the shock could be interpreted as due to the $V \times B$ motion of the plasma the enhancement observed during the shock crossing must be due to the processes occurring within the shock front itself. In laboratory experiments, where the conditions are not exactly the same as in space plasmas, Eselevich et al. (1971) reported that the major change in potential across the shock occurs within the magnetic ramp region.

Using data of numerical simulations, Lembège et al. (1999) analysed the scale size of both the magnetic ramp region L_{Br} and the scale of the major change in potential L_ϕ inside and around the ramp. They concluded that the scale lengths were of the same order, i.e. $L_{Br} \approx L_\phi$. This view is supported by the simulations of Scholer et al. (2003). The latter authors also show that during the shock reformation process, the main potential drop occurs over several ion inertial scales in the foot region and they noticed that the steepened magnetic ramp region also contributes a significant fraction of the change in total potential over much smaller scales, typically 5–10 Debye lengths. Despite the simulation shocks parameters are still rather far from observations (see Sect. 3 for more details) the tendencies in majority of simulations are well pronounced and are similar to those in laboratory experiments. Hereafter we report the observations of electric field spikes observed onboard Cluster satellites first reported by Walker et al. (2004) and the statistical study of their characteristics. The third important problem of quasiperpendicular shock physics addressed in this Review is the problem of electron heating. In contrast to the ion heating problem which has been well advanced through to detailed studies of data from ISEE, the electron heating problem has remained controversial. The action of shock quasistatic electric and magnetic fields on the electron population (which can have thermal speeds far in excess of the shock speed) is to inflate and open up a hole in the phase space distribution by accelerating (decelerating) incoming (escaping) electrons (Scudder et al. 1986c; Feldman et al. 1982). This inflation in itself is reversible thus it is not dissipation or heating if other processes would not be involved. Irreversibility may be imposed if additional scattering would take place infilling the hole. If adiabatic invariant of electrons is conserved while electrons cross the shock front it can not happen. One should conclude that some non-adiabatic process should occur inside the shock front. One of the possibilities can be related to Debye-scale electric fields (Bale et al. 1998). Another possibility is to suggest that the phase space inflation is indeed accompanied by instabilities which could scatter the electrons. Demagnetisation of the electrons due to the strong gradients in the electric field as it was mentioned above (Balikhin and Gedalin 1994) or nonlinear wave phenomena (Krasnoselskikh et al. 2002) combined with wave particle interactions can offer an alternative scattering processes. Thus the partition of energy between ions and electrons is a complex, self-consistent multi-scale interplay between electron heating, magnetic/electric field profile, shock potential, and ion reflection. This interplay remains poorly understood despite 40 years of research. That research has included detailed case studies (Scudder et al. 1986b), statistics of the inferred potential and electric field structures (Schwartz et al. 1988; Walker et al. 2004), theoretical studies (Galeev et al. 1988; Gedalin 1997; Krasnoselskikh et al. 2002) and increasingly sophisticated numerical simulations (Lembège et al. 2004; Scholer and Burgess 2006). Direct measurements of the thickness of the shock transition layer combined with the rapid simultaneous measurements of the electron distribution function can allow solving this long standing opened problem in shock physics. If the electron heating can be attributed to kinetic instabilities, the shock thickness will be measured in ion inertial lengths (c/ω_{pi}) (Papadopoulos 1985b; Matsukiyo and Scholer 2006). If such instabilities prove ineffective, above a second critical Mach number the shock steepening is expected to be limited by whistler dispersion and/or be unstable to shock reformation (Krasnoselskikh et al. 2002). Recent studies of the shock thickness (Hobara et al. 2010; Mazelle et al. 2010) do show scales comparable to whistler wavelengths. These contrasted an earlier study (Bale et al. 2003) reporting scalings that matched the gyro-scales of reflected ions. To date, studies have relied on the high temporal cadence available from magnetic or electric field experiments. However, field profiles provide only indirect evidence of the shock dissipation scales. A recent study (Lefebvre et al. 2007) used sub-populations of electrons to determine the electrostatic potential profile at one shock, suggesting that it rose in concert with the magnetic field. In the work reported here, first published in Schwartz et al. (2011), the electron distribution function major characteristics are measured at sufficient cadence to reveal directly for the first time the scale of the electron temperature profile. Many shock crossings by Cluster satellites take place on the flanks of the magnetosphere that creates quite favourable conditions for the studies of the relatively narrow shock transitions allowing one to have many measurements on small spatial scale. Hereafter we present unprecedently rapid measurements of electron distribution moments that allow to shed new light on electron heating problem and its scales.

The fourth problem presented in the Review, which is closely related to the problem of magnetic and electric field scales, is determination of the source of waves forming upstream precursor wave train. The presence of whistler/fast magnetosonic precursor wave trains in supercritical shocks was experimentally established by Balikhin et al. (1997a), Krasnoselskikh et al. (1991), Oka et al. (2006). These whistler waves have rather large amplitudes and their role in energy transformation and redistribution between different particle populations and in the formation of the shock front structure is still an open question. The energy source responsible for the generation of these waves is the subject of active debate in shock physics (see Galeev et al. 1988, 1989; Krasnoselskikh et al. 2002; Matsukiyo and Scholer 2006; Comişel et al. 2011). Often the precursor waves are almost phase-standing in the shock frame. However, if they are generated by the ramp region as the dispersive precursor their group velocity can still be greater than zero in the shock reference frame, which would allow energy flow in the form of Poynting flux to be emitted towards the upstream of the shock transition. On the other hand, if the waves are generated by instabilities related to reflected ions, their energy flux will be directed from the upstream region towards the shock ramp. The goal of Sect. 4 is to address this problem, to present the direct measurement of the Poynting flux of the upstream whistler waves aiming to establish the direction of the Poynting flux. There are two different points of view on this subject also. It has been suggested that the shock front structure of quasi-perpendicular supercritical shocks is formed in a way similar to that of subcritical shocks (Galeev et al. 1988, 1989; Krasnoselskikh et al. 2002). In such scenario the precursor wave train is a part of the shock front structure emitted by the ramp region upstream due to positive dispersion of whistler waves. The observed dynamic features of shocks have also been studied extensively using computer PIC or hybrid simulations, often with focus on the precursor wave activity and reflected ions (Hellinger and Mangeney 1997; Hellinger et al. 2007; Matsukiyo and Scholer 2006). From a kinetic viewpoint, however, it may be argued that the shock-reflected ions change the physical picture and that the principal scales, temporal and spatial, could be determined by the characteristics of the reflected ion population (Bale et al. 2003). Upstream waves can then be generated due to counterstreaming ions and electrons in the shock front region, forming unstable particle distributions with respect to some wave modes (Papadopoulos 1985b; Hellinger et al. 2007; Scholer and Matsukiyo 2004; Matsukiyo and Scholer 2006). While this is probably the case for some higher frequency waves, we present here an analysis that leads to the conclusion that the source of the upstream low frequency whistler waves is indeed related to the presence of the nonlinear ramp transition, emitting smaller scale dispersive waves towards the upstream flow. The existence of phase-standing upstream whistler waves depends on the value of the upstream flow speed Mach number relative to the phase velocity. If the Mach number of the shock does not exceed the nonlinear whistler critical Mach number $M_w = V_{w,max}/V_A = 1/2\sqrt{m_i/m_e}\cos\theta_{Bn}$, where $V_{w,max}$ represents the highest possible velocity of nonlinear whistler wave, then phase-standing (nonlinear) whistler wave trains can exist upstream of the shock (Galeev et al. 1988, 1989; Krasnoselskikh et al. 2002). The results we report here were first published by Sundkvist et al. (2012). Similar results were reported making use the Time Domain Sampling instrument (TDS) onboard Wind satellite (Wilson et al. 2012) for three of four crossings of interplanetary shocks. In one case the polarization of waves was found to be different from whistler wave. Unfortunately, single satellite measurements do not allow one to establish unambiguously the reason of this anomaly, it could be associated with some particular perturbation in the solar wind.

The fifth problem intimately related to previous is the problem of nonstationary dynamics of high Mach number shocks. Shock waves are usually considered to be nonlinear waves that cause irreversible changes of state of the media and from macroscopic point of view they are stationary (for a review, see, e.g. Tidman and Krall 1971). However, in the very beginning of the collisionless shock physics Paul et al. (1967) hypothesized that high-Mach-number shocks can be nonstationary, and the first unambiguous evidence of the nonstationarity was obtained by Morse et al. (1972) in laboratory experiments. New evidence of shock front nonstationarity was found in the 1980s. In particular, Vaisberg et al. (1984) reported low frequency oscillations of the ion flux in the Earth's bow shock. Later Begenal et al. (1987) observed a similar phenomenon in the Uranian bow shock. In the very beginning of computer simulations of the collisionless shocks Biskamp and Welter (1972) have observed the process of shock dynamic behaviour. The inflowing ions formed vortices in the phase space and dynamics of the front was definitely nonstationary. Later, numerical simulations performed by Leroy et al. (1982) using 1-D hybrid code showed that the front structure of perpendicular shocks varies with time, for instance, the maximum value of the magnetic field exhibits temporal variations with a characteristic time of the order of ion gyroperiod, the magnitude of these variations being about of 20 % if the parameters are typical for the Earth bow shock ($M_A = 8$ and $\beta_{e,i} = 0.6$, where M_A is the Alfvén Mach number, $\beta_{e,i}$ is the ratio of the thermal electron/ion and magnetic pressures). They also found that for $M_A = 10$ and $\beta_{e,i} = 0.1$ the ion reflection was bursty, oscillating between 0 and 70–75 %. Hybrid simulation of perpendicular shocks with very high Mach numbers carried out for the first time by Quest (1986) have shown that the ion reflection in the shocks can be periodic, the stages with 100 % ion reflection alternating with the stages of 100 % ion transmission. As a result, instead of a stationary structure, he observed a periodic wave breaking and shock front reformation. Later Hellinger et al. (2002) reexamined the properties of perpendicular shocks with the use of the 1-D hybrid code and observed the front reformation for a wide range of parameters if upstream protons are cold and/or Mach number is high. Scholer et al. (2003) and Scholer and Matsukiyo (2004), Matsukiyo and Scholer (2006) in their 1-D full-particle simulations with the physical ion to electron mass ratio reproduced the reformation of exactly and approximately perpendicular high-Mach-number shocks in plasmas with $\beta_i = 0.4$ and demonstrated an importance of modified two-stream instability for the reformation process. Krasnoselskikh (1985) and Galeev et al. (1988, 1989) proposed models describing the shock front instability due to domination of nonlinearity over dispersion and dissipation. This instability results in a gradient catastrophe within a finite time interval. Several aspects of the model, including the role of nonlinear whistler oscillations and existence of a critical Mach number above which a nonstationarity appears, were developed in further detail and more rigorously by Krasnoselskikh et al. (2002) and complemented by numerical simulations with the use of the 1-D full particle electromagnetic code with a relatively small ratio of electron and ion masses, $m_e/m_i = 0.005$. It was also shown that the transition to nonstationarity is always accompanied by disappearance and re-appearance of the phase-standing whistler wave train upstream of the shock front. Moreover, for large Mach numbers the nonstationarity manifests itself as a periodic ramp reformation, which influences considerably the ion reflection, in particular, the reflection becomes bursty and sometimes the ions are reflected from both old and new ramps simultaneously. The four-spacecraft Cluster mission gave much more new opportunities for experimental studies of the shocks. The first examples of some aspects of shock nonstationarity were presented by Horbury et al. (2001). These authors analyzed magnetic field data for two quasiperpendicular shocks with moderate and high Alfvén Mach number. While for moderate M_A the shock profiles measured by different spacecraft were approximately the same, with the exception of a small-amplitude wave activity in the foot, for high M_A the amplitude of the fluctuations attains 10 nT, making profiles considerably different for different spacecraft. However, the authors argued that these fluctuations stop before the ramp and do not appear to disrupt the shock structure; on the other hand, they didn't reject an opportunity that the fluctuations observed may represent the signatures of the unsteady shock reformation. Hereafter we report the first direct observation that clearly evidence the shock front reformation observed onboard Cluster mission on 24th of January 2001. This material was first published by Lobzin et al. (2007).

The sixth problem we discuss in this Review is important for the definition of the relative role of dissipative and dispersive effects, namely the problem of anomalous resistivity. The problem of electron heating mentioned above is considered for many years to be 'solved' for subcritical shocks and conventional solution proposed and widely accepted is formulated in terms of magic words 'anomalous resistivity'. This notion was introduced first by Sagdeev (1965b) and then analyzed in more detail by Galeev (1976), who made estimates of the characteristic scale of the shock transition relying on ion-sound instability. Papadopoulos (1985a) has noticed that in case of quasiperpendicular shocks the most important instabilities should be related to lower hybrid waves and has revised the model taking these effects into account. However there were no measurements that might be used to confirm or reject theoretical models of the dissipation due to anomalous resistivity. It is worth noting that this problem is very important for the determination of energy redistribution between electrons and ions, especially for the electron heating and electron acceleration. We can not present here theoretical studies of anomalous resistivity, an interested reader can find general ideas in the review papers by Galeev and Sudan (1989) and Papadopoulos (1985a). The second important problem for which short scale length waves are important is an energization of electrons within a collisionless shock. It requires the transfer of a portion of the energy associated with the incoming upstream plasma flow to the electron population. In order for this energy transfer to occur, there has to be some media that can channel energy from the incoming ion population to the electrons. One mechanism that has commonly been proposed, both for solar systems and particularly for astrophysical applications is based on excitation of lower-hybrid waves (Papadopoulos 1981; Laming 2001; Krasnoselskikh et al. 1985). The increased level of electric field fluctuations in the frequency range 10^2–10^3 Hz observed in the vicinity of a quasiperpendicular shock front is usually attributed to either ion-sound or whistler waves. One of the most comprehensive studies of the plasma waves in this frequency range was conducted by Gurnett (1985). His main conclusion was that waves observed above local electron cyclotron frequency are Doppler shifted ion-sound waves whilst those below are whistler mode waves.

The main reason of the lack of measurements of ion sound and lower hybrid waves is related to technical difficulties of small scale electric field measurements. Recently two papers were published that represent first attempts to create the experimental basis for the anomalous resistivity studies in collisionless shocks. We present in this section a short summary of the results obtained by Balikhin et al. (2005), Walker et al. (2008). The rapid changes that are observed in the plasma at the front of a supercritical, quasi-perpendicular shock and described in previous sections lead to the creation of multiple free energy sources for various plasma instabilities. Twin satellite missions, such as ISEE or AMPTE, have provided data for a number of comprehensive surveys of the waves observed in the frequency range (10^{-2}–10^{1} Hz) of the plasma turbulence encountered at the shock front. The use of multisatellite data for wave identification and turbulence studies is limited to the analysis of those waves whose coherence lengths are of the same order of magnitude as the satellite separation distance. Plasma wave modes such as the ion-sound or lower-hybrid, that are supposed to play an important role at the shock front, possess coherence lengths that are very short in comparison with any realistic satellite separation distance (Smirnov and Vaisberg 1987). For majority of these waves the coherence length is either comparable to or a few times greater than their wavelength. In such cases the waves observed by different satellites in a multisatellite mission will be uncorrelated. This will make it impossible to apply wave identification methods based on intersatellite phase delays (Balikhin et al. 1997a; Balikhin et al. 2003) or k-filtering (Pinçon and Glassmeier 2008). Nevertheless the identification of waves with short wavelengths and study of their dynamics remains very important because of their potential role in the transfer of energy associated with the upstream directed motion into other degrees of freedom. In the classical model of a quasiperpendicular low β shock anomalous resistivity occurs due to ion-sound turbulence in the shock front (Galeev 1976). Lower hybrid waves also may play an important role at the shock front since they also can be involved in resonance interactions both with electrons and ions and so may be extremely efficient at channelling the energy exchange between the two species. In order to assess the importance of ion-sound, lower hybrid and other waves with relatively short wavelengths within the plasma dynamics of the shock front the mode of the observed waves should first be identified. The strong Doppler shift that results from the large values of wavevector $|k|$ precludes the reliable use of the observed frequency for correct identification as was done in many previous studies. Here we show that the data from a single spacecraft can be used to determine propagation modes of waves observed in the frequency range 10^{2}–10^{4} Hz at the front of the terrestrial bow shock. A similar approach has been used by Tjulin et al. (2003) in a study of lower-hybrid waves in the inner magnetosphere.

The lower-hybrid wave is an electrostatic plasma wave mode whose plasma frame frequency is in the vicinity of the lower-hybrid resonance frequency $\omega_{lh} \sim \sqrt{\Omega_{ci}\Omega_{ce}}$, where Ω_{ci} and Ω_{ce} are the ion and electron gyrofrequencies respectively. The wave has linear polarisation and propagates almost perpendicular to the magnetic field ($\cos(\theta_{kB}) \sim \sqrt{m_e/m_i} \sim 89°$). The maximum growth rate γ_{max} occurs when $k_{\|}/k \sim \omega_{pi}/\omega_{pe}$, where $k_{\|}$ is parallel component of the wavevector. Since the waves are propagating in a plasma that is moving with respect to the satellite, their frequencies will be Doppler shifted in the spacecraft frame. The magnitude of this shift can be estimated using the resonance condition of the Modified Two Stream Instability (MTSI) $2V_A M_A k = \omega_{lh}$. This gives a maximum estimate for the correction in observed wave frequency due to the Doppler shift, $kV_{sw} \sim \omega_{lh}/2$, here V_{sw} is the velocity of the solar wind supposed to be equal to normal component of the upstream flow velocity.

Current models of wave turbulence that determines anomalous resistivity in the front of quasiperpendicular shocks are based on the occurrence of lower hybrid waves at a shock front being generated due to counter-streaming populations of ions and relative motion of reflected ions (Leroy et al. 1982) and bulk electrons and ions at the front via the modulational two-stream instability (MTSI) (Papadopoulos 1985b; Matsukiyo and Scholer 2006) or modified Buneman instability. These models are often used to explain the electron acceleration observed at various astrophysical shocks such as supernova remnants (Laming 2001). However, there is currently no substantial experimental evidence that these waves do indeed exist in the fronts of supercritical, quasiperpendicular, collisionless shocks. The results of data analysis from the Intershock electric field experiment, in which wave activity was observed at frequencies of a few Hertz, has been used to argue for the existence of lower hybrid waves (Vaisberg et al. 1983). An alternative explanation, however, has been proposed in which Intershock may have simply observed the electric field component of whistler wave packets propagating in the foot region (Krasnoselskikh et al. 1991; Balikhin et al. 1997a; Walker et al. 1999a). Electric field observations of Comet Halley also showed evidence for waves observed in the vicinity of the lower hybrid frequency (Klimov et al. 1986). However, their exact wave mode was not determined. The natural way to differentiate these modes is to examine their polarisations. Whistler mode waves are elliptically polarised whilst lower hybrid waves, as mentioned above, are linearly polarised. We present here the summary of direct observations of the ion-sound (Balikhin et al. 2005) and lower-hybrid waves (Walker et al. 2008) that we complete by estimates of characteristic effective collision frequencies.

The paper is organized as follows. In Sect. 2 we present the statistical studies of quasiperpendicular shock ramp widths. Section 3 is dedicated to electric field scales of the ramp of quasiperpendicular shocks. In Sect. 4 the results of evaluation of the Poynting flux of oblique whistler waves upstream of the shock front are outlined. Recent results on electron heating scale at High Mach number quasiperpendicular shocks are presented in Sect. 5. In Sect. 6 we use the data of measurements of lower hybrid and ion sound waves intensities to evaluate the characteristics of anomalous resistivity aiming to determine its role in the shock front formation. In Sect. 7 we present results of direct observations onboard Cluster satellites of nonstationarity and reformation of high-Mach number quasiperpendicular shock. In Sect. 8 we summarise the results of experimental observations and discuss the conclusions that follow from Cluster and THEMIS observations. In Appendix A we present some short comments concerning comparison of computer simulations with the observations. Appendix B contains a notation table defining the notations used in this paper.

2 Statistical Studies of Quasi-Perpendicular Shocks Ramp Widths

As was discussed in the Introduction the characteristics of the major transition within the shock in which the flow deceleration and the magnetic field and electrostatic potential variations take place are determined by the interplay between nonlinearity, dispersion and dissipation. The presence of a population of reflected ions makes it difficult to construct a reliable theoretical model based on an analytical or semi-analytical description. However, the establishment of the scales of this transition allows one to determine the characteristics of the dominant physical processes in play. Single satellite missions provide very poor possibilities for the reliable identification of the shock width and evaluation of the characteristic scales of structures within it. In such cases the spatial size of the foot or overshoot have been used (Balikhin et al. 1995) to evaluate the thickness of the ramp region. The first shock crossings by two satellites were studied in the frame of ISEE and AMPTE projects. These missions provided the first insight into the thickness of the shock transition. The decrease of the shock width and consequent increase of gradients as a function of increasing Mach number was clearly demonstrated by Farris et al. (1993). In their study the Mach number was normalized to the critical Mach number that determines the transition from sub-critical to supercritical shocks. Figure 2 (from Farris et al. 1993) shows magnetic field magnitude measurements made by ISEE 1 for five low beta quasiperpendicular shocks ordered by the ratio of their Mach number to critical Mach number increasing from top to bottom. The increase in the gradient of the shock transition is clearly independent of the differences in the angles between the normal to the shock front and the magnetic field and of the value of beta.

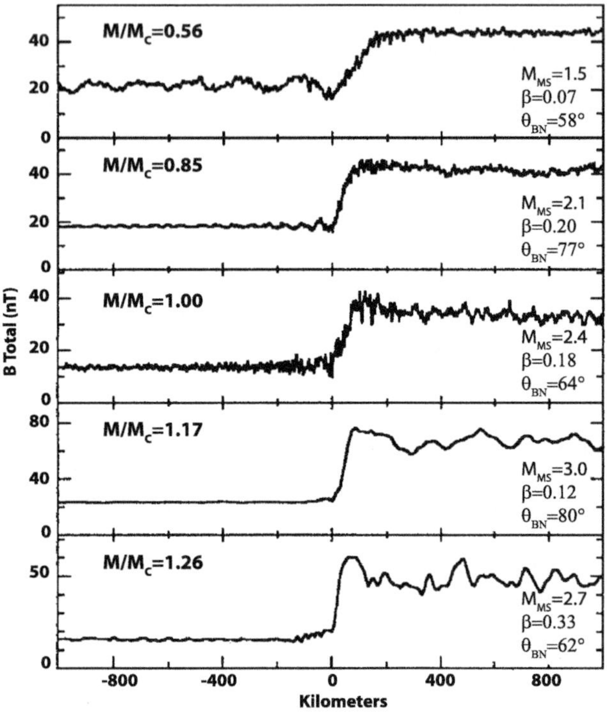

Fig. 2 The total magnetic field strength as a function of distance through five low beta, quasi-perpendicular shocks in order of increasing ratio of criticality. The Mach number, β, and θ_{Bn} for each shock are displayed. The data are shown at the highest temporal resolution available. The sampling rate for the first three shocks is 16 Hz and 4 Hz for the last two (Adapted from Farris et al. 1993)

Several studies have been dedicated to the investigation of structural elements of the shock front making use of ISEE 1,2 magnetometer data. Scudder and co-authors (Scudder et al. 1986a, 1986b, 1986c) carried out the detailed study of the shock crossing on 7th November 1977. This is presumably the most detailed study of a single event, in which all the elements of the structure were put together and compared with detailed measurements of the particle distribution functions. These authors succeeded in relating the evolution of the ion distribution function to the characteristic features of the magnetic field structure and in the determination of the major macro-features of structure of the shock front. This study concluded that the size of the magnetic field transition was determined by the dissipative process related to reflection of ions. Twin satellite measurements by ISEE provided the first indications that some shocks have quite narrow fronts (Newbury and Russell 1996; Newbury et al. 1998). Newbury and Russell (1996) reported the observation of an extremely thin, quasiperpendicular shock whose ramp width was determined to be $0.05L_i$ (where L_i is ion inertial length), i.e. of the order of electron inertial length. Figure 3 shows this particular shock crossing (bottom panel) together with a second shock, observed under similar solar wind conditions, whose ramp width was $0.89L_i$.

Cluster and THEMIS have provided new opportunities for a comprehensive study of the shock ramp scales. Recently, there have been two papers dedicated to studies of the ramp scales of the magnetic field transition (Hobara et al. 2010; Mazelle et al. 2010). They have used slightly different definition of the ramp thickness and scale and applied different methodology, however they came to very similar conclusions about statistical characteristics of the scales of the ramp transition. Hereafter we present the summary of results reported in these papers.

The article by Hobara et al. (2010) is devoted to statistical studies of the magnetic field spatial scales in the ramp region of the shock front based on Cluster and THEMIS ob-

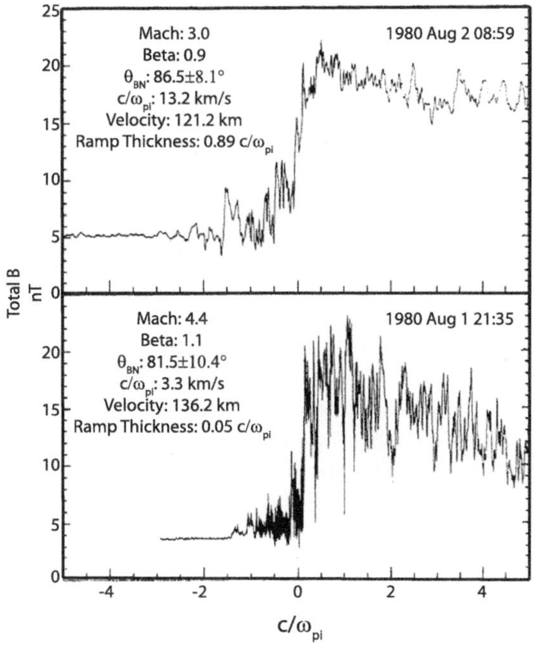

Fig. 3 These shocks formed under similar solar wind conditions, but there is great disparity between their ramp widths (Adapted from Newbury and Russell 1996)

servations. Due to their highly inclined orbit, the Cluster satellites enable the observation of shock crossings away from ecliptic plane. These shocks typically exhibit Mach numbers that are in the lower range of the whole space of terrestrial Mach numbers, since the shock normal deviates from the sunward direction. To increase the range of available Mach numbers, THEMIS shock crossings were added to the set of Cluster observations. Magnetic ramps cannot be always treated as uniform. Nonlinear substructures have been observed and reported within the ramp in several cases (e.g. Balikhin et al. 2002; Walker et al. 2004). The study of spatial-temporal characteristics of such substructures requires at least two point measurements separated by a distance that is sufficiently smaller than the inter satellite distances of both THEMIS and Cluster missions, thus the authors restricted themselves with the study of the ramp spatial scales only.

2.1 Criteria for Choosing Shocks and Definition of Notions "Size" and "Scale"

Hobara et al. (2010) have used the data from Cluster and THEMIS for a statistical study of the spatial size of the ramp. Both these missions assembled a huge stockpile of shock crossings. These data sets complement each other because of the difference in the orbits of Cluster and THEMIS satellites. The THEMIS orbit is close to equatorial plane providing an opportunity to sample the terrestrial bow shock in the vicinity of the subsolar point. Cluster crossings of the terrestrial bow shock occur mainly on the flanks. The solar wind flow in the vicinity of the terrestrial orbit is almost along Sun-Earth line, so that the Mach numbers of flank shocks are relatively low due to the greater deviation of the local shock normal from the sunward direction. Therefore, the combination of THEMIS and Cluster crossings allowed to cover a greater dynamical range of Mach numbers available for the analysis than each of these missions provides separately. Cluster crossings for two time intervals, February–April 2001 and February–March 2002, were used in the study. THEMIS shock crossings included in the study took place from the beginning of July 2007 to the end of August 2007.

The magnetic field data used in the present paper came from Cluster and THEMIS fluxgate magnetometers (FGM) (Balogh et al. 1997; Auster et al. 2008). Another reason to use the THEMIS data from the initial phase of the mission was that THEMIS C and D spacecraft separation was not very large.

The set of shocks that have been used by Hobara et al. (2010) for the study of statistical properties of the ramp width and gradient scale in the paper included 77 individual crossings of the terrestrial bow shock (30 by THEMIS and 47 by Cluster). In order to determine the spatial scale of the shock ramp by means of transformation from temporal to spatial variables an estimate of the relative shock spacecraft velocity along the shock front normal was used. These estimates are very sensitive to shock normal definition. That's why to perform reliable identification of the local normal to the shock front one of them (Hobara et al. 2010) compared normals found making use of four different methods, using the model shape of the terrestrial bow shock similar to Farris et al. (1991), using timing differences methods between the 4 Cluster spacecraft shock crossings, minimum variance and coplanarity theorem. In order to validate the results the evolution of the magnetic field component along the normal direction B_n was used. The reliability of the shock normal identification served as the only shock selection criteria. Those shock crossings, for which the calculated normal could not be considered reliable (because of the B_n evolution or due to large discrepancy in the shock normal directions found by different methods), have been excluded from consideration. The relative shock spacecraft velocity V_{ss} has been calculated using the shock normal direction, satellite separation vectors and time delay between two subsequent shock crossings.

The second recent study dedicated to the same problem (Mazelle et al. 2010) was also based on Cluster magnetic field measurements during spring seasons of 2001 and 2002 corresponding to small interspacecraft separation (100 to 600 km typically). The shock parameters (angle between upstream magnetic field and local shock normal θ_{Bn}, Alfvénic Mach number M_A and ion beta β_i) were computed from the data of Cluster and ACE.

Mazelle et al. (2010) selected the shocks for the statistical study according to following criteria. First, in order to restrict the study by almost perpendicular shocks the shocks with θ_{Bn} as close as possible to 90° were chosen. The other criteria were the existence of well-defined upstream and downstream intervals for the 4 s/c, the stability of the upstream averaged field from one s/c to another, the validity of the normal determination by checking the weak variability of the normal field component B_n around the ramp and low value of B_n upstream for θ_{Bn} to be close to 90°. Only 24 from 455 crossings of Cluster satelite quartet in 2001 and 2002 were left with all criteria validated. This means that 96 individual shock crossings were analyzed. Selected θ_{Bn} values were chosen to be in the range from nearly 90° to 75° but about 80 % of the shocks selected were above 84°. Mach numbers M_A were found to be equally distributed from 2 to 6.5 and corresponding β_i between very low values to 0.6 but with 67 of values less than 0.1.

The major difference between two works consists in using different methods of the determination of the shock ramp thickness. If the beginning of the ramp region was defined quite similarly as the beginning of the monotonous increase of the magnetic field, the end of ramp or exit from the ramp region was determined differently. Hobara et al. (2010) defined the ramp crossing duration as a time interval between the upstream edge of the ramp and the maximum of overshoot. The spatial size (width) of the magnetic ramp has been estimated as a product of V_{ss} and the duration D_t of the ramp crossing. Mazelle et al. (2010) determined at first a stationary asymptotic level of the magnetic field that might be considered as satisfying to Rankine-Hugoniot conditions. To this end a downstream interval where the magnetic field magnitude is quite steady was used. It is then considered as giving an approximate estimate of the value of the magnetic field corresponding to exit from the ramp/entry

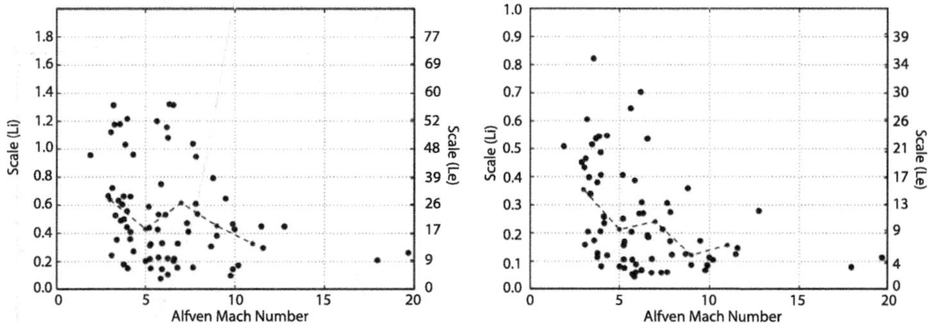

Fig. 4 Scatterplot of experimentally derived shock size (*left panel*) and shock spatial scale (*right panel*) normalized to the ion inertial scale length c/ω_{pi} (*left axis*) and electron inertial scale c/ω_{pe} (*right axis*) as a function of Alfvén Mach number. The *dashed line* represents the averaged values of shock width averaged over shocks with Alfvén Mach number in the ranges 2–4, 4–6, 6–8, 8–10, 10–12. The *vertical lines* represent the statistical *error bars* for each of these Mach number ranges. (Adapted from Hobara et al. 2010)

in the overshoot. From the initial values of the entry in and exit from the ramp a linear fit of the data points inside the estimated time interval is made and this later on is allowed to vary. The choice of a linear fit allows excluding any pollution of the ramp region by a part of the extended foot. The same analysis was repeated for all four satellites for each shock crossing. The steeper slope found for the ramp defines the 'reference satellite'. The times of the middle of the four samples of ramps for one shock crossing are then used to compute both the shock normal and velocity in the GSE frame by the 'timing-method' described in Horbury et al. (2002). This makes it possible to derive the 'apparent' width (along each satellite trajectory) and compare between the 4 s/c. Then, the velocity vector of the shock in each s/c frame is computed. Its angle with the shock normal allows reconstructing a local profile along the normal and determining the real local spatial width of each shock substructure. Readers interested in more technical details can find them in Mazelle et al. (2010) and Hobara et al. (2010).

A scatterplot of the spatial sizes of the ramp as a function of Alfvén Mach number is shown in the left hand panel of Fig. 4. The left hand panel shows the scale sizes in terms of the ion inertial length whilst the right hand panel shows the width in terms of the electron inertial length. The width of the magnetic ramp varies by about an order of magnitude from $L_r = 1.4L_i$ ($\approx 60L_e$) to $0.1L_i$ ($\approx 4L_e$). The general trend in Fig. 4 indicates that the magnetic ramp becomes thinner with increasing of Alfvénic Mach number. This trend is evident even without taking into account two shock crossings with peculiarly high Mach numbers M in the range 17–20 that correspond to the two markers in the bottom right corner of the scatterplot. The figure also shows a distinct decrease in the maximum width of the shock front with increasing Mach number. To make these tendencies more clear the characteristic width of the magnetic ramp averaged for the shocks with Alfvénic Mach numbers in ranges 2–4, 4–6, 6–8, 8–10 and 10–12 (dashed curve) are presented.

The vertical lines on Fig. 4 (left panel) represent the statistical error bars for each range of Mach numbers 2–4, 4–6, 6–8, 8–10 and 10–12. The decrease of statistical errors with the Mach number is in complete agreement with the significant decrease of the maximum shock width, while the minimum shock width undergoes much smaller changes.

The right hand panel of Fig. 4 shows the scatterplot of the spatial gradient scale of the magnetic ramp. It clearly demonstrates the same characteristic features as were evidenced in the left hand panel for the ramp width, namely a quite wide range of values, especially

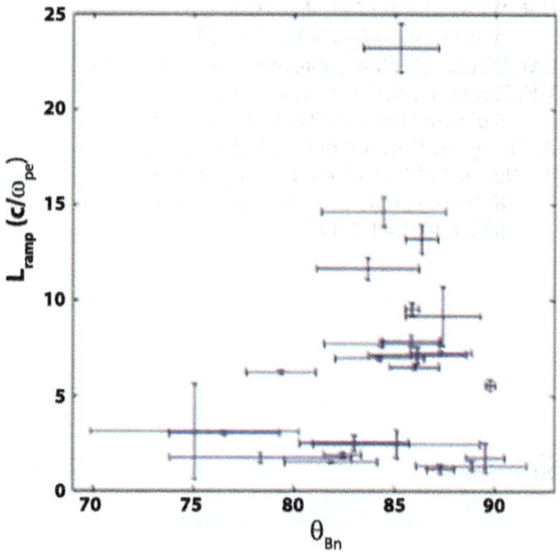

Fig. 5 Thinnest ramps observed versus shock θ_{Bn}. Reprinted with permission from Mazelle et al. (2010). Copyright 2010, American Institute of Physics

for low Mach number shocks and the trend toward shorter scales with the increase of the Mach number as well as the decrease of the maximum gradient scale while Mach number increases. As the change of the magnetic field for all chosen shocks exceeds upstream magnetic field B_0 (for many of them quite significantly) the values of the gradient scale are smaller then the width of the corresponding shocks. The ramp gradient spatial scale varies in the range 0.05–0.82L_i (2–35L_e).

2.2 Statistical Analysis (Mazelle et al. 2010)

Figure 5 displays the thinnest ramp found among each quartet of crossings for each individual shock versus θ_{Bn}. There were no simple relation found. However, it is unambiguously established that many observed thinnest ramps are less than 5 c/ω_{pe} thick and there was found an apparent trend for lower values as $\theta_{Bn} \to 90°$.

The histogram of all ramp thicknesses in Fig. 6 reveals the predominance of narrow ramp width with a Gaussian-like regular decrease towards an asymptotic limit that is still less than c/ω_{pi}.

The authors came to conclusion that their analysis confirms statistically that the magnetic field ramp of the supercritical quasi-perpendicular shock often reaches a few c/ω_{pe}.

So, the results of two independent studies by two different groups come to the same conclusion, the ramp width for quasiperpendicular high Mach number shocks as seen in magnetic field is of the order of several c/ω_{pe} and is in a perfect agreement with estimates determined from the dispersive scale of a nonlinear whistler wave, modified by the presence of reflected ions (Galeev et al. 1988, 1989; Krasnoselskikh et al. 2002).

3 Electric Field Scales of the Ramp of Quasiperpendicular Shocks

As mentioned in the introduction, there have only been a few reports regarding the scales and structure of the electric field transition at quasi-perpendicular shocks. Based on laboratory experiments, in which the conditions are not exactly the same as in space plasmas, Eselevich et al. (1971) reported that the major change in potential across the shock occurs within the magnetic ramp region.

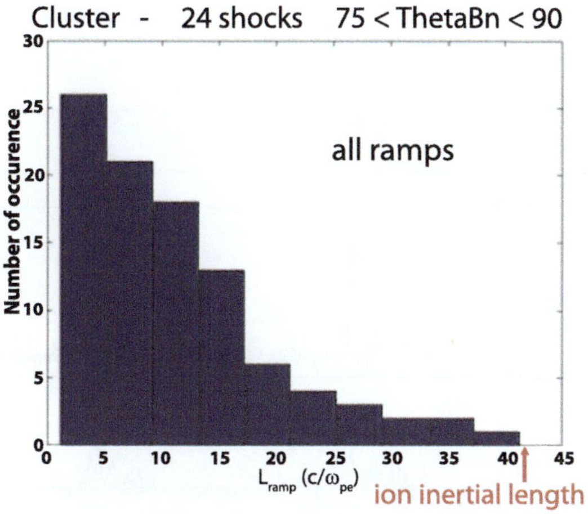

Fig. 6 Histogram of the 96 shock ramp thicknesses. Reprinted with permission from Mazelle et al. (2010). Copyright 2010, American Institute of Physics

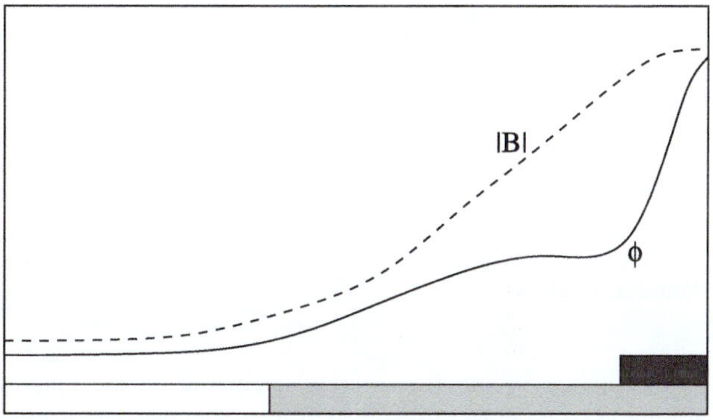

Fig. 7 Sketch of the changes observed in the magnetic field and electrostatic potential during the crossing of a quasi-perpendicular shock (based upon the experimental results of Eselevich et al. 1971)

Figure 7 is a sketch (based on the results of Eselevich et al. 1971) of the change in the magnitude of the magnetic field and the accompanying change in the electrostatic potential. These authors interpreted it as a viscous subshock similar to isomagnetic jump. From Fig. 7 it can be seen that there are two different length scales that may be associated with the change in the electrostatic potential as the shock is crossed. The first, indicated by the lightly shaded bar at the foot of the plot, shows that overall the potential changes on scales similar to that of the magnetic ramp region in agreement with the results of Lembège et al. (1999). This corresponds to an enhancement of the electric field observed as the shock is crossed. The second scale, indicated by the darkly shaded bar, corresponds to a region within the shock front in which a large increase in the potential is observed over a small spatial scale. Such changes in the potential result from large amplitude spike like features in the electric field.

The results reported here present a study of the large amplitude, short duration features in the electric field observed by the Cluster satellites during a number of crossings of the quasi-perpendicular bow shock published in Walker et al. (2004). These features contribute significantly to the overall change in potential observed at a shock crossing but their short duration implies that they are very localised. The aim of the study was to determine their scale size and amplitudes. These parameters were studied in relation to the upstream shock parameters.

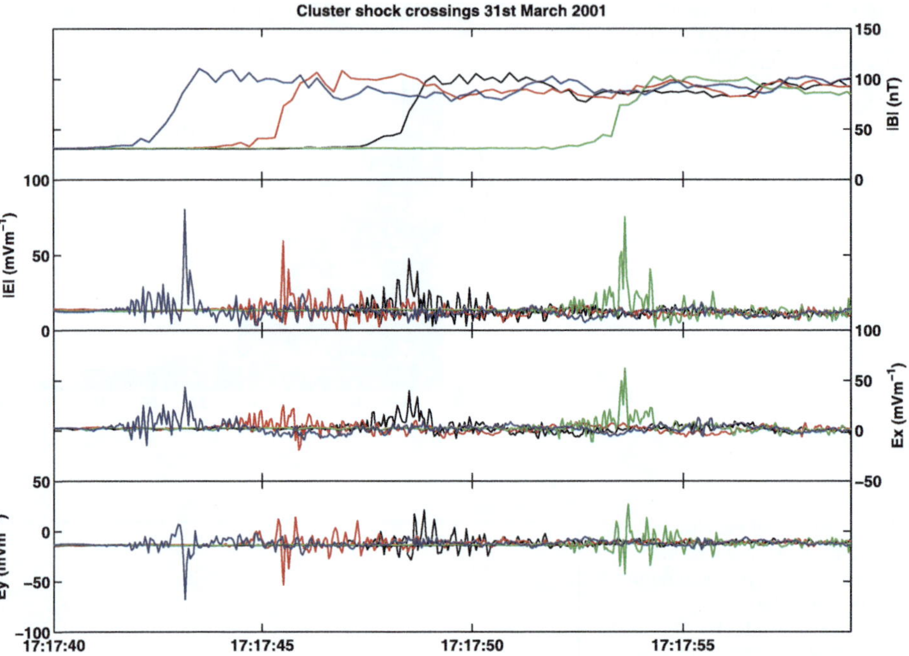

Fig. 8 Overview of the shock crossing on March 31st, 2001 at 17:18 UT. The *top panel* shows the magnitude of the magnetic field measured by FGM. The *second panel* shows the magnitude of the electric field measured in the satellites spin plane. The *lower two panels* show the spin plane components E_x and E_y (Adapted from Walker et al. 2004)

A total of 54 shock crossings, occurring on 11 separate days were investigated but not all could be analysed fully for various reasons such as unavailability of certain data sets, or the accuracy of the shock normal determination. In this section we present a case study of the electric field within the shock front, namely the crossing that occurred on 31st of March 2001 at around 17:18 UT. On this day the conditions in the solar wind were some what abnormal due to the passage of a CME. Measurements in the solar wind by Cluster indicated that the magnitude of the magnetic field was of the order of 30 nT, the normal for this shock (based upon FGM crossing times) is $n_B = (0.94, -0.17, 0.293)$ (in the GSE frame), and the shock velocity was determined to be 48.92 km s^{-1}. Other relevant parameters are $\theta_{Bn} \approx 87°$- and a density $n \sim 19$ cm^{-3}. The high value of the field resulted in an unusually small $\beta \sim 0.07$. The Alfvén Mach number for this shock ($M_A \sim 3.6$) lies close to the First Critical Mach number and to the whistler critical Mach number so the conditions of the solar wind are quite favourable for the formation of quasi-electrostatic sub-shocks at the shock front (Balikhin et al. 2002). Figure 8 (adopted from Walker et al. 2004) shows an overview of the magnetic and electric field measurements made by FGM and EFW respectively during this shock crossing. The top panel shows the magnitude of the magnetic field measured by FGM. Initially, all four Cluster spacecraft are in the solar wind just upstream of the outward moving bow shock which subsequently swept over the satellites in the order C4 (17:17:43.5), C2 (17:17:45.5), C1 (17:17:48.5), and finally C3 (17:17:53.5). The magnetic field profiles show a set of clean shock crossings that possess clearly discernible foot, ramp and overshoot regions. The second panel shows the magnitude of the electric field measured by EFW in the spin plane of each satellite ($|E|^2 = E_x^2 + E_y^2$). In the solar wind, the typical magnitude

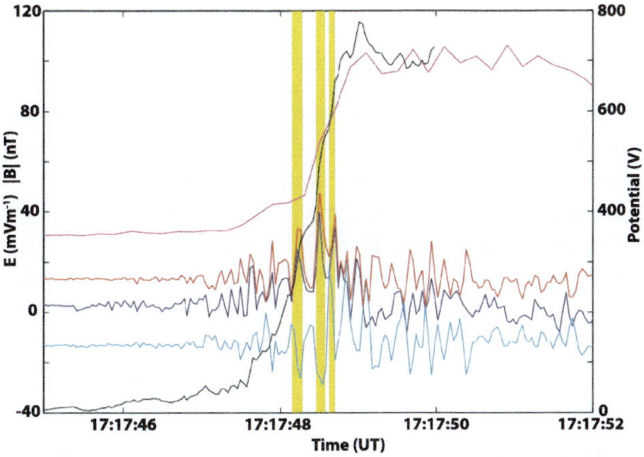

Fig. 9 The FGM magnetic and EFW electric fields measured by Cluster 1 on 31st of March 2001 around 17:18 UT. The magnetic field magnitude is shown by the *magenta line*. The spin plane electric field magnitude, and E_x and E_y components are shown in *red*, *blue* and *cyan* respectively. The *yellow regions* highlight the periods when large amplitude short duration spikes in the electric field are observed. The *black line* (Y scale of RHS) represents the change in potential within the shock (Adapted from Walker et al. 2004)

of the electric field is around 14 mV m^{-1} in the satellite spin plane. It is possible to estimate the E_z component of the upstream electric field assuming that $\mathbf{E} \cdot \mathbf{B} = 0$. This assumption is valid for estimating the field upstream and downstream of the shock but not within the shock region itself. Upstream of the shock, $E_z \approx 5$ mV m^{-1}. This value is higher than the measured E_x component ($\sim$2.5 mV m^{-1}) and less than the E_y component (-13 mV m^{-1}). Comparing the top two panels it can be seen that the disturbances measured in the electric field begin in the foot region of the shock and continue until the satellites are downstream of the overshoot/undershoot. These general disturbances have amplitudes generally in the range 5–30 mV m^{-1}. During their crossings, each of the satellites recorded a number of large amplitude, short duration features in the electric field. The largest of these spikes have maximum amplitudes of approximately 30, 40, 60, and 65 mV m^{-1} for satellites 1, 2, 3, and 4 respectively above the field measured in the solar wind just upstream of the shock front. These values represent lower limits of the strength of the electric field since the component perpendicular to the spin plane is not considered. They are seen to occur within the ramp region but there is no strong feature within the FGM data with which they correlate. It is also observed that the largest electric field peaks observed on each satellite appear to occur in pairs which may suggest field rotation. The two lower panels show the components of the electric field measured in the satellite spin planes. Both panels show that the components of the field exhibit a twin peaked structure, similar to that observed in the field magnitude and that the direction of rotation is the same for both peaks. Thus the overall structure is not due to a single rotation of the field. Our goal is a statistical study of the widths of these short living, large amplitude features.

Using the four point measurements one can determine the occurrence time of these peaks in the electric field and hence compute a normal. Examining the E_x component, the time differences between the observations of the first peak in the electric field are $\Delta t_{12} = -3.01$ s, $\Delta t_{13} = 5.03$ s, and $\Delta t_{14} = -5.35$ s. When coupled with the respective positions of the satellites this yields a normal direction $n_E = (0.946, -0.155, 0.283)$ and a velocity of $\sim$50 km s^{-1}. The difference between this normal n_E and that determined from the magnetic field (n_B) is less than a degree. Thus it appears that the electric field spikes correspond to layers within the overall shock structure. Figure 9 shows the results from Cluster 1 in greater detail. The magenta line shows the magnitude of the magnetic field. The foot region was entered around 17:17:47.3 UT whilst the ramp was crossed between 17:17:48.3 and 17:17:48.9 UT. Several large spikes in the electric field are observed in the region of the foot and shock ramp. The three largest occur around 17:17:48.2 (20 mV m^{-1}),

17:17:48.5 (30 mV m^{-1}), and 17:17:48.6 (15 mV m^{-1}). Their short duration implies that their scale size is of the order 3–5c/ω_{pe}. The black line in Fig. 9 represents an estimation of the electrostatic potential measured in the normal direction. This was calculated by removing an average of the field measured in the region just upstream of the shock from the field measured within the shock region and then integrating the projection of this electric field along the normal direction. Whilst the actual potential cannot be calculated due to the incomplete vector measurements, it can be estimated by assuming that the field perpendicular to the spin plane $E_z = 0$. This assumption is valid because for this particular shock, the normal lies very close to the spin plane. This calculation can be used to show that the largest jumps in the potential coincide with the spikes observed in the electric field and that these occurrences contribute a significant fraction of the total potential change observed at the shock. During this period, the electric field enhancements contribute around 40 % of the total change.

3.1 Scale Size

The preceding sections have presented evidence for localized increases in the electric field strength measured as the satellite traverses a quasi-perpendicular bow shock. All shocks analyzed show evidence for an enhancement in the background electric field. In most cases, the region in which this field enhancement occurs lasts longer than the crossing of the magnetic ramp. The field typically increases of the order 1–3 mV m^{-1} above that measured in the solar wind. However, as has been noted above, the turbulence in this region is dominated by spike-like fluctuations lasting a few milliseconds and with magnitudes of typically 4–20 mV m^{-1} with a maximum magnitude of the order of 70 mV m^{-1}. This existence of large gradients in the electric field has repercussions for processes involved in the heating of electrons. In the presence of strong electric field gradients the electron gyration frequency can deviate from its classically calculated value (Cole 1976; Balikhin et al. 1998), leading to an increase in its Larmor radius and the possibility of a breakdown in adiabaticity (Cole 1976; Balikhin et al. 1998). Having shown that the spikes observed in the electric field at the front of a quasiperpendicular shock appear to be physical structures that form a layer within the shock front as opposed to being the result of noise in the data or motion of the shock a statistical study of these features was performed to investigate their relationship to the properties of the shock front. Now we shall present statistical study of the data collected from a number of such spike-like features.

Figure 10 shows the distribution of the scale sizes determined from the event duration and the shock velocity evaluation of these features in terms of the electron inertial length. The scale size of these events will be unaffected by the incomplete vector measurements of the electric field. The vast majority of crossings have scale sizes of the order of 1–5c/ω_{pe}. The data that form tail of the distribution at longer scale sizes typically comprise events that have a multi-peak structure. This type of event represents an upper limit to the scale size of these short-lived events. In comparison, the typical scale of the magnetic ramp is characterized by the ion inertial length (Newbury and Russell 1996) although these authors also report one particular shock as having a ramp scale as small as $0.05c/\omega_{pi}$ or $2c/\omega_{pe}$.

Figure 11 shows the relationship between the Mach number and scale size of the spikes observed in the electric field. From the figure, it is clear that the scale size has a lower limit that decreases as the Mach number increases. One should notice that these results represent the tendency rather than the proof, the number of points at high Mach numbers is not sufficient for valuable statistical study.

Figure 12 shows a scatter plot of the relationship between θ_{Bn} and the scale size of the electric field enhancements. In general there appears to be a broad range of scales. However,

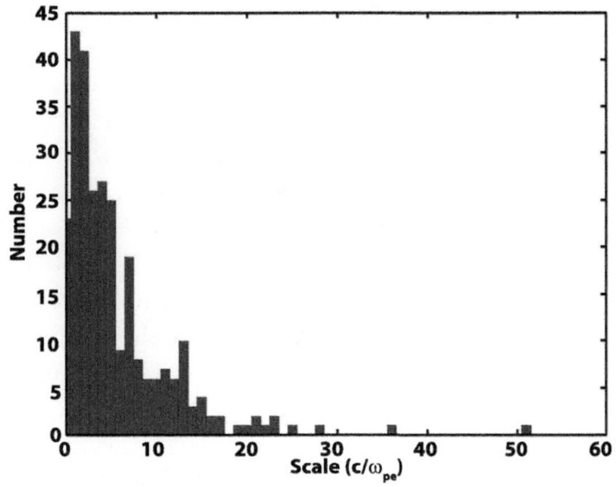

Fig. 10 Histogram of the scale sizes for the spike-like enhancements observed during a number of crossings of the quasi-perpendicular bow shock (Adapted from Walker et al. 2004)

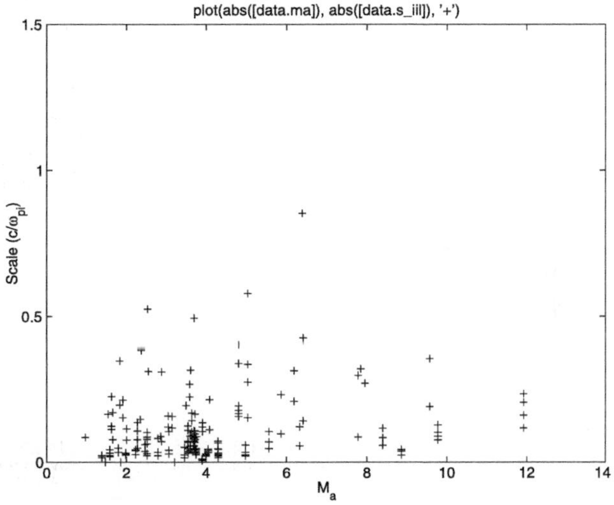

Fig. 11 Dependence of scale size on upstream Mach number (From Walker et al. 2004)

as θ_{Bn} approaches 90°—the scale length decreases. For the shocks analysed with θ_{Bn} close to 90°—the scale lengths are of the order of $2c/\omega_{pe}$. This compares favourably with theoretical estimates that for shocks close to perpendicular the scale width is estimated to be of the order of the electron inertial length as proposed by Karpman (1964), Galeev et al. (1988). This tendency corresponds exactly to dispersion dependence upon the angle θ_{Bn}.

3.1.1 Amplitude

The examples presented above show that the increase in the electric field ($\Delta E = E_{spike} - E_{upstream}$) observed during encounters with these spike-like structures varies between 4 and 70 mV m^{-1} above the average field that is measured in the solar wind just upstream of the shock. In this section the relationship between this change (ΔE) and the shock Mach number, and the angle θ_{Bn} is presented. Figure 13 shows a scatter plot of the peak amplitude observed in the electric field spike event (ΔE) as a function of the shock Mach number M_A.

Fig. 12 Dependence of scale size on θ_{Bn} (From Walker et al. 2004)

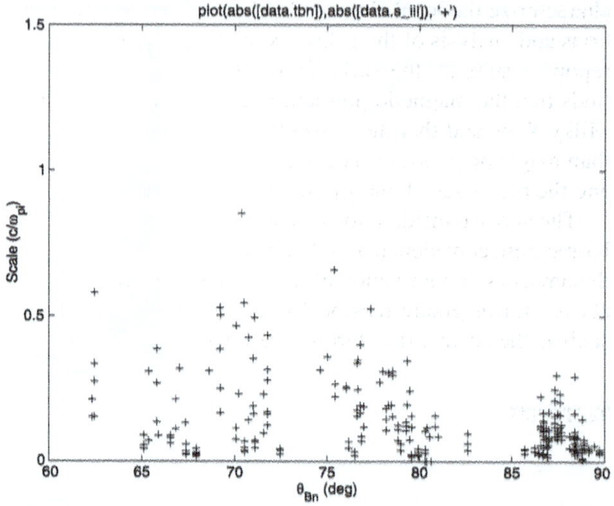

Fig. 13 Scatter plot showing the relationship between the amplitude of the electric field spikes as a function of Mach number. The *red crosses* are used to highlight the data for the shocks that occurred on 31st of March 2001 (Adapted from Walker et al. 2004)

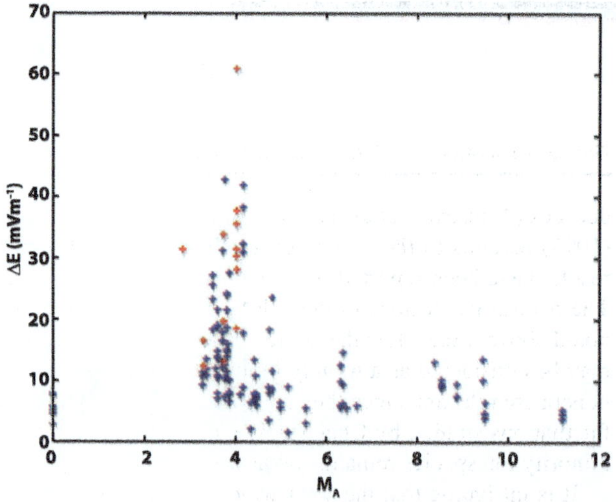

For shocks whose Mach number $M_A > 5$ there is a fairly constant trend in which $\Delta E < 15$ mV m^{-1}. In the Mach number range $3 < M_A < 5$ the range of observed amplitudes varies between 5 and 60 mV m^{-1}. It appears that in this Mach number range the electric field potential becomes more important than for low and high Mach number shocks that corresponds to dependence of the electrostatic potential upon the Mach number (Dimmock et al. 2012). The red crosses highlight the shocks observed on 31st of March 2001. All of these shocks fall into this range of Mach numbers. This set of shocks seems to possess Mach numbers corresponding to supercritical range and having large number density, at about 7 %, of alpha particles (Maksimovic, private communication). Their structure seems to resemble that of electrostatic sub-shocks similar to those observed in laboratory plasmas (Eselevich et al. 1971). A characteristic signature of sub-shocks is the occurrence of small scale electrostatic fluctuations such as those observed on this particular day. Ion sound sub-shocks have been observed in laboratory plasmas with scales of the order of 100 Debye

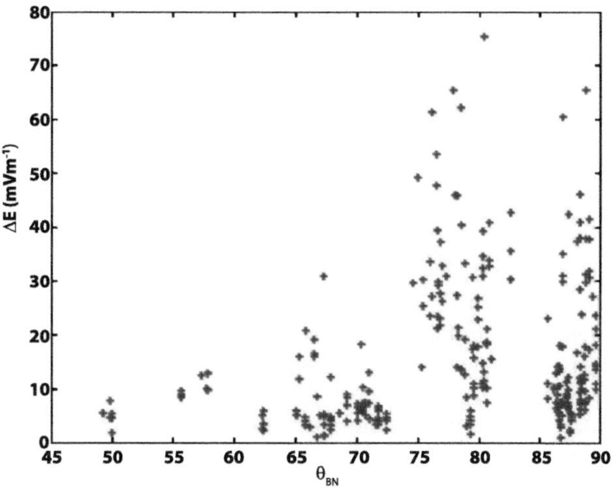

Fig. 14 The relationship between ΔE and θ_{Bn} (Adapted from Walker et al. 2004)

lengths. For the shocks observed in 31st of March 2001, the scale is closer to characteristic scale of the fast magnetosonic mode (Balikhin et al. 2002).

The relationship between ΔE and θ_{Bn} is shown in Fig. 14. It clearly shows that as θ_{Bn} approaches 90°—the range of the observed amplitudes of the electric field spikes increases.

3.2 Conclusions

In this section we presented the changes observed in the electric field during the crossing of a number of quasi-perpendicular bow shocks. It is shown that the electric field is enhanced during the crossing of the shock and that the scale size over which this enhancement is observed is larger than that of the macroscopic magnetic ramp region. Within the whole shock region, short lived electrostatic structures are observed that are intensified in the ramp region. The scale size of these structures is of the order of a few c/ω_{pe}: and was shown to decrease as θ_{Bn} approaches 90°—which corresponds to the dependencies following from theoretical model based on consideration of the shock as mainly dispersive nonlinear structure (Galeev et al. 1988; Krasnoselskikh et al. 2002). The amplitudes of these structures are typically of the order of 5–20 mV m^{-1}: but under special circumstances may reach as high as 70 mV m^{-1}. The highest amplitudes appear to be observed for shocks whose Mach number is in the range 3 to 5. This may be an indication that such shocks have quasi-electrostatic sub-shocks inside the main ramp transition. It was also demonstrated that these small scale structures make a substantial contribution to the overall change in potential observed across the shock and that the potential change is not linear.

4 Dispersive Nature of High Mach Number Shocks: Poynting Flux of Oblique Whistler Waves

It is well known that a subcritical shock has a nonlinear whistler wave train upstream of its front (Sagdeev 1966; Mellott 1985). The major transition of such a dispersive shock, the ramp, behaves as the largest peak of the whistler precursor wave package (Karpman et al. 1973; Kennel et al. 1985; Galeev et al. 1989; Krasnoselskikh et al. 2002). The presence of whistler/fast magnetosonic precursor wave trains in supercritical shocks was experimentally established in Balikhin et al. (1997b), Krasnoselskikh et al. (1991), Oka et al. (2006). These whistler waves have rather large amplitudes and their role in energy transformation and redistribution between different particle populations and in the formation of the shock front structure is still an open question. The energy source responsible for the generation of these waves is the subject of active debate in shock physics (see Galeev et al. 1988, 1989; Krasnoselskikh et al. 2002; Matsukiyo and Scholer 2006; Comişel et al. 2011). Often the precursor waves are almost phase-standing in the shock frame. However, if they are generated by the ramp region as the dispersive precursor their group velocity can still be greater than zero in the shock reference frame, which would allow energy flow in the form of Poynting flux to be emitted towards the upstream of the shock transition. On the other hand, if the waves are generated by instabilities related to reflected ions their energy flux will be directed from the upstream region towards the shock ramp. The goal of this section is to address this problem, to present the direct measurement of the Poynting flux of the upstream whistler waves aiming to establish the direction of the Poynting flux.

Below we establish the energy source of the waves by calculating the Poynting flux of the waves in the Normal Incidence Frame (NIF) of the shock, using multi-satellite Cluster data from crossings of the Earth's bowshock (Escoubet et al. 1997; Bale et al. 2005). Two events with supercritical Alfvénic Mach numbers are analyzed. In both cases it is found that the shocks show dispersive behaviour with the Poynting flux directed in upstream direction.

Poynting flux is not a Lorentz invariant and therefore depends on the frame of reference. To evaluate the value and direction of the Poynting flux with respect to the the shock we transform the electric field to the Normal Incidence Frame (NIF). The normal $\hat{\mathbf{n}} = +\hat{\mathbf{x}}$ which also serves as the x-coordinate direction in the NIF system is obtained by four-spacecraft timing, $\hat{\mathbf{z}}$ is the direction of maximum varying magnetic field obtained from a minimum variance analysis, and $\hat{\mathbf{y}}$ is the direction of the convection electric field which completes the right-handed system.

The transformation from the spacecraft frame to the NIF is given by $\mathbf{E}_{NIF} = \mathbf{E}_{sc} + \mathbf{v} \times \mathbf{B}$. The total velocity required for this transformation is defined by $\mathbf{v} = \mathbf{v}_{sh} + \mathbf{v}_{NIF}$ where $\mathbf{v}_{sh} = v_{sh}\hat{\mathbf{n}}$ is the shock velocity, $\mathbf{v}_{NIF} = \hat{\mathbf{n}} \times (\mathbf{v}_u \times \hat{\mathbf{n}})$ is the NIF velocity and $\mathbf{v}_u$ is the solar wind velocity.

A general shift of reference frame, coordinate transformation, and evaluation of the complete Poynting vector requires knowledge of the full six-dimensional electromagnetic field (three electric and three magnetic components). The Cluster spacecraft, however, only measures the two components of the elecric field in the spin-plane of the spacecraft, while the third component normal to the spin-plane is not measured. To reconstruct the third component we use the assumption that for the wave electric and magnetic fields the condition $\mathbf{E} \cdot \mathbf{B} = 0$ holds. While this is a true condition for the cross-shock (DC) electric field, it holds well for whistler wave electric fields at lower frequencies.

We study two quasi-perpendicular high Mach number shocks encountered by the Cluster multi-spacecraft mission (Escoubet et al. 1997). The first shock was observed around 04:53:40 Universal Time (UT) on 20th of January 2003, and the second around 07:07:00 UT on 24th of January 2001. We use data from the EFW (electric field) (Gustafsson et al. 1997), FGM (DC magnetic field) (Balogh et al. 1997) and STAFF (wave magnetic field) instruments (Décréau et al. 1997) from spacecraft 2 (for the 2003 shock) and spacecraft 3 (for the 2001 shock). The shock normal $\hat{\mathbf{n}}$ is established by assuming a planar shock and using the time of crossing of the four spacecraft and their relative positions (Paschmann and Daly 1998).

The first shock analyzed had an upstream $\theta_{Bn} \sim 85^\circ$ and an Alfvénic Mach number $M_A \sim 5.5$. The electric and magnetic fields in the shock front region are characterized by

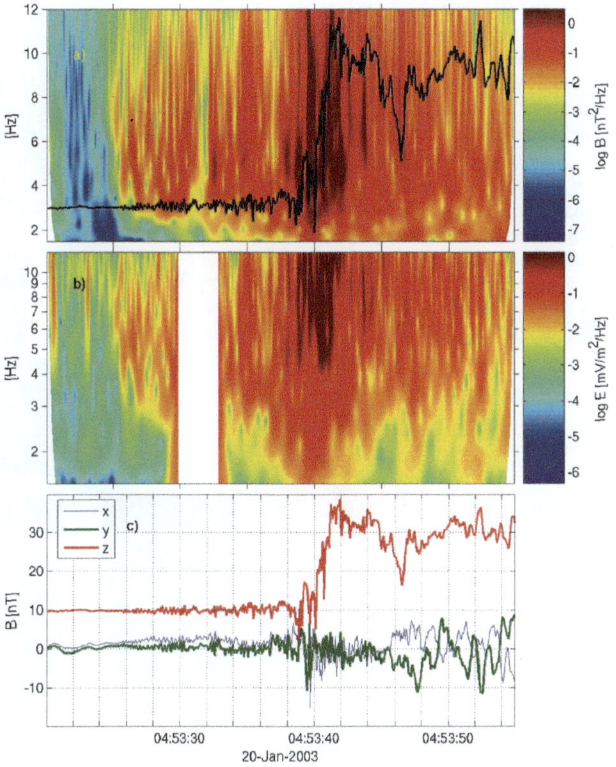

Fig. 15 Magnetic and electric fields in the Normal Incidence Frame (NIF) of a high Mach number shock. (**a**) Power spectra of the magnetic field (STAFF). The *black line* is the DC total magnetic field, included to show the waves in relation to the shock ramp structure. (**b**) Power spectra of the electric field (EFW). The data gap is due to instrumental interference. (**c**) The magnetic field in NIF coordinates B_{NIF} (Adapted from Sundkvist et al. 2012)

waves, with stronger amplitudes closer to the ramp, see Fig. 15. The waves have frequencies $f_{cp} < f$, where $f_{cp} \sim 0.1$ Hz is the proton gyrofrequency, and right-handed polarization looking along the magnetic field vector and thus belong to the magnetosonic/whistler mode. The direction of the wave-vector $\hat{\mathbf{k}}$ was determined by the Means method (Means 1972), which uses the imaginary part of the three-dimensional magnetic field spectral matrix. The angle θ_{kB} between the wave vector and the local ambient magnetic field is shown as a function of frequency in Fig. 16(b). The average value $\langle\theta_{kB}\rangle$ in the shock front region is $\sim$10–50° (right-hand scale). The whistler waves are thus oblique with respect to the local magnetic field, as well as to the shock normal. The angle increases continously as the shock front is approached and $\theta_{kB} \to 90°$ at the ramp, reflecting the quasi-perpendicular nature of the shock. This smooth transition stresses the nature of the shock as a dispersive nonlinear whistler wave.

Since Poynting flux is a second-order quantity the electric and magnetic fields in the NIF were wavelet transformed (Morlet width 5.36) and the cross-product $\mathbf{S}_f = 1/\mu_0 \mathbf{E}_f \times \mathbf{B}_f$ formed in frequency space. The calculated Poynting flux is therefore distributed in both time and frequency. The projection of the Poynting flux distribution along the magnetic field $S_\parallel = \mathbf{S}_f \cdot \mathbf{B}_0/|\mathbf{B}_0|$ using an instantaneous value of $\mathbf{B}_0$ is plotted in Fig. 16(a), where the colors red (upstream) and blue (downstream) show the direction of the flux. We note that in the front region of the shock the Poynting flux is everywhere directed upstream (red), away from the shock. In the downstream area there is a mixture of blue, green and red, where there is more turbulence and the waves are no longer coherent. The upstream and slightly oblique direction of the Poynting flux is further quantified in the instantaneous angle $\theta_{S,B}$ between the Poynting flux and the ambient magnetic field, plotted in Fig. 16(c). Figure 17

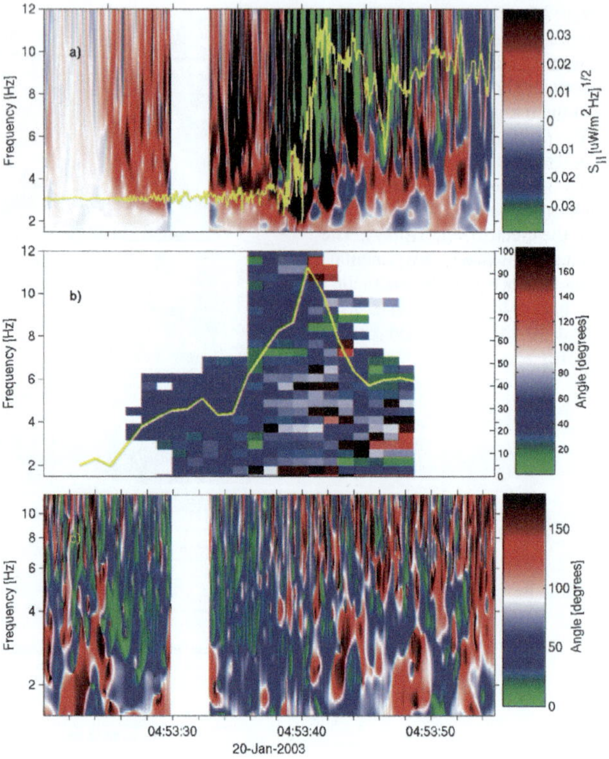

Fig. 16 Poynting flux in the Normal Incidence Frame (NIF) of the same shock as in Fig. 15. (**a**) Poynting flux $S_{\|}$ projected on the local $\mathbf{B}_0$ in the NIF, where red corresponds to the upstream flux away from the shock. (**b**) Angle $\theta_{k,B}$ between $\hat{\mathbf{k}}$ and magnetic field $\mathbf{B}_0$. The *yellow line* represents the average over all frequencies (*right scale*). (**c**) Angle between Poynting flux $\mathbf{S}$ and $\mathbf{B}_0$ (Adapted from Sundkvist et al. 2012)

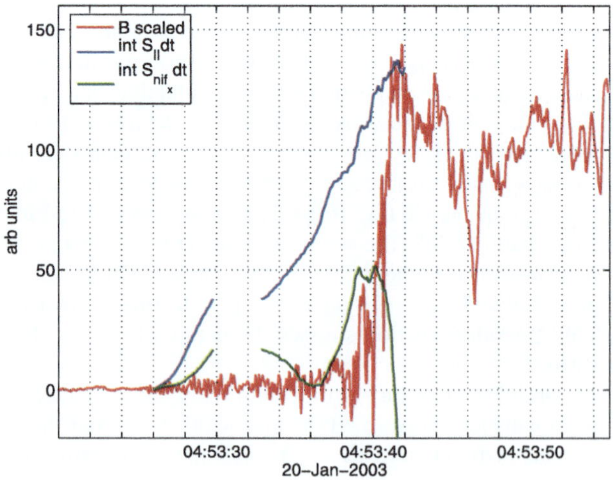

Fig. 17 Poynting flux integrated along the spacecraft trajectory. The *blue line* is a projection along the ambient magnetic field $\int S_{\|} dt$ and the *green line* is a projection on the shock normal $\int \mathbf{S} \cdot \hat{\mathbf{n}} dt$. The *red line* shows the scaled magnetic field B_0 for reference (Adapted from Sundkvist et al. 2012)

shows the Poynting flux along the spacecraft trajectory, with integrated power over frequencies corresponding to the waves in Fig. 16(a), $2 < f < 10$ Hz. In this figure the slope is the important characteristic. Positive slope means Poynting flux carried upstream, and negative slope downstream. From the figure it is evident that the source of the Poynting flux is associated with the shock ramp. The data gap and associated plateau are due to instrumental interference.

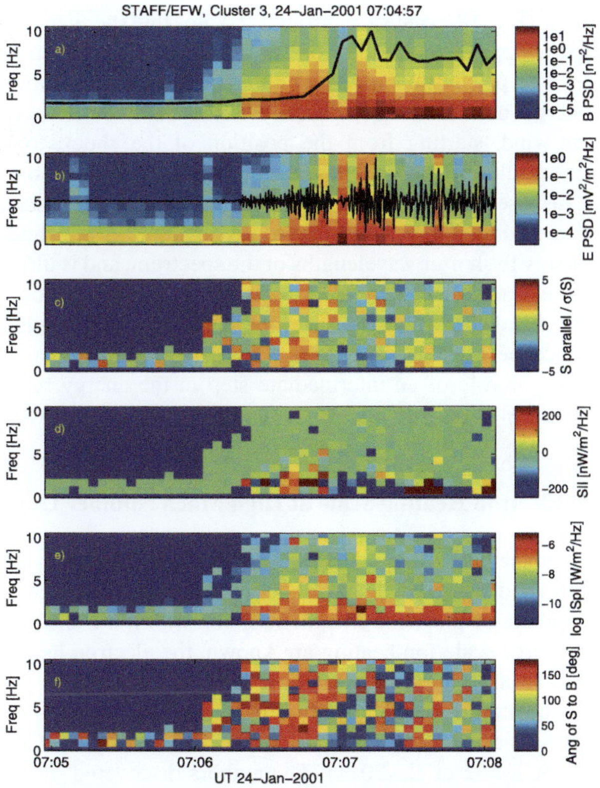

Fig. 18 Poynting flux derived from electric and magnetic fields for a high Mach number shock. (**a**) Wave magnetic field and averaged B_0. (**b**) Wave electric field. (**c**) $S_\parallel$ normalized by its standard deviation (*yellow* and *red* corresponds to upstream flux). (**d**) $S_\parallel$. (**e**) $\log_{10} S_\parallel$. (**f**) Angle of S to $\mathbf{B}_0$ (*red* meaning upstream) (Adapted from Sundkvist et al. 2012)

Another important characteristic established is that the Poynting flux direction is oblique with respect to the shock normal as well as the background magnetic field. This can be explained by analyzing how the phase velocity for whistler waves depends on this angle. The phase velocity of a wave propagating in the plane of the shock normal $\hat{\mathbf{n}}$ and background magnetic field $\mathbf{B}_0$, having an angle α with respect to the shock normal is $V_{ph} = \frac{1}{2}\sqrt{\frac{m_i}{m_e}}\cos(\theta_{Bn} - \alpha)$. Its projection on the direction of the shock normal is $V_{ph,\hat{\mathbf{n}}} = V_{ph}\cos\alpha = \frac{1}{2}\sqrt{\frac{m_i}{m_e}}\cos(\theta_{Bn} - \alpha)\cos\alpha$. Its maximum value can be found to be equal to $\max(V_{ph,\hat{\mathbf{n}}}) = \frac{1}{4}\sqrt{\frac{m_i}{m_e}}(1 + \cos\theta_{Bn})$, thus the projected phase speed can be larger than the whistler critical velocity given above. The above analysis also explains the observation of oblique whistler wave trains found in computer simulations of purely perpendicular shocks (Hellinger et al. 2007). So even in the case of shocks having Mach numbers larger than the whistler critical Mach number, whistler waves oblique with respect to the shock normal can remain quasi-standing.

The second analyzed shock crossing on 24th of January 2001 is shown in Fig. 18. This is a reforming high Mach number shock ($M_A \sim 11$) and has been analyzed in detail in Lobzin et al. (2007). Both of the shocks discussed by were analyzed using wavelet as well as Fast Fourier Transform (FFT) dynamic spectra techniques. We present the second shock using the FFT analysis, to show that the conclusions are not technique dependent. The upstream whistler waves, Figs. 18(a), (b), again have an overall Poynting flux upstream, away from the shock in the normal incidence frame, evident from the red and yellow (upstream) colors of $S_{\|}$ (Fig. 18, panels (c) through (f)). For this shock the ambient magnetic field was directed in the opposite direction, so that 180° (red) means upstream in Fig. 18.

The power flux given by the Poynting vector shows unambiguously that they carry energy over a broad frequency range from the shock ramp towards the upstream solar wind, starting from the position of the shock front. This leads to conclusion that the results of the analysis are consistent with a theoretical model (Galeev et al. 1988; Krasnoselskikh et al. 2002) that considers the shock steepening to be balanced by the effect of dispersion in addition to dissipation. As the shock steepens, nonlinearities transfer energy to shorter wavelengths of the spectrum, and is ultimately carried away from the shock as dispersive whistler wave trains. This analysis demonstrates that for high Mach number shocks, dispersive effects are dominant for the formation and stability of the shock front. Since the whistler waves are strongly damped upstream of the shock, we infer that they can play the role of an intermediate step in the energy re-partition problem, with the energy ultimately being dissipated through wave-particle interaction.

5 Electron Heating Scale at High Mach Number Quasiperpendicular Shocks

From the discussion in the previous sections the energy repartition amongst particle populations in quasiperpendicular shocks is a multi-scale process related to the spatial and temporal structure of the electromagnetic fields within the shock layer. While the major features of the large scale ion heating are known, the electron heating and smaller scale fields remain poorly understood and controversial. In this section we will discuss the scale of the electron temperature gradient based on the possibility of obtaining unprecedented high time resolution electron distributions measured in situ by the Cluster spacecraft recently discussed by Schwartz et al. (2011). The authors discovered that approximately half of the electron heating coincides with a narrow dispersive layer several electron inertial lengths (c/ω_{pe}) thick. Consequently, it gives one more argument that the nonlinear steepening is limited by wave dispersion. The DC electric field associated with the electron pressure gradient must also vary over these small scales, strongly influencing the efficiency of shocks as cosmic ray accelerators.

The 4 Cluster spacecraft (Escoubet et al. 1997) are unique in their ability to remove the time-space ambiguity in time series data taken by in situ space plasma instrumentation. By timing the passage of an event at each corner of the tetrahedron formed by the 4 spacecraft, the planar orientation and speed of the event can be determined. We employ this technique to convert the time series of data to distance along the shock normal Schwartz (1998). Figure 19 illustrates the identification of the steep shock ramp that we use as event times.

The electron instrument on Cluster measures fluxes at several energies in a half-plane containing the spacecraft spin axis. These measurements form an azimuthal wedge divided

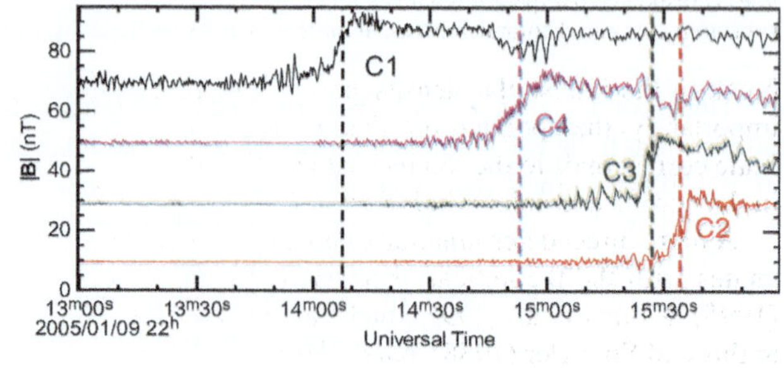

Fig. 19 Magnetic field data at a crossing of the Earth's bow shock by the 4 Cluster spacecraft on 9th of January 2005. Traces have been shifted by 20 nT for clarity. The *dashed lines* show the times of the steep ramp (Adapted from Schwartz et al. 2011)

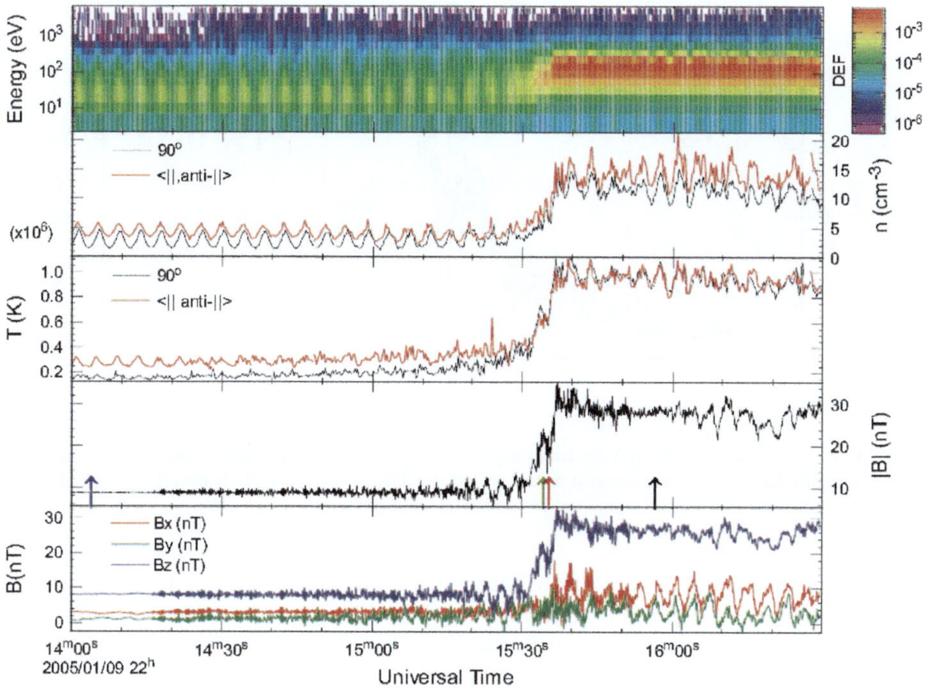

Fig. 20 Overview of data from Cluster 2 on 9th of January 2005. From top to bottom: Omni-directional electron energy-time spectrogram @ 250 ms resolution, electron pseudo-density, electron pseudo-temperatures (see Method), magnetic field magnitude, and field components. *Arrows* in the fourth panel show locations of the cuts presented in Fig. 22 (Adapted from Schwartz et al. 2011)

into 12 polar directions from aligned to anti-aligned with the spin axis, and are repeated at 125–250 ms intervals. A full 3D distribution covering all azimuths is thus built up over 1 spin ($\sim$4 s). However, when the magnetic field is roughly aligned with the spin axis, each wedge contains a full set of pitch angles from 0° to 180°. Under these circumstances, and assuming gyrotropy, the full pitch angle distribution function is available at $\leq$250 ms resolution.

We rebin the raw electron data into pitch angles α relative to the instantaneous magnetic field. We calculate pseudo-densities and temperatures for each pitch angle bin as if the distribution were isotropic, e.g., $n(90°) = 4\pi \int f(v, \alpha = 90°) v^2 dv$. These pseudo-moments better characterise the phase space distributions in the $\|, \perp$ directions than the full $T_{\|,\perp}$ moments (cf. Fig. 9 of Mitchell et al. 2012).

5.1 Results and Conclusions

An overview of the data for 9th of January 2005 is shown in Fig. 20. The transition from unshocked solar wind plasma to the shocked magnetosheath occurs around 22:15:30. Although the solar wind flow is a factor of 10 slower than the electron thermal speed, some residual modulation at the spin period is evident in the data. We have averaged the parallel and anti-parallel ($\alpha = 0, 180°$) moments so that the second and third panels of Fig. 20 reveal the pseudo-parallel and perpendicular moments. Note that the pseudo-densities $n(\alpha)$ are not, and from their definition above need not be, equal. The bottom two panels show increasing oscillations and a gradual "foot" ahead of a steeper magnetic "ramp" region. The dominant

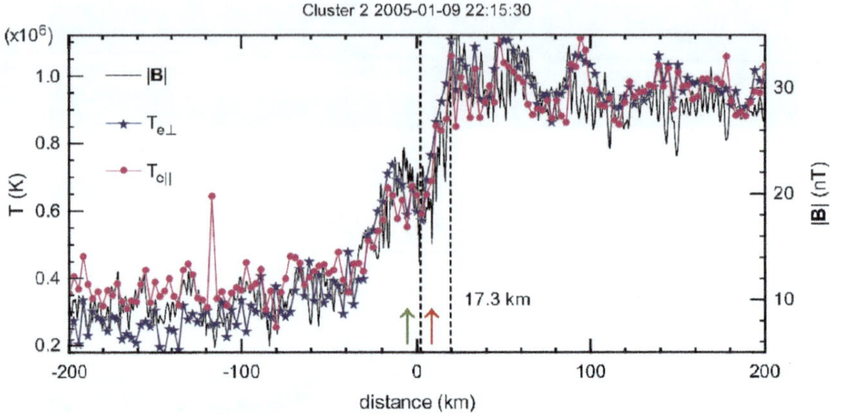

Fig. 21 Magnetic field (*solid*) and electron temperature (*symbols*) as a function of distance from the shock ramp. Roughly half the temperature rise occurs within the region 17.3 km wide between the *dashed vertical lines* corresponding to 6.4 electron inertial lengths (c/ω_{pe}) (Adapted from Schwartz et al. 2011)

$\hat{\mathbf{z}}$ magnetic field component is nearly aligned with the spin axis, enabling the parallel and perpendicular moments to be available in every 0.25 s wedge as described above. Figure 20 already suggests the main result namely that the rise in electron temperature closely follows even the steepest ramp of the magnetic field.

Figure 21 shows that both the parallel and perpendicular electron temperatures closely track the steep rise in magnetic field, with half the electron heating taking place on a scale of 17.3 km, corresponding to 6.4 electron inertial lengths and a small fraction (0.15) of an ion inertial length. Although much of the electron dynamics is linked to the DC electric and magnetic fields within the ramp (Feldman et al. 1983; Goodrich and Scudder 1984; Scudder 1995; Lefebvre et al. 2007) and is therefore reversible (the distribution function in this limiting case might be dependent upon energy and adiabatic invariant in de Hoffmann-Teller reference frame), the fact that both $T_{e\|}$ and $T_{e\perp}$ rise together suggests an inflation of the particle phase space distribution that is not reversible, due primarily to the filling in and/or entrapment of electrons in regions of phase space that would otherwise be inaccessible.

This infilling can be seen in the cuts of the distributions shown in Fig. 22. Within the steep ramp, the inflated distribution is evident, with the flat-topped infilled region already at its downstream level. This supports the notion that the temperature profiles shown in Fig. 21 really do represent irreversible heating. Interestingly, Fig. 22 shows that features previously reported with the ramp, e.g., the beam vestige of the solar wind peak (Feldman et al. 1983), are present only in the more gradual initial rise that precedes the steep ramp. That beam has been totally eroded by the time this electron scale ramp is encountered.

Thus the electron heating occurs over scales that are significantly smaller than the convected proton gyro-scale V_n/Ω_{ci} invoked in Bale et al. (2003) and also smaller than the ion inertial length that might be anticipated due to micro-instabilities within the shock current layer (Papadopoulos 1985b; Matsukiyo and Scholer 2006).

Recent statistical studies (Hobara et al. 2010) argued that previous fits to a proxy of the plasma density profile (Bale et al. 2003) mixed contributions from the more extended foot region governed by reflected gyrating ions. Restricting the measurements to just the steep ramp, they report widths in the range 3–55 c/ω_{pe} with a decreasing trend as the Mach number increases. They interpreted their work in terms of shock steepening limited by the

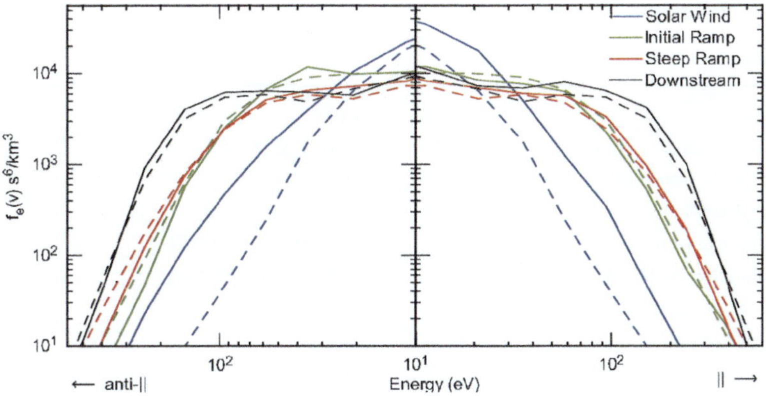

Fig. 22 Cuts of the electron distribution functions in the solar wind, initial ramp, steep ramp, and downstream along (*solid*) and perpendicular (*dashed*) to the magnetic field. The locations of the cuts are indicated along the axes in Figs. 20 and 21. Note the solar wind halo drift evident in the anti-aligned direction and the absence of features within the steep ramp (Adapted from Schwartz et al. 2011)

dispersion of electron whistler waves, with dispersion relation $\omega = \Omega_{ce} \cos\theta_{Bn} (k^2c^2/\omega_{pe}^2)$. The limiting case of a wave capable of phase standing in the incident flow has a wavelength that can be written

$$\frac{\lambda}{c/\omega_{pe}} = 2\pi \frac{\cos\theta_{Bn}}{M_A} \sqrt{\frac{m_i}{m_e}}$$

The results from Table 1 yield a value of 9.2 for this ratio, comparable to the 6.4 electron inertial lengths given above. The fact that supercritical shocks steepen to this whistler limit suggests that dissipation processes are insufficient to broaden the transition further.

It should come as no surprise that the steepening of a fast mode (right-handed) wave results in a right-handed whistler signature. Indeed, the non-coplanar component of the magnetic field (Thomsen et al. 1987), responsible for the difference in the shock electrostatic potential when viewed in different shock rest frames (Goodrich and Scudder 1984), is right-handed. There is new evidence (Sundkvist et al. 2012) that the wave Poynting flux is directed away from the ramp region upstream as expected for dispersion-limited steepening.

The present study measures directly the actual temperature profile of the electrons. The result confirms that nonlinear steepening proceeds down to scales limited by whistler dispersion. We have argued that this represents irreversible heating, implying that dissipation is operative on this, or probably smaller, scales.

We have attempted a similar analysis on other shock crossings observed by Cluster, with consistent findings. Suitable events are rare, since they require the combination of a slowly moving shock and favorable magnetic field orientations. Future space missions need to be proposed to target electron physics and hence should provide numerous examples for statistical studies.

What process(es) are actually responsible for (sub-)whistler-scale dissipation? The overall inflation in phase space is linked to the action of the cross-shock electrostatic potential in concert with the magnetic mirror forces. Some or all of the potential may be concentrated in intense spikes (Bale and Mozer 2007) that may break the adiabaticity of electron phase space trajectories despite a ramp thickness which, in our example, is 20 times the local electron gyroradius. It is worth noting that the localized spikes of the electric field are present inside the ramp region. Figure 23 represents 10 second interval of electric field measurements

Table 1 Shock parameters 2005 Jan 9 @ 22:15

Parameter	Value	
V_{shock}	+10.8	$\mathrm{km\,s^{-1}}$
Unshocked magnetic field $\mathbf{B}_u{}^{\dagger}$	(3.07, 1.35, 8.14)	nT
Unshocked electron density	4.0	$\mathrm{cm^{-3}}$
Location (Earth radii)	(12.3, 13.3, −6.7)	R_e
$\hat{\mathbf{n}}$ shock normal (timing)	(0.855, 0.418, −0.307)	
$\hat{\mathbf{n}}$ (model) Schwartz (1998)	(0.904, 0.383, −0.189)	
$V_n \equiv \mathbf{V} \cdot \hat{\mathbf{n}}$ (shock rest frame)	373	$\mathrm{km\,s^{-1}}$
Alfvén Mach no. M_A	3.8	
Magnetosonic Mach no. M_{ms}	3.0	
$\theta_{Bnu} \equiv \angle \mathbf{B}, \hat{\mathbf{n}}$	83	°
Plasma ion β_i	0.4	
Plasma electron β_e	0.34	
Electron inertial length c/ω_{pe}	2.7	km
Ion inertial length c/ω_{pi}	117	km
$V_n/\Omega_{ciu}{}^{\dagger\dagger}$	443	km
V_n/Ω_{cis}	139	km
Whistler wavelength λ	24.8	km
Electron Larmor radius r_{Leu}	1.01	km

†All vectors are in the GSE frame of reference. Subscripts "s" ("u") denote quantities in the (un)shocked region

$^{\dagger\dagger}\Omega_{ci} \equiv eB/m_p$ is the proton gyrofrequency

inside the ramp by Cluster 2, from 22:15:30 to 22:15:40 that is relatively short time with respect to time of the shock crossing but corresponds to ramp region (courtesy of F. Mozer). One can clearly see quite intense bursts of the electric field having amplitudes as large as 20–30 $\mathrm{mV\,m^{-1}}$. These bursts are very similar to those reported in Sect. 2. Such electric field bursts can be one of the possible sources of electron heating and scattering.

Another candidate processes (e.g. Balikhin and Gedalin 1994) responsible for in-filling regions of phase space, in some of which electrons are trapped, include wave scattering (Scudder et al. 1986c; Veltri and Zimbardo 1993) and demagnetization (Balikhin and Gedalin 1994); these will require further analysis and simulations.

Our discovery of short scale electron heating has an important consequence for electron and ion acceleration. Gradient drift and surfing mechanisms are sensitive to the scale of the field transitions (Zank et al. 1996), becoming very efficient at scales comparable to those reported here.

6 Can Anomalous Resistivity Account for Energy Dissipation and Electron Heating

The major results reported in this section were first published in Balikhin et al. (2005), Walker et al. (2008).

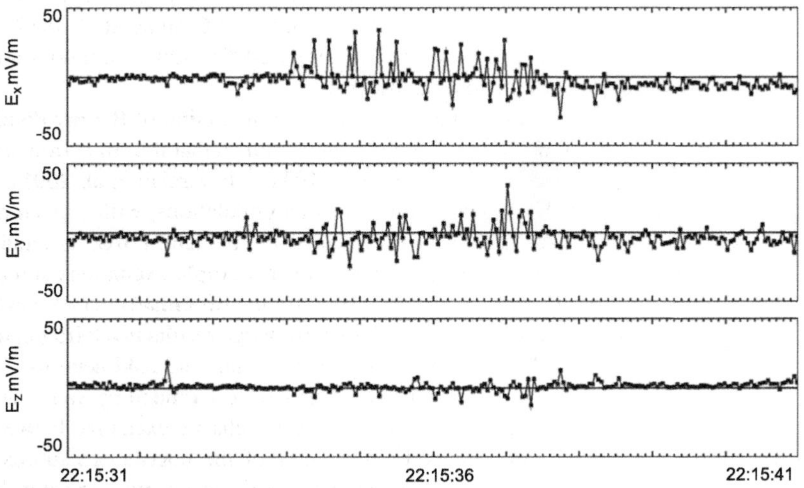

Fig. 23 Electric field measured onboard Cluster 2 satellite from 22:15:31 to 22:15:41. Electric field bursty spikes having amplitudes of 20–40 mV m^{-1} having duration of the order of 0.1 s are clearly seen during ramp crossing. Figure is provided by F. Mozer

6.1 Ion Sound Wave Packets at the Quasiperpendicular Shock Front

The data used in both articles presenting observations of short scale waves were collected by the EFW instrument on board Cluster satellites. EFW uses two pairs of spherical probes in the satellite spin plane situated on the ends of wire booms whose length is 44 m as shown in the left hand panel of Fig. 24. Thus, the distance between probes adjacent/opposite to one another is ∼62/88 m respectfully. Normally, the EFW instruments return the electric field calculated as the difference in probe potentials between probes 1 and 2 (E_{12}) and 3 and 4 (E_{34}) with a sampling rate of either 25 Hz (normal science mode) or 450 Hz (burst science mode). The individual probe potentials are also available with a time resolution of 5 Hz. In addition to these standard modes, there is a triggered internal burst mode. Using this mode, data for a short time period may be captured with a much higher sampling rate. The EFW data that has been analysed in this study consists of internal burst mode data comprising the four individual probe potentials sampled at 9 kHz for periods of around 10 seconds. Since the internal burst data is captured and stored depending upon some criteria, it may be that although the shock region was targeted for data collection, the waveforms returned may not have been captured in the shock front itself. To this end, a search was made to find possible candidate events by cross referencing the list of Cluster shock crossings for 2002 with the list of periods for which internal burst data are available. This resulted in a list of 10 possible events. Of these events, a comparison between the FGM magnetic field measurements and the time periods for which EFW internal burst data were available showed that there were only two shocks for which the period of internal burst data lay solely in the foot region of the shock. Of these, one shock possessed a magnetic profile that was highly turbulent and difficult to interpret and was also eliminated from further analysis. This left just one clean shock on which to perform the analysis.

The internal burst data sets are the only ones generated by EFW that contain the individual probe potentials at a high enough sampling rate to investigate waves and turbulence at frequencies around the lower-hybrid frequency in the vicinity of the terrestrial bow shock

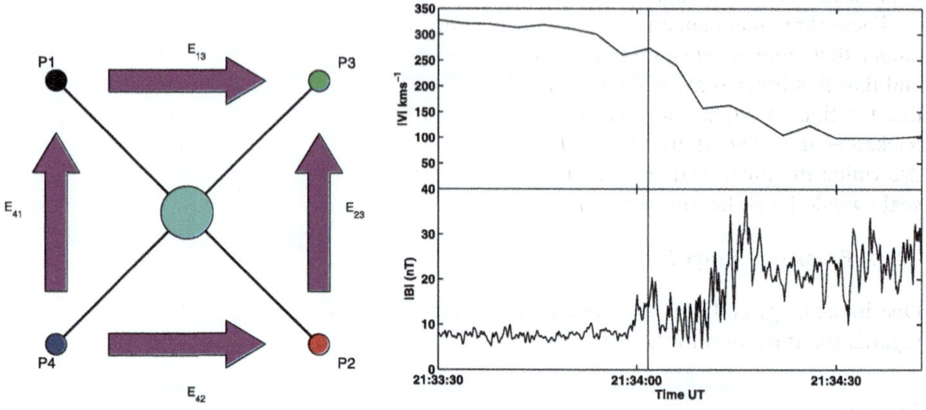

Fig. 24 The *left panel* shows the configuration of the EFW electric field probes and illustrates the electric fields calculated from them. The *right hand panel* shows the magnitude of the magnetic field (*bottom*) and ion bulk flow (*top*) measured during the shock crossing that occurred on 26th of February 2002 at 21:34 UT (Adapted from Balikhin et al. 2005)

(10–30 Hz). By using the individual probe potentials it is possible to compute two parallel electric field components one on either side of the satellite. For example, the probe pairs 1, 3 and 4, 2 maybe used to compute electric field components E_{13} and E_{42} whose directions are parallel and are spatially separated by a distance of ~62.2 m. This technique has previously been used by Balikhin et al. (2005) and Tjulin et al. (2003) to study small scale electric field structures and waves and is similar to the short baseline interferometry techniques employed in the analysis of data from sounding rockets (Pinçon et al. 1997). Since the probe potentials can be used to calculate two parallel electric field vectors it is possible to use the phase differencing technique to determine the wave vector **k**. This method may also be used to examine the polarisation characteristics of the wave in question. In this case, the phase differencing algorithm is applied to a pair of perpendicular components of the electric field (as opposed to the parallel field components mentioned above). The resulting histogram of the phase difference as a function of frequency yields a vertical line of constant phase difference with respect to frequency at a phase difference of zero for a linearly polarised wave and $\pm\pi/2$ for a circularly polarised wave. Thus, this technique may be used to help distinguish between a linearly polarised lower hybrid wave and a circularly polarised whistler mode wave, both of which have been observed at these frequencies. This method is used in preference to an examination of the coherency (see for example Krasnoselskikh et al. 1991) due to the short duration of the wave packets.

The magnetometer data, used to put the electric field measurements into context within the shock front and compute the lower hybrid resonance frequency, come from the FGM instruments (Balogh et al. 1997) and made publically available through the Cluster Active Archive. These measurements typically have a sampling rate of 22 Hz.

All the data presented in this section were recorded during one shock crossing on February 26th, 2002 at around 21:34 UT during the time intervals marked on Fig. 25 by vertical lines. Red lines mark the periods of registration of ion sound waves, green lines the periods of registration of lower hybrid and whistler waves.

During this period the Cluster satellites were situated in the foot region of a quasiperpendicular shock ($\theta_{Bn} \sim 55°$, $M_A \sim 4.3$). The EFW instrument onboard Cluster 3 was triggered to operate in internal burst mode for a few seconds.

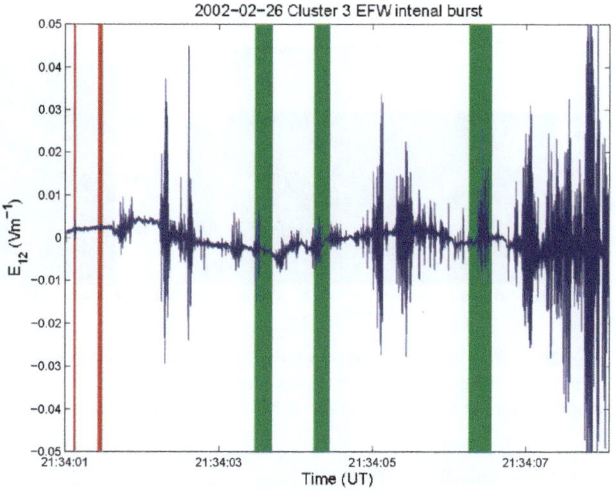

Fig. 25 Waveforms of electric field measurements during shock front crossing on 26th of February 2002 at 21:34 UT. The *vertical lines* mark periods where the waves were registered. As it will be shown later *red vertical lines* mark periods where the waves were identified as ion-sound, two *first green columns* as lower-hybrid electrostatic waves, *third green column* as whistler waves in lower hybrid frequency range

Two parallel electric field vectors of these electric field measurements lie in the same direction and have a perpendicular separation of ~62.2 m in the direction P_2 to P_3. The availability of two closely spaced, simultaneous measurements enables the use of phase differencing techniques (Balikhin et al. 1997a) for the identification of propagation modes for waves with coherence lengths down to a few Debye lengths based upon single satellite measurements. Since there is no component measured normal to the spin plane, the separation between temporal and spatial variations is possible only in the spacecraft spin plane. As a consequence, phase differencing methods are limited to the determination of the projection of the **k**-vector in the spin plane. In most cases, however, this can provide enough information to identify the plasma wave mode. This approach was implemented in these studies. Plasma measurements were obtained from the CIS HIA (ions) and PEACE (electron) instruments. Magnetic field data were obtained from FGM. It should be noted that the spin vector of the Cluster satellites is almost coincident (to within 5°) with the z GSE axis.

The ion bulk velocity (top panel) and the magnitude of the magnetic field (lower panel) as measured by Cluster 3 spacecraft are plotted in the right hand panel of Fig. 24. Initially, the spacecraft was in the solar wind. The foot region was encountered just before 21:34 UT and the shock ramp was crossed around 21:34:12.5 UT. The plasma bulk velocity began to decrease around 21:33:50. Shortly before 21:34 low frequency oscillations were observed in the magnetic field, a feature commonly observed in the foot region of supercritical shocks. The beginning of the foot region is characterised by a large amplitude, nonlinear structure similar to those previously reported by Walker et al. (1999a). A comparison of magnetic field and plasma data show that this structure is not a partial penetration of the ramp. The present study is limited to the short interval at the beginning of the internal burst mode indicated by the vertical line and coincides with the foot region.

The electric field component E_{31} as measured during the initial part of the internal burst mode interval is shown in the upper left panel of Fig. 26 and its Morlet wavelet spectra is shown in the lower left panel. The electric field fluctuations show a pair of well defined wave packets centered around 21:34:01.6 and 21:34:02.05 UT. Their frequency ranges are 100–800 Hz and 250–2000 Hz respectively. We present here the results of the identification of these wave packets to illustrate the use of the technique and its results.

The f–k_{23} spectrum, as shown in the center and right hand panels of Fig. 26, is a histogram representation of the distribution of wave energy in frequency-k space for the first

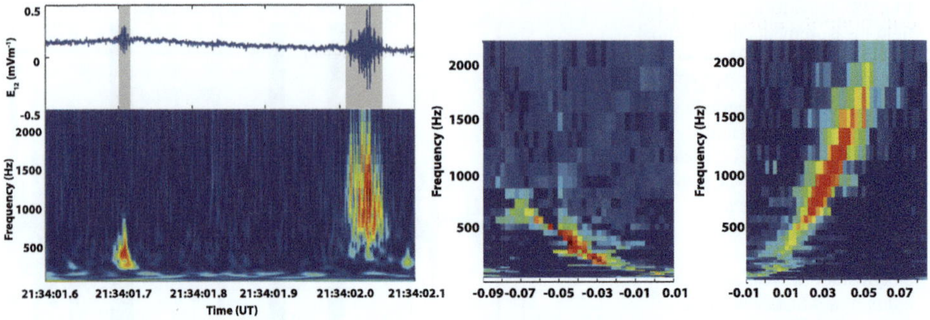

Fig. 26 *Left*: The waveform (*top panel*) and wavelet spectrogram (*bottom panel*) of the electric field computed from the difference in potential between probes 3 and 1. *Centre* and *right*: Examples of the f–k spectrograms for the first (*left*) and second (*right*) wave packets (Adapted from Balikhin et al. 2005)

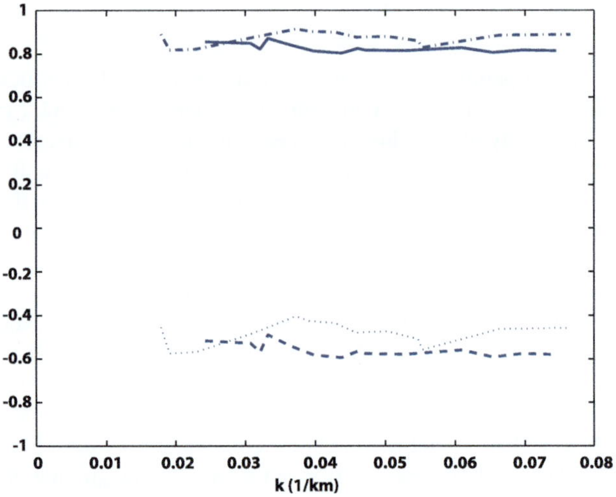

Fig. 27 A comparison of the wave vector directions for the two wave packets. The *dotted* and *dashed lines* represent the X component of wave vector for events 1 and 2 respectively. The corresponding Y components are shown by the *dash-dotted* and *solid lines* (Adapted from Balikhin et al. 2005)

wave packet (Balikhin et al. 1997a). The f–k spectrum shows a well developed ridge like maxima, the shape of which indicates the wave dispersion relation projected along the k_{23} direction. This result may be combined with a similar dispersion along the k_{13} to yield the wave vector projection in the satellite spin plane.

Since the angle between the spacecraft spin plane and the GSE XY plane is small, we will consider that the projection into the spin plane is the same as that into the GSE XY plane. The projection of the dispersion relation into the GSE XY plane is shown as the solid line in Fig. 27 for the first (centre anel in Fig. 26) and second (right on previous figure) wave packets. The observed frequency range of the first wave packet (100–800 Hz) corresponds to approximately 0.25–$1.9\Omega_{ce}$, and the magnitude of wave vector projection is in the range $0.015 < k_1 < 0.075\ \mathrm{m}^{-1}$. For this interval the electron temperature is $T_e \sim 17$ eV and plasma density $n_i = 9.7\ \mathrm{cm}^{-3}$. This leads to an estimate for the Debye length of $\lambda_d \approx 10$ m. Thus the observed values of for the projection of k correspond to $\approx$8–$40\lambda_d$.

The satellite frame dispersion relation in the satellite spin plane is shown by the solid line in Fig. 27. It's phase velocity is in the range 40–70 km s^{-1}. The Doppler shift can be estimated as the scalar product of the solar wind velocity and spin plane wave vector component. This estimation of the Doppler shift term is shown as a dashed line. It has the same sign as the phase velocity and is always greater than the observed wave dispersion indicating that in the plasma frame the waves propagate in the direction opposite to that of the solar wind, but are convected Earthwards by the plasma flow. This convection reverses the direction of propagation in the satellite frame. The average angle between the spin plane projections of wave vector and the magnetic field is about 20°.

The second wave packet analysed was observed $\approx$0.3 seconds after the first. The electric field waveforms (not shown) again indicate a good correlation between the corresponding electric field components measured by different probe pairs. The f–k_{23} spectrum calculated for this wave packet is shown in the right panel of Fig. 26. The ridge like maxima in these spectra correspond to the projections of the wave dispersion relation in the direction k_{23}. The resulting dispersion relation is shown as the solid line in Fig. 27. Its frequency range is 250–2000 Hz ($\approx$0.6–4.9Ω_{ce}), and the magnitude of wave vector projections is in the range $\approx$0.018–0.075 m^{-1}. For this wave packet, the satellite frame phase velocity is in the range 150–160 km s^{-1}. The range of wave vectors and angle of propagation with respect to the magnetic field for the second wave packet coincide with those determined for the first. Even more surprising is the fact that the angle between the two wave vector projections is less than 5°. The dashed line in Fig. 27 shows the estimation of the Doppler shift. It can be seen that the Doppler shift term for the second wave packet is less than that of the observed frequency and so the second wave packet propagates in the same direction in both the satellite and plasma frames. Therefore the first and second wave packets propagate in opposite directions in the plasma frame. While for the second wave packet the satellite frame phase velocity is the sum of of its plasma frame velocity and the solar wind convection speed for the first wave packet it is their difference. That explains why in the satellite frame the second wave packet propagates faster than the first one. The use of multipoint measurements enables one to separate temporal and spatial variations. In the current study it is possible to distinguish which of these two wave modes was observed. Thus we have a method that is independent of using the observed frequency criterion formulated by Gurnett (1985). For this interval $|B| \sim 14.8$ nT and hence the local electron cyclotron frequency $f_e = \Omega_{ce}/2\pi \sim 415$ Hz. As can be seen from the f–k spectra shown in Fig. 26 that the maximum wave energy of the first wave packet occurs at a frequency lower than f_e. According to the classification used by Gurnett (1985) this should be a whistler wave packet whose dispersion relation may be written as (neglecting thermal corrections) $\omega^2 = \Omega_{ce}^2 \cos^2\theta_{Bk} k^2c^2/(k^2c^2 + \omega_{pe}^2)$, where θ_{Bk}, Ω_e, ω_{pe} are the angle between the wave vector and the magnetic field, the electron cyclotron and electron plasma frequencies respectively. The wave vectors for the first wave packet lie in the range $(kc/\omega)^2 \sim 30$–150 and therefore correspond to the electrostatic limit of the mode for which the plasma frame frequency should be $\sim \omega_e \cos\theta_{Bk}$. If we estimate the angle θ_{Bk} using the angle between the projections of wave vector and the magnetic field in the spin plane, the plasma frame frequency can be estimated as $f_e \cos\theta_{Bk} \sim 280$ Hz. For the wavevectors found, $0.015 < k < 0.0075$ m^{-1} the electrostatic whistler phase velocity varies in the range $24 < v_{ph} < 112$ km s^{-1} in the plasma rest frame. In the spacecraft frame the slowest waves would reverse their direction of propagation, so that waves propagating in both directions would be observed. However, it has been shown earlier that all waves are propagating in the same direction. Moreover, for the strongly dispersive electrostatic whistler the phase velocity should vary by a factor of two or more over the observed range of wavevectors, while the actual variation is within 20 % only. These arguments exclude the possibility that the observed mode is the whistler in the electrostatic regime.

The other possibility is the ion-sound mode. Since we are limited to spin plane measurements of wave vectors only order of magnitude estimations of the wave parameters can be made. For such crude calculations we will disregard the factor $\theta_{Bk} \sim 18^{\circ}$ in dispersion of ion-sound waves and use the simplified form $\omega = kv_{is}/\sqrt{1+k^2\lambda_D^2}$ where $v_{is} = \sqrt{k_b T_e/m_i}$ is the ion-sound velocity and k_b is Boltzmann's constant. During the time interval in which both waves packets were observed $v_{is} \approx 40\ \mathrm{km\,s^{-1}}$. Thus, in the plasma rest frame the wave phase velocity should be in the range $0.80v_{is} < v_{ph.pf} < 0.99v_{is}$. This velocity dispersion is very close to the observations. If observed waves are indeed ion-sound waves their plasma frame frequency should be in the range ≈75–100 Hz, much lower than the observed frequency. This disagreement can be attributed to the Doppler shifts estimated as $\frac{|k|}{2\pi} V_{sw} \sim 600$–3000 Hz. In reality the Doppler shift is smaller due to the angle between the wave vector and the solar wind velocity.

The above arguments indicate that the first wave packet consists of ion-sound waves. As previously mentioned, the wave vectors for the second packet have exactly the same range as the first. Therefore, all arguments used above to deduce the wave mode of the first wave packet are valid for the second. The main difference between these two wave packets is in the sign of the Doppler shift. For the first wave packet, the observed frequency is the difference between the Doppler shift and the plasma frame frequency whilst for the second it is their sum. It can be seen that they almost coincide for the whole range of observed waves. The angle between the averaged propagation directions of these wave packets is <5°. This coincidence in the parameters for these two wave packets, observed at clearly distinct periods of time can only be explained by their simultaneous generation at the same location. The generation of ion-sound waves at the shock front are usually attributed either to electric currents or the strong electron temperature gradients in the ramp. Both waves packets were observed upstream of the ramp and carried by the solar wind flow towards it. Since there appear to be no strong gradients in the electron temperature in the foot these waves are probably generated by electric currents. The very short duration of these waves indicates that the current layer might be localized in space and time. Such small scale current layers have been predicted by a nonstationary model of the shock front (Krasnoselskikh 1985, Galeev et al. 1988, 1989; Balikhin et al. 1997b; Walker et al. 1999a). In this model quasiperiodic steepening of and overturning of the shock ramp takes place leading to the ejection of a nonlinear whistler wave into the upstream region. The amplitude of these nonlinear structures can be comparable to the $|B|$ changes in the ramp itself (Walker et al. 1999a) and will be associated with localised currents responsible for the ion-sound waves.

6.2 Observations of Lower-Hybrid Waves

The data set used in this study was collected by the EFW instrument in the same burst mode regime as in previous case onboard the Cluster satellites using onboard timing provided by the DWP instrument (Woolliscroft et al. 1997).

Figure 28(1) shows an overview of the magnetic profile of the shock encountered on February 26th at 21:34 UT. From Fig. 28(1) it can be seen that Cluster 3 first encountered the foot region of the shock just before 21:34 UT, finally crossing the ramp and entering the downstream region at approximately 21:34:15 UT. Here the EFW internal burst data selection was triggered at 21:34:01.922 UT and lasted for a period of 10.47 seconds as indicated by the shaded region in the figure.

The analysis presented here was performed on data recorded on 26th of February 2002 just after 21:34 UT on spacecraft 3. This quasi-perpendicular shock crossing took place on an inbound pass at a position (12.0, −1.60, 8.07) R_E. As can be seen from Fig. 28(1) the whole period of internal burst data was collected in the foot region of the shock. The initial

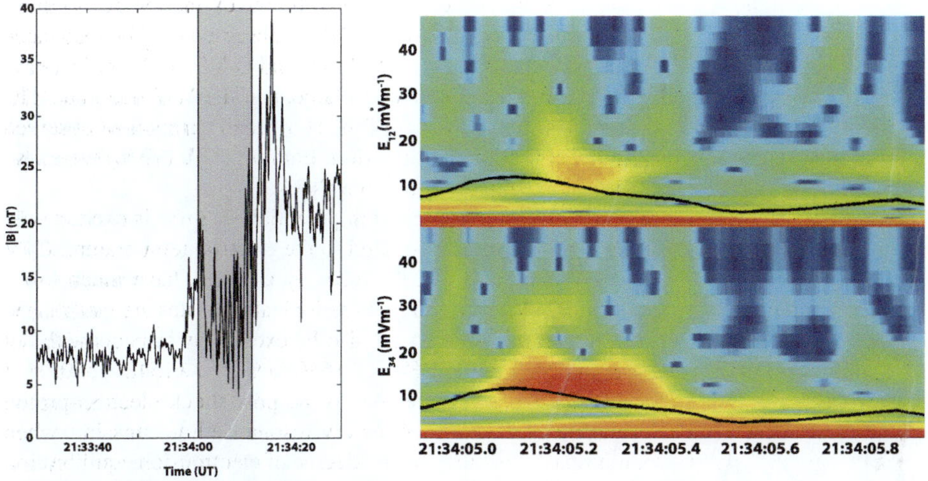

Fig. 28 Panel (*1*): The magnetic profile of the bow shock crossing observed by Cluster 3 on 26th of February 2002 just after 21:34 UT. The period for which EFW internal burst mode data is available is indicated by the *shaded region*. Panel (*2*): The wavelet dynamic spectrogram of electric fields E_{12} (*top*) and E_{34} showing a the occurrence of oscillations just above the lower-hybrid resonance frequency (*black line*) for event 1 (Adapted from Walker et al. 2008)

increase in the magnetic field profile at around 21:34 UT as has been shown above and published in Balikhin et al. (2005) to be part of the foot region rather than a partial ramp crossing.

During the 10.5 second period for which EFW internal burst data is available there were several short periods when the electric field measurements indicated that there were oscillations occurring at or just above the local lower-hybrid resonance frequency. In the following subsections the properties of the waves observed are discussed.

The first event occurred just after 21:34:05 UT. Figure 28(2) shows a dynamic spectrogram of the electric fields E_{12} (top) and E_{34} measured between probes P_1 and P_2 and probes P_3 and P_4 respectively calculated using a Morlet wavelet transform. The black line represents the lower-hybrid resonance frequency. It is clearly seen that at around 21:34:05.2 and there is a packet of waves at a frequency between 10–20 Hz, whose lower edge is just above the lower-hybrid resonance frequency. The duration of this wave packet is around 3 ms corresponding to a few wave periods. Having identified a possible occurrence of lower hybrid waves, the phase differencing technique was applied to parallel electric field vectors in an attempt to compute the dispersion relation of the waves and hence provide an unambiguous identification of the wave mode. However in this case using the spin plane electric field components E_{13} and E_{42} in the frequency range of interest (10–20 Hz) no measurable dispersion of the waves was observed on scales of the separation distance of the probe pairs (62.2 m), see Fig. 29(1).

This implies that the wave travels over the spacecraft at rather high speed so that there is virtually no difference in the phase of the wave measured at the two points on either side of the satellite. This was also evident in the waveform of the electric field signals. A comparison of the waveforms (Fig. 29(2)) shows that the two measurements which are observed to vary in phase which indicates that whatever passed over the satellite has a scale much larger than the individual probe separation distances.

So, one should conclude that the phase differencing method appears to be unable to show any dispersion in the parallel electric field vectors that means that this method cannot be used

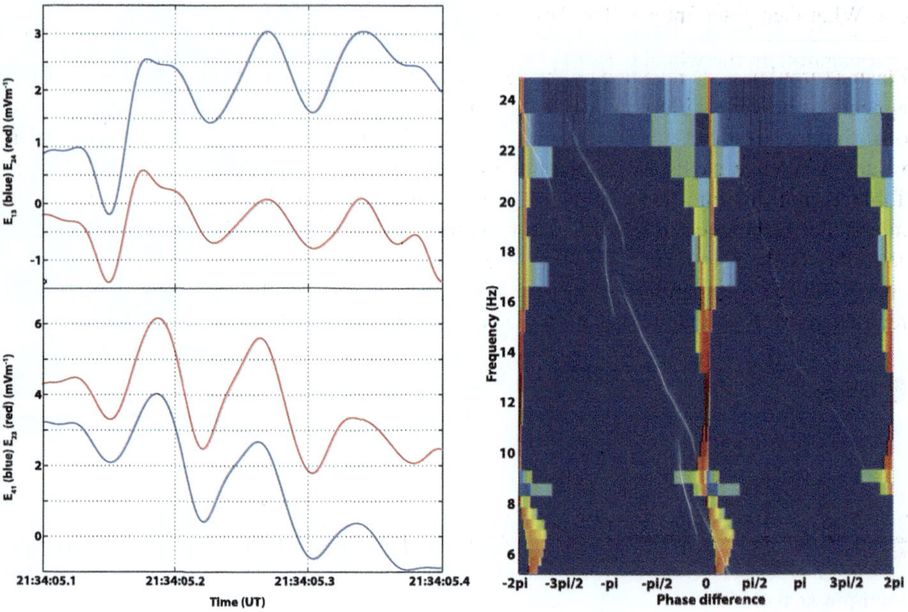

Fig. 29 Panel (*1*): The ω–k joint spectrum calculated from the phase differences measured between the two parallel electric field components E_{13} and E_{42} for event 1. Panel (*2*): The electric field waveforms E_{13} and E_{42} (*top panel*) and E_{41} and E_{23} (*lower panel*) for the first event (Adapted from Walker et al. 2008)

to reliably identify the wave packet as being lower hybrid. As a result, one needs to investigate some other wave properties of the wave packet to determine if they are compatible with the lower hybrid mode. As it was mentioned above that by applying the phase differencing method to perpendicular components of the electric field it should be possible to determine whether the wave packet is linearly or circularly polarised. To this end one should calculate the phase difference between two pairs of probes. The result of this calculation gives the estimate that in the frequency range 10–20 Hz the value of the phase difference is around zero. This result indicates that the wave possesses linear polarisation. This leads to the conclusion that the wave mode that is observed in this case corresponds to a lower-hybrid mode.

Similar analysis of the second event occurred around 21:34:04.5 on 26th of February 2002 shown the very same result, namely, it shows the phase difference of zero which again indicates that the wave is propagating with a large phase speed over the satellite and the wave packet possesses a linear polarisation and it exhibits properties that are consistent with propagation in in the lower hybrid mode.

The third event highlighted by the authors of the paper Walker et al. (2008) occurred between 21:34:07.3 and 21:34:07.45 UT on 26th of February 2002. The wavelet dynamic spectrogram analysis showed a wave packet in the frequency range 10–15 Hz that lies just above the lower hybrid resonance frequency. This wave packet was observed to drift in frequency as time increases. This change in frequency mirrors the change in the lower hybrid resonance frequency as calculated from the magnetic field. Analysis of parallel electric field components using the same phase differencing method again indicated zero phase difference between the components. However, a comparison of perpendicular electric field components indicated that a phase difference between two signals is equal to $\pi/2$. This leads to conclusion that the wave packet possesses circular polarisation and is thus not propagating in the lower hybrid mode. The circular polarisation indicates that this particular wave packet is propagating in the whistler mode.

6.3 Estimates of Efficient Collision Frequency Using Direct Measurements of Ion-Sound and Lower-Hybrid Waves

A definition of the problem of conductivity relies on exchange of momentum between electrons and waves assuming the current is mainly carried by electrons. The conventional formula for plasma conductivity reads

$$\sigma = \frac{ne^2}{m_e \nu}$$

where n is the plasma number density and ν is the collision frequency of electrons with scattering centers, usually ions or neutrals with respect to momentum loss. When electrons excite some oscillations or waves as a result of instability development they also loose the momentum and this loss is referred as the anomalous momentim loss. In order to find effective collision frequency ν_{eff} one has to use the momentum conservation law in the system consisting of electrons and waves. In the case of instability this momentum exchange can be written as follows

$$\nu_{eff} m_e n_0 \vec{u}_{ed} = \frac{2}{(2\pi)^3} \int d^3k \gamma_k W_k \left(\frac{\vec{k}}{\omega_k} \right)$$

where $\vec{u}_{ed}$ is the relative velocity of electrons carrying current, γ_k is the instability increment, W_k is the wave energy density that is defined as $W_k = \frac{\varepsilon_0 |E|^2}{2}$, $|E|$ is the turbulent electric field amplitude. We have in the left hand side the rate of the electron momentum loss per unit time, and in the right hand side we have the momentum gain by waves due to instability. It follows then that

$$\nu_{eff} = \frac{2}{(2\pi)^3 m_e n |\vec{u}_{ed}|} \left| \int d^3k \gamma_{ke} W_k \left(\frac{\vec{k}}{\omega_k} \right) \right|$$

If one would like to evaluate the efficient collision frequency from direct measurements it is necessary to have an estimate of the wave energy. Using this estimate one can evaluate the shock thickness that shock might have if it would be determined by the anomalous collisions using the characteristic length of the momentum loss:

$$L_{an} = \frac{V_{sw}}{\nu_{eff}}.$$

The comparison of the thickness obtained from this estimate with the real shock thickness can be used to evaluate the relative role of the efficient collisions.

The standard estimate of the efficient collision frequency for ion sound mode (Galeev and Sagdeev 1984) reads

$$\nu_{eff} = \omega_{pe} \frac{W}{n_0 k_B T_e}$$

Taking the estimate of the averaged electric field intensity $\langle E \rangle^2 \approx 10^{-5}$ calculated making use the data of measurements (it varies from % 1×10^{-5} to 2×10^{-5} V/m) and density and temperature from observations $n = 9.7\ \mathrm{cm}^{-3}$, $T_e = 17$ eV, $\omega_p = 1.7 \times 10^5\ \mathrm{s}^{-1}$, one can find $W = 4.5 \times 10^{-17}$, $W/(nk_B T_e) = 1.8 \times 10^{-6}$,

$$\nu_{eff}^{is} = 0.3\ \mathrm{s}^{-1}, \qquad L_{an} = \frac{V_{sw}}{\nu_{eff}^{is}} \approx 1200\ \mathrm{km}$$

where ν_{eff}^{is} is anomalous collision frequency due to ion-sound wave activity, L_{an} is the characteristic scale of anomalous energy exchange between electrons and ions. It is sufficiently larger than the electron inertial scale $c/\omega_{pe} = 1.76$ km, and comparable with the thickness of the foot region.

Another group of waves, namely lower hybrid has maximum amplitudes of the order of $10\ \mathrm{mV\,m}^{-1}$ and average electric field energy density of the same order of magnitude as ion sound waves (from 1×10^{-5} to 4×10^{-5} V/m). In order to evaluate the efficient collision frequency for these waves one should take into account the properties of lower-hybrid drift waves. To this end we shall rely on the study of lower hybrid drift instability published by Davidson et al. (1977). The maximum growth rate for these waves can be estimated as

$$\gamma_{LH} \approx \alpha \Omega_{LH}$$

where Ω_{LH} is lower hybrid frequency, coefficient $\alpha < 1$, typically $\alpha \sim 0.1$, and can reach values up to 0.3. Taking maximum of the linear growth rate we can evaluate the upper limit of the effective collision frequency. The phase velocity of waves around the maximum of increment is of the order of ion thermal velocity of ions, and the drift velocity of electrons that carry the current can be estimated evaluating current velocity from macroscopic gradient of the magnetic field. This estimation gives the value comparable with ion thermal velocity. Thus the estimate of the efficient collision frequency in this case can be written as follows:

$$\nu_{eff} \approx \alpha \Omega_{LH} \frac{m_i}{m_e} \frac{W}{nk_B T_i}$$

In the region of observations where $B = 14$ nT and lower hybrid frequency is approximately equal to $56\ \mathrm{s}^{-1}$, thus the efficient collision frequency for these waves is found to be of the of the order of

$$\nu_{eff} \approx \alpha \Omega_{LH} \frac{m_i}{m_e} \frac{W}{nk_B T_i} \sim 0.1 \times 2 \times 10^3 \times 56 \times 10^{-6} \simeq 10^{-2}\ \mathrm{s}^{-1}$$

that is sufficiently smaller than the efficient collision frequency for ion sound waves. The characteristic dissipation scale

$$L_{an} \approx \frac{350}{10^{-2}}\ \mathrm{km} \approx 3.5 \times 10^4\ \mathrm{km}$$

and is sufficiently larger than the major characteristic scales of the shock front.

This leads to the conclusion that the anomalous resistivity observed can not account for the important dissipation rate. The characteristic scales of the dissipation are too large compared to the shock transition features observed.

7 Nonstationarity and Reformation of High Mach Number Quasiperpendicular Shocks: Cluster Observations

In this section, using Cluster observations, we provide convincing evidence that high-Mach-number quasiperpendicular shocks are indeed nonstationary, and moreover, quasi-periodic shock front reformation takes place. Most of the material of this section was first published in Lobzin et al. (2007).

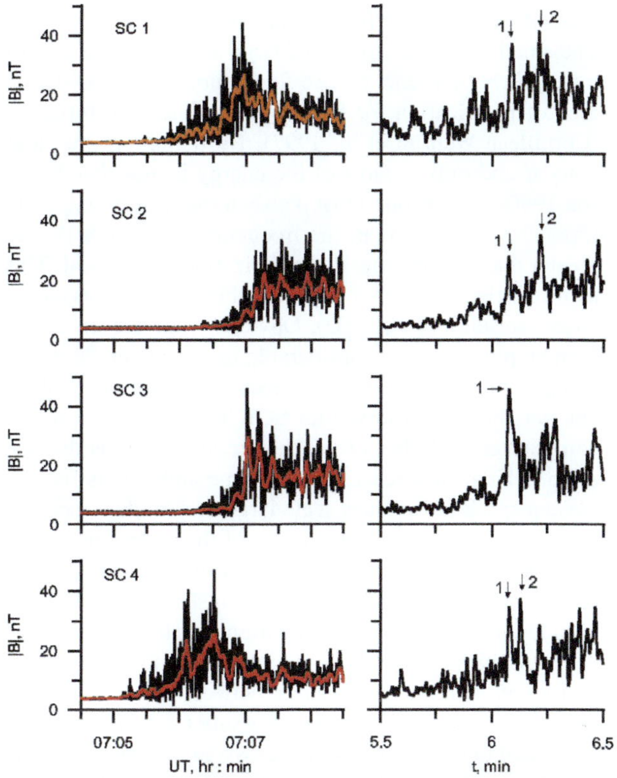

Fig. 30 The magnetic field profiles obtained by FGM experiments aboard four Cluster spacecraft during the Earth's bow shock crossing on 24 January 2001. (*left*) High-resolution magnetic field data (*black line*) and the data obtained by sliding averaging over 4 s time intervals (*red line*). (*right*) Vicinity of overshoots, with large peaks in the magnetic field magnitude. Oscillations with frequencies higher than 2 Hz were removed. To emphasize the similarity and differences of the profiles, the data for the first 3 spacecraft are shifted with respect to that for the 4th one (Adapted from Lobzin et al. 2007)

7.1 An Example of a Typical Crossing of Nonstationary Quasiperpendicular Shock Wave

A number of magnetic field profiles of the quasiperpendicular terrestrial bow shock observed by Cluster triaxial flux gate magnetometers (FGM) (Balogh et al. 1997) in the period January–May 2001 were studied. It was found that nonstationarity seems to be typical for shocks with relatively high Mach numbers. Both from numerical simulations and experiments it follows that the details of this nonstationary behaviour of the shock front may depend strongly not only on the fast magnetosonic Mach number, M_f, but also on the upstream $\beta_{e,i}$ and the angle between the upstream magnetic field and the shock normal, θ_{Bn}. For a detailed case study, a shock was chosen that could be considered as a typical quasiperpendicular, supercritical, high-Mach-number shock wave, namely the shock crossing that occurred on 24th of January 2001 at 07:05:00–07:09:00. Indeed, from the available experimental data and with the use of the multi-spacecraft timing algorithm described by Schwartz (1998) the following estimates were obtained: $\beta_e = 1.7$, $\beta_i = 2.0$, $\theta_{Bn} = 81^\circ$, $M_A = 10$, and $M_f = 5$.

Figure 30 shows the magnetic field profiles measured by the Cluster FGM instruments on 24th of January 2001. The panels on the left show the full resolution data, sampled at 67 Hz (black line) and the result of averaging this data using a 4 second sliding window (red line). The panels on the right show the result of low pass filtering the data at 2 Hz. This process enhances any large peaks in the magnetic field measurements. All the profiles can be considered as quite typical for high-Mach-number quasiperpendicular shock waves. From the averaged data shown by the red lines we observe that the shock front consists of a foot, a ramp, and at least one overshoot-undershoot cycle, i.e. large amplitude peak of the magnetic field at the end of the ramp region and following after it minimum. The small-scale oscillations of large amplitude are superimposed on this large-scale structure. To check whether these fluctuations are consistent with plane wave activity, the degree of polarization for the magnetic field waveforms obtained from STAFF experiment (Cornilleau-Wehrlin et al. 1997). By definition, the degree of polarization approaches a unity if and only if most of the energy is associated with a plane wave (Samson and Olson 1980). It was found that between the forward edge of the shock and the magnetic overshoot the oscillations in the frequency range 3–8 Hz have a high degree of polarization greater than 0.7 and that this polarization is elliptical. This wave activity can be considered as a whistler wave train nested within the shock (Galeev et al. 1988; Galeev et al. 1989; Krasnoselskikh et al. 2002). Obviously, the presence of whistler oscillations, due to their high amplitude, has a considerable impact on the large-scale shock structure. Indeed, averaging of magnetic field data reveals two regions resembling overshoots for SC4 whilst only one maximum is observed for SC1. The profiles for the other spacecraft appear to be more complicated. It follows from these considerations that the concepts of both overshoot and ramp, which must precede it, become ambiguous for such nonstationary shocks. Instead, we can speak about short scale large-amplitude structures embedded into the shock transition, with the forward edge of one of these structures playing a role of the ramp. Figure 30 also shows that the magnetic field profiles measured onboard the different spacecraft differ considerably from each other. Obviously, the number of large-amplitude peaks, their amplitudes, as well the positions within the shock front, are different. The waves observed by different spacecraft in the foot region are also different. In particular, from Fig. 30 (left) it is easily seen that the time interval between the beginning of the wave activity at the forward edge of the shock and the ramp crossing may differ by 10–20 s. This difference is substantial compared to the duration of the crossing of typical elements of the shock structure. The distinctions found between observations from the different spacecraft are related to temporal rather than spatial variations in the structure of the shock front because the spacecraft separation is comparable with shock front thickness. Indeed, the distances between spacecraft lie within the range 380–980 km. The foot thickness estimated with the use of the theoretical formula derived by Schwartz et al. (1983) is equal to 550 km, in reasonable agreement with the observations, while the total shock front thickness is considerably larger. On the other hand, the maximum time lag between the crossings is about $3T_{Bi}$, where T_{Bi} is the ion gyroperiod $T_{Bi} = 15.5$ s. This time lag is larger than the period of the shock reformation. Relying on theoretical considerations and results of numerical simulations, Krasnoselskikh et al. (2002) argue that this type of nonstationarity is closely related to nonlinear whistler wave trains embedded into the shock front and that this is a typical property of quasiperpendicular high-Mach-number shocks. Further evidence for the existence of whistler waves embedded within the shock front can be seen from the rotational features of the magnetic field observed in the vicinity of the peaks, as shown in Fig. 31, that are typical of whistler mode waves.

The large-amplitude structures seen in the magnetic field profiles within the overshoot region and its vicinity have a characteristic time of about 2 s. To examine both the similarities and differences of these profiles, oscillations with frequencies higher than 2 Hz were removed by low pass filtering the data. The filtered data were then used to calculate a set of optimal cross-correlation coefficients for profile fragments that last 35 s and include a portion of foot and the entire overshoot region. The highest correlation was found between SC1 and SC2, while the lowest one was between SC3 and SC4, a result that is in accordance with visual observations of the shifted profiles shown in Fig. 30 (right). An additional analysis of the relative position of the spacecraft tetrahedron and the shock reveals that the

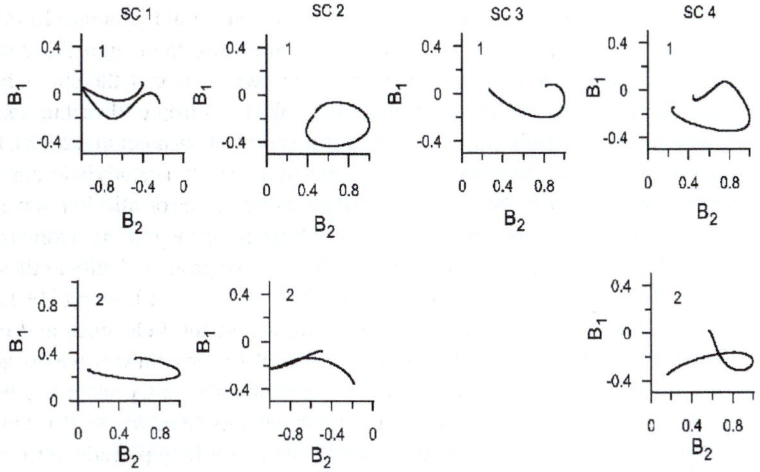

Fig. 31 Hodograms of magnetic field around their maxima for different satellites in the vicinity of supposed ramps

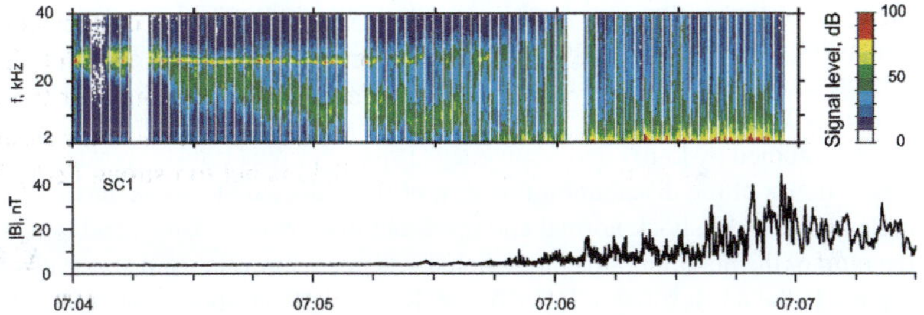

Fig. 32 (*top*) Electric field spectra and (*bottom*) magnetic field profile obtained during the Earth's bow shock crossing on 24th of January 2001 aboard SC1. The frequency-time spectrogram is measured by the Whisper experiment. The *vertical white bands* correspond to the time intervals when no data were obtained in the natural wave mode. The wave intensity is colour coded with the reference level of $10^{-7} V_{rms}/\mathrm{Hz}^{1/2}$, where rms is the root mean square to notify the averaged level of the fluctuating electric field variance. The magnetic field profile is obtained by FGM experiment. The time scales for the both panels are the same (Adapted from Lobzin et al. 2007)

similarity of the shock profiles seems to depend mainly on the time interval between the shock crossings and/or the spacecraft separation measured along the shock normal rather than on the distance along the shock surface which is in accordance with the interpretation that the observed variations are temporal rather than spatial.

Further evidence favoring the nonstationarity of this bow shock crossing comes from WHISPER measurements. In passive mode this experiment provides electric field spectra of natural emissions in the frequency range 2–80 kHz (Décréau et al. 1997). The frequency-time spectrogram obtained by WHISPER experiment aboard SC1 is shown in Fig. 32, together with the magnetic field profile with the same time scale. The bow shock crossing can be identified by a substantial enhancement in the electric field fluctuations within the frequency range 2–5 kHz. For SC1, maximum intensity for these oscillations is observed at 07:06:48 UT. One of the most obvious features of these spectra is the presence of intense waves in the vicinity of the plasma frequency, $f_{pe} = 27$ kHz, together with downshifted oscillations. The most intense feature is a narrow-band Langmuir emission with a frequency in the vicinity of f_{pe}. As compared with Langmuir waves, the power density of downshifted oscillations is usually smaller, while the frequency band they occupy is considerably wider and can be as large as 15–20 % of the central frequency. Both the plasma waves and downshifted oscillations are considered to be typical of the electron foreshock region. It is commonly believed that Langmuir waves are generated by a plasma-beam instability, while for the downshifted oscillations two different mechanisms have been proposed, namely, the plasma-beam interaction, see (see Lacombe et al. 1985; Fuselier et al. 1985) and the loss-cone instability of electron cyclotron modes (Lobzin et al. 2005). The mean frequency of the downshifted oscillations is not constant but varies within the range 0.2–1.0 f_{pe}. In addition, there exists a tendency for a large shift to occur in the vicinity of the shock front, while near the edge of the electron foreshock the shifts are considerably smaller. However, this tendency exists only on large time scales of about 1.0–1.5 min. For smaller scales, 10–15 s, there are the large-amplitude variations of the mean frequency of the downshifted oscillations. The peculiarities of the spectra described above can be explained as follows. The downshifted oscillations are produced by energetic electrons, which are reflected by the bow shock and move almost along the magnetic field lines. Because the solar wind is quiet during the time interval considered (indeed, Figs. 30 and 31 show that there are no significant variations of the magnetic field; the plasma bulk velocity is also approximately constant in the foreshock), the observed evolution of the wave spectra can only be attributed to variations of the suprathermal electron fluxes which are reflected from the bow shock and form the "rabbit ears" in the electron distributions upstream of the shock as was shown by Lobzin et al. (2005). The reflection of electrons by a nearly perpendicular bow shock was studied by Leroy and Mangeney (1984) and Wu (1984). They argued that the main characteristics of the distribution function of the reflected electrons depend first of all on the angle between the shock normal and upstream magnetic field, θ_{B_n}, and to a lesser extent on the ratio of the maximum magnetic field to its upstream value and on the electrostatic potential jump in the de Hoffmann-Teller frame. Resulting from shock front nonstationarity, slow variations of the effective normal of the reflecting part of the shock will lead to considerable variations of number density, energy of reflected electrons, and/or loss-cone angle, thereby producing the observed variations of the downshifted wave spectra. Both theoretical considerations and numerical modeling show that a characteristic time of the shock front oscillations or reformation is comparable with the ion gyroperiod (see Leroy et al. 1982; Krasnoselskikh et al. 2002; Scholer et al. 2003). The time scale of the spectra variations is also comparable with ion gyroperiod T_{Bi}, in accordance with our interpretation.

7.2 Evidence for Shock Front Reformation

As noted above, the magnetic field profiles for the shock under consideration have several nonstationary features. In this section, we consider large-amplitude structures, with a characteristic time of about 1–2 s and present the arguments in favor of front reformation for this particular bow shock crossing. Figure 30 (right) shows the magnetic field profiles obtained after low-pass filtering and shifting the data in time to clearly show the correspondence between the elements observed aboard different spacecraft. For three spacecraft there are two large narrow peaks in the overshoot region and its vicinity, while for SC3 there is only one peak in the corresponding region (see Fig. 30 (right), where these peaks are shown by arrows and numbered). The amplitudes of these peaks, both absolute and relative, differ for different spacecraft. In addition, the distance between two adjacent peaks also varies, being the smallest for SC4 and the largest for SC2. Moreover, the single peak observed by

SC3, which largest amplitude and relatively large width, may be formed due to the coalescence of two separate peaks. The observed peaks in the overshoot region can be considered as a part of the nonstationary whistler wave packets since their rotational properties are clearly evident in Fig. 31. These properties were argued to be an intrinsic element of the quasiperpendicular supercritical shock front structure (Krasnoselskikh et al. 2002). In order to investigate these features further, an analysis of their polarization was performed using the minimum variance technique. The results provide additional evidence in favor of shock front nonstationarity. Indeed, the corresponding elements have different hodograms, which can be rather complicated. However, some of the elements have approximately circular polarization typical for large-amplitude whistlers as was stated in theoretical papers (Galeev et al. 1988, 1989; Krasnoselskikh et al. 2002) and is evidenced on Fig. 31. A comparison of the magnetic field profiles, shown in Fig. 30 with the results of numerical simulations of high-Mach-number shock reformation (Krasnoselskikh et al. 2002) reveals a doubtless resemblance between them. Indeed, for large Mach numbers, quasiperiodic reformation of the shock front was observed in the simulations, with whistler wave packets playing a crucial role. In the first stage of the reformation cycle, a small-amplitude whistler perturbation upstream of the ramp is formed This perturbation grows and moves towards the ramp. When its amplitude exceeds that of the ramp, this disturbance begins to play the role of a new ramp, while the old one moves away downstream. The experimental results shown in Figs. 30 and 31 resemble 4 different snapshots for the same shock undergoing the reformation. The strongest evidence favouring the shock reformation comes from the CIS experiment, which measures the ion composition and full three-dimensional distributions for major ions with energies up to 40 keV/e (Reme et al. 1997). The time resolution of these measurements is about one spacecraft spin, 4 s. Figure 33 shows 8 snapshots obtained at the upstream edge of the shock foot, where the disturbances of the solar wind magnetic field are still small. The figure shows the number of counts vs a function of V_x and V_y in the GSE coordinate system; with the data being integratedin the V_z direction. Reflected ions are observed for the first time at 07:05:16 (see the maximum of the number of counts in the quadrant corresponding to $V_x < 0$ and $V_y < 0$ in the first snapshot). In the time interval from 07:05:16 to 07:05:44, the position of this maximum in the velocity space does not change considerably. In addition, there exists a second population of reflected ions in the quadrant corresponding to $V_x < 0$ and $V_y > 0$. From the snapshots it is easily seen that the numbers of counts corresponding to the reflected ions show approximately periodic variations with a very large modulation depth and a period of about 8 s which corresponds to half of the proton gyroperiod T_{Bi}. To confirm this statement, we performed a summation of the number of counts corresponding to these populations, the results are approximately proportional to the corresponding number densities, n_r. The temporal evolution of these number densities normalized with respect to the corresponding maximum values for the time interval considered is shown in Fig. 33 (bottom). The quasiperiodic variations seem to be more pronounced for the first population (blue line), with the minimum-to-maximum ratio being as low as $\sim$3 %. The number of counts for the second population also varies with approximately the same period, in phase with that for the first one. It is worth noting that the minimum number of counts corresponding to the reflected ions in this region is greater than the 'background noise' by a factor of 5, far beyond experimental errors, while for the maximum number of counts this factor is as large as 30 if the 'noise' level is estimated in the unperturbed solar wind just before the shock crossing. The observed peculiarities of the ion dynamics resemble the features found in the numerical simulations of Krasnoselskikh et al. (2002), where a quasiperiodic front reformation was observed for quasiperpendicular shocks with high Mach numbers. In particular, when the leading wave train before the ramp attained a large enough amplitude, a new population of reflected ions appeared upstream of the precursor. In other words,

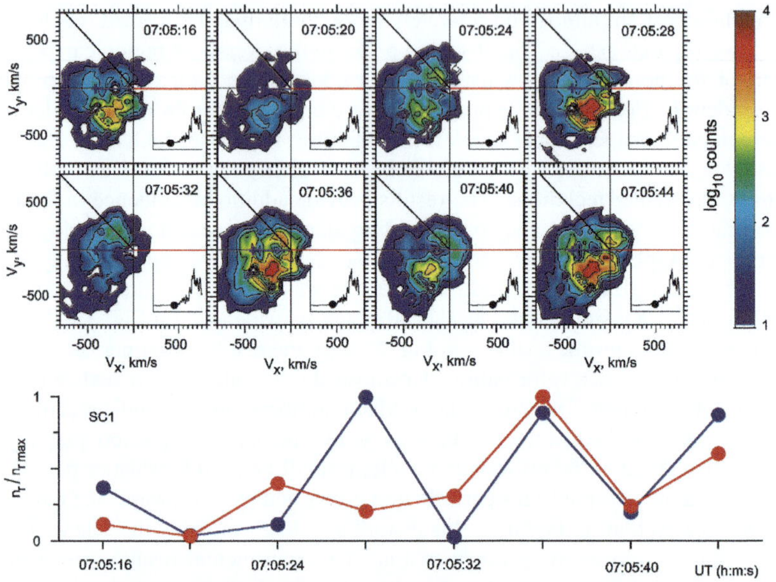

Fig. 33 (*Top*) Ion velocity distributions obtained from CIS measurements within the forward part of the foot for the Earth's bow shock crossing on 24th of January 2001 aboard SC1 and (*bottom*) temporal variations for the relative number of counts corresponding to reflected ions. The distributions were calculated in the GSE coordinates. In the *bottom panel*, a *blue line* corresponds to ions with $V_y < 0$, while a *red line* shows the data for $V_y > 0$. Strong variations of n_r, especially for $V_y < 0$, show that the reflection of ions is bursty. The relative positions where the measurements were made are indicated by the *dots* on the magnetic field profiles shown as *inserts* (Adapted from Lobzin et al. 2007)

the reflection of ions is not stationary. It is quasiperiodically modified during the reformation process. In this case a spacecraft that moves slowly across the shock, will observe the quasiperiodic appearance/disappearance of reflected ions, in accordance with experimental results outlined above.

7.3 Conclusions

In this section we have presented a set of experimental results for a high-Mach-number ($M_f = 5$) quasiperpendicular ($\theta_{Bn} = 81°$) bow shock crossing observed by Cluster spacecraft on 24th of January 2001 at 07:05–07:09 UT. The structure of this shock gives a clear evidence of its nonstationary behavior. In particular, the magnetic field profiles measured by FGM experiments onboard different spacecraft differ considerably from each other. This difference is clearly seen for large-amplitude oscillations, which have relatively short scales of about 1–2 s and resemble nonlinear whistler soliton-like structures that is confirmed by analysis of their hodograms. WHISPER measurements reveal the presence downshifted oscillations within the electron foreshock, with nonmonotonic variations of their central frequency, the characteristic time for these variations is comparable with the proton gyroperiod, $T_{Bi} = 15.5$ s. From the analysis of data from CIS experiment it follows that the reflection of ions from the shock are also highly nonstationary. Moreover, it is shown that the reflection is bursty and the characteristic time for this process is also comparable with the ion gyroperiod. From numerous numerical simulations of quasiperpendicular shocks it is well-known that for high Mach numbers the shock becomes nonstationary. Moreover, front reformation can take place with a characteristic time comparable with the ion gyroperiod. The combination of the features outlined above for the bow shock crossing under consideration is the first convincing experimental evidence favoring the shock front reformation.

8 Conclusions

There exist several models of quasiperpendicular high Mach number shocks. Theoretical considerations and computer simulations on today's level are not capable to describe correctly all physical process that determine different aspects of shock physics. The only possibility to ensure that the theory or modelling correctly capture major physical effects is to rely on analysis of experimental data of direct in situ measurements onboard satellites. The best adapted for this goal are Cluster satellites since they allow one to distinguish spatial and temporal variations and during the mission they had different intersatellite distance that allows one to probe the shock on different scales. The difficult task in such investigation program consists in formulation of the right questions to be addressed to data and to their analysis. Our aim was to determine major physical processes that define characteristics of the most important part of the shock front, its ramp and wave activity around it. The problems closely related to this major problem are electron heating mechanisms and transition of shock behaviour from stationary to nonstationary. We left beyond the scope of our Review many questions. One can mention ripples, remote sensing of the shock by field aligned beam, instabilities behind the shock front. We restricted ourselves by the analysis of scales of magnetic and electric fields, the scale of electron heating, determination of the source and generation mechanism of precursor whistler wave train and direct observation of the shock front reformation by Cluster satellites. Huge collection of data of statistical analysis and of the studies of individual events (case studies) leads to the conclusion that the ramp region of supercritical quasiperpendicular shock is nothing more than an intrinsic element of nonlinear dispersive wave structure slightly modified by reflected ions. This interpretation allows one to explain in a natural way the whole collection of data that we reported in this Review and to understand the transition from stationary to nonstationary shock behaviour when the Mach number exceeds nonlinear whistler critical Mach number. The role of anomalous resistivity is shown to be relatively weak with respect to effects of dispersion. Our study also points out several important opened questions. Presumably the most important is what is detailed mechanism of the electron heating and isotropization. Certainly, the evaluation of the role of anomalous resistivity can not be considered as the solved problem, the data set used is too poor to come to definite conclusions, thus this study still waits new measurements. We did not address the problem of particle acceleration and new results presented here will certainly have an impact on re-consideration of this important problem.

Acknowledgements V.K. acknowledges ISSI for supporting team activities of the project High Mach Number Shocks, CNES for financial support of this work through the series of grants Cluster Co-I DWP. V.K. is also grateful to A. Spitkovsky for providing a figure and useful discussions of astrophysical shocks, A. Artemyev for assistance in preparation of the manuscript, and to R.Z. Sagdeev and A.A. Galeev, teachers who supported the interest to this topic for many years. The authors are grateful to A. Balogh for his kind support and encouragement during the compilation of this review. It is also our great pleasure to acknowledge the contribution of all Cluster PI's whose tireless efforts have provided the scientific community with a readily accessible, high quality, unique data collection without which this work could not have been completed. The authors are grateful to Referee who helped us to improve the manuscript.

Appendix A: Remark on Comparison of Computer Simulation Results with Experimental Data

Recently Comişel et al. (2011) made an attempt to perform computer simulations that can properly reproduce the realistic physical conditions corresponding to observations. They modelled the shock dynamics using 1D PIC code with the realistic ion to electron mass ratio under conditions corresponding to shock conditions on 24th of January 2001 that was observed by Lobzin et al. (2007). The only difference between the model and real plasma parameters is an unrealistic ratio of $(\omega_{pe}/\Omega_{ce})$. The modelling results clearly showed that the shock indeed is nonstationary. However, it was found that there are some important differences between the results of the simulations and observations. The major differences can be summarized as follows. The electric fields observed in simulations in the close vicinity of the shock front were much higher than the electric fields experimentally registered. The energy flux of waves observed in the foot region upstream of the shock front was found to be directed toward downstream that clearly indicates that the waves observed in simulations are generated by the beam of the reflected ions and not by the ramp region as the dispersive mechanism predicts. This gives an indication that the properties of waves observed are much closer to short scale lower hybrid or lower hybrid drift waves described in Sect. 6 and not to those described in Sects. 2 and 3. The question arises where does this difference come from. To answer this question one should consider some scaling properties of equations describing dynamics of the shock. In order to do that let us re-write our equations in dimensionless form making use of natural variables

$$\widetilde{v}=v/V_A,\qquad \widetilde{t}=t\Omega_{ci},\qquad \widetilde{r}=r\Omega_{ci}/V_A,\qquad b=B/B_0,\qquad e=qE/\Omega_{ci}M_AV_A,$$
$$\widetilde{n}=n/n_0$$

The system have several dimensionless parameters that remain and should be taken into account. These are $\eta=\frac{m_e}{m_i}$ (the authors would prefer the letter μ but it is already used for magnetic permeability), $\chi=\frac{\omega_p}{\Omega_{ce}}$, and certainly $\beta_{e,i}$.

To account for the principal difference between the real physical conditions and simulations let us consider where the parameter χ may play an important role. In dimensionless variables it appears in two Maxwell equations

$$\operatorname{div}\overrightarrow{E}=\frac{\eta}{\chi^2}(\tilde{n}_i-\tilde{n}_e)$$

$$\operatorname{rot}\overrightarrow{B}=(\tilde{n}_i\overrightarrow{V_i}-\tilde{n}_e\overrightarrow{\tilde{v}_e})+\frac{\eta}{\chi^2}\frac{\partial\overrightarrow{E}}{\partial\tilde{t}}.$$

To clarify its role one can consider the properties of linear waves. One can note that the ratio of electric to magnetic field is determined by the refractive index of waves. By definition it is

$$N=\frac{kc}{\omega}$$

and it is easy to see that it is exactly this ratio is used in determination of the electric to magnetic field ratio. In SI the system in dimensional variables reads

$$\operatorname{rot}\overrightarrow{E}=-\frac{\partial\overrightarrow{B}}{\partial t}$$

that leads to following estimate for linear waves:

$$[\overrightarrow{k}\times\overrightarrow{E}]=\omega\overrightarrow{B}.$$

This can be re-written as follows:

$$\frac{cB}{E_{k\perp}} = \frac{kc}{\omega} = N.$$

It can also be expressed in terms of phase velocity

$$N = \frac{c}{V_{ph}}.$$

If we take the waves having velocities close to the shock front velocity (approximately standing whistlers in a shock front reference frame) the velocity in the plasma reference frame is $V_{up} = V_{sw} = M_A V_A$, thus the refraction index is

$$N = \frac{c}{V_{sw}} = \frac{c}{M_A V_A} = \frac{\omega_{pi}}{M_A \Omega_{ci}} = \frac{\omega_{pe}}{M_A \Omega_{ce}} \frac{1}{\sqrt{\eta}}$$

$$N_{\text{exp}} = \frac{Bc}{E_{k\perp}} = \frac{c}{V_{sw}} \left(\frac{\omega_p}{\Omega_{ce}} = \frac{2.7 \times 10^4}{1.2 \times 10^2} \sim 230 \right)$$

where $E_{k\perp}$ is the electric field component perpendicular to the k-vector.

On the 24th of January 2001 the solar wind velocity was $V_{swSW} = 440\ \text{km s}^{-1}$.

$$N_{\text{exp}} \approx 700.$$

The maximum value of ratio $\frac{\omega_p}{\Omega_c}$ in simulations is 8 thus

$$N_{sim} = 23$$

it is approximately 30 times smaller than in experiment, that means that for the same level of fluctuations of the magnetic field the electric field fluctuations are 30 times stronger than in experiment.

According to our analysis of dimensionless parameters another important difference consists in similar overestimate of electric fields due to even small deviations from quasi-neutrallity. One can suppose that this can lead to artificial increase of the role of quasi-electrostatic instabilities of short scale lower hybrid waves. As a result the dominant waves observed in simulations are similar to those reported in the section “Anomalous resistivity”, namely drift lower hybrid type waves. Presumably, the overestimation of the role of the electric field and consequently of short scale oscillations and consequently underestimation of the role of lower frequency standing precursor whistler waves results in the difference between observations and simulation results. If so, the simulated shock is really resistive while the observed one is certainly dispersive. To evaluate the influence of this overestimate of the electric field let us evaluate the electric field needed to reflect upstream ion flow assuming that for efficient reflection the potential should be of the order of half energy of the incident ion. Reflecting potential in nonlinear wave on the scale about

$$L = 5\frac{c}{\omega_{pe}} = 0.9 \times 10^6\ \text{cm}, \qquad E_{ions} = \frac{m_i V^2}{2} = 0.5 \times 10^9\ \text{eV}\frac{V^2}{c^2} = 1\ \text{keV}$$

This corresponds to the value of the electric field

$$E = \frac{0.5 \times 10^3 V}{2 \cdot 0.9 \times 10^4\ \text{m}} = 30\ \text{mV m}^{-1}$$

$$\delta B_{\text{exp}} = \frac{N_{\text{exp}} E}{c} = \frac{0.7 \times 10^3 \cdot 60\ \text{V/m} \times 10^{-3}}{2 \cdot 2 \times 10^8} = 1.5 \times 10^{-4} \sim 15\ \text{nT}$$

where δB_{exp} gives the idea of the magnetic field fluctuations really observed and obtained from the comparison with the electric field measurements. These effects are illustrated on the

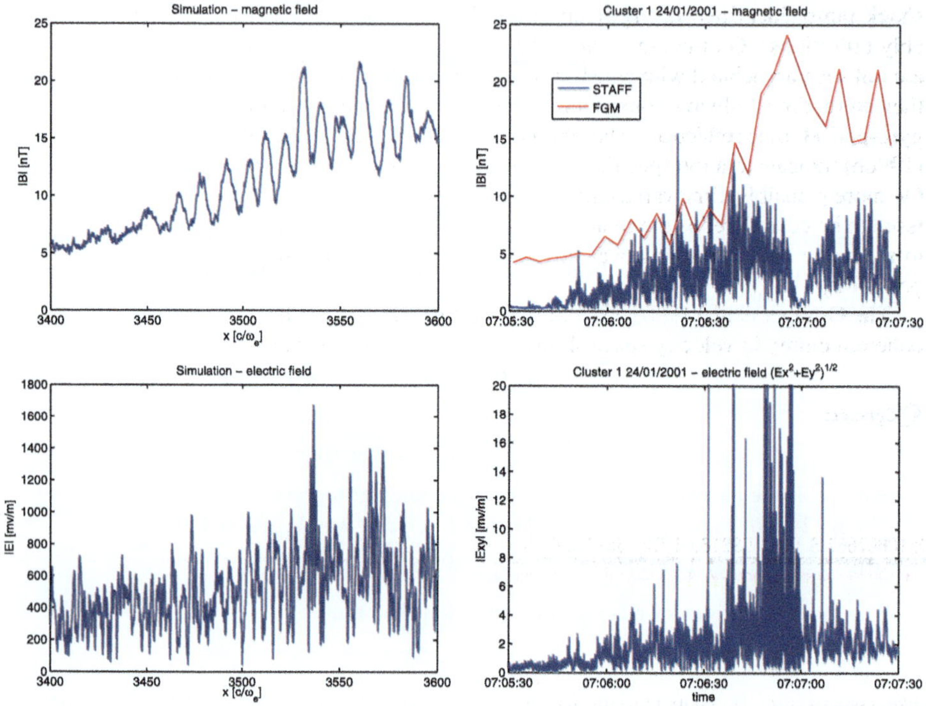

Fig. 34 Comparison of electric and magnetic fields observed on 24th of January 2001 by Cluster satellites and obtained in computer simulation by Comişel et al. (2011). *Both panels* show the fields in the vicinity of the foot/ramp regions. *Top panels* represent magnetic fields as measured by FGM instrument and STAFF search coil magnetometer (*right*), and obtained as a result of simulations (*left*). *Bottom panels* represent measured and simulated electric fields. One can see huge difference in amplitude of electric fields between measurements and simulations that results from artificial ratio of plasma to gyrofrequency and consequently unrealistic refractive index of waves (Figure is provided by J. Soucek and H. Comisel)

Fig. 34 where left hand panels show electric and magnetic field fluctuations in units similar to those experimentally observed, and right hand panels show the data obtained by Cluster satellites for similar parameters (Mach number, angle and β). The magnetic field fluctuations in simulations that will be associated with similar electric fields could be 30 times smaller. Thus electric fields capable to trap and reflect ions are associated with the magnetic field fluctuations that are quite small, namely, less than $\sim$1 nT. The ion trapping and reflection can occur in small amplitude oscillations in the foot region that can not happen in real shock. Crucial change of ion dynamics certainly results in change of the characteristics of the shock front and wave activity around.

The goal of this remark is not to understate the role and importance of computer simulations for the shock studies. We would like to point out that direct comparison of simulation results with the observations needs special attention and analysis of the simulation conditions to ensure that the process is properly described.

Appendix B: Table of Notations Used in the Article

Parameter	Interpretation
B	the magnitude of the magnetic field
B_0	upstream magnetic field
B_n	magnetic field component along the normal to the shock
$\beta = \frac{8\pi nT}{B^2}$	the ratio of total particle thermal pressure to the magnetic field pressure
ΔB	change of the magnetic field through inhomogeneous layer
Δt_{ij}	time difference of observation of shock front features such as electric field spikes by different satellites i and j
ΔE	an amplitude of the electric field spike feature
$E_{x,y,z}$	electric field components along corresponding axes
E_{spike}	maximum amplitude in electric field spike
E_{ij}	electric field as measured by means of probes i and j
$\mathbf{E}$	electric field vector
e_{ij}	electric potential difference between probes i and j onboard single satellite
$f = \omega/2\pi$	wave frequency
$f_e = \Omega_{ce}/2\pi$	electron gyrofrequency (in Hz)
γ_{LH}	growth rate of lower hybrid waves
$\mathbf{k}$	wave-vector of a wave
k_{ij}	an estimate of the **k**-vector component from electric field probes i and j measurements onboard one single satellite
$k_{\parallel}$	parallel to the magnetic field component of the wave-vector
l_{gr}	characteristic gradient scale inside the inhomogeneous layer
L_d	dissipative scale
L_{disp}	dispersive scale
$L_{I,e} = c/\omega_{pi,pe}$	ion, electron inertial length
L_{Br}	thickness of the ramp region of the shock as seen in magnetic field measurements
L_{ϕ}	characteristic scale of the electrostatic potential variation in the shock front
L_{an}	characteristic scale of energy exchange due to anomalous resistivity
L_f	characteristic spatial size of the magnetic foot
L_r	width of the magnetic foot
$m_{i,e}$	ion, electron mass
$M_A = V_{up}/V_A$	Alfvenic Mach number, the ratio of the normal component of the upstream flow velocity to Alfven velocity
$M_{Ms} = V_{up}/V_{Ms}$	magnetosonic Mach number
$M_w = V_{w,max}/V_A$	nonlinear critical whistler Mach number
n_E	shock front normal determined from timing of electric field spikes measured onboard four satellites
n_B	shock front normal vector from magnetic field measurements
n	plasma density
$\hat{\mathbf{n}}$	shock front normal vector
n_r	number density of reflected ions
$R_{Li} = V_{up}/\Omega_{ci}$	convective ion gyroradius
R_{Le}	electron Larmor radius
$S_{\parallel}$	Poynting flux along the magnetic field

Parameter	Interpretation
T	total plasma temperature
$T_{i,e}$	ion, electron temperature
$T_{e\parallel}$	parallel to magnetic field electron temperature
$T_{e\perp}$	electron temperature perpendicular to the magnetic field
T_{Bi}	ion gyroperiod
$\vec{u}_{ed}$	the relative velocity of electrons carrying current
V_A	Alfven velocity
V_{up}	the normal component of the upstream velocity to the shock surface in its rest frame
V_{Ms}	velocity of the magnetosonic wave propagating in the same direction as the shock to the background magnetic field
$V_{w,max}$	highest possible velocity of nonlinear whistler wave that can stay in the upstream flow
V_{sw}	solar wind velocity
V_{ss}	relative shock spacecraft velocity
v_{ph}	phase velocity of ion sound wave
V_{ph}	phase velocity of wave
v_{is}	ion sound velocity
$V_{x,y,z}$	velocity components along corresponding axes
W_k	electric field energy density
λ	wavelength of precursor whistler wave
$\lambda_D = \epsilon_0 k_B T_e / ne^2$	Debye radius
ν_{eff}	effective collision frequency due to wave particle interaction
θ_{Bn}	the angle between the magnetic field and shock front normal
θ_{kB}	angle between the magnetic field vector and the wave vector
$\omega_{lh} \sim \sqrt{\Omega_{ci}\Omega_{ce}}$	lower hybrid frequency
ω_{pi}	ion plasma frequency
$\Omega_{ci,e} = eB/m_{i,e}$	ion, electron gyrofrequency
ω_p	electron plasma frequency

References

J.H. Adlam, J.E. Allen, The structure of strong collision-free hydromagnetic waves. Philos. Mag. **3**, 448–455 (1958). doi:10.1080/14786435808236833

U. Ascoli-Bartoli, S. Martellucci, M. Martone, Formation and development of hydromagnetic disturbances during the implosion phase of a preionized theta pinch (cariddi) (cn-21/75), in *Plasma Physics and Controlled Nuclear Fusion Research*, vol. I (1966), p. 275

H.U. Auster, K.H. Glassmeier, W. Magnes, O. Aydogar, D. Constantinescu, D. Fischer, K.H. Fornacon, E. Georgescu, P. Harvey, O. Hillenmaier, R. Kroth, M. Ludlam, Y. Narita, K. Okrafka, F. Plaschke, I. Richter, H. Schwarzl, B. Stoll, A. Valavanoglu, M. Wiedemann, The THEMIS fluxgate magnetometer. Space Sci. Rev. **135** (2008). doi:10.1007/s11214-008-9365-9

W.I. Axford, E. Leer, G. Skadron, The acceleration of cosmic rays by shock waves, in *International Cosmic Ray Conference*, vol. 11 (1977), pp. 132–137

S.D. Bale, F.S. Mozer, Measurement of large sale parallel and perpendicular electric fields on electron spatial scales in the terrestrial bow shock. Phys. Rev. Lett. **98**, 205001 (2007). doi:10.1103/PhysRevLett.98.205001

S.D. Bale, F.S. Mozer, T.S. Horbury, Density-transition scale at quasiperpendicular collisionless shocks. Phys. Rev. Lett. **91**(26), 265004 (2003). doi:10.1103/PhysRevLett.91.265004

S.D. Bale, F.S. Mozer, V.V. Krasnoselskikh, Direct measurement of the cross-shock electric potential at low plasma β, quasi-perpendicular bow shocks. ArXiv e-prints (2008)
S.D. Bale, P.J. Kellogg, D.E. Larson, R.P. Lin, K. Goetz, R.P. Lepping, Bipolar electrostatic structures in the shock transition region: evidence of electron phase space holes. Geophys. Res. Lett. **25**, 2929–2932 (1998). doi:10.1029/98GL02111
S.D. Bale, M.A. Balikhin, T.S. Horbury, V.V. Krasnoselskikh, H. Kucharek, E. Möbius, S.N. Walker, A. Balogh, D. Burgess, B. Lembége, E.A. Lucek, M. Scholer, S.J. Schwartz, M.F. Thomsen, Quasi-perpendicular shock structure and processes. Space Sci. Rev. **118**(1–4), 161–203 (2005). doi:10.1007/s11214-005-3827-0
M. Balikhin, M. Gedalin, Kinematic mechanism of electron heating in shocks: theory vs observations. Geophys. Res. Lett. **21**, 841–844 (1994). doi:10.1029/94GL00371
M. Balikhin, M. Gedalin, A. Petrukovich, New mechanism for heating in shocks. Phys. Rev. Lett. **70**, 1259–1262 (1993). doi:10.1103/PhysRevLett.70.1259
M. Balikhin, V. Krasnosselskikh, M. Gedalin, The scales in quasiperpendicular shocks. Adv. Space Res. **15**(8–9), 247–260 (1995). doi:10.1016/0273-1177(94)00105-A
M.A. Balikhin, T. Dudok de Wit, L.J.C. Woolliscroft, S.N. Walker, H. Alleyne, V. Krasnoselskikh, W.A.C. Mier–Jedrzejowicz, W. Baumjohann, Experimental determination of the dispersion of waves observed upstream of a quasi-perpendicular shock. Geophys. Res. Lett. **24**, 787–790 (1997a). doi:10.1029/97GL00671
M.A. Balikhin, S.N. Walker, T. Dudok de Witt, H.S.C. Alleyne, L.J.C. Woolliscroft, W.A.C. Mier-Jedrzejowicz, W. Baumjohann, Nonstationarity and low frequency turbulence at a quasi-perpendicular shock front. Adv. Space Res. **20**, 729–734 (1997b)
M.A. Balikhin, V. Krasnoselskikh, L.J.C. Woolliscroft, M. Gedalin, A study of the dispersion of the electron distribution in the presence of e and b gradients: application to electron heating at quasi-perpendicular shocks. J. Geophys. Res., Atmos. **103**, 2029–2040 (1998). doi:10.1029/97JA02463
M.A. Balikhin, M. Nozdrachev, M. Dunlop, V. Krasnoselskikh, S.N. Walker, H.S.C.K. Alleyne, V. Formisano, M. André, A. Balogh, A. Eriksson, K. Yearby, Observation of the terrestrial bow shock in quasi-electrostatic sub-shock regime. J. Geophys. Res., Atmos. **107**(8), 10 (2002). doi:10.1029/2001JA000327
M.A. Balikhin, O.A. Pokhotelov, S.N. Walker, E. Amata, M. Andre, M. Dunlop, H.S.C.K. Alleyne, Minimum variance free wave identification: application to cluster electric field data in the magnetosheath. Geophys. Res. Lett. **30**(10) (2003). doi:10.1029/2003GL016918
M. Balikhin, S. Walker, R. Treumann, H. Alleyne, V. Krasnoselskikh, M. Gedalin, M. Andre, M. Dunlop, A. Fazakerley, Ion sound wave packets at the quasiperpendicular shock front. Geophys. Res. Lett. **32**(24), 24106 (2005). doi:10.1029/2005GL024660
A. Balogh, M.W. Dunlop, S.W.H. Cowley, D.J. Southwood, J.G. Thomlinson, K.H. Glassmeier, G. Musmann, H. Lühr, S. Buchert, M.H. Acuña, D.H. Fairfield, J.A. Slavin, W. Riedler, K. Schwingenschuh, M.G. Kivelson, The cluster magnetic field investigation. Space Sci. Rev. **79**, 65–91 (1997). doi:10.1023/A:1004970907748
A. Bamba, R. Yamazaki, M. Ueno, K. Koyama, Small-scale structure of the sn 1006 shock with chandra observations. Astrophys. J. **589**, 827–837 (2003). doi:10.1086/374687
F. Begenal, J.W. Belcher, E.C. Sittler, R.P. Lepping, The uranian bow shock—Voyager 2 inbound observations of a high Mach number shock. J. Geophys. Res. **92**, 8603–8612 (1987). doi:10.1029/JA092iA08p08603
A.R. Bell, The acceleration of cosmic rays in shock fronts, I. Mon. Not. R. Astron. Soc. **182**, 147–156 (1978a)
A.R. Bell, The acceleration of cosmic rays in shock fronts, II. Mon. Not. R. Astron. Soc. **182**, 443–455 (1978b)
D. Biskamp, H. Welter, Ion heating in high-Mach-number, oblique, collisionless shock waves. Phys. Rev. Lett. **28**, 410–413 (1972). doi:10.1103/PhysRevLett.28.410
R.D. Blandford, J.P. Ostriker, Particle acceleration by astrophysical shocks. Astrophys. J. Lett. **221**, 29–32 (1978). doi:10.1086/182658
K.D. Cole, Effects of crossed magnetic and /spatially dependent/ electric fields on charged particle motion. Planet. Space Sci. **24**, 515–518 (1976). doi:10.1016/0032-0633(76)90096-9
H. Comişel, M. Scholer, J. Soucek, S. Matsukiyo, Non-stationarity of the quasi-perpendicular bow shock: comparison between cluster observations and simulations. Ann. Geophys. **29**, 263–274 (2011). doi:10.5194/angeo-29-263-2011
N. Cornilleau-Wehrlin, P. Chauveau, S. Louis, A. Meyer, J.M. Nappa, S. Perraut, L. Rezeau, P. Robert, A. Roux, C. De Villedary, Y. de Conchy, L. Friel, C.C. Harvey, D. Hubert, C. Lacombe, R. Manning, F. Wouters, F. Lefeuvre, M. Parrot, J.L. Pinçon, B. Poirier, W. Kofman, P. Louarn, The STAFF investigator team, the cluster spatio-temporal analysis of field fluctuations (staff) experiment. Space Sci. Rev. **79**, 107–136 (1997)
F.V. Coroniti, Dissipation discontinuities in hydromagnetic shock waves. J. Plasma Phys. **4**, 265 (1970)

R.C. Davidson, N.T. Gladd, C.S. Wu, J.D. Huba, Effects of finite beta plasma on the lower hybrid instability. Phys. Fluids **20**, 301 (1977)

L. Davis, R. Lüst, A. Schlüter, The structure of hydromagnetic shock waves, I: nonlinear hydromagnetic waves in a cold plasma. Z. Naturforsch. Teil A **13**, 916 (1958)

F. de Hoffmann, E. Teller, Magneto-hydrodynamic shocks. Phys. Rev. **80**, 692–703 (1950). doi:10.1103/PhysRev.80.692

P.M.E. Décréau, P. Fergeau, V. Krannosels'kikh, M. Leveque, P. Martin, O. Randriamboarison, F.X. Sene, J.G. Trotignon, P. Canu, P.B. Mogensen, Whisper, a resonance sounder and wave analyser: performances and perspectives for the cluster mission. Space Sci. Rev. **79**, 157–193 (1997). doi:10.1023/A:1004931326404

A.P. Dimmock, M.A. Balikhin, V.V. Krasnoselskikh, S.N. Walker, S.D. Bale, Y. Hobara, A statistical study of the cross-shock electric potential at low Mach number, quasi-perpendicular bow shock crossings using cluster data. J. Geophys. Res., Space Phys. **117**(A2), 2210 (2012). doi:10.1029/2011JA017089

C.P. Escoubet, R. Schmidt, M.L. Goldstein, Cluster—science and mission overview. Space Sci. Rev. **79**, 11–32 (1997)

V.G. Eselevich, A.G. Eskov, R.C. Kurtmullaev, A.I. Malyutin, Isomagnetic discontinuity in a collisionless shock wave. Sov. Phys. JETP **33**, 1120 (1971)

M.H. Farris, S.M. Petrinec, C.T. Russell, The thickness of the magnetosheath: constraints on the polytropic index. Geophys. Res. Lett. **18**, 1821 (1991). doi:10.1029/91GL02090

M.H. Farris, C.T. Russell, M.F. Thomsen, Magnetic structure of the low beta, quasi-perpendicular shock. J. Geophys. Res., Atmos. **98**, 15285–15294 (1993). doi:10.1029/93JA00958

W.C. Feldman, R.C. Anderson, J.R. Asbridge, S.J. Bame, J.T. Gosling, R.D. Zwickl, Plasma electron signature of magnetic connection to the Earth's bow shock—isee 3. J. Geophys. Res., Atmos. **87**(A2), 632–642 (1982)

W.C. Feldman, R.C. Anderson, S.J. Bame, S.P. Gary, J.T. Gosling, D.J. McComas, M.F. Thomsen, G. Paschmann, M.M. Hoppe, Electron velocity distributions near the Earth's bow shock. J. Geophys. Res., Atmos. **88**, 96–110 (1983)

V. Formisano, Measurement of the potential drop across the Earth's collisionless bow shock. Geophys. Res. Lett. **9**, 1033 (1982)

V. Formisano, Collisionless shock waves in space and astrophysical plasmas, in *Proc. ESA Workshop on Future Missions in Solar, Heliospheric and Space Plasma Physics*, vol. ESA SP-235 (1985), p. 83

V. Formisano, R. Torbert, Ion acoustic wave forms generated by ion-ion streams at the Earth's bow shock. Geophys. Res. Lett. **9**, 207 (1982)

S.A. Fuselier, D.A. Gurnett, R.J. Fitzenreiter, The downshift of electron plasma oscillations in the electron foreshock region. J. Geophys. Res. **90**, 3935–3946 (1985). doi:10.1029/JA090iA05p03935

A.A. Galeev, Collisionless shocks, in *Physics of Solar Planetary Environment; Proceedings of the International Symposium on Solar-Terrestrial Physics*, Boulder, CO, June 7–18, 1976, ed. by D.J. Williams (AGU, Washington, 1976), pp. 464–490

A.A. Galeev, R.Z. Sagdeev, Current instabilities and anomalous resistivity of plasma, in *Basic Plasma Physics: Selected Chapters, Handbook of Plasma Physics*, vol. 1, ed. by A.A. Galeev, R.N. Sudan (1984), p. 271

A.A. Galeev, R.N. Sudan, *Basic Plasma Physics. Selected Chapters. Handbook of Plasma Physics*, vols. 1 and 2 (1989)

A.A. Galeev, V.V. Krasnoselskikh, V.V. Lobzin, Fine structure of the front of a quasi-perpendicular supercritical collisionless shock wave. Sov. J. Plasma Phys. **14**, 1192–1200 (1988)

A.A. Galeev, C.F. Kennel, V.V. Krasnoselskikh, V.V. Lobzin, The role of whistler oscillations in the formation of the structure of high Mach number collisionless shock, in *Plasma Astrophysics* (1989), pp. 165–171

M. Gedalin, Ion reflection at the shock front revisited. J. Geophys. Res. **101**, 4871 (1996). doi:10.1029/95JA03669

M. Gedalin, Ion heating in oblique low-Mach number shocks. Geophys. Res. Lett. **24**, 2511 (1997)

M. Gedalin, K. Gedalin, M. Balikhin, V. Krasnosselskikh, L.J.C. Woolliscroft, Demagnetization of electrons in inhomogeneous $E_\perp B$: implications for electron heating in shocks. J. Geophys. Res. **100**, 19911–19918 (1995a). doi:10.1029/95JA01399

M. Gedalin, K. Gedalin, M. Balikhin, V. Krasnosselskikh, Demagnetization of electrons in the electromagnetic field structure, typical for quasi-perpendicular collisionless shock front. J. Geophys. Res. **100**, 9481–9488 (1995b). doi:10.1029/94JA03369

G. Goldenbaum, E. Hintz, Experimental study of shock wave formation in an almost collisionfree plasma. Phys. Plasmas **8**, 2111–2112 (1965). doi:10.1063/1.1761164

C.C. Goodrich, J.D. Scudder, The adiabatic energy change of plasma electrons and the frame dependence of the cross shock potential at collisionless magnetosonic shock waves. J. Geophys. Res., Atmos. **89**, 6654–6662 (1984)

E.W. Greenstadt, R.W. Fredricks, Shock systems in collisionless space plasmas, in *Solar System Plasma Physics*, vol. 3, ed. by L.J. Lanzerotti, C.F. Kennel, E.N. Parker (North Holland, Amsterdam, 1979), pp. 3–43

D.A. Gurnett, Plasma waves and instabilities, in *Collisionless Shocks in the Heliosphere: Reviews of Current Research*, ed. by B.T. Tsurutani, R.G. Stone Geophysical Monograph (AGU, Washington, 1985), pp. 207–224

G. Gustafsson, R. Bostrom, B. Holback, G. Holmgren, A. Lundgren, K. Stasiewicz, L. Ahlen, F.S. Mozer, D. Pankow, P. Harvey, P. Berg, R. Ulrich, A. Pedersen, R. Schmidt, A. Butler, A.W.C. Fransen, D. Klinge, C.-G. Falthammar, P.-A. Lindqvist, S. Christenson, J. Holtet, B. Lybekk, T.A. Sten, P. Tanskanen, K. Lappalainen, J. Wygant, The electric field and wave experiment for the cluster mission. Space Sci. Rev. **79**, 137–156 (1997)

P. Hellinger, A. Mangeney, Upstream whistlers generated by protons reflected from a quasi-perpendicular shock. J. Geophys. Res., Space Phys. **102**(A5), 9809–9819 (1997). doi:10.1029/96JA03826

P. Hellinger, P. Trávnícek, H. Matsumoto, Reformation of perpendicular shocks: hybrid simulations. Geophys. Res. Lett. **29**(24), 240000 (2002). doi:10.1029/2002GL015915

P. Hellinger, P. Trávníček, B. Lembège, P. Savoini, Emission of nonlinear whistler waves at the front of perpendicular supercritical shocks: hybrid versus full particle simulations. Geophys. Res. Lett. **34**, 14109 (2007). doi:10.1029/2007GL030239

J.P. Heppner, N.C. Maynard, T.L. Aggson, Early results from isee-1 electric field measurements. Space Sci. Rev. **22**, 777–789 (1978)

Y. Hobara, S.N. Walker, M. Balikhin, O.A. Pokhotelov, M. Gedalin, V. Krasnoselskikh, M. Hayakawa, M. André, M. Dunlop, H. Rème, A. Fazakerley, Cluster observations of electrostatic solitary waves near the Earth's bow shock. J. Geophys. Res., Atmos. **113**, 05211 (2008). doi:10.1029/2007JA012789

Y. Hobara, M. Balikhin, V. Krasnoselskikh, M. Gedalin, H. Yamagishi, Statistical study of the quasi-perpendicular shock ramp widths. J. Geophys. Res. **115**, 11106 (2010). doi:10.1029/2010JA015659

T.S. Horbury, P.J. Cargill, E.A. Lucek, A. Balogh, M.W. Dunlop, T.M. Oddy, C. Carr, P. Brown, A. Szabo, K.-H. Fornaçon, Cluster magnetic field observations of the bowshock: orientation, motion and structure. Ann. Geophys. **19**, 1399–1409 (2001). doi:10.5194/angeo-19-1399-2001

T.S. Horbury, P.J. Cargill, E.A. Lucek, J. Eastwood, A. Balogh, M.W. Dunlop, K.-H. Fornaçon, E. Georgescu, Four spacecraft measurements of the quasi-perpendicular terrestrial bowshock: orientation and motion. J. Geophys. Res., Atmos. **107**(A8), 10–11 (2002). doi:10.1029/2001JA000273

V.I. Karpman, Structure of the shock front propagating at an angle of the magnetic field in a low density plasma. Sov. Phys. Tech. Phys. Engl. Transl. **8**, 715 (1964)

V.I. Karpman, J.K. Alekhin, N.D. Borisov, N.A. Rjabova, Electrostatic waves with frequencies exceeding gyrofrequency in magnetosphere. Astrophys. Space Sci. **22**(2), 267–278 (1973). doi:10.1007/BF00647426

C.F. Kennel, J.P. Edmiston, T. Hada, A quarter century of collisionless shock research, in *Collisionless Shocks in the Heliosphere: A Tutorial Review*, ed. by R.G. Stone, B.T. Tsurutani Geophysical Monograph, vol. 34 (American Geophysical Union, Washington, 1985), pp. 1–36

S. Klimov, S. Savin, Y. Aleksevich, G. Avanesova, V. Balebanov, M. Balikhin, A. Galeev, B. Gribov, M. Nozdrachev, V. Smirnov, A. Sokolov, O. Vaisberg, P. Oberc, Z. Krawczyk, S. Grzedzielski, J. Juchniewicz, K. Nowak, D. Orlowski, B. Parfianovich, D. Wozniak, Z. Zbyszynski, Y. Voita, P. Triska, Extremely-low-frequency plasma waves in the environment of comet Halley. Nature **321**, 292–293 (1986)

K. Koyama, R. Petre, E.V. Gotthelf, U. Hwang, M. Matsuura, M. Ozaki, S.S. Holt, Evidence for shock acceleration of high-energy electrons in the supernova Remnant sn1006. Nature **378**, 255–258 (1995). doi:10.1038/378255a0

V. Krasnoselskikh, Nonlinear motions of a plasma across a magnetic field. Sov. Phys. JETP **62**, 282–294 (1985)

V.V. Krasnoselskikh, E.N. Kruchina, A.S. Volokitin, G. Thejappa, Fast electron generation in quasiperpendicular shocks and type II solar radiobursts. Astron. Astrophys. **149**, 323–329 (1985)

V.V. Krasnoselskikh, M.A. Balikhin, H.S.C. Alleyne, S.I. Klimov, W.A.C. Mier-Jedrzejowicz, A.K. Pardaens, A. Petrukovich, D.J. Southwood, T. Vinogradova, L.J.C. Woolliscroft, On the nature of low frequency turbulence in the foot of strong quasi-perpendicular shocks. Adv. Space Res. **11**(9), 15–18 (1991)

V.V. Krasnoselskikh, B. Lembége, P. Savoini, V.V. Lobzin, Nonstationarity of strong collisionless quasiperpendicular shocks: theory and full particle numerical simulations. Phys. Plasmas **9**(4), 1192–1209 (2002). doi:10.1063/1.1457465

G.F. Krymskii, A regular mechanism for the acceleration of charged particles on the front of a shock wave. Dokl. Akad. Nauk SSSR **234**, 1306–1308 (1977)

R.H. Kurtmullaev, J.E. Nesterikhin, V.I. Pilsky, Shock waves in partially ionized plasma, in *Phenomena in Ionized Gases, VIII International Conference* (1967), p. 459

R.K. Kurtmullaev, V.K. Malinovskii, Y.E. Nesterikhin, A.G. Ponomarenko, Excitation of strong collisionless shock waves in a plasma. J. Appl. Mech. Tech. Phys. **6**, 68–73 (1965). doi:10.1007/BF00915616
C. Lacombe, A. Mangeney, C. Harvey, J. Scudder, Electron plasma waves upstream of the Earth's bow shock. J. Geophys. Res. **90**, 73–94 (1985). doi:10.1029/JA090iA01p00073
J.M. Laming, Accelerated electrons in Cassiopeia A: an explaination for the hard x-ray tail. Astrophys. J. **546**, 1149–1158 (2001)
B. Lefebvre, S.J. Schwartz, A.F. Fazakerley, P. Décréau, Electron dynamics and cross-shock potential at the quasi-perpendicular Earth's bow shock. J. Geophys. Res. **112**, 09212 (2007). doi:10.1029/2007JA012277
B. Lembège, S.N. Walker, P. Savoini, M.A. Balikhin, V. Krasnoselskikh, The spatial sizes of electric and magnetic field gradients in a simulated shock. Adv. Space Res. **24**, 109–112 (1999)
B. Lembège, J. Giacalone, M. Scholer, T. Hada, M. Hoshino, V. Krasnoselskikh, H. Kucharek, P. Savoini, T. Terasawa, Selected problems in collisionless-shock physics. Space Sci. Rev. **110**, 161–226 (2004). doi:10.1023/B:SPAC.0000023372.12232.b7
M.M. Leroy, A. Mangeney, A theory of energization of solar wind electrons by the Earth's bow shock. Ann. Geophys. **2**, 449–456 (1984)
M.M. Leroy, D. Winske, C.C. Goodrich, C.S. Wu, K. Papadopoulos, The structure of perpendicular bow shocks. J. Geophys. Res. **87**(A7), 5081–5094 (1982). doi:10.1029/JA087iA07p05081
P.C. Liewer, V.K. Decyk, J.M. Dawson, B. Lembège, Numerical studies of electron dynamics in oblique quasi-perpendicular collisionless shock waves. J. Geophys. Res., Atmos. **96**(A6), 9455 (1991)
W.A. Livesey, C.T. Russell, C.F. Kennel, A comparison of specularly reflected gyrating ion orbits with observed shock foot thicknesses. J. Geophys. Res., Atmos. **89**(A8), 6824–6828 (1984)
V.V. Lobzin, V.V. Krasnoselskikh, S.J. Schwartz, I. Cairns, B. Lefebvre, P. Décréau, A. Fazakerley, Generation of downshifted oscillations in the electron foreshock: a loss-cone instability. Geophys. Res. Lett. **32**, 18101 (2005). doi:10.1029/2005GL023563
V.V. Lobzin, V.V. Krasnoselskikh, J.-M. Bosqued, J.-L. Pinçon, S.J. Schwartz, M. Dunlop, Nonstationarity and reformation of high-Mach-number quasiperpendicular shocks: cluster observations. Geophys. Res. Lett. **34**, 5107 (2007). doi:10.1029/2006GL029095
E. Mach, J. Arbes, Einige versuche über totale Reflexion und anomale Dispersion. Ann. Phys. **263**, 436–444 (1886). doi:10.1002/andp.18862630307
E. Mach, P. Salcher, Photographische fixirung der durch Projectile in der Luft eingeleiteten Vorgänge. Ann. Phys. **268**, 277–291 (1887). doi:10.1002/andp.18872681008
S. Matsukiyo, M. Scholer, On microinstabilities in the foot of high Mach number perpendicular shocks. J. Geophys. Res., Space Phys. **111**(A6), 6104 (2006). doi:10.1029/2005JA011409
C. Mazelle, B. Lembége, A. Morgenthaler, K. Meziane, T.S. Horbury, V. Génot, E.A. Lucek, I. Dandouras, Self-reformation of the quasi-perpendicular shock: cluster observations, in *Twelfth International Solar Wind Conference*, vol. 1216 (2010), pp. 471–474. doi:10.1063/1.3395905
J.D. Means, Use of the three-dimensional covariance matrix in analyzing the polarization properties of plane waves. J. Geophys. Res. **77**, 5551–5559 (1972)
M.M. Mellott, Subcritical collisionless shock waves, in *Washington DC American Geophysical Union Geophysical Monograph Series*, vol. 35 (1985), pp. 131–140
J.J. Mitchell, S.J. Schwartz, U. Auster, Electron cross talk and asymmetric electron distributions near the Earth's bow shock. Ann. Geophys. **30**, 503–513 (2012). doi:10.5194/angeo-30-503-2012
D.L. Morse, W.W. Destler, P.L. Auer, Nonstationary behavior of collisionless shocks. Phys. Rev. Lett. **28**, 13–16 (1972). doi:10.1103/PhysRevLett.28.13
N.F. Ness, K.W. Behannon, R.P. Lepping, Y.C. Whang, K.H. Schatten, Magnetic field observations near venus: preliminary results from Mariner 10. Science **183**, 1301–1306 (1974). doi:10.1126/science.183.4131.1301
N.F. Ness, M.H. Acuna, R.P. Lepping, J.E.P. Connerney, K.W. Behannon, L.F. Burlaga, F.M. Neubauer, Magnetic field studies by Voyager 1—preliminary results at saturn. Science **212**, 211–217 (1981). doi:10.1126/science.212.4491.211
J.A. Newbury, C.T. Russell, Observations of a very thin collisionless shock. Geophys. Res. Lett. **23**, 781 (1996). doi:10.1029/96GL00700
J.A. Newbury, C.T. Russell, M. Gedalin, The ramp widths of high-Mach-number, quasi-perpendicular collisionless shocks. J. Geophys. Res., Atmos. **103**(A12), 29581–29593 (1998). doi:10.1029/1998JA900024
M. Oka, T. Terasawa, Y. Seki, M. Fujimoto, Y. Kasaba, H. Kojima, I. Shinohara, H. Matsui, H. Matsumoto, Y. Saito, T. Mukai, Whistler critical Mach number and electron acceleration at the bow shock: geotail observation. Geophys. Res. Lett. **33**, 24104 (2006). doi:10.1029/2006GL028156
K. Papadopoulos, Electron acceleration in magnetosonic shock fronts. Technical report, 1981
K. Papadopoulos, Microinstabilities and anomalous transport in collisionless shocks, in *Advances in Space Plasma Physics*, ed. by W. Grossmann, E.M. Campbell, B. Buti (1985a), p. 289

K. Papadopoulos, Microinstabilities and anomolous transport, in *Collisionless Shocks in the Heliosphere: A Tutorial Review*, ed. by R.G. Stone, B.T. Tsurutani Geophysical Monograph, vol. 34 (American Geophysical Union, Washington, 1985b), pp. 59–90
G. Paschmann, P.W. Daly, *Analysis Methods for Multi-spacecraft Data.* ISSI Scientific Reports Series sr-001, esa/issi, vol. 1 (1998). ISSN:1608-280x
J.W.M. Paul, L.S. Holmes, M.J. Parkinson, J. Sheffield, Experimental observations on the structure of collisionless shock waves in a magnetized plasma. Nature **208**, 133–135 (1965). doi:10.1038/208133a0
J.W.M. Paul, G.C. Goldenbaum, A. Iiyoshi, L.S. Holmes, R.A. Hardcastle, Measurement of electron temperatures produced by collisionless shock waves in a magnetized plasma. Nature **216**, 363–364 (1967). doi:10.1038/216363a0
J.-L. Pinçon, K.-H. Glassmeier, Multi-spacecraft methods of wave field characterisation, in *ISSI Scientific Reports Series*, vol. 8 (2008), pp. 47–54
J.-L. Pinçon, P.M. Kintner, P.W. Schuck, C.E. Seyler, Observation and analysis of lower hybrid solitary structures as rotating eigenmodes. J. Geophys. Res., Atmos. **102**, 17283–17296 (1997)
T. Piran, Magnetic fields in gamma-ray bursts: a short overview. in *American Institute of Physics Conference Series*, vol. 784, ed. by E.M. de Gouveia dal Pino, G. Lugones, A. Lazarian, 2005, pp. 164–174. doi:10.1063/1.2077181
K.B. Quest, Simulations of high mach number perpendicular shocks with resistive electrons. J. Geophys. Res., Atmos. **91**(A8), 8805–8815 (1986)
H. Reme, J.M. Bosqued, J.A. Sauvaud, A. Cros, J. Dandouras, C. Aoustin, J. Bouyssou, T. Camus, J. Cuvilo, C. Martz, J.L. Médale, H. Perrier, D. Romefort, J. Rouzaud, D. D'Uston, E. Möbius, K. Crocker, M. Granoff, L.M. Kistler, M. Popecki, D. Hovestadt, B. Klecker, G. Paschmann, M. Scholer, C.W. Carlson, D.W. Curtis, R.P. Lin, J.P. McFadden, V. Formisano, E. Amata, M.B. Bavassano-Cattaneo, P. Baldetti, G. Belluci, R. Bruno, G. Chionchio, A. di Lellis, E.G. Shelley, A.G. Ghielmetti, W. Lennartsson, A. Korth, U. Rosenbauer, R. Lundin, S. Olsen, G.K. Parks, M. McCarthy, H. Balsiger, The cluster ion spectrometry (cis) experiment. Space Sci. Rev. **79**, 303–350 (1997)
C.T. Russell, On the relative locations of the bow shocks of the terrestrial planets. Geophys. Res. Lett. **4**, 387–390 (1977). doi:10.1029/GL004i010p00387
C.T. Russell, Planetary bow shocks, in *Washington DC American Geophysical Union Geophysical Monograph Series*, vol. 35 (1985), pp. 109–130
C.T. Russell, M.M. Mellot, E.J. Smith, J.H. King, Multiple spacecraft observations on interplanetary shocks: four spacecraft determination of shock normals. J. Geophys. Res., Atmos. **88**, 4739–4748 (1983)
R.Z. Sagdeev, Asymptotic methods in the hydrodynamic theory of stability, in *Lectures Presented at the Trieste Seminar on Plasma Physics* (1965a), p. 625
R.Z. Sagdeev, Landau damping and finite resistivity instability in plasmas, in *Lectures Presented at the Trieste Seminar on Plasma Physics* (1965b), p. 555
R.Z. Sagdeev, Cooperative phenomena and shock waves in collisionless plasmas. Rev. Plasma Phys. **4**, 23 (1966)
R.Z. Sagdeyev, "Shock" waves in rarefied plasma, in *Ionization Phenomena in Gases*, vol. II, ed. by N.R. Nilsson (1960), p. 1081
J.C. Samson, J.V. Olson, Some comments on the descriptions of the polarization states of waves. Geophys. J. R. Astron. Soc. **61**, 115–129 (1980). doi:10.1111/j.1365-246X.1980.tb04308.x
M. Scholer, D. Burgess, Transition scale at quasiperpendicular collisionless shocks: full particle electromagnetic simulations. Phys. Plasmas **13**(6), 062101 (2006). doi:10.1063/1.2207126
M. Scholer, S. Matsukiyo, Nonstationarity of quasi-perpendicular shocks: a comparison of full particle simulations with different ion to electron mass ratio. Ann. Geophys. **22**, 2345–2353 (2004)
M. Scholer, I. Shinohara, S. Matsukiyo, Quasi-perpendicular shocks: length scale of the cross-shock potential, shock reformation, and implication for shock surfing. J. Geophys. Res., Space Phys. **108**, 1014 (2003). doi:10.1029/2002JA009515
S.J. Schwartz, Shock and discontinuity normals, mach numbers, and related parameters, in *ISSI Scientific Reports Series*, vol. 1 (1998), pp. 249–270
S.J. Schwartz, M.F. Thomsen, J.T. Gosling, Ions upstream of the Earth's bow shock—a theoretical comparison of alternative source populations. J. Geophys. Res. **88**, 2039–2047 (1983). doi:10.1029/JA088iA03p02039
S.J. Schwartz, M.F. Thomsen, S.J. Bame, J. Stansberry, Electron heating and the potential jump across fast mode shocks. J. Geophys. Res. **93**(A11), 12923–12931 (1988). doi:10.1029/JA093iA11p12923
S.J. Schwartz, E. Henley, J. Mitchell, V. Krasnoselskikh, Electron temperature gradient scale at collisionless shocks. Phys. Rev. Lett. **107**(21), 215002 (2011). doi:10.1103/PhysRevLett.107.215002
N. Sckopke, G. Paschmann, S.J. Bame, J.T. Gosling, C.T. Russell, Evolution of ion distributions across the nearly perpendicular bow shock—specularly and non-specularly reflected-gyrating ions. J. Geophys. Res., Atmos. **88**(A8), 6121–6136 (1983)

J.D. Scudder, A review of the physics of electron heating at collisionless shocks. Adv. Space Res. **15**, 181 (1995)

J.D. Scudder, A. Mangeney, C. Lacombe, C.C. Harvey, T.L. Aggson, R.R. Anderson, J.T. Gosling, G. Paschmann, C.T. Russell, The resolved layer of a collisionless high β supercritical quasi-perpendicular shock wave, 1: Rankine–Hugoniot geometry currents and stationarity. J. Geophys. Res., Atmos. **91**, 11019–11052 (1986a)

J.D. Scudder, A. Mangeney, C. Lacombe, C.C. Harvey, T.L. Aggson, The resolved layer of a collisionless high β supercritical quasi-perpendicular shock wave, 2: dissipative fluid electrodynamics. J. Geophys. Res., Atmos. **91**, 11053 (1986b)

J.D. Scudder, A. Mangeney, C. Lacombe, C.C. Harvey, C.S. Wu, R.R. Anderson, The resolved layer of a collisionless high β supercritical quasi-perpendicular shock wave, 3: Vlasov electrodynamics. J. Geophys. Res., Atmos. **91**, 11075 (1986c)

V. Smirnov, O. Vaisberg, Evidence of the nonlinear structure at the bow shock front, in *Collisionless Shocks*, Budapest, ed. by K. Szego (1987), pp. 70–76

D. Sundkvist, V. Krasnoselskikh, S.D. Bale, S.J. Schwartz, J. Soucek, F. Mozer, Dispersive nature of high mach number collisionless plasma shocks: Poynting flux of oblique whistler waves. Phys. Rev. Lett. **108**(2), 025002 (2012). doi:10.1103/PhysRevLett.108.025002

M.F. Thomsen, J.T. Gosling, S.J. Bame, K.B. Quest, D. Winske, On the noncoplanarity of the magnetic field within a fast collisionless shock. J. Geophys. Res., Atmos. **92**(A3), 2305–2314 (1987)

D.A. Tidman, N.A. Krall, *Shockwaves in Collisionless Plasmas* (Wiley, New York, 1971)

A. Tjulin, A.I. Eriksson, M. André, Lower hybrid cavities in the inner magnetosphere. Geophys. Res. Lett. **30**, 17–21 (2003). doi:10.1029/2003GL016915

O.L. Vaisberg, A.A. Galeev, G.N. Zastenker, S.I. Klimov, M.N. Nozdrachev, R.Z. Sagdeev, A.I. Sokolov, V.D. Shapiro, Electron acceleration at the front of strong collisionless shock waves. Zh. Èksp. Teor. Fiz. **85**, 1232–1243 (1983)

O. Vaisberg, S. Klimov, G. Zastenker, M. Nozdrachev, A. Sokolov, V. Smirnov, S. Savin, L. Avanov, Relaxation of plasma at the shock front. Adv. Space Res. **4**, 265–275 (1984). doi:10.1016/0273-1177(84)90320-X

P. Veltri, G. Zimbardo, Electron whistler interaction at the Earths bow shock, 2: electron pitch-angle diffusion. J. Geophys. Res. **98**(A8), 13335–13346 (1993)

S.N. Walker, M.A. Balikhin, M.N. Nozdrachev, Ramp nonstationarity and the generation of whistler waves upstream of a strong quasiperpendicular shock. Geophys. Res. Lett. **26**(10), 1357–1360 (1999a). doi:10.1029/1999GL900210

S.N. Walker, M.A. Balikhin, H.S.C.K. Alleyne, W. Baumjohann, M. Dunlop, Observations of a very thin shock. Adv. Space Res. **24**, 47–50 (1999b)

S.N. Walker, M.A. Balikhin, H.S.C.K. Alleyne, Y. Hobara, M. André, M.W. Dunlop, Lower hybrid waves at the shock front: a reassessment. Ann. Geophys. **26**, 699–707 (2008). http://www.ann-geophys.net/26/699/2008/

S. Walker, H. Alleyne, M. Balikhin, M. André, T. Horbury, Electric field scales at quasi-perpendicular shocks. Ann. Geophys. **22**(7), 2291–2300 (2004). http://www.ann-geophys.net/22/2291/2004/

L.B. Wilson III, A. Koval, A. Szabo, A. Breneman, C.A. Cattell, K. Goetz, P.J. Kellogg, K. Kersten, J.C. Kasper, B.A. Maruca, M. Pulupa, Observations of electromagnetic whistler precursors at supercritical interplanetary shocks. Geophys. Res. Lett. **39**, 8109 (2012). doi:10.1029/2012GL051581

L.C. Woods, On the structure of collisionless magneto-plasma shock waves at super-critical Alfvèn-Mach numbers. J. Plasma Phys. **3**(3), 435–447 (1969)

L.C. Woods, On double-structured, perpendicular, magneto-plasma shock waves. Plasma Phys. **13**, 289–302 (1971). doi:10.1088/0032-1028/13/4/302

L.J.C. Woolliscroft, H.S.C. Alleyne, C.M. Dunford, A. Sumner, J.A. Thompson, S.N. Walker, K.H. Yearby, A. Buckley, S. Chapman, M.P. Gough, The digital wave-processing experiment on cluster. Space Sci. Rev. **79**, 209–231 (1997)

C.S. Wu, A fast Fermi process—energetic electrons accelerated by a nearly perpendicular bow shock. J. Geophys. Res. **89**, 8857–8862 (1984). doi:10.1029/JA089iA10p08857

J.R. Wygant, M. Bensadoun, F.S. Mozer, Electric field measurements at subcritical, oblique bow shock crossings. J. Geophys. Res., Atmos. **92**(A10), 11109–11121 (1987)

G. Zank, H. Pauls, I. Cairns, G. Webb, Interstellar pickup ions and quasi-perpendicular shocks: implications for the termination shock and interplanetary shocks. J. Geophys. Res. **101**, 457–477 (1996)

DOI 10.1007/978-1-4899-7413-6_19
Reprinted from *Space Science Reviews* Journal, DOI 10.1007/s11214-013-9984-7

Collisionless Shocks in Partly Ionized Plasma with Cosmic Rays: Microphysics of Non-thermal Components

A.M. Bykov · M.A. Malkov · J.C. Raymond · A.M. Krassilchtchikov · A.E. Vladimirov

Received: 2 January 2013 / Accepted: 3 April 2013 / Published online: 9 May 2013

Abstract In this review we discuss some observational aspects and theoretical models of astrophysical collisionless shocks in partly ionized plasma with the presence of non-thermal components. A specific feature of fast strong collisionless shocks is their ability to accelerate energetic particles that can modify the shock upstream flow and form the shock precursors. We discuss the effects of energetic particle acceleration and associated magnetic field amplification and decay in the extended shock precursors on the line and continuum multi-wavelength emission spectra of the shocks. Both Balmer-type and radiative astrophysical shocks are discussed in connection to supernova remnants interacting with partially neutral clouds. Quantitative models described in the review predict a number of observable line-like emission features that can be used to reveal the physical state of the matter in the shock precursors and the character of nonthermal processes in the shocks. Implications of recent progress of gamma-ray observations of supernova remnants in molecular clouds are highlighted.

Keywords Collisionless shocks · Radiative shocks · Supernova remnants · Gamma-rays

A.M. Bykov (✉) · A.M. Krassilchtchikov
A.F. Ioffe Institute for Physics and Technology, 194021 St. Petersburg, Russia
e-mail: byk@astro.ioffe.ru

A.M. Krassilchtchikov
e-mail: kra@astro.ioffe.ru

M.A. Malkov
University of California, San Diego, La Jolla, CA 92093, USA
e-mail: mmalkov@ucsd.edu

J.C. Raymond
Center for Astrophysics, 60 Garden St, Cambridge, MA 02138, USA
e-mail: jraymond@cfa.harvard.edu

A.E. Vladimirov
Stanford University, Stanford, CA 94305, USA
e-mail: avladim@stanford.edu

1 Introduction

Plasma flows with collisionless shocks are found in a number of energetic space objects, such as supernova remnants (SNRs) interacting with atomic or molecular clouds (McKee and Hollenbach 1980; Draine and McKee 1993; Heng 2010), Herbig-Haro objects (Hartigan et al. 1987; Giannini et al. 2008; Teşileanu et al. 2009), winds from protostars and young stellar objects (Massi et al. 2008), recurrent novae (Evans et al. 2007; Bode et al. 2007; Tatischeff and Hernanz 2007), and also in accretion flows in the vicinities of galactic nuclei (Farage et al. 2010).

Already in the early studies of Raymond (1979) and Shull and McKee (1979) the ionization state of the pre-shock flow was recognized as an important feedback parameter that influences the dynamics of the flow, and hence, the spectrum of continuum and line emission produced in the post-shock zone. While Raymond (1979) took the pre-shock ionization state as a free parameter, Shull and McKee (1979) included the ionizing flux from the post-shock to obtain a pre-shock ionization state self-consistently with the temperature and ionization profiles in the post-shock. Dopita and Sutherland (1996) furthered the calculation of the precursor photoionization, and presented emission line spectra assuming an equilibrium with the ionizing flux from the shock.

Recent observations of forward shocks in galactic SNRs indicate a substantial role of nonthermal components in the energy budgets of post-shock flows (e.g., Helder et al. 2012). While the fraction of the energy dissipated in the shock itself may be smaller for slower radiative shocks, the enhanced radiative cooling in their post-shocks leads to stronger compression so that cosmic ray and magnetic pressure can dominate over thermal pressure in the zones where most of the observed optical and IR emission is produced. There is also a growing consensus on that in both parallel and perpendicular shocks the diffusing cosmic rays may generate unstable plasma currents and turbulence in the pre-shock zone, thus allowing for efficient particle acceleration, shock modification, and substantial enhancement of the total compression ratio (Blandford and Eichler 1987; Jones and Ellison 1991; Malkov and Drury 2001; Bell 2004; Bykov et al. 2012). Raymond et al. (1988b) analyzed a shock in the Cygnus Loop SNR and found that the nonthermal pressure exceeds the thermal pressure by an order of magnitude in the zones where the [S II] lines are formed, though they were unable to distinguish between magnetic and cosmic ray contributions. Neutral particles could affect the processes of particle acceleration (e.g., Draine and McKee 1993; O'C Drury et al. 1996; Malkov et al. 2005; Blasi et al. 2012; Inoue et al. 2012; Ohira 2012; Helder et al. 2012; Morlino et al. 2012a), magnetic field amplification and plasma heating in the upstream region (e.g., Bykov and Toptygin 2005; Reville et al. 2007).

In this paper we review physical mechanisms governing evolution of supersonic and superalfvenic flows with collisionless shocks in partially ionized plasmas with non-thermal components. Section 2 describes the photoionization precursors that help determine the neutral fraction of the pre-shock gas. In Sect. 3 Hα signatures of interaction of neutral particles with fast shocks are depicted. In Sect. 4 we discuss specific plasma heating processes due to dissipation of CR-driven magnetic fluctuations in the shock precursor. A consistent pre-shock ionization structure is given in Sect. 5 with an aim to emphasize processes that critically influence optical and IR spectra emitted from the flows. The escape of accelerated energetic particles may affect the gas compression in the shock upstream and plasma temperature in the downstream resulting in formation of radiative shocks of velocities that may well exceed that of the adiabatic shocks. We discuss the line spectra of radiative shocks modified by CR acceleration effect in Sect. 5. Specific features of CR acceleration in supernova remnants interacting with molecular clouds are discussed in Sect. 6 and their gamma-ray emission spectra are outlined in Sect. 7. A brief summary is given in Sect. 8.

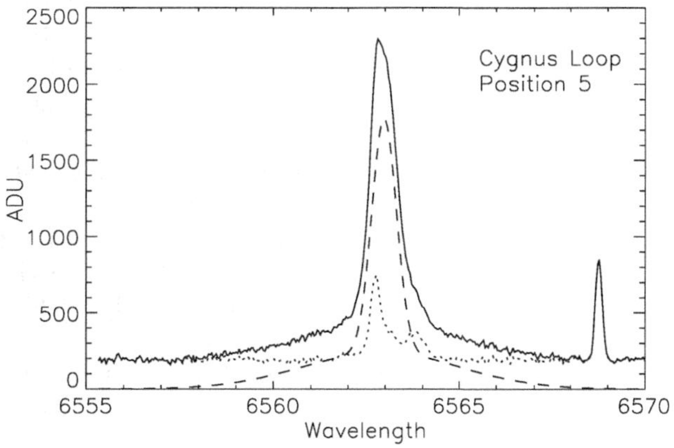

Fig. 1 Hα profile of a 350 km/s shock at the outer edge of the Cygnus Loop, obtained with the HECTOCHELLE spectrograph on the MMT. The narrow component is 30 km/s wide (FWHM), and the broad component is 225 km/s wide. *Full line* is the observation, *dashed line* is the Hα from the filament, and *dotted line* is the sky background. The feature near 6569 Å is an airglow line

2 Photoionization Precursors

The neutral fraction in the pre-shock gas often depends on photoionization by radiation from the shock itself. There are two regimes, non-radiative shocks are those in which gas is heated at the shock but does not have time to radiatively cool, while radiative shocks radiatively cool from the post-shock temperature, efficiently converting their thermal energy into radiation. Naturally, the latter produce more ionizing photons if the post-shock temperature exceeds about 150,000 K. The scale of the photoionization precursor is given by the photoabsorption cross section of hydrogen and the neutral density in the pre-shock gas, so it is typically 10^{18} or 10^{19} cm. In supernova remnants, shocks faster than 300 km/s are typically nonradiative, while slower shocks are generally radiative.

Non-radiative shocks produce some ionizing photons because some of the atoms passing through the shock are excited before they are ionized in the hot post-shock plasma. In particular, at high temperatures each He atom produces a few photons in the He I 58.4 nm and He II 30.4 nm lines. Since He makes up about 10 % of the gas, the pre-shock region will be by 30 % to 40 % ionized by these photons. EUV emission from the hot post-shock gas can further ionize the plasma. The photons are relatively energetic, so they heat the gas to around 16,000 K. Emission from such precursors outside Tycho's SNR and N132D were reported by Ghavamian et al. (2000) and Morse et al. (1996).

As mentioned above, and as described more fully in Sect. 5, radiative shocks faster than about 100 km/s produce large fluxes of ionizing photons, and those above about 120 km/s produce enough photons to fully ionize H in the pre-shock gas. However, it should be kept in mind that the equilibrium between pre-shock ionization state and ionizing flux from the shock, as assumed in Shull and McKee (1979) and Dopita and Sutherland (1996), does not always hold. In particular, an SNR shock can evolve more rapidly than the recombination time in the pre-shock gas, or the transverse scale of the shock may be smaller than the upstream photon mean free path, as it is in a Herbig-Haro object (Raymond et al. 1988a).

3 Interaction of Neutrals with Fast Shocks: Hα Signatures

Neutral H atoms are not affected by the electromagnetic fields or plasma turbulence in a collisionless shock. They pass freely through the shock transition, whose thickness is given by the ion inertial length or the proton gyroradius, which are typically $\sim 10^8$ cm in the interstellar medium. However, they become ionized in a hot downstream plasma but, because of the low ISM densities, they travel $\sim 10^{14}$–10^{15} cm downstream before it happens. This is smaller than electron-proton thermal equilibration scales, and therefore, the emission from neutrals may contain information about the post-shock conditions of electrons and ions before relaxation.

The hydrogen atoms are ionized either by collisions with protons and electrons or by a charge exchange with protons. The latter process produces a population of neutrals with a velocity distribution similar to that of the post-shock protons and, when they are excited, they produce broad Lyman and Balmer emission lines whose velocity widths are comparable to the shock speed (Chevalier and Raymond 1978; Raymond 1991; Heng 2010; France et al. 2011). Neutrals that are excited before the charge transfer produce a narrow component whose width is given by the pre-shock temperature (see Fig. 1). The broad-to-narrow intensity ratio indicates the electron temperature, while the broad component width directly measures the proton temperature. Thus one can determine the electron-ion temperature ratio at the shock (Ghavamian and Schwarz this volume). Consequently, the Hα profile provides both the post-shock temperature and T_e/T_p ratio. These quantities can be used to determine the shock speed (Smith et al. 1994; Hester et al. 1994; Ghavamian et al. 2001; van Adelsberg et al. 2008). Under the assumption that only a small fraction of the shock energy goes into cosmic rays, the shock speed can be determined from the post-shock proton and electron temperature using momentum and energy conservation laws. However, if cosmic ray acceleration is efficient, then T_e and T_p combined with a model of efficient cosmic ray acceleration can yield a shock velocity (Helder et al. 2009) though there is some ambiguity due to a contribution to the narrow component from a shock precursor (see Sect. 3.2).

The Balmer line profiles can also indicate more exotic plasma processes. First, if the post-shock proton velocity distribution is not Maxwellian, it will leave an imprint on the broad component line profile. Second, the strength and width of the narrow component can indicate heating and excitation in a shock precursor. We discuss signatures of these departures from a simple shock picture.

3.1 Non-Maxwellian Velocity Distributions

If cosmic rays are accelerated in SNR shocks, the particle velocity distribution is manifestly non-Maxwellian, taking the form of a Maxwellian core with a power law tail. Such distributions are conveniently parameterized with the κ function (Pierrard and Lazar 2010). The κ distribution approaches a Maxwellian as κ approaches infinity, while the power law tail becomes harder and includes a larger fraction of the particles as κ approaches its lower limit of 1.5.

A different sort of departure from a Maxwellian can arise because of the neutral atoms themselves. When a neutral becomes ionized, it is moving at 3/4 the shock speed with respect to the downstream plasma and magnetic field (assuming a shock compression factor of 4). It therefore behaves like a pickup ion in the solar wind (Gloeckler and Geiss 2001). Initially, all these newly-formed protons have the same velocity with respect to plasma in which they are immersed. This velocity separates into a gyration speed around the magnetic field plus a component along the field, and since all the particles have the same initial velocity (assuming that the shock speed is large compared to the pre-shock thermal speed), they form a ring beam in velocity space. That distribution is unstable, and it rapidly evolves into a bispherical shell in velocity space, a lens-like shape that depends on the ratio of the Alfvén speed to the initial particle speed (Williams and Zank 1994). The protons in the bispherical shell can then experience charge transfer to produce observable neutrals. The process of ionization, interaction with the magnetic field, and subsequent re-neutralization is the same sequence of events that produce the "IBEX ribbon" seen in H atoms from beyond the heliopause (McComas et al. 2009).

The Hα profiles produced by the pickup ion process were computed by Raymond et al. (2008). The neutral fraction in the pre-shock gas must be substantial in order to produce the Balmer line filaments seen at the edges of some supernova remnants and pulsar-wind nebulae, and therefore the neutrals can have a significant effect on the shock structure. Heng et al. (2007) computed the gradual transition over the charge transfer length scale, but found little effect on the overall dynamics. Ohira et al. (2009) examined the streaming instabilities created by the relative motion of the post-shock plasma and the protons created downstream (this effect is stronger in parallel shocks) and predicted substantial amplification of the magnetic field. Raymond et al. (2008) calculated the wave energy produced as the ring beam relaxes to a bispherical distribution, and showed that it could significantly affect the electron temperature if the waves couple efficiently to electrons. Ohira and Takahara (2010) considered the modification of the shock structure, in particular weakening of the subshock, and the effects on the cosmic ray spectrum.

The observational evidence for any departure of the proton distribution from a Maxwellian is still ambiguous. For nearly all the observed Hα broad components Gaussian fits are adequate, but in general the lines are so faint that even the determination of the line width is subject to large statistical uncertainty. A very deep exposure of a bright knot in Tycho's SNR yielded the only profile that shows a clear non-Gaussian broad component, but there are several possible interpretations (Raymond et al. 2010). The departure could be attributed to a power law tail, however, the slope of this tail is very hard, which indicates extremely efficient particle acceleration. The profile could also be matched by the combination of a pickup ion component with an ordinary Maxwellian component, though this requires a fairly high pre-shock neutral fraction. A third possibility is that some of the neutrals in the shock precursor acquire a kinetic temperature about $1/2$ of the post-shock temperature. Clearly, this implies a precursor to be both hot and thick. Finally, one could produce the observed profile by adding contributions from separate shocks along the line of sight. However, this possibility would require quite a large velocity difference, and therefore a significant density contrast between the two regions. At higher shock speeds, the velocity dependence of the charge transfer cross section can distort the profile (Heng and McCray 2007; Heng et al. 2007), but that is unlikely to account for the Tycho observations.

3.2 Shock Precursors

The precursor of a collisionless shock wave is a region upstream of the shock transition in which the plasma conditions (velocity, density, temperature, magnetic fields, ionization state) are affected by photons or superthermal particles streaming ahead of the shock front.

Shock wave precursors can be produced by ionizing photons emerging from the downstream region or by broad component neutrals that leak back through the shock. Additionally, in shocks that produce accelerated super-thermal charged particles, precursors can be produced as a consequence of this acceleration process. We first consider the observational evidence for precursors, then the physics of the three mechanisms and the implications for system parameters.

The first evidence for precursors to SNR shocks came from the observation that the widths of the Hα narrow components in several shocks were unexpectedly large. Smith et al. (1994) measured narrow component widths of 25–58 km/s in four LMC Balmer-dominated remnants, and Hester et al. (1994) found a 28–35 km/s width in a Cygnus Loop shock. These line widths correspond to temperatures of about (2–7) $\times 10^4$ K, and in static equilibrium, hydrogen is fully ionized at those temperatures. Thus Smith et al. (1994) and Hester et al. (1994) concluded that the gas must be heated in a narrow region ahead of the shock. The region must be thick enough so that charge transfer can heat the neutrals, but thin enough that the neutrals do not become ionized. Subsequent observations have revealed line widths of 44 km/s in Tycho's SNR (Ghavamian et al. 2000) and 30–42 km/s in RCW 86 and Kepler's SNR, with only SN 1006 showing a narrow component width compatible with a temperature of 10,000 K and a significant neutral fraction (Sollerman et al. 2003). Nikolić et al. (2013) have presented Integral Field Unit spectra of a section of the Hα filament in SN 1006, and they have found that a precursor makes a substantial contribution to the narrow component of Hα.

The precursor thicknesses must be about $1''$ in order for the neutrals to be heated, and they have been spatially resolved in the Cygnus Loop (Fesen and Itoh 1985; Hester et al. 1994) and in Tycho's SNR (Lee et al. 2007, 2010).

3.3 Physical Interpretation of Shock Precursors

Three ideas have been put forward to explain observed shock precursors, a photoionization precursor, a cosmic ray precursor, or a precursor created by fast neutrals leaking from the post-shock region out ahead of the shock.

The photoionization precursor heats the electrons, and it has a very large length scale. In non-radiative shocks, each He atom passing through the shock produces about 2 He II λ304 photons and 1 to 2 He I λ584 photons. These photons can ionize about 30 % of the pre-shock hydrogen, and because the photons are relatively energetic they deposit considerable heat in the electrons. Such a precursor has been reported ahead of Tycho's supernova remnant (Ghavamian et al. 2000). However, because they heat the electrons and extend over large length scales, photoionization precursors do not explain the observations discussed in Sect. 3.2.

A precursor is an integral part of diffusive shock acceleration models, in which an MHD turbulence scatters energetic particles moving away from the shock back towards it (Blandford and Eichler 1987). Its thickness is given by κ_{CR}/V_s, where κ_{CR} is the cosmic ray diffusion coefficient. While κ_{CR} is of order 10^{28} cm^2/s in the ISM, it must be somewhere closer to the Bohm limit near the shock in order for particles to reach high energies, perhaps 10^{24} cm^2/s. That implies a precursor thickness of order 10^{16} cm, or about $1''$ for nearby SNRs. The heating in such a precursor can occur by dissipation of the turbulence itself. It will be especially strong if neutrals are present to provide ion-neutral wave damping, which can limit the wave intensity and therefore the highest particle energies reached (Draine and McKee 1993; O'C Drury et al. 1996). The heating will also be strongest in shocks that accelerate cosmic rays efficiently. If the precursor heating is strong, it can reduce the Mach number of the subshock, changing the cosmic ray spectrum and pressure (see, e.g., Vladimirov et al. 2008; Wagner et al. 2009; Vink et al. 2010).

Several theoretical models have considered the effects of neutral-ion collisions on the dynamics and temperature of the precursor. Boulares and Cox (1988) computed the heating and ionization in cosmic ray precursors for the modest shock speeds in the Cygnus Loop. Ohira and Takahara (2010) considered neutrals interacting in the precursor, treating them as pickup ions. The pickup ions can strongly heat the plasma and affect the jump conditions at the subshock, and because of their high velocities they are preferentially injected into the diffusive acceleration process. Raymond et al. (2011) computed the temperature and density structures of precursors and the resulting Hα emission. If cosmic ray acceleration is efficient, the Hα profiles are strongly modified, both in broadening the narrow component and decreasing the broad-to-narrow intensity ratio. The Hα profiles reported so far are consistent with moderate ratios of cosmic ray pressure to gas pressure, but not with ratios above about 0.4. Most recently, Morlino et al. (2012b) have included a parameterized form of wave damping along with wave growth due to the cosmic ray pressure gradient for a more self-consistent treatment of the cosmic ray spectrum in the presence of neutrals and the resulting Hα profiles. They also find that an intermediate velocity width component can arise, similar to the complex profiles predicted in Raymond et al. (2011).

The other suggested mechanism for producing a precursor is leakage of some of the broad component neutrals out through the shock front. These neutrals have the thermal speeds of post-shock protons and a bulk speed of $V_S/4$ away from the shock. A significant fraction of them, of order 10 %, can overtake the shock, and of course they pass through (Smith et al. 1994; Hester et al. 1994). They carry a substantial amount of energy, but it is unclear how much of that energy heats the precursor plasma. Lim and Raga (1996) found that they became ionized to form a beam of fast protons, but that the protons are swept back through the shock without transferring much energy to the bulk plasma. Ohira (2012) constructed four-fluid models and showed that the leaking particles decelerate the upstream flow and affect the subshock compression ratio. Blasi et al. (2012) computed the energy and momentum exchange between neutrals and ions on both sides of the subshock, using a procedure like that of Heng et al. (2007) to compute the neutral velocity distribution and the flux of neutrals from the post-shock region to the precursor. Then they used the resulting precursor structure to compute spectral slopes of accelerated particles. The authors accounted for neutrals ionization by protons only, though photo ionization processes can affect the ionization states for some shock parameter space. Morlino et al. (2012a) used the kinetic simulations of Blasi et al. (2012) to calculate the Hα emission produced in the precursor. Its intensity depends on the poorly constrained efficiency of electron-ion equilibration, since the heat is deposited in the ions, but the electrons excite the line. They find that the profiles can deviate strongly from Gaussian and that in some circumstances an intermediate width component can arise. They suggest that this could explain profiles observed in Tycho's SNR. Morlino et al. (2012b) have extended this work by including the cosmic rays along with the neutrals that escape upstream through the shock.

4 CR Precursor Heating and the Post-shock Temperature

Models of collisionless shocks with large sonic and Alfvénic Mach numbers ($\mathcal{M}_s \gg 1$ and Alfvénic $\mathcal{M}_a \gg 1$) show that, through the first-order Fermi acceleration mechanism, a small minority of particles could gain a disproportionate share of the energy and populate the high energy tail of particle distribution. The energetic particles can penetrate far into the shock upstream gas, to create an extended shock precursor illustrated in Fig. 2. The cold gas in the shock precursor is decelerated and pre-heated by fluctuating magnetic field dissipation on a scale that is about c/v_{sh} times larger than a mean free path of an energetic particle λ_*. Shocks in collisionless supersonic flows produce a complex multi-scale structure of the relaxation region with an extended precursor and sub-shock of a modest sonic Mach number $\mathcal{M}_{sub} \sim 3$.

This section discusses theoretical models of strong collisionless shocks and their implications for the observations of shock precursors discussed in Sect. 3.2.

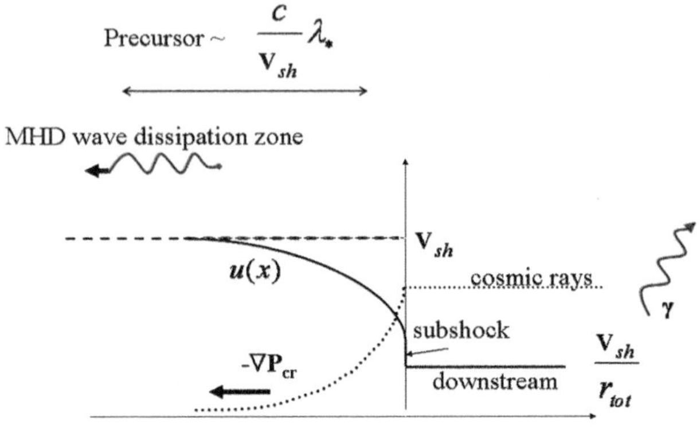

Fig. 2 A schematic view of a CR-modified shock propagating to the right

4.1 The Structure of a Non-linear Shock Precursor in the Presence of Self-generated MHD Turbulence

Consider a strong, plane-parallel collisionless shock in a plasma of finite $\beta = \mathcal{M}_a^2/\mathcal{M}_s^2$ (β is the ratio of plasma pressure to magnetic pressure). The distribution function of non-thermal particles and the bulk flow profile in the shock upstream region are sensitive to both the total upstream compression ratio r_{tot} and the subshock Mach number $\mathcal{M}_{sub}$. Direct numerical simulations of the CR-modified shock by particle-in-cell technique are nonfeasible by now because of the wide dynamical range of the simulation that requires extreme computing resources. Nevertheless, an approximate iterative approach (e.g., within the Monte Carlo model discussed in Vladimirov et al. (2008) or semi-analytical kinetic models developed by Malkov (1997) and Amato and Blasi (2006)) can be used to derive the steady-state distribution function consistent with the shock compression. These approximate models assume some diffusion model and parameterize the microphysical processes of magnetic field amplification and plasma heating. The exact calculation of the CR escape flux Q_{esc} that determines the total upstream compression ratio r_{tot} can be performed only in fully nonlinear self-consistent simulations. The Monte Carlo model of Vladimirov et al. (2008) describes the escape of particles from the shock with an assumed free escape boundary far upstream of the shock. The distance to the free escape boundary is a free parameter of the simulation that controls the maximum energy of accelerated particles and the escaping CR flux (see Ellison et al. 1996 for more details of this method).

The role of accelerated particles in the shock precursor depends on the poorly known mechanisms with which energetic particles can transfer energy to the thermal pool. However, in order to numerically estimate the impact of such a process, Vladimirov et al. (2008) parameterize the rate of plasma heating upstream. In order to do that, they assume that CR particles produce, through the resonant CR streaming instability, strong fluctuations of magnetic fields upstream of the shock. Another assumption in the model is that these fluctuations are immediately dissipated into the heat of the thermal gas, and that the rate of turbulence dissipation is proportional to the rate of turbulence generation. The dimensionless parameter α_H in the model is the ratio of dissipation to generation rate. Figure 3 adopted from Vladimirov et al. (2008) shows the self-consistently calculated flow speed, effective magnetic field and the temperature in the precursor of a strong shock with turbulence dissipation for several values of α_H. This work demonstrated that even a small rate of turbulence dissipation can significantly increase the gas temperature in the precursor, and thus increase the rate of injection of thermal particles. At the same time, the spectrum of high energy particles is only modestly affected by dissipation in these simulations.

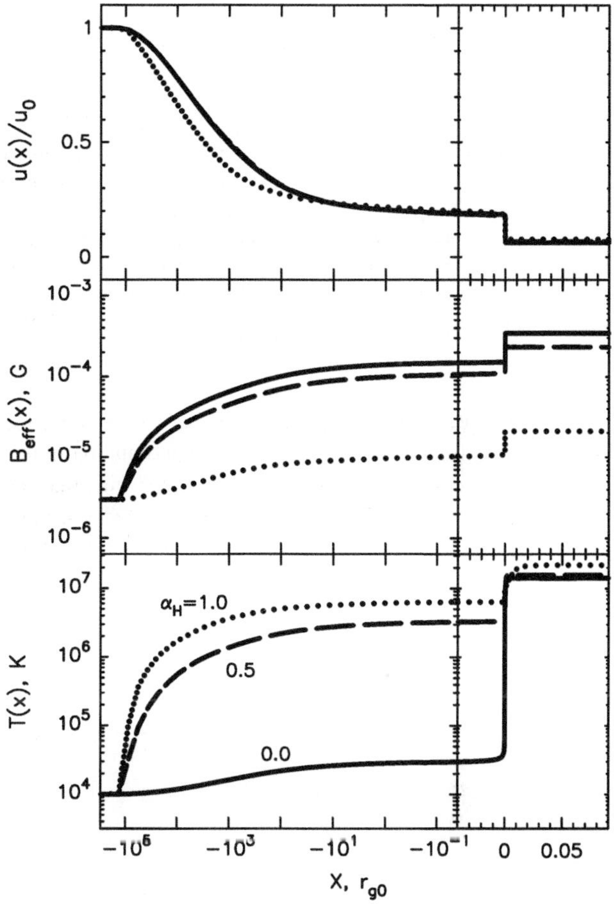

Fig. 3 The profile of a CR-modified shock simulated with Monte-Carlo nonlinear DSA model by Vladimirov et al. (2008). The *solid*, *dashed* and *dotted lines* correspond, respectively, to a fraction, $\alpha_H = 0$, 0.5 and 1.0, of the magnetic energy that goes into heating of the thermal gas. The plotted quantities are the bulk flow speed $u(x)$, the effective amplified magnetic field $B_{\rm eff}(x)$ and the thermal gas temperature $T(x)$. The shock is located at $x = 0$

In a follow-up work of Vladimirov et al. (2009) the effect of the microphysics of the fluctuating magnetic fields in the shock precursor was modeled in a more physically realistic manner. That work used the balance equation of the turbulence spectral energy density

$$\frac{\partial W}{\partial t} + \mathbf{u}\nabla_{\mathbf{r}} W + \nabla_{\mathbf{k}}\boldsymbol{\Pi} = S(x, \mathbf{k}) - L_{\rm turb}. \tag{1}$$

Here $L_{\rm turb}(x, \mathbf{k})$ represents the dissipation of turbulence, the turbulence energy injection rate is $S(x, \mathbf{k}) = \gamma(x, \mathbf{k}) W(x, \mathbf{k}, t)$, and $\gamma(x, \mathbf{k})$ is the rate of wave energy amplification by the CR-driven instabilities (see for a discussion Bykov et al. 2012; Schure et al. 2012). The fastest short-wavelength CR-driven instability studied by Bell (2004) has the following growth rate for the wavevector along the mean magnetic field:

$$\gamma_{\rm CR} = 2 v_A k_\parallel \sqrt{\frac{k_c}{k_\parallel} - 1}, \quad \text{for } 1/r_{\rm g1} < k < k_c, \tag{2}$$

where $v_A(x) = B_0/\sqrt{4\pi\rho(x)}$ is the Alfvén speed, c is the speed of light, B_0 is the far upstream magnetic field directed towards the shock normal, $\rho(x)$ is the thermal plasma mass density, $r_{\rm g1}(x)$ is the gyroradius of the least energetic current generating CR, the critical wavenumber $k_c(x) = 4\pi j_d(x)/(cB_0)$, and the local diffusive electric current of CRs responsible for the instability, $j_d(x)$, is determined via a Monte Carlo simulation. The turbulence energy density flux $\boldsymbol{\Pi}(\mathbf{r}, \mathbf{k}, t)$ describes anisotropic cascading, i.e., the transfer of turbulence energy from long to short wavelengths (Kolmogorov type cascade) as well as the inverse cascade (see, e.g., Monin and Iaglom 1975; Zhou and Matthaeus 1990; Verma 2004; Zhou 2010).

Modeling the balance, spectral distribution, and dissipation of turbulence energy with Eq. (1) allows us to construct more physically realistic models of shock precursors and account for the Balmer line observations discussed in Sect. 3. This equation must be supplemented by theoretical models of the growth, dissipation and spectral flux of MHD turbulence produced by CRs. Sections 4.2 and 4.3 discuss some of the ongoing work in this area.

4.2 MHD Turbulence Damping in Partially Ionized Plasmas

Ion-neutral collisions may dominate the frictional damping of strong magnetic oscillations in a cold photo-ionized plasma ($T \sim 2 \times 10^4$ K) of a precursor of a CR-modified shock (Draine and McKee 1993; O'C Drury et al. 1996). In a cold partially ionized plasma the generalized Ohm's law (see, e.g., Braginskii 1965; Cowling 1976)

$$\mathbf{E} + \frac{1}{c}\mathbf{u} \times \mathbf{B} = \frac{\mathbf{j}}{\sigma} + \frac{1}{n_i e c}\mathbf{j} \times \mathbf{B} + \frac{F^2 \tau_{ia}}{n_i m_i c^2}\mathbf{B} \times (\mathbf{j} \times \mathbf{B}) \tag{3}$$

results in enhanced effective magnetic diffusion ν_{ef} in the induction equation

$$\nu_{ef} = \frac{c^2}{4\pi \sigma_\perp^{ef}}, \tag{4}$$

where

$$\frac{1}{\sigma_\perp^{ef}} = \frac{1}{\sigma} + \frac{F^2 B^2}{n_i m_i c^2}. \tag{5}$$

Here F is the mass fraction of the neutrals, τ_{ia} is the mean time between the ion-neutral collisions. The neutrals in cold magnetized plasma of supernova shock precursors may strongly affect both the growth and the dissipation of CR-driven waves (see, e.g., Bykov and Toptygin 2005; Marcowith et al. 2006; Reville et al. 2007). In a precursor of a quasi-parallel shock the CR-driven turbulence is basically incompressible and the energy dissipation rate Γ^{in} in the low β limit can be estimated as

$$\Gamma^{in} = \sigma_\perp^{ef} k_\parallel^2,$$

while the ion-neutral dissipation of the compressible fluctuations that are associated with long-wavelength CR-driven oblique wave instabilities (see, e.g., Bykov et al. 2011; Schure et al. 2012) are determined by

$$\Gamma^{in} = \sigma_\perp^{ef} k^2.$$

The magnetic turbulence dissipation term in Eq. (1) due to effective magnetic diffusivity (Joule dissipation) is determined by

$$L_{\text{turb}}^{in}(x, \mathbf{k}) = \nu_{ef} k^2 W(x, \mathbf{k}, t). \tag{6}$$

To characterize a fraction of the magnetic energy dissipated in the precursor of a CR-modified shock it is instructive to estimate the minimal wavenumber k_m of the fluctuations to be strongly dissipated by ion-neutral friction while advecting through the CR-shock precursor of the scale length $L_{CR} \sim c/v_{\rm sh}\,\lambda_*$ as it is illustrated in Fig. 2, where

$$\nu_{ef}\,k_m^2 \times \frac{L_{CR}}{\mathrm{v_{sh}}} = 1. \tag{7}$$

Then the maximal wavelength to be damped is

$$\lambda_m = 2\pi/k_m \approx 2\pi\,F\mathcal{M}_{\rm a}^{-1} \times \sqrt{c\tau_{ia}\lambda_*}. \tag{8}$$

For a strong supernova shock of $\mathcal{M}_{\rm a} \approx 100$ and $F \approx 0.1$ one may get $\lambda_m \lesssim 10^{15}$ cm. Note, however, that the length λ_m is comparable to the ion-neutral collision length that is $\sim [F\,n_i\,\sigma_{in}]^{-1}$, if the ion number density in the precursor $n_i \lesssim 1\mathrm{cm}^{-3}$. Depending on the spectral shape of the CR-driven magnetic turbulence in the shock precursor, the dissipated fraction ranges from a few percent for a Kolmogorov-type spectrum to about 10 % for the flatter spectra of strong turbulence. We neglected here the MHD wave dissipation due to the thermal conduction and viscosity assuming a cold plasma case (see, e.g., Braginskii 1965, for a thorough discussion). The magnetic power dissipated by the ion-neutral collisions in the low β plasma mainly heats ions. The collisional damping discussed above results in a true irreversible conversion of the magnetic turbulence free energy into the thermal plasma energy. To the contrary, the collisionless turbulence dissipation processes that we are going to discuss below can heat electrons, thus being not completely irreversible. The collisionless heating process just increases the energy of quasi-thermal plasma components, but in general some collisionality (e.g., Ramos 2011) is required to increase the entropy and to reach the equilibrium distributions of plasma species.

4.3 Collisionless Heating of Ions and Electrons by Magnetic Turbulence

The CR-driven turbulence source $S(x, \mathbf{k})$ in the shock precursor is expected to be anisotropic in $\mathbf{k}$. The fast CR-current driven instability of Bell (2004) as well as the long-wavelength instability of Bykov et. al. (2011) in the Bohm diffusion regime are mainly amplifying the modes with the wavevectors along the local magnetic field. The acoustic instability of Drury and Falle (1986) driven by the CR-pressure gradient is also anisotropic. The strong anisotropy of the magnetic turbulence is observed in the solar wind where the outward flux significantly exceeds the ingoing one.

The solar wind is one of the best laboratories to study anisotropic magnetic turbulent dissipation and collisionless plasma heating (see, e.g., Leamon et al. 1998; Sahraoui et al. 2009; Petrosyan et al. 2010; Alexandrova et al. 2011). Recently, Sahraoui et al. (2009) reported *Cluster* spacecraft measurement providing direct evidence that the dissipation range of magnetofluid turbulence in the solar wind extends down to the electron scales. Namely, they found two distinct breakpoints in the magnetic spectrum at frequencies $f_p = 0.4$ Hz and $f_e = 35$ Hz, which correspond, respectively, to the Doppler-shifted proton and electron gyroscales. Below f_p, the spectrum follows a Kolmogorov-type scaling of power-law index about -1.62. For $f_p < f < f_e$ a second inertial range with a scaling index -2.3 was established. Above f_e the spectrum has a steeper power law -4.1 down to the noise level of the *Cluster* detectors. The authors advocated a good agreement of the results with theoretical predictions of a quasi-two-dimensional cascade into Kinetic Alfvén Waves. Chen et al. (2012) presented a measurement of the scale-dependent, three-dimensional structure of the magnetic field fluctuations in the inertial range of the solar wind turbulence. The Alfvén-type fluctuations are three-dimensionally anisotropic, with the sense of this anisotropy varying from large to small scales. At the outer scale, the magnetic field correlations are longest in the local fluctuation direction. At the proton gyroscale, they are longest along the local mean field direction and shortest in the direction perpendicular to the local mean field and the local field fluctuation. The compressive fluctuations are highly elongated along the local mean field direction, although axially symmetric in the perpendicular direction. Their large anisotropy may explain why they are not heavily damped in the solar wind by the Landau damping.

Anisotropic wavenumber spectra that are broader in wavenumber perpendicular to the background magnetic field than in the parallel one are expected even in the symmetric MHD case studied by Goldreich and Sridhar (1995) where the oppositely directed waves carry equal energy fluxes. The main features of the symmetric (but anisotropic) MHD turbulence model by Goldreich and Sridhar (1995) are:

(i) a critical balance between the linear wave mode periods and their nonlinear turnover timescales in the inertial-interval energy spectrum;
(ii) the 'eddies' are elongated in the direction of the field on small spatial scales with the scaling $k_{\|} \propto k_0^{1/3} k_{\perp}^{2/3}$, where k_0 is the wavenumber corresponding to the outer scale of the turbulence.

The three-dimensional simulations by Cho and Lazarian (2004) revealed the basic features of the Goldreich and Sridhar (1995) model in the electron magnetohydrodynamic turbulence. The kinetic Alfvén wave and whistler fluctuations are likely to make an important contribution to the turbulence below the proton gyroscale (see, e.g., Gary et al. 2012; Saito and Gary 2012; Boldyrev and Perez 2012; Mithaiwala et al. 2012). Particle-in-cell simulations show that the anisotropic whistler turbulence heats the electrons in the parallel direction as predicted by the linear theory and that in the low β plasmas the magnetic wavenumber spectrum becomes strongly anisotropic with spectral index in the perpendicular direction close to -4 (see, e.g., Gary et al. 2012; Saito and Gary 2012). Microscopic two-dimensional particle-in-cell simulations of whistler turbulence were carried out by Saito and Gary (2012) in a collisionless homogeneous magnetized plasma to study the electron and ion heating dependence on the plasma magnetization parameter β. They demonstrated that at higher values of β the magnetic energy cascade in the perpendicular direction becomes weaker and leads to more isotropic wavenumber spectra. The electron energy ratio between parallel and perpendicular components becomes closer to unity at higher β. Three-dimensional particle-in-cell plasma simulations of decay of initial long wavelength whistler fluctuations into a broadband, anisotropic, turbulent spectrum at shorter wavelengths via a forward cascade were performed by Gary et al. (2012). The simulations demonstrated a picture qualitatively similar to that in 2D but somewhat stronger anisotropy of the resulting 3D turbulence comparing to the similar 2D runs. They showed a clear break in the perpendicular wavenumber spectra qualitatively similar to that measured in the electron scale fluctuations in the solar wind. Earlier Quataert (1998) and Quataert and Gruzinov (1999) discussed the beta-dependence of particle heating by turbulence in advection-dominated accretion flows. They found that for $\beta \sim 1$, i.e. approximately equipartition magnetic fields, the turbulence primarily heats the electrons. For weaker magnetic fields, the protons are primarily heated. The division between electron and proton heating occurs between $5 < \beta < 100$, depending on unknown details of how Alfvén waves are converted into whistlers at the proton gyroscales.

The cascade of Alfvén waves, which are weakly damped down to the scale of the proton gyro-radius $k_{\perp}\rho_i \sim 1$ is a subject of gyro-kinetic models (see for a review Schekochihin et al. 2009). The models can be of interest for the parallel shock precursor heating assuming an efficient cascading of the CR-driven magnetic fluctuations down to the $k_{\perp}\rho_i \sim 1$ regime.

The cascading still remains to be demonstrated since the CR-current driven modes are very different from the standard Alfvén waves. If the cascading occurs, then the continuity equation (1) can be reduced to the equation for $b_k^2(k_\perp) = k_\perp^2 \int dk_\parallel W(\mathbf{k})$—the energy density of the anisotropic magnetic turbulence as a function of the perpendicular wavenumber (see, e.g., Howes 2010; Cranmer and van Ballegooijen 2012):

$$\frac{\partial b_k^2}{\partial t} + k_\perp \frac{\partial \epsilon(k_\perp)}{\partial k_\perp} = S(k_\perp) - \Gamma(k_\perp) b_k^2, \tag{9}$$

where the energy injection rate is S (non-zero only at the driving scale $k_\perp = k_0$), the linear energy damping rate is Γ, and $\epsilon(k_\perp)$ is the energy cascade rate. To specify the energy cascade rate both advection and diffusion in the wavenumber space models are used (see, e.g., Cranmer and van Ballegooijen 2012 for a recent discussion). Assuming critical balance at all scales and using the energy cascade rate in the form of the advection in the wavenumber space:

$$\epsilon(k_\perp) = C_1^{-3/2} k_\perp \overline{\omega} b_k^3.$$

Howes (2010) obtained the steady state solution for the energy cascade rate as

$$\epsilon(k_\perp) = \epsilon_0 \exp\left\{ - \int_{k_0}^{k_\perp} C_1^{3/2} C_2 \frac{\overline{\Gamma}(k'_\perp)}{\overline{\omega}(k'_\perp)} \frac{dk'_\perp}{k'_\perp} \right\}, \tag{10}$$

where C_1 and C_2 are the dimensionless Kolmogorov constants ($C_1 = 1.96$ and $C_2 = 1.09$) and ϵ_0 is the rate of energy input at k_0. Howes (2010) used the normalized eigenfrequencies $\overline{\omega}(k_\perp)$ from the linear gyrokinetic dispersion relations and the damping rates $\overline{\Gamma}_s$ due to different plasma species from Howes et al. (2006) and Eq. (10) (where $s = i, e$), to calculate the spectrum of heating by species

$$Q_s(k_\perp) = C_1^{3/2} C_2 (\overline{\Gamma}_s / \overline{\omega}) \epsilon(k_\perp) / k_\perp.$$

The ion damping peaks at $k_\perp \rho_i \sim 1$, while the electron damping peaks at $k_\perp \rho_i \gg 1$ unless $T_i/T_e \lesssim m_e/m_i$. The energy that passes through the peak of the ion damping at $k_\perp \rho_i \sim 1$ would lead to electron heating assuming both the cascading and the damping times are shorter than the advection time through the shock precursor. Then the total (integrated over $k_\perp \rho_i \gtrsim 1$) ion-to-electron heating rate due to the kinetic dissipation of the turbulent cascade, $Q_i/Q_e(\beta_i, T_i/T_e)$ can be approximately fitted with

$$Q_i/Q_e = c_1 \frac{c_2^2 + \beta_i^p}{c_3^2 + \beta_i^p} \sqrt{\frac{m_i T_i}{m_e T_e}} e^{-1/\beta_i}, \tag{11}$$

where $c_1 = 0.92$, $c_2 = 1.6/(T_i/T_e)$, $c_3 = 18 + 5\log(T_i/T_e)$, and $p = 2 - 0.2\log(T_i/T_e)$. A slightly better fit for $T_i/T_e < 1$ occurs with the coefficients $c_2 = 1.2/(T_i/T_e)$ and $c_3 = 18$ (Howes 2010). The model is valid for the parameter range $0.01 \le \beta_i \le 100$ and $0.2 \le T_i/T_e \le 100$. The heating rate Q_i/Q_e is an approximately monotonic function of β_i that is only weakly dependent of T_i/T_e.

The simplified model of the electron and ion heating discussed above assumed the nonlinear collisionless cascading from the energy containing scale of wavenumber $k_* R_g \sim 1$ where the amplitude of amplified magnetic field is δB_* that is determined by the gyroscale R_g of the energy containing accelerated particles to the thermal ion gyroscale $k_\perp \rho_i \sim 1$. In the shock frame the advection time of the incoming plasma through the CR precursor $\tau_{\rm adv} \sim \lambda_* c/v_{\rm sh}^2$. Therefore for efficient heating of the plasma species the cascading time τ_c must be shorter than $\tau_{\rm adv}$. If the cascading time is determined by the turn-over time of the energy containing magnetic "vortex" $\tau_c^{-1} \sim k_* \delta B_* / \sqrt{4\pi\rho}$ then the condition of efficient plasma heating in the shock precursor by the CR-driven turbulence can be written as

$$\frac{\tau_{\rm adv}}{\tau_c} = \eta \frac{c}{v_{\rm sh}^2} \frac{\delta B_*}{\sqrt{4\pi\rho}} > 1, \tag{12}$$

where $\eta = \lambda_* / R_g > 1$. The amplitude of amplified magnetic field likely scales with the shock velocity as $\delta B_* \propto v_{\rm sh}^{\rm b}$, where $1 \le {\rm b} \le 1.5$ (see, e.g., Vink 2012). The condition (12) predicts a less efficient plasma heating in the CR-precursor of the faster shocks.

Heat conduction that we did not discuss here may play a role in the electron temperature distribution in the shock precursor (cf. Breech et al. 2009). The nonlinear dynamics of the CR-driven magnetic fluctuations in the shock precursor deserves thorough modeling. Malkov et al. (2012) have obtained fully nonlinear exact solutions of the *ideal* 1D-MHD supported by the CR return current. The solutions occur as localized spikes of circularly polarized Alfvén envelopes (solitons or breathers). The sufficiently strong solitons in the model run ahead of the main shock and stand in the precursor, being supported by the return current. The CR-shock precursor in the model is dissipationless.

The electron and ion temperatures in the shock precursor determine the injection of particles into the CR acceleration regime. The temperatures can be tested by optical spectroscopy of supernova shocks. CR-modification of shocks with modest speed of a few hundred km s^{-1} may yield lower post-shock temperatures and thus make the post-shock flow switch to a radiative regime.

5 Spectroscopy of a CR-Modified Radiative Shock

Consider a one-dimensional flow around a collisionless shock consisting of three zones: (a) the pre-shock, where the unperturbed interstellar matter is preionized and preheated by the radiation (and energetic particles) generated in the downstream and where strong fluctuations of magnetic field may be generated by the CR anisotropy, (b) the thin shock front (a "viscous jump") where a substantial part of the kinetic energy of the bulk upstream flow is converted into energy of thermal motions, and (c) the post-shock, where the hot flow cools down, radiating continuum and line emission. We discuss here a class of shock flows in partly ionized media, the so-called *radiative shocks*, where the power radiated away from the post-shock flow is a sizeable fraction of the total kinetic and magnetic power dissipated at the shock (see, e.g., Spitzer 1978; Draine and McKee 1993).

Without a significant impact on accuracy, one may assume that the pre-shock is isothermal, and the flow is in a steady state (see also a comment about the equilibrium between the pre-shock ionization state and the ionizing flux at the end of Sect. 2). However, the ionization state of the plasma is non-uniform: the ionization level increases toward the subshock, as the gas absorbs the ionizing radiation flux emerging from the hot downstream region. This photoionization does not have a significant impact on the thermal state of the gas, because the hot electrons produced by photoionization do not have

[1]This section uses some results of numerical code SHELLS currently developed by A.M. Bykov, A.E. Vladimirov, and A.M. Krassilchtchikov. This code is used to model the steady-state structure and broadband continuum and line emission spectra of radiative shocks. The model accounts for the impact of efficient CR acceleration on the shock compression ratio and the postshock flow. The code incorporates modern atomic data and allows us to consider in detail the non-equilibrium microscopic, thermodynamical and radiative processes that determine the plasma flow.

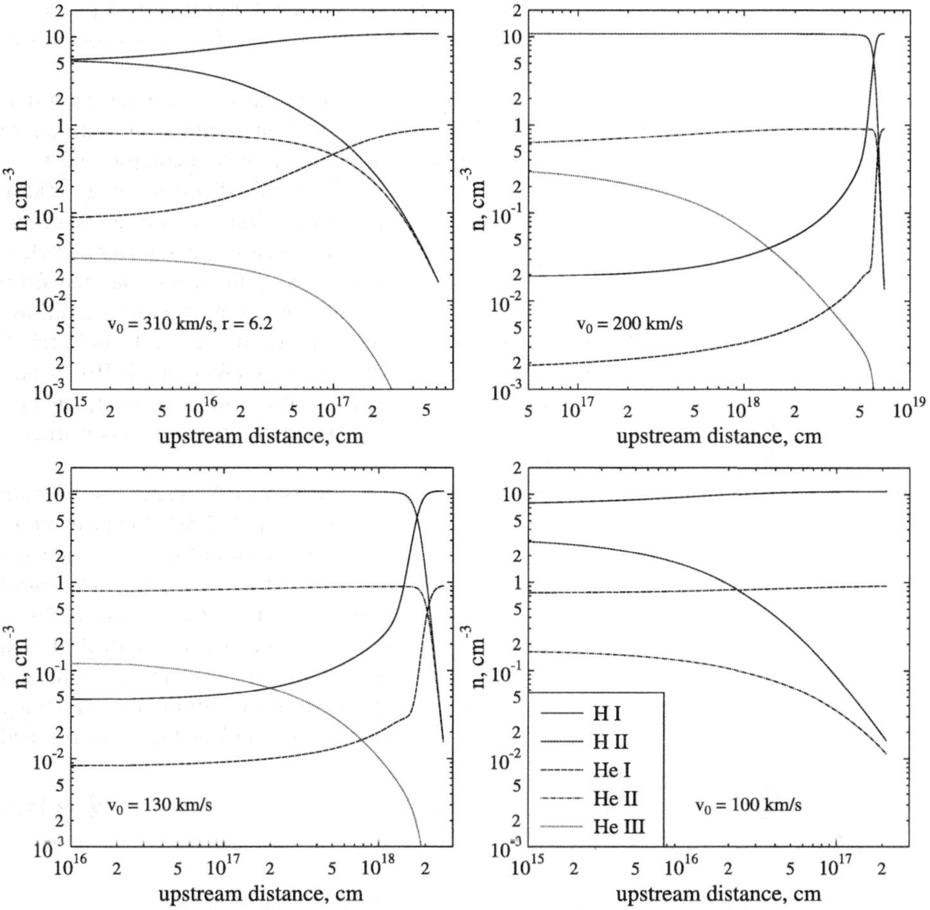

Fig. 4 Hydrogen and helium pre-shock (photo)ionization in the shock upstream for four sets of shock parameters simulated with the model of radiative shock accounting for the effect of CR escape on the shock compression r (the CR-modified case is shown in the top left panel). Within the presented model $r = 6.2$ corresponds to 30 % of the flow energy being converted into CRs

sufficient time to collisionally equilibrate with the atoms and ions. CR particles may contribute to gas ionization and heating in the interstellar clouds and shells (e.g., Spitzer 1978; Bozhokin and Bykov 1994). The effect of CRs may be important in the vicinity of fast Balmer-type interstellar shocks as it has been demonstrated by Morlino et al. (2012a), who did not account for the photo processes, though. Radiative shocks are expected to occur in relatively dense environment and, therefore, have velocities typically well below 1,000 $\mathrm{km\,s^{-1}}$. In such an environment, photo processes can provide a high ionization degree of the upstream gas for sufficiently high shock speeds. If CR acceleration is weak, the photoionization becomes significant for shock velocities above 100 $\mathrm{km\,s^{-1}}$. However, if CR acceleration is efficient, the downstream temperature is reduced and the shock compression is increased, which means that strong photoionization in the precursor occurs at much greater shock speeds. See Fig. 4 for an illustration of the effect of CR acceleration on precursor ionization. We demonstrate below the effect of the CR fluid on the shock compression and spectra of radiative shocks.

The jump conditions at the shock are either given by the Rankine-Hugoniot equations, or, if the shock structure is assumed to be modified by accelerated particles (e.g., Bykov 2004; Vladimirov et al. 2008, and references therein), the compression ratio can be parameterized by the fraction $Q_{\rm esc}$ of bulk flow energy carried away by these particles. The subshock is the standard gas viscous shock of a Mach number $\mathcal{M}_{\rm sub}$. For that simplified *two-fluid* model of a strong CR-modified shock the effective ion temperature in the downstream $T_{\rm i}^{(2)}$ can be estimated for a shock of a given velocity, if $r_{\rm tot}$ and $r_{\rm sub}$ are known:

$$T_{\rm i}^{(2)} \approx \phi(\mathcal{M}_{\rm sub}) \cdot \frac{\mu v_{\rm sh}^2}{\gamma_{\rm g} r_{\rm tot}^2(v_{\rm sh})}, \quad \text{where } \phi(\mathcal{M}_{\rm sub}) = \frac{2\gamma_{\rm g}\mathcal{M}_{\rm sub}^2 - (\gamma_{\rm g} - 1)}{(\gamma_{\rm g} - 1)\mathcal{M}_{\rm sub}^2 + 2}. \tag{13}$$

Single fluid strong shock heating represents the limit $\mathcal{M}_{\rm sub} = \mathcal{M}_{\rm s} \gg 1$, since there is no precursor in that case the temperature behind a strong shock is determined by the standard scaling

$$T^{(2)} \approx 2 \cdot \frac{(\gamma_{\rm g} - 1)}{(\gamma_{\rm g} + 1)^2} \mu v_{\rm sh}^2 = 1.38 \cdot 10^7 \, v_{s8}^2 \text{ (K)}, \tag{14}$$

In single-fluid systems the compression ratio $r_{\rm tot} = r_{\rm sub} \to (\gamma_{\rm g} + 1)/(\gamma_{\rm g} - 1)$ does not depend on the shock velocity and Eq. (13) reduces to Eq. (14). However, in multi-fluid shocks the total compression ratio depends on the shock velocity and could be substantially higher than that in the single-fluid case. Consequently, the post-shock temperature in a multi-fluid shock is lower than the post-shock temperature in a single-fluid shock of the same speed. This allows us to determine the CR acceleration efficiency using observations of post-shock temperatures and shock speeds, or the entropy profiles in the accretion shocks of clusters of galaxies (Bykov 2005; Bykov et al. 2008; Brüggen et al. 2012; Fujita et al. 2013). It is convenient to introduce the scaling $r_{\rm tot}(v_{\rm sh}) \propto v_{\rm sh}^{\xi}$ to describe the different cases of strong shock heating (Bykov et al. 2008). Then from Eq. (13) $T_{\rm i}^{(2)} \propto \phi(\mathcal{M}_{\rm sub}) \cdot v_{\rm sh}^{2(1-\xi)}$. The subshock Mach number $\mathcal{M}_{\rm sub}$ depends, in general, on $\mathcal{M}_{\rm s}$ and $\mathcal{M}_{\rm a}$. Thus, the index σ approximates the velocity dependence of $\phi(\mathcal{M}_{\rm sub}) \propto v_{\rm sh}^{\sigma}$. Finally, if $T_{\rm i}^{(2)} \propto v_{\rm sh}^{a}$, then the index $a = 2(1 - \xi) + \sigma$. For the case of shock precursor heating by CR generated Alfvén waves, the index $a \approx 1.25$ (Bykov 2005). The effects of neutrals due to charge exchange in the shock downstream with heated ions that results in a flux of high-velocity neutrals that return upstream were studied by Blasi et al. (2012) and Ohira (2012). They found that the return flux of neutrals may result in the reduced shock compression ratio and spectral steepening of test particles accelerated at the shocks slower than about 3000 km s^{-1}. The return flux of neutral atoms may also affect the radiation spectrum of the post-shock flow that we are modeling.

The ratio of the electron to ion temperature immediately after the subshock, $T_e/T_i = \delta_e$, is considered as a free parameter in our model. It can be estimated from observations as discussed in Sect. 3. It is varied in the range from $\sqrt{m_e/m_p} \approx 0.023$ to 1. Observations typically show low values of δ_e in shocks faster than 1000 km s^{-1}, and values closer to 1 in slower shocks (e.g., Ghavamian et al. 2001, 2007; Rakowski et al. 2008; Helder et al. 2010).

The post-shock plasma is treated as a stationary two-temperature single fluid flow consisting of ions, electrons, and neutral atoms that are ideal nonrelativistic gases. The neutral and ion temperatures may actually differ (Heng et al. 2007; van Adelsberg et al. 2008), but that is only important just behind a very fast shock in partially neutral gas, and it matters mainly for diagnostics based on the Hα line profile. For the considered here ranges of shock speeds and matter densities the temperature equilibration scales are below 10^{15} cm^{-2}, while the line emission zone typically lies well above 10^{16} cm^{-2} downstream.

Let z and μ be the average charge and mass of an ion, $\zeta \equiv m_e/\mu$—the average electron to ion mass ratio. Then

$$\begin{aligned} n_e &= z n_i, \quad \rho_e = m_e n_e = z m_e n_i, \\ \rho_i &= \mu n_i, \\ \rho &= n_i(z m_e + \mu) = n_i \mu (z\zeta + 1) \equiv n_i \mu / M(z), \\ M(z) &= 1/(z\zeta + 1). \end{aligned} \tag{15}$$

Let ρ_0, v_0, and B_0 be the values of density, velocity, and frozen-in transverse magnetic field just before the shock, and v_a, T_i^a be the values of flow velocity and ion temperature in the immediate post-shock defined by jump conditions.

Then, the flow evolution downstream can be described by the following system of equations.

$$\rho v = \text{const} = \rho_0 v_0, \tag{16}$$

$$\rho v^2 + p_e + p_i + p_m + p_{CR} = \text{const} \equiv \Pi \rho_0 v_0^2, \tag{17}$$

$$\frac{3}{2} n_i v \frac{dT_i}{dx} = -n_i T_i \frac{dv}{dx} - \frac{3 m_e}{\mu} \frac{n_e}{\tau_{ei}} (T_i - T_e), \tag{18}$$

$$\frac{3}{2} n_e v \frac{dT_e}{dx} = -n_e T_e \frac{dv}{dx} + \frac{3 m_e}{\mu} \frac{n_e}{\tau_{ei}} (T_i - T_e) - \Lambda - \frac{3}{2} n_i T_e v \frac{dz}{dx}, \tag{19}$$

where $p_e + p_i = n_i k_B (z T_e + T_i) = \rho k_B M (z T_e + T_i)/\mu$, the magnetic pressure $p_m = B_0^2 \rho^2 / (8\pi \rho_0^2)$, $\Pi = v_a/v_0 + B_0^2/(8\pi \rho_0 v_0 v_a) + k_B M T_i^a (z\delta_e + 1)/(v_0 v_a \mu) + Q_{\rm esc}/2$, $p_{CR} = (Q_{\rm esc}/2) \cdot \rho_0 v_0^2 (v_a/v)^{4/3}$, Λ is the cooling function, τ_{ei} is the electron-ion equilibration time, and the last term of Eq. (19) denotes electron cooling due to ionization.

The cooling term Λ is calculated as

$$\begin{aligned} \Lambda = {} & n^2 \Lambda_{coll} + n_e \sum_{i,j} (n_{i,j} \alpha_{i,j} E_{i,j}^{rec} + n_{i,j} C_{i,j} h\nu_j) \\ & - 4\pi \sum_{i,j} \int_{\nu_j}^{\infty} d\nu \cdot \sigma_{i,j}^{ph} (1 - \nu_j/\nu) n_{i,j} J_\nu, \end{aligned} \tag{20}$$

where i denotes the chemical element and j denotes the ionization state of an ion ($j = 0$ corresponds to a neutral atom). Here Λ_{coll} is due to electron-ion collisions including electron bremsstrahlung and line emission of the ions excited by electron impact, though it does not include emission due to radiative recombination; $\alpha_{i,j}$ is the recombination rate $\{j+1\} \to \{j\}$; $E_{i,j}^{rec}$ is the average energy on the recombining electrons; $C_{i,j}$ is the rate of collisional ionization; $\sigma_{i,j}^{ph}$ is the photoionization cross-section of the ion state j; J_ν is the angle-averaged density of ionizing radiation at frequency ν:

$$J_\nu(x) = \frac{1}{4\pi} \int_{-1}^{1} 2\pi \, I_\nu(\mu, x) \, d\mu, \tag{21}$$

where $\mu = \cos(\theta)$, θ is the angle between the normal to the shock front and the direction of emitted photons.

The evolution of the ionization state j of an ion i in the downstream flow is determined as

$$\begin{aligned} v \frac{dn_{i,j}}{dx} = {} & n_e (n_{i,j-1} \tilde{C}_{i,j-1} - n_{i,j} \tilde{C}_{i,j} - n_{i,j} \tilde{\alpha}_{i,j} + n_{i,j+1} \tilde{\alpha}_{i,j+1}) \\ & + n_{i,j-1} R_{i,j-1} - n_{i,j} R_{i,j} \end{aligned}$$

$$+ \sum_{s=\mathrm{H,He,He^+}} n_s\left[n_{i,j-1}V^{ion}_{s,i,j-1} - n_{i,j}\left(V^{ion}_{s,i,j} + V^{rec}_{s,i,j}\right) + n_{i,j+1}V^{rec}_{s,i,j+1}\right], \quad (22)$$

where $R_{i,j}$ is the photoionization rate, V_s^{ion} and V_s^{ion} are the rates of ionization and recombination via charge exchange reactions with the ion s, $\tilde{C}_{i,j} = C_{i,j} + C^{\mathrm{auto}}_{i,j}$, where $C^{\mathrm{auto}}_{i,j}$ is the autoionization rate, $\tilde{\alpha}_{i,j} = \alpha_{i,j} + \alpha^{2e}_{i,j}$, where $\alpha^{2e}_{i,j}$ is the dielectronic recombination rate.

To obtain the ionizing radiation field a transfer equation can be solved both in the downstream and in the upstream:

$$\cos\theta \cdot \frac{dI_\nu}{dx} = j_\nu - \kappa_\nu I_\nu, \quad (23)$$

where θ is the angle between the normal to the shock front and the direction of emitted photons. In this equation, the absorption coefficient $\kappa_\nu = \sum_{i,j} n_{i,j}\sigma^{ph}_{i,j}$ is determined by bound-free transitions in all ion species. The ionizing emission is generated as (a) permitted ultraviolet and optical lines excited by an electronic impact, (b) recombination line cascades of hydrogen and helium, (c) free-free continuum of electrons scattering at ions, (d) 2-photon continuum emission of H- and He-like ions where metastable levels are collisionally populated, (e) recombination continuum emission.

The relative contribution of each of the mechanisms to the ionizing photon field is illustrated in Fig. 5. The dominating 2-photon continuum emission comes mainly from He I and He II and the dominating ultraviolet line emission comes mainly from He I, He II, and oxygen ions upto O V. It should be noted that in the presented model the electron temperatures just after the shock are assumed to be low ($\delta_e \sim 10^{-2}$) and equilibrate with the ion temperatures downstream via Coulomb collisions.

All the ionizing lines except Lyα were considered optically thin. The optically thick case of Lyα was treated according to the standard 2-level formalism adopted from Mihalas and Mihalas (1984).

Figures 4 through 8 and Tables 1 and 2 demonstrate the results of modeling with SHELLS. The impact of efficient CR acceleration on the shock is accounted for in all calculations by using a reduced downstream temperature and the CR pressure in the shock downstream.

The steady-state pre-shock ionization structure is shown in Fig. 4 for shocks with speeds of 100, 130, 200, and 310 $\mathrm{km\,s^{-1}}$ illustrating also an effect of the CR fluid on the pre-shock ionization. The plasma temperature and gas compression just behind the CR modified shock affect the ionization structure. One can see in Fig. 4 that the ionization structure of 310 $\mathrm{km\,s^{-1}}$ shock modified by CRs may be similar to that of 100 $\mathrm{km\,s^{-1}}$ shock without CRs fluid effect. This Figure shows the ionization level of hydrogen and helium; however, all abundant chemical species up to Fe are tracked in the calculation. At speeds of approximately 130 km/s and above, hydrogen upstream becomes completely ionized, and the extent of the radiative precursor is determined by the competition between hydrogen and helium recombination and photoionization.

Figure 6 illustrates the downstream cooling region of a 200 km/s shock. Three lines correspond to three models with different assumptions regarding the efficiency of CR acceleration. For efficient acceleration, energy conservation requires greater compression ratio and lower downstream temperature. The three cases shown here illustrate shocks with compression ratios of 4.6, 5.3 and 6.2.

Shown in Fig. 7 is the result of a calculation with shock and ISM parameters plausible for the region of interaction of the north-eastern shell of the SNR IC 443 with an interstellar cloud. Unperturbed gas density is $n_0 = 15\ \mathrm{cm^{-3}}$, shock velocity $v_0 = 130$ km/s, and perpendicular magnetic field $B_0 = 3\ \mu\mathrm{G}$ are assumed. The temperature and total density of the

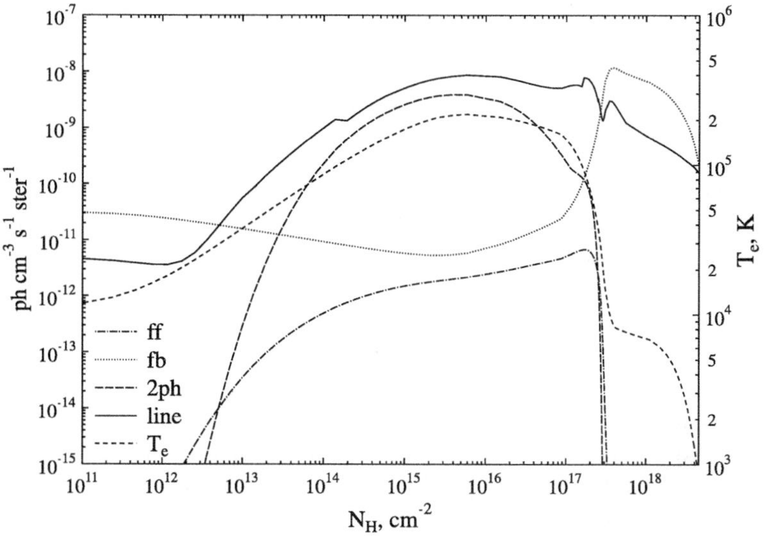

Fig. 5 Production rate of ionizing photons ($h\nu > 1$ Ry) in the downstream of a 130 km/s shock penetrating into a 10 cm^{-3} medium of solar abundance. Here "fb" denotes recombination continuum emission, "ff" denotes free-free continuum of electrons scattering at ions, "2ph" stands for 2-photon continuum emission of H- and He-like ions, and "line" denotes permitted ultraviolet lines excited by an electronic impact. The *dashed line* illustrates the simulated electron temperature profile in the downstream. The photons produced in the optically thin part of the downstream are ionizing the upstream flow

Fig. 6 Gas temperature (*solid lines*) and density (*dotted lines*) in the downstream cooling region of three 200 km/s shocks. Three lines correspond to three models with different assumptions regarding the efficiency of CR acceleration. Within the presented model $r = 6.2$ corresponds to $Q_{esc} = 30\ \%$ of the flow energy being converted into CRs, $r = 5.3$ corresponds to $Q_{esc} = 20\ \%$, and $r = 4.6$ corresponds to $Q_{esc} = 10\ \%$. See Table 1 for the corresponding line strengths

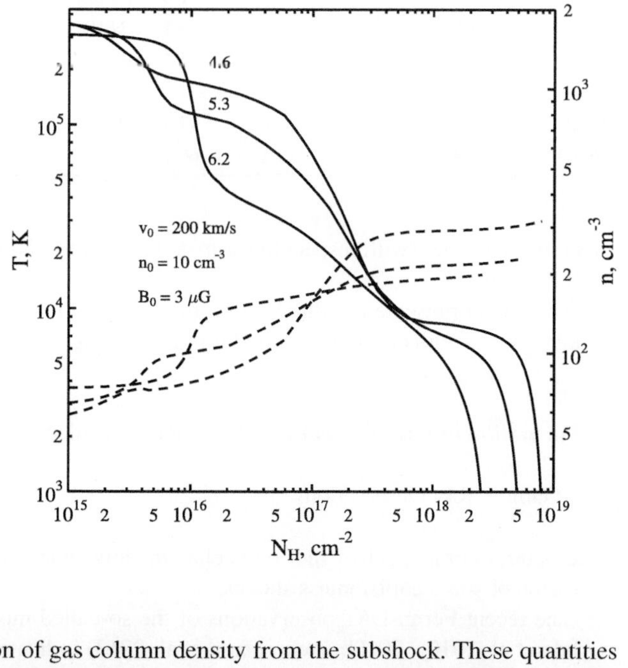

plasma are plotted as a function of gas column density from the subshock. These quantities are then used to calculate the radiative transfer of emission lines in the cooling flow, which can be used for shock diagnostics. Figure 8 shows the profiles of two of the most promi-

Table 1 Model line strengths in percent of Hβ flux for $v_0 = 200$ km/s at CR-modified compression ratios $r = 4.6, 5.3, 6.2$ (see Fig. 6). The last line presents absolute flux in Hβ 4861 Å line in 10^{-6} erg cm^{-2} s^{-1} sr^{-1} polariz^{-1}

Line	λ	$r=4.6$	$r=5.3$	$r=6.2$	Line	λ	$r=4.6$	$r=5.3$	$r=6.2$
He II	304 Å	2640	750	1.1	[S II]	6718 Å	67.1	27.8	16.1
O IV	554 Å	201	14.7	...	[S II]	6733 Å	43.9	18.9	10.6
He I	584 Å	72.5	212	103	[Fe II]	1.28 μm	126	85.1	51.1
O V	630 Å	24.5	...	...	[Fe II]	1.62 μm	29.0	19.4	11.6
O III	833 Å	98.4	142	0.5	[Fe II]	5.30 μm	95.2	78.2	49.4
C III	977 Å	637	770	52.1	[S IV]	10.51 μm	2.1	2.0	0.1
O VI	1034 Å	...	...	...	[Ne II]	12.81 μm	40.3	26.6	8.1
Lyα	1216 Å	5500	10000	17000	[Ne III]	15.55 μm	29.7	12.6	...
N V	1240 Å	2.9	0.1	...	[Fe II]	17.94 μm	10.1	8.1	5.1
Si IV	1397 Å	114	122	2.1	[S III]	18.71 μm	1.3	1.8	1.1
C IV	1549 Å	858	190	0.1	[Fe II]	24.52 μm	1.8	1.4	0.9
He II	1640 Å	58.5	12.6	...	[O IV]	25.91 μm	12.4	2.9	...
[C III]	1908 Å	219	290	50.9	[Fe II]	26.00 μm	47.5	43.5	29.6
[O II]	3729 Å	236	221	289	[S III]	33.50 μm	1.7	2.4	1.6
[Ne III]	3870 Å	23.6	14.9	...	[Si II]	34.82 μm	218	122	51.8
[Ne III]	3969 Å	7.1	4.5	...	[Fe II]	35.35 μm	10.2	9.2	6.2
He II	4687 Å	6.2	1.1	...	[Ne III]	36.02 μm	2.6	1.1	...
Hβ	4861 Å	100	100	100	[N III]	57.34 μm	1.7	1.6	0.1
[O III]	4960 Å	48.3	49.0	0.4	[O I]	63.19 μm	28.7	12.7	4.9
[O III]	5008 Å	144	146	1.0	[O III]	88.36 μm	9.8	8.2	0.1
[O I]	6300 Å	27.4	7.0	3.1	[N II]	121.8 μm	6.2	3.8	2.0
[O I]	6363 Å	27.3	6.9	3.1	[O I]	145.5 μm	3.8	1.9	0.8
Hα	6565 Å	302	310	321	[C II]	157.7 μm	18.3	13.5	7.9
[N II]	6550 Å	46.6	24.6	19.9	Hβ	4863 Å	6.38	5.61	5.05
[N II]	6585 Å	142	75.1	60.7					

nent infrared lines (with respect to estimated backgrounds) for this case: C II 157.7 μm and N II 205.3 μm.

The model presented in this Section may be used to diagnose various parameters of radiative shocks with observed infrared lines, including the cosmic ray acceleration efficiency.

6 Formation of CR Spectral Breaks in Partly Ionized Shock Precursors

6.1 Gamma-Ray Observations

The neutral component of the interstellar medium may manifest itself in the gamma-ray emission of some collisionless shocks.

The recent Fermi-LAT observations of the so-called molecular SNRs W44 and IC 443 (Abdo et al. 2010a, 2010b; Ackermann et al. 2013) indicate that the spectra of the gamma-ray producing protons (integrated over the emission region) are typically steeper than the DSA predictions for the spectra of the CRs confined in the acceleration region. The steep

Table 2 Model line intensities expressed in percent of the 4861 Å Hβ line intensity, which amounts here to $7.87 \times 10^{-6}\ \mathrm{erg\,cm^{-2}\,s^{-1}\,sr^{-1}\,polariz^{-1}}$ (see Fig. 7 for the corresponding flow profiles)

He II	304 Å	2600	[N II]	6585 Å	139
O IV	554 Å	597	[S II]	6718 Å	71.6
He I	584 Å	45.1	[S II]	6733 Å	45.1
O V	630 Å	509	[Fe II]	1.28 μm	288
O III	833 Å	179	[Fe II]	1.62 μm	69.4
C III	977 Å	438	[Fe II]	5.30 μm	204
O VI	1034 Å	0.6	[S IV]	10.51 μm	2.3
Lyα	1216 Å	641	[Ne II]	12.81 μm	61.2
N V	1240 Å	14.3	[Ne III]	15.55 μm	24.4
Si IV	1397 Å	138	[Fe II]	17.94 μm	36.4
C IV	1549 Å	1080	[S III]	18.71 μm	1.9
He II	1640 Å	74.9	[Fe II]	24.52 μm	6.7
[C III]	1908 Å	152	[O IV]	25.91 μm	15.0
[O II]	3729 Å	254	[Fe II]	26.00 μm	114
[Ne III]	3870 Å	23.6	[S III]	33.50 μm	1.8
[Ne III]	3969 Å	7.1	[Si II]	34.82 μm	143
He II	4687 Å	8.9	[Fe II]	35.35 μm	25.2
Hβ	4861 Å	100	[Ne III]	36.02 μm	2.1
[O III]	4960 Å	52.2	[N III]	57.34 μm	1.0
[O III]	5008 Å	156	[O I]	63.19 μm	47.4
[O I]	6300 Å	35.7	[O III]	88.36 μm	4.0
[O I]	6363 Å	35.5	[N II]	121.8 μm	1.9
Hα	6565 Å	302	[O I]	145.5 μm	5.6
[N II]	6550 Å	45.5	[C II]	157.7 μm	7.3

photon spectra has been found in the high energy gamma-ray spectra of some other remnants measured by, e.g., the CANGAROO (Enomoto et al. 2002), H.E.S.S (Aharonian et al. 2006) and MAGIC (Carmona 2011) atmospheric Cherenkov telescopes.

Particles accelerated by DSA mechanism generally comprised two distinct populations—CRs confined in the accelerator and the CRs escaping the system and these two populations have very different spectral shapes. The observed gamma-ray emission is a sum of the two contributions (see, e.g., Ellison and Bykov 2011). The common feature of the molecular SNRs is a significant amount of dense molecular gas in their surroundings. It has been argued on this ground (Malkov et al. 2005), that when a SNR interacts with a dense molecular cloud complex, the conditions for particle confinement to the shock are different from those adopted in conventional DSA modeling. Since the propagation of resonant Alfvén waves is inhibited by ion-neutral collisions, some particles are not confined and so escape the emission volume (e.g. Drury et al. 1996; Bykov et al. 2000; Malkov et al. 2005). The accelerated particle partial escape should result in a spectral break in their spectrum and thus, in that of the observed gamma emission. In a clear-cut case of such a limited CR confinement, the spectral index at the break should change by exactly one power $\Delta q = 1$ due to an effective reduction of the particle momentum space dimensionality by one, since particles are confined in coordinate space only when they are within a slab in momentum space oriented perpendicular to the local magnetic field.

The most convincing evidence for the breaks of index one are the recent Fermi-LAT and AGILE observations of W44 (Abdo et al. 2010a; Uchiyama et al. 2010; Giuliani et al. 2011)

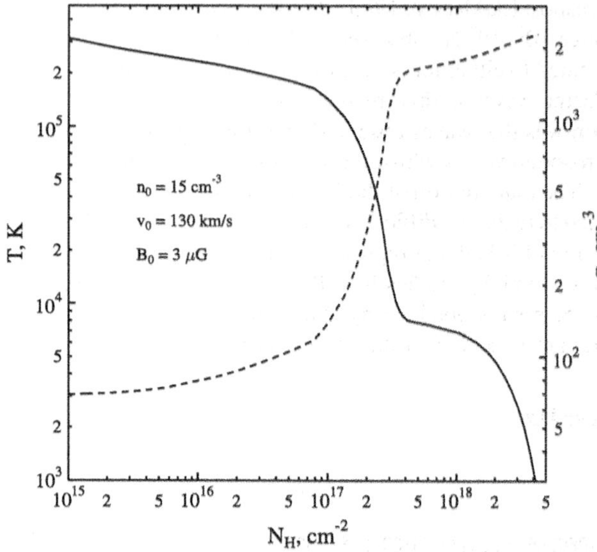

Fig. 7 Post-shock flow profile for shock and ISM parameters plausible for the region of interaction of the north-eastern shell of IC 443 SNR with an interstellar cloud (see Table 2 for the corresponding line strengths). Here the upstream density $n_0 = 15\ \text{cm}^{-3}$, the shock velocity $v_0 = 130$ km/s, and the perpendicular magnetic field $B_0 = 3\ \mu\text{G}$

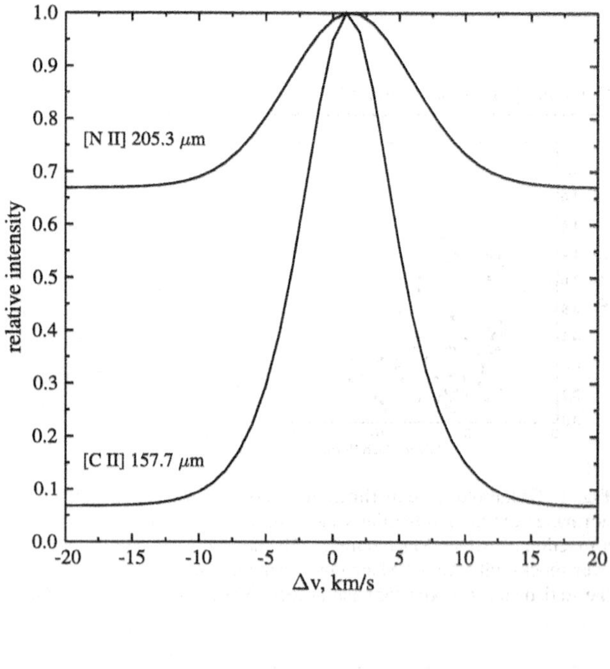

Fig. 8 [C II] 157.7 μm and [N II] 205.3 μm line profiles for the backgrounds of 140 MJy/sr and 130 MJy/sr, respectively, estimated for the north-eastern shell of IC 443 SNR with Herschel Spot (for the date 22/09/2010)

(re-analyzed in Malkov et al. 2011) and MAGIC observations of the SNR W51C (Carmona 2011; Aleksić et al. 2012). These observations are encouraging in that they unambiguously confirm the breaks. They demonstrate departure from the traditional DSA models in fully ionized plasma where the proton spectrum is a single power law with an exponential cutoff. A possible explanation to the spectral break in the gamma-ray emission of these objects is that neutral atoms modify the nature of wave-particle interactions, leading to a spectral break. This idea is discussed below.

6.2 The DSA Mechanism

A great deal of the success of the diffusive shock acceleration mechanism (DSA) is due to its ability to reproduce the observed power law spectra associated with possible CR accelerators, such as SNR shocks. For a classical step-like shock, propagating into an ionized medium that supports sufficient CR scattering on both sides of the shock, the mechanism predicts an almost universal power-law energy spectrum for the accelerated CRs, $\propto E^{-q}$, with an index $q = (r+2)/(r-1)$. It thus depends only on the shock compression r, that is, however, close to four for strong shocks. The backreaction of the accelerated CRs on the shock structure may change this result noticeably, but not dramatically. For strong shocks, $\mathcal{M} \gg 1$, the spectrum hardens from $q = 2$ to $q \simeq 1.5$, at most. This change is largely due to an enhanced compression of the shock that results from the CR escape flux and the reduction of the adiabatic index of the CR/thermal-plasma mixture compared to the ordinary plasma (i.e., $\gamma = 5/3 \to 4/3$). More problematic is to soften the CR spectrum by their backreaction effects on the shock environment. This is intuitively understandable, since spectrum steepening should diminish the backreaction coming from the softening of the equation of state ($\gamma \to 4/3$) and, especially, from the CR escape flux.

The recent Fermi-LAT observations of the SNRs W44 and IC 443 (Abdo et al. 2010a, 2010b) also urge a second look at the DSA mechanism (see Sect. 6.1).

6.3 Mechanism for a Spectral Break

When a SNR shock approaches a molecular cloud (MC) or a pre-supernova swept-up shell with a significant amount of neutrals, confinement of accelerated particles deteriorates. Due to the particle interaction with magnetic fluctuations, confinement requires their scales to be similar to the particle gyroradii (Drury 1983; Blandford and Eichler 1987). However, strong ion-neutral collisions make the wave-particle interactions more sensitive to the particle pitch angle, which can be understood from the following consideration.

While the waves are in a strongly ionized (e.g., closer to the shock) medium they propagate freely in a broad frequency range at the Alfvén speed $V_A = B/\sqrt{4\pi\rho_i}$ with the frequencies $\omega = kV_A$. Here k is the wave number (assumed parallel to the local field $\mathbf{B}$) and ρ_i is the ion mass density. As long as the Alfvén wave frequency is higher than the ion-neutral collision frequency ν_{in}, the waves are only weakly damped. When, on the other hand, the ion-neutral collision frequency is higher (deeper into the cloud), neutrals are entrained by the oscillating plasma and the Alfvén waves are also able to propagate, albeit with a factor $\sqrt{\rho_i/\rho_0} < 1$ lower speed, where ρ_0 is the neutral density. The propagation speed reduction occurs because every ion is now "loaded" with ρ_0/ρ_i neutrals. Now, between these two regimes Alfvén waves are heavily damped and even disappear altogether for sufficiently small $\rho_i/\rho_0 \ll 0.1$. The evanescence wave number range is then bounded by $k_1 = \nu_{in}/2V_A$ and $k_2 = 2\sqrt{\rho_i/\rho_0}\nu_{in}/V_A$. These phenomena have been studied in detail in Kulsrud and Pearce (1969) and Zweibel and Shull (1982), and specifically in the context of the DSA in Völk et al. (1981), Drury et al. (1996), Bykov et al. (2000) and Reville et al. (2008). Now we turn to their impact on the particle confinement and emissivity.

In the framework of a quasilinear wave-particle interaction, the wave number k is approximately related to the parallel (to the magnetic field) component of the particle momentum $p_\parallel$ by the cyclotron resonance condition $kp_\parallel/m = \pm\omega_c$, where the (non-relativistic) gyrofrequency $\omega_c = eB/mc$. Note that the appearance of $p_\parallel = p\mu$, where μ is the cosine of the pitch angle instead of the often used "sharpened" (Skilling 1975) resonance condition

$kp/m = \pm\omega_c$ is absolutely critical for the break mechanism. The frequency range where the waves cannot propagate may be conveniently translated into the parallel momentum range

$$p_1 < |p_\parallel| < p_2, \tag{24}$$

with

$$p_1 = 2V_A m\omega_c/\nu_{in}, \qquad p_2 = \frac{p_1}{4}\sqrt{\rho_0/\rho_i} > p_1. \tag{25}$$

That a spectral break must form at the photon energy corresponding to the particle momentum $p = p_1 = p_{br}$, can be readily explained as follows. The 'dead zones' $p_1 < |p_\parallel| < p_2$ imply that particles with $|p_\parallel| > p_1$ do not turn around (while moving along the magnetic field) and escape from the region of CR-dense gas collisions at a $p_\parallel/p$ fraction of the speed of light. More specifically, particles with $p_1 < |p_\parallel| < p_2$ escape because they are not scattered, whereas particles with $|p_\parallel| > p_2$, because they maintain the sign of $p_\parallel$, even though they scatter but cannot jump over the gap $p_1 < |p_\parallel| < p_2$. An exception to this rule are particles with sufficiently large $p_\perp$ that can be mirrored across the gap or overcome it via the resonance broadening.

The break can also be explained in terms of the confinement times of different groups of particles introduced above if we assume a low density pre-shock medium with clumps of dense, partially neutral, gas. Particles with $|p_\parallel| > p_1$ spend only short time $\tau_{\rm esc} \sim L_{\rm c}/c$ (where $L_{\rm c}$ is the size of the clump) inside the gas clumps. They propagate ballistically and their scattering time is assumed to be infinite, as there are no waves they can interact with resonantly ($p_1 < |p_\parallel| < p_2$) or they cannot change their propagation direction ($|p_\parallel| > p_2$). Particles with $|p_\parallel| < p_1$ are, on the contrary, scattered intensively in pitch angle, they frequently change their direction, and so sit in the clump for $\tau_{\rm conf} \sim L_{\rm c}^2/\kappa \sim L_{\rm c}^2/c^2\tau_{\rm sc}$. Here $\tau_{\rm sc}$ is their pitch-angle scattering time and κ is the associated diffusion coefficient. Not only $\tau_{\rm conf} \gg \tau_{\rm esc}$ is required, i.e., $\tau_{\rm sc} \ll L_{\rm c}/c$, but also $\tau_{\rm conf} > L_{\rm c}/U_{\rm sh}$, which means that the shock precursor is shorter than the clump $\kappa/U_{\rm sh} \lesssim L_{\rm CR} < L_{\rm c}$ (here $U_{\rm sh}$ is the shock velocity, and $L_{\rm CR}$ is the thickness of the CR front near the shock). The last condition ensures that particles with $p_\parallel > p_1$ that escape through the clump after having entered it from the shock side, will not interact with the shock after they exit through the opposite side of the clump, thus escaping upstream, Fig. 9. The reason for that is a low level of Alfvén wave turbulence ahead of the CR precursor. We also assume that the ambient magnetic field does not deviate strongly from the shock normal, in order to allow these particles to escape through the far side of the clump.

While particles with $p > p_1$ escape from the regions of enhanced gamma radiation (high gas density), an initially isotropic distribution of accelerated particles is maintained only in a slab in momentum space $|p_\parallel| < p_1$ and becomes thus highly anisotropic (a 'pancake' distribution). What matters for the integral emission, however, is a locally isotropic component $\overline{f}$ of this new proton distribution. It can be introduced by re-averaging the 'pancake' ($|p_\parallel| < p_1$) distribution in pitch angle, $\overline{f}(p) \equiv \int_0^1 f(p,\mu)d\mu$, and is readily obtained assuming that particles remaining in the dense gas (those with $|p_\parallel| < p_1$) maintain the flat pitch-angle distribution, i.e.

$$\overline{f}(p) = \int_0^{\mu_1} f_0(p)d\mu = \begin{cases} (p_1/p) f_0(p), & p \ge p_1 \\ f_0(p), & p < p_1 \end{cases} \tag{26}$$

where $f_0(p)$ is the initial (isotropic) distribution function and $\mu_1 = \min\{p_1/p, 1\}$. Thus, the slope of the particle momentum distribution becomes steeper by exactly one power above $p = p_1 \equiv p_{\rm br}$. In particular, any power-law distribution $\propto p^{-q}$, upon entering an MC, turns into p^{-q-1} at $p \ge p_{\rm br}$, and preserves its form at $p < p_{\rm br}$.

Fig. 9 SNR shock propagating into dense gas environment. The filling factor of the gas clumps is small, while some of them may be larger than the thickness of the CR layer near the shock front

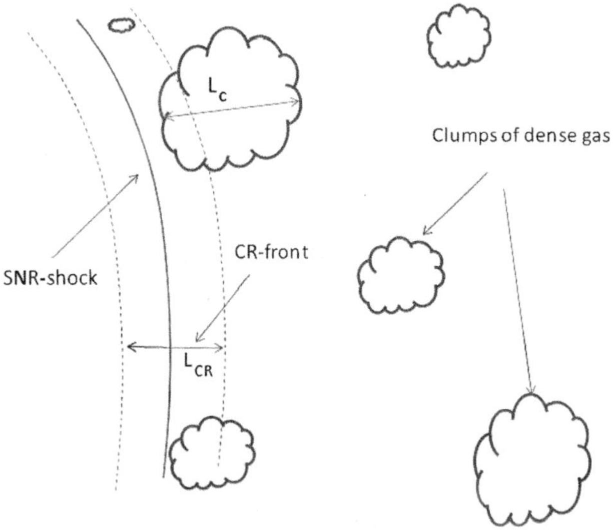

Note that the broken power-law spectrum can only be maintained if the filling factor f_{gas} of the dense gas with the significant wave evanescence interval (p_1, p_2) is relatively small, $f_{\text{gas}} \ll 1$, so that the overall particle confinement and thus the acceleration are not strongly affected. If, on the contrary, $f_{\text{gas}} \sim 1$, the resonant particles would leak into the (p_1, p_2) gap and escape from the accelerator in large amounts, thus suppressing the acceleration. Note that particles with sufficiently high momenta $p > p_2 B_0/\delta B$, where $\delta B/B_0$ is the effective mirror ratio of magnetic perturbations, can "jump" over the gap. The primary p^{-q} slope should then be restored for such particles. Recent MAGIC observations of the SNR W51C (Carmona 2011; Aleksić et al. 2012) indeed point at such spectrum recovery at higher energies. It should also be noted, that the $\Delta q = 1$ break index is a limiting case of identical gas clumps. The integrated emission from an ensemble of clumps with different p_1 and p_2 should result in a more complex spectrum.

6.4 Break Momentum

While the one power spectral break in the pitch-angle averaged particle distribution seems to be a robust environmental signature of a weakly ionized medium into which the accelerated particles propagate, the break momentum remains uncertain. According to Eq. (25), p_{br} $(\equiv p_1)$ depends on the magnetic field strength and ion density as well as on the ion-neutral collision rate $\nu_{in} = n_0 \langle \sigma V \rangle$. Here $\langle \sigma V \rangle$ is the product of the collision cross-section and collision velocity averaged over the thermal distribution. Using an approximation of Draine and McKee (1993) and Drury et al. (1996) for $\langle \sigma V \rangle$, p_{br} can be estimated as

$$p_{br}/mc \simeq 10 B_\mu^2 T_4^{-0.4} n_0^{-1} n_i^{-1/2}. \tag{27}$$

Here the gas temperature T_4 is measured in the units of 10^4 K, magnetic field B_μ—in microgauss, n_0 and n_i (number densities corresponding to the neutral/ion mass densities ρ_0 and ρ_i)—in cm^{-3}. Note that the numerical coefficient in the last expression may vary depending on the average ion and neutral masses and can be higher by a factor of a few (Kulsrud and Pearce 1969; Nakano 1984) than the estimate in Eq. (27) suggests. The remaining quantities in the last formula are also known too poorly to make an accurate independent prediction of the position of the break in the gamma ray emission region. Those are the regions near the blast wave where complicated physical processes unfold, as discussed earlier (Shull and McKee 1979; Draine and McKee 1993; Bykov et al. 2000). Also important may be the ionization by the low energy CRs accelerated at the blast wave. However, as their diffusion length is shorter than that of the particles with $p \gtrsim p_{br}$, we may assume that they do not reach the MC. Pre-ionization by the UV photons can also be ignored for the column density $N > 10^{19}$ cm^{-2} ahead of the shock beyond which they are absorbed (Uchiyama et al. 2010). Uchiyama et al. (2010), using the earlier data from Reach et al. (2005) have also analyzed the parameters involved in Eq. (27) and found the above estimate of p_{br} to be in a good agreement with the spectral break position measured by the Fermi-LAT. Nevertheless, we may run the argument in reverse and use the Fermi observations (Abdo et al. 2010a) of the gamma-ray spectrum of SNR W44 to determine the break momentum in the parent particle spectrum and constrain the parameters in Eq. (27). Since we also know the amount of the slope variation Δq, we can calculate the full spectrum up to the cut-off energy.

It should also be noted that in reality the break at $p = p_{\rm br}$ is not infinitely sharp for the following reasons. The break momentum may change in space due to variations of the gas parameters (Eq. (27)), the resonance broadening (Dupree 1966; Achterberg 1981) near $p = p_1 = p_{br}$ (so that particles with $p \sim p_1$ are still scattered, albeit weakly) and other factors, such as the contribution of small gas clumps with $L_c \ll L_{\rm CR}$, Fig. 9. The small clumps are submerged in the CR front and the CRs that escape from them are readily replenished. Note that this effect may decrease the break index Δq. However, the conversion of the parent proton spectrum into the observable gamma emission introduces a significant smoothing of the break, so that even a sharply broken proton spectrum produces a smooth gamma spectrum. It provides an excellent fit to the Fermi gamma-ray data without an ad hoc proton break smoothing adopted by the Fermi-team (Abdo et al. 2010a) to fit the data.

7 Particle and Gamma-Photon Spectra in Molecular Cloud SNRs

To calculate the particle spectra, we need to determine the degree of nonlinear modification of the shock structure. In principle, it can be calculated consistently, given the shock parameters and the particle maximum momentum, p_{max}. In the case of a broken spectrum, p_{br} likely plays the role of p_{max}, as a momentum where the dominant contribution to the pressure of accelerated particles comes from, thus setting the scale of the modified shock precursor. Note that in the conventional nonlinear (NL) acceleration theory, the cut-off momentum p_{max} plays this role, because the nonlinear spectra are sufficiently flat so as to make the pressure diverge with momentum, unlike the broken spectra.

One of the best documented gamma emission spectra comes from the SNR W44, so we fit these data using the above mechanism of the spectral break. The break in the photon spectrum is observed at about 2 GeV, which places the break in the proton distribution at about $p_{br} \simeq 7$ GeV$/c$ (Abdo et al. 2010a). For the strength of the break $\Delta q = 1$, the spectrum above it is clearly pressure converging, so that the shock structure and the spectrum may be calculated using this break momentum as the point of the maximum in the CR partial pressure. More specifically, once the break momentum is set, one can use an analytic approach (Malkov and Drury 2001) for a stationary nonlinear acceleration problem using p_{br} as an input parameter.

Apart from p_{br}, the nonlinear solution depends on a number of other parameters, such as the injection rate of thermal particles into acceleration, Mach number, the precursor heating rate and the shock velocity V_s. Of these parameters the latter is known reasonably well, $V_s \approx 300$ km/s, the injection rate can be either calculated analytically for the parallel shock geometry (Malkov and Drury 2001), or inferred from the simulations (Gargaté and Spitkovsky 2012), while the other parameters are still difficult to ascertain. Fortunately, in sufficiently strong shocks the solution either stays close to the test particle (TP) solution (leaving the shock structure only weakly modified) or else it transitions to a strongly modified NL-solution regime. The TP regime typically establishes in cases of moderate Mach numbers, low injection rates and low p_{max} (now probably closer to p_{br}), while the NL regime is unavoidable in the opposite part of the parameter space.

In the TP regime the spectrum is close to a power-law with the spectral index 2 throughout the supra-thermal energy range. In the NL regime, however, the spectrum develops a concave form, starting from a softer spectrum at the injection energy, with the index $q \simeq (r_s + 2)/(r_s - 1) > 2$, where $r_s < 4$ is the sub-shock compression ratio. Then it hardens, primarily in the region $p \sim mc$, where both the partial pressure and diffusivity of protons change their momentum dependence. The slope reaches its minimum at the cut-off (break) energy, which, depending on the degree of nonlinearity, can be as low as 1.5 or even somewhat lower if the cut-off is abrupt. The question now is into which of these two categories the W44 spectrum falls? Generally, in cases of low maximum (or, equivalently, low spectral break $p_{br} \lesssim 10$) momentum, the shock modification is weak, so the spectrum is more likely to be in a slightly nonlinear, almost TP regime. On the other hand, there is a putative indication from the electron radio emission that their spectrum may be close to $q_e \approx 1.75$, which could be the signature of a moderately nonlinear acceleration process. It should be remembered, however, that this is a global index across the W44 remnant. There are resolved bright filaments where a canonical $\alpha = -0.5$ spectrum, corresponding precisely to the TP parent electron spectrum with $q_e = 2$, is observed (Castelletti et al. 2007). Moreover, there are regions with the positive indices $\alpha \lesssim 0.4$ which cannot be indicative of a DSA process without corrections for subsequent spectral transformations such as an absorption by thermal electrons. These regions may contribute to the overall spectral hardening discussed above, thus mimicking the acceleration nonlinearity. Finally, secondary electrons give rise to the flattening of the radio spectrum as well (Uchiyama et al. 2010).

The above considerations somewhat weaken the radio data as a probe for the slope of the electron and (more importantly) for the proton spectrum. Therefore, the exact degree of nonlinearity of the acceleration remains unknown and one can consider both the TP and weakly NL regimes in calculations of the photon spectra, generated in $p - p$ collisions. Specifically, Malkov et al. (2005, 2011) calculate the π^0 production rate and the gamma-ray emissivity. In so doing, they adopt numerical recipe described in detail in Kamae et al. (2006) and Karlsson and Kamae (2008). The physical processes behind these calculations are (i) collisions of accelerated protons with the protons of the ambient gas resulting in $pp \rightarrow \pi^0$ reaction (ii) decay of π^0-mesons to generate an observable gamma emission spectrum.

An example of such calculations is shown in Fig. 10. The best fit to the Fermi and AGILE data is provided by a TP energy distribution ($\propto E^{-2}$) below $p_{br} \simeq 7$ GeV$/c$ with the spectrum steepening by exactly one power above it. The spectrum steepening is perfectly consistent with the proton partial escape described above (with no parameters involved). For comparison, a weakly NL spectrum (guided by the inferred electron spectrum with $q_e \approx 1.75$, is also used for these calculations (dashed line in Fig. 10), but its fit would require a somewhat stronger break ($\Delta q \gtrsim 1$) or a low momentum cut-off, i.e. at least one additional free parameter. It is seen that the mechanism for a break in the spectrum of shock accelerated protons suggested in Malkov et al. (2005) provides a good fit to the recent (Abdo et al. 2010a) Fermi-LAT and AGILE (Giuliani et al. 2011) observations of the SNR W44.

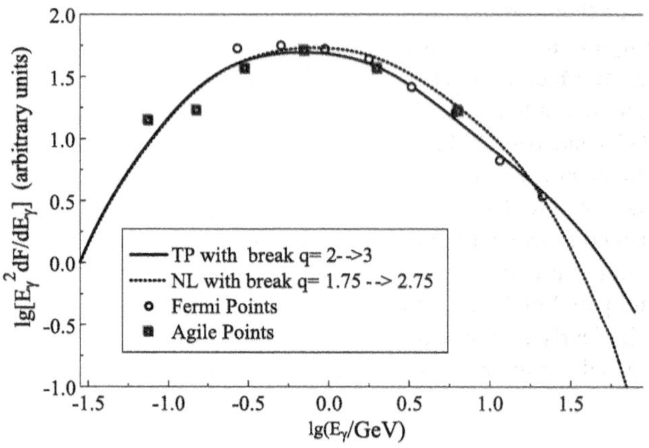

Fig. 10 Gamma radiation spectra. Photon spectra resulting from π^0 decay and calculated for two different parent proton spectra compared against the Fermi (*circles*) and AGILE (*squares*) data. *Solid line*: a test particle acceleration regime with the spectral index $q = 2$ below the break and $q = 3$ above the break at $p_{br} = 7$ GeV$/c$. *Dashed line*: a moderately nonlinear acceleration regime corresponding to the spectrum with $q \simeq 1.75$ and $q \simeq 2.75$ below and above the break, respectively. Cut-offs are placed at 300 GeV for TP- and 100 GeV, for NL-spectrum. Fermi and AGILE data are adopted from Abdo et al. (2010a) and Giuliani et al. (2011), respectively. Both curves are well within the error bars of Fermi and AGILE (not shown for clarity), which, in turn, overlap Giuliani et al. (2011)

Of course, in assessing consistency of the suggested spectral break phenomenon with the observed spectrum, the errors in the data must be taken into account. The vertical error bars near the break at 2 GeV, are fairly small (comparable to the size of the symbols used to represent the data in Fig. 10). More significant appears to be the energy dispersion. However, in the most recent Fermi-LAT publication (see Ackermann et al. 2013 including the Supplementary Online Materials) the energy dispersion is estimated to be less than 5 % for these energies, so that the broken power law is indeed consistent with the data.)

Generally, spectral breaks offer a possible resolution to the well known problem that some nonlinear DSA models produce spectra which are considerably harder than a simple test particle spectrum, and these are not consistent with the gamma-ray observations of some of supernova remnants. However, the nonlinear spectrum—i.e., diverging in energy—exhausts the shock energy available for the acceleration as the cut-off momentum grows, so that a broken spectrum should form. Broken spectra are commonly observed and the old paradigm of a single power-law with an exponential upper cut-off is maladapted to the recent, greatly improved observations (Abdo et al. 2010a, 2010b). Note, that the spectrum of the RX J1713.7-3946 (Aharonian et al. 2006) is also formally consistent with the environmental break mechanism presumably operating in W44 surrounding but with a higher $p_{br} \sim 10^3$ GeV$/c$ and thus with stronger acceleration nonlinearity (Malkov et al. 2005). However, this remnant expands into a rather complicated environment, so it is difficult to make the case for hadronic origin of the gamma-ray emission (Aharonian et al. 2004, 2006; Katz and Waxman 2008; Ellison et al. 2012). The important role of the W44 remnant for the problem of CR origin is that this particular remnant seems to be unlikely dominated by the lepton emission due to Bremsstrahlung and inverse Compton scattering (Abdo et al. 2010a; Uchiyama et al. 2010) thus favoring the hadronic origin of the gamma emission and bolstering the case for the SNR origin of galactic CRs.

8 Summary

Collisionless shocks are ubiquitous in astrophysical objects and are observed at all scales starting from the heliosphere and up to cosmological scale shocks observed in clusters of galaxies. Theoretical modeling of these shocks is challenging because shock relaxation process involves collective plasma oscillations producing long-lived highly nonequilibrium components. Straightforward numerical simulations are an extremely resource demanding task because of the very wide dynamical range of scales and time required to the resolve all of the long-lived components. Such task is even more difficult if neutral atoms and molecules are significant in the partially ionized media.

In this brief review we have discussed some observational appearance of collisionless shocks in partially ionized plasma, described the most important physical processes operating in the shocks and outlined the observational perspective of nonthermal components diagnostics of astrophysical collisionless shocks via multiwavelength observations.

Nonthermal components, i.e., energetic charged and neutral particles and fluctuating magnetic fields, can drastically modify the structure of the shock upstream providing both deceleration of the plasma flow and also efficient heating of ions and electrons. We argue that the processes of turbulence amplification and damping observed in the heliosphere may help to understand the microphysics of ion and electron heating by cosmic ray driven turbulence in the upstream regions of large scale collisionless shocks observed in the Galaxy and in clusters of galaxies. In turn, radiative signatures of the astrophysical shocks can be a unique way to study microscopic phenomena that can not be studied in the laboratory plasma on the Earth.

Hα line diagnostics of Balmer shocks provides valuable information on charge-exchange processes of the neutrals with both thermal and non-Maxwellian plasma components in the shock downstream. Nonthermal components may reduce the plasma temperature comparing to what is expected from the standard single-fluid Rankine-Hugoniot jump conditions for a particular shock velocity. In the case of radiative shocks this reduction would substantially modify the emission line spectrum coming from shock downstream and thus may serve as a valuable diagnostic tool for fast shocks interacting with clouds. The effect of neutrals on the MHD wave damping in the upstream of supernova shocks interacting with a molecular cloud may explain some spectral features in GeV energy regime, recently revealed by *Fermi* observations.

Acknowledgements A.M.B., J.C.R. and M.A.M. thank Andre Balogh and the ISSI staff for providing an inspiring atmosphere at the International Space Science Institute Workshop in Bern in 2012, which has led to the new collaboration. The authors thank the referee for a constructive report. A.M.B. and A.M.K. acknowledge support from the RAS Programs P 21 and OFN 16, and from the Ministry of Education and Science of Russian Federation (Agreement No. 8409, 2012). They performed the simulations at the Joint Supercomputing Centre (JSCC RAS) and the Supercomputing Centre at Ioffe Institute, St. Petersburg. M.M. acknowledges support by the Department of Energy, Grant No. DE-FG02-04ER54738. A.E.V. acknowledges support by NASA grants NNX09AC15G and NNX12AO73G.

References

A.A. Abdo, M. Ackermann, M. Ajello, L. Baldini et al., Gamma-ray emission from the shell of supernova remnant W44 revealed by the Fermi LAT. Science **327**, 1103–1106 (2010a)

A.A. Abdo, M. Ackermann, M. Ajello, L. Baldini et al., Observation of supernova remnant IC 443 with the Fermi Large Area Telescope. Astrophys. J. **712**, 459–468 (2010b). arXiv:1002.2198

A. Achterberg, On the propagation of relativistic particles in a high beta plasma. Astron. Astrophys. **98**, 161–172 (1981)

M. Ackermann, M. Ajello, A. Allafort, L. Baldini et al., Detection of the characteristic pion-decay signature in supernova remnants. Science **339**, 807–811 (2013)

F.A. Aharonian, A.G. Akhperjanian, K.M. Aye, A.R. Bazer-Bachi et al., High-energy particle acceleration in the shell of a supernova remnant. Nature **432**, 75–77 (2004). arXiv:astro-ph/0411533

F. Aharonian, A.G. Akhperjanian, A.R. Bazer-Bachi, M. Beilicke et al., A detailed spectral and morphological study of the gamma-ray supernova remnant RX J1713.7-3946 with HESS. Astron. Astrophys. **449**, 223–242 (2006)

J. Aleksić, E.A. Alvarez, L.A. Antonelli, P. Antoranz et al., Morphological and spectral properties of the W51 region measured with the MAGIC telescopes. Astron. Astrophys. **541**, A13 (2012). arXiv:1201.4074

O. Alexandrova, C. Lacombe, A. Mangeney, R. Grappin, Fluid-like dissipation of magnetic turbulence at electron scales in the solar wind (2011). arXiv:1111.5649

E. Amato, P. Blasi, Non-linear particle acceleration at non-relativistic shock waves in the presence of self-generated turbulence. Mon. Not. R. Astron. Soc. **371**, 1251–1258 (2006). arXiv:astro-ph/0606592

A.R. Bell, Turbulent amplification of magnetic field and diffusive shock acceleration of cosmic rays. Mon. Not. R. Astron. Soc. **353**, 550–558 (2004)

R. Blandford, D. Eichler, Particle acceleration at astrophysical shocks: a theory of cosmic ray origin. Phys. Rep. **154**, 1–75 (1987)

P. Blasi, G. Morlino, R. Bandiera, E. Amato et al., Collisionless shocks in a partially ionized medium. I. Neutral return flux and its effects on acceleration of test particles. Astrophys. J. **755**, 121 (2012). arXiv:1202.3080

M.F. Bode, D.J. Harman, T.J. O'Brien, H.E. Bond et al., Hubble space telescope imaging of the expanding nebular remnant of the 2006 outburst of the recurrent Nova RS Ophiuchi. Astrophys. J. **665**, L63–L66 (2007). arXiv:0706.2745

S. Boldyrev, J.C. Perez, Spectrum of kinetic-Alfvén turbulence. Astrophys. J. **758**, L44 (2012). arXiv: 1204.5809

A. Boulares, D.P. Cox, Application of cosmic-ray shock theories to the Cygnus Loop—an alternative model. Astrophys. J. **333**, 198–218 (1988)

S.V. Bozhokin, A.M. Bykov, Nonthermal particles in H I shells and supershells. Astron. Lett. **20**, 503–507 (1994)

S.I. Braginskii, Transport processes in a plasma. Rev. Plasma Phys. **1**, 205 (1965)

B. Breech, W.H. Matthaeus, S.R. Cranmer, J.C. Kasper et al., Electron and proton heating by solar wind turbulence. J. Geophys. Res. **114**(A13), A09103 (2009). arXiv:0907.4074

M. Brüggen, A. Bykov, D. Ryu, H. Röttgering, Magnetic fields, relativistic particles, and shock waves in cluster outskirts. Space Sci. Rev. **166**, 187–213 (2012). arXiv:1107.5223

A.M. Bykov, Shocks and particle acceleration in SNRs: theoretical aspects. Adv. Space Res. **33**, 366–375 (2004)

A.M. Bykov, Multi-fluid shocks in clusters of galaxies: entropy, σ_v T, M T and L_X T scalings. Adv. Space Res. **36**, 738–746 (2005). arXiv:astro-ph/0501575

A.M. Bykov, I.N. Toptygin, Generation of magnetic fluctuations near a shock front in a partially ionized medium. Astron. Lett. **31**, 748–754 (2005)

A.M. Bykov, R.A. Chevalier, D.C. Ellison, Y.A. Uvarov, Nonthermal emission from a supernova remnant in a molecular cloud. Astrophys. J. **538**, 203–216 (2000)

A.M. Bykov, K. Dolag, F. Durret, Cosmological shock waves. Space Sci. Rev. **134**, 119–140 (2008). arXiv: 0801.0995

A.M. Bykov, S.M. Osipov, D.C. Ellison, Cosmic ray current driven turbulence in shocks with efficient particle acceleration: the oblique, long-wavelength mode instability. Mon. Not. R. Astron. Soc. **410**, 39–52 (2011). arXiv:1010.0408

A.M. Bykov, D.C. Ellison, M. Renaud, Magnetic fields in cosmic particle acceleration sources. Space Sci. Rev. **166**, 71–95 (2012). arXiv:1105.0130

E. Carmona, Probing proton acceleration in W51C with MAGIC, in *International Cosmic Ray Conference*, vol. 7 (2011), p. 114. arXiv:1110.0950

G. Castelletti, G. Dubner, C. Brogan, N.E. Kassim, The low-frequency radio emission and spectrum of the extended SNR W44: new VLA observations at 74 and 324 MHz. Astron. Astrophys. **471**, 537–549 (2007)

C.H.K. Chen, A. Mallet, A.A. Schekochihin, T.S. Horbury et al., Three-dimensional structure of solar wind turbulence. Astrophys. J. **758**, 120 (2012). arXiv:1109.2558

R.A. Chevalier, J.C. Raymond, Optical emission from a fast shock wave—the remnants of Tycho's supernova and SN 1006. Astrophys. J. **225**, L27–L30 (1978)

J. Cho, A. Lazarian, The anisotropy of electron magnetohydrodynamic turbulence. Astrophys. J. **615**, L41–L44 (2004). arXiv:astro-ph/0406595

T.G. Cowling, *Magnetohydrodynamics* (1976)

S.R. Cranmer, A.A. van Ballegooijen, Proton, electron, and ion heating in the fast solar wind from nonlinear coupling between Alfvénic and fast-mode turbulence. Astrophys. J. **754**, 92 (2012). arXiv:1205.4613
M.A. Dopita, R.S. Sutherland, Spectral signatures of fast shocks. I. Low-density model grid. Astrophys. J. Suppl. Ser. **102**, 161 (1996)
B.T. Draine, C.F. McKee, Theory of interstellar shocks. Annu. Rev. Astron. Astrophys. **31**, 373–432 (1993)
L.O. Drury, An introduction to the theory of diffusive shock acceleration of energetic particles in tenuous plasmas. Rep. Prog. Phys. **46**, 973–1027 (1983)
L.O. Drury, S.A.E.G. Falle, On the stability of shocks modified by particle acceleration. Mon. Not. R. Astron. Soc. **223**, 353 (1986)
L.O.C. Drury, P. Duffy, J.G. Kirk, Limits on diffusive shock acceleration in dense and incompletely ionised media. Astron. Astrophys. **309**, 1002–1010 (1996)
T.H. Dupree, A perturbation theory for strong plasma turbulence. Phys. Fluids **9**(9), 1773–1782 (1966)
D.C. Ellison, M.G. Baring, F.C. Jones, Nonlinear particle acceleration in oblique shocks. Astrophys. J. **473**, 1029 (1996). arXiv:astro-ph/9609182
D.C. Ellison, A.M. Bykov, Gamma-ray emission of accelerated particles escaping a supernova remnant in a molecular cloud. Astrophys. J. **731**, 87 (2011). arXiv:1102.3885
D.C. Ellison, P. Slane, D.J. Patnaude, A.M. Bykov, Core-collapse model of broadband emission from SNR RX J1713.7-3946 with thermal X-rays and gamma rays from escaping cosmic rays. Astrophys. J. **744**, 39 (2012). arXiv:1109.0874
R. Enomoto, T. Tanimori, T. Naito, T. Yoshida et al., The acceleration of cosmic-ray protons in the supernova remnant RX J1713.7-3946. Nature **416**, 823–826 (2002)
A. Evans, C.E. Woodward, L.A. Helton, R.D. Gehrz et al., Spitzer and ground-based infrared observations of the 2006 eruption of RS Ophiuchi. Astrophys. J. **663**, L29–L32 (2007). arXiv:0705.2414
C.L. Farage, P.J. McGregor, M.A. Dopita, G.V. Bicknell, Optical IFU observations of the brightest cluster galaxy NGC 4696: the case for a minor merger and shock-excited filaments. Astrophys. J. **724**, 267–284 (2010). arXiv:1009.3070
R.A. Fesen, H. Itoh, A two-dimensional spectrum of a nonradiative shock filament in the Cygnus Loop. Astrophys. J. **295**, 43–50 (1985)
K. France, R. McCray, S.V. Penton, R.P. Kirshner et al., HST-COS observations of hydrogen, helium, carbon, and nitrogen emission from the SN 1987A reverse shock. Astrophys. J. **743**, 186 (2011). arXiv:1111.1735
Y. Fujita, Y. Ohira, R. Yamazaki, Entropy at the outskirts of galaxy clusters as implications for cosmological cosmic-ray acceleration. Astrophys. J. **767**, L4 (2013). arXiv:1303.1191
L. Gargaté, A. Spitkovsky, Ion acceleration in non-relativistic astrophysical shocks. Astrophys. J. **744**, 67 (2012). arXiv:1107.0762
S.P. Gary, O. Chang, W.J. Forward, Cascade of Whistler turbulence: three-dimensional particle-in-cell simulations. Astrophys. J. **755**, 142 (2012)
P. Ghavamian, J.M. Laming, C.E. Rakowski, A physical relationship between electron-proton temperature equilibration and Mach number in fast collisionless shocks. Astrophys. J. **654**, L69–L72 (2007). arXiv:astro-ph/0611306
P. Ghavamian, J. Raymond, P. Hartigan, W.P. Blair, Evidence for shock precursors in Tycho's supernova remnant. Astrophys. J. **535**, 266–274 (2000)
P. Ghavamian, J. Raymond, R.C. Smith, P. Hartigan, Balmer-dominated spectra of nonradiative shocks in the Cygnus Loop, RCW 86, and Tycho supernova remnants. Astrophys. J. **547**, 995–1009 (2001). arXiv: astro-ph/0010496
T. Giannini, L. Calzoletti, B. Nisini, C.J. Davis et al., Near-infrared, IFU spectroscopy unravels the bow-shock HH99B. Astron. Astrophys. **481**, 123–139 (2008). arXiv:0801.1633
A. Giuliani, M. Cardillo, M. Tavani, Y. Fukui et al., Neutral pion emission from accelerated protons in the supernova remnant W44. Astrophys. J. **742**, L30 (2011). arXiv:1111.4868
G. Gloeckler, J. Geiss, Heliospheric and interstellar phenomena deduced from pickup ion observations. Space Sci. Rev. **97**, 169–181 (2001)
P. Goldreich, S. Sridhar, Toward a theory of interstellar turbulence. 2: Strong Alfvenic turbulence. Astrophys. J. **438**, 763–775 (1995)
P. Hartigan, J. Raymond, L. Hartmann, Radiative bow shock models of Herbig-Haro objects. Astrophys. J. **316**, 323–348 (1987)
E.A. Helder, J. Vink, C.G. Bassa, A. Bamba et al., Measuring the cosmic-ray acceleration efficiency of a supernova remnant. Science **325**, 719 (2009). arXiv:0906.4553
E.A. Helder, D. Kosenko, J. Vink, Cosmic-ray acceleration efficiency versus temperature equilibration: the case of SNR 0509-67.5. Astrophys. J. **719**, L140–L144 (2010). arXiv:1007.3138
E.A. Helder, J. Vink, A.M. Bykov, Y. Ohira et al., Observational signatures of particle acceleration in supernova remnants. Space Sci. Rev. **173**, 369–431 (2012). arXiv:1206.1593

K. Heng, Balmer-dominated shocks: a concise review. Publ. Astron. Soc. Aust. **27**, 23–44 (2010). arXiv: 0908.4080

K. Heng, R. McCray, Balmer-dominated shocks revisited. Astrophys. J. **654**, 923–937 (2007). arXiv:astro-ph/0609331

K. Heng, M. van Adelsberg, R. McCray, J.C. Raymond, The transition zone in Balmer-dominated shocks. Astrophys. J. **668**, 275–284 (2007). arXiv:0705.2619

J.J. Hester, J.C. Raymond, W.P. Blair, The Balmer-dominated northeast limb of the Cygnus Loop supernova remnant. Astrophys. J. **420**, 721–745 (1994)

G.G. Howes, A prescription for the turbulent heating of astrophysical plasmas. Mon. Not. R. Astron. Soc. **409**, L104–L108 (2010). arXiv:1009.4212

G.G. Howes, S.C. Cowley, W. Dorland, G.W. Hammett et al., Astrophysical gyrokinetics: basic equations and linear theory. Astrophys. J. **651**, 590–614 (2006). arXiv:astro-ph/0511812

T. Inoue, R. Yamazaki, S. Inutsuka, Y. Fukui, Toward understanding the cosmic-ray acceleration at young supernova remnants interacting with interstellar clouds: possible applications to RX J1713.7-3946. Astrophys. J. **744**, 71 (2012). arXiv:1106.3080

F.C. Jones, D.C. Ellison, The plasma physics of shock acceleration. Space Sci. Rev. **58**, 259–346 (1991)

T. Kamae, N. Karlsson, T. Mizuno, T. Abe et al., Parameterization of gamma, e+/−, and neutrino spectra produced by p-p interaction in astronomical environments. Astrophys. J. **647**, 692–708 (2006)

N. Karlsson, T. Kamae, Parameterization of the angular distribution of gamma rays produced by p-p interaction in astronomical environments. Astrophys. J. **674**, 278–285 (2008)

B. Katz, E. Waxman, In which shell-type SNRs should we look for gamma-rays and neutrinos from p p collisions? J. Cosmol. Astropart. Phys. **2008**(1), 018 (2008)

R. Kulsrud, W.P. Pearce, The effect of wave-particle interactions on the propagation of cosmic rays. Astrophys. J. **156**, 445–469 (1969)

R.J. Leamon, C.W. Smith, N.F. Ness, W.H. Matthaeus et al., Observational constraints on the dynamics of the interplanetary magnetic field dissipation range. J. Geophys. Res. **103**, 4775 (1998)

J.J. Lee, B.C. Koo, J. Raymond, P. Ghavamian et al., Subaru HDS observations of a Balmer-dominated shock in Tycho's supernova remnant. Astrophys. J. **659**, L133–L136 (2007). arXiv:0704.1094

J.J. Lee, J.C. Raymond, S. Park, W.P. Blair et al., Resolved shock structure of the Balmer-dominated filaments in Tycho's supernova remnant: cosmic-ray precursor? Astrophys. J. **715**, L146–L149 (2010). arXiv:1005.3296

A.J. Lim, A.C. Raga, A distribution function calculation of the Hα profiles of high-velocity shocks—II. The broad component neutral precursor. Mon. Not. R. Astron. Soc. **280**, 103–114 (1996)

M.A. Malkov, Analytic solution for nonlinear shock acceleration in the Bohm limit. Astrophys. J. **485**, 638 (1997). arXiv:astro-ph/9707152

M.A. Malkov, L.O. Drury, Nonlinear theory of diffusive acceleration of particles by shock waves. Rep. Prog. Phys. **64**, 429–481 (2001)

M.A. Malkov, P.H. Diamond, R.Z. Sagdeev, On the gamma-ray spectra radiated by protons accelerated in supernova remnant shocks near molecular clouds: the case of supernova remnant RX J1713.7-3946. Astrophys. J. **624**, L37–L40 (2005)

M.A. Malkov, P.H. Diamond, R.Z. Sagdeev, Mechanism for spectral break in cosmic ray proton spectrum of supernova remnant W44. Nat. Commun. **2**, 1195 (2011). arXiv:1004.4714

M.A. Malkov, R.Z. Sagdeev, P.H. Diamond, Magnetic and density spikes in cosmic-ray shock precursors. Astrophys. J. **748**, L32 (2012). arXiv:1110.0257

A. Marcowith, M. Lemoine, G. Pelletier, Turbulence and particle acceleration in collisionless supernovae remnant shocks. II. Cosmic-ray transport. Astron. Astrophys. **453**, 193–202 (2006). arXiv:astro-ph/0603462

F. Massi, C. Codella, J. Brand, L. di Fabrizio et al., The low-mass YSO CB230-A: investigating the protostar and its jet with NIR spectroscopy and Spitzer observations. Astron. Astrophys. **490**, 1079–1091 (2008). arXiv:0809.1591

D.J. McComas, F. Allegrini, P. Bochsler, M. Bzowski et al., Global observations of the interstellar interaction from the Interstellar Boundary Explorer (IBEX). Science **326**, 959 (2009)

C.F. McKee, D.J. Hollenbach, Interstellar shock waves. Annu. Rev. Astron. Astrophys. **18**, 219–262 (1980)

D. Mihalas, B.W. Mihalas, *Foundations of Radiation Hydrodynamics* (Oxford University Press, New York, 1984), 731 p.

M. Mithaiwala, L. Rudakov, C. Crabtree, G. Ganguli, Co-existence of Whistler waves with kinetic Alfven wave turbulence for the high-beta solar wind plasma. Phys. Plasmas **19**(10), 102902 (2012). arXiv:1208.0623

A.S. Monin, A.M. Iaglom, *Statistical Fluid Mechanics: Mechanics of Turbulence*, vol. 2 (1975), revised and enlarged edition

G. Morlino, R. Bandiera, P. Blasi, E. Amato, Collisionless shocks in a partially ionized medium. II. Balmer emission. Astrophys. J. **760**, 137 (2012a). arXiv:1210.4296
G. Morlino, P. Blasi, R. Bandiera, E. Amato et al., Collisionless shocks in a partially ionized medium: III. Efficient cosmic ray acceleration (2012b). arXiv:1211.6148
J.A. Morse, W.P. Blair, M.A. Dopita, J.P. Hughes et al., Hubble space telescope observations of oxygen-rich supernova remnants in the Magellanic cloud. I. Narrow-band imaging of N132D in the LMC. Astron. ž. **112**, 509 (1996)
T. Nakano, Contraction of magnetic interstellar clouds. Fundam. Cosm. Phys. **9**, 139–231 (1984)
S. Nikolić, G. van de Ven, K. Heng, D. Kupko et al., An integral view of fast shocks around Supernova 1006 (2013). arXiv:1302.4328
L. O'C Drury, P. Duffy, J.G. Kirk, Limits on diffusive shock acceleration in dense and incompletely ionised media. Astron. Astrophys. **309**, 1002–1010 (1996). arXiv:astro-ph/9510066
Y. Ohira, Effects of leakage neutral particles on shocks. Astrophys. J. **758**, 97 (2012). arXiv:1202.4620
Y. Ohira, F. Takahara, Effects of neutral particles on modified shocks at supernova remnants. Astrophys. J. **721**, L43–L47 (2010). arXiv:0912.2859
Y. Ohira, T. Terasawa, F. Takahara, Plasma instabilities as a result of charge exchange in the downstream region of supernova remnant shocks. Astrophys. J. **703**, L59–L62 (2009). arXiv:0908.3369
A. Petrosyan, A. Balogh, M.L. Goldstein, J. Léorat et al., Turbulence in the solar atmosphere and solar wind. Space Sci. Rev. **156**, 135–238 (2010)
V. Pierrard, M. Lazar, Kappa distributions: theory and applications in space plasmas. Sol. Phys. **267**, 153–174 (2010). arXiv:1003.3532
E. Quataert, Particle heating by Alfvenic turbulence in hot accretion flows. Astrophys. J. **500**, 978 (1998). arXiv:astro-ph/9710127
E. Quataert, A. Gruzinov, Turbulence and particle heating in advection-dominated accretion flows. Astrophys. J. **520**, 248–255 (1999). arXiv:astro-ph/9803112
C.E. Rakowski, J.M. Laming, P. Ghavamian, The heating of thermal electrons in fast collisionless shocks: the integral role of cosmic rays. Astrophys. J. **684**, 348–357 (2008). arXiv:0805.3084
J.J. Ramos, Fluid and drift-kinetic description of a magnetized plasma with low collisionality and slow dynamics orderings. II. Ion theory. Phys. Plasmas **18**(10), 102–506 (2011)
J.C. Raymond, Shock waves in the interstellar medium. Astrophys. J. Suppl. Ser. **39**, 1–27 (1979)
J.C. Raymond, Supernova-remnant shock waves close up. Publ. Astron. Soc. Pac. **103**, 781–786 (1991)
J.C. Raymond, L. Hartmann, P. Hartigan, Improved bow shock models for Herbig-Haro objects—application to HH 2A-prime. Astrophys. J. **326**, 323–333 (1988a)
J.C. Raymond, J.J. Hester, D. Cox, W.P. Blair et al., Spatial and spectral interpretation of a bright filament in the Cygnus Loop. Astrophys. J. **324**, 869–892 (1988b)
J.C. Raymond, P.A. Isenberg, J.M. Laming, Non-Maxwellian proton velocity distributions in nonradiative shocks. Astrophys. J. **682**, 408–415 (2008). arXiv:0804.3808
J.C. Raymond, P.F. Winkler, W.P. Blair, J.J. Lee et al., Non-Maxwellian Hα profiles in Tycho's supernova remnant. Astrophys. J. **712**, 901–907 (2010)
J.C. Raymond, J. Vink, E.A. Helder, A. de Laat, Effects of neutral hydrogen on cosmic-ray precursors in supernova remnant shock waves. Astrophys. J. **731**, L14 (2011). arXiv:1103.3211
W.T. Reach, J. Rho, T.H. Jarrett, Shocked molecular gas in the supernova remnants W28 and W44: near-infrared and millimeter-wave observations. Astrophys. J. **618**, 297–320 (2005)
B. Reville, J.G. Kirk, P. Duffy, S. O'Sullivan, A cosmic ray current-driven instability in partially ionised media. Astron. Astrophys. **475**, 435–439 (2007). arXiv:0707.3743
B. Reville, J.G. Kirk, P. Duffy, S. O'Sullivan, Environmental limits on the nonresonant cosmic-ray current-driven instability. Int. J. Mod. Phys. D **17**, 1795–1801 (2008)
F. Sahraoui, M.L. Goldstein, P. Robert, Y.V. Khotyaintsev, Evidence of a cascade and dissipation of solar-wind turbulence at the electron gyroscale. Phys. Rev. Lett. **102**(23), 231102 (2009)
S. Saito, S.P. Gary, Beta dependence of electron heating in decaying Whistler turbulence: particle-in-cell simulations. Phys. Plasmas **19**, 012312 (2012)
A.A. Schekochihin, S.C. Cowley, W. Dorland, G.W. Hammett et al., Astrophysical gyrokinetics: kinetic and fluid turbulent cascades in magnetized weakly collisional plasmas. Astrophys. J. Suppl. Ser. **182**, 310–377 (2009). arXiv:0704.0044
K.M. Schure, A.R. Bell, L. O'C Drury, A.M. Bykov, Diffusive shock acceleration and magnetic field amplification. Space Sci. Rev. **173**, 491–519 (2012). arXiv:1203.1637
J.M. Shull, C.F. McKee, Theoretical models of interstellar shocks. I—Radiative transfer and UV precursors. Astrophys. J. **227**, 131–149 (1979)
J. Skilling, Cosmic ray streaming. I—Effect of Alfven waves on particles. Mon. Not. R. Astron. Soc. **172**, 557–566 (1975)

R.C. Smith, J.C. Raymond, J.M. Laming, High-resolution spectroscopy of Balmer-dominated shocks in the Large Magellanic Cloud. Astrophys. J. **420**, 286–293 (1994)

J. Sollerman, P. Ghavamian, P. Lundqvist, R.C. Smith, High resolution spectroscopy of Balmer-dominated shocks in the RCW 86, Kepler and SN 1006 supernova remnants. Astron. Astrophys. **407**, 249–257 (2003). arXiv:astro-ph/0306196

L. Spitzer, *Physical Processes in the Interstellar Medium* (1978)

V. Tatischeff, M. Hernanz, Evidence for nonlinear diffusive shock acceleration of cosmic rays in the 2006 outburst of the recurrent Nova RS Ophiuchi. Astrophys. J. **663**, L101–L104 (2007). arXiv:0705.4422

O. Teşileanu, S. Massaglia, A. Mignone, G. Bodo et al., Time-dependent MHD shocks and line intensity ratios in the HH 30 jet: a focus on cooling function and numerical resolution. Astron. Astrophys. **507**, 581–588 (2009). arXiv:0910.1225

Y. Uchiyama, R.D. Blandford, S. Funk, H. Tajima et al., Gamma-ray emission from crushed clouds in supernova remnants. Astrophys. J. **723**, L122–L126 (2010)

M. van Adelsberg, K. Heng, R. McCray, J.C. Raymond, Spatial structure and collisionless electron heating in Balmer-dominated shocks. Astrophys. J. **689**, 1089–1104 (2008). arXiv:0803.2521

M.K. Verma, Statistical theory of magnetohydrodynamic turbulence: recent results. Phys. Rep. **401**, 229–380 (2004). arXiv:nlin/0404043

J. Vink, Supernova remnants: the X-ray perspective. Astron. Astrophys. Rev. **20**, 49 (2012). arXiv: 1112.0576

J. Vink, R. Yamazaki, E.A. Helder, K.M. Schure, The relation between post-shock temperature, cosmic-ray pressure, and cosmic-ray escape for non-relativistic shocks. Astrophys. J. **722**, 1727–1734 (2010). arXiv:1008.4367

A.E. Vladimirov, A.M. Bykov, D.C. Ellison, Turbulence dissipation and particle injection in nonlinear diffusive shock acceleration with magnetic field amplification. Astrophys. J. **688**, 1084–1101 (2008). arXiv:0807.1321

A.E. Vladimirov, A.M. Bykov, D.C. Ellison, Spectra of magnetic fluctuations and relativistic particles produced by a nonresonant wave instability in supernova remnant shocks. Astrophys. J. **703**, L29–L32 (2009). arXiv:0908.2602

H.J. Völk, G.E. Morfill, M.A. Forman, The effect of losses on acceleration of energetic particles by diffusive scattering through shock waves. Astrophys. J. **249**, 161–175 (1981)

A.Y. Wagner, J.J. Lee, J.C. Raymond, T.W. Hartquist et al., A cosmic-ray precursor model for a Balmer-dominated shock in Tycho's supernova remnant. Astrophys. J. **690**, 1412–1423 (2009). arXiv:0809.2504

L.L. Williams, G.P. Zank, Effect of magnetic field geometry on the wave signature of the pickup of interstellar neutrals. J. Geophys. Res. **99**, 19229 (1994)

Y. Zhou, Renormalization group theory for fluid and plasma turbulence. Phys. Rep. **488**, 1–49 (2010)

Y. Zhou, W.H. Matthaeus, Models of inertial range spectra of interplanetary magnetohydrodynamic turbulence. J. Geophys. Res. **95**, 14881–14892 (1990)

E.G. Zweibel, J.M. Shull, Confinement of cosmic rays in molecular clouds. Astrophys. J. **259**, 859–868 (1982)

DOI 10.1007/978-1-4899-7413-6_20
Reprinted from *Space Science Reviews* Journal, DOI 10.1007/s11214-013-9999-0

Electron-Ion Temperature Equilibration in Collisionless Shocks: The Supernova Remnant-Solar Wind Connection

Parviz Ghavamian · Steven J. Schwartz · Jeremy Mitchell · Adam Masters · J. Martin Laming

Received: 17 January 2013 / Accepted: 27 May 2013 / Published online: 23 July 2013

Abstract Collisionless shocks are loosely defined as shocks where the transition between pre-and post-shock states happens on a length scale much shorter than the collisional mean free path. In the absence of collision to enforce thermal equilibrium post-shock, electrons and ions need not have the same temperatures. While the acceleration of electrons for injection into shock acceleration processes to produce cosmic rays has received considerable attention, the related problem of the shock heating of quasi-thermal electrons has been relatively neglected.

In this paper we review the state of our knowledge of electron heating in astrophysical shocks, mainly associated with supernova remnants (SNRs), shocks in the solar wind associated with the terrestrial and Saturnian bowshocks, and galaxy cluster shocks. The solar wind and SNR samples indicate that the ratio of electron temperature, (T_e) to ion temperature (T_p) declining with increasing shock speed or Alfvén Mach number. We discuss the extent to which such behavior can be understood on the basis of waves generated by cosmic

P. Ghavamian (✉)
Department of Physics, Astronomy and Geosciences, Towson University, Towson, MD 21252, USA
e-mail: pghavamian@towson.edu

S.J. Schwartz · J. Mitchell
Department of Physics, Imperial College London, London, UK

S.J. Schwartz
e-mail: s.schwartz@imperial.ac.uk

J. Mitchell
e-mail: j.mitchell@imperial.ac.uk

A. Masters
Institute of Space and Astronautical Science, Japan Aerospace Exploration Agency, Kanagawa 252-5210, Japan
e-mail: a.masters@stp.isas.jaxa.jp

J.M. Laming
Space Science Division, Naval Research Laboratory, Code 7684, Washington, DC 20375, USA
e-mail: j.laming@nrl.navy.mil

rays in a shock precursor, which then subsequently damp by heating electrons, and speculate that a similar explanation may work for both solar wind and SNR shocks.

Keywords Collisionless shocks · Supernova remnants · Solar wind · Cosmic ray acceleration

1 Introduction

Shock waves have been observed in a wide range of environments outside the Earth, from the solar wind to the hot gas in galaxy clusters. However, the mechanism whereby the gas in these environments is shocked has been poorly understood. While shock transitions in the Earth's atmosphere are mediated by molecular viscosity (and hence direct particle collisions), those in interstellar space and the solar wind are too dilute to form in this way. In non-relativistic shocks, the role of collisions is effectively played by collective interactions of the plasma with the magnetic field. This results in a multi-scale shock transition having sub-structure at ion kinetic length scales (Larmor radius or inertial length) and potentially electron kinetic scales (inertial lengths or whistler mode) (e.g., Schwartz et al. 2013, this issue; Treumann 2009). Such plasmas are termed collisionless. The magnetic fields threading through the charged particle plasmas in space endow the plasmas with elastic properties, much like a fluid. The kinetic energy of the inflowing gas is dissipated within this fluid via collective interactions between the particles and magnetic field, transferring energy from the magnetic field to the particles. The collective processes are the result of the DC electromagnetic fields present in the shock transition layer, kinematic phase mixing, and also plasma instabilities; the last give rise to a rich range of plasma waves and turbulent interactions.

It has long been known that these processes may heat the electrons beyond the mass-proportional value predicted by the Rankine-Hugoniot jump conditions—ample evidence is found in spacecraft studies of solar wind shocks (Schwartz et al. 1988) and multi wavelength spectroscopy of supernova remnants (Ghavamian et al. 2001, 2002, 2003, 2007; Laming et al. 1996; Rakowski et al. 2008) and galaxy cluster gas (Markevitch et al. 2005; Markevitch and Vikhlinin 2007; Russell et al. 2012) understanding how this process depends on such shock parameters as shock speed, preshock magnetic field orientation and plasma beta has been slow.

In collisionless plasmas, the downstream state of the plasma cannot be uniquely determined from the upstream parameters because the Rankine-Hugoniot jump conditions only predict the *total* pressure downstream, not the individual contributions from the electrons and ions: $P = n_i kT_i + n_e kT_e$. At the limit of a strong shock, n_e and n_i are each 4 times their preshock values, so the relative values of T_e and T_i immediately behind the shock are wholly dependent upon the nature of the collisionless heating processes occurring at the shock transition. Although an MHD description can be used to describe the behavior of the gas far upstream and far downstream of the shock, a more detailed kinetic approach is required for understanding how the dissipation at the shock front transfers energy from plasma waves and turbulence to the electrons and ions.

Non-relativistic collisionless shocks can be broadly sorted into three categories: slow, intermediate and fast. The three types are defined according to the angle between the shock velocity and upstream magnetic field, as well as the relative value of the shock speed compared to the upstream sound speed ($c_s \equiv \sqrt{\gamma P/\rho}$) and Alfvén speed ($v_A \equiv B/\sqrt{4\pi\rho_i}$). Most astrophysical shocks are quasi-perpendicular (i.e., they propagate at a nearly right angle to the preshock magnetic field), allowing only for fast-mode propagation. In that case, the relevant quantity is the magnetosonic Mach number, M_{ms} ($\equiv v_{sh}/\sqrt{v_A^2 + c_s^2}$). Collisionless shocks have also been classified according to whether the flow speed exceeds the sound speed in the downstream plasma (Kennel et al. 1985). Above the critical Mach number where the flow is subsonic, the dissipation of flow energy into thermal energy can no longer be maintained by electrical resistivity, and plasma wave turbulence (the cause of which are instabilities generated when the electron and ion distribution functions are distorted at the shock transition) is required (Kennel et al. 1985). Shocks above the critical Mach number are termed supercritical, while those below are termed subcritical. Note that even for subcritical shocks, observations suggests that kinetic processes other than resistivity and turbulence contribute to the shock dissipation (Greenstadt and Mellott 1987).There are also other, higher critical Mach numbers related to the formation of subshocks (Kennel et al. 1985) and non-steady cyclic shock reformation beyond the whistler critical Mach number (Krasnoselskikh et al. 2002). The critical Mach number for quasi-perpendicular shocks is estimated to be rather low ~2.8 (Edmiston and Kennel 1984). The Mach numbers of most SNR shocks are expected to be well in excess of this value, meaning that they are both fast-mode and supercritical.

As one approaches the high Mach numbers expected for astrophysical shocks ($M_A \sim 20$–100 for an ambient magnetic field of 3 μG), a greater and greater fraction of the incoming ions become reflected back upstream from the shock front, corresponding to an increasingly turbulent and disordered shock transition. Physical parameters such as B, T and n no longer jump in an ordered manner (i.e., the transitions are no longer laminar). In addition to the reflected particles, the hot ions from downstream become hot enough to escape upstream, further enhancing the population of ions in front of the shock. These ions, which form a precursor, are now believed to play an essential role in the dissipation of high Mach number collisionless shocks. Aside from providing the seed population for the acceleration of cosmic rays, the precursor ions are likely generate a variety of plasma waves capable of selectively heating the electrons over the ions, thereby providing an important mechanism for raising T_e/T_p above the mass-proportional value of m_e/m_p.

2 Formalism: Equilibration Timescales

Taken at face value, the Rankine-Hugoniot jump conditions predict that electrons and ions will be heated in proportion to their masses:

$$kT_{e,i} = \frac{3}{16} m_{e,i} V_{sh}^2 \tag{1}$$

In collisionless shocks, there are three relevant timescales to consider: the time required for Coulomb collisions to isotropize a distribution of electrons, t_{ee}, the time required for Coulomb collisions to isotropize distribution of ions, t_{ii}, and finally the time required for the electrons and ions to equilibrate to a common temperature, t_{ei} (Spitzer 1964). After a time scale t_{ee} or t_{ii}, the particles in question attain a Maxwellian velocity distribution. The self-collision time, $t_{c,ee}$ for electrons of density n_e and temperature T_e is (Spitzer 1964):

$$t_{c,ee} = \frac{0.266 T_e^{3/2}}{n_e \ln \Lambda_e} \text{ sec} \approx \frac{0.0116 V_s(1000)^3}{n_e \ln \Lambda_e} \text{ yr} \tag{2}$$

where $\ln \Lambda$ is the Coulomb logarithm, Eq. (1) has been used to write the relaxation time in terms of the shock speed and $V_s(1000)$ is the shock speed in units of 1000 km s^{-1}. For a young SNR having $V_s \sim 1000\ \mathrm{km\,s^{-1}}$, postshock density $n \sim 1\ \mathrm{cm^{-3}}$ and $\ln\Lambda \sim 30$ the time required to establish a Maxwellian distribution at the electron temperature given by Eq. (1) is $t_{c,ee} \sim 0.004$ yrs. The electrons are isotropized by self-collisions first, then the protons, and finally over a longer timescale the electrons and protons equilibrate to a common temperature. For this reason, Coulomb collisions alone are unable to establish equilibration electron and proton distributions at the shock front. This equilibration is described by the relation

$$\frac{dT_e}{dt} = \frac{T_p - T_e}{t_{c,ep}} \tag{3}$$

where

$$t_{c,ep} = \sqrt{\frac{m_p}{m_e}}\, t_{c,pp} = \frac{m_p}{m_e} t_{c,ee} \tag{4}$$

The temperatures T_e and T_i equilibrate to a common density-weighted average temperature T_{av}, given by $T_{av} = \frac{3}{16}\mu m_p V_s^2$, where μ is the mean mass per particle ($= (1.4/2.3) = 0.6$ for cosmic abundances). For mass-proportional heating, the timescale given by Eq. (4) is $\sim$2000 yrs for $V_s \geq 1000\ \mathrm{km\,s^{-1}}$, of similar order but longer than the proton-proton isotropization timescale, $t_{c,pp}$. These are longer than the age of the SNR, substantially so at higher V_s indicating that for minimal heating (i.e., $T_e/T_p = \frac{m_e}{m_p} \sim 1/1836$) the electrons and ions will not equilibrate to T_{av} during the lifetime of the SNR (Itoh 1978; Draine and McKee 1993).

The arguments above indicate that Coulomb collisions are ineffective at both isotropizing the heavy ion particle distributions and equilibrating the electron and ion temperatures at the transitions of collisionless shocks. The emission spectra of non-radiative SNRs (dominated mostly by X-ray and ultraviolet emission) should therefore be sensitive probes of the collisionless heating processes at the shock transition. We consider the observational constraints of these processes below.

3 Observational Constraints from SNRs

The most useful shocks for studying collisionless equilibration processes are those exhibiting detectable emission from the immediate postshock gas. To be diagnostically useful, the emission should arise from the region close to the shock front, where temperature disequilibrium between electrons, protons and heavy ions is substantial enough to affect both the relatives fluxes and relative velocity widths of emission lines. The shape and extent of the spatial profile for the different emission species behind the shock is also a useful diagnostic, especially for the UV resonance lines of He II λ1640, C IV λλ1548, 1550, N V λλ1238, 1243, and O VI λ1032, 1038. For a given shock speed, the distance behind the shock where the emission peaks depends strongly on the initial electron temperature, and electron temperature immediately behind the shock.

The ubiquity of non-radiative SNRs, as well as their relatively simple geometry and very high shock speeds, makes these objects the most important laboratories for investigating the efficiency and nature of electron-ion and ion-ion equilibration. Other non-radiative shocks available for study are those occurring in stellar wind bubbles (for example, Wolf-Rayet bubbles) (Gosset et al. 2011) and in galaxy cluster shocks (e.g., the 'Bullet Cluster' (Markevitch et al. 2005) and Abell 2146 (Russell et al. 2012)). However, these shocks are far less frequently observed than those in SNRs, leaving the latter as the most important objects providing both a broad range of shock speeds and corresponding diagnostic information (such as proper motions and spatially resolved structure). Shocks in star forming regions, such as HH objects and their associated bow shocks, tend to be clumpy and complicated, and are usually located in dense environments ($n \sim 100$–$1000\ \mathrm{cm}^{-3}$) with low enough shock speeds ($V_s \sim 100$–$200\ \mathrm{km\,s}^{-1}$) to be radiative. In those cases, the optical and UV emission (where the most valuable shock diagnostic line emission arises) is dominated by emission from the cooling and recombination zones far downstream from the shock. In those regions the cooling of the gas below 100,000 K and the accompanying compression of the gas result in a collisional plasma with $T_e = T_p = T_i$. This erases any 'memory' of the initial electron-ion equilibration. Furthermore, a significant fraction of the Ly α continuum produced in the recombination zone is expected to pass upstream and ionize the preshock gas. In strong shocks ($V_s > 80\ \mathrm{km\,s}^{-1}$), this results in complete ionization of hydrogen (Shull and McKee 1979; Cox and Raymond 1985), thus precluding the use of collisionally excited Balmer line emission from neutral H (described in the next section) as a temperature equilibration diagnostic.

3.1 Optical Spectroscopy of Balmer-Dominated Shocks: The $T_e/T_p \propto V_{sh}^{-2}$ Relation

In the late 1970s it was discovered that very fast ($\sim$2000 $\mathrm{km\,s}^{-1}$) shocks in young SNRs could generate detectable optical emission very close to the shock transition (Chevalier and Raymond 1978; Chevalier et al. 1980), providing a valuable diagnostic of physical conditions at the shock before Coulomb collisions or cooling could alter them. This emission is produced by collisional excitation of H I as it flows into the shock front. The cold neutral component does not interact directly with the plasma waves and turbulence at the shock, while the ionized component is strongly heated and compressed by a factor of four (when the shock is strong). Some of the cold H is destroyed by collisional ionization; however, the rest of the cold H undergoes charge exchange with hot ions behind the shock, generating a separate population of hot H. Approximately 1 in every 5 collisions results in collisional excitation to the $n = 3$ level of H, producing Hα and Ly β emission. The Hα line from the cold neutrals is narrow and reflects the preshock temperature ($\leq$30,000 K), while that from the hot neutrals is broad (typically $\geq$500 $\mathrm{km\,s}^{-1}$), and reflects the postchock temperature (and to a large extent, the velocity distribution) of the protons (Chevalier and Raymond 1978; Chevalier et al. 1980; Smith et al. 1991; Ghavamian et al. 2001) (Fig. 1). In Balmer-dominated shocks,the broad to narrow Hα flux ratio is proportional to the ratio of the charge exchange rate to the ionization rate, with the latter being highly sensitive to the electron and proton temperatures. This makes the broad to narrow ratio, I_b/I_n, very sensitive to T_e/T_p. There is only weak dependence of I_b/I_n on the preshock H I fraction and preshock temperature, mainly due to differences in the amount of Ly β converted into Hα in the narrow component (Ghavamian et al. 2001; van Adelsberg et al. 2008).

The first systematic attempt to use Balmer-dominated SNRs to infer T_e/T_p for collisionless shocks was attempted by Ghavamian et al. (2001). Using the broad Hα line widths measured from a sample of Balmer-dominated shocks, they constrained the range of plausible shock speeds between the limits of minimal equilibration and full equilibration. They then predicted the broad-to-narrow ratios for a grid of shock models over this range of V_s and T_e/T_p, allowing T_e/T_p to be constrained. The range of broad component Hα widths observed in SNRs ranges from $\sim$250 $\mathrm{km\,s}^{-1}$ for the slowest Balmer-dominated shocks (Cgynus Loop), to $\sim$500 $\mathrm{km\,s}^{-1}$ for intermediate-velocity shocks (RCW 86) and finally $\sim$2600 $\mathrm{km\,s}^{-1}$ for the fastest shocks (SNR 0509–67.5). This corresponds to a well-sampled range of shock speeds: nearly a factor of 10. The primary uncertainty in measurement of the broad component width at low shock speeds ($\lesssim$200 $\mathrm{km\,s}^{-1}$, as seen in the Northeastern

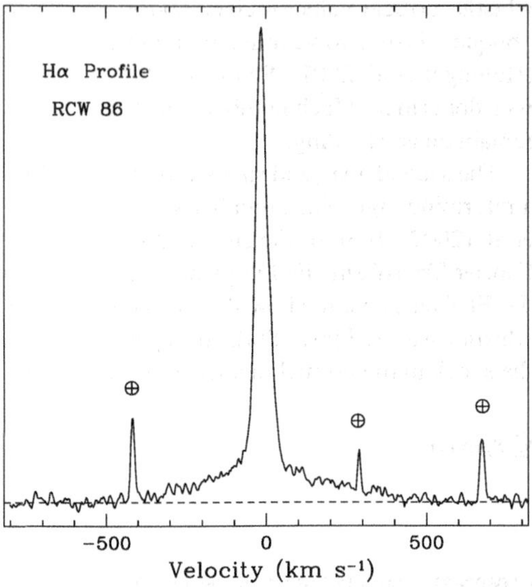

Fig. 1 The Hα line profile of a Balmer-dominated shock in RCW 86, acquired with a high dispersion (resolution $\sim$10 km s^{-1}) spectrograph (Sollerman et al. 2003). The broad ($\sim$500 km s^{-1}) and narrow ($\sim$30 km s^{-1}) Hα lines are both resolved. Night sky lines (marked by the *circles*) have been left in to demonstrate their noticeably narrower width compared to the Hα lines

Cygnus Loop; Hester et al. 1994) is disentangling the broad and narrow components when they are of comparable width. At high shock speeds ($\gtrsim$2000 km s^{-1}) the main difficulty is the baseline uncertainty of the surrounding continuum: if the peak of the broad line is low and the width very large, errors in ascertaining where the broad line merges into the background can lead to underestimates of the broad component width. The range of I_b/I_n for this sample of Balmer-dominated shocks lies between 0.4 and 1.2 (Kirshner et al. 1987; Smith et al. 1991; Ghavamian et al. 2001, 2003; Rakowski et al. 2009). However, it does not vary with broad component width in a monotonic fashion.

Recently there has been substantial improvement in the modeling of Balmer-dominated shocks. Earlier calculations of the broad Hα line profile assumed that it formed from a single charge exchange (Chevalier et al. 1980; Smith et al. 1991; Ghavamian et al. 2001), and it treated the hot and cold neutrals as two separate, distinct populations, with a given neutral belonging to either one or the other. However, in reality an interaction 'tree' is required to track the number of photons emitted by each neutral over multiple excitations and charge exchanges. These effects were first incorporated in the Balmer-dominated shock models of Heng and McCray (2007), who also found that charge exchange results in a third population of neutrals having velocity widths intermediate between the hot and cold neutrals. Further improvements in modeling of Balmer-dominated shocks were included by van Adelsberg et al. (2008), who included the momentum transferred by charge exchange between the hot neutrals and protons. This allowed the bulk velocity of the postshock neutrals to be calculated separately from those of the protons. Inclusion of the momentum transfer showed that for shock speeds $\lesssim$1000 km s^{-1}, charge exchange effectively couples the fast neutral and thermal proton distributions, while for high shock speeds ($\gtrsim$5000 km s^{-1}), it does so far less effectively. This results in a fast neutral distribution that is skewed relative to the protons in velocity space and an average velocity that is much higher for the fast neutrals than the protons (van Adelsberg et al. 2008). Together, the inclusion of all these effects has enhanced the ability of the models to match the observed broad-to-narrow ratios (and hence predict T_e/T_p). In particular, the newer models can now match the low broad-to-narrow ratio ($I_b/I_n \approx 0.67$) observed in Knot g of Tycho's SNR, yielding $T_e/T_p \approx 0.05$,

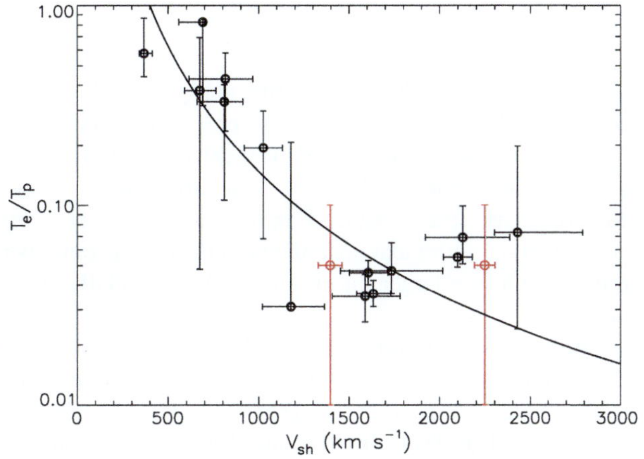

Fig. 2 The correlation between β ($\equiv T_e/T_p$) and shock speed for Balmer-dominated SNR shocks is shown, using results from van Adelsberg et al. (2008). The *solid curve* shows the V_S^{-2} dependence inferred by Ghavamian et al. (2007) for Balmer-dominated SNR shocks. New results quoted from the NE and NW portions of Tycho's SNR (this paper) are marked in *red*. The apparent upturn at $V_s \gtrsim 2000$ km s^{-1} and its associated error bars are discussed more fully in the text

$V_s \approx 1600$ km s^{-1}. However, even after all the additional physics is included, the Balmer-dominated shock models are still unable to reach the low broad-to-narrow ratios measured along the rims of DEM L71 ($I_b/I_n \approx 0.2$–0.7; Ghavamian et al. 2003; Rakowski et al. 2009). The electron-ion equilibration in DEM L 71 was instead determined via comparison of broad Hα line FWHM with postshock electron temperatures measured from *Chandra* observations (Rakowski et al. 2003). The most promising explanation advanced for the I_b/I_n discrepancy has been added narrow component flux from the shock precursor (Raymond et al. 2011; Morlino et al. 2012a), hitherto not included in the earlier shock models (Ghavamian et al. 2003; Rakowski et al. 2009). These developments are described in more detail in the next section.

The plot of T_e/T_p versus V_s for the available sample of Balmer-dominated shocks shows a declining trend of equilibration with shock speed (Ghavamian et al. 2007; Heng et al. 2007; van Adelsberg et al. 2008). The trend is described by Ghavamian et al. (2007) as full equilibration for shock speeds up to and including 400 km s^{-1}, and a declining equilibration proportional to the inverse square of the shock speed above 400 km s^{-1}. This description can be characterized in the following way:

$$\frac{T_e}{T_p} = \left\{ \begin{array}{ll} 1 & \text{if } V_s < 400 \text{ km s}^{-1} \\ \frac{m_e}{m_p} + (1 - \frac{m_e}{m_p})(\frac{V_s}{400})^{-2} & \text{if } V_s \geq 400 \text{ km s}^{-1} \end{array} \right\} \tag{5}$$

where the functional form of the T_e/T_p relation is designed to asymptotically transition to $T_e/T_p = m_e/m_p$ at very high shock velocities.

The most up to date plot of T_e/T_p versus shock speed, reproduced from van Adelsberg et al. (2008), is shown in Fig. 2 (Note that the shock models used in producing these plots do not include contribution from the shock precursor). Although the inverse correlation between T_e/T_p and V_s was largely confirmed by van Adelsberg et al. (2008), there may be some evidence of departure from the $T_e/T_p \propto V_s^{-2}$ relation at shock speeds exceeding 2000 km s^{-1}. van Adelsberg et al. find that when all three measured broad component widths and broad-to-narrow ratios from SN 1006 are included in the plot ($V_s \sim 2200$–2500 km s^{-1}), a slight upturn in the T_e/T_p–V_s relation appears. The T_e/T_p ratios for those cases are found to be $\sim$0.03, superficially similar to $\sqrt{m_e/m_p}$, rather than m_e/m_p. The reason for the discrepancy is not clear, nor whether this indicates a breakdown in the V_s^{-2} dependence at shock speeds exceeding 2000 km s^{-1}. Some caveats to consider when interpreting the upturn in T_e/T_p seen in Fig. 2 are as follows: For shock speeds $\gtrsim$2000 km s^{-1}

collisional ionization and excitation of H are primarily caused by proton (and to a lesser degree, alpha particle) impact (Laming et al. 1996; Ghavamian 1999; Ghavamian et al. 2001; Tseliakhovich et al. 2012). Experimentally measured cross sections for these interactions have still not been available to high precision (uncertainties ~20 %–30 % still exist), although more sophisticated theoretical calculations are now becoming available (see, for example, Tseliakhovich et al. 2012) . In addition, as the broad component width increases, the Hα profiles are spread out over an increasing number of pixels, resulting in noisier spectra and greater measurement uncertainty in the broad component width. These larger error bars in turn result in a larger uncertainty in V_s, especially at shock speeds of 2000 km s^{-1} and higher.

The best way to further constrain the equilibration-shock speed relation is to add new data points to the curve shown in Fig. 2. To this end, we present Balmer-dominated Hα profiles for two additional positions in Tycho's SNR (different from those of Knot g presented by Kirshner et al. (1987), Smith et al. (1991) and Ghavamian et al. (2001)). These profiles, shown in Fig. 3, were acquired with a moderate resolution spectrograph in 1998 (for details on the observational setup, see Ghavamian 1999 and Ghavamian et al. 2001). The profile marked 'NE' was obtained from a clump of Hα emission located along the northeastern edge of Tycho's SNR, approximately 1′ northward of Knot g. The clump appears behind the main body of the Balmer filaments and exhibits a broad component that is substantially

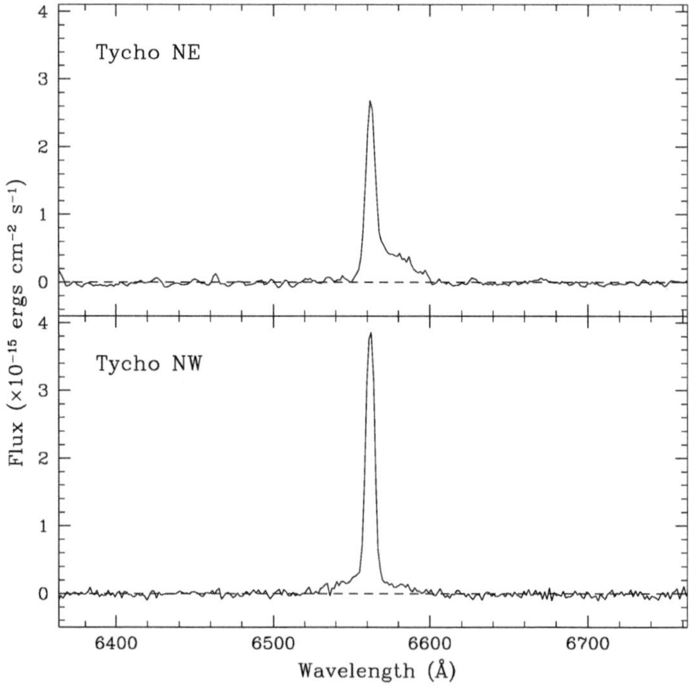

Fig. 3 Hα profiles from two locations along the Balmer-dominated rim of Tycho's SNR (Ghavamian 1999). *Top*: the spectrum of a clump immediately to the north of Knot g ($I_b/I_n = 0.85 \pm 0.04$; FWHM $= 1300 \pm 65$ km s^{-1}). The broad component is substantially redshifted from the center of the narrow component ($+10.7 \pm 1.2$ Å, or 490 ± 55 km s^{-1}), indicating that the clump is located on the far side of the shell, projected ~20° inside of the main Balmer-dominated rim. *Bottom*: profile acquired from the northernmost filament in Tycho, $I_b/I_n = 0.45 \pm 0.15$, FWHM $= 2040 \pm 55$ km s^{-1}

Doppler shifted to the red (10.7 ± 1.2 Å, or about 490 km s^{-1}). The Doppler shift reflects the bulk velocity of the hot postshock proton distribution, so the significant velocity shift of the broad component centroid indicates that the shock in the NE has a substantial velocity component into the plane of the sky, i.e., that the NE shock is located on the far side of the blast wave shell.

The broad Hα width of the NE shock is 1300 ± 65 km s^{-1} (the smallest broad component width measured in Tycho's SNR so far), with $I_b/I_n = 0.85 \pm 0.04$. The NW shock, on the other hand, has a broad Hα width of 2040 ± 55 km s^{-1} (the largest broad component width measured in this SNR so far), with $I_b/I_n = 0.45 \pm 0.15$. Neither of these broad-to-narrow ratios is strictly reproduced by the latest models of van Adelsberg et al. (2008), with the lowest predicted ratios being for Case B (the assumption of optically thick conditions for Ly β photons in the narrow component) and for low equilibrations ($T_e/T_p \lesssim 0.1$). For these low equilibrations, the corresponding shock speeds for the NE and NW shocks in Tycho are approximately 1400 km s^{-1} and 2250 km s^{-1} (using Figs. 5 and 10 of van Adelsberg et al. 2008). In Fig. 2 we have added data from the NE and NW shocks in Tycho's SNR to the $T_e/T_p - V_s$ plot. The two data points help fill in a portion of the plot where the data are sparse: the region between approximately 1200 km s^{-1} and 1500 km s^{-1}, as well as the region beyond 2000 km s^{-1}, where the existing data are taken entirely from SN 1006. The added point near 1400 km s^{-1} is fully consistent with the V_s^{-2} relation, while T_e/T_p is not well constrained enough for the point at 2250 km s^{-1} to contradict the appearance of an upturn at the highest shock velocities. It is clear that proper characterization of the equilibration-shock velocity curve above 2000 km s^{-1} will require both higher signal-to-noise spectra on existing Balmer-dominated shocks, as well as new data points beyond a broad component width of 2500 km s^{-1}.

3.2 Exceptions to $T_e/T_p \propto V_{sh}^{-2}$

Although the inverse squared relation between equilibration and shock velocity appears has been the most salient result of the study of Balmer-dominated shocks, there have been discrepant results reported in a small subset of cases. Recently the broad Hα component in the LMC SNR 0509–67.5 was detected for the first time (Helder et al. 2010). Broad emission was observed along both the northeastern rim (FWHM 3900 ± 800 km s^{-1}) and southwestern rim (FWHM 2680 ± 70 km s^{-1}), with the former being the fastest Balmer-dominated shock detected to date having the characteristic broad and narrow component Hα emission. Interestingly, the broad-to-narrow ratios for both shocks are exceptionally low, $I_b/I_n = 0.08 \pm 0.02$ and 0.29 ± 0.01 along the NE and SW rims, respectively. As noted by Helder et al. (2010) these ratios are nearly twice as large as the smallest ones predicted by the models of van Adelsberg et al. (2008), precluding measurement of T_e/T_p from the Balmer-dominated spectra and implicating excess narrow component Hα emission in a cosmic ray precursor once again. Using RGS spectra of SNR 0509–67.5 acquired with *XMM-Newton*, they obtained a forward shock speed of approximately 5000 km s^{-1} in the SW. This implies a broad Hα width of 3600 km s^{-1}, substantially smaller than the measured width of 2680 km s^{-1}. Helder et al. (2010) suggest that this indicates some thermal energy loss (~20 %) to cosmic ray acceleration. This picture is supported by the presence of nonthermal X-ray emission in their fitted RGS spectra of SNR 0509–67.5.

The result described above relies upon an accurate disentangling of the bulk Doppler broadening from the thermal Doppler broadening in the X-ray lines. The disentangling depends on parameters such as the ratio of reverse shock to forward shock velocities, as well as the ratio of the gradients in the reverse and forward shock velocities, which in turn had to be assumed from evolutionary SNR models. The result, while intriguing, is still significantly uncertain. On the other hand, Helder et al. (2010) found that the X-ray shock velocities from the NE could only be reconciled with the observed broad Hα width there if $T_e/T_p \approx 0.2$, which clearly which predicts $T_e/T_p \sim m_e/m_p$ predicted from Eq. (5). If this result were to be confirmed by future observations, it would present a new challenge in understanding how electron-ion equilibration occurs in fast collisionless shocks. One possibility, given the presence of nonthermal X-ray emission in the spectra of 0509–67.5 (Warren and Hughes 2004; Helder et al. 2010) is that the moderate loss of thermal energy to cosmic ray acceleration may have slightly increased the compression and reduced the temperature at the shock front compared to the case with no acceleration (Decourchelle and Ellison 2000; Ellison et al. 2007). Both of these effects would tend to render the plasma more collisional, possibly explaining the $T_e/T_p \approx 0.2$ result. However, it is also worth noting that the shock velocity used for obtaining T_e/T_p in the NE is subject to the same model dependence and uncertainties as the SW measurements, so similar caution is required in its interpretation.

A similar combined optical and X-ray study of RCW 86 was performed by Helder et al. (2009, 2011). There, the broad component Hα widths were supplemented with electron temperatures measured from *XMM-Newton* RGS spectra from the same projected locations along the rim. One of the main results of this study was that the slower shocks (broad Hα FWHM $\sim$ 500–600 $\mathrm{km\,s^{-1}}$) showed $T_e/T_p \sim 1$, agreeing with earlier results from similar shocks observed both in RCW 86 and elsewhere (Ghavamian et al. 1999, 2001, 2007). However, Helder et al. (2011) found while the results were indeed consistent with low equilibration at the shock front for fast shocks ($T_e/T_p \approx 0.02$ for broad FWHM $\sim$ 1100 $\mathrm{km\,s^{-1}}$)) and higher equilibration for the slower shocks ($T_e/T_p \approx 1$ for broad FWHM $\sim$ 650 $\mathrm{km\,s^{-1}}$), their X-ray derived electron temperatures were inconsistent with $T_e = 0.3$ keV at the shock front contradicting the suggestion of Ghavamian et al. (2007) that shocks above 400 $\mathrm{km\,s^{-1}}$ may all heat electrons to roughly 0.3 keV. However, a major caveat of these results is that the forward shock in RCW 86 is believed to be impacting the walls of a wind-blown bubble (Williams et al. 2011), resulting in substantial localized variations in shock speed around the rim. These variations occur as different parts of the forward shock impact the cavity wall at different times. While the broad component Hα widths closely trace the current position of the shock front, the X-ray emission behind that shock arises over a much more extended spatial scale, and is sensitive to the history of the forward shock interaction with the cavity wall. Furthermore, narrowband Hα imagery of RCW 86 with the ESO Very Large Telescope (Helder et al. 2009, 2011) shows a complex morphology of filaments, especially along the eastern side of the SNR. The broad Hα components of these filaments exhibit substantial, localized variations in line width, ranging from $\sim$ 600 $\mathrm{km\,s^{-1}}$ FWHM to 1100 $\mathrm{km\,s^{-1}}$ (Ghavamian 1999; Helder et al. 2009, 2011). These variations reflect localized changes in density and viewing geometry along the line of sight. As such, uniquely mapping the observed Balmer filaments to their corresponding X-ray emission in *XMM-Newton* data (especially given the somewhat coarse 10″ spatial resolution of that instrument) is fraught with uncertainty. Additional corroboration for these results would be desirable.

3.3 Imprint of the Shock Precursor on the Hα Line Profile

Perhaps the most important and exciting recent development in our understanding of Balmer-dominated shocks has come with the development of new kinetic-based shock models. Blasi et al. (2012) have introduced a kinetic model for following the momentum and energy exchange between neutrals and ions, along with the back-reaction of those neutrals when they pass back upstream and form a fast neutral precursor. Rather than assume a Maxwellian velocity distribution for the neutrals (as had been done in previous models, despite the lack of thermal contact between neutrals needed to justify such an assumption), both the ion and neutral distributions are computed from their appropriate Boltzmann equations. Building on these models, Morlino et al. (2012a, 2012b) have confirmed what had been suspected earlier (Smith et al. 1994; Hester et al. 1994; Sollerman et al. 2003), namely that the broadening of the narrow component beyond the expected ISM value ($\sim$25–30 km s^{-1} instead of 10 km s^{-1}) is most likely due to heating in a cosmic ray precursor. In particular, Morlino et al. (2012a) found that the characteristic charge exchange length of the incoming neutrals exceeds that of the neutrals crossing back upstream, so that the narrow component width is impacted not by the neutral return flux, but rather by heating in the cosmic ray precursor. Raymond et al. (2011) predicted a similar broadening of the narrow component, though they focused mainly on the contribution of collisional excitation in the precursor to the flux in the narrow Hα component. Spatially resolved line broadening of the narrow Hα component was detected in ground-based longslit spectroscopy of Knot g in Tycho (Lee et al. 2007). In addition, a small ramp-up in Hα emission was observed ahead of Knot g in HST imagery Tycho's SNR (Lee et al. 2010). The results, taken together, are strong evidence for the presence of cosmic ray precursors in Balmer-dominated SNRs.

One prediction of the new Balmer-dominated shock models is the existence of a third component of the Hα emission (Morlino et al. 2012a, 2012b). When the hot neutrals escape upstream, they undergo charge exchange with the colder preshock protons. This results in fast protons with cold neutrals, with the former rapidly equilibrating with preshock protons and pre-heating the gas. The temperature of the equilibrated protons in the precursor lies between the temperature of the far upstream protons ($\sim$5000 K) and the far downstream protons ($\sim 10^6$–10^8 K), typically $\sim 10^5$ K. Further charge exchange between these warm protons and the preshock neutrals gives rise to a third, 'warm' neutral component (neither fast nor slow) having velocity widths of hundreds of km s^{-1} (Morlino et al. 2012a). Interestingly, the presence of a third Hα component was first observationally reported by Smith et al. (1994) in their high resolution echelle spectroscopy of Balmer-dominated SNRs in the Large Magellanic Cloud (an example of one of the spectra from Smith et al. (1994) is reproduced in Fig. 4). A third Hα component was also reported in high resolution spectra of Knot g in Tycho's SNR (Ghavamian et al. 2000). In their models of Balmer-dominated shock emission, Morlino et al. (2012a) found that the importance of the third component relative to that of the broad and narrow components depends strongly on the preshock neutral fraction and T_e/T_p, in line with earlier theoretical predictions on properties of a fast neutral precursor (Smith et al. 1994). The fact that the third component has been detected in Tycho's SNR (width $\sim$150 km s^{-1}) is consistent with the high preshock neutral fraction ($f_{H\ I} \sim 0.9$) inferred from the broad-to-narrow ratio of Knot g by Ghavamian et al. (2001). A similar third component may have been detected in high resolution spectra of SNR 0509–67.5, where measurement of the narrow component width required the inclusion of an additional component of width 75 km s^{-1} (Smith et al. 1994). The new fast neutral precursor models predict that a substantial fraction of the Hα excitation in Balmer-dominated shocks can arise ahead of the shock, where warm neutrals are excited by electron impact. Interestingly, the relative contribution of the preshock Hα to the total (upstream + downstream) is sensitive to T_e/T_p behind the shock. Morlino et al. (2012a) found that up to 40 % of the total Hα flux from a Balmer-dominated shock can arise from the fast neutral precursor when $V_s \sim 2500$ km s^{-1} and $T_e/T_p = 1$ both upstream and downstream. In these models the preshock contribution to the total flux drops substantially for lower downstream equilibrations for $V_s \gtrsim 1000$ km s^{-1} (the slowest shocks considered by Morlino et al. 2012a). This is generally consistent with

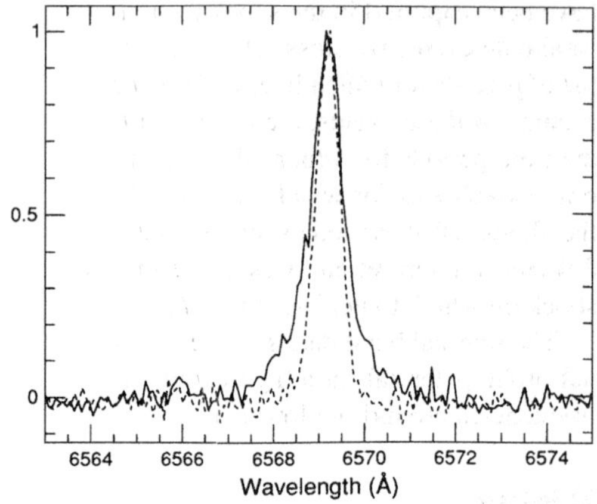

Fig. 4 High dispersion Hα profiles of the Balmer-dominated SNR 0509–67.5 in the Large Magellanic Cloud (Smith et al. 1994). The broad component is not detected due to the high dispersion. The narrow component from the east rim of the SNR (*solid line*) shows intermediate velocity wings of width ∼100 km s^{-1}, while that of the center (*dotted line*) does not. The intermediate component is now believed to arise within a fast neutral precursor (Morlino et al. 2012a)

the fact that in most cases, shock models not including the precursor Hα emission have been able to match the observed broad-to-narrow ratios. In other words, if the postshock temperature equilibration were not low for such fast shocks, the agreement between the observed and predicted I_b/I_n would have been substantially worse for such remnants as Tycho's SNR and SN 1006.

3.4 Ultraviolet and X-ray Studies of Balmer-Dominated Shocks

SN 1006 is an example of a SNR accessible to UV spectroscopy due to its galactic location, 450 pc up from the galactic plane, and therefore with relatively low extinction due to intervening dust and gas. The Hopkins Ultraviolet Telescope (HUT) observed the UV resonance lines of He II λ1640, C IV λλ1548, 1550, N V λλ1238, 1243, and O VI λ1032, 1038 (Raymond et al. 1995) emitted from the Balmer dominated filament in the NW quadrant. The line showed Doppler broadening consistent with that of the H α broad component observed in the optical, indicating insignificant ion-ion equilibration. Laming et al. (1996) were also able to infer the degree of electron-ion equilibration at the shock.

He II λ1640, by virtue of its relatively high excitation potential ∼48 eV, is excited only by electrons, and its intensity is therefore directly related to the electron temperature. C IV λλ1548, 1550, N V λλ1238, 1243, and O VI λ1032, 1038, by contrast, have much lower excitation potentials of ∼8, 10 and 12 eV, so although these ionization states are established by electrons, the line emission in these transitions can also be excited by impacts with hot protons and α particles, and the intensity ratio of He II λ1640 to C IV λλ1548, 1550, N V λλ1238, 1243, and O VI λ1032, 1038 can be sensitive to the post-shock electron-proton temperature equilibration. The spatial distribution of the UV resonance line emission, when spatially resolved, provides additional constraints on the degree of electron-ion equilibration at the shock front. For shocks slower than 1500 km s^{-1}, $T_e = T_p$ at the shock front results in both a more rapid rise and higher maximum in emissivity of the UV resonance lines with distance behind the shock (see Fig. 5).

Laming et al. (1996) calculated impact excitation cross sections for protons and α particles colliding with Li-like ions, using a partial wave expansion with the Coulomb-Bethe approximation, and applying a unitarization procedure following Seaton (1964). They found a degree of equilibration of order $T_e/T_p \sim 0.05$ or less, which implied for a 2250 km s^{-1} shock an electron temperature immediately postshock of $<5 \times 10^6$ K, in very good agreement (and actually predating) the optical results discussed above in Sect. 4.1.

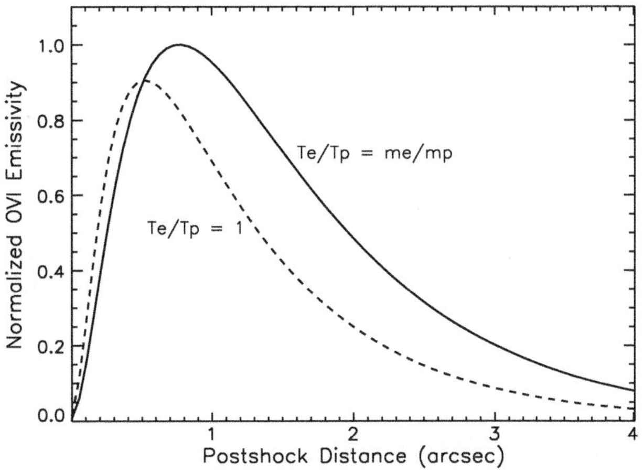

Fig. 5 The spatial distribution of the O VI(1032 + 1038) emissivity behind a 350 km s^{-1} planar shock at a distance of 500 pc (parameters appropriate to the fastest Balmer-dominated shocks in the Cygnus Loop), shown for the cases of full, instant equilibration at the shock front ($T_e/T_p = 1$) and minimal equilibration ($T_e/T_p = m_e/m_p$). Analysis and modeling of the Hα profiles, as well as the O VI and X-ray emission indicate that $T_e/T_p \approx 1$ for the Cygnus Loop

The ion-ion equilibration in SN 1006 was revisited by Korreck et al. (2004), using higher spectral resolution FUSE data comprising O VI λ1032, 1038 and the broad Ly $\beta\lambda$1025 emission lines. They found a slightly broader line profile in Ly β, implying less than mass-proportional heating and possibly a small degree of ion-ion equilibration.

SN 1987A represents another SN/SNR in a region of the sky accessible to UV observations. HST COS observed the He II λ1640, C IV $\lambda\lambda$1548, 1550, N V $\lambda\lambda$1238, 1243 and N IV $\lambda\lambda$1486 lines emitted from the reverse shock (France et al. 2011). When combined with optical spectroscopy of H α, the T_e/T_p ratio at the shock is determined to be in the range 0.14–0.35, significantly higher than similar ratios coming from Balmer dominated forward shocks. France et al. (2011) argued that a different equilibration mechanism is likely at work. Considering the relative youth of SN 1987A, and the fact that the reverse shock is the origin of the emission, significant populations of cosmic rays and associated magnetic field amplification are unlikely. In fact, in the expanding ejecta the magnetic field is likely to be very weak, leading to a very high Aflvén Mach number shock. As will be discussed below in connection with shocks in galaxy clusters, electron heating in such a case is likely to be due to acceleration in the cross-shock potential. The cross-shock potential is effective at heating electrons, and so may explain the higher T_e/T_p in SN 1987A.

The forward shock of SN 1987A has also been observed in X-rays with the grating instruments on Chandra (e.g. Zhekov et al. 2009). In general electron heating well below complete equilibration is seen, though precise interpretation is difficult because one observation sees emission from shocks at a variety of different velocities, due to irregularities in the density of the surrounding medium.

3.5 Do Results from the Balmer-Dominated Shocks Apply to Fully Ionized Shocks?

The inverse relationship between the temperature equilibration and shock speed is an interesting result from studies of Balmer-dominated SNRs. However, the applicability of this result to both fully ionized shocks and shocks undergoing efficient CR acceleration ($\gtrsim$50 % of their energy transferred to CRs) remains unsettled. Recently, Vink et al. (2010) used a two-fluid-model for cosmic rays and thermal gas to simulate the effect of cosmic ray acceleration on the temperature and ionization structure of fast, non-relativistic shocks. They found that if 5 % of the shock energy were to be channeled into cosmic rays (the minimum needed if SNRs are the dominant source of cosmic rays) then approximately 30 % of the postshock pressure must reside in cosmic rays (corresponding to a ratio of cosmic ray to total postshock pressure, w, of 0.3). For $w = 0.3$, Vink et al. (2010) predicted a lowering of the average temperature of the postshock gas to $\sim$70 % of the value when the cosmic ray contribution is ignored. This is a significant alteration of the postshock temperature profile, and should result in much more rapid equilibration of electrons and protons close to the shock.

However, do the effects described above actually occur in SNR shocks? One of the principle lines of evidence cited by Vink et al. (2010) in support of this picture was the result found in RCW 86 by Helder et al. (2009). In that SNR, the broad Hα widths of Balmer-dominated filaments were found to be nearly 50 % smaller than the minimum allowed given their X-ray proper motions. Filaments in the NE of the SNR exhibited broad Hα widths of 1000 km s^{-1}, but their apparent X-ray counterparts, which exhibited strong X-ray synchrotron (nonthermal) emission, exhibited proper motions indicating shock speeds of 3000–6000 km s^{-1}. This result, along with the theoretical prediction that X-ray synchrotron emission requires shock speeds of at least 2000 km s^{-1} (Aharonian et al. 1999) was taken by Helder et al. (2009) and Vink et al. (2010) as evidence for substantial energy loss ($w \sim 0.5$) from the Balmer-dominated shocks to cosmic rays. However, this association has now been refuted by subsequent multi-epoch optical imagery of the Hα filaments, which have failed to show the kind of high proper motions seen in the nonthermal X-ray filaments (Helder et al. 2011, in preparation). Instead, they show proper motions consistent with shock speeds predicted by the broad Hα widths without energy loss to cosmic rays($\sim$600–1200 km s^{-1}), implying that w < 0.2. The association between the Balmer-dominated shocks studied spectroscopically by Helder et al. (2009) and the X-ray filaments was due either to coincidental spatial alignment, or due to sudden deceleration of the outer shock in RCW 86 during its encounter with the surrounding cavity wall (Williams et al. 2011; Helder et al. 2011, in preparation).

The lack of association between the Balmer-dominated filaments and the non-thermal X-ray filaments in RCW 86 raises some important questions about the feasibility of using Balmer-dominated shocks to study electron-proton equilibration in cases where 50 % or more of the thermal energy is diverted to cosmic rays. In SNRs such as SN 1006 (Koyama et al. 1995; Katsuda et al. 2010a), Tycho's SNR (Warren et al. 2005; Katsuda et al. 2010b), Cas A (Vink and Laming 2003) and RX J1713.7–3946 (Koyama et al. 1997; Slane et al. 1999; Tanaka et al. 2008) the presence of strong synchrotron X-ray filaments has been interpreted as evidence for highly efficient cosmic ray acceleration. The narrowness of the synchrotron filaments most likely reflects the short emitting lifetimes of the ultra high energy electrons (energies $\sim$10–100 TeV) as they spiral in the postshock magnetic field (Vink and Laming 2003). The detection of γ-ray emission from the shells of SN 1006 (Acero et al. 2010) and RX J1713.7–3946 (Aharonian et al. 2006; Abdo et al. 2011) has shown that cosmic rays are accelerated to energies as high as 100 TeV in these SNRs. In all cases thermal X-ray emission has been exceptionally faint due to the very low inferred preshock densities ($\lesssim$0.1 cm^{-3}), making it more likely that the overall X-ray emission will be dominated by synchrotron radiation from the most energetic cosmic rays. SNRs expanding into such low density media can propagate at the high shock speeds required for cosmic ray acceleration ($\gtrsim$2000 km s^{-1}; Aharonian et al. 1999) for a longer time, allowing their structure to become modified by the back pressure from the cosmic rays. Investigating the temperature and ionization structure of such shocks with Balmer line spectroscopy requires finding Balmer-dominated shocks exhibiting X-ray synchrotron radiation.

All of the known SNRs exhibiting Balmer-dominated shocks have also been imaged at X-ray wavelengths with *Chandra* or XMM, allowing reasonably detailed searches for

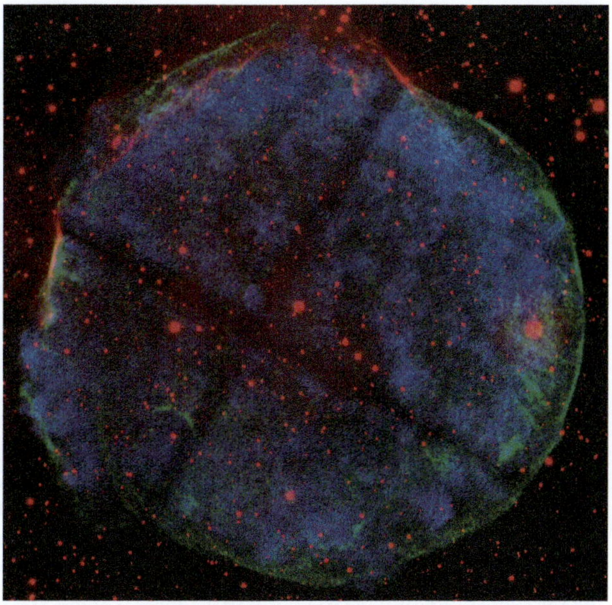

Fig. 6 An Hα narrowband image of Tycho's SNR (*red*; courtesy P.F. Winkler), overlaid onto the *Chandra* 3–8 keV hard X-rays (*green*) and *Chandra* 0.5–3 keV soft X-rays (*blue*). Both optical and X-ray images were acquired during the same epoch (2007). Note the near mutual exclusivity of the Balmer emission and hard X-ray emission

shocks emitting both Hα and synchrotron X-ray emission (the latter producing hard continuum that is dominant at energies of 2 keV and higher). A detailed comparison for all Balmer-dominated SNRs has not yet been published. However, even a cursory comparison between the narrowband Hα and hard X-ray images of these SNRs shows a distinct *anti-correlation* between shocks emitting in these two bands. For example, overlaying Hα and *Chandra* images of Tycho's SNR acquired during the same epoch (2007) (Fig. 6) shows little or no correlation between the prominent Balmer-dominated filaments on the eastern and northeastern edges and the non-thermal X-ray filaments ($E \geq 3$ keV) circling the remnant. The Balmer-dominated filaments (shown in red in Fig. 6) on the eastern side of Tycho's SNR are seen projected $30''$–$1'$ inside the edge of the nonthermal X-ray filaments (marked in green), an indication that portion of the shell along the line of sight has significantly decelerated. The Balmer filaments are seen at the outermost edge of the thermally emitting X-ray ejecta (marked in blue), but only at locations where little or no nonthermal X-ray emission is present. The bright optical filament known as Knot g (at the far left edge of Fig. 6) is the only location where Balmer filaments and X-ray synchrotron emission appear coincident. However, upon closer inspection the anticorrelation between the Balmer line and synchrotron emission can be seen in Knot g as well: the upper half of the filament, where Balmer line emission is strongest, exhibits minimal synchrotron emission, while the opposite is true in the lower half of the filament. The enhanced nonthermal emission inside of Knot g may be due to the strong recent deceleration of Knot g, where the SNR is currently propagating into a strong density gradient at the outermost edge of an H I cloud (Ghavamian et al. 2000). The lack of optical/X-ray synchrotron correlation is especially striking given that the Balmer-dominated filaments in Tycho's SNR have a high enough shock velocity ($\sim$1800–2100 km s^{-1}) to accelerate particles to TeV energies.

The anticorrelation between the optical and nonthermal X-ray emission can be observed in other SNRs as well, including SN 1006, where recently X-ray proper motions have been measured along the entire rim by Katsuda et al. (2012). As with Tycho's SNR, the locations of the Balmer-dominated filaments and the nonthermal X-ray filaments along the NW rim of SN 1006 are mutually exclusive. Instead, the Balmer-dominated shocks are closely associated with thermal X-ray filaments having a proper motion consistent with a shock velocity of $3300 \pm 200 \pm 300\ \mathrm{km\,s^{-1}}$ (statistical and registrational uncertainties, respectively) for a distance of 2.2 kpc. This result is in excellent agreement with the shock velocity of $2890 \pm 100\ \mathrm{km\,s^{-1}}$ determined from the broad Hα width and broad-to-narrow ratio of the NW filament by Ghavamian et al. (2002). Such close agreement is a strong indication that little substantial energy has been lost from the thermal plasma to cosmic ray acceleration, similar to optical proper motion studies from RCW 86 (Helder et al., in preparation).

From the above discussion it appears that Balmer-dominated SNRs, while offering powerful diagnostics of T_e/T_p and V_s, are not useful for investigating equilibration the extreme cases of strongly cosmic-ray modified shocks. In fact, the very condition allowing for the detection of the Balmer line emission—presence of neutral gas ahead of the shock—is also responsible for limiting the fraction of shock energy lost to cosmic ray acceleration. Quantitative evaluations of this effect by Drury et al. (1996) and Reville et al. (2007) show that when the preshock gas is significantly neutral, Alfvén waves driven by the cosmic rays ahead of the shock are dissipated by ion-neutral damping. As long as the charge exchange frequency, ω_{cx} ($\equiv n_{HI}\langle\sigma_{cx}v\rangle$) is larger than the Alfvén wave frequency, ω_A ($\equiv kv_A$) the ions and neutrals oscillate coherently, and ion-neutral damping is not important. However, when $\omega_{cx} < \omega_A$ the neutrals are left behind by the ions in the Alfvén wave motion, and during the incoherent oscillation between the two, charge exchange exerts a drag on the Alfvén waves, damping them. The condition required for Alfvén waves to not be strongly damped in the precursor can be written out as

$$n_{HI}\langle\sigma_{cx}v\rangle > kv_A \tag{6}$$

where $v_A \equiv \frac{B}{(4\pi m_i n_i)^{1/2}}$ is the Alfvén speed of the ions ahead of the shock and n_{HI} is the preshock neutral density. Given that cosmic rays resonantly scatter off Alfvén waves having Doppler shifted frequencies comparable to their gyrofrequency, and that the cosmic ray gyrofrequency is related to its energy via $\omega_{cr} \equiv ecB/E$, the inequality above can be cast in terms of the energy, E_{crit}, below which a significant fraction of the cosmic ray flux out of the shock is reduced by ion-neutral damping:

$$E_{crit}(\mathrm{TeV}) = 0.07\frac{B_3^2 T_4^{-0.4}}{x_{HI}(1-x_{HI})^{1/2}n^{3/2}} \tag{7}$$

where we have set $\langle\sigma_{cx}v\rangle \approx 8.4\times10^{-9}T_4^{0.4}\ \mathrm{cm^3\,s^{-1}}$ (Kulsrud and Cesarsky 1971), B_3 is the preshock magnetic field strength in units of 3 μG, n and x_{HI} are the total preshock density and neutral fraction, and T_4 is the preshock temperature in units of 10^4 K. For Balmer-dominated SNRs, where recent models have required moderate amplification of the preshock magnetic field ($\Delta B/B \sim 3$–5; Ghavamian et al. 2007) and where the preshock temperature may exceed 20,000 K (Raymond et al. 2011), $E_{crit} \sim 4$ TeV for the typical case where $x_{HI} = 0.5\ \mathrm{cm^{-3}}$. SNRs exhibiting nonthermal X-ray emission are believed to contain cosmic rays with energies of tens of TeV, so $E_{crit} \sim 4$ TeV is certainly high enough reduce the effectiveness of Balmer-dominated shocks in producing nonthermal X-ray emission.

However, as noted earlier, a modest back pressure from cosmic rays is required to explain the width of the Hα narrow component line, as well as the low broad-to-narrow ratios seen in some SNRs (Rakowski et al. 2009; Raymond et al. 2011). In fact, one model for electron heating in fast collisionless shocks requires at least some feedback from the cosmic rays in order to explain the moderate heating of electrons in SNRs, as well as the inverse squared relationship between T_e/T_p and V_s (Ghavamian et al. 2007, described in the next section).

Furthermore, as pointed out by Drury et al. (1996), the ion-neutral damping of Alfvén waves in the precursor is unimportant for cosmic rays which have already exceeded E_{crit}. Since the acceleration time for cosmic rays shortens considerably with shock speed ($\tau_{acc} \approx \kappa_{CR}/V_s^2$; (Malkov and Drury 2001), the fastest Balmer-dominated shocks are more likely to have accelerated particles beyond E_{crit} and hence will begin to exhibit nonthermal X-ray emission and cosmic-ray modified shock structure. A good example is the aforementioned SNR 0509–67.5, where the shock speeds exceed 5000 km s^{-1} (Helder et al. 2010) and nonthermal X-ray emission from cosmic ray accelerated electrons is detected from the forward shock. The forward shocks in more evolved Balmer-dominated SNRs (such as SN 1006 and Tycho's SNR) will have swept up more mass and slowed down to speeds $\lesssim$2000 km s^{-1}, by which point τ_{acc} will have lengthened and the shocks will be less cosmic-ray dominated.

4 Models for Electron Heating in SNRs

Given the lack of in situ measurements of the particle distributions in SNRs, the electron heating mechanisms in these shocks have been studied primarily via numerical methods. On one hand, a number of studies have focused on electron heating in relativistic shocks, with the aim of modeling high energy emission from gamma-ray bursts (e.g., Gedalin et al. 2008; Sironi and Spitkovsky 2011). These shocks are in a different area of parameter space than the SNR shocks discussed here, and the physics governing the electron heating in relativistic shocks is substantially different. At the very high Alfvénic Mach numbers characteristic of gamma-ray bursts, the shock transition becomes very thin (less than an electron gyroradius). Electrons in this case may be once again accelerated by the cross shock potential, similar to the very low Mach number case. On the other hand, a number of other studies consider non-relativistic shocks relevant to SNRs ($\lesssim$0.01c), where accelerated particles such as cosmic rays or solar energetic particles (SEPs) may play an important role in establishing their shock structure. These studies have sought to identify plasma waves capable of boosting electrons to mildly relativistic energies (e.g., Amano and Hoshino 2010; Riquelme and Spitkovsky 2011), with the objective of understanding how electrons are injected into the cosmic ray acceleration process. This is a different (though related) question from what we consider here, namely how electrons are promptly heated to temperatures $\sim 5\times10^6$ K at the shock front (Ghavamian et al. 2007; Rakowski et al. 2008). This limits our consideration of the work done so far to two broad scenarios of electron heating in fast, non-relativistic collisionless shocks. One scenario is based on lower hybrid wave heating in the cosmic ray precursor (Laming 2000; Ghavamian et al. 2007; Rakowski et al. 2008) while the other is based on counterstreaming instabilities ahead of the shock (e.g., the Buneman instability, Cargill and Papadopoulos (1988), Matsukiyo 2010; Dieckmann et al. 2012). We discuss these mechanisms below in turn.

4.1 Lower Hybrid Wave Heating

The most significant result of the Balmer-dominated shock studies, the inverse squared relation between T_e/T_p and V_s, places a useful constraint on the range of plausible equilibration mechanisms at the shock front. The simplest way to obtain $T_e/T_p \propto V_s^{-2}$ is to set $\Delta T_e \approx const$ at shock speeds of 400 km s^{-1} and higher, while allowing T_p to rise according to the Rankine-Hugoniot jump conditions, $k\Delta T_p \approx \frac{3}{16} m_p V_s^2$. The requirement that $T_e/T_p = 1$ at $V_s = 400$ km s^{-1} gives $\Delta T_e \approx 0.3$ keV for $V_s \geq 400$ km s^{-1}, independent of shock velocity (Ghavamian et al. 2007). Although there may be marginal evidence of a departure from this relation at shock speeds exceeding 2000 km s^{-1} (van Adelsberg et al. 2008), a velocity-independent heating of electrons in SNR shocks is an important clue to the nature of plasma heating processes in fast collisionless shocks. It suggests that plasma processes ahead of the shock front are an important (if not dominant) source of electron heating in SNRs (Ghavamian et al. 2007; Rakowski et al. 2009).

As mentioned earlier, strong interstellar shocks are expected to form a precursor where cosmic rays crossing upstream give rise to Alfvén waves and turbulence (Blandford and Eichler 1987; Jones and Ellison 1991), compressing and pre-heating the gas before it enters the shock. As long as the shock is strong ($v_{downstream}/V_s \approx 1/4$) and cosmic ray pressure does not dominate the postshock pressure (ΔB/B does not greatly exceed unity, with $\lesssim$20 % of the postshock pressure provided by cosmic rays) the thermal heating within the precursor does not depend strongly on shock velocity. The limited range of narrow component Hα widths observed in Balmer-dominated SNRs over a wide range in shock speeds (Sollerman et al. 2003; Raymond et al. 2011) is consistent with the relative insensitivity of the preshock heating to shock speed. Since the widening of the Hα narrow component line is now believed to arise in a precursor where the gas is heated by the damping of cosmic-ray driven waves (Wagner et al. 2009; Raymond et al. 2011; Morlino et al. 2012a), it stands to reason that perhaps the physical processes generating a constant electron heating with shock speed ($\Delta T_e \approx 0.3$ keV) also originate within the cosmic ray precursor.

The above argument was used by Ghavamian et al. (2007) and Rakowski et al. (2008) to advocate for a heating model where lower hybrid waves within the cosmic ray precursor preheat electrons to a constant temperature before they enter the shock front. This model was based on the conception of McClements et al. (1997), who suggested that lower hybrid waves driven by the reflected population of nonthermal ions could generate lower hybrid waves ahead of the shock, pre-heating electrons and injecting them into the cosmic ray acceleration process. The condition for generating such waves is that the shock be quasi-perpendicular, and that the reflected ions form a beam-like (gyrotropic) distribution. A similar scenario was suggested by Ghavamian et al. (2007) and Rakowski et al. (2008), but with one crucial difference: the reflected particles considered are ultra-relativistic cosmic rays rather than suprathermal ions. The lower hybrid waves are electrostatic ion waves which propagate perpendicular to the magnetic field and whose frequency is the geometric mean of the electron and ion geofrequencies, $\omega_{LH} = (\Omega_e \Omega_i)^{1/2}$. The group velocity of these waves is directed primarily along the magnetic field lines ($k_{||}^2/k_{\perp}^2 = m_e/m_p$; Laming 2001) and the waves can simultaneously resonate with ions moving across the field lines and electrons moving along the field lines. Although the growth rate of lower hybrid waves is generally small (Rakowski et al. 2008), their group velocity perpendicular to the magnetic field (and hence the shock front) can be on the order of the shock velocity ($\partial\omega/\partial k_{\perp} \approx V_s$). This allows the lower hybrid waves to remain in contact with the shock for long periods of time, attaining high intensities capable of effectively heating the electrons (McClements et al. 1997; Ghavamian et al. 2007).

In the case of cosmic rays, the time spent by the electrons in the precursor is $t \sim \kappa_{CR}/v_{sh}^2$. The kinetic energy acquired by the electrons in the precursor is $\Delta E_e \propto D_{||\,||} t$, where $D_{||\,||}$ is the momentum diffusion coefficient of electrons (Ghavamian et al. 2007). For lower hybrid wave turbulence, $D_{||\,||} \propto V_s^2$ (Karney 1978; Ghavamian et al. 2007; Rakowski et al. 2008), so that $\Delta E_e \approx \frac{1}{16}(\frac{m_e}{m_p})^{1/2} m_e \Omega_e \kappa_{CR} \propto B\kappa_{CR} \sim const$, as needed to account for the inverse squared relationship between equilibration and shock speed. Note that under the assumption that nonlinear amplification of the preshock magnetic field is not too strong ($\Delta B/B \sim 1$),

κ_{CR} is that of Bohm diffusion, which scales as $1/B$, so that ΔE_e is also approximately independent of B.

During the past decade more refined models of cosmic ray acceleration have shown that a non-resonant mode of Alfvén waves, having a higher growth rate than the previously considered resonant mode (Skilling 1975), can be excited by cosmic rays in the precursor (Bell and Lucek 2001; Bell 2004, 2005). Unlike for the resonant case, the non-resonant amplification allows for $\Delta B/B \gg 1$, driving preshock magnetic fields to values as high as 1 mG (Vink and Laming 2003; Berezhko et al. 2003; Bamba et al. 2005; Ballet 2006). Such magnetic fields are hundreds of times stronger than the canonical preshock magnetic field of 3 μG and high enough to account for the observed narrowness of X-ray synchrotron-emitting rims in such SNRs as SN 1006 (assuming the narrowness is due to rapid cooling of high energy electrons behind the shock; see Ballet 2006 and Morlino et al. 2012a). Additional studies have suggested that non-resonant amplification may dominate early in the life of the SNR, while resonant amplification may take over during the Sedov-Taylor stage of evolution (Amato and Blasi 2009; Schure et al. 2012), though in either case, $\Delta B/B > 10$ is readily attained. Such a strong magnetic field effectively reduces the acceleration time for particles, and is very well suited for explaining how cosmic rays can reach the knee in the cosmic ray spectrum near 10^{15} eV (Bell and Lucek 2001; Eriksen et al. 2011).

An important factor influencing the effectiveness of lower hybrid wave heating of electrons is the orientation of the preshock magnetic field relative to the shock front. Lower hybrid wave heating is only effective in perpendicular shocks (Vink and Laming 2003; Ghavamian et al. 2007; Rakowski et al. 2008). Given their spherical global geometry, SNR blast waves generally propagate at a range of angles to the interstellar magnetic field. X-ray observations and models of such SNRs as SN 1006 (Orlando et al. 2007; Petruk et al. 2008) have indicated that perpendicular shocks are far more effective at accelerating cosmic rays than parallel shocks. Although the detailed implications of such differences have not yet been worked out for the lower hybrid wave heating model, Rakowski et al. (2008) argue that even for quasi-parallel shock geometries the cosmic ray current driving the nonresonant Alfvén waves will generate a significant perpendicular magnetic field ahead of the shock (such a possibility has also been inferred from numerical simulations; Riquelme and Spitkovsky 2011). This would allow lower hybrid wave growth to overtake modified Alfvén wave growth for arbitrary orientations of the far upstream magnetic field, and allow for a more ubiquitous role for lower hybrid wave heating of electrons.

The amplification of the preshock magnetic field well beyond its far upstream value introduces an interesting possibility: effective lowering of the Alfvénic (and hence magnetosonic) Mach number of the shock. For the Balmer-dominated shocks, where analysis of the optical spectra has shown that at best only a moderate fraction ($\lesssim$20 %) of the shock energy has likely been channeled into cosmic rays, the widening of the Hα narrow component has been interpreted as nonthermal broadening caused by the lowest frequency waves in the precursor (Ghavamian et al. 2007; though see Raymond et al. 2011 for a thermal interpretation). To explain the 30–50 $\mathrm{km\,s^{-1}}$ widths of the Hα narrow component, the preshock magnetic field must be enhanced by a factor of a few. For the non-resonant Alfvén waves in the Bell (2004) mechanism, the magnetic field energy density immediately behind the shock is given by (Schure et al. 2012)

$$\frac{B^2}{4\pi} \approx \frac{1}{4}\phi^2 \rho v_{sh}^2 \tag{8}$$

where $\phi \equiv P_{CR}/\rho v_{sh}^2$ is the fraction of the shock ram pressure channeled into cosmic rays. Solving this expression for B gives B (μG) $\approx 228.7\phi n^{1/2} V_{1000}$, where V_{1000} is the shock speed in units of 1000 km s^{-1}. For $\phi \sim 0.1$–0.2, $n \sim 1$ cm^{-3}, a postshock compression factor of 4 and Balmer-dominated shock speeds ∼2000 km s^{-1}, this gives $\Delta B/B \sim 4$–10 ahead of the shock. Correspondingly, v_A can increase by nearly the same factor, so that M_A can be reduced by as much as an order of magnitude. Treumann and Jaroschek (2008) describe the physical picture in this case as that of the shock having to prevent an increasing number of ions from crossing the shock jump by deflecting an increasing number of them at higher and higher Mach numbers. This deflection is necessary so that the ability of the shock to dissipate the inflowing energy is not overwhelmed. By deflecting these ions back upstream into a precursor, the net inflow of momentum and energy density into the shock is reduced, reducing the net difference in velocity between the inflowing and outflowing ions. This effectively reduces the Mach number in the frame of the upstream medium.

Note that the ions in the precursor are only mildly compressed (Wagner et al. 2009; Morlino et al. 2012a), which only weakly counteracts the rise in B. In addition, v_A only scales as $n^{-1/2}$, but scales directly as B. The result is that *given the compelling evidence for enhanced preshock magnetic fields in SNRs shocks, the Mach numbers of these shocks may be overestimated by as much as an order of magnitude.* As we describe in Sect. 4, a unified description of solar wind and SNR shocks, where the physics of electron-ion temperature equilibration occurs over a similar range in Mach numbers and involves a similar range of physical processes, may be possible.

4.2 Plasma Wave Heating from the Buneman Instability

Similar to the lower hybrid wave model, the Buneman instability-driven wave model focuses on the region immediately ahead of a quasi-perpendicular shock. However, unlike the lower hybrid wave model, the Buneman instability models consider the reflected nonthermal ion distribution, rather than ultrarelativistic cosmic rays. In the latter model, ∼20 % of the ions are reflected backstream against the incoming electron and ion plasma (Cargill 1988; Cargill and Papadopoulos 1988). Here the upstream plasma is not electrically neutral due to the positive charge of the reflected ion distribution. In such cases, a drift is induced between the electrons and ions. The microinstabilities excited by this configuration depend upon the relative size of the electron thermal speed relative to the electron-ion drift velocity. The Buneman instability occurs when the drift velocity of the reflected ions relative to the upstream electrons exceeds the thermal speed of the upstream electrons ($2v_s > (kT_e/m_e)^{1/2}$)(Cargill and Papadopoulos 1988), a condition which occurs for very high Mach number ($M_A \gtrsim 50$) shocks. If the reflected ion current upstream is strong enough, the electron current generated to counteract it may produce a large enough drift between the preshock ions and electrons to cause a secondary Buneman instability when the ion speed exceeds the electron thermal speed (Dieckmann et al. 2012). The Buneman instability generates electrostatic plasma waves which damp by rapidly heating the preshock electrons to $k\Delta T_e \approx 2m_e v_s^2 \approx 0.01 v_{1000}^2$ keV, where v_{1000} is the shock speed in units of 1000 km s^{-1}. until their thermal speed matches the electron-ion drift speed, at which point the instability saturates. The rapid heating of the electrons perpendicular to the magnetic field results in $T_e/T_i \gg 1$ and makes it possible for an ion acoustic instability to occur between the preshock electrons and either the reflected ions or preshock ions (Cargill and Papadopoulos 1988). The waves generated by the ion acoustic instability can then transfer a substantial fraction of the shock energy (tens of percent) into electron thermal energy. This process occurs over a length scale of v_s/Ω_i (as opposed to κ_{CR}/v_s for the cosmic ray precursor), resulting in a $T_e/T_p \approx 0.2$, independent of shock speed. *This is in strong disagreement with the equilibrations obtained for the Balmer-dominated shocks.*

A number of other electron heating mechanisms, such as the modified two-stream instability and electron-cyclotron drift instability, have been proposed for collisionless shocks based on results from particle in cell (PIC) simulations (Umeda et al. 2012a, 2012b). A unified picture proposed by Matsukiyo (2010) suggests that electrons can be strongly energized at low Mach numbers ($M_A \leq 10$) via a modified two-stream instability, where the velocities of the reflected/incoming ions and the electrons are lower than the thermal speed of the electrons and the electrons are able to damp out the Buneman instability. In this case, obliquely propagating whistler mode waves are excited, having frequencies between the electron cyclotron frequency and the lower hybrid wave frequency. When the electron-ion drift velocity and electron thermal speed become nearly equal, the electron-cyclotron drift instability becomes important (Umeda et al. 2012a), exciting waves with frequencies that are multiples of the electron cyclotron frequency. At higher Mach numbers the electron thermal speed is lower than that of the ions, and the Buneman instability/ion acoustic wave process described earlier is predicted to take over.

The amount of electron heating predicted by the Buneman instability/ion acoustic wave model scales as M_A^2 (Cargill and Papadopoulos 1988; Matsukiyo 2010), so that for shocks in the 2000 $\mathrm{km\,s^{-1}}$–10,000 $\mathrm{km\,s^{-1}}$ range, $\Delta E_e \sim 2$–50 keV. This is clearly at odds with $\Delta E_e = 0.3$ keV observed between 400 $\mathrm{km\,s^{-1}}$ and 2000 $\mathrm{km\,s^{-1}}$ for Balmer-dominated shocks. One explanation for this discrepancy is that growth of the Buneman-like and two-stream instabilities described above requires that the reflected ions form a distribution function with a positive gradient at some velocity (Laming 2000). This distribution forms when specularly reflected ions have a mostly monoenergetic, beamlike configuration. At the low Mach numbers in the solar wind ($\lesssim 10$), where the shock structure is laminar, the reflected ions closely resemble a monoenergetic beam. However, at the higher Mach numbers, where the shock front is more turbulent and disordered, the reflected ions are likely to have a greater spread in energy and are probably less beamlike (Laming 2000). This would lead to suppression of Buneman-like instabilities. However, this line of reasoning is still speculative, and the real explanation for the lack of agreement between the observed T_e/T_p and those predicted by models in this section remains to be explored. Cosmic-ray driven processes may ultimately provide a better explanation for electron heating at SNR shocks than those involving reflected suprathermal ions.

The cross shock potential arises from the charge separation produced by the different gyroradii for ions and electrons as they cross the shock transition. It accelerates electrons into the post shock layer, and can be a means of electron heating at subcritical shocks with an approximately laminar structure. At supercritical shocks, the time dependence and nonlocality introduced reduces the degree of electron heating. However at sufficiently Alfvén Mach number (where the shock transition becomes thin (on the order of the electron convective gyroradius or inertial length), significant electron heating may again occur. In the absence of magnetic field amplification by cosmic rays, this might be expected to happen at SNR shocks. However it is much more likely in environments where the plasma beta is low, such as galaxy clusters. It may also occur in cases where a significant population of cosmic rays is unlikely due to the low age of the shock, as in gamma-ray bursts.

5 Constraints from Solar Wind Studies

From the beginning, a detailed study of the fastest collisionless shocks has been hampered by one inherent limitation: they occur in objects which are too remote for in situ study. Although some collisionless shocks in our solar system reach Alfvénic Mach numbers as high as 30 (such as those around Saturn; Achilleos et al. 2006; Masters et al. 2011), there are no physical phenomena in our solar system energetic enough to produce the type of shocks seen in SNRs (Mach numbers $\sim$100 or more if no preshock magnetic field enhancement in the CR is assumed). In addition, the range of plasma betas attainable in the solar system is larger than the range of plasma betas attained in the interstellar medium.

Another fundamental difference between solar wind and SNR shocks is the fact that the former are short-lived phenomena confined to small spatial scales (millions of km) in a curved (bow shock) geometry, whereas the latter are sustained for thousands of years, on spatial scales of parsecs, often well-described by a planar geometry. This results in an irreducible difference between the two types of shocks: particles crossing back and forth between upstream and downstream can remain in contact with SNR shocks for long periods of time, allowing accelerated CRs to acquire much more energy in SNR shocks than in solar wind shocks. This potentially allows the CRs to create shock precursors with properties needed to heat electrons and influence T_e/T_p.

The heating of electrons at the Earth's and other planetary bow shocks has been the subject of much theoretical and observational work. Typical features of the electron temperature change, ΔT_e, observed at the bow shock as observed by Schwartz et al. (1988) include (a) an approximate relationship between heating and the incident solar wind energy $\Delta T_e \propto U^2$, where U is the component of the solar wind's velocity incident upon the shock, and (b) a relationship between the change in temperature normalized by the incident energy and the fast magnetosonic Mach number, $\Delta T_e/(m_p U^2/2) \propto M_{ms}^{-1}$. A similar approximate relationship holds between the normalized electron temperature change and the Alfvén Mach number M_A, especially for shocks with a low plasma β, which is the ratio between thermal and magnetic pressures. Recent work by Masters et al. (2011) shows that this relationship with M_A holds well at Saturn's bow shock. This is particularly interesting as M_A at Saturn is often much larger than at Earth.

In Fig. 7 we plot the ratio of electron and ion temperatures downstream of the shock against the magnetosonic, and Alfvén Mach numbers, as well as the upstream flow velocity relative to the shock. The data in these figures are taken from bow shock crossings of the ISEE spacecraft, and is a subset of those listed in Schwartz et al. (1988), consisting of 61 crossings for which all the necessary data is available.

It is well known that quantities other than those displayed here may be more appropriate, specifically the change in electron temperature over the change in total temperature $\Delta T_e/\Delta(T_e + T_i)$ or even $\Delta T_e/\Delta T_i$ are better correlated with inverse Mach numbers than T_e/T_i (Schwartz et al. 1988). Nevertheless, we use the latter quantity here as it enables a comparison with data from extra-solar system and outer planetary shocks at where less data are available. Furthermore, the approximate inverse dependence upon M_{ms}, M_A, and V_s is still quite apparent in this data. It is interesting to note that the relationship with the Mach numbers is much more favourable than the dependence upon V_s, indicating that the Mach number is the more relevant quantity for organizing the relationship between T_e/T_p and shock strength.

Many mechanisms may be involved in the heating of electrons in solar system shocks. Proposed mechanisms involve acceleration of electrons by a cross-shock potential (Goodrich and Scudder 1984; Scudder et al. 1986a, 1986b), wave turbulence (Galeev 1976), microinstabilities (Wu et al. 1984), and electron trajectory scattering (Balikhin et al. 1993). The existence of a cross-shock potential may be deduced from the generalized Ohm's law in which a gradient in electron thermal pressure gives rise to an electric field. Examining the energetics of electrons crossing the shock may simplified by working in the de Hoffmann-Teller frame of reference (de Hoffmann and Teller 1950), defined as the frame

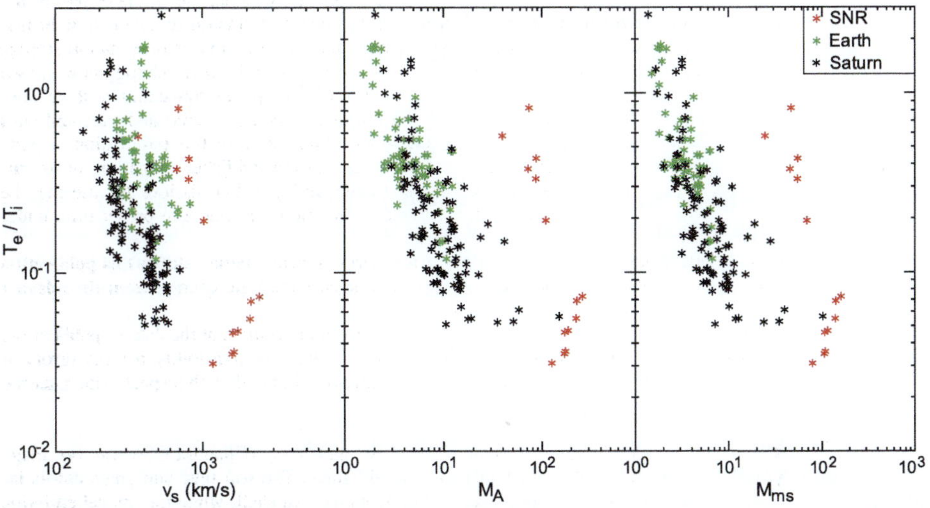

Fig. 7 Collected electron-ion equilibration data from both the solar wind bow shocks and supernova remnant shocks. T_e/T_p is plotted versus shock speed (*left*), Alfvénic Mach number (*center*) and magnetosonic Mach number (*right*). *Green symbols* show data from crossings of Earth's bow shock (Schwartz et al. 1988), while the *black symbols* show data from crossings of Saturn's bow shock (Masters et al. 2011). Shock speeds for the Saturnian bow shock are based on a solar wind model and an assumed shock speed with respect to the spacecraft of 100 $\mathrm{km\,s^{-1}}$, and ion temperatures are based on electron distribution measurements and the application of the Rankine-Hugoniot conditions (see Masters et al. 2011 for a full discussion of shock parameter derivations at Saturn's bow shock). *Red symbols* show data acquired from Balmer-dominated SNR shocks (van Adelsberg et al. 2008), and assume $v_A = 9\ \mathrm{km\,s^{-1}}$, $c_s = 11\ \mathrm{km\,s^{-1}}$

in which the shock is at rest, and in which the magnetic field and plasma flow velocity are (anti-)parallel. In this case the electric field is dominated by that generated by the electron pressure gradient, and the work done on electrons crossing the shock is determined by the cross-shock potential. Additional mechanisms are required to scatter electrons to pitch angles that are more perpendicular, and to flatten the distribution so that empty regions of phase space are filled. This results in a distribution whose temperature is controlled to a large extent by the de Hoffmann-Teller frame cross-shock potential and downstream density. In addition to direct measurement of the electric fields within the shock, (Bale and Mozer 2007; Dimmock et al. 2011), comparison of upstream and downstream electron phase space distributions have shown that these are consistent with electron acceleration by a cross-shock potential in the de Hoffmann-Teller frame (Lefebvre et al. 2007).

6 Connecting the Solar Wind Results to Those in SNRs

Figure 7 may indicate that similar mechanism(s) heat the electrons in solar wind and in SNR shocks. *This is especially appealing when we remember that M_A in SNRs may be overestimated due to preshock amplification of magnetic field by cosmic rays.* In their study of the terrestrial bow shock and interplanetary shocks, Schwartz et al. (1988) found that the electron-ion temperature equilibration organizes best by $T_e/T_i \propto 1/M_A$. Given the difficulty in determining the Mach numbers of SNR shocks, the equilibration dependence on shock strength has been characterized via the shock speed instead, and found to obey $T_e/T_i \propto 1/V_s^2$ (V_s is much more accurately known than the Mach numbers). Section 4.1 outlined a model for SNR electron heating, where the cosmic ray diffusion coefficient κ_{CR} is assumed independent of V_s. From quasi-linear theory (Blandford and Eichler 1987),

$$\kappa_{CR} = \frac{p^2 c^2 v}{3\pi e^2 U} \tag{9}$$

where U is the energy density of turbulence ($\equiv \langle \Delta B^2 \rangle / 8\pi$) and v is the cosmic ray velocity. For resonant amplification, we evaluate U at $k_{\parallel} = \Omega / v_{\parallel}$, where Ω is the gyroradius and $v_{\parallel}$ the parallel component of the cosmic ray velocity. For relativistic cosmic rays, where $v = c$, this results in $\kappa_{CR} \propto p^2 / U$, assumed constant with V_s. However, for nonrelativistic suprathermal particles, v will most likely be proportional to the shock velocity V_s, which with the same assumptions leads to $T_e / T_i \propto 1/V_s$ for solar wind shocks (as opposed to $1/V_s^2$ in SNRs). This argument is admittedly loose, and should not be viewed as much more than a hypothesis to motivate further work.

In our arguments above we have made considerable assumptions about κ_{CR}. The most obvious one is that κ_{CR} as written above applies to parallel shocks, whereas we are most likely dealing with quasi-perpendicular cases. This may reduce the difference anticipated between solar wind and SMR shocks, depending on the turbulence spectrum (summarized in Appendix A of Rakowski et al. 2008).

The degree of cosmic ray magnetic field amplification at SNR shocks required to bring SNR data points in Fig. 7 into alignment with solar wind data points is approximately an order of magnitude or less at the highest velocities considered. Given the degree of magnetic field amplification expected from cosmic ray acceleration, this appears to be highly plausible. In the case of saturated nonresonant instability (Bell 2004, 2005) by resonant scattering (Luo and Melrose 2009), $\langle \Delta B \rangle^2 / B^2 \sim 10$–$100$ is expected. In the case of nonresonant saturation, higher values, but strongly dependent on k, are predicted. Saturation by electron heating (i.e. the M_A where growth of lower hybrid waves becomes greater than the growth rate of magnetic field) leads to similar magnetic field enhancement, with $\Delta B^2 / B \sim 200$ (Rakowski et al. 2008). Such magnetic field amplification is less likely at solar wind shocks. The suprathermal particle densities are lower in solar wind shocks, and the ambient magnetic fields are higher, much closer to where the nonresonant instability would saturate (if not already beyond it).

7 Observational Constraints from Galaxy Cluster Shocks

Collisionless shocks occur over a vast range of length scales, with those in galaxy clusters being among the largest. While the shock speeds in the galaxy cluster shocks are similar to those in supernova remnants (up to 4000 km s^{-1}; Markevitch et al. 2005; Markevitch and Vikhlinin 2007; Russell et al. 2012), they occur in environments that are substantially different from both the solar wind and the ISM. These differences can be encapsulated via the plasma beta, defined as the ratio of the thermal pressure to the magnetic pressure of a plasma ($\beta \equiv nkT / (B^2 / 8\pi)$). Utilizing solar wind parameters listed by Bruno and Carbone (2005), this ratio ranges from around unity at 1 AU under fast solar wind conditions (wind velocity $\sim$900 km s^{-1}) to around 20 for the quiescent wind (wind velocity $\sim$300 km s^{-1}). The plasma β of the ISM is close to that of the fast solar wind, $\beta_{ISM} \sim 1$–4 (assuming a $B = 3$ μG, $n = 1$ cm^{-3} and $T = 10^4$ K). On the other hand, the electron temperature of the intracluster medium (ICM) ranges from $\sim 10^7$ K–10^8 K (1–10 keV), with number densities steeply declining from $\sim 10^{-2}$ cm^{-3} near the cluster centers to $\sim 10^{-4}$ cm^{-3} at the outer edges. The corresponding sound speed is close to 1000 $\mathrm{km\,s^{-1}}$, nearly two orders of magnitude higher than in the general ISM. The magnetic fields measured in galaxy clusters are actually close that of the ISM, typically on the order of a microgauss (Carilli and Taylor 2002). Therefore, $v_A \sim 50\ \mathrm{km\,s^{-1}}$ in galaxy clusters, so that $\beta_e \gg 1$ and the magnetic field pressure has negligible contribution to the dynamics of shocks in galaxy clusters. This puts galaxy cluster shocks in a different region of parameter space than solar wind and SNR shocks.

Clusters such as 1E0657–56 (the 'Bullet Cluster') and A520 show strongly enhanced X-ray emission from collisionless shocks, formed during major mergers when gas from one cluster plunges through gas from the other (Markevitch et al. 2005; Markevitch and Vikhlinin 2007; Russell et al. 2012). Shocks moving mostly along the plane of the sky have a favorable viewing geometry and appear as giant curved structures hundreds of kiloparsecs in length. The large clumps of infalling gas drive bow shocks into the cluster gas, which has already been heated to at least 1 keV, and produces thermal Bremsstrahlung emission peaking close to that energy. This is another major difference between collisionless shocks in the ISM and those in the ICM. While the Alfvénic and magnetosonic Mach numbers of SNR shocks are very difficult to determine due to the lack of available observational constraints on magnetohydrodynamic quantities upstream (such as T, B and n), those in galaxy clusters can readily be measured by spectral analysis of X-ray emission upstream. Comparison of this emission to that of the enhanced postshock region gives the density contrast between the downstream and upstream gas (i.e., n_2/n_1). This density contrast yields the sonic Mach number, M, via solving the Rankine-Hugoniot jump conditions (Russell et al. 2012):

$$M = \left(\frac{2n_2/n_1}{\gamma + 1 - \frac{n_2}{n_1}(\gamma - 1)} \right)^{1/2} \tag{10}$$

where γ is the ratio of specific heats of the cluster gas. Measurement of these jumps from X-ray observations have yielded $M \sim 1.5$–3 for the Bullet Cluster (Markevitch et al. 2005), Abell 520 and Abell 2146 (Russell et al. 2012). Using these estimated Mach numbers, the shock velocity itself can be calculated via $V_s = Mc_s$, where c_s is the upstream sound speed as inferred from the X-ray spectra. This yields shock speeds ranging between 2500 $\mathrm{km\,s^{-1}}$ and 4000 $\mathrm{km\,s^{-1}}$, similar to the fastest known Balmer-dominated shocks. However, fits to the X-ray spectra behind these shocks show prompt electron-ion equilibration at the shock front, consistent with $T_e/T_p = 1$ (Markevitch et al. 2005; Markevitch and Vikhlinin 2007), despite the extremely high shock speeds involved. This result can only be reconciled with equilibration measurements from the solar wind and SNRs if the equilibration depends on the Mach number, rather than V_s. Furthermore, given the low Mach numbers found in the galaxy clusters, it is plausible that the shock transitions in these cases are laminar (as opposed to turbulent like the SNR and fastest solar wind shocks), with electron heating occurring efficiently at the shock front via the same type of cross-shock potential as seen in the slowest solar wind and slowest SNR ($V_s \leq 400\ \mathrm{km\,s^{-1}}$) shocks. This is of course a speculation; further insight into collisionless cluster shocks may be obtained via numerical simulations for the appropriate conditions.

8 Summary and Future Work

There have been exciting advances in the study of electron-ion temperature equilibration in collisionless shocks during the past few years. Perhaps most notable has been the growing realization that temperature equilibration and cosmic ray acceleration may be intertwined processes. Optical studies of collisionless shocks in partially neutral gas (Balmer-dominated shocks) have shown that the electron-ion temperature equilibration is a declining function of shock speed, well characterized as $T_e/T_p \propto V_s^{-2}$. This relationship most likely arises due to electron heating ahead of the shock that is nearly independent of shock speed above 400 km s^{-1}. Cosmic ray precursors, with moderately amplified preshock magnetic field and density, are the most logical sites for electron heating in SNR collisionless shocks. The transition to fully equilibrated SNR shocks at speeds below 400 km s^{-1} may be due to a less turbulent and more laminar structure at low shock speeds and Mach numbers. This allows the electrons to experience a more uniform cross-shock potential, and hence higher energization compared to higher shock speeds and Mach numbers, where the shock jump is more turbulent and disordered. The magnetosonic Mach numbers of SNR shocks may match those in solar wind shocks if there is a approximately an order of magnitude increase in the Alfvén speed of the preshock gas in SNRs compared to the average ISM value. This is possible if there is a moderately amplified preshock magnetic field ($\sim$10), and is readily provided by compression and heating in a cosmic ray precursor. In solar wind shocks, the precursor is due to suprathermal, non-relativistic ions, resulting in a $T_e/T_p \propto 1/V_s$ seen in the solar wind.

While Balmer-dominated shocks have allowed us to elucidate some of the physics of electron-ion temperature equilibration, ion-neutral damping limits most of those observed to cases where the shock structure has not become strongly modified by cosmic ray acceleration. Given this limitation, *electron-ion equilibration studies of fast, collisionless shocks in fully pre-ionized gas would be highly desirable*. Such a sample would allow the equilibration to be studied over a range of speeds where shocks are increasingly affected by feedback from the accelerated cosmic rays. In such circumstances it is unclear what would happen to the T_e/T_p versus V_s relation. If, as predicted by Amato and Blasi (2009), Bell's non-resonant cosmic ray instability takes over from the resonant instability at the highest shock speeds, then additional electron heating may occur in the fastest shocks ($V_s \gtrsim 5000$ km s^{-1}), resulting in substantial deviation from the $T_e/T_p \propto V_s^{-2}$ relation. Such deviations may already have been seen in the fastest ($V_s \gtrsim 2000$ km s^{-1}) shocks, where there is evidence that T_e/T_p does not settle down to m_e/m_p, but rather $\sim$0.03. Other deviations may have been detected in SNR 0509–67.5, where T_e/T_p for a 5000 km s^{-1} shock has been estimated to be $\sim$0.2, substantially higher than predicted by the inverse squared relation. However, the study of such shocks would be challenging. Without an Hα broad component to constrain the range of V_s, shock speeds would have to be determined by other means, such as proper motion studies. That would require X-ray and or UV imagery of SNRs with well-constrained distances (such as those in the LMC or SMC), over multiple epochs. It would also require spectroscopy of the forward shocks in these SNRs, in order to constrain both the electron temperature (via X-ray continuum fits and UV emission line ratios) and the ion temperature (e.g., via He II, C IV, N V and O VI resonance lines).

An important test of our ideas concerning electron heating by cosmic ray generated waves in a shock precursor would be to measure electron temperatures at SNR shocks exhibiting strong cosmic ray modification and substantial magnetic field amplification ($\Delta B/B \gtrsim 100$). Several SNRs show X-ray filaments produced by synchrotron radiation from cosmic ray electrons, and are generally distinct from those shocks with strong Balmer emission. The absence of neutral material ahead of these shock means that optical and UV emission is weak, and electron temperatures will have to be measured from X-ray spectra. Difficulties arise in distinguishing thermal electron bremsstrahlung from cosmic ray electron synchrotron emission, requiring data of high signal to noise. Further complications in some SNRS (e.g.

Cas A) stem from scattering of bright X-ray emission from the ejecta such that it coincides spatially with emission from the forward shock. Such scattering may either be local, due to SNR dust, or instrumental, due to telescope imperfections. This leaves SN 1006 as the best likely target for such an observation, since due to its evolutionary state, only the outer layers of ejecta have encountered the reverse shock.

The development of missions like Solar Orbiter and Solar Probe Plus will allow *in situ* measurements of shocks in the solar wind, most likely associated with coronal mass ejections, much closer to the Sun. These will probe a different parameter regime, where the magnetic field pressure dominates over the gas pressure (low β, similar to ISM shocks). As such, measurements here might yield insights into the properties of similar plasma in the precursors of SNR shocks where the magnetic field has been amplified by cosmic rays.

Acknowledgements P.G. acknowledges support by HST grant HST-GO-11184.07-A to Towson University. JML acknowledges support by grant NNH10A009I from the NASA Astrophysics Data Analysis Program, and by basic research funds of the Office of Naval Research.

References

A.A. Abdo, M. Ackermann, M. Ajello et al., Observations of the Young supernova remnant RX J1713.7-3946 with the Fermi large area telescope. Astrophys. J. **734**, 28 (2011). doi:10.1088/0004-637X/734/1/28

F. Acero, F. Aharonian, A.G. Akhperjanian et al., First detection of VHE γ-rays from SN 1006 by HESS. Astron. Astrophys. **516**, A62 (2010). doi:10.1051/0004-6361/200913916

N. Achilleos, C. Bertucci, C.T. Russell et al., Orientation, location and velocity of Saturn's bow shock: initial results from the Cassini spacecraft. J. Geophys. Res. **111**, A03201 (2006). doi:10.1029/2005JA011297

F. Aharonian et al., On the origin of TeV radiation of SN 1006. Astron. Astrophys. **351**, 330 (1999)

F. Aharonian, A.G. Akhperjanian, A.R. Bazer-Bachi et al., A detailed spectral and morphological study of the gamma-ray supernova remnant RX J1713.7-3946 with HESS. Astron. Astrophys. **449**, 223–242 (2006). doi:10.1051/0004-6361:20054279

T. Amano, M. Hoshino, A critical mach number for electron injection in collisionless shocks. Phys. Rev. Lett. **104**(20), 181102 (2010). doi:10.1103/PhysRevLett.104.181102

E. Amato, P. Blasi, A kinetic approach to cosmic-ray-induced streaming instability at supernova shocks. Mon. Not. R. Astron. Soc. **392**, 1591–1600 (2009). doi:10.1111/j.1365-2966.2008.14200.x

S.D. Bale, F.S. Mozer, Measurement of large parallel and perpendicular electric fields on electron spatial scales in the terrestrial bow shock. Phys. Rev. Lett. **98**(20), 205001 (2007). doi:10.1103/PhysRevLett.98.205001

M. Balikhin, M. Gedalin, A. Petrukovich, New mechanism for electron heating in shocks. Phys. Rev. Lett. **70**, 1259–1262 (1993). doi:10.1103/PhysRevLett.70.1259

J. Ballet, X-ray synchrotron emission from supernova remnants. Adv. Space Res. **37**, 1902–1908 (2006). doi:10.1016/j.asr.2005.03.047

A. Bamba, R. Yamazaki, T. Yoshida, T. Terasawa, K. Koyama, A spatial and spectral study of nonthermal filaments in historical supernova remnants: observational results with Chandra. Astrophys. J. **621**, 793–802 (2005). doi:10.1086/427620

A.R. Bell, The interaction of cosmic rays and magnetized plasma. Mon. Not. R. Astron. Soc. **358**, 181–187 (2004). doi:10.1111/j.1365-2966.2005.08774.x

A.R. Bell, Turbulent amplification of magnetic field and diffusive shock acceleration of cosmic rays. Mon. Not. R. Astron. Soc. **353**, 550–558 (2005). doi:10.1111/j.1365-2966.2004.08097.x

A.R. Bell, S.G. Lucek, Cosmic ray acceleration to very high energy through the non-linear amplification by cosmic rays of the seed magnetic field. Mon. Not. R. Astron. Soc. **321**, 433–438 (2001). doi:10.1046/j.1365-8711.2001.04063.x

E.G. Berezhko, L.T. Ksenofontov, H.J. Völk, Confirmation of strong magnetic field amplification and nuclear cosmic ray acceleration in SN 1006. Astron. Astrophys. **412**, L11–L14 (2003). doi:10.1051/0004-6361:20031667

R. Blandford, D. Eichler, Particle acceleration at astrophysical shocks: a theory of cosmic ray origin. Phys. Rep. **154**, 1–75 (1987). doi:10.1016/0370-1573(87)90134-7

P. Blasi, G. Morlino, R. Bandiera, E. Amato, D. Caprioli, Collisionless shocks in a partially ionized medium. I. Neutral return flux and its effects on acceleration of test particles. Astrophys. J. **755**, 121 (2012). doi:10.1088/0004-637X/755/2/121

R. Bruno, V. Carbone, The Solar wind as a turbulence laboratory. Living Rev. Sol. Phys. **2**, 4 (2005)
P. Cargill, Electron heating in superhigh Mach number shocks. Astrophys. Space Sci. **144**, 535–547 (1988)
P.J. Cargill, K. Papadopoulos, A mechanism for strong shock electron heating in supernova remnants. Astrophys. J. **329**, L29–L32 (1988). doi:10.1086/185170
C.L. Carilli, G.B. Taylor, Cluster magnetic fields. Annu. Rev. Astron. Astrophys. **40**, 319–348 (2002). doi:10.1146/annurev.astro.40.060401.093852
R.A. Chevalier, J.C. Raymond, Optical emission from a fast shock wave—the remnants of Tycho's supernova and SN 1006. Astrophys. J. **225**, L27–L30 (1978). doi:10.1086/182785
R.A. Chevalier, R.P. Kirshner, J.C. Raymond, The optical emission from a fast shock wave with application to supernova remnants. Astrophys. J. **235**, 186–195 (1980). doi:10.1086/157623
D.P. Cox, J.C. Raymond, Preionization-dependent families of radiative shock waves. Astrophys. J. **298**, 651–659 (1985). doi:10.1086/163649
F. de Hoffmann, E. Teller, Magneto-hydrodynamic shocks. Phys. Rev. **80**, 692–703 (1950). doi:10.1103/PhysRev.80.692
A. Decourchelle, D. Ellison, Thermal x-ray emission and cosmic-ray production in Young supernova remnants. Astrophys. J. Lett. **543**, L57–L60 (2000). doi:10.1086/318167
M.E. Dieckmann, A. Bret, G. Sari, E. Perez Alvaro, I. Kourakis, M. Borghesi, Particle simulation study of electron heating by counter-streaming ion beams ahead of supernova remnant shocks. Plasma Phys. Control. Fusion **54**, 085015 (2012). doi:10.1088/0741-3335/54/8/085015
A.P. Dimmock, M.A. Balikhin, Y. Hobara, Comparison of three methods for the estimation of cross-shock electric potential using cluster data. Ann. Geophys. **29**, 815–822 (2011). doi:10.5194/angeo-29-815-2011
B.T. Draine, C.F. McKee, Theory of interstellar shocks. Annu. Rev. Astron. Astrophys. **31**, 373–432 (1993). doi:10.1146/annurev.aa.31.090193.002105
L.O.C. Drury, P. Duffy, J.G. Kirk, Limits on diffusive shock acceleration in dense and incompletely ionised media. Astron. Astrophys. **309**, 1002–1010 (1996)
J.P. Edmiston, C.F. Kennel, A parametric survey of the first critical Mach number for a fast MHD shock. J. Plasma Phys. **32**, 429–441 (1984). doi:10.1017/S002237780000218X
D.C. Ellison, D.J. Patnaude, P. Slane, P. Blasi, S. Gabici, Particle acceleration in supernova remnants and the production of thermal and nonthermal radiation. Astrophys. J. **661**, 879–891 (2007). doi:10.1086/517518
K.A. Eriksen, J.P. Hughes, C. Badenes, R. Fesen, P. Ghavamian, D. Moffett, P.P. Plucinsky, C.E. Rakowski, E.M. Reynoso, P. Slane, Evidence for particle acceleration to the knee of the cosmic ray spectrum in Tycho's supernova remnant. Astrophys. J. **728**, L28 (2011). doi:10.1088/2041-8205/728/2/L28
K. France, R. McCray, S.V. Penton, R.P. Kirshner, P. Challis, J.M. Laming, P. Bouchet, R. Chevalier, P.M. Garnavich, C. Fransson, K. Heng, J. Larsson, S. Lawrence, P. Lundqvist, N. Panagia, C.S.J. Pun, N. Smith, J. Sollerman, G. Sonneborn, B. Sugerman, J.C. Wheeler, HST-COS observations of hydrogen, helium, carbon, and nitrogen emission from the SN 1987A reverse shock. Astrophys. J. **743**, 186 (2011). doi:10.1088/0004-637X/743/2/186
A.A. Galeev, Collisionless shocks, in *Physics of Solar Planetary Environments*, ed. by D.J. Williams (1976), pp. 464–490
M. Gedalin, M.A. Balikhin, D. Eichler, Efficient electron heating in relativistic shocks and gamma-ray-burst afterglow. Phys. Rev. E **77**, 026403 (2008). doi:10.1103/PhysRevE.77.026403
P. Ghavamian, Optical spectroscopy and numerical models of nonradiative shocks in supernova remnants. PhD Thesis, Rice University, 1999
P. Ghavamian, J.C. Raymond, P. Hartigan, W.P. Blair, Evidence for shock precursors in Tycho's supernova remnant. Astrophys. J. **535**, 266–274 (2000). doi:10.1086/308811
P. Ghavamian, J.C. Raymond, R.C. Smith, P. Hartigan, Balmer-dominated spectra of nonradiative shocks in the Cygnus loop, RCW 86, and Tycho supernova remnants. Astrophys. J. **547**, 995–1009 (2001). doi:10.1086/318408
P. Ghavamian, P.F. Winkler, J.C. Raymond, K.S. Long, The optical spectrum of the SN 1006 supernova remnant revisited. Astrophys. J. **572**, 888–896 (2002). doi:10.1086/340437
P. Ghavamian, C.E. Rakowski, J.P. Hughes, T.B. Williams, The physics of supernova blast waves. I. kinematics of DEM L71 in the large Magellanic cloud. Astrophys. J. **590**, 833–845 (2003). doi:10.1086/375161
P. Ghavamian, J.M. Laming, C.E. Rakowski, A physical relationship between electron-proton temperature equilibration and Mach number in fast collisionless shocks. Astrophys. J. **654**, L69–L72 (2007). doi:10.1086/510740
C.C. Goodrich, J.D. Scudder, The adiabatic energy change of plasma electrons and the frame dependence of the cross-shock potential at collisionless magnetosonic shock waves. J. Geophys. Res. **89**, 6654–6662 (1984). doi:10.1029/JA089iA08p06654

E. Gosset, M. De Becker, Y. Nazé, S. Carpano, G. Rauw, I.I. Antokhin, J.-M. Vreuz, A.M.T. Pollock, XMM-Newton observation of the enigmatic object WR 46. Astron. Astrophys. **527**, A66 (2011). doi:10.1051/0004-6361/200912510

E.W. Greenstadt, M.M. Mellott, Plasma wave evidence for reflected ions in front of subcritical shocks—ISEE 1 and 2 observations. J. Geophys. Res. **92**, 4730–4734 (1987). doi:10.1029/JA092iA05p04730

E.A. Helder, J. Vink, C.G.H. Bassa, A. Bamba, J.A.M. Bleeker, S. Funk, P. Ghavamian, K.J. van der Heyden, F. Verbunt, R. Yamazaki, Measuring the cosmic-ray acceleration efficiency of a supernova remnant. Science **325**, 719 (2009). doi:10.1126/science.1173383

E.A. Helder, D. Kosenko, J. Vink, Cosmic ray acceleration efficiency versus temperature equilibration: the case of SNR 0509-67.5. Astrophys. J. Lett. **719**, L140 (2010). doi:10.1088/2041-8205/719/2/L140

E.A. Helder, J. Vink, C.G. Bassa, Temperature equilibration behind the shock front: an optical and X-ray study of RCW 86. Astrophys. J. **737**, 85 (2011). doi:10.1088/0004-637X/737/2/85

K. Heng, R. McCray, Balmer dominated shocks revisited. Astrophys. J. **654**, 923–937 (2007). doi:10.1086/509601

K. Heng, M. van Adellsberg, R. McCray, J.C. Raymond, The transition zone in Balmer dominated shocks. Astrophys. J. **668**, 275–284 (2007). doi:10.1086/521298

J.J. Hester, J.C. Raymond, W.P. Blair, The Balmer-dominated northeast limb of the Cygnus loop supernova remnant. Astrophys. J. **420**, 721–745 (1994). doi:10.1086/173598

H. Itoh, Two-fluid blast-wave model for supernova remnants. Publ. Astron. Soc. Jpn. **30**, 489–498 (1978)

F.C. Jones, D.C. Ellison, The plasma physics of shock acceleration. Space Sci. Rev. **58**, 259–346 (1991). doi:10.1007/BF01206003

C.F.F. Karney, Stochastic ion heating by a lower hybrid wave. Phys. Fluids **21**, 1584–1599 (1978). doi:10.1063/1.862406

S. Katsuda, R. Petre, J.P. Hughes, U. Hwang, H. Yamagauchi, A. Hayato, K. Mori, H. Tsunemi, X-ray measured dynamics of Tycho's supernova remnant. Astrophys. J. **709**, 1387–1395 (2010b). doi:10.1088/0004-637X/709/2/1387

S. Katsuda, R. Petre, K. Mori, S.P. Reynolds, K.S. Long, P.F. Winkler, H. Tsunemi, Steady X-ray synchrotron emission in the northeastern limb of SN 1006. Astrophys. J. **723**, 383–392 (2010a). doi:10.1088/0004-637X/723/1/383

S. Katsuda, K.S. Long, R. Petre, S.P. Reynolds, B.J. Williams, P.F. Winkler, X-ray proper motions and shock speeds along the northwest rim of SN 1006 (2012). arXiv:1211.6443

C.F. Kennel, J.P. Edmiston, T. Hada, A quarter century of collisionless shock research. Am. Geophys. Union Monograph Ser. **34**, 1–36 (1985)

R. Kirshner, P.F. Winkler, R.A. Chevalier, High-velocity emission in young supernova remnants: SN 1006 and SN 1572. Astrophys. J. **315**, L135–L139 (1987). doi:10.1086/184875

K.E. Korreck, J.C. Raymond, T.H. Zurbuchen, P. Ghavamian, Far ultraviolet spectroscopic explorer observation of the nonradiative collisionless shock in the remnant of SN 1006. Astrophys. J. **615**, 280–285 (2004). doi:10.1086/424481

K. Koyama, R. Petre, E.V. Gotthelf, U. Hwang, M. Matsuura, M. Ozaki, S.S. Holt, Evidence for shock acceleration of high-energy electrons in the supernova remnant SN1006. Nature **378**, 255–258 (1995). doi:10.1038/378255a0

K. Koyama, K. Kinugasa, K. Matsuzaki, M. Nishiuchi, M. Sugizaki, K. Torii, S. Yamauchi, B. Aschenbach, Discovery of non-thermal X-rays from the northwest shell of the new SNR RX J1713.7-3946: the second SN 1006? Publ. Astron. Soc. Jpn. **49**, L7–L11 (1997)

V.V. Krasnoselskikh, B. Lembège, P. Savoini, V.V. Lobzin, Nonstationarity of strong collisionless quasiperpendicular shocks: theory and full particle numerical simulations. Phys. Plasmas **9**, 1192–1209 (2002). doi:10.1063/1.1457465

R.M. Kulsrud, C.J. Cesarsky, The effectiveness of instabilities for the confinement of high energy cosmic rays in the galactic disk. Astrophys. J. Lett. **8**, 189 (1971)

J.M. Laming, Electron heating at SNR collisionless shocks. Astrophys. J. Suppl. Ser. **127**, 409–413 (2000). doi:10.1086/313325

J.M. Laming, Accelerated electrons in Cassiopeia A: an explanation for the hard X-ray tail. Astrophys. J. **546**, 1149–1158 (2001). doi:10.1086/318317

J.M. Laming, J.C. Raymond, B.M. McLaughlin, W.P. Blair, Electron-ion equilibration in nonradiative shocks associated with SN 1006. Astrophys. J. **472**, 267–274 (1996). doi:10.1086/178061

J.J. Lee, B.-C. Koo, J.C. Raymond, P. Ghavamian, T.-S. Pyo, A. Tajitsu, M. Hayashi, Subaru HDS observations of a Balmer-dominated shock in Tycho's supernova remnant. Astrophys. J. Lett. **659**, L133–L136 (2007). doi:10.1086/517520

J.J. Lee, J.C. Raymond, S. Park, W.P. Blair, P. Ghavamian, P.F. Winkler, K. Korreck, Resolved shock structure of the Balmer-dominated filaments in Tycho's supernova remnant: cosmic-ray precursor? Astrophys. J. Lett. **715**, L146–L149 (2010). doi:10.1088/2041-8205/715/2/L146

B. Lefebvre, S.J. Schwartz, A.F. Fazakerley, P. Décréau, Electron dynamics and cross-shock potential at the quasi-perpendicular Earth's bow shock. J. Geophys. Res. **112**, A09212 (2007). doi:10.1029/2007JA012277

Q. Luo, D. Melrose, Saturated magnetic field amplification at supernova shocks. Mon. Not. R. Astron. Soc. **397**, 1402–1409 (2009). doi:10.1111/j.1365-2966.2009.14872.x

M.A. Malkov, L.O.C. Drury, Nonlinear theory of diffusive acceleration of particles by shock waves. Rep. Prog. Phys. **64**, 429–481 (2001)

M. Markevitch, A. Vikhlinin, Shocks and cold fronts in galaxy clusters. Phys. Rep. **443**, 1–53 (2007). doi:10.1016/j.physrep.2007.01.001

M. Markevitch, F. Govoni, G. Brunetti, D. Jerius, Bow shock and radio halo in the merging cluster A520. Astrophys. J. **627**, 733–738 (2005). doi:10.1086/430695

A. Masters, S.J. Schwartz, E.M. Henley, M.F. Thomsen, B. Zieger, A.J. Coates, N. Achilleos, J. Mitchell, K.C. Hansen, M.K. Dougherty, Electron heating at Saturn's bow shock. J. Geophys. Res. **116**, A10107 (2011). doi:10.1029/2011JA016941

S. Matsukiyo, Mach number dependence of electron heating in high Mach number quasiperpendicular shocks. Phys. Plasmas **17**, 042901 (2010). doi:10.1063/1.3372137

K.G. McClements, R.O. Dendy, R. Bingham, J.G. Kirk, L.O.C. Drury, Acceleration of cosmic ray electrons by ion-excited waves at quasi-perpendicular shocks. Mon. Not. R. Astron. Soc. **291**, 241–249 (1997)

G. Morlino, R. Bandiera, P. Blasi, E. Amato, Collisionless shocks in a partially ionized medium. II. Balmer emission. Astrophys. J. **760**, 137 (2012a). doi:10.1088/0004-637X/760/2/137

G. Morlino, P. Blasi, R. Bandiera, E. Amato, D. Caprioli, Collisionless shocks in a partially ionized medium: III. Efficient cosmic ray acceleration (2012b). arXiv:1211.6148

S. Orlando, F. Bocchino, F. Reale, F. Peres, O. Petruk, On the origin of asymmetries in bilateral supernova remnants. Astron. Astrophys. **470**, 927–939 (2007). doi:10.1051/0004-6361:20066045

O. Petruk, F. Bocchino, G. Castelletti, G. Dubner, D. Lakubovskyi, M. Kirsch, M. Miceli, I. Telezhinsky, X-ray emission of the shock of SN1006. Constraints on electron kinetics, in *Proc. "The X-ray Universe 2008"*, vol. 109, Granada, Spain (2008)

C.E. Rakowski, P. Ghavamian, J.P. Hughes, The physics of supernova remnant blast waves. II. Electron-ion equilibration in DEM L71 in the large Magellanic cloud. Astrophys. J. **590**, 846–857 (2003). doi:10.1086/375162

C.E. Rakowski, J.M. Laming, P. Ghavamian, The heating of thermal electron in fast collisionless shocks: the integral role of cosmic rays. Astrophys. J. **684**, 348–357 (2008). doi:10.1086/590245

C.E. Rakowski, P. Ghavamian, J.M. Laming, The Hα diagnostic of electron heating: the case of DEM L71. Astrophys. J. **696**, 2195–2205 (2009). doi:10.1088/0004-637X/696/2/2195

J.C. Raymond, W.P. Blair, K.S. Long, Detection of ultraviolet emission lines in SN 1006 with the Hopkins ultraviolet telescope. Astrophys. J. **454**, L31–L34 (1995). doi:10.1086/309772

J.C. Raymond, J. Vink, E.A. Helder, A. de, Laat Effects of neutral hydrogen on cosmic-ray precursors in supernova remnant shock waves. Astrophys. J. **731**, L14 (2011). doi:10.1088/2041-8205/731/1/L14

B. Reville, J.G. Kirk, P. Duffy, S. O'Sullivan, A cosmic ray current-driven instability in partially ionised media. Astron. Astrophys. **475**, 435–439 (2007). doi:10.1051/0004-6361:20078336

M.A. Riquelme, A. Spitkovsky, Electron injection by Whistler waves in non-relativistic shocks. Astrophys. J. **733**, 63 (2011). doi:10.1088/0004-637X/733/1/63

H.R. Russell, B.R. McNamara, J.S. Sanders, A.C. Fabian, P.E.J. Nulsen, R.E.A. Canning, S.A. Baum, M. Donahue, A. Edge, L.J. King, C.P. O'Dea, Shock fronts, electron-ion equilibration and ICM transport processes in the merging cluster Abell 2146. Mon. Not. R. Astron. Soc. **423**, 236–255 (2012). doi:10.1111/j.1365-2966.2012.20808.x

K.M. Schure, A.R. Bell, L.O.C. Drury, A.M. Bykov, Diffusive shock acceleration and magnetic field amplification. Space Sci. Rev. **173**, 491–519 (2012). doi:10.1007/s11214-012-9871-7

S.J. Schwartz, M.F. Thomsen, S.J. Bame, J. Stansberry, Electron heating and the potential jump across fast mode shocks. J. Geophys. Res. **93**, 12923–12931 (1988). doi:10.1029/JA093iA11p12923

S.J. Schwartz, E.G. Zweibel, M. Goldman, Microphysics in astrophysical plasmas. Space Sci. Rev. (2013 in press). doi:10.1007/s11214-013-9975-8 (this issue)

J.D. Scudder, T.L. Aggson, A. Mangeney, C. Lacombe, C.C. Harvey, The resolved layer of a collisionless, high beta, supercritical, quasi-perpendicular shock wave. I—Rankine-Hugoniot geometry, currents, and stationarity. J. Geophys. Res. **91**, 11019–11052 (1986a). doi:10.1029/JA091iA10p11019

J.D. Scudder, A. Mangeney, C. Lacombe, C.C. Harvey, C.S. Wu, The resolved layer of a collisionless, high beta, supercritical, quasi-perpendicular shock wave. III—Vlasov electrodynamics. J. Geophys. Res. **91**, 11075–11097 (1986b). doi:10.1029/JA091iA10p11075

M.J. Seaton, Excitation of coronal lines by proton impact. Mon. Not. R. Astron. Soc. **127**, 191–194 (1964)

J.M. Shull, C.F. McKee, Theoretical models of interstellar shocks. I—Radiative transfer and UV precursors. Astrophys. J. **327**, 149–191 (1979). doi:10.1086/156712

L. Sironi, A. Spitkovsky, Particle acceleration in relativistic magnetized collisionless electron-ion shocks. Astrophys. J. **726**, 75 (2011). doi:10.1088/0004-637X/726/2/75

J. Skilling, Cosmic ray streaming. Nature **258**, 687–688 (1975). doi:10.1038/258687a0

P. Slane, B.M. Gaensler, T.M. Dame, J.P. Hughes, P. Plucinsky, A. Green, Nonthermal X-ray emission from the shell-type supernova remnant G347.3-0.5. Astrophys. J. **525**, 357–367 (1999). doi:10.1086/307893

R.C. Smith, R.P. Kirshner, W.P. Blair, P.F. Winkler, Six Balmer dominated supernova remnants. Astrophys. J. **375**, 652–662 (1991). doi:10.1086/170228

R.C. Smith, J.C. Raymond, J.M. Laming, High-resolution spectroscopy of Balmer-dominated shocks in the large Magellanic cloud. Astrophys. J. **4220**, 286–293 (1994). doi:10.1086/1735581

J. Sollerman, P. Ghavamian, P. Lundqvist, R.C. Smith, High resolution spectroscopy of Balmer-dominated shocks in the RCW 86, Kepler and SN 1006 supernova remnants. Astron. Astrophys. **407**, 249–257 (2003). doi:10.1051/0004-6361:20030839

L. Spitzer, *Physics of Fully Ionized Gases* (Interscience, New York, 1964)

T. Tanaka, Y. Uchiyama, F.A. Aharonian, T. Takahashi, A. Bamba, J.S. Hiraka, J. Kataoka, T. Kishishita, M. Kokubun, K. Mori, K. Nakazawa, R. Petre, H. Tajima, S. Watanabe, Study of nonthermal emission from SNR RX J1713.7-3946 with Suzaku. Astrophys. J. **685**, 988–1004 (2008). doi:10.1086/591020

R.A. Treumann, Fundamentals of collisionless shocks for astrophysical application. I. Non-relativistic shocks. Astron. Astrophys. Rev. **17**, 409–535 (2009). doi:10.1007/s00159-009-0024-2

R. A. Treumann, Jaroschek, Fundamentals of non-relativistic shock physics: I. The shock problem (2008). arXiv:0805.2132

D. Tseliakhovich, C.M. Hirata, K. Heng, Excitation and charge transfer in H–H^+ collisions at 5–80 keV and application to astrophysical shocks. Mon. Not. R. Astron. Soc. **422**, 2357–2371 (2012). doi:10.1111/j.1365-2966.2012.20787.x

T. Umeda, Y. Kidani, S. Matsukiyo, R. Yamazaki, Modified two-stream instability at perpendicular collisionless shocks: full particle simulations. J. Geophys. Res. **117**, A03206 (2012a). doi:10.1029/2011JA017182

T. Umeda, Y. Kidani, S. Matsukiyo, R. Yamazaki, Microinstabilities at perpendicular collisionless shocks: a comparison of full particle simulations with different ion to electron mass ratio. Phys. Plasmas **19**, 042109 (2012b). doi:10.1063/1.3703319

M. van Adelsberg, K. Heng, R. McCray, J.C. Raymond, Spatial structure and collisionless electron heating in Balmer-dominated shocks. Astrophys. J. **689**, 1089–1104 (2008). doi:10.1086/592680

J. Vink, J.M. Laming, On the magnetic fields and particle acceleration in Cassiopeia A. Astrophys. J. **584**, 758–769 (2003). doi:10.1086/345832

J. Vink, R. Yamazaki, E.A. Helder, K.M. Schure, The relation between post-shock temperature, cosmic-ray pressure, and cosmic-ray escape for non-relativistic shocks. Astrophys. J. **722**, 1727–1734 (2010). doi:10.1088/0004-637X/722/2/1727

A.Y. Wagner, J.-J. Lee, J.C. Raymond, T.W. Hartquist, S.A.E.G. Falle, A cosmic-ray precursor model for a Balmer-dominated shock in Tycho's supernova remnant. Astrophys. J. **690**, 1412–1423 (2009). doi:10.1088/0004-637X/690/2/1412

J. Warren, J.P. Hughes, Raising the dead: clues to type Ia supernova physics from the remnant 0509–67.5. Astrophys. J. **608**, 261–273 (2004). doi:10.1086/392528

J.S. Warren, J.P. Hughes, C. Badenes, P. Ghavamian, C.F. McKee, D. Moffett, P. Plucinsky, C.E. Rakowski, E. Reynoso, P. Slane, Cosmic-ray acceleration at the forward shock in Tycho's supernova remnant: evidence from Chandra X-ray observations. Astrophys. J. **634**, 376–389 (2005). doi:10.1086/496941

B.J. Williams, W.P. Blair, J.M. Blondin, K.J. Borkowski, P. Ghavamian, K.S. Long, J.C. Raymond, S.P. Reynolds, J. Rho, P.F. Winkler, RCW 86: a type Ia supernova in a wind-blown bubble. Astrophys. J. **741**, 96 (2011). doi:10.1088/0004-637X/741/2/96

C.S. Wu, D. Winske, M. Tanaka, K. Papadopoulos, K. Akimoto, C.C. Goodrich, Y.M. Zhou, S.T. Tsai, P. Rodriguez, C.S. Lin, Microinstabilities associated with a high Mach number, perpendicular bow shock. Space Sci. Rev. **37**, 63–109 (1984). doi:10.1007/BF00213958

S.A. Zhekov, R. McCray, D. Dewey, C.R. Canizares, K.J. Borkowski, D.N. Burrows, S. Park, High-resolution X-ray spectroscopy of SNR 1987A: Chandra Letg and HETG observations in 2007. Astrophys. J. **692**, 1190–1204 (2009). doi:10.1088/0004-637X/692/2/1190

DOI 10.1007/978-1-4899-7413-6_21
Reprinted from *Space Science Reviews* Journal, DOI 10.1007/s11214-013-9974-9

Methods for Characterising Microphysical Processes in Plasmas

T. Dudok de Wit · O. Alexandrova · I. Furno · L. Sorriso-Valvo · G. Zimbardo

Received: 18 November 2012 / Accepted: 7 March 2013 / Published online: 9 May 2013

Abstract Advanced spectral and statistical data analysis techniques have greatly contributed to shaping our understanding of microphysical processes in plasmas. We review some of the main techniques that allow for characterising fluctuation phenomena in geospace and in laboratory plasma observations. Special emphasis is given to the commonalities between different disciplines, which have witnessed the development of similar tools, often with differing terminologies. The review is phrased in terms of few important concepts: self-similarity, deviation from self-similarity (*i.e.* intermittency and coherent structures), wave-turbulence, and anomalous transport.

T. Dudok de Wit (✉)
LPC2E, CNRS and University of Orléans, 3A avenue de la Recherche Scientifique, 45071 Orléans cedex 2, France
e-mail: ddwit@cnrs-orleans.fr

O. Alexandrova
LESIA, Observatoire de Paris, 5, place Jules Janssen, 92190 Meudon, France
e-mail: olga.alexandrova@obspm.fr

I. Furno
EPFL SB CRPP, Station 13, 1015 Lausanne, Switzerland
e-mail: ivo.furno@epfl.ch

L. Sorriso-Valvo
Dipartimento di Fisica, IPCF-CNR, UOS di Cosenza, Ponte Pietro Bucci Cubo 31C, 87036 Arcavacata di Rende (CS), Italy
e-mail: sorriso@fis.unical.it

Present address:
L. Sorriso-Valvo
Space Sciences Laboratory, University of California, Berkeley, 7 Gauss Way, Berkeley, CA 94720-7450, USA

G. Zimbardo
Dipartimento di Fisica, Università della Calabria, Ponte Pietro Bucci Cubo 31C, 87036 Arcavacata di Rende (CS), Italy
e-mail: zimbardo@fis.unical.it

Keywords Plasma turbulence · Plasma fluctuations · Data analysis methods

1 Introduction

Space and laboratory plasmas are fundamentally governed by both couplings across different scales and by nonlinear processes. This applies to a wide range of scales: from the fluid regime, where the magnetohydrodynamic (MHD) approximation can be used (Biskamp 2003), down to small scales, where kinetic effects should be taken into account (Akhiezer et al. 1975a, 1975b). Progress toward their physical understanding has been closely associated with our ability to infer pertinent information by means of appropriate data analysis techniques. Usually, physical space is very sparsely sampled in the sense that one has access only to few observables (magnetic field, electron density, plasma velocity and temperature, ...), often with poor temporal and/or spatial coverage. The challenge then consists in recovering the information of interest from highly scattered observations.

For many decades, the analysis of plasma data has been dominated by techniques that implicitly assume linearity or at best rely on second order moments only (*e.g.* correlation functions and power spectra). Our prime objective here is to show how more advanced techniques can often provide deeper physical insight by extracting quantitative information that might otherwise have gone unnoticed. For example, two similarly-looking turbulent wavefields with identical power spectral densities may actually exhibit quite different properties in their higher order spectra, which in turn has direct implications on the existence of nonlinear wave couplings (or structures) that transfer energy between scales.

Many new techniques have been developed in the two last decades, thanks to numerous advances in neighbouring fields such as dynamical systems (Bohr et al. 2005; Kantz and Schreiber 2000) or complex systems (Badii and Politi 1999). Interestingly, most of these techniques are remarkably universal, and are used almost equivalently in laboratory and in astrophysical plasmas, and also in neutral fluids. Here, we discuss some of these concepts in the context of turbulence and large-scale structures in plasmas. Special emphasis will be given to commonalities between different disciplines, which are often concealed by differing nomenclatures. The focus will be on time series from geospace and from laboratory plasmas. Regrettably, the literature is almost devoid of reviews that cover more than one discipline only. For laboratory plasmas, some relevant references are the works by Ritz et al. (1988), Škorić and Rajković (2008), Tynan et al. (2009), Fujisawa (2010). In geospace plasmas, besides a few general reviews (Wernik 1996; Dudok de Wit 2003; Bruno and Carbone 2005; Vassiliadis 2006; Zimbardo et al. 2010) a large effort has been directed towards multipoint data analysis (Paschmann and Daly 2000, 2008).

This overview, rather than being exhaustive, will be phrased in terms of three frequently encountered concepts, namely self-similarity, intermittency and coherent structures (Sect. 2), measurements of wave-turbulence in space and time (Sect. 3) and anomalous transport (Sect. 4). Unfortunately we had to leave out several relevant and timely topics such as techniques for dynamical systems, or for image processing.

2 Self-Similarity in Turbulent Plasmas

Turbulence is ubiquitous in geospace and in laboratory plasmas, and occurs whenever energy injection occurs in the system and the dissipation mechanisms are weak (*i.e.* efficient at much smaller scales than the scale of injection). Turbulent states are characterised by wavefield fluctuations covering all scales from the energy injection scale L down to the scale ℓ_d at which energy dissipates.

In neutral fluids, these fluctuations exhibit a power law spectrum $E(k) \sim k^{-5/3}$ in Fourier space (k being a wavenumber) between $k_L = 1/L$ and $k_d = 1/\ell_d$, called the *inertial range.* Observation of this law for any turbulent flow indicates its universality and supports the idea of a self-similar behaviour of the turbulent velocity fluctuations, whose amplitude at scale l varies as $\delta u_l \sim \delta u_L l^\alpha$. This is to be expected if the physical processes are not scale-dependent, namely if the dynamic equations (*e.g.* Navier-Stokes for neutral fluids or MHD for magnetised plasmas) are invariant under scale transformation. This idea was used by Kolmogorov in 1941 to explain the ubiquitous $E(k) \sim k^{-5/3}$ spectrum (Kolmogorov 1941).

Some time later, it was pointed out that the idea of self-similarity would imply spatial homogeneity of the process (Kolmogorov 1962). However, studies of higher order statistics show that this property is not verified for all turbulent flows. Departure from self-similarity is called intermittency in fully developed turbulence, and is another universal property of most turbulent systems. Intermittency is attributed to the inhomogeneous energy flux through the scales, and results in the presence of small-scale coherent structures in the turbulent flow, in which most of the energy dissipation is concentrated.

In this section, we shall discuss first how to determine energy spectra and related spectral indices from fluctuating wavefields in geospace and in fusion plasmas. Next we shall introduce techniques for analysing intermittency in time series.

2.1 The Solar Wind as Turbulence Laboratory: Power Law Determinations

With its wide range of scales and parameters conditions, the solar wind is the best laboratory for studying plasma turbulence, thanks also to the availability of *in situ* measurements that are provided by dedicated space missions (*e.g.*, Bruno and Carbone 2005). Therefore, most of the following will refer to solar wind turbulence, unless stated otherwise.

Because of the large number of degrees of freedom in the dynamics of a turbulent flow, most of the results about turbulence refer to statistical quantities. As for any statistical study, in plasma turbulence it is important to check that ergodicity and stationarity apply. Ergodicity, *i.e.* the equivalence between time and space averaging, is normally ensured if the sample size of the time interval T is much larger than a typical correlation time, $T \gg t_c$. The correlation time t_c can be estimated from the data, for example as the time lag at which the autocorrelation function of the field vanishes to noise level, or reaches the value $1/e$. In turbulent flows, this should correspond to the integral scale, at which energy is injected in the system (Frisch 1995). Ergodicity cannot be formally proven on the basis of observations, and yet, various studies suggest that it is a reasonable working hypothesis. For example, ensemble averages of the solar wind magnetic fields follow the asymptotic behaviour predicted by the ergodic theorem (Matthaeus and Goldstein 1982).

Stationarity is usually supported by observations from laboratory plasmas when the discharge can be controlled, while for geospace plasmas it is often more difficult to extract stable and homogeneous records. Large scale, externally driven structures can indeed affect the statistical properties of the turbulent fields. For example, solar wind plasma properties change for fast and slow wind populations, each originating from specific regions of the solar atmosphere (Tu and Marsch 1995; Bruno and Carbone 2005). Large-scale boundaries between these different regions may mix the phases and destroy the correlations that are present in nonlinear energy cascades. This is why the first important step towards the analysis of a turbulent signal is the choice of an appropriate time interval that excludes large scale boundaries, magnetic connection to planetary shocks, *etc.* Meanwhile, many authors have studied long periods of solar wind data that included a mix of different regimes. Although this kind of analysis can give insight into the statistics of the field fluctuations, its interpretation in terms of phenomenological or theoretical modelling can be misleading, and should be carefully discussed.

Figure 1(a) shows the temporal evolution of the $B_x(t)$ component of the magnetic field along the solar wind flow direction, as measured by the CLUSTER-4 spacecraft during the time interval studied by Bale et al. (2005). Here, the velocity of the slow solar wind is $V_{sw} \simeq 350$ km/s, and $\beta_p \simeq 5$. Figure 1(b) gives the power spectral density $PSD(f)$ of $B_x(t)$, using (i) the Fourier transform, and (ii) the Morlet wavelet transform (Farge 1992; Torrence and Compo 1998; Eriksson 1998). The latter spectrum is defined as

$$PSD_x\left[\mathrm{nT}^2/\mathrm{Hz}\right] = \frac{2\delta t}{N} \sum_{j=0}^{N-1} \left|\mathcal{W}_x(\tau, t_j)\right|^2, \tag{1}$$

where

$$\mathcal{W}_x(\tau, t) = \sum_{j=0}^{N-1} B_x(t_j)\psi\left(\frac{t_j - t}{\tau}\right) \tag{2}$$

is the wavelet transform of B_x. We assume that the latter is regularly sampled at $t_j = t_0 + j\delta t$, with $j = 0, \ldots, N-1$. Here, the dilation parameter τ sets the time scale, while

$$\psi(u) = \pi^{-\frac{1}{4}} e^{-j\omega_0 u} e^{-\frac{1}{2}u^2} \tag{3}$$

is the unnormalised Morlet wavelet. Setting $\omega_0 = 2\pi$ allows the characteristic scale to be inferred from the angular frequency directly with $\tau = 2\pi/\omega$. Wavelet transforms allow us to resolve a non-stationary signal in time and scale (or frequency). Moreover, the analysing time window is scale-dependent and therefore better tailored to each scale. This is in contrast to the windowed Fourier transform, which keeps the same analysing window at all scales. Below (in Sect. 2.3), we shall exploit the time-scale resolution of wavelets; here we just use them for power spectral estimation purposes.

In Fig. 1(b) the wavelet spectrum is multiplied by a factor 10 for ease of visualisation. Strong fluctuations in the Fourier spectrum hinder the visualisation of linear slopes. The wavelet spectrum, in contrast, is smooth and clearly exhibits a power law with a $-5/3$ slope, within the $\Delta f = [10^{-3}, 0.2]$ Hz frequency range.

To verify this spectral law and help better determine the frequency range, it is useful to compensate the observed spectrum with an inverse of the observed law: if the spectrum follows indeed a power law $\sim f^{-5/3}$, the compensation by $f^{5/3}$–function should give a constant. Figure 1(c) gives the compensated spectrum with $f^{5/3}$ function (solid line). The resulting function is flat over the $\Delta f = [10^{-3}, 0.2]$ Hz frequency range (bounded by the vertical dotted lines in panels (b)–(d)). This compensated spectrum shows the ion break point at 0.3 Hz, which was not so immediate from the previous spectrum.

To check the comparison with a Kolmogorov spectrum, we compensate the observed wavelet spectrum with the inverse of the Iroshnikov-Kraichnan (IK) law $f^{3/2}$ (dashed-dotted line). The result is not constant, indicating that the IK–model is not correct. Note that $5/3 - 0.1(5/3) = 3/2$ and so, if the precision on the spectral index is close or superior to 10 %, one would not be able to distinguish between these two scalings. Figure 1(d) shows the Fourier spectrum compensated by the inverse of the Kolmogorov law. Strong fluctuations, especially for $f > 10^{-2}$ Hz, preclude the robust estimation of the spectral index. We conclude that the

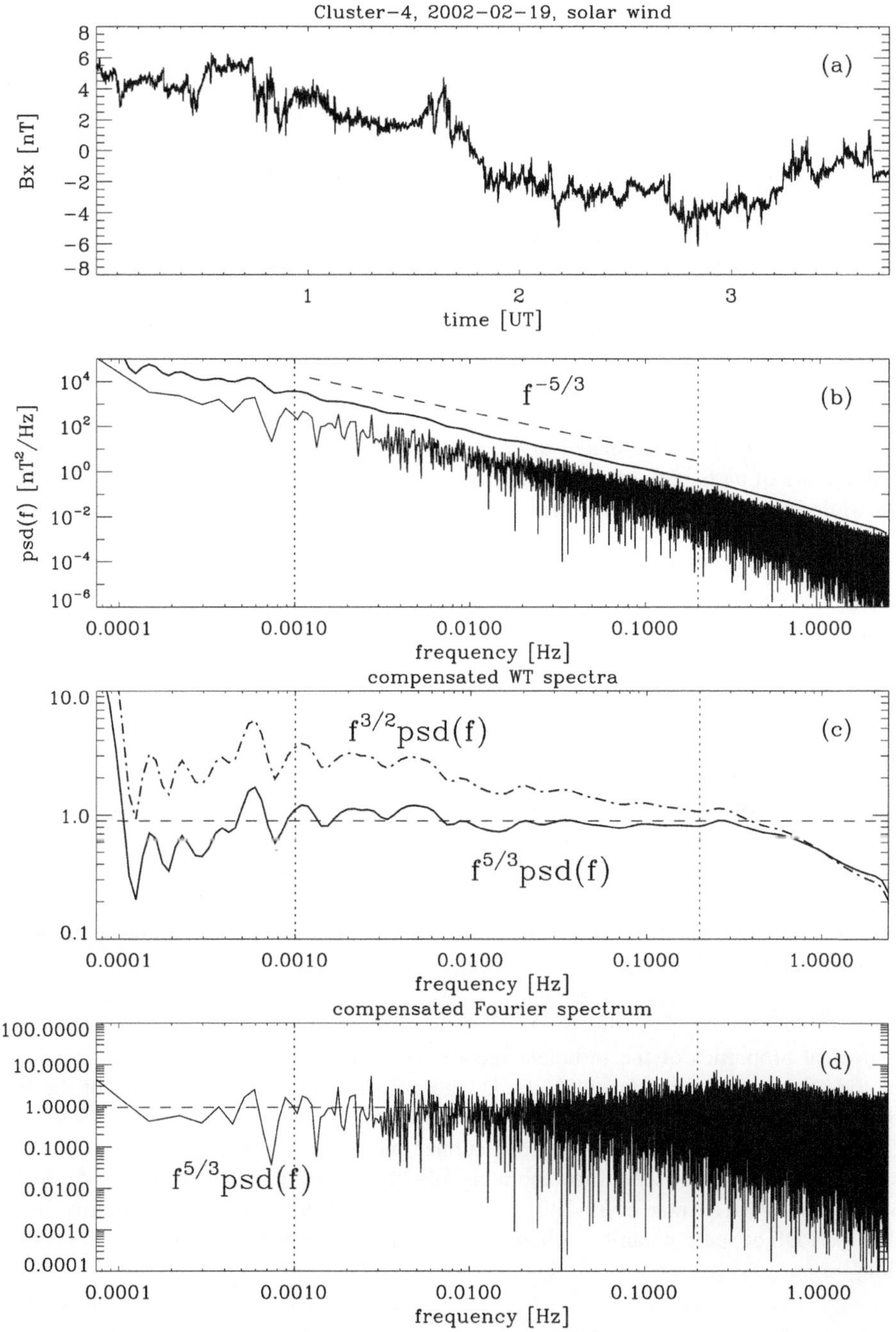

Fig. 1 Example of solar wind observations by CLUSTER. This time interval was used to study turbulence by Bale et al. (2005). The plots show: (**a**) $B_x(t)$; (**b**) Fourier and Morlet wavelet spectra of $B_x(t)$, $f^{-5/3}$ is indicated by a *dashed line*, frequency range Δf where the fitting was done is delimitated by two vertical *dotted lines*; (**c**) Compensated wavelet spectrum by $f^{5/3}$–function (*solid line*) and by $f^{3/2}$–function (*dashed-dotted line*); (**d**) Compensated Fourier spectrum by $f^{5/3}$–function

wavelet spectra help us in determining the spectral index and the frequency range of interest, where the power law is valid.

Numerous examples of the use of the wavelet spectra can be found in geospace plasmas (Eriksson 1998; Bale et al. 2005; Alexandrova et al. 2008, 2009; Kiyani et al. 2013) and in laboratory plasmas (Stroth et al. 2004; van den Berg 2004; Zweben et al. 2007).

The practical problem of power law identification and spectral index estimation is of considerable importance and has also led to a large number of erroneous interpretations. Quite often, spectral indices are derived from ranges of frequencies (or wavenumbers) that span less than a decade, which is extremely risky. Another frequent mistake is the fitting of power laws to spectra that show no compelling evidence for a linear slope in logarithmic coordinates. Rigorous frameworks such as the maximal likelihood approach, as discussed by Clauset et al. (2009), ought to be a standard for any power law study. Another issue is the unbiased estimation of spectral indices, for which discrete wavelet transforms are preferable to continuous wavelet (*e.g.* Morlet) or Fourier transforms because they are unbiased (Abry et al. 1995). Undecimated discrete wavelet transforms (also known as the *à trous* method) are of particular interest because they provide an orthonormal scale-dependent basis while keeping the timing information that is important for visualisation (Kiyani et al. 2013).

Power spectra and their spectral indices have received much attention so far because they can be conveniently estimated from time series or computed directly by means of spectral analysers. The spectrum, however, only describes a second order moment of the turbulent wavefield and as such gives a very incomplete picture of the system. Higher order moments and their properties will be addressed in the next section.

2.2 How to Measure Departure from Self-Similarity: PDFs and Structure Functions

The most complete tool for describing the statistical properties of the field fluctuations at different scales, including departure from self-similarity, is the probability density function (PDF) $\mathcal{P}(y)$ of the random variable y, defined as $\mathcal{P}(y)\,du = P(u < y < u + du)$, where P stands for the probability.

For a given turbulent record, fluctuations at different time scales τ can be approximated by increments, defined as

$$\Delta y_\tau(t) = y(t+\tau) - y(t). \tag{4}$$

Statistical properties of the turbulent record at different scales τ can be investigated by plotting the probability density $\mathcal{P}(\Delta y_\tau)$. Figure 2 shows a typical behaviour of the PDFs of magnetic fluctuations in the solar wind inertial range (Sorriso-Valvo et al. 1999, 2001; Carbone et al. 2004; Hnat et al. 2003; Leubner and Vörös 2005; Kiyani et al. 2007). The same properties are observed in laboratory plasmas (Carbone et al. 2000, 2004; Sorriso-Valvo et al. 2001; Antar et al. 2001; Hnat et al. 2008). Statistical self-similarity implies that the PDF at scale τ can be collapsed onto a unique PDF $\mathcal{P}_0$ by following transformation

$$\mathcal{P}(\Delta y_\tau) = \tau^{-H}\mathcal{P}_0(\Delta y). \tag{5}$$

where H is the Hurst exponent. Equation (5) implies that the increments are self-affine, namely $\Delta y_{a\tau} = a^H \Delta y_\tau$, where a is a scaling parameter.

The departure of the PDFs from a Gaussian with decreasing time scale is usually a signature of intermittency, and indicates that turbulent fluctuations are not self-similar at different scales.

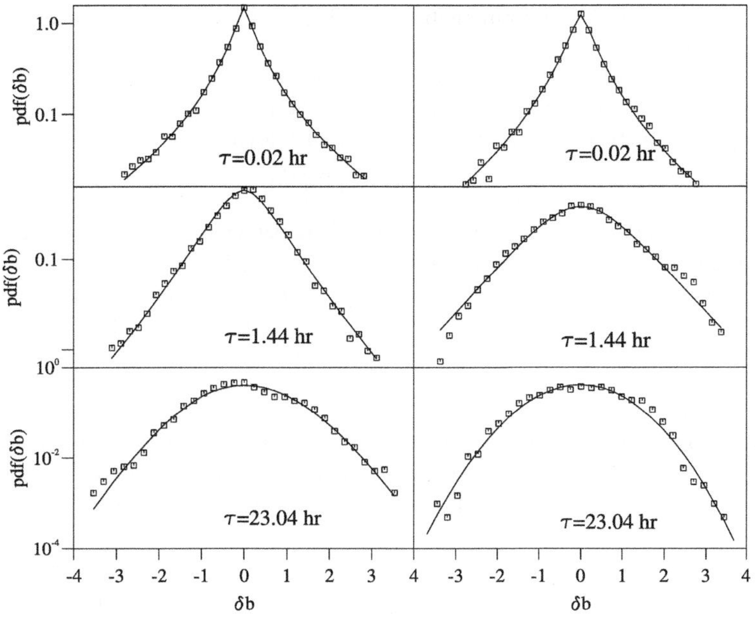

Fig. 2 Deviation of the PDFs from Gaussian statistics with scale: signature of intermittency in the inertial scale of the solar wind magnetic field (Sorriso-Valvo et al. 1999). *Left panels* for fast solar wind, *right panels* for slow solar wind

The tails of the PDF are of particular interest, because the distribution of rare events is indicative of the nature of underlying physical process. However, the practical assessment of such tails is a delicate task, and so moments of the PDF often receive more interest than the PDF itself. The moments of $\mathcal{P}(|\Delta y_\tau|)$ are called structure functions and can be estimated directly from the time series as

$$S_p(\tau) = \int_{-\infty}^{+\infty} \mathcal{P}\big(|\Delta y_\tau|\big)\big|\Delta y_\tau(t)\big|^p \, dt = \big\langle\big|\Delta y_\tau(t)\big|^p\big\rangle, \tag{6}$$

where $\langle\cdots\rangle$ denotes ensemble averaging. Equation (5) implies that the structure functions should scale with τ as

$$S_p(\tau) \propto \tau^{\zeta_p}. \tag{7}$$

For statistically self-similar processes, the scaling exponents ζ_p are a linear function of the order p; deviations from this linear behaviour can thus be used as a quantitative measure of departure from self-similarity. There is considerable experimental evidence that turbulent flows deviate from this behaviour (Frisch 1995).

Solar wind and laboratory data have been extensively studied by structure function analysis, showing the presence of intermittency (Carbone 1994; Tu and Marsch 1995; Carbone et al. 2000; Antar et al. 2001; Bruno and Carbone 2005; Matthaeus and Velli 2011). The evaluation of structure functions is straightforward, but there are pitfalls. The main danger is the increasing sensitivity of structure functions to rare and large events when the order p increases, until finite sample effects completely dominate. This often goes unnoticed as the structure function increases smoothly with order. As a rule of thumb, it is considered safe to compute structure functions up to order

$$p_{\max} = \log N - 1, \tag{8}$$

where N is the number of samples in the dataset. A more detailed check relies on the convergence of the integrand in Eq. (6) (Dudok de Wit 2004). Recursive convergence techniques have also been applied to solar wind data (Chapman and Nicol 2009).

Another common measure of intermittency is the normalised fourth order moment of PDF, also known as flatness (*e.g.*, Bruno and Carbone 2005)

$$F(\tau) = S_4(\tau)/S_2^2(\tau). \tag{9}$$

For Gaussian fluctuations $F(\tau) = 3$, so that a scale-dependent departure from this value can be used as an indication for intermittency.

Note that the increment defined in Eq. (4) is merely the convolution of the turbulent wavefield $y(t)$

$$\Delta y_\tau(t) = \int y(t') W_\tau(t - t')\, dt', \tag{10}$$

with the coarsest wavelet one could imagine, namely $W_\tau(t) = \delta(t + \tau) - \delta(t)$. The statistical properties of turbulent eddies can also be captured with increments that involve more elaborate wavelets (in terms of continuity, number of vanishing moments, *etc.*). This generalisation establishes an connection with a much broader framework wherein wavelet coefficients and their higher order moments are the prime quantities of interest (Muzy et al. 1993; Farge et al. 2004). Interestingly, this framework has intimate connections to the thermodynamic formalism.

Finally, in those cases where the estimation of the structure functions is handicapped by a short inertial range, the Extended Self-Similarity (ESS) technique (Benzi et al. 1993) can help extract relative scaling exponents (Carbone et al. 1996). Let us stress again, however, the importance of having long records for properly accessing high order moments, in particular with heavy-tailed processes (Kiyani et al. 2009).

2.3 Coherent Structure Identification

The origin of the intermittency in plasmas is still an open question and is usually ascribed to the presence of coherent structures whose typical lifetime exceeds that of incoherent fluctuations in the background. Typical examples are current sheets, shocks and vortices in space plasmas (Veltri and Mangeney 1999; Veltri 1999; Sorriso-Valvo et al. 1999; Mangeney et al. 2001; Alexandrova et al. 2006; Alexandrova 2008; Alexandrova and Saur 2008; Carbone et al. 2000; Greco et al. 2009a, 2012), and blobs, vortices, clumps, avaloids, and similar structures in laboratory plasmas (Zweben 1985; Huld et al. 1991; Antar 2003; Krasheninnikov et al. 2008; Fujisawa 2010).

A simple definition of a coherent structure is a structure whose phases are coupled for a finite range of scales, or at all measured scales. For example, coherent structures in fluid turbulence often are localized vortex filaments whose length is of the order of the energy injection scale L, and whose cross-section is of the order of the dissipation scale ℓ_d (Frisch 1995); in Fourier space, these structures occupy all scales from L to ℓ_d. Phase coherence is an essential ingredient, which can be tested, if not observed in Fourier space. One can indeed build a test statistic for the existence of coherent structures by randomising the phases of the Fourier transform, while keeping the amplitudes unchanged. This operation preserves the power spectral density and second order quantities such as the autocorrelation function, but destroys phase synchronisations. This surrogate and its statistics can be compared to the observed signal and its statistics. Surrogates are widely used in dynamical systems (Schreiber and Schmitz 2000) to test the null hypothesis against a linear, Gaussian, and stationary stochastic process; some applications to plasmas have been reported as well (Hada et al. 2003; Sahraoui 2008; Chian and Miranda 2009).

While Fourier transforms are routinely used for making surrogates and testing phase coherence, it often makes more sense to detect phase coherence locally in time. Multiresolution (*i.e.* wavelet) techniques again stand out as the most powerful tool for dealing with such problems. Here, the quantity of interest is the energy distribution of the time series in scale τ *and* in time t; notice that the latter had been ignored in the determination of frequency spectra in Sect. 2.1.

We already defined the wavelet transform in Eq. (2). Ideally, the mother wavelet function should be tuned to the kind of coherent structure one wishes to analyse. For example, Haar (or 0th order Daubechies) wavelets

$$\psi(t) = \begin{cases} 0, & \text{for } |t| \geq 1/2 \\ 1, & \text{for } 0 \leq t < 1/2 \\ -1, & \text{for } -1/2 < t < 0 \end{cases} \tag{11}$$

are appropriate for analysing sharp discontinuities (non-differentiable functions). First order Daubechies are better suited for handling smoother structures, whose first derivative exists but not the second one, *etc.* Haar wavelets have been used in various contexts to identify shocks and current sheets in the solar wind inertial range of scales (Veltri and Mangeney 1999), to identify events in fusion plasmas (Vega et al. 2008), and to de-noise plasma simulation data (Nguyen van yen et al. 2010).

Note that the Haar wavelet transform is equivalent to a set of increments at scales defined as powers of 2 of the smallest scale: $\tau_m = 2^m \tau_0$. Using these increments, a different method based on variance threshold, referred to as Partial Variance of Increments (PVI), was recently used by Greco et al. to identify magnetic discontinuities in numerical simulations and in solar wind time series (Greco et al. 2008, 2009a, 2009b, 2012).

Other wavelets have been used as well, such as Morlet wavelets for detecting magnetic Alfvén vortices in the Earth's and in Saturn's magnetosheath. Figure 3 illustrates how the structures were identified at Saturn during the Cassini orbit insertion in 2004 (Alexandrova and Saur 2008).

Let us stress that Haar wavelets, and low order wavelets in general, formally are relevant only for sharp discontinuities. These wavelets cannot properly characterise more continuous structures and fail to provide the spectral index of time series whose power law is steeper than a given threshold value (Farge et al. 2006). For that reason, higher order wavelets such as 4th or 6th order Daubechies should systematically be preferred when it comes to analysing coherent structures. Numerous other examples of coherent structure identification and extraction can be found in neutral fluid simulations (Farge 1992; Farge et al. 2001).

So far we only discussed methods that are applicable to single time series. In Sect. 3 we shall elaborate on spatio-temporal observations, which are more relevant for analysing coherent structures, but also more challenging for the observer.

3 Spatio-Temporal Observations of Turbulent Wavefields

In many observations, the temporal and spatial dimensions are intimately mixed because of plasma motion, and also because the probes often are not at rest in the plasma frame.

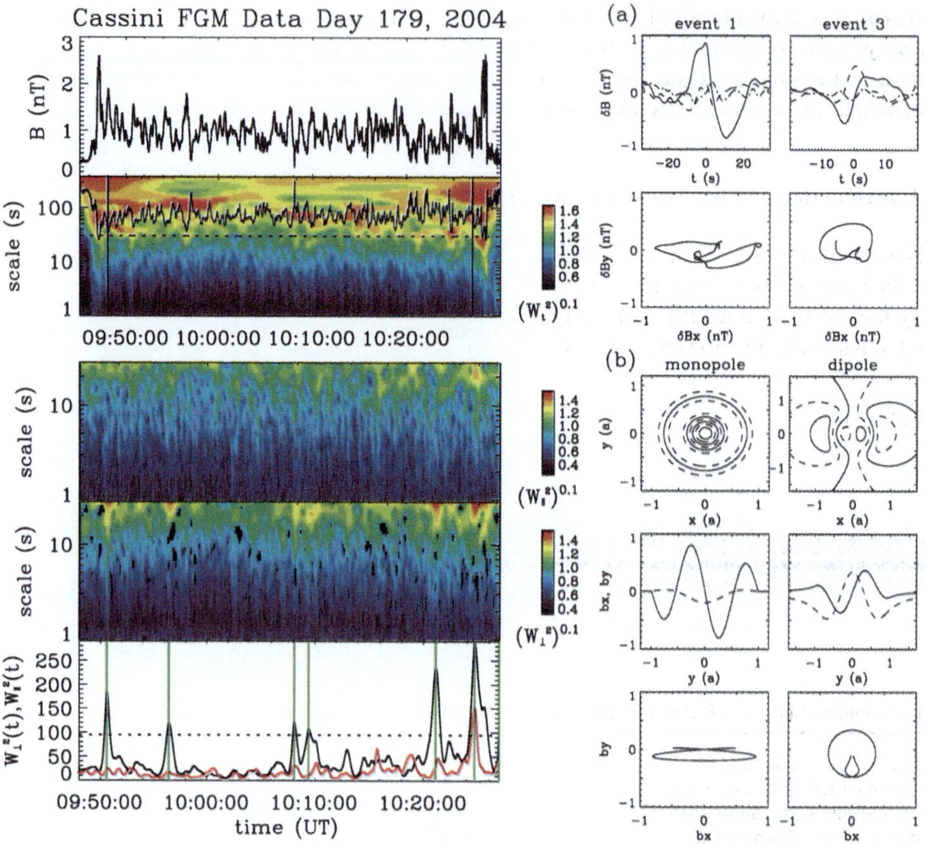

Fig. 3 *Left*, from the top to the bottom: Magnetic field intensity; Morlet wavelet scalogram, which shows how the total energy of magnetic fluctuations (*in colour*) is distributed in time-scale plane; *below* we separately plot the energy in compressible fluctuations $W_{\parallel}^2(t,\tau)$ and the distribution of the energy of Alfvénic fluctuations $W_{\perp}^2(t,\tau)$; the *bottom panel* gives the integration of $W_{\parallel}^2(t,\tau)$ (*red line*) and $W_{\perp}^2(t,\tau)$ (*black line*) over scales. The peaks of the energy of Alfvénic fluctuations (see *green vertical lines*) correspond to coherent structures, two of them are shown on the right: (**a**) 3 components of magnetic fluctuations in the local magnetic field frame (*top*), polarisation of the fluctuations in the plane perpendicular to the magnetic field (2nd line); (**b**) comparison with models of monopole and dipole Alfvén vortices. Here, a is the vortex radius. For more details, see (Alexandrova and Saur 2008)

As a consequence, the distinction between spatial structure and temporal dynamics is often elusive, except in the case where at least two simultaneous observations of the same variable can be made. The different methodologies that have been developed for that purpose depend very much on the number of simultaneous observations, which will be here our guiding thread. Here we address mostly wave-like structures but also structures of arbitrary shape.

3.1 Two-Point Measurements

Single-point measurements are formally inappropriate for disentangling space and time in plasmas. However, in the particular case where the plasma flow is steady, dispersionless, and fast with respect to its fluctuations, Taylor's hypothesis allows to infer spatial structures from temporal variations. This hypothesis is frequently used in the solar wind to infer wavenumber spectra and spatial structure functions from single time series, see for example Horbury et al. (2011) and Marsch and Tu (1997).

Two closely spaced measurements (as compared to the characteristic spatial scales of the medium) give access to a wealth of new information. The prime quantity of interest is the joint frequency-wavenumber spectrum $S(k, f)$, which is the key to the identification of plasma waves. The first step towards the extraction of that spectrum is the estimation of the cross spectrum

$$S_{XY}(f) = \langle Y(f)X^*(f)\rangle = |S_{XY}(f)|e^{j\phi_{XY}(f)}, \tag{12}$$

where $X(f)$ and $Y(f)$ are respectively the Fourier transforms of the two simultaneously measured scalar observables $x(t)$ and $y(t)$. The phase $\phi_{XY}(f)$ of this complex quantity is related to the wavenumber $\mathbf{k}$ projected along the separation vector $\mathbf{d}$ of the two measurements by

$$\mathbf{k} \cdot \mathbf{d} = \phi_{XY}(f). \tag{13}$$

This property has been widely used to infer wavenumbers in simple configurations wherein the direction of propagation is known; the latter is usually determined by minimum variance analysis (Sonnerup and Scheible 1998). An early example is the identification of whistler waves in the solar wind using two the nearby ISEE spacecraft (Hoppe and Russell 1980).

Classical cross-spectral analysis, however, has several limitations. First, it cannot properly distinguish multiple waves that have different dispersion relations. To overcome that limitation, Beall et al. (1982) proposed to use instead the local joint frequency-wavenumber spectrum, defined as

$$S_L(k, f) = \sum_i \frac{1}{2}\big[X_i(f)X_i^*(f) + Y_i(f)Y_i^*(f)\big]\delta\big(k_i(f) - k\big), \tag{14}$$

where the local wavenumber $k_i(f) = |\mathbf{k}_i(f)|$ is estimated from the cross-spectrum, for different ensembles that are indexed by i. According to this equation, the frequency-wavenumber spectrum is obtained by unfolding the frequency spectrum, conditioned by the wavenumber defined in Eq. (13). This approach has been widely used in laboratory plasmas, and in particular for drift turbulence studies, using Langmuir probe pairs (Poli et al. 2006; Tynan et al. 2009).

There is a second limitation, however, which has often been overlooked: all these Fourier-based techniques are formally applicable only to plane waves with an infinite spatial extent. Such conditions are rarely met in turbulent plasmas, which are often non-stationary and rather consist of a superposition of wave-packets. This is even more so in strong turbulence, in which the size of the wave packets becomes comparable to their characteristic period or wavelength. The generalisation of Eq. (14) to multiscale methods allows to overcome this limitation (Dudok de Wit et al. 1995). The idea simply consists in replacing the Fourier transform in Eq. (14) by a continuous wavelet transform.

Figure 4 shows a typical example taken from the solar wind magnetically connected to the Earth's bow shock (i.e. foreshock region). One single linear branch is observed in the dispersion relation when all wave packets are considered, regardless of their amplitude. According to this picture, the apparent motion amounts to simple dispersionless advection. However, the local frequency-wavenumber spectrum, in which each wave packet is now weighted by its amplitude, clearly reveals the existence of two branches, with different group velocities. The slowest branch (in the satellite rest frame) is associated with dispersive

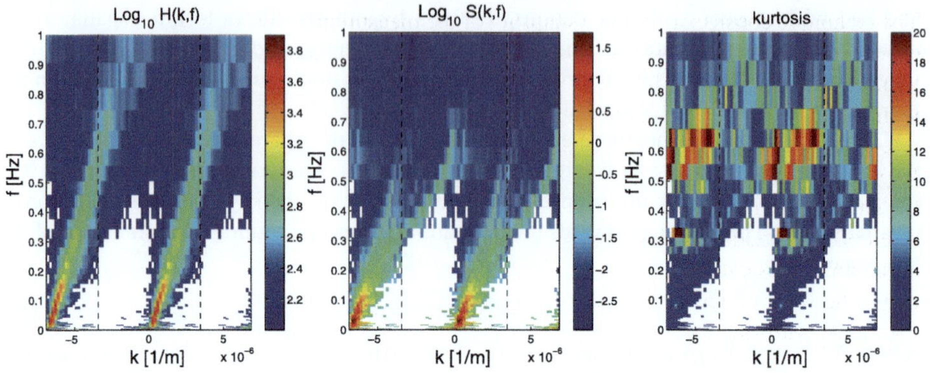

Fig. 4 Wavefield properties as a function of frequency and wavenumber, inferred from two-point magnetic field measurements in the foreshock region, from the AMPTE-UKS and AMPTE-IRM spacecraft. Details are given in Dudok de Wit et al. (1995). The *left plot* displays the number of wave packets, regardless of their amplitude; the *middle plot* shows the local frequency-wavenumber spectrum $S_L(k, f)$ estimated with wavelets; the *right plot* shows the flatness, as a signature of temporal burstiness. The *dashed rectangle* represents the principal domain, which is bounded by the spatial Nyquist number, and has been artificially unfolded. The only physically meaningful branch starts at $(f = 0, k = 0)$

whistler wave packets whose bursty nature is attested by their high flatness, see Eq. (9). Surprisingly, the use of wavelets for estimating dispersion relations is well accepted in geospace plasmas (*e.g.* Eriksson 1998; Lundberg et al. 2012), whereas they are still exceptional in laboratory plasmas (Lazurenko et al. 2008). This example shows how long it may sometimes take for a ideas to spread.

The concept of cross-spectrum can be generalised to higher order spectra in order to describe nonlinear interactions, see for example (Kim and Powers 1979; Kravtchenko-Berejnoi et al. 1995; Dudok de Wit 2003). Higher order spectra are relevant for weak turbulence, because nonlinearities can be adequately described by three-wave and four-wave interactions. Interestingly, in the Hamiltonian framework for weak turbulence (Musher et al. 1995), there is a direct connection between the physical model and higher order spectra. Langmuir turbulence is one of the few examples in which empirical statistical quantities have such an immediate physical meaning.

The seminal work of Ritz and Powers (1986), who showed how to derive spectral energy transfers from higher order spectra using double probe measurements, opened the way to several applications. Indeed, energy transfers give deep insight into the nonlinear interactions. They allow, for example, to determine whether coherent structures in a turbulent wavefield are dynamically evolving or are static remnants of a nonlinear process that took place earlier. There have been surprisingly few applications so far; some can be found in density fluctuations in tokamak edge turbulence (Kim et al. 1997; Zweben et al. 2007), in magnetic field fluctuations in the solar wind upstream the Earth's quasi-parallel bow-shock (Dudok de Wit et al. 1999), and in Langmuir turbulence simulations (Souček et al. 2003). Let us stress that they require long records, stationarity, and need careful validation to avoid misinterpretation. This may explain why they have been mostly applied to laboratory plasmas.

For fully developed turbulence, or when the spectral description is less appropriate, other techniques may be more relevant. The most elementary one is the cross-correlation function, which informs about temporal lags and about the decorrelation time. More interesting is the conditional averaging technique, also know as superposed epoch analysis in space science.

The technique consists in taking a snapshot of the measurements whenever a reference signal exceeds a threshold. Averaging these events allows the coherent patterns, if there are any, to emerge from the sea random fluctuations. Conditional averaging has been widely used to identify and characterise spatially coherent structures (often referred to as blobs or clumps) in electrostatic turbulence (Johnsen et al. 1987; Windisch et al. 2006; Diallo et al. 2008). In geospace plasmas, conditional sampling has been more frequently used for identifying the shape of transients (Baker 1986; Borovsky and Denton 2010). Laboratory applications often offer the additional advantage of allowing the probes to be moved, so that a full range of spatial scales can be investigated with two probes only, just by repeating the experiment under the same conditions (*e.g.*, Furno et al. 2008; Fattorini et al. 2012). These techniques, combined with smart detection criteria, are gaining again a lot of interest.

3.2 Three-Point Measurements and More

The transition from two-point measurements to three-point (and more) is relatively minor in terms of gain of information. Many multi-probe studies actually are made by processing the probe signals pairwise. A beautiful exception in space research is given by the 4 satellites from the CLUSTER mission, which form a tetrahedron, thereby providing access to full 3D resolution in space. With such a configuration, the determination of gradients, of the three components of wave-vectors, and normals to natural boundaries becomes possible. This has led to important improvements in the study of microphysical processes in space plasmas, both qualitatively and quantitatively. Most of these techniques and the first results obtained with them are detailed in the two books by Paschmann and Daly (2000, 2008), which we refer the reader to.

Some of the physical properties that have been inferred from CLUSTER observations, are: the estimation of the current density from the curl of the magnetic field (Haaland et al. 2004), the analysis of the geometrical structure of discontinuities such as shocks (Shen et al. 2003), the identification of a co-moving de Hoffmann-Teller frame using wavefield measurements (Khotyaintsev et al. 2004), and Alfvén vortices in the Earth's environment (Sundkvist et al. 2005; Alexandrova et al. 2006). Several successful attempts have been made for reconstructing the three-dimensional spatial spectrum of turbulence over at least one decade of scales (Tjulin et al. 2005; Sahraoui et al. 2006, 2011). These reconstructions are based on the k-filtering technique (Pinçon and Lefeuvre 1991), which allows to overcome the limitations of the Taylor hypothesis. Many of these studies have been instrumental in unveiling key physical processes such as reconnection events at the Earth's bow shock (Retinò et al. 2007) and at providing a microscopic look at the processes allowing particles from the Sun direct entry into the magnetosphere (Hwang et al. 2012). Cross-Scale, the follow-on of CLUSTER, was designed to probe both ion and electron scales, and would thereby have opened new perspectives in multipoint data analysis (Dunlop et al. 2011). To finish, let us mention a particular case wherein these techniques have been adapted to three-point measurements (Vogt et al. 2009).

Here we would like to stress again the importance of understanding the limitations behind spectral techniques such as k-filtering, which assume a superposition of uncorrelated plane waves with random phases. These assumptions are rarely met in practice and great care must be taken in analysing cases such as spatially coherent structures that are convected with the flow.

Another conceptually interesting approach involves multipoint observations of systems that exhibit "global" modes. The latter encompass here anything going from traveling waves to perturbations affecting the system globally or partly. In such spatio-temporal systems, it often makes sense to look for separable solutions of the observable $f(\mathbf{x}, t)$:

$$f(\mathbf{x}, t) = \sum_k \phi_k(t)\, \psi_k(\mathbf{x}). \tag{15}$$

This decomposition can be unique if constraints are imposed on the modes $\phi_k(t)$ and $\psi_k(\mathbf{x})$. Various techniques have been developed in what has become a highly multidisciplinary field of research that is also known as *blind source separation*: the Singular Value Decomposition or Biorthogonal Decomposition are used when the modes are orthonormal, Independent Component Analysis, when they are independent, *etc.* (Dudok de Wit 2011). In all these techniques, the heuristic idea is to concentrate the salient features of the spatio-temporal wavefield in the smallest possible number of modes. These techniques excel in separating oscillations with different poloidal mode numbers in laboratory devices (Nardone 1992; Dudok de Wit et al. 1994; Madon and Klinger 1996; Kim et al. 1999; Niedner et al. 1999; Dinklage et al. 2000) and, more generally, in extracting variations that are spatially coherent (Manini et al. 2003). Such variations may be travelling waves but also non wave-like structures. Mode decomposition techniques are also useful for handling inverse problems such as those occurring in plasma tomography (Anton et al. 1996). Applications to geospace plasmas have been slow to come because multichannel observations are so few. A notable exception is the analysis of solar and stellar spectra, which are just another kind of bivariate data (Amblard et al. 2008; Christlieb et al. 2002).

There was initially some hope that such techniques might help identifying and extracting coherent structures from turbulent wavefields, as suggested by Benkadda et al. (1994). Few attempts turned out to successful. Indeed, techniques such as the Singular Value Decomposition rely on second order moments only and so formally cannot properly extract structures whose main hallmark is their departure from a Gaussian PDF. Multiscale techniques are better suited for that, as discussed, for example, by Farge et al. (2006) and by D'Ippolito et al. (2011).

3.3 Imaging

Ideally, a complete characterisation of turbulence and meso-scale structures in plasmas, which we generally refer to as "imaging" in this section, would require full spatio-temporal measurements of the relevant fields (i.e electron and ion density, temperature, plasma potential, *etc.*) with adequate spatial and temporal resolution, and without perturbing the plasma. From the experimental point of view, this task is hampered by the intrinsic difficulty in diagnosing the region of interest with adequate temporal and spatial resolution. The seminal study by Zweben (1985), with a grid of 8×8 Langmuir probes, has remained unmatched for many years until new techniques have enabled more quantitative information to be extracted from such data sets.

We will not address here the remote sensing of solar and heliospheric plasmas, whose breathtaking images have stimulated the development of a wide variety of techniques, some of which are very specific to the structures of interest, such as coronal loops, or the motion of granules in the photosphere. For some references, see Georgoulis (2005), Meunier and Zhao (2009), Aschwanden (2010, 2011). The next conceptual step will involve data cubes such as two dimensional (2D) images taken in different wavelengths by spectro-imagers or, more generally, hyperspectral images.

In recent years, progress has been made in basic plasma physics devices, where arrays comprising a large number (of the order of 10^2) of electrostatic probes have been commonly

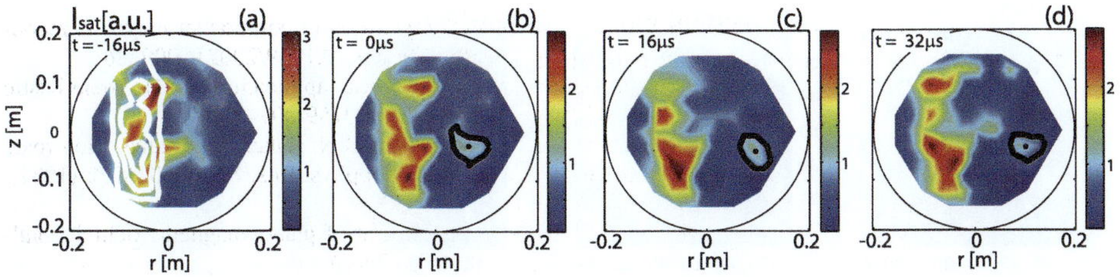

Fig. 5 Example of fast imaging using 2D profiles of ion saturation current from a 2D Langmuir probe array on the TORPEX device. An intermittent turbulence-generated structure, *i.e.* a blob (highlighted by the *thick black line*) is radially propagating in the poloidal plane of the device

introduced to typically measure ion saturation and floating potential signals (Katz et al. 2008; Müller et al. 2006). Two diagnostic issues are the quantitative interpretation of probe results, especially in magnetised plasmas where the theory of electrostatic probes is still incomplete, and the possible perturbing effects of the probes on the physics under study. Relatively few papers discuss these issues in detail, and normally some checks are made to verify that the probe does not affect the results from existing probes, but it is often not clear whether the structure and motion of blobs is unaffected by these probes.

As an example, Fig. 5 presents the application of 2D imaging to visualise intermittent meso-scale structures, or blobs, in the TORPEX device (Fasoli et al. 2010). Blob propagation is investigated using a 2D array of 86 Langmuir probes on a hexagonal pattern, the HEXTIP diagnostic described in detail, see (Müller et al. 2006). HEXTIP measures ion saturation current signals with a temporal resolution of 4 μs, which is smaller than the typical auto-correlation time of the structures under study ($\approx$100 μs), and 1.75 cm spatial resolution. The first step, common to all diagnostics and techniques providing real space temporal imaging of plasmas, is the definition of the meaning of structures. To this aim, pattern recognition techniques assuming more or less complicated definitions can be used. In this example, blobs are identified by adopting a threshold segmentation approach, which selects positive structures as regions where the ion saturation signal exceeds an appropriate value, usually expressed in terms of the standard deviation of the signals. The identified structures are bound by polygonal structures, shown in black contour in Fig. 5, which allow computing different observables linked to the structures. Examples of observables are the occupied area, the position of the center of mass of a structure, *etc.* By considering subsequent time frames, structure trajectories can be identified together with splitting events, when a single structure breaks apart, and merging events, when multiple structures merge together. The observables defined for static structures can now be computed along the trajectory of the structure itself, and new observables can be defined, such as for example the structure speed. This process allows to visualise, *i.e.* image, the structure dynamics in turbulent plasmas. Furthermore, the strength of the real space imaging is that it not only allows to identify individual structures, but also to build a statistical ensemble over which to compute the statistical/probabilistic properties of the turbulent field under investigation (Müller et al. 2006). This can be used to verify in a statistical sense predictions from theory, such as for examples scaling laws (Theiler et al. 2009).

The use of insertable probes for a full imaging of the quantities of interest for plasma turbulence studies suffers from two interlinked limitations, even in low temperature plasma experiments. A small number of probes only marginally perturbs the plasma dynamical behaviour, but would lead to either only local measurements or to insufficient spatial resolution to investigate the turbulence multiple scales. To obtain adequate spatial resolution, one

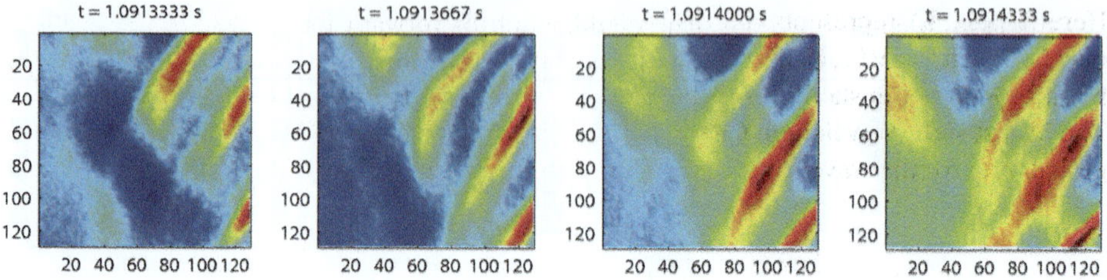

Fig. 6 Example of filament evolution obtained using a fast framing camera at the edge of TCV tokamak. The axes are in units of pixel number

would need to insert a large number of measuring tips, which would significantly perturb the phenomena under investigation. In recent years, to overcome these limitations, fast framing cameras have been adopted to monitor light emission from magnetically confined plasmas and also in basic plasma physics devices. In particular the field of fast imaging has been pushed forward by the need to investigate intermittent transport events associated with the filaments in fusion and basic plasma physics devices (Kirk et al. 2006; Ben Ayed et al. 2009; Grulke et al. 2006; Iraji et al. 2010).

Fast cameras are today used to track filamentary structures with typical spatial scales >1 mm and lifetime >1 μs in the scrape-off layer (SOL), and to compute their relevant quantities, such as average velocities and sizes. An example from the TCV tokamak (Goodman et al. 2003) is shown in Fig. 6. Fast framing cameras now offer $\approx 10^6$ pixels and acquisition frequencies up to 10^6 images per second (1 μs time resolution), which are adequate to resolve most turbulent phenomena in magnetised plasmas. These large arrays of spatially distributed measurements open new possibilities of identifying turbulent structures in two-dimensional or even three-dimensional space and of following their evolution in time.

Two main difficulties are encountered when using fast cameras to image plasma turbulence. First, the signal is usually integrated along a line-of-sight resulting in multi-chord images, which need to be inverted by tomography, using more or less complicated techniques, to reconstruct the local plasma emissivity, see for example (Svensson et al. 2008). The line integration problem is common to other diagnostic techniques which can be generally classified as "imaging" techniques. For example, measurements of line integrated electron density along satellite-to-ground path are obtained by using radio transmission from polar orbiting satellites (Kersley et al. 1997). Tomographic inversion of these data allows mapping large portion of the ionosphere in a height versus latitude plane (Bust and Mitchell 2008). The same approach can be applied to the solar corona (Butala et al. 2010). Tomographical reconstructions of heliospheric perturbations from observations of interplanetary scintillations (IPS) offer the ultimate example of how 3D structures can be reconstructed from line-of-sight observations (Jackson et al. 2011).

In laboratory devices, the line integration problem can be solved by using fast framing cameras together with gas puffing (Agostini et al. 2009; Terry et al. 2003; Maqueda et al. 2010; Zweben et al. 2011), which result in atomic line emission from the locally injected neutral atoms. In these cases, precautions must be taken to avoid that the addition of a neutral source affects the local plasma parameters and turbulence features. The second problem is related to the intrinsic origin of the quantity measured by imaging, *i.e.* optical plasma emission, which, in most cases, is a complex convolution of local density, temperature, and impurities concentration. This hampers a direct comparison of results from visible imaging with, for instance, numerical simulations of turbulent transport, unless these are coupled with radiation emission codes. The combination of forward modelling with MHD codes and direct IPS observations has been shown to enhance the reconstruction of solar wind structures (Hayashi et al. 2003).

Tomography inevitably leads to ill-posed problems, for which prior assumptions are important, and solutions are sensitive to the underlying assumptions. For such problems, the Bayesian framework (von Toussaint 2011) has proven to be remarkably powerful and will surely inspire in the decade to come new techniques that can handle the imperfections of the data more properly.

4 Anomalous Transport in Plasmas

The transport properties of suprathermal particles are important both for laboratory plasmas, where good confinement is essential in order to reach the goal of fusion, and in geospace plasmas, where understanding transport is important for predicting the arrival of solar energetic particles (SEPs) and for the acceleration processes like first order acceleration at shocks (Reames 1999; Giacalone 2011; Perri and Zimbardo 2012b). Here, an important issue that has challenged the analysis of microphysical processes is the characterisation of the transport processes.

In the classical regime of normal diffusion, the mean square displacement of particles along one dimension can be written as

$$\langle \Delta x^2 \rangle = 2Dt, \tag{16}$$

where D is the diffusion coefficient, and where $\langle \Delta x^2 \rangle$ grows linearly in time. However, in the last years it has become clear that anomalous transport regimes can be found in a large variety of physical systems, such that $\langle \Delta x^2 \rangle$ grows nonlinearly in time,

$$\langle \Delta x^2 \rangle = 2D_\gamma t^\gamma, \tag{17}$$

with $\gamma < 1$ in the case of subdiffusion, $1 < \gamma < 2$ in the case of superdiffusion, and with $\gamma = 2$ representing the ballistic regime (*e.g.*, Zimbardo 2005; Zimbardo et al. 2012).

Anomalous transport has been studied extensively in laboratory plasmas, due to the importance of understanding and controlling the plasma losses, while it is comparatively less well known in geospace plasmas. In laboratory plasmas, transport was found to be nondiffusive already two decades ago, when it became clear that the random motion of particles in fluctuating electric and magnetic fields was analogous to motion in periodic and quasi-periodic potentials (Zaslavskii et al. 1989), and that long "ballistic" displacements were possible in between the structure of nested magnetic surfaces which are characterised by broken surfaces called "cantori" (Zaslavsky et al. 1993). The combination of temporal trapping and long displacements leads to anomalous diffusion which can be both subdiffusive and superdiffusive, depending on control parameters like the turbulence level and the particle energy (Shlesinger et al. 1993).

Many studies of anomalous transport in fluids and plasmas have been performed; for instance, Benkadda et al. (1997) have shown that transport regimes including subdiffusion and superdiffusion can be obtained for passive tracer particles in a flow undergoing the transition to turbulence. Carreras et al. (2001) have similarly explored the dynamics of tracer particles in turbulence models with avalanche transport. Numerical simulations of fast ion transport in simple magnetized toroidal plasmas were performed by Gustafson et al. (2012) and the fast ion transport perpendicular to magnetic field lines was quantified. Despite the simplicity of this system, the entire spectrum of suprathermal ion dynamics, from subdiffusion to superdiffusion, depending on beam energy and turbulence amplitude was observed. Interestingly, numerical simulation of particle transport in the solar wind turbulence find that perpendicular transport is either subdiffusive or normal, while parallel transport can be either superdiffusive or normal (Zimbardo et al. 2006; Shalchi and Kourakis 2007).

An overview of the theoretical approach to anomalous transport, including the use of non Gaussian statistics, the Hurst exponent, and continuos time random walks, plus the results of numerical simulations for laboratory and geospace plasmas is given by Perrone et al. (2012). Here we discuss how to detect anomalous diffusion from *in situ* measurements in space, with a focus on superdiffusion.

In numerical simulations, one can follow a particle trajectory and numerically compute the mean square displacement; after that, a fit of $\langle \Delta x^2 \rangle$ versus time allows to distinguish between normal diffusion, subdiffusion, or superdiffusion. Another approach is based on the determination of the Lagrangian velocity autocorrelation function, since "thick" power law tails are also indicative of superdiffusion (*e.g.*, Mier et al. 2008). Again, this approach is feasible in numerical simulations but not in space measurements, where usually only the Eulerian correlation functions are available. Then, how to detect anomalous diffusion in space?

Already in 1974, by analysing the time profiles of nonrelativistic solar electron events, Lin (1974) pointed out that nonrelativistic electrons exhibit a wide range of transport regimes, going from diffusive to ballistic. Indeed, an essential property of superdiffusive transport is that it is characterised by a non Gaussian statistics, both in the PDF of the diffusing particle (the *propagator*) and in the probability of making a free path of a given length. Such a non Gaussian statistics influences the spatial distribution of superdiffusing particles, and hence the observed energetic particle time profile. In particular, superdiffusion corresponds to a Levy statistics, which implies that the propagator has the following power law shape, appropriate at some distance from the particle source at x' (Zumofen and Klafter 1993),

$$P(x - x', t - t') = \frac{A_0}{(t - t')^{1/\alpha}} \left[\frac{(t - t')^{1/\alpha}}{|x - x'|} \right]^{\alpha+1}, \tag{18}$$

where $1 < \alpha < 2$, while a different expression holds for small values of $|x - x'|$ (Zumofen and Klafter 1993). Above, A_0 is a normalisation constant. As can be seen, the long distance propagator has a power law form, sharply different from the normal Gaussian propagator. The density of energetic particles upstream of a shock can be obtained as the superposition of the particles accelerated at the shock during its propagation, and using the above propagator yields, for $t < 0$ (*i.e.*, upstream of the shock) (Perri and Zimbardo 2007, 2008)

$$f(E, t) \propto (-t)^{1-\alpha} \equiv (-t)^{-\nu}, \tag{19}$$

where $f(E, t)$ is the omnidirectional distribution function of particles of energy E. On the contrary, in the case of normal diffusion characterised by a Gaussian propagator, an exponential decay is obtained for $f(E, t)$ upstream of the shock. In other words, a power law time profile for energetic particles with slope $\nu = \alpha - 1 < 1$ is the signature of superdiffusion with anomalous diffusion exponent $\gamma = 3 - \alpha = 2 - \nu > 1$ (Perri and Zimbardo 2007, 2008; Perrone et al. 2012).

In order to extract information on the transport of energetic particles accelerated at the shock, one has to identify a "clean" shock crossing in the spacecraft data (Perri and Balogh 2010a, 2010b), then plot $f(E, \Delta t)$ upstream of the shock versus the time distance to the shock, $\Delta t = |t - t_{sh}|$. A fit of $f(E, \Delta t)$ versus Δt will reveal normal diffusion when the

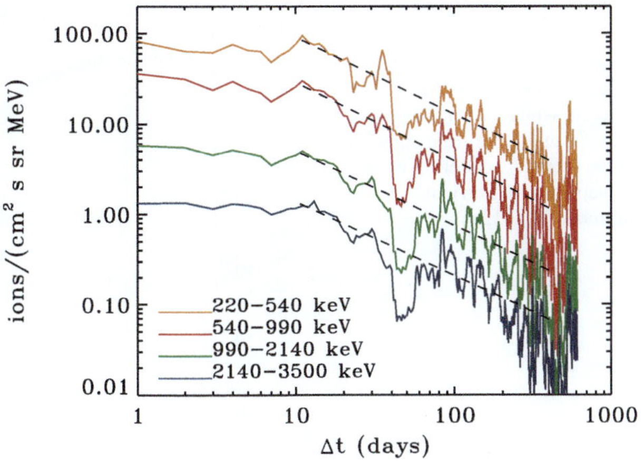

Fig. 7 Energetic ion fluxes measured by the LECP instrument onboard Voyager 2 upstream of the termination shock (energy as indicated). The *dashed black lines* represent the power law fit. Adapted from Zimbardo et al. (2012)

best fit is an exponential, and superdiffusion when the best fit is a power law. Using this technique as a diagnostic tool, Perri and Zimbardo (2007, 2008, 2009a) have shown that electron transport upstream of the shocks associated with corotating interaction regions (CIRs) detected by the Ulysses spacecraft in the solar wind at 4–5 AU is superdiffusive, with $\gamma \simeq 1.1$–1.7. On the other hand, analysing the Voyager 2 data of low energy particles, Perri and Zimbardo (2009b) have shown that ion transport upstream of the solar wind termination shock at 84 AU is superdiffusive, too, with $\gamma \simeq 1.3$, see Fig. 7. Let us stress again that a correct identification of a power law and its slope require great care, see Sect. 2.1.

We point out that the power law decay found in the energetic particle profiles upstream of CIR shocks is different from that envisaged by the numerical study by Giacalone (2004). In that study, the power law-like decay of $f(x, E, t)$ is due to the decrease of the magnetic fluctuations with increasing distance from the shock, so that pitch angle scattering decreases, too, and parallel transport becomes faster. However, Perri and Zimbardo (2012a) have shown that the variance of the magnetic field components, $\sigma_i^2 = \langle (B_i - \langle B_i \rangle)^2 \rangle_T$ being T the timescale for the average computation, measured by Ulysses at the electron resonance scale, which causes pitch angle scattering, do not vary with the distance from the shock. Therefore a constant level of magnetic fluctuations implies a constant transport regime, so that the energetic particle spatial profile, either an exponential decay or a power law with $\nu < 1$, can be used to discriminate between normal diffusion and superdiffusion.

On the other hand, once superdiffusive transport is identified as explained above, one question arises: what it the physical origin of superdiffusion in the solar wind? A first answer can be given by considering that superdiffusion is due to a Levy statistics for the distribution $\psi(\ell)$ of free path lengths ℓ. More precisely, the Levy random walk assumes $\psi(\ell) \sim |\ell|^{-1-\alpha}$ for large $|\ell|$; for constant velocity particles, this corresponds to a power law distribution of free path times, $\psi(\tau) \sim |\tau|^{-1-\alpha}$. For displacements along the magnetic field, one can relate the free path time to the pitch angle scattering time. In other words, a long pitch angle scattering time means that particles keep the same parallel velocity for a long time, thus performing the long displacements corresponding to a Levy random walk. The pitch angle scattering time is obtained from the normalised magnetic variance σ^2/B_0^2 at the gyroresonant scale as $\tau = (\sigma^2/B_0^2)^{-1}\Omega^{-1}$ (Kennel and Petschek 1966), where σ^2 is the total magnetic variance near resonance and Ω is the particle gyrofrequency. Perri and Zimbardo

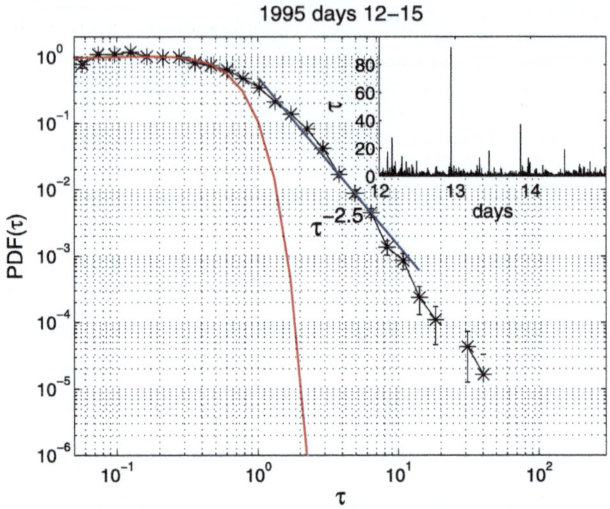

Fig. 8 PDF of scattering times measured by Ulysses in the fast solar wind in 1995 (asterisks). The *blue line* shows the power law fit, while the *red line* the Gaussian fit. The *inset* shows the time series of scattering times. Adapted from Perri and Zimbardo (2012a)

(2012a) have shown that in the case of superdiffusive events in the solar wind, the distribution of pitch angle scattering times is a power law, see Fig. 8, implying that long scattering times τ have a non negligible probability. Further, the power law slopes are between 2.4 and 3.8, values which are consistent, at least partly, with those required by the Levy random walk to have superdiffusion, *i.e.*, $2 < 1 + \alpha < 3$.

Therefore the analysis of the magnetic variances at the gyroresonant scale and the finding of a power law distribution of pitch angle scattering times with slope between 2 and 3 gives information on the microphysical processes leading to superdiffusion. On the other hand, this new data analysis technique calls for further investigations, since the scattering times determined by a spacecraft allow to compute the "Eulerian distribution function", while the random walk of the particles is determined by the scattering times seen along the particle trajectory, that is by the "Lagrangian distribution function".

A different signature of anomalous transport, which is also intimately related to the concept of self-similarity, is the occurrence of long-range correlations. Models based on self-organized criticality (Newman et al. 1996) and also various turbulence models predict the existence of such long-range correlations. By this, we mean a slow decay of the autocorrelation

$$\lim_{\tau \to \infty} C(\tau) \sim \tau^{-\gamma}, \tag{20}$$

with $\gamma < 1$. For such scalings, the integral time scale $T = \int_0^{+\infty} C(\tau)\, d\tau$ diverges and fundamental properties such as stationarity are lost. The main difficulty then consists in inferring such properties from finite and non-stationary records. Different approaches have been developed for that purpose, sometimes giving differing responses. One is the structure function approach, see Sect. 2.2. A different approach is rescaled range (R/S) analysis, in which the Hurst exponent quantifies the degree of persistence in a signal. We met this exponent before in the context of statistical self-similarity, see Eq. (5).

For a record of N regularly sampled values with zero mean $\{y_1, y_2, \dots, y_N\}$, we first create cumulative deviates $w_t = \sum_{i=1}^{t} y_i$ for $t = 1, 2, \dots, N$. From this we define the range as $R_t = \max(w_1, w_2, \dots, w_t) - \min(w_1, w_2, \dots, w_t)$. The rescaled range series then equals

$(R/S)_t = R_t/\sigma_t$, where σ_t is the standard deviation of the sequence $\{y_1, y_2, \ldots, y_t\}$. The following self-similar scaling holds

$$\lim_{\tau\to\infty} (R/S)_\tau \sim \tau^{-H}, \tag{21}$$

where H is the Hurst exponent (Carreras et al. 1998). Persistence occurs when $0.5 < H < 1$, while $H = 0.5$ indicates an uncorrelated process and $0 < H < 0.5$ indicates anti-persistence. For specific classes of processes, the value of H is connected to that of the spectral index, and of the scaling exponent of the structure function (Gilmore et al. 2002).

There have been several applications of the rescaled range analysis to laboratory plasmas, see for example (Yu et al. 2003; Tynan et al. 2009). The value of the Hurst exponent itself is not so informative; more interesting is the way it changes under different regimes, such as in the transition of from L to H-mode (Dudson et al. 2005). There have also been some applications to geospace plasmas. Kiyani et al. (2007), for example, used it to relate properties of the solar wind to the structure of the corona under different conditions of solar activity.

In many instances, H has been found to be larger than 0.5, which is often considered as signature of self-organized criticality, or rather, driven self-organized criticality. However, as with all higher order statistical methods, there are also a number of pitfalls. In particular, H must be determined within the appropriate range of time lags: τ must exceed local correlation time scales while being much smaller than the meso-scales of the system (Gilmore et al. 2002).

5 Conclusions

This brief overview of some advanced methods for geospace and laboratory plasmas reveals the power of such methods, provided that they are used in connection with plasma physics theory, and that their limitations and pitfalls are properly understood. One important, and yet missing, topic here is systems theory, which can give deep insight into the nonlinear dynamics of plasmas (Vassiliadis 2006). One of the reasons for its relevance is the growing interest for system approaches, in which the coupling between different layers or regions becomes of comparable interest as the individual regions themselves.

Many time series analysis methods are relatively mature whereas spatio-temporal analysis methods often still are in their infancy. The CLUSTER multi-satellite mission is one of the rare examples wherein a coordinated effort toward the preparation of an experiment has led to new methodological developments. In most cases, however, innovation has occurred instead by strokes of serendipity or by knowledge transfer from nearby fields. This transfer has often been slower between the geospace and laboratory plasma communities than between the plasma and neutral fluid communities. Clearly, there remains a considerable issue in fostering interactions to the point where new ideas can spread more rapidly.

Acknowledgements We all thank the International Space Science Institute (ISSI, Bern) for hospitality. In preparing this review we made extensive use of NASA's Astrophysics Data System.

References

P. Abry, P. Goncalves, P. Flandrin, Wavelets, spectrum analysis and 1/f processes, in *Wavelets in Statistics*, ed. by A. Antoniadis, G. Oppenheim. Lecture Notes in Statistics, vol. 103 (Springer, Berlin, 1995), pp. 15–30

M. Agostini, P. Scarin, R. Cavazzana, F. Sattin, G. Serianni, M. Spolaore, N. Vianello, Edge turbulence characterization in RFX-mod with optical diagnostics. Plasma Phys. Control. Fusion **51**(10), 105003 (2009). doi:10.1088/0741-3335/51/10/105003
A.I. Akhiezer, I.A. Akhiezer, R.V. Polovin, A.G. Sitenko, K.N. Stepanov, *Plasma Electrodynamics. Volume 1—Linear Theory*. Oxford Pergamon Press International Series on Natural Philosophy, vol. 1 (1975)
A.I. Akhiezer, I.A. Akhiezer, R.V. Polovin, A.G. Sitenko, K.N. Stepanov, *Plasma Electrodynamics. Volume 2—Non-Linear Theory and Fluctuations*. Oxford Pergamon Press International Series on Natural Philosophy, vol. 1 (1975)
O. Alexandrova, Solar wind vs magnetosheath turbulence and Alfvén vortices. Nonlinear Process. Geophys. **15**, 95–108 (2008)
O. Alexandrova, J. Saur, Alfvén vortices in Saturn's magnetosheath: cassini observations. Geophys. Res. Lett. **35**, 15102 (2008). doi:10.1029/2008GL034411
O. Alexandrova, C. Lacombe, A. Mangeney, Spectra and anisotropy of magnetic fluctuations in the Earth's magnetosheath: cluster observations. Ann. Geophys. **26**, 3585–3596 (2008). doi:10.5194/angeo-26-3585-2008
O. Alexandrova, A. Mangeney, M. Maksimovic, N. Cornilleau-Wehrlin, J.-M. Bosqued, M. André, Alfvén vortex filaments observed in magnetosheath downstream of a quasi-perpendicular bow shock. J. Geophys. Res. (Space Phys.) **111**(A10), 12208 (2006). doi:10.1029/2006JA011934
O. Alexandrova, J. Saur, C. Lacombe, A. Mangeney, J. Mitchell, S.J. Schwartz, P. Robert, Universality of solar-wind turbulent spectrum from MHD to electron scales. Phys. Rev. Lett. **103**(16), 165003 (2009). doi:10.1103/PhysRevLett.103.165003
P. Amblard, S. Moussaoui, T. Dudok de Wit, J. Aboudarham, M. Kretzschmar, J. Lilensten, F. Auchère, The EUV Sun as the superposition of elementary Suns. Astron. Astrophys. **487**, 13–16 (2008). doi:10.1051/0004-6361:200809588
G.Y. Antar, On the origin of "intermittency" in the scrape-off layer of linear magnetic confinement devices. Phys. Plasmas **10**, 3629–3634 (2003). doi:10.1063/1.1599855
G.Y. Antar, P. Devynck, X. Garbet, S.C. Luckhardt, Turbulence intermittency and burst properties in tokamak scrape-off layer. Phys. Plasmas **8**, 1612–1624 (2001). doi:10.1063/1.1363663
M. Anton, H. Weisen, M.J. Dutch, W. von der Linden, F. Buhlmann, R. Chavan, B. Marlétaz, P. Marmillod, P. Paris, X-ray tomography on the TCV tokamak. Plasma Phys. Control. Fusion **38**, 1849–1878 (1996). doi:10.1088/0741-3335/38/11/001
M.J. Aschwanden, Solar image processing techniques with automated feature recognition. Sol. Phys. **262**(2), 235–275 (2010). doi:10.1007/s11207-009-9474-y
M.J. Aschwanden, Solar stereoscopy and tomography. Living Rev. Sol. Phys. 8(5) (2011). http://www.livingreviews.org/lrsp-2011-5
R. Badii, A. Politi, *Complexity*. Cambridge Nonlinear Science Series, vol. 6 (Cambridge Univeristy Press, Cambridge, 1999)
D.N. Baker, Statistical analyses in the study of solar wind/magnetosphere coupling, in *Astrophysics and Space Science Library*, ed. by Y. Kamide, J.A. Slavin. Astrophysics and Space Science Library, vol. 126 (1986), pp. 17–38
S.D. Bale, P.J. Kellogg, F.S. Mozer, T.S. Horbury, H. Rème, Measurement of the electric fluctuation spectrum of magnetohydrodynamic turbulence. Phys. Rev. Lett. **94**(21), 215002 (2005). doi:10.1103/PhysRevLett.94.215002
J.M. Beall, Y.C. Kim, E.J. Powers, Estimation of wavenumber and frequency spectra using fixed probe pairs. J. Appl. Phys. **53**, 3933–3940 (1982). doi:10.1063/1.331279
N. Ben Ayed, A. Kirk, B. Dudson, S. Tallents, R.G.L. Vann, H.R. Wilson, M. Team, Inter-ELM filaments and turbulent transport in the mega-amp spherical tokamak. Plasma Phys. Control. Fusion **51**(3), 035016 (2009). doi:10.1088/0741-3335/51/3/035016
S. Benkadda, P. Gabbai, G.M. Zaslavsky, Passive particle dynamics in a flow exhibiting transition to turbulence. Phys. Plasmas **4**, 2864–2870 (1997). doi:10.1063/1.872577
S. Benkadda, T. Dudok de Wit, A. Verga, A. Sen, X. Garbet, Characterization of coherent structures in tokamak edge turbulence. Phys. Rev. Lett. **73**, 3403–3406 (1994). doi:10.1103/PhysRevLett.73.3403
R. Benzi, S. Ciliberto, R. Tripiccione, C. Baudet, F. Massaioli, S. Succi, Extended self-similarity in turbulent flows. Phys. Rev. E **48**, 29 (1993). doi:10.1103/PhysRevE.48.R29
D. Biskamp, *Magnetohydrodynamic Turbulence* (Cambridge University Press, Cambridge, 2003)
T. Bohr, M.H. Jensen, G. Paladin, A. Vulpiani, *Dynamical Systems Approach to Turbulence*. Cambridge Nonlinear Science Series, vol. 8 (Cambridge Univeristy Press, Cambridge, 2005)
J.E. Borovsky, M.H. Denton, Solar wind turbulence and shear: a superposed-epoch analysis of corotating interaction regions at 1 AU. J. Geophys. Res. (Space Phys.) **115**, 10101 (2010). doi:10.1029/2009JA014966

R. Bruno, V. Carbone, The solar wind as a turbulence laboratory. Living Rev. Sol. Phys. **2**, 4 (2005)

G.S. Bust, C.N. Mitchell, History, current state, and future directions of ionospheric imaging. Rev. Geophys. **46**, 1003 (2008). doi:10.1029/2006RG000212

M.D. Butala, R.J. Hewett, R.A. Frazin, F. Kamalabadi, Dynamic three-dimensional tomography of the solar corona. Sol. Phys. **262**, 495–509 (2010). doi:10.1007/s11207-010-9536-1

V. Carbone, Scaling exponents of the velocity structure functions in the interplanetary medium. Ann. Geophys. **12**(7), 585–590 (1994). doi:10.1007/s00585-994-0585-3

V. Carbone, R. Bruno, P. Veltri, Evidences for extended self-similarity in hydromagnetic turbulence. Geophys. Res. Lett. **23**, 121–124 (1996). doi:10.1029/95GL03777

V. Carbone, L. Sorriso-Valvo, E. Martines, V. Antoni, P. Veltri, Intermittency and turbulence in a magnetically confined fusion plasma. Phys. Rev. E **62**, 49–56 (2000)

V. Carbone, F. Lepreti, L. Sorriso-Valvo, P. Veltri, V. Antoni, R. Bruno, Scaling laws in plasma turbulence. Nuovo Cimento Riv. Ser. **27**(8), 1–108 (2004). doi:10.1393/ncr/i2005-10003-1

B.A. Carreras, V.E. Lynch, G.M. Zaslavsky, Anomalous diffusion and exit time distribution of particle tracers in plasma turbulence model. Phys. Plasmas **8**, 5096–5103 (2001). doi:10.1063/1.1416180

B.A. Carreras, B.P. van Milligen, M.A. Pedrosa, R. Balbín, C. Hidalgo, D.E. Newman, E. Sánchez, M. Frances, I. García-Cortés, J. Bleuel, M. Endler, C. Riccardi, S. Davies, G.F. Matthews, E. Martines, V. Antoni, A. Latten, T. Klinger, Self-similarity of the plasma edge fluctuations. Phys. Plasmas **5**, 3632–3643 (1998). doi:10.1063/1.873081

S.C. Chapman, R.M. Nicol, Generalized similarity in finite range solar wind magnetohydrodynamic turbulence. Phys. Rev. Lett. **103**(24), 241101 (2009). doi:10.1103/PhysRevLett.103.241101

A.C.-L. Chian, R.A. Miranda, Cluster and ACE observations of phase synchronization in intermittent magnetic field turbulence: a comparative study of shocked and unshocked solar wind. Ann. Geophys. **27**, 1789–1801 (2009). doi:10.5194/angeo-27-1789-2009

N. Christlieb, L. Wisotzki, G. Graßhoff, Statistical methods of automatic spectral classification and their application to the Hamburg/ESO survey. Astron. Astrophys. **391**, 397–406 (2002). doi:10.1051/0004-6361:20020830

A. Clauset, C. Rohilla Shalizi, M.E.J. Newman, Power-law distributions in empirical data. SIAM Rev. **51**, 661–703 (2009). doi:10.1137/070710111

A. Diallo, A. Fasoli, I. Furno, B. Labit, M. Podestà, C. Theiler, Dynamics of plasma blobs in a shear flow. Phys. Rev. Lett. **101**(11), 115005 (2008). doi:10.1103/PhysRevLett.101.115005

A. Dinklage, C. Wilke, G. Bonhomme, A. Atipo, Internally driven spatiotemporal irregularity in a dc glow discharge. Phys. Rev. E **62**, 7219–7226 (2000). doi:10.1103/PhysRevE.62.7219

D.A. D'Ippolito, J.R. Myra, S.J. Zweben, Convective transport by intermittent blob-filaments: comparison of theory and experiment. Phys. Plasmas **18**(6), 060501 (2011). doi:10.1063/1.3594609

T. Dudok de Wit, Spectral and statistical analysis of plasma turbulence: beyond linear techniques, in *Space Plasma Simulation*, ed. by J. Büchner, C. Dum, M. Scholer. Lecture Notes in Physics, vol. 615 (Springer, Berlin, 2003), pp. 315–343

T. Dudok de Wit, Can high-order moments be meaningfully estimated from experimental turbulence measurements? Phys. Rev. E **70**(5), 055302 (2004). doi:10.1103/PhysRevE.70.055302

T. Dudok de Wit, Extracting individual contributions from their mixture: a blind source separation approach. Contrib. Plasma Phys. **51**(2–3), 143–151 (2011). doi:10.1002/ctpp.201000052

T. Dudok de Wit, A. Pecquet, J. Vallet, R. Lima, The biorthogonal decomposition as a tool for investigating fluctuations in plasmas. Phys. Plasmas **1**, 3288–3300 (1994). doi:10.1063/1.870481

T. Dudok de Wit, V.V. Krasnosel'skikh, S.D. Bale, M.W. Dunlop, H. Lühr, S.J. Schwartz, L.J.C. Woolliscroft, Determination of dispersion relations in quasi-stationary plasma turbulence using dual satellite data. Geophys. Res. Lett. **22**, 2653–2656 (1995). doi:10.1029/95GL02543

T. Dudok de Wit, V.V. Krasnosel'skikh, M. Dunlop, H. Lühr, Identifying nonlinear wave interactions in plasmas using two-point measurements: a case study of short large amplitude magnetic structures (SLAMS). J. Geophys. Res. **104**, 17079–17090 (1999). doi:10.1029/1999JA900134

B.D. Dudson, R.O. Dendy, A. Kirk, H. Meyer, G.F. Counsell, Comparison of L- and H-mode plasma edge fluctuations in MAST. Plasma Phys. Control. Fusion **47**, 885–901 (2005). doi:10.1088/0741-3335/47/6/010

M.W. Dunlop, R. Bingham, S. Chapman, P. Escoubet, Q. Zhang, C. Shen, J. Shi, R. Trines, R. Wicks, Z. Pu, J. de-Keyser, S. Schwartz, Z. Liu, Use of multi-point analysis and modelling to address cross-scale coupling in space plasmas: lessons from cluster. Planet. Space Sci. **59**, 630–638 (2011). doi:10.1016/j.pss.2010.06.014

A.I. Eriksson, Spectral analysis, in *Analysis Methods for Multi-Spacecraft Data*, ed. by G. Paschmann, P.W. Daly. ISSI Scientific Report SR-01, vol. 1, Amsterdam (1998), pp. 5–42

M. Farge, Wavelet transforms and their applications to turbulence. Annu. Rev. Fluid Mech. **24**, 395–457 (1992). doi:10.1146/annurev.fl.24.010192.002143

M. Farge, G. Pellegrino, K. Schneider, Coherent vortex extraction in 3D turbulent flows using orthogonal wavelets. Phys. Rev. Lett. **87**(5), 054501 (2001). doi:10.1103/PhysRevLett.87.054501

M. Farge, K. Schneider, P. Devynck, Extraction of coherent bursts from turbulent edge plasma in magnetic fusion devices using orthogonal wavelets. Phys. Plasmas **13**(4), 042304 (2006). doi:10.1063/1.2172350

M. Farge, N.K.-R. Kevlahan, V. Perrier, K. Schneider, Turbulence analysis, modelling and computing using wavelets, in *Wavelets in Physics*, ed. by J.C. van den Berg (Cambridge University Press, Cambridge, 2004), p. 117. Chap. 4

A. Fasoli, A. Burckel, L. Federspiel, I. Furno, K. Gustafson, D. Iraji, B. Labit, J. Loizu, G. Plyushchev, P. Ricci, C. Theiler, A. Diallo, S.H. Mueller, M. Podesta, F. Poli, Electrostatic instabilities, turbulence and fast ion interactions in the TORPEX device. Plasma Phys. Control. Fusion **52**(12, Part 2), 124020 (2010). doi:10.1088/0741-3335/52/12/124020

L. Fattorini, Å. Fredriksen, H.L. Pécseli, C. Riccardi, J.K. Trulsen, Turbulent transport in a toroidal magnetized plasma. Plasma Phys. Control. Fusion **54**(8), 085017 (2012). doi:10.1088/0741-3335/54/8/085017

U. Frisch, *Turbulence, the Legacy of A.N. Kolmogorov* (Cambridge University Press, Cambridge, 1995)

A. Fujisawa, A review—observations of turbulence and structure in magnetized plasmas. J. Plasma Fusion Res. **5**, 46 (2010). doi:10.1585/pfr.5.046

I. Furno, B. Labit, M. Podestà, A. Fasoli, S.H. Müller, F.M. Poli, P. Ricci, C. Theiler, S. Brunner, A. Diallo, J. Graves, Experimental observation of the blob-generation mechanism from interchange waves in a plasma. Phys. Rev. Lett. **100**(5), 055004 (2008). doi:10.1103/PhysRevLett.100.055004

M.K. Georgoulis, Turbulence in the solar atmosphere: manifestations and diagnostics via solar image processing. Sol. Phys. **228**, 5–27 (2005). doi:10.1007/s11207-005-2513-4

J. Giacalone, Large-scale hybrid simulations of particle acceleration at a parallel shock. Astrophys. J. **609**, 452–458 (2004). doi:10.1086/421043

J. Giacalone, Cosmic-ray transport and interaction with shocks. Space Sci. Rev., 117 (2011). doi:10.1007/s11214-011-9763-2

M. Gilmore, C.X. Yu, T.L. Rhodes, W.A. Peebles, Investigation of rescaled range analysis, the Hurst exponent, and long-time correlations in plasma turbulence. Phys. Plasmas **9**, 1312–1317 (2002). doi:10.1063/1.1459707

T. Goodman, S. Ahmed, S. Alberti, Y. Andrebe, C. Angioni, K. Appert, G. Arnoux, R. Belm, P. Blanchard, P. Bosshard, Y. Camenen, R. Chavan, S. Coda, I. Condrea, A. Degeling, B. Duval, P. Etienne, D. Fasel, A. Fasoli, J. Favez, I. Furno, M. Henderson, F. Hofmann, J. Hogge, J. Horacek, P. Isoz, B. Joye, A. Karpushov, I. Klimanov, P. Lavanchy, J. Lister, X. Llobet, J. Magnin, A. Manini, B. Marletaz, P. Marmillod, Y. Martin, A. Martynov, J. Mayor, J. Mlynar, J. Moret, E. Nelson-Melby, P. Nikkola, P. Paris, A. Perez, Y. Peysson, R. Pitts, A. Pochelon, L. Porte, D. Raju, H. Reimerdes, O. Sauter, A. Scarabosio, E. Scavino, S. Seo, U. Siravo, A. Sushkov, G. Tonetti, M. Tran, H. Weisen, M. Wischmeier, A. Zabolotsky, G. Zhuang, An overview of results from the TCV tokamak. Nucl. Fusion **43**(12), 1619–1631 (2003). doi:10.1088/0029-5515/43/12/008

A. Greco, P. Chuychai, W.H. Matthaeus, S. Servidio, P. Dmitruk, Intermittent MHD structures and classical discontinuities. Geophys. Res. Lett. **35**, 19111 (2008). doi:10.1029/2008GL035454

A. Greco, W.H. Matthaeus, S. Servidio, P. Chuychai, P. Dmitruk, Statistical analysis of discontinuities in solar wind ACE data and comparison with intermittent MHD turbulence. Astrophys. J. Lett. **691**, 111–114 (2009a). doi:10.1088/0004-637X/691/2/L111

A. Greco, W.H. Matthaeus, S. Servidio, P. Dmitruk, Waiting-time distributions of magnetic discontinuities: clustering or Poisson process? Phys. Rev. E **80**(4), 046401 (2009b). doi:10.1103/PhysRevE.80.046401

A. Greco, W.H. Matthaeus, R. D'Amicis, S. Servidio, P. Dmitruk, Evidence for nonlinear development of magnetohydrodynamic scale intermittency in the inner heliosphere. Astrophys. J. **749**, 105 (2012). doi:10.1088/0004-637X/749/2/105

O. Grulke, J. Terry, B. LaBombard, S. Zweben, Radially propagating fluctuation structures in the scrape-off layer of Alcator C-Mod. Phys. Plasmas **13**(1), 012306 (2006). doi:10.1063/1.2164991

K. Gustafson, P. Ricci, I. Furno, A. Fasoli, Nondiffusive suprathermal ion transport in simple magnetized toroidal plasmas. Phys. Rev. Lett. **108**(3), 035006 (2012). doi:10.1103/PhysRevLett.108.035006

S. Haaland, B.U.Ö. Sonnerup, M.W. Dunlop, E. Georgescu, G. Paschmann, B. Klecker, A. Vaivads, Orientation and motion of a discontinuity from cluster curlometer capability: minimum variance of current density. Geophys. Res. Lett. **31**, 10804 (2004). doi:10.1029/2004GL020001

T. Hada, D. Koga, E. Yamamoto, Phase coherence of MHD waves in the solar wind. Space Sci. Rev. **107**, 463–466 (2003). doi:10.1023/A:1025506124402

K. Hayashi, M. Kojima, M. Tokumaru, K. Fujiki, MHD tomography using interplanetary scintillation measurement. J. Geophys. Res. (Space Phys.) **108**, 1102 (2003). doi:10.1029/2002JA009567

B. Hnat, S.C. Chapman, G. Rowlands, Intermittency, scaling, and the Fokker-Planck approach to fluctuations of the solar wind bulk plasma parameters as seen by the WIND spacecraft. Phys. Rev. E **67**(5), 056404 (2003). doi:10.1103/PhysRevE.67.056404

B. Hnat, B.D. Dudson, R.O. Dendy, G.F. Counsell, A. Kirk, MAST Team, Characterization of edge turbulence in relation to edge magnetic field configuration in ohmic L-mode plasmas in the mega amp spherical tokamak. Nucl. Fusion **48**(8), 085009 (2008). doi:10.1088/0029-5515/48/8/085009

M. Hoppe, C.T. Russell, Whistler mode wave packets in the Earth's foreshock region. Nature **287**, 417–420 (1980). doi:10.1038/287417a0

T. Horbury, R. Wicks, C. Chen, Anisotropy in space plasma turbulence: solar wind observations. Space Sci. Rev. 172, 325–342 (2012). doi:10.1007/s11214-011-9821-9

T. Huld, A.H. Nielsen, H.L. Pécseli, J.J. Rasmussen, Coherent structures in two-dimensional plasma turbulence. Phys. Fluids B **3**, 1609–1625 (1991). doi:10.1063/1.859680

K.-J. Hwang, M.L. Goldstein, M.M. Kuznetsova, Y. Wang, A.F. Viñas, D.G. Sibeck, The first in situ observation of Kelvin-Helmholtz waves at high-latitude magnetopause during strongly dawnward interplanetary magnetic field conditions. J. Geophys. Res. (Space Phys.) **117**, 8233 (2012). doi:10.1029/2011JA017256

D. Iraji, I. Furno, A. Fasoli, C. Theiler, Imaging of turbulent structures and tomographic reconstruction of TORPEX plasma emissivity. Phys. Plasmas **17**, 122304 (2010). doi:10.1063/1.3523052

B.V. Jackson, P.P. Hick, A. Buffington, M.M. Bisi, J.M. Clover, M. Tokumaru, M. Kojima, K. Fujiki, Three-dimensional reconstruction of heliospheric structure using iterative tomography: a review. J. Atmos. Sol.-Terr. Phys. **73**, 1214–1227 (2011). doi:10.1016/j.jastp.2010.10.007

H. Johnsen, H.L. Pécseli, J. Trulsen, Conditional eddies in plasma turbulence. Phys. Fluids **30**, 2239–2254 (1987). doi:10.1063/1.866158

H. Kantz, T. Schreiber, *Nonlinear Time Series Analysis*, 2nd edn. Cambridge Nonlinear Science Series, vol. 7 (Cambridge University Press, Cambridge, 2000)

N. Katz, J. Egedal, W. Fox, A. Le, M. Porkolab, Experiments on the propagation of plasma filaments. Phys. Rev. Lett. **101**(1), 015003 (2008). doi:10.1103/PhysRevLett.101.015003

C.F. Kennel, H.E. Petschek, Limit on stably trapped particle fluxes. J. Geophys. Res. **71**, 1 (1966)

L. Kersley, S.E. Pryse, I.K. Walker, J.A.T. Heaton, C.N. Mitchell, M.J. Williams, C.A. Willson, Imaging of electron density troughs by tomographic techniques. Radio Sci. **32**(4), 1607–1621 (1997). doi:10.1029/97RS00310

Y. Khotyaintsev, S. Buchert, K. Stasiewicz, A. Vaivads, S. Savin, V.O. Papitashvili, C.J. Farrugia, B. Popielawska, Y.-K. Tung, Transient reconnection in the cusp during strongly negative IMF B_y. J. Geophys. Res. (Space Phys.) **109**, 4204 (2004). doi:10.1029/2003JA009908

J.S. Kim, R.J. Fonck, R.D. Durst, E. Fernandez, P.W. Terry, S.F. Paul, M.C. Zarnstorff, Measurements of nonlinear energy transfer in turbulence in the tokamak fusion test reactor. Phys. Rev. Lett. **79**, 841–844 (1997). doi:10.1103/PhysRevLett.79.841

J.S. Kim, D.H. Edgell, J.M. Greene, E.J. Strait, M.S. Chance, MHD mode identification of tokamak plasmas from Mirnov signals. Plasma Phys. Control. Fusion **41**, 1399–1420 (1999). doi:10.1088/0741-3335/41/11/307

Y.C. Kim, E.J. Powers, Digital bispectral analysis and its applications to nonlinear wave interactions. IEEE Trans. Plasma Sci. **PS-7**, 120–131 (1979)

A. Kirk, N. Ben Ayed, G. Counsell, B. Dudson, T. Eich, A. Herrmann, B. Koch, R. Martin, A. Meakins, S. Saarelma, R. Scannell, S. Tallents, M. Walsh, H.R. Wilson, M. Team, Filament structures at the plasma edge on MAST. Plasma Phys. Control. Fusion **48**(12B, SI), 433–441 (2006). doi:10.1088/0741-3335/48/12B/S41

K.H. Kiyani, S.C. Chapman, N.W. Watkins, Pseudononstationarity in the scaling exponents of finite-interval time series. Phys. Rev. E **79**(3), 036109 (2009). doi:10.1103/PhysRevE.79.036109

K.H. Kiyani, S.C. Chapman, F. Sahraoui, B. Hnat, O. Fauvarque, Y.V. Khotyaintsev, Enhanced magnetic compressibility and isotropic scale invariance at sub-ion larmor scales in solar wind turbulence. Astrophys. J. **763**, 10 (2013). doi:10.1088/0004-637X/763/1/10

K. Kiyani, S.C. Chapman, B. Hnat, R.M. Nicol, Self-similar signature of the active solar corona within the inertial range of solar-wind turbulence. Phys. Rev. Lett. **98**(21), 211101 (2007). doi:10.1103/PhysRevLett.98.211101

A.N. Kolmogorov, The local structure of turbulence in incompressible viscous fluid for very large Reynolds numbers. Dokl. Akad. Nauk SSSR **30**, 301–305 (1941). Reprinted in Proc. R. Soc. Lond., Ser. A **434**, 9–13 (1991)

A.N. Kolmogorov, A refinement of previous hypotheses concerning the local structure of turbulence in a viscous incompressible fluid at high Reynolds number. J. Fluid Mech. **13**, 82–85 (1962). doi:10.1017/S0022112062000518

S.I. Krasheninnikov, D.A. D'Ippolito, J.R. Myra, Recent theoretical progress in understanding coherent structures in edge and SOL turbulence. J. Plasma Phys. **74**, 679–717 (2008)

V. Kravtchenko-Berejnoi, F. Lefeuvre, V. Krasnoselskikh, D. Lagoutte, On the use of tricoherent analysis to detect non-linear wave-wave interactions. Signal Process. **42**(3), 291–309 (1995). doi:10.1016/0165-1684(94)00136-N

A. Lazurenko, G. Coduti, S. Mazouffre, G. Bonhomme, Dispersion relation of high-frequency plasma oscillations in hall thrusters. Phys. Plasmas **15**(3), 034502 (2008). doi:10.1063/1.2889424

M.P. Leubner, Z. Vörös, A nonextensive entropy approach to solar wind intermittency. Astrophys. J. **618**(1), 547 (2005). doi:10.1086/425893

R.P. Lin, Non-relativistic solar electrons. Space Sci. Rev. **16**, 189–256 (1974). doi:10.1007/BF00240886

E.T. Lundberg, P.M. Kintner, S.P. Powell, K.A. Lynch, Multipayload interferometric wave vector determination of auroral hiss. J. Geophys. Res. (Space Phys.) **117**, 2306 (2012). doi:10.1029/2011JA017037

A. Madon, T. Klinger, Spatio-temporal bifurcations in plasma drift-waves. Physica D **91**(3), 301–316 (1996). doi:10.1016/0167-2789(95)00266-9

A. Mangeney, C. Salem, P.L. Veltri, B. Cecconi, Intermittency in the solar wind turbulence and the haar wavelet transform, in *Sheffield Space Plasma Meeting: Multipoint Measurements Versus Theory*, ed. by B. Warmbein. ESA Special Pub., vol. 492 (2001), p. 53

A. Manini, J. Moret, F. Ryter, the ASDEX Upgrade Team, Signal processing techniques based on singular value decomposition applied to modulated ECH experiments. Nucl. Fusion **43**, 490–511 (2003). doi:10.1088/0029-5515/43/6/312

R.J. Maqueda, D.P. Stotler, N. Team, Intermittent divertor filaments in the national spherical torus experiment and their relation to midplane blobs. Nucl. Fusion **50**(7), 075002 (2010). doi:10.1088/0029-5515/50/7/075002

E. Marsch, C.-Y. Tu, Intermittency, non-Gaussian statistics and fractal scaling of MHD fluctuations in the solar wind. Nonlinear Process. Geophys. **4**, 101–124 (1997)

W.H. Matthaeus, M.L. Goldstein, Stationarity of magnetohydrodynamic fluctuations in the solar wind. J. Geophys. Res. (Space Phys.) **87**, 10347–10354 (1982). doi:10.1029/JA087iA12p10347

W.H. Matthaeus, M. Velli, Who needs turbulence? Space Sci. Rev. **160**(1–4), 145–168 (2011). doi:10.1007/s11214-011-9793-9

N. Meunier, J. Zhao, Observations of photospheric dynamics and magnetic fields: from large-scale to small-scale flows. Space Sci. Rev. **144**, 127–149 (2009). doi:10.1007/s11214-008-9472-7

J.A. Mier, R. Sánchez, L. García, B.A. Carreras, D.E. Newman, Characterization of nondiffusive transport in plasma turbulence via a novel Lagrangian method. Phys. Rev. Lett. **101**, 165001 (2008). doi:10.1103/PhysRevLett.101.165001

S.H. Müller, A. Diallo, A. Fasoli, I. Furno, B. Labit, G. Plyushchev, M. Podestà, F.M. Poli, Probabilistic analysis of turbulent structures from two-dimensional plasma imaging. Phys. Plasmas **13**(10), 100701 (2006). doi:10.1063/1.2351960

S.L. Musher, A.M. Rubenchik, V.E. Zakharov, Weak langmuir turbulence. Phys. Rep. **252**, 177–274 (1995). doi:10.1016/0370-1573(94)00071-A

J.F. Muzy, E. Bacry, A. Arneodo, Multifractal formalism for fractal signals: the structure-function approach versus the wavelet-transform modulus-maxima method. Phys. Rev. E **47**, 875–884 (1993). doi:10.1103/PhysRevE.47.875

C. Nardone, Multichannel fluctuation data analysis by the singular value decomposition method. Application to MHD modes in JET. Plasma Phys. Control. Fusion **34**(9), 1447 (1992). doi:10.1088/0741-3335/34/9/001

D.E. Newman, B.A. Carreras, P.H. Diamond, T.S. Hahm, The dynamics of marginality and self-organized criticality as a paradigm for turbulent transport. Phys. Plasmas **3**, 1858–1866 (1996). doi:10.1063/1.871681

R. Nguyen van yen, D. del-Castillo-Negrete, K. Schneider, M. Farge, G. Chen, Wavelet-based density estimation for noise reduction in plasma simulations using particles. J. Comput. Phys. **229**(8), 2821–2839 (2010). doi:10.1016/j.jcp.2009.12.010

S. Niedner, H.-G. Schuster, T. Klinger, G. Bonhomme, Symmetry breaking in ionization wave turbulence. Phys. Rev. E **59**, 52–59 (1999). doi:10.1103/PhysRevE.59.52

G. Paschmann, P.W. Daly (eds.), *Analysis Methods for Multi-spacecraft Data*, 2nd edn. ISSI Scientific Report SR-001 (Springer, Amsterdam, 2000). http://www.issibern.ch/publications/sr.html

G. Paschmann, P.W. Daly (eds.), *Multi-Spacecraft Analysis Methods Revisited*. ISSI Scientific Report SR-008 (Springer, Amsterdam, 2008). http://www.issibern.ch/publications/sr.html

S. Perri, A. Balogh, Characterization of transitions in the solar wind parameters. Astrophys. J. **710**, 1286–1294 (2010a). doi:10.1088/0004-637X/710/2/1286

S. Perri, A. Balogh, Stationarity in solar wind flows. Astrophys. J. **714**, 937–943 (2010b). doi:10.1088/0004-637X/714/1/937

S. Perri, G. Zimbardo, Evidence of superdiffusive transport of electrons accelerated at interplanetary shocks. Astrophys. J. Lett. **671**, 177–180 (2007). doi:10.1086/525523

S. Perri, G. Zimbardo, Superdiffusive transport of electrons accelerated at corotating interaction regions. J. Geophys. Res. (Space Phys.) **113**, 3107 (2008). doi:10.1029/2007JA012695

S. Perri, G. Zimbardo, Ion and electron superdiffusive transport in the interplanetary space. Adv. Space Res. **44**, 465–470 (2009a). doi:10.1016/j.asr.2009.04.017

S. Perri, G. Zimbardo, Ion superdiffusion at the solar wind termination shock. Astrophys. J. Lett. **693**, 118–121 (2009b). doi:10.1088/0004-637X/693/2/L118

S. Perri, G. Zimbardo, Magnetic variances and pitch-angle scattering times upstream of interplanetary shocks. Astrophys. J. **754**, 8 (2012a). doi:10.1088/0004-637X/754/1/8

S. Perri, G. Zimbardo, Superdiffusive shock acceleration. Astrophys. J. **750**, 87 (2012b)

D. Perrone, R.O. Dendy, I. Furno, G. Zimbardo, Nonclassical transport and particle-field coupling: from laboratory plasmas to the solar wind. Space Sci. Rev. (2013, accepted)

J.L. Pinçon, F. Lefeuvre, Local characterization of homogeneous turbulence in a space plasma from simultaneous measurements of field components at several points in space. J. Geophys. Res. **96**, 1789–1802 (1991)

F.M. Poli, S. Brunner, A. Diallo, A. Fasoli, I. Furno, B. Labit, S.H. Müller, G. Plyushchev, M. Podestà, Experimental characterization of drift-interchange instabilities in a simple toroidal plasma. Phys. Plasmas **13**(10), 102104 (2006). doi:10.1063/1.2356483

D.V. Reames, Particle acceleration at the Sun and in the heliosphere. Space Sci. Rev. **90**, 413–491 (1999). doi:10.1023/A:1005105831781

A. Retinò, D. Sundkvist, A. Vaivads, F. Mozer, M. André, C.J. Owen, In situ evidence of magnetic reconnection in turbulent plasma. Nat. Phys. **3**, 236–238 (2007). doi:10.1038/nphys574

C.P. Ritz, E.J. Powers, Estimation of nonlinear transfer functions for fully developed turbulence. Physica D Nonlinear Phenom. **20**, 320–334 (1986). doi:10.1016/0167-2789(86)90036-9

C.P. Ritz, E.J. Powers, T.L. Rhodes, R.D. Bengtson, K.W. Gentle, H. Lin, P.E. Phillips, A.J. Wootton, D.L. Brower, N.C. Luhmann Jr., W.A. Peebles, P.M. Schoch, R.L. Hickok, Advanced plasma fluctuation analysis techniques and their impact on fusion research. Rev. Sci. Instrum. **59**, 1739–1744 (1988). doi:10.1063/1.1140098

F. Sahraoui, Diagnosis of magnetic structures and intermittency in space-plasma turbulence using the technique of surrogate data. Phys. Rev. E **78**(2), 026402 (2008). doi:10.1103/PhysRevE.78.026402

F. Sahraoui, G. Belmont, L. Rezeau, N. Cornilleau-Wehrlin, J.L. Pinçon, A. Balogh, Anisotropic turbulent spectra in the terrestrial magnetosheath as seen by the cluster spacecraft. Phys. Rev. Lett. **96**(7), 075002 (2006). doi:10.1103/PhysRevLett.96.075002

F. Sahraoui, M.L. Goldstein, G. Belmont, A. Roux, L. Rezeau, P. Canu, P. Robert, N. Cornilleau-Wehrlin, O. Le Contel, T. Dudok de Wit, J. Pinçon, K. Kiyani, Multi-spacecraft investigation of space turbulence: lessons from cluster and input to the cross-scale mission. Planet. Space Sci. **59**, 585–591 (2011). doi:10.1016/j.pss.2010.06.001

T. Schreiber, A. Schmitz, Surrogate time series. Physica D **142**, 346–382 (2000). doi:10.1016/S0167-2789(00)00043-9

A. Shalchi, I. Kourakis, A new theory for perpendicular transport of cosmic rays. Astron. Astrophys. **470**, 405–409 (2007). doi:10.1051/0004-6361:20077260

C. Shen, X. Li, M. Dunlop, Z.X. Liu, A. Balogh, D.N. Baker, M. Hapgood, X. Wang, Analyses on the geometrical structure of magnetic field in the current sheet based on cluster measurements. J. Geophys. Res. (Space Phys.) **108**, 1168 (2003). doi:10.1029/2002JA009612

M.F. Shlesinger, G.M. Zaslavsky, J. Klafter, Strange kinetics. Nature **363**, 31–37 (1993). doi:10.1038/363031a0

B.U.Ö. Sonnerup, M. Scheible, Minimum and maximum variance analysis. In: *ISSI Scientific Reports Series* vol. 1 (1998), pp. 185–220

L. Sorriso-Valvo, V. Carbone, P. Veltri, G. Consolini, R. Bruno, Intermittency in the solar wind turbulence through probability distribution functions of fluctuations. Geophys. Res. Lett. **26**, 1801–1804 (1999). doi:10.1029/1999GL900270

L. Sorriso-Valvo, V. Carbone, P. Giuliani, P. Veltri, R. Bruno, V. Antoni, E. Martines, Intermittency in plasma turbulence. Planet. Space Sci. **49**, 1193–1200 (2001). doi:10.1016/S0032-0633(01)00060-5

J. Souček, T. Dudok de Wit, V. Krasnoselskikh, A. Volokitin, Statistical analysis of nonlinear wave interactions in simulated langmuir turbulence data. Ann. Geophys. **21**, 681–692 (2003)

U. Stroth, F. Greiner, C. Lechte, N. Mahdizadeh, K. Rahbarnia, M. Ramisch, Study of edge turbulence in dimensionally similar laboratory plasmas. Phys. Plasmas **11**, 2558–2564 (2004). doi:10.1063/1.1688789

D. Sundkvist, V. Krasnoselskikh, P.K. Shukla, A. Vaivads, M. André, S. Buchert, H. Rème, In situ multi-satellite detection of coherent vortices as a manifestation of Alfvénic turbulence. Nature **436**, 825–828 (2005). doi:10.1038/nature03931

J. Svensson, A. Werner, J. Contributors, Current tomography for axisymmetric plasmas. Plasma Phys. Control. Fusion **50**(8), 085002 (2008). doi:10.1088/0741-3335/50/8/085002

J. Terry, S. Zweben, K. Hallatschek, B. LaBombard, R. Maqueda, B. Bai, C. Boswell, M. Greenwald, D. Kopon, W. Nevins, C. Pitcher, B. Rogers, D. Stotler, X. Xu, Observations of the turbulence in the scrape-off-layer of Alcator C-Mod and comparisons with simulation. Phys. Plasmas **10**(5, Part 2), 1739–1747 (2003). doi:10.1063/1.1564090

C. Theiler, I. Furno, P. Ricci, A. Fasoli, B. Labit, S.H. Mueller, G. Plyushchev, Cross-field motion of plasma blobs in an open magnetic field line configuration. Phys. Rev. Lett. **103**, 065001 (2009). doi:10.1103/PhysRevLett.103.065001

A. Tjulin, J.-L. Pinçon, F. Sahraoui, M. André, N. Cornilleau-Wehrlin, The k-filtering technique applied to wave electric and magnetic field measurements from the cluster satellites. J. Geophys. Res. (Space Phys.) **110**, 11224 (2005). doi:10.1029/2005JA011125

C. Torrence, G.P. Compo, A practical guide to wavelet analysis. Bull. Am. Meteorol. Soc. **79**, 61–78 (1998). doi:10.1175/1520-0477(1998)079<0061:APGTWA>2.0.CO;2

C.-Y. Tu, E. Marsch, MHD structures, waves and turbulence in the solar wind: observations and theories. Space Sci. Rev. **73**, 1–2 (1995)

G.R. Tynan, A. Fujisawa, G. McKee, A review of experimental drift turbulence studies. Plasma Phys. Control. Fusion **51**(11), 113001 (2009). doi:10.1088/0741-3335/51/11/113001

M. Škorić, M. Rajković, Characterization of intermittency in plasma edge turbulence. Contrib. Plasma Phys. **48**, 37–41 (2008). doi:10.1002/ctpp.200810006

J.C. van den Berg (ed.), *Wavelets in Physics* (Cambridge University Press, Cambridge, 2004)

D. Vassiliadis, Systems theory for geospace plasma dynamics. Rev. Geophys. **44**, 2002 (2006). doi:10.1029/2004RG000161

J. Vega, A. Pereira, A. Portas, S. Dormido-Canto, G. Farias, R. Dormido, J. Sánchez, N. Duro, M. Santos, E. Sanchez, G. Pajares, Data mining technique for fast retrieval of similar waveforms in fusion massive databases. Fusion Eng. Des. **83**(1), 132–139 (2008). doi:10.1016/j.fusengdes.2007.09.011

P. Veltri, MHD turbulence in the solar wind: self-similarity, intermittency and coherent structures. Plasma Phys. Control. Fusion **41**, 787–795 (1999). doi:10.1088/0741-3335/41/3A/071

P. Veltri, A. Mangeney, Scaling laws and intermittent structures in solar wind MHD turbulence, in *Solar Wind Nine*, ed. by S.R. Habbal, R. Esser, J.V. Hollweg, P.A. Isenberg. American Institute of Physics Conference Series, vol. 471 (1999), p. 543

J. Vogt, A. Albert, O. Marghitu, Analysis of three-spacecraft data using planar reciprocal vectors: methodological framework and spatial gradient estimation. Ann. Geophys. **27**, 3249–3273 (2009)

U. von Toussaint, Bayesian inference in physics. Rev. Mod. Phys. **83**, 943–999 (2011). doi:10.1103/RevModPhys.83.943

A. Wernik, Methods of data analysis for resolving nonlinear phenomena, in *Modern Ionospheric Science*, ed. by H. Kohl, R. Rüster, K. Schlegel (European Geophysical Society, Katlenburg-Lindau, 1996), pp. 322–345

T. Windisch, O. Grulke, T. Klinger, Radial propagation of structures in drift wave turbulence. Phys. Plasmas **13**(12), 122303 (2006). doi:10.1063/1.2400845

C.X. Yu, M. Gilmore, W.A. Peebles, T.L. Rhodes, Structure function analysis of long-range correlations in plasma turbulence. Phys. Plasmas **10**, 2772–2779 (2003). doi:10.1063/1.1583711

G.M. Zaslavskii, R.Z. Sagdeev, D.K. Chaikovskii, A.A. Chernikov, Chaos and two-dimensional random walk in periodic and quasi-periodic fields. J. Exp. Theor. Phys. **95**, 1723–1733 (1989)

G.M. Zaslavsky, D. Stevens, H. Weitzner, Self-similar transport in incomplete chaos. Phys. Rev. E **48**, 1683–1694 (1993). doi:10.1103/PhysRevE.48.1683

G. Zimbardo, Anomalous particle diffusion and Lévy random walk of magnetic field lines in three-dimensional solar wind turbulence. Plasma Phys. Control. Fusion **47**, 755–767 (2005). doi:10.1088/0741-3335/47/12B/S57

G. Zimbardo, P. Pommois, P. Veltri, Superdiffusive and subdiffusive transport of energetic particles in solar wind anisotropic magnetic turbulence. Astrophys. J. **639**, 91–94 (2006). doi:10.1086/502676

G. Zimbardo, A. Greco, L. Sorriso-Valvo, S. Perri, Z. Vörös, G. Aburjania, K. Chargazia, O. Alexandrova, Magnetic turbulence in the geospace environment. Space Sci. Rev. **156**, 89–134 (2010). doi:10.1007/s11214-010-9692-5

G. Zimbardo, S. Perri, P. Pommois, P. Veltri, Anomalous particle transport in the heliosphere. Adv. Space Res. **49**, 1633–1642 (2012). doi:10.1016/j.asr.2011.10.022

G. Zumofen, J. Klafter, Scale-invariant motion in intermittent chaotic systems. Phys. Rev. E **47**, 851–863 (1993). doi:10.1103/PhysRevE.47.851

S.J. Zweben, Search for coherent structure within tokamak plasma turbulence. Phys. Fluids **28**, 974–982 (1985). doi:10.1063/1.865069

S.J. Zweben, J.A. Boedo, O. Grulke, C. Hidalgo, B. La Bombard, R.J. Maqueda, P. Scarin, J.L. Terry, Edge turbulence measurements in toroidal fusion devices. Plasma Phys. Control. Fusion **49**, 1–23 (2007). doi:10.1088/0741-3335/49/7/S01

S.J. Zweben, J.L. Terry, B. LaBombard, M. Agostini, M. Greenwald, O. Grulke, J.W. Hughes, D.A. D'Ippolito, S.I. Krasheninnikov, J.R. Myra, D.A. Russell, D.P. Stotler, M. Umansky, Estimate of convective radial transport due to SOL turbulence as measured by GPI in Alcator C-Mod. J. Nucl. Mater. **415**(1, S), 463–466 (2011). doi:10.1016/j.jnucmat.2010.08.018

DOI 10.1007/978-1-4899-7413-6_22
Reprinted from *Space Science Reviews* Journal, DOI 10.1007/s11214-013-9963-z

Laboratory astrophysics: Investigation of planetary and astrophysical maser emission

R. Bingham · D.C. Speirs · B.J. Kellett · I. Vorgul · S.L. McConville · R.A. Cairns · A.W. Cross · A.D.R. Phelps · K. Ronald

Received: 29 November 2012 / Accepted: 28 January 2013 / Published online: 8 March 2013

Abstract This paper describes a model for cyclotron maser emission applicable to planetary auroral radio emission, the stars UV Ceti and CU Virginus, blazar jets and astrophysical shocks. These emissions may be attributed to energetic electrons moving into convergent magnetic fields that are typically found in association with dipole like planetary magnetospheres or shocks. It is found that magnetic compression leads to the formation of a velocity distribution having a horseshoe shape as a result of conservation of the electron magnetic moment. Under certain plasma conditions where the local electron plasma frequency ω_{pe} is much less than the cyclotron frequency ω_{ce} the distribution is found to be unstable to maser type radiation emission. We have established a laboratory-based facility that has verified many of the details of our original theoretical description and agrees well with numerical simulations. The experiment has demonstrated that the horseshoe distribution produces cyclotron emission at a frequency just below the local electron cyclotron frequency, with polarisation close to X-mode and propagating nearly perpendicularly to the electron beam motion. We discuss recent developments in the theory and simulation of the instability including addressing radiation escape problems, and relate these to the laboratory, space, and astrophysical observations. The experiments showed strong narrow band EM emissions at frequencies just below the cold-plasma cyclotron frequency as predicted by the theory. Measurements of the conversion efficiency, mode and spectral content were in close agreement with the predictions of numerical simulations undertaken using a particle-in-cell code and also with satellite observations confirming the horseshoe maser as an important emission mechanism in geophysical/astrophysical plasmas. In each case we address how the radiation can escape the plasma without suffering strong absorption at the second harmonic layer.

R. Bingham · B.J. Kellett
Central Laser Facility, STFC Rutherford Appleton Laboratory, Chilton, OX11 0QX, UK

R. Bingham · D.C. Speirs (✉) · S.L. McConville · A.W. Cross · A.D.R. Phelps · K. Ronald
Department of Physics, SUPA, University of Strathclyde, Glasgow, G4 0NG, UK
e-mail: david.c.speirs@strath.ac.uk

I. Vorgul · R.A. Cairns
School of Mathematics and Statistics, University of St. Andrews, St. Andrews, KY16 9SS, UK

Keywords Auroral kilometric radiation · Cyclotron maser radiation · Plasma instabilities · Blazar jets

1 Introduction

At least five sources of intense, non-thermal planetary radio emission are known in our solar system. The Earth, Jupiter, Saturn, Uranus and Neptune all produce strong maser emission in the kilometre to decametre wavelength range (Zarka 1992). In astrophysical environments the cyclotron maser instability is an important mechanism that provides an explanation for observations of intense microwave radiation (Treumann 2006). Radio emission from stars with a dipole magnetic field UV Ceti (Benz et al. 1998; Kellett et al. 2002) and Cu Virginus (Trigilio et al. 2000; Hatzes 1997; Kellett et al. 2007; Lo et al. 2012), blazar jets (Begelman et al. 2005; Treumann 2006) and shocks (Bingham et al. 2003) are also thought to be associated with maser emission. Cyclotron maser emission from astrophysical objects was first investigated by Twiss (1958). The radiation is associated with electron beams accelerated into increasing magnetic fields where the local electron plasma frequency ω_{pe} is much less than the cyclotron frequency ω_{ce} (see Fig. 1 for the terrestrial auroral case). In the low density regime, the growth rate of the R–X mode is inversely proportional to the density, while the growth rate of the ordinary mode is proportional to the density. This explains why the RX mode is the dominant feature of radiation in regions where $\omega_{pe} \ll \omega_{ce}$ (Melrose and Dulk 1982). In the case of the Earth's auroral radio emission region, the ratio $\omega_{pe}/\omega_{ce} \sim 0.01$, for blazar jets this ratio can be even smaller (Begelman et al. 2005; Treumann 2006). The idea that planetary radio emission is the result of a maser instability has been current for many years (Wu and Lee 1979; Winglee and Pritchett 1986; Melrose 1986; Louarn et al. 1990; Louarn and Le Queau 1996; Delory et al. 1998; Ergun et al. 2000) with several schemes having been proposed. The

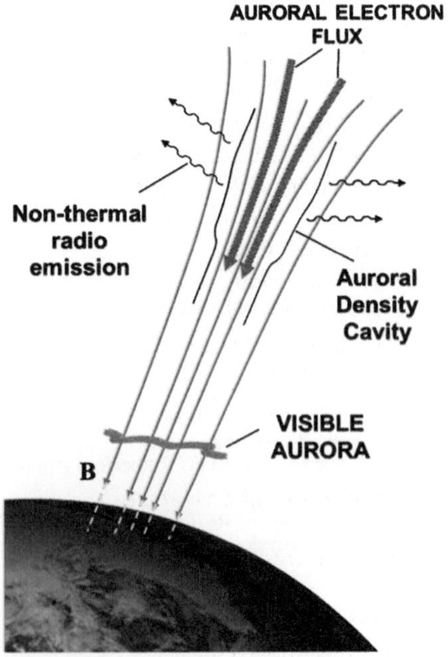

Fig. 1 Diagrammatic representation of the terrestrial auroral process

electron loss-cone distribution (Wu and Lee 1979; Winglee and Pritchett 1986) was suggested as the population inversion required for maser emission. However there is little or no evidence (Delory et al. 1998; Ergun et al. 2000) of the loss-cone distribution in the satellite data to support this assertion and as pointed out by Melrose (1998), for the emission to be explained by a loss-cone instability it has to be supposed that all observations are of a saturated state after the loss-cone has filled by pitch angle diffusion. The loss-cone instability was also questioned as the source of terrestrial auroral kilometric radiation (AKR) by the results of particle-in-cell simulations conducted by Pritchett (Pritchett 1986; Pritchett et al. 1999), based on observed electron distribution functions. These simulations show that the loss-cone instability is not a particularly efficient process. Observations from the Viking spacecraft (Louarn et al. 1990; Roux et al. 1993) in the Earth's auroral magnetosphere led to the suggestion that the energy source for AKR is instead a population of downwards accelerated electrons with a large perpendicular velocity produced by a combination of parallel accelerating electric field and converging magnetic field lines. The electron distribution that is formed by beam electrons moving into an increasing magnetic field takes the form of a horseshoe shape. These earlier observations have been backed up by more recent data from the Fast Auroral SnapshoT (FAST) satellite (Delory et al. 1998; Ergun et al. 2000; Pritchett et al. 2002) that clearly demonstrate a strong correlation between the electromagnetic wave emission and the occurrence of the horseshoe distribution. A number of authors (Delory et al. 1998; Ergun et al. 2000; Pritchett et al. 1999, 2002; Bingham and Cairns 2000; Bingham et al. 2004; Vorgul et al. 2005; Cairns et al. 2005; Speirs et al. 2005, 2008; Ronald et al. 2008a, 2008b; McConville et al. 2008; Gillespie et al. 2008) have recently focused on the horseshoe distribution as the main source of AKR and demonstrated a robust cyclotron maser instability with a large spatial growth rate (Bingham and Cairns 2000; Bingham et al. 2004; Vorgul et al. 2005), much larger than for a loss-cone distribution.

Other characteristics that are associated with the horseshoe maser but not the loss-cone maser are high brightness temperatures $>10^{14}$ K, continuous emission and high efficiency of order 1 % (Gurnett 1974). Given the prevalence of converging magnetic fields in astrophysics it has been suggested that the horseshoe maser can explain radio emission from stars with a dipole magnetic field (Benz et al. 1998; Kellett et al. 2002, 2007; Trigilio et al. 2000; Lo et al. 2012), blazar jets (Begelman et al. 2005) and shocks (Bingham et al. 2003). Since the dispersion relation describing the electromagnetic wave only depends on the factor by which the magnetic field increases and on ratios of the plasma and cyclotron frequencies, it is possible to scale the effect to laboratory dimensions. An experiment was designed and built to test the different cyclotron maser instabilities driven by loss cone and horseshoe distributions—a difficult task in the space environment. The experimental apparatus devised consisted of an electron beam injected into the convergent/fringing field of a magnetic solenoid configuration, creating a horseshoe distribution in velocity space similar to those observed by spacecraft (Delory et al. 1998; Ergun et al. 2000; Louarn et al. 1990; Roux et al. 1993; Pritchett 1986) in the auroral region, Fig. 2. An important issue for consideration is how the radiation can escape from each of the objects discussed in this paper. For each case we outline possible escape mechanisms.

2 Observations of Cyclotron-Maser Emission

2.1 Satellite Observations in the Auroral Magnetospheric Region

The most convenient source for observation is Earth, and terrestrial AKR has been observed in detail by numerous satellite missions. Such observations from within the source

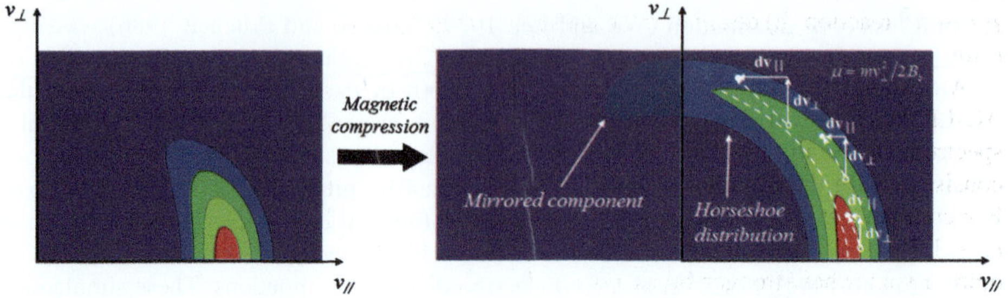

Fig. 2 Illustrative diagram showing the formation of a horseshoe shaped velocity distribution in an electron beam subject to magnetic compression

region indicate that AKR is generated at high altitudes ($\sim$1.5 $\rightarrow$ 3 Earth radii) in cavities of low background plasma density aligned along the auroral magnetic field. From the first images of the terrestrial auroral electron distribution function (DE-1 satellite), a definite crescent, or what we call a horseshoe, can be seen (Menietti and Burch 1985). Later observations by the Viking and FAST satellites provided much clearer images where one can see an obvious, highly populated horseshoe (Delory et al. 1998; Ergun et al. 2000; Pritchett et al. 2002). The main results of the satellite observations relevant to cyclotron maser emission are: the existence of fast electrons of approximately the same energy moving into a convergent magnetic field, an observed horseshoe shaped electron momentum distribution, intense kilometric radiation nearly perpendicular to the magnetostatic field, a well-defined spectral output at around the electron cyclotron frequency, X-mode polarisation of the emissions within the source region and an estimated emission efficiency of about 1 % (Menietti and Burch 1985; Gurnett 1974; Mutel et al. 2008; Menietti et al. 2011).

An important issue for consideration is how the radiation can escape from the auroral magnetospheric plasma into space. The cyclotron maser mechanism generates radiation propagating almost perpendicular to the magnetic field in the extraordinary mode. Melrose and Dulk (1982) and Melrose (1999) argue that strong absorption will take place when the radiation encounters the second harmonic absorption layer. Second harmonic absorption is maximum when the radiation propagates perpendicular to the magnetic field and goes to zero as θ approaches 0°, inferring that for radiation propagating at small angles to the magnetic field attenuation is small (Melrose 1986). Recent work by Mutel et al. (2008) using data obtained by the four spacecraft Cluster mission show that the radiation from the auroral region undergoes strong refraction, caused by the density inhomogeneity in the magnetospheric plasma with altitude (Mutel et al. 2008; Menietti et al. 2011). Mutel et al. have shown that the ray paths due to refraction are propagating at angles with respect to the magnetic field of between 10° and 20° well before they encounter the second harmonic layer, with much reduced attenuation at these angles. Both simulations (Speirs et al. 2010) and observations (Menietti et al. 2011; Mutel et al. 2008) suggest that the radiation is generated at a backward angle of a few degrees to the beam direction and not strictly perpendicular. This serves as a suitable precursor to upward refraction with respect to the decreasing background plasma density with increasing altitude as observed by Menietti et al. (2011) and Mutel et al. (2008).

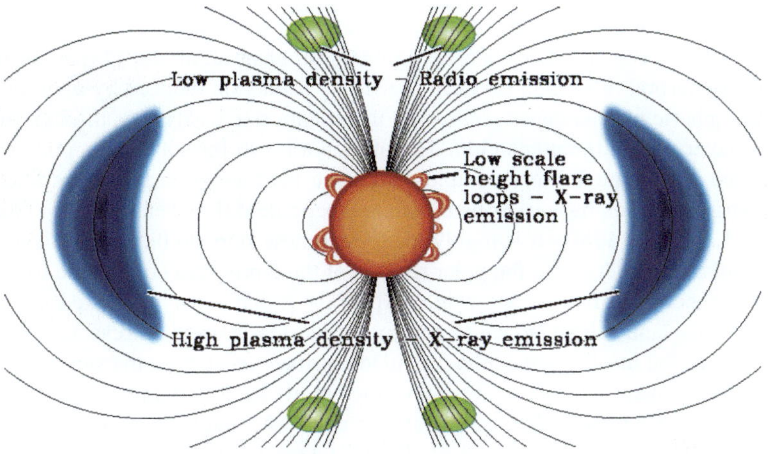

Fig. 3 Diagrammatic representation of radiation emission from the flare star UV-Ceti (Kellett et al. 2002)

2.2 Stellar Observations

Prompted by the first radio resolved image of the star UV Ceti (Benz et al. 1998), we undertook a review of the X-ray and radio observations of active late-type stars (in general), and UV Ceti, in particular. One of the aspects of the observations that we particularly noted was the highly polarised (nearly 100 % right-hand circular polarised along the direction of propagation) radio outbursts of UV Ceti. We also noted the high degree of correlation observed between the X-ray and radio luminosity of active late-type stars. Taken together, these observations strongly suggested that a new radio emission mechanism was responsible for most (and possibly all) of the radio emission of active late-type stars.

The first resolved image from UV Ceti (Benz et al. 1998) showed that the radio emission was emerging from the poles of the star. In the analysis by Benz et al. they estimated that the plasma density was $<10^8$ cm^{-3} giving a plasma frequency of $<10^8$ Hz, i.e. much less than the observed emission frequency of 8.33 GHz. Combined with a typical magnetic flux density of order 1000 Gauss (Vogt 1980), we were prompted to develop a cyclotron maser emission model driven by an electron beam entering the polar converging magnetic region of the star (Kellett et al. 2002; Bingham et al. 2001). The fact that this mechanism could easily generate the observed radio flux from these stars with keV energy electrons could then also help explain the X-ray/radio correlation. X-ray emission also implies keV electrons, so it was possible or even likely that the same electrons were responsible for the X-ray and radio emission—therefore easily explaining the observed correlation (Kellett et al. 2002). Our model for UV Ceti (Bingham et al. 2001) is summarised in Fig. 3. We adopt a dipole field geometry as suggested by the observations. Similar to the planetary case, a plasma density gradient would refract the radiation out of the plane perpendicular to the magnetic field resulting in propagation close to the magnetic field direction before encountering the second harmonic absorption layer, allowing the radiation to escape with little absorption. A similar dipole model is also invoked for CU Virginis where we see pulsed radiation signatures that very much suggest a continuous radiation source (Kellett et al. 2007; Lo et al. 2012) but which appears pulsed due to the rotation of the source across the Earth's plane. This has been coined by Trigilio the "lighthouse effect" reminiscent of pulsars but which we could call pulstars (Trigilio et al. 2000).

We conclude that maser emission from young active stars with a strong dipole-like magnetic field (Benz et al. 1998; Kellett et al. 2002; Bingham et al. 2001) may be similar to the planetary auroral models in terms of the strong refraction induced by altitude variation in magnetospheric plasma density, allowing the radiation to easily pass through the second harmonic layer with little attenuation. This may not be so for solar radio emission which is associated with a much more complex magnetic field structure than stars like UV-Ceti and CU Virginus which have a strong dipole magnetic field (Benz et al. 1998; Hatzes 1997; Borra and Landstreet 1980). It is highly likely that these stars do indeed emit cyclotron radiation similar to planets from the polar regions of the dipole that correspond to the auroral regions of planets. In this case we would expect the radiation to be emitted initially close to perpendicular and refract upwards due to the plasma density decreasing with altitude. A plasma emission mechanism cannot be ruled out for stars such as the sun that have more complicated magnetic field structures (Thejappa et al. 2012; Hillan et al. 2012), however stars such as UV-Ceti and CU-Virginus rotate much faster and have strong dipole magnetic fields of order 1000 Gauss (Benz et al. 1998; Hatzes 1997; Borra and Landstreet 1980; Vogt 1980).

Stellar and planetary cyclotron maser emission is a significant source of information on planetary/stellar magnetospheres. For offset magnetic/rotational poles, it can also allow us to study the angular momentum evolution of stars and planets in order to detect variations in rotational period (Pyper et al. 1998; Adelman et al. 2001). The common features of such astrophysical radio sources at any scale are the presence of a strong dipole or converging magnetic field, the existence of energetic electrons and observed, highly circularly polarised radio emissions at a range of frequencies correlating with the source region range in electron cyclotron frequency (Treumann 2006; Zarka 1992)

2.3 Blazar Jets and Shocks

Relativistic jets (or blazar jets when directed towards the Earth) are beam-like linear features observable over a broad range of frequencies and generated perpendicular to the accretion disc of super massive black holes (Meier et al. 2001; Nemmen et al. 2012). They can extend over intergalactic distances (many thousands of light years) and have been observed to generate highly non-thermal radio emission at frequencies ranging from a few to 100's of GHz. It has been suggested that a cyclotron-maser instability may be responsible for the generation of these emissions within the low density ($\omega_{ce} > \omega_{pe}$), magnetised relativistic electron population of the jet (Bingham et al. 2003; Begelman et al. 2005; Treumann 2006). It is believed that small scale magnetic mirrors/convergent flux tubes may be formed within the jet via hydrodynamic instabilities or shocks, providing the means of generating the required electron velocity distribution (Bingham et al. 2003; Begelman et al. 2005).

The magnetic guide fields within a blazar jet are believed to be very strong (O'Sullivan and Gabuzda 2009) and for the highly energetic electron population within the jet this can lead to large relativistic electron cyclotron frequencies in the 100's of GHz range such that $\omega_{ce} \gg \omega_{pe}$ (Begelman et al. 2005). The values used by Begelman, Ergun and Rees for the ratio of cyclotron frequency to plasma frequency indicate that the efficiency of the emission process for blazar jets could be an order of magnitude greater than for the planetary auroral case. This forms part of the argument used by Begelmann, Ergun and Rees to conclude that the radiation could indeed be due to the cyclotron-maser instability driven by an electron horseshoe/ring distribution. Previous work has shown how a suitable ring-type distribution may be generated directly by energisation of particles perpendicular the magnetostatic field via the surfatron mechanism (Katsouleas and Dawson 1983). We have also previously proposed a scheme by which electrons may be accelerated and subject to magnetic compression within a magnetised collisionless shock, generating a suitable crescent or horseshoe-type velocity distribution (Bingham et al. 2003). Both such mechanisms are believed to be viable within the highly magnetised, turbulent plasma of a blazar jet, with small-scale magnetic mirrors and field aligned currents expected to occur in association with quasi-perpendicular collisionless shocks (Begelman et al. 2005). Counterstreaming ion and ring distributions generated at such shocks can excite lower hybrid waves via the modified two-stream instability that are capable of accelerating electrons to high energies, parallel to the magnetic field (McClements et al. 1993). These high energy tail distributions can then be subject to magnetic compression when moving from the upstream to downstream region of the magnetised collisionless shock, forming a horseshoe-type distribution suitable for driving the cyclotron-maser instability.

Propagation and escape of cyclotron-maser radiation from a blazar jet has been considered in some detail (Begelman et al. 2005; Treumann 2006). Among the various factors debated, second harmonic cyclotron absorption and synchrotron absorption are potentially the most significant impediments. These can however be suitably accounted for (Begelman et al. 2005) in the case of a blazar jet, where the second harmonic absorption layer is considered to occur transversely with respect to the magnetostatic field (Begelman et al. 2005). Begelman et al. (2005) have shown as the generated X-mode radiation propagates radially outwards, where the field can be approximated as dropping off linearly as $B = B_m r_m / r$ with B_m and r_m the magnetic flux density and radius at the maser emission source respectively. Begelman et al. (2005) show that the thickness of the second harmonic layer can be approximated as $\sim r_m \omega_{pe}/\omega_{ce}$ which is relatively thin and results in attenuation of $\sim$10 %. Another factor not previously considered is the potential for refraction of the generated radiation due to the plasma density gradient associated with quasi-perpendicular shock itself. Akin to our previous consideration for the terrestrial auroral case, this could result in R-mode like radiation propagation and an associated reduction in cyclotron-wave coupling efficiency for second harmonic absorption in a layer parallel to the magnetic field.

3 Theory and Simulations

3.1 Kinetic Analysis

An analytic form of the electron distribution function with different opening angles, energy spread and density ratios between the hot component making up the horseshoe part and a background Maxwellian component is used in a dispersion relation for the R–X mode which is easily obtainable from the susceptibility tensor given by Stix (1992). We shall assume that the frequency is close to the electron cyclotron frequency, and also assume that the Larmor radius is much less than the wavelength for typical electron velocities. This latter condition means that we need only consider the susceptibility to lowest order in $k_\perp v_\perp/\Omega_e$. If we neglect all but the zero order terms we get the cold plasma result. To a first approximation we need only take account of the velocity distribution of the electrons in the resonant integral which involves $1/(\omega - \Omega_e)$ where Ω_e is the relativistic electron cyclotron frequency $eB/\gamma m_e$ with e the electron charge, B the magnetic field γ the Lorenz factor and m_e the electron rest mass. In this resonant term we must take account of the relativistic shift of the cyclotron frequency, since this picks out a particular group of resonant electrons and produces damping or growth of the wave. In terms of momentum p we have

$$\Omega_e = \Omega_{e0}\left(1+\frac{p^2}{m^2c^2}\right)^{-\frac{1}{2}} \approx \Omega_{e0}\left(1-\frac{1}{2}\frac{p^2}{m^2c^2}\right) \tag{1}$$

where Ω_{e0} is the nonrelativistic electron cyclotron frequency. For the real part of the resonant integral we can simply take the cold plasma value. Although this goes as $1/(\omega - \Omega_{e0})$ and appears to be near singular at the resonance, the $1/(\omega - \Omega_{e0})$ factors in the real part of the dispersion relation cancel out and it behaves quite smoothly in the vicinity of the cyclotron frequency. It is not crucial to include small corrections to the cyclotron frequency in the real part of the dispersion relation. The refractive index n for the R–X mode is given by

$$n^2 = \varepsilon_\perp - \frac{\varepsilon_{xy}^2}{\varepsilon_\perp} \tag{2}$$

and considering the dielectric tensor elements $\varepsilon_\perp$ and ε_{xy} with the approximations we have described, an expression for the perturbation in refractive index η due to the imaginary component can be obtained (Pritchett et al. 2002)

$$2n\delta n \approx -i\alpha\frac{(\omega-\Omega_{e0})^2(2\Omega_{e0}^2-\omega_p^2)^2}{\omega_p^4\Omega_{e0}^2} \tag{3}$$

where

$$\alpha = \frac{1}{4}\frac{\omega_{pe}^2}{\Omega_{e0}^2}2\pi^2m^2c^2\int_{-1}^{1}d\mu(1-\mu^2)p^2\left(\frac{\partial f_e}{\partial p}-\frac{\mu}{p}\frac{\partial f_e}{\partial\mu}\right)|_{p=p_0}$$

is represented in spherical polar coordinates (p,μ,ϕ) and $\mu = \cos\vartheta = p_{//}/p$ replaces the usual angle θ. $p_0 = mc(2(\Omega_{e0}-\omega)/\Omega_{e0})^{1/2}$ is the resonant momentum.

Substituting the general form of the horseshoe distribution function $f_0(p,\mu) = F(p)$ $g(\mu)$ into α gives

$$\alpha = \frac{1}{4}\frac{\omega_p^2}{\Omega_{e0}^2}2\pi^2m^2c^2p^2\left(\underset{(1)}{P\frac{\partial F}{\partial p}} + \underset{(2)}{Q\frac{F}{p}}\right)\Bigg|_{p=p_0}$$

$$\text{where}\begin{cases} P = \int_{-1}^{1}(1-\mu^2)g(\mu)d\mu, \\ Q = \int_{-1}^{1}(1-3\mu^2)g(\mu)d\mu \end{cases} \tag{4}$$

Term (1) is destabilizing, resulting in emission of waves if the gradient of F is positive at the resonant momentum. The second term is negative if g is strongly localised around $\mu = 1$ and goes to zero if g becomes uniform on the interval $[-1, 1]$. We might, therefore, have a scenario in which a beam moving down the field line with small perpendicular spread is stable. As it moves into a region of higher magnetic field and spreads into a wider horseshoe it can then become unstable and trigger the emission of AKR.

To illustrate the behaviour we consider the case of a horseshoe distribution centred at $p_\| = 0.1m_ec$, typical of the primary auroral electrons, moving down the field lines within a lower density Maxwellian plasma with a thermal temperature of 312 eV. Some results suggest that the background Maxwellian may even be absent giving rise to the auroral density cavity. The absence of a low density background Maxwellian distribution does not change the characteristics of the instability, in fact it promotes it. We solve Eq. (3) to obtain the

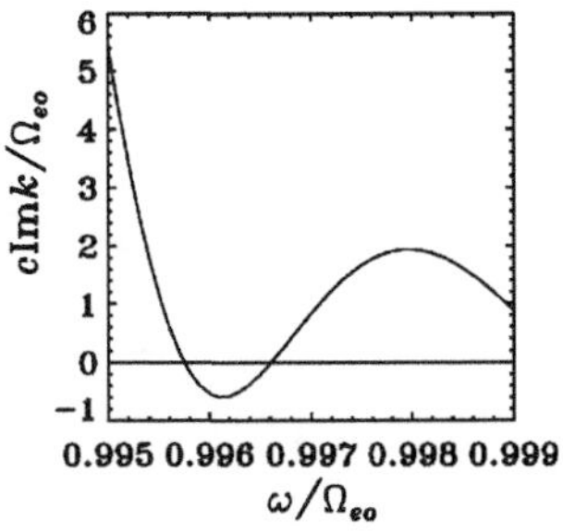

Fig. 4 Spatial growth rate $\mathrm{Im}\,k_\perp = \gamma_\perp$ normalised to Ω_{e0}/c as a function of frequency

perpendicular spatial growth rate $\gamma_\perp$ shown in Fig. 4. A typical convective growth length across the magnetic field $L_c = 2\pi/\mathrm{Im}\,k_\perp$ is 10λ for $\mu_0 = 0.5$, thermal width $= 0.02 m_e c$ and $n_h = 0.66 n_e$ and 6λ for $\mu_0 = 0.75$, thermal width $= 0.05 m_e c$ and $n_h = 0.9 n_e$ both for a cyclotron frequency of 440 kHz. These convective growth distances are 6.8 km and 4 km, respectively, allowing for many e-foldings within the auroral density cavity, which has a latitudinal width of about 100 km (Strangeway et al. 1998). We find that the growth rate decreases for increasing μ_0 and increasing thermal width of the horseshoe distribution and increasing density. The bandwidth is also extremely narrow, estimated from Fig. 4 to be of order 0.5 % or around 200 Hz, also in agreement with observations.

In a previous analysis we considered the problem of a ring distribution, unstable to a cyclotron maser instability, in a plasma with a magnetic field gradient (Cairns et al. 2008). This is reminiscent of electron distributions found at shocks and blazar jets (Bingham et al. 2003; Begelman et al. 2005). We have shown that it is possible to have a localised unstable region around the cyclotron resonance with waves radiating outwards from the beam.

3.2 PiC Simulations

For the purpose of simulating the horseshoe distribution formation and any subsequent electromagnetic interaction, the 2D axisymmetric version of the finite-difference time domain PiC (Particle-in-cell) code KARAT was used. KARAT represents the electric and magnetic fields in a simulated geometry as a summation of a static component and a time-varying component (Tarakanov 1992). The time varying electromagnetic fields generated by charges and currents within the simulation are governed by Maxwell's equations (specifically, Ampere's law and Faraday's law). The motion of PiC particles within these electromagnetic fields are governed by the relativistic Lorentz force equation, with the PiC particle current density at any point within the simulation geometry determined by the PiC method (Birdsall and Langdon 1985). The 2D axisymmetric version of KARAT in particular allows indirect observation of the distribution of particles in the transverse plane of motion. The code calculates the PiC electrons radial and azimuthal velocities and plots each particle location in transverse velocity space. Particles moving in circular orbits transition from v_θ and v_r to $-v_\theta$ and $-v_r$ periodically. The formation of spatial bunches in a cyclotron instability corresponds to modulation of electron gyrational velocities ($v_\perp$) and in the limit of small modulation can be perceived in the 2D plots of v_θ vs v_r as density variations. This is due to the bunch forming in a specific rotational phase with respect to the local AC field and thus at some instant in time having a "location" in a plot of v_θ vs v_r. In addition, energy modulation can be perceived through changes in the magnitude of $v_\perp = \sqrt{v_\vartheta^2 + v_r^2}$, illustrated by the area spanned by PiC particles in plots of v_θ vs v_r. A PiC simulation was conducted of unbounded electron cyclotron-maser emission. The parameters of this simulation included an axial length (beam

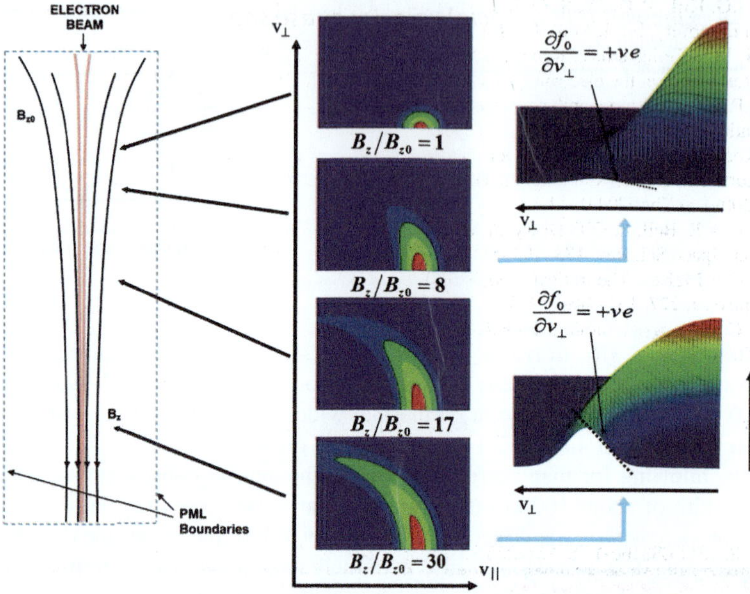

Fig. 5 Diagramatic representation of the PiC simulation geometry, with electromagnetically absorbent boundaries representative of free space

propagation path) of $36\lambda_{ce}$ (where λ_{ce} is the electron cyclotron free-space wavelength) transverse (radial) dimensions of $5\lambda_{ce}$ and radiation absorbent boundaries at the radial and axial limits. A grid resolution of 10 divisions per wavelength at the second harmonic of the electron cyclotron frequency was used along with a PiC merging factor of 3×10^6 electrons/PiC particle. Figure 5 contains a diagrammatic overview with electron beam trajectory of the unbounded simulation geometry defined within the PiC code KARAT. The electron beam was injected into the simulation with a predefined horseshoe distribution, comprising a pitch spread $\alpha = v_\perp/v_\parallel$ of $0 \rightarrow 9.5$, beam energy of 20 keV $\pm$ 5 % and beam current of 14 A. A uniform axial magnetic flux density of 0.1 T was also used in the unbounded simulation as no magnetic compression was necessary.

PiC particle velocity distributions were plotted after a 200 ns run time at two axial positions within the simulation geometry. The corresponding data is presented in Fig. 6. The injected beam distribution at $z = 0.36\lambda_{ce}$ shows a well-defined pitch spread in the $v_\perp$ vs v_z plot covering the complete pitch range from an axial electron beam to the point of magnetic mirroring (zero axial velocity). The corresponding v_θ vs v_r plot also shows a uniform spread in relative orbital phase with no evidence of coherent bunching effects. At $z = 11.7\lambda_{ce}$ the picture is very different however, with clear evidence of azimuthal bunching in the v_θ vs v_r plot (Chu 2004) and in the corresponding $v_\perp$ vs v_z plot there is spreading in the transverse velocity profile of high pitch factor electrons. Looking at $z = 18\lambda_{ce}$, evidence of phase trapping is now present with a concentration of PiC particles extending to the origin of the v_θ vs v_r plot. The corresponding $v_\perp$ vs v_z plot shows a saturated state, with minimum $\partial f/\partial v_\perp$. All of these characteristics are strongly indicative of resonant particle-wave energy transfer via a cyclotron-maser instability.

Figure 7a contains a plot of the radial Poynting flux measured in a plane at $r = 0.32\lambda_{ce}$, over the entire length of the simulation geometry. A DC offset is present in the measurement due to low frequency electromagnetic field components associated with the electron beam

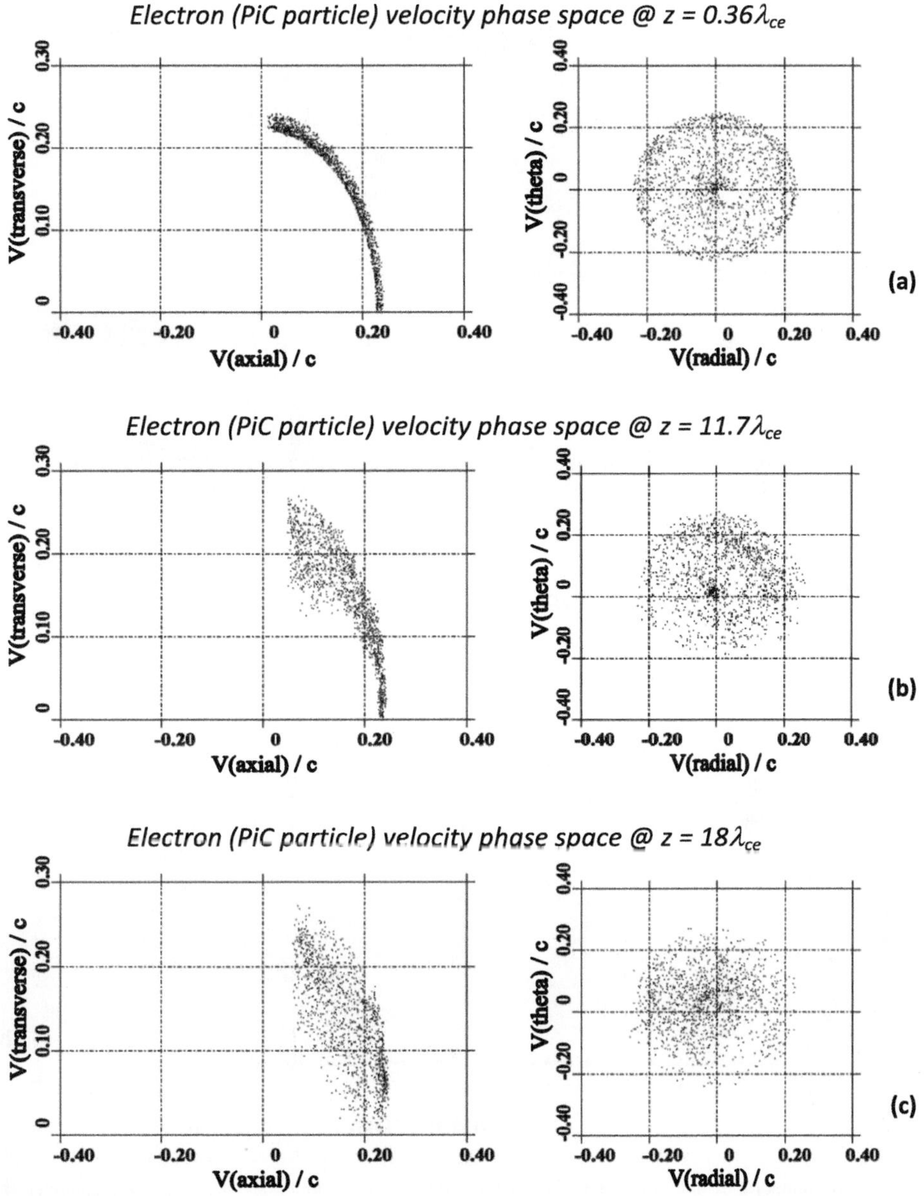

Fig. 6 PiC particle velocity distributions measured on transverse planes within the simulation geometry at (**a**) $z = 0.36\lambda_{ce}$, (**b**) $z = 11.7\lambda_{ce}$ and (**c**) $z = 18\lambda_{ce}$

propagation. The rf output power may therefore be obtained from the amplitude of the AC signal superimposed on this DC offset. Looking at Fig. 7a, there is rather sporadic growth in the electromagnetic output from $54t/t_{ce}$ onwards (where t_{ce} is the electron cyclotron period), with an increased steady growth observed after $432t/t_{ce}$ saturating at ~2.7 kW and corresponding to an rf conversion efficiency of 0.96 %. This is comparable to the 1.3 % efficiency obtained from earlier waveguide bounded simulations (Speirs et al. 2008) and

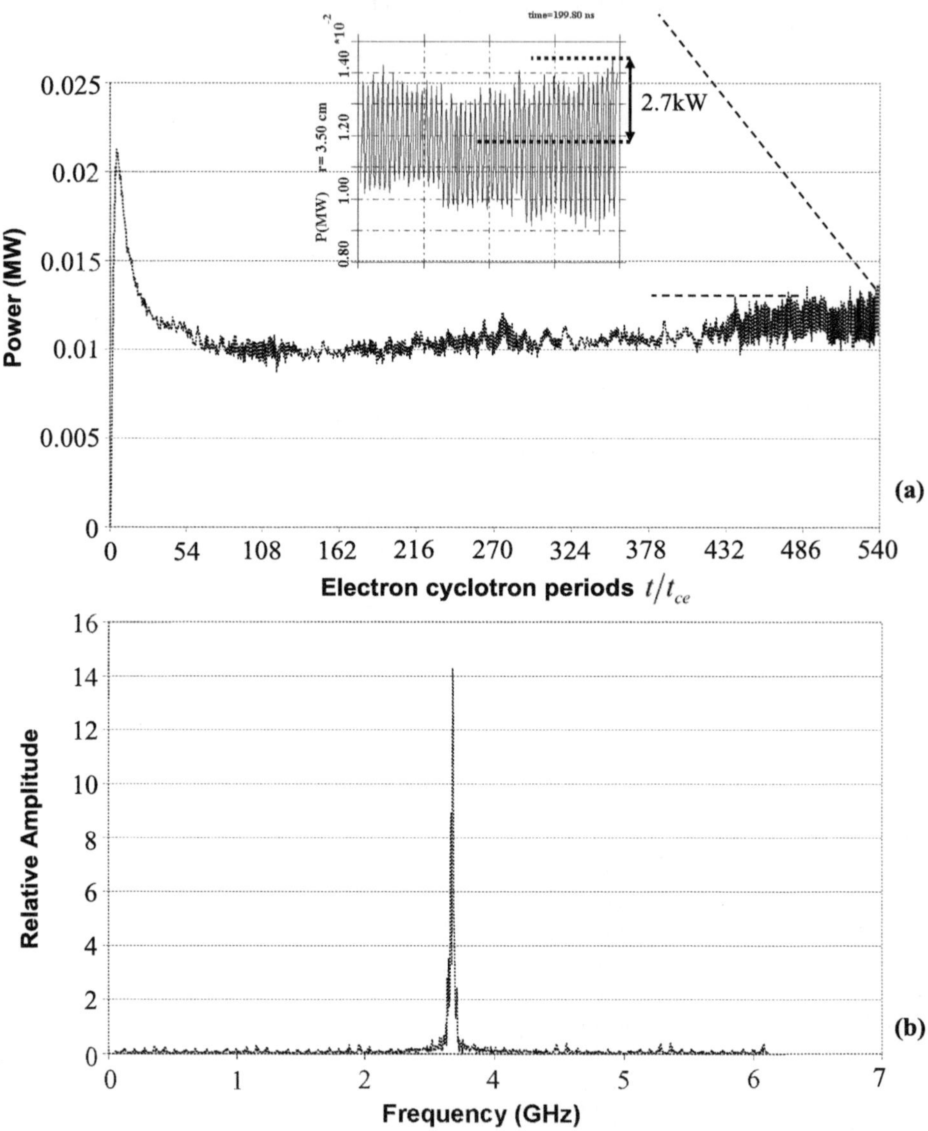

Fig. 7 (**a**) Temporal evolution of the radial Poynting flux measured in a plane at $r = 0.32\lambda_{ce}$ spanning the length of the simulation. (**b**) Fourier transform of E_{theta} from $t/t_{ce} = 0 \rightarrow 540$ at $z = 17\lambda_{ce}$

consistent with the generally accepted estimate of $\sim$1 % for the AKR generation efficiency (Gurnett 1974; Pritchett and Strangeway 1985). The corresponding output spectra is presented in Fig. 7b, showing a well-defined spectral component at 2.68 GHz. This represents a 1.1 % downshift from the relativistic electron cyclotron frequency of 2.71 GHz.

Figure 8 contains a 3D contour plot of E_{theta} mapped over the simulation geometry after a 200 ns run time. An electromagnetic wave sourced at $z \sim 13.5\lambda_{ce}$ is evident propagating near perpendicular to the electron beam with a slight backward wave character—consistent

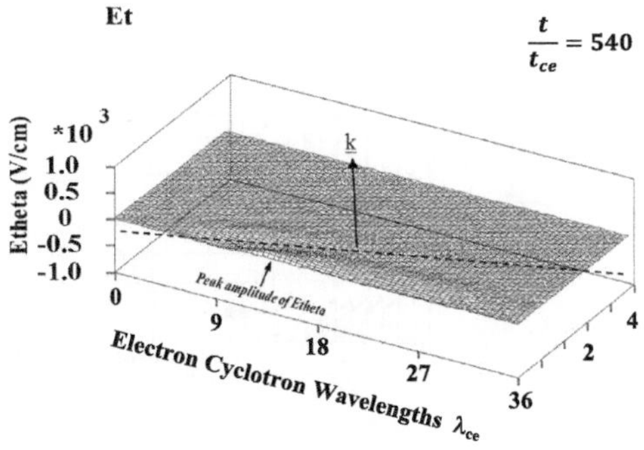

Fig. 8 3D contour plot of E_{theta} within the simulation geometry. RF emission is evident centred around an axial position of $18\lambda_{ce}$ and appears to have a slight backward wave character

with the observed 1.1 % downshift in the spectral output. The axial coordinate range over which the wave is generated corresponds to the position at which a spreading in the PiC particle velocity distribution is first observed in Fig. 6, and represents a significant number of Larmour steps from the point of beam injection for efficient cyclotron-wave coupling to be observed. This represents a reduction in spatial growth rate compared with earlier waveguide-bounded simulations (Speirs et al. 2008), although a comparable saturated RF conversion efficiency was obtained.

4 Laboratory Experiment

An experiment was conducted at the University of Strathclyde's beam-plasma laboratory (Speirs et al. 2005; Ronald et al. 2008a, 2008b; McConville et al. 2008) to simulate the magnetic compression of an electron beam and subsequent evolution/stability of the associated electron velocity distribution. The experimental apparatus was based on the use of an electron gun to inject particles into an increasing magnetic field produced by a system of electromagnets. The overall layout of the experiment is illustrated in Fig. 9, highlighting the electromagnet configuration formed by six distinct coils. Each coil was wound from OFHC copper tubing coated with a thin plastic sheath for electrical insulation. The windings were core cooled by water and were able to carry a current of up to 450 A. The first coil (solenoid 1) was half a metre in length and formed of two layers, surrounding the electron accelerator and defining the magnetic flux density experienced by the electrons as they were injected into the 16 cm diameter beam tunnel. Solenoid 2, also half a metre in length and formed of four layers, confined and transported the electrons to the interaction region and acted as a transition between the low-field electron gun region and the high-field interaction region. As the electron beam diameter reduced with increased flux density, it was possible to reduce the diameter of the interaction region to 8 cm, which also provided a useful reduction in power consumption of the DC solenoid arrangement. Solenoids 3–5 provide the maximum plateau flux density (up to 0.5 T) in the apparatus, and were wound as a 10 layer main coil with a pair of two layer balancing shim coils, respectively. Precise control of the magnetic flux density in this 'interaction' region allowed the efficiency of the cyclotron instability to be investigated as a function of cyclotron resonant detuning.

Electron emission into the apparatus was achieved by placing a cathode electrode covered with an annular region (3 cm mean radius, 1 cm radial width) of velvet emitter (Denisov et al. 1998; Speirs et al. 2005; Ronald et al. 2008a, 2008b) within 2 cm of a sparse mesh anode, Fig. 9. Design of this injector was undertaken with the aid of the 2D PiC code KARAT (Speirs et al 2005, 2008). Application of an accelerating potential of 75 kV to the cathode

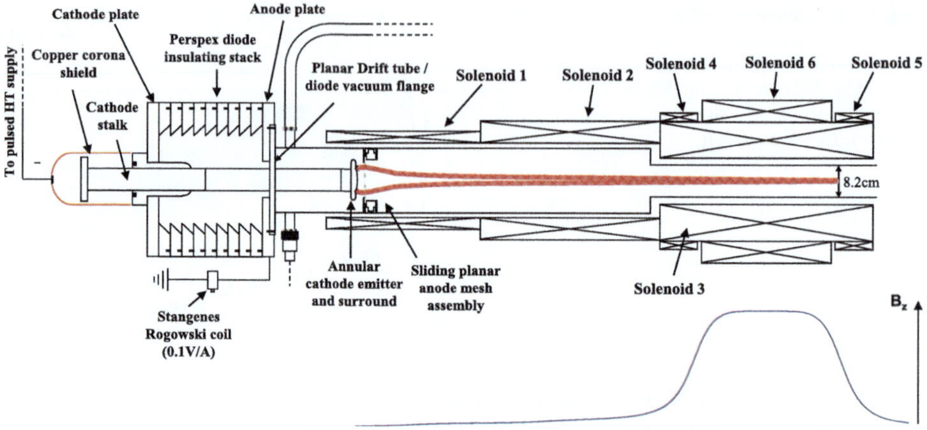

Fig. 9 Schematic diagram of the experimental setup highlighting the magnetic coil configuration and convergent axial magnetic field with peak-plateau region for cyclotron resonant energy transfer

led to field-enhanced electron emission at the discontinuities of the dielectric fibres. The field emission current was sustained by a tunnelling current and enhanced by an avalanche process within the valance electrons of the fibres and a surface flashover avalanche along the outside of the fibres (Noer 1982; Xu and Latham 1992). A high current density was emitted from each fibre tip leading to explosive vaporisation of the underlying bulk material resulting in the formation of a cathode plasma flare (Mesyats 1991) travelling at a velocity of $\sim$2 cm μs^{-1} and supporting the space-charge-limited emission of electrons into the diode gap forming a vacuum spark discharge.

Most of the current traversed the anode mesh and entered the 16 cm anode can region of the experiment, propagating into the increasing magnetic field. The electron gun was placed in the fringing field of solenoid 1, Fig. 9. As the emission surface was normal to the axis of the solenoids this ensured a variation in the magnitude of magnetic compression as a function of radial position on the annular emitter. This induced pitch variation in combination with a 12° domed electrode in the centre of the emitter annulus ensured the electrons were emitted with a finite mean pitch factor and pitch spread sufficient to ensure the formation of a horseshoe shaped velocity distribution when subject to significant magnetic compression.

The resultant velocity distribution obtained in the experiment had a population inversion in the perpendicular direction and so would be expected to be unstable to cyclotron-maser emission, since cyclotron resonance produces diffusion predominantly in the perpendicular direction in momentum space. The electron cyclotron resonance condition was defined at the fundamental frequency, with which we shall be concerned here,

$$\omega = \frac{\omega_{ce}}{\gamma} + k_{\parallel} v_{\parallel} \tag{5}$$

Here ω_{ce} is the electron cyclotron frequency, calculated using the rest mass, and γ is the usual Lorentz factor. If we consider propagation almost perpendicular to the steady magnetic field, so that the contribution from the Doppler shift is negligible, the locus of resonant particles in $(p_\perp, p_\parallel)$ is then just a circle in velocity space. If this circle lies around the inside of the horseshoe, then we may expect instability and thus a radiation frequency just below ω_{ce}.

Near cut-off TE modes were investigated within the interaction waveguide, as these most closely replicate the propagation ($\perp$ static B) and polarisation ($\perp$ static B) properties of the

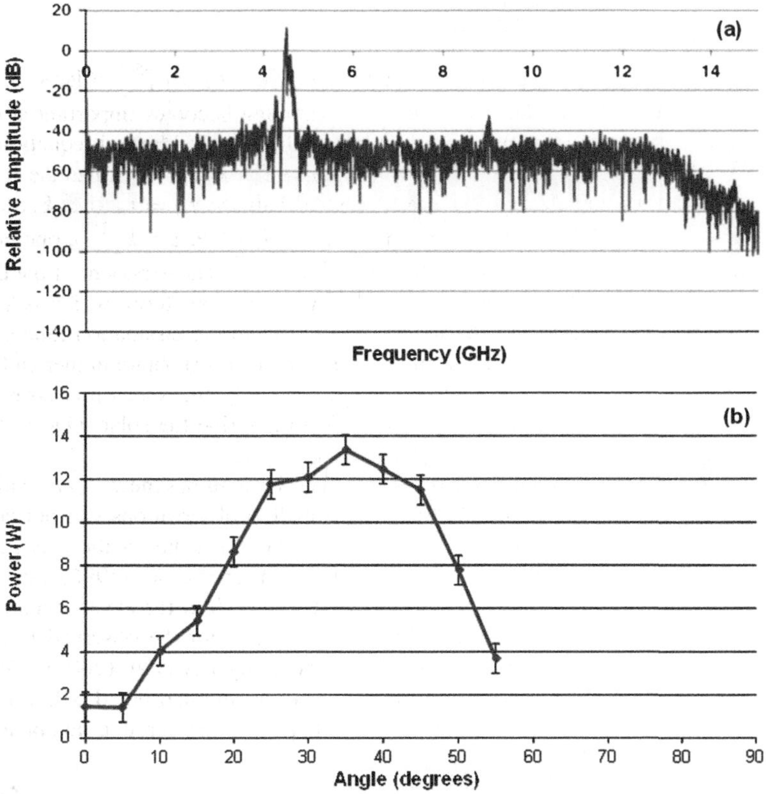

Fig. 10 Experimental measurements for the 4.42 GHz resonance, illustrating (**a**) the spectrum of the output signal, displaying a strong resonance close to the electron cyclotron frequency, 4.42 GHz and (**b**) the antenna pattern confirming excitation of the near cut-off $TE_{0,1}$ mode and yielding the efficiency by integration

X-mode, minimising the Doppler broadening of the resonance (Speirs et al. 2005; Ronald et al. 2008a, 2008b). Two interaction regimes were investigated: one comprising a peak (plateau) axial magnetic flux density of 0.18 T for resonance with the $TE_{0,1}$ mode, and a second with a peak (plateau) axial magnetic flux density of 0.48 T for resonance with the $TE_{0,3}$ mode. The spectra illustrated in Figs. 10a and 11a were obtained by Fourier transform of a directly measured AC waveform acquired by a 12 GHz single-sequence, digital oscilloscope. The first (lower) spectral peak in Fig. 10a corresponds to the $TE_{0,1}$ mode being excited at 4.42 GHz, whilst the second harmonic may also be observed at 8.9 GHz. The corresponding spectral output for the 11.7 GHz resonance regime is presented in Fig. 11a, with a well-defined spectral peak present at 11.7 GHz corresponding to a near cut-off resonance with the $TE_{0,3}$ mode. On closer inspection some minor splitting is present in the spectral peak, indicative of some slight mode competition with the $TE_{2,3}$ mode.

Mode coupling within the experiment was diagnosed by scanning single mode rectangular waveguide receivers in the output antenna pattern of the circular interaction waveguide. Both the 4.42 GHz and 11.7 GHz interaction regimes were studied, with Fig. 10b showing the azimuthally polarised antenna pattern measured for the 4.42 GHz resonance experiments. The single cycle pattern observed, peaking at ~35 degrees is indicative of the $TE_{0,1}$ mode, and consistent with expectations from the beam-wave dispersion characteristics for

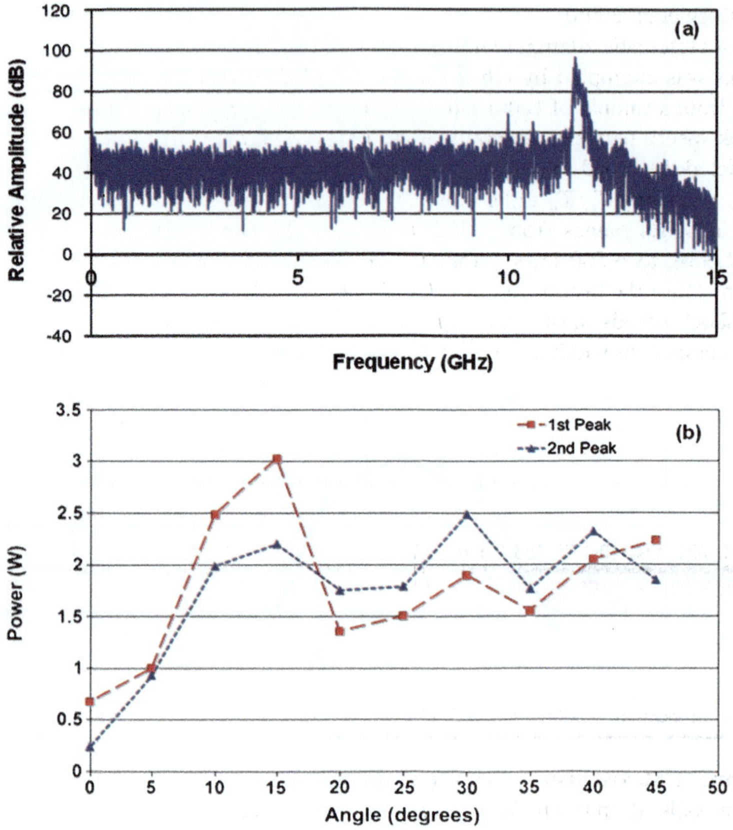

Fig. 11 Experimental measurements for the 11.7 GHz resonance, illustrating (**a**) the spectrum of the output signal, displaying a strong resonance close to the electron cyclotron frequency, 11.7 GHz and (**b**) the antenna patterns confirming excitation of the near cut-off $TE_{0,3}$ mode and yielding the efficiency by integration

the experimental parameters (Ronald et al. 2008a, 2008b; Speirs et al. 2008). Integration over the radial and azimuthal antenna patterns yielded a total output power of 19 kW corresponding to a beam-wave conversion efficiency of 2 % at optimum cyclotron detuning and a compression ratio of 18. The compression ratio of 9 yielded 35 kW and 1 % efficiency, consistent with expectations of the lower compression ratio yielding a reduction in electron population density at higher pitch factors.

For the 11.7 GHz resonance regime, multimode excitation was observed with some evidence of the $TE_{2,3}$ mode. The emission was also bursty, with two distinct temporal peaks in the output pulse envelope. Figure 11b shows the corresponding azimuthal antenna pattern for each of these temporal peaks at a magnetic compression ratio of 16. The triple peak structure observed is characteristic of both the $TE_{0,3}$ and $TE_{2,3}$ modes. Integrating over the corresponding radial and azimuthal antenna patterns yielded an output power of 30 kW and beam-wave conversion efficiency of $\sim$1 %. Overall, the efficiencies obtained for both the $TE_{0,1}$ and $TE_{0,3}$ regimes were in close agreement with the predictions of numerical simulations and are consistent with efficiency estimates for the generation of AKR (Gurnett 1974).

5 Conclusions

This paper describes an experiment created to reproduce an electron distribution function similar to those observed to be correlated with the phenomenon of cyclotron maser radiation commonly found in space and astrophysical plasmas such as planetary aurora, young active stars with a dipole-like magnetic field, blazar jets and shocks. The experiment successfully produced the required electron velocity space distribution. The radiation is associated with electron beams accelerated downwards into the increasing magnetic dipole field of the auroral regions where the local electron plasma frequency ω_{pe} is much less than the cyclotron frequency ω_{ce}. Magnetic compression leads to the formation of a velocity distribution having a horseshoe shape as a result of conservation of magnetic moment. Good agreement on both the efficiency and spectrum of the emitted radiation were obtained between experiments and numerical simulations at 4.45 and 11.7 GHz, just below ω_{ce}. The evolution of the electron velocity distribution predicted by the simulations and the narrow bandwidth of the radiation emissions appears consistent with the expectation of the theoretical models. Numerical and experimental efficiencies are comparable with the satellite observations of auroral kilometric radiation. The maser radiation is initially beamed perpendicular to the magnetic field—an important factor that has bearing on how the radiation can escape. This has been shown to be possible in the planetary case by the introduction of a plasma density profile in altitude that refracts the radiation sufficiently resulting in the beam propagating with small angles to the magnetic field, reducing the attenuation at the second harmonic layer (Mutel et al. 2008). We adopted a similar model for the stars UV Ceti and CU Virginus which both have strong dipole fields. In the case of blaser jets, Begelman et al. (2005) have demonstrated that second harmonic absorption should be low and the radiation can escape if the source region is close to the boundary in a sufficiently thin layer.

Acknowledgements This work was supported by the EPSRC and the STFC Centre for Fundamental Physics. The authors would like to thank W. He, D. Melrose and V.L. Tarakanov for helpful discussions.

References

S.K. Adelman et al., On the rotation of the chemically peculiar magnetic star 56 Arietis. Astron. Astrophys. **375**, 982–988 (2001)

M.C. Begelman, R.E. Ergun, M.J. Rees, Cyclotron maser emission from blazar jets? ApJ **625**, 51–59 (2005)

A.O. Benz et al., First VLBI images of a main-sequence star. Astron. Astrophys. **331**, 596 (1998)

R. Bingham, R.A. Cairns, Generation of auroral kilometric radiation by electron horseshoe distributions. Phys. Plasmas **7**, 3089–3092 (2000)

R. Bingham, R.A. Cairns, B.J. Kellett, Coherent cyclotron maser radiation from UV Ceti. Astron. Astrophys. **370**, 1000–1003 (2001)

R. Bingham, B.J. Kellett, R.A. Cairns, J. Tonge, J.T. Mendonca, Cyclotron maser radiation from astrophysical shocks. Astrophysical Journal **595**, 279–284 (2003)

R. Bingham, B.J. Kellett, R.A. Cairns et al., Cyclotron maser radiation in space and laboratory plasmas. Contrib. Plasma Physics **44**, 382–387 (2004)

C.K. Birdsall, A.B. Langdon, *Plasma Physics Via Computer Simulation* (McGraw-Hill, New York, 1985)

E.F. Borra, J.D. Landstreet, The magnetic fields of the Ap stars. ApJS **42**, 421–445 (1980)

R.A. Cairns et al., A cyclotron maser instability with application to space and laboratory plasmas. Phys. Scr. T **116**, 23 (2005)

R.A. Cairns et al., Cyclotron maser radiation from an inhomogeneous plasma. Phys. Rev. Lett. **101**, 215003 (2008)

K.R. Chu, The electron cyclotron maser. Rev. Mod. Phys. **76**, 489–540 (2004)

G.T. Delory et al., Fast observations of electron distributions within AKR source regions. Geophys. Res. Lett. **25**, 2069–2072 (1998)

G.G. Denisov et al., Gyrotron traveling wave amplifier with a helical interaction waveguide. Phys. Rev. Lett. **81**, 5680–5683 (1998)
R.E. Ergun et al., Electron-cyclotron maser driven by charged-particle acceleration from magnetic field-aligned electric fields. Astrophysical Journal **538**, 456–466 (2000)
K.M. Gillespie et al., 3D PiC code simulations for a laboratory experimental investigation of auroral kilometric radiation mechanisms. Plasma Phys. Controlled Fusion **50**, 124038 (2008)
D.A. Gurnett, The earth as a radio source: terrestrial kilometric radiation. J. Geophys. Res. **79**, 4227 (1974)
A.P. Hatzes, Doppler imaging of the silicon distribution on CU Vir: evidence for a decentred magnetic dipole? Mon. Not. R. Astron. Soc. **288**, 153–160 (1997)
D.S. Hillan, I.H. Cairns, P.A. Robinson, Type II solar radio bursts. 2. Detailed comparison of theory with observations. J. Geophys. Res. [Space Phys.] **115**, A06105 (2012)
T. Katsouleas, J.M. Dawson, Unlimited electron acceleration in laser-driven plasma waves. Phys. Rev. Lett. **51**, 392–395 (1983)
B.J. Kellett et al., Can late-type active stars be explained by a dipole magnetic trap? Mon. Not. R. Astron. Soc. **329**, 102–108 (2002)
B.J. Kellett et al., CU Virginis—the first stellar pulsar. e-print (2007). astro-ph/0701214
K.K. Lo et al., Observations and modelling of pulsed radio emission from CU Virginis. Mon. Not. R. Astron. Soc. **421**, 3316–3324 (2012)
P. Louarn, D. Le Queau, Generation of the auroral kilometric radiation in plasma cavities. II. The cyclotron maser instability in small size sources. Planet. Space Sci. **44**, 211–224 (1996)
P. Louarn et al., Trapped electrons as a free energy source for the auroral kilometric radiation. J. Geophys. Res. **95**, 5983–5995 (1990)
K.G. McClements et al., Lower hybrid resonance acceleration of electrons and ions in solar flares and the associated microwave emission. ApJ **409**, 465–475 (1993)
S.L. McConville et al., Demonstration of auroral radio emission mechanisms by laboratory experiment. Plasma Phys. Controlled Fusion **50**, 074010 (2008)
D.L. Meier et al., Magnetohydrodynamic production of relativistic jets. Science **291**, 84–92 (2001)
D.B. Melrose, *Instabilities in Space and Laboratory Plasmas* (Cambridge University Press, Cambridge, 1986), p. 202
D.B. Melrose, Radiation from instabilities in space plasmas. Astrophys. Space Sci. **264**, 401–410 (1998)
D.B. Melrose, Coherent emission in astrophysics: a critique. Astrophys. Space Sci. **264**, 391–400 (1999)
D.B. Melrose, G.A. Dulk, Electron-cyclotron masers as the source of certain solar and stellar radio bursts. ApJ **259**, 844–858 (1982)
D. Menietti, J.L. Burch, "Electron conic" signatures observed in the nightside auroral zone and over the polar cap. J. Geophys. Res. **90**, 5345 (1985)
J.D. Menietti et al., Simultaneous radio and optical observations of auroral structures: implications for AKR beaming. J. Geophys. Res. **116**, A12219 (2011)
G.A. Mesyats, Vacuum discharge effects in the diodes of high-current electron accelerators. IEEE Trans. Plasma Sci. **19**, 683–689 (1991)
R.L. Mutel et al., Cluster multispacecraft determination of AKR angular beaming. Geophys. Res. Letters **35**, L07104 (2008)
R.S. Nemmen et al., A universal scaling for the energetics of relativistic jets from black hole systems. Science **338**, 1445–1448 (2012)
R.J. Noer, Electron field-emission from broad-area electrodes. Appl. Phys. A **28**, 1–24 (1982)
S.P. O'Sullivan, D.C. Gabuzda, Magnetic field strength and spectral distribution of six parsec-scale active galactic nuclei jets. Mon. Not. R. Astron. Soc. **400**, 26–42 (2009)
P.L. Pritchett, Electron-cyclotron maser instability in relativistic plasmas. Phys. Fluids **29**, 2919–2930 (1986)
P.L. Pritchett, R.J. Strangeway, A simulation study of kilometric radiation generation along an auroral field line. J. Geophys. Res. [Space Physics] **90**, 9650–9662 (1985)
P.L. Pritchett et al., Free energy sources and frequency bandwidth for the auroral kilometric radiation. J. Geophys. Res. **104**, 10317–10326 (1999)
P.L. Pritchett, R.J. Strangeway, R.E. Ergun, C.W. Carlson, Generation and propagation of cyclotron maser emissions in the finite auroral kilometric radiation source cavity. J. Geophys. Res. **107**, 1437 (2002)
D.M. Pyper et al., An abrupt decrease in the rotational period of the chemically peculiar magnetic star CU Virginis. Astron. Astrophys. **339**, 822–830 (1998)
K. Ronald et al., Radio frequency resonator structure and diagnostic measurements for a laboratory simulation of auroral kilometric radiation. Phys. Plasmas **15**, 056503 (2008a)
K. Ronald et al., Electron beam measurements for a laboratory simulation of auroral kilometric radiation. Plasma Sources Sci. Technol. **17**, 035011 (2008b)
A. Roux et al., Auroral kilometric radiation sources: in situ and remote observations from Viking. J. Geophys. Res. **98**, 11657–11670 (1993)

D.C. Speirs et al., A laboratory experiment to investigate auroral kilometric radiation emission mechanisms. J. Plasma. Phys. **71**, 665–674 (2005)
D.C. Speirs et al., Numerical simulation of auroral cyclotron maser processes. Plasma Phys. Controlled Fusion **50**, 074011 (2008)
D.C. Speirs et al., Numerical investigation of auroral cyclotron maser processes. Phys. Plasmas **17**, 056501 (2010)
T.H. Stix, *Waves in Plasmas* (American Institute of Physics, New York, 1992)
R.J. Strangeway et al., FAST observations of VLF waves in the auroral zone: evidence of very low plasma densities. Geophys. Res. Lett. **25**, 2065 (1998)
V.L. Tarakanov, KARAT Code User Manual, Berkeley Research Associates Inc., VA, USA (1992)
G. Thejappa et al., Evidence for the Oscillating Two Stream Instability and Spatial Collapse of Langmuir Waves in a Solar Type III Radio Burst (2012)
R.A. Treumann, The electron–cyclotron maser for astrophysical application. Astron. Astrophys. Rev. **13**, 229–315 (2006)
C. Trigilio, P. Leto, F. Leone, G. Umana, C. Buemi, Coherent radio emission from the magnetic chemically peculiar star CU Virginis. Astron. Astrophys. **362**, 281–288 (2000)
R.Q. Twiss, Radiation transfer and the possibility of negative absorption in radio astronomy. Australian Journal of Physics **11**, 564–579 (1958)
S.S. Vogt, A magnetic study of spotted UV Ceti flair stars and related late-type dwarfs. ApJ **240**, 567–584 (1980)
I. Vorgul, R.A. Cairns, R. Bingham, Analysis of a cyclotron maser instability in cylindrical geometry. Phys. Plasmas **12**, 122903 (2005)
R.M. Winglee, P.L. Pritchett, The generation of low-frequency electrostatic waves in association with auroral kilometric radiation. J. Geophys. Res. **91**, 13531–13541 (1986)
C.S. Wu, L.C. Lee, A theory of the terrestrial kilometric radiation. ApJ. **230**, 621–626 (1979)
N.S. Xu, R.V. Latham, The application of an energy-selective imaging technique to a study of field-induced hot electrons from broad area high voltage electrodes. Surf. Sci. **274**, 147–160 (1992)
P. Zarka, The auroral radio emissions from planetary magnetospheres—what do we know, what don't we know, what do we learn from them? Adv. Space Res. **12**, 99–115 (1992)

GPSR Compliance

The European Union's (EU) General Product Safety Regulation (GPSR) is a set of rules that requires consumer products to be safe and our obligations to ensure this.

If you have any concerns about our products, you can contact us on ProductSafety@springernature.com

In case Publisher is established outside the EU, the EU authorized representative is:

Springer Nature Customer Service Center GmbH
Europaplatz 3
69115 Heidelberg, Germany

Batch number: 10607667

Printed by Printforce, the Netherlands